4th International Workshop DICE 2008: From Quantum Mechanics through Complexity to Spacetime: The Role of Emergent Dynamical Structures 2008

Journal of Physics: Conference Series Volume 174

Tuscany, Italy
22-26 September 2008

ISBN: 978-1-61738-309-0

Printed from e-media with permission by:

Curran Associates, Inc.
57 Morehouse Lane
Red Hook, NY 12571
www.proceedings.com

Some format issues inherent in the e-media version may also appear in this print version.

Printed by Curran Associates, Inc. (2010)

Institute of Physics
Dirac House, Temple Back
Bristol BS1 6BE UK

Phone: 44 1 17 929 7481
Fax: 44 1 17 920 0979

techtracking@iop.org

4th International Workshop DICE 2008: From Quantum Mechanics through Complexity to Spacetime: The Role of Emergent Dynamical Structures 2008

Journal of Physics: Conference Series Volume 174

Tuscany, Italy
22-26 September 2008

ISBN: 978-1-61738-309-0

IOP Institute of Physics

TABLE OF CONTENTS

INVITED LECTURES

BLACK HOLES, QUANTUM THEORY AND COSMOLOGY 1
Roger Penrose
DOES WAVE FUNCTION COLLAPSE CAUSE GRAVITY? 16
Lajos Diósi
THE MONTEVIDEO INTERPRETATION OF QUANTUM MECHANICS: FREQUENTLY ASKED QUESTIONS 22
Rodolfo Gambini, Jorge Pullin
MEASUREMENTS OF NON LOCAL WEAK VALUES 29
Aharon Brodutch, Lev Vaidman
EMERGENT CLASSICALITY VIA COMMUTING POSITION AND MOMENTUM OPERATORS 39
J J Halliwell
DE SITTER EQUILIBRIUM AS A FUNDAMENTAL FRAMEWORK FOR COSMOLOGY 58
Andreas Albrecht
ON SPECTRUM OF VACUUM ENERGY 69
G E Volovik
EMERGENCE OF QUANTUM MECHANICS FROM CLASSICAL STATISTICS 79
C Wetterich
DOES QUANTUM MECHANICS TELL AN ATOMISTIC SPACETIME? 99
Hans-Thomas Elze
THE PHYSICS OF MENTAL ACTS: COHERENCE AND CREATIVITY 108
F Tito Arecchi
DISSIPATIVE NEURODYNAMICS IN PERCEPTION FORMS CORTICAL PATTERNS THAT ARE STABILIZED BY VORTICES 125
Walter J Freeman, Giuseppe Vitiello
MULTIVALUED FIELDS 149
H Kleinert
THE INTERPRETATION OF QUANTUM MECHANICS: WHERE DO WE STAND? 159
Giancarlo Ghirardi
RECENT TESTS OF REALISTIC MODELS 175
Giorgio Brida, Ivo Pietro Degiovanni, Marco Genovese, Marco Gramegna, Fabrizio Piacentini, Valentina Schettini, Paolo Traina
EMERGENT/QUANTUM GRAVITY: MACRO/MICRO STRUCTURES OF SPACETIME 187
B L Hu
HIGH-ENERGY γ-RAY ASTRONOMY AND STRING THEORY 203
Nick E Mavromatos
EMERGENCE OF SPATIAL STRUCTURE FROM CAUSAL SETS 237
D Rideout, P Wallden
LIGHT, LINKS AND CAUSAL SETS 258
Rafael D Sorkin
DYNAMICS FOR CAUSAL SETS WITH MATTER FIELDS: A LAGRANGIAN-BASED APPROACH 268
Roman Sverdlov, Luca Bombelli
QUANTUM HISTORIES AND QUANTUM GRAVITY 293
Joe Henson
CONCEPT OF TIME IN CANONICAL QUANTUM GRAVITY AND STRING THEORY 316
Claus Kiefer
SOURCE OF THE OBSERVED THERMODYNAMIC ARROW 327
L S Schulman

CONTRIBUTED PAPERS

NONCOMMUTATIVITY AND QUANTUM STRUCTURE OF SPACETIME 332
B P Dolan, Kumar S Gupta, A Stern

QUANTUM MEASUREMENT AND QUANTUM GRAVITY: MANY-WORLDS OR COLLAPSE OF THE WAVEFUNCTION? ... 338
T P Singh
CONSTRAINED QUANTUM DYNAMICS ... 357
Anna C T Gustavsson
CENTER OF MASS IN SPECIAL AND GENERAL RELATIVITY AND ITS ROLE IN AN EFFECTIVE DESCRIPTION OF SPACETIME ... 365
C Chryssomalakos, H Hernandez-Coronado, E Okon
AND WHAT IF GRAVITY IS INTRINSICALLY QUANTIC? ... 369
Houri Ziaeepour
POSSIBILITY OF REALIZING BUSSEY'S THOUGHT EXPERIMENT ON COLLAPSE OF THE WAVE FUNCTION AT THE MICROSCOPIC LEVEL ... 376
Kouichi Toyoshima, Takasi Endo, Yutaka Hirayoshi
VIRTUAL STATES AND DECOHERENCE IN SINGLE IMPURITY ATOM SOLID STATE QUANTUM DEVICE ... 381
Enrico Prati
PURITY SIEVE FOR MODELS WITH FACTORIZABLE INTERACTIONS ... 387
Oleg Lychkovskiy
A FOUNDATION OF QUANTUM MECHANICS BASED ON FOCUSING AND SYMMETRY ... 394
Inge S Helland
SYMMETRIES OF NONRELATIVISTIC PHASE SPACE AND THE STRUCTURE OF QUARK-LEPTON GENERATION ... 400
Piotr Zenczykowski
ON THE RICCI FLOW AND EMERGENT QUANTUM MECHANICS ... 408
José M Isidro, J L G Santander, P Fernández De Córdoba
CAN QUANTUM MECHANICS BE AN EMERGENT PHENOMENON? ... 415
Massimo Blasone, Petr Jizba, Fabio Scardigli
BORN'S RULE FOR MOMENTUM MEASUREMENT FROM PREQUANTUM FIELD THEORY ... 425
Andrei Khrennikov
ON SCHRÖDINGER'S QUEST FOR NEW PHYSICS FOR LIFE ... 433
Alfredo Iorio
NOISE STABILIZATION EFFECTS IN MODELS OF INTERDISCIPLINARY PHYSICS ... 442
B Spagnolo, G Augello, P Caldara, A Fiasconaro, A La Cognata, N Pizzolato, D Valenti, A A Dubkov, A L Pankratov
QUANTUM-CLASSICAL INTERACTIONS AND MEASUREMENT: A CONSISTENT DESCRIPTION USING STATISTICAL ENSEMBLES ON CONFIGURATION SPACE ... 455
M Reginatto, M J W Hall
BOHMIAN MECHANICS AND THE EMERGENCE OF CLASSICALITY ... 463
A Matzkin
QUANTUM PHASES DUE TO COSMIC GRAVITY AND THE SPIN-STATISTICS CONNECTION ... 471
Aditya Gilra
RANDOM HAMILTONIAN IN THERMAL EQUILIBRIUM ... 475
Dorje C Brody, David C P Ellis, Darryl D Holm
COMPLEX NONLINEAR RELATIONS IN CLASSICAL AND QUANTUM PHYSICS ... 483
Dieter Schuch
PROPERTIES OF SOME NONLINEAR SCHRÖDINGER EQUATIONS MOTIVATED THROUGH INFORMATION THEORY ... 491
Liew Ding Yuan, Rajesh R Parwani
MATRIX MODELS, EMERGENT GRAVITY AND GAUGE THEORY ... 496
Harold Steinacker
THE EMERGENCE OF GRAVITY AS A RETRO-CAUSAL POST-INFLATION MACRO-QUANTUM-COHERENT HOLOGRAPHIC VACUUM HIGGS-GOLDSTONE FIELD ... 504
Jack Sarfatti, Creon Levit
PATH INTEGRALS ON CAUSAL SETS ... 512
Steven Johnston
EMERGENT NON-COMMUTATIVE MATTER FIELDS FROM GROUP FIELD THEORY MODELS OF QUANTUM SPACETIME ... 520
Daniele Oriti
MASSLESS PARTICLE DIFFUSION IN CAUSAL SET THEORY ... 534
Lydia Philpott, Fay Dowker, Rafael Sorkin
A RESULT IN 2D CAUSAL SET THEORY: THE EMERGENCE OF SPACETIME ... 540
Graham Brightwell, Joe Henson, Sumati Surya

QUANTUM TETRAHEDRA: A TOY MODEL FOR CLASSICAL LIMIT OF THE QUANTUM GRAVITY 546
Daniel R Terno
MATTER IN TOY DYNAMICAL GEOMETRIES 555
Tomasz Konopka
THE EMERGENCE OF SPACETIME 563
B G Sidharth
NONCOMMUTATIVE QUANTUM COSMOLOGY 571
C Bastos, O Bertolami, N C Dias, J N Prata
THE EMERGENCE OF PROBABILITIES IN ANHOMOMORPHIC LOGIC 579
Yousef Ghazi-Tabatabai, Petros Wallden
ARITHMETIC FOR THE UNIFICATION OF QUANTUM MECHANICS AND GENERAL RELATIVITY 588
Victor Korotkikh
KREIN SPACE QUANTIZATION OF LINEAR GRAVITY IN DE SITTER SPACETIME AND GENERALIZATION TO QFT 596
Farrin Payandeh
TWO-POINT FUNCTION OF MASSLESS SPIN-2 FIELD IN DE SITTER SPACE-TIME 603
A C H Shirazi
GRAVITATIONAL RENORMALIZATION OF QUANTUM FIELD THEORY: A "CONSERVATIVE" APPROACH 609
Roberto Casadio
A WINDOW TO QUANTUM GRAVITY PHENOMENA IN THE EMERGENCE OF THE SEEDS OF COSMIC STRUCTURE 616
Daniel Sudarsky, Adolfo De Unanue
COUPLING DARK ENERGY WITH STANDARD MODEL STATES 629
M C Bento, A E Bernardini, O Bertolami
EMERGENT GRAVITY AND ENTANGLEMENT ENTROPY OF BLACK HOLES 645
Mariano Cadoni
MULTIPARTITE ENTANGLEMENT IN NEUTRINO OSCILLATIONS 652
Massimo Blasone, Fabio Dell'Anno, Silvio De Siena, Fabrizio Illuminati
PARTICLE MIXING AS POSSIBLE EXPLANATION OF THE DARK ENERGY CONUNDRUM 658
Antonio Capolupo, Giuseppe Vitiello
MULTIPARTITE GEOMETRIC ENTANGLEMENT IN FINITE SIZE XY MODEL 664
Massimo Blasone, Fabio Dell'Anno, Silvio De Siena, Salvatore Marco Giampaolo, Fabrizio Illuminati
THE VIP EXPERIMENT 671
S Bartalucci, S Bertolucci, M Bragadireanu, M Catitti, M Cargnelli, C Curceanu, S Di Matteo, J-P Egger, C Guaraldo, M Iliescu, T Ishiwatari, M Laubenstein, J Marton, E Milotti, D Pietreanu, T Ponta, D L Sirghi, F Sirghi, L Sperandio, O Vasquez Doce, E Widmann, J Zmeskal
NAKED SINGULARITY IN MODIFIED GRAVITY THEORY 675
Remo Garattini
TOWARDS THE SOLUTION OF COSMOLOGICAL CONSTANT AND ZERO POINT ENERGY PROBLEMS THROUGH METRIC REVERSAL SYMMETRY 682
Recai Erdem
ON THE EXISTENCE OF TIME IN NON-SINGULAR SPACETIMES 690
E Minguzzi
NATURE OF TRACING DIMENSION, IMAGINARY ORDER OF FREEDOM AND OUR OBSERVED TIME PASSING AT CONSTANT SPEED 694
S Nagao
EVENT-DRIVEN POWER-LAW RELAXATION IN WEAK TURBULENCE: A LIQUID CRYSTALS MESOSCOPIC EXPERIMENT BRIDGING QUANTUM DOTS AND THE INTEGRATION THEORY FOR THE BRAIN 702
Ludovico Silvestri, Leone Fronzoni, Angelo Gemignani, Paolo Grigolini, Danilo Menicucci, Paolo Allegrini
QUANTUM MECHANICS IN Q-DEFORMED CALCULUS 712
A Lavagno, G Gervino
THE PROPAGATOR FOR THE STEP POTENTIAL USING THE PATH DECOMPOSITION EXPANSION 720
James M Yearsley
Author Index

Journal of Physics: Conference Series **174** (2009) 012001 doi:10.1088/1742-6596/174/1/012001

Black holes, quantum theory and cosmology

Roger Penrose
Mathematical Institute, 24-29 St Giles, Oxford OX1 3LB, U.K.
E-mail: rouse@maths.ox.ac.uk

Abstract. Some reasons are given for believing that the rules of quantum (field) theory must be changed when general relativity becomes seriously involved. If full quantum mechanical respect is paid to the principle of equivalence, we find that a superposition of gravitational fields leads to an illegal superposition of different vacua, giving support to a proposal for spontaneous quantum state reduction made earlier by Diósi, and then independently by the author. A different line of attack involves the over-riding role of black holes in the total entropy content of the universe, and in the operation of the 2nd Law of thermodynamics. The author's proposal of conformal cyclic cosmology is reviewed in order to highlight a seeming paradox, according to which the entropy of the universe of the remote future seems to return to the small kind of value that it had at the big bang. The paradox is resolved when we take into account the information loss that, from this perspective, necessarily occurs in Hawking's black-hole evaporation, with the accompanying loss of unitarity.

1. Quantum state reduction as a gravitational effect

In various places elsewhere I have expressed some of my reasons for believing that, when Einstein's theory of gravity is brought into the picture, the standard rules of quantum theory must change – with the hope that when the appropriate changes are made, the quantum *measurement paradox* will be resolved. For an account of most of these arguments, see Chapter 30 of my book *The Road to Reality.* [1] It is suggested that, in accordance with a proposal put forward initially by Diósi [2] and later, but with different motivations, by myself, [3] a quantum superposition of two states, where each on its own would be stationary, should spontaneously decay to one or the other in a time scale of the order $\sim \hbar/E_G$, where E_G is the gravitational self-energy of the mass distribution which is the *difference* between the mass distributions of the two states.

The most compelling motivation, [4] in my opinion, comes from an argument that regards Einstein's principle of equivalence as an essential ingredient in the description of quantum processes taking place in a gravitational field. Two ways of treating a uniform gravitational field may be envisaged. One is the "Newtonian" way, in which gravitation is considered as just another force, and so it merely provides a term in the Hamiltonian in addition to those describing all the other forces of relevance to the problem at hand. The other is the "Einsteinian", in which we work in a frame in free fall, so that the gravitational force disappears from the Hamiltonian, and we can then transform back for comparison with the Newtonian frame that we used before. It is found that the Newtonian wavefunction Ψ_N is simply a *phase* multiple of the Einsteinian one Ψ_E. In a sense this justifies a viewpoint that quantum theory is consistent with the principle of equivalence, a fact that has found experimental confirmation in the well-known "COW"

Journal of Physics: Conference Series **174** (2009) 012001 doi:10.1088/1742-6596/174/1/012001

experiment. [5] However, the phase factor (where $\mathbf{a}$ is the 3-vector of the acceleration, $\bar{\mathbf{x}}$ the position 3-vector of the mass centre, m the total mass, and t the time) appearing in the formula

$$\Psi_E = e^{i(\frac{1}{6}t^3\mathbf{a}\cdot\mathbf{a}+t\bar{\mathbf{x}}\cdot\mathbf{a})m/\hbar}\Psi_N$$

relating these two wavefunctions is *non-linear in the time*, because of the term $\frac{i}{6}t^3\mathbf{a}\cdot\mathbf{a}$ in the exponent, which has the effect that the two wavefunctions belong to *different vacua*. This is actually the Galilean (i.e. $c \to \infty$) limit of the *Unruh* effect, [6] where an accelerating observer (i.e. not in Einsteinian free fall) experiences a different vacuum – in fact a *thermal* vacuum – from a freely falling one at the same location. In the fully relativistic Unruh effect, the acceleration of the observer gives rise to a temperature $\mathbf{T}$ proportional to a/c, experienced by the observer. But this gives $\mathbf{T}= 0$ in the Galilean limit $c \to \infty$. Nevertheless, the two vacua are still different because of the above phase factor, which survives in the limit. [7]

I am adopting the viewpoint that the Einsteinian treatment of the gravitational field is the more fundamental one, in the quantum (as well as the classical) context. This has no real consequence if we just doing quantum mechanics within a single classically treatable gravitational field. But it does have an important implication that if we try to consider doing quantum physics in the presence of a quantum *superposition* between two gravitational fields, where the two free-fall frames differ in the neighbourhood of some point P. For then we have two distinct vacua around P involved in the superposition, so that, strictly speaking, quantum superpositions between states that are entangled with the gravitational field become "illegal" – since scalar products between states belonging to the different vacua become divergent. Since these scalar products involve integrations out to infinity, I take the view that these infinities only begin to make their mark after a finite time-scale T, and that T can be estimated by use of Heisenberg's time-energy uncertainty relation, where the energy uncertainty E_G is estimated from the coefficient of the t^3 term in the phase factor relating the two vacua, integrated over all space. Taking T to represent the rough maximum time that our quantum superposition can persist, before "decaying" to one or the other of the two gravity states under superposition, we obtain the above-mentioned ("Diósi-Penrose") proposal for the lifetime of a quantum superposition. Although current experiments have not reached the degree of sensitivity needed to test this proposal, there are projected schemes under development that might eventually do so. [8]

2. Space-time singularities and the 2nd Law

Among the other reasons for believing that the rules of quantum (field theory) must become seriously modified in the context of the curved space-time of (Einstein's) general relativity, are those relating to black holes and cosmology. Most particularly, it is normally anticipated that there ought to be a physically correct theory of "quantum gravity" that should account for the actual physical behaviour that takes place at those space-time locations where, when we attempt to describe them according to classical Einstein theory, we encounter *space-time singularities*. Yet, what we find when we try to understand the structure of the classical space-time in the neighbourhood of such singularities is something that is grossly asymmetrical in time, a feature which is hard to understand if we insist on playing according to the rules of standard quantum field theory. This time-asymmetry is deeply related to the second law of thermodynamics (which I here refer to simply as the "2nd Law"). This is not a time-asymmetry that can be explained *by* the 2nd Law; rather it is the 2nd Law (in the particular form that we find in our universe) that arises *because* of this gross time-asymmetry in space-time-singularity structure.

This point needs illumination. The 2nd Law roughly tells us that the entropy S of an isolated system should increase with time (or at least should not decrease, apart from, perhaps, occasional fluctuations). Boltzmann's definition of entropy is

$$S = k \log V,$$

k being Boltzmann's constant, where V is the volume of the coarse-graining region $\mathcal{V}$, in phase space $\mathcal{P}$, of all the possible micro-states of the system which could underlie the given macro-state under consideration. These coarse-graining volumes differ enormously from one another, a fact that is somewhat disguised by the logarithm in Boltzmann's formula (and by the smallness of Boltzmann's constant k in ordinary terms), so that a tiny entropy increase signals a vast increase of the volume of the relevant coarse-graining region. Thus, the huge majority of time-evolution trajectories within $\mathcal{P}$, which pass through a given coarse-graining region $\mathcal{V}$, will *next* enter a vastly larger region, and this corresponds to a significant entropy increase in the future time direction, as is indeed the behaviour that comes about according to the 2nd Law. However, this reasoning does not agree with observed physical behaviour in the *past* time direction, because the vast majority of time-evolution trajectories entering $\mathcal{V}$ will also have *come from* a vastly larger region, seeming to suggest that a larger entropy in the past is also to be expected. We cannot explain why the 2nd Law held in the past, simply from general considerations of this kind. Instead, we need an understanding of why it was that the entropy of the *initial* state of our universe was so extraordinarily small. If we accept this as an initial constraint, and then allow ourselves to consider *only* those trajectories originating within that utterly tiny coarse-graining region $\mathcal{B}$ relevant to our *actual* Big Bang, then we do expect to find that the vast majority of time evolutions constrained in this way will encounter increasing entropies – which is the central content of the 2nd Law.

3. The extraordinary specialness of the Big Bang

The fact that $\mathcal{B}$ does indeed have to be an extremely tiny region within $\mathcal{P}$, is a manifest implication of the 2nd Law, yet there is a seeming paradox here, in view of the most direct piece of observational evidence for an actual "primordial fireball" encompassing our entire universe some 1.4×10^{10} years ago. This evidence is provided by the *cosmic microwave background* (CMB), coming to us from all directions in space. Its frequency spectrum accords extremely closely with the Planck formula, for a temperature of 2.725K, so what we appear to be observing is an adiabatically expanding universe in a very closely thermal state, where the expansion of the universe has cooled the radiation from an extremely high temperature (of around 3000K) when the universe was a mere 3.8×10^5 years old (about 1/37000 of its present age) – this being the time of "decoupling", when the universe became fully transparent to electromagnetic radiation. This has the appearance of paradox, because the Planck spectrum is indicative of a state of *thermal equilibrium*, in other words, of *maximum* rather than minimum of entropy. The resolution of this paradox has nothing to do with the universe in its early stages being "small". There would have been as many degrees of freedom potentially available to it in its early highly condensed stages as in its later expanded stages, for the simple reason that the phase space $\mathcal{P}$, is just a *given* thing which does not itself "grow" with time (time evolution being described by trajectories *within* $\mathcal{P}$).

An issue arises, with a spatially infinite universe, that the total volumes of $\mathcal{P}$ and $\mathcal{B}$ would normally be infinite. This is not really problematic, however, in the context of standard cosmologies because, adequately for the present purposes, we can consider that our descriptions apply to a very large *co-moving* spatial volume, rather than to the entire universe. It will be sufficient to take this co-moving volume to be that originally lying within the *particle horizon* of our present space-time location, and we then find that this volume has a baryon number of around 10^{80}. I shall consider our phase space $\mathcal{P}$ and its Big-Bang-subspace $\mathcal{B}$ to refer to this co-moving space-time volume only, and then we find a finite value for the volume P of $\mathcal{P}$ and therefore also for the smaller volume B of $\mathcal{B}$.

In order understand how the 2nd Law is set on its course, we need to appreciate how B can be enormously smaller than P, despite the appearance of thermal equilibrium of the primordial fireball. A clue lies in another striking aspect of the CMB, namely its extremely closely isotropic nature, with variations of temperature in different directions in the sky being only at the level of around 10^{-5}, once the Doppler correction is made for the proper motion of the Earth. This provides a strong indication that the initial Big-Bang singularity must itself have been extremely closely isotropic, and if we assume that there is nothing significant in our particular spatial location [9] in the universe ("Copernican principle") then we conclude that it must have been closely homogeneous also. This is all in accordance with the picture that our overall universe, with its Big-Bang singularity, is rather closely approximated by a standard Friedmann-Lemaître-Robertson-Walker (FLRW) cosmological model, as is standard practice in modern cosmology.

However, it is this very closeness to an FLRW-symmetrical structure, that we apparently find in the Big Bang, that characterizes it as having an extraordinarily low entropy. The FLRW models are all very special in that there can be no gravitational contribution to the entropy, this arising only when departures from complete homogeneity or isotropy occur. In general terms, it is *gravitational clumping* that represents entropy increase. We see this most particularly, from the perspective of life here on Earth, in the formation of the Sun by gravitational clumping from a previously rather uniform distribution of gas, this being the ultimate reason for the manifest departure from thermal equilibrium that the contrast between the hot sun and the dark sky provides, and which our continued well-being on Earth depends upon. The most extreme gravitational clumping is the formation of *black holes*, this resulting in an enormous increase in entropy. In fact, the totality of large black holes residing in galactic centres provides easily the largest (known) contribution to the entropy of the universe. This is despite the fact that the contribution from the CMB itself constitutes a larger entropy per baryon than all known physical processes other than these.

Now, if we want to get some feeling for how much larger the entropy of the universe *might have been* had the Big Bang been of a completely generic nature, we can first consider the time-reversed situation of a *collapsing* universe, which acts in accordance with the 2nd Law, and in which there are some initially fairly small deviations from FLRW symmetry that represent a certain amount of gravitational clumping. As the entropy of the universe increases with this universe's collapse, matter accumulates into black holes, and as the collapse continues, the deviations from FLRW increase relentlessly. Many black holes merge together, swallowing each other up at an increasing rate, forming a great mess of a final space-time singularity, with little left of the symmetry or uniformity of the FLRW model that it initially resembled. From the perspective of strong cosmic censorship, it is not unreasonable to assume that this final ("Big-Crunch") singularity has an essentially *spacelike* character, [10] so that with a suitable choice of time parameter, it may be regarded as a simultaneous event. The entropy of the universe – or at least of the co-moving part which lies within the particle horizon of our present space-time location – may be fairly well estimated by use of the Bekenstein-Hawking formula for the entropy of the (effective) black hole which encompasses all the matter in the (relevant co-moving) universe. Taking the baryonic part of this matter to consist of 10^{80} baryons, this formula gives us an entropy of around 10^{123} in absolute units (i.e. taking $c = \hbar = k = 1$, although this choice makes little difference to this figure). In fact, baryonic matter appears to account for only $\sim 15\%$ of the dynamical material content of the universe, the remaining $\sim 85\%$ being some unknown substance referred to as "dark matter". Taking this into account, and assuming that this dark material could also contribute to the total possible black-hole entropy, we get a figure for the largest possible entropy value of perhaps 10^{125}, but owing to the uncertainties about the nature of this material, I shall take the figure 10^{123} as representing a reasonable lower limit to the possible entropy value. Inverting Boltzmann's entropy formula, we find the value $e^{10^{123}}$ as a reasonable lower limit for the volume P of the phase space $\mathcal{P}$:

$$\mathrm{e}^{10^{123}} \leq P.$$

With regard to the volume B, there is more uncertainty, since all we directly see is the CMB, and the primordial fireball which created it may well have been involved in some significant entropy-raising processes in its $\sim 3.8\times10^5$ preceding years, following the Big Bang singularity. Figures of 10^8 or 10^9 have been quoted for the entropy per baryon of this radiation, so we may take 10^{89} as a rough upper bound for the entropy in the CMB, and therefore as an upper bound for the entropy in the Big Bang itself (which might actually have been significantly lower):

$$B \leq e^{10^{89}}.$$

This tells us that the ratio of the volume of B to that of the whole phase space P is no greater than

$$e^{10^{89}} \div e^{10^{123}} = e^{10^{89}-10^{123}} = e^{-10^{123}}$$

to within the accuracy of the upper exponents, so we see that the $e^{10^{89}}$ makes no significant impression on the far more enormous $e^{10^{123}}$. The thermal nature, or otherwise, of the primordial fireball, would only affect the $e^{10^{89}}$ and is thus essentially irrelevant to the far greater problem of the utter minuteness of $e^{-10^{123}}$. On the other hand, the great uniformity of the CMB temperature across the sky is a reflection of the uniformity of the universe itself – a powerful confirmation of the fact that that the initial singularity *was* pretty closely of FLRW type, with gravitational degrees of freedom essentially inactivated and utterly *un*like the time-reverse of the maximum-entropy collapsing universe full of congealing black holes that was considered in the previous paragraph. That time reverse of a generic "Big Crunch" final singularity would have been a Big Bang of stupendous entropy ($\geq e^{10^{123}}$) riddled with white holes (the time reverses of black holes) and it was the complete absence of such white-hole monstrosities that provides the way in which our universe was so enormously special.

4. Cosmic inflation

The foregoing discussion must be compared with the alternative viewpoint of *cosmic inflation*, [11] central to most modern discussions of the very early universe, and which is commonly claimed to provide a quite different perspective, and even an "explanation" of the uniformity of the early universe. It is argued that during a period roughly between the times 10^{-35}s and 10^{-32}s following the initial Big Bang singularity, the universe exponentially expanded by a factor of perhaps between 10^{30} and 10^{60}. Without such cosmic inflation, two fairly widely separated points on the CMB sky could never have been in causal contact with each other, at any time since the Big Bang. The view is taken that such causal contact would have been *necessary* in order that the primordial fireball could have reached its global state of thermal equilibrium. It is true that the presumed inflationary phase of the very early universe indeed serves the purpose of bringing all the points on our observed CMB sky into causal contact with one another. Accordingly, it is argued that the resulting inflationary thermalization of the primordial fireball would explain the very close equality of CMB temperatures over the whole sky.

However this argument goes no way towards explaining the very tiny entropy that, as we have seen, has to be a feature of the primordial fireball. [12] Indeed, the thermalization argument – *any* thermalization argument – acts in the opposite direction. We note that the very idea of bringing distant points on the CMB sky into causal contact is in order that initially differing temperatures at such places would become equal through the *thermalization* that is enabled by this causal contact. The problem with this kind of argument, if we are trying to explain the very special nature of the early universe, is that all thermalization processes are *entropy raising* effects, so that the situation before thermalization would have to have been even more special (i.e. even lower entropy) than after thermalization. The problem of the specialness of the Big Bang is simply made *worse* by arguments of this kind. We recall, from the previous section, that the nature of this specialness could be understood if we can find some reason for there being a closely FLRW structure for the Big Bang singularity. If we are already accepting that this singularity had this structure very closely, then the close uniformity of the CMB temperature over the sky is a *consequence.* (The initial FLRW symmetry implies FLRW symmetry at all later times, if we assume deterministic evolution equations.) This uniformity of temperature is not helped by a thermalization argument. Indeed, it should be borne in mind that explicit calculations for deriving implications of inflation are almost invariably carried out within a FLRW background, which completely begs the question of *why* the initial singularity should be closely of this form.

A similar kind of objection can be raised against the common argument that inflation serves to iron out spatial irregularities that might have been present in the early universe. To see why inflation by itself cannot iron out generic irregularities in a messy Big Bang, we need to take into account the fact that the dynamical equations that are to govern inflation (these involving a special "inflaton field" φ that is introduced) are all *time symmetrical*, the only time asymmetry arising in the dynamical behaviour (e.g. symmetry breaking) coming about through *consequences* of the 2nd Law. Suppose that it is actually true that, with a generic starting point, the inflationary processes will almost inevitably lead to a smoothed-out expanding universe after inflation has finished. Clearly there will be many states that, if simply conjured up at that later time, would *not* be smooth (and if that were not the case, there would have been no need for the inflation to get rid of them). Let us reverse the direction of time for such a state and let the reversed-time dynamical evolution (with equations still allowing for the possibility of inflation, with φ-field, etc.) take over. This must lead us somewhere. Where we are led would generally be some very complicated state which is not at all like a FLRW model, but of the black-hole congealing variety that we considered in the collapsing irregular universe of the previous section. Reversing time back again we find an initial (far from FLRW) white-hole ridden initial singularity that inflation would have been powerless to smooth.

Despite the inadequacy of the arguments from cosmic inflation to explain the uniformity and smoothness of the universe (and hence the particular way in which the Big Bang was of extraordinarily low entropy), there have been observational aspects of the CMB which inflation accounts for in a fairly natural way, and which appear to be difficult to explain without invoking inflation. One of these is the closely *scale-invariant* (Harrison-Zel'dovich) nature of the deviations from uniformity in the primordial matter distribution and also in the temperature distribution in the CMB sky. [13] Inflationary cosmology accounts for this scale invariance essentially for the reason that inflation involves an *exponential* expansion, which has a self-similar character. Another observational fact is that there are correlations between the temperatures at points on the CMB sky which, according to the standard cosmological (Friedmann-Tolman) expansions rates, would never have been in causal contact. This appears to indicate that inflation is required in order for these correlations to be explained.

5. Conformal space-time geometry

Among the alternatives to inflation for accounting for such correlations are "pre-Big-Bang" models for which the correlations in the CMB sky arise because of a kind of "causal contact" which could have occurred in the previous phase of the universe, prior to the Big Bang. Likewise this previous phase could have undergone some kind of exponential behaviour, leading to a scale-invariant input into the structure of the Big Bang. [14,15] The usual idea of a pre-Big-Bang model had been to involve a previous universe phase in which *collapse* to a singularity occurred, followed by a quantum-gravity "bounce" that became our own Big Bang. [16] But a suggestion that I have myself been developing for more than three years now is that of *Conformal Cyclic Cosmology* (CCC) [17] which involves a quite different idea.

The driving geometrical notion underlying CCC is that of *conformal* space-time geometry. This geometry takes the null-cone structure, together with temporal orientation, as being more fundamental than the scale factor that provides an actual measure of time (and space) intervals. Standard general relativity is based on the full *metric tensor* **g** but, in conformal geometry, **g** is determined only up to a local scale factor, so that two metrics $\mathbf{g}$ and $\hat{\mathbf{g}}$ are deemed to be equivalent to each other if there is a smooth scalar field $\boldsymbol{\Omega}$ for which

$$\hat{\mathbf{g}} = \boldsymbol{\Omega}^2\mathbf{g}.$$

It is normally taken that $\boldsymbol{\Omega}$ is positive, but I shall also allow $\boldsymbol{\Omega}$ to become zero or negative in some regions, or even to become infinite (where $\boldsymbol{\Omega}^{-1}$ is smooth). Basically, conformal geometry is the geometry where infinitesimal shapes (or angles) have meaning, but not the infinitesimal notion of *size* that is fundamental to Riemannian geometry. Here we are concerned with *space-time*, so our $\mathbf{g}$ (or, equivalently $\hat{\mathbf{g}}$) has a Lorentzian signature, which I take in the form $(+ - - -)$. The conformal structure is equivalent to the *null-cone* structure, but I also fix a time orientation, so that the future and past null cones are distinguished. This time-oriented Lorentzian conformal structure is equivalent, locally, to the *causal* structure [18] of space-time.

It should be noted that this conformal geometry, in 4 space-time dimensions is defined by 9 numbers per point, these being the 9 independent ratios of the 10 components of $\mathbf{g}$ so, in a sense, the conformal structure is *most* of the metric. Now, there are certain aspects of physics that depend only on the conformal structure of space-time and do not need the scale factor that determines the full metric. Maxwellian electromagnetism is the most important example, the source-free Maxwell equations being invariant [19] under conformal rescalings of the form $\mathbf{g} \to \hat{\mathbf{g}} = \boldsymbol{\Omega}^2\mathbf{g}$, and this invariance holds just as well for a curved space-time metric as for flat space. Conformal invariance extends also to situations when there are (given) charged sources present, so the electromagnetic *interaction* is conformally invariant as well as are the free-field equations. Moreover, this conformal invariance extends to the wave equation governing the quantum description of a massless particle [20] of any spin $\frac{n}{2}$, although when $n \geq 3$ there are matters of consistency, in curved space-time, that complicate things. The case of the gravitational field ($n = 4$) raises particular issues, but we can meaningfully say that the free field is conformally invariant whereas the interactions are not. In rough terms, the key point seems to be that it is *mass* and the interactions that are specifically to do with mass that break conformal invariance.

Another way of addressing the relation between mass and conformal structure is to think of a classical particle picture. A massive particle has a timelike world-line along which the metric $\mathbf{g}$ determines a "length" notion, which is actually the measure of the passage of time for that particle. If we borrow from quantum mechanics Planck's formula $E = h\nu$ and combine it with Einstein's $E = mc^2$ we see that any individual particle is a kind of clock, with frequency $\nu = m(c^2/h)$, or $2\pi\nu = m$ in absolute units, showing us that it is *mass* that determines clock rates, and therefore the scale that fixes a particular metric $\mathbf{g}$ consistent with the given conformal structure. Of course, to make a reliable clock (such as an atomic or nuclear clock – which, indeed, can now be extraordinarily accurate) we need a system involving a great many particles, but the central issue is the same: we need *mass* in order to build a clock, and therefore to pick out a particular $\mathbf{g}$ consistent with the conformal structure.

If we work our way back in time, approaching the physical situation close to the Big Bang, we find that temperatures get so great that the masses of individual particles become irrelevant by comparison (or because we find energies far in excess of the Higgs mass) so it is a not unreasonable assumption that we can treat all the particles in the very early universe as being *massless*. It seems, also, that it is a plausible assumption that all relevant interactions can be regarded as being conformally invariant. Accordingly, we may take the view that, as far as the material contents of the universe were concerned, at the vicinity of the Big Bang, the real geometry of relevance there was *conformal* geometry, rather than Einstein's slightly more restrictive metric geometry. [21] In fact, if we adopt an FLRW model for the initial state of the universe (particularly if we use Tolman's radiation for the energy-tensor source – or even Friedmann's pressure-free "dust"), then we find that the conformal structure extends smoothly to a region *prior* to what we refer to as the Big Bang, where the space-time singularity that was the Big Bang is now stretched out to a smooth hypersurface $\mathscr{Z}$. One might even say that the material contents of the universe could not even clearly "notice" this singular event – for they could have been dynamically propagated across $\mathscr{Z}$ in a consistent way, owing to this conformal invariance.

The basic exception is *gravity*, which interacts with mass-energy in a non-conformally invariant way. But if we wish to express the particular condition on the Big Bang which appears actually to have held pretty closely – and which ensures a very tiny initial entropy for the universe, compared with what it might have been, whence the 2nd Law was set on its course, then we need to ensure that gravitational degrees of freedom were not initially activated. For many years, I have tried to formulate this condition as what I have referred to as the "Weyl curvature hypothesis". This asserts, in rough terms, that the *Weyl curvature* – which is the entirely trace-free part of the Riemann curvature, and which measures the *conformal* space-time curvature (and the free gravitational field) – must be zero at initial-type space-time singularities, though it is unconstrained at final-type singularities (inside black holes).

However, there are various difficulties and ambiguities about the precise mathematical statement of the needed condition if it is formulated in this particular way, so it is fortunate that Paul Tod [22] has developed an alternative approach which is geometrically clear and unambiguous. According to this procedure, the required condition is that the conformal geometry can be extended smoothly to a "pre-Big-Bang" Lorentzian manifold. This is seen through the introduction of a local metric $\hat{\mathbf{g}}$, with $\hat{\mathbf{g}} = \boldsymbol{\Omega}^2\mathbf{g}$, where $\mathbf{g}$ is the conventional Einstein metric, but where $\hat{\mathbf{g}}$ is finite and smooth at the Big Bang (where $\boldsymbol{\Omega} \to \infty$, with $\boldsymbol{\Omega}^{-1}$ smooth) so the singularity is expanded out infinitely, providing us with a *smooth* past boundary $\mathscr{Z}$, which is a spacelike hypersurface instead of a singularity. Just as with the standard FLRW models, the metric $\hat{\mathbf{g}}$ extends smoothly to the other side of $\mathscr{Z}$, so we can say that the *conformal* metric of space-time extends across $\mathscr{Z}$.

The very possibility of performing this operation provides us with a powerful restriction on the conformal geometry at the Big Bang, this restriction being of an appropriate character for a good formulation of the "Weyl curvature hypothesis". Yet, it should be noted that Tod's procedure was introduced just as a kind of "mathematical trick", providing merely a plausible and mathematically elegant way of characterizing the sort of restriction on the Big Bang that could precisely characterize the starting region $\mathcal{B}$, within the phase space $\mathcal{P}$ of possible universes. No *physical* reality had been intended for this earlier phase of the universe.

6. Conformal cyclic cosmology

Here is where CCC adopts a fundamentally different attitude, and the proposal is indeed to assign an actual reality to the postulated space-time prior to the Big Bang. This would appear to provide a situation resembling earlier ideas [23] in which the Big Bang would be replaced by a kind of "bounce" from a previous phase of the universe which was in *collapse* – but in which the 2nd Law had been presumably severely violated, in order for the entropy to get down to the tiny value that it had when emerging from the bounce. The idea behind CCC is more subtle than this, however, and to understand what is involved, we must pass to the other end of the scale of time for our universe and consider, instead, what is to be expected in the very remote *future*.

I assume that the so-called "dark energy" of our universe is simply a *cosmological constant* Λ (as introduced by Einstein in 1927), where $\Lambda > 0$. Then the accelerated expansion that appears to be going on now will continue indefinitely. Galaxies and galactic clusters may be expected to hang together, although much larger conglomerations such as superclusters are likely to disperse. Large black holes in galactic centres will swallow great numbers of stars and grow larger. Many stars will escape from their galaxies, but will remain within the cluster to which its galaxy belongs, eventually to be swallowed by the huge black hole that results from the congealing of those in the various galaxies in the cluster. Other "rogue" stars will, however, escape altogether from the cluster, and will evolve to dead black dwarfs or to individual black holes of a few solar masses. Presumably, most of the dust and accompanying dark matter in the galaxies will eventually be swallowed by the black holes. Ultimately – and here we are thinking in terms of time scales of the order of 10^{100} years, or the like – even the largest of the black holes will evaporate away through mass loss due to Hawking radiation, where almost the entire black holes' mass-energy would ultimately be converted to electromagnetic radiation. A considerable fraction of the black holes' mass-energy would, however, have been converted to *gravitational* radiation during violent encounters between different black holes residing in galaxies of the cluster.

A picture is thereby presented of a final configuration of an expanding universe inhabited largely by massless radiation, and the issue arises as to whether one can adequately treat this ultimate "asymptotic" state of the universe as something which is inhabited only by massless entities. For the purposes of argument, I am going to assume this to be the case, although the status of rogue material such as the aforementioned black dwarfs, or stray individual particles, such as protons, electrons, and neutrinos, must be considered. The strict philosophy of CCC does indeed require that nothing massive is left in the future asymptotic limit, for then we can consider that conformal geometry again becomes the relevant space-time structure. For the purposes of the present discussion, it will be necessary merely to assume that the major part of the universe's contents can be modelled accurately in this way, though a plausible case can be made that the stronger assumption that, in the very far limit, mass will itself die away so that the *strict* CCC philosophy may well ultimately hold true.

A mathematical device for treating the future asymptotic behaviour of an expanding universe, with conformally invariant massless contents, has been around for some 45 years. [24] The idea is to introduce a conformal factor $\mathbf{\Omega}$ which appropriately goes to zero in the future asymptotic limit, so that the metric $\hat{\mathbf{g}} = \mathbf{\Omega}^2\mathbf{g}$ (where $\mathbf{g}$ is the conventional Einstein metric) remains finite and smooth in the asymptotic limit (where now $\mathbf{\Omega} \to 0$). The conformal space-time then acquires a future boundary $\mathscr{I}$ which represents the future "infinity" of the Einsteinian space-time. Here we have a positive cosmological constant, whence this future hypersurface boundary $\mathscr{I}$ turns out to be *spacelike*. (This is advantageous, from the strict mathematical point of view, [25] since problematic issues that arise when $\Lambda = 0$ with $\mathscr{I}$ null, are here avoided.) Both the electromagnetic and the gravitational radiation, escaping ultimately to infinity, will make their significant marks on $\mathscr{I}$, but they do not spoil the smoothness of this bounding hypersurface. Moreover, the conformal geometry of the space-time, in its asymptotic future, extends smoothly to a conformal space-time manifold on the other side of $\mathscr{I}$. As with Tod's formulation of the Weyl curvature hypothesis, no physical reality is intended to be attached to this region of "conformal space-time" on the other side of $\mathscr{I}$, but it is introduced simply as a mathematical convenience.

The central idea of CCC, however, is to take seriously both the space-time to the future of $\mathscr{I}$ obtained here, and that to the past of the Big-bang hypersurface $\mathscr{B}$ that occurs in Tod's proposal. According to CCC, the history of the universe consists of a (perhaps infinite) succession of *aeons*, where the indefinitely expanding remote future of each aeon can be joined smoothly as a conformal manifold to a big bang for the succeeding one. Thus, the "$\mathscr{I}$" of each aeon is identified with the "$\mathscr{B}$" of the next. The conformal geometry is taken to be smooth across the join. The entire succession of aeons taken together provides a conformal manifold which is non-singular in *past* directions, although there will be black-hole space-time regions of divergent Weyl curvature that are singularities in the conformal geometry, and which can be encountered as *future* end-points of particles' timelike world-lines.

Journal of Physics: Conference Series **174** (2009) 012001 doi:10.1088/1742-6596/174/1/012001

7. The 2nd Law and black-hole information loss

Finally, let us come to an important issue which is made particularly manifest in relation to CCC, but it also arises as a fundamental issue for the universe as we know it, quite independently of CCC. This is the question of how the 2nd Law can remain true when we compare the situation close to the Big Bang with that in the very remote future. In CCC we require a very strong correspondence indeed, between the two, since as we pass across the $\mathcal{I}/\mathcal{Z}$ hypersurface – which I shall henceforth call the *crossover surface* – via which each aeon joins to the next, all the relevant fields and conformal geometry of space-time must match exactly, as we pass from $\mathcal{I}$ to $\mathcal{Z}$. We must ask how the entropy at $\mathcal{Z}$ can be so enormously different from the entropy at $\mathcal{I}$, while the two are physically almost identical, so long as we remain in the realm of conformally invariant physics.

It should first be realized that a conformal rescaling of the metric, according to $\hat{\mathbf{g}} = \boldsymbol{\Omega}^2\mathbf{g}$, does *not* affect the volume measure of phase space $\mathcal{P}$. The position coordinates scale in the opposite way from the momentum coordinates, so that the 2-form surface elements $dx \wedge dp$, of which the full (Liouville) volume measure is composed, are *invariant* under the rescaling. Thus, entropy measures are not affected by the conformal rescalings.

On the other hand, the temperature does undergo an enormous change, approaching zero just before crossover and infinity just following it. This is because energy certainly undergoes rescaling, whereas the degrees of freedom do not. Photon goes to photon at the crossover, but each individual photon's energy (and therefore frequency) is completely rescaled at crossover. If we think in terms of a local metric $\hat{\mathbf{g}}$ which is taken so as to be smooth over crossover, then the photon's energy (frequency), with respect to the metric $\hat{\mathbf{g}}$, remains unchanged as it passes through the crossover surface. But in the scaling back to the Einsteinian metric $\mathbf{g}$, we have an enormous stretching just prior to crossover (so we get physical photons which have very low energy) and an enormous squashing just following crossover (so we get physical photons of very high energy).

The situation with gravity is technically a little different, on account of the fact that there is a difference between the conformal scaling factors for the "graviton field" and for the Weyl conformal tensor, [26] the latter passing smoothly through zero at crossover. However, the graviton information also carries through the crossover surface without loss of information (and without change in entropy), where in CCC the information is converted into that carried by a *scalar field* which must be introduced at crossover. This scalar field arises initially simply from the conformal factor $\boldsymbol{\Omega}$, but it is then interpreted as the origin of new *dark matter* for the succeeding aeon (compensating for dark matter swallowed by black holes in the previous aeon).

If we do not adopt CCC, there is still a huge issue to be faced, since although we need not expect such a close correspondence between degrees of freedom at $\mathcal{I}$ and $\mathcal{Z}$ as we do in CCC, it is still very difficult to see how we can account for the enormous entropy difference at these two places. This difference would be far more than 10^{101} which is a very rough estimate of the entropy in black holes that are within our particle horizon now, and we must expect this discrepancy to be greatly increased in the future. The conformal geometry and physical content of the universe at $\mathcal{I}$ and $\mathcal{Z}$ are still rather similar at these two places, irrespective of CCC, and it would appear that there can be very little difference in the entropies at $\mathcal{I}$ and $\mathcal{Z}$ and certainly nothing like 10^{101} or more.

One issue that needs to be brought up at this stage, however, is the huge entropy that is often attributed to the presence of the *cosmological event horizons* that come about from a positive Λ. If we try to think of such a cosmological horizon as being similar to a black-hole horizon (and in certain respects they are indeed similar), then we might want to attribute an entropy to such a horizon in accordance with the Bekenstein-Hawking formula. This would provide us with a value of about 3×10^{122}, if the argument for this entropy is accepted. However, as we can see from the considerations of conformal rescaling just given (telling us that the entropy at crossover is not affected by the conformal scale), this "entropy" does not actually correspond to any actual degrees of freedom – certainly if we take Λ to be a *constant*, as it should be in standard Einstein general relativity. Some cosmologists prefer to take the view that we should take Λ to represent some kind of "dark-energy field" which could be varying in time, and accordingly subject to spatial variation also. This presents us with various theoretical difficulties, however – especially in connection with the "weak energy condition", which is only marginally satisfied by a cosmological constant and would be likely to be violated as soon as $\Lambda\mathbf{g}$ is replaced by a varying field.

There are additional difficulties involved in attempting to attribute a Bekenstein-Hawking entropy to the whole universe on the basis of a particle-horizon area. Although this horizon does indeed have a finite cross-sectional area $12\pi\Lambda^{-1}$ (suggesting a Bekenstein-Hawking entropy of $3\pi\Lambda^{-1}$ in absolute units), it is not at all clear how this "entropy" could actually be interpreted. If it indeed applies to the *whole* universe, and if the universe is spatially infinite, then the single value "3×10^{122}" would provide only an *infinitesimal* contribution to the *finite* part of the universe that has been relevant to our discussion.

However, we need to take into account the fact that cosmological horizons (unlike black-hole horizons) are highly dependent on an observer's world-line, so if we consider that it is "our" world-line that is the relevant one, we might hope to find an entropy value that is relevant just to the ($\sim 10^{80}$-baryon) matter content that lies within our particle horizon. In fact we would need to extend our world-line to infinity, in order to arrive at the cosmological horizon, and this would lead (basing calculations on a constant Λ) to a matter content M within our ultimate particle horizon that would be somewhat over 3 times the volume that is within our present particle horizon. We might expect a cosmological entropy that would roughly correspond to the collapse of M to a black hole. This would be about an order of magnitude greater than the maximum entropy figure of 10^{124}, or so (i.e. $\sim 10^{125}$), for the matter (including dark matter) within our present particle horizon.

This figure is perhaps tantalizingly close (given various uncertainties) to the entropy figure of 3×10^{122} that we arrive at from the cosmological horizon size, although appearing to be somewhat *larger*. A first impression might be that an *equality* between these figures should somehow be a logical necessity. The limiting past light cone of points moving progressively up our world-line is indeed our cosmological event horizon, and it is tempting to think that this is actually a black-hole horizon for all the material M. However, this is not a correct identification, as the null cones along this horizon point in the wrong direction. A more accurate identification would be that this is the black-hole horizon for all the material in the universe which does *not* lie within our ultimate particle horizon, which would be far far larger, so this identification seems to make little sense. Accordingly, my own position (for these and other reasons which I hope to explain elsewhere) is to take the cosmological event-horizon entropy as spurious, or at least as having no dynamical content. The remainder of the discussion given here is in accordance with this standpoint.

In that case, we must look elsewhere for the seeming enormous discrepancy between, on the one hand, what the 2nd Law tells us coupled with the vastness of the entropy increase engendered by the formation of black holes and, on the other, the remarkably close (conformal) similarity between $\mathcal{Z}$ and $\mathcal{I}$. The clue indeed lies with the black holes. Why is this? We might say that the 2nd Law is beautifully exemplified by the entire history of a black hole's evolution, from its initial creation in gravitational collapse – the epitome of an irreversible process – through to its final (presumed) disappearance through Hawking evaporation. The relentless increase of the hole's surface area is seen as a close reflection, within classical general relativity, of the 2nd Law itself, once we have appreciated the Bekenstein-Hawking attribution of entropy to the hole's surface area, this increase being reversed only when the ambient temperature reduces below the Hawking temperature of the hole, so that it becomes thermodynamically favourable for the hole to lose (mass-)energy to the outside world.

Yet all this refers to activity that is proceeding according to the well-understood physics of the world *outside* the black holes' horizons. If we ask what takes place deep within the holes' interiors, we get a different story. What we find is a space-time singularity with none of the simplicity of structure that we found at $\mathcal{Z}$, where it had been not unreasonable to suppose that the conformal structure of space-time could be extended beyond it. Instead, we find a singularity of such a violent nature that its conformal geometry cannot be extended beyond it and, as far as we can tell, all information is destroyed at this singularity.

Although not a great deal is known for sure about the structure of the space-time singularity within a generic collapse to a black hole, there are three things which we must take note of, all of which are certainly plausible. The first is that, on the basis of the conjecture of *strong cosmic censorship*, [10] the singularity should generally have a spacelike character, and all timelike world-lines originating within the horizon and extended as far as possible into the future, will encounter this singularity. The second is that, in accordance with the proposal of Belinskii, Khalatnikov, and Lifshitz, [27] the Weyl curvature will oscillate in an extremely wild and complicated manner, whilst diverging to infinity. Finally, in his original analysis, Hawking [28] deduced the temperature and entropy of a black hole using quantum field theory in a background describing a collapse to a black hole, where he explicitly assumed that all information falling into the hole was destroyed (at the singularity). This information loss is what gives rise to the thermal state of the black hole.

It must be mentioned that in more recent discussions, this information loss has now become a matter of considerable controversy, one of the main reasons being that it implies a breakdown of unitarity. However, I find this argument unconvincing, particularly since the very process of quantum measurement, in standard quantum mechanics, violates unitarity, and since my own view (as expressed in the first section of this article) is that the physical process of quantum state reduction is a gravitationally induced phenomenon, it is reasonable to expect violations of unitarity in other quantum processes involving gravity. Other arguments in favour of information loss [29] have also been put forward in recent years.

In fact, this information loss is, in my view, the crucial ingredient which resolves the conundrum that I am addressing in this section, whereby the 2nd Law is, in effect, *not* violated, even though the physical situation can return to a low-entropy state following the process of black-hole evaporation. To understand how this can be, we must return to Boltzmann's formula $S = k\log V$, and ask what volume this "V" is supposed to be measuring. When a black hole is present, there will be some degrees of freedom internal to the black hole that, sooner or later, will get destroyed by the hole's singularity, but *exactly* when this destruction is considered to have taken place will have no real concern for us. We must certainly consider that it is gone by the time that the black hole finally evaporates away, via Hawking's runaway process at the end of the hole's life. But we should probably not think of this as a "sudden" process which "takes place" at a particular time. It is just that this information gets lost at some stage, and has become irrelevant during the course of the black hole's existence. It has finally gone by the time the hole disappears. The volume "V" (occurring in Boltzmann's formula) that is relevant to physical processes that are *not* caught behind a black hole's horizon may be calculated either by taking into account this interior information or not. If we do, then the volume measure V_{ext}, of relevance to the external processes that concern us and which we would normally use for calculating the entropy of those processes, must be multiplied by the total volume V_{bh} of the phase space describing all the degrees of freedom in possible information caught in the black hole (and due, eventually, for destruction), giving a phase-space volume $V_{\text{ext}} \times V_{\text{bh}}$. But if we do *not* take these degrees of freedom lost inside the black hole into consideration, then we would just use V_{ext}. The difference between these choices merely amounts to whether or not we add the quantity $k\log V_{\text{bh}}$ into our measure of entropy for the physics external to the black hole.

I would guess that a rough estimate for the internal entropy $k\log V_{\mathrm{bh}}$ that disappears, in effect, when the black hole is gone, would generally be not hugely more than the Bekenstein-Hawking entropy of the black hole when it was at its maximum size, but it would be somewhat more than this value, by an amount that would depend upon the hole's thermal history. (A good estimate of the internal entropy, it seems to me, would be to take that of the hypothetical black hole that would have resulted, had there been no loss through Hawking evaporation.) The sum of all these contributions $k\log V_{\mathrm{bh}}$, arising from all separate black holes, could easily account the discrepancy that we are looking for. If we add these contributions in, then we get an entropy value consistent with the 2nd Law holding for any particular aeon (or for the history of the entire universe, if we do not accept CCC). In CCC, these contributions must be removed – an effective "renormalization" of entropy – by the time all black holes have evaporated, so we can consider the entropy at the crossover surface, joining one aeon to the next, to be completely continuous.

My final comment is that however we look at this, we have another powerful reason for believing that the unitary rules of standard quantum mechanics (or quantum field theory) must be transcended when general relativity is involved.

Acknowledgments
I am grateful to many of my colleagues and friends (particularly Paul Tod and James Peebles) for useful comments and to NSF for support under grant PHY 00-90091.

Notes

1. Penrose (2004).
2. Diósi (1984, 1987, 1989).
3. Penrose (1993, 1996).
4. See Penrose (2004), Ex.[21.6] and note 30.37.
5. Colella, Overhauser, and Werner (1975), Bonse and Wroblewski (1983).
6. Unruh (1976).
7. B.S.Kay (private communication).
8. Marshall *et al.* (2003).
9. Compare the discussion given in Ellis (2009), Section 5, for an alternative viewpoint.
10. Penrose (1978).
11. Guth (1981, 1997).
12. Penrose (1990).
13. See, for example, Kolb and Turner (1994).
14. Veneziano (1998).
15. Steinhardt and Turok (2007).
16. See, for example, Ashtekar, Pawlowski, and Singh (2006), Bojowald (2007).
17. Penrose (2006, 2008, 2009).
18. Kronheimer and Penrose (1967).
19. MacLennan (1956).
20. Penrose (1965), Penrose and Rindler (1984).
21. Rugh and Zinkernagel (2007).
22. Tod (2003).
23. See, for example, notes 14, 15, and 16.
24. Penrose (1965), Penrose and Rindler (1986).
25. Friedrich (1998).
26. Penrose (1965), Penrose and Rindler (1984, 1986).
27. Belinskii, Lifshitz, and Khalatnikov (1970, 1972).
28. Hawking (1975, 1976).
29. Braunstein and Pati (2007).

References

Ashtekar A, Pawlowski T and Singh P 2006 Quantum nature of the Big Bang *Phys. Rev. Lett.* **96** 141301

Belinskii V A, Khalatnikov I M and Lifshitz E M 1970 Oscillatory approach to a singular point in the relativistic cosmology *Usp. Fiz. Nauk* **102** 463-500. Engl. transl. in *Adv. in Phys.* **19** 525-73

Belinskii, V.A., Khalatnikov, I.M. and Lifshitz, E.M. 1972 *Soviet Phys. JETP* **62** 1606

Bonse U and Wroblewski T 1983 *Phys. Rev. Lett.* **51** 1401

Bojowald M 2007 What happened before the Big Bang? *Nature Physics* **3**(8) 523-5

Braunstein S L and Pati A K 2007 Quantum information cannot be completely hidden in correlations: implications for the black-hole information paradox *Phys. Rev. Lett.* **98** 080502

Colella R, Overhauser A W and Werner S A 1975 *Phys. Rev. Lett.* **34** 1472

Diósi L 1984 Gravitation and quantum mechanical localization of macro-objects *Phys. Lett.* A **105** 199-202

Diósi L 1987 *Phys. Lett.* A **120** 377

Diósi L 1989 Models for universal reduction of macroscopic quantum fluctuations *Phys. Rev.* A **40** 1165-74

Ellis G F R 2009 Dark energy and inhomogeneity (paper for NEB XIII conference, Thessaloniki 2008)

Friedrich H 1998 Einsteins equation and conformal structure, in *The Geometric Universe; Science, Geometry, and the Work of Roger Penrose* eds S A Huggett, L J Mason, K P Tod, S T Tsou and N M J Woodhouse (Oxford: Oxford U. Press)

Guth A H 1981 Inflationary universe: A possible solution to the horizon and flatness problems *Phys. Rev.* D **23** 347-56

Guth A H 1997 *The Inflationary Universe* (London: Jonathan Cape)

Hawking S W 1975 Particle creation by black holes *Comm. Math. Phys.* **43** 199-220

Hawking S W 1976 Black holes and thermodynamics *Phys. Rev.* D **13**(2) 191

Kolb E W and Turner M S 1994 *The Early Universe* (New York: Perseus Publishing)

Kronheimer E H and Penrose R 1967 On the structure of causal spaces *Proc. Camb. Phil. Soc.* **63** 481-501

Marshall W, Simon C, Penrose R and Bouwmeester D 2003 Towards quantum superpositions of a mirror *Phys. Rev. Lett. 91* 130401; http://link.aps.org/abstract/PRL/v91/e130401

McLennan Jr. J A 1956 Conformal invariance and conservation laws for relativistic wave equations for zero rest mass *Nuovo Cim.* **3** 1360-79

Penrose R 1965 Zero rest-mass fields including gravitation: asymptotic behaviour *Proc. Roy. Soc. London* A **284** 159-203

Penrose R 1978 Singularities of Space-Time, in *Theoretical Principles in Astrophysics and Relativity* eds N R Liebowitz, W H Reid and P O Vandervoort (Chicago: Chicago U. Press)

Penrose R 1990 Difficulties with inflationary cosmology, in *Proceedings of the 14th Texas Symposium on Relativistic Astrophysics* ed E Fenves (New York: N.Y. Acad. Sci.)

Penrose R 1993 Gravity and quantum mechanics, in *General Relativity and Gravitation 13, Cordoba, Argentina. Part 1: Plenary Lectures 1992* eds R J Gleiser, C N Kozameh and O M Moreschi (Bristol and Philadelphia: Inst. of Phys. Publ.) pp 179-89

Penrose R 1996 On gravitys role in quantum state reduction *Gen. Rel. Grav.* **28** 581-600

Penrose R 2004 *The Road to Reality: A Complete Guide to the Laws of the Universe* (London: Jonathan Cape)

Penrose R 2006 Before the big bang: an outrageous new perspective and its implications for particle physics, in *EPAC 2006 Proceedings, Edinburgh, Scotland* ed C R Prior (European Physical Society Accelerator Group, EPS-AG) pp 2759-62 http://accelconf.web.cern.ch/AccelConf/e06/PAPERS/THESPA01.PDF

Penrose R 2008 Causality, quantum theory and cosmology,in *On Space and Time* ed Shahn Majid (Cambridge, UK: Cambridge U. Press) pp 141-195 (ISBN 978-0-521-88926-1)

Penrose R 2009 The Basic Ideas of Conformal Cyclic Cosmology, in *Death And Anti-Death, Volume 6: Thirty Years After Kurt Gödel (1906-1978)* ed C Tandy (Stanford and Palo Alto, Calif.: Ria U. Press) Ch 7 pp 223-242 (ISBN 978-1-934297-03-2)

Penrose R and Rindler W 1984 *Spinors and Space-Time, Vol. 1: Two-Spinor Calculus and Relativistic Fields* (Cambridge, UK: Cambridge U. Press)

Penrose R and Rindler W 1986 *Spinors and Space-Time, Vol. 2: Spinor and Twistor Methods in Space-Time Geometry* (Cambridge, UK: Cambridge U. Press)

Rugh S E and Zinkernagel H (2007) Cosmology and the meaning of time (Symposium, 'The Socrates Spirit' Section for Philosophy and the Foundations of Physics, Hellabaekgade 27, Copenhagen N, Denmark)

Steinhardt P J and Turok N 2007 *Endless Universe: Beyond the Big Bang* (New York: Doubleday)

Tod K P 2003 Isotropic cosmological singularities: other matter models *Class. Quant. Grav.* **20** 521-534 doi:10.1088/0264-9381/20/3/309

Unruh W G 1976 Notes on black hole evaporation *Phys. Rev.* D **14** 870

Veneziano G 1998 A Simple/Short Introduction to Pre-Big-Bang Physics/Cosmology *Preprint* arXiv: hep-th/9802057v2

Fourth International Workshop DICE2008 IOP Publishing
Journal of Physics: Conference Series **174** (2009) 012002 doi:10.1088/1742-6596/174/1/012002

Does wave function collapse cause gravity?

Lajos Diósi

Research Institute for Particle and Nuclear Physics, H-1525 Budapest 114, P.O.Box 49, Hungary

E-mail: diosi@rmki.kfki.hu

Abstract. We give a twist to the assumption - discussed in various earlier works - that gravity plays a role in the collapse of the wave function. This time we discuss the contrary assumption that the collapse of the wave function plays a role in the emergence of the gravitational field. We start from the mathematical framework of a particular Newtonian gravitational collapse theory proposed by the author longtime ago, and we reconciliate it with the classical equivalence principle.

1. Introduction

For a suitable theory of quantum-gravity, one has to revise the standard concept of gravity or, alternatively, the standard concept of the quantum, or both. Let us, e.g., consider the Wheeler-DeWitt [1, 2, 3] equation of quantum-gravity:

$$H\left(g, \frac{\partial}{\partial g}, q, \frac{\partial}{\partial q}\right)\Psi(g,q) = 0 , \tag{1}$$

where H and Ψ are the quantized Hamilton function and the wave function of the universe, resp., in function of the 3-geometry g and the matter fields q. The generic solutions $\Psi(g,q)$ do not adapt any 4-geometry, only the semi-classical approximate solutions do. The wave function $\Psi(g,q)$ does not possess the standard statistical interpretation because there is no room for the concept of measurement in a fully quantized universe. These difficulties need a quantum mechanical remedy. We might suppose a - yet hypothetic - universal decoherence mechanism to make the wave function $\Psi(g,q)$ collapse into 'pointer states' which are, like the semi-classical states, localized in g. The relativistic implementation of this project is completely missing. The non-relativistic task is tractable. We introduce the following Euclidean distance between two mass densities $f(\mathbf{r})$ and $f'(\mathbf{r})$:

$$\|f - f'\|_G^2 = G\int\int[f(\mathbf{r}) - f'(\mathbf{r})][f(\mathbf{s}) - f'(\mathbf{s})]\frac{\mathrm{d}\mathbf{r}\mathrm{d}\mathbf{s}}{|\mathbf{r}-\mathbf{s}|} , \tag{2}$$

where G is the Newton constant. Based on this, a dynamical collapse theory was born long ago [4, 5, 6, 7, 8, 9]. Each collapse violates Einstein's classical equivalence principle. I discuss how its restoration may lead to an emergent Newtonian interaction.

2. Collapse from gravity

At the heart of the collapse theory is the master equation [4, 5] to describe decoherence governed by the distance (2):

$$\frac{d\hat{\rho}}{dt} = -\frac{i}{\hbar}[\hat{H},\hat{\rho}] - \lambda\frac{G}{2\hbar}\int\int[\hat{f}(\mathbf{r}),[\hat{f}(\mathbf{s}),\hat{\rho}]]\frac{d\mathbf{r}d\mathbf{s}}{|\mathbf{r}-\mathbf{s}|} \equiv -\frac{i}{\hbar}[\hat{H},\hat{\rho}] + \lambda\mathcal{D}\hat{\rho}\ , \tag{3}$$

where $\hat{\rho}$ is the density matrix, $\hat{H}$ is the Hamiltonian of the system, and $\mathcal{D}$ is the decoherence term whose coupling will be tuned by the number λ. By adding a suitable stochastic mechanism to the r.h.s., we get the collapse equation [6]:

$$\frac{d\hat{\rho}}{dt} = -\frac{i}{\hbar}[\hat{H},\hat{\rho}] + \lambda\mathcal{D}\hat{\rho} - \frac{1}{\hbar}\int\{\hat{f}-\langle\hat{f}\rangle,\hat{\rho}\}\delta\phi d\mathbf{r}\ , \tag{4}$$

where the auxiliary random field $\delta\phi$ has vanishing mean and a white-noise correlation:

$$\mathbf{M}[\delta\phi(\mathbf{r},t)\delta\phi(\mathbf{s},u)] = \lambda\frac{\hbar G}{|\mathbf{r}-\mathbf{s}|}\delta(t-u)\ . \tag{5}$$

We do not include the Newtonian interaction

$$\hat{H}_G = -\frac{G}{2}\int\int\frac{\hat{f}(\mathbf{r})\hat{f}(\mathbf{s})d\mathbf{r}d\mathbf{s}}{|\mathbf{r}-\mathbf{s}|} \tag{6}$$

into $\hat{H}$ because we want to see if this term can - in some way - emerge from the collapse mechanism (4).

We may, for simplicity, consider the free motion of two (or more) rigid homogeneous balls of mass M and radius R, with c.o.m. canonical variables are $\hat{\mathbf{x}}_1, \hat{\mathbf{p}}_1$ and $\hat{\mathbf{x}}_2, \hat{\mathbf{p}}_2$, resp. Then $\hat{H} = (\hat{\mathbf{p}}_1 + \hat{\mathbf{p}}_2)/2M$ and

$$\hat{f}(\mathbf{r}) = \frac{M}{V_R}\left[\theta(R-|\mathbf{r}-\hat{\mathbf{x}}_1|) + \theta(R-|\mathbf{r}-\hat{\mathbf{x}}_2|)\right]\ , \tag{7}$$

where $V_R = 4\pi R^3/3$. The collapse eq. (4) yields pure states for long time, then the collapse and the kinetic terms balance each other, the wave function becomes a product state of localized pointer states:

$$\psi(\mathbf{x}_1,\mathbf{x}_2) = \psi_1(\mathbf{x}_1)\psi_2(\mathbf{x}_2)\ . \tag{8}$$

3. Pointer states, probe trajectories

The dynamics of the pointer states ψ_1, ψ_2 is independent and only statistically correlated. Therefore we consider a single probe first. In the limit where the spread σ_ψ of $\hat{\mathbf{x}}$ in the given pure state $|\psi\rangle$ satisfies $\sigma_\psi \ll R$, the collapse eqs. (4,5) reduce [6]:

$$\frac{d|\psi\rangle}{dt} = -\frac{i}{\hbar}\frac{\hat{\mathbf{p}}^2}{2M}|\psi\rangle - \frac{\lambda}{2\hbar}M\omega_G^2(\hat{\mathbf{x}}-\langle\hat{\mathbf{x}}\rangle)^2|\psi\rangle + \frac{1}{\hbar}\mathbf{F}(\hat{\mathbf{x}}-\langle\hat{\mathbf{x}}\rangle)|\psi\rangle\ , \tag{9}$$

where $\omega_G = GM/R^3$ and $\mathbf{F}$ is a random auxiliary force of vanishing mean and of white-noise isotropic correlation:

$$\mathbf{M}[\mathbf{F}(t)\circ\mathbf{F}(u)] = \lambda\hbar M\omega_G^2\delta(t-u)\mathbf{1}\ . \tag{10}$$

The latter equation follows from eq. (5) if one compares the stochastic terms of both eqs. (4) and (9) which yields the relationship:

$$\mathbf{F} = -\frac{M}{V_R}\int\nabla\delta\phi(\langle\hat{\mathbf{x}}\rangle + \mathbf{r})\theta(R-r)d\mathbf{r}\ . \tag{11}$$

Observe, however, that if $\mathbf{F}$ and $\delta\phi$ were real external force and field, resp., they would then contribute to the Hermitian terms $(i/\hbar)\mathbf{F}\hat{\mathbf{x}}|\psi\rangle$ and $(-i/\hbar)\int\delta\phi\hat{f}\mathrm{d}\mathbf{r}|\psi\rangle$. But they stand there without the i: the collapse needs the non-unitary mechanism. Yet, this is not the last word about the 'missing i'.

The longtime asymptotic solution of the collapse eq. (9) - when collapse and dynamics reach an equilibrium - is well known [10]. The equilibrium wave packet becomes a complex Gaussian pointer state

$$\psi(\mathbf{x}) = \frac{1}{\sqrt[4]{2\pi\sigma_\infty^2}} \exp\left(i\langle\hat{\mathbf{p}}\rangle\mathbf{x} - \sqrt{-2i}\left(\frac{\mathbf{x}-\langle\hat{\mathbf{x}}\rangle}{2\sigma_\infty}\right)^2\right) \tag{12}$$

of stationary (squared) width $\sigma_\infty^2 = 2(\hbar/M\omega_G\sqrt{\lambda})$. The c.o.m. of the wave packet will perform a sort of 'Brownian' motion:

$$\frac{\mathrm{d}\langle\hat{\mathbf{p}}\rangle}{\mathrm{d}t} = \mathbf{F}\,, \tag{13}$$

$$\frac{\mathrm{d}\langle\hat{\mathbf{x}}\rangle}{\mathrm{d}t} = \frac{\langle\hat{\mathbf{p}}\rangle}{M} + \frac{2\sigma_\infty^2}{\hbar}\mathbf{F}\,. \tag{14}$$

The trajectory of the c.o.m. is the classical inertial one apart from the 'Brownian' motion which becomes ignorable for macroscopic masses. However, our 'Brownian' motion is not the standard one because the relationship $\mathrm{d}\langle\hat{\mathbf{x}}\rangle/\mathrm{d}t = \langle\hat{\mathbf{p}}\rangle/M$ is violated by a diffusion term in (14) therefore the strict continuity of the spatial trajectory is lost. In classical physics this would be a fatal inconsistency, in quantum physics we must cope with it. As for the eq. (13), it yields standard Brownian momentum diffusion (friction is ignored).

We make a crucial observation related to the 'missing i'. The imaginary force $i\mathbf{F}$ in the collapse eq. (9) yields a real force $\mathbf{F}$ in eq. (13). Such miraculous transition $i\mathbf{F} \to \mathbf{F}$ does not hold in general but for the equilibrium shape (12) of the wave function. This will nevertheless allow us to argue that the 'Brownian' trajectories may - in a subtle way - induce the central attractive Newtonian force $g = -GM/r^2$ in the c.o.m. system of the object.

4. Excursion: Ambient pressure from 'Brownian' trajectory

As a very remote analogue, I would like to present the model of another non-standard 'Brownian' motion where the presence and the average value of a central compressing force can exactly be derived from the 'Brownian' motion. Suppose our classical rigid ball of radius R and mass M is put into a thin gas of small rigid ball molecules that will repeatedly hit the 'Brownian' ball whose trajectory becomes a random broken line. You would not think that this broken line encodes the value P of the ambient pressure. Yet it does!

Imagine that you are sitting inside the 'Brownian' ball and experiencing the long sequence $\Delta\mathbf{v}_1, \Delta\mathbf{v}_2, \dots, \Delta\mathbf{v}_k, \dots$ of velocity jumps during a long period T. This knowledge is not yet sufficient to conclude that there must be an ambient pressure. You have to know that you sit inside a rigid ball of mass M and surface $4\pi R^2$ and the velocity jumps are caused by forces perpendicular to the surface, which is indeed the case in all collisions, see Fig. 1. Then you argue that the absolute value $M|\Delta\mathbf{v}_k|$ of each momentum change is caused by a force perpendicular to the surface. The average of these perpendicular forces over a long period T is thus equal to the average of the experienced momentum jumps $M|\Delta\mathbf{v}_k|$. On the other hand, the average of the perpendicular forces is, by definition, equal to the ambient pressure times the surface, which leads you to the exact relationship for the ambient pressure:

$$P = \lim_{T\to\infty} \frac{M}{4\pi R^2} \sum_{k=1}^{n} |\Delta\mathbf{v}_k|\,. \tag{15}$$

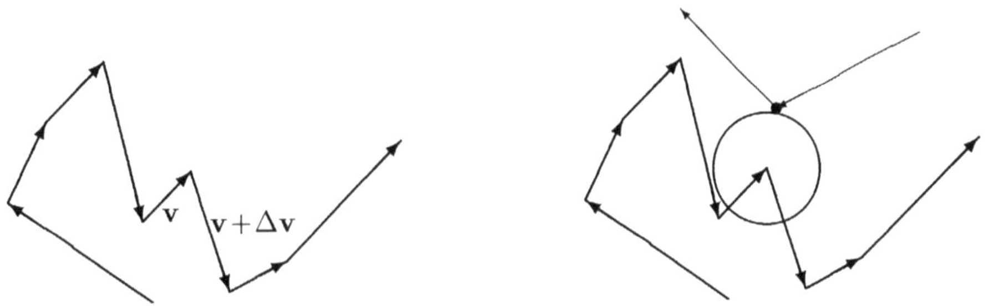

Figure 1. The trajectory of a Brownian mass M is visualized. The lines correspond to free inter-collision motion of the center of mass, break points are due to random separate collisions with the ambient molecules. On the r.h.s., one of the collisions is detailed: an ambient molecule hits the surface of the Brownian mass to cause the momentum transfer $M\Delta\mathbf{v}$.

Can't we use the lessons of the above analogue to reach our original goal? The first complete version [4] of the collapse theory [6] used jump-like randomness instead of the diffusive term of the collapse eq. (4), derived a collapse rule different from (9) in the special case of a rigid ball. Accordingly, the state satisfies the frictional Schrödinger equation:

$$\frac{\mathrm{d}|\psi\rangle}{\mathrm{d}t} = -\frac{i}{\hbar}\frac{\hat{\mathbf{p}}^2}{2M}|\psi\rangle - \frac{\lambda}{2\hbar}M\omega_G^2(\hat{\mathbf{x}} - \langle\hat{\mathbf{x}}\rangle)^2|\psi\rangle + \frac{\lambda}{2\hbar}M\omega_G^2\sigma_\psi^2|\psi\rangle \tag{16}$$

apart from random jumps

$$|\psi\rangle \longrightarrow \frac{\hat{\mathbf{x}} - \langle\hat{\mathbf{x}}\rangle}{\sigma_\psi}|\psi\rangle \tag{17}$$

which occur at the time dependent rate $\lambda M\omega_G^2\sigma_\psi^2$. These random jumps would yield a broken line trajectory (with tiny discontinuities) similar to that on Fig. 1. Unfortunately, the equations of the jump-process have not yet been solved. Therefore we return to the diffusive theory discussed in secs. 2 and 3.

5. Gravity from collapse?

We wish to derive the presence of an average attractive Newtonian field $g = -GM/r^2$ from the features of the random trajectory of the c.o.m. described by eqs. (13,14). Again, we could imagine that we are sitting inside the massive ball and experiencing the random accelerations which we shall attribute to the random force $\mathbf{F}$. This force is not yet the emergent Newtonian interaction force we are looking for. Because of quantum features, the situation is more involved.

Einstein's equivalence principle, that free falling objects move along geodesics, is violated by the collapse mechanism. Can our diffusive trajectories (13,14) become geodesics in a suitable curved space-time? No, they can not since the coordinate diffusion makes the spatial path discontinuous and this can never happen in any space-time continuum. Fortunately, the momentum diffusion of the trajectories can be attributed to curved space-time - in the Newtonian limit - and this will lead us to the emergence of the Newtonian interaction.

But we must learn some quantum subtleties first. In fact, the c.o.m. trajectory $\{\langle\hat{\mathbf{x}}\rangle, \langle\hat{\mathbf{p}}\rangle\}$ of the probe is a mathematical fiction in a sense that it is not observable in a single run, only on a large statistics is $\{\langle\hat{\mathbf{x}}\rangle, \langle\hat{\mathbf{p}}\rangle\}$ testable. Fortunately enough, quantum mechanics tells us the *unique observable.* Consider the following field:

$$\phi(\mathbf{r},t) = \delta\phi(\mathbf{r},t) - 2\lambda G \int \frac{\langle \hat{f}(\mathbf{s})\rangle_t}{|\mathbf{r}-\mathbf{s}|} \mathrm{d}\mathbf{s} \equiv \delta\phi(\mathbf{r},t) + 2\lambda\phi^{scl}(\mathbf{r},t) \ , \tag{18}$$

where we have formally introduced the notation ϕ^{scl} for the $|\psi\rangle$-dependent semi-classical Newtonian field (mean-field) which does not yet play any dynamical role, which is still to play it after our arguments below. Without going into the details of the continuous measurement interpretation [6, 11, 12] of collapse, we invoke the fact that the above field ϕ is the only one available to the observer. Apart from $|\psi\rangle$-independent functionals of the $|\psi\rangle$-dependent ϕ, no other quantities are testable. Testability (observability) has a perfect mathematical criterion. If we completed the r.h.s. of the collapse eq. (4) by a deliberate $|\psi\rangle$-dependent feed-back Hamiltonian, it would not allow for the existence of the master equation like (3) while quantum mechanics requires its existence [13]. The only guarantee of the existence is if we make the feed-back Hamiltonian depend on $|\psi\rangle$ through the above ϕ.

To fulfill or, at least, to approximate Einstein's classical equivalence principle, we are looking for a curved space-time where the equilibrium c.o.m. behaviour of the probes (13,14) will be as close to the the classical geodesics as possible. In the Newtonian limit, our search for the space-time geometry means our search for a certain Newtonian potential $\phi^?$ that we apply to the corresponding classical dynamics of the c.o.m.:

$$\frac{\mathrm{d}\langle\hat{\mathbf{p}}\rangle}{\mathrm{d}t} = -\frac{M}{V_R}\int \nabla\phi^?(\langle\hat{\mathbf{x}}\rangle + \mathbf{r})\theta(R-r)\mathrm{d}\mathbf{r} \ , \tag{19}$$

$$\frac{\mathrm{d}\langle\hat{\mathbf{x}}\rangle}{\mathrm{d}t} = \frac{\langle\hat{\mathbf{p}}\rangle}{M} \ , \tag{20}$$

and we expect it yields similar dynamics to (13,14) obtained from the collapse eq. (4). We cannot expect the coincidence between the eqs. (14) and (20) because coordinate diffusion is a typical quantum effect. We expect the coincidence of the eqs. (13) and (19). Let us compare the forces on the r.h.s. of both, invoke eq. (11), and inspect that the choice $\phi^? = \delta\phi$ would match the forces. Since, however, the field $\phi^?$ describes our space-time geometry therefore $\phi^?$ must be a testable field. The only choice is that $\phi^?$ is a $|\psi\rangle$-independent functional of ϕ. The choice is uniqe:

$$\phi^? = \phi = \delta\phi + 2\lambda\phi^{scl} \ , \tag{21}$$

otherwise the momentum diffusion in eq. (13) remains unmatched. For a single probe, the contribution of ϕ^{scl} cancels from the classical eq. (19) so that it will, as desired, coincide with the collapse result (13). However, in the presence of two (or more) probes, ϕ^{scl} creates interaction between them:

$$\frac{\mathrm{d}\langle\hat{\mathbf{p}}\rangle}{\mathrm{d}t} = \mathbf{F} - 2\lambda\frac{M}{V_R}\int \nabla\phi^{scl}(\langle\hat{\mathbf{x}}\rangle + \mathbf{r})\theta(R-r)\mathrm{d}\mathbf{r} \ . \tag{22}$$

This corresponds to the classical Newtonian interaction, the standard strength will, however, require one half ($\lambda = 1/2$) of the decoherence strength proposed by the author [4, 5, 6]; it corresponds to the choice by Penrose [7, 8, 9].

We are reaching our goal now. The result (22) is still at variance with the collapse dynamics (13) because of the Newtonian interaction term. One has to pay a price for the (partial) fulfillment of the equivalence principle, i.e., for the gravitational explanation of the momentum diffusion caused by the collapse mechanism. We are happy to pay the price: we add the Newtonian interaction $\hat{H}_G$ to the Hamiltonian part of the collapse eq. (4) from the beginning. As a result, the equilibrium (collapse vs. kinetics) c.o.m. trajectories of our quantum probes become classical geodesics - apart from a tiny coordinate diffusion.

6. Discussion, outlook

Ever since the birth of the gravity-related collapse models [4, 6], the author can't stop thinking about an emergent Newtonian interaction. The key concept has always been the re-enforcement of the Einstein equivalence principle for the observable (testable) consequences of the quantum collapse. The present work is a first tentative proposal, using heuristic as well as sophisticated theoretical arguments. The classical corner stone of our construction is Einstein's equivalence principle known to everybody. The quantum corner stone is known to few, that is the uniqueness of the $|\psi\rangle$-dependent testable field in the theory of time-continuous quantum measurement. The present work may leave concerns as to its ultimate consistence, it is nonetheless a useful first attempt to outline the concept and some perspectives.

Some remarks are in order. First, it is obvious that without a topologically fluctuating space-time the coordinate diffusion can never be interpreted classically. In this respect, our success is partial: the emergent space-time structure restores momentum conservation rather than the full principle of equivalence. Second, the construction is sensitive to the numeric strength factor λ of decoherence. The value $\lambda = 1$ allows for an underlying frictional Schrödinger-Newton equation [14, 15], the value 1/2 does not. We desire this innocent looking discrepancy be relaxed by future refinements. Third, we imposed the classical equivalence principle on the c.o.m. trajectories $\{\langle\hat{\mathbf{x}}\rangle, \langle\hat{\mathbf{p}}\rangle\}$ which are not testable, whereas one corner stone of the construction is that the space-time should not depend but on testable quantity ϕ. This is a conceptual flaw. An improved construction shall impose the classical equivalence principle to the testable quantities. Fourth, the emergent Newtonian interaction assumes equilibrium pointer states for the probes. One might think that we predict non-standard gravity for non-equilibrium wave functions which would be a basis for characteristic new phenomena to test. This is not yet the case. If in a simple universe, dominated by rigid ball constituents, most of them are in equilibrium pointer states than we predict standard Newton interaction everywhere.

Acknowledgments

This work was supported by the Hungarian OTKA Grant No. 49384 and No. 75129.

[1] Wheeler J A 1968 in *Battelle Rencontres: 1967 Lectures on Mathematical Physics* (New York: Benjamin)
[2] DeWitt B S 1967 *Phys. Rev.* **160** 1113
[3] Peres A 1968 *Phys. Rev.* **171** 1335
[4] Diósi L 1986 *A quantum-stochastic gravitation model and the reduction of the wavefunction* (in Hungarian) http://www.rmki.kfki.hu/ diosi/thesis1986.pdf
[5] Diósi L 1987 *Phys. Lett.* A **120** 377
[6] Diósi L 1989 *Phys. Rev.* A **40** 1165
[7] Penrose R 1996 *Gen. Relativ. Gravit.* **28** 581;
[8] Penrose R 1998 *Phil. Trans. R. Soc. Lond.* A **356** 1927
[9] Penrose R 2004 *The Road to Reality* (London: Jonathan Cape Publishers)
[10] Diósi L 1988 *Phys. Lett.* A **132** 233
[11] Diósi L 1988 *Phys. Lett.* A **129** 419
[12] Diósi L 1990 *Phys. Rev.* A **42** 5086
[13] Gisin N 1984 *Phys. Rev. Lett.* A **53** 1776
[14] Diósi L 2007 *J. Phys. Conf. Ser.* **67** 011024
[15] van Wezel J and van den Brink J 2008 *Philos. Mag.* **88** 1659

Journal of Physics: Conference Series **174** (2009) 012003 doi:10.1088/1742-6596/174/1/012003

The Montevideo interpretation of quantum mechanics: frequently asked questions

Rodolfo Gambini[1] and Jorge Pullin[2]

[1] Instituto de Física, Facultad de Ciencias, Iguá 4225, esq. Mataojo, Montevideo, Uruguay
[2] Department of Physics and Astronomy, Louisiana State University, Baton Rouge, LA 70803-4001

E-mail: `rgambini@phys.lsu.edu`, `pullin@lsu.edu`

Abstract. In a series of recent papers we have introduced a new interpretation of quantum mechanics, which for brevity we will call the Montevideo interpretation. In it, the quantum to classical transition is achieved via a phenomenon called "undecidability" which stems from environmental decoherence supplemented with a fundamental mechanism of loss of coherence due to gravity. Due to the fact that the interpretation grew from several results that are dispersed in the literature, we put together this straightforward-to-read article addressing some of the main points that may confuse readers.

In a series of recent papers we have introduced a new interpretation of quantum mechanics, which for brevity we will call the Montevideo interpretation. The most complete description published up to date is the essay [1]. Here we clarify some points about the interpretation in the form of a set of questions and answers.

- **Briefly, what is the Montevideo interpretation of quantum mechanics?**
  This interpretation explains the emergence of the classical world via decoherence through the interaction with the environment plus loss of coherence of the quantum theory when studied with real clocks and measuring rods (a pedagogical discussion of why there is loss of coherence when one considers real rods and clocks is in [2]). The combined effect of both losses of coherence implies that all information about quantum coherence in the system plus apparatus plus environment becomes inaccessible.
  After a while, there is no experimental arrangement that is able to decide if the evolution of the state of the complete quantum system was unitary or suffered a collapse. Whenever such situation of *undecidability* is reached, the interpretation assumes that an event (measurement) takes place [1].
- **Is the interpretation realist?**
  This interpretation is realist in the sense that it attempts to yield a description of reality based on available physical theories. We assume that physical reality is constituted by two fundamental elements: the states and the events. The latter include observable events (the observable phenomena). From the latter we define the physical properties of the system, which are all quantitative expressions of phenomena. The states characterize the disposition of the system to produce events with certain probabilities. We admit that certain elements of reality are undecidable, that is, the laws of physics may not describe reality completely (see below).

- **What is this loss of coherence due to real clocks and rods?**
Ordinary quantum theory evolves unitarily in terms of a time parameter t. Such parameter is assumed to be a completely classical variable, which we know with arbitrary accuracy. In reality, no variable in nature is purely classical, so when we use a real clock, we are not measuring t but a new variable T. Even if one chooses a very classical clock, trying to correlate T with t as well as possible, such a variable has some level of quantum fluctuations and other inaccuracies. Our inability to measure time accurately implies that pure states evolve into mixed states, the superposition being that "a state at time T" really corresponds, due to the clock's inaccuracies, to a probabilistic distribution of several possible values of t upon evolution. The spread of the distribution grows as the system evolves. Similar effects arise in quantum field theories when one measures the spatial coordinates. A pedagogical introduction can be seen in [2].
- **Couldn't one eliminate the effects by using better clocks?**
At some point one will encounter fundamental limitations to the accuracy of a clock, so the effect will always be there. Some authors have argued that the fundamental limits are surprisingly large (in Planck units), by noting that every measurement requires the expenditure of energy and if you expend too much energy trying to be accurate your clock becomes a black hole. This is the rough argument, there are variations on it in the literature and they all seem to come to remarkably similar limits [3]. This particular limit states that the inaccuracy of a clock that measures a time T is given by $\delta T \sim T_{\text{Planck}}^{2/3} T^{1/3}$, where $T_{\text{Planck}} \sim 10^{-44}s$. One sees that for laboratory typical times of a few hours, $\delta T \sim 10^{16} T_{\text{Planck}} \sim 10^{-28}s$.
- **But aren't those fundamental limits quite questioned in the literature?**
Indeed (see [4]). We have, however, carried out detailed calculations [5] with much more conservative estimates on the clock inaccuracies, for instance that the fluctuations in the clock grow with any power of the time measured —even close to zero—. We find that the effects we are discussing are still present even for systems having a number of degrees of freedom several orders of magnitude smaller than what we would consider "classical world".
- **Does one need to involve quantum gravity for all this?**
Not really. It is true that in quantum gravity the theory is generally covariant and one naturally accepts that there is no absolute time t (the so-called "problem of time" [6]). A favored solution to this problem is to build the theory relationally by choosing a clock from among the physical variables [7] (for readers with more interest in how the fundamental decoherence relates to the problem of time in quantum gravity, see the next to last question). Gravity also plays a role in setting a fundamental limit to how accurate a clock can be. Up to now detailed models of this from quantum gravity are lacking, the bounds placed are phenomenological semi-classical estimates. That in part justifies why they are criticized in the literature.
- **But ultimately, you are just talking about loss of coherence. In addition to the degrees of freedom of the environment you now just added the degrees of freedom of the clock. How could that change things so much?**
What makes things different is that a clock is a very special set of degrees of freedom (in practice it is also a quite large set of degrees of freedom, at least if one wants it to remain accurate for a while). These degrees of freedom are being used to establish a property of the rest of the system. This is very different from the degrees of freedom of the environment. A clock is also special in that it is supposed to stay accurate during the lifetime of the experiment one studies (and better yet, beyond). Real clocks set limits on what is experimentally accessible, in particular from a pure state after it has evolved for a certain amount of time.

- **One of the objections to environmental decoherence is that the evolution of the "system plus measuring device plus environment" viewed as a closed system is unitary and one could recover quantum coherence eventually, by waiting a long time**
The loss of coherence due to the loss of the clocks' accuracy is such that it gets worse the longer one waits. So waiting longer does not help. In fact it guarantees that the "revivals" of coherence several authors have studied for closed macroscopic systems are killed off by the loss of coherence due to the clock. A concrete calculation was worked out in [5, 8].
- **d'Espagnat [9] has argued in models like Zurek's [10] that one could in principle measure global observables of the system that would indicate if collapse has occurred, even without the need of observing "revivals". Can't the observables be used to decide if the system collapsed or not?**
It is obviously difficult to attempt to give a general answer to such a question since global observables require including quantum measurements of properties of the environment, which are in practice virtually impossible to carry out. We have tried to model the situation in a modification of Zurek's example where one has some control of the environment. This allows a chance at least in principle of measuring the global observable proposed by d'Espagnat. We have found that fundamental limitations of quantum mechanics and the use of real clocks disallow its measurement [5]. Notice that, at least in this example, these limitations are not practical in nature, but fundamental, resulting from applying the laws of quantum mechanics.
- **But there are other problems with the solution to the measurement problem by decoherence. What about the and/or problem as Bell [11] called it, or "the problem of outcomes"?**
The "problem of outcomes" is that although decoherence may have taken place, taking the system plus measurement apparatus to a state in which their density matrix is quasi-diagonal, the system remains in a superposition of states, each of them associated with different macroscopic behaviors. States corresponding to the diagonal of the density matrix, coexist with each other. How does one go from this situation to the classical world, where one of those alternatives and one only is realized? How does one go from a situation where one has alternatives "A *and* B" to "*either* A *or* B"? In our approach we argue that there exist situations where the decoherence due to the environment and the use of real rods and clocks implies that the "A *and* B" provided by unitary evolution and the "*either* A *or* B" result attributed to reduction are indistinguishable. When one reaches this state of *undecidability*, an event has taken place, without affecting any experimental outcome.
- **What happens to the quantum state, does it collapse or not?**
Since we claim an event (measurement) has happened when one cannot decide if the state has collapsed or has evolved unitarily, one clearly cannot decide if it collapsed or not. We have proposed that a point of view is that both alternatives be accepted as possible for the system. Entering the realm of philosophy, this means we are taking a "regularist" point of view. In philosophy there are different attitudes that have been taken towards the physical laws of nature (see for instance [12]). One of them is the "regularity theory", many times attributed to Hume [13]; in it, the laws of physics are statements about uniformities or regularities of the world and therefore are just "convenient descriptions" of the world. Ernest Nagel in *The Structure of Science* [14] describes this position in the following terms: *"...according to Hume [physical laws consist] in certain habits of expectation that have been developed as a consequence of the uniform but de facto conjunctions of [properties]."* The laws of physics are dictated by our experience of a preexisting world and are a representation of our ability to describe the world but they do not exhaust the content of the physical world. A second point of view sometimes taken is the "necessitarian theory" [12], which

states that laws of nature are "principles" which govern the natural phenomena, that is, the world "necessarily obeys" the laws of nature. The laws are the cornerstone of the physical world and nothing exists without a law. The presence of the undecidability we point out suggests strongly that the "regularity theory" point of view is more satisfactory since the laws do not dictate entirely the behavior of nature. Notice that the freedom of the system to collapse or not is not governed by any probabilistic rule. This introduces a notion of free will into our description of nature [1].

- **Can the distinction between collapse or unitary evolution be relevant in spite of the fact that it is not observable?**
  Undecidability does not imply that the difference between these two scenarios is irrelevant. For instance, there may exist complex systems (an intriguing example are living organisms) for which it is impossible to prepare their initial state or consider ensembles. For such systems, undecidability may occur among widely different states. No matter what is the outcome, reduction or unitary evolution, the choice between these alternatives could produce observable phenomena later on. The specific outcome may have important consequences on the occurrence of future events.
- **Does your point of view argue against the many-worlds interpretation?**
  Not necessarily. The many-worlds interpretation is compatible with what we are claiming. However, it becomes less compelling. The many-worlds interpretation was introduced in part to address the "and/or problem" by assigning different worlds to different alternatives. Since we are given a criterion for the passage from and to or, the other alternatives (and other worlds) are therefore not really necessary. That does not mean that they are precluded by our point of view.
- **Does your point of view argue against the "modal" interpretations?**
  Well, let's see. Dieks [15] lists the following set of properties as shared by all modal interpretations:
  ◇ The interpretation is based on the standard formalism of quantum mechanics. [OK, plus quantum gravity].
  ◇ The interpretation is realist, that is, it aims at describing how reality would be if quantum mechanics were true. [OK].
  ◇ Quantum mechanics is a fundamental theory, which must describe not only elementary particles but also macroscopic objects. [OK].
  ◇ Quantum mechanics describes single systems: the quantum state refers to a single system, not to an ensemble of systems. [OK].
  ◇ The quantum state of the system (pure state or mixture) describes the possible properties of the system and their corresponding probabilities, and not the actual properties. The relationship between the quantum state and the actual properties of the system is probabilistic. [OK].
  ◇ Systems possess actual properties at all times, whether or not a measurement is performed on them. [NO, actual properties only appear when the system becomes undecidable].
  ◇ A quantum measurement is an ordinary physical interaction. There is no collapse: the quantum state always evolves unitarily according to the Schrödinger equation. [NO, this cannot be decided]
  ◇ The Schrödinger equation gives the time evolution of probabilities, not of actual properties. [OK].
  As we see, there are a lot of points in common, but not all properties of a "modal" interpretation are satisfied by ours. In modal interpretations a central problem is the determination of when events occur. There have been many proposals for such a point and there has not emerged a unanimous consensus on this (see [16] and references therein)

- **A friend of mine works in string theory and claims the theory is unitary and your effect therefore is wrong.**
String theory has been studied in many situations and in all cases where it makes sense to talk about an evolution as a function of a parameter, such evolution is unitary. In these situations however, we are just like in ordinary quantum mechanics, when formulated in terms of a classical parameter t (ok, the story is slightly more complicated since we are talking about a field theory but we mentioned the effect is present there too, supplemented with additional spatial decoherence effects). Our point is that such classical parameters are not accessible in reality. So if one were to formulate string theory in terms of a real clock T one would have the same types of effects we discuss here. Ask your friend if she believes that string theory solves the "problem of time" of quantum gravity. The reply will likely be that the problem persists, it is just that string theorists tend to concentrate on other issues and situations, where the problem of time is not relevant.
- **This loss of coherence of yours could be made arbitrarily large by choosing a very inaccurate clock. Has this been observed experimentally?**
Actually yes, although no experimentalist has deliberately used a bad clock to observe the loss of coherence, certain correlations in experiments involving Rabi oscillations can be interpreted as using some of the atoms as "bad clocks" and indeed one sees loss of coherence like the one we mention [17].
- **Will the fundamental loss of coherence arise naturally in a technical treatment of the problem of time in quantum gravity, or is it a choice?**
It does. In general relativity there is no notion of absolute time. In fact, there is no absolute notion. All physical predictions have to be formulated as relations between physical quantities. This has been recognized since the early days of general relativity through Einstein's "hole argument" [18]. In particular the notion of time has to emerge "relationally". One possible way of attacking this was introduced by Page and Wootters [19]. In their proposal one takes any physical quantity one is interested in studying and chooses another physical variable that will act as "clock". One then studies how the first variable "evolves as a function of the second one". In this view, time does not play any preferred role among other physical variables. This is in contrast to ordinary quantum mechanics where one has to unnaturally assume that time is supposed to be the only variable in the universe not subject to quantum fluctuations. In spite of the simplicity and naturalness of this proposal to tackle the problem of time in quantum gravity, technical problems arise. The problems are related with what one considers to be physical quantities in a theory like general relativity. Usual things that one may consider physical quantities, like "the scalar curvature of space-time at a given point" are not well defined objects in general relativity. The problem is what is "a given point"? Points in space have to be defined physically in general relativity. One can characterize a point as a "place where something physical happens" (for instance, a set of physical fields takes certain values). Then one could ask "how much is the curvature at that point". The end result is indeed physical. But it is again a relation between the values of curvatures and fields. Such relation is given and immutable. How could one construct a clock out of something immutable? It appears that the only things that are physical are immutable relations and the only things that evolve are the members of the relations, like the curvatures and fields. In technical terms, what one can consider as physical observable in general relativity is a quantity that is left invariant under the symmetries of the theory, or in the canonical language, that commutes with the constraints. Since one of the constraints is the Hamiltonian, physical quantities do not evolve. Therefore they cannot work as clocks. This created problems [6] for the Page–Wootters proposal. A way out was sought by trying to establish relations not between physical quantities but between mathematical quantities one uses to describe the

theory that are not directly measurable (like for instance, the components of the metric at a point). Far from helping, this led to significant technical problems since one ends formulating the theory in terms of unobservable quantities. Ultimately it was shown in model systems that the proposal cannot be used to compute elementary things, for instance quantum probabilities of transition [6].
The observation we have recently made [7] is that the Page–Wootters construction can be rescued by using Rovelli's proposal of "evolving constants of the motion" [20], a concept that can be traced back to DeWitt, Bergmann and Einstein himself. This idea is to introduce genuinely observable physical quantities, i.e. relations between magnitudes as we highlighted above, but that depend on a continuous parameter. If one imagines evolution as changes in such a parameter, one can actually construct the relational description of Page and Wootters and show that it actually leads to the correct quantum probabilities of transition, at least in model systems [7]. The beauty of the complete construction is that the continuous parameter in the evolving constants completely drops out at the end of the day and the formulation remains entirely written in terms of truly observable physical quantities, even in the extreme situations that can develop in physics when quantum gravity effects become important. The calculation of the propagator with this construction, at least in model systems, yields the usual propagator at leading order, but one has corrections to it due to the mechanisms of loss of coherence that we are referring to in this paper.

- **What is a good reference for all this?**
  We have an Fqxi essay where these ideas are presented for non-experts [1]. A pedagogical detailed discussion of the fundamental loss of coherence is in [2].

Acknowledgments

This work was supported in part by grants NSF-PHY-0650715, FQXi, CCT-LSU, Pedeciba, ANII PDT63/076 and the Horace Hearne Jr. Institute for Theoretical Physics.

References

[1] Gambini R and Pullin J 2009 Free will, undecidability, and the problem of time in quantum gravity – FQXi essay *Preprint* arXiv:0903.1859
[2] Gambini R, Porto R and Pullin J 2007 *Gen. Rel. Grav.* **39** 1143
[3] Wigner E 1957 *Rev. Mod. Phys.* **29** 255; Ng Y J and van Dam H 1995 *Annals N.Y. Acad. Sci.* **755** 579 (*Preprint* arXiv:hep-th/9406110); *do.* 1994 *Mod. Phys. Lett.* A **9** 335; Károlhyázy F, Frenkel A and Lukács B 1986 *Quantum concepts in space and time* ed R Penrose and C Isham (Oxford: Oxford U. Press); Lloyd S and Ng J 2004 *Scientific American* November 2004
[4] Requardt M 2008 *Preprint* arXiv:0807.3619
[5] Gambini R, García Pintos L and Pullin J 2009 Undecidability and the problem of outcomes in quantum measurements *Preprint* arXiv:0905.4222
[6] Kuchař K 1992 *Proceedings of the 4th Canadian conference on general relativity and relativistic astrophysics* ed G Kunstatter, D Vincent and J Williams (Singapore: World Scientific) – online version http://www.phys.lsu.edu/faculty/pullin/kvk.pdf
[7] Gambini R, Porto R A, Pullin J and Torterolo S 2009 *Phys. Rev.* D **79** 041501 (*Preprint* arXiv:0809.4235)
[8] Gambini R and Pullin J 2007 *Found. Phys.* **37** 1074 (*Preprint* arXiv:quant-ph/0608243)
[9] d'Espagnat B 1995 *Veiled reality* (New York: Addison Wesley)
[10] Zurek W 1982 *Phys. Rev.* D **26** 1862
[11] Bell J 1988 *Speakable and unspeakable in quantum mechanics* (Cambridge, UK: Cambridge U. Press)
[12] See for instance: *Laws of Nature* The Stanford encyclopedia of philosophy (http://plato.stanford.edu/entries/laws-of-nature/) and references therein. Also N. Swartz N 2003 *The concept of physical law* second edition (available online at http://www.sfu.ca/philosophy/physical-law/)
[13] Hume D 1999 *An Enquiry concerning Human Understanding* (USA: Oxford U. Press)
[14] Nagel E 1979 *The structure of science: problems in the logic of scientific explanation* (New York: Hackett)
[15] Dieks D 2007 *Stud. Hist. Phil. Mod. Phys.* **38** 292-310
[16] Lombardi O and Castagnino M 2008 *Stud. Hist. Phil. Mod. Phys.* **39** 380 (*Preprint* arXiv:quant-ph/0610121)

Journal of Physics: Conference Series **174** (2009) 012003 doi:10.1088/1742-6596/174/1/012003

[17] Bonifacio R, Olivares S, Tombesi P and Vitali D 2000 *Phys. Rev.* A **61** 053802 (*Preprint* arXiv:quant-ph/9911100)
[18] Iftime M and Stachel J 2006 *Gen. Rel. Grav.* **38** 1241 (*Preprint* arXiv:gr-qc/0512021)
[19] Page D N and Wootters W K 1983 *Phys. Rev.* D **27** 2885
[20] Rovelli C 1991 *Phys. Rev.* D **43** 442

Measurements of non local weak values

Aharon Brodutch[1,2] **and Lev Vaidman**[1]
[1]Raymond and Beverly Sackler School of Physics and Astronomy, Tel Aviv University, Tel Aviv, Israel
[2]Center for Quantum Computer Technology, Faculty of Science, Macquarie University, Sydney NSW 2113, Australia
E-mail: abrodutc@ics.mq.edu.au

Abstract. Some recent attempts at measuring non local weak values via local measurements are discussed and shown to be less robust than standard weak measurements. A method for measuring some non local weak values via non local measurements (non local weak measurements) is introduced. The meaning of *non local weak values* is discussed.

1. Introduction

The concept of *observable* or *variable* in quantum mechanics is not as simple and intuitive as in classical physics and it requires definition. While in classical physics every observable has a value for any state of the system (which might be known or not known to the observer), in quantum mechanics we cannot associate a value for every observable. Instead, a measurement procedure is defined and every variable is associated with a set of *eigenvalues*, the possible outcomes of its measurement. Every state of a quantum system is associated with probability distributions for outcomes of measurements of every variable. It might happen that for some variables the probability distribution is singular, i.e., a particular eigenvalue is obtained with certainty and in this case, this eigenvalue can be associated with the variable as in the classical case. Otherwise, a statistical expectation value can be associated with a variable.

The definition of a quantum measurement procedure is, therefore, crucial for the concept of the value of a variable in quantum mechanics. The standard model of quantum measurement is the Von Neumann procedure [1] which consists of a short interaction between the system and a measuring device. The quantum measurement of a variable O is described by the Hamiltonian :

$$H = g(t)PO, \tag{1}$$

where P is the momentum conjugate to the pointer variable of the measuring device Q, and the normalized coupling function $g(t)$ specifies the time of the measurement interaction. The outcome of the measurement is the shift of the pointer variable during the interaction. In an ideal measurement the function $g(t)$ is nonzero only during a very short period of time, and the free Hamiltonian during this period of time can be neglected.

A discussion of ideal and non ideal measurements requires a few definitions. An *ideal, nondemilition, instantaneous*, measurement is defined as a Von Neumann measurement that leaves the state of a system, which initially was in an eigenstate of the measured observable, unchanged. A faithful demolition measurement is one that gives the required result but disturbs the system by changing its state. Nonlocality too needs definition in this context. A nonlocal system is one that has two or more parts placed in separate locations. A local system is one that exists in a single location. Local interactions are those that can be created between two systems in the same place.

Formally, in non relativistic quantum mechanics one can consider any Hamiltonian and thus one can measure any variable. However, relativistic quantum mechanics limits us to local interactions and thus we cannot construct the Hamiltonian (1) for a variable O related to a composite system with parts placed in separate locations. Nevertheless, measurements of some nonlocal variables are possible [2] in the sense that at the end of an instantaneous measurement procedure the composite system ends up in an eigenstate of this variable and the information about the eigenstate is written down (although in separate locations). We call this procedure a *nonlocal measurement.*

For local measurements we can consider the standard Von Neumann procedure with weakened coupling. In such measurements the pointer does not point sharply to zero before the interaction, and it does not point sharply to an eigenvalue after the interaction. The probability distribution of the pointer after the interaction points at the expectation value. Such measurements are called *weak measurements.* Particularly interesting are weak measurements performed on pre and post-selected quantum systems. For a quantum system pre-selected in a state $|\Psi\rangle$ and post-selected in a state $|\Phi\rangle$ the probability distribution of the pointer variable points to (a real part of) the *weak value* [3]

$$O_w \equiv \frac{\langle\Phi|O|\Psi\rangle}{\langle\Phi|\Psi\rangle}. \quad (2)$$

In some cases weak values might be much larger than the eigenvalues of O. This amplification is sometimes called the Aharonov-Albert-Vaidman (AAV) effect.

The simple formula above defines the weak value of any variable, including non local variables. In this paper we consider a possibility of combining the ideas of weak and nonlocal measurements to measure such non local weak values. This work was inspired by recently introduced concept of *joint weak measurements* [4]. In this procedure a new analysis of the readings on local measuring devices which performed local weak measurements allows to calculate the weak values of the product of two variables related to separate parts of the system. We analyze this procedure and show that it lacks some of the fundamental features of weak measurements and requires much larger resources than standard weak measurements. We show that some non local weak values can be measured using *non local weak measurements* and introduce a method for making such measurements. We argue that only values (local or non local) that can be measured directly using a weak measurement can be thought of as weak values, while those that are measured indirectly are just the result of a calculation leading to the result of (2).

2. Nonlocal measurements

The special theory of relativity limits us to local interaction, so if A is a nonlocal variable related to separate locations, the interaction described by (1) does not exist in nature. Nevertheless there are some nonlocal measurements which can be performed. A composite measuring device is used, with parts near every location of the variable A.

For example, a nonlocal variable $A = \sum B_i$ where each variable B_i is related to location i can be measured with the Hamiltonian

$$H = g(t) \sum P_i B_i, \quad (3)$$

where P_i is the momentum conjugate to the pointer variables Q_i located in location i.

If initially all pointer variables are well localized around zero, we can learn the outcome of a measurement of A from the sum of readings of all local measuring devices $A = \sum Q_i$. This procedure is a faithful measurement of A, but it is not an ideal measurement of A. In an ideal measurement a system initially in an eigenstate of A should remain unchanged after the measurement. This is usually not the case for the method described above. For example, if our system consists of two spin$-\frac{1}{2}$ particles in a singlet state $\frac{1}{\sqrt{2}}(|\uparrow\rangle|\downarrow\rangle - |\downarrow\rangle|\uparrow\rangle)$ and $A = \sigma_{1z} + \sigma_{1z}$, then we will learn from our measurement that $A = Q_1 + Q_2 = 0$, but the singlet state will be changed either to $|\uparrow\rangle|\downarrow\rangle$ or to $|\downarrow\rangle|\uparrow\rangle$.

In order to perform the nonlocal measurement we should do the following [2]. Instead of localizing all pointers Q_i we start with the following entangled state of the measuring device:

$$|\Psi\rangle_{in}^{MD} = |\sum Q_i = 0, P_1 = P_2 = ... = P_N\rangle \tag{4}$$

After the interaction, we can read the value of A as before: $A = \sum Q_i$, but now the eigenstates of A remain unchanged.

It has been shown [2] that beyond the measurement of a sum of local variables separated in space one can also perform a measurement of a modular sum of local variables. But there are nonlocal variables which cannot be measured in a non-demolition way [2]. In particular, consider a product, $\sigma_z^A\sigma_z^B$, of spin components of two separate spin-1 particles, one in Alice's hand and another in Bob's hand.We now prove that the possibility of an ideal non-demolition measurement of that variable contradicts causality, and thus, such a measurement does not exist.

Before time $t = 0$ we prepare the two particles in the initial state

$$|\Psi\rangle_{in} = \frac{1}{\sqrt{2}}(|-1\rangle_A + |0\rangle_A)\ |0\rangle_B \tag{5}$$

We assume that at time $t = 0$ somebody performs a non-demolition measurement of $\sigma_z^A\sigma_z^B$. Immediately after $t = 0$ Alice performs her projective local measurement of the state $\frac{1}{\sqrt{2}}(|-1\rangle_A + |0\rangle_A)$. Bob, who has access to particle B can send a superluminal signal to Alice in the following way. Just before $t = 0$ he decides to change the state of his spin to $|1\rangle_B$ or to leave it as it is $|0\rangle_B$. If he decides to do nothing, then a nonlocal measurement of $\sigma_z^A\sigma_z^B$ will not change the state of the particles because state (5) is an eigenstate of $\sigma_z^A\sigma_z^B$. Therefore, Alice, in her local projective measurement will find the state $\frac{1}{\sqrt{2}}(|-1\rangle_A + |0\rangle_A)$ with certainty. However, if Bob decides to change the state of his spin to $|1\rangle_B$ the initial state before the nonlocal measurement at $t = 0$ will be changed to

$$|\Psi'\rangle_{in} = \frac{1}{\sqrt{2}}(|-1\rangle_A + |0\rangle_A)\ |1\rangle_B \tag{6}$$

It is not an eigenstate of $\sigma_z^A\sigma_z^B$ and thus after the measurement we will end up with equal probability with the state $|-1\rangle_A\ |1\rangle_B$ or $|0\rangle_A\ |1\rangle_B$. In both cases the probability of obtaining a positive outcome in Alice's's projective measurement on the state $\frac{1}{\sqrt{2}}(|-1\rangle_A + |0\rangle_A)$ is just one half. Instantaneous change of probability of a measurement performed by Alice breaks causality, therefore instantaneous measurement of $\sigma_z^A\sigma_z^B$ is impossible.

Note that not any product is unmeasurable. If instead of two spin-1 particles we consider two spin-$\frac{1}{2}$ particles then the product $\sigma_z^A\sigma_z^B$ is measurable. Indeed, we can express this product as a modular sum: $\sigma_z^A\sigma_z^B = (\sigma_z^A\sigma_z^B)mod4 - 1$ and every modular sum is measurable.

If we relax the requirement of an ideal measurement that it should be a non-demolition measurement, (which is a very rare property of real quantum measurements) and require only that it gives us a faithful result, then, conceptually, there are no constraints on measuring nonlocal variables. Given a large enough resource of entanglement we can "teleport" the quantum states of all separate parts of the system to one location and perform a measurement of any variable [5]. This is not a real teleportation which requires sending classical bits, this procedure can be performed instantaneously. Of course, the result of measurement can only be read later, when the results of the local measurements will be brought together. If the system is in addition pre and post-selected, the procedure should be slightly modified, although (somewhat surprisingly) we need not add a lot of entanglement resources [6].

3. Weak measurements

A weak measurement is a standard Von Neumann measurement with weakened interaction. One of the ways to weaken the interaction is to prepare the state of the measuring device in such a way that P is very small. A good model of weak measurement is given by the coupling (1) with the initial state of the pointer variable a given by a Gaussian centered around zero:

$$\Psi_{in}^{MD}(Q) = (\Delta^2\pi)^{-1/4}e^{-Q^2/4\Delta^2}. \tag{7}$$

with the position uncertainty Δ large ensuring a small P. We will henceforth use this model to describe weak measurements.

Weak values might lie very far from the range of the eigenvalues. For example, a spin half particle prepared in the x direction and post-selected in an almost orthogonal state $\frac{1}{2}(\cos(\frac{\pi}{4}+\epsilon)|\uparrow\rangle - \sin(\frac{\pi}{4}+\epsilon)|\uparrow\rangle$ will yield a very large weak value for σ_z [8]. Even in these cases there is an obvious shift of the pointer variable by the weak value (2). This shift gives weak values their significance [9].

While the meaning of weak values remains controversial [10, 11, 12], the justification of considering weak values as a description of the pre and post-selected quantum systems relies on the universality of the influence of the coupling to a variable in the limit of its weakness. The pointer variable prepared in a natural way (see Jozsa for some limitations [13]) shifts due to a weak measurement coupling as if it were coupled to a classical variable with the value equal to the weak value.

There have been numerous experiments showing weak values [14, 15, 16, 17, 18], mostly of photon polarization and the AAV effect has been well confirmed. Hosten and Kwiat [19] applied weak measurement procedure for measuring spin Hall effect in light. This effect is so tiny that it can not be observed without the amplification.

We must not forget that there is a method for performing any (demolition) measurement [5]. Thus, given large enough ensemble we can, in particular, measure the two-state vector at the time of the measurement. This is a complete description of a pre and post-selected quantum system, so it allows to calculate any function of the pre and post-selected state, and among others, the weak value or rather the result of (2).

4. Local and nonlocal measurements of the sum of local variables

We will start with the simplest example of non local variables, the sum of two local variables. Using the expression (2) we get

$$(A+B)_w = \frac{\langle\Phi|A+B|\Psi\rangle}{\langle\Phi|\Psi\rangle} \tag{8}$$

which in turn gives us the simple relation:

$$(A+B)_w = A_w + B_w \tag{9}$$

Thus, in order to find the value of $(A+B)_w$ we can measure A_w and B_w locally and just add the two numbers. This fact is somewhat surprising because the analogous relation for an expectation values of the measuring devices performing strong measurements on pre and post-selected systems (which we signify as $\langle \quad \rangle_{\Phi\Psi}$) does not hold. In general

$$\langle A+B\rangle_{\Phi\Psi} \neq \langle A\rangle_{\Phi\Psi} + \langle B\rangle_{\Phi\Psi} \tag{10}$$

Consider the following example . At time t_1 two spin$-\frac{1}{2}$ particles are prepared in a state

$$|\Psi\rangle = \sqrt{\frac{1}{2+\epsilon^2}}(|\uparrow\rangle_A|\downarrow\rangle_B + |\downarrow\rangle_A|\uparrow\rangle_B) + \epsilon|\uparrow\rangle_A|\uparrow\rangle_B \tag{11}$$

Later, at time t_2, the particles are found in a state

$$|\Phi\rangle = \sqrt{\frac{1}{2+\epsilon^2}}(|\uparrow\rangle_A|\downarrow\rangle_B - |\downarrow\rangle_A|\uparrow\rangle_B) + \epsilon|\uparrow\rangle_A|\uparrow\rangle_B \tag{12}$$

We can use the Aharonov-Bergmann-Lebowitz (ABL)[20] formula for calculating the probabilities for the outcomes of strong intermediate measurements of a variable C given the pre-selection $|\Psi\rangle$ and post-selection $|\Phi\rangle$

$$\mathrm{Prob}(c_n) = \frac{|\langle\Phi|\mathbf{P}_{C=c_n}|\Psi\rangle|^2}{\sum_j |\langle\Phi|\mathbf{P}_{C=c_j}|\Psi\rangle|^2} \tag{13}$$

where $P_{c=c_j}$ is a projection operator for the eigenstate(s) with the eigenvalue c_j. Thus, for an ideal measurement of the nonlocal variable $\sigma_z^A + \sigma_z^B$ we obtain:

$$\langle\sigma_z^A + \sigma_z^B\rangle_{\Phi\Psi} = 2p(\uparrow\uparrow) + 0p(\sigma_z^A + \sigma_z^B = 0) - 2p(\downarrow\downarrow) = \tag{14}$$

$$\frac{2|\langle\Phi|\mathbf{P}_{\uparrow\uparrow}|\Psi\rangle|^2 - 2|\langle\Phi|\mathbf{P}_{\downarrow\downarrow}|\Psi\rangle|^2}{|\langle\Phi|\mathbf{P}_{\uparrow\uparrow}|\Psi\rangle|^2 + |\langle\Phi|\mathbf{P}_{\downarrow\downarrow}|\Psi\rangle|^2 + |\langle\Phi|\mathbf{P}_{\sigma_z^A+\sigma_z^B=0}|\Psi\rangle|^2} = \frac{2|\epsilon^2\frac{1}{2+\epsilon^2}|^2}{|\epsilon^2\frac{1}{2+\epsilon^2}]| + |0|^2 + |0|^2} = 2.$$

For an ideal measurement of a local variable σ_z^A, given that it is the only intermediate measurement that has been performed we have:

$$\langle\sigma_z^A\rangle_{\Phi\Psi} = p(\uparrow) - p(\downarrow) = \frac{|\langle\Phi|P_{\sigma_z^A=\uparrow}|\Psi\rangle|^2 - |\langle\Phi|P_{\sigma_z^A=\downarrow}|\Psi\rangle|^2}{|\langle\Phi|P_{\sigma_z^A=\uparrow}|\Psi\rangle|^2 + |\langle\Phi|P_{\sigma_z^A=\downarrow}|\Psi\rangle|^2} = \tag{15}$$

$$= \frac{|(\epsilon^2+1)\frac{1}{2+\epsilon^2}|^2 - |(-1)\frac{1}{2+\epsilon^2}|^2}{|(\epsilon^2+1)\frac{1}{2+\epsilon^2}|^2 + |(-1)\frac{1}{2+\epsilon^2}|^2} = \frac{2\epsilon^2+\epsilon^4}{2+\epsilon^4+2\epsilon^2}$$

The expectation value of σ_z^B measured alone is

$$\langle\sigma_z^B\rangle_{\Phi\Psi} = p(\uparrow) - p(\downarrow) = \frac{|\langle\Phi|P_{\sigma_z^B=\uparrow}|\Psi\rangle|^2 - |\langle\Phi|P_{\sigma_z^B=\downarrow}|\Psi\rangle|^2}{|\langle\Phi|P_{\sigma_z^B=\uparrow}|\Psi\rangle|^2 + |\langle\Phi|P_{\sigma_z^B=\downarrow}|\Psi\rangle|^2} = \tag{16}$$

$$\frac{|(\epsilon^2-1)\frac{1}{2+\epsilon^2}|^2 - |(+1)\frac{1}{2+\epsilon^2}|^2}{|(\epsilon^2+1)\frac{1}{2+\epsilon^2}|^2 + |(+1)\frac{1}{2+\epsilon^2}|^2} = \frac{-2\epsilon^2+\epsilon^4}{2+\epsilon^4-2\epsilon^2}$$

It is easy to see that the expectation value of the nonlocal variable (14) is very different from the sum of (15) and (16). However, it is more reasonable to compare (14) with the sum of expectation values of the outcomes of local measurements of σ_z^A and σ_z^B performed simultaneously. This corresponds to the measurement of a variable with non-degenerate eigenstates $|\uparrow\rangle_A|\downarrow\rangle_B, |\downarrow\rangle_A|\uparrow\rangle_B, |\uparrow\rangle_A|\uparrow\rangle_B, |\downarrow\rangle_A|\downarrow\rangle_B$. In this case we have

$$\langle\{\sigma_z^A\}+\{\sigma_z^B\}\rangle_{\Phi\Psi} = 2p(\uparrow\uparrow)+0p(\uparrow\downarrow)+0p(\downarrow\uparrow) = \frac{2|\langle\Phi|\mathbf{P}_{\uparrow\uparrow}|\Psi\rangle|^2}{|\langle\Phi|\mathbf{P}_{\uparrow\uparrow}|\Psi\rangle|^2+|\langle\Phi|\mathbf{P}_{\uparrow\downarrow}|\Psi\rangle|^2+|\langle\Phi|\mathbf{P}_{\downarrow\uparrow}|\Psi\rangle|^2} = \tag{17}$$

$$\frac{2|\epsilon^2\frac{1}{2+\epsilon^2}|^2}{|\epsilon^2\frac{1}{2+\epsilon^2}|^2+|\frac{1}{2+\epsilon^2}|^2+|(-1)\frac{1}{2+\epsilon^2}|^2} = \frac{2\epsilon^2}{2+\epsilon^4}$$

the brackets {} signify a separate measurement of each observable.

We see that expectation value of the strong measurements of the sum of two variables related to separate parts is not equal to the sum of the expectation values of local measurements, even if they are performed simultaneously.

$$\langle\sigma_z^A+\sigma_z^B\rangle_{\Phi\Psi} \neq \langle\sigma_z^A\rangle_{\Phi\Psi}+\langle\sigma_z^B\rangle_{\Phi\Psi} \neq \langle\{\sigma_z^A\}+\{\sigma_z^B\}\rangle_{\Phi\Psi} \tag{18}$$

For weak measurements, however, the equality (9) holds, both for separate and joint local measurements. If we assume the existence of nonlocal interactions, then we can directly couple to the sum $\sigma_z^A+\sigma_z^B$.

$$H = g(t)P(\sigma_z^A+\sigma_z^B). \tag{19}$$

Since a strong measurement yields the outcome 2 with certainty, the weak value should also equal 2 according to the theorem proved in [21]. In this particular example, the coupling need not be weak to find the weak value 2. Since there are no nonlocal interactions in nature we have to apply local weak measurements of σ_z^A and σ_z^B

$$H = g(t)(P_A\sigma_z^A+P_B\sigma_z^B) \tag{20}$$

and add the outcomes. We use a model in which the initial wave functions of the measuring devices are Gaussians around zero

$$\Psi_{in}^{MD}(Q_A,Q_B) = Ne^{\frac{-Q_A^2}{2\Delta^2}}e^{\frac{-Q_B^2}{2\Delta^2}} \tag{21}$$

The measurement interaction leads to a shift of the pointer wave function by the eigenvalue of σ_z so that after the measurement, the measuring device will be described by the state

$$\begin{gathered}\Psi_{fin}^{MD}(Q_A,Q_B) = \\ \Psi_{in}^{MD}(Q_A-1,Q_B+1)-\Psi_{in}^{MD}(Q_A+1,Q_B-1))+\epsilon^2\Psi_{in}^{MD}(Q_A-1,Q_B-1) = \\ Ne^{\frac{-(Q_A^2+Q_B^2)}{2\Delta^2}}[e^{\frac{-(-2Q_A+2Q_B+2)}{2\Delta^2}}-e^{\frac{-(2Q_A-2Q_B+2)}{2\Delta^2}}+\epsilon^2e^{\frac{-(-2Q_A-2Q_B+2)}{2\Delta^2}}]\end{gathered} \tag{22}$$

with N being a normalization constant.

The expectation value of the measurement outcome is:

$$\begin{gathered}\langle Q_A+Q_B\rangle = \int(Q_A+Q_B)|\Psi_{fin}^{MD}(Q_A,Q_B)|^2dQ_AdQ_B = \\ \frac{2\epsilon^4}{\epsilon^4+(2-2e^{\frac{-2}{\Delta^2}})} \approx 2-\frac{8}{\Delta^2\epsilon^4}\end{gathered} \tag{23}$$

for large Δ. The statistical measurement error is the width Δ. At the limit of weak measurements, i.e. large Δ it indeed yields the weak value $(\sigma_z^A+\sigma_z^B)_w = 2$. However, we can go close to the weak value only for very large Δ, (very weak measurement) and thus very large uncertainty in the final reading. For example, in the case of $\epsilon = 0.1$ we will need to have

$\Delta > 600$ in order to measure the weak value with a deviation of 10%. Of course, if we make the same measurement on a large ensemble the statistical error will be made smaller according to $\Delta_n = \frac{\Delta}{\sqrt{n}}$ requiring us to use an ensemble of about 3.6×10^5 such pre and post selected systems just to be within the right order of magnitude. This number will be increased if we want to make the deviation or the statistical error smaller. Since for a true nonlocal measurement we can get the required result at the strong limit, we only need one such system to get the correct expectation value. We see that local measurements allow us to find nonlocal weak values, but it is a very inefficient procedure. The nonlocal weak value is the result of a calculation made on the readings of two pointer variables rather than the direct result of the reading of a single pointer variable.

Let us now try to combine the techniques of nonlocal measurements based on a measuring device with entangled parts and local interactions, with the weak measurement techniques. The measuring device in an ideal strong nonlocal measurement has an initial state (4). To make it weak we have to prepare the conjugate momenta to be centered around 0 which requires the pointer (centered around $Q_A + Q_B = 0$) to have a large uncertainty. In this case, the initial state of the measuring device (using the usual Gaussian model) will be

$$\Psi_{in}^{MD}(Q_A + Q_B) = Ne^{\frac{(Q_A+Q_B)^2}{2\Delta^2}} \tag{24}$$

and the final state

$$\Psi_{fin}^{MD}(Q_A + Q_B) = Ne^{\frac{(Q_A+Q_B-2)^2}{2\Delta^2}} \tag{25}$$

This is a Gaussian around the weak value $(\sigma_z^A + \sigma_z^B)_w = 2$. In fact for this example, the measurement need not be weak to get it right due to an accidental fact: the strong measurement also yields the eigenvalue "2" with certainty.

Let us consider another example to compare various measurement methods. The system is described by the two state vector

$$\langle\uparrow\downarrow + \downarrow\uparrow + \uparrow\uparrow + \downarrow\downarrow| \quad |0.95 \uparrow\downarrow -1.05 \downarrow\uparrow +0.11 \uparrow\uparrow\rangle \tag{26}$$

so that the weak value is

$$(\sigma_z^A + \sigma_z^B)_w = \frac{\langle\uparrow\downarrow + \downarrow\uparrow + \uparrow\uparrow + \downarrow\downarrow|\, \sigma_z^A + \sigma_z^B\, |-1.05 \uparrow\downarrow +0.95 \downarrow\uparrow +0.11 \uparrow\uparrow\rangle}{\langle\uparrow\downarrow + \downarrow\uparrow + \uparrow\uparrow\, | -1.05 \uparrow\downarrow +0.95 \downarrow\uparrow +0.11 \uparrow\uparrow\rangle} = 22, \tag{27}$$

while the local weak values are

$$(\sigma_z^A)_w = 211 \tag{28}$$
$$(\sigma_z^B)_w = -189. \tag{29}$$

Using the measuring device with the initial state (24) we obtain the final state

$$\Psi_{fin}^{MD}(Q_A, Q_B) = N[-1.05e^{\frac{-(Q_A+Q_B)^2}{2\Delta^2}} + 0.95e^{\frac{-(Q_A+Q_B)^2}{2\Delta^2}} + 0.11e^{\frac{-(Q_A+Q_B-2)^2}{2\Delta^2}}], \tag{30}$$

and the expectation value of the pointer variable:

$$\langle Q_A + Q_B\rangle = \frac{22.0\left(11.0e^{2.0/\Delta^2} - 10.0\right)}{221.0e^{2.0/\Delta^2} - 220.0} \approx 22 - \frac{2360}{\Delta^2}. \tag{31}$$

A deviation of 1% and an uncertainty of 10% will require an ensemble of about 2.2×10^3 particles see fig 1. If on the other hand we have only local weak measurements, i.e. two local measuring devices, we will get an expectation value of $\langle Q_A + Q_B\rangle \approx 22 - \frac{8.8\times10^5}{\Delta^2}$. Here a deviation of 1% and an uncertainty of 10 % will require an ensemble of about 8.2×10^5 particles. In this example we can see that entanglement in the measuring device provides an improvement of more than two orders of magnitude.

Journal of Physics: Conference Series **174** (2009) 012004 doi:10.1088/1742-6596/174/1/012004

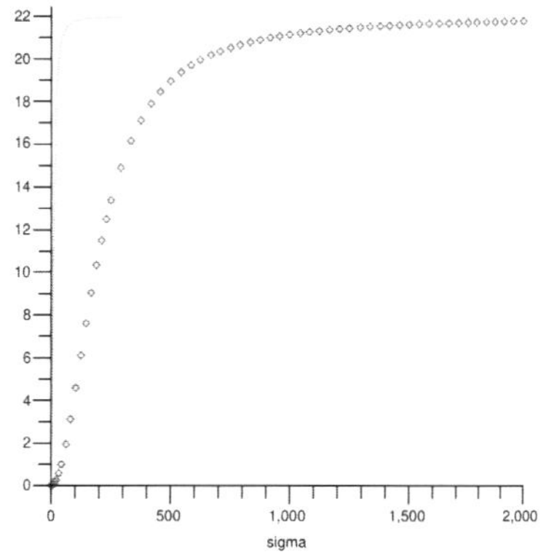

Figure 1. Comparison of local and entangled measuring devices. Expectation values of the pointer variable as a function of Δ. Deviations of 10% are at values of $\Delta > 30$ for the nonlocal measuring device (full line) and $\Delta > 600$ for the local one (dots) . While Deviations of 1% are at values of $\Delta > 100$ for the nonlocal measuring device and $\Delta > 2,000$ for the local one.

5. Local and nonlocal weak measurements of the product of local variables

We now have enough background to analyze recent results about the measurements of the product of separate local variables named "joint weak values" [4]. If we are given nonlocal (unphysical) interactions, then a weak measurement of the product is not different from any other weak measurement and all theory of local weak measurements is applicable. However, if we consider only local interactions, the situation is very different. The product rule does not hold, not only for expectation values of strong measurements

$$\langle AB\rangle_{\Phi\Psi} \neq \langle A\rangle_{\Phi\Psi}\langle B\rangle_{\Phi\Psi} \tag{32}$$

but also for weak values:

$$(AB)_w \neq A_w B_w. \tag{33}$$

At first glance it seems that local weak measurements cannot help us find the weak value of the product. Here is an example of two pre- and post-selected states which yield different values, of $(\sigma_z^A\sigma_z^B)_w$, but have the same *joint* probability for the pointer variables of local weak measurements. In both cases the initial state is the product $|\uparrow_x\rangle_A|\uparrow_x\rangle_B$, but the example is more transparent if we write it in spin z basis

$$|\Psi\rangle = \frac{1}{2}(|\uparrow\rangle_A|\uparrow\rangle_B + |\downarrow\rangle_A|\downarrow\rangle_B + |\uparrow\rangle_A|\downarrow\rangle_B + |\downarrow\rangle_A|\uparrow\rangle_B) \tag{34}$$

The first post-selected stated state is

$$|\Phi\rangle = \frac{1}{2}[|\uparrow\rangle_A|\uparrow\rangle_B + |\downarrow\rangle_A|\downarrow\rangle_B + i(|\uparrow\rangle_A|\downarrow\rangle_B - |\downarrow\rangle_A|\uparrow\rangle_B)] \tag{35}$$

and the second is

$$|\Phi'\rangle = \frac{1}{2}[|\uparrow\rangle_A|\uparrow\rangle_B - |\downarrow\rangle_A|\downarrow\rangle_B + i(|\uparrow\rangle_A|\downarrow\rangle_B + |\downarrow\rangle_A|\uparrow\rangle_B)] \tag{36}$$

It is easy to see that in the first case $(\sigma_z^A\sigma_z^B)_w = 1$ while in the second case $(\sigma_z^A\sigma_z^B)_w = -1$. In fact, it is a special case in which the weak value is equal to the result which is obtained with certainty in a strong nonlocal measurement.

For local measurement, we use the same model as before and write down the state of the two local measuring devices as a product of Gaussians

$$\Psi_{in}^{MD}(Q_A, Q_B) = e^{-\frac{Q_A^2}{4\Delta^2}} e^{-\frac{Q_B^2}{4\Delta^2}} \tag{37}$$

After the interaction and post-selection, the joint distribution for the pointer variables Q_a, Q_B given by $\Psi^\dagger\Psi$ turns out to be the same in both cases, so that a measurement of Q_A, Q_B or any combination of the two will not provide us with a method for distinguishing between the two initial states.

Nevertheless, Resch and Steinberg showed that one can find the weak value of the product by looking at the local measuring device. They proved the following formula (for a measuring device initially centered around zero)

$$Re(AB)_w = 2\langle Q_A Q_B\rangle - Re(A_w^* B_w) \tag{38}$$

Resch and Laudeen provided another expression

$$Re(AB)_w = \langle Q_A Q_B\rangle - \frac{4\Delta^4}{\hbar^2}\langle P_A P_B\rangle \tag{39}$$

Now it is clear why there is no contradiction between the fact that the probability distribution of the pointer variables of the measuring devices are identical, while the joint weak values of the product are different. It is not enough to look at the pointer variables, we have to look at their conjugate momenta as well (a different type of measurement). It is explicit in formula (39). (In formula (38) we have the complex local weak variables and in order to see their imaginary parts we need to observe the conjugated momenta of the pointer variables.)

An error analysis of this method for the example above shows that a set of measurements resulting in a deviation of< 1% and a statistical error of 10% requires an ensemble of about 2×10^6 such pre and post selected systems. We obtain the weak value of the product from local

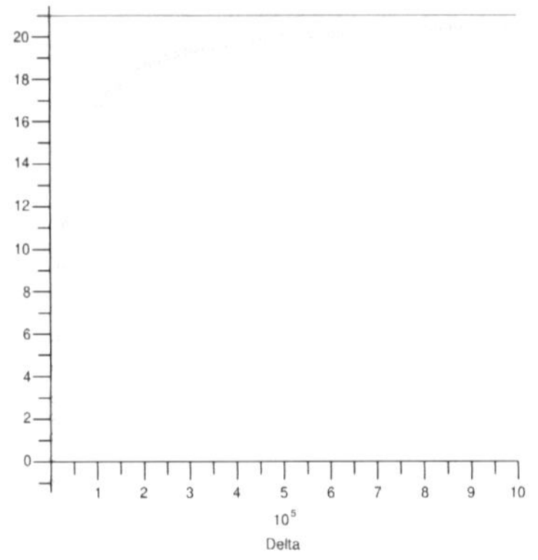

Figure 2. Comparison of local and nonlocal (unphysical) measuring devices for the measurement of a product. Expectation values of the pointer variable as a function of Δ. Deviations of 1% are at values of $\Delta > 100$ for the nonlocal measuring device (full line) and $\Delta > 5 \times 10^6$ for the local one (dots) .

weak measurements but at the expense of extreme errors and with a requirement of two different readings (one of Q and one of P).

For a comparison of this method and a non-local method with the (unphysical) coupling term $g(t)\sigma_z^A\sigma_z^B P$ we look at the same pre and post-selected state as before (26). The weak value for the product should be

$$A_w \equiv \frac{\langle\Phi|\sigma_z^A\sigma_Z^B|\Psi\rangle}{\langle\Phi|\Psi\rangle} = \frac{2-\epsilon^2}{2+\epsilon^2} = 21 \tag{40}$$

As can be seen in fig 2, the non local method converges much faster then the local one. A deviation of 1% with an uncertainty of 10% would require an ensemble of about 2×10^3 for the non-local method and about $10^{12}$for the local method described above. It is not surprising that this method is even less practical then the one for the measurement of a sum.

Journal of Physics: Conference Series **174** (2009) 012004 doi:10.1088/1742-6596/174/1/012004

6. Conclusions

Like eigenvalues and expectation values, weak values of a non local system can also be obtained using local methods. Such methods require larger resources, and have no pointer pointing at the desired result. Calculations using results from different types of measurement are required to arrive at the final result. We showed that the weak value of some non local variables (a sum of two or more local observables) can be measured directly. We have not found a direct way for weak measurement of a modular sum of nonlocal variables in spite of the existence of the method for strong measurement of nonlocal modular sum.

If we try to give an interpretation of weak values as *elements of reality* [22], one of the strengths of weak measurements is that it corresponds to a shift of the pointer variable by the weak value. This is not the case when making local measurements for calculating non-local values. Such methods require us to look at different pointers and use a formula for reaching the desired result. Lundeen and Steinberg [23] measured non local weak values using the method of joint weak values to measure non local weak values in an optical experiment. Their results had large deviations (more than 25%). The question of interpretation still remains open but unlike local weak values which have been measured precisely in the lab, some non-local weak values might still be thought of as accounting artifacts rather then physical observables.

Acknowledgments

This work has been supported in part by the European Commission under the Integrated Project Qubit Applications (QAP) funded by the IST directorate as Contract Number 015848 and by grant 990/06 of the Israel Science Foundation. We would like to thank Daniel Terno and Judie Kupferman for their helpful comments.

References

[1] Von Neumann J 1932 *Mathematische Grundlagen der Quantenmechanik*
[2] Aharonov Y, Albert D Z and Vaidman L 1986 *Phys. Rev.* D **34(6)** 1805-13
[3] Aharonov Y and Vaidman L 2002 *Time in Quantum Mechanics* ed J G Muga, R Sala Mayato and I L Egusquiza
[4] Resch K J and Steinberg A M 2004 *Phys. Rev. Lett.* **92(13)** 130402
[5] Vaidman L 2001 *Phys. Rev. Lett.* **90(1)** 010402
[6] Vaidman L and Nevo I 2006 *Int. J. Mod. Phys.* B **20** 1528-1535
[7] Aharonov Y and Vaidman L 1990 *Phys. Rev.* A **41(1)** 11-20
[8] Aharonov Y, Albert D Z and Vaidman L 1988 *Phys. Rev. Lett.* **60(14)** 1351-1354
[9] Vaidman L 1996 *Found. Phys.* **26** 895
[10] Leggett A J 1989 *Phys. Rev. Lett.* **62(19)** 2325
[11] Peres A 1989 *Phys. Rev. Lett.* **62(19)** 2326
[12] Aharonov Y and Vaidman L *Phys. Rev. Lett.* **62(19)** 2327
[13] Jozsa R 2007 *Phys Rev* A **76(4)** 044103
[14] Ritchie N W M, Story J G and Hulet R G 1991 *Phys. Rev. Lett.* **66** 1107
[15] Parks A D, Cullin D W and Stoudt D C 1998 *Proc. Roy. Soc. Lon.* A **454** 2997
[16] Brunner N, Acin A, Collins D, Gisin N and Scarani V 2003 *Phys. Rev. Lett.* **93** 180402
[17] Resch K J, Lundeen J S and Steinberg A M 2004 *Phys. Lett.* A **324** 125
[18] Pryde G L, O'Brien J L, White A G, Ralph T C and Wiseman H M 2005 *Phys. Rev. Lett.* **94** 220405
[19] Hosten O and Kwiat P 2008 *Science* **319** 787
[20] Aharonov Y, Bergmann P G and Lebowitz J L 1964 *Phys. Rev* **134(6B)** B1410-B1416
[21] Aharonov Y and Vaidman L 1991 *Journal of Physics* A **24** 2315-2328
[22] Vaidman L 1996 *Found. Phys* **26** 895-906
[23] Lundeen J and Steinberg A 2009 *Phys. Rev. Lett* **102** 020404

Emergent classicality via commuting position and momentum operators

J. J. Halliwell
Blackett Laboratory, Imperial College, London SW7 2BZ, UK
E-mail: j.halliwell@ic.ac.uk

Abstract. Any account of the emergence of classicality from quantum theory must address the fact that the quantum operators representing positions and momenta do not commute, whereas their classical counterparts suffer no such restrictions. To address this, we revive an old idea of von Neumann, and seek a pair of commuting operators X, P which are, in a specific sense, "close" to the canonical non-commuting position and momentum operators, x, p. The construction of such operators is related to the problem of finding complete sets of orthonormal phase space localized states, a problem severely limited by the Balian-Low theorem. Here these limitations are avoided by restricting attention to situations in which the density matrix is reasonably decohered (i.e., spread out in phase space).

1. Introduction

1.1. Opening Remarks

The canonical commutation relation between position and momentum operators,

$$[\hat{x}, \hat{p}] = i\hbar \tag{1}$$

lies at the heart of all quantum phenomena. Physically, this relation corresponds to the fact that measurements of position and momentum depend on the order of the measurements. It is this relation, in essence, that is responsible for the key differences between classical and quantum mechanics, since in classical mechanics, measurements of position and momentum can be made in such a way that the order makes no difference. It follows that any account of the emergence of classical behaviour from quantum theory must reconcile these two very different aspects of classical and quantum theory. A typical point of view is that classical mechanics emerges only at a very coarse-grained level and for sufficiently coarse-grained samplings of position and momentum, their non-commutativity makes little difference [1].

Still, one wonders whether there is a deeper or more precise way of reconciling the non-commuting quantum operators with their commuting classical counterparts. Indeed, this clearly troubled the founders of quantum theory, since von Neumann addressed the issue in his 1932 book, Mathematical Foundations of Quantum Mechanics [2]. He noted that when we make observations of a macroscopic system, we are in fact able to make observations of position and momentum simultaneously (although imprecisely, of course). This suggested to him that the measurements we make do not in fact correspond directly to the usual operators $\hat{x}$ and $\hat{p}$, but to some other operators, $\hat{X}$ and $\hat{P}$, say, which commute

$$[\hat{X}, \hat{P}] = 0 \tag{2}$$

and which must in some sense be "close" to the original operators $\hat{x}$, $\hat{p}$. Such a pair of operators could be particularly useful in bringing a degree of precision to discussions of emergent classicality. The aim of this paper is to discuss the construction and utility of such operators.

1.2. The Imprecision of Classical Physics

To what degree do macroscropic measurements fix the position and momentum operators in quantum mechanics? When we make macroscopic measurements of, say, a particle, there will be imprecisions Δx, Δp in the specifications of position and momentum. These imprecisions may be "small" to classical eyes but they will typically be very large compared to the quantum scale. It will be useful for what follows to get a quantitative idea of this. Suppose that the position is measured to a precision of $10^{-8}m$ and the velocity to within $10^{-8}m/s$ (both extremely precise specifications, by macroscopic standards). For a mass of, say, $10^{-6}kg$, we then have,

$$\frac{\Delta p \Delta x}{\hbar} \sim 10^{12} \tag{3}$$

This means that when we specify the phase space location of a classical system in a way that is very precise to classical eyes, at the quantum scale it could be in any one of about 10^{12} phase space cells. There is therefore a considerable amount of freedom at the quantum level to redefine the position and momentum operators without making any noticeable difference to macroscopic observations. Perhaps within this freedom there is the possibility to find position and momentum operators which commute.

1.3. Motivations

What are the motivations for constructing commuting operators which are close to the position and momentum operators? An important indicator of emergent classicality is approximate diagonality of the density operator ρ (often referred to as decoherence) [3]. For example, for a point particle interacting with a thermal environment, it may be shown that after a short time scale, the density operator approaches the form

$$\rho = \int dpdq \ f(p,q) \ |\psi_{pq}\rangle\langle\psi_{pq}| \tag{4}$$

where $f(p,q)$ is a non-negative function and $|\psi_{pq}\rangle$ are a set of phase space localized states (such as generalized coherent states) [4]. This indicates that it is approximately diagonal in both position or momentum. (It cannot be exactly diagonal in both, since they do not commute.) This means that there is negligible interference between different values of position and momenta and, loosely speaking, they may then be treated as if they are classical. The statement is imprecise, however, since there is still *some* interference, so the variables are only imprecisely defined.

One would like to be able to make a more exact statement about diagonality of ρ and this is possible in terms of the commuting operators X, P. For example, using the eigenstates $|\psi_{nm}\rangle$ of X, P one could construct a density operator of the form

$$\rho' = \sum_{nm} f_{nm} |\psi_{nm}\rangle\langle\psi_{nm}| \tag{5}$$

which is exactly diagonal in X and P, indicating there is exactly *zero* interference between different values of these quantities. One can then choose the coefficients f_{nm} to make ρ' as close as possible to ρ. The question of the degree of approximate diagonality is therefore shifted to the question of the closeness of the density operator to a pseudo-classical density operator ρ'.

A more comprehensive approach to emergent classicality is the decoherent histories approach [5, 6, 7, 8, 9, 10]. There, the central object of interest is the decoherence functional,

$$D(\underline{\alpha}, \underline{\alpha}') = \mathrm{Tr}\left(P_{\alpha_n}(t_n)\cdots P_{\alpha_1}(t_1)\rho P_{\alpha_1'}(t_1)\cdots P_{\alpha_n'}(t_n)\right) \tag{6}$$

The histories are characterized by the initial state ρ and by the strings of projection operators $P_{\alpha}(t)$ (in the Heisenberg picture) at times t_1 to t_n (and $\underline{\alpha}$ denotes the string of alternatives $\alpha_1 \cdots \alpha_n$). Intuitively, the decoherence functional is a measure of the interference between pairs of histories $\underline{\alpha}$, $\underline{\alpha}'$. When

$$\mathrm{Re}D(\underline{\alpha}, \underline{\alpha}') = 0 \tag{7}$$

for $\underline{\alpha} \neq \underline{\alpha}'$, we say that the histories are consistent and probabilities $p(\underline{\alpha}) = D(\underline{\alpha}, \underline{\alpha})$ obeying the usual probability sum rules may be assigned to them. (Typically, the physical mechanisms producing consistency actually cause the stronger condition $D(\underline{\alpha}, \underline{\alpha}') = 0$ for $\underline{\alpha} \neq \underline{\alpha}'$ to be satisfied, which is referred to as decoherence.) One can then ask whether these probabilities are strongly peaked about trajectories obeying classical equations of motion, and if they are, we can say that the system is emergently classical.

For histories in which the projections $P_{\alpha_k}(t_k)$ are onto positions at different times of a point particle interacting with a thermal environment, the decoherence functional, like the density operator, is *approximately* diagonal [6, 10]. The approximation is typically exceptionally good, but still only approximate. Again one wonders whether more exact statements can be made. Indeed, it has been conjectured that approximately consistent histories can be in some sense distorted into exactly consistent ones [11]. There are a number of ways in which a set of histories could be distorted: one can change the initial state, the times of the projections, the widths of the projections or the operators who spectrum is projected onto. The possible existence of the commuting variables $\hat{P}$, $\hat{X}$ suggests a particular way of distorting the histories so as to make them exactly decoherent.

Each position at time t is, in the Heisenberg picture, a function of the canonical pair, $\hat{x}, \hat{p}$, so $\hat{x}_t = f_t(\hat{x}, \hat{p})$. The projections onto $\hat{x}_t$ at different times do not commute, since $[\hat{x}_t, \hat{x}_t'] \neq 0$ in general. The decoherence functional is therefore not diagonal in general, but can be approximately diagonal if the system is coupled to a thermal environment.

Now suppose we consider projections onto the variables $\hat{X}_t = f_t(\hat{X}, \hat{P})$ at different times. Under reasonable dynamics, $\hat{X}_t$ will be close to $\hat{x}_t$ as long as $\hat{x}, \hat{p}$ are close to $\hat{X}, \hat{P}$. The operators $\hat{X}_t$ *do* commute at different times so all the projections commute and, as is easy to see, the decoherence functional will be exactly diagonal. So exact decoherence is achieved by shifting the operators $\hat{x}, \hat{p}$ to the commuting pair $\hat{X}, \hat{P}$. Furthermore, it is known that the probabilities for histories of positions $\hat{x}_t$ are typically peaked about classical evolution. The probabilities for the commuting variables $\hat{X}_t$ will therefore have the same property if the $\hat{X}_t$ are close to $\hat{x}_t$. The question of approximate decoherence and emergent classicality is therefore shifted to the question of the closeness of the old and new operators.

These then, at least in outline, are the reasons why a commuting pair of position and momentum-like operators may be useful for discussing emergent classicality. We turn now to the construction of such operators.

1.4. Von Neumann's Construction

Von Neumann outlined a prescription whereby the commuting operators $\hat{X}$ and $\hat{P}$ may be constructed [2]. This involved first taking a discrete subset $|m, n\rangle$ of the coherent states $|p, q\rangle$, with one state per cell of size $2\pi\hbar$ (a von Neumann lattice). He alleged that these states are complete (this was later proved [12, 13, 14]). He then stated that they may be orthogonalized using the Schmidt process to produce an orthonormal set $|\psi_{nm}\rangle$. From these, he constructed position and momentum-like operators

$$\hat{X} = \sum_{nm} na|\psi_{nm}\rangle\langle\psi_{nm}|, \qquad \hat{P} = \sum_{nm} m\frac{2\pi\hbar}{a}|\psi_{nm}\rangle\langle\psi_{nm}| \tag{8}$$

(where a is a constant with the dimension of length). These operators clearly commute. He argued that these new operators are indeed "close" to the old ones, in the sense that

$$\langle\psi_{nm}|(\hat{x}-\hat{X})^2|\psi_{nm}\rangle\ \langle\psi_{nm}|(\hat{p}-\hat{P})^2|\psi_{nm}\rangle \leq K^2\hbar^2 \tag{9}$$

Von Neumann's calculations imply that the constant K is about $1,800$ (but he thought more detailed calculations could give a smaller value).

However, von Neumann's prescription is at best at sketch of how this works and he certainly did not give full details (such as the explicit form of the $|\psi_{nm}\rangle$). Furthermore, we now know a lot more about phase space localized states than was known in 1932, and, as will be described below, there are obstructions to constructing such states. These obstructions do not necessarily apply to what von Neumann did, but nevertheless, it is still interesting to revisit his ideas from a more modern perspective.

1.5. A General Approach and the Balian-Low Theorem

We start with a more general statement of the problem. We consider a set of states of the form

$$|\psi_{nm}\rangle = U_{nm}|\psi\rangle \tag{10}$$

where $|\psi\rangle$ is a fiducial state and U_{nm} is the unitary shift operator,

$$U_{nm} = \exp\left(\frac{i}{\hbar}na\hat{p} - \frac{i}{\hbar}mb\hat{x}\right) \tag{11}$$

A particularly interesting case is that of a von Neumann lattice, in which case $b = 2\pi\hbar/a$ and the translations in the p and x directions then commute, and we have

$$U_{nm} = (-1)^{mn}\ \exp\left(\frac{i}{\hbar}na\hat{p}\right)\ \exp\left(-i\frac{2\pi m}{a}\hat{x}\right) \tag{12}$$

There is then one state per cell of size $2\pi\hbar$, as in von Neumann's case. It is of interest to find a fiducial state $|\psi\rangle$ in Eq.(10) such that the states are complete and orthonormal, that is,

$$\sum_{nm}|\psi_{nm}\rangle\langle\psi_{nm}| = 1 \tag{13}$$

$$\langle\psi_{nm}|\psi_{n'm'}\rangle = \delta_{nn'}\delta_{mm'} \tag{14}$$

Fiducial states leading to states satisfying these properties are easily found and we will exhibit a set below. There is, however, a crucial difficulty. According to the theorem of Balian and Low [15, 16, 17], if the three properties (10), (13) and (14) are satisfied, then the fiducial state $|\psi\rangle$ has either $(\Delta x)^2$ or $(\Delta p)^2$ infinite, so is not phase space localized. If we used such states to construct commuting operators $\hat{X}, \hat{P}$ as von Neumann did, then at least one of the averages in Eq.(9) would diverge, and there would be no sense in which the new operators are close to the old. (Note that von Neumann claims to have used the Schmidt procedure to construct his orthonormal set, which one would not expect to produce states satisfying Eq.(10), so his construction does not necessarily fall foul of the Balian-Low theorem).

The problem of constructing orthonormal phase space localized states is one of great interest in a number of fields so some effort has been expended in finding ways around the Balian-Low theorem. Zak has proved some interesting results in this area. In Ref [18], he showed that the coherent states restricted to a von Neumann lattice $|m,n\rangle$ obey a sort of orthogonality relation if the usual inner product $\langle m,n|m',n'\rangle$ is averaged over a single phase space cell. It is not yet clear if this result can be used to produce commuting position and momentum operators.

He has also considered complete orthonormal sets of states which are localized in position, but double-peaked in momentum (so localized in p^2, but not in p. From these one can construct commuting operators $\hat{X}$ and $\hat{P}^2$, which are "close" to $\hat{x}$ and $\hat{p}^2$ [19]. This is tantalizing close to the goal of this paper, but not quite there (and also suggests that the $p \rightarrow -p$ transformation plays a crucial role in the Balian-Low theorem). Many of these and similar results are proved using the so-called kq representation, a technique which is particularly well-adapted to these problems [20].

To avoid the Balian-Low theorem, one has to drop one of the three requirements (10), (13) and (14) in order to get phase space localization. For the purposes of this paper, which is to construct useful commuting position and momentum operators, the orthogonality Eq.(14) is essential. We will therefore explore the possibility of dropping the other two requirements. Eq.(10), the requirement that the states be obtained by translation of a single fiducial state is mainly for practical convenience, so there is no harm in relaxing this as long as the resulting states are not unmanageable.

More significantly, we will relax the requirement of completeness, Eq.(13). The motive behind this is as follows. We are interested in using commuting operators to discuss emergent classicality. In practice, this means that we are only concerned with the physical situation in which the density matrix of the system has undergone a degree of decoherence. This means that it is approximately diagonal in both position and momentum, or equivalently, its Wigner function is reasonably spread out in phase space. The density matrix is therefore insensitive to the fine structure of phase space and it seems reasonable to suppose that the physics could be well-described by a less than complete set of states, if they are carefully chosen. Of course, this is a quantitative matter that needs to be checked in detail and we will do this.

For the purposes of constructing commuting position and momentum operators, we will actually use a construction slightly more general than the one indicated by von Neumann in Eq.(8). In particular, we will look for commuting operators $\hat{X}$, $\hat{P}$ of the form

$$\hat{X} = \sum_{nm} X_n E_{nm} \tag{15}$$

$$\hat{P} = \sum_{nm} P_m E_{nm} \tag{16}$$

Here, X_n and P_m are c-numbers and E_{nm} are projection operators localized onto a region of phase space labelled by n, m which will consist of *more than one* $2\pi\hbar$-sized cell. They are exclusive

$$E_{nm} E_{n'm'} = E_{nm} \delta_{nn'} \delta_{mm'} \tag{17}$$

and exhaustive

$$\sum_{nm} E_{nm} = 1 \tag{18}$$

We will also insist that they are obtained from unitary shifts of a single projector E,

$$E_{nm} = U_{nm} E \left(U_{nm}\right)^\dagger \tag{19}$$

Here, U_{nm} shifts from one cell to the next, so is of the form Eq.(11), with a, b chosen so that the translation in the position and momentum directions commute, but $ab > 2\pi\hbar$. These three conditions are the natural generalization for projection operators of the requirements (10), (13) and (14), and the original case is obtained with the choice

$$E_{nm} = |\psi_{nm}\rangle\langle\psi_{nm}| \tag{20}$$

The quantity $\mathrm{Tr}E$ is a measure of the number of $2\pi\hbar$-sized cells projected onto, so clearly $\mathrm{Tr}E = 1$ in the pure state case, but $\mathrm{Tr}E \gg 1$ in our case, as we will see.

One might have thought that this more general construction could avoid the Balian-Low theorem, since the restrictions on E_{nm} are in fact weaker than the restrictions (10), (13) and (14) for pure states. This is not in fact the case. We will prove a modest extension of the Balian-Low theorem which shows that there is no phase space localized projector E satisfying the three requirements Eqs.(17), (18) and (19). However, what we will do is find an "almost" complete set of phase space localized pure states from which we can construct a projector E that satisfies the exhaustivity condition Eq.(18) to a good approximation when acting on sufficiently decohered density operators. From this we can construct commuting operators $\hat{X}$ and $\hat{P}$ may with useful properties.

1.6. Earlier Work

Finally, we briefly mention a related approach. An earlier attempt to construct commuting position and momentum operators was considered in Ref.[21]. This construction involved doubling the original Hilbert space and then using operators defined on this enlarged space. (See also Ref.[22] for similar ideas). The resulting theory is essentially the same as 't Hooft's deterministic quantum theory [23]. However, this is no longer standard quantum theory. In the present work, by contrast, we stay within the framework of standard quantum theory.

1.7. This Paper

In Section 2, we briefly summarize some known properties of the Wigner function which help to make precise the idea that the state is sufficiently spread out in phase space. We also briefly note that the Wigner function naturally suggests an alternative method of defining commuting position and momentum operators. In Section 3, we prove a modest generalization of the Balian-Low theorem for projector operators. In Section 4 we introduce an orthornormal set of phase space localized states that are "almost" complete. We then use them to construct a set of phase space localized projection operators E_{nm} which are almost exhaustive. In Section 5 it is shown that the incompleteness does not matter if the density operator of the system is reasonably spread out in phase space. In Section 6, we use the projection operators E_{nm} to construct commuting position and momentum operators and compute the "distance" between these operators and the usual canonical pair, as in Eq.(9). In Section 7, we compare the probabilities for X and P with those for the usual position and momentum operators and find them to be close. We summarize and conclude in Section 8. This contribution is based on the published work Ref.[24].

2. Some Properties of the Wigner Function

It will be useful for the rest of the paper to briefly summarize here some aspects of quantum mechanics in phase space and the Wigner function. Most of this is standard material and may be skipped by the informed reader, except the brief comments at the end of this section.

The Wigner representation for a density operator ρ (or indeed a wide class of operators) is defined by

$$W(p,q) = \frac{1}{2\pi\hbar}\int d\xi\ e^{-\frac{i}{\hbar}p\xi}\ \rho(q+\frac{1}{2}\xi, q-\frac{1}{2}\xi) \tag{21}$$

with inverse

$$\rho(x,y) = \int dp\ e^{\frac{i}{\hbar}p(x-y)}\ W(p, \frac{x+y}{2}) \tag{22}$$

Many calculations involving operators are usefully expressed in the Wigner representation. For example,

$$\mathrm{Tr}\,(AB) = 2\pi\hbar \int dpdq\ W_A(p,q) W_B(p,q) \tag{23}$$

where $W_A(p,q)$ and $W_B(p,q)$ are the Wigner functions of A and B.

We will be interested in later sections in the behaviour of the density operator or Wigner function in a simple model of the decoherence process. We take the simplest case of a single particle coupled to a thermal environment in the limit of high temperature and negligible dissipation, with no external potential. The master equation for the density matrix $\rho(x,y)$ is,

$$\frac{\partial\rho}{\partial t} = \frac{i\hbar}{2m}\left(\frac{\partial^2\rho}{\partial x^2} - \frac{\partial^2\rho}{\partial y^2}\right) - \frac{D}{\hbar^2}(x-y)^2\rho \tag{24}$$

where $D = 2m\gamma kT$. In the Wigner representation, the corresponding Wigner function obeys the equation,

$$\frac{\partial W}{\partial t} = -\frac{p}{m}\frac{\partial W}{\partial x} + D\frac{\partial^2 W}{\partial p^2} \tag{25}$$

Evolution according to the master equation Eq.(24) tends to produce approximate diagonality in position and momentum. In the Wigner representation, this appears as a spreading out phase space. Indeed, using Eq.(25) it may be shown that

$$(\Delta p)_t^2 = 2Dt + (\Delta p)_0^2 \tag{26}$$

$$(\Delta x)_t^2 = \frac{2}{3}\frac{Dt^3}{m^2} + (\Delta p)_0^2\frac{t^2}{m^2} + \frac{2}{m}\sigma(x,p) + (\Delta x)_0^2 \tag{27}$$

where

$$\sigma(x,p) = \frac{1}{2}\langle\hat{x}\hat{p} + \hat{p}\hat{x}\rangle - \langle\hat{x}\rangle\langle\hat{p}\rangle \tag{28}$$

evaluated in the initial state. In particular, for long times, the phase space spreading behaves according to

$$\frac{(\Delta p)_t(\Delta x)_t}{\hbar} \sim \left(\frac{\gamma kT}{\hbar}\right)t^2 \tag{29}$$

This means that any initial state becomes spread out in phase space on the (typically very short) timescale $(\hbar/\gamma kT)^{1/2}$. (See Ref.[25] for similar calculations).

In the case above, a free particle, the spreading continues indefinitely. However, for a bound system with dissipation, equilibrium is eventually reached. For a simple harmonic oscillator at thermal equilibrium, for example, the ratio of thermal to quantum fluctuations is

$$\frac{(\Delta p)(\Delta x)}{\hbar} \approx \frac{kT}{\hbar\omega} \tag{30}$$

At typical laboratory temperatures and frequencies that are not unrealistically fast (for macroscopic systems), this number can be very large, of order 10^{10} say. These estimates will be relevant later on. In brief, they show that the density operator becomes very spread out in phase space (and hence slowly varying) very readily.

This much is known material and will be used later. However, we now note that the existence of the Wigner representation suggests an alternative method for constructing commuting position and momentum operators. It is well-known that when the Wigner function is sufficiently spread out, it becomes positive and may be regarded as a probability distribution for the variables p and q (which clearly commute, since they are numbers, not operators). The variables p and q do not correspond exactly to the operators $\hat{p}$ and $\hat{x}$, but they are close when the Wigner function is spread out. In fact, it is easy to see that we have the following correspondences:

$$qW(p,q) \leftrightarrow \frac{1}{2}(\hat{x}\rho + \rho\hat{x}) \tag{31}$$

$$pW(p,q) \leftrightarrow \frac{1}{2}(\hat{p}\rho + \rho\hat{p}) \tag{32}$$

That is, multiplication by q or p in the Wigner representation corresponds to the operation of anticommutation with $\hat{x}$ or $\hat{p}$ on the density operator (and it is easy to show that these two operations on ρ commute). The point here is that these operators on ρ are operations on the space of density operators which have no counterpart in terms of operations on pure states. This is therefore not a route to producing the desired pair of commuting operators envisage by von Neumann.

3. An Extension of the Balian-Low Theorem

We first show that there is no projection operator satisfying the three properties Eqs.(18), (19) and (20). This is an almost trivial extension of the proof of the Balian-Low theorem given by Battle [17].

We consider the object $\mathrm{Tr}(Exp)$, which we assume exists. (If it does not, i.e., is infinite, this means that E has infinite dispersion in either x or p, so is not phase space localized). We have, from the three properties of E_{mn}, Eqs.(17)-(19),

$$\begin{aligned}
\mathrm{Tr}(Exp) &= \sum_{mn} \mathrm{Tr}\left(ExE_{mn}p\right) \\
&= \sum_{mn} \mathrm{Tr}\left(ExU_{mn}EU_{mn}^{\dagger}p\right) \\
&= \sum_{mn} \mathrm{Tr}\left(EU_{mn}(x+na)E(p+bm)U_{mn}^{\dagger}\right) \\
&= \sum_{mn} \mathrm{Tr}\left(E_{-m,-n}(x+na)E(p+bm)\right) \\
&= \sum_{mn} \mathrm{Tr}\left(E_{-m,-n}xEp\right) \\
&= \mathrm{Tr}\left(xEp\right) \\
&= \mathrm{Tr}\left(Epx\right)
\end{aligned} \tag{33}$$

Or in other words,

$$\mathrm{Tr}\left(E[x,p]\right) = 0 \tag{34}$$

which, via the commutation relations, implies that $\mathrm{Tr}E = 0$. Since $E \geq 0$, this means that $E = 0$. So no projector satisfying the three properties exists, if one insists that they be satisfied exactly. Hence, as indicated we will relax the conditions in what follows.

We note in passing that this simple result has implications for the Balian-Low theorem in the pure state case. It may seem that one could avoid the Balian-Low theorem by dropping the requirement Eq.(10) and requiring that the states $|\psi_{nm}\rangle$ are generated from more than one fiducial state. However, one could then use those fiducial states (assuming they are orthogonal) to construct a projection operator E satisfying the properties Eqs.(17)-(19) and the above result shows that the Balian-Low theorem is not in fact avoided.

4. An Almost Complete Set of Orthonormal Phase Space Localized States

We now show how to construct an almost complete orthonormal set of phase space localized states, from which we can construct the projector E. Our starting point is the set of states considered by Low [16],

$$\psi_{nm}(x) = a^{-\frac{1}{2}}h(x-na)\ e^{2\pi i m x/a} \tag{35}$$

where $h(x)$ a window function on $[-\frac{1}{2}a, \frac{1}{2}a]$. These clearly satisfy the three conditions, Eqs.(10), (13) and (14). The Fourier transform of these wave function is

$$\tilde{\psi}_{nm}(p) = \left(\frac{2a}{\pi\hbar}\right)^{\frac{1}{2}} (-1)^m\ \frac{\sin(pa/2\hbar)}{(pa/\hbar - 2\pi m)}\ e^{-\frac{i}{\hbar}nap} \tag{36}$$

from which it is easy to see that $(\Delta p)^2$ diverges because $\tilde{\psi}_{nm}(p)$ goes to zero like $1/p$ for large p which is not fast enough. Differently put, the window function in Eq.(35) causes the derivative of the wave function to involve the δ-functions $\delta(x - na \pm \frac{1}{2}a)$, which are not square-integrable. These properties are fully in line with the Balian-Low theorem.

However, inspired by a suggestion of Zak [26], it is easy to see from Eq.(36) that we may take linear combinations of these states which fall off like $1/p^2$ and therefore have finite $(\Delta p)^2$. In particular, the states

$$\psi^{(1)}_{nm}(x) = \frac{1}{\sqrt{2}} \left(\psi_{n,2m}(x) + \psi_{n,2m+1}(x)\right) \tag{37}$$

have this property. One can also see this in configuration space: they vanish at $x = na \pm \frac{1}{2}a$, and the offending δ-function appearing in their derivative therefore causes no problems. They are not complete, consisting of just "half" the states in the momentum direction, and indeed the states missed out are the states

$$\chi^{(1)}_{nm}(x) = \frac{1}{\sqrt{2}} \left(\psi_{n,2m}(x) - \psi_{n,2m+1}(x)\right) \tag{38}$$

and these have infinite dispersion in p.

But it is not hard to see that the "remainder" states $\chi^{(1)}_{nm}$ decay like $1/p$ in momentum space, so a further process of "halving" is possible to produce more states with finite dispersion. That is, we define

$$\psi^{(2)}_{nm} = \frac{1}{\sqrt{2}} \left(\chi^{(1)}_{n,2m}(x) - \chi^{(1)}_{n,2m+1}(x)\right) \tag{39}$$

with new remainder states

$$\chi^{(2)}_{nm} = \frac{1}{\sqrt{2}} \left(\chi^{(1)}_{n,2m}(x) + \chi^{(1)}_{n,2m+1}(x)\right) \tag{40}$$

So the set of states $\psi^{(1)}_{nm}$, $\psi^{(2)}_{nm}$, $\chi^{(2)}_{nm}$ is complete and orthonormal, but the states $\psi^{(1)}_{nm}$ have infinite dispersion.

We can continue in this way to define the sequence of states

$$\psi^{(K+1)}_{nm} = \frac{1}{\sqrt{2}} \left(\chi^{(K)}_{n,2m+1} - \chi^{(K)}_{n,2m}\right) \tag{41}$$

and

$$\chi^{(K+1)}_{nm} = \frac{1}{\sqrt{2}} \left(\chi^{(K)}_{n,2m+1} + \chi^{(K)}_{n,2m}\right) \tag{42}$$

for $K = 1, 2, 3, \cdots$. If we truncate the sequence at some finite value of K, $K = N$, say, then the set of states $\psi^{(1)}_{nm}, \psi^{(2)}_{nm} \cdots \psi^{(N)}_{nm}$ together with the remainder states $\chi^{(N)}_{nm}$ are orthonormal and complete. We therefore have the completeness relation

$$\sum_{K=1}^{N} \sum_{n,m} \psi^{(K)}_{nm}(x)\psi^{(K)}_{nm}(y)^* + \sum_{n,m} \chi^{(N)}_{nm}(x)\chi^{(N)}_{nm}(y)^* = \delta(x-y) \tag{43}$$

In some sense, "most" of the states, namely the $\psi^{(K)}_{nm}$, have finite dispersion, and "some" of them, namely the $\chi^{(N)}_{nm}$ have infinite dispersion. In this way, as N increases, the infinite dispersion anticipated from the Balian-Low theorem is pushed into a progressively smaller set of states. (An interesting question is whether the limit $N \to \infty$ may be taken in any meaningful or useful way, but we will not pursue that here).

Since each state is a linear combination of the ψ_{nm}, one may also derive the following general formula,

$$\psi_{nm}^{(K)} = 2^{-K/2} \sum_{j=0}^{2^K-1} c_j^K \; \psi_{n,2^K m+j} \tag{44}$$

where

$$c_j^K = (-1)^j \;\; \text{if} \;\; 0 \leq j \leq 2^{K-1}-1 \tag{45}$$

and

$$c_j^K = -(-1)^j \;\; \text{if} \; 2^{K-1} \leq j \leq 2^K - 1 \tag{46}$$

We also have

$$\chi_{nm}^{(K)} = 2^{-K/2} \sum_{j=0}^{2^K-1} (-1)^j \; \psi_{n,2^K m+j} \tag{47}$$

This may also be written,

$$\chi_{nm}^{(K)} = 2^{-K/2} \frac{(1-e^{2^{K+1}\pi i x/a})}{(1+e^{2\pi i x/a})} \psi_{n,2^{K+1}m} \tag{48}$$

The states $|\psi_{nm}^{(K)}\rangle$ are not all obtained from a single fiducial state, since we have

$$|\psi_{nm}^{(K)}\rangle = U_{nm}^{(K)} |\psi^K\rangle \tag{49}$$

where

$$U_{nm}^{(K)} = U_{n,2^K m} \tag{50}$$

and

$$|\psi^K\rangle = |\psi_{00}^K\rangle \tag{51}$$

There are therefore N fiducial states for the set $\psi_{nm}^{(K)}$, $K = 1, \cdots N$.

At some length, one can compute the averages and dispersions of p and x in the fiducial states $|\psi^K\rangle$. One obtains

$$\langle p\rangle_K = \frac{2\pi\hbar}{a}\left(2^{K-1} - \frac{1}{2}\right) \tag{52}$$

$$\langle x\rangle_K = 0 \tag{53}$$

$$(\Delta p)_K^2 = \left(\frac{2\pi\hbar}{a}\right)^2 \frac{(2^{2K}-1)}{12} \tag{54}$$

$$(\Delta x)_K^2 = \frac{a^2}{12} \tag{55}$$

(Note that $\langle p\rangle_K$ is not zero. Because there is more than one fiducial state, it does not appear to be possible to shift the states so that $\langle p\rangle_K = 0$ in the fiducial states without spoiling orthogonality.)

The construction described above is concisely summarized as follows: For each set of 2^N lattice points in the momentum direction, there are $2^N - 1$ states with finite dispersion and just 1 state (the remainder state), with infinite dispersion.

Consider the following simple example to illustrate the construction. Consider the 8 lattice points $m = 0, 1 \cdots 7$ in the momentum direction, so $N = 3$. Then the 8 states which depend only on these points are

$$\begin{aligned}\psi^{(1)}_{n,0} &= \frac{1}{\sqrt{2}}(\psi_{n,0}+\psi_{n,1})\\ \psi^{(1)}_{n,1} &= \frac{1}{\sqrt{2}}(\psi_{n,2}+\psi_{n,3})\\ \psi^{(1)}_{n,2} &= \frac{1}{\sqrt{2}}(\psi_{n,4}+\psi_{n,5})\\ \psi^{(1)}_{n,3} &= \frac{1}{\sqrt{2}}(\psi_{n,6}+\psi_{n,7})\end{aligned} \tag{56}$$

$$\begin{aligned}\psi^{(2)}_{n,0} &= \frac{1}{2}(\psi_{n,0}-\psi_{n,1}-\psi_{n,2}+\psi_{n,3})\\ \psi^{(2)}_{n,1} &= \frac{1}{2}(\psi_{n,4}-\psi_{n,5}-\psi_{n,6}+\psi_{n,7})\end{aligned} \tag{57}$$

$$\psi^{(3)}_{n,0} = \frac{1}{2^{3/2}}(\psi_{n,0}-\psi_{n,1}+\psi_{n,2}-\psi_{n,3}-\psi_{n,4}+\psi_{n,5}-\psi_{n,6}+\psi_{n,7}) \tag{58}$$

$$\chi^{(3)}_{n,0} = \frac{1}{2^{3/2}}(\psi_{n,0}-\psi_{n,1}+\psi_{n,2}-\psi_{n,3}+\psi_{n,4}-\psi_{n,5}+\psi_{n,6}-\psi_{n,7}) \tag{59}$$

There are 7 finite dispersion states and one infinite dispersion state. It is easy to see that they are orthonormal and form a complete set on these 8 lattice points. Note also that widths of the states $\psi^{(K)}_{nm}$ increases with K, but this is in some sense offset by the fact that the states become progressively sparser.

So far we have been working with a complete set of states. We now come to the specific form of the proposal to relax the requirement of using a complete set of states (in the construction of commuting operators $\hat{P}$ and $\hat{X}$): we quite simply drop the remainder states $\chi^{(N)}_{nm}$, which have infinite dispersion, and use only the incomplete set $\psi^{(K)}_{nm}$, $K = 1, 2\cdots N$, which have finite dispersion. That is, in each set of 2^N lattice points in the momentum direction, we use $2^N - 1$ out of the 2^N states.

It seems likely that this approximation will be valid for sufficiently large N, for suitable density matrices. We will show in the next section that reasonable results can be obtained using the $\psi^{(K)}_{nm}$ alone. Here, we briefly look at the remainder states $\chi^{(N)}_{nm}$ to see under what conditions they may be dropped. Note first that the states $\psi^{(K)}_{nm}(x)$ have the property that they vanish at the end-points of the intervals, $x = na \pm a/2$ (as they must, so that their derivative does not have a δ-function). Not surprisingly therefore, the remainder states $\chi^{(N)}_{nm}(x)$ become narrower and progressively more concentrated about the end-points as $N \to \infty$, as one can see from Eq.(48) (since they are in some sense making up for the fact that the $\psi^{(K)}_{nm}(x)$ vanish at the end-points, and the whole set of states is complete). This suggests that, under suitably coarse grained conditions, the behaviour of theses states at single points will become insignificant. We will see this more explicitly below.

We may now use the almost complete set of states to construct the desired projection operators E_{nm} localized on phase space cells. We choose the phase space cells to have 2^N lattice points in the p-direction and 1 lattice point in the x-direction. We have found $2^N - 1$ states with finite dispersion in each of those cells. We therefore take the projector E to be

$$E = \sum_{K=1}^{N}\sum_{m=0}^{2^{N-K}-1} |\psi^{(K)}_{0m}\rangle\langle\psi^{(K)}_{0m}| \tag{60}$$

It is a sum over all states depending only on the lattice points 0 to $2^N - 1$ (in the momentum direction). Importantly, this is an *exact* projection operator which localizes onto a region of phase space – it satisfies

$$E^2 = E \tag{61}$$

exactly. The projection operator for any other cell is easily obtained by unitary displacement, using steps of size 2^N lattice points in the momentum direction (and single steps in the x-direction):

$$E_{nm} = U_{nm}^{(N)} E \left(U_{nm}^{(N)} \right)^\dagger \tag{62}$$

These projectors clearly satisfy the exclusivity condition, Eq.(17), exactly, but do not satisfy the exhaustivity condition Eq.(18) since we have

$$\sum_{nm} E_{nm} + \sum_{n,m} |\chi_{nm}^{(N)}\rangle\langle\chi_{nm}^{(N)}| = 1 \tag{63}$$

But, as we have argued, we expect the χ terms to be negligible under suitable conditions so the projectors E_{nm} should be almost exhaustive

$$\sum_{nm} E_{nm} \approx 1 \tag{64}$$

The approximate nature of this property may not in fact matter for many practical applications. The phase space projector onto a large cell Γ in phase space is defined by

$$E_\Gamma = \sum_{n,m\in\Gamma} E_{nm} \tag{65}$$

The projector onto the region outside Γ is then defined to be

$$\bar{E}_\Gamma = 1 - E_\Gamma \tag{66}$$

so they are trivially exhaustive, $E_\Gamma + \bar{E}_\Gamma = 1$. The key point here is that the remainder states dropped in the construction of E have infinite dispersion so they do not naturally belong in the construction of a phase space projector for a finite region of phase space.

We will need some further properties of E. We have

$$\mathrm{Tr} E = 2^N - 1 \tag{67}$$

which means it projects onto a phase space region of size $(2^N - 1)(2\pi\hbar)$. The object $E/\mathrm{Tr}E$ may be thought of as a density operator and we can compute averages and variances to see the properties of E. We have

$$\langle\hat{p}\rangle_E = \frac{\mathrm{Tr}(pE)}{\mathrm{Tr}E} \tag{68}$$

$$= \frac{2\pi\hbar}{a}\left(2^{N-1} - \frac{1}{2}\right) \tag{69}$$

and

$$(\Delta p)_E^2 \approx \sum_{K=1}^{N} \frac{(\Delta p)_K^2}{2^K} \tag{70}$$

$$\approx \frac{2^{N+1}\pi^2\hbar^2}{3a^2} \tag{71}$$

where we show only the leading terms of large N. We also have

$$\langle \hat{x} \rangle_E = 0 \tag{72}$$

$$(\Delta x)^2_E = \frac{a^2}{12} \tag{73}$$

Since $\langle \hat{p} \rangle_E \neq 0$, it is in fact useful to perform a simple translation in momentum and define a related projector E' with all the same properties as E except that $\mathrm{Tr}(pE') = 0$. We will use this in what follows.

The construction of an exact projector with the above properties is the main achievement of this section and will be used to construct the commuting $\hat{X}$ and $\hat{P}$ operators below.

Finally, we make two minor remarks. First, note that position and momentum enter in the construction in very different ways. However, the constant a is arbitrary, so may be tuned to make the width of the projector E arbitrarily small in either the x or p direction. Also, the states Eq.(35) and (36) are a Fourier transform pair, so we could easily interchange them and start with a set of states that have perfect localization in p, instead of x. It would of course be of interest to find a construction in which x and p entered on an equal footing.

Second, we note for comparison that Omnès has made extensive use of phase quasi-projectors of the form

$$P_\Gamma = \int_\Gamma dpdq \ |p,q\rangle\langle p,q| \tag{74}$$

where $|p,q\rangle$ are phase space localized states, such as the coherent states [9]. These have proved very useful for discussing emergent classicality in quantum theory. However, in contrast to the projectors constructed here, these are not exact projectors, since they obey the approximate relation

$$P_\Gamma^2 \approx P_\Gamma \tag{75}$$

(and so P_Γ and $1 - P_\Gamma$ are only approximately exclusive). It would be of interest to revisit some of Omnès results using the exact projectors constructed here.

5. Validity of Approximate Completeness

Now we come to a crucial check of our approach, which is to determine the conditions under which working with an approximately complete set of states gives reasonable results. The density matrix ρ of the system satisfies $\mathrm{Tr}\rho = 1$. If the states $|\psi^{(K)}_{nm}\rangle$ for $K = 1, \cdots N$ are approximately complete, then we should have

$$\sum_{K=1}^{N} \sum_{nm} \langle \psi^{(K)}_{nm} | \rho | \psi^{(K)}_{nm} \rangle \approx 1 \tag{76}$$

which is the same as

$$\sum_{nm} \mathrm{Tr}\left(E_{nm}\rho\right) \approx 0 \tag{77}$$

We check this. It is most useful to exploit the Wigner representation, Eqs.(21),(22) together with the property Eq.(23), so we have

$$\sum_{K=1}^{N} \sum_{nm} \langle \psi^{(K)}_{nm} | \rho | \psi^{(K)}_{nm} \rangle = 2\pi\hbar \sum_{K=1}^{N} \sum_{nm} \int dpdq \ W^{(K)}(p,q) \ W_\rho(p + \frac{2^{K+1}\pi\hbar}{a} m, q + na) \tag{78}$$

where W_ρ is the Wigner function of ρ and $W^{(K)}$ is the Wigner function of $|\psi^K\rangle$. Now the crucial step is to approximate the discrete sum over n, m with an integral over continuous variables,

$\bar{p} = 2^{K+1}\pi\hbar m/a$, $\bar{q} = na$ and we have

$$\sum_{nm} W_\rho(p + \frac{2^{K+1}\pi\hbar}{a}m, q + na) \approx \int \frac{d\bar{p}d\bar{q}}{2^{K+1}\pi\hbar} W_\rho(p+\bar{p}, q+\bar{q}) \tag{79}$$

$$= \frac{1}{2^{K+1}\pi\hbar} \tag{80}$$

This approximation is valid as long as the Wigner function of ρ is slowly varying over phase space volumes of size $2^K(2\pi\hbar)$. Since K runs up to N, we require slow variation on a scale of size $2^N(2\pi\hbar)$. This is typically the case for density matrices that are sufficiently decohered, as we saw in Section 2. We now have

$$\sum_{K=1}^{N}\sum_{nm} \langle\psi_{nm}^{(K)}|\rho|\psi_{nm}^{(K)}\rangle \approx \sum_{K=1}^{N} \frac{1}{2^K} \tag{81}$$

$$= 1 - \frac{1}{2^N} \tag{82}$$

Hence we do indeed get a result close to 1 as long as N is sufficiently large.

One other related result is worth recording here since it will be used in the next section. Any density operator satisfies the relation

$$\int \frac{dkdq}{2\pi\hbar} U^\dagger(q,k)\rho U(q,k) = 1 \tag{83}$$

where $U(q,k)$ is the unitary shift operator in phase space. This is easily proved using the Wigner representation above, since we have

$$\langle x|U^\dagger(q,k)\rho U(q,k)|y\rangle = \int dp\ e^{\frac{i}{\hbar}p(x-y)}\ W_\rho(p-k, \frac{x+y}{2} - q) \tag{84}$$

and integrating the right-hand side over k and q yields $2\pi\hbar\delta(x-y)$.

One would expect that, for sufficiently slowly varying Wigner functions, a discrete version of the result Eq.(83) would hold. So suppose instead of the operator $U(q,k)$, we use the operators $U_{nm}^{(K)}$ defined in Eq.(50). We then have

$$\langle x|(U_{nm}^{(K)})^\dagger \rho U_{nm}^{(K)}|y\rangle = \int dp\ e^{\frac{i}{\hbar}p(x-y)}\ W_\rho(p - \frac{2^{K+1}\pi\hbar}{a}m, \frac{x+y}{2} - na) \tag{85}$$

Using the same approximation and same steps as in Eq.(79) to do the sum over n, m, it follows that

$$\sum_{nm} (U_{nm}^{(K)})^\dagger \rho U_{nm}^{(K)} \approx \frac{1}{2^K} \tag{86}$$

and therefore

$$\sum_{K=1}^{N}\sum_{nm} (U_{nm}^{(K)})^\dagger \rho U_{nm}^{(K)} \approx 1 - \frac{1}{2^N} \tag{87}$$

6. Construction of Commuting Position and Momentum Operators

We now use the projection operator E'_{nm} to construct commuting position and momentum operators, as outlined in the introduction. They are,

$$\hat{X} = \sum_{nm} X_n\ E'_{nm} \tag{88}$$

$$\hat{P} = \sum_{nm} P_m\ E'_{nm} \tag{89}$$

where $X_n = na$ and $P_m = m \times 2^N\,(2\pi\hbar/a)$. Clearly

$$[\hat{X}, \hat{P}] = 0 \tag{90}$$

as required.

We need to determine whether these operators are close to the original canonical pair, $\hat{x}, \hat{p}$. To do this we need some measure of distance, $\| \hat{P} - \hat{p} \|$. We will define this by

$$\| \hat{P} - \hat{p} \|^2 = \mathrm{Tr}\left((\hat{P} - \hat{p})^2 \rho \right) \tag{91}$$

(and similarly for $\hat{X}$) where ρ is the density operator of the system, which we will assume is reasonably spread out in phase space. A more general measure of distance would include some sort of optimization over ρ, but, for the class of reasonably decohered density operators, the result depends only weakly on ρ.

We have

$$\| \hat{P} - \hat{p} \|^2 = \mathrm{Tr}\left((\hat{P}^2 - 2\hat{P}\hat{p} + \hat{p}^2)\rho \right) + \mathrm{Tr}\left(\hat{P}[\hat{p}, \rho] \right) \tag{92}$$

where the operators $\hat{P}$ are moved to the left in the first term in preparation for inserting the explicit form for $\hat{P}$ below (since $[\hat{P}, \hat{p}] \neq 0$). It is useful to introduce the notation $\mathrm{tr}(\cdots)$ to denote a trace over the incomplete set of states $|\psi^{(K)}_{nm}\rangle$. We may then write

$$\| \hat{P} - \hat{p} \|^2 = \mathrm{tr}\left((\hat{P}^2 - 2\hat{P}\hat{p} + \hat{p}^2)\rho \right) + d_p^2 + \mathrm{Tr}\left(\hat{P}[\hat{p}, \rho] \right) \tag{93}$$

where

$$d_p^2 = \mathrm{Tr}(\hat{p}^2 \rho) - \mathrm{tr}(\hat{p}^2 \rho) \tag{94}$$

In the previous section we saw that $\mathrm{Tr}\rho$ and $\mathrm{tr}\rho$ are both very close (to order 2^{-N}) and it is reasonable to expect the same if ρ is replaced with $\hat{p}^2\rho$, so d_p^2 will negligible and we drop it.

We also expect the term $\mathrm{Tr}\left(\hat{P}[\hat{p}, \rho] \right)$ to be small, since a decohered ρ will be approximately diagonal in momentum. Inserting the explicit expression for $\hat{P}$, it is easily shown that

$$\mathrm{Tr}\left(\hat{P}[\hat{p}, \rho] \right) = \sum_{nm} P_m \mathrm{Tr}\left([\hat{p}, (U^{(N)}_{nm})^\dagger \rho U^{(N)}_{nm}] \right) \tag{95}$$

Since we are working in the approximation in which ρ is sufficiently slowly varying that the sum over n becomes an integral. It is then easily seen (using Eq.(83, for example) that, in this approximation, the object

$$\sum_n (U^{(N)}_{nm})^\dagger \rho U^{(N)}_{nm} \tag{96}$$

is diagonal in momentum. Therefore the term Eq.(95) is zero to the approximation we are using.

Consider the one remaining term in Eq.(93). Inserting the explicit for for $\hat{P}$, we have

$$\| \hat{P} - \hat{p} \|^2 = \sum_{nm} \mathrm{tr}\left(E'_{nm} (P_m - \hat{p})^2 \rho \right) \tag{97}$$

$$= \sum_{nm} \mathrm{tr}\left(E' \hat{p}^2 (U^{(N)}_{nm})^\dagger \rho U^{(N)}_{nm} \right) \tag{98}$$

From Eq.(86), we have

$$\sum_{nm} (U^{(N)}_{nm})^\dagger \rho U^{(N)}_{nm} \approx \frac{1}{2^N} \tag{99}$$

and so

$$\| \hat{P} - \hat{p} \|^2 \approx (\Delta p)^2_E \tag{100}$$

$$\approx \frac{2^{N+1}\pi^2\hbar^2}{3a^2} \tag{101}$$

where only the leading term for large N is given.

A similar (and simpler) calculation shows that

$$\| \hat{X} - \hat{x} \|^2 \approx (\Delta x)^2_E \tag{102}$$

$$= \frac{a^2}{12} \tag{103}$$

Putting all these results together we obtain

$$\| \hat{P} - \hat{p} \| \, . \, \| \hat{X} - \hat{x} \| \approx C\hbar \tag{104}$$

where

$$C = 2^{N/2}\frac{\pi}{3\sqrt{2}} \tag{105}$$

This result is valid in the approximation of large N and for slowly varying density operators.

Eqs. (82) and (104) show that N needs to be chosen to be "large" to get approximate completeness, yet "small" for the commuting operators to be close to the original canonical pair. It seems likely, however, that there is a range of intermediate values that will meet both of these requirements. For example, take $N = 20$. Then $C \sim 10^3$, safely within the limit estimated in Eq.(3), so the difference between the old and new operators will be completely invisible to macroscropic observations. The error due to approximate completeness is of order 2^{-N}, which is about 10^{-6}. So there appears to be a large regime in which the approach works well.

7. Probabilities for Position and Momentum

The results Eq.(82) and Eq.(104) are only indicative, and the true test of closeness of the old and new operators is a comparison of the probabilities for x, p and X, P.

Consider the probability that the variable X lies in the range Δ_X, where Δ_X is the interval $[n_1 a, n_2 a]$, for some integers n_1, n_2. The probability is

$$p(\Delta_X) = \mathrm{Tr}\,(\rho E_{\Delta_X}) \tag{106}$$

where the projector E_{Δ_X} is defined by

$$E_{\Delta_X} = \sum_{n \in \Delta_X} \sum_m E'_{nm} \tag{107}$$

(and the loose notation $n \in \Delta_X$ means $n \in [n_1, n_2]$). The probability for lying outside the region Δ_X is defined using the projector $1 - E_{\Delta_X}$, so that the probabilities add to 1 exactly, and the approximate completeness discussed earlier poses no problems. The probability for P is similarly defined, in terms of a projector E_{Δ_P} defined by

$$E_{\Delta_P} = \sum_{m \in \Delta_P} \sum_n E'_{nm} \tag{108}$$

Both of these probabilities therefore involve the probability p_{nm} associated with our basic phase space cell of size $2^N(2\pi\hbar)$, given by

$$p_{nm} = \mathrm{Tr}(E'_{nm}\rho) = \mathrm{Tr}\left(E' \; (U^{(N)}_{nm})^\dagger \rho U^{(N)}_{nm}\right) \tag{109}$$

Journal of Physics: Conference Series **174** (2009) 012005 doi:10.1088/1742-6596/174/1/012005

It is more usefully written in the Wigner representation,

$$p_{mn} = 2\pi\hbar \int dpdq \; W_{E'}(p,q) \; W_\rho(p+P_m, q+X_n) \tag{110}$$

where, recall, $X_n = na$ and $P_m = m \times 2^N(2\pi\hbar/a)$.

The probability for the variable X is

$$p(\Delta_X) = \sum_{n\in\Delta_X} \sum_m p_{nm} \tag{111}$$

Since the Wigner function of ρ is assumed slowly varying, the sum over m may be approximated by an integral over P_m regarded as a continuous variable. The p in $p + P_m$ is absorbed into the integration and we obtain,

$$p(\Delta_X) \approx \frac{a}{2^N} \sum_{n\in\Delta_X} \int dpdq \; W_{E'}(p,q) \; \rho(q+X_n, q+X_n) \tag{112}$$

Now the p integral may be carried out with the result

$$p(\Delta_X) \approx a \sum_{n\in\Delta_X} \int dq \; \frac{\langle q|E'|q\rangle}{2^N} \; \rho(q+X_n, q+X_n) \tag{113}$$

Since E' is phase space localized, one can see that the first part of the integrand is a smearing function peaked about $q=0$ and with width $(\Delta q)^2 \approx a^2/12$. It is normalized to 1 (for large N) when integrated over q since $\mathrm{Tr}E' = 2^N - 1$. If we assume that

$$\Delta_X^2 \gg \frac{a^2}{12} \tag{114}$$

then the presence of the smearing function makes no difference and we obtain

$$p(\Delta_X) \approx a \sum_{n\in\Delta_X} \rho(na, na) \tag{115}$$

If, as we assume, the density operator is sufficiently slowly varying for the discrete sum to become an integral, we obtain

$$p(\Delta_X) \;\approx\; \int_{\Delta_X} dX \; \rho(X,X) \tag{116}$$

the usual probability for position.

Similarly for P, with analogous approximations, we obtain

$$\begin{aligned} p(\Delta_P) &= 2^N\left(\frac{2\pi\hbar}{a}\right) \sum_{m\in\Delta_P} \int dp \; \frac{\langle p|E'|p\rangle}{2^N} \; \tilde\rho(p+P_m, p+P_m) \\ &\approx 2^N\left(\frac{2\pi\hbar}{a}\right) \sum_{m\in\Delta_P} \tilde\rho(P_m, P_m) \end{aligned} \tag{117}$$

where $\tilde\rho(p,p')$ is the Fourier transform of the density matrix $\rho(x,y)$. Again, for slowly varying density operators the discrete sum becomes an integral and we obtain

$$p(\Delta_P) \;\approx\; \int_{\Delta_P} dP \; \tilde\rho(P,P) \tag{118}$$

which coincides with the usual probability for p.

We therefore find that the probabilities for X, P coincide with those for x, p as long as the following conditions hold:

(i) N is sufficiently large that the errors $1/2^N$ are tolerably small
(ii) The density operator ρ is slowly varying on scales of size $2^N(2\pi\hbar)$.
(iii) The widths of the projections satisfy $\Delta_X\Delta_P \gg 2^N(2\pi\hbar)$.

The key restriction is (ii), the restriction on the density operator. Eqs.(29) and (30) indicate that the density operator can easily become sufficiently broad for (ii) to be satisfied. For example, for a time-evolving state, (ii) will be satisfied for

$$t \gg \left(\frac{\hbar}{\gamma kT}\right)^{1/2} 2^{N/2} \tag{119}$$

This can easily be extremely short, even for large N. Similarly, for a state close to thermal equilibrium, (ii) is satisfied for

$$\frac{kT}{\hbar\omega} \gg 2^N \tag{120}$$

which is again easily satisfied.

8. Summary and Discussion

We have constructed a pair of commuting operators $\hat{X}, \hat{P}$ which, at sufficiently coarse grained scales, are close (in a variety of ways) to the canonical position and momentum operators $\hat{x}, \hat{p}$. These commuting operators offer a new way of defining the relationship between approximate and exact decoherence.

There are two ways in which this programme could be developed and improved. First of all, this paper has concentrated on the technicalities of constructing $\hat{X}$ and $\hat{P}$. It would be useful to develop more details of the conceptual framework in which they are used to discuss emergent classicality.

Secondly, the present approach works at the coarse grained scales of about 2^N phase space cells. Although arguably small to classical eyes, there are some ways in which this scale is quite large, and there are indications that some version of the present approach should work at finer scales. For example, it is known that the Wigner function (and also the P-function) become positive when coarse grained over just one or two phase space cells (rather than 2^N cells, as here), a feature that is often taken as indicator of approximate decoherence and emergent classicality [27]. This suggests that there might be another way to construct commuting position and momentum operators which does not require such a large amount of phase space coarse graining. It would, for example, be of interest to see if von Neumann's original suggestion (involving an explicit orthogonalization of the coherent states) can actually be made to work.

These and related questions will be pursued in future publications.

Acknowledgments

I am particularly grateful to Joshua Zak for many useful conversations over a long period of time. I would also like to thank Jeremy Butterfield and John Cardy for many useful comments, and Keith Hannabus for bringing Ref.[22] to my attention. I would particularly like to thank Thomas Elze for putting together such an interesting meeting.

References

[1] For general discussions of the emergence of classical behaviour from quantum theory, see: Halliwell J J (2005) *Contemp.Phys.* **46** 93, and Hartle J B (1994) *Proceedings of the Cornelius Lanczos International Centenary Confererence* ed J D Brown, M T Chu, D C Ellison and R J Plemmons (Philadelphia: SIAM)
[2] Von Neumann J 1955 *Mathematical Foundations of Quantum Mechanics* (Princeton: Princeton U. Press)
[3] Joos E and Zeh H D 1985 *Z. Phys.* B **59** 223
[4] Halliwell J J and Zoupas A 1995 *Phys.Rev.* D **52** 7294; *do.* 1997 *Phys. Rev.* D **55** 4697

[5] Gell-Mann M and Hartle J B 1990 *Complexity, Entropy and the Physics of Information, SFI Studies in the Sciences of Complexity*, Vol. **VIII** ed W Zurek (Reading, MA: Addison Wesley)
[6] Gell-Mann M and Hartle J B 1993 *Phys. Rev.* D **47** 3345
[7] Griffiths R B 1984 *J.Stat.Phys.* **36** 219; *do.* 1993 *Phys. Rev. Lett.* **70** 2201; *do.* 1996 *Phys. Rev.* A **54** 2759; *do.* 1998 *Phys. Rev.* A **57** 1604
[8] Omnès R 1992 *Rev. Mod. Phys.* **64** 339 and references therein
[9] Omnès R 1997 *J. Math. Phys.* **38** 697
[10] Halliwell J J 1996 *Fundamental Problems in Quantum Theory* ed D Greenberger and A Zeilinger *Annals of the New York Academy of Sciences* **775** 726
[11] Dowker H F and Kent A 1996 *J. Stat. Phys.* **82** 1575; *do.* 1995 *Phys. Rev. Lett.* **75** 3038
[12] Perelomov A M 1971 *Teor. Mat. Fiz* **6** 213
[13] Bargmann V, Butera P, Giradello L and Klauder J R 1971 *Rep. Math. Phys.* **2** 221
[14] Bacry H, Grossmann A and Zak J 1975 *Phys. Rev.* B **12** 1118
[15] Balian R 1981 *C. R. Acad. Sci. (Paris)* **292** 1357
[16] Low F 1985 *Passion for Physics – Essays in Honor of Geoffrey Chew* ed C DeTar, J Finkelstein and Chung-I Tan (Singapore: World Scientific)
[17] Battle G 1988 *Lett. Math. Phys.* **15** 175
[18] Zak J 2001 *J. Phys.* A **34** 1063
[19] Zak J 2003 *J. Phys.* A **36** L553
[20] Zak J 1967 *Phys. Rev. Lett.* **19** 1385
[21] Halliwell J J 2001 *Phys. Rev.* D **63** 085013
[22] Haagersup U and Rordam M 1995 *Duke Math. J.* **77** 627
[23] 't Hooft G 1999 *Class. Quant. Grav.* **16** 3263
[24] Halliwell J J 2005 *Phys. Rev.* A **72** 042109
[25] Anastopoulos C and Halliwell J J 1995 *Phys. Rev.* D **51** 6870
[26] Zak J, private communication.
[27] Diósi L and Kiefer C 2002 *J. Math. Phys.* A **35** 2675

de Sitter equilibrium as a fundamental framework for cosmology

Andreas Albrecht
Department of Physics, One Shields Avenue, University of California, Davis, CA 95616

Abstract. Cosmology might turn out to be the study of fluctuations around a "de Sitter equilibrium" state. In this article I review the basic ideas and the attractive features of this framework, and respond to a number of common questions raised about the de Sitter equilibrium picture. I show that this framework does not suffer from the "Boltzmann Brain" problem, and relate this cosmological picture to recent work on the "clock ambiguity".

1. Introduction

Most approaches to theoretical cosmology ultimately require a framework which can quantify the relative probabilities of different cosmological scenarios. For example, the widely held belief that cosmic inflation gives an account of the initial state of the big bang which is better or "more natural" than models without inflation can only be substantiated within such a framework. One could hope that such a framework would allow the various features of inflation that seem intuitively appealing to directly enhance the probability assigned to cosmologies with inflation in a quantifiable way.

A great many models of cosmic inflation yield a picture called "eternal inflation" [1] where inflation continues forever in (exponentially) increasingly large regions, periodically reheating in isolated regions known as "pocket universes" which can look something like the big bang cosmology we observe (for some reviews and a sample of recent work see [2, 3, 4, 5, 6, 7, 8, 9, 10, 11, 12]). It has long been hoped that the eternal nature of eternal inflation would allow the predictions to be independent of any initial conditions at the start of inflation, and thus yield stable results that could be regarded as the robust predictions of the inflationary cosmology.

So far things have not turned out this way. Eternal inflation typically leads to a variety of types of pocket universes, each produced in infinite quantities. To predict which type of pocket universes are most likely one must regulate these infinities in some way. To date, there is no physically determined (or even universally agreed upon) way to regulate the infinities, and discussions of this matter usually invert the question to: "Which of the various possible regularization schemes yield predictions that actually match what we observe?" (Many do not.) Posing the question this way may be realistic and even interesting in the absence of something better. But as long as one is framing the problem in these terms one is, at least for the time being, abandoning the prospect of actually making concrete predictions from inflation. If one embraces the string theory landscape picture the problem is only made greater by an unspecified multitude of possible inflatons experiencing eternal inflation. One might hope that a better understanding of the underlying physics would eventually regulate the infinities for us and lead to robust predictions (perhaps holographic considerations [13, 14, 15, 12] and/or other approaches to carefully developing the formalism [16, 17, 18, 19, 20] will ultimately help with this). Another possibility is that the current predicament is a reflection of serious problems with whole framework of eternal inflation which will prevent it from ever gaining predictive power.

This paper is about an alternative framework for assigning probabilities in cosmology which I currently find more promsing than eternal inflation. This framework, which I call the "de Sitter Equilibrium" (dSE) cosmology has been examined in two previous papers [21, 22]. In this article I briefly review the dSE picture and call out the key reasons I find this framework attractive. I then comment on some of the questions that have come up regarding dSE since the earlier work appeared. I also offer a quantitative estimate that illustrates how the dSE picture can evade the so-called "Boltzmann Brain" problem. I then discuss how the dSE framework might relate to my recent work on the "clock ambiguity" which at least superficially might appear to be taking things in an orthogonal direction.

2. The dSE framework

2.1. General features

Intuitively, one can think of equilibrium for a system as the state a system achieves if left without external intervention for an arbitrarily long time. By this standard, de Sitter space can be thought of as the equilibrium state achieved by any system which obeys Einstein gravity with a positive cosmological constant Λ. This point requires the cosmological constant to be truly constant, not the energy of a false vacuum that would ultimately decay through tunneling processes [23] or other instabilities [24, 25].

If our universe does have a true (perfectly stable) positive cosmological constant the correct long-term description of the universe would be a state closely approximating de Sitter space. The Hawking temperature of de Sitter space would create fluctuations which could (very rarely) be sufficiently large to cause a temporary deviation into a state that does not look at all like a classical de Sitter space.

Under these conditions the field of theoretical cosmology become the study of the full range of fluctuations out of a de Sitter background, and the interpretation of these different fluctuations as possible cosmological scenarios. One very attractive feature of this picture is that probabilities are assigned to fluctuations in an equilibrium state without any reference to "initial conditions", a concept which is meaningless in a system eternally in equilibrium. For fluctuations in an equilibrium state, it is just the laws of physics (the Hamiltonian) that determine the probabilities of various fluctuations. Another attractive feature of the dSE picture is that the presence of a horizon surrounding any observer in de Sitter space leads to quantitative treatments that appears much more naturally finite, in contrast to the infinities which seem to fundamentally plague eternal inflation.

Of course, the main prediction of the dSE picture might be taken to be that the universe should be observed in a pure de Sitter equilibrium state. The viability of dSE as a framework for theoretical cosmology depends crucially on one's willingness to impose the thermodynamic arrow of time as a condition, rather than something that must predicted as a universal and eternal feature of one's theory. Many authors (although not all [26, 27]) have indeed found it reasonable to use the arrow of time as a condition [28, 29, 30, 31]. This could come about through some sort of anthropic argument related to how critical the arrow of time is for the functioning of observers like us, or it could be a much more narrowly formulated choice to use the fact that we observe an arrow of time as a condition to pose conditional probability questions about the universe.

2.2. de Sitter entropy
An observer in a de Sitter space with positive cosmological constant Λ is surrounded by an event horizon of radius R_Λ given by

$$R_\Lambda^2 = \frac{\rho_\Lambda}{3m_P^2} = \frac{\Lambda}{3}. \tag{1}$$

Throughout we use $\hbar = c = k_B = 1$ and $m_P^2 \equiv l_P^2 \equiv 1/8\pi G$. As with the black hole case, the horizon is associated with an entropy

$$S_\Lambda = \pi \frac{R_\Lambda^2}{l_P^2} \tag{2}$$

Gibbons and Hawking (who were the first to propose and study de Sitter entropy [32]) showed that when other objects (in particular black holes) are put in a de Sitter space the overall entropy goes down. Specifically, when a black hole with entropy S_{BH} is place inside a de Sitter space with entropy S_Λ the horizon of the de Sitter space shrinks so that the entropy of the combined system is

$$S_\Lambda \to S_\Lambda - \sqrt{S_\Lambda S_{BH}} + S_{BH} \approx S_\Lambda - \sqrt{S_\Lambda S_{BH}}. \tag{3}$$

Even though the black hole entropy has added to the total, the decrease due to horizon shrinkage (2nd term) is greater than the increase due to the additional entropy of the black hole (for black holes small enough to fit in the de Sitter horizon in the first place), leading to a net entropy decrease. Any localized matter will look like a black hole sufficiently far away and will decrease the horizon entropy accordingly while adding less entropy than would be the case for the true black hole. Similar arguments can be made about adding a more uniform radiation field to de Sitter space, and thus the statement that the de Sitter entropy is the maximum possible entropy for a system with a stable cosmological constant appears to be quite robust. This maximal entropy feature is one of the reasons de Sitter space appears fit to be regarded as an equilibrium state.

2.3. Recurrences
Both previous papers on dSE cosmology have made use of the following picture: The de Sitter space is regarded as a finite system in a Hilbert space of dimension

$$N_\Lambda \equiv e^{S_\Lambda} . \tag{4}$$

One can then make the ergodic argument that a particular fluctuation which is consistent with N_F microstates occurs with probability

$$P_F = \frac{N_F}{N_\Lambda} \equiv \frac{t_F}{t_R} . \tag{5}$$

The finite system is expected to experience recurrences on a time $t_R \propto N_\Lambda$ and the system spends a time $t_F \propto N_F$ in the fluctuation, which is the origin of the last equality in Eqn. (5). Furthermore, one can infer N_F from the minimum entropy S_F exhibited by the entire system during the fluctuation, and thus write

$$P_F = \frac{N_F}{N_\Lambda} \equiv \frac{e^{S_F}}{e^{S_\Lambda}} . \tag{6}$$

If the fluctuation is a small localized mass concentration in the de Sitter background, then (using Eqn. (3))

$$P_F = \exp\left\{-\sqrt{S_\Lambda S_{BH}}\right\} \tag{7}$$

where S_{BH} is the entropy of the black hole with equivalent ADM mass to the localized fluctuation.

2.4. Probabilities for fluctuation into our observed universe

The formalism being outlined here should in principle serve as a means to calculate the probability of any fluctuation out of the de Sitter equilibrium. Of particular interest are the probabilities for a fluctuation into an inflating state, a fluctuation into a standard big bang cosmology without inflation, and a fluctuation into a "Boltzmann Brain" [22]. If the probability for a fluctuation to enter cosmic inflation is much higher than the others, the formalism supports the prevailing beliefs in theoretical cosmology. Otherwise, the formalism is in conflict with these beliefs. In what follows, we will simply assume that the physical degrees of freedom include an inflaton and other fields suitably chosen and coupled to allow the standard cosmological picture: *inflation* $\to$ *reheating* $\to$ *standard big bang* (SBB) to be a possible behavior of the system.

2.4.1. Probabilities for the Farhi Guth Guven process In [22] Sorbo and I (AS) studied the formation of an inflating universe form dSE via the following process: A small seed fluctuation of localized matter forms with probability P_c which then has probability P_q of fluctuating further into an inflating state. We used Eqns. 3 and 7 to calculate

$$P_c = \exp\left\{-\sqrt{S_\Lambda S_S}\right\} \tag{8}$$

where S_S is the entropy of the black hole with the same ADM mass as the fluctuation. The (necessarily quantum) probability that the local fluctuation excites the inflaton field and tunnels into an inflating state is given by

$$P_q = \exp\left\{-\pi\left(\frac{R_I}{l_P}\right)^2\right\} \approx exp\{-S_I\} \tag{9}$$

where R_I is the de Sitter radius during inflation

$$R_I^2 \equiv \frac{\rho_I}{3m_P^2} \tag{10}$$

and ρ_I is the energy density during inflation. This two-stage process is known as the Farhi Guth Guven process [33, 34]. Combining Eqns. 8 and 9 gives a probability

$$P_I \equiv P_c P_q = \exp\left\{-\sqrt{S_\Lambda S_S} - S_I\right\} \tag{11}$$

for entering into an inflating state.

One should compare this with the probability of fluctuating into the standard big bang without inflation. One can think of the formation of the SBB without inflation as the time reverse of the process of "heat death" of the SBB into de Sitter space. Seen that way the probability of fluctuating straight into the SBB is given by equating the entropy of the fluctuation (S_F in Eqn. (6)) to the entropy of the observed universe $S_{SBB} \approx 10^{100}$ (or perhaps one should use $S_{SBB} \approx 10^{85}$, the entropy the observed universe had in the radiation era). Either way, the probability for fluctuating into the SBB without inflation (from Eqn. (6)) is exponentially lower than getting there via inflation (Eqn. (11))[1]

$$\frac{P_I}{P_{SBB}} = \exp\left\{S_\Lambda - \sqrt{S_\Lambda S_{SBB}}\right\} \gg 1 \tag{12}$$

As discussed at length in [22], this calculation appears to validate the standard picture of modern cosmology, where inflation is highly favored over other scenarios.

[1] This calculation of P_{SBB} follows the calculation of this quantity in [21]. In [22] we considered a different expression which has a less straightforward motivation (and one which I find less compelling at the time I write this) One could use either expression for P_{SBB} in what follows without changing any of the key points.

2.4.2. DKS probabilities In [21] Dyson, Kleban and Susskind (DKS) argued that a universe undergoing inflation looks a lot like a de Sitter space, but with $\rho_\Lambda = \rho_I$, the energy density of the inflaton field during inflation. In parallel with Eqn. (2) one can define

$$S_I = \pi \frac{R_I^2}{l_P^2} \equiv \frac{\rho_I}{3m_P^2} \tag{13}$$

and based on holographic considerations DKS argued that Eqn. (13) should give the entropy of the *entire universe* during inflation. When seen in that light, the probability for a fluctuation leading to inflation is given by

$$P_I = \exp\{-(S_\Lambda - S_I)\} \tag{14}$$

giving

$$\frac{P_I}{P_{SBB}} = \exp\{-(S_{SBB} - S_I)\} \ll 1 \tag{15}$$

Since inflation is disfavored in this picture it runs totally against the modern picture of cosmology. Furthermore DKS argued that, even ignoring inflation there are lots of cosmologies that would be exponentially favored over the observed one (for example, cosmologies with a slightly higher cosmic microwave background temperature and thus a larger value $S'_{SBB} > S_{SBB}$).

3. Issues and discussion

In this section I will comment on some issues that have been raised in various informal conversations that have taken place since the publication of [22]. Many are related to the differences between the AS and DKS calculations outlined in section 2.4.1 and 2.4.2, so that is a good place to start.

3.1. dSE as a heat bath

In the AS approach the de Sitter space is seen as a large system for which the fluctuation involves only a small fraction of the entropy. When the seed forms and fluctuates into an inflating universe, the full physical system describes *both* the de Sitter space (with the fluctuation present, which just looks like a small black hole to an observer elsewhere in the de Sitter space) *and* separately the inflating state, destined to reheat and form the SBB.

This picture depends critically on the belief (not universally shared [35]) that standard features of quantum mechanics should extend to quantum gravity so as to allow a wavefunction that describes one semiclassical spacetime to fluctuate into one that describes more than one semiclassical spacetime represented by different physical degrees of freedom. Thus one would expect the wavefunction following the FGG process to be given by a combination of two wavefunctions, one describing the perturbed background de Sitter space (expressed in one subspace) and one describing the inflating state (expressed in another subspace). As the inflating state reheats, goes through the SBB phase and eventually achieves "heat death" in the final approach to the background de Sitter space, the two parts of the wavefunctions would represent the same physical state and the system would again describe only one semiclassical spacetime, the equilibrium de Sitter space.

To track the entropy of this process, one must evaluate the entropy of the entire system (including both semiclassical spacetimes). From that perspective, the fact that the actual fluctuation that starts inflating has a small entropy makes such a fluctuation more likely, because it removes less entropy from the background de Sitter space than a larger fluctuation would. In other words, the entropy of the entire system makes a smaller drop to start inflation than it would to fluctuate directly into the SBB cosmology, and that makes inflation more likely.

By contrast, the DKS calculation assigns an entropy S_I to the entire universe once inflation starts. In that approach, the smaller S_I, the greater the entropic cost of fluctuating into an inflating state, and generally the cost is much greater to fluctuate into inflation rather than directly into the SBB.

At this point whether one chooses the FGG analysis or DKS depends on what one believes about the way a fundamental theory of quantum gravity should work (whether, for example, the holographic interpretation of inflation used by DKS is appropriate). In this paper I focus on the FGG process, which results in a cosmological picture which I find quite attractive.

Interestingly, the holographic analysis that leads to Eqn. (14) in the DKS approach can be duplicated by the Coleman De Luccia [23] (CDL) tunneling process. The CDL process leads to the same quantitative tunneling probability for entering an inflating state from the background or "fundamental" de Sitter space. However, if both the CDL and FGG processes are allowed, FGG will win because it is much faster [36]. Some have expressed skepticism that the FGG process is physically allowed. If FGG could be eliminated, then the tunneling analysis will be dominated by CDL and thus give the same result as the DKS analysis.

3.1.1. Issues with the FGG process Here are a few issues some have with the FGG process, along with my thoughts in response:

Ill-defined path integral: The path integral methods used by FGG have some curious properties, including a Euclidean interpolating solution that is not a manifold. This has led to some skepticism that perhaps the FGG process is not valid. However, Fischler, Morgan and Polchinski (FMP) [37, 38] used a Hamiltonian approach to calculate the same process which does not have any similar technical peculiarities, and the results were the same (this is why the FMP approach was used in [22]).

The $m \to 0$ limit: In the limit where the mass of the seed fluctuation that initiates the FGG process is taken to zero, the process still proceeds at a finite rate. Some cite this as a problematic feature that suggests the FGG process in unphysical. Personally, I would expect the $m \to 0$ behavior is an artifact of the thin wall approximation (used by both FGG and and FMP) which is likely to break down at some finite value of the seed mass (I consider a lower bound on the seed mass based on such considerations in section 4 below).

AdS/CFT calculations: Freivogel *et al.* [39] analyzed something like the FGG process using AdS/CFT techniques. They showed that the FGG process they considered violates unitarity and thus they argued the FGG process is not physically allowed. However, AdS/CFT analysis is not able to describe processes that allow the boundary of the bulk to fluctuate. While some believe this is a reflection of the fundamental nature of quantum gravity, another interpretation is that this means that an AdS/CFT analysis cannot give a full account of quantum gravity. If one takes that view, then the results in [39] might a reflection of the limitations of the AdS/CFT analysis.

4. dSE and Boltzmann Brains

I now turn to the question of "Boltzmann Brains" in dSE cosmology. The "Boltzmann Brain problem" (first posed in [40] and recently receiving renewed attention [22, 41, 35, 9, 42]) refers to the situation where one's cosmological framework assigns a higher probability to isolated "observers" fluctuating briefly out of equilibrium versus observers which are correlated with large cosmological (non-equilibrium) states of matter the way we are. Boltzmann Brain observers may have (false) memories of planet Earth, the solar system, galaxies, cosmic microwave background maps etc., but their destiny is to be immediately reabsorbed in the background equilibrium state. Since this is not what we experience, cosmological frameworks which strongly favor Boltzmann Brain observers are typically considered to be failures.

Journal of Physics: Conference Series **174** (2009) 012006 doi:10.1088/1742-6596/174/1/012006

In dSE, the probability assigned to a single Boltzmann Brain observer (BB) is given by using

$$S_{BH} = S^S_{Br} \equiv \left(\frac{m_{Br}}{m_P}\right)^2 \tag{16}$$

in Eqn. (7), where S^S_{Br} is the entropy of a black hole with the same ADM mass, m_{Br} as the BB, giving

$$P_{Br} = \exp\left\{-\sqrt{S_\Lambda S_{Br}}\right\} \tag{17}$$

To compare this with Eqn. (11) for the probability for inflation one must consider the value of S_S, the entropy of a black hole with ADM mass equal to the seed mass m_S for the FGG process. While the $m_S \to 0$ gives the dominant process, as discussed above and in [22], the thin wall approximation should break down as one takes this limit. Here we assume the scale of the inflaton potential gives a lower bound on m_S given by

$$m_S = \rho_I R_I^3 = 0.03kg \left(\frac{(10^{16}GeV)^4}{\rho_I}\right)^{1/2} . \tag{18}$$

Keeping only the dominant terms gives

$$\frac{P_{Br}}{P_I} = \frac{\exp\{-\sqrt{S_\Lambda S_{Br}}\}}{\exp\{-\sqrt{S_\Lambda S_S}\}} . \tag{19}$$

Inflation is favored when $S_S < S_{Br}$ which gives

$$\left(\frac{m_{Br}}{0.03kg}\right) > \left(\frac{(10^{16}GeV)^4}{\rho_I}\right)^{1/2} \tag{20}$$

which is a condition that is very easily met. Thus the dSE framework (using the FGG process) does not appear to suffer from the BB problem.

5. dSE, entropy and time symmetry

5.1. The low entropy of inflation-produced SBB's

From a sufficiently fine-grained point of view, an SBB cosmology that started with reheating at the end of an inflationary era will always have much lower entropy than other SBB cosmologies that don't have that special starting point. This can be illustrated by the fact that the time-reverse of an SBB cosmology is highly unlikely to suck all the energy of the universe into the coherent (and extremely low entropy form) of a homogeneous rolling scalar field.

Although this feature is sometimes cited as a problem for inflation [43, 44, 45], in the context of dSE cosmology things are just as they should be. It is a remarkable property of inflation that it can take a low entropy (inflating) state and reheat it into an state that appears sufficiently high in entropy (from a suitably coarse-grained point of view) to look like an SBB cosmology. But the fact that the inflationary path to the SBB actually has much lower fine grained entropy than the "typical" SBB allows it to leave more entropy in the "heat bath", making it a much more likely fluctuation out of dSE.

5.2. What about the time reversed process?

As discussed already in section 3.1, the late time evolution of our observed universe will eventually re-equilibrate with the eternal de Sitter space. Also, since there is only some small probability of tunneling into inflation in the first place, there will be different parts of the wavefunction representing the part that tunneled into inflation and the part that did not. These different parts will "re-cohere" as they all approach (or remain in) the dSE state.

It seems reasonable to expect that this late-time re-equilibration would look very much the same whether it was an inflation-produced SBB or a "regular" SBB doing the equilibration. Furthermore, with nothing defining a universal time direction for the dSE state, one could equally well expect the "tunneling out" and "equilibrating back" events to happen in either order (a meta-observer with a universal time arrow would equally often see the time reversed process, but of course a resident of the corresponding SBB phase would just see time in the direction of increasing entropy).

The dSE framework implies that the process of an SBB "decohering out" of the dSE (the time reverse of re-equilibration) should be much more likely if the SBB state in question is destined (due to extremely subtle microscopic features) to eventually back into an inflating state, than if it were a "regular" SBB with no period of inflation at the "beginning". But it seems extremely hard to imagine how that could be the case, since it seems natural for the re-coherence process to look the same in terms of features that matter to the dynamical properties at that stage (i.e. localized mass concentrations that could change the de Sitter horizon).

My response to this point is that indeed it is hard to see how the probabilities would work out to be so different for the two cases, but we should be used to that sort of thing being hard. A box of gas has a much lower probability of fluctuating in and out of a gold watch state than it does fluctuating in and out of a pile of dirt state. But one would expect the final stages of re-cohering to the equilibrium state (at both ends of the fluctuation) would look very much the same for the cases of the gold watch and the dirt. It is probably very hard to tell at that stage of the evolution what makes one fluctuation much more probable than the other. So I believe if one can tolerate that being a difficult problem for the box of gas, one should be able to tolerate the equivalent problem for the dSE picture being at least as hard.

One can also ask whether there is something fundamentally wrong with the FGG process since it looks so time-asymmetrical (tunneling at one end and equilibration at the other). What place is there for such an asymmetric process in system with so much time symmetry? Even the "watch" and "dirt" fluctuations discussed directly above would be expected to be fully time-symmetric.

To further consider this question I propose the following analogy: Consider a box of gas divided in two by a barrier that can only be penetrated by quantum tunneling, as depicted in Fig. 1. The gas on both sides of the barrier is in equilibrium, with the same equilibrium properties on both sides. Figure 2 depicts a zoomed-in view of a gas particle tunneling through the barrier. The wavefunction of the incoming particle divides into two pieces, a coherent sum of a piece that tunnels through and a piece that bounces back. The time reverse of this process seems very odd: A particle coherently matched on both sides of the barrier in order to pull onto just one side after striking the barrier. But the box of gas is perfectly time-symmetrical and the tunneling process occurring in both directions is equally likely. I suspect our preference for viewing this tunneling event in a particular time direction is a reflection of our own experience living in a very time-asymmetric world, and I believe a similar assessment applies to concerns about the apparent time-asymmetry of the FGG process in the dSE picture.

6. dSE and the clock ambiguity

The dSE cosmological framework appears to be built on laws of physics that are truly eternal and stable. One has the picture of a universe evolving though infinitely many recurrence times

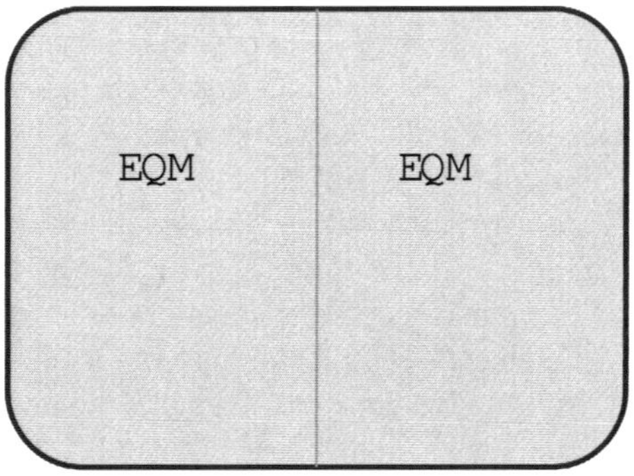

Figure 1. A box of gas with a barrier that can only be penetrated by quantum tunneling. The gas on both sides is in an equilibrium state with the same properties.

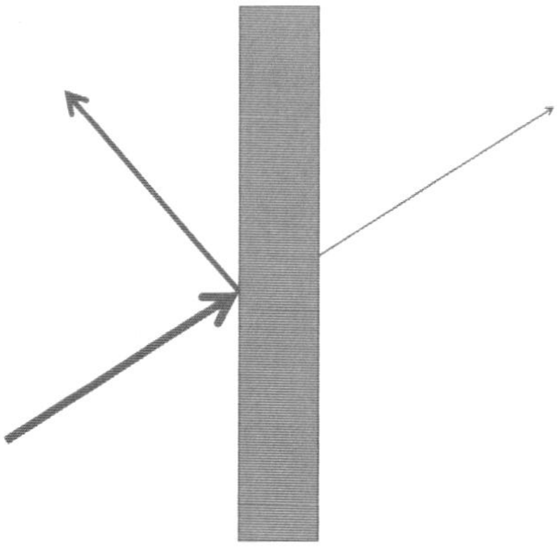

Figure 2. A zoomed-in cartoon of a tunneling event at the barrier depicted in Fig. 1. We find it convenient to think of the tunneling in this highly time-asymmetric way, even though the system as a whole is time-symmetric.

and building up statistics for even the rarest of fluctuations. Apparently all this would require perfectly stable physical laws.

In another line of investigation, Iglesias and I [46, 47, 48] have studied a picture motivated by the "clock ambiguity" in which the laws of physics are emergent and are only stable for a finite period of time. Using arguments developed in [48] (especially Eqn. (8) in that paper) one can find lower bounds on N_H, the number of states in the Hilbert space describing our physical world. One gets values for the lower bound around $\exp\{10^{60}\}$ or even $\exp\{10^{100}\}$, but typically not higher. It the context of that work it makes sense to regard the lower bound as the typical or possibly even the predicted value of N_H. From that point of view, N_H from the clock ambiguity work would be exponentially smaller than the value $N_H \geq \exp\{S_\Lambda\} \approx \exp\{10^{120}\}$ required to have stable physical laws even for one recurrence time.

Thus it would seem that these two ways of thinking about the cosmos and physical laws (the clock ambiguity work and the dSE cosmology) are deeply at odds with one another. That in itself is not necessarily a problem, since all these ideas lie in such a speculative domain that this is probably not the right time to expect them to fit together. But still, I find it interesting to reflect a bit further on the possible connections between these ways of thinking.

In the dSE cosmology, the universe spends by far most of the time simply sitting in equilibrium, subject to tiny fluctuations that are completely uninteresting for cosmology. How much stability is required of the laws of physics in order to sustain this this equilibrium picture? Perhaps not a lot[2]. Perhaps these two pictures can coexist as long as the laws of physics can be stable over cosmologically interesting timescales, a requirement that we showed in [48] is quite easy to meet.

7. Conclusions

Viewing cosmology as the study of fluctuations around a de Sitter equilibrium state has a number of appealing features. It seems more naturally finite than the picture that emerges from "chaotic inflation" (for which the regulation of infinities is a problem), and does not depend on what one

[2] This is related to, but not quite the same as a point made in [49] about the stability of quantum measurements over a recurrence time

assumes for "initial conditions" (it does not have any). When viewed from the point of view of the FGG process (as in [22]), this framework shows that inflation is favored over big bang cosmologies without inflation because inflation presents an entropically "cheaper" way to form cosmological fluctuations. In this paper I have also quantified the claim (first made in [29]) that one of the important roles of inflation can be to evade the "Boltzmann Brain" problem. I've also sought to address various questions that have been raised about this picture, in the hopes of moving those discussions forward.

There are many open questions: Which (if either) interpretation of the dSE cosmology (DKS [21] or AS [22]) is correct? (They give totally different answers.) How can one make the treatment more rigorous? (A new formalism for correlation functions in de Sitter space [50] or other explorations [51] may hold some clues.) How can the predictive power be more fully developed? I find these questions interesting in their own right, but especially so given the many attractive features of de Sitter equilibrium cosmology. Progress on these open questions will help us see if these ideas can ultimately make good on their promise to provide deep theoretical foundations for cosmology.

Acknowledgments
I thank Anthony Aguirre, Raphael Bousso, Alan Guth, Damien Martin and Daniel Phillips for discussions that have influenced this work. Also I thank the organizers, especially Thomas Elze, for a really excellent conference. This work was supported in part by DOE grant DE-FG03-91ER40674.

Bibliography
[1] Linde A D 1986 *Phys. Lett.* B **175** 395–400
[2] Linde A 2007 *JCAP* **0706** 017 (*Preprint* arXiv:0705.1160)
[3] Guth A H 2007 *J. Phys.* A **40** 6811–6826 (*Preprint* arXiv:hep-th/0702178)
[4] Garriga J and Vilenkin A 2008 *Phys. Rev.* D **77** 043526 (*Preprint* arXiv:0711.2559)
[5] Linde A 2008 *Lect. Notes Phys.* **738** 1–54 (*Preprint* arXiv:0705.0164)
[6] Aguirre A, Gratton S and Johnson M C 2007 *Phys. Rev.* D **75** 123501 (*Preprint* arXiv:hep-th/0611221)
[7] Bousso R, Freivogel B and Yang I S 2008 *Phys. Rev.* D **77** 103514 (*Preprint* arXiv:0712.3324)
[8] Linde A, Vanchurin V and Winitzki S 2009 *JCAP* **0901** 031 (*Preprint* arXiv:0812.0005)
[9] De Simone A *et al* 2008 (*Preprint* arXiv:0808.3778)
[10] De Simone A, Guth A H, Salem M P and Vilenkin A 2008 *Phys. Rev.* D **78** 063520 (*Preprint* arXiv:0805.2173)
[11] Bousso R, Freivogel B and Yang I S 2009 *Phys. Rev.* D **79** 063513 (*Preprint* arXiv:0808.3770)
[12] Garriga J and Vilenkin A 2009 *JCAP* **0901** 021 (*Preprint* arXiv:0809.4257)
[13] Albrecht A J, Kaloper N and Song Y S 2002 *Preprint* arXiv:hep-th/0211221
[14] Arkani-Hamed N, Dubovsky S, Nicolis A, Trincherini E and Villadoro G 2007 *JHEP* **05** 055 (*Preprint* arXiv:0704.1814)
[15] Dubovsky S, Senatore L and Villadoro G 2009 *JHEP* **04** 118 (*Preprint* arXiv:0812.2246)
[16] Bousso R 2006 *Phys. Rev. Lett.* **97** 191302 (*Preprint* arXiv:hep-th/0605263)
[17] Linde A 2007 *JCAP* **0701** 022 (*Preprint* arXiv:hep-th/0611043)
[18] Hartle J B, Hawking S W and Hertog T 2008 *Phys. Rev. Lett.* **100** 201301 (*Preprint* arXiv:0711.4630)
[19] Hartle J B, Hawking S W and Hertog T 2008 *Phys. Rev.* D **77** 123537 (*Preprint* arXiv:0803.1663)
[20] Bousso R and Yang I S 2009 *Preprint* arXiv:0904.2386
[21] Dyson L, Kleban M and Susskind L 2002 *JHEP* **10** 011 (*Preprint* arXiv:hep-th/0208013)
[22] Albrecht A and Sorbo L 2004 *Phys. Rev.* D **70** 063528 (*Preprint* arXiv:hep-th/0405270)
[23] Coleman S R and De Luccia F 1980 *Phys. Rev.* D **21** 3305
[24] Mottola E 1985 *Phys. Rev.* D **31** 754
[25] Tsamis N C and Woodard R P 1995 *Ann. Phys.* **238** 1–82
[26] Feynman R P 1965 *The character of physical law [by] Richard Feynman* (Cambridge, Mass.: M.I.T. Press)
[27] Carroll S M and Chen J 2004 *Preprint* arXiv:hep-th/0410270
[28] Page D N 1983 *Nature* **304** 39–41
[29] Albrecht A 2004 In *Science and Ultimate Reality: From Quantum to Cosmos* – honoring John Wheeler's 90th birthday, eds J D Barrow, P C W Davies and C L Harper (Cambridge, UK: Cambridge U. Press) (*Preprint* arXiv:astro-ph/0210527)

[30] Banks T and Fischler W 2005 *Phys. Scripta* T **117** 56–63 (*Preprint* arXiv:hep-th/0310288)
[31] Bousso R, Harnik R, Kribs G D and Perez G 2007 *Phys. Rev.* D **76** 043513 (*Preprint* arXiv:hep-th/0702115)
[32] Gibbons G W and Hawking S W 1977 *Phys. Rev.* D **15** 2738–2751
[33] Farhi E and Guth A H 1987 *Phys. Lett.* B **183** 149
[34] Farhi E, Guth A H and Guven J 1990 *Nucl. Phys.* B **339** 417–490
[35] Banks T 2007 *Preprint* arXiv:hep-th/0701146
[36] Aguirre A and Johnson M C 2006 *Phys. Rev.* D **73** 123529 (*Preprint* arXiv:gr-qc/0512034)
[37] Fischler W, Morgan D and Polchinski J 1990 *Phys. Rev.* D **41** 2638
[38] Fischler W, Morgan D and Polchinski J 1990 *Phys. Rev.* D **42** 4042–4055
[39] Freivogel B *et al.* 2006 *JHEP* **03** 007 (*Preprint* arXiv:hep-th/0510046)
[40] Boltzmann L 1895 *Nature* **51** 413–415
[41] Page D N 2006 *Preprint* arXiv:hep-th/0611158
[42] Mersini-Houghton L and Adams F C 2008 *Class. Quant. Grav.* **25** 165002 (*Preprint* arXiv:0810.4914)
[43] Penrose R 1989 *Annals N. Y. Acad. Sci.* **571** 249–264
[44] Unruh W G 1997 In *Critical Dialogs in Cosmology* ed N Turok (Singapore: World Scientific) p249
[45] Hollands S and Wald R M 2002 *Preprint* arXiv:hep-th/0210001
[46] Albrecht A 1994 *Preprint* arXiv:gr-qc/9408023
[47] Albrecht A J and Iglesias A 2008 *Phys. Rev.* D **77** 063506 (*Preprint* arXiv:0708.2743)
[48] Albrecht A J and Iglesias A 2008 *Preprint* arXiv:0805.4452
[49] Banks T, Fischler W and Paban S 2002 *JHEP* **12** 062 (*Preprint* arXiv:hep-th/0210160)
[50] Freivogel B and Kleban M 2009 *Preprint* arXiv:0903.2048
[51] Hertog T and Horowitz G T 2005 *JHEP* **04** 005 (*Preprint* arXiv:hep-th/0503071)

Journal of Physics: Conference Series **174** (2009) 012007 doi:10.1088/1742-6596/174/1/012007

On spectrum of vacuum energy

G.E. Volovik

Low Temperature Laboratory, Helsinki University of Technology, Finland
L.D. Landau Institute for Theoretical Physics, Moscow, Russia

E-mail: volovik@boojum.hut.fi

Abstract. We discuss the problem of the spectral function of vacuum energy. In traditional approach the ultraviolet divergencies of the vacuum energy are cancelled by imposing relations between different quantum fields and their masses. The emergent theories suggest that the microscopic degrees of the underlying quantum vacuum add to the spectral function and their contribution cancels the diverging zero point energy of quantum fields. Examples of the spectral function of the vacuum energy in the condensed-matter systems with relativity emerging at low energy are presented. In the Sakharov induced gravity situation may be even more dramatic: only microscopic (Planck scale) constituent fields contribute to the vacuum energy, while the diverging zero-point energy of emergent quantum field (gravitational field) is missing. On the other hand consideration of the fermionic condensed matter systems suggests that emergent relativistic fermionic quasiparticles contribute in conventional way as Dirac vacuum.

1. Introduction

The thermodynamic approach [1, 2, 3] to the cosmological constant problems assumes that the quantum vacuum is a stable self-sustained equilibrium state, which is described by compressibility and other characteristics of the response to external perturbations. In this approach, the vacuum energy density appears in two forms. First, there is the microscopic vacuum energy density $\epsilon \sim E_{\rm UV}^4$ characterized by an ultraviolet energy scale $E_{\rm UV}$. Second, there is the macroscopic vacuum energy density which is determined by a particular thermodynamic potential, $\tilde{\epsilon}_{\rm vac}(q) \equiv \epsilon - q\,d\epsilon/dq$, where q is a microscopic variable describing the physics of the ultraviolet vacuum. Thermodynamics and dynamics of q are described by macroscopic equations, because q is a conserved quantity similar to the particle density in liquids, which describes a microscopic quantity – the density of atoms – but obeys the macroscopic equations of hydrodynamics, because of particle-number conservation. Different from known liquids, the quantum vacuum is Lorentz invariant, and the quantity q is Lorentz invariant. For a self-sustained vacuum in full thermodynamic equilibrium and in the absence of matter, the vacuum energy density $\tilde{\epsilon}_{\rm vac}$ is automatically nullified (without fine tuning) by the spontaneous adjustment of the vacuum variable q to its equilibrium value q_0, so that $\tilde{\epsilon}_{\rm vac}(q_0) = 0$. It is important that $\tilde{\epsilon}_{\rm vac}$ contributes to the effective gravitational field equations as cosmological constant Λ. This implies that the effective cosmological constant Λ of a perfect quantum vacuum is strictly zero. This result follows from the stability of the quantum vacuum in the underlying microscopic theory, but it does not depend on details microscopic structure of the vacuum from which gravity emerges.

The problem which we discuss here is the fine structure of the vacuum: how different degrees of freedom, macroscopic (sub-Planckian) and microscopic (super-Planckian), contribute

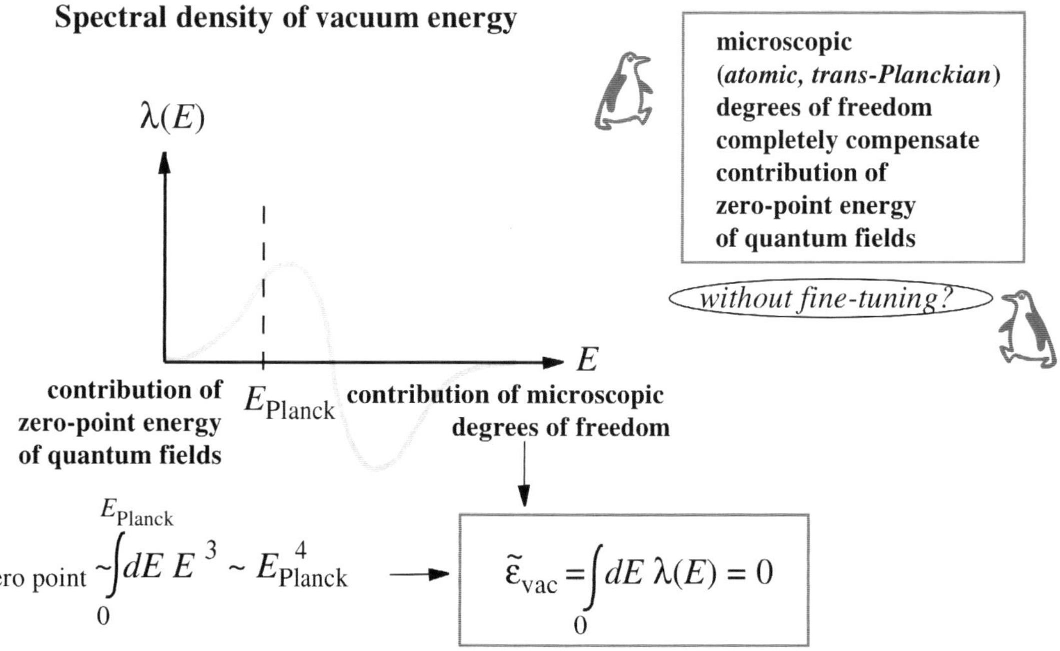

Figure 1. Spectral density of Λ suggested for the vacuum in which bosonic quantum fields dominate in the sub-Planckian region (from the talk in Ref. [5]). Zero point energy of bosonic fields gives rise to the contribution to the energy of quantum vacuum which diverges as E^4 and is estimated as $E_{\rm P}^4$, where $E_{\rm P}$ is Planck energy scale. In the equilibrium self-sustained quantum vacuum [1, 2, 3] this contribution is fully compensated by microscopic (trans-Planckian) degrees of freedom, so that the resulting $\Lambda = \int dE\ \lambda(E) = 0$.

to the vacuum energy. For that one may introduce the spectral density $\lambda(E)$ which describes contributions of different energy scales E to the vacuum energy:

$$\Lambda = \int_0^\infty dE\ \lambda(E)\ , \tag{1}$$

Fig. 1 shows the spectral density $\lambda(E)$ suggested in Ref. [5].

According to this suggestion, at low energy the dominating contribution to the vacuum energy comes from zero point energy of macroscopic bosonic quantum fields and/or from the occupied negative energy levels of fermionic fields. This contribution quartically diverges and approaching the Planck scale produces the huge vacuum energy:

$$\int_0^{\rm P} dE\ \lambda(E) \sim \int_0^{\rm P} dE\ E^3 \sim E_{\rm P}^4\ . \tag{2}$$

However, at higher energy scale $E \sim E_{\rm P}$, the microscopic degrees of the quantum vacuum intervene and finally compensate the contribution of quantum fields. As a result, the spectral density satisfies the sum rule

$$\int_0^\infty dE\ \lambda(E) = 0\ . \tag{3}$$

Such a symmetry between sub- and super-Planckian degrees of freedom [6] is generated by thermodynamics of the equilibrium self-sustained vacuum.

In Refs. [7, 8] the other the sum rules were introduced, which impose the relations between masses of fermionic and bosonic particles in the Standard Model. We discuss these sum rules using our experience with condensed matter systems, and also the Frolov-Fursaem scheme of Sakharov induced gravity [9, 10] which uses the relation between the masses of constituent fields.

2. Contribution of macroscopic quantum fields to Λ

In Refs. [7, 8] the spectral density of the relativistic vacuum has been introduced in a different form which takes into account zero point energy of different quantum fields:

$$\Lambda = \int \frac{d^3p}{(2\pi)^3} \int_0^\infty dx\ \rho(x)\sqrt{x+p^2}\ . \tag{4}$$

Here $\rho(x)$ is a kind of density of states of modes with mass $M = x^{1/2}$. The same spectral function was introduced earlier by Zeldovich [11]. The contribution of a bosonic scalar field with mass M_b to the spectral function is $\rho_b(x) = (1/2)\delta(x - M_b^2)$, while the contribution of one spin component of the fermionic Dirac field with mass M_f is $\rho_f(x) = -\delta(x - M_f^2)$. Summation over all bosonic and fermionic fields gives the spectral function

$$\rho_{\rm macro}(x) = \frac{1}{2}\sum_b \delta(x - M_b^2) - \sum_f \delta(x - M_f^2)\ . \tag{5}$$

In this representation the spectral function $\lambda(E)$ in Eq.(1) is expressed via the spectral density $\rho(x)$ as:

$$\lambda(E) = \frac{E^2}{\pi^2}\int_0^E dp\ p^2\rho(E^2 - p^2)\ . \tag{6}$$

In the energy range $M_{b,f} \ll E \ll E_{\rm UV}$ one has the conventional zero-point contribution to the vacuum energy:

$$\lambda_{\rm macro}(E) = \left(\frac{1}{2}N_b - N_F\right)\frac{E^3}{2\pi^2}\ ,\quad M \ll E \ll E_{\rm UV}\ , \tag{7}$$

where N_b and N_F denote the number of bosonic and fermionic species (which properly includes summation over spin components). In this representation, the nullification of Λ imposes several constraints (sum rules) for $\rho(x)$:

$$\int_0^\infty dx\rho(x) = \int_0^\infty dx\ x\rho(x) = \int_0^\infty dx\ x^2\rho(x) = \int_0^\infty dx\ x^2 \ln x\ \rho(x) = 0\ . \tag{8}$$

While there is only one condition for $\lambda(E)$ in Eq.(3), there are four conditions for $\rho(x)$. The first three conditions in (8) nullify the quartically, quadratically and logarithmically divergent contributions to the vacuum energy respectively [11, 7], while the last condition nullifies the rest finite contribution, which only depends on masses of fields.

These sum rules (13) give the following relation between bosonic and fermionic species and their masses:

$$\frac{1}{2}N_b = N_F\ , \tag{9}$$

$$\sum_b M_b^2 = 2\sum_f M_f^2\ , \tag{10}$$

$$\sum_b M_b^4 = 2\sum_f M_f^4\ , \tag{11}$$

$$\sum_b M_b^4 \ln M_b = 2\sum_f M_f^4 \ln M_f\ . \tag{12}$$

As before, the first three conditions (9-11) remove the diverging terms in vacuum energy [7], while the last condition (12) nullifies the cosmological constant. It was suggested [7, 8] that these sum rules impose relations between the masses of the fermionic and bosonic fields in Standard Model. This is probably valid if all these fields are fundamenta. However this suggestion is not supported in emergent theories.

3. Contribution of microscopic degrees of deep vacuum

In the emergent theory [4, 5], the bosonic and fermionic fields are considered as effective low-energy fields, and in addition there are microscopic degrees of freedom of the deep vacuum which should also contribute to the spectral function of the vacuum energy:

$$\rho(x) = \frac{1}{2}\sum_b \delta(x - M_b^2) - \sum_f \delta(x - M_f^2) + \rho_{\rm micro}(x) \ . \tag{13}$$

The microscopic contribution $\rho_{\rm micro}(x)$ is determined by physics of extreme ultraviolet and thus involves the ultraviolet scale $E_{\rm UV}$, which in principle can be larger that the Planck scale. Let us consider for example $\rho_{\rm micro}(x)$ as continuous function of x. Then the dimensional analysis gives the following estimates: $\rho_{\rm micro}(0) \sim \rho_{\rm micro}(E_{\rm UV}^2) \sim E_{\rm UV}^{-2}$.

At low energy $E \ll E_{\rm UV}$, the contribution of microscopic degrees to the spectral function is:

$$\lambda_{\rm micro}(E) \approx \frac{E^5}{3\pi^2}\rho_{\rm micro}(0) \sim \frac{E^5}{E_{\rm UV}^2} \quad , \quad E \ll E_{\rm UV} \ . \tag{14}$$

This contribution is small compared to the contribution of zero point energy of macroscopic quantum fields in Eq.(7). But at the microscopic energy scale the contribution of the microscopic physics becomes comparable with the macroscopic contribution and finally it should compensate the zero point energy nullifying Λ as required by thermodynamics. The sum rules in Eq.(8) become

$$\frac{1}{2}N_b - N_F + \int_0^\infty dx \rho_{\rm micro}(x) = 0 \ , \tag{15}$$

$$\frac{1}{2}\sum_b M_b^2 - \sum_f M_f^2 + \int_0^\infty dx \ x\rho_{\rm micro}(x) = 0 \ , \tag{16}$$

$$\frac{1}{2}\sum_b M_b^4 - \sum_f M_f^4 + \int_0^\infty dx \ x^2\rho_{\rm micro}(x) = 0 \ , \tag{17}$$

$$\frac{1}{2}\sum_b M_b^4 \ln M_b - \sum_f M_f^4 \ln M_f + \int_0^\infty dx \ x^2 \ln x \ \rho_{\rm micro}(x) = 0 \ . \tag{18}$$

Contrary to conditions (9-12), they do not require any relations between the masses for masses of fields.

Let us take into account that the magnitude of $\rho_{\rm micro}$ is $|\rho_{\rm micro}| \sim E_{\rm UV}^{-2}$ and the characteristic scale of the argument of $\rho_{\rm micro}(x)$ is $x \sim E_{\rm UV}^2$. Then since $(M_b, M_f) \ll E_{\rm UV}$, the masses of fields can be neglected in Eqs.(16-18) and in the main approximation (i.e. in the limit of zero masses) the conditions become:

$$\frac{1}{2}N_b - N_F + \int_0^\infty dx \rho_{\rm micro}^{(0)}(x) = 0 \ , \tag{19}$$

$$\int_0^\infty dx \ x\rho_{\rm micro}^{(0)}(x) = \int_0^\infty dx \ x^2\rho_{\rm micro}^{(0)}(x) = \int_0^\infty dx \ x^2 \ln x \ \rho_{\rm micro}^{(0)}(x) = 0 \ . \tag{20}$$

Correction to the microscopic function $\rho_{\rm micro}(x)$ due to masses of quantum fields can be estimated comparing Eq.(16) with Eq.(20):

$$\rho_{\rm micro} = \rho^{(0)}_{\rm micro} + \delta\rho_{\rm micro} \ , \quad \frac{\delta\rho_{\rm micro}}{\rho^{(0)}_{\rm micro}} \sim \frac{M^2}{E^2_{\rm UV}} \ll 1 \ . \tag{21}$$

This estimate is consistent with the variation of another microscopic parameter – the Newton constant, $\delta G/G \sim M^2/E^2_{\rm UV}$ (see discussion in Ref. [1]).

4. Examples of microscopic contributions

Let us consider several examples of the microscopic contribution $\rho_{\rm micro}(x)$ to the spectral function. The first two examples represent some simple choices of the continuous function $\rho_{\rm micro}(x)$, which explicitly demonstrate the magnitudes of parameters involved. The third one is the induced gravity discussed by Frolov and Fursaev [9].

Let us choose $\rho_{\rm micro}(x)$ as a sum of exponents:

$$\rho_{\rm micro}(x) = \frac{1}{E^2_{\rm UV}} \sum_n a_n \gamma_n \exp(-\gamma_n \frac{x}{E^2_{\rm UV}}) \ , \tag{22}$$

with parameters $|a_n| \sim \gamma_n \sim 1$. The factor $\rho_{\rm micro}(0)$ which enters the low-energy part of the microscopic contribution in Eq.(14) is $\rho_{\rm micro}(0) = E^{-2}_{\rm UV} \sum_n a_n \gamma_n$.

Sum rules in Eqs.(15-18) are

$$\frac{1}{2}N_b - N_F + \sum_n a_n = 0 \ , \tag{23}$$

$$\frac{1}{2}\sum_b M^2_b - \sum_f M^2_f + E^2_{\rm UV} \sum_n \frac{a_n}{\gamma_n} = 0 \ , \tag{24}$$

$$\frac{1}{2}\sum_b M^4_b - \sum_f M^4_f + 2E^4_{\rm UV} \sum_n \frac{a_n}{\gamma^2_n} = 0 \ , \tag{25}$$

$$\frac{1}{2}\sum_b M^4_b \ln M_b - \sum_f M^4_f \ln M_f + 2E^4_{\rm UV} \sum_n \frac{a_n}{\gamma^2_n}\left(\frac{3}{2} - C + \ln\frac{1}{\gamma_n}\right) = 0 \ , \tag{26}$$

where C is Euler's constant. In the main approximation $(M^2_b, M^2_f) \ll E^2_{\rm UV}$ one has

$$\frac{1}{2}N_b - N_F + \sum_n a^{(0)}_n = 0 \ , \tag{27}$$

$$\sum_n \frac{a^{(0)}_n}{\gamma^{(0)}_n} = \sum_n \frac{a^{(0)}_n}{\left(\gamma^{(0)}_n\right)^2} = \sum_n \frac{a^{(0)}_n}{\left(\gamma^{(0)}_n\right)^2} \ln\frac{1}{\gamma^{(0)}_n} = 0 \ . \tag{28}$$

Solutions of Eqs.(27,28) exist only for the number of exponents $n > 3$.

The corrections to the parameters a_n and γ_n due to masses are consistent with Eq.(21): from Eq.(24) it follows that $\delta a_n/a^{(0)}_n \sim \delta\gamma_n/\gamma^{(0)}_n \sim M^2/E^2_{\rm UV}$. The same properties are shared by another simple choice for $\rho_{\rm micro}(x)$ as a polynomial with argument restricted by the microscopic energy scale:

$$\rho_{\rm micro}(x) = \frac{N_F - \frac{1}{2}N_b}{E^2_{\rm UV}} \theta(E^2_{\rm UV} - x) \sum_{n=0}^{3} a_n \frac{x^n}{E^{2n}_{\rm UV}} \ . \tag{29}$$

5. Induced gravity: absence of zero-point energy for graviton?

The third example is the Frolov-Fursaev scheme of Sakharov induced gravity [9, 10, 12]. In this theory, the microscopic degrees of freedom are represented by constituent fermionic and bosonic quantum fields, whose masses are of order of Planck energy scale. The macroscopic physics contains only one effective field – gravity. The action for the induced gravitational field is obtained by integration over the constituent quantum fields. Masses of constituent fields satisfy 6 equations. Four of them, Eqs.(2.16) in Ref. [9], nullify divergent and finite contributions to the cosmological constant, and the other two in Eq.(2.17) in Ref. [9] are needed for nullification of the ultraviolet divergences in the induced Newton constant G. The first four equations coincide with conditions (9-12). This would indicate that in the Frolov-Fursaev scheme the spectral function for the vacuum energy is

$$\rho(x) = \rho_{\rm micro}(x) = \frac{1}{2}\sum_b \delta(x - M_{cb}^2) - \sum_f \delta(x - M_{cf}^2) \, , \tag{30}$$

where M_{cb} and M_{cf} are masses of constituent bosons and constituent fermions respectively. Note that the other two conditions in Eq.(2.17) of Ref. [9], which are determined by Seeley-DeWitt coefficients, discriminate between vector and scalar bosonic fields. The resulting six conditions require existence of the all three types of constituent fields in the quantum vacuum: vector, spinor and scalar fields (see Eq.(2.18) in Ref. [9]). But this is not important when only the vacuum energy is considered.

What is important for us is that all the masses of microscopic constituent fields are of Planck scale, $M_{cb} \sim M_{cf} \sim E_{\rm P}$. As a result, the spectral function $\rho(x)$ is identically zero in the sub-Planckian region of energies: $\rho(x \ll E_{\rm P}^2) \equiv 0$. This would mean that at low energy the spectral function $\lambda(E)$ is also identically zero: $\lambda(E \ll E_{\rm P}) \equiv 0$. At first glance, this is a rather surprising result: there is no contribution $\rho_{\rm macro}(x)$ of the low-energy gravitational field to the vacuum energy, i.e. zero point energy of gravitons is missing contrary to expectation that the contribution of graviton should be the same as that of two massless scalar fields corresposponding to two values of the helicity. On the other hand, in induced theory, graviton is a composite object made of microscopic fields. It would be strange if, say, hydrogen atom composed of 'fundamental' particles – quarks and leptons – will produce diverging terms to the vacuum energy: such a composite object exists only below the scale of compositeness (atomic scale), and thus cannot contribute to the vacuum energy at the scale of the ultraviolet cut-off.

What should one expect from composite objects in a more realistic approach? There are several possibilities. One of them is that zero point energy of graviton does not contribute to the vacuum energy (which does not mean, however, that the Casimir effect is absent for the gravitational waves, since we considered the energy of a homogeneous vacuum). In this case, the low-energy part of $\lambda(E)$ in Eq.(14) starts with $E^5/E_{\rm UV}^2$ rather than with E^3. Alternatively, the zero point part E^3 is present, but is cancelled by the contribution at the compositeness scale $M_{cb} \sim M_{cf}$. This would mean that there is a natural sum rule which nullifies the contribution of a composite object leaving only contribution of microscopic degrees of freedom to the vacuum energy (unless the Higgs condensate is formed).

6. Vacuum spectral function in condensed matter systems

6.1. Relativistic fermions emerging near Fermi surface

It is instructive to consider the spectrum of vacuum energy in such effective relativistic theories which emerge in the known microscopic background. For example the relativistic 1+1 dimensional fermions emerging near the Fermi surface of interacting non-relativistic bare fermions in 1+1 dimension. For illustration, we consider the simplified case of a mean-field model of Fermi gas with attractive interaction, which can exist as isolated system without environment. This model ignores possible instability of the Fermi gas towards the Cooper pairing and other subtle phenomena occurring in 1+1 dimension, which also simplifies the consideration. We choose the non-relativistic fermionic energy spectrum $p^2/2m$, and the energy density as a function of number density n (non-relativistic analog of q in [1, 2, 3]) in the form:

$$\epsilon(n) = \epsilon_0(n) + \frac{1}{2\pi}\int_{-p_F}^{p_F} dp\, \frac{p^2}{2m} = \epsilon_0(n) + \frac{\pi^2}{6m}n^3 \quad , \quad n = \frac{1}{2\pi}\int_{-p_F}^{p_F} dp = \frac{p_F}{\pi} \ , \tag{31}$$

where $\epsilon_0(n)$ is the dominating term in the vacuum energy, while the kinetic part is considered as small perturbation. We need the system, which can live without the external environment, and thus has zero vacuum pressure in equilibrium. Condition of stability of equilibrium vacuum state in the absence of external pressure reads [1]:

$$P_{\rm vac} = -\tilde{\epsilon}_{\rm vac} = -\epsilon(n) + n\frac{d\epsilon}{dn} = 0 \ . \tag{32}$$

The equilibrium chemical potential is

$$\mu = \frac{d\epsilon}{dn} = \frac{p_F^2}{2m} + \frac{d\epsilon_0}{dn} \ . \tag{33}$$

The proper thermodynamic potential which corresponds to cosmological constant is

$$\Lambda = \tilde{\epsilon}_{\rm vac} = -P_{\rm vac} = \epsilon(n) - \mu n = \frac{1}{2\pi}\int_{-p_F}^{p_F} dp\, \frac{p^2 - p_F^2}{2m} + \epsilon_0(n) - n\frac{d\epsilon_0}{dn}. \tag{34}$$

This vacuum energy is the sum of the interaction term and energies of individual particles. Only the latter are relevant for the spectrum of vacuum energy. Let us now rewrite this in terms of the emergent relativistic fermions. Relativistic fermionic quasiparticles live near two Fermi surfaces (Fermi points) at $p = \pm p_F$. Close to Fermi surface at $p = p_F$, the energy spectrum of fermions is linear and relativistic:

$$E = c\tilde{p} \ , \tag{35}$$

where $c = v_F$ and $\tilde{p} = p - p_F$. In terms of the momentum $\tilde{p}$ counted from the Fermi surface, equation (34) reads

$$\Lambda = \Lambda_0 + \frac{N_F}{2\pi}\int_0^{p_F} d\tilde{p}\left(-v_F\tilde{p} + \frac{\tilde{p}^2}{2m}\right) = \Lambda_0 + \frac{c}{\pi}\int_0^{p_F} d\tilde{p}\left(-\tilde{p} + \frac{\tilde{p}^2}{2p_F}\right) \ , \tag{36}$$

where $N_F = 2$ is the number of emergent relativistic massless fermionic species, emerging in the vicinity of the Fermi surface. The parameter Λ_0 plays the role of the bare cosmological constant:

$$\Lambda_0 = \epsilon_0(n) - n\frac{d\epsilon_0}{dn} \ . \tag{37}$$

This contribution to vacuum energy solely comes from the microscopic physics of deep vacuum, and thus its energy spectrum does not make much sense. Introducing the Planck energy $E_{\rm P} = cp_F$ in equation (36), one obtains the following spectrum of Λ:

$$\lambda(E) = \frac{1}{\pi c}\left(-E + \frac{E^2}{2E_{\rm P}}\right)\Theta(E_{\rm P} - E) \quad , \quad \Lambda = \Lambda_0 + \int_0^\infty dE\ \lambda(E) = 0 \ . \tag{38}$$

The first term in brackets in (36) is the contribution of the Dirac sea of two species of relativistic fermions with the spectrum $E = c\tilde{p}$. At low E this is the leading term in the spectrum. After

Journal of Physics: Conference Series **174** (2009) 012007 doi:10.1088/1742-6596/174/1/012007

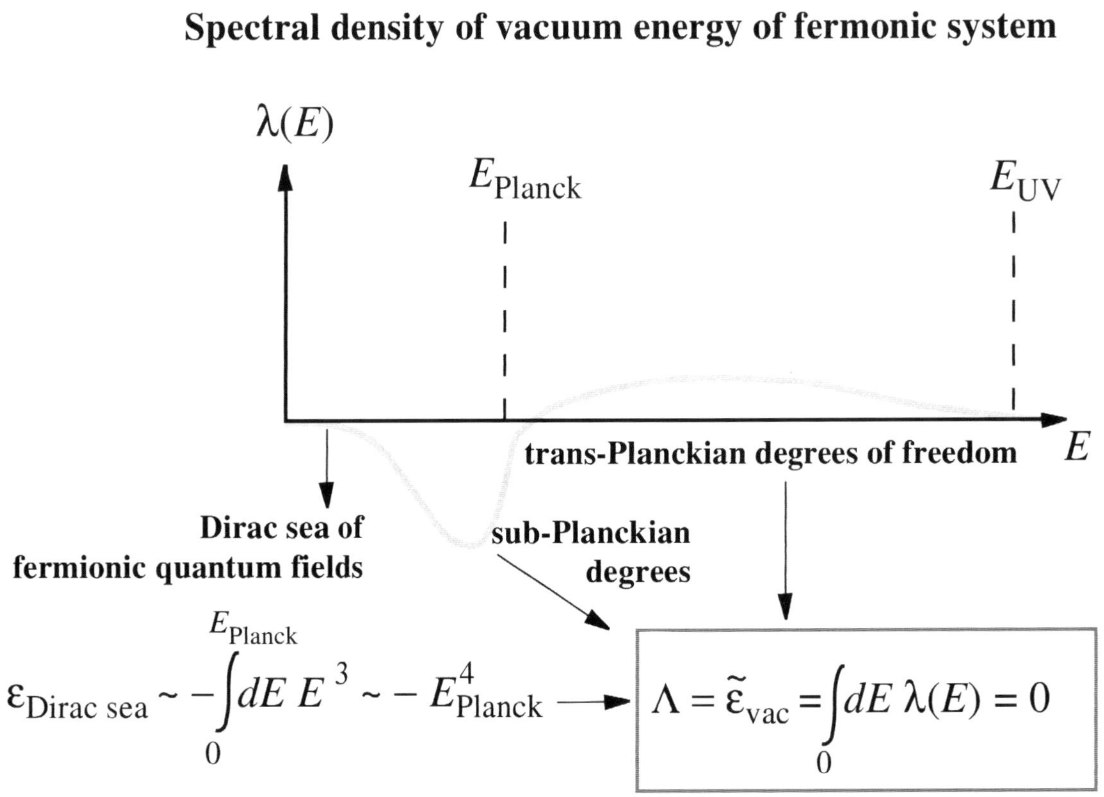

Figure 2. Spectral density of Λ suggested by fermionic condensed matter. The Dirac sea energy of relativistic fermions dominates at $E \ll E_{\rm P}$. It diverges as $-E^4$ producing contribution $\sim -E_{\rm P}^4$ to the vacuum energy. The contribution of the sub-Planckian region is fully compensated by a tiny response of the deep vacuum, whose characteristic energy scale $E \sim E_{\rm UV} \gg E_{\rm P}$.

integration over all energy scales this terms gives the traditional estimate for the Dirac sea energy in 1+1 dimesnions:

$$\epsilon_{\rm Dirac\ vacuum} = -2\int_0^{E_{\rm P}} \frac{dp}{2\pi} E = -\frac{E_{\rm P}^2}{2\pi c} \ . \tag{39}$$

The second term in brackets in (38) comes from the non-linear correction to the Lorentz invariant spectrum (35). At low energy $E \ll E_{\rm P}$ it is the subleading term in the spectrum, but at high energy this term gives the same order of magnitude as the Dirac sea, $\epsilon_{\rm non-linear} = E_{\rm P}^2/6\pi c$. In equilibrium vacuum both these contributions are compensated by the term Λ_0.

The bare cosmological constant can be considered as the vacuum energy density originating from the deep vacuum. To justify that, let us consider the situation, when the term $\epsilon_0(n)$ in (31) has characteristic energy scale $E_{\rm UV}$ much larger than the Planck energy, i.e. $|\epsilon_0(n)| \sim E_{\rm UV}^2/\hbar c \gg E_{\rm P}^2/\hbar c$. In this case, the kinetic part of $\epsilon(n)$ – the second term in (31) – can be considered as small perturbation. Let us find the response of the deep vacuum to this perturbation which leads to the full compensation of vacuum energy. Let n_0 and μ_0 be the equilibrium values of n and μ of the quantum vacuum in zero approximation, when the kinetic term is absent. The obey equations

$$-P_{\rm vac} = \epsilon(n_0) - \mu_0 n_0 = 0 \ , \quad \left.\frac{d\epsilon}{dn}\right|_{n=n_0} = \mu_0 \ . \tag{40}$$

Then from (32) it follows that in the first order approximation one has the following correction to the equilibrium density:

$$n = n_0(1+\alpha) \quad , \quad \alpha = -\frac{\pi^2}{3m} n_0^3 \chi = -\frac{E_{\rm P}^2}{3\pi c}\chi \ . \tag{41}$$

where χ is compressibility of the vacuum state [1, 2, 3]

$$\chi^{-1} = n_0^2 \frac{d^2\epsilon}{dn_0^2} > 0 \ , \tag{42}$$

Since $1/\chi \sim E_{\rm UV}^2/c$, the perturbation of the deep vacuum is small: $|\alpha| \sim E_{\rm P}^2/E_{\rm UV}^2 \ll 1$. The ultraviolet vacuum is so strong that even a tiny response of this vacuum fully compensates the contribution of sub-Planckian region. Adding these ultraviolet degrees of freedom to $\lambda(E)$ one obtains the spectrum of Λ illustrated in Fig. 2.

6.2. Relativistic fermions emerging near Fermi point
In a similar manner one can calculate the spectrum $\lambda(E)$ coming from the 3+1 relativistic Weyl fermions emerging in condensed matter systems with topologically protected Fermi point. In case of p-wave superfluid (see Chapter 7 in Ref. [4]), we present omitting the derivation the leading contributions to $\lambda(E)$ coming in the low-energy region in vicinity of Fermi points:

$$\lambda(E) \approx -N_F \frac{E^3}{2\pi^2}\left(1 - \frac{E}{2E_{\rm P}}\right) \quad , \quad E \ll E_{\rm P} \ . \tag{43}$$

Here N_F is number of emergent relativistic massless fermionic species (for a p-wave superfluid it is $N_F = 2$); and we used such parameters of the system that all three Planck energy scales in Eq.(7.31) of Ref. [4] coincide. The leading E^3 term has the correct form, while the subleading term has power 4 instead of 5 in (14); it has the same relative magnitude as the subleading term in (38) and is normalized to the Planck scale instead of the $E_{\rm UV}$ scale in (14). This is because in both cases the underlying vacuum – the quantum liquid – is non-relativistic even below the Planck scale, and becomes relativistic only in the vicinity of Fermi points, in the limit $E \ll E_{\rm P}$. On the contrary, our vacuum remains Lorentz invariant even above the Planck scale, i.e. even in the region $E_{\rm P} \ll E \ll E_{\rm UV}$ [1, 2].

7. Discussion
The consideration of condensed matter systems suggests that it is the trans-Planckian degrees of freedom of deep ultraviolet vacuum, which compensate the divergences of the vacuum energy occurring in the sub-Planckian region. In equilibrium, the back reaction of the deep vacuum leads to complete nullification of the vacuum energy responsible for the cosmological constant. This suggests that deep vacuum contributes also to the sum rules introduced by Zeldovich [11] for the spectrum of the vacuum energy, as we discussed in Sec. 3. That is why the emergent theories do not support the idea [7, 8] that the sum rules impose relations between the masses of the fermionic and bosonic fields in Standard Model.

The considered examples demonstrate a non-trival behavior of the spectrum of vacuum energy. Massless relativistic fermionic quasiparticles in condensed matter systems and graviton in the Frolov-Fursaev scheme of induced gravity both emerge form the constituent microscopic fields. However, we found that in the fermionic systems, the contribution of the emergent fermions to Λ has a traditional form at low energy, corresponding to the contribution of Dirac sea, Eqs.(43) and (38). On the other hand, the Frolov-Fursaev scheme suggests that induced gravitons do not contribute its zero point energy to Λ. This contradiction requires the further investigation of the spectrum of vacuum energy, in particular the contribution to the vacuum energy from the composite bosons and fermions must be studied.

Acknowledgments
I thank Dmitri Fursaev, Alexander Kamenshchik and Andrei Zelnikov for discussion.

References
[1] Klinkhamer F R and Volovik G E 2008 Self-tuning vacuum variable and cosmological constant *Phys. Rev.* D **77** 085015 (*Preprint* arXiv:0711.3170)
[2] Klinkhamer F R and Volovik G E 2008 Dynamic vacuum variable and equilibrium approach in cosmology *Phys. Rev.* D **78** 063528 (*Preprint* arXiv:0806.2805)
[3] Klinkhamer F R and Volovik G E 2008 $f(R)$ cosmology from q–theory *JETP Lett.* **88** 289 (*Preprint* arXiv:0807.3896)
[4] Volovik G E 2003 *The Universe in a helium droplet* (Oxford: Clarendon Press)
[5] Volovik G E 2006 Emergent physics on vacuum energy and cosmological constant *AIP Conference Proceedings* **850** 26 (*Preprint* arXiv:cond-mat/0507454); Fig. 1 reproduces slide 19 of the oral presentation in: http://ltl.tkk.fi/wiki/images/7/77/Volovik-LT24talk.pdf
[6] Nobbenhuis S 2006 Categorizing different approaches to the cosmological constant problem *Found. Phys.* **36** 613–680 (*Preprint* arXiv:gr-qc/0411093)
[7] Kamenshchik A Y, Tronconi A, Vacca G P and Venturi G 2007 Vacuum energy and spectral function sum rules *Phys. Rev.* D **75** 083514 (*Preprint* arXiv:hep-th/0612206)
[8] Alberghi G L, Kamenshchik A Y, Tronconi A, Vacca G P and Venturi G 2008 Vacuum energy, cosmological constant and Standard Model physics *Pis'ma ZhETF* **88** 819–824 (*Preprint* arXiv:0804.4782)
[9] Frolov V P and Fursaev D 1998 Black hole entropy in induced gravity: Reduction to 2D quantum field theory on the horizon *Phys. Rev.* D **58** 124009 (*Preprint* arXiv:hep-th/9806078)
[10] Frolov V, Fursaev D and Zelnikov A 2003 CFT and black hole entropy in induced gravity *JHEP* **0303** 038 (*Preprint* arXiv:hep-th/0302207)
[11] Zel'dovich Ya B 1968 *Usp. Fiz. Nauk* **95** 209–230 [*Sov. Phys.-Uspekhi* **11** 381 (1968)]
[12] Visser M 2002 Sakharov's induced gravity: a modern perspective *Mod. Phys. Lett.* A **17** 977–992 (*Preprint* arXiv:gr-qc/0204062)

Emergence of quantum mechanics from classical statistics

C. Wetterich
Institut für Theoretische Physik
Universität Heidelberg
Philosophenweg 16, D-69120 Heidelberg
E-mail: c.wetterich@thphys.uni-heidelberg.de

Abstract. The conceptual setting of quantum mechanics is subject to an ongoing debate from its beginnings until now. The consequences of the apparent differences between quantum statistics and classical statistics range from the philosophical interpretations to practical issues as quantum computing. In this note we demonstrate how quantum mechanics can emerge from classical statistical systems. We discuss conditions and circumstances for this to happen. Quantum systems describe isolated subsystems of classical statistical systems with infinitely many states. While infinitely many classical observables "measure" properties of the subsystem and its environment, the state of the subsystem can be characterized by the expectation values of only a few probabilistic observables. They define a density matrix, and all the usual laws of quantum mechanics follow. No concepts beyond classical statistics are needed for quantum physics - the differences are only apparent and result from the particularities of those classical statistical systems which admit a quantum mechanical description. In particular, we show how the non-commuting properties of quantum operators are associated to the use of conditional probabilities within the classical system, and how a unitary time evolution reflects the isolation of the subsystem.

A realization of quantum mechanics as a classical statistical system may shed new light on the conceptual interpretation of experiments based on entanglement, as teleportation or quantum cryptography [1]. One may even speculate that steps in quantum computing [2] could be realized by exploiting classical statistics. Recently, classical statistical ensembles that are equivalent to four-state and two-state quantum mechanics have been constructed explicitely [3, 4]. This constitutes a proof of equivalence of few-state quantum statistics and classical statistics for infinitely many degrees of freedom. In view of the particular manifolds of classical states for these examples one may wonder, however, if quantum statistical systems are very special cases of classical statistics, or if they arise rather genuinely under certain conditions. In this note we argue that quantum statistics can indeed emerge rather generally if one describes small "isolated" subsystems of classical ensembles with an infinity of states.

An atom in quantum mechanics is an isolated system with a few degrees of freedom. This contrasts with quantum field theory, where an atom is described as a particular excitation of the vacuum. The vacuum in quantum field theory is a complicated object, involving infinitely many degrees of freedom. In a fundamental theory of particle physics, which underlies the description of atoms, collective effects as spontaneous symmetry breaking are crucial for its understanding. Our treatment of atoms in the context of classical statistics is similar to the conceptual setting of quantum field theory. A classical statistical system with infinitely many states describes the atom and its environment or the atom and the vacuum. The quantum statistical features become apparent if one concentrates on a subsystem that describes the isolated atom.

Our classical statistical description of quantum systems has three crucial ingredients. (1) *Probabilistic observables* have in a given quantum state only a probabilistic distribution of possible measurement values, rather than a fixed value as for classical observables in a given state of the classical ensemble. Probabilistic observables obtain from classical observables by "integrating out" the environment degrees of freedom. The fact that this map is not invertible avoids conflicts with the Kochen-Specker theorem [5, 6]. The probabilistic nature of the observables can be understood as a result of "coarse graining of the information", starting from deterministic observables on a suitable level. Alternatively, this concept may be used as a basic definition of observables [7].

(2) *Conditional probabilities* are used for a computation of the probabilities for the possible outcomes of two measurements of observables A and B. In particular, if B is measured after A, the outcome of the measurement of B depends on the previous measurement of A. Conditional probabilities induce the concept of *quantum correlations*, and we propose that quantum correlations rather than classical correlations should be used for a description of two measurements. This avoids conflicts with Bell's theorem [9, 10]. The classical correlations are actually not uniquely defined for the quantum system - they depend on properties of the environment.

(3) *The unitary time evolution* of quantum mechanics is a special case of a more general classical evolution which can also describe the phenomena of decoherence and syncoherence. We propose that the unitary time evolution reflects the isolation of the subsystem and corresponds to a partial fixed point (or better "fixed manifold") of the more general classical time evolution.

A classical statistical description of quantum mechanics should not be confounded with a deterministic description. The no go theorems for large classes of local deterministic "hidden variable theories" remain valid. We rather take the attitude that a probabilistic approach is appropriate for the basic setting of both the quantum and the classical world. Deterministic behavior is a particular case (albeit rather genuine) which originates from the collective properties of many degrees of freedom, as the laws of thermodynamics, or the motion of planets which are composed of many atoms.

Probabilistic observables

The most crucial effect of the embedding of the subsystem into classical statistics for infinitely many states is the appearance of probabilistic observables for the description of the subsystem. For a given state of the subsystem - which will be associated with a quantum state - they have a probability distribution of values rather than a fixed value as for the standard classical observables. For subsystems that are equivalent to M-state quantum mechanics the spectrum of the possible outcomes of measurements for the probabilistic observables contains at most M different real values γ_a. In a given quantum state the probabilistic observable is characterized by probabilities w_a to find γ_a, where w_a depends on the state. The simplest example are two-level observables, which can resolve only one bit, such that $\gamma_1 = 1$, $\gamma_2 = -1$.

Let us consider a subsystem that can be described by n discrete classical two-level-observables $A^{(k)}$. The subsystem is embedded into a classical statistical ensemble with infinitely many states labeled by τ. On this level the observables $A^{(k)}$ are standard classical or deterministic observables. They can only take fixed values $A_\tau^{(k)} = \pm 1$ for any state τ of the classical statistical ensemble. We assume that these observables form a basis in a sense to be specified later. The simplest quantum mechanical analogue for $n = 3$ is two-state quantum mechanics, with $A^{(k)}$ corresponding to three orthogonal "spins" in an appropriate normalization. This may be viewed as an atom with spin one half where only the spin degree of freedom is resolved, as for example in Stern-Gerlach type experiments. For arbitrary n, we denote the expectation value or average of $A^{(k)}$ by ρ_k,

$$\rho_k = \langle A^{(k)} \rangle = \sum_\tau p_\tau A_\tau^{(k)} \ , \ -1 \le \rho_k \le 1. \tag{1}$$

Here we have formulated $A^{(k)}$ as classical observables with classical probabilities $p_\tau \ge 0$ for the states τ of the classical statistical ensemble. As usual, one has $\sum_\tau p_\tau = 1$. (Not all classical probability distributions $\{p_\tau\}$ correspond to quantum systems, and we will discuss restrictions below.) The probabilistic observables associated to these classical observables have probabilities $w_\pm^{(k)} = (1 \pm \rho_k)/2$ to find $\gamma_{1,2} = \pm 1$. We can compute $w_+^{(k)}$ by summing the probabilities p_τ over all classical states for which $A_\tau^{(k)} = 1$, and similar for $w_-^{(k)}$. This maps the classical observable, as characterized by the values $A_\tau^{(k)}$ in every classical state τ, to a probabilistic observable characterized by $\gamma_a^{(k)}$ and $w_a^{(k)}$.

The map from the classical observables to the probabilistic observables is not invertible. A different classical observable $A'^{(k)}$, with different values $A'^{(k)}_\tau$ in the classical states, may be mapped to the same probabilistic observable $A^{(k)}$. For this it is sufficient that the probabilities $w_\pm^{(k)}$ are the same for every probability distribution $\{p_\tau\}$ which corresponds to a quantum system. This lack of invertibility avoids conflicts of our classical statistical description with the Kochen-Specker theorem [5], as we we will discuss at the end of this note. Since the classical observables $A_\tau^{(k)}$ contain much more information than the associated probabilistic observables, they "measure" properties of both the quantum system and its environment. The transition to probabilistic observables "integrates out" the environment degrees of freedom [4].

Quantum system and system observables

We will assume that the n numbers ρ_k are the only information that is needed for a computation of expectation values for the "system observables" of the subsystem. In this sense, the state of the subsystem is characterized by the n expectation values of the basis observables $A^{(k)}$. Only a very limited amount of the information contained in the probability distribution $\{p_\tau\}$ for the total system is needed for the subsystem. Quantum mechanics is "incomplete statistics" in the sense of ref. [11]. More precisely, system observables are those classical observables that lead to probabilistic observables A for which the probabilities w_a can be computed in terms of $\{\rho_k\}$. Then the expectation values of functions $f(A)$ can also be computed from $\{\rho_k\}$, $\langle f(A) \rangle = \sum_a f(\gamma_a) w_a(\rho_k)$. We will assume that the relation between w_a and ρ_k is linear.

Our first question concerns a classification of possible system observables for the subsystem. It is straightforward to define rescaled observables $cA^{(k)}$ by $(cA^{(k)})_\tau = cA_\tau^{(k)}, \langle cA^{(k)} \rangle = c\rho_k$. Furthermore, we can trivially shift the observable by a piece e_0 proportional to the unit observable. The rescaled and shifted observables A obey

$$\langle A \rangle = \rho_k e_k^{(A)} + e_0^{(A)}, \tag{2}$$

where repeated indices are summed. Here we associate to each $A^{(k)}$ an n-dimensional unit vector $e^{(k)}$ with components $e_m^{(k)} = \delta_m^k$. Then the vector $e^{(A)}$ reads $e^{(A)} = ce^{(k)}$ if $A = cA^{(k)}$. One may use $c = \hbar/2$ if A describes a spin with standard units of angular momentum. Other units may be employed for alternative interpretations, as for example occupation number $n = (1 + A^{(3)})/2$ which equals one for occupied and zero for empty. (Contrary to widespread belief, $\hbar$ is not a genuine property of quantum mechanics, but rather an issue of units.) In the following we mainly discuss observables with $e_0^{(A)} = 0$.

For an arbitrary system observable A we may write the expectation value $\langle A \rangle = \sum_a \gamma_a w_a(\rho_k)$, with w_a depending linearly on ρ_k, as a linear combination $\langle A \rangle = \sum_k c_k \langle A^{(k)} \rangle$. (If necessary, we substract an appropriate constant shift.) This should hold for all probability distributions

$\{p_\tau\}$ which describe the quantum system. (Such observables may correspond to rotated spins in the quantum mechanical analogue.) We can associate to each such observable the vector $e^{(A)} = \sum_k c_k e^{(k)}$, $e_k^{(A)} = c_k$, such that eq. (2) remains valid. Explicit constructions of such probabilistic observables can be found in [4]. The possibility to write the expectation value in the form (2) is a necessary, albeit not sufficient condition for a classical observable to be a system observable.

Probabilistic observables are characterized by the spectrum $\{\gamma_a\}$ of possible measurement values, and the associated probabilities w_a. The multiplication of the observable by a constant c and the addition of a piece proportional to the unit observable are always defined by $\gamma_a \to c\gamma_a + e_0$. However, the sum and the product of two probabilistic observables A, B are, in general, not defined. If system observables exist for arbitrary real c_k this allows a definition of linear combinations of probabilistic system observables as represented by linear combinations of the associated vectors e. For $C = c_A A + c_B B$ one has $e^{(C)} = c_A e^{(A)} + c_B e^{(B)}$, $\langle C\rangle = c_A\langle A\rangle + c_B\langle B\rangle$.

Next we are interested in some general properties of the basis observables. For example, one typical question may ask if two of them can have simultaneously a sharp value. A "classical eigenstate" of a probabilistic observable A is an ensemble for which A has a "sharp value" with vanishing dispersion, $\langle A^2\rangle - \langle A\rangle^2 = 0$. For example, the eigenstate of the observable $A^{(k)}$ with "classical eigenvalue" $\langle A^{(k)}\rangle = 1$ is characterized by $p_\tau = 0$ whenever $A_\tau^{(k)} = -1$. The maximal number of sharp "basis observables" $A^{(k)}$ can be characterized by the "purity" P of the ensemble,

$$P = \rho_k \rho_k. \tag{3}$$

(Note that P depends on the set of basis observables that characterize the subsystem.) For $P = 0$ one finds equipartition with $\langle A\rangle = 0$ for all two level observables. Obviously, $\tilde{M}$ sharp observables require $P \geq \tilde{M}$, since at least for $\tilde{M}$ values of k one needs $\rho_k = \pm 1$. For an ensemble with $P = 1$ at most one observable $A^{(k)}$ can be sharp. Typical classical ensembles that describe isolated quantum systems will have a maximal purity smaller than n, such that not all $A^{(k)}$ can have sharp values simultaneously. We recall that the purity (3) is a statistical property involving expectation values. For a given classical state τ all observables $A^{(k)}$ have a sharp value.

Density matrix and wave function

For M an integer obeying $M \geq P + 1$ we may represent the ρ_k by an $M \times M$ hermitean "density matrix" $\rho_{\alpha\beta}$:

$$\rho = \frac{1}{M}(1 + \rho_k L_k). \tag{4}$$

The matrices L_k [3] are $SU(M)$-generators $k = 1..M^2 - 1$, obeying

$$\begin{aligned} \mathrm{tr}L_k &= 0 \ , \ L_k^2 = 1 \ , \ \mathrm{tr}(L_k L_l) = M\delta_{kl}, \\ \{L_k, L_l\} &= 2\delta_{kl} + 2d_{klm}L_m \ , \ [L_k, L_l] = 2if_{klm}L_m. \end{aligned} \tag{5}$$

The matrix ρ has indeed the properties of a density matrix,

$$\mathrm{tr}\rho = 1 \ , \ \rho_{\alpha\alpha} \geq 0 \ , \ \mathrm{tr}\rho^2 = \frac{1}{M}(1 + P) \leq 1. \tag{6}$$

In analogy to quantum mechanics, a "classical pure state" obeys $\rho^2 = \rho$ and therefore requires $P = M - 1$. We can also associate to any system observable A an operator $\hat{A}$ such that the quantum mechanical rule for the computation of expectation values holds ($e_k \equiv e_k^{(A)} = c_k$)

$$\hat{A} = e_k L_k \ , \ \langle A\rangle = \rho_k e_k = \mathrm{tr}(\rho \hat{A}). \tag{7}$$

We will concentrate on the minimal M needed for a given maximal purity of the ensemble. For $n = M^2 - 1$ the operators for the basis variables $A^{(k)}$ are given by L_k. If $n < M^2 - 1$ only part of the L_k are used as a basis for the observables.

Many characteristic features of the system observables A can now be inferred from standard quantum mechanics, as demonstrated by a few examples. For $M = 2$ at most one of the three possible two-level-observables $A^{(k)}$ can have a sharp value. This occurs for an ensemble where the density matrix describes a quantum mechanical pure state, $\rho = \frac{1}{2}(1 \pm \hat{A})$, $\mathrm{tr}\rho^2 = \frac{1}{2} + \frac{1}{4}\mathrm{tr}\hat{A}^2 = 1$, with $\hat{A}^{(k)} = L_k = \tau_k$. For such an ensemble the expectation value of the two orthogonal two-level-observables must vanish, $\langle A^{(l)} \rangle = 0$ for $l \neq k$. Thus, whenever one basis observable is sharp, the two others have maximal uncertainty, as for the spin one-half system in quantum mechanics.

Another example for $M = 4$ describes two different two-level-observables (say the z-direction of two spins S_z^1, S_z^2) by $L_1 = \mathrm{diag}\ (1, 1, -1, -1)$ and $L_2 = \mathrm{diag}\ (1, -1, 1, -1)$. The product of the two spins is represented by $L_3 = \mathrm{diag}\ (1, -1, -1, 1)$. Consider an ensemble characterized by $\rho_3 = -1$, $\rho_1 = \rho_2 = 0$. For this ensemble one has $\langle A^{(1)} \rangle = \langle A^{(2)} \rangle = 0$ such that for both two-level-observables the values $+1$ and -1 are randomly distributed in the ensemble. Nevertheless, $\langle A^{(3)} \rangle = -1$ indicates that the two "spins" are maximally anticorrelated. Whenever the first spin takes the value $+1$, the second one necessarily assumes -1 and vice versa. At the end of this note we discuss that pure states of this type show the characteristics of an entangled quantum state. Our third example considers the observable S corresponding to the sum $\hat{S} = L_1 + L_2$. For the particular pure state density matrices $(\hat{\rho}_m)_{\alpha\beta} = \delta_{m\alpha}\delta_{m\beta}$ one has $\langle S \rangle = 2$ (for $m = 1$), $\langle S \rangle = 0$ (for $m = 2, 3$) and $\langle S \rangle = -2$ (for $m = 4$). Thus S has the properties of a total spin, composed of two half integer spins (say $S_z = S_z^1 + S_z^2$).

The density matrix can be diagonalized by a unitary transformation. In consequence, any pure state density matrix can be written in the form $\rho = U\hat{\rho}_m U^\dagger$ for a suitable U, with $UU^\dagger = 1$. This allows us to "take the root" of a pure state density matrix by introducing the quantum mechanical wave function ψ_α as an M-component complex normalized vector, $\psi^\dagger\psi = 1$,

$$\begin{aligned} \rho_{\alpha\beta} &= \psi_\alpha\psi_\beta^* \ , \ \psi_\alpha = U_{\alpha\beta}(\hat{\psi}_m)_\beta \ , \\ (\hat{\psi}_m)_\beta &= \delta_{m\beta} \ , \ \langle A \rangle = \psi^\dagger \hat{A}\psi. \end{aligned} \tag{8}$$

All the usual rules for expectation values in quantum mechanical pure states apply.

Pure states play a special role since they describe classical ensembles with minimal uncertainty for a given integer P. For $M = 4$ a pure state has purity $P = 3$ and three different observables can have sharp values, corresponding to the maximum number of three commuting quantum mechanical operators. For $P > 1, M = P + 1$ and $\{L_k, L_l\} = 2\delta_{kl} + 2d_{klm}L_m$ the condition for a pure state, $\rho^2 = \rho$ or $\rho_k[\rho_l d_{klm} - (M-2)\delta_{km}] = 0$, is not automatically obeyed for all ρ_k with $\rho_k\rho_k = P$. For a pure state, the "copurity" $C = \mathrm{tr}[(\rho^2 - \rho)^2]$ must also vanish. While the purity P is conserved by all orthogonal $SO(n)$ transformations of the vector (ρ_k), pure states are transformed into pure states only by the subgroup of $SU(M)$ transformations. The $SU(M)$ transformations are realized as unitary transformations of the wave function ψ, where the overall phase of ψ remains unobservable since it does not affect ρ in eq. (8). In our classical statistical description of quantum phenomena, the particular role of the classical pure states constitutes the basic origin for the unitary transformations in quantum mechanics. Just as in quantum mechanics, we can write the density matrix ρ for an arbitrary ensemble as a linear combination of appropriate pure state density matrices.

Algebra of probabilistic quantum observables

The vector $e^{(A)}$ is sufficient for a determination of the expectation value of A in any state of the subsystem (characterized by ρ_k). However, the typical observables for the subsystem are probabilistic observables and we further have to specify the probability distribution for a possible outcome of measurements for every state $\{\rho_k\}$. This is needed for a computation of expectation values $\langle A^p\rangle$ of powers of A. We will next concentrate on "quantum observables". They constitute a subclass of the system observables for which the operators associated to A^p are given by $\hat{A}^p$. For a given M we consider observables with a spectrum of at most M different values γ_a. We first consider a non-degenerate spectrum of M different γ_a and identify $a = \alpha$. The probabilities to find γ_α in the state ρ_k of the subsystem are denoted by $w_\alpha(\rho_k) \geq 0$, $\sum_\alpha w_\alpha(\rho_k) = 1$, $\langle A\rangle = \sum_\alpha w_\alpha(\rho_k)\gamma_\alpha = \rho_k e_k$. The expectation value $\langle A\rangle = \text{tr}(\rho\hat{A})$ is invariant under a change of basis by unitary transformations, $\rho \to \rho' = U\rho U^\dagger$, $\hat{A} \to \hat{A}' = U\hat{A}U^\dagger$. We may choose a basis with diagonal $\hat{A}' = diag(\lambda_1, \ldots, \lambda_M)$, $\langle A\rangle = \sum_\alpha \rho'_{\alpha\alpha}\lambda_\alpha$, suggesting that the spectrum γ_α can be identified with the eigenvalues λ_α of the operator $\hat{A}$, and $w_\alpha(\rho_k) = \rho'_{\alpha\alpha}$.

Among the classical observables we therefore consider the subclass of quantum observables which generate probabilistic observables with the property

$$\langle A\rangle = \sum_\alpha w_\alpha(\rho_k)\gamma_\alpha \ , \ w_\alpha(\rho_k) = \rho'_{\alpha\alpha}. \tag{9}$$

To each quantum observable, we can associate a hermitean quantum operator, with a spectrum of possible measurement values given by the eigenvalues $\lambda_\alpha = \gamma_\alpha$ of the operator. The classical probability for the outcome of the measurement in a given state is the corresponding diagonal element of the density matrix in a basis where $\hat{A}$ is diagonal. We describe the mathematical structures related to quantum observables in more detail at the end of this note, where we also deal with degenerate spectra of less than M different γ_a. There we will construct an explicit realization for the quantum observables.

We can now consider powers of the probabilistic observable A, $\langle A^p\rangle = \sum_\alpha w_\alpha(\rho_k)\gamma_\alpha^p$. The observable A^p should belong to the observables of the subsystem, since it can be associated with p measurements of A, multiplying the p measurement results that must be identical. We can therefore associate an operator $\tilde{A}_p$ to the observable A^p , $\langle A^p\rangle = \text{tr}(\rho\tilde{A}_p)$. For quantum observables this is realized by $\tilde{A}_p = \hat{A}^p$ and we conclude $\langle A^p\rangle = \text{tr}(\rho\hat{A}^p)$. For $\langle A^2\rangle = \text{tr}(\rho\hat{A}^2)$ a classical eigenstate of A obeys $[\text{tr}(\rho\hat{A})]^2 = \text{tr}(\rho\hat{A}^2)$ and the possible classical eigenvalues are the eigenvalues of the operator $\hat{A}$. If a pure state is an eigenstate of A one has $\hat{A}\psi = \lambda\psi$ with $\lambda \equiv \lambda_\alpha = \gamma_\alpha$ one of the eigenvalues of $\hat{A}$.

Consider now the classical ensemble of states τ which describe the subsystem together with its environment. It is straightforward to characterize the properties of the classical observables A_τ that are mapped to the probabilistic quantum observables. First, the spectrum of possible outcomes of individual measurements equals the (sharpe) values of the classical observable A_τ in the classical states τ. It consists of the eigenvalues of $\hat{A}$. Second, the probabilities for the states p_τ must be distributed such that w_α can be written as a linear combination of $\rho_k\langle A^{(k)}\rangle$ for all states with $P \leq M - 1$. At first sight this second requirement for a classical observable to be a quantum observable may seem rather special. However, in many circumstances the quantum observables are related to the basis observables by simple "physical operations". For example, the spin observable in an arbitrary direction obtains from the three basis observables $A^{(k)}(M = 2)$ by rotation, such that the relation (9) arises naturally [4]. Other simple operations are the addition and multiplication of "commuting observables", as we will explain below. At this point it may be worthwhile to pause. We have selected a set of classical observables with a remarkable property: whenever the ensemble obeys a simple constraint on the purity, $P \leq M - 1$, all the laws of quantum mechanics apply for these classical observables, as for example Heisenberg's uncertainty relation based on the commutator of the associated operators.

In summary, we define the quantum observables as a subclass of the classical observables with a spectrum of at most M different possible measurement values γ_a. Furthermore, the associated probabilities w_a should be given by the relation (9) which is linear in ρ_k, with coefficients depending only on the observable. This implies the relation (2) and defines $e_k^{(A)}$. (We may add a piece proportional to the unit observable.) It is then also possible to compute the expectation values $\langle A^p \rangle$ in terms of the information contained in the quantum state through an expression linear in the ρ_k. Therefore $\langle A^p \rangle$ does not involve any properties of the environment - these expectation values are completely determined by the subsystem. To any quantum observable we can associate a unique vector (e_k) and therefore a unique quantum operator. In turn, to each quantum operator we can also associate a unique probabilistic observable, with w_α given by eq. (9). On the level of probabilistic observables the quantum observables can therefore be fully characterized by the vector (e_k).

The correspondence between classical probabilistic observables and quantum operators allows for the introduction of a "quantum product" AB between the probabilistic observables, which is associated to the operator product $\hat{A}\hat{B}$. Together with linear combinations, this defines an algebra for the probabilistic quantum observables. The algebra of probabilistic quantum observables is isomorphic to the algebra of quantum operators. For both, it can be expressed as operations in the space of (e_k). In particular, we observe $A^2 = AA$. We will see below that a particular use of the "quantum product" AB arises from an investigation of conditional probabilities.

In general, a probabilistic quantum observable A describes an equivalence class of classical observables that all lead to the same quantum operator $\hat{A}$. The product $\hat{A}\hat{B}$ then induces a product structure between equivalence classes. If a representative for each class is selected one can also define AB on the level of classical observables [11]. We emphasize that the product AB is *not* the classical or pointwise product $A \cdot B$ where $(A \cdot B)_\tau = A_\tau B_\tau$.

Quantum correlations

Beyond a rule for the computation of expectation values of observables, any theory must provide a prediction for the outcome of two consecutive measurements. After a first measurement of the observable A the result of a subsequent measurement of another observable B is, in general, influenced by the first measurement. The measurement of A has changed the ensemble and the knowledge of the observer. For simplicity we concentrate on two-level-observables, $\langle A^2 \rangle = \langle B^2 \rangle = 1$, $\hat{A}^2 = \hat{B}^2 = 1$. The probability of finding $B = 1$ after a measurement $A = 1$ amounts to the *conditional probability* $(w_+^B)_+^A$. After the measurement $A = 1$ the ensemble must be an eigenstate to the eigenvalue $\bar{A} = 1$ - otherwise a subsequent measurement of A would not necessarily yield the same value as the first one. Then B is measured under this condition.

We take here the attitude that there is only one given reality, but physicists can at best give a statistical description of it. The "fundamental laws" are genuinely of a statistical nature [12] and only establish relations within different possibilities for the history of the real world. Measuring for an observable A in a given state the value $\gamma_{\bar{\alpha}}$ simply eliminates the other possible alternatives (which may have nonvanishing probabilities $w_{\alpha \neq \bar{\alpha}}$). After the measurement of A it makes only sense to ask what are the outcomes of other measurements under the condition that A has been measured to have the value $\gamma_{\bar{\alpha}}$.

Correspondingly, the *measurement correlation* or *conditional correlation* $\langle BA \rangle_m$ multiplies the measured values of A and B, weighed with the probabilities that they occur

$$\begin{aligned} \langle BA \rangle_m &= (w_+^B)_+^A w_{+,s}^A - (w_-^B)_+^A w_{+,s}^A \\ &\quad -(w_+^B)_-^A w_{-,s}^A + (w_-^B)_-^A w_{-,s}^A, \end{aligned} \tag{10}$$

with $w_{\pm,s}^A$ the probability that A is measured as ± 1 in the state s and $\langle A \rangle = w_{+,s}^A - w_{-,s}^A = \mathrm{tr}$ $(\rho \hat{A})$, $w_{+,s}^A + w_{-,s}^A = 1$, $(w_+^B)_\pm^A + (w_-^B)_\pm^A = 1$. The conditional correlation needs a specification of the conditional probabilities as $(w_\pm^B)_+^A$. For their computation we use the prescription that after the first measurement $A = 1$ the density matrix ρ_{A+} must describe an eigenstate of A,

$\mathrm{tr}(\hat{A}\rho_{A+}) = 1$. This effect of a first measurement may be called *state reduction*. The subsequent measurement of B then involves this state,

$$(w_+^B)_+^A - (w_-^B)_+^A = \mathrm{tr}(\hat{B}\rho_{A+}). \tag{11}$$

The relation (11) is based on the property that the quantum observable B obeys eq. (2) for an arbitrary quantum state of the subsystem. Our assumption is therefore that after the measurement of A the classical ensemble still describes a quantum system. This seems reasonable for appropriate measurements since otherwise the first measurement destroys the isolation of the subsystem instead of only changing its state. This assumption has far reaching consequences, however. It necessarily implies eq. (11) and excludes the option of using the classical correlation for the general description of subsequent measurements, as we will see below.

For $M = 2$ the matrix ρ_{A+} is unique, $\rho_{A+} = \frac{1}{2}(1 + \hat{A})$, such that

$$(w_\pm^B)_+^A = \frac{1}{2} \pm \frac{1}{4}\mathrm{tr}(\hat{B}\hat{A}) \ , \ (w_\pm^B)_-^A = \frac{1}{2} \mp \frac{1}{4}\mathrm{tr}(\hat{B}\hat{A}). \tag{12}$$

However, for $M > 2$ one has $\mathrm{tr}(\hat{A}\rho_{A+}) = 1$ for

$$\begin{aligned} \rho_{A+} &= \frac{1}{M}(1 + \hat{A} + X) \ , \ \mathrm{tr}(\hat{A}X) = 0 \ , \\ \mathrm{tr}X &= 0 \ , \ \mathrm{tr}X^2 = M(P-1), \end{aligned} \tag{13}$$

with

$$\rho_{A+}^2 - \rho_{A+} = \frac{1}{M^2}(X^2 + \{\hat{A}, X\}) - \left(1 - \frac{2}{M}\right)\rho_{A+}. \tag{14}$$

A necessary condition for ρ_{A+} describing a pure state is $\mathrm{tr}X^2 = M(M-2)$, $P = M - 1$, which implies $X = 0$ only for $M = 2$.

We may distinguish between a "maximally destructive measurement" where all information about the original ensemble except for the value of A is lost, and a "minimally destructive measurement" for which an original pure state remains a pure state after the measurement. A maximally destructive measurement is described by $X = 0$ in eq. (13), leading to

$$\langle B \rangle_{A+} = (w_+^B)_+^A - (w_-^B)_+^A = \frac{1}{M}\mathrm{tr}(\hat{B}\hat{A}) = \langle BA \rangle_{\max}. \tag{15}$$

Here we denote by $\langle BA \rangle_{\max}$ the conditional correlation for maximally destructive measurements and use that ρ_{A-} obtains from ρ_{A+} by changing the sign of $\hat{A}$ in eq. (13) (with $X = 0$). We can use $\langle BA \rangle_{\max}$ for the definition of a scalar product between the observables B and A, since it does not depend on the initial ensemble. The two-level observables $A^{(k)}$ form an orthogonal basis in this sense, $\langle A^{(k)} A^{(l)} \rangle_{\max} = \delta_{kl}$.

A minimally destructive measurement of $A = 1$ projects out all states with $A = -1$, without further changes of the original ensemble and associated density matrix ρ,

$$\rho_{A\pm} = \frac{1}{2(1 \pm \langle A \rangle)}(1 \pm \hat{A})\rho(1 \pm \hat{A}). \tag{16}$$

With

$$\langle BA \rangle_m = \mathrm{tr}(\hat{B}\rho_{A+})w_{+,s}^A - \mathrm{tr}(\hat{B}\rho_{A-})w_{-,s}^A, \tag{17}$$

and

$$w_{\pm,s}^A = (1 \pm \langle A \rangle)/2 \tag{18}$$

this yields

$$\langle BA\rangle_m = \frac{1}{2}\mathrm{tr}(\{\hat{A},\hat{B}\}\rho). \tag{19}$$

The conditional correlation for minimally destructive measurements in the classical statistical ensemble corresponds precisely to the expression of this correlation in quantum mechanics. It involves the anticommutator and is therefore related to the quantum mechanical operator product. On the level of probabilistic observables we can express the conditional correlation $\langle BA\rangle$ in terms of expectation values of the quantum product AB

$$\langle BA\rangle_m = \frac{1}{2}(\langle BA\rangle + \langle AB\rangle), \tag{20}$$

demonstrating the close connection between the quantum product and the conditional correlation.

The two point correlation is commutative, $\langle BA\rangle_m = \langle AB\rangle_m$. We will postulate that the two point correlation (20) describes in general the correlation between two measurements for quantum systems and call it "*quantum correlation*". We have motivated its use by two subsequent measurements, but the order of the measurements does actually not matter. Any rule for the computation of a correlation between two measurements needs some choice of an appropriate product of observables. At first sight, a possible alternative choice may be the classical correlation which is based on the classical product $A \cdot B$, as defined on the level of the classical ensemble, $(A \cdot B)_\tau = A_\tau B_\tau$. We will see below, however, that $A \cdot B$ is usually not a quantum observable and can therefore not be determined from the information characterizing a quantum state, i.e. $\{\rho_k\}$. Its use would therefore need information which relates to the environment, but not only to the subsystem. In other words, the use of the classical product corresponds to a state reduction after the first measurement where substantial information about the relation between the subsystem and the environment is retained. This is not what a usual measurement in a quantum system does and the classical correlation can therefore not serve for the description of such measurements. For any measurement where the outcome (including the state reduction) can be expressed in terms of information available for the subsystem the choice of the quantum product seems natural. It retains a maximum of the information which is available in the subsystem.

Since the correct choice of the correlation for a description of two measurements is crucial we may describe the issue in some more detail. In any statistical setting one should distinguish the probability w_{++} that the two level observables A and B are *measured* with values $A = 1, B = 1$ from the probability p_{++} that they "have" the values $A = B = 1$ in the classical statistical ensemble. On the classical statistical level one can express

$$p_{++} = \frac{1}{4}(1 + \langle A\rangle + \langle B\rangle + \langle A \cdot B\rangle) \tag{21}$$

in terms of the "classical" correlation

$$\langle A \cdot B\rangle = \sum_\tau p_\tau A_\tau B_\tau. \tag{22}$$

The probability p_{++} does not specify a priori the conditional information relating two subsequent measurements, which is necessary for w_{++}. In general, one needs a separate prescription how w_{++} should be computed from the available statistical information. Only under particular circumstances, one may be able to identify w_{++} with p_{++}. In other words, $w_{++} = p_{++}$ is an additional basic assumption which does not hold true in general. This contrasts to the case of a single measurement for A, where $w_+ = p_+ = (1 + \langle A\rangle)/2$ by definition.

If we try the identification $w_{++} = p_{++}$ we run into severe problems. In general, the classical correlation $\langle A \cdot B \rangle$ is not computable in the subsystem. It is a property of the system and its environment and cannot be obtained from the information characterizing the quantum state, i.e. from ρ_k. This problem is closely linked to the fact that the mapping from the classical observables described by A_τ to the probabilistic observables described by $\gamma_a^{(A)} = \pm 1$ and $w_a^{(A)} = w_\pm^{(A)}$ is not invertible. The observable A_τ describes properties of the subsystem and its environment, while the characterization of the environment is only lost on the level of the probabilistic observable A. The use $w_{++} = p_{++}$ corresponds to an implicit definition of the conditional probability for two measurements where after the first measurement $A = 1$ all classical states τ for which $A_\tau = -1$ are eliminated. This elimination process depends, however, on the particular observable A_τ and therefore also reflects properties of the environment, not only of the subsystem. Two observables A_τ and A'_τ, which lead to the *same* probabilistic observable A, produce, in general, *different* classical products $\langle A \cdot B \rangle \neq \langle A' \cdot B \rangle$. (These issues are discussed in more detail at the end of this note.)

If the state of an isolated subsystem can be described by $\{\rho_k\}$, this information must also be sufficient for a prediction of the outcome of two measurements. The probability w_{++} must be computable in terms of $\{\rho_k\}$. For this reason we employ the quantum correlation $\langle AB \rangle$ and postulate

$$w_{++} = \frac{1}{4}(1 + \langle A \rangle + \langle B \rangle + \langle AB \rangle_m), \tag{23}$$

as advocated already before for subsequent measurements. (A different motivation for the use of the quantum correlation is given in [11].) Indeed, now w_{++} can be expressed in terms of ρ_k

$$w_{++} = \frac{1}{4}\Big(1 + e_k^{(A)} e_k^{(B)} + \rho_k [e_k^{(A)} + e_k^{(B)} + d_{mlk} e_m^{(A)} e_l^{(B)}]\Big) \tag{24}$$

The other probabilities w_{+-} etc. obtain from eqs. (23), (24) by appropriate changes of relative signs.

The prescription for the probabilities of the outcome of two measurements influences strongly the statistical properties of correlations. For example, one may ask if a "hidden variable theory" is possible, where there exist discrete functions $\tilde{A}(v) = \pm 1$, $\tilde{B}(v) = \pm 1$ such that $\langle AB \rangle_m = \int dv \tilde{p}(v) \tilde{A}(v) \tilde{B}(v)$ with some probability distribution $\tilde{p}(v)$. Just as in the quantum mechanics, this can be excluded by the use of Bell's inequalities [9]. In our classical statistical setting the correlation function (19) or the probabilities for the outcome of two measurements are exactly the same as in quantum mechanics. On the other hand, Bell's inequalities apply to the classical correlation $\langle A \cdot B \rangle$. Besides theoretical arguments we have therefore also experimental evidence that in general the classical correlation function should not be used for the description of the outcome of two measurements.

Quantum time evolution

We have seen how quantum structures can arise from the description of subsystems where the "state of the system" is described by n expectation values of "basis observables". For $P < n$ the appearance of "non-commuting structures" is mandatory. The question remains why such quantum systems are omnipresent in nature, in contrast to "commuting structures" for $P = n$. The answer may be rooted in stability properties of the time evolution.

Let us consider some continuous time evolution of the classical probability distribution $\{p_\tau\}$. It relates the ensemble at time t_2 to the ensemble at some earlier time t_1, and induces a transition from $\rho_k(t_1)$ to $\rho_k(t_2)$,

$$p_\tau(t_2) = \tilde{S}_{\tau\rho}(t_2, t_1) p_\rho(t_1) \ , \ \rho_k(t_2) = S_{kl}(t_2, t_1) \rho_l(t_1). \tag{25}$$

We may decompose the transition matrix S_{kl} into the product of an orthogonal matrix $\hat{S}_{kl}$, which preserves the length of the vector $(\rho_1 \ldots, \rho_n)$ and therefore the purity, and a scaling d, $S_{kl} = \hat{S}_{kl} d$. For an infinitesimal evolution step this implies

$$\begin{aligned} \partial_t \rho_k(t) &= T_{kl}\rho_l(t) + D\rho_k(t)\ ,\ D = \partial_t \ln d(t, t_1), \\ T_{kl} &= -T_{lk} = \partial_t \hat{S}_{km}(t, t_1)\hat{S}_{lm}(t, t_1). \end{aligned} \tag{26}$$

For a given maximal purity during the evolution, eq. (26) can be rewritten as an equation for the density matrix ρ,

$$\begin{aligned} \partial_t \rho_{\alpha\beta} = &- i[H, \rho]_{\alpha\beta} + R_{\alpha\beta\gamma\delta}\left(\rho_{\gamma\delta} - \frac{1}{M}\delta_{\gamma\delta}\right) \\ &+ D(\rho_{\alpha\beta} - \frac{1}{M}\delta_{\alpha\beta}). \end{aligned} \tag{27}$$

This corresponds to a split of the infinitesimal $SO(n)$ transformation $\delta\rho_k = T_{kl}\rho_l$ into a unitary part corresponding to the subgroup $SU(M)$ and represented by the hermitean Hamiltonian $H = H_k L_k + H_0$, and remaining rotations of $SO(n)/SU(M)$ represented by R or $\tilde{T}_{kl}$,

$$\begin{aligned} T_{kl} &= -2f_{klm}H_m + \tilde{T}_{kl}\ ,\ [L_k, L_l] = 2if_{klm}L_m, \\ \tilde{T}_{kl} &= \frac{1}{M}R_{\alpha\beta\gamma\delta}(L_k)_{\beta\alpha}(L_l)_{\gamma\delta}. \end{aligned} \tag{28}$$

In general, H, R and D may depend on ρ_k.

We are interested in possible partial fixed points of the evolution for which $R = 0$ and $D = 0$, while H is independent of ρ_k. (Partial fixed points of this type have been found explicitly in the classical time evolution of non-relativistic boson fields [13].) Then eq. (27) reduces to the linear von-Neumann equation for the density matrix. In case of a pure state density matrix this implies the Schrödinger equation $i\partial_t\psi = H\psi$. One recovers the unitary time evolution of quantum mechanics. The more general evolution equation away from the "unitary partial fixed point" can describe "decoherence" [14] as a decrease of purity for $D < 0$, or "syncoherence" as the approach to the pure state partial fixed point with increasing purity for $D > 0$. The latter typically accounts for a situation where the subsystem described by the observables $A^{(k)}$ can exchange energy with the environment. An example is the evolution from a mixed state of atoms in different energy states to a pure state of atoms in the ground state by virtue of radiative decay of the excited states. A static pure state density matrix obtains as usual as a solution of the quantum mechanical eigenvalue problem $H\psi = E_j\psi$. We suggest that the omnipresence of quantum systems in nature is due to the existence of such partial fixed points which reflect the isolation of the subsystem.

If H is independent of ρ_k it can be considered as an observable of the subsystem, $H = H_k L_k + H_0$, with fixed coefficients H_k, H_0. By Noether's theorem it is associated with the energy of the subsystem, where E_j denotes the possible energy eigenvalues. (If one wants to use standard energy units one replaces $H \to H/\hbar$.) On the other hand, R and D account for the interactions of the subsystem with its environment. They vanish in the limit of "perfect isolation" of the subsystem. If the interactions with the environment are strong enough, the subsystem is typically not evolving towards the equipartition fixed point, $\rho_{\alpha\beta} = \frac{1}{M}\delta_{\alpha\beta}$, but rather towards a Boltzmann type density matrix $\rho \sim \exp[-\beta(H + \mu_i N_i)]$ (for conserved quantities N_i and chemical potentials μ_i), which may be close to a pure state density matrix if the temperature $T = \beta^{-1}$ is small as compared to the typical separation of the two lowest energy eigenvalues E_j. In contrast, if the isolation from the environment becomes efficient fast enough, the subsystem follows subsequently its own unitary time evolution, as well known from quantum mechanics.

At this point we may recapitulate what we have achieved. Starting form a classical statistical ensemble we have identified a class of classical observables which have the same expectation values and admit the same algebra as the operators in a corresponding M-state quantum system. This holds provided the "purity condition" $P \leq M-1$ is obeyed. Also the conditional correlations which describe measurements of two such observables are the same as in the quantum system. Furthermore, we have found the criteria for the time evolution of the classical probability distribution that ensure a unitary evolution of the density matrix for the corresponding quantum system. Obviously, these properties are related to the specific subset of classical observables that can describe the subsystem, in the sense that no further information about the environment is needed for a prediction of the outcome of measurements in the subsystem. In the remaining part of this paper we will discuss the properties of these specific quantum observables in some detail.

Quantum observables

The quantum structures for probabilistic observables discussed so far do not need any specification of the representation as classical observables. In eq. (1) many classical systems with different states τ, classical probabilities p_τ and classical values of the observable in a given state, $A_\tau^{(C)}$, may describe the same quantum properties of the subsystem. Nevertheless, the implementation of quantum observables implies certain restrictions on the possible classical realizations. We will discuss those in the following, mainly for the purpose of conceptual foundations.

Classical observables are maps from the set of probability distributions Ω, with elements $\{p_\tau\} = (p_1 \ldots, p_z)$, to real numbers, $\Omega \overset{A^{(C)}}{\to} \mathbb{R}$, with $\{p_\tau\} \to \langle A^{(C)} \rangle = \sum_\tau p_\tau A_\tau^{(C)}$. (For simplicity we employ a language with a finite number z of classical states, which can be extended to an infinite set at the end in some specified limiting procedure.) We will restrict the discussion to those elements of Ω which correspond to "quantum states", i.e. which obey the bound for the purity of the ensemble. In general, the classical observables $A^{(C)}$ describe the system and its environment. (We use "system" for the (isolated) subsystem or quantum system from now on.)

We are interested in the subclass of quantum observables $A^{(Q)}$ whose expectation values and quantum correlations can be computed in the system. Quantum observables are classical observables with special properties. First, for a M-state quantum system, a quantum observable has at most M different classical values $A_\tau^{(Q)} = \gamma_a$. (For a non-degenerate spectrum we can identify $a = \alpha = 1 \ldots M$.) For any given quantum observable we can classify the classical states τ according to the value of $A_\tau^{(Q)}$, $\tau = (\gamma_a, \sigma_{\gamma_a})$, where for each given γ_a one typically has a large degeneracy of classical states, labeled by σ_{γ_a}. We can define the probability w_a for the occurrence of a possible measurement value γ_a as

$$w_a = \sum_{\sigma_{\gamma_a}} p_{(\gamma_a, \sigma_{\gamma_a})} \;,\; \langle A^{(Q)} \rangle = \sum_a \gamma_a w_a. \tag{29}$$

As a crucial ingredient, w_a must be computable from the quantities which specify the quantum state, i.e. from the expectation values of the "basis observables" $\langle A^{(k)} \rangle = \rho_k$. We assume a linear relation

$$w_a(\rho_k) = \sum_k c_{ak} \rho_k + c_{a0}. \tag{30}$$

The probabilities must be normalized, $\sum_a w_a(\rho_k) = 1$, for arbitrary ρ_k, which implies the conditions

$$\sum_a c_{ak} = 0 \;,\; \sum_a c_{a0} = 1. \tag{31}$$

Journal of Physics: Conference Series **174** (2009) 012008 doi:10.1088/1742-6596/174/1/012008

For quantum observables the coefficients c_{ak}, c_{a0} are restricted further since $w_a(\rho_k)$ must obey eq. (9). For non-degenerate eigenvalues one needs

$$c_{\alpha k} = \frac{1}{M}(UL_kU^\dagger)_{\alpha\alpha}\ ,\ c_{\alpha 0} = \frac{1}{M}, \tag{32}$$

for some suitable unitary matrix U.

Thus quantum observables are defined by two properties:
(i) the restriction of the spectrum to at most M different values, (ii) the "quantum determination" of probabilities $w_a(\rho_k)$. While (i) only involves a property of the classical observable, the second restriction (ii) depends on the selection of possible quantum states out of the most general probability distributions $\{p_\tau\}$ and on relations to the basis observables. For the specification of a quantum observable we need at least γ_a and c_{ak}, c_{a0}. A quantum observable $A^{(Q)}$ has the important property that its classical product $A^{(Q)} \cdot A^{(Q)}$ (defined by $(A^{(Q)} \cdot A^{(Q)})_\tau = (A_\tau^{(Q)})^2$) is again a quantum observable, with spectrum (γ_a^2) and the same w_a as for $A^{(Q)}$. This extends to higher polynomials and arbitrary functions $f(A^{(Q)})$.

We can now associate to any classical quantum observable $A^{(Q)}$ a probabilistic quantum observable, $A^{(Q)} \to A$, which is characterized by the spectrum of possible measurement values (γ_a) and the associated probabilities w_a. Only this information will be needed for a computation of expectation values $\langle (A^{(Q)})^p \rangle$ in a "quantum state" of the system, while the detailed form of $\{p_\tau\}$ is not relevant. On the level of classical observables the quantum observables are characterized by a distribution of values $A_\tau^{(Q)} = \gamma_a$ for the classical states τ. This distribution still contains much more information than the spectrum γ_a and the associated probabilities w_a. Therefore $A_\tau^{(Q)}$ still describes the system and partly the environment. Only on the level of probabilistic observables A the parts of $A_\tau^{(Q)}$ relevant for the environment are projected out, such that A only "measures" properties of the system.

We also can associate to every $A^{(Q)}$ a quantum operator $\hat{A}$ by a map $A^{(Q)} \to \hat{A}$. It is constructed from eq. (7) by observing

$$\begin{aligned}
\langle A^{(Q)} \rangle &= \sum_a \gamma_a (c_{a,0} + \sum_k c_{a,k}\rho_k) = e_0^{(A)} + \sum_k e_k^{(A)} \rho_k, \\
e_0^{(A)} &= \sum_a \gamma_a c_{a,0}\ ,\ e_k^{(A)} = \sum_a \gamma_a c_{a,k}, \\
\hat{A} &= e_0^{(A)} + \sum_k e_k^{(A)} L_k\ ,\ \langle A^{(Q)} \rangle = \mathrm{tr}(\hat{A}\rho).
\end{aligned} \tag{33}$$

We note that the map (33) is possible for arbitrary probabilistic system observables obeying eq. (30). Without the restriction of the type (32) for quantum observables, however, $(A^{(Q)})^2$ will, in general, not be mapped to $\hat{A}^2$. A simple example is the random two-level observable R, with $\gamma_1 = 1\ ,\ \gamma_2 = -1\ ,\ c_{ak} = 0\ ,\ c_{10} = c_{20} = 1/2$. It is mapped to $\hat{A} = 0$, while $R^2 = 1$.

In turn, we have a map from the space of quantum operators $\mathcal{O}$ to the space of probabilistic observables $\mathcal{P}$, since for every operator $\hat{A}$ the spectrum $\{\gamma_a\}$ is defined, and the probabilities w_a can be computed for all quantum states $\{\rho_k\}$ or density matrices ρ. The latter obtain from the diagonal elements $(U\rho U^\dagger)_{\alpha\alpha}$, with U the unitary matrix used for the diagonalization of $\hat{A}$.

However, the map from the ensemble of classical quantum observables $\mathcal{Q}$ to the space of quantum operators $\mathcal{O}$ is not invertible. The classical observables $A^{(Q)}$ involve a specification of $A_\tau^{(Q)}$ for every classical state τ, which is much more information than contained in the coefficients $c_{a,k}\ ,\ c_{a,0}$. We may encounter situations where a quantum observable B is mapped to an operator $\hat{B}$, while also a function $f(A)$ of a different quantum observable A is mapped to the same operator, $f(\hat{A}) = \hat{B}$. (Here $f(\hat{A})$ is an operator valued function, while $f(A)$ is based on the classical product $A \cdot A$.) Such a situation does not imply an identification of the quantum observables, i.e. in general one has $B \neq f(A)$. This lack of invertibility of the map $\mathcal{Q} \to \mathcal{O}$ constitutes an important difference between our approach and many alternative attempts of a "classical formulation of quantum mechanics", which associate to each $\hat{A}$ a unique classical observable. For example, this is typically assumed for "hidden variable theories". Also for the Kochen-Specker theorem [5] the existence of a map $\hat{A} \to A^{(Q)}$ is a crucial hypothesis, which is not obeyed in our setting.

On the level of classical observables we always can define a linear combination, $C = \lambda_A A^{(Q)} + \lambda_B B^{(Q)}$, and the pointwise product, $D = A^{(Q)} \cdot B^{(Q)}$, of two quantum observables $A^{(Q)}, B^{(Q)}$, where $C_\tau = \lambda_A A_\tau^{(Q)} + \lambda_B B_\tau^{(Q)}, D_\tau = A_\tau^{(Q)} B_\tau^{(Q)}$. However, in general neither C nor D are quantum observables. Consider the simplest case, $M = 2$, and the two basis observables $A^{(1)}$ and $A^{(2)}$ with spectrum $\gamma_\alpha^{(1)} = \gamma_\alpha^{(2)} = \pm 1$. A linear combination $C = \cos\vartheta A^{(1)} + \sin\vartheta A^{(2)}$ has a spectrum $\gamma_\alpha^{(C)} = \pm\cos\vartheta \pm \sin\vartheta$. This observable has four different possible measurement values. It can therefore not be a quantum observable of the system with $M = 2$, even though $\langle C \rangle$ can be computed in terms of $\rho_{1,2} = \langle A^{(1),(2)} \rangle$. We conclude that the "rotated spin", which corresponds to the operator $\hat{A}(\vartheta) = \cos\vartheta \hat{A}^{(1)} + \sin\vartheta \hat{A}^{(2)}$, has to be described by a quantum observable $A_{(\vartheta)}^{(Q)}$ that is again a two level observable with spectrum $\gamma_\alpha = \pm 1$, rather than by a linear combination of $A^{(1)}$ and $A^{(2)}$ of the type C. This necessity arises for each value of the angle ϑ and we have discussed in detail in [4] that this needs a classical ensemble with infinitely many classical states τ. The reader should note that one can define two types of linear combinations. On the level of classical observables one can define combinations of the type C, while on the level of operators or the associated quantum observables a natural definition is $\hat{A}(\vartheta)$ or $A^{(Q)}(\vartheta)$. In general, a projection on the subsystem does not map C to $\hat{A}(\vartheta)$ or the probabilistic observable $A^{(Q)}(\vartheta)$.

The classical product $D = A^{(1)} \cdot A^{(2)}$ has a spectrum $\gamma_\alpha^{(D)} = \pm 1$. The condition (i) for a quantum observable is obeyed by D. However, the probability $w_\alpha^{(D)}$ for finding $\gamma_\alpha^{(D)} = 1$ needs knowledge of the joint probability to find $A^{(1)} = 1, A^{(2)} = 1$ or $A^{(1)} = -1$, $A^{(2)} = -1$. This information cannot be extracted from ρ_1 and ρ_2, which only yield the probabilities for finding $A^{(1)} = \pm 1$ (namely $w_\pm^{(1)} = (1 \pm \rho_1)/2$) or for finding $A^{(2)} = \pm 1$ (namely $w_\pm^{(2)} = (1 \pm \rho_2)/2$). Nor is it contained in the expectation value ρ_3 of the third basis variable for $M = 2$. We conclude that the classical observable D does not obey the condition (ii) for a quantum observable.

Representation as classical observables

After these general remarks we now present an explicit classical ensemble and quantum observables for a system with given M. We recall that the classical quantum observables $A^{(Q)}$ which are mapped to a given $\hat{A}$ are not unique. Also the specification of the classical states τ and the corresponding construction of classical observables is not supposed to be unique. At the end all measurable information of the system can be expressed in terms of the expectation values of quantum operators, such that the details of the classical observables do not matter. Only the existence of the classical observables in a setting free of contradictions is therefore needed in order to demonstrate a realization of quantum mechanics as a classical statistical ensemble.

It is sufficient to determine at least one classical quantum observable for every operator $\hat{A}$, i.e. for every hermitean $M \times M$ matrix. The quantum observable $\lambda A^{(Q)}$, with $\lambda \in \mathbb{R}$, is mapped to the operator $\lambda\hat{A}$. We will therefore restrict our discussion to operators with unit norm, say tr $\hat{A}^2 = M$. Also the addition of a part proportional to the unit observable translates for operators to the addition of a corresponding piece proportional to the unit operator, $A + c \to \hat{A} + c$. We can therefore restrict the discussion to traceless operators, tr$\hat{A} = 0$. We follow a simple construction principle. Consider first a single operator $\hat{A}$ with a spectrum of $m(\hat{A}) \leq M$ distinct eigenvalues $\lambda_{a(\hat{A})}(\hat{A})$. We associate to it $m(\hat{A})$ discrete classical states, labeled by $a(\hat{A}) = 1 \dots m(\hat{A})$. In these states the classical observable $A^{(Q)}$, which is mapped to $\hat{A}$, takes the values $A^{(Q)}_{a(\hat{A})} = \lambda_{a(\hat{A})}(\hat{A})$. Add now a second operator $\hat{B}$ with $m(\hat{B})$ distinct eigenvalues $\lambda_{a(\hat{B})}(\hat{B})$. If this operator is "independent" we construct the direct product space with states τ labeled by the double index $\tau = \big(a(\hat{A}), a(\hat{B})\big)$, and $A^{(Q)}_\tau = \lambda_{a(\hat{A})}(\hat{A})$, $B^{(Q)}_\tau = \lambda_{a(\hat{B})}(\hat{B})$. This is continued until all independent operators are included. As stated above, the resulting ensemble has infinitely many classical states τ, since the number of independent operators is infinite. (A well defined sequence of subsequently included operators induces a well defined limit process for the construction of the ensemble [4].) Our construction yields explicitly a classical quantum observable for every independent operator. We recall that many further classical observables that do not obey the restrictions for quantum observables can be defined in the ensemble.

Details of the construction will depend on the notion of independent operators. As a simple criterion we may call two operators independent if $\mathrm{tr}(\hat{A} - \hat{B})^2 \geq \epsilon$, and take the limiting process $\epsilon \to 0$. Other more restrictive definitions of "independent" may be possible. Different contradiction-free definitions of "independent" lead to different classical realizations, which all result in the same quantum properties of the system. The explicit construction above has demonstrated that such classical realizations exist. Of course, there are also classical ensembles with "many more" states than those used in our explicit construction. If one is only interested in the quantum observables the states $\tilde{\tau}$ of such a larger ensemble can be mapped to the states τ of the ensemble used in the construction by summing the probabilities of all states which have the same A_τ for all quantum observables.

Classical product for observables

We next turn to the conditions under which linear combinations or classical products of two quantum observables are again quantum observables. For $C^{(Q)} = \lambda_A A^{(Q)} + \lambda_B B^{(Q)}$ the expectation value can always be expressed in terms of ρ_k. We can therefore compute $e^{(C)}_k = \lambda_A e^{(A)}_k + \lambda_B e^{(B)}_k$ such that $\langle C\rangle$ obeys eq.(2). This allows for the construction of an operator $\hat{C}$ (7) which obeys for all ρ

$$\langle C\rangle = \mathrm{tr}(\hat{C}\rho) = \lambda_A\langle A\rangle + \lambda_B\langle B\rangle = \mathrm{tr}\Big[(\lambda_A\hat{A} + \lambda_B\hat{B})\rho\Big]. \tag{34}$$

We therefore can identify $\hat{C} = \lambda_A\hat{A} + \lambda_B\hat{B}$. At this step, however, the spectrum of possible measurement values for C does not necessarily coincide with the spectrum of eigenvalues of $\hat{C}$, which is a necessary condition for a quantum observable. On the classical level the spectrum of C consists of all linear combinations

$$\gamma^{(C)}_c = \gamma^{(C)}_{(a,b)} = \lambda_A\gamma^{(A)}_a + \lambda_B\gamma^{(B)}_b \tag{35}$$

for all possible pairs $(a, b) = c$. It may be reduced by those $\gamma^{(C)}_c$ for which the probability w_c vanishes for all quantum states. Even if the number $\tilde{M}^{(C)}$ of different values $\gamma^{(C)}_c$ obeys $\tilde{M}^{(C)} \leq M$ as realized, for example, if $\tilde{M}^{(A)}\tilde{M}^{(B)} \leq M$, there is no guarantee that all $\gamma^{(C)}_c$ coincide with the eigenvalues of $\hat{C}$.

If $C^{(Q)}$ is a quantum observable its spectrum must coincide with the spectrum of $\hat{C}$. Furthermore, the classical product $C^{(Q)} \cdot C^{(Q)}$ must also be a quantum observable and obey

$$\begin{aligned}\langle C\cdot C\rangle &= \lambda_A^2\langle A\cdot A\rangle + \lambda_B^2\langle B\cdot B\rangle + 2\lambda_A\lambda_B\langle A\cdot B\rangle \\ &= \mathrm{tr}(\hat{C}^2\rho) \\ &= \lambda_A^2\mathrm{tr}(\hat{A}^2\rho) + \lambda_B^2\mathrm{tr}(\hat{B}^2\rho) + \lambda_A\lambda_B\mathrm{tr}\Big(\{\hat{A},\hat{B}\}\rho\Big).\end{aligned} \tag{36}$$

We find as a necessary condition that the classical product $A\cdot B$ must be computable in terms of ρ_k

$$\langle A\cdot B\rangle = \frac{1}{2}\mathrm{tr}\Big(\{\hat{A},\hat{B}\}\rho\Big). \tag{37}$$

This has to hold for arbitrary ρ. If any (nontrivial) linear combination of $A^{(Q)}$ and $B^{(Q)}$ is a quantum observable one concludes that $A^{(Q)}\cdot B^{(Q)}$ must be a quantum observable with associated operator $A\cdot B\to\frac{1}{2}\{\hat{A},\hat{B}\}$. Similar restrictions arise for higher powers of C.

The condition (37) is nontrivial. On the classical level we can derive from $\{p_\tau\}$ the probabilities $w_{(a,b)}$ that A has the value $\gamma_a^{(A)}$ and B takes the value $\gamma_b^{(B)}$. For quantum observables $A^{(Q)},B^{(Q)}$ the probabilities $w_a^{(A)}=\sum_b w_{(a,b)}, w_b^{(B)}=\sum_a w_{(a,b)}$ can be computed from $\{\rho_k\}$. In general, the information contained in $\{\rho_k\}$ will not be sufficient to determine $w_{(a,b)}$, however. It will therefore often not be possible to express $\langle A\cdot B\rangle=\sum_{a,b}\gamma_a^{(A)}\gamma_b^{(B)}w_{(a,b)}$ in terms of $\{\rho_k\}$. Then $D=A\cdot B$ cannot be a quantum observable. On the other hand, if a linear relation between $\langle D\rangle$ and ρ_k exists, $\langle D\rangle=e_0^{(D)}+e_k^{(D)}\rho_k$, we can write $\langle D\rangle=\mathrm{tr}(\hat{D}\rho)$ and eq. (37) implies $\hat{D}=\frac{1}{2}\{\hat{A},\hat{B}\}$.

If $\lambda_A A^{(Q)}+\lambda_B B^{(Q)}$ is a quantum observable for arbitrary λ_A,λ_B the associated operators $\hat{A}$ and $\hat{B}$ must commute, $[\hat{A},\hat{B}]=0$. In order to show this, we first consider the case where a given $\gamma_{\bar{c}}^{(C)}=\lambda_A\gamma_{\bar{a}}^{(A)}+\lambda_B\gamma_{\bar{b}}^{(B)}$ corresponds to a unique combination $(\bar{a},\bar{b})$. If $C^{(Q)}$ is a quantum observable, there must exist probability distributions $\{p_\tau\}$ which are an "eigenstate" for the "eigenvalue" $\gamma_{\bar{c}}^{(C)}$. This implies $w_{\bar{c}}=1$, $w_{c\neq\bar{c}}=0$ or $w_{(\bar{a},\bar{b})}=1$, $w_{(a,b)}=0$ if $a\neq\bar{a}$ or $b\neq\bar{b}$ and therefore $w_{\bar{a}}=1$, $w_{a\neq\bar{a}}=0$, $w_{\bar{b}}=1$, $w_{b\neq\bar{b}}=0$. We conclude that this state is also a simultaneous eigenstate of the observables $A^{(Q)}$ and $B^{(Q)}$, with respective eigenvalues $\gamma_{\bar{a}}^{(A)}$ and $\gamma_{\bar{b}}^{(B)}$. In particular, we may consider a pure state $\psi_{\bar{c}}$ which is an eigenstate of $\hat{C}$ with eigenvalue $\gamma_{\bar{c}}^{(C)}$. It must obey $\hat{A}\psi_{\bar{c}}=\gamma_{\bar{a}}^{(A)}\psi_{\bar{c}}$, $\psi_{\bar{c}}^T\hat{A}=\gamma_{\bar{a}}^{(A)}\psi_{\bar{c}}^T$, $\hat{B}\psi_{\bar{c}}=\gamma_{\bar{b}}^{(B)}\psi_{\bar{c}}$, $\bar{\psi}_c^T\hat{B}=\gamma_{\bar{b}}^{(B)}\bar{\psi}_c^T$. We choose a basis where $\hat{C}$ is diagonal and $\psi_{\bar{c}}^T=\hat{\psi}_1^T=(1,0,\ldots 0)$. In this basis $\hat{A}$ and $\hat{B}$ must be block-diagonal, $\hat{A}_{\alpha 1}=\gamma_{\bar{a}}^{(A)}\delta_{\alpha 1}$, $\hat{A}_{1\beta}=\gamma_{\bar{a}}^{(A)}\delta_{1\beta}$ and similar for $\hat{B}$. We can repeat this for other eigenvalues of $\hat{C}$. If for every eigenvalue $\gamma_c^{(C)}$ of $\hat{C}$ the composition out of eigenvalues of $A^{(Q)}$ and $B^{(Q)}$ is unique, we can infer that $\hat{A}$ and $\hat{B}$ must commute. Indeed, in the basis where $\hat{C}$ is diagonal both $\hat{A}$ and $\hat{B}$ must be simultaneously diagonal, and therefore $[\hat{A},\hat{B}]=0$.

In presence of multiple possibilities of composing $\gamma_{\bar{c}}^{(C)}$ from linear combinations of $\gamma_a^{(A)}$ and $\gamma_b^{(B)}$ the discussion is more involved. This case appears, however, only for particular coefficients λ_A,λ_B. If C is a quantum observable for arbitrary λ_A and λ_B such degenerate cases can be avoided such that $\hat{A}$ and $\hat{B}$ must commute. Indeed, consider the case of a "degenerate decomposition" for a particular pair (λ_A,λ_B). This occurs if there are two solutions $\lambda_A\gamma_{a_1}^{(A)}+\lambda_B\gamma_{b_1}^{(B)}=\gamma_{c_1}^{(C)}$, $\lambda_A\gamma_{a_2}^{(A)}+\lambda_B\gamma_{b_2}^{(B)}=\gamma_{c_2}^{(C)}$, with $\gamma_{c_1}^{(C)}=\gamma_{c_2}^{(C)}$, $\gamma_{a_1}^{(A)}\neq\gamma_{a_2}^{(A)}$. Performing an infinitesimal shift $\lambda_A\to\lambda_A+\delta_A$, while keeping λ_B fixed, results in a separation of $\gamma_{c_1}^{(C)}$ and $\gamma_{c_2}^{(C)}$, $\gamma_{c_2}^{(C)}-\gamma_{c_1}^{(C)}=\delta_A(\gamma_{a_2}^{(A)}-\gamma_{a_1}^{(A)})\neq 0$. Then $\gamma_{c_1}^{(C)}$ has a unique composition from $\gamma_{a_1}^{(A)}$ and $\gamma_{b_1}^{(B)}$. (We have discussed here the case of two-fold degeneracy where other eigenvalues of C are separated from $\gamma_{c_1}^{(C)}$, $\gamma_{c_2}^{(C)}$ by a finite distance. Higher degeneracies can be treated similarly.)

Two classical quantum observables $A^{(Q)}$ and $B^{(Q)}$ are called "*commuting observables*" if arbitrary linear combinations $\lambda_A A^{(Q)}+\lambda_B B^{(Q)}$ are also quantum observables. The operators $\hat{A},\hat{B}$ associated to a pair of commuting observables must commute, $[\hat{A},\hat{B}]=0$. Furthermore, the classical product of two commuting observables is a quantum observable. The associated operators and probabilistic observables are given by the chain of maps

$$A^{(Q)}\cdot B^{(Q)}\to\frac{1}{2}\{\hat{A},\hat{B}\}\to AB. \tag{38}$$

For commuting quantum observables the classical correlation is computable from the quantum system and equals the quantum correlation.

We next consider general conditions for the classical product being a quantum observable, $D^{(Q)} = A^{(Q)} \cdot B^{(Q)}$. If for all eigenvalues of $D^{(Q)}$ the decompositions $\gamma_{\bar{c}}^{(D)} = \gamma_{\bar{a}}^{(A)}\gamma_{\bar{b}}^{(B)}$ are unique, the operators $\hat{A}$ and $\hat{B}$ must again commute. In this situation the quantum state specifies the probabilities w_a for the observable A to have the value $\gamma_a^{(A)}$, the analogue for w_b, and in addition the joint probability $w_c = w_{(a,b)}$ that a measurement of A yields $\gamma_a^{(A)}$ and a measurement B yields $\gamma_b^{(B)}$. Since the associated operators $\hat{A}$ and $\hat{B}$ commute we may choose a basis where both are diagonal. The probability for the operator $\hat{A}\hat{B}$ to take the value $\gamma_c^{(D)} = \gamma_a^{(A)}\gamma_b^{(B)}$, with $(a,b) = c$, is given by the corresponding diagonal element of the density matrix in this basis, $(\rho')_{\alpha\alpha}$. This must equal w_c, and we conclude that for all quantum states $\mathrm{Tr}(\hat{A}\hat{B}\rho) = \sum_c w_c\gamma_c^{(D)} = \langle D\rangle = \mathrm{Tr}(\hat{D}\rho)$ and therefore eq. (38) applies.

Inversely, we cannot infer that for every pair of quantum observables $A^{(Q)}, B^{(Q)}$, for which the associated operators commute, $[\hat{A}^{(Q)}, \hat{B}^{(Q)}] = 0$, the classical product $A \cdot B$ must be a quantum observable. There is simply no guarantee that the joint probabilities $w_{(a,b)}$ find an expression in terms of $\{\rho_k\}$. As an upshot of this discussion we conclude that the lack of a map $\mathcal{O} \to \mathcal{Q}$ leaves a lot of freedom in the choice and properties of the quantum observables. Generically, linear combinations and classical products of quantum observables are not quantum observables themselves.

Bit chains

Finally, we discuss the special setting of "bit chains". The simplest bit chain is a set of three commuting two level observables which we take among the set of basis observables $A^{(k)}$. For the example $M = 4$ we may consider the observables T_1, T_2, T_3 associated to the three commuting diagonal operators L_1, L_2, L_3. A bit chain arises if the expectation values of two (or several) one-bit-observables as well as their products can be determined simultaneously in a quantum system. Suppose that the first bit corresponds to T_1, the second to T_2. Each bit can take the two values $+1$ or -1. For a bit chain the classical product $T_1 \cdot T_2$ is also a quantum observable, associated with T_3. From the expectation values $\langle T_1\rangle$, $\langle T_2\rangle$, $\langle T_3\rangle$ we can determine all probabilities for the four possibilities of values $(+,+)$, $(+,-)$, $(-,-)$ and $(-,+)$ for bits one and two [3]. This can be easily generalized: for a bit chain of quantum observables with $\tilde{P}$ members T_j , $j = 1 \ldots \tilde{P}$, all mutual classical products are members of the bit chain, $T_i \cdot T_j = c_{ijk}T_k(i \neq j)$, with $c_{ijk} = c_{jik} = 1$ for one particular combination (i,j,k) and zero otherwise. All associated operators $\hat{T}_j$ mutually commute, and $\hat{T}_i\hat{T}_j = c_{ijk}\hat{T}_k$, $\hat{T}_j^2 = 1$. Also all linear combinations $\lambda_i T_i + \lambda_j T_j$ are quantum observables, represented by the operators $\lambda_i\hat{T}_i + \lambda_j\hat{T}_j$.

For given M the maximal number of members of a bit chain is $\tilde{P} = M - 1$ and we call such chains "complete bit chains". This restriction follows simply from the maximal number of mutually commuting operators. The presence of a bound for $\tilde{P}$ poses certain restrictions on the classical realizations of bit chains associated to different sets of mutually commuting operators.

As an example, consider the case $M = 8$. A possible complete three bit chain with seven members can be associated to the operators $C_1 \to (\tau_3 \otimes 1 \otimes 1)$, $C_2 \to (1 \otimes \tau_3 \otimes 1)$, $C_3 \to (1 \otimes 1 \otimes \tau_3)$, $\tilde{C}_1 \to (1 \otimes \tau_3 \otimes \tau_3)$, $\tilde{C}_2 \to (\tau_3 \otimes 1 \otimes \tau_3)$, $\tilde{C}_3 \to (\tau_3 \otimes \tau_3 \otimes 1)$, $\approx{C} \to (\tau_3 \otimes \tau_3 \otimes \tau_3)$. For this "$C$-chain" one has $C_j\tilde{C}_j = \approx{C} = C_1C_2C_3$ for all $j = 1,2,3$, $\tilde{C}_1 \cdot \tilde{C}_2 = \tilde{C}_3$. Alternative candidates for complete three bit chains are the "A-chain" where τ_3 is replaced by τ_1, or the "B-chain" which obtains from the C-chain by the replacement $\tau_3 \to \tau_2$, Further candidates are the "F-chain" $(C_1, A_2, A_3, \tilde{F}_1, \tilde{F}_2, \tilde{F}_3, \approx{F})$, "$G$-chain" $(A_1, C_2, A_3, \tilde{G}_1, \tilde{G}_2, \tilde{G}_3, \approx{G})$ or "H-chain" $(A_1, A_2, C_3, \tilde{H}_1, \tilde{H}_2, \tilde{H}_3, \approx{H})$, with analogous multiplication structures given by the order of the elements in the list, i.e. $\approx{F} \to (\tau_3 \otimes \tau_1 \otimes \tau_1)$, $\approx{G} = (\tau_1 \otimes \tau_3 \otimes \tau_1)$, $\approx{H} \to (\tau_1 \otimes \tau_1 \otimes \tau_3)$.

Finally, we may consider a possible candidate "Q-chain" $(\stackrel{\approx}{F}, \stackrel{\approx}{G}, \stackrel{\approx}{H}, \tilde{Q}_1, \tilde{Q}_2.\tilde{Q}_3, \stackrel{\approx}{Q})$, with $\stackrel{\approx}{Q} \to -(\tau_3 \otimes \tau_3 \otimes \tau_3)$. If all these sets of observables are simultaneously realized as bit chains, we run into contradiction. From the Q-chain we conclude $\stackrel{\approx}{F} \cdot \stackrel{\approx}{G} \cdot \stackrel{\approx}{H} = \stackrel{\approx}{Q}$. In turn, from the F, G, H chains we infer $C_1 \cdot A_2 \cdot A_3 = \stackrel{\approx}{F}$, $A_1 \cdot C_2 \cdot A_3 = \stackrel{\approx}{G}$, $A_1 \cdot A_2 \cdot C_3 = \stackrel{\approx}{H}$ and therefore $\stackrel{\approx}{Q} = C_1 \cdot A_2 \cdot A_3 \cdot A_1 \cdot C_2 \cdot A_3 \cdot A_1 \cdot A_2 \cdot C_3 = C_1 \cdot C_2 \cdot C_3 = \stackrel{\approx}{C}$. The operators associated to $\stackrel{\approx}{C}$ and $\stackrel{\approx}{Q}$ have opposite sign, however, showing the contradiction.

This clearly demonstrates that not every set of two level observables for which the associated operators mutually commute can be a bit chain simultaneously. In our case this only poses a consistency condition for the possibilities of classical products of quantum observables being quantum observables themselves. If we had a map $\hat{A} \to A^{(Q)}$, with $f(\hat{A}) \to f(A^{(Q)})$ and $f(A^{(Q)})$ based on the classical product, one could show that for every pair of commuting operators $\hat{A}, \hat{B}$ the map implies $\hat{A}\hat{B} \to A \cdot B$. The resulting contradiction is a proof of the Kochen-Specker-theorem [5] - actually the above chains of observables correspond precisely to the elegant proof of this theorem by N. Straumann [6].

The observation that not all "candidate chains" C, F, G, H, Q can be simultaneously bit chains does not mean that the associated sets of seven commuting operators are inequivalent. There is no problem to associate to each such operator set a bit chain. Only the bit chain associated to the set $\hat{Q}_1 = (\tau_3 \otimes \tau_1 \otimes \tau_1)$, $\hat{Q}_2 = (\tau_1 \otimes \tau_3 \otimes \tau_1)$, $\hat{Q}_3 = (\tau_1 \otimes \tau_1 \otimes \tau_3)$, $\hat{Q}_2\hat{Q}_3$, $\hat{Q}_3\hat{Q}_1$, $\hat{Q}_1\hat{Q}_2$, $\hat{Q}_1\hat{Q}_2\hat{Q}_3 = -(\tau_3 \otimes \tau_3 \otimes \tau_3)$ should be a new bit chain with quantum observables $(Q_1, Q_2, Q_3$, $Q_2 \cdot Q_3$, $Q_3 \cdot Q_1$, $Q_1 \cdot Q_2$, $Q_1 \cdot Q_2 \cdot Q_3)$ which are different from the candidate Q-chain $(\stackrel{\approx}{F}, \stackrel{\approx}{G}, \stackrel{\approx}{H}, \tilde{Q}_1, \tilde{Q}_2, \tilde{Q}_3, \stackrel{\approx}{Q})$ discussed above. This does not lead to any contradiction, since the map from quantum observables to operators is not invertible. Both the observables Q_1 and $\stackrel{\approx}{F}$ are mapped to the same operator $\hat{Q}_1$, but the classical product may be a quantum observable for $Q_1 \cdot Q_2$ and not for $\stackrel{\approx}{F} \cdot Q_2$. In our explicit construction of a classical representation of observables we should not exclude $\hat{Q}_1$ from the set of independent operators with the argument that it can be obtained as the product of two commuting observables.

Classical entanglement

Entanglement is a key feature of quantum mechanics. Its classical realization is best discussed in the context of bit chains. Consider $M = 4$ and a bit chain of observables T_1, T_2, T_3, with corresponding commuting diagonal operators $L_1, L_2, L_3 = L_1 L_2$. We associate a first bit to T_1 and a second bit to T_2. The product of the two bits is then determined by the classical product $T_1 \cdot T_2 = T_3$. Let us concentrate on a state with $\rho_3 = \langle T_3 \rangle = -1$, $\rho_1 = \langle T_1 \rangle = 0$, $\rho_2 = \langle T_2 \rangle = 0$. Depending on the other ρ_k this may be a pure or mixed state, with purity $P = 1 + \sum_{k \geq 4} \rho_k^2$. From $\rho_1 = 0$ we infer an equal probability to find for the first bit the values $+1$ and -1, and similar for the second bit from $\rho_2 = 0$. On the other hand, $\rho_3 = -1$ implies a maximal anticorrelation between bits one and two. The probabilities vanish for all classical states for which both bit one and bit two have the same value corresponding to $w_{++} = w_{--} = 0$, $w_{-+} = w_{-+} = 1/2$.

We may assume that a first apparatus measures bit one, and a second one bit two. Whenever the first apparatus shows a positive result, the second apparatus will necessarily indicate a negative result, and vice versa. By itself, this anticorrelation does not yet indicate an entangled pure state. For example, it may be realized by a mixed state with $\rho_k = 0$ for $k \geq 4$, $P = 1$, corresponding to a diagonal density matrix $\rho = (1/2)diag(0, 1, 1, 0)$.

We may compute the quantum correlation between a rotated spin observable $A(\vartheta)$ with associated operator $\hat{A}(\vartheta) = \cos\vartheta L_1 + \sin\vartheta L_8$ and a second rotated spin observable $B(\varphi)$ with

Journal of Physics: Conference Series **174** (2009) 012008 doi:10.1088/1742-6596/174/1/012008

$\hat{B}(\varphi) = \cos\varphi L_2 + \sin\varphi L_4$. One finds for arbitrary ρ_k

$$\begin{aligned} \langle A(\vartheta)B(\varphi)\rangle &= C(\vartheta,\varphi) = \frac{1}{2}\mathrm{tr}\big(\{\hat{A}(\vartheta),\hat{B}(\varphi)\}\rho\big) \\ &= \cos\vartheta\cos\varphi\rho_3 + \cos\vartheta\sin\varphi\rho_6 \\ &\quad + \sin\vartheta\cos\varphi\rho_{10} + \sin\vartheta\sin\varphi\rho_{12}, \end{aligned} \tag{39}$$

where we use the representations

$$\begin{aligned} L_1 &= (\tau_3\otimes 1)\ ,\ L_2 = (1\otimes\tau_3)\ ,\ L_3 = (\tau_3\otimes\tau_3), \\ L_8 &= (\tau_1\otimes 1)\ ,\ L_4 = (1\otimes\tau_1)\ ,\ L_{12} = (\tau_1\otimes\tau_1), \\ L_6 &= (\tau_3\otimes\tau_1)\ ,\ L_{10} = (\tau_1\otimes\tau_3)\ ,\ L_{14} = -(\tau_2\otimes\tau_2). \end{aligned} \tag{40}$$

For all states with $\rho_3 = \rho_{12} = -1\ ,\ \rho_6 = \rho_{10} = 0$, one obtains the familiar quantum result for two spins with relative rotation

$$\langle A(\vartheta)B(\varphi)\rangle = -\cos(\vartheta-\varphi) = \bar{C}(\vartheta-\varphi). \tag{41}$$

Bell's inequality for local deterministic theories reads for this situation

$$|C(\vartheta_1,0) - C(\vartheta_2,0)| \le 1 + C(\vartheta_1,\vartheta_2). \tag{42}$$

With eq. (41) this reduces to $|\bar{C}(\vartheta_1) - \bar{C}(\vartheta_2)| \le 1 + \bar{C}(\vartheta_1-\vartheta_2)$. It is violated for $\vartheta_1 = \pi/2\ ,\ \vartheta_2 = \pi/4$.

We observe that the contribution $\sim \rho_{12}$ to the quantum correlation matters. For our choice $\vartheta_1 = \pi/2\ ,\ \vartheta_2 = \pi/4$ the inequality (42) reads $|C(\pi/2,0) - C(\pi/4,0)| \le 1 + C(\pi/2,\pi/4)$. For $\rho_6 = \rho_8 = 0\ ,\ \rho_3 = -1$ and general ρ_{12} one finds $C(\pi/2,0) = 0\ ,\ C(\pi/4,0) = -1/\sqrt{2}$ and $C(\pi/2,\pi/4) = \rho_{12}/\sqrt{2}$. For $\rho_{12} = 0$ Bell's inequality is now obeyed.

We conclude that the presence of off-diagonal elements in the density matrix (in a direct product basis for the two entangled spins) plays an important role for the coexistence of different complete bit chains. For our example with $M = 4$ a second bit chain besides $T_1, T_2, T_3 = T_1\cdot T_2$ is given by $T_8, T_4, T_{12} = T_8\cdot T_4$, with associated commuting operators $L_8, L_4, L_{12} = L_8L_4$. Not only the spins in one direction are maximally anticorrelated for $\rho_3 = -1$, but also the spins in an orthogonal direction (represented by L_8, L_4) are maximally anticorrelated for $\rho_{12} = -1$. The quantum state of the subsystem allows for a specification of several correlations by independent elements as ρ_3 and ρ_{12}. This possibility is closely connected to the use of quantum correlations for the calculation of the outcome of two measurements. In a setting where only the classical correlations are available, a simultaneous implementation of the two two-bit chains (T_1, T_2, T_3) and (T_8, T_4, T_{12}) would require more than three mutually commuting objects and can therefore not be implemented for a state with purity $P \le 3$.

We finally display the classical formulation for two particular entangled pure states. They are given by

$$\rho_3 = \epsilon\rho_{12} = -\epsilon\rho_{14} = -1\ ,\ \epsilon = \pm 1. \tag{43}$$

The sign $\epsilon = +1$ corresponds to the rotation invariant spin singlet state with density matrix

$$\rho = \frac{1}{4}\big(1 - (\tau_1\otimes\tau_1) - (\tau_2\otimes\tau_2) - (\tau_3\otimes\tau_3)\big), \tag{44}$$

and wave function

$$\psi = \frac{1}{\sqrt{2}}(\hat{\psi}_2 - \hat{\psi}_3). \tag{45}$$

For $\epsilon = -1$ the relative sign between $\hat{\psi}_2$ and $\hat{\psi}_3$ is positive.

Conclusions

We have obtained all laws of quantum mechanics from classical statistics, including the concept of probability amplitudes ψ and the associated superposition of states with interference and entanglement, as well as the unitary time evolution. Our classical statistical description is genuinely probabilistic and not a local deterministic model. It allows to predict probabilities for the outcome of a chain of measurements, but not a deterministic result of a given measurement in terms of some "hidden variables". Bell's inequalities do not apply for our formulation of a correlation function which is based on conditional probabilities for a sequence of measurements. Since the mapping from classical observables to quantum operators is not invertible, no contradiction to the Kochen-Specker theorem arises.

Our setting can be extended to include observables like location and momentum by considering many two-level observables on a space-lattice and taking the limit of vanishing lattice spacing. In our statistical mechanics setting quantum mechanics describes isolated subsystems of a larger ensemble that also includes the environment. Isolation does not mean that the subsystem can be described by classical probabilities for the states of the subsystem and sharp values of the observables in these states. It rather relates to a separated time evolution of the subsystem and to observables which can be described by quantities only associated to the subsystem, without explicit reference to the environment.

We do not intend to enter here the debate if quantum mechanics or classical statistics are more fundamental -it is well known that classical statistics can be obtained as a limiting case of quantum mechanics. In our view classical statistics and quantum mechanics are two sides of the same medal. This may have far reaching consequences, as the possibility that the late time asymptotic state of a classical ensemble is given by the equilibrium ensemble of quantum statistics, or that certain steps in quantum computations can find a classical analogue. We find it remarkable that the conceptual foundations of quantum mechanics need not to go beyond the concepts of classical statistics.

[1] Bouwmeester D, Pan J W, Mattle K, Eibl M, Weinfurter H and Zeilinger A 1997 *Nature* **390** 575
[2] Feynman R 1982 *Int. J. Theor. Phys.* **21** 467; Deutsch D 1985 *Proc. R. Soc. London* A **400** 97; Cirac J I and Zoller P 1995 *Phys. Rev. Lett.* **74** 4091
[3] Wetterich C *Preprint* arXiv:0809.2671
[4] Wetterich C *Preprint* arXiv:0810.0985
[5] Kochen S, Specker E P 1967 *Journal of Mathematics and Mechanics* **17** 59; Mermin N D 1990 *Phys. Rev. Lett.* **65** 3373; Peres A 1991 *J. Phys. A: Math. Gen.* **24** L175
[6] Straumann N *Preprint* arXiv:0801.4931
[7] Holevo A S 1982 *Probabilistic and Statistical Aspects of Quantum Theory* (Amsterdam: North Holland); Ali S T and Prugovecki E 1977 *J. Math. Phys.* **18** 219; Singer M and Stulpe W 1992 *J. Math. Phys.* **33** 131
[8] Beltrametti E and Bugajski S 1995 *J. Phys. A: Math. Gen.* **28** 3329; *do.* 1995 *Int. J. Theor. Phys.* **34** 1221; Bugajski S 1996 *Int. J. Theor. Phys.* **35** 2229;
[9] Bell J S 1964 *Physica 1* 195
[10] Greenberger D, Horne M and Zeilinger A 1989 *Bell's Theorem, Quantum Theory, and Conceptions of the Universe* ed M Kafatos (Dortrecht: Kluver) 69-72 (*Preprint* arXiv:0712.0921)
[11] Wetterich C 2004 *Decoherence and Entropy in Complex Systems* ed H-T Elze (Berlin: Springer Verlag) 180 (*Preprint* arXiv:quant-ph/0212031)
[12] Wetterich C 1993 *Nucl. Phys.* B **314** 40; *do.* 1993 *Nucl. Phys.* B **397** 299
[13] Wetterich C 1997 *Phys. Lett.* B **399** 123
[14] Zeh H D 1970 *Found. Phys.* **1** 69; Joos E and Zeh H D 1985 *Z. Phys.* B **59** 273; Joos E, Zeh H D, Kiefer C, Giulini D, Kupsch J and Stamatescu I O 2003 *Decoherence and the appearance of the classical world* (Berlin: Springer Verlag); Zurek W 2003 *Rev. Mod. Phys.* **75** 715

Fourth International Workshop DICE2008 IOP Publishing
Journal of Physics: Conference Series **174** (2009) 012009 doi:10.1088/1742-6596/174/1/012009

Does quantum mechanics tell an atomistic spacetime?

Hans-Thomas Elze

Dipartimento di Fisica "Enrico Fermi", Largo Pontecorvo 3, I-56127 Pisa, Italia

E-mail: `elze@df.unipi.it`

Abstract. The canonical answer to the question posed is "Yes." – tacitly assuming that quantum theory and the concept of spacetime are to be unified by 'quantizing' a theory of gravitation. Yet, instead, one may ponder: Could quantum mechanics arise as a coarse-grained reflection of the atomistic nature of spacetime? – We speculate that this may indeed be the case. We recall the similarity between evolution of classical and quantum mechanical ensembles, according to Liouville and von Neumann equation, respectively. The classical and quantum mechanical equations are indistinguishable for objects which are free or subject to spatially constant but possibly time dependent, or harmonic forces, if represented appropriately. This result suggests a way to incorporate anharmonic interactions, including fluctuations which are tentatively related to the underlying discreteness of spacetime. Being linear and local at the quantum mechanical level, the model offers a decoherence and natural localization mechanism. However, the relation to primordial deterministic degrees of freedom is nonlocal.

1. Quantum features, information loss and spacetime discreteness

Classical ensemble theory with its Liouville equation implies the absence of a stable ground state, when rewritten as a Hilbert space theory with an analogue of the Schrödinger equation [1, 2, 3]. Correspondingly, the analogue of the von Neumann equation obtains an unusual superoperator which couples the Hilbert space and its dual [4], cf. Section 2 below.

These features have presented serious obstacles to a number of recent attempts to understand quantum mechanics as an emergent phenomenon [5, 6, 7, 8, 9, 10], see also Refs. [11, 13, 14].

Such studies are strongly motivated by issues surrounding quantum gravity, in particular, the fact that quantum theory is hardly compatible with the symmetry requirements of general relativity or of other theories describing gravity and spacetime, which are based on general coordinate invariance. Furthermore, despite the great successes in explaining statistical aspects of experiments, the intrinsically indeterministic features of quantum mechanics remain problematic. This is seen, for example, in the unresolved measurement problem, with related issues of wave function collapse or objective reduction [15, 16], and in the quantum mechanics of the universe, considered as a whole, which might be self-contradictory.

Yet there is a proof of existence of deterministic models for quantum mechanical objects, in which dissipation, i.e., a fundamental information loss mechanism, has been an essential ingredient [17, 18].

With these findings in mind, we presently discuss aspects of a dynamical transition *from* classical *to* quantum behaviour, assuming that: *Quantum features originate from a dissipative process which affects all physical objects. Dynamics and statistical interpretation of quantum states (Born rule) originate from deterministic rules, such as embodied in classical mechanics, in an ensemble theory.*

We speculate that the *atomistic* structure of spacetime itself is responsible for effects which are attributed to quantum mechanics, typically operating at length scales very much larger than the Planck length. Here "atomistic" refers to a discrete set of elements with the presence, or absence, of a certain order relation between any two elements. Furthermore, this set, in particular the number of its elements changes dynamically, possibly establishing new order relations, or erasing old ones. In this way, "time happens".

Similar ideas about the nature of spacetime have been formulated every now and then throughout the history of natural philosophy [19, 20]. However, only recently such general scenario has been elaborated in more detail in the theory of *causal sets*. In mathematical terms, these are locally finite ordered sets. Their evolution by sequential growth through random ("sprinkling") appearance of new set elements together with their order relations has been studied [20, 21, 22].

In the absence of an equally elaborate theory of matter in relation to such atomistic structure of spacetime, we can nevertheless state the following, concerning the situation of a typical object. Consider an electron, for example, an object that to highest known precision behaves according to the laws of quantum mechanics. We may term its "situation" a hypothetical complete set of its properties that are accessible by experiments.

Now, first of all, there are interactions between such object and its environment (the "rest of the universe"), last not least gravitational ones. Besides more familiar aspects, however, there must exist a continual *information loss* about its situation, since the atomistic spacetime beneath evolves. With respect to the latter, quantum theory at present deals with very coarse-grained phenomena, when describing the dynamics of matter. Loosely speaking, in order to fully characterize an electron, the evolution of its causal relations with a continually changing number of spacetime "atoms" has to come into play. Consequently, in a coarse-grained picture where this is not explicitly taken into account, information about an object degrades, due to the evolving – always and everywhere present – environment which is spacetime.

Secondly, however, common objects are characterized by a certain persistence, which makes them identifiable in experiments. Therefore, the information loss must be a delicate one. It must be compatible with the *conservation of probability*, which is a basic tenet of a reasonable ensemble theory.

Contrary to measurement processes in quantum mechanics, where information is transferred from microscopic to macroscopic objects, [1] we propose here that matter degrees of freedom are continually "measured" by the evolution of spacetime.

These heuristic considerations lead us to modify the classical ensemble theory in important ways. We will incorporate dissipation into the Liouville equation, however, in such a way that probability conservation remains intact. This provides us with a glimpse of the mechanism that might be responsible for turning the deterministic evolution of classical objects, described here by an ensemble theory, into the Schrödinger evolution of quantum objects.

2. The Liouville equation in terms of Hilbert space operators

It will be shown here how the *classical* Liouville equation can be written in the form of the *quantum mechanical* von Neumann equation, generally, incorporating a characteristic extra term.

We consider objects with a single continuous degree of freedom, for simplicity. [2] To begin with, we consider conservative forces, such that the equations of motion are determined by the generic Hamiltonian function:

$$H(x,p) := \frac{1}{2}p^2 + v(x) \quad , \tag{1}$$

[1] In a way which is not understood in all its aspects in theory. For recent discussions, which critically review related ideas, see Refs. [15, 16].

[2] The following is easily repeated for few-body systems or fields.

depending on generalized coordinate x and momentum p, and where v denotes the *true potential.* In Section 3, we will come back to the notion of the true potential and distinguish it from a related *coarse-grained potential* V.

An ensemble of such objects, for example, following trajectories with different initial conditions, is described by a distribution function f in phase space, i.e., by the probability $f(x,p;t)\mathrm{d}x\mathrm{d}p$ to find a member of the ensemble in an infinitesimal volume at point (x,p). This distribution evolves according to the Liouville equation:

$$-\partial_t f = \frac{\partial H}{\partial p}\cdot\frac{\partial f}{\partial_x} - \frac{\partial H}{\partial x}\cdot\frac{\partial f}{\partial_p} = \{p\partial_x - v'(x)\partial_p\}f \ , \tag{2}$$

with $v'(x) := \mathrm{d}v(x)/\mathrm{d}x$. – We recall that the relative minus sign in the Poisson bracket, or between terms here, reflects a symplectic phase space symmetry. This will translate into the familiar commutator structure in Eq. (5).

A Fourier transformation, $f(x,p;t) = \int \mathrm{d}y\ \mathrm{e}^{-ipy}f(x,y;t)$, replaces the Liouville equation by:

$$i\partial_t f = \{-\partial_y\partial_x + yv'(x)\}f \ , \tag{3}$$

without changing the notation for the distribution function, whenever changing variables. Thus, momentum is eliminated and a *doubled number of coordinates* results. Finally, with the transformation:

$$Q := x + y/2 \ , \quad q := x - y/2 \ , \tag{4}$$

we obtain the Liouville equation in the form:

$$i\partial_t f(Q,q;t) = \Big\{\hat{H}_Q - \hat{H}_q + \mathcal{E}(Q,q)\Big\}f(Q,q;t) \ , \tag{5}$$

$$\hat{H}_\chi := -\frac{1}{2}\partial_\chi^2 + v(\chi) \ , \quad \text{for } \chi = Q,q \ , \tag{6}$$

$$\mathcal{E}(Q,q) := (Q-q)v'(\frac{Q+q}{2}) - v(Q) + v(q) = -\mathcal{E}(q,Q) \ . \tag{7}$$

Several comments are in order here:

- The present reformulation of classical dynamics in phase space can be carried out rather independently of the number of degrees of freedom and is applicable to matrix or Grassmann valued variables as well; see, for example, Refs. [6, 23]. Gauge theories or, generally, theories with constraints have to be examined carefully.

- Most importantly, the Eq. (5) closely resembles the *von Neumann equation* for a density operator $\hat{f}(t)$, considering $f(Q,q;t)$ as its matrix elements. We automatically recover the Hamiltonian operator $\hat{H}$ related to the Hamiltonian function, Eq. (1), as in quantum theory. However, an essential difference consists in the interaction $\hat{\mathcal{E}}$ between *bra-* and *ket-* states. The Hilbert space and its dual here are coupled by a *superoperator.* [3]

- Alternatively, the Eq. (5) might be read as the *Schrödinger equation* for two identical (sets of) degrees of freedom. However, their respective Hamilton operators, $\hat{H}_{Q,q}$, contribute with opposite sign, which must be traced to the classical symplectic symmetry. Since their

[3] This superoperator is of a very specific form, which leads to the antisymmetry in Eq. (7). It differs from a Lindblad superoperator, often obtained as a symmetric double commutator structure, in the case of open quantum mechanical systems [24]; this is seen, for example, in our Eq. (23) below.

Journal of Physics: Conference Series **174** (2009) 012009 doi:10.1088/1742-6596/174/1/012009

interaction $\hat{\mathcal{E}}$ is antisymmetric under $Q \leftrightarrow q$, the complete (Liouville) operator on the right-hand side of Eq. (5) has a symmetric spectrum with respect to zero and, in general, will not be bounded below.

- Finally, the following observation will be important:

$$\hat{\mathcal{E}} \equiv 0 \iff \text{true potential } v(x) \text{ is constant, linear, or harmonic} . \tag{8}$$

The analogous vanishing of $\hat{\mathcal{E}}$ in a field theory is equivalent with having massive or massless free fields, with or without external sources, and with or without bilinear couplings. Generally, in all these cases, anharmonic forces or interactions are absent.

We emphasize that the coupling of the Hilbert space and its dual and the related lacking of a stable ground state, in general, show that our reformulation of Hamiltonian dynamics does not qualify as a proper quantum theory. However, we will motivate certain modifications from which dynamical aspects of quantum mechanics seem to emerge after all.

3. Spatiotemporal discreteness and fluctuations it may cause

Here we present a simple-minded argument indicating that the discreteness of spacetime may be relevant for the emergence of quantum mechanical phenomena, turning the Liouville equation into the von Neumann - Lindblad equation, in particular.

Given that spacetime is discrete, there must be a characteristic length scale, the Planck length, where the continuum description of all phenomena breaks down. This implies that one might overlook important traces of this atomistic structure by employing continuum quantities which allow to arbitrarily extrapolate their functional dependence to scales $l \approx l_{Pl}$, where l_{Pl} denotes the Planck length.

Instead of having a *coarse-grained potential* V, for example, the *true potential* v should become piecewise defined somehow, when approaching smaller and smaller scales in the continuum picture. Thus, the function V is an approximation to v and the difference between the two must give rise to local *fluctuations* δV. Therefore, we set:

$$v(\mathbf{x}) = V(\mathbf{x}) + \delta V(\mathbf{x}) \ , \tag{9}$$

in order to relate the short distance behaviour to its coarse-grained description.

Two sources of fluctuations can enter here: First, the spatiotemporal discreteness [25]. Second, the possibly discrete nature of interactions or matter. The latter might go beyond what is usually seen as *effects* of quantum mechanics. This has not been explored in parallel with the causal set theory of the deep structure of spacetime and will not be further discussed at present.

3.1. "Asymptotic freedom" on causal sets, piecewise linear potentials and emergence of the von Neumann equation

However, there is an important "asymptotic freedom" effect, caused by spatiotemporal discreteness: on a (background) causal set, cross sections must fall to zero when the center of mass energy of two scattering particles reaches the Planck scale [26]. [4] This is based on

[4] I thank Rafael Sorkin for the following illuminating description of this "No Interaction Theorem": *Imagine each particle or wave-packet as a "world tube" in Minkowski space. In the center of mass frame, the two tubes make an "X". The center of the X, where the tubes meet, is the interaction region. Now boost each particle to a very high energy. Because of the Lorentz contraction, the tubes become so flattened that the interaction region shrinks down to less than a Planck volume. Hence there is (very likely) no element of the causet to represent this interaction region, whence there is no interaction.*

viewing the causal set *as if* generated by a Poisson process, "sprinkling" set elements with an average density of one element per Planckian volume, $(l_{Pl})^4$ – for example, into Minkowski space, if this is considered as the continuum limit.

The "asymptotic freedom" effect suggests to consider the *true potential* function v as *piecewise linear*, with the pieces – simplices in higher dimensions – characterized by short "linearity lengths" δ_n. They must be, however, sufficiently larger than the Planck length, $\delta_n > l_{Pl}$, such that the continuum description is meaningful. [5]

With these remarks in mind, we reconsider the force term $\propto v'$ in Eq. (5). We rewrite this equation, admitting more than one spatial dimension, more simply as:

$$i\partial_t f(\mathbf{Q},\mathbf{q};t) = [\hat{H}_0,\hat{f}]_{(\mathbf{Q},\mathbf{q};t)} + (\mathbf{Q}-\mathbf{q})\cdot\nabla v(\frac{\mathbf{Q}+\mathbf{q}}{2})f(\mathbf{Q},\mathbf{q};t) \;, \tag{10}$$

where $\hat{H}_0 := -\frac{1}{2}\Delta$, with Δ denoting the Laplacian, in an obvious generalization. Now, the term involving the derivative of the potential is related to the *would-be quantum mechanical* term – the potential difference between points Q and q – as a linear mid-point approximation to an integral:

$$(\mathbf{Q}-\mathbf{q})\cdot\nabla v(\frac{\mathbf{Q}+\mathbf{q}}{2}) \;\approx\; \int_{\mathbf{q}}^{\mathbf{Q}} d\mathbf{s}\cdot\nabla v(\mathbf{s}) = v(\mathbf{Q}) - v(\mathbf{q}) \;. \tag{11}$$

The approximate relation becomes an exact equality in the cases covered by observation (8).

More generally, the integral – which yields *exactly* the quantum mechanical term – can be decomposed into a sum of integrals along straight line segments:

$$v(\mathbf{Q}) - v(\mathbf{q}) = \sum_n \int_{\mathbf{s_n}}^{\mathbf{s_{n+1}}} d\mathbf{s}\cdot\nabla v(\mathbf{s}) = \sum_n (\mathbf{s_{n+1}} - \mathbf{s_n})\cdot\nabla v(\frac{\mathbf{s_n}+\mathbf{s_{n+1}}}{2}) \;, \tag{12}$$

determined by a set of positions $\{\mathbf{s_n}\}$ which match the piecewise linearity of the true potential with the path of integration from $\mathbf{q}$ to $\mathbf{Q}$, i.e., with the linearity lengths, $\delta_n = |\mathbf{s_{n+1}} - \mathbf{s_n}|$.

We conclude that within small volumes, the linear size of which is determined by the linearity length δ of the potential there, the Liouville equation of *classical* statistical mechanics, in the form of Eq. (5) or Eq. (10), is *indistinguishable* from the corresponding *quantum mechanical* von Neumann equation, which both refer to the continuum description.

Furthermore, we learn from Eqs. (10)–(12) that for larger distances, covering several linearity lengths δ_n, the potential (commutator) term of the von Neumann operator is obtained as a *sum of classical contributions*. This summation is natural in view of the fact that each term adds an appropriate amount of energy to the potential difference $v(\mathbf{Q}) - v(\mathbf{q})$, with one contribution per interval over which the potential is linear.

In this way, we obtain the Hamilton operator for arbitrarily extended, piecewise linear potentials:

$$\hat{H} = -\frac{1}{2}\partial_{\mathbf{x}}^2 + v(\mathbf{x}) \;, \tag{13}$$

in the coordinate representation, and readily infer the von Neumann equation in operator form:

$$\partial_t\hat{f} = -i[\hat{H},\hat{f}] \;. \tag{14}$$

However, following the argument from Eq. (10) to Eq. (14), there is a subtlety behind the last two equations here, concerning the composition of the simplices.

[5] Such considerations may deviate from the strategy to adapt continuum quantities to a discrete spacetime – for example, by approximating a quartic potential by a quartic expression with support on the elements of a causal set. It seems a valuable alternative to look for a model of matter degrees of freedom and their interactions which is formulated in terms innate to a causal set, as far as possible. For free particles, work in this direction has been reported in Ref. [27].

In the continuum picture, each simplex forms a tiny "box" with a *sharply localized* boundary condition, which prohibits the leaking of probability to the outside – as long as they are sufficiently separated. Bringing two such boxes next to each other, such that a common border region arises which is less than a Planck length "thick", the confining boundary condition will dissolve into a free boundary condition, no matter how it originated in the first place. This is another manifestation of the "asymptotic freedom" property, caused by the discrete spacetime structure. Thus, we imagine that the range of variables entering the matrix elements of $\hat{f}$ can adiabatically increase to the full range over both simplexes.

While this "Gedankenexperiment" might seem plausible, a serious justification of the extension of the range of validity of Eq. (14) over more than one linearity length seems necessary. We intend to come back to this important point elsewhere.

3.2. Coarse-graining, decoherence and localization mechanism

In order to extend our considerations further to reach the scales where quantum mechanics is known to work, we have to address also the coarse-graining that must be involved. This cannot be analyzed rigorously without understanding how the common forces / interactions that we encounter relate to phenomena at much shorter distance scales, such as the linearity scale.

However, qualitatively, we expect the coordinates to *fluctuate*, which enter the continuum description, in particular, when we write down the von Neumann equation with the potential terms obtained in Eq. (12). – Indeed, let us consider, for example, an event determined in the continuum picture. The position ascribed to it in a fixed Lorentz frame cannot be exactly maintained with time, if Minkowski space is a *continuum approximation* to an underlying discrete structure: to the future of this event there likely will be Planck scale size (or larger) patches of Minkowski space which contain no element of the approximated causal set. It appears to us that the event, respectively its successors, must change position by small amounts with time ("swerve"), effectively performing a random walk within its future lightcone [27].

We can get a feeling for the linear size $\triangle r$ of these swerves by estimating the probability to have a spatial void of volume $\nu = \frac{4}{3}\pi\triangle r^3$ that lasts for a Planck time interval. For the Poisson process mentioned earlier and with $\triangle r \gg l_{Pl}$, we find this to be $e^{-(\triangle r/l_{Pl})^3}$. Therefore, uncertainties $\triangle r$ of the order of a typical distance pertaining to the Standard Model, for example, are extremely improbable. Nevertheless, at the linearity scale they may play an essential role in washing out the piecewise linear potentials. [6]

By experience, smooth polynomial interactions play an essential role, say in the Standard Model and, to be derived from there, in phenomenological forces of physics at still lower energies. Since details of the coarse-graining are unknown, in particular those leading to highly constrained gauge symmetries, we now consider the phenomenological *Ansatz* of Eq. (9), which connects *short distance behaviour at the linearity scale* with its *coarse-grained description*. In particular, this replaces a true, piecewise linear potential v by a coarse-grained potential V plus local fluctuations δV.

The fluctuations are treated as white noise with mean and correlation, respectively, given by:

$$\langle \delta V(\mathbf{x}) \rangle = 0 \ , \tag{15}$$

$$\langle \delta V(\mathbf{x}) \delta V(\mathbf{y}) \rangle = \nu^2(\mathbf{x}) \delta(\mathbf{x}-\mathbf{y})/\delta(0) \ , \tag{16}$$

where $\nu(\mathbf{x})$ describes the width of the local distribution of fluctuations. This distribution cannot be assessed without further insight into how forces arise at small spacetime scales.

[6] Similarly, there, they will have an effect on the kinetic terms of the Liouville or von Neumann operator; we simply take them as they are, since they yield the appropriate quantum mechanical operator. However, a detailed study of the emergence of such terms from a causal set perspective, of the wave operator in particular, has recently been performed [25], eventually to be combined with considerations along the lines presented here.

We now implement Eq. (9), $v(\mathbf{x}) = V(\mathbf{x}) + \delta V(\mathbf{x})$, in the von Neumann equation following from Eqs. (10)–(12), with piecewise linear potential. This yields the von Neumann equation with the coarse-grained potential V and additional fluctuating terms:

$$\partial_t \hat{f} = -i[\hat{H}_{eff}, \hat{f}] \ , \tag{17}$$

where the effective Hamilton operator is now given by:

$$\hat{H}_{eff} := \hat{H} + \delta V \tag{18}$$

$$:= -\frac{1}{2}\partial_{\mathbf{x}}^{\ 2} + V(\mathbf{x}) + \delta V(\mathbf{x}) \ , \tag{19}$$

in the coordinate representation, cf. Eq. (14).

In order to evaluate the influence of the fluctuations, we iterate Eq. (17) once:

$$\partial_t^{\ 2} \hat{f} = -i[\hat{H}, \partial_t \hat{f}] + \mathrm{O}(\delta V) - [\delta V, [\delta V, \hat{f}] \ . \tag{20}$$

Integrating once, with the initial condition:

$$\partial_t \hat{f}|_{t=0} = -i[\hat{H}, \hat{f}]_{t=0} + \mathrm{O}(\delta V) \ , \tag{21}$$

and averaging over the fluctuations, with the help of Eqs. (15)–(16), leads to:

$$\partial_t \hat{f} = -i[\hat{H}, \hat{f}] - \int_0^t \mathrm{d}t' (\{\hat{\nu}^2, \hat{f}\} - 2\hat{\nu}\hat{f}\hat{\nu}) \ , \tag{22}$$

with the matrix elements $\langle \mathbf{x}|\{\hat{\nu}^2, \hat{f}\}|\mathbf{y}\rangle := \nu^2(\mathbf{x}) f(\mathbf{x}, \mathbf{y}) + f(\mathbf{x}, \mathbf{y})\nu^2(\mathbf{y})$ and $\langle \mathbf{x}|\hat{\nu}\hat{f}\hat{\nu}|\mathbf{y}\rangle := \nu^2(\mathbf{x}) f(\mathbf{x}, \mathbf{y})\delta(\mathbf{x} - \mathbf{y})/\delta(0)$. For sufficiently short times and slowly varying $\hat{f}$, we may use the approximation:

$$\partial_t \hat{f} \approx -i[\hat{H}, \hat{f}] - t(\{\hat{\nu}^2, \hat{f}\} - 2\hat{\nu}\hat{f}\hat{\nu}) \ , \tag{23}$$

Thus, we arrive at a Markovian master equation with a dissipative *Lindblad term*, in addition to the leading commutator, which is responsible for unitary quantum evolution à la von Neumann in the absence of dissipation.

Linearity in the density matrix is an important feature of our resulting equations. In particular, in the form of Eqs. (22)–(23), the master equation preserves the normalization of $\hat{f}$, say $\mathrm{Tr}\hat{f} = 1$, which expresses the conservation of probability.

It seems worth while to point out that we arrive at the standard, i.e., in particular *local*, quantum mechanical evolution equation – even if modified by a less standard yet wellcome Lindblad term. However, we emphasize that the relation to the primordial deterministic degrees of freedom, which involves a Fourier transformation (cf. Section 2), is highly nonlocal. This may ease the tension created by Bell's theorem, when it comes to deterministic ("hidden") variables.

The Lindblad term in Eq. (23) implies an interesting *decoherence and continuous localization mechanism*, which causes the decay of spatial superpositions ("Schrödinger cat states"). While the diagonal matrix elements of $\hat{f}$ are not affected by the Lindblad term, the off-diagonal matrix elements decay:

$$f(\mathbf{x}, \mathbf{y}; t) = f(\mathbf{x}, \mathbf{y}; 0)\mathrm{e}^{-\frac{1}{2}t^2(\nu^2(\mathbf{x}) + \nu^2(\mathbf{y}))} \ , \tag{24}$$

where we neglected the effect of $\hat{H}$, for simplicity, which cannot stop the decay. Ultimately, spacetime discreteness produces this mechanism – via the induced fluctuations in interactions.

Since there is no theory yet to tell us about the correlation function ν^2, it may suffice to point out that considerations of stochastic effects on quantum mechanics – as it is – have quite a history, see, for example, Refs. [15, 16, 28, 29] and the literature cited there. This is related to attempts to solve the measurement problem and to account for the apparent absence of spatial superposition states of macroscopic objects [30, 31]. Also quantum gravity is conjectured to lead to characteristic stochastic effects [32, 33, 34]. In this context, various proposals for the analogue of our Lindblad operator $\hat{\nu}$ in Eq. (23) have been made as well and it has similarly been concluded that spatial superposition states decohere and decay.

While these issues are not settled, our heuristic arguments suggest that if quantum mechanical behaviour indeed emerges dynamically, then the resolution of the measurement problem may be linked to this. 'Prequantum' dynamics may account for an objective selection mechanism in accordance with the observed wave function collapse in measurement outcomes.

Generally, Lindblad master equations present a large class of linear Markovian master equations, which are usually derived to describe open quantum systems that interact with particular environments [24, 35, 36]. Here spacetime itself forms the universal environment that "measures" all physical objects.

We have derived this *quantum mechanical master equation*, beginning with *classical statistical mechanics*, by incorporating several assumptions about the atomistic nature of spacetime, as well as about the nature of forces acting on matter.

4. Conclusions

We have presented a heuristic discussion to the effect that quantum mechanics – together with a natural decoherence and continuous localization mechanism – is a consequence of classical statistics and concerns dynamics with respect to an atomistic spacetime that is observed with low resolution, i.e., at large distance scales. The spatiotemporal discreteness results in permanent information loss affecting all matter, when described in the corse-grained way that is appropriate with distances much larger than the Planck scale. We were motivated here by the theory of causal sets, where such a picture of spacetime does not follow from a quantization of gravity but is assumed as primary feature.

There are a number of topics for further study. Work is in progress making newly use of the formalism adopted here, of using Hilbert space operators to describe classical statistics, and going beyond the early suggestions [1, 2].

Furthermore, as pointed out in Ref. [4], the problem of negative probabilities needs to be resolved, which is met when carrying the notion of probability density from the classical phase space theory over to the emergent quantum mechanical one, as we do.

The generalization of the ideas sketched here to field theories has to face the fundamental problem to understand the nature of matter fields and the emergence of (the symmetries of) their interactions with decreasing spatiotemporal resolution. Our approach, paying special attention to piecewise linear potentials at sufficiently small scales, is in accordance with the suggestion of noninteracting fields at the Planck scale. There, the atomistic structure of spacetime becomes effective and – for an underlying causal set – induces such an "asymptotic freedom property".

Acknowledgments

I am grateful to Lajos Diósi, Jose Isidro, Andrei Khrennikov and Rafael Sorkin for discussions or correspondence.

References

[1] Koopman B O 1931 *Proc. Nat. Acad. Sci. (USA)* **17** 315
[2] von Neumann J 1932 *Ann. Math.* **33** 587, 789
[3] 't Hooft G 2007 *J. Phys.: Conf. Ser.* **67** 012015 (*Preprint* arXiv:quant-ph/0604008)
[4] Elze H-T 2009 *Int. J. Qu. Inf. (IJQI)* **7** 83-96 (*Preprint* arXiv:0806.3408); *do.* 2009 Symmetry aspects in emergent quantum mechanics *Proceedings Discrete'08, Valencia (Spain), December 11-16, 2008, J. Phys.: Conf. Ser.* to appear

[5] 't Hooft G 2003 *Int. J. Theor. Phys.* **42** 355-361; *do.* 1999 *Class. Quant. Grav.* **16** 3263-3279
[6] Elze H-T 2006 *J. Phys.: Conf. Ser.* **33** 399-404 (*Preprint* arXiv:gr-qc/0512016); *do.* 2005 *Braz. J. Phys.* **35** 343-350; *do.* 2003 *Phys. Lett.* A **310** 110-118
[7] Blasone M, Jizba P and Kleinert H 2005 *Ann. Phys.* **320** 468-486 (*Preprint* arXiv:quant-ph/0504200); *do.* 2005 *Braz. J. Phys.* **35** 497-502; *do.* 2005 *Phys. Rev.* A **71** 052507
[8] Adler S L 2005 *Quantum Mechanics as an Emergent Phenomenon* (Cambridge, UK: Cambridge U. Press)
[9] Markopoulou F and Smolin L 2004 *Phys. Rev.* D **70** 124029 (*Preprint* arXiv:gr-qc/0311059)
[10] Blasone M, Jizba P and Vitiello G 2001 *Phys. Lett.* A **287** 205-210 (*Preprint* arXiv:hep-th/0007138)
[11] Blasone M, Jizba P and Scardigli F 2009 Can quantum mechanics be an emergent phenomenon? *Preprint* arXiv:0901.3907, to appear in Ref. [12]
[12] Elze H-T, Diósi L, Fronzoni L, Halliwell J J and Vitiello G (ed) 2009 Proceedings 4th Intern. Conf. DICE 2008, Castiglioncello (Italy), September 22-26, 2008 *J. Phys.: Conf. Ser.* to appear
[13] Isidro J M, Santander J L G and Fernandez de Cordoba P 2008 A note on the quantum-mechanical Ricci flow *Preprint* arXiv:0808.2717; *do.* 2008 Ricci flow, quantum mechanics and gravity *Preprint* arXiv:0808.2351
[14] Wetterich C 2008 Emergence of quantum mechanics from classical statistics *Preprint* arXiv:0811.0927, to appear in Ref. [12]; *do.* 2008 Probabilistic observables, conditional correlations, and quantum physics *Preprint* arXiv:0810.0985; *do.* 2008 Quantum entanglement and interference from classical statistics *Preprint* arXiv:0809.2671
[15] Ghirardi GC 2009 The interpretation of quantum mechanics: where do we stand? *Preprint* arXiv:0904.0958, to appear in Ref. [12]
[16] Bassi A and Ghirardi GC 2003 *Phys. Rep.* **379** 257 (*Preprint* arXiv:quant-ph/0302164)
[17] 't Hooft G 2007 *AIP Conf. Proc.* **957** 154-163 (*Preprint* arXiv:0707.4568)
[18] Elze H-T 2008 *J. Phys. A.: Math. Theor.* **41** 304020 (*Preprint* arXiv:0710.2765)
[19] Jammer M 1954 *Concepts of Space: The History of Theories of Space in Physics* (Harvard: Harvard U. Press)
[20] Sorkin R D 2005 Causal sets: discrete gravity *Lectures on Quantum Gravity* ed A Gomberoff and D Marolf (New York: Springer) (*Preprint* arXiv:gr-qc/0309009)
[21] Dowker F 2005 Causal sets and the deep structure of spacetime *100 Years of Relativity - Space-Time Structure: Einstein and Beyond* ed A Ashtekar (Singapore: World Scientific) (*Preprint* arXiv:gr-qc/0508109)
[22] Rideout D P and Sorkin R D 2000 *Phys. Rev.* D **61** 024002 (*Preprint* arXiv:gr-qc/9904062)
[23] Elze H-T 2007 *Int. J. Theor. Phys.* **46** No 8 2063-2081 (*Preprint* arXiv:hep-th/0510267)
[24] Diósi L 2007 *A Short Course in Quantum Information Theory, Lecture Notes in Physics* **713** (Berlin: Springer)
[25] Sorkin R D 2007 *Approaches to Quantum Gravity: Toward a New Understanding of Space, Time and Matter* ed D Oriti (Cambridge, UK: Cambridge U. Press) (*Preprint* arXiv:gr-qc/0703099)
[26] Sorkin R D 2009 Light, links and causal sets *Preprint*, to appear in Ref. [12]
[27] Philpott L, Dowker F and Sorkin R 2008 Energy-momentum diffusion from spacetime discreteness *Preprint* arXiv:0810.5591v2
[28] Diósi L 1984 *Phys. Lett.* A **105** 199
[29] Diósi L 2005 *Braz. J. Phys.* **35** 260-265 (*Preprint* arXiv:quant-ph/0412154v2)
[30] Penrose R 1998 *Phil. Trans. R. Soc.* **356** 1927
[31] Adler S L 2003 *Stud. Hist. Philos. Mod. Phys.* **34** 135
[32] Ellis J, Mavromatos N E and Nanopoulos D V 1992 *Phys. Lett.* B **293** 37-48 (*Preprint* arXiv:hep-th/9207103v2)
[33] Gambini R, Porto R A and Pullin J 2008 *Phys. Lett.* A **372** 1213-1218 (*Preprint* arXiv:0708.2935v2)
[34] Hu B-L 2009 Emergent/quantum gravity: macro/micro structures of spacetime *Preprint* arXiv:0903.0878, to appear in Ref. [12]
[35] Lindblad G 1976 *Commun. Math. Phys.* **48** 119-130
[36] Gorini V, Kossakowski A and Sudarshan E C G 1976 *J. Math. Phys.* **17** 821-825

The physics of mental acts: coherence and creativity

F. Tito Arecchi

University of Firenze and INOA (Istituto Nazionale di Ottica Applicata)-Firenze, Italy

E-mail: tito.arecchi@inoa.it

Abstract. Coherence is a long range order absent at thermal equilibrium, where a system is the superposition of many uncorrelated components. To build non-trivial correlations, the system must enter a nonlinear dynamical regime.The nonlinearity leads to a multiplicity of equilibrium states, the number of which increases exponentially with the number of partners; we call complexity such a situation. Complete exploration of complexity would require a very large amount of time. On the contrary, in cognitive tasks, one reaches a decision within a few hundred milliseconds. Neuron synchronization lasting around 301 msec is the indicator of a conscious perception (Gestalt); however, the loss of information in the chaotic spike train of a single neuron takes a few msec, thus a conscious perception implies a control of chaos, whereby the information stored in a brain area survives for a time sufficient to elicit an action. Control of chaos is achieved by the interaction of a bottom-up stimulus with a top-down control (induced by the semantic memory). We call creativity this optimal control of neuronal chaos; it goes beyond the Bayesian inference, which is the way a computer operates, thus it represent a non-algorithmic step.

1. Introduction

The terms introduced in the Abstract are defined as follows:

Coherence = long range order (in space [vision] or time [music]);
Complexity = display of different coherences;
Creativity = jump from one coherence regime to another.

What are the relations among the three concepts? A detailed analysis has been given elsewhere [Arecchi 2007] and here we recall the main points. Coherence is associated with long range correlations, in space or time; complexity arises whenever an array of coupled dynamical systems displays multiple paths of coherence. Creativity corresponds to a selection of a coherence path within a complex nest. As we will see soon, it seems dynamically related to control of chaos.

Exploration of a complex situation would require a very large amount of time, in order to classify all possible coherences, i.e. long range correlations. In cognitive tasks facing a complex scenario, our strategy consists in attaining a decision within a finite short time. Any conscious perception (we define conscious as that eliciting a decision) requires a few hundred milliseconds, whereas the loss of information in the chaotic spike train of a single neuron takes a few msec.

The interaction of a bottom-up signal (external stimulus) with a top-down modification of the control parameters (induced by the semantic memory) leads to a collective synchronization lasting a sizable fraction of a second: this is the indicator of a conscious perception. The operation is a control of chaos, and it has an optimality; if it lasts less than 200msec, no decisions emerge, if it lasts much longer, there is no room for sequential cognitive tasks. We call creativity this optimal control of neuronal chaos. It amounts to selecting one among a large number of possible coherences all present in a complex situation. *The selected coherence is the meaning of the object under study.*

In Sec. 2, we survey the main steps of a perceptual process, with reference to vision. In Sec. 3, we discuss the cognitive process, starting from the most fundamental inference based on Bayes theorem; we show that it is limited to a single model (or likelihood) whereas complex situations require changes of model. This is the creative step peculiar of human decisions and impossible to a Turing machine. In Sec. 4, we review critically a quantum approach to brain processes. In Sec. 5 (conclusions), we compare two pairs of conflicting categories relevant for decision making, namely, partial truth versus global truth and hermeneutic circle versus hermeneutic spiral.

2. How we see: processing visual events

Fig. 1 shows a brain section with the path of the visual information. A given light signal impinging on the eyes retina stimulates the production of electrical pulses (spikes) high about 100 millivolt and lasting about 1 millisecond. The time sequence of these standard pulses represents a neural code, this means that different inputs are coded in different spike sequences. Spikes travel on the neuron axons , like on transmission lines At the axon end (synapse) the electrical signal releases some amount of neurotransmitters, that is, specialized chemicals which diffuse in the inter-neuron space and arrive to a next neuron, where they again stimulate electrical spikes and so on. The brain has a huge number of neurons tangled by a web of mutual couplings. A detailed account of all couplings is out of reach at the present state of neuroscientific investigation. It would require resolving each neuron separately; this can be done in laboratory animals by inserting a sparse number of microelectrodes each one probing a single neuron . Coarse grained information is captured by EEG (electro encephalo-gram) whereby we measure the electrical activity probed by some electrodes glued on the scalp. At variance with the microelectrodes probing the single neuron, the technique is non invasive; however an external electrode sums up the activity of many neurons, with a poor space resolution. The fMRI (functional magnetic resonance imaging) is sensitive to the magnetic properties of oxygenated blood; thus it records the blood intake to a brain area which is in operation and hence requires energy; again, the space resolution is poor.

Let us consider the visual system; the role of elementary feature detectors has been extensively studied [Hubel]. Thus, we know that some neurons are specialized in detecting exclusively vertical or horizontal or tilted bars, or a specific luminance contrast, etc.

However, the problem arises: how elementary detectors contribute to a holistic (Gestalt) perception? A hint is provided by [Singer]. Suppose we are exposed to a visual field containing two separate objects. Both objects are made of the same visual elements, horizontal and vertical contour bars, different degrees of luminance, etc. What are then the neural correlates of the identification of the two objects? We have one million fibers connecting the retina to the visual cortex. Each fiber results from the merging of approximately 100 retinal detectors (rods and cones) and as a result it has its own receptive field. Each receptive field isolates a specific detail of an object. We thus split an image into a mosaic of adjacent receptive fields. Now the feature binding hypothesis consists of assuming that all the cortical neurons whose receptive fields are pointing to a specific object synchronize the corresponding spikes, and as a consequence the visual cortex organizes into separate neuron groups oscillating on two distinct spike trains for the two objects. Direct experimental evidence of this synchronization is obtained by insertion of microelectrodes in the cortical tissue of animals just sensing the single neuron (Fig. 2).

The interaction of a bottom-up signal (external stimulus) with a top-down change of the control parameters (induced by the semantic memory) leads to a collective synchronization lasting 200 msec: this is the indicator of a conscious perception. The operation is a control

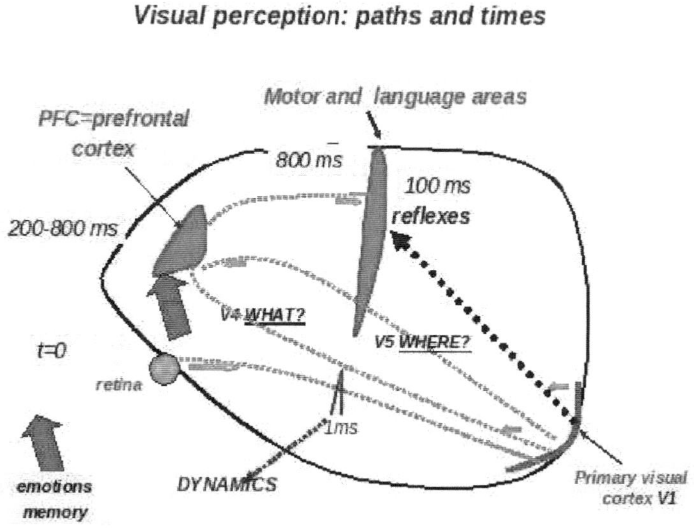

Figure 1. Time sequence of brain events following a visual stimulus on the retina at time t=0. After 200 msec, the signal coded as a train of electrical pulses of 1 msec duration(spikes) traveling on neurons axons, arrives at the primary visual cortex V1; the dynamics of spikes (interspike interval) depends on the sensory stimuli and on the receptive cell (Fig. 2). The signal from V1 is then elaborated in two distinct areas, V4 or WHAT area which recognizes shapes and color, and V5 or WHERE area, which recognizes motion and space relations. The separate information is combined in the PFC (prefrontal cortex) together with top-down signals coming from the inner brain (emotions, memory).This mixing takes about half a second starting from 200 msec; at 800 msec after the sensitization of the retina, a decision emerges activating the motor and language areas, The elaboration in PFC takes so long because the input is compared with previous memories and hence re-coded. Instead vital responses arrive directly to the motor area in even less than 100 msec (reflexes: heavy dotted line).

of chaos, and it has an optimality; if it lasts less than 200msec, no decisions emerge, on the contrary, if it lasts much longer, there is no room for sequential cognitive tasks. The addition of extra degrees of freedom implies a change of code, thus it can be seen as a new level of description of the same physical system.

Based on the neurodynamical facts reported above, we can understand how this occurs [Grossberg]. The higher cortical stages from where decision emerge have two inputs. One (bottom-up) comes from the sensory detectors via the early stages which classify elementary features. This single input is insufficient, because it would provide the same signal for e.g. horizontal bars belonging indifferently to either one of the two objects. However, as we said already, each neuron is a nonlinear system passing close to a saddle point, and the application of a suitable perturbation can stretch or shrink the interval of time spent around the saddle, and thus lengthen or shorten the interspike interval. The perturbation consists of top-down signals corresponding to conjectures made by the semantic memory (Fig. 3). ART (adaptive resonance theory) refers to a re-coding of bottom up stimuli based on previous learning. For simple stimuli which require a fast reaction, there is a direct access from V1 to the motor areas as shown in Fig. 1.

On the other hand, exploration of a complex situation would require a very large amount of

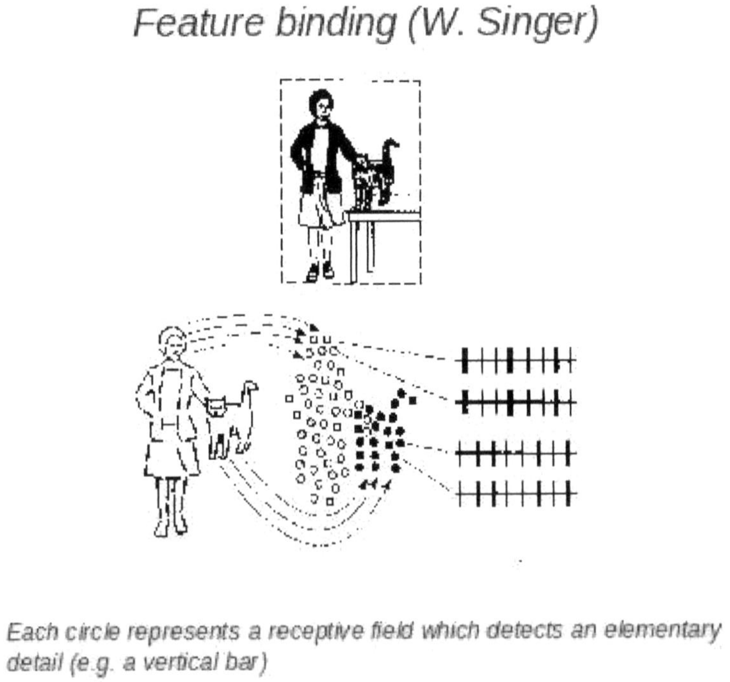

Figure 2. Feature binding: the lady and the cat are respectively represented by the mosaic of empty and filled circles, each one representing the receptive field of a neuron group in the visual cortex. Within each circle the processing refers to a specific detail (e.g. contour orientation). The relations between details are coded by the temporal correlation among neurons, as shown by the same sequences of electrical pulses for two filled circles or two empty circles. Neurons referring to the same individual (e.g. the cat) have synchronous discharges, whereas their spikes are uncorrelated with those referring to another individual (the lady) [from Singer].

time. In cognitive tasks facing a complex scenario, the cognitive strategy consists in converging to a decision within a finite time . Various experiments [Libet, Rodriguez et al.] prove that a decision is taken after a few hundred milliseconds of exposure to a sensory stimulus Thus, any conscious perception (we define conscious as that eliciting a decision) requires a few hundred msec, whereas the loss of information in a chaotic train of neural spikes takes a few msec. There must be a mechanism of chaos control which holds a given information for the time necessary to decide about.

Synchronization of a chain of chaotic lasers provides a promising model for a physics of cognition. Indeed, a peculiar type of chaos, consisting of sequences of identical spikes erratically occurring in time, has been discovered in lasers and called HC [Arecchi and Meucci 2008]. The chaotic train can be regularized by synchronization actions. An array of weakly coupled HC systems represents the simplest model for a physical realization of feature binding. The array can achieve a collective synchronized state lasting for a finite time (corresponding to the physiological 200 ms!) if there is a sparse (non global) coupling, if the input (bottom-up) is applied to just a few neurons and if the inter-neuron coupling is suitably adjusted(top-down control of chaos) [Ciszak, Montina and Arecchi 2009 a,b]. This approach is discussed in the next section.

Journal of Physics: Conference Series **174** (2009) 012010 doi:10.1088/1742-6596/174/1/012010

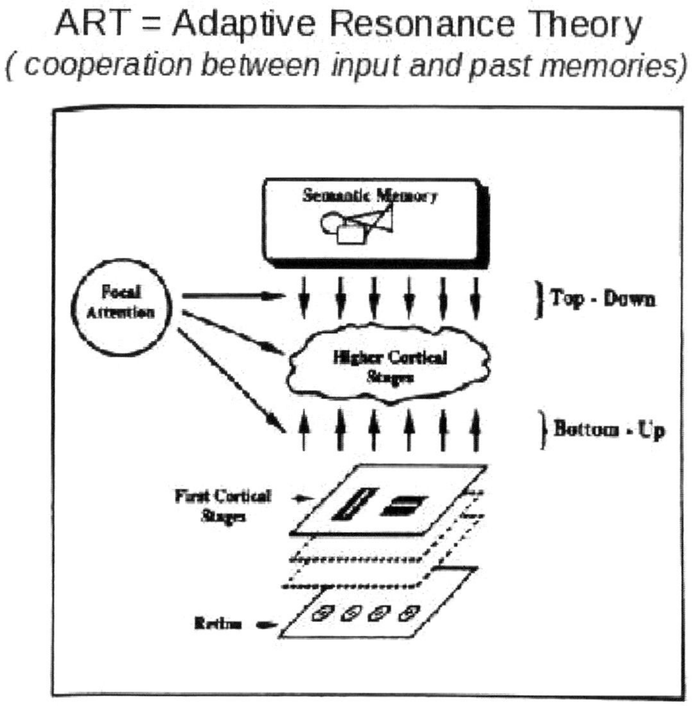

Figure 3. ART = Adaptive Resonance Theory. Role of bottom-up stimuli from the early visual stages an top-down signals due to expectations formulated by the semantic memory. The focal attention assures the matching (resonance) between the two streams.

3. The scientific insight

As Galileo was saying in a letter to Marc Welser, in 1610, a sound scientific approach means avoiding philosophical considerations on the nature of objects, and limiting oneself to measurable data coded in numbers, then connecting these numbers by a mathematical formalism The scientific method of Galileo is based on two pillars:

Sensate esperienze= experiments performed through our senses: if the case helped by equipment, e.g. the telescope and the clock;

Necessarie dimostrazioni = compelling proofs: in a theorem, if one agrees on the premises and understands the proof, than he must necessarily accept the conclusions.

Thus, in science meaning appears irrelevant and only measurable details and repeatable phenomena make sense. The Galileian innovation was to replace words with numbers. Some time later, Newton discovered that assignment of position and velocity of a particle at a given time (the initial condition) determines univocally its future trajectory. All objects in nature are made of elementary particles endowed of this Newtonian property. Thus the future of the whole object can be predicted once we know its initial condition. This was the basis of Laplace determinism (1812) disproved a few decades later by Poincaré.

3.1. Deterministic chaos and its control

For a Newtonian particle, once we know the forces ,the trajectory emerging from a given initial condition ,as the star in Fig. 4, is unique. The coordinates of the initial point are in general

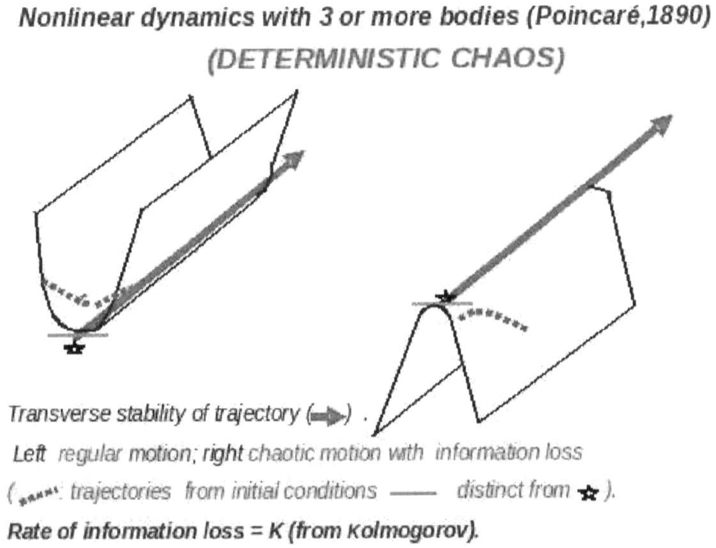

Figure 4. Deterministic chaos-the trajectory emerging from a precise initial condition, as the star, is unique. However, in general the initial condition is spread over a small patch. Points of the patch converge toward the calculated trajectory or diverge from it, depending on whether the transverse stability landscape is a valley (left) or a hill (right). From 3 coupled dynamical degrees of freedom on, the right situation is generic. We call it deterministic chaos. The information loss rate K is called Kolmogorov entropy.

real numbers truncated to a finite number of digits, thus the initial condition is spread over a small patch. Points of the patch converge toward the calculated trajectory or diverge from it, depending on whether the transverse stability yields a landscape like a valley (left) or the ridge of a hill (right). Poincar proved that from 3 coupled dynamical degrees of freedom on, the right situation is generic. Nowadays we call it deterministic chaos. It is a sensitive dependence on initial conditions and implies a loss of information of the initial preparation. The loss rate K is called Kolmogorov entropy. We can adjust its value by adding extra variables which change the slope of the downhill fall without practically perturbing the longitudinal trajectory (control of chaos), as shown in Fig. 5. Control of chaos occurs everywhere in interacting systems and it is essential to establish coherent features.

Suppose we have a N-dimensional chaotic system: adding a few, say p, extra variables we embed the dynamical problem in a $(N+p)$-dimensional space. By careful choice of the control, we can keep the longitudinal trajectory of the original dynamics, modifying however its transverse stability. In the example of Fig. 6, we consider a perceptual feature encoded by a chaotic dynamics which loses information over 2 msec. This time is insufficient to elicit a decision; thus a mechanism of chaos control must stabilize chaos for about 200 msec. The chaos control was initially devised for complete stabilization (as shown on the left of Fig. 6). This strategy however is unfit for perceptual purposes, since the agent should be able to react to a sequence of stimuli, thus it has to fix a given input just for the time necessary to decide about, and then be ready for the next input.

We thus hypothesize that transient control of chaos is the strategy whereby a cognitive agent exploits previously learned resources in order to re-code the input arriving from the early sensory stages. Re-coding over a transient interval can be done in multiple ways. The selected way is the

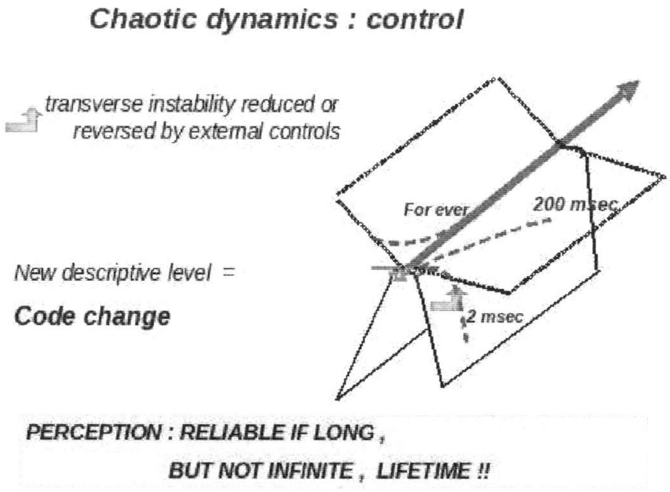

Figure 5. Chaos is controlled by adding extra-dynamic variables, which change the transverse instability without affecting the longitudinal trajectory. In the perceptual case, the most suitable top-down signals are those which provide a synchronized neuron array with an information lifetime sufficient to activate successive decisional areas (e.g. 200 ms), whereas the single HC neuron has a chaotic lifetime of 2 ms. If our attentional-emotional system is excessively cautious, it provides a top-down correction which may stabilize the transverse instability for ever, but then the perceptual area is blocked to further perceptions.

one that best fits the interaction of the agent with its environment. This hypothesis elaborated in the following sub-sections.

3.2. Stabilizing a perception (sensorial input)

From a dynamical point of view, the single neuron is a chaotic system. Even though the electrical spikes are all equal to each other, their time separation is chaotic. As a general consideration, a living being is a semiotic agent; by this we mean that the agent, embedded in an environment, changes the descriptive code in order to stabilize the cognitive dynamics and compensate for unavoidable information loss associated with the chaotic dynamics of its neurons. We have already stated that any conscious perception (we define conscious as that eliciting a decision) requires about 200 msec, whereas the loss of information in a chaotic train of neural spikes takes a few msec.

We have already seen Singer experiment on feature binding. An array of weakly coupled HC systems represents the simplest model for a physical realization of feature binding. The array can achieve a collective synchronized state lasting for a finite time (corresponding to the physiological 200 ms!) if there is a sparse (non global) coupling, if the input (bottom-up) is applied to just a few neurons and if the inter-neuron coupling is suitably adjusted(top-down control of chaos) [Arecchi 2004; Ciszak, Montina and Arecchi 2009 a,b]. The operation is a control of chaos, and it has an optimality ; if it lasts less than 200msec, no decisions emerge, on the contrary, if it lasts much longer, there is no room for sequential cognitive tasks. The addition of extra degrees of freedom implies a change of code, thus it can be seen as a new level of description of the same input.

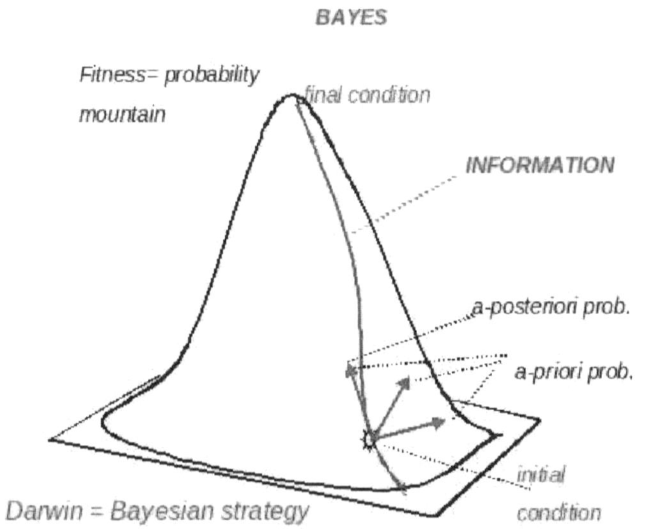

Figure 6. Successive applications of the Bayes theorem to the experiments. The procedure is an ascent of the Probability Mountain through a steepest gradient line. Each point of the line carries an information related to the local probability by Shannon formula. Notice that Darwinian evolution by mutation and successive selection of the best fit mutant is a sequential implementation of Bayes theorem.

3.3. From perception to cognition – Creativity

Considerations analogous to those developed for perception (elaboration of a sensorial input) can be made for cognition (conceptual input). We distinguish two types of cognitive task. In type I, we work within a prefixed framework and readjust the hypotheses at each new cognitive session, by a Bayes strategy. Bayes theorem [Bayes] consists of the relation:

$$P(h|data) = P(h)[P(data|h)/P(data)] \quad (1)$$

That is: the probability $P(h|data)$ of an hypothesis h, conditioned by the observed data (this is the meaning of the bar |), called a-posteriori probability of h, is the product of the a-priori probability $P(h)$ of that hypothesis (we assume to have guessed a package of convenient hypotheses with different probabilities), times the ratio of the probability $P(data|h)$ that data is generated by an hypothesis h, (this is the model) to the probability $P(data)$ of the effectively occurred data. When this ratio is largest, the a-posteriori probability is maximum. As shown in Fig. 6, starting from an initial observation and formulating a large number of different hypotheses, the one supported by the experiment suggests the most appropriate dynamical explanation. Going a step forward and repeating the Bayes procedure amounts to climbing a probability mountain along a steepest gradient line. Recent neurological studies [Beck 2008] explain the readiness of reflexes in terms of a fast coding done by a Bayesian strategy which operates very fast, in about 100 msec. Thus fast reactions are NOT the result of an interplay with previous semantic memories at the PCF level (Fig. 1) as modeled in ART. Fast indeed,but:how accurate? What confidence is provided by a Bayesian procedure? To answer the question, we must consider a new feature, namely **complexity**.

A complex problem is characterized by a probability landscape with many peaks, insofar as it can not be described by a single model (Fig. 7). Jumping from a probability hill to another

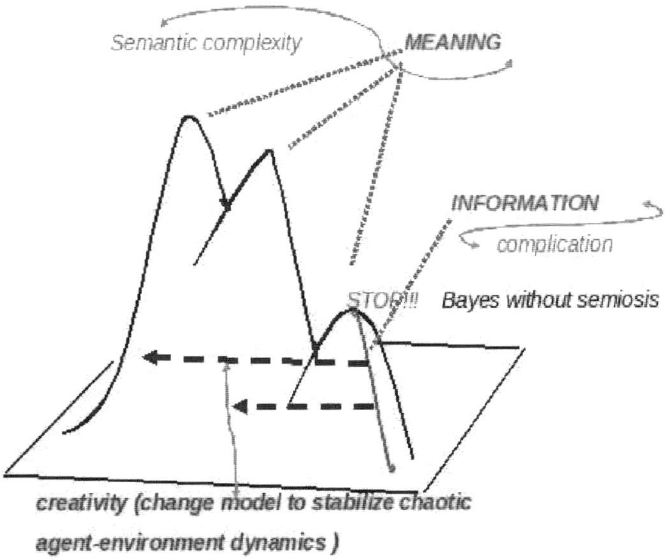

Figure 7. Semantic complexity A complex system is one with a many-peak probability landscape. The ascent to a single peak can be automatized by a steepest gradient program. On the contrary, to jump to other peaks, and thus continue the Bayes strategy elsewhere, is a creativity act, implying a holistic comprehension of the surrounding world (semiosis). We call meaning the multi-peak landscape and semantic complexity the number of peaks. It has been guessed that semiosis is the property that discriminates living beings from Turing machines [Sebeok]; here we show that a non-algorithmic procedure, that is, a non-Bayesian jump from one model to another is what we have called creativity. Semiosis is then equivalent to creativity .The difference between Bayesian strategy and creative jump is the same as the difference between normal science and paradigm shift [Kuhn]

is non- Bayesian; let us call it type II cognition.

In human cognition, type II is driven by hints suggested by the context (semiosis) yet not included in the model. Type II task is a creativity act because it implies a change of code, at variance with type I, which operates within a fixed code. The ascent to a single peak can be automatized in a steepest gradient program; once the peak has been reached, the program stops, any further step would be a downfall.

Question: can the changes of hill in Fig. 7 be handled by a computer? Answer: No. Indeed, as the computer behavior is perturbed by some added noise (as done in Monte Carlo simulations or in genetic algorithms) it explores the immediate surrounding of a Bayes uphill path. A finite step away from it would require a guideline, otherwise it would be whimsical. Thus, a non-deterministic computer cannot perform the jumps of type II, since it intrinsically lacks semiotic abilities. In order to do that, the computer must be assisted by a human operator. We call meaning the multi-peak landscape and semantic complexity the number of peaks.

This intrinsic limitation of a formalism was stated in two famous theorems, namely, the first incompleteness theorem for a consistent theory (Gödel, 1931) and its computer version (Turing, halting problem, 1936) (Fig. 8). Gödel's theorem, stating that there are propositions which appear true to the intuition (semiotic appreciation) of a mathematician endowed with the body of axioms, yet they can not be deduced via the syntax of the formalism, signed a sharp end to the formalistic dream of David Hilbert (1900), to explain all mathematics in terms of logic.

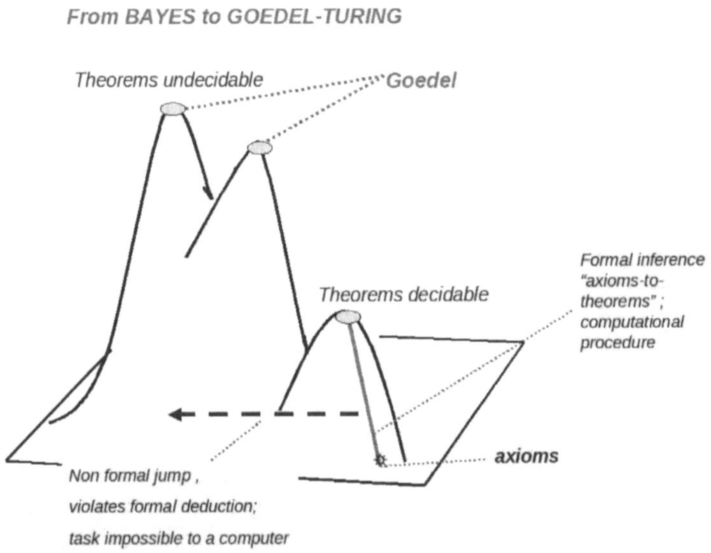

Figure 8. Gödel's first incompleteness theorem (1931): For any consistent formal, computably enumerable theory that proves basic arithmetical truths, an arithmetical statement that is true, but not provable in the theory, can be constructed. "Provable in the theory" means "derivable from the axioms and primitive notions of the theory, using standard first-order logic". Computer equivalent (Turing1936=halting problem): a universal computer, for a generic input, cannot decide to stop.

Since a universal computer is a syntactical machine which grinds the input data via its inner instructions, a similar limitation applies to a computational procedure (Turing). In the space in which we represent Bayesian procedures, the Goedel-Turing theorems are visualized as true propositions (peaks of left-ward hills), accessible from the axioms by a creative jump but not by via the formalism.

Let us discuss in detail the difference between a type I cognitive task, which implies changing hypothesis h within a model, that is, climbing a single mountain, and a type II cognitive task, which implies changing model, that is, jumping over to another mountain. We formalize a model as a set of dynamical variables x_i $(i = 1, 2, N)$, N being the number of degrees of freedom, with the equations of motion

$$\dot{x}_i = F_i(x_1 ..., x_N; \mu_1 ..., \mu_M) \tag{2}$$

Where F_i are the force laws and the M numbers represent the control parameters; The set $\{\mathbf{F}, \mathbf{x}, \mu\}$ is the model.

Changing hypotheses within a model means varying the control parameters $\{\mu\}$, as we do when exploring the transition from regular to chaotic motion within some model dynamics.

Instead, *changing code, or model*, means selecting different sets of degrees of freedom $\mathbf{y}$, control parameters μ and equations of motion $\mathbf{G}$ as follows:

$$\dot{y}_i = G_i(x_1 ..., x_R; \mu_1 ..., \mu_L) \tag{3}$$

Where R and L are different respectively from N and M. The set $\{\mathbf{G}, \mathbf{y}, \nu\}$ is the new model.

While changing hypotheses within a model is an a-semiotic procedure that can be automatized in a computerized *expert system*, changing model implies catching the meaning of the observed world, and this requires what is called *embodied cognition* [Varela]. Embodied cognition has been developed over thousands of generations of evolutionary adaptation, and we are unable so far to formalize it as an algorithm.

This no-go statement seems to be violated by a class of complex systems, which has been dealt with successfully by recursive algorithms. Let us consider a space lattice of spins, with couplings that can be either ferromagnetic or anti-ferromagnetic in a disordered, yet frozen way (spin glass at zero temperature, with quenched disorder). It will be impossible to find a unique ground state. For instance having three spins A, B, and C in a triangular lattice, if all of them have ferromagnetic interaction, then the ground state will consist of parallel spins, but if instead one (and only one) of the mutual coupling is anti-ferromagnetic, then there will be no satisfactory spin orientation compatible with the coupling (try with: A-up, B-up, C-up; it does not work; then try to reverse a single spin, it does not work either).

This model has a cognitive flavor, since a brain region can be modeled as a lattice of coupled neurons with coupling either excitatory or inhibitory, thus resembling a spin glass, [Hopfield; Amit; Toulouse et al]. We have a large number of possible ground states, all including some frustration. Trying to classify all possible configurations is a task whose computational difficulty (either, program length or execution time) diverges exponentially with the size of the system. Sequentially related changes of code have been successfully introduced to arrive at finite-time solutions. [Mezard et al; Solomon].

Can we say that the mentioned solutions realize the reductionistic dream of finding a suitable computer program that not only climbs the single probability hill, but also is able to chose the best hill? If so, the optimization problem would correspond to understanding the meaning of the object under scrutiny.

We should realize however that spin glasses are frozen objects, given once for ever. A clever search of symmetries has produced a spin glass theory [Mezard et al] that, like the Renormalization Group (RG) for critical phenomena [Wilson] discovers a recursive procedure for changing codes in an optimized way. Even though the problem has a large number of potential minima, and hence of probability peaks, a suitable insight in the topology of the abstract space embedding the dynamical system has led to an optimized trajectory across the peaks. In other words, the correlated clusters can be ordered in a hierarchical way and a formalism analogous to RG applied.

It must be stressed that this has been possible because the system under scrutiny has a structure assigned once for ever. In everyday tasks, we face a system embedded in an environment, which induces unpredictable changes in course of time. This rules out the nice symmetries of hierarchical approaches, and rather requires an adaptive approach. Furthermore, a real life context sensitive system has to be understood within a reasonably short time, in order to take vital decisions about it.

We find again a role of control of chaos in cognitive strategies, whenever we go beyond the limit of a Bayes strategy. We call creativity this optimal control of neuronal chaos. Fig. 9 sketches the reduction of complexity and chaos which results from a creative scientific step. Some examples of scientific creativity are listed in Fig. 10.

The difference here outlined between a Bayesian strategy, that can be implemented in an expert system, and a creative non-algorithmic jump is alike Kuhns distinction between normal science and paradigm shift [Kuhn].

The difference occurs also in translating a literary piece from a language to another. The translator does not trivially replace each word by the corresponding one of the new language. Rather he/she lives a given situation in the original language and then narrates it in the new language.

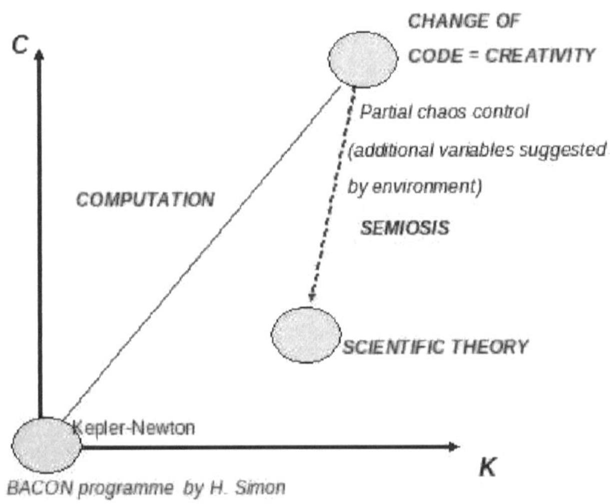

Figure 9. C-K diagram (C=computational complexity; K=information loss rate in chaotic motion): Comparison between the procedure of a computer and a semiotic cognitive agent (say: a scientist). The computer operates within a single code and C increases linearly with K. A scientist explores how adding different degrees of freedom one can reduce the high K of the single-code description. This is equivalent to the control operation of Fig. 6; it corresponds to a new model with reduced C and K. The BACON program [Simon] could retrieve automatically Keplers laws from astronomical data just because the solar system, approximated by Newton two-body interactions, is chaos-free.

1 - electricity; magnetism; optics	*Maxwell equations*
2 - Mendeleev table	*Quantum atom (Bohr,Pauli)*
3 - zoo of 100 elementary particles	*SU(3) - quarks (M Gell Mann)*
4 – scaling laws in phase transitions	*Renormalization group (K.Wilson)*

Figure 10. Reduction of complexity by code change.

4. Adaptive creative jumps vs. quantum approaches to mind

Quantum approaches to mind processes have a long story [Stapp]. I do not plan to survey the many different trials. I myself was puzzled by an uncertainty relation between the information handled in a brain task and the time duration of that task [Arecchi 2005]; but later I realized that the uncertainty was like a thermodynamical limitation that constrains the size of two correlated variables, without any quantum flavour. Indeed, the distinctive mark of a quantum dynamics is to be ruled by a wave-like behaviour which is responsible for interference.

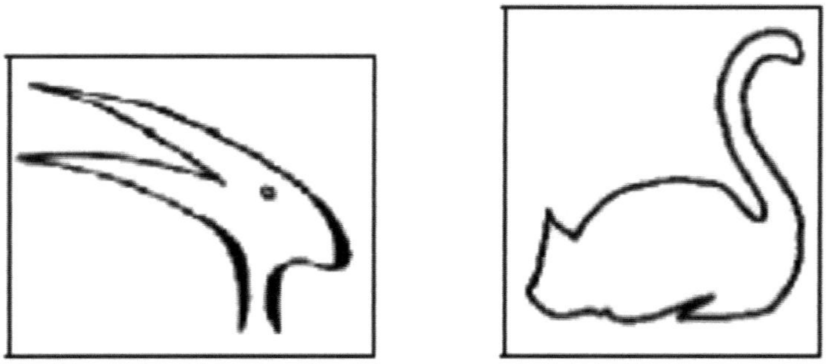

From E Conte et al. arxiv:0710.5092

Figure 11. Two ambiguous figures: left, either rabbit or stork; right, either swan or cat.

A recent series of experiments has reported the alleged evidence of interference in brain processes [Conte et al], based on the following facts. Let us take two ambiguous figures (Fig. 11). The first one can be read either as a rabbit with long ears or as a stork with along beak; the second one either as a swan with a long neck or as a cat with a long tail. In standard experiments on bistable perceptions, a subject is exposed to just one of the two pictures and he/she reports the switching time between the two perceptions. For a standard subject, the average switching time is about 2 or 3 seconds, with strong statistical fluctuations around the average. Of course, over a long run there is equal likelihood for either one of the two perceptions, unless the experimenter introduces a bias which forces , say, the rabbit over the stork. This means that $p(x_1) = p(x_2) = 0.5$, where x_1 and x_2 are the two outcomes, and the probabilities are defined as Eq.(4), where the apex a(b) denote the left or the right figure, respectively.

$$p^a(x) = \frac{\text{the number of results a} = x}{\text{the total number of observations}}, \; x \in X. \tag{4}$$

Let us now alternate the presentation of a with that of b. If we evaluate $p(x_1)$ conditional on perceiving y_1 or y_2 in the b presentation we would expect that summing over all y occurrences on the channel b we recover the previous unbiased value, as Eq.(5).

$$p^a(x) = p^b(y_1)p^{a|b}(x|y_1) + p^b(y_2)p^{a|b}(x|y_2), \; x \in X. \tag{5}$$

Instead ,the perturbation due to intercalating the other picture in the sequence of observations is expressed by an extra-term, which occurs with opposite signs for x_1 and x_2, thus unbalancing the two outcomes associated with the first figure. Eq.(5) must be replaced by Eq.(6)

$$p^a(x) = p^b(y_1)p^{a|b}(x|y_1) + p^b(y_2)p^{a|b}(x|y_2) + 2\lambda(a = x|b, C)\sqrt{p^b(y_1)p^{a|b}(x|y_1)p^b(y_2)p^{a|b}(x|y_2)} \quad (6)$$

The extra-term is like interference in QM; being limited between 0 and 1, it can be written as the cosine of an angle, hence justifying a Hilbert space picture. There is however a logical error. In fact, in physics two states of Hilbert space can be measured simultaneously by different observers; here instead, the two events are NOT synchronous.

In ambiguous patterns (bistable percepts) the two patterns do NOT form a superposition in the quantum sense, since they are NOT different states of the same Hilbert space; in fact, they appear successively in course of time, and one cannot speak of quantum interference. The extra-term in Eq.(6)implies a bias of the memory mechanisms, modifying the balance.

More generally, all authors who resort to quantum explanations are violating the Occam razor. Indeed, our previous explanation of non-algorithmic jumps as transient chaos control operations does not require any quantum assumption.

In a similar fashion, quantum decision approaches [Yukalov and Sornette] seem elegant formalisms implying extra assumptions with respect to those necessary to understand the decision process; thus they conflict with Occams economy principle.

5. Conclusions

From what we have discussed, the search for meaning, i.e. a creative decision, is NON-Bayesian.

The world is complex, that is , not grasped by a unique model .How to choose among alternative models? There are two ways:

i) arbitrarily (kind of oracle), this amounts to **relativism**;

or

ii) along some guidelines (fertility and latitude of the new explanation; useful outcomes); since any cognitive decision modifies (through our actions) the environment, we perform several trials with the aim of stabilizing the chaotic agent-environment dynamics. Thus, a **semiotic** approach to cognition discloses hard facts, it is a hint of an **ontology**.

The same re-coding freedom inherent in the transient stabilization of chaos holds for ethical decisions. But behold, our freedom is CONDITIONAL on the environmental features in which we are embedded. Let me quote two philosophers:

- Aristotle: thought and desire play a coordinate role (Nicomachean Ethics)
- Vico: poetic wisdom is the result of perception, memory and imagination
- "The New Science": our freedom is by NO means ABSOLUTE, as instead stated by
- Kant: freedom = independency from cause-effect chains = not-caused cause (Critique of Practical Reason).

The creative jump is not arbitrary, but guided by the situation in which we are embedded. We re-adjust the code until we find a satisfactory reading of the world around us.

Let us return to Fig. 7 but with a large number of peaks, to depict a complex situation (Fig. 12). The ascent to a single Bayes peak yields certainty, i.e. fidelity to a protocol, NOT truth but PAC (probably approximately correct) knowledge [Vidyasagar]. As we look for meaning, we jump non-algorithmically across several peaks, exploring a sizable region of a complex space; we speak of partial truth, from a certain point of view [Agazzi]. This is the only truth accessible to a scientific program; it is a modern formulation of the old definition of truth as adaequatio intellectus et rei [Thomas Aquinas]. Global truth is not accessible to us, as a complex universe is beyond our investigation power Would this mean a relativistic attitude? Not al all, since any starting point X has a pre-scientific motivation. It corresponds to a different choice of an investigation area, thus it corresponds to an ethical, better to say, ethological motivation, depending on personal motivations. For instance, in selecting an area of physics, I have devoted

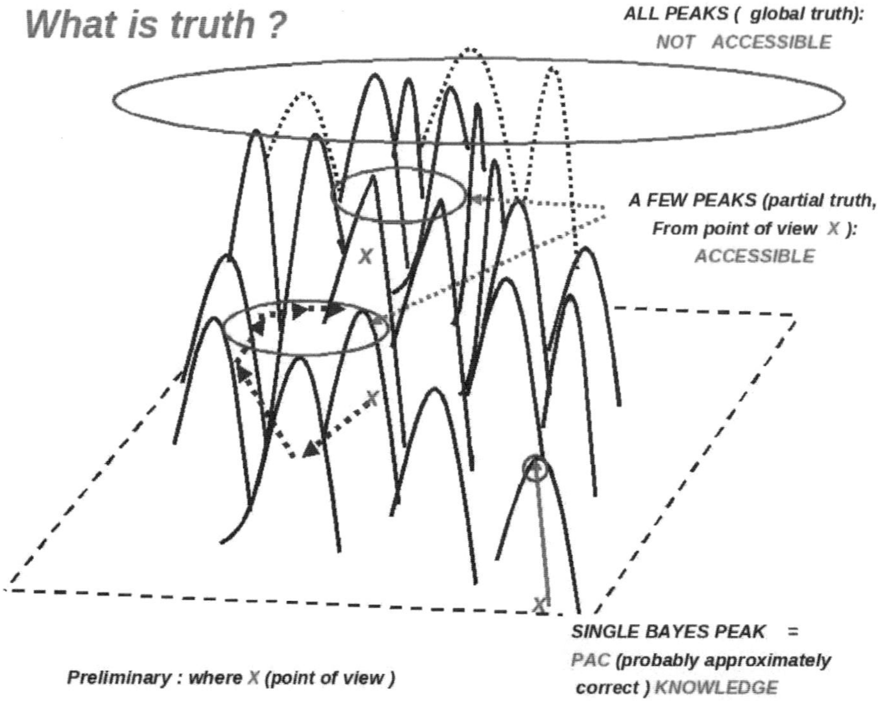

Figure 12. Comparisson of three procedures i) Climbing the single Bayes peak, leading to PAC Knowledge (what a computer does); ii) Different creative endeavors, starting from different X and exploring separate regions of a complex space = partial truth (what a creative mind does); iii) Exhausting all peaks of a complex problem, thus reaching a global truth (what nobody can do).

my efforts to quantum optics and chaotic phenomena, staying away from elementary particles and cosmology. Once made clear my X, I can compare my results (my partial truth) with that of other colleagues.

As a final remark, we compare the hermeneutic circle, peculiar of cognitive agents (animals or robots) with a limited repertoire, with what I call the hermeneutic spiral (Fig. 13) , which corresponds to the human language ability to make an infinite use (in course of time) of finite resources. The words polysemy implies that we in general attribute several meanings or connotations to the same word. If these connotations are frozen once for ever, as in a historical dictionary, then we are in the upper case. Suppose a grammatical operator takes from a sub-meaning of the word A to a sub-meaning of B; applying the inverse operator we recover the initial connotation of A, either in one or several steps. A neurobiological example is offered by the olfactory system of the locust [Laurent et al]. Exposed to a cherry odour, it responds with a specific sequence of neuronal spikes ; re-exposing the locust to the same odour at a later time, the odour is encoded by the same sequence.

If instead we do the experiment with a more sophisticated animal as a rabbit [Freeman], any successive exposure to the same odour is encoded in different spike patterns. The rabbit feels that some time has passed by and that it is experiencing the same perception but not for the first time. This spiral along time, whereby we never repeat exactly the same connotation, is peculiar of human creativity, in language (we make new poetry with the same lexicon) or music

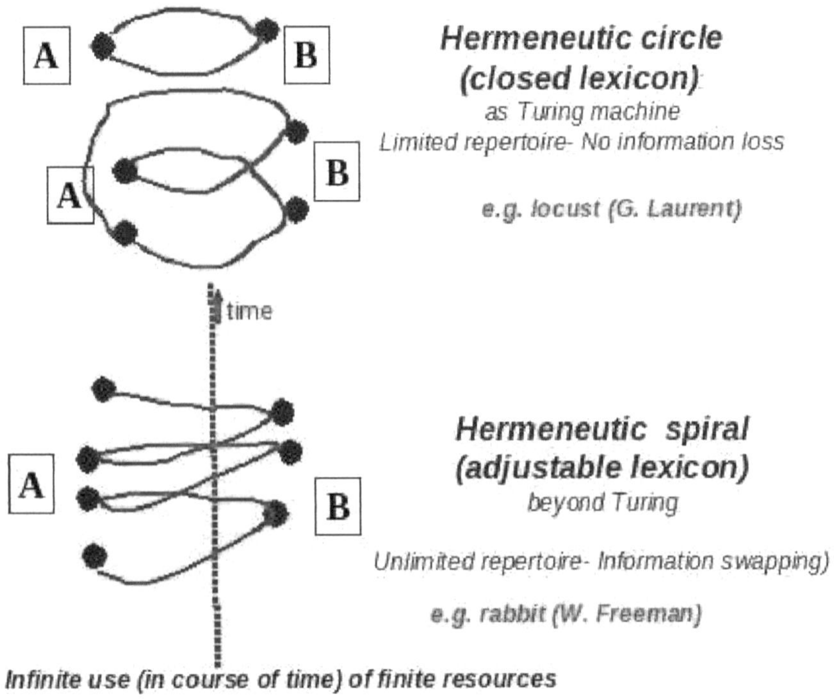

Figure 13. Difference between hermeneutic circle and hermeneutic spiral. Let us consider an algorithm which takes from an item A to an item B. It the items are words of an ordinary language, they are polysemic, that is both A and B do not have a unique connotation. Selecting a specific connotation of A the algorithm arrives at a specific connotation of B. By reversing the algorithmic procedure, one returns at the original connotation of A either in one loop (uppermost figure) or in a finite number of loops (central figure).These cases correspond to a cognitive agent with a limited repertoire and no information loss. If, as decribed in non-Bayesian cases, there information swapping, i.e. re-coding, the number of connotations of the words A and B is for all purposes unlimited. As time flows ,there is a semiotic re-adjustment between cognitive agent and environment which yields new connotations for the same lexical terms, as it occurs in poetry and music.

(we make new pieces with the same notes). Indeed ,it is our common experience that revisiting after a while a piece dear to us, e.g. a Bach Cantata, we discover new meanings, since our inner universe has grown in the meantime.

Acknowledgments

The Author acknowledges the financial support of Ente Cassa di Risparmio di Firenze under the project "dinamiche cerebrali caotiche".

References

Since the topics of this review track my research line along the years, quite a few considerations can be found scattered in my research papers. My publications can be downloaded from my homepage: www.inoa.it/home/arecchi, see: List of publications - Research papers in Physics Furthermore, these topics have been dealt with extensively in the book:

- Arecchi F T 2007 *Coerenza, Complessità, Creatività* (Roma: S Di Renzo)
All other references are listed here below in alphabetic order:
- Agazzi E 1974 *Temi e problemi di Filosofia della fisica* (Roma: Edizioni Abete)
- Amit D J, Gutfreund H and Sompolinski H 1985 *Phys. Rev* A **32** 1007
- Arecchi F T 2004 *Physica* A **338** 218-237
- Arecchi F T 2005 *Braz. J. Phys.* **35** 253
- Arecchi F T and Meucci R 2008 *Scholarpedia* **3**(9) 7066
- Bayes T 1763 *Phil. Trans. Royal Soc. (London)* **53** 370-418
- Beck J M *et al* 2008 *Neuron* **60** 1142-1152
- Chaitin G J 1987 *Algorithmic information theory* (Cambridge, UK: Cambridge U. Press)
- Ciszak M, Montina A and Arecchi F T 2009 a *Cognitive Processing* **10**(1) 33-39
- Ciszak M, Montina A and Arecchi F T 2009 b Control of transient synchronization with external stimuli *Chaos* **19** to appear
- Conte E *et al* 2008 Mind states follow quantum mechanics during perception and cognition of ambiguous figures *Preprint* arXiv:0802.1835
- Freeman W J 1991 *Scientific American* **264**(2) 78-85
- Grossberg S 1995 *The American Scientist* **83** 439
- Hopfield J J 1982 *Proc. Nat. Acad. Sci. (USA)* **79** 2554
- Hubel D H 1995 *Eye, Brain and Vision* Scientific American Library, No. **22** (New York: W. H. Freeman)
- Kuhn T S 1962 *The Structure of Scientific Revolutions* 1st ed (Chicago: U. of Chicago Press)
- Laurent G, Wehr M and Hananel Davidowitz H 1996 *Journ. Neurosci.* **16** 3837
- Libet B, Wright E W, Feinstein F and Pearl D K 1979 *Brain* **102** 193
- Mezard M, Parisi G and Virasoro M 1987 *Spin glass theory and beyond* (Singapore: World Scientific)
- Rodriguez E, George N, Lachaux J P, Martinerie J, Renault B and Varela F 1999 *Nature* **397** 340343
- Sebeok T A 2001 *Semiotica* **1341**/4 61-78
- Shilnikov A, Shilnikov L and Turaev D 2004 *Int. J. Bif. And Chaos* **14** 2143
- Simon H A 1980 *Cognitive Science* **4** 33-46
- Singer W and Gray E C 1995 *Annu. Rev. Neurosci.* **18** 555
- Solly D R *et al* 2007 *Nature* **450** 1054
- Solomon S 1995 *Ann. Rev. of Comp. Physics II* (Singapore: World Scientific) 243-294
- Stapp H 1993 *Mind, matter and quantum mechanics* (Berlin: Springer)
- Thomas Aquinas 1992 *Quaestiones de Veritate* (Bologna: Ed. Studio Domenicano)
- Toulouse G, Dehaene S and Changeux J P 1986 *Proc. Nat. Acad. Sci. (USA)* **83** 1695
- Varela F, Thompson E and Rosch E 1991 *The Embodied Mind* (Cambridge, MA: MIT Press)
- Vidyasagar M 1997 *A Theory of Learning and Generalization, with Applications to Neural Networks and Control Systems* (Berlin: Springer)
- Wilson K 1983 *Rev. Mod. Phys.* **55** 583
- Yukalov V I and Sornette D 2008 *J. Math. Phys.* to appear (*Preprint* arXiv:0808.0112)

Journal of Physics: Conference Series **174** (2009) 012011 doi:10.1088/1742-6596/174/1/012011

Dissipative neurodynamics in perception forms cortical patterns that are stabilized by vortices

Walter J Freeman[1] **and Giuseppe Vitiello**[2, 3]

[1] Department of Molecular and Cell Biology University of California, Berkeley CA 94720-3206 USA

[2] Dipartimento di Matematica e Informatica and INFN Universitá di Salerno, I-84100 Salerno, Italy

E-mail: dfreeman@berkeley.edu - http://sulcus.berkeley.edu
vitiello@sa.infn.it - http://www.sa.infn.it/giuseppe.vitiello

Abstract. In the engagement of the brain with its environment, large-scale neural interactions in brain dynamics create a mesoscopic order parameter, which is evaluated by measuring brain waves (electrocorticogram, ECoG). Such large-scale interactions emerge from the background activity of the brain that is sustained by mutual excitation in cortical populations and manifest in spatiotemporal patterns of neural activity. Band pass filtering reveals beats in ECoG power that recur at theta rates ($3-7$ Hz) as null spikes in $\log_{10}$ power. The order parameter transiently approaches zero, and the microscopic activity is both disordered and symmetric. As the null spikes terminate, the order parameter resurges and imposes a mesoscopic spatial pattern of ECoG amplitude modulation that then governs the microscopic gamma activity and retrieves the memory of a stimulus. The brain waves reveal a spatial pattern of phase modulation in the form of a cone. The dissipative many-body model of brain dynamics describes these phase cones as vortices, which are initiated by the null spikes, and which stabilize the amplitude modulated patterns embedded in the turbulent neural noise from which they emerge.

1. Introduction

In 1958 von von Neumann observed that the operations performed by the brain are not done with logic and mathematics: "Brains lack the arithmetic and logical depth that characterize our computations... We require exquisite numerical precision over many logical steps to achieve what brains accomplish in very few short steps" (p. 63 of ref. [1]). Schematically, we might identify three of such steps. The brain provides guidance for the engagement of the body with its environment, which requires constant attention and control with the various senses. The first step in perception thus consists in the transduction of microscopic energies from the surroundings at sensory receptors that expresses the information in action potentials (nerve impulses). The acquisition of information about the current status of the body and its surroundings is indeed a necessary step. Then, the rapid recognition of momentary changes to which the body must accommodate and compare at each moment with all relevant past experience in similar conditions is crucial in order to successfully engage the body with its environment. Thus the second step is refinement of the requisite information at a higher resolution in the sensory systems. The third step is abstraction and generalization. Such a step consists in the continual updating of

[3] Corresponding author

the *meaning* of the flow of information exchanged in the brain's behavioral relation with its environment. The brain thus stores *experience*, which constitutes the basis for *knowledge*. The survival of the body in the environment in which it is situated thus crucially depends on the brain *action-perception* cycle, namely the capability of the brain to be *open* on the environment and to react to it. This means that neurodynamics is intrinsically dissipative. Dissipation, i.e. the reciprocal exchange of energy (in various forms of information, matter, radiation, etc.) between the brain and its environment, is a characterizing feature of brain dynamics. Therefore, non-equilibrium processes, critical phenomena, and phase transitions necessarily enter into any realistic modeling of brain function [2, 3].

In this paper we will discuss neurophysiological data from the perspective of dissipative neurodynamics. Nonlinear ordinary differential equations (ODE) in networks called Katchalsky-sets (KI, KII, etc. sets) [4] have been implemented in an analog very large scale integration [5], putting a KII set on a chip with about 800 first order ODE to describe an 8×8 electrode array. This report extends the modeling of the third step into many-body physics [6]. We are motivated in this by the fact that there is no conventional neural network offered by neuroscientists or neuroengineers that can adequately model or simulate our data. In our approach, the cerebral cortex is described as a network of populations that organize into a state of criticality, which has many unitarily inequivalent ground states that constitute memories. Access and retrieval of memories, in the act of recognition, are shown to occur through phase transitions. Retrieved memories are expressed in spatial patterns of amplitude modulation (AM) of carrier frequencies. The AM patterns are stabilized by vortices in the turbulence of cortical background activity in the form of phase cones.

This paper is organized as follows. We will first review neurophysiological data and show how from these some features emerge which point to consider collective dynamical processes in theoretical modeling. We will then review the dissipative model and show its agreement with observations.

2. Comparison of the microscopic sensory recept with the mesoscopic percept

A large part of contemporary neuroanatomy and neurophysiology provides the basis for understanding of the first two steps mentioned above. The understanding of the third step, heretofore unknown, presents difficulties which were clearly stated over fifty years ago by Lashley: "Generalization is one of the primitive basic functions of organized nervous tissue. Here is the dilemma. Nerve impulses are transmitted ... from cell to cell through definite intercellular connections. Yet all behavior seems to be determined by masses of excitation. ... What sort of nervous organization might be capable of responding to a pattern of excitation without limited specialized paths of conduction? The problem is almost universal in the activities of the nervous system" (p. 302-306 of ref.[7]).

This third step is taken in the several sensory cortices that receive the information from receptors. The prototype for all the sensory and perceptual stems in the evolution of mammalian brains is that for olfaction combined with the hippocampal system for memory formation. The functions of the three-layered allocortex in these ancient structures illustrate the elementary mechanisms that transpose from information in sensory input to memory in perception.

The topology of the olfactory system

The topology of the olfactory system is shown schematically in three sequential surfaces: receptors, bulb and cortex. In round figures, there are 10^3 types among the 10^8 receptors giving 10^5 of each type. Any one sniff selects very few of each type, e.g., 10^2, with random variation on successive sniffs owing to nasal turbulence (Fig. 1). Assignment of the sample on each sniff to the class of odors to which it belongs requires inductive generalization. This is done in reinforcement

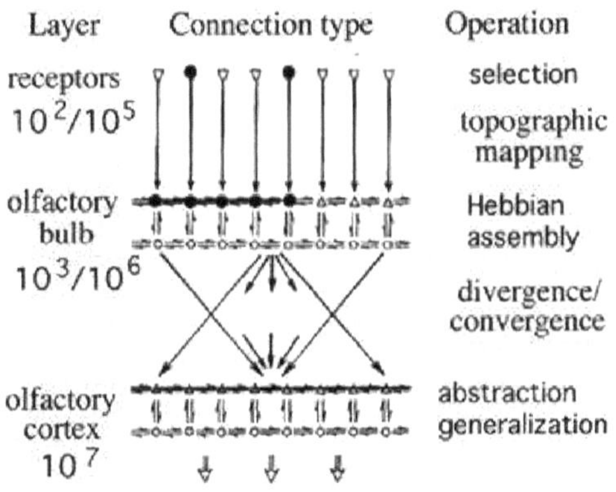

Figure 1. The topology of the olfactory system is schematized in 3 surfaces. In Step 1, receptors transduce an input of chemical energy to nerve impulses. In Step 2, after topographic mapping and convergence, a Hebbian assembly generalizes the input to an odorant class. In Step 3, a mesoscopic AM pattern (Fig. 2) is created by a phase transition. Transmission by a divergent-convergent pathway enacts an integral transformation that abstracts the output, deleting the sensory details.

learning by strengthening of synapses (increased gain, k_{ee}) among bulbar excitatory neurons ('∇' in Fig. 1) that are co-activated on each sniff by a reinforced odorant (rewarded or punished). Cumulatively, increases form a Hebbian nerve cell assembly ('$\bullet - \bullet - \bullet - \bullet$') for each learned class by pair-wise correlation on reinforced trials and decorrelation by habituation on sniffs without the reinforcement. Neurons selected by odorant sensing show membership in an assembly by vigorous firing of microscopic impulses. Excitation of any members of the assembly by a learned stimulus activates the entire assembly. However, an assembly only comprises an estimated 0.1% of the excitatory neurons in the bulb, so its activation at the microscopic level does not explain perception. Amplification is required to the entire olfactory bulb at the mesoscopic level. That is done by dendritic integration, which produces continuous fields of gamma neural oscillations that are observed and measured in the olfactory bulbar and cortical electrocorticogram (ECoG). The dynamical formation and spatial extension of these observed fields of neural oscillations are explained by the dissipative many-body model (Section 5).

The transmission from receptors to the bulb is done by topographic mapping, which is compatible with microscopic content (action potentials as point processes); transmission from the bulb to the cortex is done by divergent-convergent projection performing a spatiotemporal integration of the global output of the bulb. The act of integration enhances gamma oscillatory field potentials that share the same instantaneous analytic frequency. It attenuates activity that is spectrally dispersed, including sensory-driven activity, and expresses as output only the activity constructed by the bulb from its updated memory store. The transmission in the step from bulb to cortex is done by action potentials, but the organization differs from the prior step. In the topographic input path to the bulb, the state variable is pulse frequency (Fig. 1); in the output projection from the bulb, the state variable is pulse density. The conversion from microscopic frequency to mesoscopic density is initiated at the input synapses, where pulse frequencies are transduced to dendritic current amplitudes; the conversion is completed at the output synapses of target neurons, by the spatiotemporal integration performed by the divergent-convergent projection. The wave densities of the olfactory bulb ECoG reveal its mesoscopic output, resulting from bulbar dendritic integration prior to the global integral transformation. The cortical ECoG in the third layer reveals the cortical transformation of its bulbar input.

Journal of Physics: Conference Series **174** (2009) 012011 doi:10.1088/1742-6596/174/1/012011

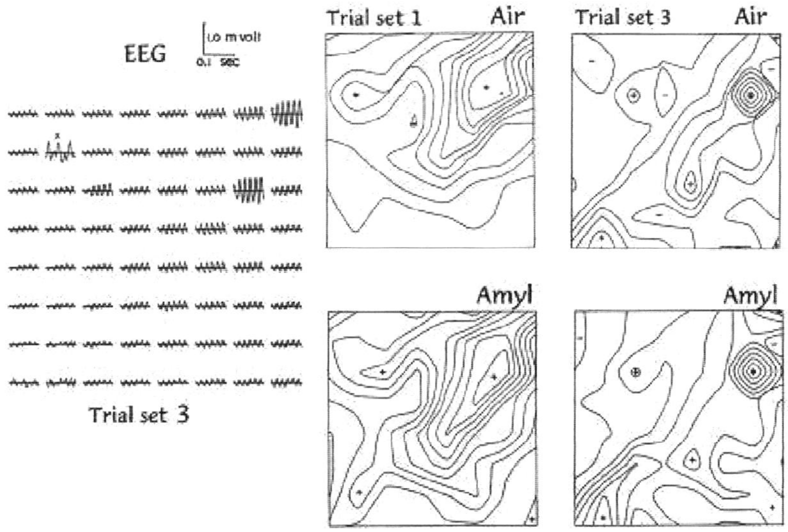

Figure 2. Left: Example of an AM carrier wave in one frame. Center: contour plots of AM patterns from 4 × 4 mm frames showing differences correlated with stimuli. Right: Demonstration of lack of invariance of AM patterns with respect to stimuli, showing that AM patterns are not representations of stimuli; they are operators containing activated memory about stimuli (center frames), which is subject to continual updating with new experience (right frames). The vertical differences reflect short-term memory changes upon the formation of an assembly. The horizontal differences reflect long-term memory changes in consolidation. From ref. [8]

ECoG spatial patterns of amplitude modulation and phase modulation

Bulbar sensation of a microscopic sensory input is expressed in high pulse frequencies of axons in the sparse network of an assembly. Perception of the input is expressed in a spatial pattern of ECoG, specifically the spatial patterns of amplitude modulation (AM) accompanied by phase modulation (PM) of the gamma carrier frequency in the band pass filtered ECoG. All neurons in the olfactory bulb contribute to the AM pattern (Fig. 2). It is mesoscopic, because it is continuous over the entire bulb, and in every local neighborhood it is the sum of dendritic current from 10^4 to 10^5 neurons. The carrier wave of the AM pattern has the same waveform and mean frequency over the entire bulb, so the AM and PM patterns are both embedded in and revealed by the band pass filtered ECoG in each stepped window. The realization of microscopic dynamical properties in AM and PM mesoscopic ordered patterns is the theoretical challenge to which the many-body model answers by using the mechanism of spontaneous breakdown of symmetry (see Section 5).

The 64 ECoG signals recorded from an 8×8 array forming a 4×4 mm window onto the bulb yield 64 amplitude values in a 64×1 feature vector, $A_{x,y}(t)$, at each time step. The tip of the vector inscribes a trajectory in 64-space. Small steps of the trajectory reveal the maintenance of a stable pattern. On repeated trials with the same stimulus the trajectory returns to the same region of 64-space. The center of gravity and surrounding standard deviation (SD) define the AM pattern for a learned stimulus and its assembly. Different stimuli have different assemblies and patterns, so that AM patterns of ECoG are classifiable with respect to learned stimuli. Unlike the microscopic sensory information, the mesoscopic classificatory information is non-local; no channel has any more or less value in classification than any other. However, the AM

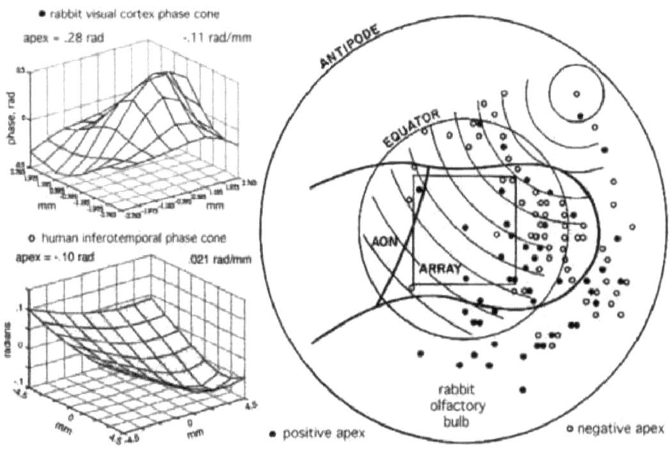

Figure 3. The silhouette shows the outline of the olfactory bulb, on which is superimposed a 4×4 mm rectangle giving the position of the surgically placed 8×8 electrode array. The two circles specify the flattened surface of the spherical bulb. The solid and open dots show the locations of conic apices. The arcs show the isophase contours of one phase cone at 0.1 rad intervals. The insets at left give examples of phase cones respectively from rabbit and human. [Adapted from ref. [9]]

patterns lack invariance with respect to invariant stimuli; instead they vary with the context and history of the individual subject. As expected for an associative memory, the AM patterns change on assimilation to any new AM pattern formation. They express the contextual meaning of the stimuli as they emerge from the memory store, which structurally is stored in the modified synaptic gains (k_{ee} in Fig. 1). These features are specifically predicted by the dissipative many-body model (see Section 5). The contextual nature of the meaning and its emergence as a product of the nonlinear relation between the brain, its *experience* (memory), and its surroundings can be traced back to formal (mathematical) aspects of the dissipative model.

Phase cones

The PM pattern, $\phi_{x,y}(t)$, of the 8×8 phase values that accompany the AM pattern of the carrier wave at its relatively fixed frequency has the shape of a cone in the cortical surface coordinates (Fig. 3). The location and sign (lag 'o' or lead '•') of the cone vary randomly from each AM pattern to the next but are fixed within AM patterns. The phase gradient in rad/mm of the cone varies randomly inversely with the carrier frequency in rad/sec; the mean phase velocity from the ratio in m/sec is invariant and equals the mean conduction velocity of cortical axons running parallel to the surface [9]. Maps of the phase cones on the cortical surface show that they last 3 to 5 cycles of the carrier oscillation. The phase cone, $\phi_{x,y}(t)$, helps to determine the location, diameter, starting time, and duration of each event containing an AM pattern, but it has no classificatory value with respect to stimuli.

Display of the filtered ECoG amplitude by frames at twice the digitizing rate in a movie (Fig. 4) shows two overlapping AM patterns with different locations, carrier frequencies, and PM patterns. One focus oscillates with the repeated expansion of alternating positive and negative peaks (+o-o+). The other focus oscillates with alternating positive and negative peaks (PNP) rotating counterclockwise about a fixed point which corresponds to the downward apex of a phase cone with maximal lag (implosion). Hence the conic phase gradients are time-dependent (Fig. 5) with six observed types of spatial PM patterns: repetitive expansion (explosion), contraction

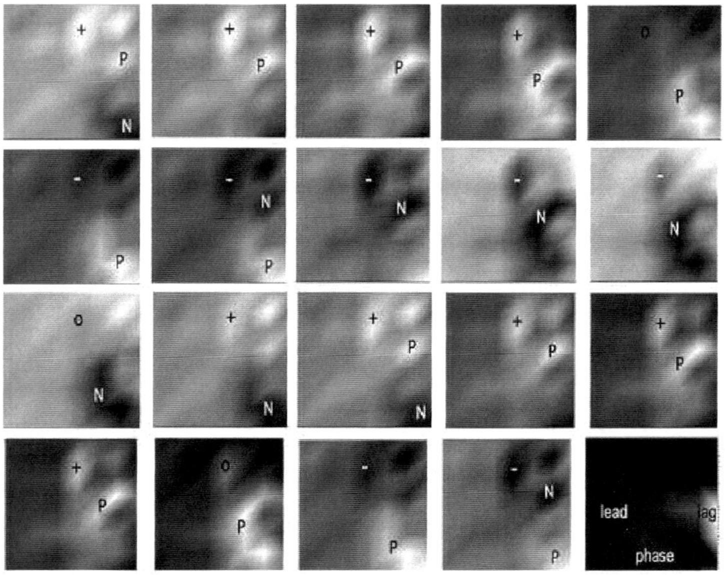

Figure 4. Frames from a 5.6 × 5.6 mm 8 × 8 electrode array were shown in time steps of 4 ms of ECoG filtered in a pass band 20 − 25 Hz. A small focus (+ ∘ − ∘ +) oscillated in place with cycle duration near 48 ms, the other focus rotated counterclockwise with cycle duration near 46 ms. Each pattern persisted for several cycles, then terminated. Four to six independent phase-locked ECoG patterns commonly overlapped, giving the appearance of a pan of boiling water.

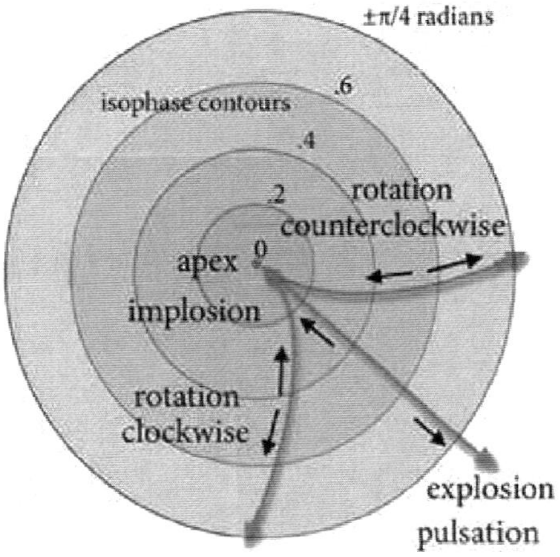

Figure 5. The conic phase gradients were either inward (implosion) or outward (explosion); they rotated clockwise or counterclockwise or pulsed without rotation.

(implosion), with or without clockwise rotation, or counterclockwise rotation. The dissipative model accounts for these observed types of spatial PM patterns and the behavior of the phase gradients.

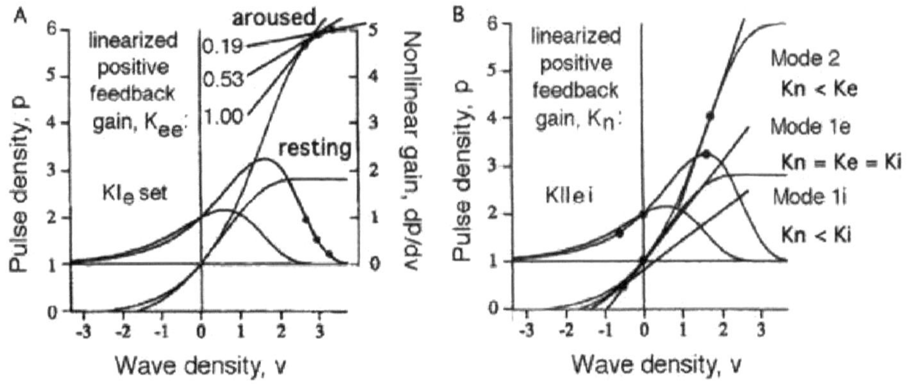

Figure 6. The linearized gain is calculated by fitting a tangent to the operating point of the nonlinear gain, dp/dv. The steady states at unity gain differ for positive (A) vs. negative (B) feedback. The asymmetric function $G(v)$ provides input-dependent gain (B, Mode 2) that is crucial for induction of phase transitions. Maximal gain, Q_m, increases with arousal (here from $Q_m = 2$ at rest to $Q_m = 5$). From [10].

3. Spontaneous background activity and oscillatory carrier waves

Cortical dynamics is expressed in networks of populations, each population representing a node. The dynamics of the network of nodes is described by a set of nonlinear integro-differential equations. Experiments have demonstrated that each population has a small-signal, near-linear domain. Thus, the equation for each neural population can be decomposed into linear time- and distance-dependent functions, $F(t)$, $H(x, y)$, describing spatial and temporal integration by dendrites, and the static nonlinear function, $G(v)$, describing the conversion (Fig. 6) of wave density v to pulse density p at trigger zones:

$$p(x, y, t) = F(t)H(x, y)G(v) \ . \tag{1}$$

The conversion of pulse density to wave density at synapses is:

$$v(x, y, t) = k_{ij}p(x, y, t) \ , \tag{2}$$

where the "gain" k_{ij} is a coefficient of proportionality from the j-th population to the i-th population. The dendritic conversion in self-regulated activity of neuronal feedback is kept within a near-linear range by the limitation in range at the trigger zone. This restriction is violated by external stimulation, such as by application of electric shocks. See Ref. [4] for details on the description of neural activity by means of ordinary linear differential equations. Here we observe that the parameters for learning and habituation are the gains k_{ee} and k_{ei} at synapses selected by input and by reinforcement under limbic control.

The operating point in the steady state is at unity gain, $dp/dv = 1$, when input and output amplitudes are equal (see Fig. 6). Demonstration of stability properties is by varying the input intensity and measuring the amplitude and rates of rise and decay of impulse responses (Fig. 7). The sequence of solutions is plotted as root loci in the complex plane. Both the amplitude and the decay rate extrapolate linearly to threshold at zero. This implies the existence of a pole at the origin of the complex plane and shows that the background activity is self-regulated by the refractory periods in every local area of cortex. The hypothesis is proposed that the pole represents a point attractor that maintains the spontaneous background activity at a set point, which is regulated by multifactorial limbic control. The control parameter Q_m is expressed by the value of the asymptotic maximum.

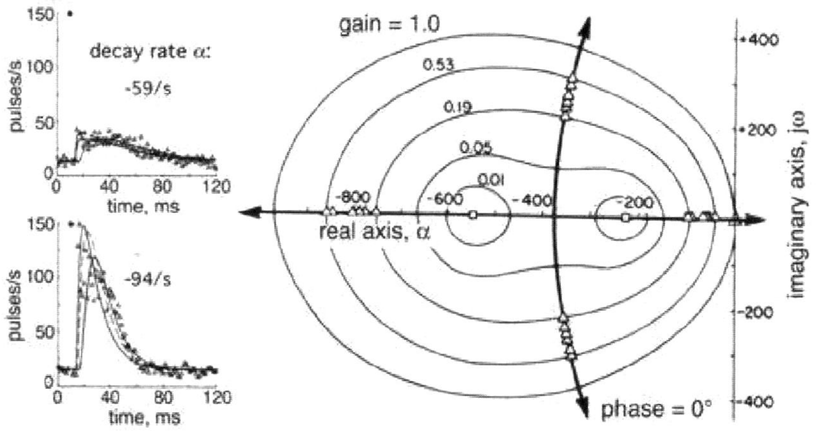

Figure 7. Left: Impulse responses of a periglomerular cell on excitation of the primary olfactory nerve at 6 intensities. Data (△) in averaged post-stimulus time histograms were fitted with solutions to a matrix of six 1st order ODE. Right: Root loci (dark lines) of the changes in poles as a function of gain (elliptical contours), k_p. At zero gain the poles (△) lie at the open loop values (□), rise rate = $-530/s$, decay rate = $-230/s$. The left-right arrows indicate the direction of change in the real rate constants as gain contours (ellipses) increase to infinity. The up-down arrows show the direction of approach of the complex conjugate roots to zeroes in the right half of the complex plane (not shown). From Fig. 5.10 p. 289, Fig. 5.13 p. 292 in [11].

The oscillatory carrier wave (inset in Fig. 2) in the gamma range ($20-80\ Hz$) is generated by negative feedback interaction of the excitatory neurons ('△') with inhibitory neurons ('∘' in Fig. 1). Each neuron interacts with 10^4 other neurons, so the ECoG is mesoscopic. Modeling shows that the synapses that undergo modification in learning are not those in the negative feedback loop, k_{ei} and k_{ie}; the modified synapses are between the co-excited excitatory neurons, k_{ee}. A 20% increase in synaptic gain is shown by simulation to give 50-fold increase in the output power (Fig. 8). Activation by unwanted input is reduced by habituation, so that a 25% decrease in synaptic gain decreases power by as much as 10^{-3} attenuation [11]. The simulation of these mesoscopic properties show that the Hebbian assembly is a powerful signal amplifier and noise attenuator, and that these operations are enhanced by Q_m in arousal (Fig. 6), but it does not explain how the neural populations undergo phase transitions by which the AM pattern corresponding to a selected assembly is actualized.

The analysis of the root loci properties indicate that a limit cycle attractor exists in cortical dynamics at or near the frequency of the complex pole pair (△). If the operating point were to shift to the right across the imaginary axis with increasing ratio of amplitude to background noise, the root loci predict that the amplitude would continue to increase without limit, as the operating point approached the limit cycle attractor, at which a singularity would occur. This prediction cannot be tested experimentally with linear analysis, because the impulse responses must converge for observation. Nonlinear analysis indicates that this trajectory does occur and is responsible for the phase transition owing to the spontaneous fluctuations. It has now become apparent that the fluctuations can be attributed to fluctuations in the background activity, because the driving input is constant, and no changes in sensitivity occur with learning. The proposed mechanism is the beats in Rayleigh noise, which are power minima in band pass filtered noise that alternate with power maxima.

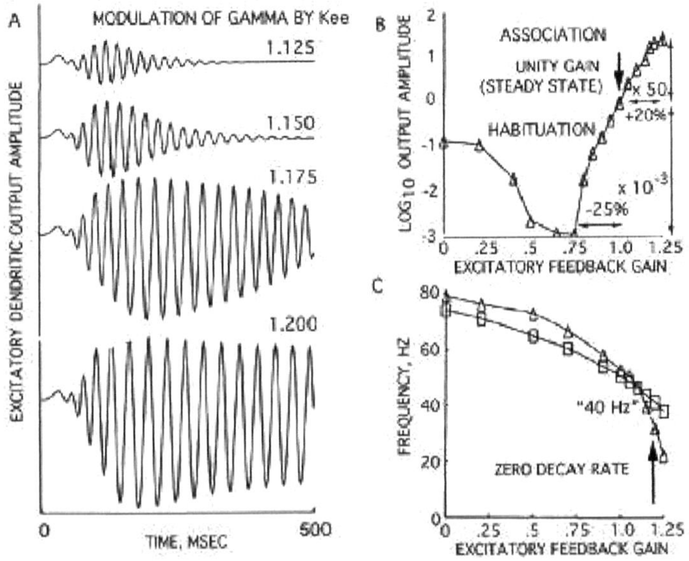

Figure 8. Negative feedback between excitatory and inhibitory neurons sustains oscillation in the gamma range (20 – 80 Hz). An assembly forms by selective strengthening of synaptic connections (A). Increased excitatory feedback gain, k_{ee}, in reinforcement learning increases the power (B) and decreases the frequency (B) of the oscillations selectively on reception of the learned input. Habituation decreases the sensitivity of the olfactory system to unreinforced stimuli [10].

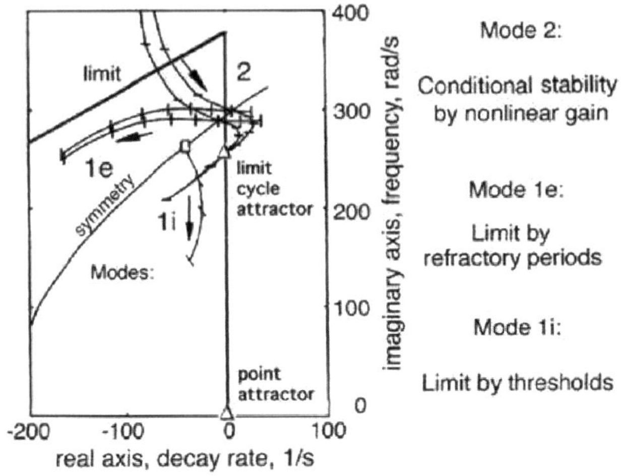

Figure 9. Root loci (solution sets for linearized gains, k_{ee}, k_{ei}, k_{ie}, k_{ii}) are shown in the upper half of the complex plane (one of a pair of complex conjugate roots) for three families of root loci. The control parameter for movement along trajectories is the output amplitude. Arrows show direction of pole movement with increasing ratio of response amplitude to background activity. The stable point attractor at the origin governs the level of the background activity. The limit cycle attractor manifests a singularity, because if the operating point is shifted into the right half of the complex plane, the increase in amplitude carries it inexorably to the attractor. Crossing the imaginary axis in phase transition requires that four conditions be met: formation of an assembly (Fig. 1), arousal (Fig. 6), input that selectively activates the assembly and spontaneous suppression of the order parameter, $\mathbf{A}^2(t)$, in a null spike (Fig. 10) from Rayleigh noise. Adapted from Fig. 6.30, p. 388 in [11].

4. Null spikes, phase transitions and scale-free dynamics in perception

In piecewise linear analysis, the operating point moves along a trajectory in Mode 2 across the imaginary axis into the right half of the complex plane by reducing the background activity. The sudden decrease in the amplitude of the band pass filtered ECoG (Fig. 10, A, C) manifests a decrease in the order parameter, $\mathbf{A}^2(t)$. The crossing of the operating point into the right half of the complex plane is revealed by a beat that precedes a phase transition.

The event repeats as down spikes in $\log_{10}$ power (D), the deepest of which are labeled null spikes. The analytic phase (B) from the Hilbert transform [12] at these points is indeterminate, as reflected in wide deviations of the analytic frequency (D) and the increased spatial standard deviation of the analytic phase, $SD_X(t)$ [13].

The neural mechanism of down spikes in ECoG stems from that of the background activity in mutual excitation. The microscopic action potentials that carry the self-regulated activity are modeled as independent Poisson processes collectively generating white noise, which is easily simulated by time series of random numbers. The cumulative sum simulates the mesoscopic ECoG background activity that conforms to brown noise ($1/f^2$) [14, 15] and more specifically to black noise ($1/f^\alpha$, $2 < \alpha < 4$) [16, 17]. These time series have continuous power spectral densities (PSD). Inhibition does not change the power-law PSD in the rest state of symmetry where $k_n = k_{ee} = k_{ei}$ (symmetry in Fig. 9). The scale-free cortical dynamics [18] gives the power-law PSD, so that band pass filtering the simulated and real ECoG gives beats (Rayleigh noise) of alternating maxima and minima (Fig. 10, A) for every pass band in the relevant spectral range ($1 - 100$ Hz). The beat frequencies are distributed; the modal beat frequency for both simulated noise and ECoG is proportional to the bandwidth by a factor of 0.641 [19] at every center frequency [20].

These coincident spikes serve as approximate indicators of temporal locations of phase transition. The precise location is indicated by a temporal discontinuity in the analytic phase [21], at which the analytic frequency jumps to a new value in what appears as *phase slip* [22] immediately preceding the appearance and stabilization of new AM patterns with phase cones [23].

Proof that the null spikes are intrinsic to the neural mechanisms governing microscopic and mesoscopic background activity is by replicating the wave forms of the analytic power and frequency of the EEG (Fig. 10, A); the power-law distributions of power spectral density (PSD) of the EEG at rest and in sleep (Fig. 11, A) [17]; the distributions of $\log_{10}$ analytic power of downward spikes (C, D) [15], and the distributions of the durations of frames between null spikes (B) [20].

These experimental data provide the evidence needed to construct a dissipative dynamic hypothesis of perception. The turbulence in the ECoG at the cortical surface illustrated in Fig. 4 holds in every pass band in the clinical range, conforming to the scale-free dynamics predicted from the power-law PSD of the resting ECoG (Fig. 11, A) and other variables [12]. The existence of a pole at the origin of the complex plane (Fig. 7) shows that cortex homeostatically holds its operating point at or very near a state of criticality, which can justifiably be called self-organized, because it is kept at a homeostatically controlled set point by randomly distributed, abortive phase transitions that are manifested in phase cones (Fig. 3) having power-law distributions of durations and diameters. Few among the phase cones have durations that exceed those expected for noise [Fig. A1.08 in [23]; Section 3.4 and Fig. 2.06, E in [21]], and these few fall into the Rice distribution with longer durations than inter-spike intervals in the Rayleigh distribution [20]. That is significant, because these frames accompany classifiable AM patterns (Fig. 2) that last long enough to transmit 3 to 5 cycles of the carrier frequency [25], and they also have the long correlation distances needed to span vast areas of primary sensory cortices. These attributes of size and persistence make them prime candidates for the neural correlates of retrieved memories.

Null spikes set the system in the symmetric (zero order parameter) ground state so that

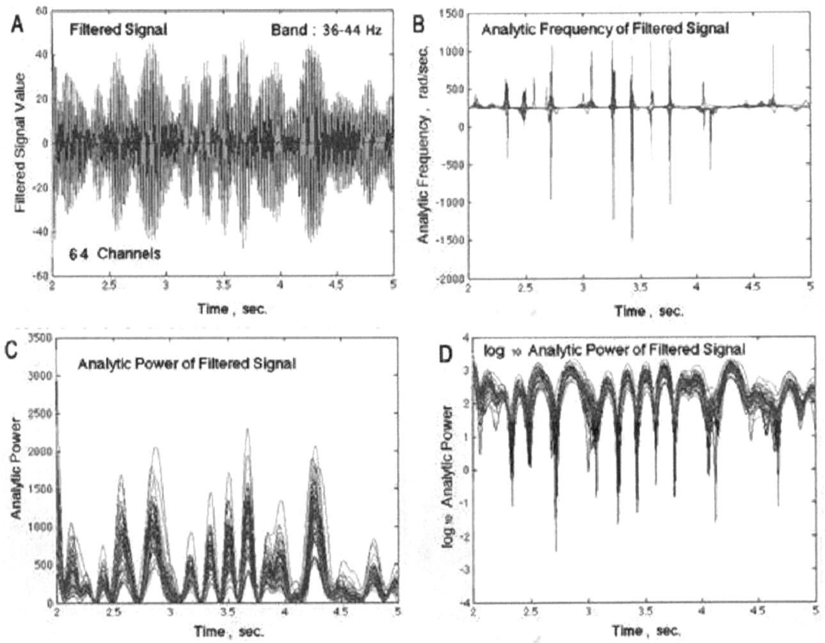

Figure 10. Null spikes are observed by band pass filtering the EEG (A), applying the Hilbert transform [24] to get the analytic power (B), and taking the logarithm (C). On each channel the downward spikes coincide with spikes in analytic frequency (D) reflecting increased analytic phase variance. The flat segment between spikes reflects the stability of the carrier frequency of AM patterns. The spikes form clusters in time but are not precisely synchronized. One or more of these null spikes coincides with phase transitions leading to emergence of AM patterns. The modal repetition rate of the null spikes in Hz is predicted to be 0.641 times the pass band width in Hz (p. 90, Equation 3.8-15 in ref. [19]).

spontaneous symmetry breaking (Mode 2) may then occur. When that happens, the sensory input that activates a Hebbian assembly already formed by learning introduces into the broken symmetry a powerful narrow-band gamma burst (Fig. 8) that is facilitated by the increased synaptic gain, k_{ee}, with learning (Fig. 1), the increased control parameter, Q_m, with arousal, and the asymmetric gain around the operating point (Fig. 6).

5. Dissipative many-body neurodynamics

In the previous Sections we have seen that the mesoscopic neural activity of neocortex appears in laboratory observations to consist of the dynamical formation of spatially extended domains in which widespread cooperation supports brief epochs of patterned oscillations. In 1967 Umezawa and Ricciardi [26, 27, 28] proposed to describe the collective neural activity which manifests in the formation of spatially extended domains by using the mechanism of spontaneous breakdown of symmetry in quantum field theory (QFT). By resorting to preceding studies on the physics of living matter [29, 30, 31] and to the QFT formalism for dissipative systems [32], the extension to the dissipative dynamics of the Umezawa and Ricciardi many-body model has been worked out [33, 3] and the comparison of the predictions of the dissipative quantum model of brain with the laboratory observations has been pursued [6, 34, 35].

It appears that the dissipative model predicts the experimentally observed coexistence of physically distinct amplitude modulated (AM) and phase modulated (PM) patterns correlated with categories of conditioned stimuli and the remarkably rapid onset of AM patterns into irreversible sequences that resemble cinematographic frames [6, 34]. The dissipative model also

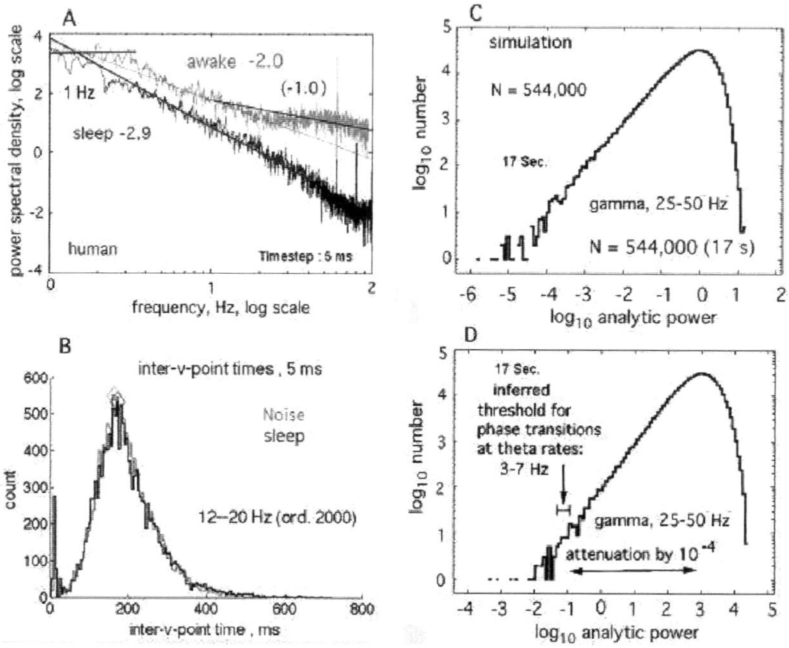

Figure 11. Evidence is summarized showing that the mesoscopic background activity conforms to scale-free, low-dimensional noise [18]. Engagement of the brain in perception and other goal-directed behaviors is accompanied by departures from randomness upon the emergence of order (A), as shown by comparing PSD in sleep, which conforms to black noise, vs. PSD in an aroused state showing excess power in the theta (3 – 7 Hz) and gamma (25 – 100 Hz) ranges. B. The distributions of time intervals between null spikes of brown noise and sleep ECoG are superimposed. C,D. The distributions are compared of $\log_{10}$ analytic power from noise and ECoG. Hypothetically the threshold for triggering a phase transition is 10^{-4} down from modal analytic power. From [15, 17]

accounts for the formation of phase cones and vortices in the transition from one AM pattern to another one [6].

The collective behavior of an ensemble of a large number of elementary components is the object of study of Statistical Mechanics. In the case of neural components, Hopfield [36] asked whether stability of memory and other macroscopic properties of neural nets are also derivable as collective phenomena and emergent properties. Classical Statistical Mechanics provides very powerful tools in answering Hopfield's question [37, 38]. However, at a classical level analysis, the electric field of the extracellular dendritic current and the magnetic fields inside the dendritic shafts appear to be much too weak and the chemical diffusion appears to be much too slow to be able to fully account for the cortical collective activity observed in laboratory [6, 14]. Molecular (neuro-)biology provides crucial tools in the discovery of many mechanisms and chemical functions in brains and living matter in general. The question is now how to put together all these data so to derive the complex behavior of the whole system. In this connection, Schrödinger has introduced the important distinction between the "two ways of producing orderliness" (p. 80 of ref. [39]), namely, ordering generated by the "statistical mechanisms" and ordering generated by "dynamical" interactions among the atoms and the molecules, which, as it is well known, are *necessarily* quantum interactions.

The functional stability in living systems is characterized by time ordering of pathways of biochemical reactions sequentially interlocked. Common laboratory experience is that even the simplest chemical reaction pathway, once embedded in a random chemical environment, soon collapses. Chemical efficiency and functional stability to the degree observed in living matter, i.e. not as "regularity only in the average" [39], appears to be out of reach of any probabilistic approach *solely* based on microscopic random kinematics. Thus, it is still a *matter of belief* that out of a purely random kinematics there may arise with high probability a unique, time ordered sequence of chemical reactions as the one required by the macroscopic history of the system. It is a fact that, although highly sophisticated probabilistic methods have been developed, there is no available computation or even abstract proof which shows how to obtain the characteristic chemical efficiency and stability of living matter by resorting uniquely to statistical concepts. In Schrödinger's words: "it needs no poetical imagination but only clear and sober scientific reflection to recognize that we are here obviously faced with events whose regular and lawful unfolding is guided by a "mechanism" entirely different from the "probability mechanism" of physics" (p.79 of ref. [39]). Classical statistical mechanics and short range forces of molecular biology, *although necessary*, do not seem to be completely adequate tools. Therefore, it appears to be necessary to supplement them with a further step so to include underlying quantum *dynamical* features. Moreover, one more motivation to follow the research path pioneered by Schrödinger is in the fact that there is no conventional neural network offered by neuroscientists or neuroengineers that can adequately model or simulate the neurophysiological data on the brain activity which are today available [35]. In another approach, random graph theory modified as neuropercolation theory shows promise in complementing the use of ordinary and partial differential equations [18].

Such a conceptual frame and the experimental observations motivated the formulation of the many-body model for the brain and its subsequent extension to the dissipative dynamics.

The model central feature rests on the fact that patterns of correlated elements (ordered patterns) are described in QFT by the mechanism of spontaneous breakdown of symmetry (SBS). The observable specifying the degree of ordering of the system vacuum is called the order parameter and acts as a macroscopic variable. It is related with the density of coherently condensed Nambu-Goldstone (NG) bosons in the system ground state and may be considered to be the *code* specifying the system macroscopically observable ordered state, namely its physical phase, among many possible degenerate ground states (vacua). In the quantum model of the brain the code of the ground state specifies its memory content: the memory recording process is depicted by the NG boson condensation in the brain ground state. The external informational input acts as the trigger of the symmetry breakdown out of which the NG bosons and their condensation are generated. The recall of the recorded information occurs under the input of a stimulus "similar" to the one responsible for the memory recording [26] (see also ref. [40]).

We stress that the external stimulus only acts as a trigger inducing the SBS, the specific ordered pattern being generated by the "internal" brain dynamics, which, except for the breakdown of the symmetry, is not conditioned by the external stimulus (*spontaneous* breakdown of the symmetry). This model feature is perfectly consistent with laboratory observations mentioned in Section 2; it accounts for the observed lack of invariance of AM patterns with invariant stimuli [34, 41, 42].

A different regime is, in contrast, obtained when symmetry is *explicitly* broken, as, for example, under the effect of an electric shock, by which the cortex dynamics is enslaved with consequent response by the so-called evoked or *event-related potential* (ERP). The explicit breakdown in cortical dynamics is observed by resort to stimulus-locked averaging across multiple presentations in order to remove or attenuate the background activity, so as to demonstrate that the location, intensity and detailed configuration of the ERP is predominantly determined by the stimulus.

The symmetry which gets broken is the rotational symmetry of the electrical dipoles of the water molecules [43, 44, 45] and the NG modes are the vibrational dipole wave quanta (DWQ) [29, 30, 31]. The whole brain dynamics is indeed embedded in a matrix of molecules carrying the quantum degree of freedom associated to the electric dipole oscillatory motion. The well known electrical properties of cell membranes and the experimental observations [46, 47] of slow fluctuations in neuronal membrane polarization (the so-called up and down states) corresponding to that of spontaneous fluctuations in the fMRI signal confirm that one cannot ignore the electrical dipole oscillatory matrix in which the neuronal electrophysical and electrochemical activity is embedded [48]. Thus, the electric dipole vibrational field dynamics at basic molecular level cannot be without effects on the rich electrochemical and biochemical activity observed at the neural classical level.

We stress that *no ambiguity should be born on the point that neurons and other brain cells are by no means considered quantum objects in the dissipative many-body model.* The dissipative many-body model differs in a substantial way from brain models formulated in the Quantum Mechanics frame, such those discussed in refs. [49, 50, 51]. In contrast with such last models, *the neurons, the glia cells and other brain cells are classical systems in the dissipative many-body model.*

Let us very briefly summarize some formal aspects of the dissipative model. For details see refs. [33, 3].

The procedure of the canonical quantization of a dissipative system requires the "doubling" of the degrees of freedom of the system [32] in order to ensures that the flow of the energy exchanged between the system and the environment is balanced.

Let A_k and $\tilde{A}_k$ denote the annihilation operators for the DWQ mode and its "doubled mode", respectively. k denotes the momentum and other specifications of the A operators ($A_k^\dagger$ and $\tilde{A}_k^\dagger$ denote the creation operators). Let $\mathcal{N}$ be the *code* imprinted in the vacuum at the initial time $t_0 = 0$ by the external input and representing the *memory record* of the input. The code $\mathcal{N}$ is the set of the numbers $\mathcal{N}_{A_k}$ of modes A_k, for any k, condensate in the vacuum state denoted by $|0>_\mathcal{N}$, which thus represents the memory state at $t_0 = 0$ [33, 52]. $\mathcal{N}_{A_k}(t)$ is given, at each t, by:

$$\mathcal{N}_{A_k}(t) \equiv {}_\mathcal{N}< 0(t)|A_k^\dagger A_k|0(t) >_\mathcal{N} = \sinh^2(\Gamma_k t - \theta_k) \quad , \tag{3}$$

and similarly for the modes $\tilde{A}_k$. Γ_k is the damping constant (related to the memory life-time of the mode k) and θ_k fixes the code value at $t_0 = 0$. The state $|0(t)>_\mathcal{N} \equiv |0(\theta, t)>$ is the time-evolved of the state $|0>_\mathcal{N}$ at $t_0 = 0$ and is given by:

$$|0(t)\rangle = \prod_\kappa \frac{1}{\cosh(\Gamma_\kappa t - \theta_k)} \exp\left(\tanh(\Gamma_\kappa t - \theta_k) A_\kappa^\dagger \tilde{A}_\kappa^\dagger\right)|0\rangle \ . \tag{4}$$

$|0>_\mathcal{N}$ and $|0(t)>_\mathcal{N}$ are normalized to 1 and in the infinite volume limit we have

$${}_\mathcal{N}< 0(t)|0>_{\mathcal{N}'} \underset{V\to\infty}{\longrightarrow} 0 \quad \forall\, t \neq t_0, \ \ \forall\, \mathcal{N}\,, \mathcal{N}' \ , \tag{5}$$

$${}_\mathcal{N}< 0(t)|0(t')>_{\mathcal{N}'} \underset{V\to\infty}{\longrightarrow} 0, \ \ \forall\, t, t' \text{ with } \ t \neq t' \ , \ \forall\, \mathcal{N}, \ \mathcal{N}' \ , \tag{6}$$

with $|0(t)>_{\mathcal{N}'} \equiv |0(\theta', t)>$. Eqs. (5) and (6) also hold for $\mathcal{N} \neq \mathcal{N}'$, $t = t_0$ and $t = t'$, respectively. Eqs. (5) and (6) show that in the infinite volume limit the vacua of the same code $\mathcal{N}$ at different times t and t', for any t and t', and, similarly, at equal times, but different $\mathcal{N}$'s, are orthogonal states. The corresponding Hilbert spaces are unitarily inequivalent spaces. The number $(\mathcal{N}_{A_k} - \mathcal{N}_{\tilde{A}_k})$ is a constant of motion for any k and θ. The physical meaning of the $\tilde{A}$ system is the one of the sink where the energy dissipated by the A system flows. The $\tilde{A}$ modes describe the thermal bath or the environment modes.

The balance of energy flow between the system and the environment is ensured by the requirement $\mathcal{N}_{A_k} - \mathcal{N}_{\tilde{A}_k} = 0$, for any k. Such a requirement, however, does not uniquely fix the code $\mathcal{N} \equiv \{\mathcal{N}_{A_k}$, for any $k\}$. Also $|0>_{\mathcal{N}'}$ with $\mathcal{N}' \equiv \{\mathcal{N}'_{A_k}; \mathcal{N}'_{A_k} - \mathcal{N}'_{\tilde{A}_k} = 0$, for any $k\}$ ensures the energy flow balance. Thus, also $|0>_{\mathcal{N}'}$ is an available memory state: it corresponds, however, to a different code number (*i.e.*$\mathcal{N}'$) and therefore to a different information than the one of code $\mathcal{N}$. In the infinite volume limit $\{|0>_{\mathcal{N}}\}$ and $\{|0>_{\mathcal{N}'}\}$ are representations of the canonical commutation relations each other unitarily inequivalent for different codes $\mathcal{N} \neq \mathcal{N}'$. Thus, infinitely many memory (vacuum) states, each one of them corresponding to a different code $\mathcal{N}$, may exist: a huge number of sequentially recorded inputs may *coexist* without destructive interference since infinitely many vacua $|0>_{\mathcal{N}}$, for all $\mathcal{N}$, are *independently* accessible in the sequential recording process.

In conclusion, the "brain (ground) state" is represented as the collection (or the superposition) of the full set of states $|0>_{\mathcal{N}}$, for all $\mathcal{N}$. The brain is thus described as a complex system with a huge number of macroscopic states (the memory states).

The degree of the coupling of the system A with the system $\tilde{A}$ can be parameterized by an index, say n, in such a way that in the limit of $n \to \infty$ the possibilities of the system A to couple to $\tilde{A}$ (the environment) are "saturated": the system A then gets *fully* coupled to $\tilde{A}$. It can be shown [52] that a higher or lower *degree of openness* (measured by n) to the external world may produce a better or worse ability in setting up neuronal correlates, respectively (different under different circumstances, and so on, e.g. during the sleep or the awake states, the childhood or the older ages). The functional or effective connectivity (here we do not consider the structural or anatomical one) is highly dynamic in the dissipative model. Once these functional connections are formed, they are not necessarily fixed. On the contrary, they may quickly change and new configurations of connections may be formed extending over a domain including a larger or a smaller number of neurons. The finite size of the correlated domain implies a non-zero effective mass of the DWQs. These propagate through the domain with a greater inertia than in the case of large (infinite) volume where they are (quasi-)massless. The domain correlations are then established with a certain time-delay, which contributes to the delay observed in the recruitment of neurons in a correlated assembly under the action of an external stimulus.

Free energy, phase transitions and attractor landscape

One can show [33] that the minimization of the free energy, $d\mathcal{F}_A = dE_A - \frac{1}{\beta}dS_A = 0$, is ensured and the change in time $d\mathcal{N}_A$ of particles condensed in the vacuum turns into heat dissipation $dQ = \frac{1}{\beta}dS$.

In the infinite volume limit Eqs. (5) and (6) also hold true for $\mathcal{N} = \mathcal{N}'$. Time evolution of the state $|0\rangle_{\mathcal{N}}$ is thus represented as the (continual) transition through the representations $\{|0(t)\rangle_{\mathcal{N}}, \ \forall\mathcal{N} \ , \ \forall t\}$, namely by the "trajectory" through the "points" $\{|0(t)\rangle_{\mathcal{N}}, \ \forall\mathcal{N} \ , \ \forall t\}$ in the space of the representations (each one minimizing the free energy functional). The trajectory initial condition at $t_0 = 0$ is specified by the $\mathcal{N}$-set. It can be shown [55, 56, 57] that: *a)* these trajectories are classical trajectories and *b)* they are chaotic trajectories. This means that the trajectories are bounded and each trajectory does not intersect itself; there are no intersections between trajectories specified by different initial conditions and trajectories of different initial conditions are diverging trajectories. The property that trajectories specified by different initial conditions ($\mathcal{N} \neq \mathcal{N}'$) never cross each other implies that no *confusion* (interference) arises among the codes of different neuronal correlates, even as time evolves. In realistic situations of finite volume, states with different codes may have non–zero overlap (the inner products Eqs. (5) and (6) are not zero). In such a case, at a "crossing" point between two, or more than two, trajectories, there can be "ambiguities" in the sense that one can switch from one of these trajectories to another one which there crosses. This may be felt indeed as an *association* of memories.

For a very small difference $\delta\theta_k \equiv \theta_k - \theta'_k$ in the initial conditions of the two initial states, the difference between originally even slightly different $\mathcal{N}_{A_k}$'s can be shown to grow as time evolves.

For large enough t, the modulus of the difference $\Delta\mathcal{N}_{A_k}(t)$ and its time derivative diverge as $\exp(2\Gamma_k t)$, for all k's. This may account for the high perceptive resolution in the recognition of the perceptual inputs. Moreover, the difference between *k–components* of the codes $\mathcal{N}$ and $\mathcal{N}'$ may become zero at a given time $t_k = \frac{\theta_k}{\Gamma_k}$. However, the difference between the codes $\mathcal{N}$ and $\mathcal{N}'$ does not necessarily become zero. The codes are different even if a finite number of their components are equal since they are made up by a large number of $\mathcal{N}_{A_k}(\theta, t)$ components (infinite in the continuum limit). Suppose that, for $\delta\theta_k \equiv \theta_k - \theta'_k$ very small, the time interval $\Delta t = \tau_{max} - \tau_{min}$, with τ_{min} and τ_{max} the minimum and the maximum, respectively, of $t_k = \frac{\theta_k}{\Gamma_k}$, for *all* k's, be very small. Then the codes are recognized to be *almost* equal in such a Δt, which then expresses the recognition (or recall) process time. This shows how it is possible that "slightly different" $\mathcal{N}_{A_k}$-patterns (or codes) are recognized to be the *same code* even if corresponding to slightly different inputs. We also remark that the vacua $\{|0(t)\rangle_{\mathcal{N}},\ \forall\mathcal{N}\ ,\ \forall t\}$, through which trajectories go, behave as nonlinear dynamical "attractors" since they are the system ground states (for each $\mathcal{N}$ at each t). Their manifold, which constitutes the space of the brain states, may be thus thought as the attractor landscape for the (nonlinear) brain dynamics. The fact that the dissipative quantum model brings to the classicality of the trajectories through such a space of brain states is one of the model major merits [6, 35].

In the engagement of the brain with the environment in the action-perception cycle, the existing attractor landscape provides the "accumulated experience", or the "knowledge", or the"context" in which the new perception happens to be situated (cf. Section 2). Under the influence of the new perception the brain reaches a new "balance" in a new vacuum, which enters the attractor landscape producing its global rearrangement. The process of the formation of the *meaning* of the newly acquired information and of its "contextualization" consists in such a global updating of the attractor landscape, which, in turn, defines and imposes the conditions to the subsequent action. The whole process involving the engagement of the brain with its environment has been referred to as the "dialog" between the brain and its *Double* (the environment is formally described by the "doubled" degrees of freedom, which express at the same time the environment and the system (brain) time-reversed image, see above). Consciousness has been proposed to be rooted in such a permanent dialog between the subject and its Double [3, 33].

Null spikes, vortices and phase cones

Time–dependence of the DWQ frequency implies that higher momentum k-components of the $\mathcal{N}$-set possess longer life–times. Momentum is proportional to the inverse distance over which the mode propagates, thus modes with a shorter range of propagation (more "localized" modes) survive longer. On the contrary, modes with a longer range of propagation decay sooner. The dissipative model thus predicts the existence of condensation domains of different finite sizes with different degrees of stability [52]. They are described by the condensation function $f(x)$, which acts as a "form factor" specific for the considered domain [53, 54], and in the presence of the e.m. field $f(x)$ has to carry some topological singularity in order for the condensation process to be physically detectable. Phase transitions driven by boson condensation are always associated with some singularity (indeterminacy) in the field phase at the phase transition point[53, 54]. The dissipative model thus accounts for the observed crucial mechanism described in Section 4: the event that initiates a perceptual phase transition is an abrupt decrease, the *null spike*, in the analytic power of the background activity to near zero. The model also accounts for the observation of vortices and propagating waves of phase gradients (the phase cones) in perceptual phase transitions. These features are summarized in the following. See details in ref. [35].

By use of the Hilbert transform, the ECoG sampled wave function $\psi(x)$ in the selected spectral pass band is represented as

$$\psi(x) = \mathbf{A}(x)e^{i\phi(x)} , \tag{7}$$

where $x \equiv (x, y, t)$ in the two surface dimensions of cortex (3 dimensions for the microscopic level of networks), and the analytic power $\mathbf{A}^2(x)$ and the analytic phase $\phi(x)$ are, in terms of the real and imaginary parts $a(x)$ and $b(x)$,

$$\mathbf{A}^2(x) = a^2(x) + b^2(x) , \tag{8}$$

$$\phi(x) = \arctan \frac{b(x)}{a(x)} , \tag{9}$$

respectively. During periods of high amplitude the spatial deviation of phase (SD_X) is low and the phase spatial mean tends to be constant within frames and to change suddenly between frames, indicating coherence and coordinated phase differences. $\mathbf{A}^2(x)$ forms a feature vector that serves as our order parameter (see Section 2 and Refs. [6, 34]).

The reduction in the amplitude of the spontaneous background activity induces a brief state of indeterminacy in which the power in significant pass band of the electrocorticogram (ECoG) is near zero and the phase of ECoG is undefined.

As already discussed in previous sections, the cortex can be driven across such a "phase transition" to a new AM pattern by the stimulus arriving at or just before the onset of a null spike. The observed velocity of spread of phase transition is finite, i.e. there is no "instantaneous" phase transition; it is determined by the conduction velocities of axons of intracortical neurons, not those of the input axons. Experimental evidence indicates that the neuronal correlation length would cover an entire cerebral hemisphere very quickly (practically without delay in the gamma activity), if measured at the critical transition. Between the null spikes the cortical dynamics is (nearly) stationary for $\sim 60 - 160\ ms$. This is called a frame. The transitions by which they form are shorter by an order of magnitude. The spatial phase gradient propagates as a phase cone that is imposed on the carrier wave of the wave packet in a frame by the propagation velocity of the largest axons having the highest velocity in a distribution. The location of the apex is a random variable across frames that is determined by the accidents of where the null spike is lowest and the background input is highest. The null spike has rotational energy at the geometric mean frequency of the pass band, so it is called a vortex. The vortex occupies the whole area of the phase-locked neural activity of the cortex for a point in time. One more observed feature is the random variation from each frame to the next of the slope of the spatial conic phase gradient, negative with explosion, positive with implosion. The negative gradient could be explained in conventional neurodynamics (e.g. in terms of a pacemaker), but not the positive gradient. Also, there is no explanation in the conventional framework of why both gradients, the positive and the negative one, occur.

The spontaneous breakdown of symmetry considered above leads to non-vanishing polarization density $\mathcal{P} = \rho\delta$, where ρ and δ are the charge density and the (average) dipole length, in the ground state. One then considers the (further) spontaneous breakdown of the phase symmetry and the charge density wave function $\sigma(x)$ written [30] as

$$\sigma(x) = \sqrt{\rho(x)}e^{i\theta(x)} , \tag{10}$$

with real $\rho(x)$ and $\theta(x)$. The "phase" $\theta(x)$ is the NG field associated with the breakdown of global phase symmetry. The boson condensation of the field $\theta(x)$ in the system ground state is formally described by the transformation

$$\theta(x) \rightarrow \theta(x) - \frac{e_0 v^2}{Z} f(x) . \tag{11}$$

The c-number condensation function $f(x)$ satisfies the same equation satisfied by the $\theta(x)$ field, i.e. $\partial^2 f(x) = 0$. The constant Z is the wave function renormalization constant, e_0 and v are the electron charge and the constant entering the symmetry breakdown condition $< 0|\rho(x)|0 >= v \neq 0$ [30, 58, 59]. Coherent domains of finite size are obtained by non-homogeneous boson condensation. The condensate is described by the function $f(x)$ which, as said above, acts as a "form factor" specific for the considered domain [52, 53, 54, 59]. One can show [30, 58] that mathematical consistency requires that the electromagnetic vector potential $a_\mu(x)$ has to satisfy the equation

$$(\partial^2 + m_V^2)\, a_\mu(x) \;=\; \frac{m_V^2}{e_0}\, \partial_\mu f(x)\;. \tag{12}$$

We adopt the gauge condition $\partial^\mu a_\mu(x) = 0$. Eq. (12) is the classical Maxwell equation for the massive vector potential a_μ (m_V is its mass). The classical ground state current $j_{\mu,cl}$ turns out to be

$$j_{\mu,cl}(x) \equiv \langle 0|j_\mu(x)|0\rangle \;=\; m_V^2 \left[a_\mu(x) - \frac{1}{e_0}\, \partial_\mu f(x) \right]\;, \tag{13}$$

and we have $\partial^\mu j_{\mu,cl}(x) = 0$. The term $m_V^2 a_\mu(x)$ is the well known *Meissner current*, while $\frac{m_V^2}{e_0}\partial_\mu f(x)$ is the *boson current*. The mesoscopic field and current are thus given in terms of the boson transformation function. It is also remarkable that the classical current is related with $\partial_\mu f$, i.e. with variations in the boson transformation function.

The important point is that such a condensation function $f(x)$ has to carry some topological singularity in order for the condensation process to be physically detectable. Eq. (12) implies $a_\mu(x) = \frac{1}{e_0}\partial_\mu f(x)$ for regular $f(x)$, which in turn implies zero classical field ($F_{\mu\nu} = \partial_\mu a_\nu - \partial_\nu a_\mu$) and zero classical current ($j_{\mu,cl} = 0$) since the Meissner and the boson current cancel each other. It is indeed well known [60] that the gauge field a_μ is expelled out of the ordered domain region (there it vanishes) where the order parameter is of course non-zero and $f(x)$ does not have singularities. On the contrary, the gauge field is non zero in the regions where $f(x)$ presents non-trivial topological singularities such as line singularities, e.g. on the line $r = 0$ in the core of a vortex: we have there the "normal" (disordered) state rather than the ordered one and the non-vanishing massive gauge field there propagates (the Anderson-Higgs-Kibble mechanism) [60, 61]. On the boundaries between the normal and the ordered regions the phase field gradients are non-zero. Instead they are zero in the normal region, i.e. in the vortex core. One can also show [53, 54, 59] that the phase transition from one state space to another (unitarily inequivalent) one can only be induced by a singular boson transformation function $f(x)$. We thus recognize that in the brain the null spike (the observed abrupt decrease in the order parameter and the concomitant increase in the phase field gradients in the phase transition from an AM amplitude to another one) is indeed characterized by the topological singularity of the function $f(x)$. In the case of phase symmetry summarized above, the stationary function $f(x)$ solution of our problem carries a vortex singularity and is given by

$$f(x) = \arctan\left(\frac{x_2}{x_1}\right)\;. \tag{14}$$

Eq. (14) shows that the phase is undefined on the line $r = 0$, with $r^2 = x_1^2 + x_2^2$, consistently with the observed phase indeterminacy in the process of transition between two AM pattern frames. As a result of the single-valuedness of $\sigma(x)$ the topological singularity is characterized by the winding number n: $\oint \nabla f \cdot dl = 2\pi n$, $n = 0, \pm 1, \pm 2, ...$, when the integration is performed along the closed circle $(0, 2\pi)$ (flux quantization). We also remark that clockwise rotation and counterclockwise rotation are both allowed.

As already remarked above, the dissipative model predicts that the response amplitude depends not on the input amplitude, but on the intrinsic state of the cortex, specifically the degree of reduction in the power and order of the background brown noise.

We also observe that the initial site where non-homogeneous condensation starts (the phase cone apex) is not conditioned by the incoming stimulus, but is randomly determined by the concurrence of a number of local conditions, such as where the null spike is lowest and the background input is highest, in which the cortex finds itself at the transition process time. The apex is never initiated within frames (in the broken symmetry phase or ordered region), but between frames (during phase transitions), as it is indeed predicted by the dissipative model (vortices occur during the critical regime of phase transitions). The null spike appears in the band pass filtered brown noise activity and can be conceived as a *shutter* that blanks the intrinsic background ECoG. When the order parameter goes to zero the microscopic activity (of the background state) does not decrease but, consistently with the model description, it becomes disordered (fully symmetric). In such a state of very low analytic amplitude, the analytic phase is undefined, as it is indeed at the center line of the vortex core, and the system, under the incoming weak sensory input, may re-set the background activity in a new AM frame, if any, formed by reorganizing the existing activity. The analytic amplitude decrease repeats in the theta or alpha range, independently of the repetitive sampling of the environment by limbic input. Consistently with observations, in the dissipative model the reduction in activity constitutes a singularity in the dynamics at which the phase is undefined. The aperiodic shutter allows opportunities for phase transitions.

Imploding and exploding regime of the phase cone

In the phase transition processes occurring in a finite span of time, in which the formation of defects (e.g. vortices) occurs, a maximally stable new configuration is attained after a certain lapse of time since the transition has started. The system is said to be in the critical or Ginzburg regime during such a lapse of time. Reliable information on the critical regime behavior is provided by using the harmonic approximation for the evolution of the space-time dependent order parameter v [53, 54, 62]. By resorting to such an approximation, we expand the v field into partial waves:

$$v(x,t) = \sum_{\mathbf{k}} \{u_{\mathbf{k}}(t)e^{i\mathbf{k}\cdot\mathbf{x}} + u_{\mathbf{k}}^{\dagger}(t)e^{-i\mathbf{k}\cdot\mathbf{x}}\} \ . \tag{15}$$

Here we omit the dependence on temperature since this does not affect our discussion. In the Ginzburg-Landau (GL) formalism, for each k-mode ($k \equiv \sqrt{\mathbf{k}^2}$), the equations for the parametric oscillators $u_{\mathbf{k}}$ is [63] (see also [53, 54]):

$$\ddot{u}_{\mathbf{k}}\,(t) + (\mathbf{k}^2 - m^2)\,u_{\mathbf{k}} = 0 \ . \tag{16}$$

The oscillator frequency is

$$M_k(t) \;=\; \sqrt{\mathbf{k}^2 - m^2(t)}, \tag{17}$$

In full generality, we are assuming that m^2 may depend on time. At each t, $M_k(t)$ is real for each k provided, during the critical regime time interval, it is

$$\mathbf{k}^2 \geq m^2(t) \ , \tag{18}$$

for each k-mode. This turns out to be a condition on the k-modes propagation. Let $t = 0$ and $t = \tau$ denote the times at which the critical regime starts and ends, respectively. For a given $\mathbf{k}$, Eq. (18) holds up to a time τ_k after which $m^2(t)$, for $t > \tau_k$, is larger than $\mathbf{k}^2$. The corresponding k-mode can propagate in a span of time $0 \leq\ t\ \leq \tau_k$. Therefore, the "effective causal horizon" [64, 65] can happen to be inside the system (possible formation of more than a domain) or outside (single domain formation) according to whether the time occurring for reaching the boundaries of the system is longer or shorter than the allowed propagation time. This determines the dimensions to which the domains can expand. The explicit form of $m^2(t)$ determines the value of τ_k.

The time dependence of $m(t)$ may be modeled in a way to allow defect (i.e. vortex) formation [54]. We put:

$$m^2(t) = m_0^2 \, e^{2h(t)} \,, \tag{19}$$

where the function $h(t)$ has to be chosen in a specific model. The correlation propagation time is given by:

$$h(\tau_k) = \ln\left(\frac{k}{m_0}\right) \propto \ln\left(\frac{L}{\xi}\right) \,, \tag{20}$$

where ξ is the correlation length corresponding to the k-mode propagation and $L \propto m_0^{-1}$. L acts as an intrinsic infrared cut-off. Due to Eq. (18), small k values are excluded by the non-zero minimum value of $m^2(t)$. Long wave-lengths are thus precluded, which means that only domains of finite size can be obtained. At the end of the critical regime the correlation may extend over domains of linear size of the order of $\lambda_k \propto m^{-1}(\tau)$. $h(\tau_k)$ resembles the commonly called string tension [65].

In order to further specify our model, the explicit analytic expression for $h(t)$ is chosen to be [54]:

$$h(t) = \pm \frac{at}{bt^2 + c} \,, \tag{21}$$

where a, b, c are (positive) arbitrary parameters. We put $c/a\lambda \equiv \tau_Q$, $a\lambda/b \equiv \tau_0$, with λ an arbitrary constant. Note that $h(\tau_Q) = h(\tau_0)$. The time derivative of $h(t)$, and thus of $m^2(t)$, is zero at $t = \tau = \pm\sqrt{\tau_Q \tau_0}$. τ thus plays the role of the equilibrium time scale. We observe that

$$h(t) = \pm \frac{1}{\lambda \tau_Q} \frac{1}{1 + \frac{t^2}{\tau^2}} \, t \approx \pm \frac{\Gamma}{2} \, t \,, \tag{22}$$

for $t^2/\tau^2 \approx 1$, with $\Gamma \equiv 1/\lambda\tau_Q$. Since the size of the vortex core is given by $(m(t))^{-1}$, Eqs. (19) and (22) show that such a size evolves in time as $e^{\mp\Gamma \, t}$, $t < \tau$ ($t < \tau_k$ for the k-mode). This means that we have converging (imploding) and diverging (exploding) regimes, as indeed found in laboratory observations of the phase cone behaviors. We remark that Eq. (22) shows that the $\pm$ signs in Eq. (21) amount to working with both elements of the basis $(e^{+\Gamma/2 \, t}, e^{-\Gamma/2 \, t})$, as indeed required by mathematical correctness. In this sense, the $\pm$ double sign cannot be avoided in the model choice of $h(t)$. From a physical point of view, it is equivalent to working with time evolution pointing in one given time direction (say the $t > 0$ arrow of time) and with its "time-reversed" copy or image. This is perfectly consistent with one of the main features of the dissipative model where time-reversed excitations are introduced, thus "doubling" the system degrees of freedom. Moreover, in ref. [66] clockwise and counterclockwise (loop and anti–loop) rotations have been shown to be related by time–reversal transformations.

We observe that since the disordered ("normal") state is confined to the vortex core, in the imploding regime (the shrinking of the vortex core) long range correlation, i.e. ordering, is prevailing (the vortex is "squeezed out"); in the exploding regime (enlargement of the vortex core), local correlations (disorder) prevail.

Finally, the number of defects (of vortices) n_{def} possibly appearing during the critical regime is given in the linear approximation by [54, 65]:

$$n_{def} \propto m^2(\tau) \approx m_0^2 \, |\tau/\lambda\tau_Q| \,. \tag{23}$$

In conclusion, in the process of non-instantaneous phase transitions (as those observed in brain) the dissipative model predicts the existence of a vortex singularity associated (at the vortex core) with the abrupt decrease (null spike) of the order parameter (the analytic amplitude) and the concomitant increase of spatial variance of the phase field (the analytic phase). The resulting phase cones present both phase gradients, the positive and the negative one, as in the observations. The formation of imploding and exploding phase cones is shown to be allowed, as indeed deduced from observations.

Let us close this Section with a final comment. We have seen that coherent states play a relevant role in the dissipative many-body model. For example, the transformation (11) induces boson condensation and generates a (Glauber) coherent state. In ref. [67] it has been discussed the self-similarity property of coherent states in connection with fractal self-similarity property [68, 69]. The dissipative model thus predicts self-similarity as a characterizing property of brain dynamics [67, 70]. As observed in the previous Sections, laboratory observations show indeed that brain background activity is characterized by self-similarity. Measurements of the durations, recurrence intervals and diameters of neocortical EEG phase patterns have power-law distributions with no detectable minima. The power spectral densities in time and space of ECoGs from surface arrays conform to power-law distributions [21, 23, 71, 72, 73]. This suggests that the activity patterns generated by neocortical neuropil might be scale-free [74, 75, 76] with self-similarity in ECoGs patterns over distances ranging from hypercolumns to an entire cerebral hemisphere [34].

6. Conclusions

The discussion presented in this paper leads us to conclude that the crucial step in perception is the transition from the excited microscopic assembly to the large-scale mesoscopic AM pattern. That possibility occurs when a null spike (Fig. 10, D) of the necessary depth (Fig. 11, D) supervenes in the appropriate frequency range, and the order parameter manifested in the ECoG vanishes. Any pre-existing dissipative structure dissolves, leaving the microscopic neurons in maximal disorder, which is a state of symmetry. The null spike sets the conditions allowing spontaneous breakdown of symmetry, which opens the possibility for the activated assembly to capture the population by directing the trajectory into its basin of attraction. The precise location of the phase transition in time, space and spectrum vary randomly (Fig. 3, Fig. 10). When the null spike terminates and the order parameter returns, it sustains the new self-organized dissipative structure. The stability of the AM pattern is enhanced by the vortex (Fig. 5) that is frozen into the conic phase gradient (Fig. 4).

These basic features have been described by the dissipative many-body model. Several predictions have been derived and compared with experiments. It can be useful to list few of these experimentally confirmed model predictions. As already mentioned, the dissipative many-body model accounts for the observed dynamical formation of spatially extended domains of neuronal synchronized oscillations and of their rapid sequencing. The model explains indeed two main features of the ECoG data:

- the textured patterns of AM in distinct frequency bands correlated with categories of conditioned stimuli, i.e. *coexistence* of physically distinct AM patterns, and
- the remarkably rapid onset of AM patterns into (irreversible) sequences that resemble cinematographic frames.

Each spatial AM pattern is described to be consequent to spontaneous breakdown of symmetry triggered by external stimulus and is associated with one of the emerging unitarily inequivalent ground states. Their sequencing is associated to the non-unitary time evolution implied by dissipation. Moreover, consistently with experimental observations the model predicts that [6, 34]

- very low energy is required to excite AM correlated neuronal patterns,

- AM patterns have large diameters, with respect to the small sizes of the component neurons,
- duration, size and power of AM patterns are decreasing functions of their carrier wave number
- there is lack of invariance of AM patterns with invariant stimuli.
- there is heat dissipation at (almost) constant in time temperature (see also [6, 34]),
- null spikes occur in the process of phase transitions and the whole behavior of phase gradients and phase singularities in the vortices formation agrees with observations,
- that the phase field is constant within the frames,
- the insurgence of a phase singularity is associated with the abrupt decrease of the order parameter and the concomitant increase of spatial variance of the phase field,
- the onsets of vortices occurs *between* frames, not *within* them,
- the phase cones occur with random variation of sign (implosive and explosive) at the apex,
- that the phase cone apices occur at random spatial locations,
- that the apex is never initiated *within* frames, but *between* frames (during phase transitions).

The model leads to the classicality (not derived as the classical limit, but as a dynamical output) of functionally self-regulated and self-organized background activity of the brain.

Finally, self-similarity and scale-free dynamics suggested by observation finds its correspondence in the coherent state structure of brain dynamics as described by the model.

Acknowledgments
The authors express their appreciation for programming by Brian C. Burke and William Redfearn.

References

[1] von Neumann J 1958 *The Computer and the Brain* (New Haven CT: Yale UP)
[2] Freeman W J 2007 *Intentionality. Entry for Encyclopedia for Computational Neuroscience* ed E Izhikevich http://www.scholarpedia.org/article/Intentionality
[3] Vitiello G 2001 *My Double Unveiled* (Amsterdam: John Benjamins)
[4] Freeman W J and Erwin H 2008 *Freeman K-set. Entry for Encyclopedia for Computational Neuroscience* ed E Izhikevich
http://www.scholarpedia.org/article/Freeman_K − set
[5] Principe J C, Tavares V G, Harris J G and Freeman W J 2001 *Proceedings IEEE* **89** 1030
[6] Freeman W J and Vitiello G 2008 *J. Phys. A: Math. Theor.* **41** 304042 http://Select.iop.org (*Preprint* arXiv:q-bio.NC/0701053v1)
[7] Lashley K 1948 *The Mechanism of Vision, XVIII, Effects of Destroying the Visual "Associative Areas" of the Monkey* (Provincetown MA:Journal Press)
[8] Freeman W J 2001 *How Brains Make Up Their Minds* (New York: Columbia UP)
[9] Freeman W J and Baird B 1987 *Behavioral Neuroscience* **101** 393
[10] Freeman W J 1979 *Biol. Cybern.* **33** 237
[11] Freeman W J 1975/2004 *Mass Action in the Nervous System* (New York: Academic Press)
[12] Freeman W J 2007 *Scale-free neocortical dynamics. Encyclopedia for Computational Neuroscience* ed E Izhikevich
http://www.scholarpedia.org/article/Scale − free_neocortical_dynamics
[13] Freeman W J, Burke B C and Holmes M D 2003 *Human Brain Mapping* **19** 248
[14] Freeman W J 2006 *Clin. Neurophysiol.* **117** 572 http://repositories.cdlib.org/postprints/1480/
[15] Freeman W J, O'Nuillain S and Rodriguez J 2008 *J. Integr. Neurosci.* **7** 337
[16] Schroeder M 1991 *Fractals, Chaos, Power Laws* (San Francisco: WH Freeman)
[17] Freeman W J and Zhai J 2008 *Cognitive Neurodynamics* **3** in press
[18] Freeman W J, Kozma R, Bollobás B and Riordan O 2009 Scale-Free Cortical planar Graphs *Handbook of Large-Scale Random Networks. Bolyai-Springer Series on Advanced Combinatorics* ch 7 ed B Bollobás, R Kozma, D Mikls (New York: Springer)
[19] Rice S O 1950 *Mathematical Analysis of Random Noise - and Appendixes. Technical Publications Monograph B-1589* (New York: Bell Telephone Labs Inc)
[20] Freeman W J 2009 *Cognitive Neurodynamics* **3** in press

[21] Freeman W J 2004 *Clin. Neurophysiol.* **115** 2089
http://repositories.cdlib.org/postprints/1486
[22] Pikovsky A, Rosenblum M and Kurths J 2001 *Synchronization - A Universal Concept in Non-linear Sciences* (Cambridge UK: Cambridge UP)
[23] Freeman W J 2004 *Clin. Neurophysiol.* **115** 2077
[24] Freeman W J 2007 *Hilbert transform for brain waves. Entry for Encyclopedia for Computational Neuroscience* ed E Izhikevich *http : //www.scholarpedia.org/article/Hilbert_transform_forbrain_waves*
[25] Freeman W J 2005 *Clin. Neurophysiol.* **116** 1118
http://authors.elsevier.com/sd/article/S1388245705000064 http://repositories.cdlib.org/postprints/2134/
[26] Ricciardi L M and Umezawa H 1967 *Kibernetik* **4** 44; Reprint in *Brain and Being* 2004 eds G G Globus, K H Pribram and G Vitiello (Amsterdam: John Benjamins) 255
[27] Stuart C I J, Takahashi Y and Umezawa H 1978 *J.Theor. Biol.* **71** 605
[28] Stuart C I J, Takahashi Y and Umezawa H 1979 *Found. Phys.* **9** 301
[29] Del Giudice E, Doglia S, Milani M and Vitiello G 1985 *Nucl. Phys.* B **251** (FS 13) 375
[30] Del Giudice E, Doglia S, Milani M and Vitiello G 1986 *Nucl. Phys.* B **275** (FS 17) 185
[31] Del Giudice E, Preparata G and Vitiello G 1988 *Phys. Rev. Lett.* **61** 1085
[32] Celeghini E, Rasetti M and Vitiello G 1992 *Annals of Phys.* **215** 156
[33] Vitiello G 1995 *Int. J. Mod. Phys.* B **9** 973
[34] Freeman W J and Vitiello G 2006 *Phys. of Life Reviews* **3** 93
http://www.arxiv.org/find [Freeman] (*Preprint* arXiv:q-bio.OT/0511037)
[35] Freeman W J and Vitiello G 2008 Vortices in brain waves, forthcoming
[36] Hopfield J J 1982 *Proc. of Nat. Acad. Sc. USA* **79** 2554
[37] Amit D J 1989 *Modeling Brain Function: The World of Attractor Neural Networks* (Cambridge UK: Cambridge University Press)
[38] Mezard M, Parisi G and Virasoro M 1987 *Spin glass theory and beyond* (Singapore: World Sci.)
[39] Schrödinger E 1944. *What is life?* [1967 reprint] (Cambridge: Cambridge University Press)
[40] Sivakami S and Srinivasan V 1983 *J. Theor. Biol.* **102** 287
[41] Freeman W J, Burke B C and Holmes M D 2003 *Human Brain Mapping* **19** 248
[42] Freeman W J, Burke B C, Holmes M D and Vanhatalo S 2003 *Clin. Neurophysiol.* **114** 1055
[43] Jibu M and Yasue K 1992 *Cybernetics and System Research* ed R Trappl (Singapore: World Scientific) 797
[44] Jibu M and Yasue K 1995 *Quantum brain dynamics and consciousness* (Amsterdam: John Benjamins)
[45] Jibu M, Pribram K H and Yasue K 1996 *Int. J. Mod. Phys.* B **10** 1735
[46] Haider B, Duque A, Hasenstaub A R and McCormick D A 2006 *J. Neurosci.* **26** 4535
[47] Peterson C C, Hahn T T G, Mehta M, Grinvald A and Sakmann B 2003 *Proc. Natl. Acad. Sci. U.S.A.* **100** 13638
[48] Raichle M E 2006 *Science* **314** 1249
[49] Stapp H P 1993/2003 *Mind, matter and quantum mechanics* (New York: Springer-Verlag)
[50] Hameroff S and Penrose R 1996 *J. Conscious. Stud.* **3** 36
[51] Penrose R 1994 *Shadows of the mind* (Oxford: Oxford University Press)
[52] Alfinito E and Vitiello G 2000 *Int. J. Mod. Phys.* B **14** 853 [Erratum 2000 ibid. B **14** 1613]
[53] Alfinito E, Romei O and Vitiello G 2002 *Mod. Phys. Lett.* B **16** 93
[54] Alfinito E and Vitiello G 2002 *Phys. Rev.* B **65** 054105
[55] Vitiello G 2004 *Int. J. Mod. Phys.* B **18** 785
[56] Pessa E and Vitiello G 2003 *Mind and Matter* **1** 59
[57] Pessa E and Vitiello G 2004 *Int. J. Mod. Phys.* B **18** 841
[58] Matsumoto H, Papastamatiou N J, Umezawa H and Vitiello G 1975 *Nucl. Phys.* B **97** 61
Matsumoto H, Papastamatiou N J and Umezawa H 1975 *Nucl. Phys.* B **97** 90
[59] Umezawa H 1993 *Advanced field theory: micro, macro and thermal concepts* (New York: AIP)
[60] Anderson P W 1984 *Basic Notions of Condensed Matter Physics* (Menlo Park: Benjamin)
[61] Higgs P 1996 *Phys. Rev.* **145** 1156
Kibble T W B 1967 *Phys. Rev.* **155** 1554
[62] Bettencourt L M A, Antunes N A and Zurek W H 2000 *Phys. Rev.* D **62** 065005
Antunes N A, Bettencourt L M A and Zurek W H 2000 *Phys. Rev. Lett.* **82** 2824
[63] Perelomov A M 1986 *Generalized Coherent States and their Applications* (Berlin: Springer)
[64] Kibble T W B 2000 *Topological defects and the non-equilibrium dynamics of symmetry breaking phase transitions. NATO Science Series C 549* ed Y M Bunkov and H Godfrin (Dordrecht: Kluwer Acad.) 7
Volovik G E 2000 *Topological defects and the non-equilibrium dynamics of symmetry breaking phase transitions. NATO Science Series C 549* ed Y M Bunkov and H Godfrin (Dordrecht: Kluwer Acad.)

353
[65] Zurek W H 1997 *Phys. Rep.* **276** 177 and refs. therein quoted
[66] Alfinito E and Vitiello G 2003 *Mod. Phys. Lett.* B **17** 1207
[67] Vitiello G 2009 *New Mathematics and Natural Computation* **5** 245
[68] Peitgen H O, Jürgens H and Saupe D 1986 *Chaos and fractals. New frontiers of Science* (Berlin: Springer-Verlag)
[69] Bunde A and Havlin S (eds.) 1995 *Fractals in Science* (Berlin: Springer-Verlag)
[70] Vitiello G 2009 Fractals and the Fock-Bargmann representation of coherent states, in press
[71] Braitenberg V and Schüz A 1991 *Anatomy of the Cortex: Statistics and Geometry* (Berlin: Springer-Verlag)
[72] Linkenkaer-Hansen K, Nikouline V M, Palva J M and Iimoniemi R J 2001 *J. Neurosci.* **15** 1370
[73] Hwa R C and Ferree T 2002 *Phys. Rev.* E **66** 021901
[74] Wang X F and Chen G R 2003 *IEEE Trans. Circuits Syst.* **31** 6
[75] Freeman W J 2005 *J. Integrative Neuroscience* **4** 407
[76] Bassett D S, Meyer-Lindenberg A, Achard S, Duke T and Bullmore E 2006 *Proc. Natl. Acad. Sci. U.S.A.* **103** 19518

Multivalued Fields

H. Kleinert

Institut für Theoretische Physik, Arnimallee 14, D14195 Berlin, Germany

E-mail: `kleinert@physik.fu-berlin.de`

Abstract. The statistical mechanics of line-like excitations of many-body systems can most easily be formulated with the help of multivalued fields which do not satisfy the Schwarz integrability conditions. These fields must be included into the hydrodynamic description of such systems in order to allow for a proper understanding of the observed phase transitions. We sketch the construction of the associated quantum field theory and its application to the phase transitions of superfluids, Bose-Einstein condensates, superconductors, crystals, and magnetic monopole gases.

1. The simplest systems in which multivalued fields play an important role are superfluid helium and Bose-Einstein condensates (BECs). The Landau-Pitaevski theory of such systems is based on the energy density

$$\mathcal{H}_{\rm LP}(\psi, \boldsymbol{\nabla}\psi) \;=\; \frac{1}{2}\Big\{|\boldsymbol{\nabla}\psi|^2 + \tau|\psi|^2 + \frac{g}{2}|\psi|^4\Big\}, \tag{1}$$

which in a BEC is applicable for all temperatures, in superfluid helium only near the critical point. The parameter $\tau \equiv T/T_c^{\rm MF} - 1$ is a reduced temperature measuring the distance from the characteristic temperature $T_c^{\rm MF}$ at which the $|\psi|^2$-term changes sign.

The partition function is given by the functional integral [1]

$$Z_{\rm LP} \;\equiv\; \int \mathcal{D}\psi\mathcal{D}\psi^* \exp\left[-\int d^3x\, \mathcal{H}_{\rm LP}(\psi, \boldsymbol{\nabla}\psi)\right]. \tag{2}$$

Under a field transformation $\psi(x) \to \rho(x)e^{i\theta(x)}$, and using the naive Leibnitz rule

$$D\psi = e^{i\theta}(i\rho\boldsymbol{\nabla}\theta + \boldsymbol{\nabla}\rho), \tag{3}$$

the energy density is replaced by

$$\mathcal{H}_1 \;=\; \frac{\rho^2}{2}(\boldsymbol{\nabla}\theta)^2 + \frac{1}{2}(\boldsymbol{\nabla}\rho)^2 + \frac{\tau}{2}\rho^2 + \frac{g}{4}\rho^4. \tag{4}$$

and the partition function becomes

$$Z_1 \;\approx\; \int \mathcal{D}\theta\mathcal{D}\rho\, \rho \exp\left[-\int d^3x\, \mathcal{H}_1\right]. \tag{5}$$

The phase angle appears harmonically in $\mathcal{H}_1$ and can be integrated out to yield

$$Z_2 \;\approx\; {\rm Det}^{-1/2}(-\boldsymbol{\nabla}^2) \int \mathcal{D}\rho \exp\left[-\int d^3x\, \mathcal{H}_2\right]. \tag{6}$$

where

$$\mathcal{H}_2 = \frac{1}{2}(\boldsymbol{\nabla}\rho)^2 + \frac{\tau}{2}\rho^2 + \frac{g}{4}\rho^4. \tag{7}$$

At the mean-field level, this partition function (6) possesses a second-order phase transition. It takes place when τ drops below zero where the pair field $\psi(x)$ acquires the nonzero expectation value $\langle\psi(x)\rangle = \rho_0 = \sqrt{-\tau/g}$, the order parameter of the system. The ρ-fluctuations around this value have a *coherence length* $\xi = 1/\sqrt{-2\tau}$.

This phase transition is often identified as the famous λ-transition observed in superfluid helium at 2.18 K. This, however, is a false identification. Indeed, the universality class of the transition of (6) is *Ising* and has different critical exponents than that of the λ-transition. For instance, the critical exponent $\alpha \equiv 2 - D\nu$ of the singularity of the specific heat $C \propto |1 - T/T_c|^{-\alpha}$ following from (6) would be $\alpha \approx 0.112$ [2]. The transition of the partition function (2), on the other hand, can be calculated to be $\alpha \approx -0.0129$ [3], which agrees perfectly with the experimental result for the λ-transition $\alpha \approx -0.0127$ measured in a satellite orbit [4] to minimize pressure differences inside the sample.

The culprit for this discrepancy is the illegal application of the Leibnitz rule in (3), which must be corrected. Since $\theta(\mathbf{x})$ and $\theta(\mathbf{x}) + 2\pi$ correspond to the same complex field $\psi(\mathbf{x})$, the corrected Leibnitz rule reads

$$D\psi = e^{i\theta}[i\rho(\boldsymbol{\nabla}\theta - \boldsymbol{\theta}^v) + \boldsymbol{\nabla}\rho], \tag{8}$$

The cyclic nature of the scalar field $\theta(\mathbf{x})$ manifests itself in the presence of a vector field $\boldsymbol{\theta}^v(\mathbf{x})$ called *vortex gauge field* [5]. This field is a sum of δ-functions on Volterra surfaces $\boldsymbol{\delta}(S;\mathbf{x}) \equiv \int_S d^2\bar{\mathbf{x}}\, \delta(\mathbf{x} - \bar{\mathbf{x}})$, across which $\theta(\mathbf{x})$ has jumps by 2π. For brevity, we have defined $\eth(\mathbf{x}) \equiv 2\pi\delta(\mathbf{x})$.

The boundary lines of the surfaces are vortex lines. Their shape is found from the vortex gauge field $\boldsymbol{\theta}^v(\mathbf{x})$ by forming the curl

$$\boldsymbol{\nabla} \times \boldsymbol{\theta}^v(\mathbf{x}) = \mathbf{j}^v(\mathbf{x}), \tag{9}$$

where $\mathbf{j}^v(\mathbf{x})$ is the *vortex density*, a sum over δ-functions $\boldsymbol{\delta}(L;\mathbf{x}) \equiv \int_L d\bar{\mathbf{x}}\, \delta(\mathbf{x} - \bar{\mathbf{x}})$ along the vortex lines L. Here we have used the differential version of Stokes' integral theorem

$$\boldsymbol{\nabla} \times \boldsymbol{\delta}(S;\mathbf{x}) = \boldsymbol{\delta}(L;\mathbf{x}). \tag{10}$$

Vortex gauge transformations correspond to deformations of the surfaces at fixed boundary lines which add to $\boldsymbol{\theta}^v(\mathbf{x})$ pure gradients of the form $\boldsymbol{\nabla}\eth(V;\mathbf{x})$, where $\eth(V;\mathbf{x}) \equiv \int_V d^3\bar{x}\, \eth(\mathbf{x} - \bar{\mathbf{x}})$ are δ-functions on the volumes V over which the surfaces have swept. The theory of these fields has been developed in the textbook [5], the Cambridge lectures [6], and in the recent monograph [7]. Being a gauge field, $\boldsymbol{\theta}^v(\mathbf{x})$ may be modified by a further gradient of a smooth function to make it purely transverse, $\boldsymbol{\nabla} \cdot \boldsymbol{\theta}^v_T(\mathbf{x}) = 0$, as indicated by the subscript T.

Since the vortex gauge field is not a gradient, it cannot be absorbed into the vector potential by a gauge transformation. Hence it survives in the first term of the functional (4), and the correct partition function is

$$Z_{\rm LP} = \int \mathcal{D}\boldsymbol{\theta}^v_T \int \mathcal{D}\theta\, \mathcal{D}\rho\, \rho \exp\left\{-\int d^3x \left[\frac{\rho^2}{2}(\boldsymbol{\nabla}\theta)^2 + \frac{1}{2}(\boldsymbol{\nabla}\rho)^2 + \frac{\tau}{2}\rho^2 + \frac{g}{4}\rho^4 + \frac{\rho^2}{2}\boldsymbol{\theta}^{v\,2}_T\right]\right\}. \tag{11}$$

The symbol $\int \mathcal{D}\boldsymbol{\theta}^v_T$ does not denote an ordinary functional integral. It is defined as a sum over any number and all shapes of Volterra surfaces S in $\boldsymbol{\theta}^v(\mathbf{x})$, across which the phase jumps by 2π

[6]. Each configuration has to be subjected to a gauge transformation to achieve the transverse form with $\boldsymbol{\nabla}\cdot\boldsymbol{\theta}^v_T(\mathbf{x})=0$.

The phase angle can again be integrated out so that (11) reduces to

$$Z_{\rm LP} = {\rm Det}^{-1/2}(-\boldsymbol{\nabla}^2)\int\mathcal{D}\boldsymbol{\theta}^v_T\int\mathcal{D}\rho\,\exp\left\{-\int d^3x\left[\frac{1}{2}(\boldsymbol{\nabla}\rho)^2+\frac{\tau}{2}\rho^2+\frac{g}{4}\rho^4+\frac{\rho^2}{2}\boldsymbol{\theta}^{v\,2}_T\right]\right\}. \quad (12)$$

This differs from the wrong partition function (6) with the Ising universality class by the last term.

The important observation is now that due to the last term, the partition function (12) possesses a phase transition for *negative* τ, i.e., when the expectation value of the field ρ is still larger than zero. It is caused by the fluctuations of the vortex gauge field in the last term of (12). Let us exhibit the vortex lines in this term. First we take account of the transversality of $\boldsymbol{\theta}^v_T$ by writing the last term as

$$\begin{aligned}\int d^3x\,\frac{\rho^2}{2}\boldsymbol{\theta}^{v\,2}_T &= \int d^3x\left[\boldsymbol{\theta}^{v2}(\mathbf{x})-\boldsymbol{\nabla}\cdot\boldsymbol{\theta}^v(\mathbf{x})\frac{1}{-\boldsymbol{\nabla}^2}\boldsymbol{\nabla}\cdot\boldsymbol{\theta}^v(\mathbf{x})\right]\\ &= \int d^3x\,[\boldsymbol{\nabla}\times\boldsymbol{\theta}^v(\mathbf{x})]\frac{1}{-\boldsymbol{\nabla}^2}[\boldsymbol{\nabla}\times\boldsymbol{\theta}^v(\mathbf{x})]=\int d^3x\,\mathbf{j}^v(\mathbf{x})\frac{1}{-\boldsymbol{\nabla}^2}\mathbf{j}^v(\mathbf{x}). \quad (13)\end{aligned}$$

This is a Biot-Savart-like energy between vortex lines described by the δ-functions in $\mathbf{j}^v(\mathbf{x})$. Being the curl of the gauge-field $\boldsymbol{\theta}^v$, these lines satisfy $\boldsymbol{\nabla}\cdot\mathbf{j}^v=0$ and are therefore closed.

The fluctuations of the vortex lines show a proliferation transition when the entropy becomes larger than the energy of the lines. This can be simply shown as follows. A long line of length l with an energy per length ϵ is suppressed strongly by a Boltzmann factor $e^{-\epsilon l/T}$. This suppression is, however, counteracted by configurational entropy. Suppose the line can bend easily on a length scale ξ which is of the order of the coherence length of the system. Hence it can occur in approximately $(2D)^{l/\xi}$ possible configurations, where D is the space dimension. A rough approximation to the partition function of a single loop of arbitrary length is given by the integral

$$Z_1\approx\oint\frac{dl}{l}\,(2D)^{l/\xi}e^{-\epsilon l/T}. \quad (14)$$

The factor $1/l$ in the integrand accounts for the cyclic invariance of the loop. By exponentiating this one-loop expression we obtain the partition function of a grand-canonical ensemble of loops of arbitrary length l, whose free energy is therefore $F=-Z_1/\beta$.

The integral (14) converges only below a critical temperature

$$T<T_c=\epsilon\xi/\log(2D). \quad (15)$$

Above T_c, the integral diverges and the ensemble undergoes a phase transition in which the loops proliferate and become infinitely long. This process will also be called *loop condensation*.

This rough argument can be made more quantitative by approximating the partition function (12) by a lattice version. For this purpose we observe that the Hamiltonian in (11) possesses classical solutions in the form of vortex lines. These have a core radius of the order of the coherence length $\xi=\sqrt{-\tau}$. Let us replace the partition function (11) by a model partition on a simple cubic lattice of distance $a=\xi$ in which the fluctuations of the field ρ are ignored, since they are irrelevant to the universality class of the phase transition. Thus we replace ρ by their mean-field value $\rho_0=\sqrt{-\tau/g}$ and approximate the last term in (12) by its lattice expression

$$\int d^3x\,\frac{\rho_0^2}{2}\boldsymbol{\theta}^{v\,2}_T\to a^3\rho_0\sum_{\mathbf{x}}\boldsymbol{\theta}^{v\,2}_T\to\rho_0^2a\,4\pi^2\sum_{\mathbf{x}}\mathbf{n}^2_T(S;\mathbf{x})=\rho_0^2a\,4\pi^2\sum_{\mathbf{x}}\mathbf{l}(L;\mathbf{x})\frac{1}{-\boldsymbol{\nabla}\overline{\boldsymbol{\nabla}}}\mathbf{l}(L;\mathbf{x}). \quad (16)$$

Journal of Physics: Conference Series **174** (2009) 012012 doi:10.1088/1742-6596/174/1/012012

Here the symbol $\boldsymbol{\nabla}$ denotes the *lattice derivative* whose components ∇_i act on an arbitrary function $f(\mathbf{x})$ as

$$\nabla_i f(\mathbf{x}) \equiv a^{-1}[f(\mathbf{x}+a\mathbf{e}_i) - f(\mathbf{x})], \tag{17}$$

where $\mathbf{e}_i$ are the unit vectors to the nearest neighbors in the plane, and a is the lattice spacing. There exists also a conjugate lattice derivative

$$\overline{\nabla}_i f(\mathbf{x}) \equiv a^{-1}[f(\mathbf{x}) - f(\mathbf{x}-a\mathbf{e}_i)]. \tag{18}$$

It arises naturally in the lattice version of partial integration

$$\sum_{\mathbf{x}} f(\mathbf{x})\nabla_i g(\mathbf{x}) = -\sum_{\mathbf{x}}[\overline{\nabla}_i f(\mathbf{x})]g(\mathbf{x}), \tag{19}$$

which holds for functions $f(\mathbf{x})$, $g(\mathbf{x})$ vanishing on the surface of the system, or satisfying periodic boundary conditions. This will always be assumed in the remainder of this discussion. In Fourier space, the eigenvalues of ∇_i, $\overline{\nabla}_i$ are $K_i = (e^{ik_i a} - 1)/a$, $\overline{K}_i = (1 - e^{-ik_i a})/a$, respectively, where k_i are the wave numbers in the i-direction.

The operator $\boldsymbol{\nabla}\overline{\boldsymbol{\nabla}} = \overline{\boldsymbol{\nabla}}\boldsymbol{\nabla}$ in (16) is the *lattice Laplacian*, the lattice version of the Laplace operator $\boldsymbol{\nabla}^2$ in the continuum. Its eigenvalues in D dimensions are [8]

$$\overline{\mathbf{K}}\mathbf{K} = \frac{2}{a^2}\sum_{i=1}^{D}(1 - \cos k_i a), \tag{20}$$

where $k_i \in (-\pi/a, \pi/a)$ are the wave numbers in the Brillouin zone of the lattice. In the continuum limit $a \to 0$, both lattice derivatives reduce to the ordinary derivative ∂_i, and $\overline{\mathbf{K}}\mathbf{K}$ goes over into $\mathbf{k}^2$.

In going from the left-hand side of (16) to the lattice version on the right-hand side we have replaced $\boldsymbol{\theta}^v$ by its lattice version

$$\boldsymbol{\theta}^v = \boldsymbol{\delta}(\mathbf{x}) = 2\pi\boldsymbol{\delta}(S;\mathbf{x}) \to 2\pi\mathbf{n}(S;\mathbf{x})/a^2, \tag{21}$$

where $\mathbf{n}(S;\mathbf{x})$ is an integer-valued gauge field on the lattice sites. Similarly we have replaced $\mathbf{j}^v(\mathbf{x})$ by its lattice version

$$\mathbf{j}^v = \boldsymbol{\delta}(L;\mathbf{x}) = 2\pi\boldsymbol{\delta}(L;\mathbf{x}) \to 2\pi\mathbf{l}(L;\mathbf{x})/a^2, \tag{22}$$

where $\mathbf{l}(L;\mathbf{x}) = \boldsymbol{\nabla}\times\mathbf{n}(S;\mathbf{x})$ is an integer-valued divergenceless field on the lattice sites which displays directly the line elements.

On a lattice, the contribution of the vortex lines to the partition function coming from the last term in (12) can be summed up in a partition function associated with the lattice energy (16)

$$Z_{\text{lines}}[\rho_0] \approx \sum_{\{\mathbf{l};\boldsymbol{\nabla}\cdot\mathbf{l}=0\}} \exp\left[-\frac{4\pi^2\rho_0^2 a}{2}\sum_{\mathbf{x}}\mathbf{l}(L;\mathbf{x})\frac{1}{-\boldsymbol{\nabla}\overline{\boldsymbol{\nabla}}}\mathbf{l}(L;\mathbf{x})\right]. \tag{23}$$

The condition $\boldsymbol{\nabla}\cdot\mathbf{l} = 0$ restricts the sum over all $\mathbf{l}(\mathbf{x})$-configurations to all non-selfbacktracking integer-valued closed loops. The Biot-Savart interaction is now given by the lattice version of the Coulomb potential

$$v_0(\mathbf{x}) \equiv \frac{1}{-\boldsymbol{\nabla}\overline{\boldsymbol{\nabla}}}(\mathbf{x}), \tag{24}$$

in terms of which (23) reads more explicitly

$$Z_{\text{lines}}[\rho_0] \approx \sum_{\{\mathbf{l};\boldsymbol{\nabla}\cdot\mathbf{l}=0\}} \exp\left[-\frac{4\pi^2\rho_0^2 a}{2}\sum_{\mathbf{x},\mathbf{x}'}\mathbf{l}(L;\mathbf{x})v_0(\mathbf{x}-\mathbf{x}')\mathbf{l}(L;\mathbf{x}')\right]. \tag{25}$$

The Coulomb potential can easily be calculated explicitly. For the sake of generality we consider also the lattice version of the Yukawa potential

$$v_m(\mathbf{x}) \equiv \frac{1}{-\boldsymbol{\nabla}\overline{\boldsymbol{\nabla}}+m^2}(\mathbf{x}). \tag{26}$$

Its values at $\mathbf{x}$ are obtained from the Fourier transform of $1/(\mathbf{K}\overline{\mathbf{K}}+m^2)$ [9].

$$v_m(\mathbf{x}) = \prod_{i=1}^{3}\int\frac{d^3(ak_i)}{(2\pi)^3}\frac{e^{i(k_1x_1+k_2x_2+k_3x_3)}}{2\sum_{i=1}^3(1-\cos ak_i)+a^2m^2} = \int_0^\infty dse^{-(6+m^2)s}I_{x_1}(2s)I_{x_2}(2s)I_{x_3}(2s). \tag{27}$$

The partition function (23) is a particular representations of the partition functions of the *Villain model* [10, 13], whose partition functions are defined by

$$Z_V = \sum_{\{n_i(\mathbf{x})\}}\left[\prod_{\mathbf{x}i}\int_{-\pi}^{\pi}d\theta(\mathbf{x})\right]e^{-\beta_V H_V}, \tag{28}$$

where H_V is

$$H_V = \frac{1}{2}\sum_{\mathbf{x}}[a\boldsymbol{\nabla}\theta(\mathbf{x}) - 2\pi\mathbf{n}(\mathbf{x})]^2 \tag{29}$$

and $\beta_V = \rho_0^2 a$.

The equality of (28) and (25) follows directly from the fact that the Villain model is a lattice version of

$$Z_V[\rho] = \int\mathcal{D}\theta\int\mathcal{D}\boldsymbol{\theta}_T^v\exp\left[-\frac{b}{2}\int d^3x\,(\boldsymbol{\nabla}\theta-\boldsymbol{\theta}_T^v)^2\right]. \tag{30}$$

After integrating out $\theta(\mathbf{x})$, this becomes

$$Z_V[\rho] = \text{Det}^{-1/2}(-\boldsymbol{\nabla}^2)\int\mathcal{D}\boldsymbol{\theta}_T^v\exp\left(-\frac{b}{2}\int d^3x\,\boldsymbol{\theta}_T^{v\,2}\right), \tag{31}$$

and we can replace $\boldsymbol{\theta}_T^{v\,2}$ by $(\boldsymbol{\nabla}\times\boldsymbol{\theta}_T^v)(-\boldsymbol{\nabla}^2)^{-1}(\boldsymbol{\nabla}\times\boldsymbol{\theta}_T^v) = \mathbf{j}^v(-\boldsymbol{\nabla}^2)^{-1}\mathbf{j}^v$. By taking this expression to a simple-cubic lattice we obtain the partition function (23).

There exists an equivalent representation for this partition function which is obtained by so-called *duality transformation* [5]:

$$Z_{\text{lines}}[\rho_0] \approx \sum_{\{\mathbf{b};\boldsymbol{\nabla}\cdot\mathbf{b}=0\}}\exp\left[-\frac{1}{2\beta_V}\sum_{\mathbf{x}}\mathbf{b}^2(L;\mathbf{x})\right]. \tag{32}$$

Now it describes a grand-canonical ensembles of non-selfbacktracking closed lines along the integer-valued vectors $\mathbf{b}(L;\mathbf{x})$ *without* Biot-Savart type interactions.

The important point is now that this model is known to have a phase transition of the XY-model universality class. This can be shown either by high- and low-temperature expansions or by Monte Carlo simulations. The transition lies at

$$\beta_V = \rho_0^2 a \approx 1/3. \tag{33}$$

The effect of the long-range part of the Biot-Savart forces seems to be very weak. This can be seen as follows. Suppose we replace the partition function (25) by another one with only contact part of the Coulomb interaction between the vortex lines

$$Z_{\rm lines}[\rho_0] \approx \sum_{\{\mathbf{l};\boldsymbol{\nabla}\cdot\mathbf{l}=0\}} \exp\left[-\frac{4\pi^2\rho_0^2 a}{2} v_0(\mathbf{0}) \sum_{\mathbf{x}} \mathbf{l}^2(L;\mathbf{x})\right] \tag{34}$$

This looks precisely line the partition function (32) which has the phase transition of the XY-class at $\rho_0^2 a \equiv 1/3$. This implies that the partition function (34) has a transition at

$$4\pi^2\rho_0^2 a\, v_0(\mathbf{0}) \approx 3. \tag{35}$$

Indeed, inserting $v_0(\mathbf{0}) = 0.252\,731\ldots$ obtained from (27) we obtain $4\pi^2\rho_0^2 a\, v_0(\mathbf{0}) \approx 3.33$ which is only wrong by 10%. Thus the replacement of $v_0(\mathbf{x}-\mathbf{x}')$ in the vortex energy of (25) by the local expression $v_0(\mathbf{0})\delta_{\mathbf{x},\mathbf{x}'}$ in the partition function (32) must be done with a renormalization factor $r \approx 1.11$:

$$v_0(\mathbf{x}-\mathbf{x}') \to v_0(\mathbf{0})\delta_{\mathbf{x},\mathbf{x}'}/r. \tag{36}$$

The phase transition of the Villain model at (33) implies that the original partition function (12) has a transition of the XY-class at roughly

$$-\tau_c \approx (g/3\sqrt{2})^{2/3}. \tag{37}$$

This follows directly by inserting into (33) $a = \xi = \sqrt{-\tau}$ and $\rho_0^2 = -\tau/g$. Thus, when heating a superfluid, the transition occurs *before* it would undergo a transition of the Ising type at $\tau = 0$. This is the reason for observing the critical exponents of the XY-model type.

The alert reader may not be convinced by this argument and object that also the fluctuations of the real ρ-field in (11) will shift the transition from the mean-field value $\tau = 0$ to a negative value. Indeed, in the lattice formulation of (11), a one-loop Feynman diagram yields a temperature shift

$$\Delta\tau = 6g\langle\rho^2\rangle = 6gv_0(\mathbf{0}) \tag{38}$$

2. Having understood the nature of the phase transition in BEC and helium, we are ready to solve an old controversy whether this transition persists if a fluctuating vector potential $\mathbf{A}(\mathbf{x})$ is coupled minimally to the field $\psi(\mathbf{x})$ in (1), which then becomes the Ginzburg-Landau Hamiltonian density of superconductivity

$$\mathcal{H}_{\rm sc}(\psi,\boldsymbol{\nabla}\psi,\mathbf{A},\boldsymbol{\nabla}\mathbf{A}) = \frac{1}{2}\Big\{[(\boldsymbol{\nabla}-iq\mathbf{A})\psi]^2+\tau|\psi|^2+\frac{g}{2}|\psi|^4\Big\}+\frac{1}{2}(\boldsymbol{\nabla}\times\mathbf{A})^2. \tag{39}$$

Now $\psi(\mathbf{x})$ is the field describing Cooper pairs of charge $q = 2e$. The theory needs gauge fixing, which may be done by absorbing the gradient of the phase $\theta(\mathbf{x})$ of the field $\psi(\mathbf{x})$ in the vector potential, so that we can replace $\psi(\mathbf{x}) \to \rho(\mathbf{x})$. The transverse vortex gauge field $\boldsymbol{\theta}_T^v(\mathbf{x})$, however, cannot be absorbed and it interacts with the vector potential $A(\mathbf{x})$. This has a partial partition function

$$Z_A[\rho] \equiv \int \mathcal{D}\boldsymbol{\theta}_T^v \mathcal{D}\mathbf{A} \exp\left\{-\frac{1}{2}\int d^3x(\boldsymbol{\nabla}\times\mathbf{A})^2 - \frac{1}{2}\int d^3x\,\rho^2(e\mathbf{A}-2\pi\boldsymbol{\theta}_T^v)^2\right\}. \tag{40}$$

Without the vortex gauge field $\boldsymbol{\theta}^v_T(\mathbf{x})$, the partition function (40) describes free vector fields of space-dependent mass $e^2\rho^2(\mathbf{x})$. If we ignore $\boldsymbol{\theta}^v_T(\mathbf{x})$ and $\mathbf{A}(\mathbf{x})$, the total partition function has the same form as in (11) and describes a second order phase transition.

Let us now admit the vector partition function (40), but still ignore vortices by setting $\boldsymbol{\theta}^v_T(\mathbf{x}) \equiv 0$, and ignoring the space dependence of $\rho(\mathbf{x})$. Then the second term in (40) in the condensed phase with $\rho_0 \neq 0$ generates a Meissner-Higgs mass term. This gives rise to a finite *penetration depth* of the magnetic field $\lambda = 1/m_A = 1/\rho_0 q$. The ratio of the two length scales $\kappa \equiv \lambda/\sqrt{2}\xi$ (which for historic reasons carries a factor $\sqrt{2}$) is the Ginzburg parameter whose mean field value is $\kappa_{\rm MF} \equiv \sqrt{g/q^2}$. Type I superconductors have small values of κ, type-II superconductors have large values. At the mean-field level, the dividing line lies at $\kappa = 1/\sqrt{2}$.

Let us now allow for $\mathbf{A}(\mathbf{x})$-fluctuations while still ignoring the vortex gauge field $\boldsymbol{\theta}^v_T(\mathbf{x})$. At very smooth $\rho(\mathbf{x})$, they can be integrated out in (40) which becomes

$$Z^0_A[\rho] = \exp\left[\int d^3x \frac{e^3\rho^3}{6\pi}\right] \tag{41}$$

This adds the energy density (4) a cubic term $-e^3\rho^3/6\pi$. Such a term makes the transition first-order. The free energy has now a minimum at

$$\rho_0 = \frac{c}{2g}\left(1+\sqrt{1-\frac{4\tau g}{c^2}}\right). \tag{42}$$

If τ decreases below

$$\tau_1 = 2c^2/9g. \tag{43}$$

the new minimum lies *lower* than the one at the origin (see Fig. 1), so that the order parameter jumps from zero to

$$\rho_1 = 2c/3g \tag{44}$$

in a phase transition. At this point, the coherence length of the ρ-fluctuations $\xi = 1/\sqrt{\tau + 3g\rho^2 - 2c\rho}$ has the finite value

$$\xi_1 = \frac{3}{c}\sqrt{\frac{g}{2}}, \tag{45}$$

this being the same as at $\rho = 0$. The jump from $\rho = 0$ to ρ_1 implies a phase transition of first-order [12].

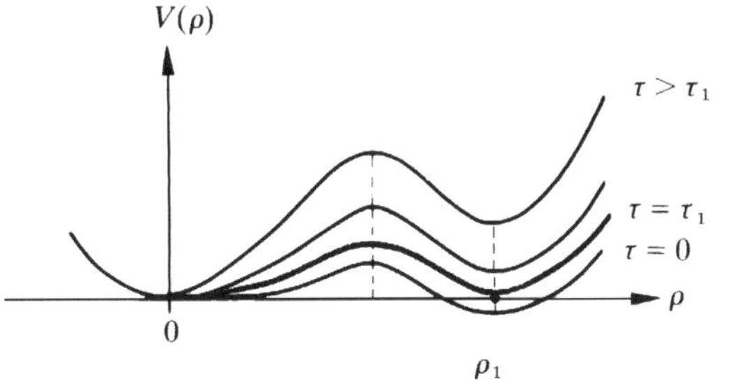

Figure 1. Potential for the order parameter ρ with cubic term. At τ_1, the order parameter jumps from $\rho = 0$ to ρ_1, corresponding to a phase transition of first-order.

3. However, this result is reliable only under the assumption of a smooth $\rho(\mathbf{x})$. It is applicable only in the type-I regime where vortex loops are strongly suppressed. In the type-II regime, such lines can more easily be excited thermally and we can no longer ignore the vortex gauge field $\boldsymbol{\theta}_T^v(\mathbf{x})$]. This invalidates the above conclusion and gives rise to a second-order transition (of the same XY-universality class as in the superfluid) if the Ginzburg parameter κ is sufficiently large. In fact, if we now integrating out the **A**-field, we obtain

$$Z_A[\rho] = \exp\left[\int d^3x \frac{e^3\rho^3}{6\pi}\right] \int \mathcal{D}\boldsymbol{\theta}_T^v \exp\left[-\frac{4\pi^2\rho^2}{2}\int d^3x \frac{1}{2}\left(\boldsymbol{\theta}_T^{v\,2} - \boldsymbol{\theta}_T^v \frac{\rho^2 q^2}{-\boldsymbol{\nabla}^2+\rho^2q^2}\boldsymbol{\theta}_T^v\right)\right],$$

rather than (41). The second integral can be simplified to

$$\frac{4\pi^2\rho^2}{2}\int d^3x \left(\boldsymbol{\theta}_T^v \frac{-\boldsymbol{\nabla}^2}{-\boldsymbol{\nabla}^2+\rho^2q^2}\boldsymbol{\theta}_T^v\right). \tag{46}$$

Integrating this by parts, and replacing $\nabla_i\boldsymbol{\theta}_T^v\,\nabla_i\boldsymbol{\theta}_T^v$ by $(\boldsymbol{\nabla}\times\boldsymbol{\theta}_T^v)^2 = \mathbf{j}^{v\,2}$, since $\boldsymbol{\nabla}\cdot\boldsymbol{\theta}_T^v = 0$, the partition function (46) takes the form

$$Z_A[\rho] \;=\; \exp\left[\int d^3x \frac{e^3\rho^3}{6\pi}\right] \int \mathcal{D}\boldsymbol{\theta}_T^v \exp\left[-\frac{4\pi^2\rho^2}{2}\int d^3x \left(\mathbf{j}^v \frac{1}{-\boldsymbol{\nabla}^2+\rho^2q^2}\mathbf{j}^v\right)\right]. \tag{47}$$

This is the partition function of a grand-canonical ensemble of closed fluctuating vortex lines L described by the δ-functions over lines in $\mathbf{j}^v(\mathbf{x})$. The interaction between them looks is similar to the Biot-Savart interaction between the vortex lines in (13), except that it has now a finite range equal to the penetration depth $\lambda = 1/\rho q$. It is well-known how to compute pair and magnetic fields of the Ginzburg-Landau theory for a single straight vortex line from the extrema of the energy density (1). In an external magnetic field, there exist triangular and various other regular arrays of vortex lattices and various phase transitions. In the core of each vortex line, the pair field ρ goes to zero over a distance ξ. If we want to sum over grand-canonical ensemble of fluctuating vortex lines of any shape in the partition function (11), the space dependence of ρ causes complications. As before, we avoid these by a lattice approximation. in which the system is placed on a simple-cubic lattice of spacing $a = \xi$, and replacing the variable $\rho(\mathbf{x})$ by a *fixed* $\rho = \rho_0$ given by Eq. (42). Thus we replace the partial partition function (47) approximately by

$$Z_2[\rho_0] \;=\; \sum_{\{\mathbf{l};\boldsymbol{\nabla}\cdot\mathbf{l}=0\}} \exp\left[-\frac{4\pi^2\rho_0^2 a}{2}\sum_{\mathbf{x}} \mathbf{l}(L;\mathbf{x})v_{\rho_0 e}(\mathbf{x}-\mathbf{x}')\mathbf{l}(L;\mathbf{x}')\right], \tag{48}$$

where $v_{\rho_0 e}(\mathbf{x}-\mathbf{x}')$ is the lattice Yukawa potential (27). As before in (23) and (25), the condition $\boldsymbol{\nabla}\cdot\mathbf{l} = 0$ restricts the sum over all $\mathbf{l}(\mathbf{x})$-configurations in (48), to all non-selfbacktracking integer-valued closed loops.

We have seen before for the longe-range Coulomb potential, that the local replacement (36) in the partition function (23) introduces only a renormalization factor $r \approx 1.11$. Thus we expect that the similar replacement for the short-range Yukawa potential

$$v_m(\mathbf{x}-\mathbf{x}') \to v_m(\mathbf{0})\delta_{\mathbf{x},\mathbf{x}'}/r_m. \tag{49}$$

to have a renormalization factor r which lies the closer to unity the larger m. For large m, r_m becomes exactly equal to unity. Thus the partition function (48) has the same phase transition as the local approximation

$$Z_2[\rho_0] \;\approx\; \sum_{\{\mathbf{l};\boldsymbol{\nabla}\cdot\mathbf{l}=0\}} \exp\left[-\frac{4\pi^2\rho_0^2 a}{2r} v_{\rho_0 e}(\mathbf{0})\sum_{\mathbf{x}} \mathbf{l}^2(L;\mathbf{x})\right]. \tag{50}$$

This partition function has an XY-class phase transition at [compare (35)].

$$4\pi^2\rho_0^2 a\, v_{\rho_0 e}(\mathbf{0}) \approx 3r_m. \tag{51}$$

It is useful to rewrite (33) as $\rho_0^2 a \approx 3.33/4\pi^2 v_0(\mathbf{0}) \approx 1/3$. Similarly we may write Eq. (51) as

$$\rho_0^2 a \frac{v_{\rho_0 e}(\mathbf{0})}{v_0(\mathbf{0})} = \frac{r_m}{3}. \tag{52}$$

The left-hand side is most easily evaluated by the so-called *hopping expansion* of (27), which is obtained from by expanding the modified Bessel functions in (27) in powers of s and doing the integral over s yielding [14]

$$v_m(\mathbf{0}) = \sum_{n=0,2,4} \frac{H_n}{(a^2m^2+6)^{n+1}}, \quad H_0 = 1, H_2 = 6, \ldots . \tag{53}$$

To lowest order, this yields the ratio $v_m(\mathbf{0})/v_0(\mathbf{0}) \equiv 1/(m^2/6+1)$. A more accurate numerical fit to the ratio $v_m(\mathbf{0})/v_0(\mathbf{0})$ which is good up to $m^2 \approx 10$ (thus comprising all interesting κ-values since m^2 is of the order of $3/\kappa^2$) is $1/(\sigma\, m^2/6+1)$ with $\sigma \approx 1.38$. Hence the transition takes place at

$$\frac{\rho_0^2 a}{(\sigma\, a^2\rho_0^2 q^2/6+1)} \approx \frac{r_m}{3} \quad \text{or} \quad \rho_0 \approx \frac{1}{\sqrt{3a}}\sqrt{\frac{r_m}{1-\sigma r_m a q^2/18}}. \tag{54}$$

The important point is now that this transition can occur only until ρ_0 reaches the value $\rho_1 = 2c/3g$ of Eq. (44). From there on, the transition will no longer be of the XY-model type but takes place discontinuously as a first-order transition.

Replacing in (54) a by $\alpha\xi_1$ of Eq. (45), and ρ_0 by ρ_1, we find the equation for the mean-field Ginzburg parameter $\kappa_{\rm MF} = \sqrt{g/q^2}$:

$$\kappa_{\rm MF}^3 + \alpha^2\sigma\frac{\kappa_{\rm MF}}{3} - \frac{\sqrt{2}\alpha}{\pi r_m} = 0. \tag{55}$$

Inserting $\sigma \approx 1.38$ and choosing $\alpha \approx r_m \approx 1$, the solution of this equation yields the tricritical value

$$\kappa_{\rm MF}^{\rm tric} \approx 0.81/\sqrt{2}. \tag{56}$$

In spite of the roughness of the approximations, this result is very close to the value $0.8/\sqrt{2}$ derived from the dual disorder field theory [11]. The approximation has three uncertainties. First, the identification of the effective lattice spacing $a = \alpha\xi$ with $\alpha \approx 1$; second the associated neglect of the $\mathbf{x}$-dependence of ρ and its fluctuations, and third the localization of the critical point of the XY-model type transition in Eq. (54).

Thus have shown the existence of a tricritical point in a superconductor directly within the fluctuating Ginzburg-Landau theory, by taking the vortex fluctuations into account. This became possible after correcting the covariant derivative (3) of $\psi = \rho e^{i\theta}$ to (8). For $\kappa > 0.81/\sqrt{2}$, vortex fluctuations give rise to an XY-model type second-order transition before the cubic term becomes relevant. This happens for $\kappa < 0.81/\sqrt{2}$ where the cubic term causes a discontinuous transition.

4. These examples show that the subtleties of functional integration over multivalued fields are crucial for understanding important physical phenomena such as phase transitions. Similar considerations are necessary in the context of elasticity theory where the energy is usually expressed in terms of the strain $u_{ij} = \partial_i u_j(\mathbf{x})$ of the displacement field $u_i(\mathbf{x})$ of the atoms from their rest position. Such a description is also false since the displacement field $u_i(\mathbf{x})$ is a multivalued field. It is defined only up to multiples of the lattice vectors. This multivaluedness must be taken into account with the help of a *defect gauge field* similar to $\boldsymbol{\theta}^v(\mathbf{x})$, one for each lattice direction. Its fluctuations give rise to the melting transition, as has been shown in the textbook [15].

5. The defects in a crystal form a Riemann-Cartan space with curvature and torsion, and it is possible to rederive all properties of the Riemann-Cartan geometry by starting from a multivalued crystal world and converting the variables to metric and affine connection. This has been shown in my Piombino lecture 2004 [16].

References

[1] Kleinert H 2006 *Path Integrals in Quantum Mechanics, Statistics, Polymer Physics, and Financial Markets* third extended edition (Singapore: World Scientific) 1–1450 (`kl/b5` where `kl` is short for `http://www.physik.fu-berlin.de/~kleinert`)

[2] Kleinert H and Schulte-Frohlinde V 2001 *Critical Phenomena in* Φ^4*-Theory* (Singapore: World Scientific) 1–487 (`http://www.physik.fu-berlin.de/~kleinert/b8`), see Table 20.1.

[3] Kleinert H 1999 *Phys. Rev.* D **60** 085001 (*Preprint* arXiv:hep-th/9812197) (`http://www.physik.fu-berlin.de/~kleinert/279`)

[4] Lipa J A, Nissen J A, Stricker D A, Swanson D R and Chui T C P 2003 *Phys. Rev.* B **68** 174518

[5] Kleinert H 1989 *Gauge Fields in Condensed Matter* Vol. I Superflow and Vortex Lines (Singapore: World Scientific) 1–742 (`kl/b1`)

[6] Kleinert H 1995 *Proceedings of NATO Advanced Study Institute on Formation and Interaction of Topological Defects* (New York: Plenum) 201–232 (`kl/227`)

[7] Kleinert H 2008 *Multivalued Fields* (Singapore: World Scientific) 1-497 (`http://www.physik.fu-berlin.de/~kleinert/b11`)

[8] See pp. 136–145 in textbook [5] (`kl/b1/gifs/v1-136s.html`)

[9] See Eq. (6.110) in Part I of Ref. [5] (`kl/b1/gifs/v1-165s.html`)

[10] Villain J 1977 *J. Phys. (Paris)* **36** 581; see also `kl/b1/gifs/v1-489s.html`

[11] Kleinert H 1982 *Lett. Nuovo Cimento* **37** 295 (`kl/108`). The tricritical value $\kappa \approx 0.8/\sqrt{2}$ derived in this paper was confirmed only recently by Monte Carlo simulations: Hove J, Mo S and Sudbø A 2002 *Phys. Rev.* B **66** 64524

[12] See Eq. (3.111) in Part II of Ref. [5] (`kl/b1/gifs/v1-337s.html`)

[13] See the figure on p. 501 of Ref. [5] (`kl/b1/gifs/v1-501s.html`)

[14] See Eq. (6.122) in Part I of Ref. [5] (`kl/b1/gifs/v1-168s.html`)

[15] Kleinert H 1989 *Gauge Fields in Condensed Matter* Vol. II Stresses and Defects (Singapore: World Scientific) 743-1443 (`kl/b1`)

[16] Kleinert H 2005 *Braz. J. Phys.* **35** 359 – *Proceedings of the conference DICE 2004, Piombino, 1-4 Sept. 2004* ed H-T Elze (`kl/359/braz.pdf`).

The interpretation of quantum mechanics: where do we stand?

GianCarlo Ghirardi
Department of Theoretical Physics, University of Trieste,
The Abdus Salam International Center for Theoretical Physics
and
The Istituto Nazionale di Fisica Nucleare, Trieste, Italy
E-mail: ghirardi@ts.infn.it

Abstract. We reconsider some important foundational problems of quantum mechanics. After reviewing the measurement problem and discussing its unavoidability, we analyze some proposals to overcome it. This analysis leads us to reconsider the current debate on our best theory, i.e. quantum mechanics itself. We stress that, after the remarkable interest and the many efforts which have lead, in the last years of the past century, to a revival of the subject, and, more important, to new interesting results, we are now witnessing a re-emergence of the vague and unprofessional positions which have characterized the debate in the second quarter of the XXth century. In particular we consider as extremely serious the fact that a completely mistaken position concerning the real meaning of Bell's theorem seems to have been taken by many scientists in the field.

1. Introduction

As well known, almost one century after its formulation and in spite of its unprecedented predictive successes in accounting for physical processes, quantum mechanics is still at the centre of a lively debate. Everybody knows very well the disputes between some of the great founding fathers of the theory, from N Bohr to A Einstein to E Schrödinger to W Heisenberg, which have marked the twenties and the thirties of the past century. The debate has been less heated in the subsequent years, perhaps because, as stated by M Gell-Mann in his Nobel acceptance speech in 1976:

> *Niels Bohr brainwashed a whole generation of physicists into believing that the problem (of the interpretation of the theory) had been solved fifty years ago.*

Actually I remember well that, at the beginning of the sixties when I started my scientific carreer, to work on foundational issues was considered by a great part of the scientific community a loss of time, a choice to pay more attention to (irrelevant) philosophical issues than to precise scientific problems. Luckily enough, the sixties were also the years in which another deep thinker, John S Bell, by deriving the celebrated inequality that bears his name [1], has given a tremendous imput to our understanding of reality by making clear that nonlocal features characterize most natural processes. However, this great scientist did not limit his considerations to the peculiar aspects of the theory stemming from one of its most surprising features, entanglement, but has devoted many important papers to give voice to his unsatisfaction with the conceptual and logical status of the theory, in particular with the so-called measurement or macro-objectification problem. His immense prestige has pushed many physicists to reconsider such a problem and during the fall of the past century there has been a re-flourishing of foundational investigations.

In this paper we will first of all concentrate our attention on the macro-objectification problem and on some recent proposals to overcome it, and then we will briefly review and analyze the positions which characterize the present debate on this problem.

2. The macro-objectification problem

The problem we are interested in is usually presented by making reference to the so-called von Neumann ideal measurement scheme. Synthetically it goes as follows.

Suppose one is interested in *measuring* a physical observable F of a microscopic system S. Let us denote as $\mathcal{F}$ the (self-adjoint) operator associated to it, as f_i its eigenvalues and as $|\phi_i\rangle$ the corresponding eigenvectors. Obviously, since S is microscopic we have not direct access to it and therefore, to actually perform the measurement, we assume, with von Neumann, that we have at our disposal a macroscopic quantum system in a "ready state" $|A_R\rangle$ which we put into interaction with S and that the system apparatus interaction which lasts for a certain time interval is such that the initial state $|\phi_i\rangle \otimes |A_R\rangle$ evolves into the final state $|\phi_i\rangle \otimes |A_i\rangle$. Here the final states of the apparatus $|A_i\rangle$ are orthogonal and macroscopically different. In this way, by looking at the final state of the apparatus we can deduce in which of the eigenstates of $\mathcal{F}$ the microsystem was before the measurement.

Now comes the problem: quantum mechanics is a linear theory, implying that if various states are possible states for a system S then also any normalized linear combination of them is a possible state for it. Moreover, the evolution law is linear, so that a linear superposition of states evolves into the same linear superposition of the evolved of such states. In formulae and with reference to our example:

$$|\phi_i\rangle \otimes |A_R\rangle \rightarrow |\phi_i\rangle \otimes |A_i\rangle \Rightarrow \sum_i c_i|\phi_i\rangle \otimes |A_R\rangle \rightarrow \sum_i c_i|\phi_i\rangle \otimes |A_i\rangle, \tag{1}$$

where the simple arrow denotes the time evolution and the double one denotes the implication.

As Eq. (1) clearly shows, in the final state a superposition of macroscopically different states appears. Now, what meaning whatsoever can we attribute to a state which is a superposition (and not a statistical mixture) of such states? Actually the states $|A_i\rangle$ of the apparatus might correspond to different locations of a macroscopic pointer. How can then a macroscopic object be in a state which does not correspond to a macroscopically definite position?

We all know the answer that the so-called ortodox interpretation of quantum mechanics gives to our question: wave packet reduction takes place during the process and the final state is not the one appearing at the end of Eq.(1), but, with a probability given by $|c_j|^2$, the state $|\phi_j\rangle \otimes |A_j\rangle$ of the superposition.

This assumption is fundamentally inconsistent since it contemplates two contradictory evolution laws for physical systems: one, which is linear and deterministic, governing the evolution of microsystems, and one, which is nonlinear and stochastic, holding when macroscopic systems enter into play. Why should macrosystems not to be governed by quantum mechanics? Are they not built up of elementary constituents (electrons, protons, neutrons) which are subjected to the laws of quantum mechanics? But this is not the whole story. In fact, on the one hand, we know very well that there are macroscopic systems whose properties can be understood only in terms of quantum theory, while, on the other, even if we would be inclined to accept two different evolution principles, we cannot avoid to recognize that there is nothing, absolutely nothing, in the theory which allows us to locate the split identifying which systems obey the linear equation and which ones do not. As lucidly summarized by Bell [2]:

There is a fundamental ambiguity in quantum mechanics, in that nobody knows exactly what it says about any particular situation, for nobody knows exactly where the boundary between the wavy quantum world and the world of particular events is located.

To conclude this section we remark that the above ideal scheme of measurement (such measurement processes are usually denoted as 'of the first kind') is an extremely idealized one: in practice, the final apparatus states are not strictly orthogonal, we cannot control all degrees of freedom characterizing them, a macroscopic object cannot be isolated from the environment, no apparatus is exempt from malfunctioning, etc. Some scientists maintain that it is just resorting to such an idealized description as the one of Eq.(1) which leads to the inconsistencies we have just discussed. As a paradigmatic example we will mention a sentence [3] by H. Primas:

If you really want to discuss measurements of the first kind you may try your luck. But remember that measurements of the first kind are unrealistic and completely irrelevant for experimental science. Replace in all books and philosophical treatises on quantum mechanics the word 'measurement' by the proper expression 'measurement of the first kind' and add a footnote: measurements of the first kind are idealizations which never play any role in experimental sciences.

Our next step is to prove that this position is plainly wrong and that the macro-objectification problem cannot be avoided. To this purpose we will briefly recall the argument of a recent paper [4] in which the problem has been tackled in its full generality.

3. A completely general measurement scheme

To exhibit an explicit proof of the last statement of the previous section, let us consider a microsystem which we assume can be prepared in two orthogonal states $|u\rangle$ and $|d\rangle$ (typically we can think of the two spin states of a spin 1/2 particle corresponding to a definite value of σ_z), and in their equal amplitudes superposition $|u+d\rangle = \frac{1}{\sqrt{2}}[|u\rangle + |d\rangle]$. We consider also a statistical ensemble $\mathcal{E}\{|A_R, \alpha\rangle, p(\alpha)\}$ of apparata, devised to measure precisely the observable σ_z. We have resorted to an ensemble of apparata just because we cannot have full control of all their degrees of freedom. The state $|A_R, \alpha\rangle$ corresponds to the 'ready state' of the apparatus, let us say one for which the macroscopic pointer of the apparatus 'points, at 0', while the index α distinguishes situations differing for uncontrollable degrees of freedom such as the precise state of the molecules of the pointer, and so on[1]. The quantities $p(\alpha)$ are the weights of the various subensembles of the statistical ensemble characterized by the macroproperty A_R. Completely in general, when we will resort to expressions of the type $|X, \mu\rangle$ we intend to indicate a situation of the macro-apparatus corresponding to what we perceive[2] as 'the pointer point at X' and all other relevant variables are summarized by the symbol μ. Let us denote as $\mathcal{X}$ the set of all such states for all possible values of μ. We will not require orthogonality of states corresponding to macroscopically different situations, but, obviously, we have to impose that two such states are, to a large extent, "distinguishable", a fact that we will characterize by imposing:

$$inf_{\mu\in\mathcal{X},\nu\in\mathcal{Y}} \parallel |X,\mu\rangle - |Y,\nu\rangle \parallel \geq \sqrt{2} - \eta, \quad \eta \ll 1. \tag{2}$$

Let us go on with our assumptions. We put the system in one of the two states $|u\rangle$ or $|d\rangle$ into interaction with the specific apparatus in the state $|A_R, \alpha\rangle$, and we assume that the quantum

[1] Note that we can include in α also degrees of freedom which are external to the apparatus, typically, those referring to the environment.

[2] It goes without saying that all statements concerning the position of the pointer do not make reference to a situation in which the position of the pointer is a geometrical point, but to one in which it lays in a quite narrow interval on a graduated scale.

evolution during the measurement is governed by a unitary operator $U(t_i, t_f)$. Let us denote as $|F, u, \alpha\rangle$ and $|F, d, \alpha\rangle$ the final states obtained when we trigger the apparatus in the state $|A_R, \alpha\rangle$ by the microstates $|u\rangle$ and $|d\rangle$, respectively:

$$U(t_i, t_f)[|u\rangle \otimes |A_R, \alpha\rangle] = |F, u, \alpha\rangle; \quad U(t_i, t_f)[|d\rangle \otimes |A_R, \alpha\rangle] = |F, d, \alpha\rangle. \tag{3}$$

Let us now consider some physically meaningful sets:

$$J_{\mathcal{U}}^{-} = \{\alpha : |F, u, \alpha\rangle \notin \mathcal{U}\}; \quad J_{\mathcal{D}}^{-} = \{\alpha : |F, d, \alpha\rangle \notin \mathcal{D}\}, \tag{4}$$

and their complements $J_{\mathcal{U}}^{+} = \mathcal{C}J_{\mathcal{U}}^{-}$, $J_{\mathcal{D}}^{+} = \mathcal{C}J_{\mathcal{D}}^{-}$. The physical meaning of these sets should be clear. For instance $J_{\mathcal{U}}^{-}$ represents the subset of those apparata (actually the set of the α's) for which, in spite of having being triggered by the microstate $|u\rangle$, their final statevector does not belong to the set $\mathcal{U}$, i.e. the one for which the pointer 'points at U'. For α in such a set the apparata did not register correctly the fact that the system we are interested in is in the state $|u\rangle$. The meaning of the other three sets, is obvious. We can now formulate our final assumption. By taking advantage of the fact that the function $p(\alpha)$ is a probability measure on the set on which the index α runs, we can define a measure by the formula:

$$\mu\{\mathcal{X}\} = \int_{\alpha \in \mathcal{X}} p(\alpha) d\alpha. \tag{5}$$

We now make a natural reliability assumption concerning our ensemble of apparatuses:

$$\mu\{J_{\mathcal{U}}^{-}\} < \epsilon, \quad \mu\{J_{\mathcal{D}}^{-}\} < \epsilon \quad i.e. \quad \mu\{J_{\mathcal{U}}^{+}\} > 1 - \epsilon, \quad \mu\{J_{\mathcal{D}}^{+}\} > 1 - \epsilon. \tag{6}$$

Note that these relations imply that the set of all α's for which the apparatuses give the right answer both when triggered by $|u\rangle$ or by $|d\rangle$ has a measure:

$$\mu\{J_{\mathcal{U}}^{+} \cap J_{\mathcal{D}}^{+}\} > 1 - 2\epsilon, \quad \epsilon \ll 1. \tag{7}$$

which is near to 1.

Having made precise our assumptions and notation we can now proceed as follows. Let us trigger those apparatuses for which α belongs to $J_{\mathcal{U}}^{+} \cap J_{\mathcal{D}}^{+}$ (i.e. those which register correctly both the state $|u\rangle$ as well as the state $|d\rangle$) by the state $|u + d\rangle$. The linear nature of quantum mechanics implies, from Eq. (3):

$$U(t_i, t_f)\frac{1}{\sqrt{2}}[|u + d\rangle \otimes |A_R, \alpha\rangle \equiv \frac{1}{\sqrt{2}}]U(t_i, t_f)[|u\rangle + |d\rangle] \otimes |A_R, \alpha\rangle = \frac{1}{\sqrt{2}}[|F, u, \alpha\rangle + |F, d, \alpha\rangle]. \tag{8}$$

Let us now raise the question: does the final state of Eq.(8) belong to the set $\mathcal{U}$ of states? The answer is no, because its distance from the state $|F, d, \alpha\rangle$ is:

$$\| \, |F, u + d, \alpha\rangle - |F, d, \alpha\rangle \, \| = \| \, \frac{1}{\sqrt{2}}|F, u, \alpha\rangle + (1 - \frac{1}{\sqrt{2}})|F, d, \alpha\rangle \, \| \leq \frac{1}{\sqrt{2}} + 1 - \frac{1}{\sqrt{2}} = 1, \tag{9}$$

while, according to Eq.(2) a state belonging to $\mathcal{U}$ must have a distance almost equal to $\sqrt{2}$ from any state belonging to the set $\mathcal{D}$ of states corresponding to a macroscopically and perceptively different situation from the one which is, by assumption, associated to the perception 'the pointer points at U'. The same argument can be followed to prove that the final state cannot belong to $\mathcal{D}$ or to any other set of states corresponding to a definite position of the pointer.

The conclusion should be obvious. Having assumed that we have an ensemble $\mathcal{E}\{|A_R, \alpha\rangle, p(\alpha)\}$ of macro-apparata which allows us reliably (but not perfectly) to know whether the microstate triggering them is either $|u\rangle$ or $|d\rangle$, then the large majority of such apparata (i.e. all those characterized by an α belonging to the set $J^+_{\mathcal{U}} \cap J^+_{\mathcal{D}}$, whose measure is near to 1), when triggered by the superposition $|u+d\rangle$ ends up in a state which cannot correspond to any definite position-perception concerning the macroscopic pointer. The macro-objectification problem cannot be overcome invoking measurement processes more realistic than the von Neumann one: it is the very possibility of measuring an observable with an acceptable level of reliability and the assumption of the unrestricted validity of quantum mechanics which imply that the same micro-macro interaction leads almost always to a perceptually nondefinite situation when the triggering state is the superposition of different eigenstates of the observed quantity.

4. Decoherence

A quite popular proposal which has been put forward to overcome the macro-objectification problem is based on the consideration of the unavoidable and uncontrollable interaction of any macroscopic body with its environment. A sketchy way of summarizing this proposal is the following. Let us consider a state which is a linear superposition of two macroscopically different states of a macro-object S:

$$|\Psi^{(sys)}\rangle = \alpha|M\rangle + \beta|N\rangle. \tag{10}$$

One then takes into account the above mentioned unavoidable interactions of S with the environment, which we suppose to be described by the state $|E_0\rangle$ at the initial time. Such interactions and the subsequent evolution lead to a state of the system+environment which we will express as:

$$|\Psi^{(sys+env)}\rangle = \alpha|M\rangle \otimes |E_M\rangle + \beta|N\rangle \otimes |E_N\rangle. \tag{11}$$

Now we have to take into account two important facts. The states $|E_M\rangle$ and $|E_N\rangle$ appearing in the above equation turn out to be practically orthogonal and they are essentially out of the control of the experimenter. The second fact is obvious, the first is made understandable by taking into account that, e.g., the interaction of a macrosystem which is, let us say, located in a certain position (when the state is $|M\rangle$) interacting with a single molecule of the environment can easily throw it into an excited state which is orthogonal to the state of the same molecule when it has not interacted with the macrosystem since it is differently located (as, we suppose, is the case for the state $|N\rangle$). But this is not the whole story; even if for the two macroscopically different states of the macrosystem no one of the microsystems of the environment is in one of a pair of orthogonal states, for sure many microsystems will be in different states in the two cases. Such states, being different, have a scalar product which is smaller than 1. If one then takes into account that we can easily have, let us say, an Avogadro number of such states (corresponding to the many particles which have interacted with the macrosystem), one easily realizes that the two states $|E_M\rangle$ and $|E_N\rangle$ are, *de facto*, orthogonal.

As already stated, however, the experimenter has no control over the states of the particles of the environment: he is only interested in the physics of the macrosystem he is studying. Now, quantum mechanics teaches us that to describe the properties of a subsystem of a composite system, the most efficient and formally direct way is the one of resorting to the statistical operator formalism and to take the partial trace on the degrees of freedom we are disregarding. Let me summarize the steps which are involved using both the language of the statevectors and of the statistical operators:

$$\begin{aligned} |\Psi^{(sys)}\rangle &\rightarrow |\Psi^{(sys+env)}\rangle \\ \rho^{(sys)} = |\Psi^{(sys)}\rangle\langle\Psi^{(sys)}| &\rightarrow \rho^{(sys+env)} = |\Psi^{(sys+env)}\rangle\langle\Psi^{(sys+env)}| \end{aligned} \tag{12}$$

We recall now the explicit form of $\rho^{(sys+env)}$ following from Eq.(11):

$$\begin{aligned}\rho^{(sys+env)} = |\alpha|^2|M\rangle\otimes|E_M\rangle\langle E_M|\otimes\langle M| \;&+\; |\beta|^2|N\rangle\otimes|E_N\rangle\langle E_N|\otimes\langle N|+ \\ +\alpha\beta^*|M\rangle\otimes|E_M\rangle\langle E_N|\otimes\langle N| \;&+\; \alpha^*\beta|N\rangle\otimes|E_N\rangle\langle E_M|\otimes\langle M|.\end{aligned} \tag{13}$$

If we now evaluate the reduced statistical operator, taking into account the orthogonality of the states of the environment we have:

$$\rho^{(Red,sys)} \equiv Tr^{(env)}\rho^{(sys+env)} = |\alpha|^2|M\rangle\otimes|E_M\rangle\langle E_M|\otimes\langle M| + |\beta|^2|N\rangle\otimes|E_N\rangle\langle E_N|\otimes\langle N|. \tag{14}$$

Such a statistical operator is the one associated to the statistical mixture of the states $|M\rangle$ and $|N\rangle$ with statistical weights $|\alpha|^2$ and $|\beta|^2$, a fact which would render legitimate the replacement of the original pure state of the system with the corresponding mixture:

$$|\Psi^{(sys)}\rangle = \alpha|M\rangle + \beta|N\rangle \Rightarrow \mathcal{E}\{|M\rangle,|N\rangle;|\alpha|^2,|\beta|^2\}, \tag{15}$$

with obvious meaning of the symbols.

The argument seems appealing and actually puts into evidence one relevant fact, i.e., that to experimentally distinguish the pure state from the statistical mixture is an extremely difficult task. But if one looks at it from the conceptual point of view one immediately understands that it is seriously misleading for at least two reasons:

- In the previous section we have proved that actually linear superpositions of macroscopically different states occur and that their occurrence leads to a deadlock. To replace them with a statistical mixture is a trick which can have only a FAPP (for all practical purposes) validity - to use a specification due to Bell [5].
- There is also a much more serious objection to the now described game. Most of the proponents of this solution to the puzzling aspects of linear superpositions of macroscopically distinguishable states seem to ignore a fundamental fact, i.e. that, within quantum mechanics, the correspondence between statistical ensembles and statistical operators is infinitely many to one. Thus, even if one would accept that the statistical operator which must be used is the one of Eq.(14), one has no reason to interpret it as describing the statistical ensemble at the r.h.s. of Eq.(15). In fact, just to mention a trivial example, one immediately realizes that, e.g., also the following statistical mixture of three pure quantum states:

$$\mathcal{E}\{\frac{1}{\sqrt{2}}[|M\rangle+|N\rangle],\frac{1}{\sqrt{2}}[|M\rangle-|N\rangle],|M\rangle;|\beta|^2,|\beta|^2,(|\alpha|^2-|\beta|^2)\}, \tag{16}$$

corresponds precisely to the same statistical operator of Eq.(14), but it still contains embarassing superpositions of macroscopically distinguishable states.

Actually, the decoherence approach to the measurement problem has been repeatedly criticized (see, e.g. [6]) and, moreover, the most serious proponents of this solution have plainly recognized that the situation which we have outlined is puzzling, and have proposed (vague) ways out from the puzzle. The most paradigmatic example can be found in an important paper [7] by Joos and Zeh, which clearly admidts the arbitrariness of choosing one statistical mixture (the most reasonable from a perceptual point of view and the one corresponding to wave packet reduction), among the infinitely many which are possible:

- *The local description is assumed and the specific choice of a basis can perhaps be justified by a fundamental underivable assumption about the local nature of the observer ... and his way of perceiving,*

- *No unitary treatment of the time dependence can explain why only one of these dynamically independent components is experienced,*
- *The difficulty in giving a complete derivation of classical concepts may as well signal the need of entirely novel concepts.*

As a final remark we would like to stress that the most recent attempts to achieve important technological improvements based on quantum mechanics - such as Quantum Cryptography, Quantum Teleportation and Quantum Computation - deal unavoidably with individual physical systems and they make a systematic use of wave packet reduction at the individual level as an important resource. Unless one ignores these fundamental aspects of modern research one has to face the necessity of breaking the linear nature of quantum mechanics at an appropriate stage.

5. The primitive ontology of a theory

Many scientists maintain that the purely technical, formal and logical aspects of a theory represent all what deserves attention. We share with J S Bell and many others the opinion that further requirements must be imposed to any theoretical scheme to be considered as a fundamental account of natural processes. We do not want to spend many words on this point; for a deep analysis we refer the reader to a recent lucid paper [8] in which the demand for a richer elaboration of the meaning of the formal scheme one is considering has been put forward. The authors of this paper have stressed the necessity of equipping any theory with what they call 'the Primitive Ontology' (PO) of the formalism. In brief, the PO consists in the clear and precise specification of what the theory is fundamentally about.

As already stated we will go quickly through this point, limiting ourselves to illustrate it with reference to two theories, two proposals for the solution of the macro-objectification problem which, at least at the nonrelativistic level, are fully consistent. We have in mind Bohmian Mechanics and the Dynamical Reduction Models, in particular the so-called GRW theory. Their different status deserves to be stressed. In fact while Bohmian Mechanics is, by assumption, predictively identical to quantum mechanics the GRW theory represents a modification of the standard theory. They correspond to the two alternatives which Bell [9] has indicated as the only two possible ways out of the macro-objectification problem:

> *Either the wavefunction, as given by the Schrödinger equation, is not everything, or it is not right.*

Let us come to analyze these theories and to discuss their primitive ontologies.

5.1. Bohmian Mechanics

The formal aspects of the proposal put forward in his celebrated papers [10] by D Bohm in 1952 can be summarized as follows:

- The quantum description of the state of a physical system is incomplete; for a complete specification one has to add to the wavefunction also the positions of all particles which constitute the system under examination (the positions are the hidden variables of the theory because we are able to prepare any desired wavefunction, but we have no way to control at which point of the support of the wavefunction the particle actually is - and the theory assumes that it has a precise position),
- The initial state of a system is fully specified by the assignment of the initial wavefunction in configuration space $\Psi(q_1, q_2, ..., q_n, 0) \equiv \Psi(q_i, 0)$, and the initial positions $Q_i(0), (i = 1, 2, ..., n)$ actually occupied by the constituents particles,
- There are two linked evolution equations in the theory. The wavefunction evolves according to the standard quantum equation. Then one defines, in terms of the wave function, a

velocity field $v_{B,k}(Q_j,t)$ for each particle and considers a first order (in time) differential equation for the positions themselves. In formulae:

$$\begin{aligned} i\hbar\frac{\partial\Psi(Q_i,t)}{\partial t} &= H\Psi(Q_i,t); \;\; \frac{dQ_k(t)}{dt} = v_{B,k}(Q_j,t); \\ v_{B,k}(Q_j,t) &= \frac{\hbar}{m_k}\frac{Im[\Psi^*(Q_j,t)\nabla_k\Psi(Q_j,t)]}{|\Psi(Q_j,t)|^2}. \end{aligned} \tag{17}$$

The theory exhibits some nice features. First of all, in virtue of the continuity equation satisfied by Schrödinger's wavefunction one immediately proves that if one defines the position density distribution $\rho(Q_k,t) = |\Psi(Q_k,t)|^2$ one has the so called equivariance property:

$$\rho(Q_k,0) = |\Psi(Q_k,0)|^2 \Rightarrow \rho(Q_k,t) = |\Psi(Q_k,t)|^2. \tag{18}$$

This equation tells us that if one has an ensemble of particles whose initial positions are distributed according to the modulus square of the initial wavefunction, then, by letting the particles evolve along the uniquely defined trajectories ensuing from the differential equations for the positions, one recovers at any time the position density distributions predicted by quantum mechanics.

The theory is purposedly equivalent to standard quantum mechanics but it can be shown to imply that macroscopic objects end up, with the correct probabilities, in those positions that are assigned to them by the standard theory enriched with the reduction postulate. In short the theory solves the measurement problem in a perfectly consistent way.

The primitive ontology of the theory emerges clearly from the given picture: all particles of the Universe have at all times perfectly defined positions, they move along precise trajectories in such a way to reproduce the position density distribution of standard quantum mechanics. In the case of micro-macro interactions, macroscopic systems end up, in turn, in the precise positions predicted by quantum mechanics plus the wave packet reduction postulate. Obviously, one has to assume that all measurements reduce essentially to position measurements, just as it happens when we consider a system with a macroscopic pointer, from the position of which we can infer the outcome of the measurement. What the theory is about are the positions and they match perfectly our perceptions concerning macroscopic systems. Using Bell's terminology, the positions are *the beables* of the theory, and everything is built up from them.

5.2. The dynamical reduction program

This attempt to account for all natural processes in terms of a unique dynamical principle is based on two important remarks: first of all the characteristic trait which makes the standard evolution and wave packet reduction radically different is that while the first is linear and deterministic, the second one is nonlinear and stochastic. Secondly, if one wants to have a mechanism leading to the macro-objectification of some variable, one has to make a precise choice about it since different quantities are associated to noncommuting operators and one cannot ask that incompatible observables become simultaneously objective. The most natural candidate, if one takes into account our definite perceptions, are the positions, in accordance with the remark by Einstein [11]:

> *A macro-body must always have a quasi-sharply defined position in the objective description of reality.*

Accordingly, in the first model of this kind which has been proposed [12] (see also the comprehensive review[13]) one adds to Schrödinger's equation nonlinear and stochastic terms affecting all elementary constituents of any system and involving a universal localization mechanism. Formally one assumes that, at random times, with an appropriate frequency $\lambda \approx 10^{-16} sec^{-1}$, the wavefunction suffers a process corresponding to a localization of one of the constituents. When this process involves the $i-th$ particle the wavefunction changes according to:

$$\begin{aligned} \Psi(q_1, q_2, ..., q_n) &\Rightarrow \frac{\Lambda_i(x)\Psi(q_1, q_2, ..., q_n)}{\| \Lambda_i(x)\Psi(q_1, q_2, ..., q_n) \|} \\ \Lambda_i(x) &= \frac{1}{(2\pi\sigma^2)^{3/2}} \cdot e^{-\frac{(q_i - x)^2}{2\sigma^2}}; \quad \sigma \approx 10^{-5} cm. \end{aligned} \tag{19}$$

A last specification concerns the positions at which collapses occur. It is embodied in the assumption that the position x is chosen at random with a probability distribution:

$$dP(x \in [x, x+dx] | \Psi(t), i) = \| \Lambda_i(x)\Psi(q_1, q_2, ..., q_n) \|^2 \, dx. \tag{20}$$

The model exhibits some physically interesting features:

- The spatial probability density of the localizations, according to Eq.(20), practically coincides with the one which the standard theory attributes to position measurements,
- It is immediately proved, by passing to the centre-of-mass and relative coordinates, that the theory entails the so called *Trigger Mechanism*: the localizations of the c.o.m. are amplified with the number of particles, actually of nucleons (because the frequency λ is assumed to be proportional to the mass of the constituents), the value given above referring to nucleons,
- According to this prescription a nucleon (or an atom) suffers a localization about every 10^8 years, and, consequently, microscopic systems are practically unaffected, while a macroscopic object, containing an Avogadro's number of particles, suffers a localization about every 10^{-7}sec.,
- The statistical operator obeys an equation of the Quantum Dynamical Semigroup type,
- The theory is, in principle, testable against quantum mechanics.

The conclusion should be obvious: the proposed universal dynamics leaves all quantum predictions for microsystems unaltered but it accounts for wave packet reduction with probabilities in agreement with the quantum ones and for the classical behaviour of macroscopic systems, as well as for our definite perceptions with them.

In the simplified version we have presented the theory suffers of a serious limitation: the localizations break the symmetry character of the wavefunction for identical particles. This feature is easily corrected as firstly shown in[14]. Before coming to consider the ontology of the model we mention that much more formally elegant (even though physically equivalent) variants of the theory have been presented [14, 15]. They are based on the formalism of stochastic differential equations of the Ito or Stratonovich type.

Let us now come to discuss the PO of the theory. A first proposal has been put forward by Bell himself in his presentation of the GRW theory at the Imperial College meeting celebrating the centenary of Schrödinger's [9]:

> *There is nothing in the theory but the wavefunction. ... However, the GRW jumps (which are part of the wavefunction, not something else) are well localized in ordinary space. Indeed each is centred on a particular spacetime point* ($\mathbf{x}$,t)*. So we can propose these events as the basis of the 'local beables' of the theory. These are the mathematical counterparts in the theory to real events at definite times and places in the real world. ... A piece of matter then is a galaxy of such events.*

This assumption that 'what the theory is fundamentally about' are the localizations themselves has been characterized as 'The flashes ontology' in refs.[8] and [16, 17, 18]. An alternative ontology has been proposed in [19] and has been subsequently characterized as 'the mass density ontology'. This assumes that what the theory is about, what is real 'out there', is the mass density in 3-dimensional space:

$$m(\mathbf{x},t) = \sum_{i=1}^{N} \int_{R^{3N}} d\mathbf{r}_1, ..., d\mathbf{r}_N \delta(\mathbf{r}_i - \mathbf{x}) |\Psi(\mathbf{r}_1, ..., \mathbf{r}_N, t)|^2. \tag{21}$$

This concludes our description of two ways out from the macro-objectification problem. As already mentioned there have been many other interesting attempts aiming at the same result. We have concentrated our attention on these two because they are precise models which, at the nonrelativistic level, account on the basis of a unique universal dynamics both for the behavior of microscopic systems, as well as for the reduction process which takes place during measurement procedures. Bell has qualified them [5] as 'exact', a term by which he intended *which are precisely formulated and which neither require nor are embarrassed by a conscious observer.* Bohmian Mechanics has been discussed many times and there are also entire books devoted to it. Concerning the GRW theory there is an extensive literature (which can be found, e.g., in [13]) in which many physically interesting aspects of the theory are analyzed.

6. The recent debate

In what follows we will be specifically interested in reviewing the new trends in the foundational investigations about quantum mechanics. In spite of many new proposals, we witness a revival of the old positions taken by the Copenhagen school in the thirties. In particular some scientists have expressed various criticisms concerning the two approaches we have just discussed and their consideration will allow us also to call attention to the new trends in the field.

6.1. The case of Bohm theory

A quite widespread critical attitude towards Bohmian Mechanics makes reference to the fact that since this theory is empirically indistinguishable from the standard theory, it should be considered an example of 'bad science', a 'degenerate research program' in the sense of I Lakatos. To stress this position I cannot do better than quoting a sentence from a letter written in 1966 by S Weinberg to S Goldstein which can be found in ref.[20]:

> *At the regular weekly luncheon meeting today ot our Theory Group, I asked my colleagues what they think of Bohm's version of quantum mechanics. The answers were pretty uniform and much what I would have said myself. First, as we understand it, Bohm's quantum mechanics uses the same formalism as ordinary quantum mechanics, including a wavefunction that satisfies the Schrödinger equation, but adds an extra element, the particle trajectory. The predictions of the theory are the same as for ordinary quantum mechanics, so, there seems little point in the extra complication, except to satisfy some a priori ideas about what a physical theory should be like... In any case, the basic reason for not paying attention to the Bohm approach is not some sort of ideological rigidity, but much simpler – it is just that we are all too busy with our own work to spend time on something that doesn' t seem likely to help us make progress with our real problems.*

In our opinion it is remarkable that various (presumably) high level scientists and even a giant like S Weinberg do not consider important at least to mention the puzzling aspects of the standard interpretation of the theory, the aspects for which Bohm has been led to work out his proposal. Without paying the due attention to this fundamental point one cannot appreciate the great merit of the approach, i.e., the fact that it consistently solves the measurement problem and leads to the classical behaviour of macroscopic objects, a feature which is by no means trivial.

At any rate, if the predictive equivalence of the two theories is considered as a drawback, people interested in foundational issues should not ignore that the other 'exact' theory we have presented qualifies itself as a rival theory of quantum mechanics which, in principle, can be tested against it. So, let us pass to analyze some of the critical remarks adressing the dynamical reduction program.

6.2. The dynamical reduction models

As already remarked, it is often claimed that one can very well adopt the orthodox interpretation of quantum theory provided one takes into account the decoherence due to the environment. Typically, J Bub, in describing what he calls 'the new orthodoxy', claims [21] that the superpositions of macroscopically different states actually occur but we do not see them due to the coupling with the environment. We have already discussed this aspect and we believe to have proved that, even though one might adopt it FAPP, one cannot avoid facing it when one is interested in the internal consistency of the theory. Moreover, invoking decoherence leads to consider practically only the statistical aspects of the theory: in a sense one claims that, when dealing with ensembles, one can use the reduced statistical operator obtained by disregarding the environmental degrees of freedom. But, as we have already remarked, modern technology deals more and more with individual physical systems and only if one resorts to individual reduction processes one can go on. We have the impression that the attitude of many of the adherents to the 'decoherence' approach make an a priori and vague choice aimed to render nonrelativistic quantum mechanics a theory which, from an experimental point of view, cannot be caught in telling a lie[3].

Such, in our opinion, not well pondered attitude emerges also from a formal aspect to which these scientists make often reference: the so-called 'ancilla argument'. It is well known that any quantum evolution equation of the quantum dynamical semigroup type for the statistical operator is physically equivalent to a quantum mechanical theory with a unitary dynamics provided one considers an enlarged Hilbert space. The game is quite simple: one introduces a new quantum ancilla whose degrees of freedom are unaccessible. Then one cooks up a unitary dynamics in the Hilbert space $System \otimes Ancilla$ such that by partial tracing on the degrees of freedom of the ancilla one recovers the evolution equation for the statistical operator of a dynamical reduction model of the type we have considered. A natural question arises: what is the purpose of this position besides the desire to protect the standard formalism come what may? The ancilla has, by construction, no observable effect and the procedure amounts simply to introduce hidden elements whose only role is to save the formal structure of the standard theory.

In view of the fact that dynamical reduction theories qualify themselves as rival theories of quantum mechanics, would it not be more serious, scientifically, to try, as R Penrose [22, 23, 24], S Adler [25, 26, 27] and many others do, to investigate the possibility of performing some crucial test? I do not intend to suggest that the GRW theory has to be taken seriously in its present formulation as a fundamental theory of natural processes, but I cannot avoid calling attention to the fact that, being a precise theory, it can give some indications about where to look for possible violations of the linearity of quantum mechanics.

[3] For a brilliant discussion of the arguments presented in this subsection we refer the reader to ref.[20].

7. Relativistic generalizations

Bell in one of his last papers, after having discussed Bohmian Mechanics and the GRW theory has stated [5]:

> *The real problem now is which one of these two exact theories admits a relativistic generalization*

Concerning this fundamental problem it is useful to mention that Bohmian Mechanics admits relativistic generalizations of various kinds, from the original attempts of Bohm and Hiley [28] based on a preferred Lorentz frame, to the recent investigations by Dürr *et al.* [29] and Salmos [30], resorting to an arbitrary preferred space-time slicing. Berndl *et al.* [31] and Dewdney and Horton [32] suggest a preferred joint parametrization of the world lines. We are not interested here in the technical details of these approaches, what is relevant is that they are characterized by a general feature which turns out to be unavoidable: they are not 'genuinely invariant' in the precise sense that they require a (hidden) preferred reference frame. This is a consequence of the fact that [33] any theory which violates Bell's locality condition by violating the parameter independence requirement, does not admit a 'genuine' relativistic generalization.

There have been also various attempts at a relativistic generalization of dynamical reduction models. The first is due to P Pearle [34] and it has been discussed in detail and proved to be formally perfectly Lorentz invariant in ref.[35]. The idea is quite simple but deep: one considers a fermion field which is coupled to a meson field and introduces a reduction mechanism which forbids the superpositions of different mesonic states. Since the mesonic clouds associated to different positions of a fermion differ, one induces in this way an indirect localization of the fermions. However, a new problem arises: the introduction of stochastic processes in a relativistic context leads to the emergence of untractable divergences.

Other attempts should be mentioned, among them those of Dove and Squires [36], of Dowker *et al.* [37, 38, 39], which are formulated on a discrete space-time. Up to very recent times no real step forward has been done. In the year 2006, R Tumulka has presented [16, 17, 18] a relativistic generalization of the GRW theory for N noninteracting distinguishable fermions based on the consideration of a multi-time Dirac equation. The theory sticks to what we have called the flashes ontology[4]. It is particularly interesting to mention Tumulka's conclusions which he has reinforced recently [17]:

> *A somewhat surprising feature of the present situation is that we seem to arrive at the following alternative: Bohmian mechanics shows that one can explain quantum mechanics, exactly and completely, if one is willing to pay with using a preferred slicing of space-time; our model suggests that one should be able to avoid a preferred slicing if one is willing to pay with a certain deviation from quantum mechanics.*

8. Quantum computation, nonlocality and realism

Quite recently, an attitude which we see as an attempt to revive the Copenhagen position emerged in connection with two quite different subjects, i.e., on the one side, the investigations concerning the promising and interesting field of quantum computation and, on the other, some theoretical and experimental analyses of quantum nonlocality. Let us proceed to discuss them.

People working in quantum computation have repeatedly put forward the idea that what quantum mechanics is about is not 'something existing out there', but only and exclusively 'our information'. A paradigmatic example of this attitude appears in a paper by A Zeilinger [40]:

- *The distinction between reality and our knowledge of reality, between reality and information, cannot be made,*

[4] Models based on the mass-density ontology require a serious reconsideration before one can resort to them for relativistic generalizations.

- Bell's result suggests *that the concept of reality itself is at stake,*
- *There is no way to refer to reality without using the information we have about it.*

Leaving apart the tautological structure of the last sentence, I believe that the best response to the first has been given in a recent paper [41]:

> *The distinction between reality and our knowledge of reality, not only can be made; it must be made if the notions of knowledge and information are to have any meaning in the first place.*

To further illustrate the discussion about the recent positions taken by some scientists we consider it appropriate to mention a recent exchange of views between D Mermin and the present author [42, 43]. Mermin asserts that the present status of quantum computation is completely satisfactory on the basis of the fact that:

- *... all gates, including the measurement gate, alter the state associated with the incoming Q-bits in a well defined, generally discontinuous manner which is precisely defined by the state,*
- *... in spite of the fact that there remains the distinction between linear, invertible, unitary gates with output state fully determined by the imput, and nonlinear, irreversible, measurement gates, with output state stochastically determined by the input, ..., the action of both types of gates are fully determined, with nothing left to the discretion of the theoretical physicist.*

We perfectly agree on these statements, but we stress that they amount simply to a way of evading the real problem. In fact Mermin himself stresses that:

> *it is through the readings of 1-Qbit measurement gates, and only through such readings, that one can extrect information from a quantum computer ... measurement is essential at both ends of a quantum computation. ... the 'user' which looks at a visual display or a print out* [plays the fundamental role in the process].

Here one can see, on the one side, the plain recognition that the very practice of quantum information and computation requires to accept the validity of two different and incompatible dynamical principles. On the other side, one ignores completely the fact that while we know everything about the working of the unitary gates used to implement our algoritms, we have an extremely vague idea of why and when individual physical processes occur which are not governed by the (usually) assumed universal laws of quantum mechanics and actually contradict it. This is, in essence the criticism to Mermin that we have raised. He has replied [44] and, besides making various remarks he has made quite clear his position:

> *I stress that the state and the unitary transformations it is subject to are mathematical abstractions that enable one to compute ... the probabilities of the readings of the final measurement gates. Mathematical abstractions do not require stochastic 'hits' originating in unknown physical processes or interactions with gravitons to be reset; they are reset by us, when we acquire more information and want to calculate what we expect to experience next.*

This sentence makes explicit reference to the GRW theory and to the proposals by Penrose of relating wave packet reduction to gravity. We believe that one cannot ignore that these alternative theories represent serious attempts to account consistently, in physical terms and not by vague verbal assumptions, for what *'we will experience ... next'*.

To conclude this first part of our comments we would like to mention that the idea that quantum mechanics deals simply with 'information', is not new. For instance it was put forward during the discussions concerning the decoherent histories approach to quantum mechanics when it has been suggested to look at them exclusivively with reference to IGUSes (*information gathering and using systems*). Bell has considered this position and he has made clear [5] that he was inclined to reject any reference to information unless one would, first of all, answer to the following basic questions: *Whose information?, Information about what?*

This remark allows to focus in a precise way the two incompatible positions which characterize the present debate which, in our opinion, echoes the one of more than 60 years ago. In fact, e.g., Mermin himself, has recently taken a quite precise position [45] concerning the issue under discussion:

> *The question of 'information about what' is a fundamentally metaphysical question that ought not to distract tough-minded physicists.*

This concludes our first set of remarks.

We come now to the second point, i.e., to the recent debate concerning nonlocality. An anticipation of this attitude is already contained [40] in the second sentence reported at the beginning of this section. The story goes as follows: A Leggett has considered [46] a class of theories, referred to as *nonlocal realistic theories*, which exhibit nonlocal features but satisfy a particularly natural (in his opinion) set of conditions. Leggett himself has shown that, in spite of the fact that one could not derive a Bell's type of inequality for such theories, they were at variance with the predictions of quantum mechanics and had to be abandoned. Subsequently various authors [47, 48, 49, 50] have considered the same problem both with reference to Leggett's proposal as well as for a somewhat extended class of nonlocal theories and have proved once more their incompatibility with quantum mechanics. Even experimental tests of Leggett's model have been performed, confirming the crash with quantum predictions.

This line of research is, in our opinion, quite interesting and deserves to be pursued. However, the above facts - i.e. the supposed 'naturalness' of the requests and the persisting incompatibility with quantum mechanics - have given rise to a very peculiar type of argument. In a sketchy way it can be summarized as follows: the derivation of Bell's inequality is based on two assumptions: Locality *and* Realism. Accordingly, the experimental violation of Bell's inequality requires to abandon at least one of these assumptions. Since giving up locality along the 'natural' way suggested by Leggett does not eliminate the crash of such proposals with quantum theory, the experimentally tested incompatibility gives a clear indication that one has to give up the assumption of realism. This argument, combined with the already emerging opinion, coming from the quantum computational community, i.e. that *the concept of reality itself is at stake*, has seen the adoption of this point of view by many, even extremely brilliant physicists [40, 47, 51].

We would like to make some remarks on this point. First of all we will call attention on the factual irrelevance of the argument. Secondly we will stress, and we consider this an extremely important point, that the just mentioned papers take a completely mistaken attitude towards the real meaning of Bell's work. Let us proceed to present the just mentioned criticisms.

- The class of models which have been considered is quite restricted, in particular it embodies a sort of weak locality request, which is in no way compelling. This point has been already raised in an important paper [50].
- There is no doubt that there exist theories which can legitimately be qualified as 'realistic' and fully agree with the standard theory. We mention first of all Bohmian Mechanics and, secondly, a quite elementary toy example dealing with a system of two entangled spin 1/2 particles in the singlet state which has been devised in 1964 by Bell in his fundamental paper [1].

In agreement with these remarks one cannot claim that the program of building up a 'fully realistic nonlocal theory' equivalent to quantum mechanics is not viable, an important counterexample of what has been suggested in the above mentioned papers. But this remark is by no way the most relevant one, actually it turns out to be rather marginal. What really matters is the fact that the derivation of Bell's inequality in no way whatsoever needs an assumption of realism. In spite of this fundamental fact, which everybody can verify by going carefully through the proof, a large part of the scientific community shares the completely wrong opinion that realism is among the basic assumption needed for the derivation of Bell's result. Bell himself has stressed this aspect and has remarked [52] that it is extremely difficult to eradicate this prejudice[5]:

> *My own first paper (Physics* **1***, 195 (1965.) on this subject starts with a summary of the EPR argument* **from locality to** *deterministic hidden variables. But the commentators have almost universally reported that it begins with deterministic hidden variables.*

This being the situation we must conclude that in no way whatsoever Bell's inequality has something to do with realism. It simply identifies in a straightforward and lucid way that what quantum phenomena impose to us is to accept the unescapable fact that natural processes involving entangled states of composite and far-away systems turn out to be unavoidably non-local. If one is keen to consider the possibility of giving up realism he can very well try his luck, but for sure such a big price is not necessary, and even not suggested, by the quantum nonlocal correlations.

The question whether Bell's theorem involves some request of realism has been discussed and proven once more to be unjustified by various authors [53, 54, 55].

We would like to conclude this analysis of some of the most relevant issues of the foundational investigations on quantum mechanics by expressing our worries concerning the fact that there are some indications that 80 years of lively debate on the conceptual problems of our best theory by the most brilliant physicists of the last century are facing a serious risk of being cancelled by those scientists who Mermin has qualified as the Q-computation crowd or by those who derive from experimental results inspired by not strictly convincing theoretical models unjustified conclusions concerning such an important issue as the one of the reality of the world around us.

References

[1] Bell J S 1964 *Physics* **1** 195
[2] Bell J S 1989 *Themes in Contemporary Physics II* ed S. Deser and R.J. Finkelstein (Singapore: World Scientific) p 1
[3] Primas H 1991 *Symposium on the foundations of modern physics 1990, Discussion Sections* ed K.V. Laurikainen and J. Viri (University of Turku, Finland)
[4] Bassi A and Ghirardi G C 2000 *Phys. Lett.* A **275** 373
[5] Bell J S 1990 *Phys. World* **3** 33
[6] Adler S 2003 *Stud. Hist. Phil. Mod. Phys.* **34** 135
[7] Joos E and Zeh H D 1985 *Z. Phys.* B **59**, 223
[8] Allori V, Goldstein S, Tumulka R and Zanghí N 2008 *British J. Philos. Sci.* **59** 353
[9] Bell J S 1987 *Schrödinger–Centenary Celebration of a Polymath* ed C W Kilmister (Cambridge: Cambridge University Press)
[10] Bohm D 1952 *Phys. Rev.* **85** 166, 1952 *Phys. Rev.* **85** 180
[11] Pauli W in Born M 1971 *The Born Einstein Letters* (New York: Walter and Co)
[12] Ghirardi G C, Rimini A and Weber T 1986 *Phys. Rev.* D **34** 470
[13] Ghirardi G C and Bassi A 2003 *Phys. Rep.* **379** 257
[14] Pearle P 1999 *Phys. Rev.* A **59** 80
[15] Ghirardi G C, Pearle P and Rimini A 1990 *Phys. Rev.* A **42** 78
[16] Tumulka R 2006 *AIP Conference Proceedings 844* ed A. Bassi *et al* (Melville, New York: AIP) p 340
[17] Tumulka R 2006 *Proc. Roy. Soc. London* A **462** 1897

[5] It has to be remarked that deterministic hidden variable theories assume that the complete specification of the state of the system implies that all physical properties are actually possessed by the systems prior to any measurement process. This is equivalent to the request of realism discussed by the above mentioned authors

[18] Tumulka R 2006 *J. Stat. Phys.* **125** 821
[19] Ghirardi G C, Grassi R and Benatti F 1995 *Found. Phys.* **25** 5
[20] Hagar A 2007 *Stud. Hist. Philos. Mod. Phys.* **38** 906
[21] Bub J 1997 *Interpreting the Quantum world* (Cambridge: Cambridge University Press)
[22] Penrose R 1996 *Gen. Relativ. Gravit.* **28** 581
[23] Penrose 2000 *Mathematical Physics 2000* (London: Imperial College)
[24] Marshall W, Simon C, Penrose R and Bouwmeester D 2003 *Phys. Rev. Lett.* **91** 130401
[25] Adler S 2005 *J. Phys.* A **38** 2729
[26] Adler S, Bassi A and Ippoliti E 2005 *J. Phys.* A **38** 2715
[27] Bassi A, Adler S and Ippoliti E 2005 *Phys. Rev. Lett.* **9** 030401
[28] Bohm D and Hiley B J 1993 *The Undivided Universe: an Ontological Interpretation of Quantum Theory* (London: Routledge)
[29] Dürr D, Goldstein S, Münch-Berndl K and Zanghí N 1999 *Phys. Rev.* A **60** 2729
[30] Salmos T M 1995 *J. Stat. Phys.* **80** 793, 1996 *Bohmian Mechanics and Quantum Theory: an Appraisal* ed J.T. Cushing *et al.* (Dordrecht: Kluwer) p 191
[31] Berndl K, Dürr D, Goldstein S and Zanghí N 1996 *Phys. Rev.* A **53** 2062
[32] Dewdney C and Horton G 2001 *J. Phys.* A **34** 9871
[33] Ghirardi G C and Grassi R 1996 *Bohmian Mechanics and Quantum Theory: an Appraisal* ed J.T. Cushing *et al.* (Dordrecht: Kluwer) p 353
[34] Pearle P 1990 *Sixty-two Years of Uncertainty* ed A. Miller (New York: Plenum)
[35] Ghirardi G C, Grassi R and Pearle P 1990 *Found. Phys.* **20** 1271
[36] Dove C and Squires E 1995 *Found. Phys.* **25** 1267
[37] Dowker F and Henson J 2004 *J. Stat. Phys.* **115** 1327
[38] Dowker F and Herbauts I 2004 *Classical Quant. Grav.* **21**, 2963
[39] Dowker F and Herbauts I 2005 *Found. Phys. Lett.* **18**, 499
[40] Zeilinger A 2005 *Nature* **438** 743
[41] Daumer M, Dürr D , Goldstein S, Maudlin T and Tumulka R 2006 *AIP Conference Proceedings 844* ed A. Bassi *et al* (Melville, New York: AIP) p 129
[42] Mermin D 2006 *Quantum Inf. Processes* **35** 239
[43] Ghirardi G C 2008 *Found. Phys.* **38** 1011
[44] Mermin D 2008 arXiv:0808.1582
[45] Mermin D 2002 *Quantum [Un]speakables* ed R.A. Bertlmann and A. Zeilinger (Berlin: Springer) p 271
[46] Leggett A J 2003 *Found. Phys.***33** 1496
[47] Gröblacher S, Paterek T, Kaltenbaek R, Bruckner C, Zukowski M, Aspelmeyer M and Zeilinger A 2007 *Nature* **446** 871
[48] Branciard C, Ling A, Gisin N, Kurtsiefer C, Lamas-Linares A and Scarani V 2008 *Nature Physics* **4** 681
[49] Paterek T, Fedrizzi A, Gröblacher S, Jennewein T, Zukovski M, Aspelmeyer M and Zeilinger A 2007 *Phys. Rev. Lett.* **99** 210406
[50] Colbeck R and Renner R 2008 *Phys. Rev. Lett.* **101** 050403
[51] Aspect A 2005 *Nature* **438** 745
[52] Bell J S 1981 *Journal di Physique* C2 **42** 41
[53] Norsen T 2006 *Found. Phys. Lett,* **19** 633, 2007 *Found. Phys.* **37** 311
[54] Tausk D V 2008 *arXiv:0809.4000*
[55] Laudisa 2008 *Found. Phys.* **38** 1110

Fourth International Workshop DICE2008 IOP Publishing
Journal of Physics: Conference Series **174** (2009) 012014 doi:10.1088/1742-6596/174/1/012014

Recent tests of realistic models

Giorgio Brida, Ivo Pietro Degiovanni, Marco Genovese, Marco Gramegna, Fabrizio Piacentini, Valentina Schettini and Paolo Traina
Istituto Nazionale di Ricerca Metrologica, Strada delle Cacce 91, 10135 Torino, Italy
E-mail: m.genovese@inrim.it

Abstract. In this article we present recent activity of our laboratories on testing specific hidden variable models and in particular we discuss the realizations of Alicki – van Ryn test and tests of SED and of Santos' models.

1. A short summary of a long history

The research concerning the possible existence of a classical theory (i.e. a theory where all the values of observables are deterministically fixed by some hidden variables, a Hidden Variable Theory, HVT) able to reproduce the results of Quantum Mechanics (QM) has a pluridecennial history, stemming from the 1935 Einstein-Podolsky-Rosen paper [1], where the completeness of QM was questioned.

In 1964 Bell showed that for every Local Hidden Variable Theory (LHVT) [2], correlations among certain observables must satisfy a set of inequalities (the Bell's Inequalities, BIs) while for QM they can be violated. In practice, even though many experiments, mainly realized with entangled photons (see list [3]), have shown violation of BIs, their interpretation always calls for additional hypothesis due to experimental limitations. In particular most of those results rely on the accessory assumption that the events observed, due to the finite efficiency of real detection apparatuses, are a faithful statistical sample of the whole ensemble (this is often called *detection loophole* or *fair sampling assumption*) [4, 5]. No definitive experimental test of local realism has yet been performed where all loopholes are closed simultaneously. In the last decade the study of QM vs. HVT has attracted much interest, fueled by the development of Quantum Information science [4]. Furthermore, the tests of certain realistic models are receiving much current attention (see list [6]). In particular, some classes of HVTs have not been excluded by Bell's inequality experiments because of experimentally induced loophole(s). Experiments specifically aimed at testing these HVTs are the focus of recent interest. Other differences between quantum and classical treatments have also been discovered and pointed out [7]. Finally, it is worth to mention that a classical non-unitary evolution could eventually be restored at high energy scales (Planck scale); recent works [8] have paved the road to build a theory of this kind with the hope of a possible future experimental test.

In this article we review some recent activity in this field of our laboratories at INRIM.

2. The Alicki – van Ryn test

In 2008 a simple test of nonclassicality at the single qubit level was proposed [9, 10] to show that some quantum states in a two dimensional Hilbert space cannot have a classical description.

This test is appealing for its simplicity compared to other tests of quantumness (see [4] and references therein) and could represent a useful tool for various applications in the fields of quantum technologies (as quantum information, metrology, imaging, ...), fundamental quantum optics, foundations of quantum mechanics, etc.

As it pertains single particle states, this test is not relevant to the question of locality, rather it is a more fundamental test of nonclassicality with respect to the possibility of an underlying HVT. Furthermore, we underline that it does not apply to every conceivable HVT like Bell's inequalities, but only to a restricted class of HVTs [11]. We also report that Zukowski [12] suggests that the Alicki – van Ryn classicality criterium is equivalent to the von Neumann theorem.

2.1. Theoretical model

The proposal [9] stems from the fact that given any two positive real functions $\mathcal{A}, \mathcal{B}$ obeying the relation:

$$0 \leq \mathcal{A}(x) \leq \mathcal{B}(x) \tag{1}$$

for any probability distribution $\rho(x)$ it must be true that:

$$\langle \mathcal{A}^2 \rangle \equiv \int \mathcal{A}^2(x)\rho(x)dx \leq \int \mathcal{B}^2(x)\rho(x)dx \equiv \langle \mathcal{B}^2 \rangle \tag{2}$$

On the other hand for quantum systems, one can find pairs of observables $\widehat{A}, \widehat{B}$ such that the minimum eigenvalue of $\widehat{B} - \widehat{A}$ is greater than zero which we refer to as the inequality:

$$0 \leq \widehat{A} \leq \widehat{B} \tag{3}$$

The commutation relations deriving from the classical approach leading to Eq. (2), prescribe that for all systems (described by the density matrix $\widehat{\rho}$):

$$\langle \widehat{A}^2 \rangle \leq \langle \widehat{B}^2 \rangle \tag{4}$$

where $\langle \widehat{O} \rangle \equiv \mathrm{Tr}[\widehat{O}\ \widehat{\rho}]$, while, to the contrary, quantum theory allows that for certain quantum states:

$$\langle \widehat{A}^2 \rangle > \langle \widehat{B}^2 \rangle \tag{5}$$

This sharp difference between classical (in the sense discussed above) and quantum theory predictions at a single qubit level can be tested experimentally on an ensemble of single particles.

One possible pair of operators $\widehat{A}$ and $\widehat{B}$ is:

$$\widehat{A} = a\frac{\mathbf{1} + \widehat{Z}}{2} \tag{6}$$

$$\widehat{B} = b\frac{\mathbf{1} + r\cos\beta\widehat{Z} + r\sin\beta\widehat{X}}{2} \tag{7}$$

where $\widehat{Z}$, $\widehat{X}$ are Pauli matrices. To ensure the positivity of $\widehat{A}$ and $\widehat{B}$, we assume $a > 0$, $b > 0$, and $0 \leq r \leq 1$ for a, b, and r real.

In order to have that, given $\langle \widehat{B} \rangle > \langle \widehat{A} \rangle$, at least one state can be found such that $\langle \widehat{B}^2 \rangle < \langle \widehat{A}^2 \rangle$ (*i.e.* the condition for Alicki – van Ryn nonclassicality), the minimum of the eigenvalue of $\widehat{B} - \widehat{A}$ should be positive while $\widehat{B}^2 - \widehat{A}^2$ should be negative for at least one eigenvalue, requiring that:

$$\frac{1 - r^2}{2\sqrt{1 + r^2 - 2r\cos\beta}} < \frac{a}{b} < \frac{1 - r^2}{2(1 - r\cos\beta)} \tag{8}$$

The test proposed in Ref. [9] concerns the class of HVTs satisfying Eq. (3) only.

The quantum objects we use to implement this test are linearly polarized single photons:

$$|\Psi\rangle = \cos\psi|H\rangle + \sin\psi|V\rangle \tag{9}$$

If we choose the parameter values $a = 0.74$, $b = 1.2987$, $r = 3/5$, $\beta = 2/9\ \pi$ (we note that, with this choice, the condition in Eq. (8) is satisfied), and $\psi = -11/36\ \pi$ the results for $\langle\widehat{B}^2\rangle - \langle\widehat{A}^2\rangle$, and $\langle\widehat{B}\rangle - \langle\widehat{A}\rangle$ predicted by quantum theory are those presented in Table 1, while the minimum eigenvalue of $\widehat{B} - \widehat{A}$ is $d_- = 0.0189$ (satisfying the $\widehat{B} - \widehat{A} > 0$ requirement for the Alicki – van Ryn test), where:

$$d_- \equiv \frac{1}{2}\left\{b - a - \sqrt{a^2 + b^2\, r^2 - 2\, a\, b\, r\, \cos\beta}\right\} \tag{10}$$

The simplest example of a HVT that does not satisfy this condition and can mimic QM, relies on a hidden (or simply unmeasured) variable x uniformly distributed between 0 and 1 ($\rho(x) = 1$ when $0 \le x \le 1$), and classical quantities $\mathcal{A}(x) = a_0\theta(X_A - x)$, and $\mathcal{B}(x) = b_0[p_1\theta(X_B - x) + (1 - p_1)\theta(x - X_B)]$, where $\theta(\xi)$ is the step function (1 for $\xi \ge 0$, and 0 elsewhere). By choosing $X_A = a\cos^2\psi$ and $X_B = \frac{b}{2}(1 + r\cos(\beta - 2\psi))$ and using the experimental parameters defined above, we obtain the quantum mechanical predictions, $\langle\mathcal{B}\rangle - \langle\mathcal{A}\rangle = 0.0685$ and $\langle\mathcal{B}^2\rangle - \langle\mathcal{A}^2\rangle = -0.0449$. It is easy to verify that this model does not belong to the class of hidden variable models falsified by this test, as the condition given in Eq. (1) is not satisfied for some x. In particular, for $X_B < x < X_A$ we have $\mathcal{A}(x) > \mathcal{B}(x)$.

2.2. *First experimental test: heralded photon at 1550 nm*

A first experimental test was performed with the state (9) obtained by a heralded single photon source based on PDC (the experimental setup is presented in Fig. 1); the PDC was generated by a PPLN MgO-doped bulk crystal pumped by a continuous wave (CW) laser at 532 nm, producing pairs of correlated photons at 810 nm and 1550 nm.

A cutoff filter blocked the pump laser after the crystal and a dichroic mirror separated the 810 nm and 1550 nm photons. Extra interference filters at 810 and 1550 nm with a full width at half-maximum (FWHM) of 10 nm and 30 nm, respectively, further suppressed fluorescence from the PPLN crystal reducing background counts. The collection geometry on the heralding arm restricted the visible bandwidth to ≈2 nm FWHM.

In the experiment, photons in the heralding arm were routed by a single-mode fiber (SMF) to a silicon single photon avalanche diode (Si-SPAD) operating in Geiger mode, while photons in the heralded arm, coupled into a second SMF, were sent to the apparatus giving the probabilistic projections according to the parameter values determined above (necessary to measure $\langle\widehat{B}\rangle$ and $\langle\widehat{B}^2\rangle$). These projections were implemented by means of an all-fiber variable beam splitter and polarizers.

The variable beam splitter was obtained by means of an optical switch routing heralded photons with an adjustable splitting ratio into two different optical paths. The input polarization state in each optical path after the beam splitter was controlled with a three paddle SMF polarization rotator followed by a rotating polarizer (POL). The paddles were able to control the polarization state of the beam inside the fiber by using the mechanical stress, able to locally change the birefringence of the fiber.

The presence of the switch with adjustable splitting ratio allowed us to experimentally set the parameter r and estimate it.

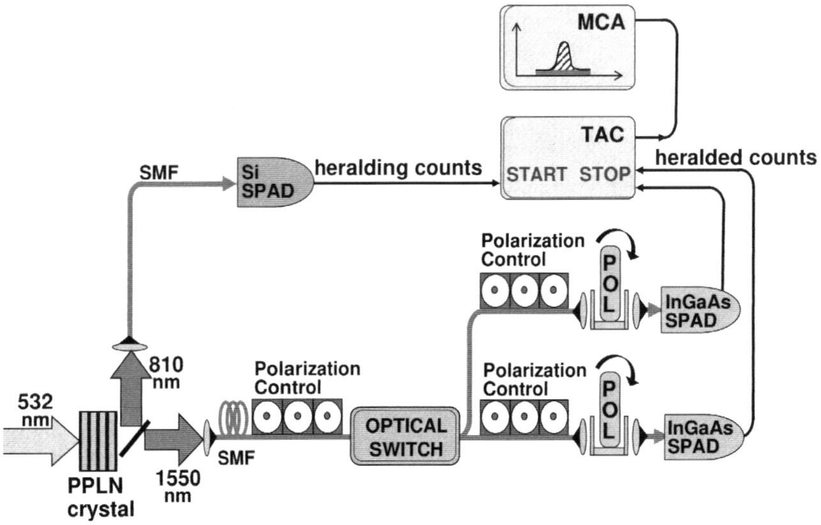

Figure 1. Setup of the first Alicki - van Ryn test experimental scheme. A PDC heralded single photon source generates pairs of photons at 810 nm (heralding) and 1550 nm (heralded) in a PPLN crystal pumped by a 532 nm laser. The heralded photons are sent to the measurement apparatus designed to evaluate the observables $\langle\widehat{A}\rangle$, $\langle\widehat{A}^2\rangle$, $\langle\widehat{B}\rangle$, and $\langle\widehat{B}^2\rangle$.

In fact, the expectation values of $\langle\widehat{A}\rangle$, $\langle\widehat{B}\rangle$, $\langle\widehat{A}^2\rangle$, and $\langle\widehat{B}^2\rangle$ can be obtained experimentally by projecting the heralded photons onto the linear polarizations states as the operators $\widehat{A}$ and $\widehat{B}$ can be rewritten as:

$$\widehat{A} = a\widehat{P}_0 \tag{11}$$

and:

$$\widehat{B} = b\left[p_1\widehat{P}_{\frac{\beta}{2}} + (1-p_1)\widehat{P}_{\frac{\beta+\pi}{2}}\right] \tag{12}$$

where $\widehat{P}_\phi$ is the projection operator on the state $|s(\phi)\rangle = \cos\phi|H\rangle + \sin\phi|V\rangle$.

Thus p_1 is the splitting ratio to set on the optical switch to measure $\langle\widehat{B}\rangle$, and in terms of r is $p_1 = \frac{1+r}{2}$.

In the measurement of $\langle\widehat{B}^2\rangle$ the splitting ratio p_2 should be $p_2 = \frac{p_1^2}{p_1^2+(1-p_1)^2}$, thus the operator $\langle\widehat{B}^2\rangle$ is:

$$\widehat{B}^2 = b_0^2\frac{1-2\sqrt{(1-p_2)p_2}}{(1-2p_2)^2}\left[p_2\widehat{P}_{\frac{\beta}{2}} + (1-p_2)\widehat{P}_{\frac{\beta+\pi}{2}}\right] \tag{13}$$

The experimental scheme allowed us to properly set the polarizers to perform projections on $\widehat{P}_{\frac{\beta}{2}}$, and $\widehat{P}_{\frac{\beta+\pi}{2}}$, and set the splitting probability of the optical switch to p_1, or p_2, to make necessary measurements of the observables. After passing through the polarizers that performed the projections, the heralded photons were finally sent to InGaAs-SPADs gated by the Si-SPAD heralding counts.

We determined the true coincidence probability for each gate, rather than using the raw measured counts, in order to eliminate the contribution of accidental coincidences, detector

Table 1. Measurement results obtained with the first experimental scheme with statistical and total uncertainties and theoretical predictions. The total uncertainty (in parentheses) accounts for both statistical and systematic effects.

Quantity	Measurement	QM theory
$\mathcal{E}[\langle\widehat{B}\rangle - \langle\widehat{A}\rangle]$	0.0581 ± 0.0049 (± 0.0112)	0.0685
$\mathcal{E}[\langle\widehat{B}^2\rangle - \langle\widehat{A}^2\rangle]$	-0.0403 ± 0.0043 (± 0.0066)	-0.0449
		(> 0 HVTs)
$\mathcal{E}[p_1]$	0.80 ± 0.01	0.800
$\mathcal{E}[p_2]$	0.94 ± 0.01	0.941

deadtimes, and drifts. The probability for each measurement i was evaluated according to:

$$\eta_i(\theta, p) = \frac{N_i(\theta, p)}{M_{g,i}} \tag{14}$$

where $N_i(\theta, p)$ is the number of coincidences sent with probability p towards the detection system with the polarizer projecting photons onto the state $|s(\theta)\rangle$, and $M_{g,i}$ is the number of heralding gate counts.

Thus, in each experimental configuration, $\langle\widehat{P}(\theta)\rangle$ was estimated to be:

$$\mathcal{E}[\langle\widehat{P}(\theta)\rangle] = \frac{\sum_i \eta_i(\theta, p)}{\sum_i[\eta_i(\theta, p) + \eta_i(\theta + \pi/2, p)]} \tag{15}$$

while the probability of sending a photon towards a detection system (whose nominal value is p) was:

$$\mathcal{E}[p] = \frac{\sum_i[\eta_i(\theta, p) + \eta_i(\theta + \pi/2, p)]}{\sum_i \begin{bmatrix} \eta_i(\theta, p) + \eta_i(\theta, 1-p) + \\ \eta_i(\theta + \pi/2, p) + \eta_i(\theta + \pi/2, 1-p) \end{bmatrix}} \tag{16}$$

Using Eqs. (14) and (15) we computed the experimental values of $\langle\widehat{A}\rangle$, $\langle\widehat{A}^2\rangle$, $\langle\widehat{B}\rangle$, and $\langle\widehat{B}^2\rangle$ as seen in Table 1. From the same experimental results we obtained an indirect evaluation of the minimum eigenvalue of $\widehat{B} - \widehat{A}$ as (0.0101 ± 0.0065), showing that we have met requirement (8) (we point out that the high relative uncertainty of this evaluation is due to its indirect determination).

The value of $\mathcal{E}[\langle\widehat{B}^2\rangle - \langle\widehat{A}^2\rangle]$ shows a violation of the classical limit (for appropriate HVTs) by more than 6 standard deviations.

Table 1 presents both the statistical and total uncertainties. Statistical uncertainties include those due to Poisson counting statistics as well as those due to random misalignment of the polarizers (we estimate an angular uncertainty of 2.5°), while the total uncertainties also include systematic effects such as the uncertainty on the setting of the optical switch voltage bias used to obtain the required splitting ratio. As an additional test, we measured $\mathcal{E}[p_1]$ and $\mathcal{E}[p_2]$ and found them consistent with the intended settings.

A necessary requirement for a convincing realization of the Alicki - Van Ryn proposal [9] is a demonstration that our source in fact produces single photon states, thus we need to verify that the possibility of having more than one photon in the heralded arm after detecting the heralding photon is low. To check the source performance we used the same setup as for the main experiment (Fig. 1), but with the polarizers removed and the splitting factor of the switch

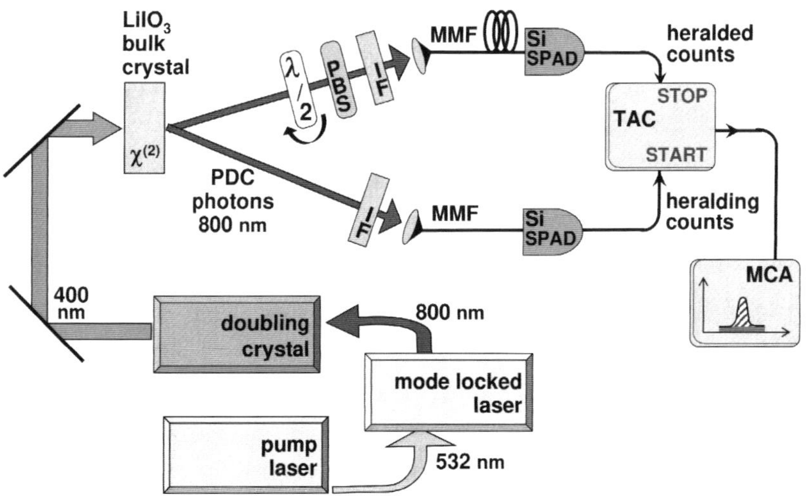

Figure 2. Setup of the second Alicki - van Ryn test experimental scheme. A PDC heralded single photon source generates pairs of correlated photons at 800 nm in a $LiIO_3$ bulk crystal pumped by a 400 nm mode-locked laser. The counted photon on the heralding arm announces the presence of its twin on the heralded arm. The heralded photons are sent to the measurement apparatus designed to evaluate the observables $\langle\widehat{A}\rangle$, $\langle\widehat{A}^2\rangle$, $\langle\widehat{B}\rangle$, and $\langle\widehat{B}^2\rangle$.

set to $p = (0.50 \pm 0.01)$. The complete analysis on how the non-ideal (multi-photon) behavior of our single photon source might have affected the experimental results is reported in [11]; in summary, we found this non-ideal effects to be negligible, being more than an order of magnitude below the listed uncertainties.

2.3. Second experimental test: heralded photons at 800 nm

A second scheme was then realized with the purpose to reduce systematic uncertainties [13]. The main differences between this experimental scheme and the previous one were that a different configuration for polarization projection was adopted and that both the heralded and the heralding photons were at 800 nm, while in the previous case the heralded photon was at a telecom wavelength [11]. This allowed us to use more efficient and less noisy detection systems, reducing the experimental uncertainty contributions.

The experimental apparatus is sketched in Fig. 2. The PDC source is a 5 mm long bulk $LiIO_3$ crystal, pumped by a 400 nm laser, that produces pairs of correlated photons at 800 nm. The pump light is obtained by doubling the frequency of a mode-locked laser (with a repetition rate of $\approx$76 MHz) pumped by a 532 nm CW laser. Two interference filters (IF) with FWHM of 20 nm were placed in both the heralded and heralding arms to reduce background light. Microscope objectives (20x) collected the light into multimode fibers (MMF), and then the photons were finally counted by Si-SPAD operating in Geiger mode. A half-wave plate ($\lambda/2$) and a polarizing beam splitter (PBS) were used for our polarization projective measurements.

Again, the spectrum of the operators $\widehat{B}$, $\widehat{B}^2$ (containing r) was simply obtained by proper polarization projection of the single photon state. Nevertheless, the random splitting of the

Table 2. Measurement results obtained with the second experimental scheme, with statistical and total uncertainties and theoretical predictions.

Quantity	Measurement	QM theory
$\mathcal{E}[\langle\widehat{B}\rangle - \langle\widehat{A}\rangle]$	0.0701 ± 0.0015	0.0685
$\mathcal{E}[\langle\widehat{B}^2\rangle - \langle\widehat{A}^2\rangle]$	-0.0461 ± 0.0010	-0.0449
		(> 0 HVTs)

measured photons in this last experiment was not performed. The expectation values of $\langle\widehat{A}\rangle$, $\langle\widehat{A}^2\rangle$, $\langle\widehat{B}\rangle$, and $\langle\widehat{B}^2\rangle$ were obtained experimentally by projecting the heralded photons onto the linear polarizations states, as the operators $\widehat{A}$ and $\widehat{B}$ can be rewritten as:

$$\widehat{A} = a\widehat{P}_0 \tag{17}$$

and:

$$\widehat{B} = b\left(\frac{1+r}{2}\widehat{P}_{\frac{\beta}{2}} + \frac{1-r}{2}\widehat{P}_{\frac{\beta+\pi}{2}}\right) \tag{18}$$

The experimental results are presented in Table 2: from the experimental value of $\langle\widehat{B}^2\rangle - \langle\widehat{A}^2\rangle$ we observe a very large violation (≈ 46 standard deviations) of the theoretical classical limit $\langle\widehat{B}^2\rangle - \langle\widehat{A}^2\rangle > 0$.

Also in this experiment the single photon nature of our source was verified, showing that it was sufficiently good for not affecting our conclusions [13].

3. Stochastics electrodynamics and Santos' HVT models

Other recent experiments, performed in our laboratories, pertain testing more restrict HVT specifically built for escaping Bell inequalities tests with today loopholes.

One of the most elaborated model of this kind is Stochastic Electro-Dynamics (SED), where the electrodynamics is described classically, but with the presence of a vacuum field that allows to reproduce many quantum mechanical features [14]-[25]. Specific predictions of this theory suited to be tested experimentally were proposed in [20]-[25]. In particular it was suggested that Bell inequalities would not be violated under a certain intensity level of the impinging field and a phenomenon of up-conversion was predicted.

This allowed an experimental falsification of this theory [26, 27, 28]. Indeed no up conversion was observed well beyond the prediction of the theory, whilst a clear violation (more than 20 standard deviations) of Clauser–Horn inequality was observed at an intensity many orders of magnitude smaller than the limit of SED.

These results induced a further theoretical elaboration, which, abandoned the project of an elaborated alternative to Quantum Electro-Dynamics, tried to develop specific models [29, 30] not excluded by all the performed experiments, with the purpose of obtaining some clue for building a more refined theory.

In a little more detail, Refs. [29, 30] considered a set-up based on polarization entangled photons, where one measures the single count rates R_1 and R_2 on the idler and signal arm, the coincidence rate R_{12} and the total pair production rate R_0. According to QM one has:

$$R_1 = R_2 = \frac{1}{2}\eta R_0 \qquad R_{12} = \frac{1}{4}\eta^2 R_0[1 + V\cos(2\phi)] \tag{19}$$

being ϕ the angle between the two polarization axes, V the observed visibility and η the detection efficiency (considered identical for the two arms).

The results of such an experiment can be reproduced by a LHV model constituted by the functions [29]:

$$\rho(|\lambda_1 - \lambda_2|) \geq 0 \qquad 0 \leq P_j(|\lambda_j - \phi_j|) \leq 1 \qquad \int \rho(|\lambda_1 - \lambda_2|) d\lambda_1 d\lambda_2 = 1 \tag{20}$$

where λ_1, λ_2 are the hidden (angular) variables and ϕ_1, ϕ_2 the angles of the polarization planes with respect to an axis perpendicular to pump direction and polarization. From this model, introducing the visibility parameters:

$$V_A = \frac{R_{12}(0) - R_{12}(\pi/2)}{R_{12}(0) + R_{12}(\pi/2)} \qquad V_B = \sqrt{2}\frac{R_{12}(\pi/8) - R_{12}(3\pi/8)}{R_{12}(\pi/8) + R_{12}(3\pi/8)} \tag{21}$$

one derives with some algebra the inequality [29]:

$$\frac{V_B}{V_A} \geq F = 1 + \cos^2\left(\frac{\pi\eta}{2}\right)\left[V_B - \frac{\sin^2(\pi\eta/2)}{(\pi\eta/2)^2}\right] \tag{22}$$

that can be violated in quantum mechanics (where, in the ideal case, $V_B/V_A = 1$): this was the first inequality checked in our experiment.

Then we studied an inequality of the same type, but valid for a larger family of LHVT [30], taking into account a set of four hidden variables $\lambda \equiv \{\chi_1, \chi_2, \mu_1, \mu_2\}$ (of two different types) instead of two. The equations of this local realistic model are [30]:

$$\rho(\lambda) = \rho_\chi(|\chi_1 - \chi_2|) g_1(\mu_1) g_2(\mu_2) \geq 0 \tag{23}$$

$$\int \rho_\chi(|\chi_1 - \chi_2|) d\chi_1 d\chi_2 = \int g_j(\mu_j) d\mu_j = 1 \tag{24}$$

$$0 \leq P_j(|\chi_j - \phi_j|) = \int Q_j(\mu_j, |\chi_j - \phi_j|) g_j(\mu_j) d\mu_j \leq 1 \tag{25}$$

giving raise, after some approximations and algebraic manipulations, to the inequality:

$$\Delta_{min} = \left[\frac{n\sum_{j=1}^n R_{12}^2(\phi_j)}{\left(\sum_{j=1}^n R_{12}(\phi_j)\right)^2} - \frac{\left(\sum_{j=1}^n R_{12}(\phi_j)\cos(2\phi_j)\right)^2}{\left(\sum_{j=1}^n R_{12}(\phi_j)\right)^2} - 1\right]^{\frac{1}{2}} \geq D(\eta) \tag{26}$$

where $\phi_j = \pi \cdot (j-1)/n$ and the parameters V and $D(\eta)$ are defined as follows:

$$V = 2\frac{\sum_{j=1}^n R_{12}(\phi_j)\cos(2\phi_j)}{\sum_{j=1}^n R_{12}(\phi_j)} \tag{27}$$

$$D(\eta) \equiv \frac{4}{3\pi}\sqrt{\frac{2}{3\eta} - \frac{1}{2} - \frac{\sin^4(\pi\eta/2)}{(\pi\eta/2)^4}}\left(V - \frac{\sin^2(\pi\eta/2)}{(\pi\eta/2)^2}\right)_+^{\frac{3}{2}} \tag{28}$$

where $()_+$ means putting zero if the quantity inside the bracket is negative. Testing this second inequality will be the second purpose of the present work.

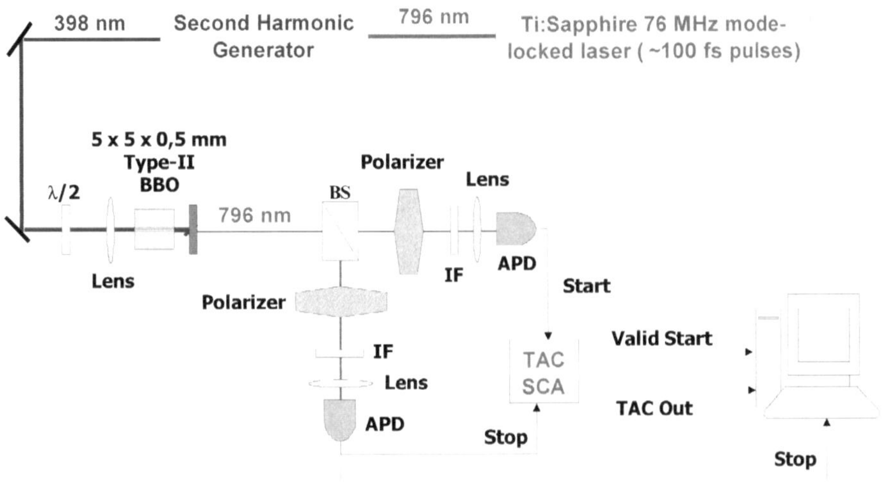

Figure 3. Santos' HVT test experimental setup: a 76 MHz mode-locked laser (doubled in frequency) pumps a 0,5 mm thick type-II BBO crystal, producing degenerate collinear PDC at 796 nm. The PDC emission is split by the BS, and each branch goes passes through a polarizer, an IF and a collection lens before being addressed to an APD.

3.1. Experimental set-up and results

In order to test these inequalities, we used the experimental setup reported in Fig. 3, with a 0.2 W, 398 nm pulsed (with 200 fs pulses) laser beam, obtained by second harmonic generation from a titanium-sapphire beam at 796 nm, pumping a type-II BBO $5 \times 5 \times 0,5$ mm crystal. With proper phase-matching conditions, we obtained the collinear degenerate (796 nm) PDC and sent it to a non-polarizing 50%-50% beam splitter (BS). On the transmission (t) arm we put a Glan-Thomson polarizer, a pinhole, an interference filter (FWHM 10 nm, λ_{peak} 795 nm) and an optical fiber coupler connected with a multi-mode fiber to the detector, a Perkin-Elmer SPCM-AQR-14-FC-9491 silicon avalanche photo-diode (quantum efficiency at 796 nm: 62%); the reflection (r) arm featured a Glan-Thomson polarizer, a large band interference filter (FWHM 40 nm, λ_{peak} 800 nm) and the same fiber coupling and detection devices as in the other arm. Taking the t photons as trigger, a coincidence window was opened on the r detector to reveal the correlated pairs: this was obtained by sending the output of the t detector as start to a Time-to-Amplitude Converter (TAC) that received the r detector signal as stop. The 20 ns coincidence window was set properly to exclude spurious coincidences with PDC photons of the following pulse. The background has been properly evaluated and subtracted.

By fixing the t polarizer at an angle of $\frac{\pi}{4}$ and rotating the r one by $\frac{\pi}{8}$ per step, we obtained the coincidence rates data shown in Fig. 4. As a first test for our entangled state, we checked (of course within the fair sampling hypothesis) the Clauser–Horne inequality [4], written in the form:

$$\frac{|R(\frac{\pi}{8}) - R(\frac{3\pi}{8})|}{R_{tot}} \leq \frac{1}{4} \tag{29}$$

The collected data, with a measured visibility 97.8%, led to the following violation:

CH violation	σ	violation/σ
0,1026	0,0021	47,8

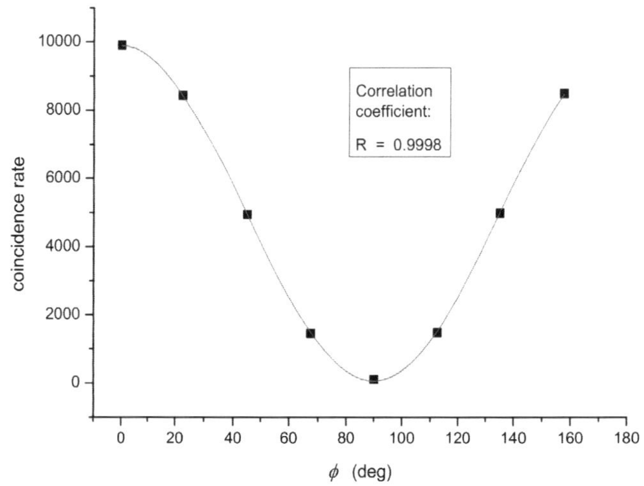

Figure 4. Coincidence rates versus ϕ (angle between the axes of t and r polarizers): each step is the average of 30 different acquisitions of 30 s each. The agreement between experimental data and theoretical predictions are very good (correlation coefficient R = 0.9998). Uncertainty bars are too small to be seen.

Then, to violate the inequality (22), we evaluated the quantities F, V_B and V_A, obtaining:

$F - V_B/V_A$	σ	violation/σ
0,0671	0,0057	11,7

Finally, the results for the inequality (26) were:

$D(\eta) - \Delta_{min}$	σ	violation/σ
0,0073	0,0022	3,3

We like to stress that both the inequalities (22),(26) still were violated even without background subtraction (of 11.6 and 1.7 standard deviations respectively). Thus, we have clearly shown a violation of the predictions made upon the LHVT models presented in [29] and [30], whilst perfectly agreeing with SQM.

In summary, we have experimentally demonstrated, following the theoretical predictions of [29, 30], that the specific families of hidden variable theories considered there fail to properly reproduce the observed correlations between measurements of bipartite polarization photon entangled states, and are therefore falsified by our experiment.

4. The Croca's wavelet model

Before concluding we would like to mention a work in progress in our laboratories. Recently Croca et al. proposed a modified quantum mechanics based on wavelets [31]. The main prediction of this model for an experimental test derives from the fact that in this scheme a wave packet has a width σ that does not depend only on the frequency bandwidth σ_k only, but also on the size of the basic mother Morlet wavelet σ_0, being altogether:

$$\sigma^2 = \frac{1}{\sigma_k^2 + 1/\sigma_0^2} \tag{30}$$

This result leads to the prediction that in a single-particle (e.g. photon) interferometric experiment the visibility drops to zero when the bandwidth is restricted under $1/\sigma_0^2$. We are now studying the implementation of such a set up. However, even without a search in the huge literature, a first limit, $\sigma_0 > 500m$, is posed by an experiment where we obtained high visibility fringes with a He-Ne laser beam attenuated to photo-counting regime addressed to a Mach-Zender interferometer.

5. Conclusion

In conclusion, we have presented recent activity of our laboratories on testing specific hidden variable models and in particular we have discussed realizations of Alicki – van Ryn test, tests of SED and of Santos' models.

Acknowledgments

This work has been supported by Compagnia di San Paolo, by PRIN 2007FYETBY (CCQOTS) and by Regione Piemonte (E14). We thank R. Alicki, J. Croca and E. Santos for helpful discussions, and also N. Gisin, A. Migdall and S. Polyakov for a fruitful and pleasant collaborative work.

References

[1] Einstein A, Podolsky B and Rosen N 1935 *Phys. Rev.* **47** 777
[2] Bell J S 1964 *Physics* **1** 195
[3] Aspect A *et al* 1982 *Phys. Rev. Lett.* **49** 1804; Franson J P 1989 *Phys. Rev. Lett.* **62** 2205; Rarity J G and Tapster P R 1990 *Phys. Rev. Lett.* **64** 2495; Brendel J *et al* 1992 *Eur. Phys. Lett.* **20** 275; Kwiat P G *et al* 1990 *Phys. Rev.* A **41** 2910; Kiess T E *et al* 1993 *Phys. Rev. Lett.* **71** 3893; Kwiat P G *et al* 1995 *Phys. Rev. Lett.* **75** 4337; Ou Z J and Mandel L 1988 *Phys. Rev. Lett.* **61** 50; Shih Y H *et al* 1993 *Phys. Rev.* A **47** 1288; Torgerson J R *et al* 1995 *Phys. Lett.* A **204** 323; Pan J W *et al* 2000 *Nature* **403** 515; Brida G *et al Phys. Lett.* A **268** 12; Tittel W *et al* 1998 *Phys. Rev. lett.* **81** 3563; Weihs G *et al* 1998 *Phys. Rev. Lett.* **81** 5039; Rowe M A *et al* 2001 *Nature* **409** 791
[4] Genovese M 2005 *Phys. Rep.* **413** 319-396
[5] De Caro L and Garuccio A 1996 *Phys. Rev.* A **54** 174; Pearle P M 1970 *Phys. Rev.* D **2** 1418
[6] Gröblacher S *et al* 2007 *Nature* **446** 871; Branciard C *et al* 2007 *Phys. Rev. Lett.* **99** 210407; Paterek T *et al* 2007 *Phys. Rev. Lett.* **99** 210406; Eisaman M D *et al* 2008 *Phys. Rev.* A **77** 032339; Leggett A 2003 *Found. Phys.* **33** 1469; Khrennikov A 2007 *Preprint* arXiv:0709.3909
[7] Auletta G 2000 *Foundations and Interpretation of Quantum Mechanics* (Singapore: World Scientific)
[8] 't Hooft G 2002 "Quo Vadis Quantum Mechanics" Conf. Proceedings (Philadelphia) *Preprint* arXiv:quant-ph 0212095; *do.* 2003 *Int. J. Theor. Phys.* **42** 355-362; Blasone M *et al* 2001 *Phys. Lett.* A **287** 205-210; Biró T S *et al* 2001 *Found. Phys. Lett.* **14** 471-485; van De Bruck C 2000 *Preprint* arXiv:gr-qc/0001048; Blasone M *et al* 2004 *Preprint* arXiv:quant-ph/0409021; *do.* 2003 *J. Phys. Soc. Jap. Suppl.* C **72** 50-53; Biró T S and Müller B 2003 *Preprint* arXiv:hep-lat/0307028; Elze H-T 2003 *Phys. Lett.* A **310** 110-118; *do.* 2007 *J. Phys.: Conf. Ser.* **67** 012016; *do.* 2008 *J. Phys. A: Math. Theor.* **41** 304020; *do.* 2008 *Int. J. Theor. Phys.* **47** 455-467; *do.* 2009 *Int. J. Qu. Inf. (IJQI)* **7** 83-96
[9] Alicki R and Van Ryn N 2008 *J. Phys. A: Math. Theor.* **41** 062001
[10] Alicki R, Piani M and Van Ryn N 2008 Quantumness Witnesses *Preprint* arXiv:0807.2615v1
[11] Brida G, Degiovanni I P, Genovese M, Schettini V, Polyakov S V and Migdall A 2008 *Opt. Express* **16** 11750-11758
[12] Zukowski M 2008 Unexpected reemergence of von Neumann theorem *Preprint* arXiv:0809.0115v1
[13] Brida G, Degiovanni I P, Genovese M, Piacentini F, Schettini V, Gisin N, Polyakov S V and Migdall A 2008 Improved implementation of nonclassicality test for a single particle *Preprint* arXiv: 0811.3376v1
[14] Marshall T W 1965 *Proc. Cambridge Philos. Soc.* **61** 537-546

[15] Boyer T H 1969 *Phys. Rev.* **182** 1374-1383
[16] Santos E 1972 *Lett. Nuovo Cimento* **4** 497-501
[17] Surdin M 1971 *Ann. Inst. Henri Poincaré* **15** 203-241
[18] de la Pena L and Cetto A M 2000 *The quantum dice: an introduction to stochastic electrodynamics* (Kluwer)
[19] Casado A *et al* 1997 *J. Opt. Soc. Am.* B **14** 494-502
[20] Casado A *et al* 1997 *Phys. Rev.* A **55** 3879-3890
[21] Casado A *et al* 1997 *Phys. Rev.* A **56** 2477-2480
[22] Casado A *et al* 1998 *J. Opt. Soc. Am.* B **15** 1572-1577
[23] Casado A *et al* 2000 *Eur. Phys. Journ.* D **11** 465-472
[24] Casado A *et al* 2001 *Eur. Phys. Journ.* D **13** 109-119
[25] Dechoum K, Marshall T W and Santos E 2000 *Journ. Mod. Opt.* **47** 1273-1287
[26] Brida G *et al* 2002 *Phys. Lett.* A **299** 121-124
[27] Brida G *et al* 2003 *Journ. Mod. Opt.* **50** 1757-1762
[28] Brida G *et al* 2007 *J. Phys.: Conf. Ser.* **67** 012047
[29] Santos E 2004 *Phys. Lett.* A **327** 33
[30] Santos E 2007 *Eur. Phys. Journ.* D **42** 501-509
[31] Croca J R 2007 *Int. J. Qu. Inf. (IJQI)* **5** 1; Croca J R, private communication

Journal of Physics: Conference Series **174** (2009) 012015 doi:10.1088/1742-6596/174/1/012015

Emergent/quantum gravity: macro/micro structures of spacetime[1]

B. L. Hu

Department of Physics, University of Maryland, College Park, Maryland 20742-4111 U.S.A

E-mail: blhu@umd.edu

Abstract. Emergent gravity views spacetime as an entity emergent from a more complete theory of interacting fundamental constituents valid at much finer resolution or higher energies, usually assumed to be above the Planck energy. In this view general relativity is an effective theory valid only at long wavelengths and low energies. We describe the tasks of emergent gravity from any ('top-down') candidate theory for the microscopic structure of spacetime (quantum gravity), namely, identifying the conditions and processes or mechanisms whereby the familiar macroscopic spacetime described by general relativity and matter content described by quantum field theory both emerge with high probability and reasonable robustness. We point out that this task may not be so easy as commonly conjured (as implied in the 'theory of everything') because there are emergent phenomena which cannot simply be deduced from a given micro-theory. Going in the opposite direction ('bottom-up') is the task of quantum gravity, i.e., finding a theory for the microscopic structure of spacetime, which, in this new view, cannot come from quantizing the metric or connection forms because they are the collective variables which are meaningful only for the macroscopic theory (valid below the Planck energy). This task looks very difficult or almost impossible because it entails reconstructing lost information. We point out that the situation may not be so hopeless if we ask the right questions and have the proper tools for what we want to look for. We suggest pathways to move 'up' (in energy) from the given macroscopic conditions of classical gravity and quantum field theory to the domain closer to the micro-macro interface where spacetime emerged and places to look for clues or tell-tale signs at low energy where one could infer indirectly some salient features of the micro-structure of spacetime.

1. Emergent / Quantum Gravity

Beneath the diversity of theories and proposals [1] there is at least general agreement from all camps on the definition of quantum gravity: It is a theory for the microscopic structure of spacetime. The major disagreement lies in whether such a theory may be obtained by quantizing general relativity (GR), a highly successful theory for the macroscopic structure of spacetime we are familiar with. The alternative viewpoint is that GR is a low energy effective theory, and

[1] *This is a descriptive summary of the main themes presented in invited talks in the following meetings: Workshop "From Quantum to Emergent Gravity: Theory and Phenomenology", Trieste, Italy. June 11-15, 2007; Peyresq Physics Meeting, June 16-22, 2007; Loops '07 Meeting, Morelia, Mexico, June 25-30, 2007; Workshop on "Condensed Matter meets Gravity", Lorentz Center, University of Leiden, Holland, August 27-31, 2007; Symposium on Foundations of Physics, University of Maryland, April 24-27, 2008; Workshop on "Emergent Gravity", MIT, August 25-29, 2008; 4th International Conference DICE2008, Castiglioncello, Italy, September 22-26, 2008.*

the metric and connection forms are the collective or hydrodynamic variables of some unknown microscopic theory. These variables will lose their meaning at shorter wavelengths and higher energies. Classical gravity in this view is emergent from quantum gravity (definition above) but not the classical limit or correspondence of a theory obtained by quantizing GR. Taking this view the effort in the past half a century in seeking suitable ways of quantizing this classical theory is considered largely misdirected, in that doing so will not lead to a microscopic theory of spacetime. In the analogy of a crystal made of atoms quantizing the vibrational modes yields phonons, not atoms. Finding the atomic structure of matter does not come from simply quantizing its collective degrees of freedom, but takes a very different path. We will discuss these two routes, namely, finding features in the microscopic structure of spacetime (quantum gravity) from the given macroscopic spacetime described by GR (we shall refer to this as the 'bottom up' approach – up here refers to energy scale), and recovering the emergent properties of the familiar macroscopic theory of spacetime from candidate theories of quantum gravity ('top-down') by studying the mechanisms and processes of emergence. Thus the title of this essay.

The emphasis and approaches, the goals and methodology of these two routes are very different. To motivate why this new view makes better sense I have given examples before from atomic - condensed matter physics, molecular and hydro-dynamics, quantum fluids and critical phenomena (see below). Instead of finding better ways to quantize GR the new challenge is to infer the microscopic structure from the known macroscopic phenomena, not an easy task as it entails reconstructing lost information. However, this is not a new demand, it has been the task for physicists for centuries. For this purpose, concepts and methods from nonequilibrium statistical mechanics and examples from strongly correlated many-body systems will probably play an essential role hitherto largely ignored in quantum gravity.

In the top-down direction one may think that deducing the macro theory from a *known* micro theory is easy. This is the attitude behind a 'theory of everything'. One can invoke familiar procedures such as taking the hydrodynamic or thermodynamics limits in classical physics or using the correspondence principle or applying effective field theory in quantum physics. But as we shall see, this may not be so straightforward as imagined because there are different types of emergence. If we don't know the underlying constituents we can learn from the characteristics of many emergent theories in the description of nature to understand and formulate the basic rules of emergence. We could then apply the insights gained to describe the conditions and processes or mechanisms for the emergence of spacetime. These are the new tasks for emergent gravity in the 'top-down' direction.

1.1. Emergence

The view that gravity is emergent can be reached in a number of ways. My own path of search and discovery is illustrated by the following observations: a) *Cosmology as 'condensed matter' physics* (1988) [2], pointing to the importance of recognizing that processes like cosmological phase transition, structure formation or even 'birth of the universe' scenarios are not only governed by the underlying general relativity and quantum field theories, but also follow the rules more often found in condensed matter physics. b) *Semiclassical (mean field) and stochastic (fluctuations) gravity as mesoscopic physics* (1994) [3], highlighting the key role played by noise and fluctuations in many important processes in cosmology and black hole physics. c) *Decoherence and the emergence of classical spacetime* from wavefunctions of the universe. We point out that the outcome of decoherence (see, e.g., [4]) and the existence of relatively stable quasi-classical domains resembling our universe depends critically on how coarse-grainings in the environment or histories are chosen and implemented, including considerations of the balance between variability and robustness in the advent of these structures [5]. And, in an open system [6] conceptual framework it is always important to make judicious choices of collective variables for the best description of macroscopic phenomena at every level of structure and their interactions (see also [7]). d) *General relativity as hydrodynamics* (1996) [8]: conservation laws aids in the decoherence of long wavelength modes [9, 10] and the appearance of quasiclassical domains, and e) *From stochastic gravity to quantum gravity* (1999) [11, 12]: How to retrieve quantum coherence information in the gravity sector from metric fluctuations induced by (higher order) correlations in the quantum field f) *Kinetic theory approach to quantum gravity* (2002) [13], via a new Einstein-Boltzmann hierarchy of equations relating the correlations in the matter to that in the gravity sector, with the Einstein equation as the lowest rung of this hierarchy. g)*Spacetime as condensate* (2005) [14]: Why it is not total nonsense to view the ultracold spacetime *today* as a quantum entity? What benefits can one derive from taking this view? What can we learn about quantum gravity from Bose-Einstein condensate (BEC)? The overarching goal is to find the microscopic constituents and their interactions from the collective dynamics of the macroscopic variables. (The last three themes will be summarily described in Sec. 4 below.)

1.1.1. Emergent Theories Emerging In the last three years this viewpoint is increasingly embraced by researchers from the string theory [15, 16], particle and nuclear physics [17, 18, 19, 20] communities in the light of string-gravity-gauge dualities [21, 22, 23, 16], and pursued by an increasing number of young researchers in the quantum gravity community as witnessed in the talks at the present meeting. Two major proponents of this view before it became popular are Volovik [24] and Wen [25].

Emergence is not a strange new concept. Originated in biology and sociology but also prevalent in condensed matter physics, emergence phenomena are more commonly encountered in nature than constructs deduced from the reductionist paradigm based on linear logical inductions. An extreme version of the latter is the theory of everything (TOE) which says that if we knew all the elementary constituents and their interactions we could deduce the structure and dynamics of everything in the physical world.

In reality, there are new emergent rules or laws governing the organization and dynamics of the basic constituents and new modes of interaction at every level of structure. This counterpoint is cannonized in Anderson's 1971 outcry (at a time when elementary particle physics and its philosophy seemed to overshadow other subfields): "More is different". Nonetheless it is well-known that progress in particle physics required the injection of ideas from condensed matter physics, symmetry breaking being a well- acknowledged example.

In truth these two aspects are always present in our quest to comprehend the physical world: Something assumed to be 'elementary' at a lower energy or with a coarser resolution at one time will later be found to be a composite at a higher energy or under a finer resolution. Though their emphasis and approaches are very different these two paradigms are equally important and functional in the development of our conception and understanding of nature. The degree of difficulty may vary. Constructing the macroscopic structure and dynamics from known microscopic constituents is often easier, such as getting hydrodynamics from molecular dynamics. Going the other way, i.e., deducing molecular features from hydrodynamics, is more difficult. For spacetime structures, the former is the task of emergent gravity, the latter is the goal of quantum gravity. Note, however, that even if we have the precise knowledge of the fundamental theory, seeing the emergent behavior at lower energies is not as straightforward as often presumed.

1.1.2. Deducible and non-deducible Emergence To understand the characteristics of emergent theories, and the rules of emergence one should distinguish in the start those which can be logically or methodically deduced from a microscopic theory and those which cannot, at least not without the benefit of knowing certain attributes of the macroscopic theory. Hydrodynamics from molecular dynamics or nuclear physics from QCD are examples of the former and quantum

Hall effect is often quoted as an example of the latter: One can construct useful theories only after such an effect was observed, but not (or highly unlikely) before. One can appreciate how nontrivial this task is by looking at the challenges of string theory or loop quantum gravity in reproducing the physical world of low energy phenomena.

1.2. Emergent Gravity: Macro from Micro

If spacetime is an emergent entity then its large scale structures should be deductible from the underlying constituents and their interactions, like a crystal structure or a condensate constructed from atoms. Collective (macro) features such as elasticity (in gravity, metric elasticity [26]) associated with such a (micro) structure can lend themselves to a nice geometric depiction but these features will cease to make sense at scales close to or smaller than an atom. In these cases, the task of emergent gravity is similar to the derivation of hydrodynamic equations of motion and thermodynamic laws from molecular dynamics.

1.2.1. Top-down: Micro to Macro Spacetime is the resultant large scale structure formed by the interaction of sub-level constituent particles. Even when the micro constituents are believed to be known – strings [27], loops [28], spin-foams [29], simplicials [30] – one still needs to show the macro limit exists and how it is attained from their interactions. It is a necessary criterion any candidate theory of quantum gravity needs to meet. For deductive emergent behavior, path could be tortuous but in principle attainable.

As is known from many examples of condensed matter physics, it is often a nontrivial task to deduce mesoscopic behavior from micro-dynamics: one usually encounters nonlinear or even nonlocal interactions in strongly correlated systems, and needs some ingenuity to identify collective variables at successive levels of structure to make the analysis simpler. Using the molecular to hydro dynamics example, this intermediate scale corresponds to finding the appropriate kinetic variables and the associated dynamics (e.g., use of maximal entropy laws at successive stages of complexity).

In addition to nonlinearity, nonlocal properties can emerge from every level of structure and dynamics. This makes the transition to macroscopic limit much more involved. It requires not just a linear progression of deductions from one single level, but the injection of new ideas at many levels of construction.

The above assumed that we are dealing with deductive-predictive theories. But we should be mindful that this macro manifestation could be via a nondeductive emergent behavior from the micro structures, in which case even micro- to macro-, sub- to super- structures would not be easy, some cases could even be impossible.

1.2.2. Emergence from existing theories of quantum gravity We can use some major candidate theories of quantum gravity (micro structure of spacetime) to examine the key issues listed above. I see the causal dynamical triangulation program pursued by Ambjorn and Loll et al (see [30] and earlier references therein) as the more intuitive and operationally accessible option. Regge's original conception (of Regge Calculus and Regge-Ponsano calculus) is to reduce the essential ingredients of spacetime to its rudimentary constructs, i.e., the simplices, such as having the curvature of spacetime residing on the vertices. One can see how piecing together these simplices give rise to different spacetime structures, many of them have fractal dimensions and few grow to the size of our universe. This scheme is useful to see how the macroscopic structure emerges, a welcoming recent result is the 4-dimensionality of our spacetime. However, since the simplices are chosen to contain the key features of the macroscopic spacetime (e.g., curvature), they are good building blocks [2] yet may not contain readily decipherable information about the true degrees of freedom in the microscopic substructure (like strings or loops, if they are proven so). Spin-foam theory [29] is aesthetically appealing and rich in mathematical structure. But I still cannot see how a macro variable such as the connection form when quantized would turn into a micro-variable (such as the basic differences between a phonon and an atom). As for string theory, the AdS/CFT correspondence [21, 22] is often cited as an example of emergent gravity. This to me is not a real answer. One needs to show that both gauge theory and gravity emerge from the same more basic theory, presumably the M theory, but that is a hope short of a proof [31]. For recent developments of these theories see [32]. For other proposals, especially the causal set theory, see [33] and [1]. To claim emergence each of the candidate theories of microscopic structure of spacetime needs to identify at least the principal mechanisms responsible for generating the macroscopic spacetime with distinct identifiable properties. Gauge-gravity duality relation [16] has been applied to many branches of physics from gauge fluid viscosity in RHIC [18, 34] to quantum phase transition in condensed matter (in terms of dyonic black holes) [35] and superconductivity [36] with surprising results. I will have a short comment on the implications of these recent developments in the last section.

1.3. Quest for Quantum Gravity: A theory for the microscopic structures of spacetime

Compared to working out the macroscopic limits of a known microscopic theory, going the reverse way is always more difficult because of the missing information. One has only the macroscopic features at hand which are often grossly coarse-grained from the microscopic. From the degraded information one hopes to catch a glimpse of the micro structure. The task is daunting. But this is how physics has progressed through the centuries.

1.3.1. Bottom-up: Macro to Micro To proceed from bottom up we think that focusing on two key elements may help:

1) *Topological structures*, because they are more resilient to evolutionary or environmental changes. For examples see the works of Volovik [24] (He^3 analog, Fermi surface) and Wen [25] (string-nets, emergent light and fermions; see also [37]).

2) *Noise-fluctuations.* Fluctuations could reveal some sub-structural contents and behavior, such as deducing molecular interaction from hydrodynamical fluctuations, or finding the universality classes of systems from the critical exponents in a phase transition.

1.3.2. Common features of collective states built from different constituents In the depiction of the structure and properties of matter there are two almost orthogonal perspectives: One is by way of its constituents and interactions, the other according to its collective behavior. If we regard this chain of QED - QCD - GUT - QG as a vertical progression depicting the hierarchy of basic constituents, there is also a horizontal progression in terms of the stochastic - statistical - kinetic - thermodynamic/hydrodynamic depiction of the collective states. There exist similarities between matters in the same collective state (e.g., hydrodynamics) but made from different constituents. Macroscopic behavior of electron plasmas are similar in many respects to the quark-gluon plasma. Indeed, one talks about magneto-hydrodynamics from Maxwell's theory as well as magneto-chromo-hydrodynamics from QCD. In this long wavelength, collision-dominated regime, their behavior can both be adequately depicted by the laws of hydrodynamics. The underlying micro-theories are different, but the hydrodynamic behavior of these constituents are similar. The macroscopic, hydrodynamic equations and their conservation laws are all based

[2] It is like the fine powders ground from a diamond, even though much smaller in scale, each speckle still possesses the crystal structure of the macroscopic object. Without a drastic increase in resolution it cannot reveal the underlying atomic structure.

on the dynamical and conservation laws of microphysics (e.g., Newtonian mechanics) but now in terms of a new set of collective (hydrodynamic) variables, this new theory of the macroscopic behavior emerges with a simple yet distinct life of its own.

2. Nonlocality and Stochasticity in quantum and statistical mechanics

With increased attention to quantum information in recent years one hears more about 'nonlocality' in quantum mechanics, referring to the issues raised in EPR [38] and Bell's inequalities. Even though predictable correlation exists between spatially-separated quantum states and quantum entanglement can exist between objects outside each other's light-cones [39], quantum mechanics is unambiguously local [40], as is quantum field theory. [3]

Nonlocality in statistical mechanical is an older subject, perhaps less conspicuous than that which is mis-conjured in quantum mechanics. We will start with a simple observation which illustrates two main themes of this exposition: The first theme is that locality or nonlocality depends sensitively on the level of structure and dynamics it pertains to or the degree of precision in one's measurements. Micro-locality in general has nothing to do with macro-locality. The second theme is that nonlocality (correlations) and stochasticity (fluctuations) often appear together, and one could use noise and fluctuations to probe into the nonlocality of structures and dynamics.

2.1. Collective Variables constructed from constituent variables

At the risk of appearing naive and pedantic, I'll give two household examples to illustrate this point. Perhaps the first collective behavior we learned in physics is wave phenomena. There, many individual particles each undergoing a simple harmonic motion in a well-coordinated manner give rise to waves propagating in the medium of these particles. Already here we see the relation between the micro dynamics and the macro behavior. The rules of wave mechanics is different from that of the constituent particles. The waves obey a wave equation while the molecules obey Newton's second law. Surely one can deduce the macro behavior from the micro dynamics but the reverse relation is many to one: transverse mechanical and electromagnetic waves or longitudinal sound waves share the same waves mechanics description. One feature to observe relevant to our theme is the relation of locality to nonlocality. Waves can propagate from one point in space to another (snapshot graph) which is correlated with the evolution in time at any one point in space (history graph). However each individual particle only undergoes a simple harmonic motion around its point of origin and there is nothing nonlocal (used here in a non-technical sense) about it. Another feature to observe is that quantizing the wave degrees of freedom (sound modes becomes phonon) neither yields the dynamics of the constituent particles nor reveal any microscopic information about the molecules. The conception of wave phenomena simply ceases to exist and the theoretical constructs of wave mechanics irrelevant at the microscopic level. For the description of atomic motion we need to use the quantum wave functions for the atom but that is a far cry from the mechanical waves.

As another commonly encountered example, consider two depictions of gas or fluid dynamics, one at the level of molecular dynamics, the other of hydrodynamics. A molecule will collide with many others at very high speed for a long time before diffusing a short distance. That molecule's locality in a collective entity as a gas or fluid element is very different from that induced by following its own history at the micro scale where multiple collisions take place along the trajectory. Other molecules arriving at the same time in this gas or fluid element bear no locality relation with this particular molecule before or after the instant of collision (assuming a short-ranged or contact potential). The smoothness and continuity of the fluid elements'

[3] We thus advise against the use of the word 'nonlocal' in this context because it leads to unnecessary confusion.

movement described by the Euler equation as reported by a macroscopic observer belie the stochastic nature of molecular dynamics [4].

2.2. Locality is specific to the choice of observables

These two elementary examples show that the notion of locality is very specific to the observables chosen for the description of a particular physical phenomenon of interest (e.g., molecular dynamics or wave propagation). Nonlocality often appears when one tries to translate physics expressed in one set of collective variables suitable for one level of structure to another set.

2.3. Nonlocality and Stochasticity in statistical mechanics

I shall comment here only on temporal nonlocality, i.e., processes with memory (non-Markov). This is common in open system dynamics (see, e.g., [42] and references therein). Nonlocality in space or spacetime is more involved. An example of how this issue is addressed can be found in the causal sets approach of Sorkin [43]. For two interacting subsystems the two ordinary differential equations governing each subsystem can be written as an integro-differential equation governing one such subsystem, thus rendering its dynamics non-Markovian, with the memory of the other subsystem's dynamics carried by the non-local kernels. All this is happening in a closed system except now one has shifted the attention to one of its subsystems. Should the other subsystem possess a much greater number of degrees of freedom (called environment) and are coarse-grained in some specified way, the nonlocal kernels are then responsible for the appearance of dissipation and noise. It is the act of coarse-graining which renders open this particular subsystem of interest. (For a general discussion, see, e.g., [44, 45].)

Nonlocality is linked to stochasticity in nonequilibrium dynamics. Nonlocal dissipation and nonlocal fluctuations (colored noise) arise naturally in the open-system dynamics of Langevin and the effectively-open system dynamics of Boltzmann, or the dynamics of correlation hierarchy. These features originate from the choice of *coarse-graining* measures and *backreaction* processes. NonMarkovian dynamics is an example of nonlocality in time, meaning that prior history or past memories enter in the determination of future time development.

Another angle towards nonlocality and stochasticity is correlation. Correlation functions measure nonlocality and can be related to fluctuations. (e.g., one form of fluctuation-dissipation theorem can be phrased in terms of correlations). Strongly interacting and correlated systems offer an excellent arena to see how these two aspects are played out: locality at one level versus nonlocality at a different level, and how coarse-graining leads to stochasticity. Many condensed matter physics models offer these insights [46].

3. Nonlocality and Stochasticity in Quantum Gravity

There is probably more confusion in our understanding and use of the notion of nonlocality in quantum gravity. An example in loop quantum gravity is how to relate a weave state to spacetime. This was addressed in a classic paper by Ashtekar, Rovelli and Smolin [47]: A weave state is a kinematical state designed to match a given slowly varying classical spatial metric. The concept of quantum threads (say, from a spin-network) weaved into a fabric (manifold) of classical spacetime already tacitly assumes (wrongly, as is now known) a particular kind of micro to macro transition, where there is a simple correspondence or even equivalence between locality at the micro and the macro levels. An improvement over the original suggestion was the so-called nonlocal weave states which leave marks on the fabrics of spacetime [48]. Bombelli, Corichi and Winker [49] use combinatorial methods to turn random weave (micro) states into (macro) states of spacetime manifold. But this remains an oversimplification, as the low-brow example of emergence from molecular to hydro- dynamics has illustrated. Markopoulou and

[4] Quantum correlations and entanglement have added layers of complexity [41].

Smolin [50] recently pointed out that most weave states, including Bombelli et al's, assume that these weave states all satisfy an unstated condition of locality (edges connect two nodes of Planck length metric distance), and that there are plenty of weave states which do not satisfy this condition. The question is: How does one weave from these nonlocal micro states a fabric (manifold) with the familiar macroscopic locality? From our view, (non) locality at one level may have little to do with (non) locality at other levels.

To unravel these two aspects in quantum gravity one can first try to understand how these issues play out in known physical models, then apply to spacetime, incorporating its special features such as diffeomorphism invariance (which, in the emergent gravity view, is associated only with the emergence of spacetime manifold and is thus no longer an issue at the deeper level of its sub-structures). Many symmetries we are familiar with associated with the macroscopic spacetime will become meaningless at the microscopic level (quantum gravity). One should also use the various candidate theories of quantum gravity, such as strings, loops, spin-nets, spin foams and causal sets, to examine the sense of locality at different levels of structure, and see how they could be related to locality in our macroscopic spacetime.

3.1. Metric Fluctuations and Stochastic Gravity

Keeping gravity classical but allowing for the averaged value of quantum matter field as source constitutes the theory of *semiclassical gravity*. Including the consideration of the fluctuations of quantum matter fields as source constitutes the theory of stochastic semiclassical gravity (see [12] for a review). Metric fluctuations, both intrinsic (in the gravitational field) and induced (by quantum matter fields) are the central figures in this relatively new theory. With a wider acceptance of the view that gravity is emergent and that quantizing general relativity does not yield a theory of the microscopic structure of spacetime, stochastic gravity's theoretical significance and practical applications are better recognized and utilized. For instance, the influence of induced metric fluctuations on cosmological structure formation is not easily considered in the traditional methods [51]. The Einstein-Langevin (E-L) equation has also been applied recently to the analysis of backreaction of Hawking radiation and metric fluctuations near the event horizon of an evaporating black hole [52].

Stochastic gravity can be derived in the quantum open system conceptual framework [12]. As with all open systems nonlocality and stochasticity in their dynamics are innate features. They appear in the nonlocal dissipation and colored noise kernels which give rise to history-dependent and stochastic trajectories. It is anticipated that these features are prevalent as one traverses any two adjacent levels of structure in any of the macro, meso and micro domains.

3.2. Stochastic in relation to Semiclassical and Quantum Gravity

Intuitively, the difference between quantum and semiclassical gravity is that the latter loses all the coherence in the quantum gravity sector. Stochastic gravity improves on semiclassical gravity in that partial information related to the coherence in the gravity sector is preserved. This is reflected in the backreaction from the quantum fields and manifests as induced metric fluctuations. The noise terms carry quantum information absent in semiclassical gravity. The coherence in gravity is related to the coherence in the matter field, as a complete quantum description should be given by a coherent wave function of the combined matter and gravity sectors. Since the degree of coherence can be measured in terms of correlations the strategy of this approach is to examine the higher correlations of the matter field, starting with the two point function of the energy momentum tensor in order to probe into the partial quantum coherence remaining in the gravity sector. Note that in retrieving this quantum coherence of the gravity sector, we do not pretend to be constructing a microscopic theory of spacetime. As long as one uses metric or connection as a dynamical variable, the theory remains macroscopic. For our purpose here metric fluctuations can provide precious information about how spacetime behaves

Journal of Physics: Conference Series **174** (2009) 012015 doi:10.1088/1742-6596/174/1/012015

at the interface (phase transition or cross-over) between the micro and macro domains, not unlike fluctuations (in, e.g., heat capacity at constant volume) near a critical point contains valuable information about the universality class of the system undergoing such a phase transition.

If we view classical gravity as an effective theory, i.e., the metric or connection functions as collective variables of some fundamental constituents which make up spacetime in the large and general relativity as the hydrodynamic limit, we can also ask if there is a mid-way weighing station like kinetic theory from molecular dynamics, from quantum micro-dynamics to classical hydrodynamics. This transition involves both the micro to macro transition and the quantum to classical transition, characteristics of the mesoscopic regime. Stochastic gravity is of the nature of a mesoscopic theory: the noise in the Einstein-Langevin (E-L) equation contains the 4th order correlation of the quantum field (or gravitons when considered as matter source). One can work out higher correlation functions of the quantum matter field and obtain from a generalized E-L equation the higher order induced metric fluctuations.

In our research program several routes have been proposed to move from the bottom up, that is, starting from the well-established foundation of semiclassical gravity, adding the considerations of fluctuations and trying to inch our way towards the macro-micro interface from below, with the hope of unraveling some microscopic features of spacetime. We describe a few below: a) a kinetic theory approach to quantum gravity, b) universal 'metric conductance' fluctuations; and c) spacetime as condensate[5].

4. Routes from bottom up

4.1. Spacetime Correlations and Fluctuations: Einstein-Boltzmann hierarchy of equations

To better appreciate the relation of this mesoscopic regime between the macro and the micro, we need to explain the representation of kinetic theory in terms of correlation dynamics. In [53] Calzetta and I introduced the master effective action for an interacting quantum field where one can obtain the hierarchy of Schwinger-Dyson equations. From this one can derive the kinetic equations [54] and an expression for the correlation noise arising from the slaving of the higher correlation functions. The resulting equation is known as the stochastic Boltzmann equation. This kinetic theory structure emerging from an interacting quantum field illustrates how the macro (hydro) dynamics is linked to the micro (molecular) dynamics. The minor difference is that there are two distinct sectors, classical gravity and quantum matter field. This enables one to adopt an open system approach in treating the relation between them, such as the Einstein-Langevin equation, while using the Boltzmann-BBGKY hierarchy to systemize the quantum matter field sector. The linkage between these two sectors exists at every level of structure and dynamics, the lowest being the semiclassical Einstein equation which involves only the expectation value of the stress energy tensor. The next level up is stochastic gravity which involves the two point function of the stress-energy tensor. The hierarchy of equations governing the higher order induced metric fluctuations from the higher correlations of the stress-energy tensor of the quantum field is what I called the Einstein-Boltzmann hierarchy of equations which provides a staircase towards the micro-structure in this kinetic theory approach to quantum gravity[6].

[5] The following section summarizes the ideas described in Sec 1.1. Readers familiar with them can skip over to the next section.

[6] This is not the Einstein-Boltzmann equation in classical general relativity and kinetic theory – it frames the classical matter in the Boltzmann style as source of the Einstein equation. Our Boltzmann-Einstein hierarchy of equation refers to the layers of structure in the gravity sector. 'Boltzmann' is to show the kinetic theory nature, and 'Einstein' to show its spacetime structure, albeit these two giants provide their respective theories only for the lowest correlation order and its dynamics: in distribution function of spacetime and in geometro-hydrodynamics respectively.

4.2. Strongly correlated systems: Spacetime Conductance Fluctuations
In earlier expositions of this subject matter, I proposed to view semiclassical and stochastic gravity as mesoscopic physics [3]. Viewing the issues of correlations and quantum coherence in the light of mesoscopic physics, the stress-energy two point function on the right hand side of the Einstein-Langevin equation is analogous to conductance which is given by the current-current two point function. What this means is that we are actually calculating the transport function of (the matter particles as depicted by) the quantum fields. Following Einstein's keen observation that spacetime dynamics is determined by (while also dictates) the matter (energy density), we expect that the transport function represented by the current correlation in the matter (fluctuations of the energy density) may also have some physical significance or even (perhaps less refined) a geometric representation at a higher energy scale (akin to the pseudo-Riemannian spacetime of GR at the low energy limit).

This is consistent with viewing general relativity as hydrodynamics: Conductivity, viscosity and other transport functions are hydrodynamic quantities. For many practical purposes we don't need to know the details of the fundamental constituents or their interactions to establish an adequate depiction of the low or medium energy physics, but can model them with semi-phenomenological concepts (like mean free path and collisional cross sections) [7]. When the interaction among the constituents gets stronger, effects associated with the higher correlation functions of the system begin to show up. Studies in strongly correlated systems are revealing in these regards [55, 56, 57]. For example, fluctuations in the conductance – from the 4 point function of the current – carry important information such as the sample specific signature and universality. Although we are not quite in a position, technically speaking, to calculate the energy momentum 4 point function, thinking about the problem in this way may open up many interesting conceptual possibilities, e.g., what does universal conductance fluctuations mean for spacetime and its underlying constituents? (In the same vein, I think studies of nonperturbative solutions of gravitational wave scattering [58] will also reveal interesting information about the underlying structure of spacetime beyond the hydrodynamic regime.) Thus, viewed in the light of mesoscopic physics, with stochastic gravity we may probe into the higher correlations of quantum matter and the associated excitations of the collective modes in geometro-hydrodynamics and the kinetic theory of spacetime micro-dynamics.

4.3. Spacetime Condensate: the Universe as the Ultimate Macroscopic Quantum Phenomenon
Working within this novel conceptual framework of geometro-hydrodynamics, we suggest a new way to look at the nature of spacetime inspired by Bose-Einstein condensate (BEC) physics. We ask the question whether spacetime could be considered a condensate, a low temperature quantum entity, and if so, its immediate implications. This is an utterly unconventional thought because the universe below the Planck energy is still at a very high temperature, and even more radically, it implies macroscopic quantum phenomena observed in today's universe. The challenge here is to see if such a view could make outstanding issues like the dark energy easier to understand. Note also that unlike BEC in atomic physics we want to describe its salient features even *without the knowledge of what the 'atom of spacetime' is.* Please refer to [14] for background discussions such as the meaning of a condensate in different context, how far this idea can sustain, its advantages and pitfalls, and its implications on the basic tenets of physics and existing programs of quantum gravity. See also [59]

4.3.1. Unconventional view 1: All sub-Planckian physics are low temperature physics We believe that the physical laws governing today's universe are valid all the way back to the GUT

[7] In the mesoscopic domain the simplest kinetic model of transport using these concepts are no longer accurate. One needs to work with system-environment models and keep the phase information of the collective electron wave functions, in so doing extending the Buttiker-Landauer formula.

(grand unification theory) and the Planck epochs, even at the Planck temperature $T_{Pl} = 10^{32}K$. Since the spacetime structure is supposed to hold for all sub-Planckian eras, if we consider spacetime as a condensate today, it should have remained so from the Planck era onwards.

In fact, one should push this concept to its limit and come to the conclusion that all known physics today, as long as a smooth manifold structure remains valid for spacetime, the arena where all physical processes take place, are low-temperature physics. Spacetime condensate began to take shape at below the Planckian temperature T_{Pl}, according to our current understanding of the physical laws. In this sense spacetime physics as we know it is low temperature hydrodynamics, and, in particular, today we are dealing with ultra-low temperature physics, similar to superfluids and BECs.

The metric or connection forms are hydrodynamic variables, and most macroscopic gravitational phenomena can be explained as collective modes and their excitations (of the underlying micro-theory): from gravitational waves in the weak regime as perturbations, to black holes in the strong regime as solitons (nonperturbative solutions). There may even be analogs of turbulence effects in geometro-hydrodynamics, possibly predictable and detectable when our numerical techniques and observational skills are improved.

4.3.2. Unconventional view 2: Spacetime is, after all, a quantum entity An even more severe difficulty in viewing spacetime as a condensate is to recognize and identify the quantum features in spacetime as it exists today, not at the Planck time. The conventional view holds that spacetime is classical below the Planck energy, but quantum above. That was the rationale for seeking a quantum version of general relativity, beginning with quantizing the metric function and the connection forms. Our view is that the universe is fundamentally a quantum phenomena, but at the mean field level the many body wave functions (of the micro-constituents, or the 'atoms' of spacetime) which we use to describe its large scale behavior (order parameter field) obey a classical-like equation, similar to the Gross-Pitaevsky equation in BEC, which has proven to be surprisingly successful in capturing the large scale collective dynamics of BEC [60], until quantum fluctuations and strong correlation effects enter into the picture [61].

Could it be that the Einstein equation depicting the collective behavior of the spacetime quantum fluid has the same footing as a Gross-Pitaevsky equation for BEC? The deeper layer of structure is ostensibly quantum, it is only at the mean field level that the many-body wave function is amenable to a classical description [8].

4.3.3. The Universe as a Macroscopic Quantum Phenomenon The obvious challenge is, if the universe is intrinsically quantum and coherent, where can one expect to see the quantum coherence phenomena of spacetime? Here again we look to analogs in BEC dynamics for inspiration, and there are a few useful ones, such as particle production in the collapse of a BEC (called bosenova in the experiment of Donley et al [62]). One obvious phenomenon staring at our face is the vacuum energy of the spacetime condensate, because if spacetime is a quantum entity, vacuum energy density exists unabated for our present day late universe, whereas its origin is somewhat mysterious for a classical spacetime in the conventional view. It is highly desirable to explore the implications of this view on the cosmological constant and coincidence problems. One encouraging fact is that in most condensed matter systems the collective excitations have energies of the same order of magnitude as the ground state, rarely with a severe discrepancy as in particle physics scales (the gauge hierarchy problem). The observed fact that the dark energy is 0.7 of the total budget and not 10^{-120} came as a big surprise. But not so much so

[8] In truth, for any quantum system which has bilinear coupling with its environment or is itself Gaussian exact (or if one is satisfied with a Gaussian approximation description) the equations of motion for the expectation values of the quantum observables have the same form as its classical counterpart. The Ehrenfest theorem interwoven between the quantum and the classical is one common example under these conditions.

from the viewpoint of spacetime as a condensate. The vacuum energy of the condensate and its excitations are similar in magnitude. One can investigate this issue from this perspective, using well-studied models of superconductivity and superfluidity to explore the implications of viewing spacetime as a condensate [63].

5. Quantum + Gravity: Look harder at low energy physics

Rather than forcing the union of quantum mechanics (QM) and general relativity (GR), such as is attempted to quantize GR, to try to get the microscopic structures of spacetime, it pays to explore more thoroughly the intersection of these two theories at today's low energy in order to find out the true nature of their conflicts and their implied meanings. Perhaps both theories lose their meaning and applicability at a higher energy. In the emergent gravity view, GR is an effective theory which is no longer valid at the Planck energy. There are also speculations that quantum mechanics is also an emergent theory which fails at a deeper level where nonlocality and stochasticity appear [64, 65, 66, 67, 68]. Our improved understanding of basic issues like quantum measurement brings forth new issues in this arena such as the effect of gravity in quantum decoherence [69] and collective effects in macroscopic quantum systems [70]. Penrose has made challenging theoretical claims [71] and proposed interesting experimental tests [72] in this realm. We share the same view in the underlying assumption of Penrose's proposal, namely, low energy physics could be useful in our search for a new theory which encompasses both GR and QM as limiting cases of it. We mention two aspects worthy of further probing and deliberations.

5.1. Quantum and gravitational low energy phenomena

To explore the interplay of quantum mechanics or field theory and gravitational phenomena at today's energy, one could start with the simplest set up of the interaction between a particle (with charge and mass) or atom (entity with internal degrees of freedom) with a gravitational field under the influence of a quantum field, including finer considerations of quantum correlation, entanglement and decoherence. This is necessary for designing laboratory experiments for high precision tests of quantum effects in gravitation and cosmology. It can also address basic issues of quantum measurement. These preparatory probes are useful for finding anomalies or 'cracks' in the existing theoretical structure which is the starting point of the 'bottom-up' approach to QG.

In a quantum field a classical particle will not move along a geodesic [73] because of the influence of quantum fluctuations. We know how these effects show up as stochastic forces acting on its mean trajectory, i.e., via the Langevin equation [74, 75]. Particle motion is affected both by the intrinsic and induced (called active and passive by Ford [76]) fluctuations of the quantum field. Studying the effects of an atom in a gravitational field (see, e.g., [77]) and the effects of a gravitational field on atom-optical or quantum fluid devices (e.g., SQUID) can also provide some channel to explore subtle effects hitherto unknown arising from the interplay between quantum systems and a classical gravitational field. One could tap into the capabilities of today's atomic-optical and superconductor / superfluid experimentations which have reached a high precision level for these purposes (many experimental schemes have been proposed, see, e.g. [78, 79]) .

From a theoretical viewpoint many of the systems of relevance to our goals are at the mesoscopic scale. In order to check against the results from prospective precision experiments our current understanding of the interaction of mesoscopic quantum systems with gravity need be improved [80]. The interaction of strongly correlated quantum systems with coherent classical gravitational fields using, e.g., the Schwinger-Keldysh effective action method for quantum transport [81]), real time description of macroscopic and dissipative quantum tunneling [82] (which can be applied to 'Birth' of the Universe scenarios), and analysis of the properties of the stochastic Schrodinger equation in the presence of gravity are just a few sample problems.

Coherence (even for classical fields) from a much extended space and time range can affect the local observations. Thus all nonlocal and memory effects should be properly accounted for. (One sees this in gravitational radiation reaction). The best way to ensure these effects are not lost is to do a quantum field calculation from the beginning and only at the end take the non-relativistic quantum limits to compare the results with experiments. The physics we are probing is very delicate, thus prompting this caution.

These land-based or space experiments in Earth's vicinity are designed to test the interaction of quantum probes (atom-optical or superconductivity devices) with weak gravity, but to really get to the heart of the problem, one needs to probe quantum effects in strong gravitational fields, such as near a black hole or in the early universe. Obviously it is not easy to do experiments near a real black hole but analog models may offer some valuable insights [84]. By contrast, observations in cosmology can provide bounds on quantum processes in the early universe, at the GUT or even the Planck scales. Investigations like this could provide some hints on how macro features may emerge from the underlying constituents. Again we do not want to imply that such studies whose backdrop assumes a manifold structure whose dynamics is described by GR will give us directly the microscopic structure, but they may reveal some traits of it at the interface. A more direct link is provided by the duality between strong gravity (black holes) and quantum gauge fields [16] in a holographic view , which we mention below with a comment.

5.2. Implications of results from gauge-gravity duality

In recent years the gauge-gravity duality relation has been expounded in application to a variety of sub-fields of physics with astounding results. Examples are viscosity of strongly interacting gauge fluids obtained from gravitational perturbations from black holes in an AdS space [17, 18], gluon-gluon scattering and deconfinement transition in QCD [85, 86] related to Hawking-Page transition [87] in hot Schwarzschild-AdS space, quantum phase transition [35] and superconductivity [36].

What do these unexpected connections suggest, and what essential elements enter into the duality? AdS-CFT is built on the holography principle [23], i.e., physics in a classical AdS space in the bulk is related to conformal field theory (CFT) on the boundary. Thus one could get information on strongly interacting quantum gauge fields (such as viscosity in a strongly interacting fluid or deconfinement transition in QCD) by examining the corresponding gravity effects in AdS space (such as quasi-normal modes of black holes and the Hawking-Page transition, respectively).

A very attractive aspect of this approach as I see it is that both parties entering in this duality relation, quantum gauge field and classical gravitational field, are familiar objects in ordinary low energy physics. Although this duality was first discovered in the context of string theory one does not need to invoke any progenitor theory at the Planck scale. This is an example of what we mean by looking more closely at the relations between quantum (here the strongly interacting gauge fields) and gravity (here the black holes, which are non-perturbative solutions in GR) at today's low energy.

Another observation I can make related to the main theme discussed here is that in all the applications of this duality principle we have seen so far only classical gravity enters, not quantum gravity. The only entity in gravity which is quantized is the linear gravitational perturbations (gravitons, or spin-2 particles, acting in the same role as phonons in solids). There is no need or place for quantum geometry in this relation which governs macroscopic phenomena between quantum gauge fields and classical gravity.

Acknowledgments

I thank the organizers of the meetings and hosts of my visits where some of these ideas were presented, in reverse chronological order, Thomas Elze, Olaf Dreyer, Fotini Markopoulou, Laurent Friedel, Lee Smolin, Rafael Sorkin, Jeffrey Bub, Pawel Mazur, Jan Zaanan, Edgard Gunzig, Alejandro Corichi, Robert Oeckl, Stefano Liberati and Stephen Adler for their invitations and hospitality. I am much obliged to many of my collaborators and colleagues for discussions on the wide range of topics over the years, notably, Charis Anastopoulos, Esteban Calzetta, Ted Jacobson, Albert Roura and Enric Verdaguer. I have also been influenced by exchanges with or reading the papers of T.K. Ng, Grisha Volovik and X.-G. Wen. This work is supported in part by NSF grants PHY-0601550 and PHY-0801368.

References

[1] Oriti D 2008 ed *Approaches to Quantum Gravity* (Cambridge, UK: Cambridge U. Press) – contains reviews of different schools of thoughts on quantum gravity.

[2] Hu B L 1989 COSMOLOGY AS 'CONDENSED MATTER' PHYSICS – invited talk at Third Asia-Pacific Physics Conference, Hong Kong, June 1988 – *Proceedings* ed K Young (Singapore: World Scientific Publ. Co.) (*Preprint* arXiv:gr-qc/9511076)

[3] Hu B L 1997 SEMICLASSICAL GRAVITY AND MESOSCOPIC PHYSICS – invited talk at International Symposium on Quantum Classical Correspondence, Drexel University, Philadelphia, September 1994 *Proceedings* ed D H Feng and B L Hu B L (Boston: International Publishers) (*Preprint* arXiv:gr-qc/9511077)

[4] Gell-Mann M and Hartle J B 1993 *Phys. Rev.* D **47** 3345; Omnes R 1994 *The Interpretation of Quantum Mechanics* (Princeton: Princeton U. Press); Giulini D *et al* 1996 *Decoherence and the Appearance of a Classical World in Quantum Theory* (Berlin: Springer); Zurek W H 2003 *Rev. Mod. Phys.* **75** 715

[5] Hu B L 1994 *The Physical Origin of Time-Asymmetry* ed J J Halliwell, J Perez-Mercader and W H Zurek (Cambridge, UK: Cambridge U. Press) (*Preprint* arXiv:gr-qc/9302021

[6] See, e.g., Davies E B 1976 *The Quantum Theory of Open Systems* (London: Academic Press); Lindenberg K and West B J 1990 *The Nonequilibrium Statistical Mechanics of Open and Closed Systems* (New York: VCH Press); Weiss U 1993 *Quantum Dissipative Systems* (Singapore: World Scientific)

[7] Giddings S B, Marolf D and Hartle J B 2006 *Phys. Rev.* D **74** 064018

[8] Hu B L 1996 General Relativity as Geometro-Hydrodynamics – invited talk at Second Sakharov Conference, Moscow, May 1996 *Preprint* arXiv:gr-qc/9607070

[9] Brun T A and Halliwell J J 1996 *Phys. Rev.* D **54** 2899; Halliwell J J 1998 *Phys. Rev.* D **58** 105015; *do.* 2003 *Phys. Rev.* D **68** 025018

[10] Hartle J B, Laflamme R and Marolf D 1995 *Phys. Rev.* D **51** 7007

[11] Hu B L 1999 *Int. J. Theor. Phys.* **38** 2987 (*Preprint* arXiv:gr-qc/9902064)

[12] Hu B L and Verdaguer E 2003 *Class. Quant. Grav.* **20** R1-R42. (*Preprint* arXiv:gr-qc/0211090); *do.* 2004 *Living Reviews in Relativity* **7** 3 (update: *Preprint* arXiv:0802.0658)

[13] Hu B L 2002 A KINETIC THEORY APPROACH TO QUANTUM GRAVITY – invited talk at 6th Peyresq Meeting, France, June 2001; *do.* 2002 *Int. J. Theor. Phys.* **41** 2111-2138 (*Preprint* arXiv:gr-qc/0204069)

[14] Hu B L 2005 *Int. J. Theor. Phys.* **44** 1785 (*Preprint* arXiv:gr-qc/0503067)

[15] Seiberg N 2005 Emergent Spacetime – rapporteur talk at 23rd Solvay Conference in Physics, December 2005 *Preprint* arXiv:hep-th/0601234)

[16] Horowitz G and Polchinsky J 2006 in [1] (*Preprint* arXiv:gr-qc/0602037)

[17] Horowitz G T and Hubeny V E 2000 *Phys. Rev.* D **62** 024027

[18] Kovtun P K, Son D T and Starinets A O 2005 *Phys. Rev. Lett.* **94** 111601

[19] Herzog C P, Kovtun P, Sachdev S and Son D T 2007 *Phys. Rev.* D **75** 085020

[20] Shuryak E 2006 Emergent Theory of Strongly Coupled Quark-Gluon Plasma – talk at Yukawa Institute meeting (YUKIS), November 2006 *Preprint* arXiv:hep-ph/0703208

[21] Maldacena J M 1998 *Adv. Theor. Math. Phys.* **2** 231; *do.* 1999 *Int. J. Theor. Phys.* **38** 1113

[22] Witten E 1998 *Adv. Theor. Math. Phys.* **2** 253; Gubser S S, Klebanov I R and Polyakov A M 1998 *Phys. Lett.* B **428** 105

[23] Brown J D and Henneaux M 1986 *Commun. Math. Phys.* **104** 207-226; 't Hooft G 1993 Dimensional Reduction in Quantum Gravity *Preprint* arXiv:gr-qc/9310026; Susskind L 1993 *Phys. Rev. Lett.* **71** 2367; Bousso R 2002 *Rev. Mod. Phys.* **74** 825-874

[24] Volovik G E 2003 *The Universe in a Helium Droplet* (Clarendon Press); *do.* Fermi-point scenario for

emergent gravity – talk at Trieste workshop "From Quantum to Emergent Gravity: Theory and Phenomenology" *Preprint* arXiv:0709.1258

[25] Wen X G 2004 *Quantum Field Theory of Many-Body Systems* (Oxford: Oxford U. Press); Levin M and Wen X G 2003 *Phys. Rev.* B **67** 245316; *do.* 2005 *Phys. Rev.* B **71** 045110

[26] Sakharov A D 1967 *Akad D. Nauk S. S. R.* **177** 70-71; *do.* 1968 *Sov. Phys. Doklady* **12** 1040-1041

[27] Green M B, Schwarz J H and Witten E 1990 *Superstring Theory* (Cambridge, UK: Cambridge U. Press); Polchinsky J 1998 *Superstring Theory* (Cambridge, UK: Cambridge U. Press)

[28] Rovelli C 2004 *Quantum Gravity* (Cambridge, UK: Cambridge U. Press); Gambini R and Pullin J 2004 *Loops, knots, gauge theories and quantum gravity* (Cambridge, UK: Cambridge U. Press 1996); Ashtekar A and Lewandowski J 2004 *Class. Quant. Grav.* **21** R53 (*Preprint* arXiv:gr-qc/0404018); Thiemann T 2008 *Introduction to Modern Canonical Quantum General Relativity* (Cambridge, UK: Cambridge U. Press) (*Preprint* arXiv:gr-qc/0110034); see also: Nicolai H, Peeters K and Zamaklar M 2005 Loop quantum gravity: An outside view *Preprint* arXiv:hep-th/0501114

[29] Friedel L and Krasnov K 2008 *Class. Quant. Grav.* **25** 125018 (*Preprint* arXiv:0708.1595); Conrady F and Freidel L 2008 *Phys. Rev.* D **78** 104023, and references therein

[30] Ambjorn J, Jurkiewicz J and Loll R 2004 *Phys. Rev. Lett.* **93** 131301; *do.* 2005 *Phys. Rev. Lett.* **95** 171301; Loll R, Ambjorn J and Jurkiewicz J 2006 *Contemp. Phys.* **47** 103-117 (*Preprint* arXiv:hep-th/0509010); Ambjorn J, Jurkiewicz J and Loll R 2006 in [1] (*Preprint* arXiv:hep-th/0604212)

[31] Kaloper N, Silverstein E and Susskind L 2001 *JHEP* 0105 031; Freidel L Reconstructing AdS/CFT *Preprint* arXiv:0804.0632

[32] See the proceedings of recent meetings in string theory http://www.stringwiki.org/wiki/Conferences; loop quantum gravity http://www.matmor.unam.mx/eventos/loops07/; for relation to foundational issues in quantum mechanics, see *Proceedings DICE Conferences* ed H T Elze *et al*: *DICE2004* – *Braz. J. Phys.* **35** (2005); *DICE2006* – *J. Phys.: Conf. Ser.* **67** (2007); *DICE2008* – this volume

[33] See, e.g., contributions by Sorkin R, Bombelli L, Rideout D, Surya S, and Hansen H – this volume

[34] Mateos D 2007 *Class. Qu. Grav.* **24** S713

[35] Hartnoll S A, Kovtun P K, Müller M and Sachdev S 2007 *Phys. Rev.* B **76** 144502

[36] Hartnoll S A, Herzog C P and Horowitz G T 2008 *Phys. Rev. Lett.* **101** 031601

[37] Hamma A, Markopoulou F, Premont-Schwartz I, Severini S 2008 *Phys. Rev. Lett.* **102** 017204

[38] Einstein A, Podolsky B and Rosen N 1935 *Phys. Rev.* **47** 777

[39] Lin S Y and Hu B L 2008 Temporal and Spatial Dependence of Quantum Entanglement: Quantum 'Nonlocality' in EPR from Field Theory Perspective *Preprint* arXiv:0812.4391 and references therein

[40] Unruh W G 1997 Is Quantum Mechanics Non-Local? *Preprint* arXiv:quant-ph/9710032

[41] I. Cirac and co-workers have demonstrated the relation between quantum entanglement and quantum correlations and showed that correlations are insufficient to represent the full content of entanglement. For atoms in optical lattices, see, e.g: Vollbrecht K G and Cirac J I 2007 *Phys. Rev. Lett.* **98** 190502

[42] Hu B L, Paz J P and Zhang Y 1992 *Phys. Rev.* D **45** 2843; *do.* 1993 *Phys. Rev.* D **47** 1576

[43] Sorkin R 2007 – invited talk at DICE2006 – proceedings *J. Phys.: Conf. Ser.* **67**

[44] Zwanzig R 2001 *Nonequilibrium Statistical Mechanics* (Oxford: Oxford U. Press)

[45] Calzetta E and Hu B L 2008 *Nonequilibrium Quantum Field Theory* (Cambridge, UK: Cambridge U. Press)

[46] Konopka T, Markopoulou F and Smolin L 2007 *Preprint* arXiv:gr-qc/0702044; Konopka T, Markopoulou F and Severini S 2008 *Phys. Rev.* D **77** 104029; Konopka T 2008 *Phys. Rev.* D **78** 044032

[47] Ashtekar A, Rovelli C and Smolin L 1992 *Phys. Rev. Lett.* **69** 237

[48] Bombelli L 2000 Statistical geometry of random weave states – proceedings of the IX Marcel Grossmann Meeting, Rome, July 2-8 *Preprint* arXiv:gr-qc/0101080

[49] Bombelli L, Corichi A and Winkler O 2005 *Annalen Phys.* **14** 499-519 (*Preprint* arXiv:gr-qc/0409006

[50] Markopoulou F and Smolin L *Preprint* arXiv:gr-qc/0702044

[51] Roura A and Verdaguer E 2008 *Phys. Rev.* D **78** 064010

[52] Hu B L and Roura A 2007 *Phys. Rev.* D **76** 124018

[53] Calzetta E and Hu B L 2000 *Phys. Rev.* D **61** 025012

[54] Calzetta E and Hu B L 1988 *Phys. Rev.* D **37** 2878

[55] E.g., Imry I 1997 *Introduction to Mesoscopic Physics* (Oxford: Oxford U. Press)

[56] E.g.: Datta S 1995 *Electronic Transport in Mesoscopic Systems* (Cambridge, UK: Cambridge U. Press); Rammer J 1998 *Quantum Transport Theory* (Reading, Mass.: Perseus Books)

[57] Akkermans E and Montambaux G 2006 *Mesoscopic physics of electrons and photons* (Paris: EDP Sciences)

[58] Becker K, Becker M, Polchinski J and Tseytlin A 1997 *Phys. Rev.* D **56** R3174 - R3178; Becker K and Becker M 1998 *Phys. Rev.* D **57** 6464

[59] See, e.g.: Elze H T 2008 Does quantum mechanics tell an atomistic spacetime? – this volume

[60] Pethick C and Smith H *Bose-Einstein Condensate* (Cambridge: Cambridge U. Press)

[61] Rey A M, Hu B L, Calzetta E, Roura A and Clark C 2004 *Phys. Rev.* A **69** 033610
[62] Donley E *et al* 2001 *Nature* **412** 295
[63] The smallness of the cosmological constant is suggested to be due to quantum fluctuations of a discrete spacetime. See: Sorkin R 1997 *Int. J. Theor. Phys.* **36** 2759; Ahmed M *et al* 2004 *Phys. Rev.* D **69** 103523; see also: Samuel J and Sinha S 2006 *Phys. Rev. Lett.* **97** 161302 – for viewing the cosmological constant as surface tension of a fluid membrane
[64] Ghirardi G C, Rimini A and Weber T 1986 *Phys. Rev.* D **34** 470; Pearle P 1976 *Phys. Rev.* D **13** 857; *do.* 1984 **29** 235; *do.* 1986 **33** 2240; Ghirardi G C, Pearle P and Rimini A 1990 *Phys. Rev.* A **42** 78; Ghirardi G C and Bassi A 2003 *Phys. Rep.* **379** 257
[65] Diosi L 1984 **Phys. Lett.** A **105** 199; *do.* 1987 *Phys. Lett.* A **120** 377; *do.* 1989 *Phys. Rev.* A **40** 1165
[66] Gisin N 1984 *Phys. Rev. Lett.* **52** 1657
[67] Penrose R 1986 *Quantum Concepts in Space and Time* ed R Penrose and C J Isham (Oxford: Clarendon Press); *do.* 1996 *Gen. Rel. Grav.* **28** 581; *do.* 1998 *Phil. Trans. Roy. Soc. (London)* A **356** 1927
[68] Adler S L 2005 *Emergent Quantum Mechanics* (Cambridge, UK: Cambridge University Press)
[69] Kok P and Yurtsever U 2003 *Phys. Rev.* D **68** 085006; Reynaud S *et al* 2002 *Int. J. Mod. Phys.* A **17** 1003; *do.* 2003 HYPER and Gravitational Decoherence *Preprint* arXiv:quant-ph/0311024; Roura A, Anastopoulos C and Hu B L, Critique on Decoherence in Planetary Motion by Stochastic Gravitons (unpublished)
[70] See, e.g.: Arndt M, Nairz O, Vos-Andreae J, Keller C, van der Zouw G and Zeilinger A 1999 *Nature* **401** 680; Brezger B, Hackermüller L, Uttenthaler S, J. Petschinka, Arndt M and Zeilinger A 2002 *Phys. Rev. Lett.* **88** 100404
[71] Penrose R 2000 *Mathematical Physics* ed A Fokas *et al* (London: Imperial College Press) 266-282
[72] Marshall W, Simon C, Penrose R and Bouwmeester D 2003 *Phys. Rev. Lett.* **91** 130401
[73] Dalvit D A R and Mazzitelli F D 1999 *Phys. Rev.* D **60** 084018
[74] Johnson P R and Hu B L 2002 *Phys. Rev.* D *65* 065015
[75] Galley C R and Hu B L 2005 *Phys. Rev.* D **72** 084023
[76] Yu H and Ford L H 2004 *Phys. Rev.* D **70** 065009; Borgman J and Ford L H 2004 *Phys. Rev.* D **70** 064032; *do.* 2004 *Phys. Rev.* D **70** 127701
[77] Pinto F 1993 *Phys. Rev. Lett.* **70** 3839-3843; Fischer U 1994 *Class. Quantum Grav.* **11** 463-474; Parker L, Vollick D and Redmount I 1997 *Phys. Rev.* D **56** 2113
[78] Lämmerzahl C L 1998 *Class. Quantum Grav.* **15** 13-27
[79] Turyshev S G *et al* 2005 Fundamental Physics with Laser Astrometric Test of Relavitity *Preprint* arXiv: gr-qc/0506104
[80] Kiefer C and Weber C 2005 *Annalen Phys.* **14** 253-278 (*Preprint* arXiv:gr-qc/0408010)
[81] Rammer J and Smith H 1986 *Rev. Mod. Phys.* **58** 323-359
[82] Takagi S 2002 *Macroscopic Quantum Tunneling* (Cambridge, UK: Cambridge U. Press)
[83] Gambini R, Porto R A and Pullin J 2004 *Phys. Rev. Lett.* **93** 240401
[84] Barcel C, Liberati S and Visser M 2005 *Living Rev. Relativity* **8** 12; Girelli F, Liberati S and Sindoni L 2008 *Phys. Rev.* D **78** 084013
[85] Bern Z, Dixon L J, Kosower D A, Roiban R, Spradlin M, Vergu C and Volovich A 2008 *Phys. Rev.* D **78** 045007
[86] Herzog C 2007 *Phys. Rev. Lett.* **98** 091601
[87] Hawking S W and Page D N 1983 *Commun. Math. Phys.* **87** 577

Journal of Physics: Conference Series **174** (2009) 012016 doi:10.1088/1742-6596/174/1/012016

High-energy γ-ray astronomy and string theory

Nick E. Mavromatos

King's College London, Department of Physics, Strand, London WC2R 2LS, UK

E-mail: nikolaos.mavromatos@kcl.ac.uk

Abstract.
There have been observations, first from the MAGIC Telescope (July 2005) and quite recently (September 2008) from the FERMI Satellite Telescope, on non-simultaneous arrival of high-energy photons from distant celestial sources. In each case, the highest energy photons were delayed, as compared to their lower-energy counterparts, by clearly observable time intervals. Although the astrophysics at the source of these energetic photons is still not understood, and such non simultaneous arrival might be due to non simultaneous emission as a result of conventional physics effects, nevertheless, rather surprisingly, the observed time delays can also fit excellently some scenarios in quantum gravity, predicting Lorentz violating space-time "foam" backgrounds with a non-trivial subluminal vacuum refractive index suppressed linearly by a quantum gravity scale of order of the reduced Planck mass ($\sim 10^{18}$ GeV). In this talk, I discuss the MAGIC and FERMI findings in this context. First, I review the high-energy astrophysics models for cosmic acceleration at celestial sources, stressing that currently there is no consensus as regards the observed delays. Then I derive estimates/bounds on the quantum gravity scale that reproduces the observed time delays on the assumption of a vacuum refractive index for photons with linear suppression, and argue on the consistency of such bounds with measurements from other Gamma Ray Telescopes, such as H.E.S.S. I then explain under which circumstances the MAGIC and FERMI findings could be accommodated in such models in agreement with all the other, currently available, astrophysics constraints of Lorentz Violation. The key features are: (i) transparency of the foam to electrons, (ii) absence of birefringence effects and (iii) a breakdown of the local effective lagrangian formalism. In contrast to other Quantum Gravity (field-theoretic) models with non-trivial optical properties available to date, a string model based on brane-worlds with the bulk space being punctured by space-time point-like D0-brane defects, that provide the seeds for Lorentz violating foamy structures, seems to respect all three requirements. The model provides an explanation for the observed photon time delays in a natural range of the string coupling and mass scale.

1. Introduction and Summary: the MAGIC observations and string theory

An alternative title for the talk could be *MAGIC and String Theory*: Usually the terminology *M-theory*, with M standing for either *Magical* or Marvelous or Mysterious, is attributed to the underlying (yet not completely understood) unifying theory of all known string theories [1], as a result of the many appealing duality and other symmetries it possesses, which result in the unification of the five known string theories, viewed as a low-energy limit of M-theory. This is a super-unification picture, which may prompt the way for a detailed understanding of the yet elusive theory of the quantum structure of space-time, otherwise termed as "Quantum Gravity" (QG). Our current knowledge/understanding of M-theory is limited. Nevertheless, for an analysis of some of the predictions of string theory that could have some relevance to observable low-energy physics this may not be an obstacle, as we shall attempt to discuss in this work.

In this review the word *MAGIC* is used for something completely different. It is an acronym (*M.A.G.I.C* = ***M****ajor* ***A****tmospheric* ***G****amma-ray* ***I****maging* ***C****herenkov telescope*), pertaining to the initials describing the full name of a Physics Instrument, specifically a Telescope based on the Canary Islands observatory (c.f. fig. 1), dedicated to the study of Cherenkov radiation emitted by highly energetic cosmic particles as they enter our atmosphere From the study of the emitted Cherenkov radiation, when a highly energetic cosmic particle enters the Earth's atmosphere, which is characterised by a non-trivial refractive index, one can deduce several important conclusions on the nature of the initial particle and through this to try to understand the mechanisms of production of such energetic cosmic particles.

Figure 1. Major Atmospheric Gamma-ray Imaging Cherenkov Telescope at the Canary Islands (Spain) Observatory (right panel). The telescope observed very high energy gamma rays from the active galactic nucleus Markarian 501 (radio image on left panel, by J.M. Wrobel and J.E. Konway, picture taken from `http://www.vlba.nrao.edu/whatis/mark.html`) with energies up to the order of 10 TeV (i.e. 10^{12} eV).

On July 9th 2005, the telescope observed [2] (c.f. fig. 1) very high energy gamma rays from the active galactic nucleus Markarian 501 (Mkn 501), which lies at red-shift $z = 0.034$ (i.e. about half a million light years away) from Earth, with energies up to the order of 10 TeV (1 TeV = 10^3 GeV = 10^{12} eV), which were delayed up to four minutes as compared with their lower-energy counterparts (in the 0.6 TeV or lower range) (c.f. fig. 2). It was the first and currently the only observation of such a distinct delay. The effect may be related to the astrophysics of the active galactic nucleus (*source* effect), which, as we shall discuss below, is not well understood at present and hence there is no consensus among the astrophysicists on the appropriate mechanism for the production of such high-energy photons at the source.

These uncertainties prompted more ambitious, although admittedly looking far-fetched, explanations [3], pertaining to new fundamental physics, affecting the photon *propagation* due, for instance, to space-time *foamy* vacuum structures that lead to modified dispersion relations for photons, that is a departure from the Lorentz invariant energy (E)-momentum ($\vec{p}$) relations of Special Relativity, $E = |\vec{p}|c$. Indeed, the reader should bear in mind that at small length scales, of the order of the Planck length, $\ell_P = \sqrt{\frac{\hbar G_N}{c^3}} = \frac{\hbar}{M_P c} \sim 10^{-35}$ m, which is the characteristic scale that quantum gravity effects are expected to be dominant, the structure of space time may be quite different from what we perceive at our (low-energy) scales, and it might be even discrete and non-commutative, that is the space-time coordinates (as we perceive them at present) might be average values of non commutative quantum operators. Moreover, one may have highly curved non-trivial fluctuations of the space-time metric, giving space time a *"foamy"* structure. In such complicated Quantum Gravity (QG) vacua, therefore, the very concept of Lorentz symmetry might break down at these short length scales, pointing towards the possibility of *spontaneous Lorentz symmetry breaking* by the QG vacuum, since if Lorentz symmetry is intact, it strictly

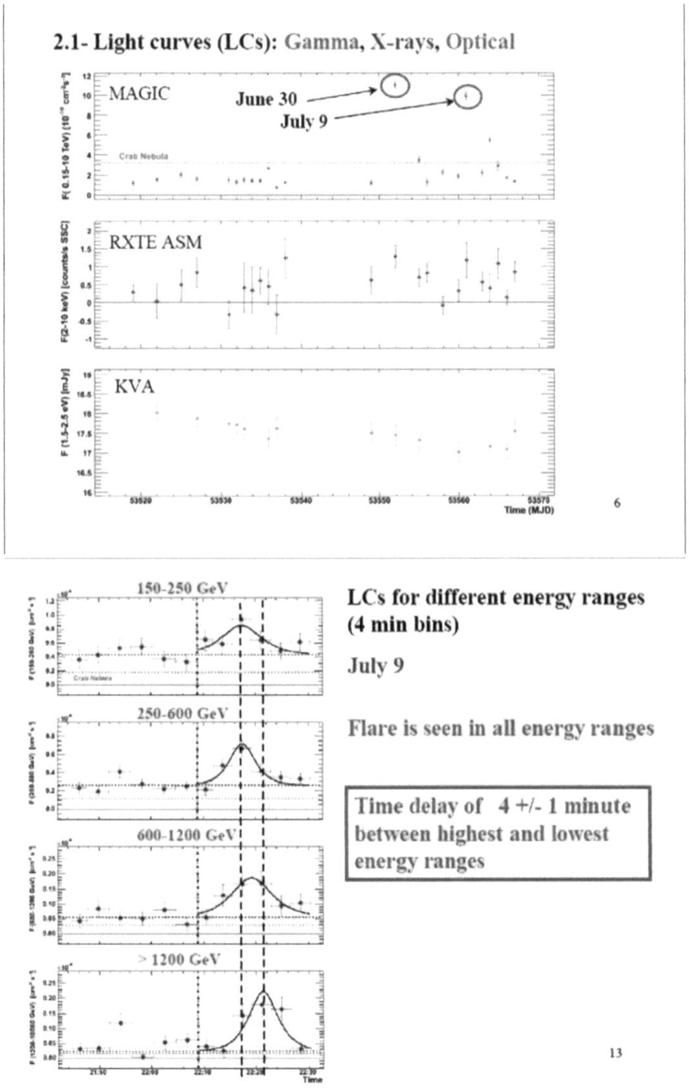

Figure 2. The observations of the MAGIC telescope [2] regarding very high energy Gamma Rays (with energies in the TeV range) showed that the most energetic photons were delayed up to four minutes as compared with their lower-energy counterparts (in the 0.6 TeV or lower range). The figures show light curves (LC), i.e. the photon flux vs. time of arrival. Observations like this may be used to prompt new fundamental physics on the structure of space-time. The lower panel shows LC at different energy ranges, demonstrating clearly the time delay (of order of 4 $\pm$1 minutes) of the more energetic photons.

prohibits such modifications in the dispersion relations. Such departures from the standard Special Relativity form of dispersion relations correspond to the generation of a momentum dependent mass gap for the photon, and hence a non trivial refractive index, as the photon propagates through the *medium* of quantum gravity. Here lies the origin of the *spontaneous breaking* of the *Lorentz symmetry* by the ground state of such foamy quantum gravity models.

It is the point of this review to touch upon such issues, through the description of of the most important physical consequences of such a breaking. As we shall see, surprisingly enough, many models of QG that entail such a breaking can already be falsified in current astrophysical experiments, which set very stringent bounds on Lorentz symmetry breaking. Some of the models we are going to discuss have experimental consequences that can be tested (or are already falsified(?)) in Nature, at least in principle. In this article we shall discuss in some detail only one class of theories of QG that can entail such a breaking, which is a subset of the modern version of string theory, including brane (domain-wall-like) defects in space time. Such defects will play the rôle of the non trivial space time structures that would be deemed responsible for the Lorentz symmetry spontaneous breaking by the ground state of these systems.

As we shall discuss here, there are very stringent conditions for such exotic explanations of the MAGIC observations to come into play in agreement with the plethora of many existing astrophysical tests of Lorentz symmetry. The space-time foam must be transparent to electrically charged probes, such as electrons, while photons should exhibit non-trivial refractive indices in this theory, but with no birefringence effects. Moreover, the local effective lagrangian description of the effects, that is a low-energy representation of the QG medium effects in terms of local higher derivative operators in flat space-time backgrounds, should break down.

Thus, although one cannot exclude the possibility that *both* effects, source and propagation due to quantum gravity, may be simultaneously responsible for the observed photon delays in the MAGIC experiment, nevertheless the available theoretical models that do the job are very limited. The thesis of this article will be that only certain models [4] of (the modern version of) string theory, including space-time defects, whose dynamics breaks Lorentz symmetry and provides a "foamy" structure of space-time, can offer an explanation for the MAGIC photon-arrival-times anomaly, which actually might be unique in interpreting these observations as a result of a stringy space-time foam situation, in a way consistent with all the current astrophysical constraints on Lorentz symmetry.

The key features of such stringy models are: (i) transparency of the foam to electrons, (ii) absence of birefringence effects and (iii) a breakdown of the local effective lagrangian formalism. In contrast to other Quantum Gravity (field-theoretic) models with non-trivial optical properties available to date, a string model based on brane-worlds with the bulk space being punctured by space-time point-like D0-brane defects, that provide the seeds for Lorentz violating foamy structures, seems to respect all three requirements. Due to electric charge conservation, only electrical neutral excitations (such as photons) interact with the D-particle defects, thereby leading to time delays proportional to the photon energies, but with no birefringence effects, in the sense that the delays are independent of the photon polarization. Moreover, the recoil and the quantum fluctuations of the D-particles, during their scattering with matter excitations, imply a breakdown of the local effective lagrangian formalism in the sense that the phenomena cannot be wholly represented by the addition of local higher-derivative operators in a flat-space time effective lagrangian. The string model thus provides an explanation for the observed photon time delays, in a natural range of the string coupling and mass scale, and avoids all the other stringent constraints of Lorentz Invariance Violation coming from non observation of birefringence effects or of very high energy photons, that would be the case if local effective lagrangian formalism were in place, as a result of the modification of the energy thresholds for pair production in the scattering of ultra-high-energy photons ($\sim 10^{19}$ eV) with infrared ones in the Universe [5].

A word of caution to the reader is due at this point. Our point is not to advocate string theory as a superior candidate to other quantum gravity models available to date, but rather to discuss situations in which string theory predictions can be falsified by experiment. And high energy photon astrophysics may be a useful arena for this purpose! Of course, it goes without saying that, at present, we are very far from reaching any conclusions on such matters. Definite falsification of these stringy quantum gravity scenarios, if at all possible, would require a plethora of further studies by means of other high-energy astrophysics processes and observations.

To understand in detail the possible explanations of the MAGIC "photon anomaly", and why string theory comes into play as a potential explanation provider, I will first recapitulate the up to date knowledge on theoretical models of QG, followed by a brief review of the scenarios for the production of very high energy gamma rays in the Universe, known to date. In this way I hope I will be able to convince the reader that the theoretical uncertainties in both fields, astrophysical production of high energy photons and possible propagation effects due to quantum gravity media, are more or less on equal footing, and presently there is no consensus in both communities. This leaves the field open to speculations and it is in this sense that the exotic explanation of the MAGIC effect we shall provide here, based on string theory, may not sound so far fetched after all! Hopefully, at the end of the day, the reader will be able to make his/her own judgement on whether such a theory is really required to explain the facts.

And such facts, at present, seem to be increasing in numbers. Indeed, three years after the MAGIC observations, in September 2008, the FERMI (formerly known as GLAST) Satellite Telescope [6], also observed time delays of the higher-energy photons, from the distant Gamma Ray Burst (GRB) 080916c. As we shall discuss in this talk, the pattern of the delay fits [4] the above-mentioned string model of quantum-gravity-induced refractive index, with the pertinent quantum gravity scale being essentially the same as that inferred from the MAGIC observations. Viewed as a lower bound, this scale is also compatible with that obtained from other Gamma-Ray data of the H.E.S.S. Collaboration [7].

The structure of the article is as follows: in the next section, 2, I discuss issues pertaining to the violation of Lorentz invariance in media or quantum electrodynamics vacua with non trivial structure that break Lorentz invariance, in the sense of inducing a non-trivial vacuum refractive index, *e.g.* the case of thermal plasma. Such cases may be thought of as analogues of the Quantum Gravity (QG) space-time foam vacuum that may also not respect Lorentz symmetry but can also characterise the source regions of the cosmic high energy Gamma Rays. In this latter sense, it is important to understand photon propagation in those cases first, so as to disentangle possible source effects from the Quantum-Gravity-induced ones. In section 3, we discuss possible interpretations of the MAGIC (and FERMI) observations, including exotic ones involving QG dispersive media. We discuss bounds and sensitivities of various astrophysical experiments, and state carefully the stringent requirements that must be met by a theoretical model of QG, in order for the observed delays in the MAGIC and FERMI Telescopes to be attributed to effects due to a space-time-foam medium, in agreement with all other current tests. All these requirements are surprisingly respected by a model in the modern version of string theory, which is discussed in section 4. The model involves membrane-like defects in a higher-than-four dimensional space time, with our world being viewed as a hyper-membrane (D(irichlet)-brane) embedded in this space time. The model is of the kind of large-extra-dimension models to be tested at LHC and future colliders, with the important ingredient of having point-like space-time (D-particle) defects, responsible for the "foamy" structure of space time. Finally, section 5 contains our concluding remarks.

A note is in order at this point concerning the *units* used in this work: throughout the article, unless otherwise stated, we shall work in *natural Planck units*, in which $\hbar = c = 1$. In these units, length and time are identified, and they are inversely proportional to mass or energy. The latter are also identified and expressed in units of multiples of eV ($=1.6 \times 10^{-19}$ Joules), in particular GeV ($=10^9$ eV) and TeV ($=10^{12}$ eV) in this work. From time to time, for concreteness, the speed of light *in vacuo* c and Planck's constant h or $\hbar = h/2\pi$ may appear explicitly in some formulae.

2. Lorentz invariance and the vacuum structure of quantum fields

One of the cornerstones of Modern Physics is Einstein's theory of Special Relativity (SR), which is based on the assumption that the speed of light in vacuo c is an invariant under all observers, and in fact this implies the Lorentz transformation in flat space times, and the nature of c as a limiting velocity for *all* particle species in SR.

The generalization (by Einstein) of SR to include curved space times, that is the theory of General Relativity (GR), encompasses SR locally in the sense of the *strong form of the equivalence principle.* According to it, *at every space-time point, in an arbitrary gravitational field, it is possible to choose a locally inertial ('free-float') coordinate frame, such that within a sufficiently small region of space and time around the point in question, the laws of Nature are described by special relativity, i.e. are of the same form as in unaccelerated Cartesian coordinate frames in the absence of Gravitation.* In other words, locally one can always make a coordinate transformation such that the space time looks *flat.* This is not true globally, of course, and this is why GR is a more general theory to describe gravitation. The equivalence principle relies on another fundamental invariance of GR, that of general coordinate, that is the invariance of the gravitational action under arbitrary changes of coordinates. This allows GR to be expressed in a generally covariant form.

In such a locally *Lorentz-invariant* vacuum, the photon dispersion relation, that is a local in space-time relation between the photon's four-wavevector components $k^\mu = (\omega, \vec{k})$ (where ω denotes the frequency, and $\vec{k}$ the momentum) reads in a covariant notation:

$$k^\mu k^\nu \eta_{\mu\nu} = 0 \tag{1}$$

where repeated indices $\mu, \nu = 0, 1, \ldots 3$, with 0 referring to temporal components, denote summation and $\eta_{\mu\nu}$ denotes the Minkowski space-time metric, with components $\eta_{00} = -1$, $\eta_{0i} = \eta_{i0} = 0$, $\eta_{ij} = \eta_{ji} = \delta_{ij}$, $i = 1, 2, 3$ with δ_{ij} the Kronecker delta symbol.

The above relation (1) implies the equality of all three kinds of photon velocities *in vacuo* that stem from its wave nature (due to the particle-wave duality relation):

$$\begin{aligned}
&\text{phase}: && v_{\rm ph} = \frac{\omega}{|\vec{k}|} \equiv \frac{c}{n(\omega)} = c \\
&\text{group}: && v_{\rm gr} = \frac{\partial \omega}{\partial |\vec{k}|} \equiv \frac{c}{n_{\rm gr}(\omega)} = c\ , \quad n_{\rm gr}(\omega) = n(\omega) + \omega \frac{\partial n(\omega)}{\partial \omega} \\
&\text{front}: && v_{\rm front} = c/n(\infty) = c
\end{aligned} \tag{2}$$

since the phase and group *refractive indices* of the trivial *vacuum* equal unity $n(\omega) = n_{\rm gr}(\omega) = 1$. For brevity we shall work from now on in units where $c = 1$.

2.1. Photon propagation in conventional media and in non-trivial field-theory vacua with a refractive index

The above results (2) changes significantly when light propagates in a material *medium*, in which its speed is different from c in vacuo. This is due to the non trivial refractive index the material has, as a result of the electromagnetic interactions of the photon with the electrons in the medium. As we shall discuss later on, this case seems to bear some quite instructive analogies with our string model of space time foam, which lies at the focus of our attention here.

The simple model of quantum oscillators has been adopted by Feynman [10] as a simplified but well motivated analogue for describing the situation in the case of photons in ordinary media. In that case, the electrons of the medium, of mass m, are represented by simple harmonic oscillators, with frequency ω_0, which provides the necessary *restoring force* during the scattering of light off the electrons in order to keep the latter oscillating around their initial position. In that problem the induced refractive index is obtained by the reduction of the phase velocity of the photon wave.

It can then be shown by means of an elementary analysis, based on quantum mechanics [10], that light propagates through the medium with a speed c/n, where n is the refractive index of the medium, which is given by the following formula—:

$$n = 1 + \frac{\rho_{e_e} e^2}{2\epsilon_0 m(\omega_0^2 - \omega^2)}. \tag{3}$$

In this relationship, ϵ_0 is the dielectric constant of the vacuum and n_e is the area density of electrons in the medium (plate in the example of [10] adopted here for concreteness), which is given by $n_e = \rho_e \Delta z$, where ρ_e is the volume density of electrons.

Thus causes a delay Δt in traversing the distance Δz, given by: $\Delta t = (n-1)\Delta z/c$. We see in (3) that the refractive index in an ordinary medium is inversely proportional to (the square of) the frequency ω of light, as long as it smaller than the oscillator frequency, where the refractive index diverges.

If the couplings of the two polarizations of the photon to the electrons in the medium are different, the phenomenon of birefringence emerges, namely different refractive indices for the two polarizations. Moreover, we see from (3) that the propagation of light is subluminal if the frequency (energy) of the photon $\omega < \omega_0$, whereas it is superluminal for higher frequencies (energies) [10]. This reflects the fact that the phase shift induced for the scattered light can be either positive or negative, but there such a superluminal refractive index causes no issue with causality, since the speed at which information may be sent is still subluminal.

As we shall see later on, this conventional model will provide us with a rather good analogue of what happens in some string models of quantum space time foam, also characterized by non-trivial refractive indices, which we shall analyse in section 4. However, as we shall see there, contrary to the conventional situation discussed in this section, in the string case the refractive index is found proportional to the photon frequency, while the effective mass scale that suppresses the effect is the quantum gravity (string) scale and not the electron mass as in (3).

The rôle of a non-trivial refractive index material can be played under certain circumstances by a *non-trivial vacuum* in which photons propagate, such as quantum electrodynamics at finite temperature plasmas [11] or the Casimir vacuum between parallel capacitor plates [12] (or other geometries, as long as the space is bounded appropriately) (c.f. fig. 3). In such cases, the loop corrections due to *vacuum polarization*, *i.e.* creation and annihilation of virtual electron-positron pairs, in quantum electrodynamics (QED) result in a modified photon propagator, and a non-trivial group velocity and refractive index $n(\omega) \neq 1$. The reason for this is the breaking of Lorentz invariance, due to either the existence of spatial boundaries (Casimir vacuum) or finite temperature (thermal vacuum in case of plasmas). Moreover one may consider quantum electrodynamics in a homogeneous and isotropic Friedman-Robertson-Walker (FRW) expanding-Universe background [13] and examine the non-trivial effects of vacuum polarization on photon propagation there. This was in fact historically the first instance where the effects of curvature induced *superluminal* propagation for low-frequency photon modes.

The situation concerning all the above cases can in fact be represented in a rather unified way, covering all cases, by the following formula giving the group velocity of (low-energy) photons in such non-trivial vacua due to vacuum polarization in four-dimensional QED [11]:

$$v_{\rm gr} = 1 - \frac{44}{135}\alpha^2 \frac{\rho}{m_{\rm e}^4} \tag{4}$$

where ρ is the energy density *relative* to the standard vacuum. This formula is valid in all cases except the Gravitational Background case of [13], where α^2 should be replaced by $\alpha m_{\rm e}^2 G_{\rm N}$, with

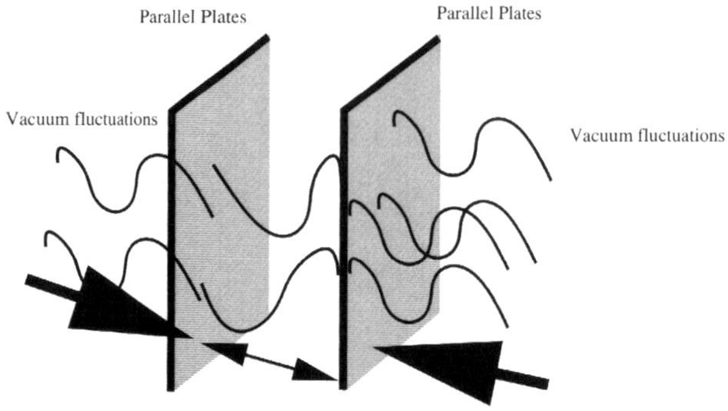

Figure 3. The Casimir Vacuum of Quantum Electrodynamics: in the compact space between the parallel plates (or more generally in other compact geometries), the quantum fluctuations of the electromagnetic field are responsible for the macroscopically measured Casimir force between the plates, scaling with the size L of the compact space as L^{-4} in four space-time dimensions. The force can be either attractive or repulsive, depending on the geometric set up. However, besides this force from "nothing", the Casimir Vacuum provides another interesting phenomenon in quantum physics. According to calculations by Scharnhorst and collaborators [12], the presence of the boundaries breaks manifestly Lorentz invariance, and results in modified dispersion relation, and hence a non-trivial refractive index, for the virtual photons of the *non-trivial Quantum Electrodynamics vacuum* in the compact region. From this latter perspective, the situation is entirely analogous to the what happens in thermal plasmas, such as the ones occurring in the interior of stars or source regions of active galactic nuclei, which we are interested in in this work. There, the periodic boundary conditions of the Casimir case are replaced by similar ones due to the finite temperature, which also break Lorentz symmetry of the ground state.

G_N the Newton gravitational constant. In the (four space-time dimensional) Casimir vacuum, for instance, the presence of boundaries in space (e.g. capacitor parallel plates at a distance L in the simplest geometry), imply loss of low-energy photon modes, and as such the energy density of the vacuum is lowered relative to the standard one (without the boundaries). In this case $\rho = -\frac{\pi^2}{720L^4} < 0$ in (4), and one has the phenomenon that the effective low-energy photon modes appear to propagate in a *superluminal way*, $v\mathrm{gr} > 1$ (we give here the formula for propagation *perpendicular* to the plates for concreteness and ease of comparison with the plasma case later on):

$$v_{\mathrm{gr}}^{\mathrm{Casimir}} = 1 + \frac{11\pi^2}{8100}\alpha^2 \frac{1}{L^4 m_{\mathrm{e}}^4} > 1 \tag{5}$$

But as explained in [11], and mentioned above, there is no contradiction with relativity here, as this result applies only to low energy photons, with energies much lower than m_{e}, and indicates the effective loss of degrees of freedom due to the spatial boundaries.

In the plasma case at finite temperature T, which can characterise the interior regions of stars or active galactic nuclei, of interest to us in this work, there is a formal analogy [11] between the rôles played by the temperature T and the plate separation L. They both break Lorentz invariance in a rather similar way. In fact, there is a correspondence between the Casimir and thermal-plasma vacuum formulae by replacing $2T$ by L^{-1}. In the plasma case, the low energy photon modes with momentum $k^2 \ll m_e^2$ ($k \equiv |\vec{k}|$) have a group velocity:

$$v_{gr}(kT \ll m_e^2) = 1 + \frac{11\pi^2}{8100}\alpha^2\left(\frac{2T}{m_e}\right)^4 > 1 \tag{6}$$

which can be directly compared with the Casimir one (5) upon the substitution $2T \to L^{-1}$.

For future reference we give the group velocity for the high-energy (as compared to the effective QED scale m_e) photon modes [11] in the finite T plasma case:

$$v_{gr}(kT \gg m_e^2) = 1 - \frac{\alpha^2}{6}\left(\frac{T}{k}\right)^2 \ln^2\left(\frac{kT}{m_e^2}\right) < 1 \tag{7}$$

which again can be compared with the corresponding Casimir vacuum case (5), for photon modes *perpendicular* to the plates [12], upon the replacement $2t \to L^{-1}$. Similar effects also characterise the curved background case of [13], where again the high-momentum photon modes are subluminal. We remark at this point that the relation (7) can characterise the source regions of active galactic nuclei, and thus such effects can be responsible for inducing time delays of photons from these regions. We shall come back to it, when discussing the MAGIC delays in section 3.3 below.

We notice here the momentum (k) dependence of the subluminal velocity for the high-energy photon modes. In such non-trivial vacua, therefore, high energy photon modes will exhibit a momentum dependent subluminal refractive index, which will affect their arrival times at the observation point, if one considers simultaneous emission of modes within a certain energy range. From (7) we observe that the higher the momentum k of the high-energy photon mode the higher the group velocity, since $\frac{\partial v_{gr}(kT \gg m_e^2)}{\partial k} = \frac{\alpha^2 T^2}{3k^3}\left(\ln(\frac{kT}{m_e^2})\right) > 0$, and thus fast modes will arrive first if emitted simultaneously in this vacuum.

At this point it is instructive to make some clarifications regarding the speed of light and causality, that is the fact that signals do not arrive before they occur. The phase, front and group velocities have all been found to exceed the value of the speed of light in vacuo. However, this is not in conflict with causality, that is the fact that signals never arrive before they occur.

Indeed, the various light velocities do not have to be subluminal, as they carry no information. Information can be transmitted by (or not) sending pulses. The information then *appears* at first sight to propagate with the group velocity, *i.e.* the velocity of the peaks of the pulses. However, as experimentally demonstrated in several instances since the 1980's, the group velocity of the photons can also be super-luminal. This is because the group velocity can be disentangled from the *signal velocity*, that is the speed by which information can be transferred, and thus it can exceed the speed of light *in vacuo*, c, without contradicting the *causality* requirements of Special Relativity. Group velocities larger than c can occur *e.g.* in tunneling experiments and appear to lead to superluminal transmission. However there is no contradiction with causality or Special Relativity in such cases. The error lies in identifying the peak of the pulse with the temporal position of the carried information. For example, a Gaussian-shaped pulse can be detected long before its peak due to the rise of intensity at earlier times. Therefore, a different kind of signal must be considered, where no information at all is sent out before a certain moment of time. For such signals, it can be proven that the earliest time at which that switching event can be observed is limited exactly by propagation with the vacuum velocity c. A so-called *precursor* is traveling with that speed, but is normally too weak to be detected, except in certain circumstances. For direct measurements of optical precursors in regions with anomalous dispersion the reader is referred, for instance, to ref. [8].

Super-luminal group photon velocities have been measured in laser pulses passing through specially prepared materials in ref. [9]. Using special set up involving optical fibers, a super-luminal group velocity of photons (in the fiber) has been measured and found different from the *signal velocity*, which is defined as the speed of the front of a square pulse. It is the signal speed that cannot exceed c, due to *causality*, as mentioned above. The precursors (or "forerunners") of the signal, mentioned above, which travel with sub-luminal speed, arrive first before the main front. This signal velocity was measured for the first time directly in an experiment in [9], and indeed was found to be less than c.

This is an important issue that the reader should have in mind, especially when we discuss non-trivial vacuum refractive indices in several non trivial ground states of quantum systems, including gravity.

In fact, as argued in ref. [15], where we refer the interested reader for details, specifically for the case of Casimir vacuum [12] but the discussion can be generalised, super-luminal group velocities are "benign" as far as causality and compatibility with the kinematics of Special Relativity are concerned. In particular, as stressed in [15], kinematics of Special Relativity requires *only* an invariant speed *not* actually a maximum one. Moreover, causality can be guaranteed in such super-luminal cases because the pertinent kinematics is equivalent to an "effective" metric in space-time, $g^{\mu\nu}_{\rm eff}$, $\mu, \nu = 0, \dots 3$, different from the Minkowski one, that describes the kinematics in the non-trivial (Casimir) vacuum. The modified dispersion relations leading to super-luminal group velocities (5) acquire the form

$$k_\mu k_\nu g^{\mu\nu}_{\rm eff} = 0 \ , \tag{8}$$

where as usual repeated indices denote summation, $k_0 = -\omega$, $k_i = (\vec{k})_i, i = 1, 2, 3$, $g^{\mu\nu}_{\rm eff} = \eta_{\mu\nu} + \xi n^\mu n^\nu$, with $\eta^{\mu\nu}$ the (inverse) Minkowski metric, n^μ a unit (space-like) vector orthogonal to the plates of fig. 3, and $\xi = \frac{11\pi^2\alpha^2}{4050 L^4 m_e^4}$, with L the distance between the Casimir plates.

The basic point of the discussion in [15], and how causality is maintained, is that the presence of the effective metric in (8) widens slightly (due to the deviations of order $\xi \ll 1$ of the effective metric from the Minkowski one) the light cone in the direction orthogonal to the plates, and hence light in that direction travels at a speed $c_{\rm light} > c$, while light in the direction parallel to the plates still travels with speed c. In a given inertial reference frame, moving with four-velocity u^μ with respect to the rest frame of the apparatus of fig. 3, the photons inside the Casimir cavity travel, at a direction perpendicular to the plates, with a definite speed $c^{(u)}_{\rm light} > c$, which has *only one value* for each observer and is not universal among observers. In this way super-luminal group velocities (5) are compatible with causality, since violation of the latter occurs *only* if signals travel with the *same* speed greater than c in two different frames. The reader should have these features in mind when studying space-time foam theories of quantum gravity.

2.2. Non-trivial optical properties of the quantum gravity vacuum?

A truly unspeakable feature on the speed of light may appear when considering the ground state of *Quantum Gravity per se* as a non-trivial vacuum with non-standard optical properties, leading to a *non-trivial* refractive index. This may characterise, for instance, certain approaches in which path-integration over microscopic singular fluctuations of the metric field, such as Planck size black holes and other topologically non-trivial configurations, implies a sort of "*foamy*" structure of the space time at small (Planck-size) length scales, over which photons can propagate, a proposal made initially by J.A. Wheeler [16]. In such situations, the concept of a *local Effective Quantum-Field Theory* Lagrangian may *break down*, and the situation resembles that of a quantum-decoherent motion of matter in open quantum mechanical systems interacting with an environment [17, 18, 19].

The ground state of such QG foam situations may behave as a (subluminal) *refractive medium*, as suggested originally in [20], which to our knowledge, constitutes the first concrete attempt to consider the non-trivial optics effects effect of such vacua on massless particle (photon) propagation, based on earlier works by the current authors on non-critical strings in black hole backgrounds [21]. We note at this stage that similar ideas on modified dispersion relations for particles in a quantum gravity medium, but on a purely phenomenological basis, without any attempt to present concrete models, have also been advocated in ref. [22]. In the quantum gravity case, the effective modified dispersion relations of the low-energy matter theory, assumes the generic form:

$$E^2 = |\vec{p}|^2 + m^2 + \sum_{n=1}^{\infty} c_n |\vec{p}|^2 \left(\frac{|\vec{p}|}{M_{\rm QG}}\right)^n \qquad (9)$$

where m is the rest mass of the probe, and c_n are constant coefficients, with signature and values that depends crucially on the type of theory considered. It is not clear whether the series converges, or is resummable, as this information depends crucially on the details of the underlying microscopic theory. The important point to notice in (9) is that the natural suppression scale of the (Lorentz-symmetry-violating) correction terms of the standard Special Relativity dispersion relation is that of Quantum Gravity itself, $M_{\rm QG}$. According to what was mentioned earlier, there seems to be no fundamental issue with violation of (micro-)causality [15] or contradiction with the principles of Special Relativity in cases where super-luminal group velocities arise from the QG anomalous dispersion relations (9). The important point is to observe that effective metrics, as in the Casimir vacuum case (8), can indeed be found in order to describe the QG anomalous dispersion (9) at least in some cases, that we shall be interested in here. Thus causality can be saved for those QG cases by applying the same logic as for the Casimir vacuum, discussed previously. Nevertheless, as we shall discuss later on in subsection 3.5, in cases where super-luminal modes exist, one has QG-induced birefringence phenomena, which are severely constrained by the current phenomenology (*i.e* by the absence of the relevant signals).

Before closing this subsection we also mention that there is another approach towards modified dispersion relations of particles, of the type (9), based on the so-called doubly special or deformed special relativity (DSR) theories [23, 24]. In such approaches, which are formulated on flat space times, the modification to the dispersion relation arises by the postulate that the local symmetry group of space time is no longer the Lorentz group but a different one. In their original version [23], DSR modified dispersion relations were obtained by the requirement that the length scale of quantum gravity ("Planck") remain invariant under transformations, which thus leads to deviations from the Lorentz group. There is no unique prescription to achieve this, however, and in this way one may even arrive at models where there is no upper limit in momentum. In their subsequent version [24], DSR models postulated the existence of upper limits in velocities of species, and in this way the modified local group was determined by appropriate combinations of dilatations and translations.

It is unclear to us whether such theories are fundamental or effective, and most likely, as fundamental there will be not consistent in their quantization. For instance, at present a known problem is the behaviour of multi-particle states in such DSR fundamental theories. We shall not discuss these theories further here, apart from mentioning that, as in the stringy model we shall consider in section 4 [20, 25, 26], they are not characterised by birefringence effects, leading only to subluminal propagation (defined appropriately [23, 24]). There is, however, an important difference from the stringy model: the action of gravity is universal among particle species in DSR theories, and therefore the resulting modification in the dispersion relation pertain to all species. In contrast, for specifically stringy reasons to be outlined in section 4, in the string model only photons and at most electrically neutral particles [27] are allowed to interact non-trivially with the stringy defects, and are thus subjected to non-trivial modification of their dispersion relations.

3. Possible interpretations of the MAGIC Effect

To understand in detail the possible explanations of the MAGIC observations, let us first recapitulate the up to date knowledge on the production of very high energy gamma rays in the Universe [28] and ways of cosmic acceleration. I must stress that there is still no consensus among the astrophysical community as regards the various ways of particle acceleration at various regions in the Universe, and in fact it is most likely that there are several mechanisms taking place, depending on the source. The MAGIC observations added to this puzzle, in particular why the more energetic photons from the Mk501 Galaxy arrived later than the lower energy ones. This is the only case up to now where such delays have been observed. The situation i n our understanding of the source mechanism for the production of high energy particles from cosmic sources will improve only by making more and more precision measurements from a variety of celestial sources, which is currently under way, through several terrestrial and extra-terrestrial facilities on high-energy astrophysics, which are either under construction or just have been put in operation.

It is these uncertainties in the conventional astrophysics of the sources that allow for speculations that fundamental physics, such as photon propagation in a quantum gravity "medium", might play a significant rôle on the MAGIC effect. However, if the latter has a chance of being true it must respect all the other stringent astrophysical constraints on Lorentz invariance that are currently available. As we shall discuss in section 4, it seems that, at present, only a specific string theory model of quantum foam, in which only photons are not transparent to the foam effects, could stand up to this chance.

3.1. Conventional astrophysics mechanisms for cosmic Very-High-Energy (VHE) gamma-ray production

Gamma rays constitute the most interesting part of the spectrum of active galactic nuclei (AGN). An AGNs is a compact region at the centre of a galaxy, with much higher than normal luminosity over some or all of the electromagnetic spectrum (in the radio, infrared, optical, ultra-violet, X-ray and/or gamma ray wavebands). The radiation from AGN is believed to be a result of accretion on to the supermassive black hole at the centre of the host galaxy. AGN are the most luminous persistent sources of electromagnetic radiation in the universe, and as such can be used as a means of discovering distant objects; their evolution as a function of cosmic time also provides constraints on cosmological models.

Gamma Rays with energies higher than 20 MeV and up to TeV have been observed today from such AGNs. It is customary (although somewhat arbitrary) to classify these Gamma Rays according to their energies as follows:

- (i) *High-Energy* Gamma rays: with energies from 20 MeV - 100 GeV ,
- (ii) *Very High-Energy* Gamma rays: with energies from 100 GeV - 30 TeV ,
- (iii) *Ultra High-Energy* Gamma rays: with energies from 30 GeV - 30 PeV ,
- (iv) *Extremely High-Energy* Gamma rays: with energies from 30 PeV - ?.

Theoretically, the last category incorporate energies up to the ultraviolet cutoff energy scale (Planck-scale energies 10^{19} GeV) that defines the structure of low-energy field theories as we know them.

The production of very high energy gamma rays is still not understood well, and constitutes a forefront of research on galactic and/or extragalactic astrophysics.

At present there are three major categories believed responsible for the production of very- and ultra- high-energy Gamma rays:

- (i) Photons from conversion of gravitational energy in Active Galactic Nuclei (AGN) (c.f. fig. 4),

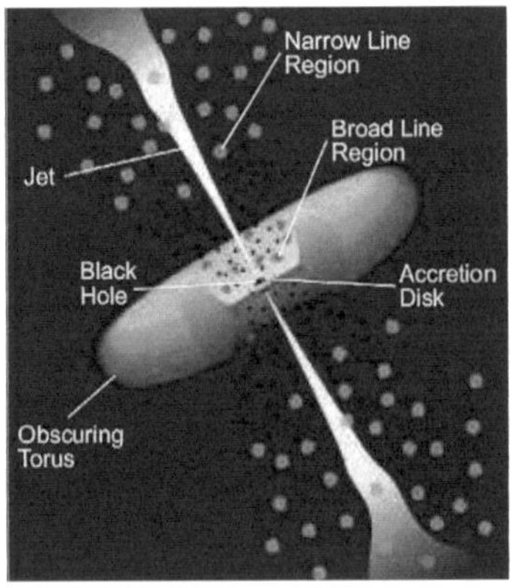

Figure 4. The basics of a cosmic accelerator model (picture taken from ref. [29]): Very high energy Photon production from Gravitational Energy conversion (Penrose process) in AGNs, believed to take place in AGN Mkn501. Relativistic matter, such as electrons are beam ejected as a result of the enormous gravitational energy available during the collapse process forming a black hole at the center of the galaxy. Then such electrons undergo synchrotron radiation due to their interaction with the magnetic fields existing in the galactic region, and eventually inverse Compton scattering (IC), i.e. interactions of these very high energy electrons with low-energy photons (say of e energies), to produce TeV photons from Mkn501 observed by MAGIC (see fig. 5). This combined process is called Synchrotron-self-Compton process.

- (ii) From self-annihilation of Dark Matter and
- (iii) From decays of exotic massive particles (with masses of order $10^{15} - 10^{16}$ GeV/c^2), appearing in Grand Unifying Models Beyond the Standard Model (such as string-theory inspired models), in the very Early Universe.

In what follows we shall restrict ourselves to (i), which most likely pertains to the production of very high energy Gamma rays observed in AGN such as Mkn 501 observed by the MAGIC Telescope. It is believed today that the centres of galaxies contain massive black holes due to matter collapse [30], with typical masses in the range $10^6 - 10^9$ solar masses. AGN therefore are celestial systems with very high mass density, and it has long been assumed that they consist of a massive black hole, of say 10^8 solar masses or more, accreting the gas and dust at the center of the galaxy. The gravitational energy liberated during accretion onto a black hole is 10% of the rest mass energy of that matter and is the most efficient mass-energy conversion process known involving normal matter (Gravitational Energy Conversion). Collapsing matter towards this massive central galactic object releases gravitational energy (Gravitational Energy Conversion) and results in spectacular relativistic material jet emissions. Since the accreting matter has in general a non-trivial angular momentum, angular momentum conservation is responsible for the matter orbiting the black hole and, through energy dissipation, the formation of a material (flat) accretion disk. This also results in material jets with ultra-relativistic particles outflowing the accretion plane (fig. 4).

In addition, since the black holes at the centres of the AGN are probably rotating (as a result of having a non-zero angular momentum (Kerr type) [30], due to angular momentum conservation in the formation (collapse) process), one might also speculate that part of the relativistic jet might be due to the so-called Penrose process, which allows the extraction of energy from a rotating black hole [31]. The extraction of energy from a rotating black hole is made possible by the existence of a region of the Kerr spacetime called the ergoregion, in which

Journal of Physics: Conference Series **174** (2009) 012016 doi:10.1088/1742-6596/174/1/012016

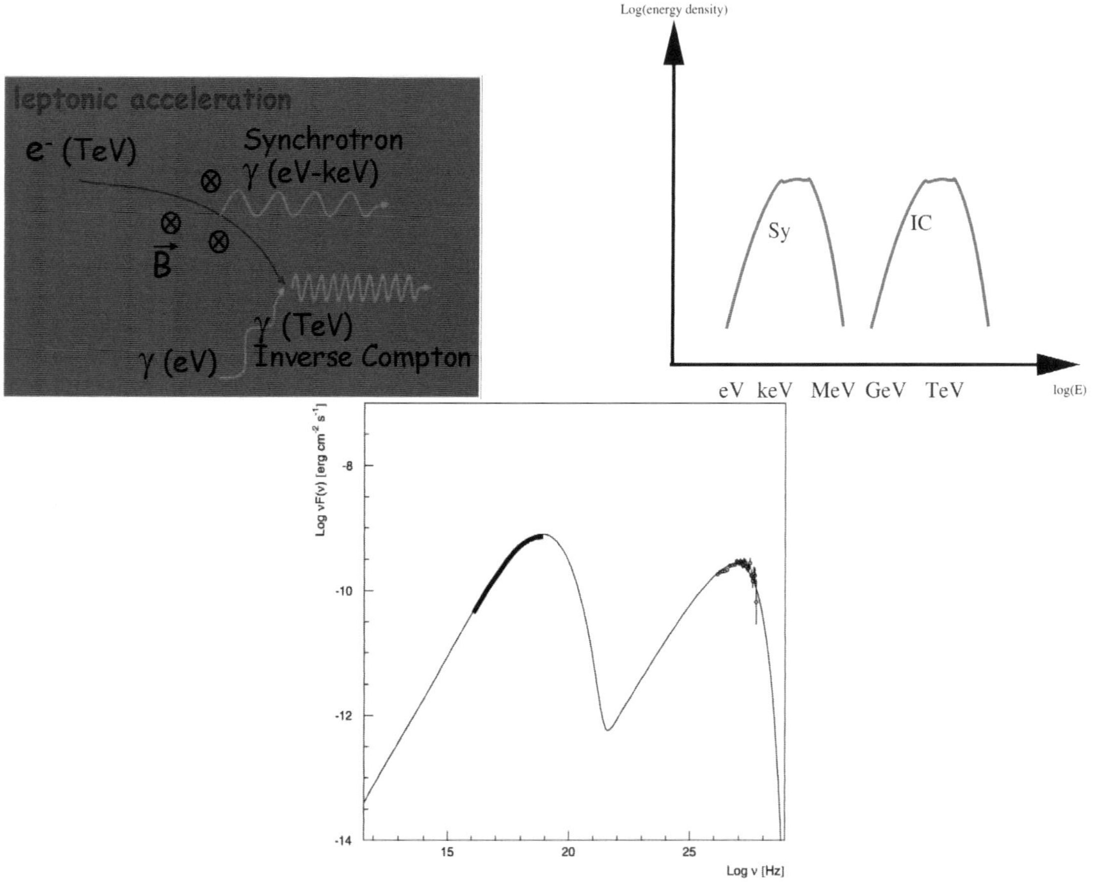

Figure 5. One of the suggested models (leptonic acceleration) for the production of very high energy gamma rays: Synchrotron-self-Compton (SSC), involving synchrotron (Sy) and inverse Compton (IC) scattering for the same electron, provides a mechanism for the production of photons with energies up to several TeV. In the middle figure the typical energy spectrum is indicated, with the characteristic double peak of the SSC mechanism, one of the synchrotron radiation and the other for the IC scattering. The Inverse Compton spectrum has a peak at TeV energies, as those observed by MAGIC when looking at the AGN Mkn 501 (lower picture, from Konopelko *et al*, ApJ. **597**, 851 (2005)).

a particle is necessarily propelled in locomotive concurrence with the rotating spacetime. In the process, a lump of matter enters into the ergoregion of the black hole and splits into two pieces, the momentum of which can be arranged so that one piece escapes to infinity, whilst the other falls past the outer event horizon into the hole (a rotating black hole has two event horizons, an outer and an inner one). The escaping piece of matter can possibly have greater mass-energy than the original infalling piece of matter. In summary, the process results in a decrease in the angular momentum of the black hole, leading to a transference of energy, whereby the momentum lost is converted to energy extracted. The process obeys the laws of black hole mechanics. A consequence of these laws is that if the process is performed repeatedly, the black hole can eventually lose all of its angular momentum, becoming rotationally stationary.

The particles in the relativistic jet undergo acceleration, but currently the pertinent mechanism is a matter of debate and active research. Most likely it depends on the source. In general there are two generic ways of cosmic acceleration.

Mechanisms for cosmic acceleration

Leptonic acceleration: Among the particles in the jet are charged electrons, whose paths are curved as a result of the existing magnetic fields in the galactic regions, which accelerate the electrons. The curved path of a charged objects implies, as well known, synchrotron radiation, as a result of energy conservation. Moreover, in AGN's like Mkn 501, the same (high-energy) electron can also undergo inverse-Compton scattering with low-energy photons (with energies of order eV, e.g. photons of the cosmic microwave background radiation that populate the Universe as remnants of the Big-Bang). The terminology inverse-Compton (IC) scattering refers here to the fact that, contrary to the conventional Compton photon-electron scattering, here it is the electron which is the high energy particle, and whose loss of energy is converted to outgoing radiation (c.f. fig. 5). This IC outgoing radiation can have very high energies. In fact, the Compton spectrum peak can be at several TeV energies, as observed for the AGN Mkn 501. This combined process, whereby the *same electron* that is responsible for *synchrotron* radiation in AGN also undergoes *IC scattering* to produce high energy photons is known as Synchrotron-self-Compton (SSC) mechanism [32], and is believed [28] -with some variations to be discussed below - that is responsible for the production of the very high energy photons observed in the AGN Mkn 501 (c.f. fig. 5). This is the so-called Leptonic acceleration mechanism, to distinguish it from a different type of acceleration, the *hadronic* one, to be discussed below, and which is believed by many as being the main mechanism for extragalactic ultra-high energy cosmic rays. Notably, AGNs are also extragalactic objects and hence hadronic acceleration mechanism may be relevant for the production of very high energy Gamma rays, as alternative scenarios to (or co-existing with) SSC-leptonic acceleration mechanism described above.

Hadronic acceleration: A prominent way of producing high energy Gamma Rays of extragalactic origin, is the scattering of very high energy protons (produced in the jet of the AGNs by means of gravitational energy conversion mentioned above (c.f. fig.4)) off protons in the *interstellar* medium. Such collisions result in pion production, of which neutral pions decay ($\pi^0 \to 2\gamma$) and give rise to very high energy photons that are detected directly. The charged pions on the other hand are converted to muons, whose decays produces electrons or positrons ($\pi^\pm \to \mu^\pm \to e^\pm$); the scattering of the latter with low-energy photons of the cosmic microwave background radiation (CMB) then results, through (inverse) Compton scattering, in photons being detected on Earth or satellites (c.f. fig. 6).

The energy spectra of hadronic acceleration models are different in shape and location of their peaks in the energy axis from those of leptonic acceleration, as can be seen in fig. 7, where various photon spectra in the Universe are superimposed for comparison. It is worthy of mentioning that in order to interpret the current high energy gamma ray data using leptonic inverse Compton measurements, one has to assume that the galactic magnetic fields are of low intensity. From measurements of high energy Gamma ray spectra from AGNs or other extragalactic sources, such as Gamma Ray Bursters, we can then soon get sufficient information into the precise way of acceleration of cosmic particles. The current experimental knowledge on cosmic high-energy gamma ray spectra can be summarised in fig. 8, from which it is clear that we need more measurements in the lower-energy part of the spectrum before conclusions can be drawn on the kind of cosmic acceleration taking place at various celestial sources.

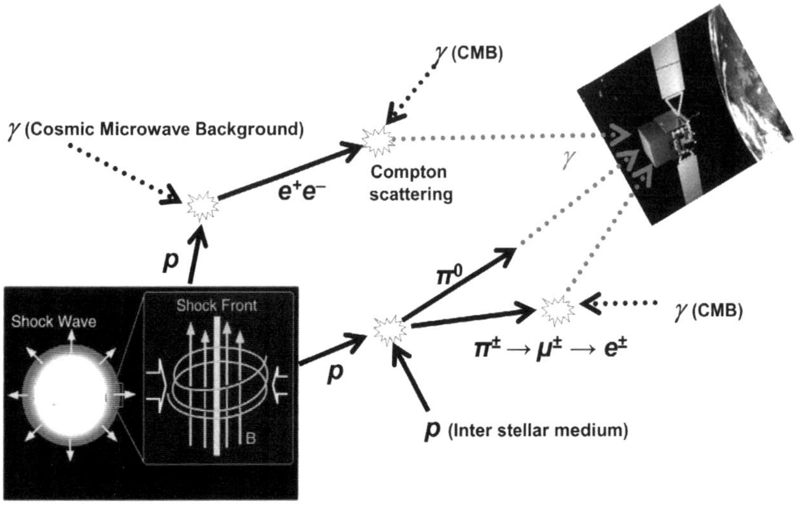

Figure 6. A hadronic acceleration model for the production of very high energy gamma rays from extragalactic sources. High-energy Protons which have been accelerated in AGNs or other sources in the presence of magnetic cosmic fields, can interact with the protons of the interstellar medium to give rise to pions (neutral π^0 and charged $\pi^{\pm}$). Photons from neutral pion decays($\pi^0 \to 2\gamma$) could be detected together with those from coming from charged-pion conversion to muon processes $\pi^{\pm} \to \mu^{\pm} \to e^{\pm}$, which eventually yield electrons or positrons that scatter á la Inverse Compton with low-energy CMB photons resulting in detectable (on Earth or satellite experiments) high-energy photons (picture taken from H. Tajima SLAC-DOE (USA) Programme review talk (June 7, 2006), (`www-conf.slac.stanford.edu/programreview/2006/Talks`)).

3.2. The MAGIC delays: adding to the uncertainties on the VHE gamma-ray production mechanisms

The observed time delays of order 4 ± 1 minutes between the most energetic (TeV) Gamma rays from AGN Mkn 501 (c.f. fig. 2), observed by the MAGIC Telescope [2, 3], lead to further uncertainties in the production mechanism of such photons.

As discussed in [2], the conventional model of SSC used to explain the origin of VHE Gamma Rays as being due to an electronic *uniform* acceleration in the AGN jet region, which finds a good application in other AGNs, such as Crab Nebula, fails to account for the observed time delay by MAGIC. The use of the acceleration parameters in the Crab Nebula AGN leads to only millisecond delays of the more energetic photons if applied to the Mkn 501 case.

This prompted speculations that the conventional SSC mechanisms involving uniform acceleration of the relativistic blob of particles in the jet of the AGN (c.f. fig. 4) need to be modified for the Mkn 501 case. In fact, several propositions along this line have been made so far:

- (i) Particles inside the emission region moving with constant Doppler factor need some time to be accelerated to energies that enable them to produce γ rays with specific high energies in the TeV region [2].
- (ii) The γ-ray emission has been captured in the initial phase of the acceleration of the relativistic material blob in the jet of the AGN (c.f. fig. 4), which at any point in time radiates up to highest γ-ray energies possible [34].
- (iii) A version of the SSC scenario (termed *one-zone SSC model*), which invokes a brief

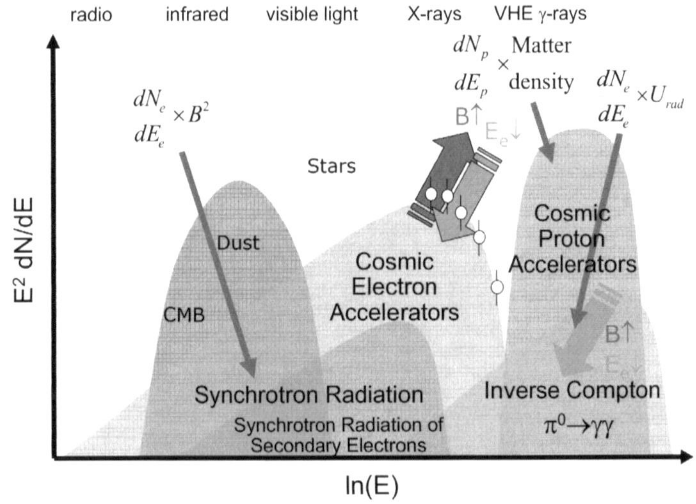

Figure 7. Comparing the spectra of very high energy gamma rays in models of leptonic and hadronic acceleration of cosmic particles. In the figure E denotes the photon energy (in some generic units) and N the observed cosmic photon number, while the suffix "e" ("p") denotes quantities pertaining to electrons (protons) and B denotes the magnetic field. From the differences in the shape and position of these spectra one can get information on the kind of acceleration that takes place (picture taken from talk of M. de Naurois, Workshop of HSSHEP, April 2008, Olympia (Greece), (`http://www.inp.demokritos.gr/conferences/HEP2008-Olympia/`)).

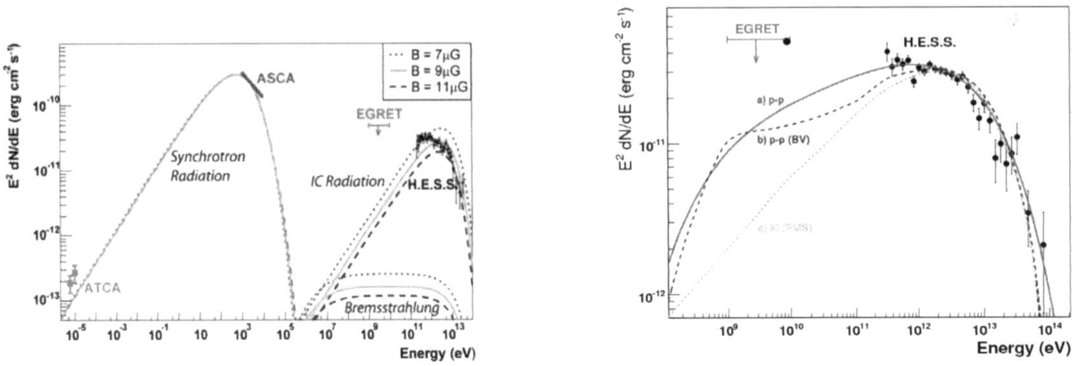

Figure 8. Recent data on Gamma ray spectra. The symbols in the axes have the same meaning as in fig. 7. To be sure on the kind of cosmic acceleration taking place (i.e. leptonic or hadronic) one needs data at the lower energy part of the spectrum. Such data can be provided, for instance, by the FERMI (GLAST) satellite, which was recently launched [33] (picture taken from talk of M. de Naurois, Workshop of HSSHEP, April 2008, Olympia (Greece), (`http://www.inp.demokritos.gr/conferences/HEP2008-Olympia/`)).

episode of increased particle injection at low energies [35]. Subsequently, the particles are accelerated to high energies, which thus accounts for the observed delays, but they also emit synchrotron and SSC, thereby loosing energy. As in scenario (ii) above, also according to this model the MAGIC observations have caught the relativistic electrons in the jet of the AGN at their acceleration phase.

To the above I would also like to add the possibilities that some hadronic mechanisms might also be in operation here, which seems not to have been discussed by the community so far. It therefore becomes clear from the above brief discussion that the situation concerning the delayed production of VHE Gamma Rays from the AGN Mkn 501 is far from being resolved by means of conventional (astro)physics at the source.

This brings us to the main topic of this article, which is a possible (albeit speculative at this stage) link of the MAGIC observation with more fundamental physics associated with the very structure of space time on which the *propagation* of the VHE Gamma Rays takes place.

3.3. Quantum gravity spacetime foam and the MAGIC delays: wild speculation or realistic scenarios?

In [3] it has been observed that, as a result of the ability of the experiment to measure individual (within the accuracy of the observations of course) photons from Mkn 501, of various energies, it should be possible to reconstruct the peak of the flare of July 9th 2005 using dispersion relations of these individual photons during their journey from emission till observation. In fact, we went one step further and assumed sub-luminal modified dispersion relations of the type expected [20, 25, 36] to be encountered in a model of quantum-gravity (QG) induced space-time foam [16] coming from string theory [4]. As we shall discuss below, when we describe in detail this model, the sub-luminality of the QG-induced refractive index in such models is guaranteed by the very nature of string theory, which respects the cornerstone of special relativity that the speed of light in vacuo is the maximal material velocity. In this respect the space-time foam in such theories leads to the *absence of birefringence*, in other words the refractive index is the same for both photon polarizations. This is an important feature, which allows the MAGIC results to be compatible with other stringent limits of Lorentz invariance from other astrophysical sources, as we shall explain below.

In [3] we examine two cases of QG-induced modified dispersion for photons, stemming from (9) upon assuming:

- Case I: Photon Refractive index *suppressed Linearly* by the QG energy scale $M_{\rm QG1}$, i.e. only the coefficient $c_1 > 0$ in the series of Eq. (9) is non zero, its positivity being required by the sub-luminal nature assumed for the propagation, as ensured by the string theory underlying model [20, 36, 4].
- Case II: Photon Refractive index *suppressed Quadratically* by the QG energy scale $M_{\rm QG2}$, i.e. only the coefficient $c_2 > 0$ in the series of Eq. (9) is non zero, its positivity again being linked to the sub-luminal nature assumed for the propagation. String theory models with this kind of quadratic suppression also exist in the modern approach to string theory, including representation of our world as a brane (domain wall hyperplane) [37], but will not be discussed here.

The method of reconstructing the peak ("most active part") of the flare by implementing modified dispersion relations for individual photons is based on the following well known fact of classical electrodynamics [38]: a pulse of electromagnetic radiation propagating through a linearly-dispersive medium, as postulated above, becomes diluted so that its power (the energy per unit time) decreases. The applicability of classical electrodynamics for estimating the low-energy behavior induced by space-time foam and the corresponding pulse-broadening effect have been discussed in the paper by J. Ellis *et al.* (2000) in ref. [36], where we refer the interested reader for further details and explicit examples.

The dilution effects for the linear or quadratic cases may easily be obtained as described in [38] by applying the dispersion laws

$$\omega(k) = k[1 - k/(2M_{\rm QG1})\ , \quad \text{or} \quad \omega(k) = k[1 - k^2/(3M_{\rm QG2}^2)]\ , \tag{10}$$

where ω denotes the frequency of the photon, with wave vector k. Any transformation of a signal to reproduce the undispersed signal tends to recover the original power of the pulse. If the parameter $M_{\rm QGn}$, $n = 1, 2$ is *chosen correctly*, the power of the recovered pulse is *maximized*.

This observation has been implemented in the analysis of [3] by appropriately choosing (using statistical-analysis techniques) a time interval $(t_1; t_2)$ containing the most active part of the flare. For the record, we mention that this procedure has been applied in [2] to 1000 Monte Carlo (MC) data samples generated by applying to the measured photon energies the (energy-dependent) Gaussian measurement errors. The results of the reconstruction of the peak ("most active part") of the flare using the linear Case I are demonstrated in fig. 9 for completeness. In a similar way one gets bounds on the quadratic Case II. These results have also been confirmed using different statistical analysis techniques, independent of the ECF. The interested reader is referred to [3] for details of the analysis.

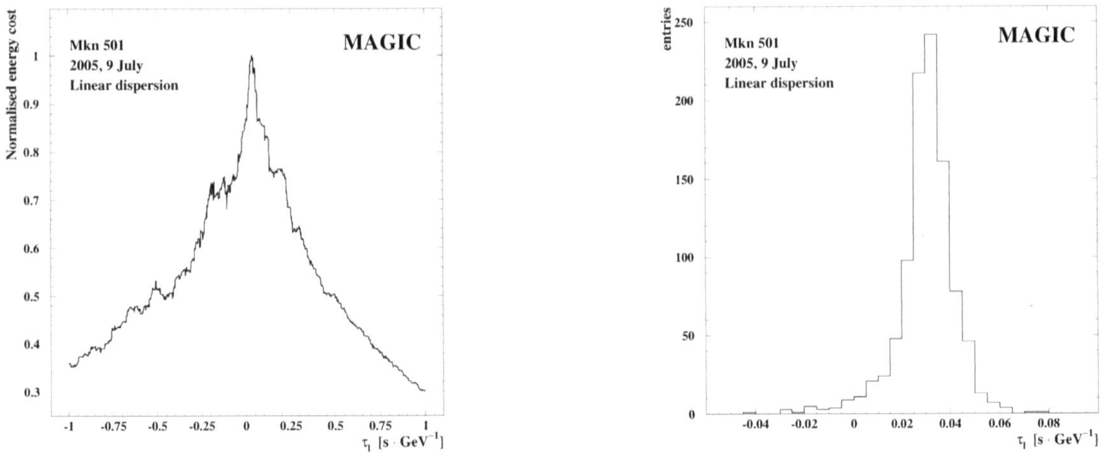

Figure 9. The *Left figure* shows the Energy Cost Functional (ECF) from one realization [3] of the MAGIC measurements with photon energies smeared by Monte Carlo, for the case of a vacuum refractive index that is linear in the photon energy (Case I). The ECF exhibits a clear maximum, whose position may be estimated by fitting it with a Gaussian profile in the peak vicinity. The *Right figure* shows the results of such fits to the ECFs with τ_l for the 1000 energy-smeared realizations of the July 9 flare. From this distribution we derive the value $\tau_l = (0.030 \pm 0.012)$ s/GeV, where $M_{\rm QG1} = 1.445 \times 10^{16}\,{\rm s}/\tau_l$, leading to a lower limit $M_{\rm QG1} > 0.21 \times 10^{18}$ GeV at the 95% C.L.

Taking into account the *uncertainties in the production mechanism* at the source, which could also contribute as we have discussed before to the observed delays, we can only place lower bounds on the *quantum gravity energy scale* from such an analysis. In fact for the linear and quadratic cases we obtain the lower bounds (c.f. fig 9):

$$\begin{aligned} M_{\rm QG1} &> 0.21 \times 10^{18}\ \text{GeV at 95\% Confidence Level (C.L.)}, \\ M_{\rm QG2} &> 0.26 \times 10^{11}\ \text{GeV at 95\% C.L..} \end{aligned} \tag{11}$$

It is important to notice that, had the mechanism at the source been understood, and the emission of the different photons been more or less simultaneous, i.e. by at least two orders of magnitude smaller as compared to the observed delays, then the above lower bounds could be turned into a real measurement of the quantum gravity scale. It is surprising, therefore, that in such a situation Case I can reproduce the observed delays, provided $M_{\rm QG1}$ is of order of the so-called *reduced Planck mass*, which is an energy scale characterising conventional string theory models.

In the analysis of [3] it was possible to exclude the possibility that the observed time delay may be due to a conventional QED plasma refraction effect induced as photons propagate through the source. From the discussion in sub-section (2.1), in particular eq. (7), it becomes clear that if the delay would be due to plasma effects at the source region, then this would induce

$$\Delta t = D(\alpha^2 T^2/6k^2)\ln^2(kT/m_e^2) \ ,$$

where α is the fine-structure constant, k is the photon momentum, T is the plasma temperature, m_e is the mass of electron, D is the size of the plasma, and we use natural units: $c, \hbar = 1$. Plausible numbers such as $T \sim 10^{-2}$ MeV and $D \sim 10^9$ km (for a review see [39]) yield a negligible effect for $k \sim 1$ TeV, which are the photon momenta relevant to the MAGIC experiment. Exclusion of other source effects, such as time evolution in the mean emitted photon energy, might be possible with the observation of more flares, e.g., of different AGNs at varying redshifts. Observations of a single flare cannot distinguish the quantum-gravity scenarios considered here from modified synchrotron-self-Compton mechanisms.

3.4. H.E.S.S. and FERMI observations and quantum gravity scale bounds

However, the above-described pioneering study demonstrates clearly the potential scientific value of an analysis of multiple flares from different sources. The most promising candidate for applying the analyses proposed here is the flare from the Active Galaxy PKS 2155-304 detected recently by H.E.S.S. (*H*igh *E*nergy *S*tereoscopic *S*ystem) Collaboration [40], another experiment involving arrays of Cherenkov Telescopes in Namibia (Africa). This galaxy lies further than Mk501 at redshift $z = 0.116$ and there is a much higher statistics of photons at energy ranges of a few TeV.

In fact, quite recently, H.E.S.S. collaboration published their measurements [7] on the arrival time of photons from PKS 2155-304. However, unlike the MAGIC observations from Mk501 Galaxy, there was no time lag found between higher- and lower-energy photons in this case. These results can thus place only bounds on the quantum gravity scale, if space-time foam is assumed to affect the photon propagation. The bounds are similar to the MAGIC case (11) above.

The H.E.S.S. result can mean several things, and certainly points towards different source mechanisms for the acceleration of photons between the two galaxies Mkn 501 and PKS 2155-304. However, this second measurement by H.E.S.S. cannot still rule out the possibility that quantum gravity plays a rôle in photon propagation. For instance, one cannot exclude the (admittedly remote) possibility that, due to a still unknown source effect, the high-energy photons in the PKS 2155-304 Galaxy are emitted first, in contrast to the Mkn 501 case, in such a way that a sub-luminal vacuum refractive index quantum-gravity effect, of a strength appropriate to produce the delays observed by the MAGIC experiment, "slowed these photons down" as compared to the lower-energy ones, so that there are no observable delays in the arrival times between the higher and lower energy photons in the H.E.S.S. experiment. Of course, one cannot exclude the possibility that the conditions of this set of measurements, for some reason, prohibited the detection of an observable time lag between high and low energy photons.

Although the available sample of AGNs is still not large enough for a robust analysis on bounds of Quantum-Gravity medium effects, nevertheless, one can at least check for consistency between the available MAGIC and HESS results, and gauge the magnitude of possible intrinsic fluctuations in the AGN time-lags. Comparing the time-lag measured by MAGIC for Mkn 501 at redshift $z = 0.034$: $\Delta t/E_\gamma = 0.030 \pm 0.012$ s/GeV, with that measured for PKS 2155-304 at $z = 0.116$: $\Delta t/E_\gamma = 0.030 \pm 0.027$ s/GeV, we see that they are compatible with a common, energy-dependent *intrinsic* time-lag at the source. On the other hand, they are also compatible with a universal redshift- and energy-dependent *propagation* effect:

$$\Delta t/E_\gamma = (0.43 \pm 0.19) \times K(z)\text{s/GeV}, \qquad K(z) \equiv \int_0^z \frac{(1+z)dz}{\sqrt{\Omega_\Lambda + \Omega_m(1+z)^3}}, \tag{12}$$

assuming an expanding Universe within the framework of the standard Cosmological-constant-Cold-Dark-Matter (ΛCDM) model. The best fit of the MAGIC and HESS data based on (12) leads to the following result for the Quantum Gravity scale, assuming that it is the dominant cause of the delay: $M_{QG1} = (0.98^{+0.77}_{-0.30}) \times 10^{18}$ GeV.

With measurements from only a few available flares from AGN it is, therefore, not possible to disentangle with any certainty source from propagation effects. For this we need statistically significant populations of available data. Unfortunately the occurrence of fast flares in AGNs is currently unpredictable, and since no correlation has yet been established with observations in other energy bands that could be used as a trigger signal, only serendipitous detections are currently possible. It seems unlikely that the relatively rare and unpredictable sharp energetic flares produced only occasionally by AGNs, which have a relatively restricted redshift range and hence a small lever arm, will soon be able to provide the desired discrimination.

On the other hand, Gamma Ray Bursts (GRBs) are observed at a relatively high rate, about one a day, and generally have considerably larger redshifts. The advent of the FERMI (née GLAST) Telescope with its large acceptance offers the possibility of achieving the required sensitivity. Indeed, the FERMI Collaboration has already made a preliminary report of GeV-range γ rays from the GRB 080916c [6]. In this GRB, there is a 4.5-second time-lag between the onsets of high- ($>$ 100 MeV) and low-energy ($<$ 100 KeV) emissions. Moreover, the highest-energy photon GRB 080916c measured by the FERMI γ-ray telescope had an energy $E = 13.2^{+0.70}_{-1.54}$ GeV, and was detected $\Delta t = 16.5$ s after the start of the burst. Spectroscopic information has been used by the GROND Collaboration [41] to estimate the redshift of GRB 080916c as $z = 4.2 \pm 0.3$ [6]. Assuming this value of the redshift, the best fit (12) would correspond to a time-lag

$$\Delta t = 25 \pm 11 \text{ s} \tag{13}$$

for a 13 GeV photon from GRB 080916c. As discussed in [4, 6], such time delays can fit excellently within the above-mentioned QG scenario for a subluminal refractive index for photons, with the following lower bound for the QG scale $M_{QG1} > 1.50 \pm 0.20 \times 10^{18}$ GeV, where the inequality is due to the ignorance of the source mechanism. This bound is consistent with the MAGIC and HESS results stated previously. The reader should also bear in mind that the 4.5-second time-lag observed for $\sim$ 100 MeV photons could not be explained by a propagation effect that depends linearly on the energy [6]. Because of ignorance of the source mechanism, the preliminary analysis of these data by the FERMI Collaboration [6] quoted a lower bound $M_{QG1} > 1.50 \pm 0.20 \times 10^{18}$ GeV, which is consistent with the MAGIC and HESS results stated previously. It is clear therefore that the FERMI Telescope has already demonstrated the sensitivity to probe a possible linearly energy-dependent propagation effect at the level reached by the available AGN data, and it is appropriate and possibly helpful to consider how such an effect could be probed in the future.

The analysis of the second reference in [4] has also demonstrated that although the three above-mentioned sets of data, from MAGIC, FERMI and HESS Collaborations, can be explained simultaneously by a linear in energy vacuum refractive index, suppressed by a single power of the quantum gravity scale, this is not the case for a refractive index scaling quadratically with the photon energy, for instance of the type encountered in some brane models with asymmetric warp factors, as in ref. [37]. Indeed, on assuming that the quadratic refractive index is the sole cause of the observed delays in the arrival of high- vs. low -energy photons in *both* the MAGIC and FERMI cases, this would imply a time delay of order 0.24 ± 0.16 s for the most energetic photon (13.22 GeV) of the GRB 080916c, for a quantum gravity scale that saturates $M_{\rm QG2}$ in (11). This is two orders of magnitude smaller than the measured time-lag (16.5 s) by the FERMI Collaboration [6].

Prompted by the MAGIC (and subsequently FERMI) results, in particular their consistency with a non-trivial linear in energy vacuum refractive index, the authors of [4] attempted to discuss space-time foam models within the framework of string theory and in particular its modern extension, involving solitonic domain-wall defects (D(irichlet) branes. We review such a model (or better a class of such models) in the next section. Before embarking on such a task, however, we consider it as essential to compare the bounds (11) on QG-induced anomalous photon dispersion (10) obtained from the MAGIC observations with similar bounds on Lorentz Violation from other astrophysical processes. This will be important for the viability of the string model as a possible explanation of the MAGIC "anomaly". Indeed, it is important to notice that, if the results of MAGIC (and FERMI) collaborations are attributed as being predominantly due to the "*medium*" of Quantum Gravity, which leads to anomalous photon dispersion, then it is imperative that this medium is transparent to electrons, and it does not imply birefringnece effects, since the latter would lead to much more stringent bounds on the anomalous terms, incompatible with the bounds (11) from photon measurements.

3.5. MAGIC and FERMI time-delays versus other (astrophysical) constraints on quantum gravity foam

The sensitivity of the MAGIC (and FERMI) observations to Planck scale physics (11), at least for linearly suppressed modified dispersion relations, calls for an immediate comparison with other sensitive probes of non-trivial optical properties of QG medium.

Indeed, from the analysis of [3], there was no microscopic model dependence of the induced modifications of the photon dispersion relations, other than the sub-luminal nature of the induced refractive index and the associated absence of birefringence, that is the independence of the refractive index on the photon polarization. The latter feature avoids the otherwise very stringent constraints on the photon dispersion relation imposed by astrophysical observations, as we now come to discuss.

We shall be very brief in our description of the complementary astrophysical tests on Lorentz invariance and quantum-gravity modified dispersion relations, to avoid large diversion from our main point of this review article which is string theory.

There are three major classes of complementary astrophysical constraints, to be considered in any attempt to interpret the MAGIC, FERMI or more general γ-ray Astrophysics results in terms of quantum-gravity induced anomalies in photon dispersion.

- *Birefringence and strong constraints on QG-induced photon dispersion:*
  In certain models of quantum gravity, with modified dispersion relations, for instance the so-called loop-quantum gravity [42], the ground state breaks reflexion symmetry (parity) and this is one of the pre-requisites for a dependence of the induced refractive index on the photon polarization, *i.e.* birefringence. We remind the reader that in birefringent materials this is caused precisely by the existence of some kind of anisotropies in the material. The velocities of the two photon polarizations (denoted by $\pm$) in such QG models may be parametrised by:

$$v_{\pm} = c\,(1 \pm \xi(\hbar\omega/M_P)^n) \tag{14}$$

where $M_P = 1.22 \times 10^{19}$ GeV is the Planck energy scale, and ξ is a parameter following from the underlying theoretical model, which is related with the modifications of the pertinent dispersion relations for photons. The order of suppression of these effects is described by n which in the models of [42] assumed the value $n = 1$, but in general one could have higher order suppression, as we have discussed in (9).
Vacuum QG birefringence should have showed up in optical measurements from remote astrophysical sources, in particular Gamma Ray Bursters (GRB). The latter are cosmic explosions of titanic proportions, due to collapsing massive stars at distant parts of the universe.
Ultraviolet (UV) radiation measurements from distant galaxies [43] and UV/optical polarization measurements of light from Gamma Ray Bursters [44] rule out birefringence unless it is induced at a scale (way) beyond the Planck mass (for linear models the lower bound on the QG scale in such models can exceed the Planck scale ($\sim 10^{19}$ GeV) by as much as *seven orders of magnitude*). Indeed, in terms of the parameter η introduced above (c.f. (14)), for the case $n = 1$ of [42], one finds from optical polarization observations that the absence of detectable birefringence effects imply the upper bound $|\xi| < 2 \times 10^{-7}$, which is incompatible with the MAGIC observed delays, saturating from below the bounds (11).
At this point, we wish to mention that, using recent polarimetric observations of the Crab Nebula in the hard X-ray band by INTEGRAL [45], the authors of [46] have demonstrated that the absence of vacuum birefringence effects constrains linearly suppressed Lorentz violation in quantum electrodynamics to the level $|\xi| < 6 \times 10^{-10}$ at 95% C.L., thereby tightening by about three orders of magnitude the above-mentioned constraint.

- *Synchrotron Radiation and further stringent constraints for electronic QG-induced anomalous dispersion in vacuo.*

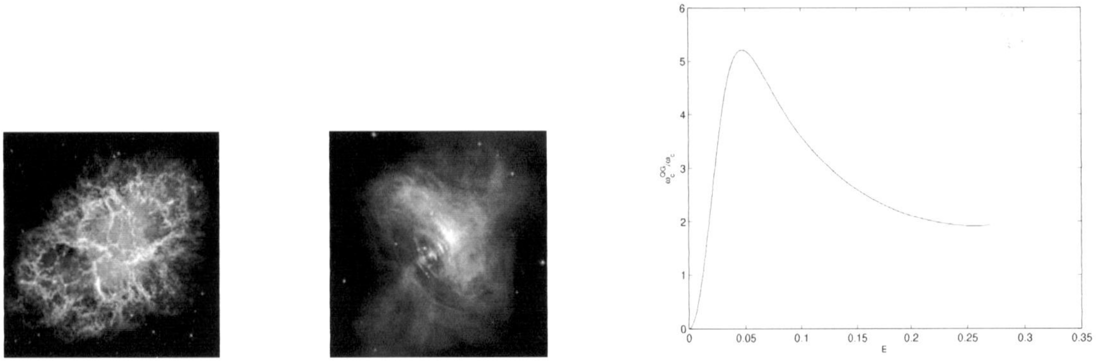

Figure 10. The Crab Nebula (left image) is a supernova remnant, with a rotating neutron star (the Crab Pulsar) at its centre (middle image). Observations of synchrotron radiation (right image, radiation spectrum (arbitrary units)) from such celestial objects places very stringent constraints on quantum gravity models with anomalous dispersion relations for electrons (images taken from NASA, `http://www.nasa.gov/` (Crab Nebula) and wikipedia, `http://en.wikipedia.org/wiki/CrabNebula` (Crab Pulsar)) and [27] (synchrotron spectrum).

Another important experimental constraint on models with QG-induced anomalous dispersion relations comes from observations of synchrotron radiation from distant galaxies [47, 27, 48], such as Crab Nebula (c.f. fig. 10). As mentioned previously, the magnetic fields at the core regions of galaxies curve the paths of (and thus accelerate) charged particles, in particularly electrons (which are stable and therefore appropriate for astrophysical observations), and thus, on account of energy conservation this results in synchrotron radiation.

In standard electrodynamics [38], electrons in an external magnetic field H, follow helical orbits transverse to the direction of H. The so-accelerated electrons in a magnetic field emit synchrotron radiation with a spectrum that cuts off sharply at a frequency ω_c (c.f. fig. 10, right panel):

$$\omega_c^{LI} = \frac{3}{2}\frac{eH}{m_0}\frac{1}{1-\beta^2}, \tag{15}$$

where e is the electron charge, m_0 its mass, and $\beta_\perp \equiv v_\perp$ is the component of the velocity of the electron perpendicular to the direction of the magnetic field. The superfix LI in (15) stresses that this formula is based on a LI approach, in which one calculates the electron trajectory in a given magnetic field H and the radiation produced by a given current, using the relativistic relation between energy and velocity.

All these assumptions are be affected by violations of Lorentz symmetry, such as those encountered in quantum-gravity space-time foam models, leading to modified dispersion relations for photons and electrons of the form:

$$\omega^2(k) = k^2 + \xi_\gamma \frac{k^{2+\alpha}}{M_P^\alpha}, \tag{16}$$

$$E^2(p) = m_0^2 + p^2 + \xi_e \frac{p^{2+\alpha}}{M_P^\alpha}, \tag{17}$$

for photons and electrons, respectively, where ω and k are the photon frequency and wave number, and E and p are the electron energy and momentum, with m_0 the electron (rest) mass. In the spirit of the MAGIC observation analysis above, we assume here linear ($\alpha = 1$) or quadratic QG ($\alpha = 2$) effects, characterized by parameters ξ_γ and ξ_e, extracting the Planck mass scale $M_P = 1.22 \times 10^{19}$ GeV. In fact one can do the analysis [27] for a general α (single power) and attempt to extract limits on this parameter by matching with observations.

A detailed analysis [47, 27], including the modifications in the electron's trajectories due to space-time foam [27], yields:

$$\omega_c^{QG} \propto \omega_c^{LI} \frac{1}{(1+\sqrt{2-1/\eta^2})^{1/2}\left(\frac{m_0^2}{E^2} + (\alpha+1)\left(\frac{E}{\mathcal{M}}\right)^\alpha\right)}, \quad \mathcal{M} \equiv M_P/|\xi_e|, \quad \eta \equiv 1-(E/\mathcal{M})^\alpha, \tag{18}$$

where ω_c^{LI} is given in (15) and the superscript "QG" indicates that the QG-modified dispersion relations (17) are used. This function is plotted schematically (for $\alpha = 1$) in fig. 10 (right panel).

In [47, 27], the above QG-modified dispersion relations have been tested using observations from Crab Nebula. It should be emphasized that the estimate of the end-point energy of the Crab synchrotron spectrum and of the magnetic field used above are indirect values based on the predictions of the Synchrotron Self-Compton (SSC) model of very-high-energy emission from Crab Nebula [49]. In [27] the choice of parameters used was the one that gives good agreement between the experimental data on high-energy emission and the predictions of the SSC model [49, 39]. Estimating the magnetic field of Crab Nebula in the region 160×10^{-6} Gauss $< H < 260 \times 10^{-6}$ Gauss, and requiring $|\xi_e| \leq 1$ (which thus sets the quantum gravity scale as at least M_P) one obtains the following bounds on the exponent α of the dispersion relations (17) [27]:

$$\alpha \geq \alpha_c \ : \qquad 1.72 < \alpha_c < 1.74 \tag{19}$$

These results imply already a sensitivity to quadratic QG corrections with Planck mass suppression M_P.
However, for photons there are no strong constraints on ξ_γ coming from synchrotron radiation studies, unless in cases where QG models entail birefringence [48], where, as we discussed above, strong constraints on the photon dispersion are expected at any rate from optical measurements on GRBs. In this sense, the result (19) *excludes the possibility* that the MAGIC observations leading to a four-minute delay of the most energetic photons are due to a quantum foam that acts *universally* among photons and electrons. However, the synchrotron radiation measurements cannot exclude anomalous photon dispersion with linear Planck-mass suppression, leading to a saturation of the lower bound (11), in models where the foam is transparent to electrons, as in the string case to be discussed in the next section.

- *Strong constraints from Ultra-High-Energy cosmic photon annihilation*
Further strong constraints on generic modified dispersion relations for photons, like the ones used in the aforementioned QG-interpretation of the MAGIC results [3], comes from processes of scattering of ultra-high-energy photons, with energies above 10^{19} eV off very-low energy cosmic photons, such as the ones of the cosmic microwave background (CMB) radiation that populates the Universe today, as a remnant from the Big-Bang epoch. In [5] it has been argued that the non-observation of such ultra-high energy (UHE) photons places very strong constraints on the parameters governing the modification of the photon dispersion relations, that are several order of magnitude smaller than the values required to reproduce the MAGIC time delays, should the effect be attributed predominantly to photon propagation in a QG dispersive medium.
The main argument relies on the fact that an ultra-high-energy photon would interact with a low-energy ("infrared") photon of the CMN background (with energies in the eV range) to produce electron prositron pairs, according to the reaction:

$$\gamma_{\rm UHE} \quad + \quad \gamma_{\rm CMB} \quad \Rightarrow \quad e^+ \, e^- \, . \tag{20}$$

The basic assumption in the analysis is the strict energy and momentum conservation in the above reaction, despite the modified dispersion relations for the photons. Such an assumption stems from the validity of the *local-effective-lagrangian formalism* to the case of an effective description of the QG foam effects on particles with energies much lower than the QG energy scale (assumed close to Planck scale $M_{\rm Pl} = 10^{19}$ GeV). In this formalism, one can represent effectively the foam *dispersive effects* as corresponding to higher-derivative *local operators* in a *flat-space-time* Lagrangian. The upshot of this is the modification of the pertinent equations of motion for the photon field (which in a Lorentz-invariant theory would be the ordinary Maxwell equations) by higher-derivative terms, suppressed by some power of the QG mass scale.
In the notation of [5] one can consider the following modified dispersion relations:

$$\begin{aligned} \omega_\pm &= k^2 + \xi_n^\pm k^2 \left(\frac{k}{M_{\rm Pl}}\right)^n , \qquad \omega_b^2 = k_b^2 , \\ E^2_{e,\pm} &= p_e^2 + m_e^2 + \eta_n^{e,\pm} p_e^2 \left(\frac{p_e}{M_{\rm Pl}}\right)^n \end{aligned} \tag{21}$$

where $(\omega, \vec{k})$ indicate four-momenta for photons, $(E, \vec{p}_e$ are the corresponding four-momentum vectors for electrons, and the suffix b indicates a low-energy CMB photon, whose dispersion relations are assumed approximately the normal ones, as any QG correction is negligible due to the low values of energy and momenta. The +(-) signs indicate left(right) polarizations (photons) or helicities (electrons). Positive (negative) ξ indicate subluminal

(superluminal) refractive indices. Upon the assumption of energy-momentum conservation in the process, one arrives at kinematic equations for the threshold of the reaction (20), that is the minimum energy of the high-energy photon required to produce the electron-positron pairs.
For the linear- or quadratic- suppression case (Cases I and II in the MAGIC analysis above, for which $n = 1, 2$ respectively (21)), for instance, one finds that for the relevant subluminal photon refractive indices corresponding to the saturation of the M_{QG1} lower bound in (11), the threshold for pair production disappears for ultra-high-energy photons, and hence such photons should have been observed. The non-observation of such photons implies constraints for the relevant parameters ξ, η which are stronger by *several orders of magnitude* than the bounds (11) inferred from the MAGIC observations.
From the analysis of [5] for the $n = 1$ case, one concludes that in the case of linear Planck-mass suppression of the sub-luminal QG-indunced modified dispersion relations for photons, of interest for the QG-foam interpretation of the MAGIC results [3], parameters with size $\xi_1 > 10^{14}$ are ruled out. This exceeds the sensitivity of the MAGIC experiment (c.f. (11)) to such Lorentz-symmetry violating effects by fifteen orders of magnitude ! Similar strong constraints are also obtained from the non observations of *photon decay* ($\gamma \to e^+e^-$), a process which, if there modified dispersion relations, is generally allowed [5].
However, the reader should recall that this type of analysis is based on exact energy momentum conservation in the process (20), stemming from the assumption of the local-effective lagrangian formalism for QG-foam. As we discussed in [50], and shall review briefly below, however, such a formalism need not be applicable in the case of quantum gravity, where the fluctuations of space-time or other defects of gravitational nature paly the rôle of an external environment, resulting in *energy fluctuations* in the reaction (20). The presence of such fluctuations does affect the relevant energy-threshold equations, for the reaction to occur, which stem from kinematics, in such a way that the above stringent limits are no longer valid.

After this necessary digression, we now come back to discuss a string theory model (or better a class of such models), which can explain the MAGIC observations in terms of vacuum-induced non-trivial refractive indices, in agreement with all the above-mentioned complementary constraints on Lorentz Violations.

4. A stringy model for space time medium with non-trivial "optical" properties: D–particle "foam"

From the above discussion it becomes clear that any model of refraction in space-time foam that exhibits effects at the level of the MAGIC sensitivity [3] (11) should be characterised by the following specific properties:

- (i) photons are *stable* (i.e. do *not* decay) but should exhibit a modified *subluminal* dispersion relation with Lorentz-violating corrections that grow linearly with $E/(M_{\mathrm{QG}\gamma}c^2)$, where $M_{\mathrm{QG}\gamma}$ is close to the Planck scale,
- (ii) the medium should not refract electrons, so as to avoid the synchrotron-radiation constraints [47, 27], and
- (iii) the coupling of the photons to the medium must be independent of photon polarization, so as not to have birefringence, thus avoiding the pertinent stringent constraints [43, 44, 48].
- (iv) The formalism of local effective lagrangians breaks down, in the sense that there are quantum fluctuations in the total energy in particle interactions, due to the presence of a quantum gravitational 'environment', such that stringent constraints, which otherwise would be imposed from the non-observation of ultra-high energy photons ($\hbar\omega > 10^{19}$ eV), are evaded.

Figure 11. A type IA string theory model of D-particle "foam". The model consists of appropriate stucks (left panel) of D(irichlet)branes, some of which are moving in a higher-dimensional Bulk space time, punctured by point-like D-brane defects (D-particles). As the brane world moves (right panel), D-particles from the bulk cross the brane world, and thus they appear to an observer on the brane as "flashing on and off" space-time foam defects ("D-particle foam"). Photons, represented as open strings on the D3 brane, interact with these defects via capture/recoil, and this leads to non-trivial refractive indices. The effect is therefore "classical" from the bulk space time point of view, but appears as an effective "quantum foam" from the D3-brane observer effective viewpoint.

A model with all these properties has been suggested by us some years ago [51, 27, 4], and is based on a stringy model for space-time foam, which we now come to discuss.

4.1. Brief review of the model and its cosmology

In [51] we have attempted to construct a brane/string-inspired model of space time foam which could have realistic cosmological properties. For this purpose we exploited the modern approach to string theory [1], involving membrane hypersurfaces (D(irichlet)-branes). Such structures are responsible for reconciliating (often via duality symmetries) certain string theories (like type I), which before were discarded as physically uninteresting, with Standard-Model phenomenology in the low-energy limit.

In particular, we considered (c.f. figure 11) a ten-dimensional bulk bounded by two eight-dimensional orientifold planes, which contains two stacks of eight-dimensional branes, compactified to three spatial dimensions. Owing to special reflective properties, the orbifolds act as boundaries of the ninth-dimension. The bulk space is punctured by point-like D0-branes (D-particles), which are allowed in type IA string theory (a T-dual of type I strings [52]) we consider in [51] and here [1]. These are massive objects in string theory [1], with masses M_s/g_s, where M_s is the string mass scale (playing the rôle of the quantum gravity scale in string theory), and $g_s < 1$ is the string coupling, assumed weak for our purposes. These objects are viewed

[1] One can extend the construction to phenomenologically realistic models [53] of type IIB strings, in which the "D-particles" are constructed out of D3-branes wrapped around appropriate three cycles. The observable universe is represented by appropriate stacks of higher dimensional Dp-branes (with some of their extra dimensions compactified). Standard model excitations are represented by strings stretched among such branes. The foamy backgrounds of brane worlds punctured by D-particles consist of intersecting Dp-D3(compactified)-brane configurations, with open strings stretching between them, which describe the capture process. The calculation of time delays from such processes is technically more involved in such cases, but the basic conclusions remain the same as in our type-IA string model. Photons and not electrons are susceptible to significant time delays, leading to observed effects.

as space-time *defects*, analogous, *e.g.* to cosmic strings, but these are point-like and electrically neutral. I have to stress at this point that, according to modern ideas in string theory [1], the scale M_s is in general different from the four-dimensional Planck-mass scale $M_P = 1.2 \times 10^{19}$ GeV/c^2, and in fact it is a free parameter in string theory to be constrained by experiment. The energy scale $M_s c^2$ can be as low as a few TeV; it cannot be lower than this, though, since in such a case we should have already seen fundamental string structures experimentally, which is not the case.

Our model assumes a collision between two branes from the original stack of branes (c.f. fig. 11) at an early epoch of the Universe, resulting in an initial cosmically catastrophic Big-Bang type event in such non-equilibrium cosmologies [54]. After the collision, the branes bounce back.

It is assumed that currently the branes are moving slowly towards the stack of branes from which they emanated. Hence populations of bulk D-particles cross the brane worlds and interact with the stringy matter on them. To an observer on the brane the space-time defects will appear to be *"flashing"* on and off. The model we are using involves eight-dimensional branes and so requires an appropriate *compactification* scheme to three spatial dimensions e.g. by using manifolds with non-trivial *fluxes* (unrelated to real magnetic fields). Different coupling of fermions and bosons to such external fields breaks target space supersymmetry. The consequent induced mass splitting [55, 56] between fermionic and bosonic excitations on the brane world is proportional to the intensity of the flux field (a string generalization of the well-known Zeeman effect of ordinary quantum mechanics, whereby the presence of an external field leads to energy splittings, which are however different between (charged) fermions and bosons. In this way one may obtain phenomenologically realistic mass splittings in the excitation spectrum (at TeV or higher energy scales) owing to *supersymmetry obstruction* rather than spontaneous breaking (this terminology, which is due to E. Witten [57], means that, although the ground state could still be characterised by zero vacuum energy, the masses of fermion and boson excitations differ and thus supersymmetry is broken at the level of the excitation-spectrum). The assumption of a population concentration of massive D-particle defects in the haloes of galaxies can lead to modified galactic dynamics [58]. Hence, we have an alternative scenario to standard cold dark matter, using vector instabilities arising from the splitting of strings and attachment of their free ends to a D-particle defect (c.f. fig. 11).

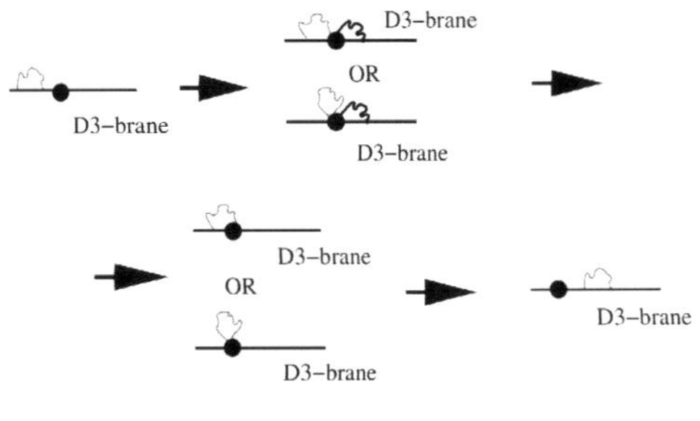

Figure 12. Schematic view of the capture process of an open string state, representing a photon, propagating on a D3-brane world by a D-particle on this world. The intermediate string state, indicated by thick wave lines, which is creating on capture of the end(s) of the photon by the D-particle, stretches between the D-particle and the brane world, oscillates in size between 0 and $\alpha' p^0$, where p^0 is the energy of the incident photon, and thus produces a series of outgoing photon waves, with attenuating amplitudes, constituting the re-emission process. The intermediate string state provides also the restoring force, necessary for keeping the D-particle roughly in its position after scattering with the photon string state.

4.2. Time delays in D-particle foam: a matter of uncertainty with...strings attached

We can now proceed to discuss a possible origin of time delays induced in the arrival time of photons, emitted simultaneously from an astrophysical object, as a result of their propagation in the above-described D-particle space-time foam model. As we shall argue, the above model is in principle capable of reproducing photon arrival time delays proportional to the photon energies, of the kind observed in the MAGIC experiment [2, 3]. Such delays occur during the encounter of photons with the D-particles defects in the foam. This microscopic phenomenon, which -as we shall see below-is essentially stringy and does not *characterise* local field theories, contributes to a sub-luminal non-trivial refractive index *in vacuo*, induced by the capture of photons or electrically neutral probes by the D-particle foam [4]. The capture process is described schematically in figure 12 and we next proceed to describe the underlying physics, which is essentially *stringy*.

An important feature of the model is that, on account of electric charge conservation, only electrically neutral excitations are subjected to capture by the D-particles. To the charged matter the D-particle foam looks *transparent*. This is because the capture process (12) entails a splitting of the open string state, representing matter excitations. Charged excitations are characterized by an electric flux flowing across the string, and when the latter is cut in two pieces as a result of its capture by the D-particle defect, the flux should go somewhere because charge is conserved. The D-particle, being neutral, cannot support this conservation, and hence only *electrically neutral* excitations, such as photons, are subjected to this splitting and the associated delays, as we shall discuss below. The reader should bear in mind of course that the D-particles carry other kinds of fluxes, unrelated to electromagnetism [1]. These are conserved separately, and it is for this reason that isolated D-particles cannot exist, but there must always be in the company of other D-branes, as in our model above, so that the relevant fluxes are carried by the stretched strings between the latter and the D-particles.

For our purposes in this work it is also important to remark that the D-particles are treated as static when compared to photons. This is because the ends of the open string representing the photon move on the D3 brane world with the speed of light in (normal) vacuo, $c = 1$ in our units, while the relative velocities of the D-particles with respect to the brane world (which propagates in the bulk space) are much lower than this, in the model of brane cosmology of [51], which we use here as our concrete example of D-particle foam. For instance, as discussed in [54], to reproduce cosmological observations in this model, in particular the spectrum of primordial density fluctuations, which are affected by the relative motion of the brane world, the speed of propagation of the D3-brane Universe should be smaller than $10^{-4}c$.

When the end(s) of the open-string photon state are attached to the D-particle, there is an intermediate string state formed, *stretched* between the D-particle and the D3-brane, which absorbs the incident energy p^0c of the photon state (where p^0 denotes the temporal component of the four-momentum), to grow in size from zero size to a length L that is determined by the requirement of energy minimization as follows. One assumes [59] that the intermediate string state needs N oscillations to achieve its maximal length L. This implies:

$$p^0 = \frac{L}{\alpha'} + \frac{N}{L} \tag{22}$$

where $\alpha' = \ell_s^2$ is the Regge slope, which is equal to the square of the string length ℓ_s, and hence it has dimensions of $[\text{length}]^2$. In (22) and throughout the article we are working in relativistic units of $\hbar = c = 1$, and in this sense p^0 is identified with the energy of the photon. In such units we also have that the string length $\ell_s = 1/M_s$, where M_s is the string mass scale.

The first term on the right-hand side of (22), proportional to the length L, is due to the tension $1/\alpha'$ of the stretched string, while the second term represents the energy storage due to oscillations. The relation (22) then expresses energy conservation in the capture process of fig. 12.

We can minimize the right hand side by varying with respect to L and demanding the derivative to vanish, which guarantees energy minimization in the process. This procedure determines N, which then is substituted back to the equation to yield the required maximal L: $L_{\rm max} = \frac{1}{2}\alpha' p^0$. Since the end of the stretched string state attached on the D3-brane moves with the sped of light in (normal) vacuo $c = 1$, this implies that the time taken (delay) for the intermediate string state to first grow to this maximal length and then shrink again to zero size is:

$$\Delta t \sim \alpha' p^0 \qquad (23)$$

This describes the time delay encountered in photon propagation in D-particle foam, due to the formation of the intermediate composite string state between D-particles and the photon (fig. 12).

Above we have discussed the situation in the Dirichlet picture, describing the attachment of the end of the strings on D-branes. The (quantum) oscillations of the intermediate string state will produce a series of outgoing wave-packets, with attenuating amplitudes, which will correspond to the re-emission process of the photon after capture. The presence of the stretched string state, which carries the characteristic flux of the D-brane interactions, provides the *restoring force*, necessary to keep the D-particle in its position after scattering with the photon. The situation may be thought of as the stringy/brany analogue of the restoring force in situations in local field theories of photons propagating in media with non trivial refractive indices, as discussed by Feynman [10] and reviewed above in section (2.1). The rôle of the electrons in that case (represented as harmonic oscillators) is played here by the D-particle defects of the space-time. The stringy situation, however, is more complicated, since the D-particles have an infinite number of oscillatory excitations, represented by the various modes of open strings with their ends attached to them. Moreover, contrary to the conventional medium case, in the string model the refractive index is found proportional to the photon frequency, while the effective mass scale that suppresses the effect (23) is the quantum gravity (string) scale M_s and not the electron mass, as in (3). The latter property can be understood qualitatively by the fact that the mass of these D-particle defects is of order [1] $1/(g_s\sqrt{\alpha'}) = M_s/g_s$, where g_s is the string coupling (in units of $\hbar = c = 1$).

The time delays (23) pertain to a single encounter of a photon with a D-particle. In case of a foam, with a linear density of defects $n^*/\sqrt{\alpha'}$, *i.e.* n^* defects per string length, the overall delay encountered in the propagation of the photon from the source to observation, corresponding to a traversed distance D, is:

$$\Delta t_{\rm total} = \alpha' p^0 n^* \frac{D}{\sqrt{\alpha'}} = \frac{p^0}{M_s} n^* D \qquad (24)$$

When the Universe's expansion is taken into account, one has to consider the appropriate red-shift-z dependent stretching factors which affect the measured delay in the propagation of two photons with different energies. Thus, from (24) we obtain in such a case a total delay in the arrival times of photons with energy difference ΔE, which has the form considered in [36], namely it is proportional to ΔE and is suppressed linearly by the quantum gravity (string) scale, M_s:

$$(\Delta t)_{\rm obs} = \frac{\Delta E}{M_s} {\rm H}_0^{-1} \int_0^z n^*(z) \frac{(1+z)dz}{\sqrt{\Omega_\Lambda + \Omega_m (1+z)^3}} \qquad (25)$$

where z is the red-shift, ${\rm H}_0$ is the (current-era) Hubble expansion rate, and we have assumed for concreteness the ΛCDM standard model of cosmology, with $\Omega_i \equiv \frac{\rho_{i(0)}}{\rho_c}$ representing the present-epoch energy densities (ρ_i) of matter (including dark matter), Ω_m, and dark vacuum energy, Ω_Λ, in units of the critical density $\rho_c \equiv \frac{3H_0^2}{8\pi G_N}$ of the Universe (G_N is the Newton's gravitational constant), that is the density required so that the Universe is spatially flat. The current astrophysical measurements of the acceleration of the Universe are all consistent with a non zero Cosmological-Constant Universe with Cold-Dark-Matter (ΛCDM Model), with $\Omega_\Lambda \sim 73\%$ and $\Omega_m \sim 27\%$.

Notice in (25) that the essentially stringy nature of the delay implies that the characteristic suppression scale is the string scale M_s, which plays the rôle of the quantum gravity scale in this case. The scale M_s is a free parameter in the modern version of string theory, and thus it can be constrained by experiment. As we have discussed in this article, the observations of delays of energetic (TeV) photons from AGN by the MAGIC telescope [2] can provide such an experimental way of constraining n^*/M_s in (25). For $\Delta E \sim 10$ TeV, for instance, the delay (25) can lead to the observed one of order of minutes, provided $M_s/n^* \sim 10^{18}$ GeV (in natural units with $c = 1$) [3]. This implies natural values for both n^* and M_s, although it must be noted that n^* is another free parameter of the bulk string cosmology model of [51], considered here. In general, $n^*(z)$ is affected by the expansion of the Universe, as it is diluted by it. This depends on the bulk model and the interactions among the D-particles themselves. For redshifts of relevance to the MAGIC experiments, $z = 0.034 \ll 1$, one may ignore the z-dependence of n^* to a good approximation.

The total delay (25) may be thought of as implying [10] an effective *subluminal* refractive index $n(E)$ of light propagating in this space time, since one may assume that the delay is equivalent to light being slowed down due to the medium effects. On account of the theoretical uncertainties in the source mechanism, however, the result of the AGN Mkn 501 observations of the MAGIC Telescope translate to *upper* bounds for the quantity n^*/M_s in (25), which determines the strength of the anomalous photon dispersion in the string/D-particle foam model.

We remark at this stage that the above time delays are a direct consequence of the *stringy uncertainty principles*, and as we have discussed they are essentially stringy effects, associated with the capture process of fig. 12. Indeed, strings are characterised – apart from the phase-space Heisenberg uncertainty relation, modified by higher order terms in α' [60] as a result of the existence of the minimal string length $\ell_s \equiv \sqrt{\alpha'}$ in target space-time,

$$\Delta X \Delta P \geq \hbar + \alpha' (\Delta P)^2 + \dots \tag{26}$$

also by the space-time uncertainty relation [61] Such space-time uncertainty relations are consistent with the corresponding space-time string uncertainty principle [61]

$$\Delta X \Delta t \geq \alpha' \tag{27}$$

Since the momentum uncertainty $\Delta P < p^0$, we have from (26), to leading order in α' (in units $\hbar = 1$):

$$\Delta X \geq \frac{1}{\Delta P} > \frac{1}{p^0} \tag{28}$$

In view of (23), we then arrive at consistency with the space-time un certainty (27),

$$\Delta X \geq \frac{\alpha'}{\alpha' p^0} \sim \frac{\alpha'}{\Delta t} \tag{29}$$

These delays are *causal*, i.e. consistent with the fact that signals never arrive before they occur. They do not characterize local field theories in non commutative space times (the analogue of having strings in constant electric fields), which suffer from *non causal effects*, due to the existence of advanced outgoing waves after the scattering. For the expert reader we note that the fact that string theory produces only retarded waves after the scattering, and not advanced ones, has been explained by a detailed calculation in [59] in the case of the scattering of two open-string modes (scalar tachyon modes, for simplicity). There are scattering phases which are such that in string theory only delays occur. This issue is related with the *maintenance of causality* by strings, which notably is violated in the case of field theories in non-commutative space times, where a similar scattering of wave packets would result to advanced outgoing wave-packets, violating causality.

In view of the above discussion, if the time delays observed by MAGIC can finally be attributed partly or wholly to this type of stringy space-time foam, then the AGN Mkn 501, and other such celestial sources of very high energy photons, may be viewed as playing the rôle of the "Heisenberg microscopes" for these string/brane uncertainties.

5. Instead of Conclusions

From the above discussion it becomes clear that the properties/requirements (i)-(iv) at the beginning of the last section (4), which the photon must satisfy, so that the MAGIC results on the time four-minute delay [3] are attributed to propagation in a stochastic quantum-gravity medium, are indeed respected by the D-particle string foam model just presented. This of course does not mean that there are no conventional astrophysics explanations for the MAGIC results, but it demonstrates clearly that string theory (or better its modern version involving D-brane defects) is capable of explaining the observed photon delays, in agreement with all the other astrophysical data currently available. This is at least amusing, since it provides a framework for experimentally testing some models of string theory at present or in the foreseeable future.

The key point of course in the approach is the existence of space-time defects in the ground state of the model, whose topologically non-trivial interactions with the string states, via string-stretching during the capture process (c.f. fig. 12), are mainly responsible for the observed delays, proportional to the incident photon energy. The peculiarity of the D-particle foam in being *transparent* to charged particles (as a result of electric charge conservation requirements), evades the stringent constraints on linear Planck scale suppression refractive indices that would be induced by electron synchrotron radiation studies from Crab Nebula [47]. Moreover, the absence of birefringence avoids the similarly stringent constraints on such models that otherwise would be imposed by galactic measurements [48]. Finally, it worths mentioning that the string-stretched linear in energy time delays (23),(25), when applied to neutrinos, can be flavour (i.e. neutrino-species) independent, thus avoiding [62] the stringent constraints that would be obtained from models of quantum gravity with flavour-dependent modifications of neutrino propagation and thus modifications in their oscillations [63].

As already mentioned, the MAGIC result needs confirmation by other experiments like H.E.S.S. [40, 7], or other photon dedicated experiments, like FERMI (formerly GLAST) [33], where photons from Gamma Ray bursts will be observed. One needs many more high-energy astrophysical photon measurements to be able to disentangle source from possible propagation effects due to fundamental physics. If a statistically significant population of data on photons from cosmic sources is collected, exhibiting refractive indices varying linearly with the distance of the source [36], as well as the photon energy, then this would be a very strong confirmation of the D-particle foam model, for reasons explained above. However, it must be noted that GRB's, which are expected to lead to statistically significant data in the next few years, will produce photons much lower in energies than the flares observed in AGN, and this could be a drawback. At any rate there are attempts to claim that observations from FERMI will have sensitivity close to the Planck scale [64] for such linear-suppression models. And, indeed, as we mentioned above, in September 2008 the FERMI collaboration claims [6] observations of a delayed arrival of high energy ($\sim$ 13 GeV) photons, as compared to lower-energy ones, from the distant GRB 080916c, which can be fitted excellently by the above-described string theory model of space time foam, with a string scale of order 10^{18} GeV, in agreement with the MAGIC results. The situation starts to becoming exciting...

From the above discussion it becomes, hopefully, clear to the reader that experimental searches for quantum gravity, if the latter is viewed as a medium, are highly model dependent as far as the sensitivity to experimental falsification of the predictions of the underlying theoretical models is concerned. However, we are entering an era where low-energy (compared to Planck scale) physics experiments can already provide valuable information on the structure of space-time at the scales where Quantum Gravity is expected to set in. Very- and Ultra- high energy Astrophysics is at the forefront of such fundamental research. We therefore expect that, for the years to come, this branch of physics will proceed in parallel with terrestrial high energy experiments, such as the Large Hadron Collider launched at CERN recently, and be able to provide us soon with complementary important information on the underlying fundamental structure of our Cosmos. Time will then show whether quantum gravity can be finally put to experimental confirmation. *Affaire à suivre...*

Acknowledgments

It is a real pleasure to thank V.A. Mitsou for discussions, especially on astro-particle physics issues. I also wish to thank the organisers of the DICE2008 Conference (Castiglioncello, Italy) for the invitation to speak, and for creating a stimulating atmosphere for discussions during the meeting. This work is partially supported by the European Union through the Marie Curie Research and Training Network *UniverseNet* (MRTN-2006-035863).

References

[1] Polchinski J 1998 *String theory* Vols. 1 and 2 (Cambridge, UK: Cambridge U. Press)
[2] Albert J *et al* [MAGIC Collaboration] 2007 *Astrophys. J.* **669** 862
[3] Albert J *et al* [MAGIC Collaboration] and Ellis J R, Mavromatos N E, Nanopoulos D V, Sakharov A S and Sarkisyan E K G 2008 *Phys. Lett.* B **668** 253
[4] Ellis J R, Mavromatos N E and Nanopoulos D V 2008 *Phys. Lett.* B **665** 412; *do.* 2009 (addendum) *Phys. Lett.* B in press *Preprint* arXiv:0901.4052
[5] Galaverni M and Sigl G 2008 *Phys. Rev.* D **78** 063003; *do.* 2008 *Phys. Rev. Lett.* **100** 021102
[6] Abdo A A *et al* [The Fermi LAT and Fermi GBM Collaborations] 2009 *DOI10.1126/science.1169101 (Science Express Research Articles)* published online 19 February 2009
[7] Aharonian F *et al* [H.E.S.S. Collaboration] 2008 *Phys. Rev. Lett.* **101** 170402
[8] Jeong H *et al* 2006 *Phys. Rev. Lett.* **96** 143901; for further references and information see: *Encyclopedia of Laser Physics and Technology*, `http://www.rp-photonics.com/superluminal_transmission.html`
[9] Brunner N *et al* 2004 *Phys. Rev. Lett.* **93** 203902
[10] Feynman R P, Leighton R B and Sands M 1977 *The Feynman Lectures on Physics* Vol. 2 (Reading, Mass.: Addison-Wesley)
[11] Latorre J I, Pascual P and Tarrach R 1995 *Nucl. Phys.* B **437** 60
[12] Scharnhorst K 1990 *Phys. Lett.* B **236** 354; *do.* 1998 *Annalen Phys.* **7** 700; Barton G and Scharnhorst K 1993 *J. Phys.* A **26** 2037
[13] Drummond I T and Hathrell S J 1980 *Phys. Rev.* D **22** 343
[14] Hollowood T J and Shore G M 2008 *Nucl. Phys.* B **795** 138 (2008); *do.* 2007 *Phys. Lett.* B **655** 67; see also: Shore G M 2007 *Nucl. Phys.* B **778** 219; *do.* 2003 *Contemp. Phys.* **44** 503
[15] Liberati S, Sonego S and Visser M 2002 *Annals Phys.* **298** 167, and references therein
[16] See for instance: Wheeler J A and Ford K 1998 *Geons, black holes and quantum foam: A life in physics* (New York: Norton), and references therein
[17] Hawking S W 1982 *Commun. Math. Phys.* **87** 395
[18] Ellis J R, Hagelin J S, Nanopoulos D V and Srednicki M 1984 *Nucl. Phys.* B **241** 381
[19] Banks T, Susskind L and Peskin M E 1984 *Nucl. Phys.* B **244** 125
[20] Amelino-Camelia G, Ellis J R, Mavromatos N E and Nanopoulos D V 1997 *Int. J. Mod. Phys.* A **12** 607
[21] Ellis J R, Mavromatos N E and Nanopoulos D V 1992 *Phys. Lett.* B **293** 37; *do.* 1999 *J. Chaos Solitons Fractals* **10** ed C Castro and M S El Naschie (Elsevier Science, Pergamon) 345–363 (*Preprint* arXiv:hep-th/9805120
[22] Gonzalez-Mestres L 1997 *Preprints* arXiv:physics/9703020, arXiv:physics/9702026, arXiv:hep-ph/9610474, arXiv:astro-ph/9610089
[23] Amelino-Camelia G 2002 *Int. J. Mod. Phys.* D **11** 35; Bruno N R, Amelino-Camelia G and Kowalski-Glikman J 2001 *Phys. Lett.* B **522** 133

[24] Magueijo J and Smolin L 2002 *Phys. Rev. Lett.* **88** 190403; *do.* 2003 *Phys. Rev.* D **67** 044017
[25] Ellis J R, Mavromatos N E and Nanopoulos D V 2000 *Phys. Rev.* D **62** 084019; *do.* 2000 *Gen. Rel. Grav.* **32** 127; Ellis J R, Kanti P, Mavromatos N E, Nanopoulos D V and Winstanley E 1998 *Mod. Phys. Lett.* A **13** 303 (1998)
[26] Kogan I I, Mavromatos N E and Wheater J F 1996 *Phys. Lett.* B **387** 483; Ellis J R, Mavromatos N E and Nanopoulos D V 1998 *Int. J. Mod. Phys.* A **13** 1059
[27] Ellis J R, Mavromatos N E and Sakharov A S 2004 *Astropart. Phys.* **20** 669; Ellis J R, Mavromatos N E, Nanopoulos D V and Sakharov A S 2004 *Int. J. Mod. Phys.* A **19** 4413; *do.* 2004 *Nature* **428** 386
[28] For a review see: De Angelis A, Mansutti O and Persic M 2008 *Riv. Nuovo Cim.* **31** 187
[29] Urry C M and Padovani P 1995 *Publ. Astron. Soc. Pac.* **107** 803
[30] Misner R, Thorne K and Wheeler J A 1973 *Gravitation* (Freeman and Company)
[31] For an explanation of the Penrose process, in particular in the presence of electromagnetic fields, as is the case of AGNs, see: Bhat M, Dhurandhar S and Dadhich N 1985 *J. Astrophys. Astr.* **6** 85, and references therein
[32] See, e.g.: Rybicki G B and Lightman A P 1979 *Radiative Processes in Astrophysics* (New York: John Wiley & Sons)
[33] See, e.g.: Morselli A [GLAST Collaboration] 2008 *J. Phys.: Conf. Ser.* **110** 062017, and references therein
[34] Bednarek W and Wagner R M 2008 *Astron. & Astrophys.* **486** 679
[35] Mastichiadis A and Moraitis K 2008 *Preprint* arXiv:0810.2420
[36] Amelino-Camelia G, Ellis J R, Mavromatos N E, Nanopoulos D V and Sarkar S 1998 *Nature* **393** 763; Ellis J R, Farakos K, Mavromatos N E, Mitsou V A and Nanopoulos D V 2000 *Astrophys. J.* **535** 139; Ellis J R, Mavromatos N E, Nanopoulos D V, Sakharov A S and Sarkisyan E K G 2006 *Astropart. Phys.* **25** 402, [*ibid.* **29**, 158 (2008)]; Jacob U and Piran T 2008 *JCAP* **0801** 031
[37] Farakos K, Mavromatos N E and Pasipoularides P 2009 *JHEP* **0901** 057
[38] Jackson J D 1975 *Classical Electrodynamics* (John Wiley & Sons, Inc.)
[39] Hillas A M *et al* 1998 *Astrophys. J.* **503** 744
[40] Aharonian F *et al* [H.E.S.S. Collaboration] 2007 *Astrophys. J.* **664** L71
[41] Greiner J *et al* 2008 *Preprint* arXiv:0801.4801
[42] Gambini R and Pullin J 1999 *Phys. Rev.* D **59** 124021
[43] Gleiser R J and Kozameh C N 2001 *Phys. Rev.* D **64** 083007
[44] Fan Y Z, Wei D M and Xu D 2006 *Mon. Not. Roy. Astron. Soc.* **376** 1857
[45] Dean A J *et al* 2008 *Science* **321** 1183
[46] Maccione L, Liberati S, Celotti A, Kirk J G and Ubertini P 2008 *Phys. Rev.* D **78** 103003
[47] Jacobson T, Liberati S and Mattingly D 2003 *Nature* **424** 1019
[48] Maccione L, Liberati S, Celotti A and Kirk J G 2007 *JCAP* **0710** 013
[49] Ong R A 1998 *Phys. Rept.* **305** 93
[50] Ellis J R, Mavromatos N E and Nanopoulos D V 2001 *Phys. Rev.* D **63** 124025
[51] Ellis J R, Mavromatos N E and Westmuckett M 2004 *Phys. Rev.* D **70** 044036; *do.* 2005 *Phys. Rev.* D **71** 106006; Ellis J R, Mavromatos N E, Nanopoulos D V and Westmuckett M 2006 *Int. J. Mod. Phys.* A **21** 1379
[52] Schwarz J H 1999 *Preprint* arXiv:hep-th/9907061
[53] Li T, Mavromatos N E, Nanopoulos D V and Xie D, to appear
[54] Ellis J R, Mavromatos N E, Nanopoulos D V and Sakharov A 2004 *New J. Phys.* **6** 171
[55] Bachas C 1995 *Preprint* arXiv:hep-th/9503030
[56] Gravanis E and Mavromatos N E 2002 *Phys. Lett.* B **547** 117; Mavromatos N E 2003 *Beyond the Desert, Oulu 2002 (Finland)* ed H V Klapdor-Kleingrothaus (IoP) 3
[57] Witten E 1995 *Int. J. Mod. Phys.* A **10** 1247
[58] Mavromatos N E and Sakellariadou M 2007 *Phys. Lett.* B **652** 97
[59] Seiberg N, Susskind L and Toumbas N 2000 *JHEP* **0006** 044
[60] Veneziano G 1986 *Europhys. Lett.* **2** 199; Amati D, Ciafaloni M and Veneziano G 1988 *Int. J. Mod. Phys.* A **3** 1615; Gross D J and Mende P F 1987 *Phys. Lett.* B **197** 129; *do.* 1988 *Nucl. Phys.* B **303** 407
[61] Yoneya T 2000 *Prog. Theor. Phys.* **103** 1081
[62] Ellis J R, Harries N, Meregaglia A, Rubbia A and Sakharov A 2008 *Phys. Rev.* D **78** 033013
[63] Brustein R, Eichler D and Foffa S 2002 *Phys. Rev.* D **65** 105006
[64] Lamon R 2008 *JCAP* **0808** 022

Journal of Physics: Conference Series **174** (2009) 012017 doi:10.1088/1742-6596/174/1/012017

Emergence of spatial structure from causal sets

D Rideout[1] and P Wallden[2]

[1] Perimeter Institute for Theoretical Physics, 31 Caroline St N, Waterloo, ON N2K 1S5, Canada

[2] Raman Research Institute, Sadashivanagar, Bangalore - 560 080, India

E-mail: `drideout@perimeterinstitute.ca`, `petros@rri.res.in`

Abstract. There are numerous indications that a discrete substratum underlies continuum spacetime. Any fundamentally discrete approach to quantum gravity must provide some prescription for how continuum properties emerge from the underlying discreteness. The causal set approach, in which the fundamental relation is based upon causality, finds it easy to reproduce timelike distances, but has a more difficult time with spatial distance, due to the unique combination of Lorentz invariance and discreteness within that approach. We describe a method to deduce spatial distances from a causal set. In addition, we sketch how one might use an important ingredient in deducing spatial distance, the 'n-link', to deduce whether a given causal set is likely to faithfully embed into a contintuum spacetime.

1. Introduction

1.1. Fundamental discreteness

There are many reasons to believe that some sort of Planck scale discreteness will be present in any theory of quantum gravity. Perhaps the most convincing evidence is the finiteness of black hole entropy, which requires a cutoff at around the Planck scale. Thus we expect quantum gravity to be described as a discrete theory of geometry, in some form or other. One can regard the continuum as *emerging* from the discrete, in a similar way in which continuum theories for fluids emerge from the underlying physics of their discrete molecules.

In the case of fluids, the discrete elements (molecules) naturally live in a continuum background, and inherit their interrelationships from their embedding in this background. The situation for gravity, however, is different. There it is natural to expect that the discrete elements do not live in a background medium. In order to recover geometry, in this absense of a background, one must impose some sort of relationship among the discrete elements, for example in the form of a graph or binary relation, which indicates which discrete elements are 'nearest neighbors' in spacetime.

It is important to note in this vein that nature appears to satisfy local Lorentz invariance, at least to the degree accessible by current observations. Since the intuitive notion of nearest neighbors is naively frame dependent, a fundamental relation for discrete spacetime must contain the nearest neighbors for every Lorentz frame. Thus one expects that the number of nearest neighbors of any element will be infinite (or if finite they should generically extend to cosmological scales). From these heuristic arguments one may expect that the discreteness of quantum gravity may be expressible, in its simplest form, as some sort of very highly connected graph or binary relation.

As suggested above, it makes sense to guess that a fundamental relation for discrete quantum gravity would be compatible with the Lorentz symmetry. Stated the other way around, one might expect that the fundamental relation is such that it allows the macroscopic Lorentz symmetry to emerge naturally from the discrete. There are two possibilities which come to mind, one which takes a microscopic causal ordering as this relation, and another which takes spatial nearest neighbors, as given by the invariant spacetime interval ds^2. It is not possible to specify both freely because the causal relation alone is sufficient to recover the metric up to a conformal factor. Thus given one relation, it must be possible to deduce the other.

A number of approaches to quantum gravity take the second approach, such as loop quantum gravity [1] [1, 2, 3], 'quantum graphity' [4], and the approach outlined in [5]. Quantum graphity, for example, has as its fundamental discrete structure a graph which is regarded as a spatial object, which evolves in time. It postulates a form for the Hamiltonian which describes its time evolution, and from this one is able to derive the presence of light cones and non-trivial causal structure [6, 5].

Here we take the alternate approach, that microscopic causal ordering is fundamental, and show that it is possible to derive from it a symmetric, spacelike relation, at least for discrete structures (causal sets) which are well approximated by Minkowski space. This approach has the advantage that one can address the question of how continuum structures emerge from the discrete, without reference to a particular dynamical law.

The ability to recognize continuum properties from the discrete relation may be an important stepping stone toward constructing a full theory of quantum gravity. In particular, most of our understanding of how to formulate gravitational dynamics is within the continuum. It would be helpful to understand how this carries over to the discrete context. This understanding may also provide crucial hints as to a fundamental origin of black hole entropy and the covariant entropy bound, by providing clues as to how to count states associated with a spacelike or null hypersurface [7].

Once we have the ability to deduce spatial distances between discrete elements in the approximating continuum, it is not difficult to use this to compose a symmetric, spatial nearest neighbor relation, which can be useful for recovering topology and geometry of curved spacetime. In addition this may allow contact with other approaches to quantum gravity, which hold the spacelike relation to be fundamental.

We address the question of how to deduce the spatial distances in an approximating continuum, given only a discrete partial order. It turns out that, while the recovery of timelike distances from the discrete causal ordering is relatively striaghtforward, recovering spacelike distances is much more difficult, due to the relatively unfamiliar nature of Lorentz invariant discreteness.

1.2. Causal sets

The causal set approach to quantum gravity is based on two observations. One is that the causal structure of a spacetime, namely a list of which events 'can causally influence' which others, is a very rich structure, enough to reconstruct the conformal metric. The other is the abundance of evidence suggesting that some sort of discrete structure underlies continuum spacetime. [8]

The intent is to approach 'quantization' via histories, i.e. to take the *histories*, rather than states defined on some 'spatial hypersurface', as the fundamental objects of the theory. The quantum theory can then be expressed in terms of some appropriate generalization of the Feynman path integral [9, 10, 11].

[1] Loop quantum gravity does not literally take this approach, however the fundamental object with which the theory is expressed is purely spatial in character, and the edges of the spin network encode (among other things) a binary spatial relation on the vertices.

In the causal sets approach, the histories are taken to be *causal sets*, which are countable sets of 'atoms of spacetime', endowed with a partial order relation $\prec$ which is transitive and irreflexive. For a causal set C, transitivity requires that if $x \prec y$ and $y \prec z$ then $x \prec z$, $\forall x, y, z \in C$. Irreflexivity is simply the condition that no element can precede itself, namely $x \not\prec x \, \forall x \in C$. To enforce discreteness, one imposes a local finiteness condition, that every *order interval* $[x, y] = \{z \in C | x \prec z \prec y\}$ has finite cardinality.

The correspondence between the discrete causal set and a continuum spacetime is described in terms of a *faithful embedding*, which is an order preserving map ϕ from a causal set C to a spacetime (M, g), which has the property that "the number of elements mapped into any region of spacetime volume V is Poisson distributed with mean V". Here *order preserving* means that two events $\phi(x)$ and $\phi(y)$ in the image of ϕ are causally related in the spacetime ($\phi(x) \in J^-(\phi(y))$) iff the corresponding elements in C are related $x \prec y$. The Poisson distribution with mean V assigns a probability of $\frac{V^n e^{-V}}{n!}$ to the event of mapping n elements to the region of volume V.

A faithful embedding is easy to realize, one simply selects points in the target spacetime (M, g) at random by a Poisson process (a 'sprinkling'), and then computes a partial order on them using the causal structure of the spacetime. This gives a causal set C, along with a faithful embedding ϕ of C into (M, g). It is useful to make use of the map ϕ implicitly when discussing the causal set, so that one can speak of "the causet elements in a region of, or at points in, (M, g)" to refer to those elements which are mapped to regions of (M, g) by ϕ.

It is important to keep in mind, however, that it is the discrete causal set that is fundamental, and the continuum into which it may faithfully embed is to be regarded only as an approximation to the discrete substructure. The faithful embedding is merely a tool with which to discuss how the continuum arises from the discrete. It is worth remarking that it may be the case that a 'physical' causal set may only faithfully embed into a continuum spacetime after some coarse graining, so the physics near the Planck scale need not be continuum-like.

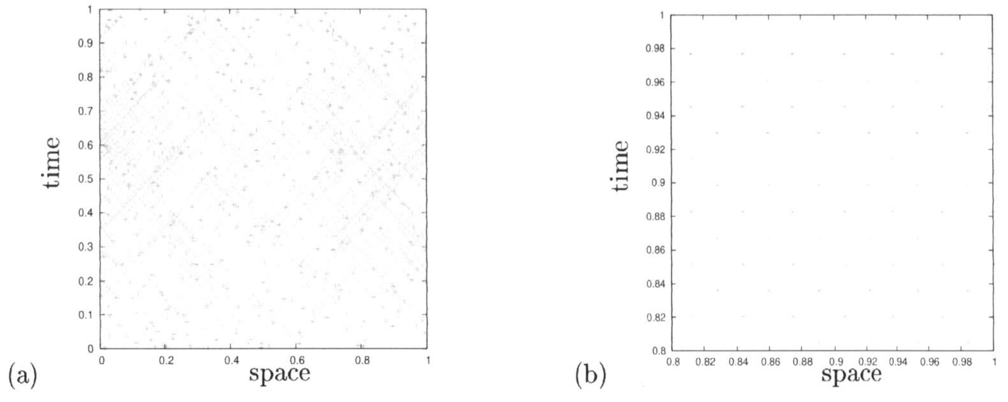

Figure 1. Two embeddings of a causal set into a spacetime. (a) The causal set has 512 elements, and is faithfully embedded into a flat $2d$ spacetime with clyindrical topology $S^1 \times I$. The links are shown in green. (b) The figure depicts a (non-faithful) embedding into a region of $2d$ Minkowski space. The red and green distributions of points are related to each other by a boost of 4/5.

Figure 1 illustrates two embeddings of causal sets into spacetimes, one faithful and one not. Some important features to note in these diagrams regard the behavior of the *links* in the causal sets, which are those relations which are 'irreducible', in that they are not implied by transitivity. For the faithful embedding on the left, one can note that (a) the number of links connected to each element is very large, and (b) the vast majority of links connect elements from very distant regions of the spacetime, in any given frame (such as that of the figure). This is because any given link only appears short in one frame, it will be long in all others. Since almost all frames are 'another frame', almost all links appear to be very long. In the embedding on the right, on the other hand (in which the links are not shown explicitly), each element is connected to only four links. In the red frame the links connect nearby elements, while in the green frame the ends of the link are becomming more distant in one direction (and extremely close in the other). This embedding fails the 'faithfulness' condition, that the expected number of elements mapped to any region is equal to its volume. There are obviously very large regularly shaped regions which fail this criterion. In this way regular lattices fail to be manifoldlike, if one does not go all the way to the continuum ($\rho \to \infty$) limit. Thus in this way the Lorentz invariance appears to in fact *emerge* from a discrete relational theory, as the only way in which the correspondence between number of discrete elements and spacetime volume can be independent of the region one is considering.

1.3. Hauptvermutung

Named after the famous conjecture in topology, that every triangulable space has a unique triangulation (which was later proved to be false), it is conjectured that a causal set which faithfully embeds into a spacetime manifold determines the manifold up to 'approximate isomorphism'. Somewhat more precicely, we can define the 'Hauptvermutung' (central conjecture) for causal sets as follows (see figure 2). Given a causal set C, and two faithful

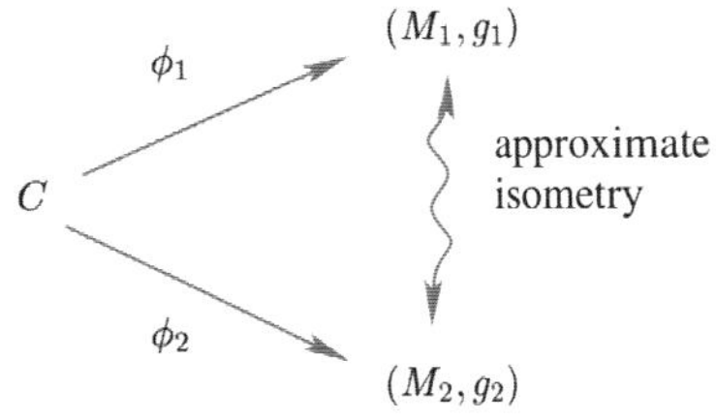

Figure 2. 'Hauptvermutung' of causal sets

embeddings $\phi_1 : C \to (M_1, g_1)$ and $\phi_2 : C \to (M_2, g_2)$ from C into spacetime manifolds, then (M_1, g_1) and (M_2, g_2) must be approximately isometric. The 'appromately' is necessary because of the discreteness — the embedding is obviously blind to continuum structures smaller than the discreteness scale. The precise mathematical statement of the Hauptvermutung remains open, however there does exist a body of work on defining a distance measure between Lorentzian geometries [12].

It may seem a lot to hope, to recover the full spacetime geometry from the discrete order. However, the expectation is not unreasonable. The microscopic order more-or-less directly encodes the macroscopic causal order. By a theorem of David Malament and others [13, 14], we know that the causal ordering of events in spacetime contains enough information to recover the topology, differential structure, and conformal metric. Thus all that remains is the conformal factor, which encodes volume information. Here the discreteness plays a crucial role, by providing the missing volume information, via the correspondence between number and spacetime volume expressed in the faithfulness condition on the embedding. In the continuum one needs to add the gravitational field to get geometry, while in the discrete the geometry arises naturally, without needing to add any additional mathematical structures. Thus it is not unreasonable to expect that one can recover the complete spacetime geometry from the discrete causal order. Note here that the Lorentzian signature arises naturally, as the only one capable of distinguishing past from future.

1.3.1. Dimension An obvious question which arises when discussing causal sets is how can one predict the dimension of an approximating spacetime, given only the causal set. There have been a number of proposals for how to do this, for example the midpoint scaling and Myrheim-Meyer dimension, which are both described in [15, 16]. The Myrheim-Meyer dimension [17, 18] works by counting the number of relations in an order interval, and comparing it with what one would get from sprinkling into $\mathbb{M}^d$.

In section 7 we sketch an alternate method for estimating the dimension of causal sets, based upon the counting of n-links (defined in section 4), and present some preliminary results.

1.3.2. Timelike distance Since the fundamental relation defining the causal set is intrinsically timelike, one can imagine that it is relatively straightforward to extract timelike distances from the causal set. The natural choice which is likely extendible to curved spacetime is to count the number of links in the longest chain connecting the two related elements $x \prec y$ [17, 19, 20]. (In general such a longest chain will be far from unique. Figure 3 shows the collection of longest chains between a pair of elements in $\mathbb{M}^2$.)

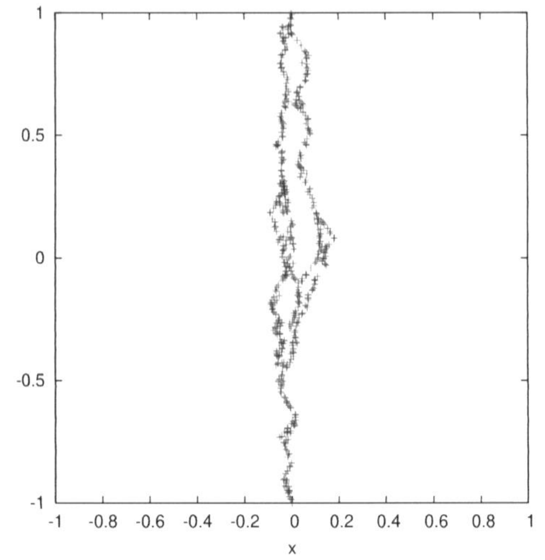

Figure 3. The collection of longest chains (*geodesics*) between a pair of elements in $\mathbb{M}^2$.

By a theorem stated in [21], we know that this length L converges to the proper time separation between x and y, in particular

$$\frac{L}{(\rho V)^{1/d}} \to m_d$$

in probability as $\rho V \to \infty$. Here ρ is the sprinkling density, V is the volume of the interval $[x, y] = J^+(x) \cap J^-(y)$, d is the spacetime dimension, and m_d is a constant depending only on the dimension. It's exact value is known only in two dimensions, $m_2 = 2$. It's value in three dimensions was measured in [22] to be $m_3 = 2.296 \pm .012$. If we write the volume of the interval as $V = D_d T^d$, with T the proper time we seek, and D_d a dimension dependent coefficient (we will need $D_3 = \frac{\pi}{12}$), then in the infinite ρ limit

$$T = \frac{L}{m_d \rho^{1/d} D_d^{1/d}} \,. \tag{1}$$

To get some feel for how well this works 'in practice', we check it on a computer, using code for the Cactus high performance computing framework [23] mentioned in [22]. We sprinkle into an interval of height $T = 2$ (in arbitrary units), count the length of the longest chain within that interval, and from it compute, from (1), an 'effective' m_d

$$m_d^{\text{eff}} = \frac{L}{T(\rho D_d)^{1/d}} \tag{2}$$

which is relevant for finite ρV. For $d = 3$ we get the results depicted in figure 4. Each datapoint

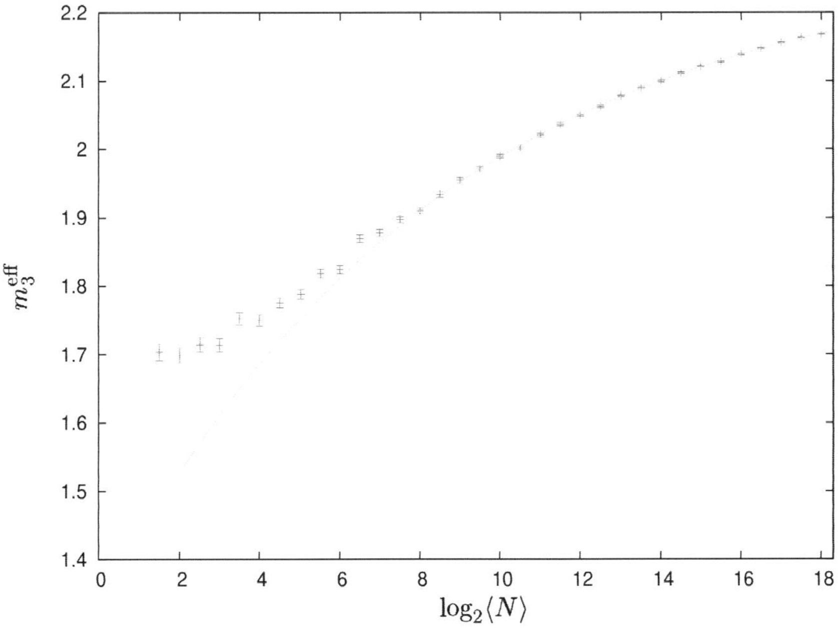

Figure 4. Convergence of length of longest chain L to proper time T for a faithful embedding into an interval of $\mathbb{M}^3$. The green fit to the asymptotic form measures the constant of proportionality m_3.

results from 1600 sprinklings into an interval of $\mathbb{M}^3$, each with a mean number of elements $\langle N\rangle$ (the mean of the Poisson distribution) as indicated. We plot the mean m_3^{eff} from (2), along with the standard estimate of its error. The fit is to the function $m_3 + ae^{b\log_2\langle N\rangle}$, for the datapoints $\langle N\rangle \geq 2^8$. We get $m_3 = 2.2856 \pm 0.0063$, which is consistent with the measurement in [22] (that employs a slightly different procedure) [2].

Note that for any finite region (ρV finite), the m_3^{eff} is less than the asymptotic value m_3, and thus a measurement of proper time using the length of the longest chain via (1) will always underestimate the continuum proper time. We call this effect *timelike underestimation.*

2. Early prescription for spacelike distance

We would like to recover the full spacetime geometry from the discrete causal ordering. We have seen above how to get dimension and timelike distances. How can one extract spatial

[2] Here we are effectively conditioning on the existence of sprinkled elements at the endpoints of the interval. Since the occupation probability at each point in a Poisson process is independent of those at every other point, this should make no difference. In [22] the authors select a pair of elements that land nearest the endpoints of the interval, and use the interval formed by those two elements. It appears that this distinction does not affect our measurement of m_3.

information? How does spatial geometry come out? We start with the simple question of spatial distance in flat spacetime. Can we recognize the spatial distance between elements of a causal set which can be faithfully embedded into Minkowski space?

An obvious first guess is to use the spacelike points to locate a timelike pair, whose distance is the same as the spacelike distance, and then use the timelike method above. In the continuum, this can be done, as follows. Consider a pair of spacelike separated elements x and y, whose spatial distance we would like to measure. Consider those pairs of points w in the common past $J^-(x) \cap J^-(y)$ and z in the common future $J^+(x) \cap J^+(y)$ which minimize the timelike distance (length of the longest chain L_{wz}) between w and z. In the continuum, every such pair is separated by exactly the spacelike distance between x and y.

In the discrete causal set context, the above construction works fine for causal sets which faithfully embed into $\mathbb{M}^d$, for $d < 3$. This is because for $d = 2$ there is a unique choice for the pair (w, z) in the continuum (w is the point at the intersection of the past light cones of x and y, and z is the intersection of their future light cones). For a causal set sprinkled into $\mathbb{M}^2$, the pairs (w, z) at minimum timelike separation (length of the longest chain) will lie close to the corresponding 'minimizing pair' in the continuum.

At larger dimension d, in the continuum, one find a $d-2$ dimensional submanifold of points (w, z) at minimum timelike separation. One way to see this is that the pair (x, y) breaks the Lorentz symmetry in one direction, but the picture is still invariant under boosts in any orthogonal direction. Starting from any arbitrary frame, in which the pair (w, z) both occur at the centroid of x and y, one can trace out the submanifold by applying all boosts which leave x and y invariant.

Unfortunately this infinity of 'minimizing pairs' causes problems in the discrete case of a random sprinkling, because it causes one to consider an infinite number of independent regions when computing the spatial distance between x and y. To understand why this simple prescription for spacelike distance fails in dimensions greater than 1+1, consider the spacelike pair (x, y) in figure 5(a). Shown are a pair (w, z) (a *minimizing pair*) which are close to the intersection of x and y's future and past light cones, and might locally minimize the length of the longest chain between w and z. However, our distance prescription seeks a global minimum, not a local minimum, so we must consider 'what happens in other frames' as well. Figure 5(b) depicts the same situation, in the frame in which x and y are simultaneous, and w and z occur at the same spatial location. (x and y are not shown explicitly in the diagram; they are displaced into and out of the page by a small amount.) Shown are three candidates for pairs (w, z), along with the order interval between them, which is the portion of the causal set which determines the length of the longest chain between them. If we compare the intervals for two pairs which are 'highly boosted with respect to each other' (meaning the boost parameter relating the frames in which the pairs (w, z) are simultaneous) have very little overlap. But each of these intervals contains the same spacetime volume of Minkowski space. The sprinklings in each of these regions is independent (save the tiny region of overlap), since the Poisson distribution is independent at each point. Since there are an infinite number of candidate pairs (w, z), to find the global minimum, we must sample the Poisson distribution an infinite number of times, to get the portion of the causal set in each of these independent regions. One possible result for this sample, which has finite probability, is that the region is empty. Thus this event (of finding a candidate pair (w, z) whose interval $[w, z]$ is empty of sprinkled elements) must occur an infinite number of times, and so the distance between x and y will always be exactly two [3], regardless of where they fall in Minkowski space.

[3] Two because by definition $w \prec x \prec z$ and $w \prec y \prec z$ (and $x \not\prec y$), each of which are chains of length 2.

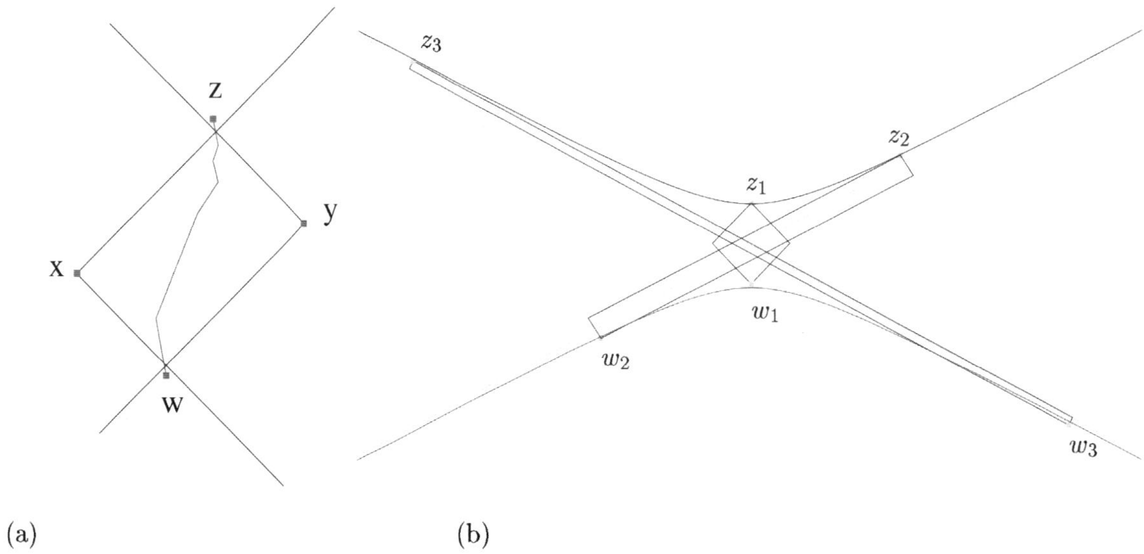

Figure 5. (a) Two spacelike elements x, y in $\mathbb{M}^3$, and a pair w, z in their common past/future. (b) The elements x and y lie at the center of the figure, but displaced out of and into the page by some finite equal amount. The hyperbolae indicate the intersection of the past and future light cones of x and y. The long straight green lines are the asymptotes of the hyperbolae. The green dots with black intervals indicate the projection of the causal intervals for minimizing pairs onto the plane of the page.

2.1. Numerical evidence

The above argument depends upon a sprinkling into the entirety of Minkowski space. What if one has a sprinkling into a finite region of Minkowski, such as might be physically realistic in a finite universe, or as can be simulated on a computer? Is it possible to see the above degeneracy in spatial distance, while considering only that portion of the causal set which is contained in a finite portion of Minkowski space? The probability to see a region of spacetime volume V which is empty of sprinkled elements is e^{-V}. In $\mathbb{M}^3$, an interval of height $T = 4$ has volume $V = \frac{\pi}{12}4^3 \approx 16.755$, and the probability to find an empty interval of height $T = 4$ is 5×10^{-8}. Thus a lower bound on the size of a region of $\mathbb{M}^3$ which would be required to see this degenerate distance is $17/5 \times 10^{-8} \approx 3 \times 10^8$.

To explore this on the computer, we sprinkle into a $T \times T \times \Delta$ box, as illustrated in figure 2.1. In this section we use units such that $\rho \equiv 1$. The spacelike pair (x, y) are placed at the small red squares, which lie in the center of the two largest faces as shown. In the simulations, we hold the distance Δ between x and y fixed, while increasing the size of the box in the lateral directions. If the distance measure works well, then obviously increasing the size of the box in the lateral directions should have no affect. We see in figure 2.1 that this is not the case. There we show the results from two simulations, one with $\Delta = 4$, and one for a more distant pair $\Delta = 8$. X is the smallest value of (1) for a pair of elements (w, z) in the common past and future of x and y. In all these simulations we use a $m_3^{\text{eff}} = 1.75815$, which we carefully measure to be the appropriate value for an interval of size $\rho V = \frac{\pi}{12}4^3$. This eliminates the effect of timelike underestimation.

At $T = \Delta$ the common future and past of (x, y) are outside of the sprinkling region, so the spatial distance is undefined. For slightly larger T, we observe a substantial overestimation of the continuum distance Δ. This occurs because it is unlikely that a sprinkled element will land close to the intersection of x and y's light cones. The pair (w, z) will always be separated by a continuum (timelike) distance $> \Delta$, hence the overestimation. This overestimation will always

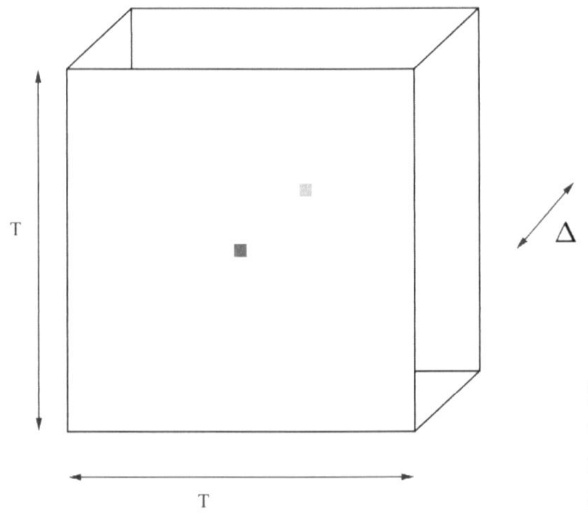

Figure 6. Growing region of $\mathbb{M}^3$ into which we sprinkle. Δ is held fixed to a small value (4 or 8), while T is increased as far as the computer allows ($2^{7.75} \approx 215$).

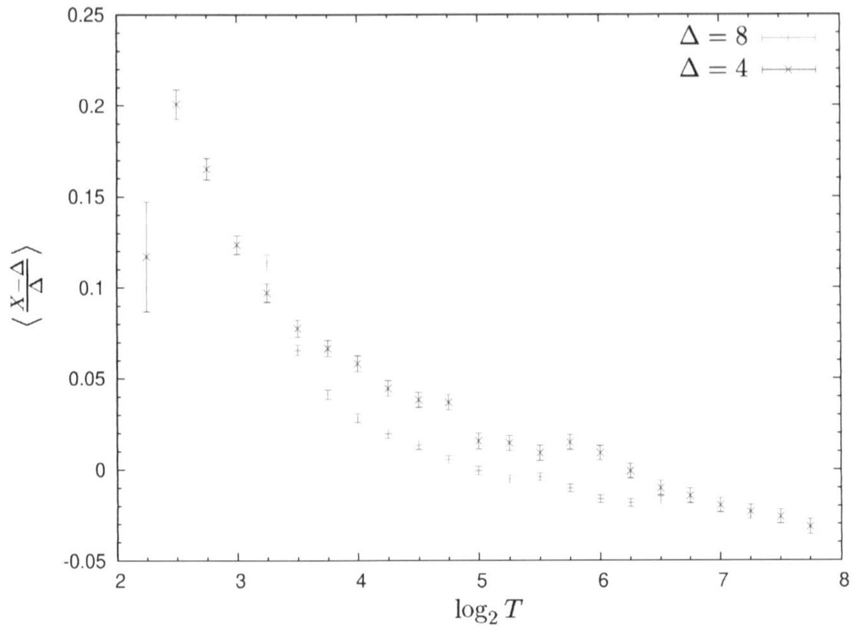

Figure 7. A demonstration of the degeneracy of the 'naive' prescription for spacelike distance of section 2. Each data point gives the mean and standard error for 900 sprinklings.

be of the order of the discreteness scale, and so, even though in this context is rather large, will be irrelevant for macroscopic distances.

At large T, we see the overestimation decrease, past zero to negative values. Since the use of m_3^{eff} already corrects for the timelike underestimation effect, this decrease must be due to the degeneracy of the spacelike distance definition of section 2. If we could expand T all the way to infinity, we expect to see the deviation decrease to $\frac{\frac{2}{D_3^{1/3} m_3^{\text{eff}}} - \Delta}{\Delta} \approx -.55544$ for $\Delta = 4$.

It appears that this naive prescription for spatial distance on a causal set fails, due to the Lorentz invariant nature of the embedding, and the random nature of the Poisson process. How might one deduce spatial distances on a causal set? One possibility is to select enough additional elements to eliminate the boost freedom mentioned above. This leads to more of a 'diameter' measure, than a distance measure, in that it can be interpreted as giving the diameters of spheres defined by the d elements. This is described in section 3.

An alternate proposal is to take advantage of the fact that links in a causal set closely track the light cones, and this can be used to locate appropriate pairs (w, z) without resorting to a global minimum. This leads to the 2-link distance proposal of section 5.

3. Diameter measures

In order to locate a relatively unique pair (w, z), one can select, instead of simply a pair of elements (x, y), a collection of d mutually unrelated elements (an *antichain*) in $\mathbb{M}^d$. One can then select a w which is in the common past of the entire antichain, and a z which is in the common future, and seek a pair which minimizes the length of the longest chain between w and z. It turns out that this measures the diameter of the smallest $d-1$-ball, which lives on the surface of simultaneity of all d elements of the antichain, and contains them all. It is called l_g in [22].

Using the same antichain, one can also construct a $d-2$-sphere, which lives on the same spatial hypersurface, and contains each of the d elements of the antichain. It is possible to write down an expression in terms of order invariants which measures the diameter of this sphere. It is called l_s in [22]. Both of these diameter measures are described there in detail, along with a number of possible applications.

4. n-links

As mentioned above, an alternate approach to measuring spatial distances in a causal set is to use the fact that the *links* of the causal set lie very close to the light cone in a faithful embedding, and effectively give the light cone structure of the causet. One way to see this is that the proper time between a pair of related elements is measured by the length of the longest chain connecting them. A null relation corresponds to a minimum proper time (zero), so the analogue in a causal set should correspond to a minimum length of the longest chain, which is a single link. Figure 8 depicts some of the links to the future of an element in Minkowski space.

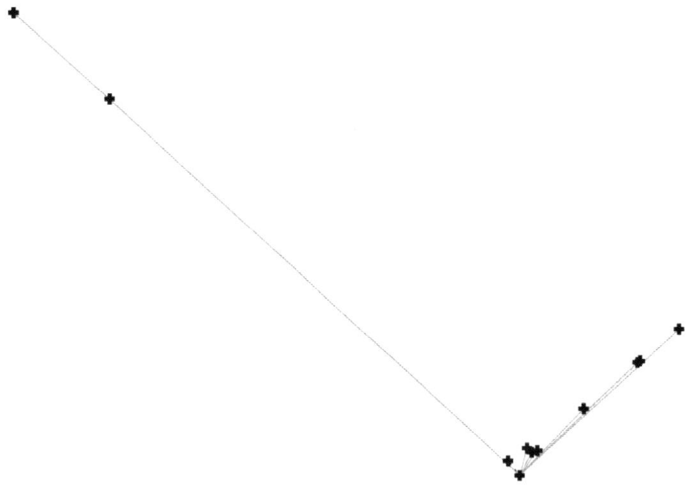

Figure 8. Links to the future of an element in Minkowski space.

We define an *n-link* as an element which is linked to each member of an n element antichain. If this element is to the future (past) it is called a future (past) n-link. Some examples are depicted in figure 9. Given the above discussion, it is fairly clear that an n-link corresponds to the intersection of light cones emanating from each element of the antichain.

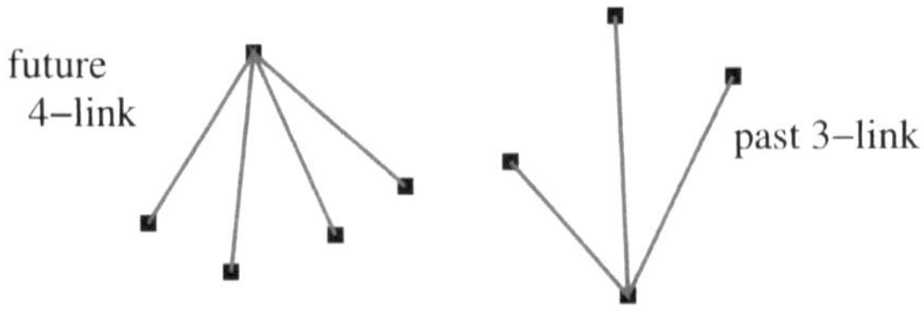

Figure 9. Illustration of n-links.

How many n-links does one expect to find, to the future of a given n-antichain (an antichain with n elements), for a causal set faithfully embedded into $\mathbb{M}^d$? For $n < d$, one expects an infinite number, for similar reasons as discussed in section 2. Roughly speaking n elements are not enough to select a surface of simultaneity, so in each of an infinite number of frames one expects to see an n-link with some finite probability. For $n > d$ there should not be any n-links, unless the n elements are arranged in some special way (e.g. as in figure 8, which depicts a 9-link within a sprinkling of $\mathbb{M}^2$). For $n = d$, the intersection of the future light cones consists of a single point. The probability that there is an element near (and to the future of) this point is moderate, but the probability that it is linked to each element of the antichain diminishes rapidly with their distance.

One application of n-links in causal sets is they can provide a definition of spatial distance which is devoid of the degeneracy from which the definition given in section 2 suffers. This will be explained in detail in the next section.

Another application may be that they can be used as in indicator of 'manifoldlikeness' in a causal set (cf. [24]). The behavior of the number of n-links in a causal set that is well approximated by Minkowski space is sketched above, which is potentially quite specific given that it should hold for all n. For example, if one does have a causal set which faithfully embeds into $\mathbb{M}^d$, then the above behavior should fairly clearly pick out the value of d. This could compared with other dimension estimators as a further test. An initial attempt at identifying manifoldlikeness in this way appears in section 7 below.

5. 2-link distance

The definition of spacelike distance given in section 2 fails because for any spacelike pair (x, y) there are an infinite number of potential 'minimizing pairs' (w, z) which lie close to the intersections of the light cones of x and y. Since one takes a global minimum over all such pairs (w, z), one gets a trivial result. If one could take an average of 'minimizing pairs' (w, z), rather than minimum, this problem could be avoided. The problem is how to locate appropriate pairs of elements close to the intersection of the light cones of x and y, however this is exactly what is provided by 2-links.

We thus define the 2-link distance between two unrelated elements x and y by the following algorithm:

(i) Compute the set of all future 2-links of x and y.
(ii) For each future 2-link z, compute the smallest distance L_i from an element $w \in \text{past}(x) \cap \text{past}(y)$ to z.
(iii) Store L_i.
(iv) Repeat the above with 'future' and 'past' interchanged.
(v) Compute the average of the L_i.

In figure 10 we contrast 2-link distance against the 'naive' spatial distance of section 2. Each data point gives the mean and its error, for 900 sprinklings into the boxes of figure 2.1. Again we see the large spacelike overestimation for the 2-links, as for the naive spatial distance. As

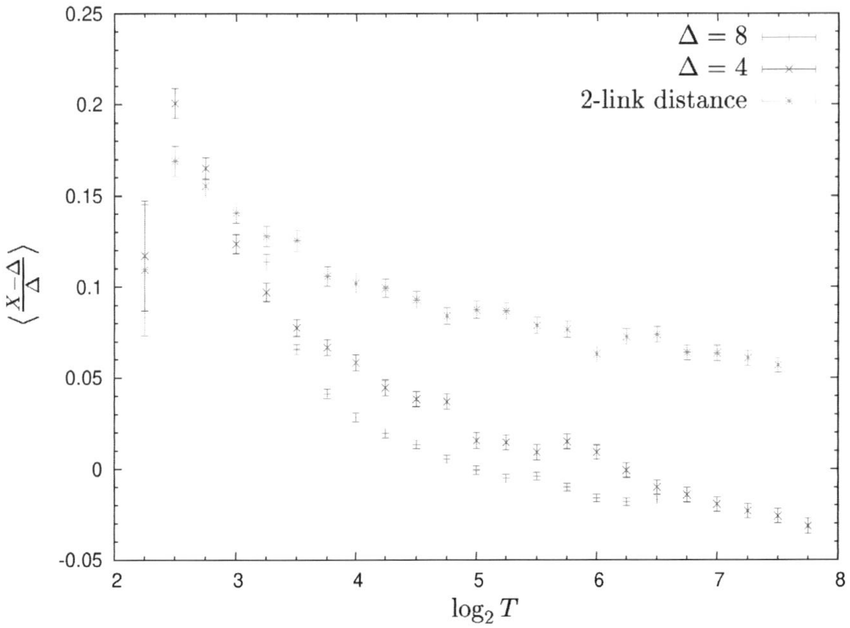

Figure 10. Comparison of 2-link distance at $\Delta = 4$ (magenta) with the 'naive' prescription for spacelike distance of section 2 (red and blue).

the size of the box is increased, its accuracy increases slightly because it finds more 2-links with which to estimate the distance. The random fluctuations in this figure are smaller than those of figure 14 of [22] because here we include past 2-links as well as future, and also condition on there being causet elements at the center of the faces of the box.

One may wonder how common are 2-links, if we are to rely on their existence in order to define spatial distance in a finite region. We expect to find an infinite number in dimensions > 2, but this assumes infinite Minkowski space. Will there exist many in a reasonably sized portion of our universe? We address this question by searching for 2-links, using the same computation as described in section 2.1, with $\Delta = 8$. In this case, however, we simply count the number of (future and past) 2-links, rather than compute the 2-link distance itself. The results are shown in figure 11. The error bars indicate the usual estimate of the standard deviation. Given that they overlap with, or are not far from extending to, zero, there will be a number of sprinkled causets which do not find any relevant 2-links. In these cases we simply discard the causet, though we do count it toward the total number generated (900). The blue curve is a fit to $a + b \ln\langle T \rangle$. In figure 12 we count the number of 2-links in a rectangular $T \times 8 \times T \times T$ box. The spacelike elements (x, y) are placed in the centers of the x-faces, as before. Here the number of 2-links grows linearly with the box size, as one might expect. The fitting function is $a + b\langle T \rangle$. Here we show error bars for both the standard deviation and the error in the mean.

6. Spatial nearest neighbors and curved spacetime

One can use the 2-link distance in the context of curved spacetime, however it has the unfortunate property that the distance between x and y depends upon portions of the causal set which are arbitrarily 'distant' from x and y. To measure a length in my lab I must in principle consider the happenings in Andromeda millions of years ago and also millions of years from now. Given that we know that spacetime is locally flat, we can truncate the search for 2-links at some intermediate mesoscale, beyond which we expect curvature effects to become relevant. It may be that the introduction of a mesoscale is inevitable in discrete quantum gravity, as it arises in a number of quite different contexts [25, 24].

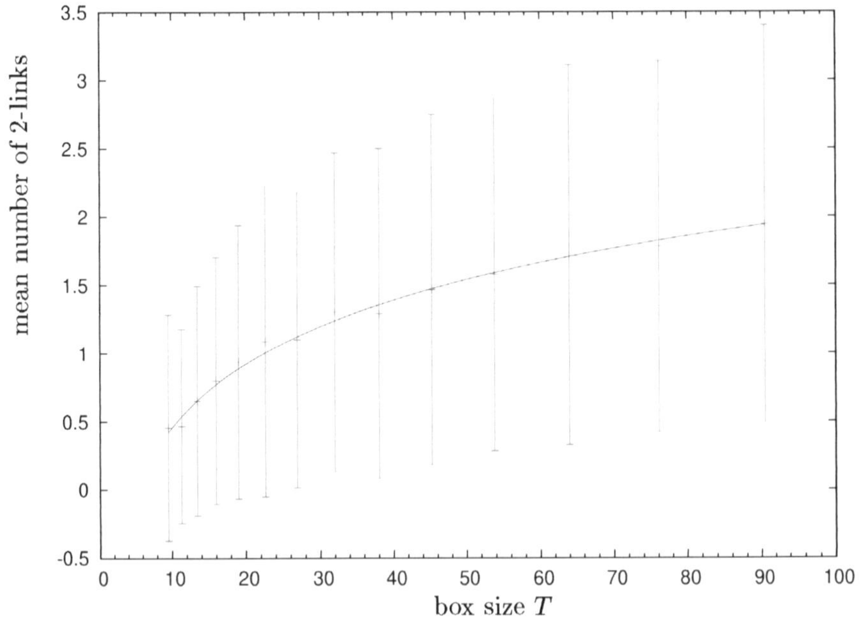

Figure 11. Expected number of 2-links in the sequence of boxes of figure 2.1. Each data point gives the mean and standard deviation from 900 sprinklings.

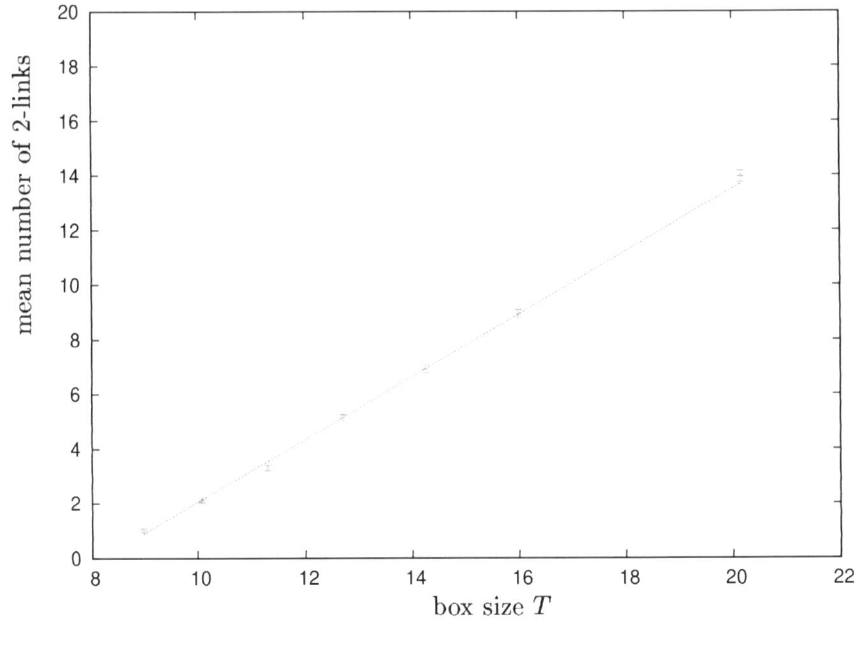

Figure 12. Expected number of 2-links in a sequence of boxes similar to figure 2.1, but with an added spatial dimension of size T. Here each data point gives the mean, standard deviation, and standard error, from 403 sprinklings.

In order to get a reasonable distance measure from the 2-link distance, one simply needs to find a 'reasonable' number of 2-links. One can thus define an 'infrared cutoff' in terms of some number of 2-links. This (subset of 2-links) of course depends upon an arbitrary choice of frame, but this is not a problem, in that the resulting 2-link distance should be independent of this choice, provided that the region of the causal set that they 'enclose' is well approximated by flat spacetime.

A proposal to address the need for the region enclosed by the '2-link minimizing pairs' to be flat is to employ the 2-link distance to locate 'spatial nearest neighbors' in the causal set. These would consist of unrelated pairs of elements whose 2-link distance is beneath some threshold. If the threshold is chosen small enough, then it can be easy to find enough 2-links with a quite modest cutoff. Such nearest neighbors of an element within a sprinkling of $\mathbb{M}^3$ is shown in figure 13.

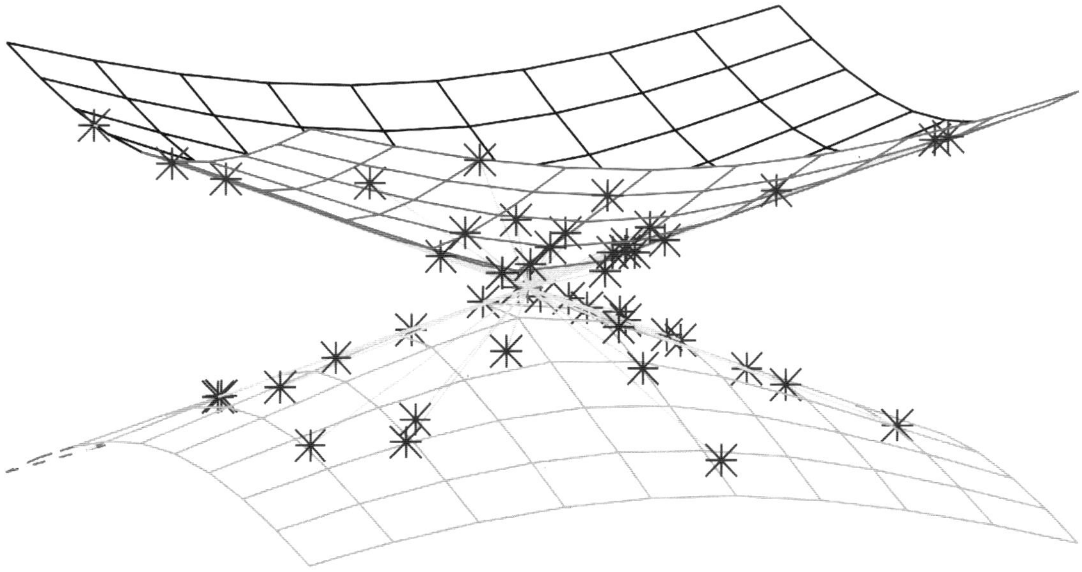

Figure 13. Spatial nearest neighbors of an element in a sprinkling into (a fixed cube in) $\mathbb{M}^3$. $\langle N \rangle = 65\,536$. The future and past light cones of the 'origin element' x are shown. The spacelike cyan lines are drawn between x and each neighbor, for emphasis.

With such a spatial nearest neighbor relation in hand, one can define distances of curves in a curved spacetime setting. See [22] for some illustrations of spatial nearest neighbors in Minskowski space, including an adjacency graph of a spatial slice, derived from a sprinkled causal set.

7. n-links as manifoldlikeness test

It is noted in [22], and also above in section 4, that the particular behavior of the number of n-links 'attached' to a given antichain for sprinklings into Minkowski space may be used as an indicator for 'manifoldlikeness'. By this we mean that, if the numbers of n-links look like what one finds for Minkowski space, with many n-links for $n < d$, few for $n = d$, and almost none for $n > d$, then the causal set may be likely to faithfully embed into $\mathbb{M}^d$.

To get some idea how this might work in practice, we preform a preliminary study, by counting 1- and 2-links in a number of (finite) causal sets. We find that the counts of 1- and 2-links are easily able to distinguish obviously non-manifoldike causal sets from sprinklings, but they have difficulty distingishing causal sets generated by the sequential growth dynamics of [26] from sprinklings into $\mathbb{M}^d$. Physically, one might think of an abundance of 1-links as indicating the potential existence of light cones, as one finds in spacetime, while an abundance of 2-links may indicate the existence of a spatial direction orthogonal to the pair of elements in question. 3-links could indicate spatial directions orthogional to a plane, and so on.

The particular computation we perform is as follows. Given a causal set C, for each element x, we count the number of links l connected to x, and then form a histogram from these counts l.

We do likewise for 2-links: for every 2-element antichain in C we count the number of attached 2-links, and form a histogram of these counts. For example, for the causal set in figure 14, there are six elements with 2 (1-)links (the minimal and maximal elements (those with an empty past and future respectively)), and three with 4 links (those in the middle layer). For 2-links, there are three pairs with 2 2-links (those consisting of elements in the middle layer), six with one 2-link (3 pairs each in the top and bottom layers), and six with zero 2-links (these come from pairs with elements in different layers). The correponding histogram is displayed in figure 15.

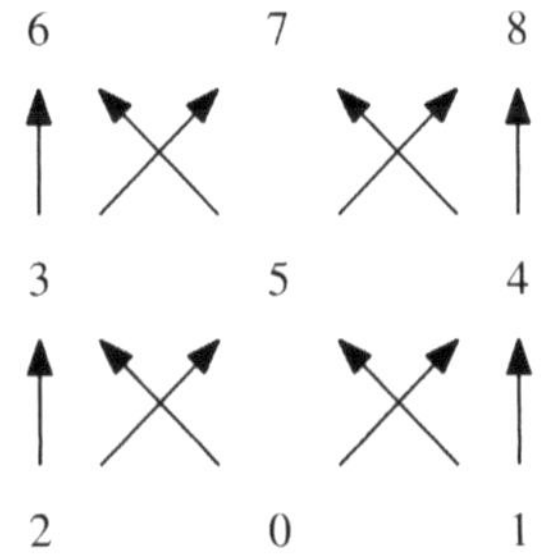

Figure 14. A tower of three 3-crowns. The numeric labels are arbitrary and have no significance.

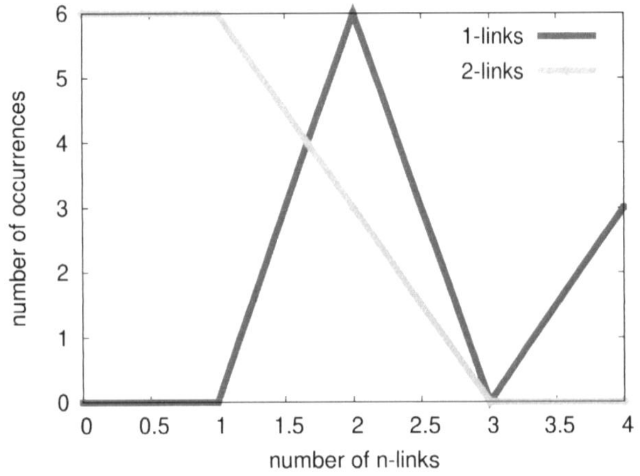

Figure 15. 1- and 2-link histograms for the causal set in figure 14. Lines are drawn between the data points for clarity.

We begin with some sprinklings into intervals of Minkowski space, to see how the counts of 1- and 2-links behave for manifoldlike causal sets. Figure 16 shows histograms of the number of links attached to a causet element, for sprinklings into six different spacetimes. Four are intervals in Minkowski space, of various dimensions. The fifth is a region of a conformally flat Friedman Robertson Walker universe, which contains the initial singularity. (The region is $\eta, x_1, x_2, x_3 \in [0, 1]$, in which coordinates the metric is $ds^2 = \eta^4(-d\eta^2 + dx_1^2 + dx_2^2 + dx_3^2)$. The spatial topology is T^3, so $x_i = 0$ is identified with $x_i = 1$ for $i = 1 \ldots 3$.) For each of these five we sprinkle with a mean number of elements $\langle N \rangle = 512$. The sixth is a $4d$ Schwarzschild black hole, where we have used the technique described in [27] to deduce the causal relations. We sprinkle $\langle N \rangle = 256$ elements into the region $0 < t < 10$, $0 < r < 3M$ in Eddington-Finkelstein coordinates. Figure 14 depicts a similar histogram, for 2-links. We expect to find an infinite number of 2-links for every 2-antichain, for spacetime dimension > 2. This manifests itself in extremely long (power law) tails, in our finite simulations. Even in two dimensions we find some 2-antichains with a large number of 2-links, though the fall off is much sharper. The fact that the results for the $4d$ Schwarzschild spacetime more closely resemble lower dimensional spacetimes may be because many of the elements get sprinkled behind the horizon, where the light cones rapidly fall into the singularity, before they have a chance to build up a large number of links.

A popular concept in physics is that we live in some sort of a product manifold, perhaps with compactified internal dimensions, such as in Kaluza-Klein or String theory, or with large extra dimensions of a braneworld scenario. To investigate how some of these ideas might play out in the causal set context, we consider sprinklings into a flat spacetime with topology $T^3 \times I$ (i.e. a box shaped region of $\mathbb{M}^3$ with opposite sides identified), in which we vary the size of the

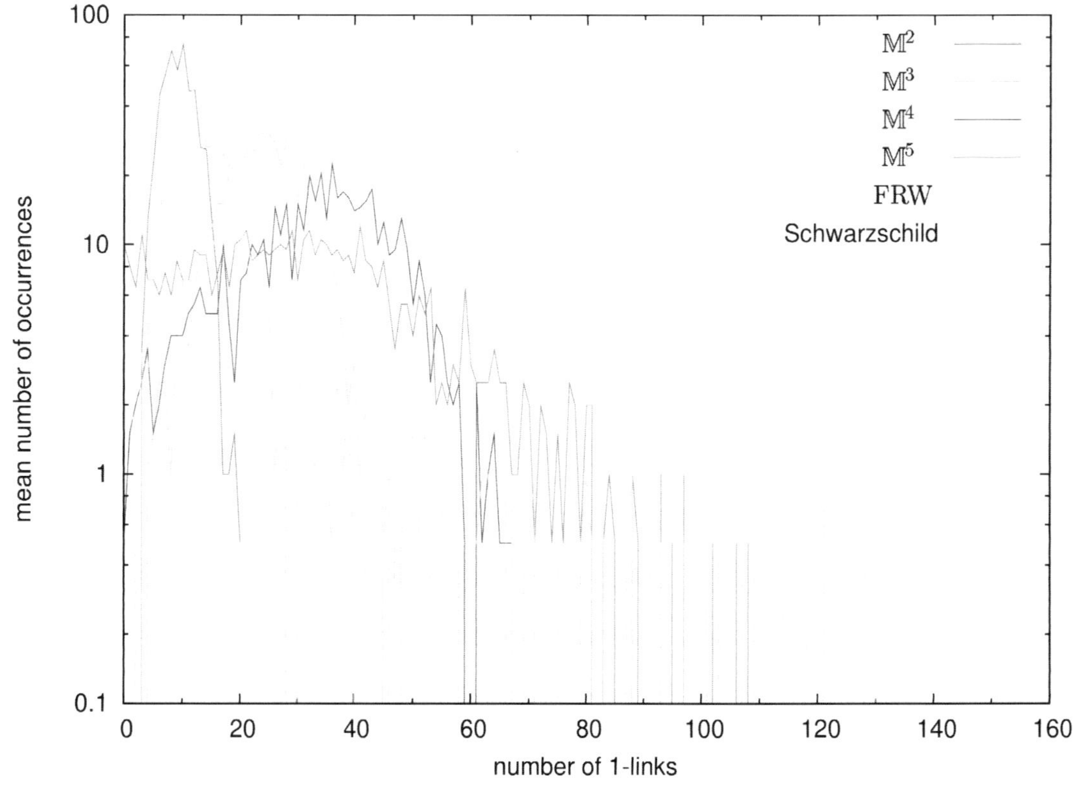

Figure 16. Abundances of (1-)links in some manifoldlike causal sets. Each histogram gives the mean from two sprinklings.

dimensions of the torus. Our results are given in figure 18. We see that for the case in which all dimensions have the same size (purple), the results closely match that of the interval in $\mathbb{M}^4$ from figure 17. As the size of the internal dimensions is decreased, the approximately 256 elements of the causal set begin to be unable to 'see' the internal dimensions, and the asymptotic behaviour of 2-link counts becomes like those of lower dimensional spacetimes. In the extreme case of the red data, with the 'internal' dimensions $\leq \frac{1}{16}$ times the size of the 'external', the causal set is not able to resolve the internal dimensions, and the 2-link counts are almost exactly what one finds for an interval in $\mathbb{M}^2$. (As an alternative to what we did here, one could instead hold the manifold fixed, and sprinkle increasingly more elements to resolve the small internal dimensions. We chose this approach of varying the manifold at fixed $\langle N \rangle$ simply to save on compute time.)

To see how this works as a measure of manifoldlikeness, we try it on two obviously non-manifoldlike causal sets: the 'tower of crowns' and the generic 'Kleitman-Rothschild orders'. A tower of m-crowns consists of a number of layers (m-antichains). If we label the elements in each layer by $0 \ldots m-1$, then element i in layer t precedes elements i and $i+1 \bmod m$ in layer $t-1$. Figure 14 depicts a tower of three 3-crowns. The Kleitman-Rothschild (KR) orders form a generic subset of the set of all finite partially ordered sets [28, 29]. They consist of three layers, the middle of which contains approximately half the elements of the causet set, and the top and bottom layer contain about a quarter each. [4] For each pair of elements in adjacent

[4] In our simulations we select the cardinality of the bottom layer from a Poisson distribution with mean $N/4$ (for an N element causet set), and likewise for the top layer. All remaining elements are placed in the middle layer.

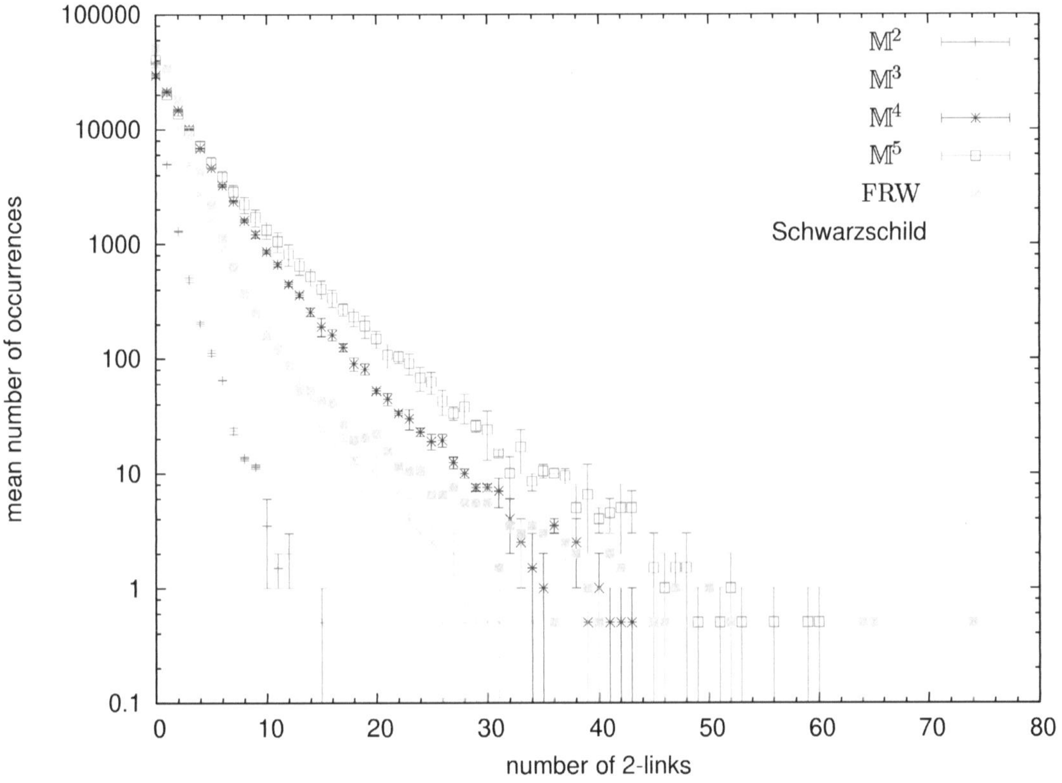

Figure 17. 2-links in some manifoldlike causal sets. Each histogram gives the mean (and standard error) from two sprinklings.

layers, we place a relation between them with probability 1/2. Every element of the bottom layer precedes every element of the top layer. The results for these causal sets are depicted in figures 19 and 20. The tower of crowns are obviously not manifoldlike, as the only non-zero bins are 2 and 4 for the 1-links, and 0, 1, and 2 for the 2-links. The 1-links histogram for the KR orders looks like it could have come from a sprinkling. The 2-links histogram, however, possess a 'discontinuity', with a huge spike at zero followed by empty bins at 1-3. The spike at zero results from unrelated pairs of elements, each from different layers. This sort of behaviour does not resemble sprinklings into spacetime.

The last class of causal sets we consider are those generated by the 'transitive percolation' dynamics, which is a simple special case of the generic class of sequential growth models derived in [26]. Here one begins with N elements, and considers every pair of elements $i < j$ in turn, introducing a relation between them with some fixed probability p. After relations are introduced in this way, one takes the transitive closure to arrive at a causal set. In figure 21 we depict the 2-links for two choices of parameters, $N = 2^{15}$, $p = 2/10$, and $N = 128$, $p = 7/10$. They fall off very quickly, but otherwise are not obviously distinct from the results that one gets from sprinkling into spacetime. To get some idea as to whether there are sprinklings into spacetime which mimic these results, we sprinkle into a rectangle of $\mathbb{M}^2$, with a height/duration 55 times the (spatial) width. Although the 2-link abundances for transitive percolation at $N = 128, p = 7/10$ fall off a bit faster than for the sprinkling into the rectangle, it remains possible that that some choice of parameters for percolation will give a 2-link histogram which can be matched by a

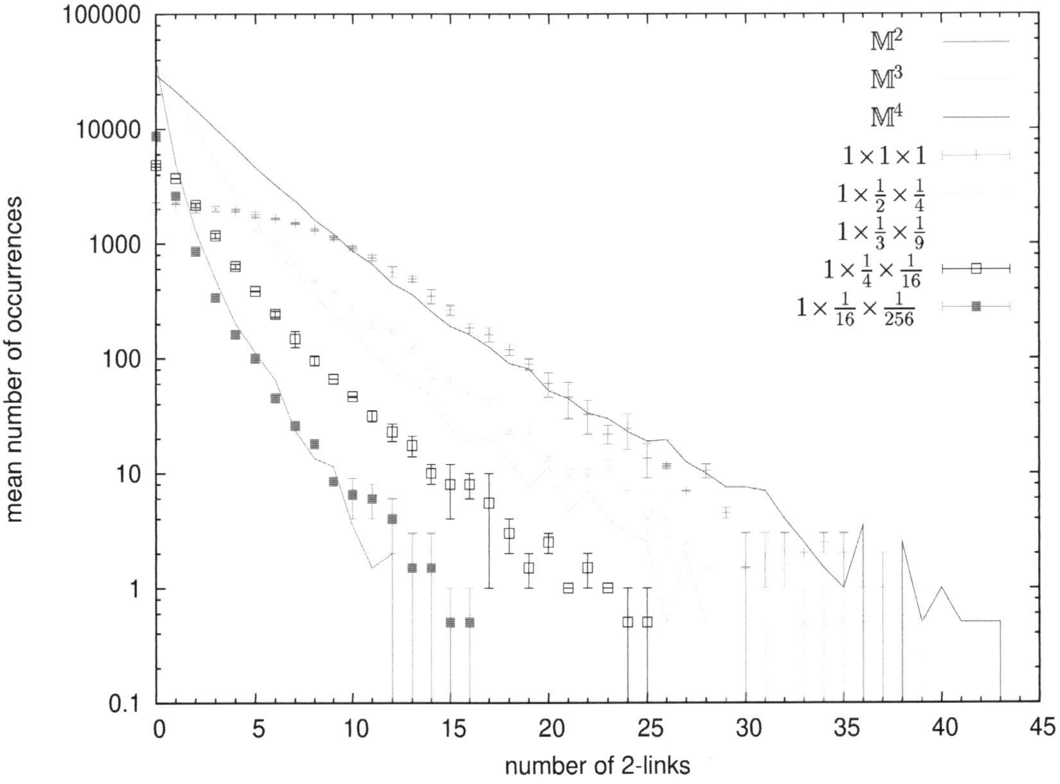

Figure 18. 2-links in $\langle N\rangle = 256$ sprinklings into $T^3 \times I$ spacetimes. The 'circumferences' of the toroidal dimensions are given in the key. Each histogram gives the mean (and standard error) from two sprinklings. The solid lines indicate the means from figure 17, for comparison.

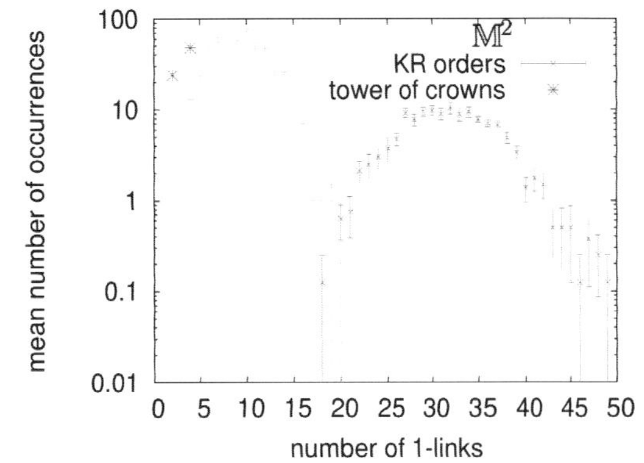

Figure 19. 1-links for some non-manifoldlike causal sets: (Two) Kleitman-Rothschild order(s) on 64 elements, and a tower of 6 12-crowns. Results from sprinkling into $\mathbb{M}^2$ is added for comparison.

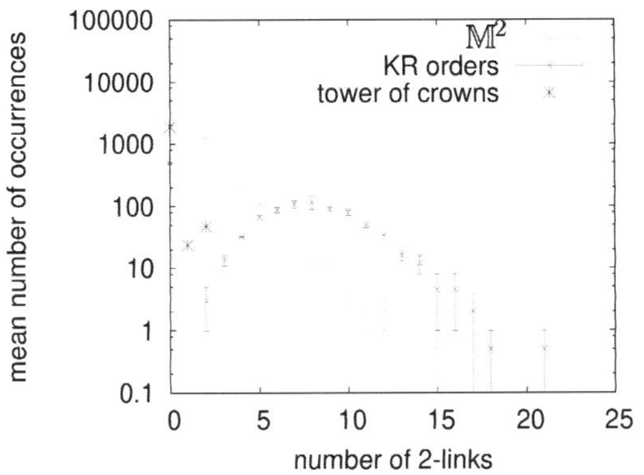

Figure 20. 2-links for some non-manifoldlike causal sets: (Two) Kleitman-Rothschild order(s) on 64 elements, and a tower of 6 12-crowns. Note the large number of pairs in the KR-order (red) without any 2-links. Results from sprinkling into $\mathbb{M}^2$ is added for comparison.

sprinkling.

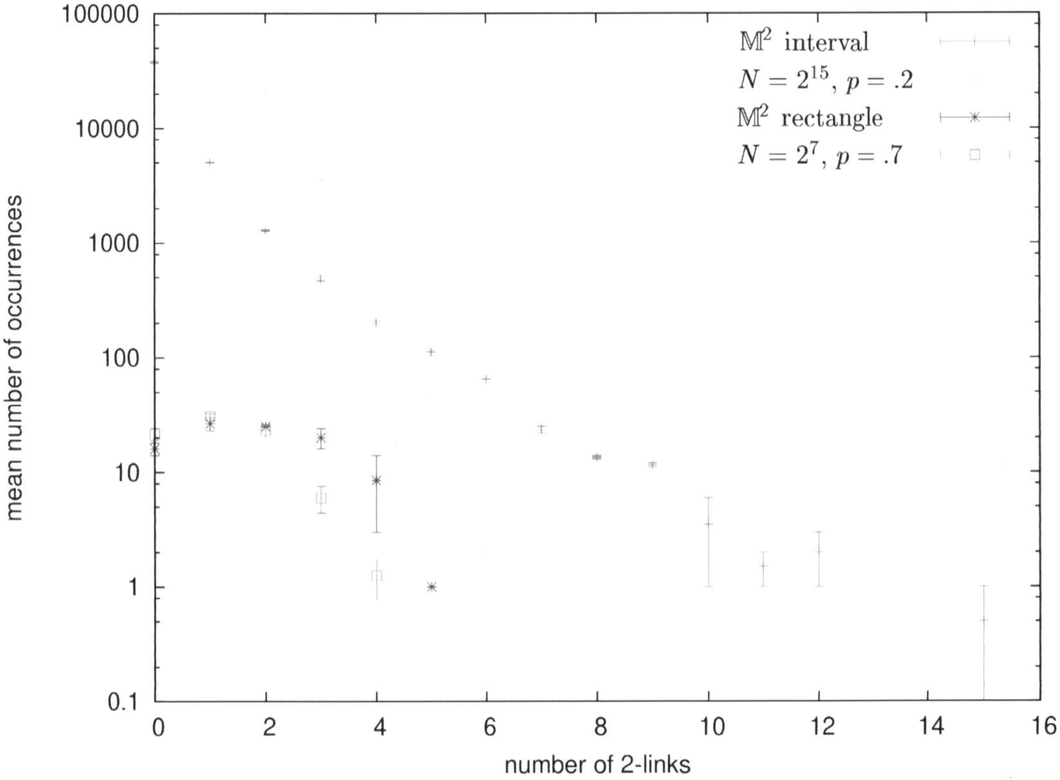

Figure 21. Comparison of transitive percolation with sprinklings into $\mathbb{M}^2$. The interval in $\mathbb{M}^2$ uses $\langle N \rangle = 512$, while the rectangle $\langle N \rangle = 128$. Each histogram gives the mean and error from two causal sets.

8. Conclusion

A fundamentally atomistic theory of quantum gravity must come with some sort of relation defined on the discrete elements, in order for nontrivial geometric structures to emerge. In the case of causal sets, this relation is regarded as a causal ordering. This association with causality makes it relatively straightforward to see how timelike structures emerge, but spatial quantities are more subtle. Here we proposed some quantities, derived from the discrete causal ordering, which extract spatial information from the causal set, such as spatial distances.

In order to define spatial distance, we introduced the notion of an n-link, which is a simple derived object which encodes information about the intersection of light cones of spatially separated elements. From this we introduced the 2-link distance, and showed that it is able to overcome the degeneracy of the former measure of spatial distance for pairs of elements on a causal set.

In order to recover curved geometry, we further proposed to use the 2-link distance within some local region, to identify spatial nearest neighbors in the causal set. This defines a symmetric, spacelike relation, which is derived from the causal order. From it we may hope to recover the length of continuous curves, and thus derive the metric geometry. It is hoped that such spacetime structures will be useful in the development of a dynamical law for causal sets.

The n-link captures some element of spatial information, which plain links (in and of themselves) do not. In [22] is was observed that their abundance depends somewhat strongly upon dimension, and it was proposed that they could be used as a dimension estimator, or even as a potentially stringent condition for a causal set to be faithfully embeddable into a spacetime manifold. Here we performed a preliminary investigation of this 'manifoldlikeness' condition, and observed somewhat sharp dependence of 2-link abundances on dimension.

In order to see the effect of compact dimensions at various scales, we explored flat $T^3 \times I$ spacetimes, with varying ratios of scales of the 'internal' dimensions. We observed that the abundance of 2-links is able to see the correct dimension of 4 at large sprinkling densities, but sees an effective, smaller dimension as the sprinkling density is reduced as compared to the ratio of circumferences of the toroidal dimensions. The results mirror those of [18] for embeddings into $S^1 \times I$ spacetime.

The question of how to compute the dimension of discrete spacetime at a range of length scales is important currently, because of a number of results from various approaches to quantum gravity which predict a scale dependent dimension, in particular a smaller fractal dimension at high energies [30] (and references therein). In addition to the older methods mentioned in the introduction, counting of n-links may provide an important alternate method to deduce which dimensions arise dynamically at various scales in causal set quantum gravity.

In section 7 we presented some preliminary results on a manifoldness test based on counting n-links. It is easily able to distinguish some obviously non-manifoldlike causal sets from those which are faithfully embeddable into spacetime, however it had some difficulty in distinguishing causal sets generated from transitive percolation from those of sprinklings into regions of Minkowski space. This may be rectified when considering n-links for $n > 2$, or by performing a careful analysis on the parameter space of the transitive percolation and sprinkling models. Of course it can only be a necessary condition for there to exist a faithful embedding into spacetime, and not sufficient, because one can always compose partial orders with a particular form for the n-link counts, which is not manifoldlike.

Acknowledgments
This research was supported primarily by the Perimeter Institute for Theoretical Physics. Research at Perimeter Institute is supported by the Government of Canada through Industry Canada and by the Province of Ontario through the Ministry of Research & Innovation. Earlier work was supprted by the Marie Curie Research and Training Network ENRAGE (MRTN-CT-2004-005616), and the Royal Society International Joint Project 2006-R2. Some numerical results were made possible by the facilities of the Shared Hierarchical Academic Research Computing Network (SHARCNET:www.sharcnet.ca).

References

[1] Thiemann T 2007 *Modern Canonical Quantum General Relativity* (Cambridge, UK: Cambridge U. Press)
[2] Ashtekar A 2004 *Class.Quant.Grav.* **21** R53 (*Preprint* gr-qc/0404018)
[3] Rovelli C 2004 *Quantum Gravity* (Cambridge University Press)
[4] Konopka T, Markopoulou F and Severini S 2008 *Phys. Rev.* D **77** 104029 (*Preprint* 0801.0861)
[5] Wolfram S 2002 *A New Kind of Science* (Wolfram Media, Inc.)
[6] Hamma A, Markopoulou F, Premont-Schwarz I and Severini S 2009 *Phys. Rev. Lett.* **102** 017204 (*Preprint* 0808.2495)
[7] Rideout D and Zohren S 2006 *Class. Quant. Grav.* **23** 6195–6213 (*Preprint* gr-qc/0606065)
[8] Bombelli L, Lee J, Meyer D and Sorkin R D 1987 *Phys. Rev. Lett.* **59** 521–524
[9] Sorkin R D 1994 *Mod. Phys. Lett.* A **9** 3119–3127 (*Preprint* gr-qc/9401003)
[10] Sorkin R D 2007 *J. Phys. A: Math. Gen.* Special Issue: Quantum Universe *Preprint* quant-ph/0610204
[11] Craig D and Hartle J B 2004 *Phys. Rev.* D **69** 123525 (*Preprint* gr-qc/0309117)

Journal of Physics: Conference Series **174** (2009) 012017 doi:10.1088/1742-6596/174/1/012017

[12] Noldus J 2004 *Class. Quant. Grav.* **21** 839–850
[13] Malament D B 1977 *J. Math. Phys.* **18** 1399–1404
[14] Hawking S W, King A R and McCarthy P M 1976 *J. Math. Phys.* **17** 174–181
[15] Sorkin R D 2005 *Lectures on Quantum Gravity, Proceedings of the Valdivia Summer School, Valdivia, Chile, January 2002* ed A Gomberoff and D Marolf (Plenum) (*Preprint* gr-qc/0309009)
[16] Reid D D 2003 *Phys. Rev.* D 024034 (*Preprint* gr-qc/0207103)
[17] Myrheim J 1978 *Preprint* CERN TH-2538
[18] Meyer D A 1988 *The Dimension of Causal Sets* Ph.D. thesis (Massachusetts Institute of Technology)
[19] Bachmat E 2008 *Contemporary Mathematics* **458** 347–359 (*Preprint* gr-qc/0702140)
[20] Henson J 2007 *Towards Quantum Gravity* ed D Oriti D (Cambridge, UK: Cambridge U. Press) (*Preprint* gr-qc/0601121)
[21] Brightwell G and Gregory R 1991 *Phys. Rev. Lett.* **66** 260–263
[22] Rideout D and Wallden P 2008 *Preprint* 0810.1768
[23] Goodale T, Allen G, Lanfermann G, Massó J, Radke T, Seidel E and Shalf J 2003 *Vector and Parallel Processing — VECPAR 2002, 5th International Conference* (Berlin: Springer) 197–227
[24] Major S, Rideout D and Surya S 2009 *Preprint* 0902.0434
[25] Sorkin R D 2007 *Towards Quantum Gravity* ed D Oriti (Cambridge, UK: Cambridge U. Press) (*Preprint* gr-qc/0703099)
[26] Rideout D P and Sorkin R D 2000 *Phys. Rev.* D **61** 024002 (*Preprint* gr-qc/9904062)
[27] He S and Rideout D 2008 *Preprint* 0811.4235
[28] Brightwell G 1993 *Surveys in Combinatorics* (*London Math. Soc. Lecture Notes Series* vol 187) ed Keith 53–83
[29] Kleitman D and Rothschild B 1975 *Trans. Amer. Math. Society* **205** 205–220
[30] Modesto L 2008 *Preprint* 0812.2214

Light, Links and Causal Sets

Rafael D Sorkin
Perimeter Institute, 31 Caroline Street North, Waterloo ON, N2L 2Y5 Canada
Department of Physics, Syracuse University, Syracuse, NY 13244-1130, U.S.A.
E-mail: sorkin@physics.syr.edu

Abstract. After sketching a context in which to seek observable signals of spatio-temporal discreteness, I briefly review the status of the causal set program for quantum gravity, concluding with a simple model for the field produced by a moving charge in a background causal set.

1. How might discreteness show up?

Riemann had a nice phrase somewhere that talked about "the reality that underlies spacetime."[1] What we know of quantum gravity suggests that that reality is more in the nature of a discrete structure than a continuous one. If so then it's important to think about phenomena that could reveal this underlying discreteness. Several such possibilities come to mind, some of them present by definition, others suggested by analogy with the discreteness of ordinary material, and still others specific to one or another of the deep structures of spacetime that have been proposed by workers on quantum gravity.

Almost by definition, discreteness implies a "cutoff" in energy and wavelength. However (unless the idea of "large extra dimensions" turns out to be correct, cf. [1]) the expected Planckian scale of this effect places it beyond *direct* observation for the present. Moreover (and this is why I included the qualification "Almost" just above) the concept of a cutoff takes on a more subtle meaning in the context of a discreteness that respects Lorentz invariance, like that of the causal set. One cannot naively translate such a granularity into the statement that an upper limit to frequency would exist in any given frame of reference. Such a claim would conflict with the fact, that, for example, a plane-wave solution of the massless wave equation can have any frequency at all, depending on which frame it is referred to.[2] This subtlety does not mean that a Lorentz-respecting cutoff must remain forever elusive, of course, only that it can be detected solely by effects whose nature is intrinsically frame-independent. For example, the notion of center of mass energy is invariant, and correspondingly, one can easily see in the context of (background) causal sets that scattering cross sections must fall to zero at Planckian energies. One could in principle accelerate particles to arbitrarily high energy (limited only by infrared cutoffs of cosmological origin), but if such "transplanckian" particles were to collide,

[1] Of course he would have said "space" rather than "spacetime"

[2] Here I am of course identifying the physical symmetry group of Minkowski spacetime with the Poincaré transformations as normally defined. This ignores the possible extension of the concept of symmetry to a Hopf algebra that is not a group algebra, as essayed in "deformed special relativity" and non-commutative geometry more generally. [2]

the associated wave packets would simply "pass through" each other. They would not "see each other" because they would be supported on disjoint sets of causet elements.

Interesting as this "asymptotic freedom" effect is in principle, it seems far removed from observability in practice, and people have sought elsewhere for the phenomenology of spatio-temporal discreteness. By way of inspiration, one can call to mind some of the analogous phenomena that testify to the atomicity of ordinary matter. Within this ambit, the effect that has received the most attention is perhaps the possible breaking of Lorentz invariance. Here the condensed-matter analogs are the partial breaking of translational and rotational invariance by a crystal (resulting for example in Umklapp scattering processes and the Mössbauer effect) and the related high frequency modification of dispersion relations for waves traveling through the crystal. However, the observational and experimental evidence to date speaks against any such failure of Lorentz invariance, suggesting that we should limit our search to phenomenology consistent with this symmetry. Indeed, the vista from causal sets suggests that this negative verdict will persist indefinitely, because it seems to be impossible to design a causet that would single out any frame of reference without coming into contradiction with the basic hypothesis that "number = volume". (More specifically, it seems that the most uniform distribution of points in Minkowski spacetime $\mathbb{M}^4$ is given by a Poisson distribution; and one knows that Poisson distributions are frame-independent in a strong sense [3].)

But, not all matter is crystalline, and dispersion relations are not the only – or even the best – messengers from the realm of atoms. In gases and liquids, another whole family of phenomena springs from the irregular arrangement and chaotic motions of the constituent molecules. Some familiar instances of these "fluctuation phenomena" are the Brownian motion of microscopic particles suspended in a medium and the scattering and extinction of light propagating through air. By way of analogy people have asked whether particles moving freely through the vacuum might deviate from geodesic motion, and whether light from cosmologically distant sources might be attenuated or deviated or otherwise affected by a Planck-scale discreteness [4].

To fully answer these questions is not possible outside of a completed theory of "quantum gravity", but that need not stop us from seeking partial answers based on plausible models of an incomplete and/or partly phenomenological nature. Below I will briefly present one such model [5] that describes the propagation of light (more precisely, a massless scalar field) through a causal set as a process of "direct transmission from source to sink." Limited though it is, this model lets us appreciate that neither modified dispersion relations nor loss of coherence is a necessary consequence of discreteness. It will also allow one to estimate the size of some of the fluctuations in signal strength that do result from the discreteness. Before getting to that, however, I'd like to recall two or three more types of phenomena that have been suggested as potential "signals" of a quantum gravity founded on an atomistic deep structure of spacetime.

The first of these is the causal set counterpart of Brownian motion, but a similar effect could be expected in any Lorentz-symmetric discrete theory. Precisely because of the absence of a distinguished frame, a direct analog of the Wiener process is ruled out in the sense that fluctuations in the position cannot not be Markovian; but fluctuations of velocity or momentum pose no such difficulty. In fact, one can define a spacetime analog of the "Ornstein-Uhlenbeck" version of Brownian motion, in which the particle's worldline is differentiable but its tangent-vector is not, the acceleration being subject to a sort of Lorentz-invariant white noise. For a massive particle the resulting stochastic process [6, 7, 8] is characterized by a single phenomenological parameter, a diffusion constant with the dimensions of inverse proper time.[3] For a massless particle, a second free parameter enters (a drift coefficient in energy), while for a photon (massless with polarization), still other parameters become relevant [8]. Since we have no estimate of any of these phenomenological parameters from first principles, the best that

[3] These are the dimensions if the diffusion is referred to velocity-space. If it is referred to momentum-space the dimensions are $mass^2/time$.

one can do is to seek observational bounds on them. Such bounds have been deduced from the undistorted Planckian shape of the cosmic microwave background (CMB) spectrum [8] and from cosmological bounds on the kinetic energy carried by relic neutrinos [9].

The second type of phenomenon stemming from the discreteness of the causet is a potential failure of locality at sufficiently high energies. At the root of this possibility is an inner contradiction among discreteness, Lorentz invariance, and locality: any two can coexist but not all three together.[4] Some models of discrete spatio-temporal structure maintain an (approximate) locality at the expense of Lorentz invariance. But with causal sets, this option is difficult or impossible to implement (even if it were desirable), and the problem becomes how to recover at least the appearance of locality at sufficiently great length scales, where a continuum spacetime offers a good approximation to the underlying causet. Work on this problem has illustrated in concrete examples how a coarse-grained locality on mesoscopic scales can emerge from microscopic interactions that are radically nonlocal, but it also suggests that the nonlocality might persist at length scales much greater than the fundamental discreteness scale (Planck scale) [10]. Were evidence of such nonlocality discovered, that would open up an entirely new phenomenological area for quantum gravity. However I don't know that anyone has proposed a clear experimental signal of nonlocality that one could easily look for.

A final phenomenon that deserves incidental mention here – concerning the cosmological constant Λ – differs from the others I've been referring to in that it pertains to full quantum gravity, as opposed to the more limited setting of the propagation of fields and particles on a fixed, non-dynamical background. In conjunction with the quantum mechanical conjugacy between Λ and spacetime volume V, Poisson fluctuations in the "density of causet elements" imply fluctuations in Λ with an order of magnitude $1/\sqrt{V}$. This heuristic argument [11] led to the prediction, which later came true, that the contemporary value of Λ would be found to be comparable in magnitude (but not necessarily in sign) to the density of baryonic and other "matter". A simple model built on the above idea can be found in [12], but a fully consistent phenomenological theory of such fluctuations remains to be devised.

2. The causal set program

In the preceding introduction, I have sketchily reviewed some possible "phenomenology" corresponding to the hypothesis that a discrete structure supersedes the Lorentzian manifold at Planckian scales. Partly to provide background for those remarks (and for the remarks which follow), and partly in obedience to a request from the conference organizer, I will try in the present section to review even more sketchily, the overall status of the causal set program itself.[5]

At the root of this program is the recognition that, when it is combined with volume information, causal structure alone suffices to reproduce fully the geometry of spacetime. In a continuum – what Riemann called a "continuous manifold" – volume information is lacking. But what Riemann called a "discrete manifold" does carry an obvious volume-measure obtained simply by counting elements. In this case "less is more, as well as less". The kinematic meaning of the causet hypothesis, then, is that macroscopic spacetime M is an effective description of (an approximation to) a causal set C. Mathematically C is a partial order or "poset". Its defining order relation $\prec$ gives rise to the light-cone structure of spacetime, while its innate volume measure gives rise to the continuum volume element $\sqrt{-g}d^4x$. In brief "Number = Volume", and thus, "Order + Number = Geometry".

A more precise notion of the causet-spacetime correspondence has been based on the concept

[4] To get a feel for this contradiction, observe that two "infinitesimally nearby" point-events are "infinitesimally nearby" in all frames, but two points at a small but finite timelike distance in a given frame will be separated by arbitrarily great times in other frames.

[5] For some general references to this program see [13].

of "faithful embedding".[6] One says that $f : C \to M$ is faithful if the point-set $f[C]$ "could have been obtained by Poisson-sprinkling points into M at unit density". It it this notion of correspondence that is employed in the phenomenological applications discussed in the previous and following sections.

What would it take to turn the causet hypothesis into a theory of quantum gravity? For a physicist, it seems natural to follow the scheme advocated by Taketani that views a theory as comprised of three components or "stages", which one may call kinematical, dynamical and phenomenological. For causets, kinematics refers first of all to the *kind* of structure one has, i.e. to the definition of a causet as a locally finite partial order, and more broadly to the development of the mathematical ideas proper to causets (the mathematics that would play the same role for causet theory that differential geometry plays for general relativity). Of special importance is the development of a "dictionary" allowing one to translate back and forth between order theoretic and geometric concepts. By dynamics I mean what one might describe as the "equations of motion" of the causet. But in addition to this (which pertains to full quantum gravity) there are also questions analogous to those belonging to the theory of quantum fields on a background spacetime. With the exception of the above remarks on Λ, all the various effects mooted above fit into this second category. The word phenomenology, finally, needs no definition and no separate discussion in this section, since the other two sections of the paper are devoted to it.[7]

2.1. Kinematics

At the level of kinematics, much is known, although some important questions still remain open. Some of the necessary concepts can be borrowed from the mathematical discipline of combinatorics (for example the "height of an interval"), others can be copied from the "global causal analysis" of continuum general relativity, while still others (like Myrheim's "ordering fraction", or "coarse-graining" by random selection of a suborder) seem not to have been studied before. As things stand now, one has a growing dictionary allowing one to "read out" geometrical and topological information from order information.[8] Corresponding to the timelike distance between two points of spacetime is the length of the longest (not shortest) chain between two causet elements.[9] (As its name suggests, a *chain* in a partial order is a sequence of causet elements, $e_1 \prec e_2 \prec e_3, \cdots \prec e_n$, each an "ancestor" of the next one in the sequence.) Corresponding to spacetime dimensionality we have a number of different estimators. Here I will just mention three. For a given interval I we have first of all the Myrheim-Meyer dimension, which seems to be the most accurate so far [15]. It is a certain function of the ordering fraction $R/\binom{N}{2}$, R being the number of related pairs of elements of I and $N = |I|$ being simply the number of elements of I. (An *interval* in a poset is the set of elements causally between two given elements a and b: $I = \{x \in C \,|\, a \prec x \prec b\}$.) A second estimator, the "midpoint scaling dimension" of an interval is analogous to a Hausdorff or fractal dimension. A third estimator suggested by Eitan Bachmat [16], can be applied directly to the causet as a whole, not just to intervals within it. It also is a kind of scaling dimension, based in this case on how the "height" of C (the length of the longest chain it contains) scales under coarse-graining. All these estimators (for timelike distance and dimension) have been tested in a range of examples,

[6] With Lorentzian geometry, the concept of "closeness" between two spacetimes M or between a spacetime and causet C turns out to be much more delicate than one would have been led to believe by the example of Euclidean signature.

[7] A "phenomenological" application of causets not mentioned in those sections concerns black hole entropy. Unfortunately, the available accounts are rather outdated by now [14].

[8] Of course this type of translation only makes sense when the causet in question is in fact well approximated by some Lorentzian manifold.

[9] This estimator turns out to be remarkably accurate. There is some evidence that its fluctuations grow only *logarithmically* with the number of elements, in dimensions $3 + 1$ and higher!

but rigorous proofs of their validity exist only in special cases, like that of Minkowski spacetime.

In principle, if one knows all the distance relations in a Lorentzian manifold C, one know everything there is to know about M, including its topology. Thus, one can expect that something similar would be true of a causet well approximated by M: one could deduce the topology approximately by counting longest chains within C. However, no technique for doing this is known. Nonetheless, there has been recent progress in recovering algebraic-topological information directly, more specifically in assigning homology groups to "slices" (the discrete analogs of spacelike hypersurfaces) [17, 18]. There also exist proposed definitions for homology groups pertaining to C as a whole (as opposed to slices within it), but they remain untested.

Probably the most exciting of the recent kinematical developments, is the prospect of an estimator for the Ricci scalar R. Were this prospect to pan out, it would furnish a causal set counterpart to the Lagrangian of continuum gravity, and therefore a direct route to setting up a "path integral" for causet dynamics. The estimator in question grew out of the attempt, referred to earlier, to formulate an approximately local analog of the scalar wave equation for a field ϕ defined on a causet [10]. The resulting "discrete D'Alembertian operator" $\Box$ does yield such an analog, and it can indeed be used to study propagation of waves in a causet. But once one has it, $\Box$ can also be combined with the "Synge world function" σ to obtain a discrete analog of R, by forming (purely at a kinematical level) the expression $\Box\,\Box\,\sigma$. In the continuum, this combination is known to reduce to (a multiple of) R in the coincidence limit. The question now is whether it will do likewise in the discrete case.

A final kinematical result that bears mention here relates back to the question of Lorentz breaking and the causet-spacetime correspondence. I asserted earlier that discreteness of the causet type respects the full symmetry group of $\mathbb{M}^4$, including the Lorentz boosts. This claim rests on the fact that the Poisson "sprinkling" process is Lorentz invariant (which invariance rests in turn on the fact that 4-volume is invariant). What this really signifies is that, given a large number of sprinklings into $\mathbb{M}^4$, one could not deduce a distinguished reference frame from the entire collection. Logically this still leaves open the possibility that even though the *process* of sprinkling is invariant an *individual* sprinkling would still determine a frame, so that invariance could return only in some sort of "sum over causets". Were things to turn out this way, the project of studying wave or particle propagation in a *single* causet, taken as typical, would be misguided. In effect, one would have to go all the way to full quantum gravity in order to analyze the effects of discreteness on propagation. Fortunately, one can prove that (with probability 1) even a single sprinkling fails to determine a frame [3]. Accordingly, it makes sense, as a first effort, to study propagation in a single sprinkled causet, and this is what we will do in Section 3.

2.2. Dynamics

In relation to causal sets, dynamics refers first of all to the laws or principles that would govern the structure of the causet itself. In this sense the dynamics would be expressed in terms of a quantum sum-over-causets, or some replacement for it. But dynamics also refers to any process of evolution of fields, particles, or other forms of "matter" within a single causet, treated as a non-dynamical background or "arena". As discussed in the introduction, most of our phenomenological insights so far (with the exception of those relating to Λ) stem from the study of dynamics in this second sense. I have already spoken about work on the motion of a classical particle and on the propagation of a classical field, and we will return to a special case of the latter after dwelling briefly on the dynamics of the causet itself. However, first I want to mention some results on *quantum* particle propagation that are described more fully in the contribution of Steven Johnston to this volume [19].

In a certain sense, the massless Green's function G of [5] (summarized below) already has a quantal character, not for fields but for particles. That is, it can be expressed as a sum over particle paths, each path contributing a single "amplitude". The only thing is that these paths verge on triviality: each is a chain of just two linked elements of the causet. That it is nonetheless appropriate to regard these "paths" as discrete particle worldlines becomes clearer when one generalizes to the massive case. In the continuum the corresponding retarded G has support within the light cone, not just on it and one finds [19] that most contributions now come from paths of high cardinality.

Let us proceed now to causet dynamics proper. Perhaps the first question one should ask in this connection is how such a dynamics could even be formulated, given that Hamiltonian evolution can get no foothold when time as well as space is discrete, and when there is no background manifold to support any sort of analog of a metric-field operator. What springs to mind, of course, is the path integral, but even this is problematic if one limits its role to furnishing a "propagator" between fixed boundaries. Rather, one must (or so it seems) replace the "path integral qua propagator" by the "path integral qua decoherence functional or quantal measure". (For definitions of these essentially equivalent objects see [20].) When this is done, quantum theory closely resembles the classical (but probabilistic) theory of stochastic processes like the random walk. In fact it appears as a generalization of those theories which replaces the Kolmogorov sum rule with a weaker sum rule that allows for pairwise interference between alternatives.

In this light, it seemed natural to seek a quantal dynamical scheme for causal sets by setting oneself the simpler task of finding a classical stochastic theory of causet structure. In fact what emerged was not a "structural law" as such, but rather a dynamics based on *growth* or accretion, in other words a birth process. The resulting models are known as classical sequential growth (CSG) models, and they follow almost uniquely from two or three general principles, including in particular a principle of "Bell causality". (Except for some exceptional solutions, all of these models belong to the so called "generalized percolation" family, and as such are closely related to what are called "random graph orders" in the mathematical literature [21].)

Although the CSG models were initially intended only as stepping stones to a full quantum dynamics – a dynamics of "*quantal* sequential growth" – one can ask to what extent any of them can produce causal sets that resemble spacetimes (i.e. that are well approximated by a Lorentzian manifold, preferably a spacetime that also satisfies the Einstein equations!) The answer seems to be mixed. On one hand there are good reasons to believe that fully manifold-like causets will never be produced by any of the CSG models (more precisely, will be produced with negligible probability). On the other hand, for certain choices of the parameters, one obtains causets that resemble manifolds in important ways, e.g. de Sitter space [22].

Thanks to this resemblance, the CSG models can bring to life possibilities inherent in the causal set hypothesis, that might be realized more fully (i.e. more realistically) in some future theory of *quantal* sequential growth. One such example concerns a "Boltzmann-Tolman universe" scenario that could solve some of the "large number riddles" of cosmology [23]. It turns out that for a large range of their parameters or "coupling constants", the CSG models lead to a qualitatively similar overall behavior of the evolving causet C. It begins life as a single element (or as the empty set if you prefer to start the growth there), expands rapidly to a large size, stabilizes there, and after some time (depending on the parameter values chosen initially) collapses back to a single element (called a "post" in the combinatorial literature). There ensues a succession of cycles of expansion and contraction separated by posts, each of which appears to its descendants as a "big bang". In each new cycle the initial expansion looks exactly the same at first, beginning with a "tree phase" in which each new element is born from a single "parent" chosen at random from the elements of the growing tree born earlier.[10] The subsequent phases of the expansion do depend on the specific values of the parameters, but what is interesting is

[10] A causet is a *tree* iff each element has exactly one parent (or none in the case of the post).

this, that the effective parameters governing the evolution change from cycle to cycle. They are modified by the action of a certain transformation, the "cosmic renormalization group" [24], that drives them toward the values characteristic of the particular CSG model known as "transitive percolation", which contains a single free parameter p that, among other things, governs the size of maximum expansion. The value of this "renormalized" p decreases from cycle to cycle, and when it is much less than unity, C expands to a very large and homogeneous configuration. Overall the picture seems remarkably lifelike, given the limitations of classical growth. In such a scenario, the large numbers of cosmology like the very great value of the spatial radius of curvature in natural units, would owe their size to the large age of the cosmos (measured in number of elements), rather than any special "fine tuning".[11]

The classical sequential growth models have also led to important progress of a conceptual nature by showing how the so called "problems of time" can be resolved in a way that applies equally well to the classical and quantum cases [25]. Without any appeal to "clocks" or other surrogates for coordinate systems (such as asymptotic regions where an S-matrix can be defined), one has identified "observables" which first of all carry a clear physical meaning and second of all are complete in the sense that all other covariant (label independent) observables can be formed from them in a well defined sense. Because this completeness rests only lightly on the details of the CSG models, there is every reason to expect that it will go through unchanged in the case of quantal sequential growth (i.e. quantum gravity à la causet).

Of course this expectation can only be tested after we have a quantum dynamics to work with. How might one arrive at such a dynamics? The example of classical gravity suggests two very different but complementary paths, one based on general principles the other on imitation of the continuum theory.

Along the first path one would seek to emulate the way in which the Einstein equation was (or might have been) derived from the twin principles of general covariance and locality. This was in fact the type of reasoning that led to the CSG models. As I stressed above, locality is not to be met with in a causal set, and covariance by itself (in the sense of labeling independence) was clearly an inadequate guide. But it turned out that a cognate condition that came to be called "Bell causality" could take over where locality left off. The resulting CSG models are almost uniquely selected by these two principles, together with a so-called principle of "internal temporality". (Full details of the derivation can be found in [26].) To arrive at a QSG dynamics in a similar manner has so far not been possible, because of the lack of a satisfactory generalization of the condition of Bell causality to the quantum case.

The more general issue lurking here has nothing to do with quantum gravity as such, and raises questions which are worthy of study in their own right. Can one formulate the requirement of relativistic causality without invoking external agents like "observers"? Can one do so in the language of histories? And if so how? The answers, if we had them [27, 28], might remove the main obstruction blocking the first path to a theory of quantal sequential growth. (The quest for a suitable criterion of relativistic causality lies in the background of the work on quantum foundations reported on in Petros Wallden's article in this volume [29].)

Along the second path one would try to discover a quantity in the causet that would go over to the scalar curvature in the continuum limit, and having found it one would try to incorporate it into a discrete analog of the gravitational functional-integral. For a long time it was difficult to get started on this path because no estimator for the curvature was known. Now that we do have a candidate expression (based on the Synge function as described earlier), the next step would be to confirm that it really does have the Ricci scalar as its continuum limit.[12]

[11] The puzzle of why the radius of curvature is so much greater than the Planck length is often called the "flatness problem".

[12] A more "implicit" way to arrive at an analog of the scalar-curvature action has recently been suggested by Bombelli and Sverdlov [30], based on a kind of "dynamical selection" of a local frame that would allow one to

3. Direct transmission along causal links

In the introduction, I promised to return in this section to the model of propagation of a scalar field by direct transmission along causal links from source to sink. In my actual talk at the DICE conference, I devoted most of the time to this model, but since the details will appear soon in another place [5], I will limit myself here to a thumbnail sketch.

In the continuum, the field emitted by a specified source can be computed in two ways, either by solving the field equations with suitable boundary conditions or (for a free field) by folding the source in with a suitable Green function. Ordinarily one adopts the "thermodynamic" boundary condition of no incoming radiation and correspondingly employs the retarded Green function. What the model in question does is to adapt the second of these two methods to the causet.

Now for a massless scalar field ϕ in $\mathbb{M}^4$, G is particularly simple, being given (up to normalization) just by the delta-function on the future light cone, i.e. by $\delta(x^\mu x_\mu)$ in Cartesian coordinates ("Huygens' principle"). In this sense, ϕ propagates strictly between pairs of points joined by a null line segment. Now let C be a causet well approximated by (or by a portion of) $\mathbb{M}^4$. The nearest discrete analog of a null segment is perhaps a *link* of C, that is, a pair of elements $x \prec y$ related as parent and child, or in other words a pair such that the interval delimited by x and y is empty.[13] If, then, we define $G(x, y) = 1$ or 0 according as x is or is not a child of y, we obtain a natural analog of the continuum Green function; and it is perhaps not too surprising that the continuum limit of $G(x, y)$ coincides (up to normalization) with the retarded Green function in $\mathbb{M}^4$.

To complete the model, we can represent the source as simply (though not as realistically) as possible as a so called saturated chain or *path* within C,[14] and we assume that the value of $\phi(x)$ due to the source is, modulo a prefactor of the source's charge, just the sum of $G(x, y)$ over all elements y of the source. Clearly this is just the number of source elements y linked to x. Finally we can imagine that the detector is active in a specified region of spacetime (it has a certain spatial volume and is turned on for a certain time) and just outputs the integral of the field over the region. Translated into causet terms, this just means (because "volume = number") that the signal is the *sum* of $\phi(x)$ over those $x \in C$ where the detector is active.

Given this setup, it becomes relatively easy to compute the expectation value of the detector's response to a given source, and also the leading correction thereto in powers of the Planck length. At leading order, one recovers exactly the continuum expression,[15] while the first correction involves an integral over the boundary of the source region and is completely negligible for any realistic detector and source. Thus as expected, there is no change to the dispersion relations, and no loss of coherence. (Nor is there any scattering, but that conclusion seems less certain to persist in the context of a more complete model that attributes degrees of freedom to ϕ itself, and then deduces the radiated field by solving the corresponding discrete field equations.)

There remains the question of the signal fluctuations induced by the discreteness of the underlying causet – the fact that the signal is actually a sum rather than an integral. The fluctuations in ϕ are less easy to compute than its expectation-value, and not every detail has been pinned down yet. However, it seems fairly clear that – unfortunately – the fluctuations are also too tiny to show up in observations of any realistic source using any realistic detector, even if the source is at cosmological distances from us.

deploy known estimators for certain components of the Ricci *tensor* in a causet [31, 32].

[13] A slightly more general analog would be a pair $x \prec y$ such that the interval between x and y is a chain. In the (flat) continuum, this precisely characterizes a light ray.

[14] A chain is saturated when no further elements can be interpolated between its endpoints. In other words, it is "made exclusively of links".

[15] In particular, (the AC part of) the field of an oscillating charge falls off, as it should, like $1/R$ at large distances, not as $1/R^2$ as one might have concluded naively from the fact that the DC part goes like $1/R$. Interestingly, it is the Doppler shift in the number of contributing source elements that invalidates the naive reasoning.

In summary, the model is one in which the field spreads via a process of direct transmission along causal links. It seems remarkable that this law of propagation, which is probably the simplest one could have imagined had one thought directly in terms of the causal set, turns out to reproduce all the main features of the relationship between the field and its source in the continuum.

Acknowledgments

Research at Perimeter Institute for Theoretical Physics is supported in part by the Government of Canada through NSERC and by the Province of Ontario through MRI. This research was partly supported by NSF grant PHY-0404646.

References

[1] Sorkin R D 2005 Big extra dimensions make Λ too small, in *Proceedings of the Second International Workshop "DICE 2004", held September 2004, Piombino, Italy* ed H-T Elze *Braz. J. Phys.* **35** 280–283 (*Preprint* arXiv:gr-qc/0503057)

[2] See for example: Balachandran A P and Martone M 2009 Space-time from Symmetry: The Moyal Plane from the Poincaré-Hopf Algebra *Preprint* arXiv:0902.3409

[3] Bombelli L, Henson J and Sorkin R D 2006 Discreteness without symmetry breaking: a theorem *Preprint* arXiv:gr-qc/0605006

[4] Lieu R and Hillman L W 2003 The phase coherence of light from extragalactic sources: Direct evidence against first order quantum gravity fluctuations in time and space *Astrophys. J.* **585** L77–L80

[5] Dowker F, Henson J and Sorkin R D Wave propagation on a causet I: direct transmission along causal links (in preparation)

[6] Dudley R M 1965 Lorentz-invariant Markov processes in relativistic phase space *Arkiv för Matematik* **6**(14) 241–268; *do.* 1967 A note on Lorentz-invariant Markov processes *Arkiv för Matematik* **6**(30) 575–581;
Schay Jr. G 1961 *The equations of diffusion in the Special Theory of Relativity* (doctoral thesis, Princeton University)

[7] Dowker F, Henson J and Sorkin R D 2004 Quantum Gravity Phenomenology, Lorentz Invariance and Discreteness *Modern Physics Letters* A **19** 1829–1840 (*Preprint* arXiv:gr-qc/0311055)
http://www.physics.syr.edu/ sorkin/some.papers/

[8] Philpott L, Dowker F and Sorkin R D 2008 Energy-momentum diffusion from spacetime discreteness *Preprint* arXiv:0810.5591; http://www.physics.syr.edu/ sorkin/some.papers/
See also the article by Lydia Philpott in this volume.

[9] Kaloper N and Mattingly D 2006 Low energy bounds on Poincaré violation in causal set theory *Phys. Rev.* D **74** 106001 (*Preprint* arXiv:astro-ph/0607485)

[10] Sorkin R D 2009 Does Locality Fail at Intermediate Length-Scales? in *Approaches to Quantum Gravity – Towards a new understanding of space and time* ed D Oriti (Cambridge, UK: Cambridge U. Press) 26–43 (ISBN: 978-0-521-86045-1) (*Preprint* arXiv:gr-qc/0703099)
http://www.physics.syr.edu/ sorkin/some.papers/
Henson J 2006 The causal set approach to quantum gravity *Preprint* arXiv:gr-qc/0601121

[11] Sorkin R D 1987 A Modified Sum-Over-Histories for Gravity, in *Highlights in gravitation and cosmology: Proceedings of the International Conference on Gravitation and Cosmology, Goa, India, 14-19 December, 1987* ed B R Iyer, A Kembhavi, J V Narlikar and C V Vishveshwara;
see pp 184–186 in the article Brill D and Smolin L 1988 in *Workshop on quantum gravity and new directions* (Cambridge, UK: Cambridge U. Press) 183–191;
Sorkin R D 1987 On the Role of Time in the Sum-over-histories Framework for Gravity paper presented to the conference *The History of Modern Gauge Theories, held Logan, Utah, July 1987* – published in *do.* 1994 *Int. J. Theor. Phys.* **33** 523–534

[12] Ahmed M, Dodelson S, Greene P and Sorkin R D 2004 Everpresent Λ *Phys. Rev.* D **69** 103523 (*Preprint* arXiv:astro-ph/0209274); http://www.physics.syr.edu/ sorkin/some.papers/
Ahmed M 2006 First indications of causal set cosmology (doctoral dissertation, Syracuse University);
Ng Y J and van Dam H 2001 A small but nonzero cosmological constant *Int. J. Mod. Phys* D **10** 49 (*Preprint* arXiv:hep-th/9911102);
Sorkin R D 2007 Is the cosmological "constant" a nonlocal quantum residue of discreteness of the causal set type? – in *Proceedings of the PASCOS-07 Conference, held July 2007, London, England – American Institute of Physics Conference Proceedings* **957** 142–153 (ISBN: 978-0-7354-0471-7) (*Preprint* arXiv:0710.1675); http://www.physics.syr.edu/ sorkin/some.papers/

[13] Bombelli L, Lee J, Meyer D and Sorkin R D 1987 Spacetime as a Causal Set *Phys. Rev. Lett.* **59** 521–524; Bombelli L 1987 *Space-time as a Causal Set* (Ph.D. thesis, Syracuse University)
Sorkin R D 2005 Causal Sets: Discrete Gravity (Notes for the Valdivia Summer School), in *Lectures on Quantum Gravity (Series of the Centro De Estudios Científicos) – proceedings of the Valdivia Summer School, held January 2002 in Valdivia, Chile* eds A Gomberoff and D Marolf (Plenum) (*Preprint* arXiv:gr-qc/0309009);
Dowker 2005 F Causal sets and the deep structure of Spacetime, in *100 Years of Relativity – Space-time Structure: Einstein and Beyond* ed A Ashtekar (Singapore: World Scientific) 445-464 (*Preprint* arXiv:gr-qc/0508109);
Sorkin R D Geometry from order: causal sets – online article in the *Spotlights on relativity* series, maintained by the Max-Planck-Institut für Gravitationsphysik (Albert-Einstein-Institut), Potsdam http://www.einstein-online.info/en/spotlights/causal_sets/index.html;
Henson J 2006 The causal set approach to quantum gravity *Preprint* arXiv:gr-qc/0601121
[14] Dou D 1999 *Causal Sets, a Possible Interpretation for the Black Hole Entropy, and Related Topics* (Ph.D. thesis, SISSA, Trieste) (*Preprint* arXiv:gr-qc/0106024);
Dou D and Sorkin R D 2003 Black Hole Entropy as Causal Links *Foundations of Physics*, to appear (*Preprint* arXiv:gr-qc/0302009)
[15] Meyer D A 1988 *The Dimension of Causal Sets* (Ph.D. thesis, M.I.T.)
[16] Bachmat E, private communication
[17] See the contributions by Sumati Surya and David Rideout in this volume for these developments and for further references.
[18] Major S, Rideout D and Surya S 2006 On Recovering Continuum Topology from a Causal Set *Preprint* arXiv:gr-qc/0604124
[19] Contribution by Steven Johnston in this volume.
[20] Hartle J B 1991 The Quantum Mechanics of Cosmology, in *Quantum Cosmology and Baby Universes: Proceedings of the 1989 Jerusalem Winter School for Theoretical Physics* eds S Coleman *et al* (Singapore: World Scientific);
Sorkin R D 1994 Quantum Mechanics as Quantum Measure Theory *Mod. Phys. Lett.* A **9** (No. 33) 3119–3127 (*Preprint* arXiv:gr-qc/9401003)
[21] Brightwell G 1993 Models of Random Partial Orders, in *Surveys in Combinatorics, 1993 – London Math. Soc. Lecture Notes Series* **187** ed K Walker (Cambridge, UK: Cambridge U. Press) 53–83
[22] Maqbool Ahmed and David P. Rideout (unpublished).
[23] Sorkin R D 2000 Indications of causal set cosmology *Int. J. Theor. Ph.* **39** (7) 1731–1736 (an issue devoted to the proceedings of the Peyresq IV conference, held June - July 1999, Peyresq, France) (*Preprint* arXiv:gr-qc/0003043); http://www.physics.syr.edu/ sorkin/some.papers/
[24] Martin X, O'Connor D, Rideout D and Sorkin R D 2001 On the 'renormalization' transformations induced by cycles of expansion and contraction in causal set cosmology *Phys. Rev.* D **63** 084026 (*Preprint* arXiv:gr-qc/0009063); http://www.physics.syr.edu/ sorkin/some.papers/;
Ash A and McDonald P 2003 Moment problems and the causal set approach to quantum gravity *J. Math. Phys.* **44** 1666–1678 (*Preprint* arXiv:gr-qc/0209020)
[25] Brightwell G, Dowker H F, García R S, Henson J and Sorkin R D 2002 General Covariance and the 'Problem of Time' in a Discrete Cosmology in *Correlations – Proceedings of the ANPA 23 conference, held August 16-21, 2001, Cambridge, England* ed K G Bowden (London: Alternative Natural Philosophy Association) 1–17 (*Preprint* arXiv:gr-qc/0202097);
Brightwell G, Dowker F, García R S, Henson J and Sorkin R D 2003 'Observables' in Causal Set Cosmology Phys. Rev. D **67** 084031 (*Preprint* arXiv:gr-qc/0210061)
[26] Rideout D P 2001 *Dynamics of Causal Sets* (Ph.D. thesis, Syracuse University);
Rideout D P and Sorkin R D 2000 A Classical Sequential Growth Dynamics for Causal Sets *Phys. Rev.* D **61** 024002 (*Preprint* arXiv:gr-qc/9904062)
[27] Henson J 2005 Comparing causality principles *Stud. Hist. Philos. Mod. Phys.* **36** 519–543 (*Preprint* arXiv:quant-ph/0410051)
[28] Craig D, Dowker F, Henson J, Major S, Rideout D and Sorkin R D 2007 A Bell Inequality Analog in Quantum Measure Theory *J. Phys. A: Math. Theor.* **40** 501–523 (*Preprint* arXiv:quant-ph/0605008) http://www.physics.syr.edu/ sorkin/some.papers/
[29] Contribution by Petros Wallden in this volume.
[30] See the contribution by Luca Bombelli and Roman Sverdlov in this volume, also *Preprint* arXiv:0905.1506
[31] Myrheim J 1978 Statistical geometry *CERN preprint TH-2538* (available from SPIRES)
[32] Gibbons G W and Solodukhin S N 2007 The Geometry of Small Causal Diamonds *Preprint* arXiv:hep-th/0703098

Dynamics for causal sets with matter fields: a Lagrangian-based approach

Roman Sverdlov[1] **and Luca Bombelli**[2,3]

[1] Physics Department, University of Michigan,
450 Church Street, Ann Arbor, MI 48109-1040, USA
[2] Department of Physics and Astronomy, University of Mississippi,
108 Lewis Hall, University, MS 38677-1848, USA
[3] Departament de Física Fonamental, Universitat de Barcelona,
Av. Diagonal 647, 08028 Barcelona, Spain

E-mail: sverdlov_roman@hotmail.com, bombelli@olemiss.edu

Abstract. We present a framework for the dynamics of causal sets and coupled matter fields, which is a simplification and generalization of an approach we recently proposed. Given a set of fields including the gravitational one, the main step in implementing our proposal consists in writing their continuum-based action using as variables for the spacetime geometry the causal order and volume element. One then discretizes the resulting expression, with a procedure designed to maintain covariance. After a discussion of the general framework, we treat in detail the case of scalar fields, Yang-Mills gauge fields and the gravitational field.

1. Introduction

Despite the fact that the causal set program for quantum gravity [1] has been around for more than 20 years [2, 3, 4], identifying an appropriate formulation of the dynamics of the theory remains an open issue. Such a formulation would allow us to recover a smooth Lorentzian geometry with matter fields in a suitable continuum approximation, and would have a semiclassical limit in which general relativity coupled to those matter fields, including possible quantum corrections, is recovered. Some proposals have been made, most notably the classical sequential growth dynamics models of Rideout and Sorkin [5], a class of stochastic growth models defined by the fact that they satisfy causal-set versions of the general covariance and causality requirements. While a number of interesting results have been obtained on these models, no way to recover the continuum geometry and dynamics from them is known, and no proposal has been made for how to include matter fields in them.

Here we will present a different approach to the dynamics of causal sets, one which is not based on general principles, but rather tries to mimick directly the continuum dynamics by using a discretized expression for the action, written in terms of variables that are meaningful for causal sets. Since a causal set is simply a locally finite partially ordered set, the only variables that characterize it are the *number* of elements in each region and the *order* in which they come. Thus, as far as gravity is concerned, our tasks are to express the Einstein-Hilbert action in terms of volumes of (finite) spacetime regions and causal relations, the continuum counterparts to the counting measure and the partial order, respectively, and then to discretize the result, making sure that we end up with a covariant expression.

But one of our main motivations for proposing this approach is that the action for other fields can be treated in the same way, as we will show below for scalar and gauge fields. The reason for including matter is not simply completeness, or even a desire to develop the phenomenology of the theory, but the belief that ultimately the gravitational field will have to be treated on the same footing as the other forms of matter from the beginning. Since all matter fields will be coupled to the geometry at the causal-set level, and most causal sets are not well approximated by smooth manifolds, it may be that not only the metric structure of spacetime, but even the possibility of defining topological and differentiable structures depends on the matter fields, that their presence is necessary in order to understand how manifoldlike causal sets emerge.

In section 2 we discuss the general framework and introduce the concept of Lagrangian generator, a building block from which suitable forms for the action of continuum fields can be written; section 3 treats the scalar field case and section 4 the case of a charged scalar field, i.e., one coupled to a gauge field; in section 5 we discuss the dynamics of the (Yang-Mills) gauge field itself, and in section 6 that of the gravitational field; the concluding section 7 contains some closing remarks. This article is based on the talk by LB at the DICE2008 conference, at which previous work along these lines [6] was presented. After the talk was given, however, RS developed an improved version of this approach, and we felt that it would be more useful for this writeup to reflect the recent developments. For this reason, the material in this paper reflects to a large extent the content of one chapter of the PhD dissertation by RS.

2. Lagrangians, pre-Lagrangians and Lagrangian generators

It turns out that the causal-set versions of the Lagrangians all look somewhat complicated, but they are all complicated in the same way. To emphasize the pattern, we introduce the concept of Lagrangian generator, together with a prescription for going from a Lagrangian generator to the corresponding Lagrangian. Our aim in doing that is to ensure that Lagrangian generators for all known fields are simple-looking and the actual Lagrangians, as complicated as they might appear, can be "read off" from the expression for the Lagrangian generator.

The key observation to guide us is the form in which one can write the scalars that appear in the usual continuum Lagrangian densities for the Klein-Gordon scalar, electromagnetic and gravitational fields. As will be shown in the next four sections, if $\alpha(p,q) := J^+(p) \cap J^-(q)$ is the Alexandrov set defined by two causally-related points $p \prec q$ in Minkowski space, then for any scalar fields and electromagnetic potentials varying linearly in inertial coordinates, respectively, the Lagrangian density at a point $x \in \alpha(p,q)$ can be written as

$$k_{\rm sc}\, \partial^\mu\phi\, \partial_\mu\phi = \frac{1}{\tau^2(p,q)} \left(\phi(q) - \phi(p)\right)^2 + \frac{E_{\rm sc}}{\tau^{2d+2}(p,q)} \int_{\alpha(p,q)} {\rm d}^d r\, {\rm d}^d s\, (\phi(r) - \phi(s))^2 \tag{2.1}$$

$$\begin{aligned} k_{\rm em}\, F^{\mu\nu} F_{\mu\nu} = {} & \frac{1}{\tau^{d+2}(p,q)} \int_{\alpha(p,q)} {\rm d}^d r\, \big(a(p,r) + a(r,q) + a(q,p)\big) + \\ & + \frac{E_{\rm em}}{\tau^{3d+2}(p,q)} \int_{\alpha(p,q)} {\rm d}^d r\, {\rm d}^d s\, {\rm d}^d t\, \big(a(r,s) + a(s,t) + a(t,r)\big)\ , \end{aligned} \tag{2.2}$$

for appropriately adjusted pairs of coefficients (k, E), while for any metric whose expansion in Riemann normal coordinates only contains quadratic non-trivial terms (the lowest possible order, so we will still refer informally to this situation as the "linear case"),

$$k_{\rm gr}\, R = \frac{V(\alpha(p,q))}{\tau^{d+2}(p,q)} + \frac{E_{\rm gr}}{\tau^{2d+2}(p,q)} \left(\int_{\alpha(p,q)} {\rm d}^d r\, \big[V(\alpha(p,r)) + V(\alpha(r,q))\big] \right), \tag{2.3}$$

for appropriately adjusted values for $(k_{\rm gr}, E_{\rm gr})$. Since $\alpha(p,q)$ is defined by causal relations alone without a direct reference to the continuum structure, and in particular observing that the right-hand sides of these integrals don't involve spacetime tensor indices, the replacement of integrals by sums makes it easy to generalize these expressions to a causal set.

A point that needs to be emphasized is that the expressions (2.1)–(2.3) are covariantly defined scalars at a point, despite the fact that they involve the use of one Alexandrov set, as long as E is chosen to have the correct, dimension-dependent value. This, however, brings up a problem when we take these expressions over to the causal-set context: in light of the fact that the theory is formulated for arbitrary causal sets and not just manifoldlike ones, if E was viewed as a constant, it would be an extremely lucky coincidence that its value in the fundamental theory happened to be precisely the one needed for the exact cancellation of the non-covariant terms in the special case of a four-dimensional manifold. This, of course, is not satisfactory.

For this reason, we will view E as a variable, subject to some physical laws that adjust its value to what it should be at any given point. The general idea is that the properties of a manifoldlike partial order determine the dimensionality of the manifold in which it can be faithfully embedded [1], which in turn determines the value of E. More specifically, while in the linear manifoldlike scenario the only field E depends on is gravity, as will be seen later in the non-linear case there is a slight dependence on the non-linear behavior of other fields, albeit very small. In the non-manifoldlike scenario, however, the dependence on non-gravitational fields might be arbitrarily large, just like the dependence on gravity would be. Consider a general theory for a set of fields of interest $F \in \mathcal{F}$, with a continuum Lagrangian density $\mathcal{L}(F, x, E)$ that can be obtained from expressions of the type (2.1)–(2.3). Then

$$E = E(F, x) \tag{2.4}$$

is a function both of the point x and of the value of the fields $F \in \mathcal{F}$ in its neighborhood. Notice that since E is not an independent field, a path integral for the field F will not include a separate integration over E. Rather, it will be of the form

$$\int \mathcal{D}F \exp\left(\mathrm{i} \int \mathrm{d}^d x\, \mathcal{L}\big(F, x, E(F, x)\big)\right). \tag{2.5}$$

Let us now discuss how $E(F, x)$ is determined. The definition of E for a general causal set requires a notion of degree of relativistic non-covariance for an expression such as (2.1)–(2.3), which we will denote generically by $\mathcal{L}(F, E, p, q)$. If we identify a reference frame with the axis of an Alexandrov set (one defined by two points on the t-axis of the reference frame), then the degree of non-covariance represents the extent to which the value of $\mathcal{L}(F, E, p, q)$ depends on the choice of Alexandrov set $\alpha(p, q)$, and can be quantified as the range of variation of the values of $\mathcal{L}(F, E, p, q)$ as p and q are allowed to vary.

In the case of linear fields, the non-covariance is zero as long as E is appropriately adjusted. In a curved Lorentzian geometry we can still have vanishing non-covariance, provided we go to the infinitesimal-$\alpha(p, q)$ limit. In the case of fields that can be seen as a discretized continuum it is not possible to select an infinitesimal region and the best one can do is constrain either the timelike length $\tau(p, q)$ or the volume $V(\alpha(p, q))$ between p and q to have a fixed (small) value, which means that some degree of non-covariance will appear, depending on the details of the discretization procedure; as long as the fields are linear or well-behaved functions that can be considered as slowly-varying, for an appropriately chosen E the non-covariance should be small.

However, in the case of a general causal set, most of the assumptions that are made for the manifold can no longer be trusted, which means that it is possible for the degree of non-covariance to be large no matter what E is selected to be. However, it is still possible to *formally* select E in such a way that it minimizes the degree of non-covariance, even though the minimum might be very large.

One feature of Lorentzian geometries that complicates this procedure is that if the manifold is not compact the set of pairs (p, q) may not be compact either, even when one imposes a condition on how far p and q are, so fields that are slowly varying in one Alexandrov set may appear to vary uncontrollably in another, highly boosted one that occupies a stretched region "near the light cone of point p". Therefore, if all possible Alexandrov sets $\alpha(p,q)$ with $\tau(p,q) = \tau_0$ were considered, the value of E might have nothing to do with the one it would have had in the linear case. For that reason, we constrain the varying Alexandrov set to vary within the boundaries of some other, larger "fiducial" Alexandrov set $\alpha(P,Q)$, with

$$\tau_1 = \tau(p,q) < \tau(P,Q) = \tau_2\,. \tag{2.6}$$

The variation, then, is defined by

$$\mathrm{var}_{\tau_1}(\mathcal{L},F,E,P,Q) := \max\{\mathcal{L}(F,E,p,q) \mid P \prec p \prec q \prec Q \;\wedge\; \tau(p,q) = \tau_1\}\,, \tag{2.7}$$

and for any point x in a causal set, $\alpha_{\tau_1,\tau_2}(F,x)$ is defined as the set of triples (E,P,Q) for which $\mathrm{var}_{\tau_1}(\mathcal{L},F,E,P,Q)$ is minimized with the constraint that $\tau(P,Q) = \tau_2$:

$$\begin{aligned}\alpha_{\tau_1,\tau_2}(F,x) = \{(E,P,Q) \mid \tau(P,Q) = \tau_2 \;\wedge\; \forall\, E',P',Q' \text{ with } \tau(P',Q') = \tau_2,\\ \mathrm{var}_{\tau_1}(\mathcal{L},F,E',P',Q') \geq \mathrm{var}_{\tau_1}(\mathcal{L},F,E,P,Q)\}\,.\end{aligned} \tag{2.8}$$

Typically, $\alpha_{\tau_1,\tau_2}(F,x)$ is a one-element set, and the Lagrangian at x can simply be taken to be $\mathcal{L}(F,E,P,Q)$, where (E,P,Q) is the unique element of that set. But in order to formally accommodate the cases where $\alpha_{\tau_1,\tau_2}(F,x)$ has more than one element, the Lagrangian at x is formally defined as the average of the above over all elements of $\alpha_{\tau_1,\tau_2}(F,x)$:

$$\mathcal{L}_{\tau_1,\tau_2}(F,x) = \frac{1}{\sharp\{\alpha_{\tau_1,\tau_2}(F,x)\}} \sum_{(E,P,Q)\in\alpha_{\tau_1,\tau_2}(F,x)} \mathcal{L}(F,E,P,Q)\,, \tag{2.9}$$

where $\sharp\{\,\}$ denotes the cardinality of a set.

Now, dropping the issue of the value of E and the selection of p and q for a second, let's switch gears and go back to the expression for $\mathcal{L}(F,E,p,q)$, which, from now on, will be referred to as a "pre-Lagrangian". As the examples in the beginning of this section illustrate, the expressions for $\mathcal{L}(F,E,p,q)$ for scalar, gauge, and gravitational fields look similar, in the sense that they are all linear combinations of a function of p and q and some form of integral of the same function over the interior of $\alpha(p,q)$. So the natural question arises: why do Lagrangians take this particular form and not some other one? That question is answered by introducing the concept of Lagrangian generator and formally defining a procedure of going from a Lagrangian generator to a pre-Lagrangian in such a way that the above mentioned linear combination arises in a natural way if one formally follows the steps of the procedure.

For simplicity, let's consider first an expression for a pre-Lagrangian of the general type (2.3),

$$\mathcal{L}_{\mathcal{J}}(F,E,p,q) = \mathcal{J}(p,q) + \frac{E}{\tau^d(p,q)} \int \mathrm{d}^d r\, \mathcal{J}(p,r)\,, \tag{2.10}$$

where $\mathcal{J}$ is a real-valued function depending on the set of fields F, which maps $(p,q) \mapsto \mathcal{J}(F,p,q) \in \mathbb{R}$. After the selection of E and q will be made, the value of the pre-Lagrangian for that specific selection will be identified with the actual Lagrangian density at p. To put (2.10) in a useful form, define two maps f, g: $S^3 \to S^2$, where here S is any set, as follows:

$$f(a,b,c) = (a,b)\,, \qquad g(a,b,c) = (a,c)\,. \tag{2.11}$$

With these definitions,

$$\int_{\alpha(p,q)} \mathrm{d}^d r\, \mathcal{J}(F,f(p,q,r)) = \int_{\alpha(p,q)} \mathrm{d}^d r\, \mathcal{J}(F,p,q) = V(\alpha(p,q))\, \mathcal{J}(F,p,q) \tag{2.12}$$

and

$$\int_{\alpha(p,q)} \mathrm{d}^d r\, \mathcal{J}(F, g(p,q,r)) = \int_{\alpha(p,q)} \mathrm{d}^d r\, \mathcal{J}(F,p,r)\,, \tag{2.13}$$

and a pre-Lagrangian equivalent to (2.10) in the continuum is given by

$$\mathcal{L}_{\mathcal{J}}(F,E,p,q) = \frac{1}{V(\alpha(p,q))} \int_{\alpha(p,q)} \mathrm{d}^d r\, \big[\mathcal{J}\big(F, f(p,q,r)\big) + E\,\mathcal{J}\big(F, g(p,q,r)\big)\big]\,, \tag{2.14}$$

because $V(\alpha(p,q))$ and τ^d are proportional, through a constant k_d (see (3.6) below) that can be absorbed into E. In fact, since $\tau(p,q) = \tau_1$ will also be a constant, $V(\alpha(p,q))$ may be omitted, and in order to preserve time-reversal symmetry, we would like to replace $\mathcal{J}(F, f(p,q,r))$ and $\mathcal{J}(F, g(p,q,r))$ by $\mathcal{J}(F, f(p,q,r)) + \mathcal{J}(F, f(q,p,r))$ and $\mathcal{J}(F, g(p,q,r)) + \mathcal{J}(F, g(q,p,r))$, respectively, up to an unimportant factor of 2. Thus, the pre-Lagrangian can be written as

$$\begin{aligned}\mathcal{L}_{\mathcal{J}}(F,E,p,q) = \int_{\alpha(p,q)} \mathrm{d}^d r\, \Big(&\mathcal{J}\big(F, f(p,q,r)\big) + \mathcal{J}\big(F, f(q,p,r)\big) + \\ &+ E\,\big[\mathcal{J}\big(F, g(p,q,r)\big) + \mathcal{J}\big(F, g(q,p,r)\big)\big]\Big)\,.\end{aligned} \tag{2.15}$$

The Alexandrov-set function $\mathcal{J}(F,p,q)$ will be called a "Lagrangian generator" for the pre-Lagrangian (2.15). It is the basic object in the dynamics of the field F, and needs to be defined without reference to spacetime tensors or a differentiable structure on the underlying set.

More generally, in order to allow $\mathcal{L}_{\mathcal{J}}$ to include multiple integrals such as in (2.1) and (2.2), $f(p,q,r)$ and $g(p,q,r)$ should be generalized to $f(p,q,r_1,...,r_n)$ and $g(p,q,r_1,...,r_n)$. For the sake of completeness and convenience to the reader, our final definition of Lagrangian generator $\mathcal{J}$ will include both the prescription for going from the Lagrangian generator to the pre-Lagrangian $\mathcal{L}_{\mathcal{J}}(F,E,p,q)$ just discussed, and the transition from the pre-Lagrangian to the actual pointwise Lagrangian $\mathcal{L}_{\mathcal{J}}(F,x)$, discussed earlier in this section:

Definition: Let $\mathcal{F}$ be the set of possible distributions of a set of field of interest. A Lagrangian generator is a triple $(\mathcal{J}, f, g)$ with $\mathcal{J}\colon \mathcal{F} \times S^n \to \mathbb{R}$, and $f,\, g\colon S^{2+m} \to S^n$, with $n \le 2+m$. The pre-Lagrangian corresponding to this Lagrangian generator is $\mathcal{L}_{\mathcal{J}}\colon \mathcal{F} \times \mathbb{R} \times S^2 \to \mathbb{R}$, given by

$$\begin{aligned}\mathcal{L}_{\mathcal{J}}(F,E,p,q) = \sum_{p \prec r_i \prec q} \Big(&\mathcal{J}\big(F, f(p,q,r_1,...,r_m)\big) + \mathcal{J}\big(F, f(q,p,r_1,...,r_m)\big) + \\ &+ E\big[\mathcal{J}\big(F, g(p,q,r_1,...,r_m)\big) + \mathcal{J}\big(F, g(q,p,r_1,...,r_m)\big)\big]\Big)\,,\end{aligned} \tag{2.16}$$

in the causal set case, with the sums replaced by integrals in the continuum case. The "variation" of this pre-Lagrangian in $\alpha(P,Q)$ is the real-valued function defined by

$$\mathrm{var}_{\tau_1}(\mathcal{J},F,E,P,Q) = \max\{\mathcal{L}_{\mathcal{J}}(F,E,p,q) \mid P \prec p \prec q \prec Q\ \wedge\ \tau(p,q) = \tau_1\}\,, \tag{2.17}$$

and for any $x \in S$, the (F,τ_1,τ_2)-based fiducial neighborhood of x is given by

$$\begin{aligned}\alpha_{\tau_1,\tau_2}(F,x) = \{(E,P,Q) \mid \tau(P,Q) = \tau_2\ \wedge\ &\forall E',P',Q' \text{ with } \tau(P',Q') = \tau_2, \\ &\mathrm{var}_{\tau_1}(\mathcal{L},F,E',P',Q') \ge \mathrm{var}_{\tau_1}(\mathcal{L},F,E,P,Q)\}\,.\end{aligned} \tag{2.18}$$

Finally, the pointwise Lagrangian density corresponding to $\mathcal{J}$ is given by

$$\mathcal{L}_{\mathcal{J},\tau_1,\tau_2}(F,x) = \frac{1}{\sharp\{\alpha_{\tau_1,\tau_2}(F,x)\}} \sum_{E,P,Q \in \alpha(x)} \mathcal{L}(F,E,P,Q)\,. \tag{2.19}$$

Let us now see what this expression becomes for some important types of fields.

3. Scalar fields

The Lagrangian generator for a scalar field ϕ is given by $(\mathcal{J}, f, g)$, where

$$\mathcal{J}(\phi, r, s) = \big(\phi(r) - \phi(s)\big)^2 - \tfrac{1}{2}\, m^2 \phi^2(r) \tag{3.1}$$

$$f(r_1, r_2, r_3, r_4) = (r_1, r_2)\,, \qquad g(r_1, r_2, r_3, r_4) = (r_3, r_4)\,. \tag{3.2}$$

For reasons that will soon become apparent, m is not the actual mass, although it is related to it. In fact, m is assumed to be very small, of order τ. The above expression implies that the pre-Lagrangian for a scalar field is given by

$$\mathcal{L}(\phi, E, p, q) = \int \mathrm{d}^d r\, \mathrm{d}^d s\, \big(\mathcal{J}(f(p,q,r,s)) + E\, \mathcal{J}(g(p,q,r,s))\big) \tag{3.3}$$

$$= \int \mathrm{d}^d r\, \mathrm{d}^d s\, \big(\mathcal{J}(\phi, p, q) + E\, \mathcal{J}(\phi, r, s)\big) = \mathcal{J}(\phi, p, q)\, V^2(\alpha(p,q)) + E \int \mathrm{d}^d r\, \mathrm{d}^d s\, \mathcal{J}(\phi, r, s)\,.$$

By remembering that the volume of an n-dimensional ball is

$$V(\text{ball}) = \frac{2\pi^{n/2}}{n\,\Gamma(n/2)}\, r^n\,, \tag{3.4}$$

we obtain, for the volume of an Alexandrov set in Minkowski space,

$$V(\alpha(p,q)) = \frac{2\pi^{(d-1)/2}}{(d-1)\,\Gamma((d-1)/2)} \int_{-\tau/2}^{\tau/2} \Big(\frac{\tau}{2} - |t|\Big)^{d-1} \mathrm{d}t = k_d\, \tau^d\,, \tag{3.5}$$

where

$$k_d = \frac{2\pi^{(d-1)/2}}{(d-1)\,\Gamma(2^{d-2} d\,(d-1))}\,. \tag{3.6}$$

Substituting the above expression for volume, along with the Lagrangian generator $\mathcal{J}$, in the expression for $\mathcal{L}$, we obtain

$$\begin{aligned}\mathcal{L}(\phi, E, p, q) &= k_d^2\, \tau^{2d}(p,q) \big[\big(\phi(q) - \phi(p)\big)^2 - \tfrac{1}{2}\, m^2 \big(\phi^2(p) + \phi^2(q)\big)\big] \\ &\quad + E \int \mathrm{d}^d r\, \mathrm{d}^d s\, \big[\big(\phi(r) - \phi(s)\big)^2 - m^2 \phi^2(r)\big]\,.\end{aligned} \tag{3.7}$$

We would now like to compute the Lagrangian density, if the spacetime is assumed to be flat Minkowskian and ϕ is assumed to be linear.

Let's start from the mass term. It is given by

$$\mathcal{L}_m(\phi, E, p, q) = \frac{m^2\, k_d^2}{2}\, \tau^{2d}(p,q)\, \big(\phi^2(p) + \phi^2(q)\big) + E\, m^2 \int \mathrm{d}^d r\, \mathrm{d}^d s\, \phi^2(r)\,. \tag{3.8}$$

The above expression tells us that the leading order of the mass term is $m^2\, \tau^{2d}$. When we will get to the kinetic term, it will be shown that the order of magnitude that we are interested in is τ^{2d+2}. Thus, if m is assumed to be of the order of τ, then the leading order is τ^{2d+2} which coincides with the leading order for the kinetic term.

This means that, as far as the mass term is concerned, we can throw away all the higher-order terms. This can be done by using the approximation $\phi \approx \phi_0$, which tells us that

$$\begin{aligned}\mathcal{L}_m(\phi, E, p) &\approx m^2\, k_d^2\, \tau^{2d}(p,q)\, \phi_0^2 + E\, m^2\, \phi_0^2\, V^2(\alpha(p,q)) \\ &= (1+E)\, m^2\, k_d^2\, \phi_0^2\, \tau^{2d}(p,q)\,.\end{aligned} \tag{3.9}$$

Now let's look at the kinetic term. The linearity assumptions imply that in the integrand

$$\big(\phi(r) - \phi(s)\big)^2 = (r^\mu - s^\mu)\,(r^\nu - s^\nu)\,\partial_\mu\phi\,\partial_\nu\phi\,. \tag{3.10}$$

Consider a coordinate system in which the t axis passes through p and q, while the origin lies midway between these points. Denoting $\tau(p,q)$ by τ,

$$p = (-\tau/2, 0, 0, 0)\,, \qquad q = (\tau/2, 0, 0, 0)\,. \tag{3.11}$$

In this coordinate system,

$$\int_{\alpha(p,q)} \mathrm{d}^d r\, r^\mu = 0\,, \tag{3.12}$$

since the above integrand is antisymmetric with respect to the center of the Alexandrov set. By slicing the Alexandrov set into $t =$ constant balls, we get

$$\int_{\alpha(p,q)} (x^0)^2\, \mathrm{d}^d r = \frac{2\pi^{(d-1)/2}}{(d-1)\,\Gamma((d-1)/2)} \int_{-\tau/2}^{\tau/2} t^2 \Big(\frac{\tau}{2} - |t|\Big)^{d-1} \mathrm{d}t = I_{d0}\, \tau^{d+2}\,, \tag{3.13}$$

where

$$I_{d0} = \frac{2\pi^{(d-1)/2}}{(d-1)\,\Gamma(2^{d-1}(d-1)\,d\,(d+1)\,(d+2))}\,. \tag{3.14}$$

Furthermore, it can be shown that

$$\int_{\alpha(p,q)} (x^k)^2\, \mathrm{d}^d r = I_{d1}\, \tau^{d+2}\,, \tag{3.15}$$

where, by cylindrical symmetry, the coefficient is the same for each k and is given as

$$I_{d1} = \frac{2\pi^{d/2-1}}{(d-2)\,\Gamma(d/2-1)}\,. \tag{3.16}$$

Substituting these expressions into the integral in $\mathcal{L}_{\rm kin}$ we obtain

$$\begin{aligned} \int \mathrm{d}^d r\, \mathrm{d}^d s\, \big(\phi(r) - \phi(s)\big)^2 &= \Big(\int \mathrm{d}^d s\Big) \sum_{\mu=0}^{d-1} \mathrm{d}^d r\, (r^\mu)^2 + \Big(\int \mathrm{d}^d r\Big) \sum_{\mu=0}^{d-1} \mathrm{d}^d s\, (s^\mu)^2 \\ &= 2\, k_d\, \tau^{2d+2} \big[\big(I_{d0} + I_{d1}(d-1)\big)\,(\partial_0\phi)^2 - I_{d1}\,(d-1)\,\partial^\mu\phi\,\partial_\mu\phi\big]\,. \end{aligned} \tag{3.17}$$

Thus, the kinetic term of the pre-Lagrangian is

$$\mathcal{L}_{\rm kin}(\phi, E, p, q) = k_d\, \tau^{2d+2} \big\{(\partial_0\phi)^2 \big[k_d + 2\, E_d\, \big(I_{d0} + I_{d1}\,(d-1)\big)\big] - 2\, E_d\, I_{d1}\,(d-1)\,\partial^\mu\phi\,\partial_\mu\phi\big\}\,. \tag{3.18}$$

Switching from the coordinate system in which the t axis passes through p and q to an arbitrary one, the result becomes

$$\begin{aligned} \mathcal{L}_{\rm kin}(\phi, E, p, q) = {}& k_d\, \tau^{2d}(p,q)\,(q^\mu - p^\mu)\,(q^\nu - p^\nu)\,\partial_\mu\phi\,\partial_\nu\phi\, \big[k_d + 2\, E_d\, \big(I_{d0} + I_{d1}\,(d-1)\big)\big] + \\ & - 2\, k_d\, E_d\, I_{d1}\, \tau^{2d+2}(d-1)\,\partial^\mu\phi\,\partial_\mu\phi\,. \end{aligned} \tag{3.19}$$

If the choice of points p and q varies with the constraints that both the midpoint 0 between p and q as well as the Lorentzian distance between the two points are fixed, then $\mathcal{L}(\phi, E, p, q)$ undergoes a variation of order τ^{2d+2} due to the $(q^\mu - p^\mu)\,(q^\nu - p^\nu)\,\partial_\mu\phi\,\partial_\nu\phi$ term. The mass term, on the other hand, only gives variations to higher orders. Thus, if variations of orders higher than τ^{2d+2} are neglected, then the variation can be "minimized", or in this case set to 0, if the first term in (3.19) (depending on the "tilting" of the axis of the Alexandrov set, while the second one is constant) vanishes, or

$$k_d + 2\, E_d \left(I_{d0} + I_{d1}(d-1)\right) = 0\,, \tag{3.20}$$

which determines the value of E_d:

$$E_d = -\frac{k_d}{2\left(I_{d0} + I_{d1}(d-1)\right)}\,. \tag{3.21}$$

Substituting this into the expression for the Lagrangian gives

$$\mathcal{L} = \frac{I_{d1}\, k_d^2\,(d-1)}{I_{d0} + I_{d1}\,(d-1)}\, \tau^{2d+2}\, \partial^\mu\phi\, \partial_\mu\phi - m^2\, \phi^2\, k_d^2\, \tau^{2d} \left(1 - \frac{k_d}{2\left(I_{d0} + I_{d1}\,(d-1)\right)}\right). \tag{3.22}$$

This Lagrangian can be rewritten as

$$\mathcal{L} = \hbar_d^2\, \frac{v_0}{2}\, \partial^\mu\phi_d\, \partial_\mu\phi_d - \frac{v_0\, m_d^2}{2}\, \phi_d^2\,, \tag{3.23}$$

where v_0 is the volume taken up by one point and

$$\hbar_d\phi_d = \frac{k_d}{v_0}\, \tau^{d+1} \sqrt{\frac{I_{d1}\,(d-1)}{I_{d0} + I_{d1}\,(d-1)}}\; \phi \tag{3.24}$$

$$m_d = \frac{k_d}{v_0}\, \tau^d \sqrt{1 - \frac{k_d}{2\left(I_{d0} + I_{d1}\,(d-1)\right)}}\; m\,. \tag{3.25}$$

In future sections other fields will be similarly scaled, but the coefficients will be different from field to field. At first this might seem wrong since the kinetic terms of all Lagrangians have the same coefficient 1 in standard quantum field theory. But it can be easily shown that this difference does not amount to anything but a change of overall factor:

$$\begin{aligned} &\int [\mathcal{D}\phi_1 ... \mathcal{D}\phi_n]\, \exp\left\{\mathrm{i}\, S(\phi_1) + ... + \mathrm{i}\, S(\phi_n)\right\} \\ &= \rho_{1d} \cdots \rho_{nd} \int [\mathcal{D}\phi_{1d} ... \mathcal{D}\phi_{nd}]\, \exp\left\{\mathrm{i}\, S(\phi_{1d}/\rho) + ... + \mathrm{i}\, S(\phi_{nd}/\rho)\right\}, \end{aligned} \tag{3.26}$$

where

$$\phi_{kd} = \rho_{kd}\, \phi_k\,. \tag{3.27}$$

4. Charged scalar field

Consider a charged spin-0 particle, described by a set of complex scalar fields $\phi = (\phi_1, ..., \phi_n)$ coupled to a $\mathrm{SU}(n)$ gauge field. (Notice that in this approach to the dynamics of matter fields in causal set theory, although it will be assumed that spacetime is discretized, the internal degrees of freedom will still have a continuous invariance group.) In the continuum, the dynamics of such a field can be described starting with the matter Lagrangian density

$$\mathcal{L}_\mathrm{m}(g_{\mu\nu}, \phi, A_\mu; x) = \tfrac{1}{2}\, |g|^{1/2} \left[g^{\mu\nu}\, (D_\mu\phi)^\dagger\, (D_\nu\phi) - m^2\, \phi^\dagger\phi\right], \tag{4.1}$$

where the gauge-covariant derivative is defined as usual by $D_\mu\phi^a := \partial_\mu\phi^a + \mathrm{i}\, e\, A_\mu{}^a{}_b\, \phi^b$, and $A_\mu = A_\mu{}^k\, T^k$ is the Lie-algebra-valued connection form representing the gauge field on a differentiable manifold. (Here, Latin indices a, b, ..., are Lie-algebra tensor indices, while k, l, ..., label elements of the basis T^k of the Lie algebra.) In the causal set context, the scalar field will be simply replaced by a corresponding field defined at each causal set element, but to write down the action it is important to specify what variables will replace A_μ.

A gauge field is defined in terms of holonomies, where the word "holonomy" refers to the group transformation corresponding to the parallel transport of a Lie-algebra-valued field such as ϕ^a between two points p and q. In a differentiable Lorentzian manifold M (of dimension d), holonomy is defined as a function $a : M \times M \to \mathrm{TSU}(n)$, where $\mathrm{TSU}(n)$ is the tangent bundle to $\mathrm{SU}(n)$, and consists therefore of all traceless $n \times n$ tensors. This map assigns to any two elements p, $q \in M$ the holonomy of A_μ along the geodesic segment $\gamma(p,q)$ connecting p and q in M, given by the path-ordered exponential

$$a(p,q) = \int_{\gamma(p,q)} A_\mu{}^k\, T^k\, \mathrm{d}x^\mu\,, \tag{4.2}$$

in terms of which the expression $D_\mu\phi(x)$ appearing in the scalar field Lagrangian arises from the leading-order term in the expansion of the expression $(1 + a(x,y))\,(\phi(y) - \phi(x))$.

This means that the causal set version of the charged scalar field Lagrangian can be obtained by making some simple substitutions in the one obtained in the previous section for the Klein-Gordon field. Thus, if the gauge field a is assumed to be fixed and not subject to any Lagrangians, then the Lagrangian generator for a matter field $\phi \in \mathbb{R}^n$ interacting with a is given by $(\mathcal{J}, f, g)$, where

$$\mathcal{J}_{\rm sc}(\phi, a, r, s) = |\phi(r) - a(r,s)\,\phi(s)|^2 + \tfrac{1}{8}\, m^2\, (\phi^*(r) + \phi^*(s))\, \big(\phi(r) + \phi(s)\big) \tag{4.3}$$

and

$$f_{\rm sc}(r_1, r_2, r_3) = (r_1, r_3)\,, \qquad g_{\rm sc}(r_1, r_2, r_3) = (r_1, r_2)\,. \tag{4.4}$$

However, as discussed in the next section, a itself is subject to a Lagrangian generator given by

$$\begin{aligned} &\mathcal{J}_{\rm YM}(a, r_1, r_2, r_3) = \mathrm{tr}[(a(r_1,r_2) + a(r_2,r_3) + a(r_3,r_1))^2]\,, \\ &f_{\rm YM}(r_1, r_2, r_3, r_4, r_5) = (r_3, r_4, r_5)\,, \\ &g_{\rm YM}(r_1, r_2, r_3, r_4, r_5) = (r_1, r_2, r_3)\,. \end{aligned} \tag{4.5}$$

In order for the theory to possess an $\mathrm{SU}(n)$ symmetry in the setup, these two Lagrangian generators are combined into one as $(\mathcal{J}_{\rm tot}, f, g)$,

$$\begin{aligned} &\mathcal{J}_{\rm tot}(a, r_1, r_2, r_3) = \mathcal{J}_{\rm sc}(\phi, a, r_1, r_2) + \mathcal{J}_{\rm YM}(a, r_1, r_2, r_3)\,, \\ &f_{\rm tot}(r_1, r_2, r_3, r_4, r_5) = f_{\rm YM}(r_1, r_2, r_3, r_4, r_5)\,, \\ &g_{\rm tot}(r_1, r_2, r_3, r_4, r_5) = g_{\rm YM}(r_1, r_2, r_3, r_4, r_5)\,. \end{aligned} \tag{4.6}$$

5. Yang-Mills field

In this section, the main goal is to express the Yang-Mills Lagrangian density,

$$\mathcal{L}_{\rm YM}(g_{\mu\nu}, A_\mu; x) = \tfrac{1}{2}\, |g|^{1/2}\, \mathrm{tr}(F_{\mu\nu} F^{\mu\nu})\,, \tag{5.1}$$

in terms of the holonomies for the gauge field introduced in the previous section, as well as variables describing the geometry that are meaningful for causal set [7]. Once this is done, the Lagrangian density can be easily rewritten in the discrete setting.

In the causal set context, a gauge field is defined as a map $a\colon S^2 \to \mathrm{TSU}(n)$. The Lagrangian generator for the gauge field is given by $(\mathcal{J}, f, g)$, where

$$\mathcal{J}(a, r_1, r_2, r_3) = \mathrm{tr}[(a(r_1, r_2) + a(r_2, r_3) + a(r_3, r_1))^2] \tag{5.2}$$

and

$$f(r_1, r_2, r_3, r_4, r_5) = (r_1, r_2, r_3)\,, \qquad g(r_1, r_2, r_3, r_4, r_5) = (r_3, r_4, r_5)\,. \tag{5.3}$$

This means that in the case of Minkowski space the pre-Lagrangian is given by

$$\begin{aligned}\mathcal{L}(a, E, p, q) &= \int_{\tau(p,q)} \mathrm{d}^d r\, \mathrm{d}^d s\, \mathrm{d}^d t \left[\mathcal{J}(a, f(p, q, r, s, t)) + \mathcal{J}(a, f(q, p, r, s, t)) + \right.\\ &\qquad\qquad \left. + E\left(\mathcal{J}(a, g(p, q, r, s, t)) + \mathcal{J}(a, g(q, p, r, s, t))\right)\right]\\ &= \int_{\tau(p,q)} \mathrm{d}^d r\, \mathrm{d}^d s\, \mathrm{d}^d t \left[\mathcal{J}(a, p, q, r) + \mathcal{J}(a, q, p, r) + E\left(\mathcal{J}(a, r, s, t) + \mathcal{J}(a, r, s, t)\right)\right]\,. \end{aligned} \tag{5.4}$$

Integrating (trivially) over s and t in the first two terms, and using the fact that $\mathcal{J}(a, p, q, r) = \mathcal{J}(a, q, p, r)$, we get that

$$\mathcal{L}(a, E, p, q) = 2\left(V^2(\alpha(p, q)) \int_{\tau(p,q)} \mathrm{d}^d r\, \mathcal{J}(a, p, q, r) + E \int_{\tau(p,q)} \mathrm{d}^d r\, \mathrm{d}^d s\, \mathrm{d}^d t\, \mathcal{J}(a, r, s, t)\right). \tag{5.5}$$

Assume that the gauge field is differentiable and reasonably well behaved. In particular, it is well-behaved-enough for ${F_{\mu\nu}}^k$ to be approximately constant in $\alpha(P, Q)$ whenever $(E, P, Q) \in \alpha_{\tau_1, \tau_2}(a, u)$ for some $u \in S$. Assume for definiteness that the three points are spacelike related. Choose a coordinate system so that r coincides with the origin, the x axis points from r to s, and the y axis is perpendicular to the x axis in the r-s-t plane. Then in this coordinate system $a = (0, 0, 0, ...)$, $b = (0, b^1, 0, ...)$, $c = (0, c^1, c^2, ...)$.

The flux of ${F_{\mu\nu}}^k$ through the interior of that triangle is expressed by the relationship

$$a(r, s) + a(s, t) + a(t, r) = \tfrac{1}{2}\, s^1\, t^2\, {F_{12}}^k\, T^k + ... \tag{5.6}$$

This result generalizes to points at arbitrary locations, and can be written covariantly as

$$a(r, s) + a(s, t) + a(t, r) = \tfrac{1}{2}\, {F_{\mu\nu}}^k\, T^k\, (s^\mu - r^\mu)\, (t^\nu - r^\nu) + ... \tag{5.7}$$

Recalling that, for SU(n), $\mathrm{tr}(T^k T^l) = C_2\, \delta_{kl}$, to leading order in the separation between points,

$$\mathrm{tr}[(a(r, s) + a(s, t) + a(t, r))^2] = \frac{C_2}{4}\, {F_{\mu\nu}}^k\, (s^\mu - r^\mu)\, (t^\nu - r^\nu)\, {F_{\rho\sigma}}^k\, (b^\rho - a^\rho)\, (c^\sigma - a^\sigma)\,. \tag{5.8}$$

Let's start from

$$\int_{\tau(p,q)} \mathrm{d}^d r\, \mathrm{d}^d s\, \mathrm{d}^d t\, \mathcal{J}(a, r, s, t)\,. \tag{5.9}$$

Expand the right-hand side of Eq (5.8), and integrate term by term. Clearly any term with an odd number of powers of any variable will integrate to 0. Thus, the only terms that may potentially survive the integration are those of the form $r^\mu\, r^\nu\, r^\rho\, r^\sigma$ or quadratic terms in two of the three points. Simple counting of terms gives

$$\begin{aligned}&\int_{p \prec r, s, t \prec q} \mathrm{d}^d r\, \mathrm{d}^d s\, \mathrm{d}^d t\, \mathrm{tr}\left[(a(r, s) + a(s, t) + a(t, r))^2\right]\\ &= \frac{C_2}{4}\, {F_{\mu\nu}}^k\, {F_{\rho\sigma}}^k \int_{p \prec a, b, c \prec q} \mathrm{d}^d a\, \mathrm{d}^d b\, \mathrm{d}^d c\, (s^\mu - r^\mu)\, (t^\nu - r^\nu)\, (s^\rho - r^\rho)\, (t^\sigma - r^\sigma)\\ &= \frac{C_2}{4}\left[3\, V \sum_{k, \mu, \nu} ({F_{\mu\nu}}^k)^2 \left(\int_{\alpha(p,q)} \mathrm{d}^d a\, (r^\mu)^2\right)\left(\int_{\alpha(p,q)} \mathrm{d}^d b\, (s^\nu)^2\right) + \right.\\ &\qquad\qquad \left. - V^2 {F_{\mu\nu}}^k\, {F_{\rho\sigma}}^k \int_{\alpha(p,q)} \mathrm{d}^d a\, r^\mu\, r^\nu\, r^\rho\, r^\sigma\right], \end{aligned} \tag{5.10}$$

where V is the volume of the Alexandrov set $\alpha(p,q)$.

The only terms of $F_{\mu\nu}{}^k\, F_{\rho\sigma}{}^k\, r^\mu\, r^\nu\, r^\rho\, r^\sigma$ that survive integration are the ones whose indices are pairwise equal. But if either $\mu=\nu$ or $\rho=\sigma$, then $F_{\mu\nu}{}^k=0$ or $F_{\rho\sigma}{}^k=0$, respectively, which would set the whole thing to 0. Thus, the only options are $\mu=\rho$, $\nu=\sigma$ and $\nu=\rho$, $\mu=\sigma$. The antisymmetry of $F_{\mu\nu}{}^k$ then implies that these two cases are opposites of each other, which in turn implies that $F_{\mu\nu}{}^k\, F_{\rho\sigma}{}^k\, r^\mu\, r^\nu\, r^\rho\, r^\sigma=0$. Thus, Eq (5.10) becomes

$$\begin{aligned}
&\int_{p\prec r,s,t\prec q} \mathrm{d}^d r\, \mathrm{d}^d s\, \mathrm{d}^d t\, \mathrm{tr}\big[(a(r,s)+a(s,t)+a(t,r))^2\big] \\
&= \frac{3\,V C_2}{4} \sum_{k,\mu,\nu} (F_{\mu\nu}{}^k)^2 \Big(\int_{\alpha(p,q)} \mathrm{d}^d r\, (r^\mu)^2\Big)\Big(\int_{\alpha(p,q)} \mathrm{d}^d s\, (s^\nu)^2\Big) \\
&= \frac{3\,k_d\, C_2\, \tau^{3d+4}}{2} \sum_k \Big(J^0 J^1 \sum_{i=1}^{d-1} (F_{i0}{}^k)^2 + (J^1)^2 \sum_{i<j} (F_{ij}{}^k)^2\Big)\,, \qquad (5.11)
\end{aligned}$$

where $J^\mu=\tau^{-d-2}\int_{\alpha(p,q)} \mathrm{d}^d x\, (x^\mu)^2$, or in other words

$$J^0 = \frac{2\pi^{(d-1)/2}}{2^d\,(d-1)\,d\,(d+1)\,(d+2)}\,\Gamma((d-1)/2) \qquad (5.12)$$

$$J^1 = \ldots = J^{d-1} = \frac{2\pi^{d/2-1}}{2^{d+1}\,(d-2)\,d\,(d+2)}\,\Gamma((d-2)/2) \int_{-\pi/2}^{\pi/2} (\cos\theta)^d\, \mathrm{d}\theta\,. \qquad (5.13)$$

Now let's move to the second integral, $\int_{\alpha(p,q)} \mathrm{d}^d x\, \mathrm{tr}[(a(p,x)+a(x,q)+a(q,p))^2]$, where $p\prec q$ are the endpoints of the Alexandrov set. Rewriting Eq (5.8) in terms of the points p, x, and q gives

$$\begin{aligned}
&\mathrm{tr}\big[(a(p,x)+a(x,q)+f(q,p))^2\big] \qquad (5.14)\\
&= \tfrac{1}{4}\, C_2\, F_{\mu\nu}{}^k\, (p^\mu-x^\mu)\,(q^\nu-x^\nu)\, F_{\rho\sigma}{}^k\, (p^\rho-x^\rho)\,(q^\sigma-x^\sigma)\,.
\end{aligned}$$

Again this can be expanded and integrated term by term. There are several conditions each term has to meet, in order for its integral not to vanish. First of all, it needs to contain an even number of factors of x. Secondly, as was shown before, for symmetry reasons

$$F_{\mu\nu}{}^k\, F_{\rho\sigma}{}^k \int_{\alpha(p,q)} \mathrm{d}^d x\, x^\mu\, x^\nu\, x^\rho\, x^\sigma = 0\,. \qquad (5.15)$$

Finally, $F_{\mu\nu}\, p^\mu p^\nu = F_{\mu\nu}\, q^\mu q^\nu = 0$, and the identities $p=(-\frac{\tau}{2},0,0,0)$ and $q=(\frac{\tau}{2},0,0,0)$ imply $F_{\mu\nu}\, p^\mu q^\nu = -F_{\mu\nu}\, p^\mu p^\nu = 0$. The only terms in Eq (5.15) that do *not* vanish for any of the above reasons are

$$\begin{aligned}
&F_{\mu\nu}{}^k\, F_{\rho\sigma}{}^k\, p^\mu\, x^\nu\, p^\rho\, x^\sigma\,, \qquad F_{\mu\nu}{}^k\, F_{\rho\sigma}{}^k\, p^\mu\, x^\nu\, x^\rho\, q^\sigma\,, \\
&F_{\mu\nu}{}^k\, F_{\rho\sigma}{}^k\, x^\mu\, q^\nu\, p^\rho\, x^\sigma\,, \qquad F_{\mu\nu}{}^k\, F_{\rho\sigma}{}^k\, x^\mu\, q^\nu\, x^\rho\, q^\sigma\,.
\end{aligned}$$

Plugging in the coordinate values of p and q, each of the above four expressions evaluates to $\frac{1}{4}\,\tau^2\, F_{\mu0}{}^k\, F_{\rho0}{}^k\, x^\mu\, x^\rho$. In order for this not to be an odd function, $\mu=\rho$ has to hold, and in order for $F_{\mu0}$ to be non-zero $\mu\neq 0$ has to hold. Thus, this becomes $\frac{1}{4}\,\tau^2\,(F_{i0}{}^k)^2\,(x^i)^2$ and, since there are four such terms, the integral becomes

$$\begin{aligned}
&\int_{\alpha(p,q)} \mathrm{d}^d x\, \mathrm{tr}\big[(a(p,x)+a(x,q)+a(q,p))^2\big] \\
&= \tfrac{1}{4}\, C_2\, \tau^2 \sum_{i=1}^{d-1} \int_{\alpha(p,q)} \mathrm{d}^d x\, (F_{i0}{}^k)^2\, (x^i)^2 = \tfrac{1}{4}\, C_2\, \tau^{d+4} J^1 \sum_{i=1}^{d-1} (F_{i0}{}^k)^2\,, \qquad (5.16)
\end{aligned}$$

where, based on rotational symmetry, $J^1 = ... = J^{d-1}$ was used.

Substituting this into the original expression for the pre-Lagrangian, and doing some basic algebra, the latter becomes

$$\mathcal{L}(a, E, p, q) = \tau^{3d+4} C_2 k_d J^1 \left((3 E J^0 + \tfrac{1}{4} k_d) \sum (F^k_{i0})^2 + 3 E J^1 \sum_{i<j} (F^k_{ij})^2 \right). \tag{5.17}$$

In order to get rid of the variation that results from different choices of Alexandrov sets, we would like the above expression to be relativistically invariant; in other words, we would like it to be proportional to $F^{\mu\nu} F_{\mu\nu}$. This means

$$3 E_d J^0 + \tfrac{1}{2} k_d = -3 E_d J^1 , \tag{5.18}$$

which implies

$$E_d = -\frac{k_d}{6 (J^0 + J^1)} . \tag{5.19}$$

Substituting this expression for E_d gives the Lagrangian

$$\mathcal{L}(a, p, q) = -\frac{k_d^2 (J^1)^2 C_2}{4 (J^0 + J^1)} \tau_2^{3d+4} F^{\mu\nu} F_{\mu\nu} , \tag{5.20}$$

which can be expressed as

$$\mathcal{L}(a, p, q) = -\frac{v_0}{4} F_d{}^{\mu\nu} F_{d\mu\nu} , \tag{5.21}$$

where v_0 is the volume taken up by a single point if $F_d{}^{\mu\nu}$ is defined as

$$F_d{}^{\mu\nu} = k_d J^1 F^{\mu\nu} \sqrt{\frac{C_2 \tau^{3d+4}}{v_0 (J^0 + J^1)}} . \tag{5.22}$$

This means that

$$A_d{}^{\mu} = k_d J^1 A^{\mu} \sqrt{\frac{C_2 \tau^{3d+4}}{v_0 (J^0 + J^1)}} . \tag{5.23}$$

From this we can determine how the charge changes from dimension to dimension. At a first glance, since we haven't yet defined a Lagrangian for fermions, we are only ready to talk about the charge of bosonic fields. However, a simple symmetry consideration allows us to overcome this barrier and include fermionic charges in the discussion. Whether a field is bosonic or fermionic, we would like to be able to say

$$\partial_\mu \to \partial_\mu + e A_\mu . \tag{5.24}$$

We would also like to be able to say

$$\partial_\mu \to \partial_\mu + e_d A_{d\mu} . \tag{5.25}$$

This means that, regardless whether the field in question is bosonic or fermionic, and regardless of any other properties of the field (such as mass) it should satisfy

$$e A^\mu = e_d A^\mu_d . \tag{5.26}$$

This immediately implies that, both for bosons and fermions,

$$e_d = \frac{e}{k_d J^1} \sqrt{\frac{v_0 (J^0 + J^1)}{C_2 \tau^{3d+4}}} . \tag{5.27}$$

Journal of Physics: Conference Series **174** (2009) 012019 doi:10.1088/1742-6596/174/1/012019

6. Gravitational field

The Lagrangian generator for gravity is the triple $(\mathcal{J}, f, g)$, defined simply by

$$\mathcal{J}(r,s,t) = \begin{cases} (8\pi G)^{-1} & \text{if } r \prec t \prec s \\ 0 & \text{otherwise} \end{cases}, \qquad f(r,s,t,u) = (r,s,t), \qquad g(r,s,t,u) = (r,t,u). \tag{6.1}$$

This means that the pre-Lagrangian is given by

$$\mathcal{L} = \int \mathrm{d}^d r\, \mathrm{d}^d s \sqrt{|\det g(r)|}\sqrt{|\det g(s)|}\left[\mathcal{J}\big(f(p,q,r,s)\big) + E\,\mathcal{J}\big(g(p,q,r,s)\big)\right]. \tag{6.2}$$

Since $\mathcal{J}$ is a constant inside a certain domain and is 0 outside, the above integrals amount to restricting r and s to certain domains. The first term is $\mathcal{J}(p,q,r)$, thus its restriction is $p \prec r \prec q$. That statement, of course, is trivial, which means that the first term has no restriction at all. On the other hand, the second term is $\mathcal{J}(p,r,s)$, thus it has a restriction $p \prec r \prec s \prec q$. Taking this into account, the pre-Lagrangian becomes

$$\begin{aligned} \mathcal{L} &= \frac{1}{8\pi G}\Big(\int \mathrm{d}^d r\, \mathrm{d}^d s \sqrt{|\det g(r)|}\sqrt{|\det g(s)|} + E\int_{p\prec r\prec s\prec q} \mathrm{d}^d r\, \mathrm{d}^d s \sqrt{|\det g(r)|}\sqrt{|\det g(s)|}\Big) \\ &= \big(V(\alpha(p,q))\big)^2 + E\int_{\alpha(p,q)} \mathrm{d}^d s \sqrt{|\det g(s)|}\, V(\alpha(p,s)). \end{aligned} \tag{6.3}$$

One can parametrize the interior of the Alexandrov set $\alpha(p,q)$ with normal geodesic coordinates around p. Suppose that Greek indices μ, ν are used for coordinates satisfying

$$g_{\mu\nu}(p) = \eta_{\mu\nu}, \tag{6.4}$$

but which are otherwise arbitrary, not necessarily geodesic. Then one can define the normal geodesic coordinates derived from the above in the following way:

$$r^{\bar{\mu}} = \eta^{\mu\nu}\, \partial_\nu \gamma_{pr}\big|_p \left(\int_{\gamma_{pr}} g_{\mu\nu} \mathrm{d}x^\mu\, \mathrm{d}x^\nu\right)^{1/2}, \tag{6.5}$$

where γ_{ab} denotes the geodesic segment connecting a and b. From this, it is straightforward to see that the following equation is satisfied exactly:

$$\tau(p,r) = \eta_{\bar{\mu}\bar{\nu}}\, r^{\bar{\mu}} r^{\bar{\nu}}. \tag{6.6}$$

It should be noticed, however, that the above is true only if one of the two points is p, while

$$\tau(r,s) \neq \eta_{\bar{\mu}\bar{\nu}}\,(s^{\bar{\mu}} - r^{\bar{\mu}})\,(s^{\bar{\nu}} - r^{\bar{\nu}}). \tag{6.7}$$

However, a different causal relation, $\prec_p$, will be introduced in addition to the already existing one, $\prec$. While $\prec$ matches flat spacetime expectations only if the pair of points of interest includes p, $\prec_p$ does so for arbitrary pairs of points:

$$a \prec_p b \quad \Leftrightarrow \quad \eta_{\bar{\mu}\bar{\nu}}\,(b^{\bar{\mu}} - a^{\bar{\mu}})\,(b^{\bar{\nu}} - a^{\bar{\nu}}) \geq 0. \tag{6.8}$$

At the same time, we will retain the original causal relation, $\prec$, for which the above is not true:

$$\prec \;\neq\; \prec_p. \tag{6.9}$$

Let's introduce the following notation:

$$J^+(a) = \{r \succ a\}\,, \qquad J_p^+(a) = \{r \succ_p a\}\,,$$
$$J^-(a) = \{r \prec a\}\,, \qquad J_p^-(a) = \{r \prec_p a\} \tag{6.10}$$
$$\alpha(a,b) = J^+(a) \cap J^-(b)\,, \qquad \alpha_p(a,b) = J_p^+(a) \cap J_p^-(b)\,, \tag{6.11}$$

and define $V_p(a,b)$, not to be confused with $V(\alpha_p(a,b))$, as follows

$$V_p(a,b) = k_d \left(\eta_{\mu\nu}\,(b^\mu - a^\mu)\,(b^\nu - a^\nu)\right)^{d/2}. \tag{6.12}$$

It is important to notice that

$$\mathrm{vol}(\alpha_p(a,b)) \neq V_p(a,b)\,, \tag{6.13}$$

because the definition of V_p neglects the $|g|^{1/2}$ factor in the volume element. But, for our purposes, V_p as defined above is the simplest one to use. On the other hand, $V(a,b)$ is defined as the actual volume of the Alexandrov set,

$$V(a,b) = \mathrm{vol}(\alpha(a,b))\,. \tag{6.14}$$

Then, the integral of the volume can be expanded as follows

$$\int_{\alpha(p,q)} \mathrm{d}^d r\, \sqrt{|\det g(r)|}\, V(\alpha(p,r))$$
$$= \int_{\alpha_p(p,q)} \mathrm{d}^d r\, V_p(p,q) + \Delta_{1d}(p,q) + \Delta_{2d}(p,q) + \Delta_{3d}(p,q) + O(\tau^{2d+4})\,, \tag{6.15}$$

where $\Delta_{1d}(p,q)$ is the correction coming from the mismatch between $\alpha(p,q)$ and $\alpha_p(p,q)$, $\Delta_{2d}(p,q)$ is the correction due to the error in $V(\alpha(p,r))$, and Δ_3 is the correction due to the use of $\mathrm{d}^d r$ instead of $\mathrm{d}^d r\sqrt{(-1)^{d-1}\det g}$ for the volume element. These corrections are formally defined as follows:

$$\Delta_{1d}(p,q) = \int_{(\alpha(p,q)\setminus\alpha_p(p,q))} \mathrm{d}^d r\, V_p(p,r) - \int_{\alpha_p(p,q)\setminus\alpha(p,q))} \mathrm{d}^d r\, V_p(p,r) \tag{6.16}$$
$$\Delta_{2d}(p,q) = \int_{\alpha_p(p,q)} \mathrm{d}^d r\, (V(p,r) - V_p(p,r)) \tag{6.17}$$
$$\Delta_{3d}(p,q) = \int_{\alpha_p(p,q)} \mathrm{d}^d r\, \left(\sqrt{(-1)^{d-1}\det g} - 1\right). \tag{6.18}$$

Whenever any of these three correction terms is computed the other two are neglected, since the "correction of the correction" is of order τ^{2d+4}, while the calculation is performed to order τ^{2d+2}. Thus, the shape of the Alexandrov set is assumed to be unchanged in the calculation of Δ_{2d} and Δ_{3d}, $\sqrt{(-1)^{d-1}\det g}$ is dropped in the calculation of Δ_{1d} and Δ_{2d}, and the correction to $V(\alpha(p,x))$ is neglected in the calculation of Δ_{1d} and Δ_{3d}.

Let's start by computing Δ_{1d}. In normal coordinates, the light cone of p is not deformed by curvature, while the light cone of q still is,

$$J^+(p) = J_p^+(p)\,, \qquad J^-(q) \neq J_p^-(q)\,. \tag{6.19}$$

Substitution of above into simple set theory algebra gives

$$\alpha_p(p,q) \setminus \alpha(p,q) = J_p^+(p) \cap (J_p^-(q) \setminus J^-(q)) \subset J_p^-(q) \setminus J^-(q) \tag{6.20}$$
$$\alpha(p,q) \setminus \alpha_p(p,q) = J_p^+(p) \cap (J^-(q) \setminus J_p^-(q)) \subset J^-(q) \setminus J_p^-(q)\,. \tag{6.21}$$

Assuming that $J^-_p(q)$ and $J^-(q)$ are very close to each other, the above implies that most of the contribution to Δ_{1d} comes from the vicinity of $J^-_p(q)$. So, let's evaluate $V_p(p,r)$ for $r \in J^-_p(q)$. In other words, assume that

$$\sum (r^{\bar{k}})^2 = (\tau - r^{\bar{0}})^2 \,. \tag{6.22}$$

This implies that

$$\eta_{\bar{\mu}\bar{\nu}}\, r^{\bar{\mu}}\, r^{\bar{\nu}} = (r^{\bar{0}})^2 - (\tau - r^{\bar{0}})^2 = \tau\,(2r^{\bar{0}} - \tau)\,, \tag{6.23}$$

which, using the fact that

$$2\,r^{\bar{0}} - \tau = r^{\bar{0}} - (\tau - r^{\bar{0}}) = r^{\bar{0}} - \sqrt{\sum (r^{\bar{k}})^2}\,, \tag{6.24}$$

becomes

$$\eta_{\bar{\mu}\bar{\nu}}\, r^{\bar{\mu}}\, r^{\bar{\nu}} = \tau_0 \left(r^{\bar{0}} - \sqrt{\sum (r^{\bar{k}})^2} \right) . \tag{6.25}$$

Now define two functions χ_1 and χ_2 as follows:

$$\chi_1(r) = \int_{\alpha(r,q)} \mathrm{d}^d s\, k_d\, (\eta_{\bar{\mu}\bar{\nu}}\, s^{\bar{\mu}}\, s^{\bar{\nu}})^{d/2} - \int_{\alpha_p(r,q)} \mathrm{d}^d s\, k_d\, (\eta_{\bar{\mu}\bar{\nu}}\, s^{\bar{\mu}}\, s^{\bar{\nu}})^{d/2} \tag{6.26}$$

$$\chi_2(r) = \int_{\alpha(r,q)} \mathrm{d}^d s\, k_d \left[\tau_0 \left(s^{\bar{0}} - \sqrt{\sum (s^{\bar{k}})^2} \right) \right]^{d/2} - \int_{\alpha_p(r,q)} \mathrm{d}^d s\, k_d \left[\tau_0 \left(s^{\bar{0}} - \sqrt{\sum (s^{\bar{k}})^2} \right) \right]^{d/2} . \tag{6.27}$$

As a consequence of the fact that $J^+(p) = J^+_p(p)$, we get

$$\chi_1(p) = \chi_2(p)\,, \tag{6.28}$$

and it is easy to see that

$$\chi_1(q) = \chi_2(q) = 0\,. \tag{6.29}$$

Therefore

$$\Delta_{1d} = \chi_1(p) = -(\chi_2(q) - \chi_2(p)) = -\int_0^{\tau(p,q)} \mathrm{d}\tau\, \frac{\mathrm{d}\chi_2(r(\tau))}{\mathrm{d}\tau}\,, \tag{6.30}$$

where $r(\tau)$ is defined as a point on γ_{pq} whose distance to r is τ:

$$\gamma_{pr(\tau)} \subset \gamma_{pq}\,, \qquad l(\gamma_{pr(\tau)}) = \tau\,. \tag{6.31}$$

This immediately implies that

$$\delta\tau > 0 \Rightarrow r(\tau) \prec r(\tau + \delta\tau)\,. \tag{6.32}$$

Furthermore, in normal coordinates

$$r^{\bar{\mu}}(\tau) = \tau\, V^{\bar{\mu}}\,, \tag{6.33}$$

where $V^{\bar{\mu}}$ is a tangent vector to γ_{pq} at p. The fact that the latter is timelike implies that

$$\delta\tau > 0 \quad \Rightarrow \quad r(\tau) \prec_p r(\tau + \delta\tau)\,. \tag{6.34}$$

In order to compute $\mathrm{d}\chi_2(r(\tau))/\mathrm{d}\tau$ notice that, schematically,

$$\begin{aligned} \delta\chi_2(r) &= \int_{S_1} f - \int_{S_2} f - \int_{S_3} f + \int_{S_4} f = \int_{S_1\backslash S_2} f - \int_{S_2\backslash S_1} f - \int_{S_3\backslash S_4} f + \int_{S_4\backslash S_3} f \\ &= \int_{(S_1\backslash S_2)\backslash(S_3\backslash S_4)} f - \int_{(S_3\backslash S_4)\backslash(S_1\backslash S_2)} f - \int_{(S_3\backslash S_4)\backslash(S_4\backslash S_3)} f + \int_{(S_4\backslash S_3)\backslash(S_3\backslash S_4)} f\,, \end{aligned} \tag{6.35}$$

where

$$S_1 = \alpha(r(\tau+\delta\tau),q)\,, \quad S_2 = \alpha_p(r(\tau+\delta\tau),q)\,, \quad S_3 = \alpha(r(\tau),q)\,, \quad S_4 = \alpha_p(r(\tau),q)\,. \tag{6.36}$$

Substituting both $r(\tau) \prec r(\tau+\delta\tau)$ and $r(\tau) \prec_p r(\tau+\delta\tau)$ and using simple set-theory algebra, one obtains

$$\begin{aligned}&\alpha(r(\tau),q) \setminus \alpha_p(r(\tau),q) \setminus (\alpha(r(\tau+\delta\tau),q) \setminus \alpha_p(r(\tau+\delta\tau),q))\\ &= (J^+(r(\tau)) \setminus J^+(r(\tau+\delta\tau))) \cap (J^-(q) \setminus \alpha_p(r(\tau,q))) \subset J^+(r(\tau)) \setminus J^+(r(\tau+\delta\tau))\end{aligned} \tag{6.37}$$

$$\begin{aligned}&\alpha_p(r(\tau),q) \setminus \alpha(r(\tau),q) \setminus (\alpha_p(r(\tau+\delta\tau),q) \setminus \alpha(r(\tau+\delta\tau),q))\\ &= (J_p^+(r(\tau)) \setminus J_p^+(r(\tau+\delta\tau))) \cap (J_p^-(q) \setminus \alpha(r(\tau,q))) \subset J_p^+(r(\tau)) \setminus J_p^+(r(\tau+\delta\tau))\end{aligned} \tag{6.38}$$

$$\begin{aligned}&(\alpha(r(\tau+\delta\tau),q) \setminus \alpha_p(r(\tau+\delta\tau),q)) \setminus (\alpha(r(\tau),q) \setminus \alpha_p(r(\tau),q))\\ &= \alpha(r(\tau+\delta\tau),q) \cap (J_p^+(r(\tau)) \setminus J_p^+(r(\tau+\delta\tau))) \subset J_p^+(r(\tau)) \setminus J_p^+(r(\tau+\delta\tau))\end{aligned} \tag{6.39}$$

$$\begin{aligned}&(\alpha_p(r(\tau+\delta\tau),q) \setminus \alpha(r(\tau+\delta\tau),q)) \setminus (\alpha_p(r(\tau),q) \setminus \alpha(r(\tau),q))\\ &= \alpha_p(r(\tau+\delta\tau),q) \cap (J^+(r(\tau)) \setminus J^+(r(\tau+\delta\tau))) \subset J^+(r(\tau)) \setminus J^+(r(\tau+\delta\tau))\,.\end{aligned} \tag{6.40}$$

Thus, all four integrals are performed either over a subset of $J^+(r(\tau)) \setminus J^+(r(\tau+\delta\tau))$ or over a subset of $J_p^+(r(\tau)) \setminus J_p^+(r(\tau+\delta\tau))$. In either case, the range of integration is in the vicinity of the lightcone of $r(\tau)$.

Now let s be an arbitrary point in that region. Remembering that τ denotes the distance from p to the point of intersection of that region with the geodesic γ_{pq}, and is not to be confused with the distance from p to a floating point, we have

$$\begin{aligned}&s \in \left[J^+(r(\tau)) \setminus J^+(r(\tau+\delta\tau))\right] \cup \left[J_p^+(r(\tau)) \setminus J_p^+(r(\tau+\delta\tau))\right]\\ &\Rightarrow \sqrt{\textstyle\sum(s^{\bar{k}})^2} = s^{\bar{0}} - \tau + O(\tau^2) \Rightarrow k_d\left[\tau_0\left(s^{\bar{0}} - \sqrt{\textstyle\sum(s^{\bar{k}})^2}\right)\right]^{d/2} = k_d\,(\tau\tau_0)^{d/2} + O(\tau^{d+2})\,.\end{aligned} \tag{6.41}$$

Thus, all integrals reduce to integrations over a constant:

$$\begin{aligned}&\int_{S_1} f\,\mathrm{d}^d x - \int_{S_2} f\,\mathrm{d}^d x - \int_{S_3} f\,\mathrm{d}^d x + \int_{S_4} f\,\mathrm{d}^d x\\ &= \Big(\int_{(S_1\setminus S_2)\setminus(S_3\setminus S_4)} - \int_{(S_3\setminus S_4)\setminus(S_1\setminus S_2)} - \int_{(S_3\setminus S_4)\setminus(S_4\setminus S_3)} + \int_{(S_4\setminus S_3)\setminus(S_3\setminus S_4)}\Big) f\,\mathrm{d}^d x\\ &= (\tau\tau_0)^{d/2}\Big(\int_{(S_1\setminus S_2)\setminus(S_3\setminus S_4)} - \int_{(S_3\setminus S_4)\setminus(S_1\setminus S_2)} - \int_{(S_3\setminus S_4)\setminus(S_4\setminus S_3)} + \int_{(S_4\setminus S_3\setminus(S_3\setminus S_4)}\Big)\,\mathrm{d}^d x\\ &= (\tau\tau_0)^{d/2}\Big(\int_{S_1} \mathrm{d}^d x - \int_{S_2} \mathrm{d}^d x - \int_{S_3} \mathrm{d}^d x + \int_{S_4} \mathrm{d}^d x\Big)\\ &= (\tau\tau_0)^{d/2}\,\big(V(S_1) - V(S_2) - V(S_3) + V(S_4)\big)\,.\end{aligned} \tag{6.42}$$

Now, $V(S_1)$, $V(S_2)$, $V(S_3)$ and $V(S_4)$ can be read off from reference [8] as follows:

$$\begin{aligned}V(S_1) &= \left[k_d + (A_d\, Rg_{\mu\nu} + B_d\, R_{\mu\nu})\,(q^\mu - r^\mu(\tau+\delta\tau))\,(q^\nu - r^\nu(\tau+\delta\tau))\right]\tau^d(r(\tau+\delta\tau),q)\\ V(S_2) &= k_d\,\tau^d(r(\tau+\delta\tau),q)\\ V(S_3) &= (k_d + (A\, Rg_{\mu\nu} + B\, R_{\mu\nu})\,(q^\mu - r^\mu(\tau))\,(q^\nu - r^\nu(\tau)))\,\tau^d(r(\tau),q)\\ V(S_4) &= k_d\,\tau^d(r(\tau),q)\,.\end{aligned} \tag{6.43}$$

Since the coordinate system is defined in terms of geodesics coming out of p, all of these geodesics, including γ_{pq} are, by definition, straight lines in the chosen coordinate system. Therefore, it can be assumed that γ_{pq} coincides with the t axis, which simplifies the above equations:

$$V(S_1) = k_d\,(\tau_1 - \tau - \delta\tau)^d + (A_d\,R + B_d\,R_{00})\,(\tau_1 - \tau - \delta\tau)^{d+2} \tag{6.44}$$

$$V(S_2) = k_d\,(\tau_1 - \tau - \delta\tau)^d \tag{6.45}$$

$$V(S_3) = k_d\,(\tau_1 - \tau)^d + (A_d\,R + B_d\,R_{00})\,(\tau_1 - \tau)^{d+2} \tag{6.46}$$

$$V(S_4) = k_d\,(\tau_1 - \tau)^d\,. \tag{6.47}$$

This implies that

$$\begin{aligned}\chi_2(r(\tau+\delta\tau)) - \chi_2(r(\tau)) &= (\tau_1\tau)^{d/2}\big(V(S_1) - V(S_2) - V(S_3) + V(S_4)\big) \qquad (6.48)\\ &= (A_d\,R + B_d\,R_{00})\,(\tau_1\tau)^{d/2}\,\frac{\mathrm{d}}{\mathrm{d}\tau}\,(\tau_1-\tau)^{d+2}\,\delta\tau + O(\tau^{d+2}\,(\delta\tau)^2)\,,\end{aligned}$$

which in turns implies that

$$\frac{\mathrm{d}\chi_2(r(\tau))}{\mathrm{d}\tau} = (A_d\,R + B_d\,R_{00})\,(\tau_1\tau)^{d/2}\,\frac{\mathrm{d}}{\mathrm{d}\tau}\,(\tau_1-\tau)^{d+2}\,. \tag{6.49}$$

Thus,

$$\Delta_{1d} = \chi_1(p) - \chi_1(q) = \chi_2(p) - \chi_2(q) = -(A_d\,R + B_d\,R_{00})\int_0^{\tau_1}\mathrm{d}\tau\,(\tau_1\tau)^{d/2}\,\frac{\mathrm{d}}{\mathrm{d}\tau}(\tau_1-\tau)^{d+2}\,. \tag{6.50}$$

The binomial expansion of $(\tau_1 - \tau)^{d+2}$ gives

$$\Delta_{1d} = -(A_d\,R + B_d\,R_{00})\int_0^{\tau_1}\mathrm{d}\tau\,(\tau_1\tau)^{d/2}\,\frac{\mathrm{d}}{\mathrm{d}\tau}\left(\sum_{k=0}^{d+2}(-1)^k\binom{d+2}{k}\,\tau_1^{d+2-k}\,\tau^k\right). \tag{6.51}$$

Evaluating the derivative and combining it with the $\tau^{d/2}$ factor we find that

$$\Delta_{1d} = -(A_d\,R + B_d\,R_{00})\int_0^{\tau_1}\mathrm{d}\tau\,\tau_1^{d/2}\left(\sum_{k=0}^{d+2}(-1)^k\binom{d+2}{k}\,\tau_1^{d+2-k}\,k\,\tau^{d/2+k-1}\right). \tag{6.52}$$

Finally, integration of this last expression gives

$$\Delta_{1d} = (A_d\,R + B_d\,R_{00})\,\tau_1^{2d+2}\sum_{k=0}^{d+2}(-1)^k\binom{d+2}{k}\frac{k}{d/2+k}\,. \tag{6.53}$$

Now let's compute Δ_{2d}. We will go back to our usual notation, where r (which will be denoted by x in order to avoid conflict with r being a radius of a ball) is an arbitrary element of $\alpha(p,q)$, and it is no longer assumed to lie on γ_{pq}. As was stated earlier, the Δ_{2d} correction is due to the error of computing the volume of $\alpha(p,x)$, where x is far away from the boundary of $\alpha(p,q)$. Since that correction is already of order τ^{d+2}, its integral over $\alpha(p,q)$ is of order τ^{2d+2}, while its integral over the "corrections" to the shape of $\alpha(p,q)$ is of order τ^{2d+4}. For this reason, the latter term will be neglected, and it will be assumed that the shape of $\alpha(p,q)$ has not been affected by curvature, while the shapes of $\alpha(p,x)$ has. Furthermore, in light of the geodesic coordinates, while the corrections to the volume of $\alpha(p,x)$ are not neglected, the flat space equation for the distance will be used in computing them. Thus, the equation for Δ_{2d} becomes

$$\Delta_{2d} = A_d\, R \int_{\alpha_p(p,q)} \mathrm{d}^d x\, (x^{\bar\mu}\, x_{\bar\mu})^{1+d/2} + B_d\, R_{\bar\mu\bar\nu} \int_{\alpha_p(p,q)} \mathrm{d}^d x\, x^{\bar\mu}\, x^{\bar\nu}\, (x^{\bar\rho}\, x_{\bar\rho})^{d/2}\,. \tag{6.54}$$

By using the cylindrical symmetry, this becomes

$$\begin{aligned}\Delta_{2d} = A_d\, R \int_{\alpha_p(p,q)} \mathrm{d}^d\overline{x}\, (x^{\bar\mu} x_{\bar\mu})^{1+d/2} + B_d\, R_{\bar0\bar0} \int_{\alpha_p(p,q)} \mathrm{d}^d\overline{x}\, (x^{\bar0})^2\, (x^{\bar\rho}\, x_{\bar\rho})^{d/2} \,+ \\ + B_d \left(\sum R_{kk}\right) \int_{\alpha_p(p,q)} \mathrm{d}^d\overline{x}\, (x^{\bar1})^2\, (x^{\bar\rho}\, x_{\bar\rho})^{d/2}\,,\end{aligned} \tag{6.55}$$

where, due to cylindrical symmetry, $(x^{\bar k})^2$ was replaced with $(x^{\bar1})^2$.

From cylindrical symmetry,

$$\int_{\alpha_p(p,q)} \mathrm{d}^d\overline{x}\, (x^{\bar\rho}\, x_{\bar\rho})^{1+d/2} = \int_{\alpha_p(p,q)} \mathrm{d}^d\overline{x}\, (x^{\bar0})^2\, (x^{\bar\rho}\, x_{\bar\rho})^{d/2} - (d-1) \int \mathrm{d}^d\overline{x}\, (x^{\bar1})^2\, (x^{\bar\rho}\, x_{\bar\rho})^{d/2}\,. \tag{6.56}$$

This can be used to get rid of the integral involving $(x^{\bar1})^2$, and get

$$\Delta_2 = \tau^{2d+2}\Big[R\Big(\Big(A_d + \frac{B_d}{d-1}\Big) H_{d,2,d/2}\Big) + \frac{R_{00}}{d-1}\big(d\, B_d\, H_{d,2,d/2} - B_d\, H_{d,0,1+d/2}\big)\Big]\,, \tag{6.57}$$

where

$$H_{d,i,j} = \frac{1}{\tau^d} \int_{\alpha_p(p,q)} \mathrm{d}^d\overline{x}\, (x^{\bar0})^i\, (x^{\bar\mu}\, x_{\bar\mu})^j \tag{6.58}$$

and

$$H_{d,i,j+1/2} = \frac{1}{\tau^d} \int_{\alpha_p(p,q)} \mathrm{d}^d\overline{x}\, (x^{\bar0})^i\, (x^{\bar\mu}\, x_{\bar\mu})^{j+d/2}\,; \tag{6.59}$$

here j can be either an integer or a half-integer.

Let us now compute these coefficients. In the calculations that follows we will treat the above integrands simply as functions, and "forget" that their source is a curvature. If we imagine slicing $\alpha_p(p,q)$ into $\overline{t}$ = constant balls and then slicing each ball into $\overline{r}$ = constant spheres, it is easy to see that

$$\begin{aligned}&\int_{\alpha_p(p,q)} \mathrm{d}^d\overline{x}\, (x^{\bar0} - p^{\bar0})^i\, \big((x^{\bar\mu} - p^{\bar\mu})\, (x_{\bar\mu} - p_{\bar\mu})\big)^{j/2} \mathrm{d}^d x \\ &= \frac{2\pi^{(d-1)/2}}{\Gamma((d-1)/2)} \Big(\int_0^{\tau/2} \mathrm{d}\overline{t}\; \overline{t}^i \int_0^t \mathrm{d}\overline{r}\; \overline{r}^{d-2}\, (\overline{t}^2 - \overline{r}^2)^{j/2} + \int_{\tau/2}^{\tau} \mathrm{d}\overline{t}\; \overline{t}^i \int_0^{\tau-\overline{t}} \mathrm{d}\overline{r}\; \overline{r}^{d-2} (\overline{t}^2 - \overline{r}^2)^{j/2}\Big)\,.\end{aligned} \tag{6.60}$$

By changing variables to

$$\overline{u} = \frac{\overline{t}}{\tau}\,, \qquad \overline{s} = \frac{\overline{r}}{\overline{t}}\,, \tag{6.61}$$

the above expression becomes

$$\begin{aligned}&\int_{\alpha_p(p,q)} \mathrm{d}^d\overline{x}\, (x^{\bar0} - p^{\bar0})^i \big((x^{\bar\mu} - p^{\bar\mu})\, (x_{\bar\mu} - p_{\bar\mu})\big)^{j/2} \\ &= \frac{2\pi^{(d-1)/2}}{\Gamma((d-1)/2)}\, \tau^{i+j+d} \Big[\int_0^{1/2} \mathrm{d}\overline{u}\, \Big(\overline{u}^{i+j+d-1} \int_0^1 \mathrm{d}\overline{s}\; \overline{s}^{d-2} (1 - \overline{s}^2)^{j/2}\Big) + \\ &\qquad + \int_{1/2}^1 \mathrm{d}\overline{u}\, \Big(\overline{u}^{i+j+d-1} \int_0^{1/u-1} \mathrm{d}\overline{s}\; \overline{s}^{d-2}\, (1 - \overline{s}^2)^{j/2}\Big)\Big]\,.\end{aligned} \tag{6.62}$$

Since the limits of integration in the first term are constants, that term can be represented as a product of two separate integrals. After evaluating the u-integral, the expression becomes

$$\int_{\alpha_p(p,q)} \mathrm{d}^d\overline{x}\,(x^{\bar{0}}-p^{\bar{0}})^i\left((x^{\bar{\mu}}-p^{\bar{\mu}})(x_{\bar{\mu}}-p_{\bar{\mu}})\right)^{j/2}$$
$$= \frac{2\pi^{(d-1)/2}}{\Gamma((d-1)/2)}\tau^{i+j+d}\left[\frac{1}{(i+j+d)\,2^{i+j+d}}\int_0^1 \mathrm{d}\overline{s}\,\overline{s}^{d-2}\,(1-\overline{s}^2)^{j/2}+\right. \tag{6.63}$$
$$\left.+\int_{1/2}^1 \mathrm{d}\overline{u}\left(\overline{u}^{i+j+d-1}\int_0^{1/u-1}\mathrm{d}\overline{s}\,\overline{s}^{d-2}\,(1-\overline{s}^2)^{j/2}\right)\right].$$

From now on the calculation splits into four cases: even and odd d and even and odd j. From the original intentions of the calculation it is clear that whenever d is odd j is also add, and viceversa. So only these two cases need be considered.

Case 1: Both d and j are even. Since j is even, denote it as

$$j = 2h\,. \tag{6.64}$$

Expanding $(1-\overline{s}^2)^h$ binomially, the integral becomes

$$\int_{\alpha_p(p,q)} \mathrm{d}^d\overline{x}\,(x^{\bar{0}}-p^{\bar{0}})^i\left((x^{\bar{\mu}}-p^{\bar{\mu}})(x_{\bar{\mu}}-p_{\bar{\mu}})\right)^{j/2}$$
$$= \frac{2\pi^{(d-1)/2}}{\Gamma((d-1)/2)}\tau^{i+j+d}\sum_{k=0}^{h}(-1)^k\binom{h}{k}\left[\frac{1}{(i+2h+d)\,2^{i+2h+d}}\int_0^1 \mathrm{d}\overline{s}\,\overline{s}^{d-2+2k}+\right. \tag{6.65}$$
$$\left.+\int_{1/2}^1 \mathrm{d}\overline{u}\left(\overline{u}^{i+2h+d-1}\int_0^{\frac{1}{\overline{u}}-1}\mathrm{d}\overline{s}\,\overline{s}^{d-2+2k}\right)\right].$$

After evaluating the $\overline{s}$ integrals this becomes

$$\int_{\alpha_p(p,q)} \mathrm{d}^d\overline{x}\,(x^{\bar{0}}-p^{\bar{0}})^i\left((x^{\bar{\mu}}-p^{\bar{\mu}})(x_{\bar{\mu}}-p_{\bar{\mu}})\right)^{j/2}$$
$$= \frac{2\pi^{(d-1)/2}}{\Gamma((d-1)/2)}\tau^{i+j+d}\sum_{k=0}^{h}(-1)^k\binom{h}{k}\left[\frac{1}{2^{i+2h+d}\,(i+2h+d)\,(d-1+2k)}+\right.$$
$$\left.+\int_{1/2}^1 \mathrm{d}\overline{u}\left(\overline{u}^{i+2h+d-1}\,\frac{(1/\overline{u}-1)^{d-1+2k}}{d-1+2k}\right)\right]. \tag{6.66}$$

After pulling out $d-1+2k$ and expanding out $(1/\overline{u}-1)^{d-1+2k}$, that becomes

$$\int_{\alpha_p(p,q)} \mathrm{d}^d\overline{x}\,(x^{\bar{0}}-p^{\bar{0}})^i\left((x^{\bar{\mu}}-p^{\bar{\mu}})(x_{\bar{\mu}}-p_{\bar{\mu}})\right)^{j/2}$$
$$= \frac{2\pi^{(d-1)/2}}{\Gamma((d-1)/2)}\tau^{i+j+d}\sum_{k=0}^{h}\frac{(-1)^k}{d-1+2k}\binom{h}{k}\left(\frac{1}{2^{i+2h+d}\,(i+2h+d)}+\right. \tag{6.67}$$
$$\left.+\sum_{l=0}^{d-1+2k}(-1)^l\binom{d-1+2k}{l}\int_{1/2}^1 \mathrm{d}\overline{u}\,\overline{u}^{i+2h-2k+l}\right).$$

Evaluating the integral that is left this becomes

$$\int \mathrm{d}^d\overline{x}\,(x^{\bar{0}}-p^{\bar{0}})^i\left((x^{\bar{\mu}}-p^{\bar{\mu}})(x_{\bar{\mu}}-p_{\bar{\mu}})\right)^{j/2} = H_{dih}\,\tau^{i+j+d}\,, \tag{6.68}$$

where

$$H_{dih} = 2\pi^{(d-1)/2} \sum_{k=0}^{h} \frac{(-1)^k}{d-1+2k} \binom{h}{k} \left(\frac{1}{2^{i+2h+d}\,(i+2h+d)} + \right. \tag{6.69}$$

$$\left. + \sum_{l=0}^{d-1+2k} (-1)^l \binom{d-1+2k}{l} \frac{1-(\frac{1}{2})^{1+2h-2k+l+1}}{i+2h-2k+l+1} \right).$$

Case 2: Both d and j are odd. Since j is odd, it will be replaced with

$$j = 2\,h + 1\,. \tag{6.70}$$

Then the binomial expansion tells us that

$$(1-\overline{s}^2)^j = \sqrt{1-\overline{s}^2} \sum_{k=0}^{j} \binom{h}{k} (-1)^k\, \overline{s}^{2k}\,, \tag{6.71}$$

which means that the original integral can be rewritten as

$$\int_{\alpha_p(p,q)} \mathrm{d}^d\overline{x}\, (x^{\bar{0}} - p^{\bar{0}})^i \left((x^{\bar{\mu}} - p^{\bar{\mu}})\,(x_{\bar{\mu}} - p_{\bar{\mu}}) \right)^{j/2} = \frac{2\pi^{(d-1)/2}}{\Gamma((d-1)/2)}\, \tau^{2h+1+i+d} \times$$

$$\times \sum_{k=0}^{h} \binom{h}{k} (-1)^k \left[\frac{1}{(i+2h+1+d)\,2^{i+2h+1+d}} \int_0^1 \mathrm{d}\overline{s}\, \overline{s}^{d-2+2k} \sqrt{1-\overline{s}^2} \; + \right.$$

$$\left. + \int_{1/2}^{1} \mathrm{d}\overline{u} \left(\overline{u}^{i+2h+d} \int_0^{\frac{1}{\overline{u}}-1} \mathrm{d}\overline{s}\, \overline{s}^{d-2+2k} \sqrt{1-\overline{s}^2} \right) \right]. \tag{6.72}$$

Since d is odd, so is $d-2+2k$. Thus,

$$d - 2 + 2k = 2a + 1\,, \tag{6.73}$$

where

$$a = \frac{d-3+2k}{2}\,. \tag{6.74}$$

Thus, the integral of interest is $\int \overline{s}^{2a+1}\sqrt{1-\overline{s}^2}\,\mathrm{d}\overline{s}$.
By using

$$\overline{s} = \sin\overline{\theta} \tag{6.75}$$

the integral becomes

$$\int \overline{s}^{2a+1}\sqrt{1-\overline{s}^2}\,\mathrm{d}\overline{s} = \int \sin^{2a+1}\overline{\theta}\, \cos\overline{\theta}\, \mathrm{d}(\sin\overline{\theta}) = \int \sin^{2a+1}\overline{\theta}\, \cos^2\overline{\theta}\, \mathrm{d}\overline{\theta}\,. \tag{6.76}$$

Combining one of the $\sin\overline{\theta}$ factors with $\mathrm{d}\overline{\theta}$ gives

$$\int \overline{s}^{2a+1}\sqrt{1-\overline{s}^2}\,\mathrm{d}\overline{s} = -\int \sin^{2a}\overline{\theta}\, \cos^2\overline{\theta}\, \mathrm{d}(\cos\overline{\theta})\,. \tag{6.77}$$

Expanding $\sin^{2a}\overline{\theta}$ as

$$\sin^{2a}\overline{\theta} = (1-\cos^2\overline{\theta})^a = \sum_{b=0}^{a} (-1)^b \binom{a}{b} \cos^{2b}\overline{\theta} \tag{6.78}$$

the above integral becomes

$$\int \mathrm{d}\overline{s}\, \overline{s}^{2a+1} \sqrt{1-\overline{s}^2} = -\sum_{b=0}^{a} (-1)^b \binom{a}{b} \int \mathrm{d}(\cos\overline{\theta}) \cos^{2b+2}\overline{\theta} \tag{6.79}$$

$$= -\sum_{b=0}^{a} (-1)^b \binom{a}{b} \frac{\cos^{2b+3}\overline{\theta}}{2b+3} \tag{6.80}$$

By substituting $s = \sin\theta$ this becomes

$$\int \mathrm{d}\overline{s}\, \overline{s}^{2a+1} \sqrt{1-\overline{s}^2} = -\sum_{b=0}^{a} \frac{(-1)^b}{2b+3} \binom{a}{b} (1-\overline{s}^2)^{b+3/2} . \tag{6.81}$$

Substituting $2a+1 = d-2+2k$ we obtain

$$\int \mathrm{d}\overline{s}\, \overline{s}^{d-2+2k} \sqrt{1-\overline{s}^2} = -\sum_{b=0}^{k+\frac{d-3}{2}} \frac{(-1)^b}{2b+3} \binom{k+\frac{d-3}{2}}{b} (1-\overline{s}^2)^{b+3/2} . \tag{6.82}$$

Thus, the original integral becomes

$$\int_{\alpha_p(p,q)} \mathrm{d}^d\overline{x}\, (x^{\bar{0}} - p^{\bar{0}})^i \left((x^{\bar{\mu}} - p^{\bar{\mu}})\, (x_{\bar{\mu}} - p_{\bar{\mu}}) \right)^{h+1/2}$$

$$= \frac{2\pi^{(d-1)/2}}{\Gamma((d-1)/2)} \sum_{k=0}^{h} \left\{ (-1)^k \binom{h}{k} \sum_{b=0}^{k+\frac{d-3}{2}} \frac{(-1)^b}{2b+3} \binom{k+\frac{d-3}{2}}{b} \times \right.$$
$$\left. \times \left[\frac{1}{(i+2h+1+d)\, 2^{i+2h+1+d}} + \int_{1/2}^{1} \mathrm{d}\overline{u}\, \overline{u}^{i+2h+d} \left(\left(\frac{2}{\overline{u}} - \frac{1}{\overline{u}^2} \right)^{b+3/2} - 1 \right) \right] \right\} . \tag{6.83}$$

By evaluating the (-1) term of the integral over $\overline{u}$, and also pulling $\overline{u}$ out of the denominators, the expression becomes

$$\int_{\alpha_p(p,q)} \mathrm{d}^d\overline{x}\, (x^{\bar{0}} - p^{\bar{0}})^i \left((x^{\bar{\mu}} - p^{\bar{\mu}})\, (x_{\bar{\mu}} - p_{\bar{\mu}}) \right)^{h+1/2}$$

$$= \frac{2\pi^{(d-1)/2}}{\Gamma((d-1)/2)} \sum_{k=0}^{h} \left[(-1)^k \binom{h}{k} \sum_{b=0}^{k+\frac{d-3}{2}} \frac{(-1)^b}{2b+3} \binom{k+\frac{d-3}{2}}{b} \left(\frac{1}{(i+2h+1+d)\, 2^{i+2h+1+d}} + \right. \right.$$
$$\left. \left. - \frac{1-(\frac{1}{2})^{i+2h+d+1}}{i+2h+d+1} + \int_{1/2}^{1} \mathrm{d}\overline{u}\, \overline{u}^{i+2h+d-2b-3} (2\,\overline{u}-1)^{b+3/2} \right) \right] . \tag{6.84}$$

By expanding $(2\,\overline{u}-1)^{b+1}$ we get

$$\int_{1/2}^{1} \mathrm{d}\overline{u}\, \overline{u}^{i+2h+d-2b-3}\, (2\,\overline{u}-1)^{b+3/2} \tag{6.85}$$

$$= \sum_{c=0}^{b} (-1)^c \binom{b+1}{c} 2^{b-c+1} \int_{1/2}^{1} \mathrm{d}\overline{u}\, \overline{u}^{i+2h+d-b-c-2} \sqrt{2\,\overline{u}-1} .$$

By setting

$$\overline{v} = \sqrt{2\,\overline{u}-1} \tag{6.86}$$

this becomes

$$\int_{1/2}^{1} \mathrm{d}\overline{u}\, \overline{u}^{i+2h+d-2b-3}\,(2\,\overline{u}-1)^{b+3/2} \tag{6.87}$$

$$= \sum_{c=0}^{b}(-1)^c \binom{b+1}{c} 2^{b-c+1} \int_0^1 \mathrm{d}\overline{v}\, \overline{v}^2 \left(\frac{\overline{v}^2+1}{2}\right)^{i+2h+d-b-2-c} . \tag{6.88}$$

We can carry out the $\overline{v}$ integration if we expand the power of the binomial $(\overline{v}^2+1)$, and we get

$$\int_{1/2}^{1} \mathrm{d}\overline{u}\, \overline{u}^{i+2h+d-2b-3}\,(2\,\overline{u}-1)^{b+3/2} \tag{6.89}$$

$$= \sum_{c=0}^{b}\left((-1)^c \binom{b+1}{c} 2^{2b-i-2h-d+3} \sum_{e=0}^{i+2h+d-b-2-c} \binom{1+2h+d-b-2-c}{e} \frac{1}{2e+3}\right).$$

Substituting these into the original integral we obtain

$$\int_{\alpha_p(p,q)} \mathrm{d}^d\overline{x}\,(x^{\bar{0}}-p^{\bar{0}})^i \left((x^{\bar{\mu}}-p^{\bar{\mu}})\,(x_{\bar{\mu}}-p_{\bar{\mu}})\right)^{h+1/2} = H_{d,i,h+1/2}\,\tau^{2h+i+d+1} , \tag{6.90}$$

where

$$H_{d,i,h+1/2} = \frac{2\pi^{(d-1)/2}}{\Gamma((d-1)/2)} \sum_{k=0}^{h}\left\{(-1)^k \binom{h}{k} \sum_{b=0}^{k+(d-3)/2} \frac{(-1)^b}{2b+3}\binom{k+\frac{d-3}{2}}{b} \times \right.$$
$$\times \left[\frac{1}{(i+2h+d+1)\,2^{i+2h+d+1}} - \frac{1-(\frac{1}{2})^{i+2h+d+1}}{i+2h+d+1} + \right. \tag{6.91}$$
$$\left.\left. + \sum_{c=0}^{b}\left((-1)^c \binom{b+1}{c} 2^{2b-i-2h-d+3} \sum_{e=0}^{i+2h+d-b-2-c} \binom{1+2h+d-b-2-c}{e}\frac{1}{2e+3}\right)\right]\right\}.$$

Finally, let's compute Δ_3. Writing the metric in normal coordinates,

$$g_{\bar{\mu}\bar{\nu}} = \eta_{\bar{\mu}\bar{\nu}} - \tfrac{1}{3}\, R_{\bar{\mu}\bar{\rho}\bar{\nu}\bar{\sigma}}\, x^{\bar{\rho}}\, x^{\bar{\sigma}} , \tag{6.92}$$

we find that

$$\sqrt{|\det g|} = 1 - \tfrac{1}{6}\, R_{\bar{\rho}\bar{\sigma}}\, x^{\bar{\rho}}\, x^{\bar{\sigma}} . \tag{6.93}$$

Thus,

$$\Delta_{3d} = -\frac{k_d}{6}\, R_{\bar{\rho}\bar{\sigma}} \int \mathrm{d}^d x\, x^{\bar{\rho}}\, x^{\bar{\sigma}} \left((x^{\bar{\alpha}}-p^{\bar{\alpha}})\,(x_{\bar{\alpha}}-p_{\bar{\alpha}})\right)^{d/2} . \tag{6.94}$$

As was done with Δ_2, the correction to the correction term will be neglected, which means that integration is performed over $\alpha_p(p,q)$ instead of $\alpha(p,q)$ and no correction term is introduced to the $V(\alpha(p,x))$ when the expression $((x^{\bar{\alpha}}-p^{\bar{\alpha}})\,(x_{\bar{\alpha}}-p_{\bar{\alpha}}))^d$ was used.

Substituting the above into the expression for Δ_{3d} gives

$$\Delta_{3d} = \frac{k_d}{6}\, R_{\bar{\rho}\bar{\sigma}} \int_{\alpha_p(p,q)} \mathrm{d}^d\overline{x}\, x^{\bar{\rho}}\, x^{\bar{\sigma}} \left((x^{\bar{\alpha}}-p^{\bar{\alpha}})\,(x_{\bar{\alpha}}-p_{\bar{\alpha}})\right)^{d/2} . \tag{6.95}$$

By cylindrical symmetry, $(x^k)^2$ in the above integral can be replaced with $(x^1)^2$ which means

$$\Delta_{3d} = \frac{k_d}{6}\left[R_{\bar{0}\bar{0}} \int_{\alpha_p(p,q)} \mathrm{d}^d\overline{x}\,(x^{\bar{0}})^2 \left((x^{\bar{\alpha}}-p^{\bar{\alpha}})\,(x_{\bar{\alpha}}-p_{\bar{\alpha}})\right)^{d/2} + \right.$$
$$\left. - \left(\sum R_{\bar{k}\bar{k}}\right) \int_{\alpha_p(p,q)} \mathrm{d}^d\overline{x}\,(x^{\bar{1}})^2 \left((x^{\bar{\alpha}}-p^{\bar{\alpha}})\,(x_{\bar{\alpha}}-p_{\bar{\alpha}})\right)^{d/2}\right] . \tag{6.96}$$

Again, by cylindrical symmetry,

$$\int_{\alpha_p(p,q)} \mathrm{d}^d\overline{x}\,\left((x^{\bar\alpha}-p^{\bar\alpha})\,(x_{\bar\alpha}-p_{\bar\alpha})\right)^{1+d/2} = \int_{\alpha_p(p,q)} \mathrm{d}^d\overline{x}\,(x^{\bar 0})^2\,\left((x^{\bar\alpha}-p^{\bar\alpha})\,(x_{\bar\alpha}-p_{\bar\alpha})\right)^{d/2} + \\ -\,(d-1)\int_{\alpha_p(p,q)} \mathrm{d}^d\overline{x}\,(x^{\bar 1})^2\,\left((x^{\bar\alpha}-p^{\bar\alpha})\,(x_{\bar\alpha}-p_{\bar\alpha})\right)^{d/2}, \tag{6.97}$$

which allows us to express the integral involving $(x^{\bar 1})^2$ in terms of integrals involving $(x^{\bar 0})^2$. Substituting this expression into the expression for Δ_{3d} and doing some simple algebra gives

$$\Delta_{3d} = -\frac{k_d}{6\,(d-1)}\,\tau^{2d+2}\left[R\,(H_{d,0,1+d/2}-H_{d,2,d/2}) + R_{00}\,(d\,H_{d,2,d/2}-H_{d,0,1+d/2})\right]. \tag{6.98}$$

Adding Δ_{1d}, Δ_{2d} and Δ_{3d} together, the total correction becomes

$$\Delta_d = \Delta_{1d}+\Delta_{2d}+\Delta_{3d} = \tau^{2d+2}\,(C_d\,R + D_d\,R_{00}), \tag{6.99}$$

where

$$C_d = A_d\sum_{k=0}^{d+2}(-1)^k\binom{d+2}{k}\frac{k}{d/2+k} + H_{d,0,d/2+1}\left(A_d+\frac{B_d}{d-1}-\frac{k_d}{6\,(d-1)}\right) + \frac{H_{d,2,d/2}}{d-1}\left(\frac{k_d}{6}-B_d\right) \tag{6.100}$$

and

$$D_d = B_d\sum_{k=0}^{d+2}(-1)^k\binom{d+2}{k}\frac{k}{d/2+k} + \frac{1}{d-1}\left[\left(B_d-\frac{k_d}{6}\right)d\,H_{d,2,d/2} + \left(\frac{d\,k_d}{6}-B_d\right)H_{d,0,1+d/2}\right]. \tag{6.101}$$

Now, as we were computing the corrections to the integral, we almost forgot the main term! Here it is:

$$\int_{\alpha_p(p,q)} \mathrm{d}^d x\, k_d\left((x^{\bar\mu}-p^{\bar\mu})\,(x_{\bar\mu}-p_{\bar\mu})\right)^{d/2} = k_d\,\tau^{2d}\,H_{d,0,d/2}\,. \tag{6.102}$$

This means that the total integral is

$$\int_{\alpha(p,q)} \mathrm{d}^d x\, V(\alpha(p,x)) = k_d\,H_{d,0,d/2}\,\tau^{2d} + (C_d\,R + D_d\,R_{00})\,\tau^{2d+2}\,. \tag{6.103}$$

At the beginning of this section it was shown that the gravitational pre-Lagrangian is given by

$$\mathcal{L}(\prec,E,p,q) = \frac{1}{4\pi G}\left(V^2(\alpha(p,q)) + E\int_{\alpha(p,q)} \mathrm{d}^d r\,\sqrt{(-1)^{d-1}\det g}\,V(p,r)\right). \tag{6.104}$$

By substituting the above expression for the integral, as well as

$$V(\alpha(p,q)) = k_d\,\tau^d + \tau^{d+2}\,(A_d\,R + B_d\,R_{00}), \tag{6.105}$$

we obtain

$$\mathcal{L}(\prec,E,p,q) = \tau^{2d}\,k_d\,(k_d + E\,H_{d,0,d/2}) \\ +\,\tau^{2d+2}\left(R\,(2\,k_d\,A_d + E\,C_d) + R_{00}\,(2\,k_d\,B_d + E\,D_d)\right). \tag{6.106}$$

Thus, in order to minimize variations, we have to get rid of the $O(\tau^{2d})$ contribution, which comes form the R_{00} term. Thus, E_d is selected in such a way that the coefficient of R_{00} vanishes:

$$2\,k_d\,B_d + E_d\,D_d = 0 \Rightarrow E_d = -2\,k_d\,B_d/D_d\,. \tag{6.107}$$

Finally, substituting this into the expression for the gravitational Lagrangian gives

$$\mathcal{L}(\prec,p,q) = \frac{1}{4\pi G}\left[\tau^{2d}\,k_d^2\Big(1-\frac{2B_d}{D_d}\,H_{d,0,d/2}\Big) + 2\,k_d\,R\,\tau^{2d+2}\Big(A_d-\frac{B_d\,C_d}{D_d}\Big)\right]. \tag{6.108}$$

As long as d is fixed, the first term in this Lagrangian is constant, and in fact has the form of a cosmological constant; its role, however, needs to be studied in more detail, since it appears to have a very large value, while a different type of argument based on causal sets by R Sorkin [9] leads to a value of the right order of magnitude for the cosmological constant. The second term in (6.108) is proportional to the scalar curvature R, as expected, and can be rewritten in the more familiar form

$$\mathcal{L}_{\rm EH} = \frac{v_0}{8\pi\,G_d}\,R\,, \tag{6.109}$$

where v_0 is the volume taken up by a single point of a causal set and G_d is given by

$$G_d = G\,\frac{v D_d}{2\,k_d\,\tau_2^{2d+2}\,(A_d\,D_d - B_d\,C_d)}\,. \tag{6.110}$$

Regarding the use of the above gravitational Lagrangian, it should be noted that variations in the partial order $\prec$ are much more general than the variations in the metric that lead to the Einstein equation in the continuum version of the theory. One can speculate that, while causal-set variations that stay within the class of manifoldlike ones will still lead to an effective continuum gravitational field equation, based on the second term of (6.108), the first term in that expression suggests that other types of variations may lead to dynamical restrictions on how manifoldlike "classical causal sets" are allowed to be, and particularly to conditions on their dimensionality. Such topics are beyond the scope of this contribution.

7. Conclusion

In this work, Lagrangians for the main types of bosonic fields [10] have been successfully defined for arbitrary causal sets. The machinery we introduced allowed us to derive otherwise complicated-looking causal-set Lagrangians from simple Alexandrov-set based Lagrangian generators, and we showed that, if the relevant fields are defined on a Lorentzian manifold and the Alexandrov sets are small enough that we can use the appropriate slow-variation approximation for the fields inside them, the Lagrangian densities coincide with the ones used in standard field theory. For similar work on fermionic fields, see reference [11]. We should point out here that, in light of the non-linearity of the procedure for selecting an Alexandrov set, while Lagrangians for different fields can be added, the same is not true for Lagrangian generators. Thus, for a set of fields with individual Lagrangian generators $\mathcal{J}_1$, ..., $\mathcal{J}_n$, we can formally write

$$\mathcal{J} = \{\mathcal{J}_1,...,\mathcal{J}_n\}\,, \qquad \mathcal{L}_{\mathcal{J}} = \sum\nolimits_i \mathcal{L}_{\mathcal{J}_i}\,. \tag{7.1}$$

This line of research inherits the same problem that other approaches to quantum field theory have to face: if we treat gravity like all other fields, then a path integral over it implies an integration over all possible topologies and geometries, thus leaving us with none of the background information that is needed to define propagators or solve any other problems in ordinary quantum field theory. While this issue has been addressed in other approaches [12], our belief is that in our context these should be addressed by an appropriate theory of quantum measurement that accommodates general relativistic covariance and includes gravity as one of the "measured" fields. A formulation of that theory is beyond the scope of this work.

Another difficulty is that even if the topology was not an issue (say, we are dealing with a toy model of propagation of non-gravitational fields in fixed gravitational background), there are simply too many degrees of freedom to integrate over. In the case of regular quantum field theory this issue is addressed by visualizing spacetime as a regular cubic lattice, which allows one to perform the integrals over "all lattice points at once". In the case of a general causal set, however, such structure is absent, which means that one has to separately integrate over the fields at every single point in spacetime, one by one. While numerical simulations on the computer might make it possible to do so in toy models of causal sets consisting of only a few points, this becomes a problem for more realistic models. There is a way to make oneself feel better about it: The Einstein equation does not have many exact solutions either. Nevertheless, in order to bring causal set theory to the level of the Einstein equation, there have to be some simple cases for which there is a solution, such as for example a field with a very large gradient which assures a specific selection of Alexandrov sets. This, however, is still work for a future.

The above problems are made much worse by the fact that causal set theory claims to be a candidate for a quantum theory of gravity, which implies that it should be precise enough to talk about the gravitational field of elementary particles: a precision much higher than the one of any other existing theory. Thus, reducing it to crude approximations might well defeat the very purpose causal set theory is there in the first place.

Finally, the fact that the selection of Alexandrov sets is non-linear implies that the superposition principle does not work on microscopic scale. This might also put under question other principles that are based on linearity, such as the law of conservation of energy. In the case of smooth differentiable behavior, since these Lagrangians do reduce to the ones observed in regular quantum field theory, all the known laws, including the law of conservation of energy, should hold. They can, however, be violated once the behavior of fields is no longer smooth (such as, for instance, near the big bang). On the positive side, the existence of a source of non-conservation might have verifiable cosmological consequences.

References

[1] Bombelli L, Lee J, Meyer D and Sorkin R D 1987 Space-time as a causal set *Phys. Rev. Lett.* **59** 521
[2] Sorkin R D 2005 Causal sets: discrete gravity *Lectures on Quantum Gravity* ed A Gomberoff and D Marolf (New York: Springer) (*Preprint* arXiv:gr-qc/0309009)
[3] Dowker F 2005 Causal sets and the deep structure of spacetime *100 Years of Relativity – Space-time Structure: Einstein and Beyond* ed A Ashtekar (Singapore: World Scientific) (*Preprint* arXiv:gr-qc/0508109)
[4] Henson J 2009 The causal set approach to quantum gravity *Approaches to Quantum Gravity: Toward a New Understanding of Space, Time and Matter* ed D Oriti (Cambridge, UK: Cambridge U. Press) (*Preprint* arXiv:gr-qc/0601121)
[5] Rideout D and Sorkin R 2000 Classical sequential growth dynamics for causal sets *Phys. Rev.* D **61** 024002 (*Preprint* arXiv:gr-qc/9904062)
[6] Sverdlov R and Bombelli L 2009 Gravity and matter in causal set theory *Class. Quantum Grav.* **26** 075011 (*Preprint* arXiv:0801.0240)
[7] Sverdlov R 2008 Gauge fields in causal set theory *Preprint* arXiv:0807.2066
[8] Gibbons G W and Solodukhin S 2007 The geometry of small causal diamonds *Phys. Lett.* B **649** 317 (*Preprint* arXiv:hep-th/0703098)
[9] Sorkin R D 2007 Is the cosmological "constant" a nonlocal quantum residue of discreteness of the causal set type? *AIP Conf. Proc.* **957** 142 (*Preprint* arXiv:0710.1675)
[10] Sverdlov R 2008 Bosonic fields in causal set theory *Preprint* arXiv:0807.4709
[11] Sverdlov R 2008 Spinor fields in causal set theory *Preprint* arXiv:0808.2956
[12] Rovelli C 2006 Graviton propagator from background-independent quantum gravity *Phys. Rev. Lett.* **97** 151301 (*Preprint* arXiv:gr-qc/0508124)

Journal of Physics: Conference Series **174** (2009) 012020 doi:10.1088/1742-6596/174/1/012020

Quantum Histories and Quantum Gravity

Joe Henson

Perimeter Institute, 31 Caroline St. N. , Waterloo, Canada N2L 2Y5.

E-mail: jhenson@perimeterinstitute.ca

Abstract. This paper reviews the histories approach to quantum mechanics. This discussion is then applied to theories of quantum gravity. It is argued that the results of all measurements in suitably defined classical situations can be identified with properties of the histories. This observation has significance for the formulation of new theories (such as quantum gravity theories) as it puts a constraint on the kinematics, if the quantum/classical correspondence principle is to be preserved. Consequences for quantum gravity, particularly for Lorentz symmetry and the idea of "emergent geometry", are discussed.

1. Introduction

Quantum histories are those entities summed over in a quantum path integral or "state sum". How much of the physical interpretation that we give to classical histories (solutions of the classical equations of motion for the theory in question) is preserved in the quantum case? Is it, at least, fair to say that some of the quantum histories must approximate, in some sense, to classical ones? The answer has important consequences. If it was true, then given a classical theory we could put some constraints on the possible spaces of quantum histories that could underlie it, even before we had written down the rest of the dynamics.

This may seem like a weak condition considering simple cases such as a single particle. However, it becomes non-trivial if we consider the idea of emergence: when the apparent properties that we (approximately) test for at the classical level are only effective descriptions of some more fundamental properties of the quantum system. In that case there may be many possibilities for the quantum history space; it becomes more important to find methods for ruling out unsuitable ones. This is relevant to quantum gravity (QG). The idea that the spacetime manifold is merely an approximation to some deeper (and perhaps discrete) structure is a prevalent one in current QG research. In that case, it is argued below, these matters have an impact on questions such as the preservation of Lorentz symmetry, and strategies for producing emergent geometry theories in general.

However, in view of the current literature – particularly in quantum gravity – the situation is unclear. Certainly, quantum histories are quite unlike the classical solutions in some respects. Many pieces of research are influenced by the fact that effective, macroscopic properties may be only cryptically related to the microscopic properties of a system. As a result, it is sometimes argued that *only* properties of quantum *sums* of histories, rather than those of any single history, can be safely related to semiclassical properties (see *e.g.* [1, 2]). Also, in a certain view of quantum mechanics, the path integral exists only to calculate the evolution of the wave-function, and so the physical interpretation of the path integral is subsidiary to that of the wave-function.

Taking on this view encourages one to downplay questions about the properties of quantum histories.

Some of these statements seem to run contrary to other viewpoints on quantum mechanics. In what might be called the "histories viewpoint", the results of measurements are directly related to properties of the histories [3] [4](see also [5] and references therein). This allows an easy treatment of measurements distributed in time, and other insights into new and standard theories that would be difficult by other means.

Some closer analysis is necessary to reconcile, or choose between, the two views. In the first part of this paper, some claims are made that explain how properties of quantum histories relate to measurements and to classical approximations. The central claim argued for is that some of the quantum histories must approximate the classical histories in some reasonable class, if the correct classical regime is to be recovered. A more precise idea of the required approximation will be given. The relationship between symmetries and this approximation is discussed (section 3), and it is explained how all this is consistent with the idea of effective descriptions (section 4). This leads to some physical conditions on effective descriptions: roughly, that the fundamental-effective correspondence should respect the symmetries of the classical theory in question. In the remaining sections, this principle is applied to quantum gravity and theories of emergent spacetime, resulting in some constraints on such theories (coming from the requirement of local Lorentz invariance in the effective continuum description).

Some of the main points raised are not at all novel. The general viewpoint on quantum mechanics is present in Feynman's seminal paper [3] and makes an appearance in other well-known discussions of quantum mechanics, e.g. the work of Bell [6]. Hopefully, therefore, the conclusions of sections 2 to 4 will be unsurprising to many researchers. However, the application to quantum gravity and emergent dynamics does call on some details of this discussion.

At the outset, it should be noted that these arguments are in no way reliant on a particular view of the measurement problem. Neither is it necessary to accept that a histories-based view of quantum mechanics, as opposed to the state-vector based one, is somehow primary. It is only necessary that the described histories view of a quantum mechanical theory is consistent; it need not be preferable on some other grounds. Consequently, the conclusions should not only be relevant to those who are committed to a histories viewpoint – it is the possibility of such a viewpoint that is argued to have important consequences.

2. Histories and measurements

The purpose of this first part of the paper is to review some basic facts about the histories approach to quantum mechanics, and to justify some claims about histories and measurements in semi-classical situations. When a specific example is called for, the quantum mechanical theory of a single non-relativistic particle will be used. Although this example is useful for explanation, the points made will mostly be relevant to other systems as well. Where it is not obvious, any important differences will be stated in discussion.

The predictions of quantum mechanics come in the form of conditional probabilities of some experimental outcomes given some preparatory observations. All of these can be computed from a joint probability distribution over the outcomes of all of the measurements. Below, unless otherwise noted, we will use the Heisenberg picture, and measurements will be projective Von Neumann measurements. We will consider making measurements at N separate times $t_1, t_2, ..., t_i, ..., t_N$ on a single non-relativistic particle. At time t_i we test whether the position of the particle has some value in the set Δ_i. Let us define the corresponding (classical) observable $P_i\big(x(t_i)\big) := 1$ when $x(t_i) \in \Delta_i$ and 0 otherwise. In the operator formalism, the joint probability for passing all of the tests takes the following form:

$$\Pr(x(t_i) \in \Delta_i \,\forall\, i) = \| \, U(t_0, T) \, \hat{P}_n \ldots \hat{P}_2 \, \hat{P}_1 \, |\psi_0\rangle \, \|^2, \tag{1}$$

where $|\psi_0\rangle$ is the initial wave vector, $U(t',t)$ is the unitary evolution operator between times t and t', and from our definition of P_i,

$$\hat{P}_i = U^\dagger(t_0,t_i)\int_{\Delta_i} dx|x\rangle\langle x|\, U(t_0,t_i). \tag{2}$$

The time evolution to the final time T is included for later convenience. Using a path integral formulation, it is possible to write down another form for the joint probability. First let us give the probability amplitude for the particle to pass all of these tests and end up at some position x_T at a final time $T \geq t_n$:

$$\Phi\big(x(t_i)\in\Delta_i\,\forall\, i; x_T, T\big) = \int_{t_0}^{x_T,T} \mathcal{D}x(t)\left(\prod_{i=1}^{N} P_i\big(x(t_i)\big)\right) e^{(i/\hbar)S[x(t)]}\psi_0(x(t_0))\ , \tag{3}$$

where the limits of the path integral mean that we sum over all paths from an initial time t_0 to final time T with a specific final position x_T, $\psi_0(x) := \langle x|\psi_0\rangle$ is the initial wave function, and $S[x(t)]$ is the action of the path or "history" $x(t)$. The joint probability to pass all the tests is

$$\Pr(x(t_i)\in\Delta_i\,\forall\, i) = \int dx_T\, \Big|\Phi\big(x(t_i)\in\Delta_i\,\forall\, i; x_T, T\big)\Big|^2. \tag{4}$$

This[1] removes any direct reference to Hilbert space, operators and state vectors from our joint probability expression. Similarly we can find the probability for any combination of passed and failed tests, and from this we can make a probability distribution on the space of all possible sequences of outcomes. For measurements in the position (generalised) basis, and for dynamics that is local and Markovian in the appropriate sense [7], this is equivalent to the state-vector expression. These results follow from repeated application of Feynman's prescription for calculating the probability of an outcome of one measurement with path integrals [8], as explained in [4] [2].

Now let us consider the space of histories summed over in eqn.(4), which will be called the *history space* Ω. Each set of experimental outcomes is codified as a proposition about the system (a particle's position in this case), such as "$x(t_i)\in\Delta_i\,\forall\, i$". This proposition corresponds to a possible property of a history $x(t)$ in Ω. Also, it corresponds to a subset of the history space Ω, namely, the set of histories that have that property. For instance, let $\Gamma_i = \{x(t)\in\Omega : x(t_i)\in\Delta_i\}$. Then $\Gamma = \bigcap_{i=1}^N \Gamma_i = \{x(t)\in\Omega : x(t_i)\in\Delta_i\,\forall\, i\}$ is the set that corresponds to passing all the tests. This is the correspondence between measurement outcomes and properties of histories that we will be interested in throughout the following.

The joint probability can be rewritten using this idea. Note that $\prod_{i=1}^N P_i\big(x(t_i)\big) = 1$ for any history $x(t)\in\Gamma$ and 0 for other histories. We now write the probability as a function of a subset of Ω rather than of the corresponding proposition. From equations (3) and (4),

$$\Phi(\Gamma; x_T, T) = \int_{t_0;\, x(t)\in\Gamma}^{x_T,T} \mathcal{D}x(t)\, e^{(i/\hbar)S[x(t)]}\psi_0(x(t_0))\ , \tag{5}$$

where $x(t)\in\Gamma$ in the limits of the integral means that only paths in Γ are summed over, and the joint probability (which has been renamed) is

$$\mu(\Gamma) = \int dx_T\, \Big|\Phi(\Gamma; x_T, T)\Big|^2. \tag{6}$$

[1] The "incoherent sum" (*i.e.* the sum outside the square) over final positions in eqn.(4) is necessary for agreement with eqn.(1) , and follows from the use of the inner product in (1) – the value of the final time T has no effect on the result as is easily seen from (1).

[2] This is a special case of Caves' prescription, which allowed amplitudes for detection as a function of position (in the operator formalism, POVM measurements rather than the projection measurements used here for simplicity).

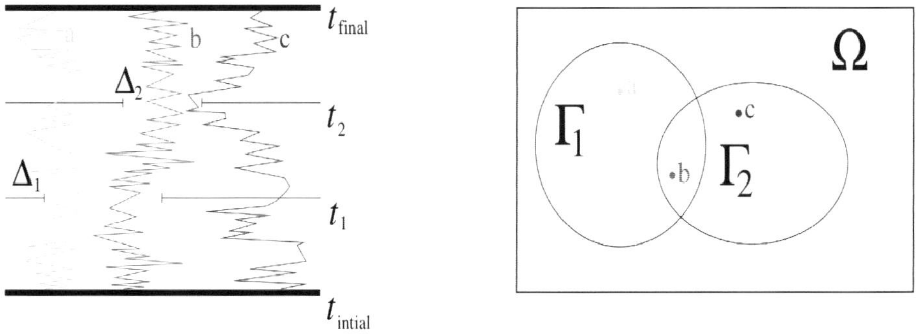

Figure 1. The left hand diagram is a space-time diagram of some paths in a typical single particle path integral, and the right hand diagram shows the space of histories as a set diagram. On the left time runs between initial and final times, and intervals Δ_1 (at time t_1) and Δ_2 (at time t_2) correspond to two measurements as described in the text. They also correspond to sets of histories Γ_1 and Γ_2, as shown on the right. Three histories are shown as examples on both sides.

This form best illustrates the connection between experimental outcomes and properties of histories, shown in figure 1. Having made this connection, we can say the following: $\mu(\Gamma)$ is the probability that the history lies in the set Γ – assuming that we make this particular set of measurements [3]. This defines the probability distribution over the set of alternative outcomes for our chosen measurements. However, while we can uniquely define $\mu(\Gamma)$ for a set of histories Γ, μ need not satisfy the conditions to be a probability measure on Ω [9]. The double slit experiment provides a simple counter-example: conditioning on the final position of the particle being in a dark line on the screen, the probability of finding the particle to have passed through some slit (had we measured for that) is not the sum of the probabilities for passing through the left slit and passing through the right slit (had we measured for that), violating the probability sum rule. That is why notation has been switched from Pr to μ, which is sometimes called a "quantum measure" (as opposed to a probability measure) on Ω [9], or the "diagonal of the decoherence functional" [10].

One advantage of this path integral way of writing the probabilities is that it can easily be extended to measurements distributed in time, unlike the standard Hilbert space/operator formalism [4]. There is nothing to stop the definition of (and, in principle, measurement for) more general properties such as "The particle was within the interval Δ at some time between times t_1 and t_2".

Similar prescriptions can be made for other systems, given a suitable choice of a basis to function as the position basis does above, which will be called the *histories basis* (it is important to note that not all bases are suitable in this regard). Such a sum-over-histories formalism can be defined with greater or lesser levels of rigour for all of the standard quantum theories, and when the history space is discrete, the path integral (or state sum) is well-defined. To summarise, we can make the following uncontroversial assertion:

Claim 1 *The outcomes of the measurements in the histories basis can be identified with properties of the quantum histories, i.e. with subsets of the history space.*

2.1. Comparing quantum and classical histories

Claim 1 can be used to defend the central claim of this section: that some quantum histories in our theory must approximate to classical histories in the corresponding classical theory.

Claim 1 does not assert that *all* measurements correspond to properties of histories. For instance, a measurement of the momentum $\hat{p}$ of our non-relativistic particle cannot always be derived from sequential measurements of its position, even on a coarse-grained level. But it is possible to claim this in certain situations, namely, those in the domain of validity of the corresponding classical theory. Otherwise, we would have no reason to give $\hat{p}$ the same name as the familiar classical momentum. Deferring a fuller explanation until the next section, the claim is the following (note again that measurements here are restricted to projective Von-Neumann measurements):

Claim 2 *In a classical situation, all measurements can be represented by projection operators in one basis*[3].

From claims 1 and 2, the next claim follows easily.

Claim 3 *In a classical situation, the outcomes of all measurements can be identified with properties of the histories (when a suitable basis is used as histories basis), i.e. with subsets of the history space.*

With the help of the correspondence principle, that quantum mechanics must agree with the predictions of classical mechanics in its domain of validity, this leads to our central claim. Firstly it is useful to restate the principle as follows:

Correspondence principle: If a quantum theory is to be consistent with its classical counterpart, then, for every classical history γ^{cl} in some reasonable set, there must exist a classical situation in that quantum theory containing enough measurements to approximately reconstruct γ^{cl}.

Combining this principle with claim 3 leads to the following:

Claim 4 *For every classical history (in some physically reasonable subset of the classical histories), there must exist some quantum history that approximates it*[4].

This claim will be a focus of the following discussion. It will rule out some history spaces as good starting points for the quantisation of a given theory. In standard cases our quantisation procedures ensure that this requirement is met, but this may no longer be true in a more general setting, as we will see. Later we will use such a strategy to give insights into quantum gravity.

So far, no details have been given to justify the claims, or even to define properly the new terms and concepts used therein. Chief amongst these are the term "classical situation", and the notion of "approximation" referred to in the statement of the correspondence principle and in claim 4. Classical situations will be investigated in the following section. This will justify claim 2, from which the others follow.

How do we decide on the level of approximation necessary to maintain the correspondence principle? For later convenience we will introduce the following notation: $\gamma \sim \gamma'$ if the two histories γ and γ' approximate to each other in the above sense. The most basic requirement comes from observation. Two histories cannot approximate to each other if they can be distinguished by an experiment, if that experiment confirms the classical theory (with high probability). Since it refers only to measurements in the domain of validity of the classical theory, this requirement can be known *prior* to the formulation of the quantum theory. In practice we might demand slightly more: that there will be similar classical situations possible

[3] The argument that projection operators (as opposed to POVMs) are general enough to describe classical measurements is made in [11].

[4] It should be possible to strengthen this claim, and to demand not one approximating quantum history but some larger set. Once the dynamics is specified, set of histories approximating the classical history should have a large quantum measure, and it seems unreasonable to expect a set composed of one history to do so. However this weak form of the claim seems sufficient for the following applications.

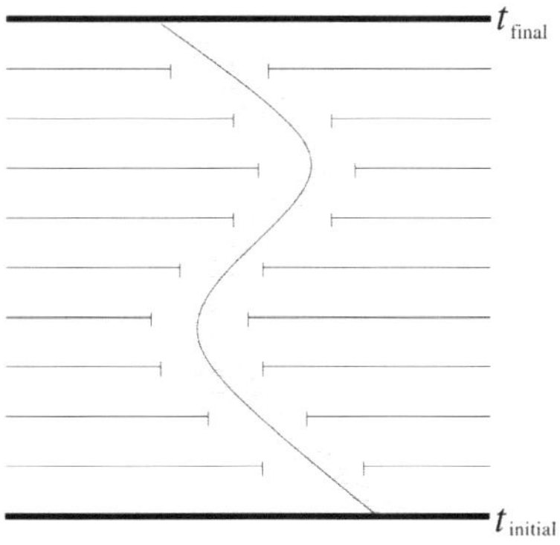

Figure 2. A sketch informally illustrating a classical situation. The sketch is a space-time diagram like the left hand side of figure 1. The black curve represents a classical history. The light-coloured path represents a typical path in the path integral. Measurements, corresponding to the spatial intervals represented by capped lines, are consistent with classical properties with high probability. Such measurements must be of finite accuracy, and be separated by finite intervals in time. However, with enough measurements the properties of the classical history are sufficiently well approximated.

in the theory for other possible classical solutions (although one must be careful not too demand too much [12]). This point is returned to in section 3, where it is argued that a condition must be put on $\sim$ if we are to preserve the symmetries of the classical theory.

First, a common objection to comparing quantum and classical histories is noted. There are seemingly some good reasons to be suspicious of comparing classical histories to their quantum counterparts. They can be shown to have several counter-intuitive and non-classical features. A "typical" path in a position path integral[5] is nowhere differentiable, and with discrete time steps the length along the path does not scale linearly with the distance between the end-points, exhibiting fractal behaviour. Because of this, in the continuum, the length along the path is typically infinite (these points are briefly reviewed in a quantum gravity context in [14]). In these ways the typical quantum histories are quite unlike classical histories.

This gives good reason to be skeptical of applying classical thinking to all aspects of the path integral. However, this argument does not speak against the claims made above. The point in question is how the properties of the quantum histories relate to *observable* properties of the classical history, at a level of approximation at which the classical description is valid. The above objections are not relevant to this: when do we ever measure the smoothness of the spacetime trajectory of a particle, or its total length? These properties are extremely fine-grained; to measure them, one would have to measure with total precision the position of the particle at all times. Measurements of these properties would be impossible, and quite unlike the measurements we perform to verify our classical theories in their domains of validity.

Claim 4 states that there must be some quantum histories that share all the (coarse-grained) properties of a classical history that we could observe in a classical situation. Paths may be unsmooth and of infinite length in some fine-grained sense, but all the appropriate coarse-grained properties can be classical, as sketched in figure 2. This is the basic meaning of "approximation" in the claim; a lot hangs on the subtleties of this approximation, which will be examined later.

Claim 4 puts a condition on the properties of individual histories in the sum-over-histories, not on "quantum sums" of histories. How are the properties of sets of histories different from those of the quantum histories themselves? At the classical level, these collective properties are

[5] This idea of typicality is an informal notion. It means a path which is a member of some reasonable high amplitude set, with "reasonable" having the same meaning as in the context of reasonable measurements [13].

nothing more than *some subset* of the properties of some quantum history: those properties common to all histories in the sum. According to the claims, there are no properties at the level of the classical approximation *additional* to those of the histories; the classical properties are not "strongly emergent" in that sense.

2.2. Classical situations

A *classical situation* is here defined as one in which no deviation from the classical theory is observed. In QM (using the Heisenberg picture), we have a state, a dynamics and a set of projection measurements M making up an "experimental situation" $\{|\Psi\rangle, U(t,t'), M\}$.

Definition 1 *A situation $\{|\Psi\rangle, U(t,t'), M\}$ is classical if, with high probability, none of the measurements in M contradict the predictions of some solution of the corresponding classical theory*[6]*, and if any subset of M has the same property*[7].

This includes measurements in all bases. A projection measurement $P_i \in M$ gives a range of possible values Δ_i for some observable O_i at some time t_i. A classical history $\gamma^{\rm cl}$ will assign one value to the observable at that time, $O_i^{\rm cl}(\gamma^{\rm cl}; t_i)$. Not contradicting the classical results means

$$\exists\, \gamma^{\rm cl} \text{ s.t. } O_i^{\rm cl}(\gamma^{\rm cl}; t_i) \in \Delta_i \quad \forall\, P_i \in M. \tag{7}$$

The most obvious property of such a situation is that the measurements included will not be of arbitrary accuracy. For instance, if M is classical then, whatever the initial state, no two sharp measurements of anything can be in M because the results would almost surely disagree with the exact values for any classical solution.

Any measurement, no matter how unsharp, does affect the state to some degree. However the choice of whether to carry out a classical measurement will not affect the results of any of the others. This can be seen from the above definition. Consider a set of projection measurements M on our system, and a larger set $M' = M \cup A$ containing some further measurement. Assume M' is classical for our system. By definition 1, all of the measurements in M' must confirm the classical predictions (with high probability). Since it is a subset of M', the same is true of M. If adding the further measurements A altered the results of any measurement in M, they would no longer confirm the classical predictions. But this cannot be true as M' is classical by assumption. Therefore the choice of whether to carry out the additional measurement, under this condition for classicality, does not affect the results.

Conversely, we can start with the following definition:

Definition 2 *A set of measurements M is classical (with respect to a particular dynamics and state) if, with high probability, none of the measurements contradict the predictions of some solution of the corresponding classical theory, and the choice of whether to carry out any subset of the measurements M does not affect the results of the other measurements.*

[6] It is relevant to ask whether giving the quantum state and dynamics approximately fixes the classical solution that will be "tracked" in any valid classical situation, or whether different classical solutions could be possible within one classical situation. For a semi-classical initial state one solution is approximately determined. In [15, 11] Kofler and Brukner argue that this need not always be the case; in fact, if one considers approximating the quantum mechanics with a probability distribution over classical states, the initial quantum state does not play a role in determining classicality at all. Even in this case, some small set of initial measurements will approximately decide the classical history to be followed, so the distinction is not critical in the following.

[7] One might worry that, in theory, the last part of the condition is not obeyed by some macroscopic systems: without external interference some macroscopic systems can diverge from classical behaviour in surprisingly short times [16, 17, 18]. Some measurements in that case could make the system behave *more* classically and it might seem that classicality need not be preserved if they are removed from the set M. However, this apparently does not lead to violation of the quantum/classical correspondence once a typical environment is added to the system, introducing decoherence effects. At least in this limited sense, decoherence effects play the role of the missing measurements.

From this we can easily infer that all subsets of M are also valid as the set of measurements in a classical situation (given the same initial state and dynamics). This gives an equivalent formulation to definition 1.

Whether a situation is classical or not will depend strongly upon the included measurements. Examples are situations in which the positions and angular momentum of solid macroscopic bodies are approximately measured; looking forward to a theory of quantum gravity, any set of measurements so far performed in confirmation of General Relativity (along with a suitable state and dynamics) would constitute a classical situation. It is interesting to ask when and how classical situations arise from more general quantum mechanical regimes, what degree of approximation is necessary for a given system, and so on. Such studies can be found in some standard textbooks [13, 19] and this is still an active area of research (see *e.g.* [15, 11, 16]). In particular [11] gives a specific and precise example in which a classical situation can be realised and the above claims are valid.

It is now straightforward to justify claim 2. It is necessary for the correspondence principle that classical situations exist in which the measurements in M give enough information to approximately reconstruct a classical history, as we often do in reality. Needless to say, given a classical history, *i.e.* the configuration variables at all times, the value of other observables such as momentum and energy can be calculated. The same is approximately true if the configuration as a function of time is only approximately known[8]. In classical situations, the results of reasonably unsharp measurements of, *e.g.*, momentum could be approximated by a series of (approximate) position measurements. It is sometimes said that "classical measurements approximately commute" in this context. We can conclude that all of the measurement outcomes that we find in a classical situation can be derived from a series of position measurements. In those situations, there is no loss of generality incurred by talking only in terms of measurements in this one basis. Similar reasoning would apply to more general systems than single particles. In other words, all the measurements in our set M can be re-expressed as a series of measurements one basis, which we can identify with the histories basis mentioned above. This justifies claim 2.

Having used it to justify the claims, we should note that definition 1 above is not as strong as it might appear. According to this requirement, it is possible for two sets of measurements A and B both to be good classical sets of measurements with respect to some state and dynamics, but for $A \cup B$ not to be. This is not a fault in the formalism. Every measurement affects the state, and it is possible that performing one extra measurement could tip the balance and prevent another measurement from verifying the classical prediction, within our standards of high probability.

2.3. The histories viewpoint

For completeness, a stronger histories viewpoint should also be mentioned, although this has less relevance to the following discussion. We can consider the properties of histories as giving a full account of all experimental outcomes, rather than just those in classical situations, albeit in a less direct fashion. In his original paper [3] Feynman makes this point:

> The [SOH measurement] postulate is limited to defining the results of position measurements. It does not say what must be done to define the result of a momentum measurement, for example. This is not a real limitation however, because in principle

[8] The approximation necessary for the values of these other variables is a function of the approximation for measurements of distance. We should not expect it to always be possible for the approximations to be small in terms of the average value, *e.g.* for contrived variables such as exponentials of the energy [12]. This is acceptable physically, as it possible to satisfy the correspondence principle without measuring such contrived variables more accurately.

the measurement of momentum of one particle can be performed in terms of position measurements of other particles, e.g. meter indicators. Thus, an analysis of such an experiment will determine what it is about the first particle which determines its momentum.

With this in mind, it is possible to hold that measurements in other bases are an unnecessary "fifth wheel" in quantum mechanics, which need never be considered when corresponding our theory to reality. In this view, concepts like momentum measurements are viewed as deriving from series of position measurements only in a classical approximation, and have no more fundamental meaning. One gives up Hilbert space basis symmetry as being of fundamental significance, perhaps in the hope of incorporating relativity more naturally into quantum mechanics. That is an expression of the histories view of quantum mechanics. It is even possible to go further, and consider interpretations in which one history (from the set corresponding to all measurement outcomes) represents reality (see [20, 21] and references therein). Such a history would indeed have all the properties needed to explain observations (although these approaches would arguably have to face the same challenges as De Broglie-Bohm pilot wave theories: does this notion of "the real history" serve all or any of the same purposes as its classical analogue, or is it itself a fifth wheel?). In developing a feel for the histories approach, it may help to know that such views are possible; however, the comments below will not make direct use of this stronger interpretation.

Having reviewed the basic claims, we can now turn to some more specific questions. The next two sections explain how symmetries and effective descriptions affect these claims, and thus show how claim 4 can be more than a trivial requirement.

3. Symmetries and classical measurements

As has been emphasised, it is necessary for the correspondence principle that there exist classical situations in our quantum theory, allowing us to approximately reconstruct parts of a classical history, and thus find consistency with our experience of the classical world. The classical theory may have symmetries, which we can verify by measurements in its domain of validity. Does this affect the level of approximation that we can demand for claim 4?

Let us consider the case of a macroscopic, rigid rotating body. In everyday life, we are used to the fact that we can approximately measure all of the orientation and angular momentum components of such an object. In other words, in a certain regime, we never find that measurement of the z-component of angular momentum of, say, a baseball, alters or precludes the measurement of the y-component. Therefore it is tempting to make the following claim:

Definition 3 *Consider measurements P and $P' = U^{-1}PU$, where U is a unitary operator representing a symmetry transformation. In a classical situation $\{|\Psi\rangle, U(t,t'), M\}$ that is "generic", if measurement P is in M, then $\{|\Psi\rangle, U(t,t'), M'\}$ is also a classical situation, where $M' = M \cup P'$.*

However, it is not impossible to imagine a classical situation, as defined previously, in which this was not the case. As noted at the end of section 2.2, it is possible that for some state and dynamics, the sets of measurements A and B can both be classical, while the combined set $A \cup B$ is not. For example, A could be a measurement of the z-component of angular momentum with some high accuracy so that the situation is "only just" classical (*i.e.* the result matches the classical one within our standards of high probability, but the probability is close to breaking those standards). Then B could be a rotated version of A, and carrying out both A and B could result in a probability to match the classical path that is outside our standards of high probability.

This situation is possible, but, as noted, not typical – hence the insertion of the word "generic" in definition 3. To engineer such measurements on a baseball, we would have to measure the angular momentum components with tremendous accuracy, and perform the two measurements in such quick succession that environmental disturbance could not return the ball to a typical semi-classical state in the meantime. It would be interesting to be more exact about this, but the main point stands without further analysis: everybody is aware that we can, and almost always do, observe classical situations in which such difficulties do not arise, and verify symmetries of the classical theory in them.

In the light of this, it is reasonable to add definition 3, as a strengthening of our definition of classical situations. Again, we leave aside the question of how and when exactly such generic classical situations arise in standard theories.

Adding this claim is important, as it preserves the meaning of the symmetries in classical situations. Take a classical theory and a symmetric state in that theory (the rotational symmetry of a region in a field, for example, or the global Lorentz symmetry of Minkowski space realised as a solution of GR). In the domain of validity of the classical theory, we should be able to test the symmetry by performing a measurement, followed by several transformed versions of the measurement. If the results of all measurements do not match, the symmetry is ruled out. We can embed this classical case as a (generic) classical situation in a quantum theory, with the help of definition 3. We start with a symmetric quantum state and we can perform all measurements $P' = U_1^{-1}PU_1$, $P'' = U_2^{-1}PU_2$, and so on. All of these can be included in a classical situation, if one of them can, according to definition 3 (as long as the classical situation remains generic after the inclusion of each measurement). Again, if results fail to match we can conclude that the symmetry is broken. This is the situation in standard quantum mechanical theories. For example, in a semi-classical state, the rotational symmetry of a region in a field could be verified in this way.

Without definition 3, the situation is unclear. Now, in the classical situation, it is possible to start with a symmetric quantum state, but obtain different results for the measurements P, P' *etc.* There is nothing wrong with the symmetry of the quantum state; it is classicality that has failed instead in this case. But what becomes of the classical test for symmetry? Following the classical rules we would say that the symmetry is broken, but it is still possible that the initial quantum state was symmetric. Indeed, performing a symmetry transformation on *all* of the measurements, and repeating the experiment with an initial symmetric state, would of course produce the same results (statistically).

Further, consider the case in which we have no control over the first measurement P – we could even substitute for it an environmental decoherence effect. In this case the failure of classicality caused by measurement would depend on the symmetry group element defined by P, something we could not even alter in different runs of the experiment. The breaking of the symmetry by the uncontrollable disturbance of P, *even though it did not affect the classical state being tracked*, could destroy the classical symmetry. Emphasising the classical or quantum side, we could say: (*i*) the notion of symmetry that we test classically is broken, and thus not properly captured by the symmetry of the quantum state, and (*ii*) we cannot not trust our usual classical way of verifying symmetry. In other words, claiming the existence of the symmetry would not be enough to predict that the usual classical tests would verify the symmetry. Fortunately in all standard theories we can apply definition 3 and so this difficulty has never arisen, so far. If it did arise, the onus would be on the inventors of the new theory to clarify matters.

The existence of these generic classical situations is also consistent with "quasi-classical" or "disturbance uncertainty" arguments. The requirement is the following: the quantum uncertainty in some measurement of an object system S should accord with the uncertainty produced by coupling a measurement system T to S, if S is treated as classical and T is granted the uncertainty relations between conjugate variables dictated by quantum mechanics. The Bohr-Rosenfeld argument [22] is a venerable example of this type. Failure of the classical symmetries by repeated measurement seems to be inconsistent with this, and might re-open some difficult conceptual problems for quantum mechanics.

In conclusion, such generic classical situations must exist in order for us to test (unambiguously) symmetries without leaving the domain of validity of the classical theory, *i.e.* to test the symmetries using classical methods. In the statement of the correspondence principle (section 2.1), the classical situation referred to must be a generic one. As in claim 4, this should allow us to approximately reconstruct a classical history, symmetries and all. So we can now demand something more from the approximation $\sim$:

Claim 5 *The level of approximation $\sim$ allowed in claim 4 should be consistent with the symmetries of the classical history being tracked.*

I.e., the approximation allowed for observable O should be the same as $U^{-1}OU$ for a symmetry transformation U. In yet other words, $\sim$ should respect the symmetries of the classical theory. This ensures that classicality can be preserved under symmetry transformed measurements.

4. Effective descriptions

The variables that give the most useful description of a system at a macroscopic scale are often not those that we need to employ at a more fundamental level. What is required is an effective (otherwise called coarse-grained or emergent) description of our fundamental theory. This is the case, for example, whenever a hydrodynamical description of many particles is used. Often, our classical theory is expressed in terms of effective properties, rather than the fundamental ones. We may then want to discuss the quantum/classical correspondence principle in this effective language. In quantum gravity, it has been hypothesised that the continua of GR may be effective descriptions of some discrete structure (see the later sections on quantum gravity). We will now discuss how this idea can be consistent with the claims made above.

Our starting point is a quantum theory as in the discussion of section 2. This is the fundamental theory, and should be complete in itself: in principle, it should supply all of our predictions, and all of the claims of the previous sections apply to it. Certain measurements correspond to subsets of the history space Ω as in claims 1 and 3. However, the fundamental description of these subsets might be too complicated to use directly; we would like to relabel them in a way that is easily related to measurement outcomes. The effective description is an efficient way of achieving this: we can express measurement outcomes in terms of subsets of some different "effective" history space Ω', and then relate these to subsets of the fundamental history space. That is, we can use some relation that, given an effective property corresponding to $\Gamma' \subset \Omega'$ (*e.g.* some value of a hydrodynamical mass density integrated over a region), outputs a fundamental property that is its expression in the fundamental theory (*e.g.* some condition on configurations of many particles). For convenience, the space Ω' should contain all of the effective histories that we need to have a good effective description (*i.e.* so that, when the effective description is valid, all of the measurement outcomes correspond well to effective history properties), and nothing else.

Let us formalise the idea of effective properties. We will use a binary relation $\rightharpoonup$ to associate fundamental histories $\gamma \in \Omega$ to effective histories $\gamma' \in \Omega'$ [9]. That is, $\gamma \rightharpoonup \gamma'$ if γ' is an effective description of the history γ. Note that $\rightharpoonup$ need not be a function. There may be many effective descriptions of one fundmental history, and some fundamental histories may have no effective description at all (*e.g.* a hydrodynamical description of a system may not cover all possible states of that system).

As required, every subset of the effective history space is related to a subset of the fundamental history space via $\rightharpoonup$. Consider an effective property P and the associated set of histories $\Gamma'_P \subset \Omega'$.

[9] In other words, there is a subset $E_{\rightharpoonup}$ of the Cartesian product $\Omega \times \Omega'$, and $x \rightharpoonup y$ means $x \times y \in E_{\rightharpoonup}$.

Let us say that a fundamental history γ "has effective property P" if and only if *all* effective descriptions of γ have the property. Thus, the subset of the fundamental history space associated to P is

$$\Gamma_P = \{\gamma \in \Omega : \gamma \rightharpoonup \gamma' \Rightarrow \gamma' \in \Gamma'_P\}. \tag{8}$$

This is the physically relevant concept: if a measurement outcome definitely corresponds to Γ'_P, this implies that the fundamental history is in set Γ_P.

Using $\rightharpoonup$, we can in principle calculate probabilities of measurement outcomes from the fundamental theory as before. In practice approximations are used, and direct reference to the fundamental theory becomes unnecessary. This effective description may be indispensable in practice, and can lead to a theory that looks, even qualitatively, quite different from the fundamental one[10]. However, the fact that it is in principle possible to relate histories in Ω and Ω' is the lynch-pin that holds the two descriptions together; without it the idea of an effective description could not be meaningful. If a histories formulation is possible for the fundamental theory, then this way of describing the effective theory must also be possible, whatever other formulations might also be possible.

This represents no fundamental change in perspective from that of the previous section. Using $\rightharpoonup$, any question about effective properties corresponds to a question about fundamental histories, and the correspondence principle can be dealt with as before. Conversely, there are no properties of the effective histories which cannot be re-expressed as properties of fundamental histories. Again, there are no "strongly emergent" observable properties with no counterpart at the fundamental level. The only difference is the extra layer of approximation introduced.

This conclusion follows from the claims of the previous section, and is not at variance with the perspectives coming from condensed state matter systems or quantum field theory. It seems that the only field in which explicitly contradictory views are sometimes expressed is quantum gravity. It is interesting to note that Wilson and Kadanoff's coarse-grainings, an essential ingredient of their renormalisation group techniques, are consistent with this picture [23]. There, a map is made between individual fine-grained lattice field configurations and individual field configurations on a coarser lattice, which gives the relation $\rightharpoonup$ in the above terminology. Roughly speaking, the quantum amplitude attributed to a coarse-grained configuration is calculated by summing over the fine-grained configurations that correspond to it. A candidate $\rightharpoonup$ is chosen on the basis that all fine-grained configurations that map to the same coarse-graining must be physically indistinguishable when probing on large scales. This is an example of a situation in which a histories view has helped in the derivation and understanding of new physical concepts.

4.1. Physical conditions on effective descriptions

There are some necessary conditions on $\rightharpoonup$, if a good effective description of our theories in terms of Ω' is to be recovered. We have already said that we require *all* members of Ω' to be possible histories, so $\rightharpoonup$ must be surjective (or "right-total"): for all $\gamma' \in \Omega'$, there must exist a $\gamma \in \Omega$ such that $\gamma \rightharpoonup \gamma'$.

Still, a fundamental history may contain *less* information than an effective one: consider approximating a discrete set of atoms with a continuous medium, for example. This can be accommodated by allowing $\rightharpoonup$ to be one-to-many[11]. This is physically acceptable: because experiments cannot be carried out to arbitrary accuracy, not all histories in Ω' need be physically distinguishable. A fundamental history γ can have more than one approximating effective

[10] Note that similar statements could be made in the case of classical statistical mechanics, where is is obvious that all measurement outcomes correspond to properties of the fundamental histories.

[11] In other terms, $\rightharpoonup$ is not functional. A binary relation R between domain X and codomain Y is functional (or right definite) if for all $x \in X$, and all $y, z \in Y$ it holds that if xRy and xRz then $y = z$. It is left-total if there exists a $y \in Y$ such that xRy for all $x \in X$. We have already noted that $\rightharpoonup$ need not be left-total.

history, as long as these effective histories are physically indistinguishable from one another. We will write $\gamma'_1 \sim \gamma'_2$ if these two effective histories are physically indistinguishable. Using this, the condition is:

$$\forall \gamma \in \Omega, \forall \gamma'_1, \gamma'_2 \in \Omega', \ \gamma \rightharpoonup \gamma'_1 \,\text{and}\, \gamma \rightharpoonup \gamma'_2 \ \Rightarrow \ \gamma'_1 \sim \gamma'_2. \tag{9}$$

Here we have deliberately identified "physical indistinguishability" with the level of approximation allowed between classical and quantum histories, $\sim$. This represents the minimal requirement that the accuracy of our measurements (*i.e.* ability to distinguish the histories) must be at least as good as that used in the domain of validity of the classical theory. If this condition were not true, we would be in the untenable situation of being able to measure for a property of a system, while denying that it has any meaning in our fundamental theory of that system; hence we would be unable to make any predictions about it from that theory. These conditions can be combined with claim 5 of section 3, which put a physical condition on the approximation relation $\sim$.

The requirements that we arrive at are the following:

Physical conditions on effective descriptions: $\rightharpoonup$ *is surjective; the condition in eqn.(9) must hold; in that condition, the relation of physical indistinguishability* $\sim$ *must respect any classical symmetries that we wish to preserve.*

It is important to note that the symmetries mentioned here operate on the *effective* histories $\gamma' \in \Omega'$ that are the arguments of $\sim$; they may be absent or even meaningless on the fundamental level. The symmetries need only have meaning at the effective level to connect with the classical theory. For example, when gases are commonly referred to as translationally and rotationally invariant, this is a statement referring to the effective continuum level. It has no counterpart at the fundamental level.

At this point the relevance to quantum gravity can be emphasised. Say we have a classical theory to quantise. Claim 4 gives a necessary condition on the history space, in order to maintain the correspondence principle. This seems like a weak condition, and in standard cases our quantisation procedures ensure that the condition is satisfied. However, say we believe that the best route is to propose some more fundamental history space which is only linked to the classical histories via an effective description $\rightharpoonup$. In this case claim 4 leads to the above conditions, and may become non-trivial. Further investigation is needed to ensure the correspondence principle can hold. This can be taken care of once something is known about Ω and $\rightharpoonup$; we need not wait for the full definition of our quantum theory.

5. Applying the histories approach to quantum gravity

This section begins the second part of the paper, in which the ideas previously discussed are applied to quantum gravity.

How would gravity fit into the histories framework? The review in section 2 started with the example of non-relativistic particles; histories as described there were functions of some fixed time co-ordinate. In GR such a time co-ordinate does not have a physical meaning independent of some co-ordinatisation. However, in a more general view, the quantum histories need only be some unsmooth generalisation of the classical histories, and the discussion above would carry over unaltered. For example, theories of the time-reparameterisation invariant particle can be made in which the histories are not functions of time, and a theory of gravity might work in a similar way [10]. It may even be possible to use the standard idea of histories as a function of time, given some gauge fixing.

Several prominent QG approaches have been built on the presupposition that a histories framework will remain valid here: Regge calculus approaches [24] including Causal Dynamical

Triangulations [25, 26, 27], as well as the related spin-foam models [28, 29] conjectured to include a spacetime formulation of loop quantum gravity [29], and causal sets [30, 31, 32] (See also [33] for a discussion of the advantages of a histories approach in quantum gravity). Any approach based on condensed matter or quantum-computational analogues, since they use more-or-less standard quantum theory, could be cast in a histories language. Whether the histories framework for gravity is, as in standard theory, equivalent to a canonical formulation, or whether it would be more fruitful in this case to go beyond the canonical form altogether, is a debated issue. The only assumption below is that a histories view is possible. Indeed, if a generalised quantum theory was made which somehow did *not* allow a histories formulation, this would itself raise the question whether too much of the standard quantum picture had been given up for the theory to be physically reasonable. The whole physical interpretation of such a theory, especially the correspondence principle, would have to be thoroughly re-evaluated before it could be accepted.

In claims 3 and 4, we have seen that there are physical constraints on the quantum history space, coming from the requirement that our quantum theory satisfies the correspondence principle. For a quantum gravity theory to have the correct classical approximation, there must be some histories in Ω which have all the properties that we expect to be able to observe in a classical situation. A classical situation here (real or hypothetical) would be one in which General Relativity is verified by experiment and observation. It could, for instance, include all measurements that have actually been made that are relevant to GR. The following is an adaptation of claim 4 to quantum gravity.

Claim 6 *Consider a quantum gravity theory that is consistent with the histories framework. To be physically reasonable, some histories of the theory must approximate to solutions of general relativity*[12].

The appropriate level of approximation should obey the same conditions as for claim 4. Using effective descriptions as in section 4, the fundamental histories do not have to be spacetimes (meaning diffeomorphism classes of Lorentzian manifolds) in order to satisfy claim 6. Also, from claim 5, we might add that *the level of approximation required in claim 6 for each observable should be consistent with the symmetries of the spacetime being tracked.*

Claim 6, though derived from fairly uncontroversial grounds in standard quantum theory, may have interesting implications, particularly for two interrelated issues: "emergent gravity", and the fate of Lorentz symmetry in quantum gravity.

6. Lorentz invariance

Local Lorentz Invariance, or more specifically the question of whether this physical principle will be maintained or broken at the Planck scale, has become a much debated topic both in the theory and phenomenology of quantum gravity. On the phenomenological side, constraints on violation of Lorentz invariance (termed "Lorentz violation" below) are becoming ever more stringent [34, 35], and strong arguments have been made that Planck scale Lorentz violation is incompatible with current observations, without (at best) additional fine-tuning being introduced [36, 37]. With this progress, it becomes increasingly important for the theoretical side to provide predictions, at least of a heuristic nature: in any particular theoretical framework, what becomes of Lorentz invariance? Is it possible to maintain it, or must it be broken at some scale?

It is well-known that waves travelling on lattices violate Lorentz invariance, and this is often used as an argument against fundamental discreteness. Some Loop Quantum Gravity (LQG)

[12] The argument leading to claim 6 requires that the histories be written in some reasonable histories basis, corresponding to some reasonable basic observable of the classical theory. This is the case for spin-foams, where the histories are supposed to correspond to triangulations of spacetimes with labelled faces. These variables are supposed to relate to those of GR in a direct way.

based arguments have been made to support this expectation [38, 39, 40], and some further discussion of the methods used is given in [41]. However, there are conflicting arguments for local Lorentz invariance in LQG [42], and also some [43, 44] for the "third way" of Doubly Special Relativity (DSR) [45] in which the Planck length or Planck energy is introduced as an invariant scale along with the speed of light, deforming the Lorentz transformations (an idea that is criticised in [46]). If the spin-foam quantum gravity program is to provide a path-integral formulation of LQG, then it might be expected that the argument which prevails in the LQG program will also prevail for these models, and *vice-versa.*

Here the question will be more closely examined. Planck-scale lattices do violate Lorentz invariance, when considered as fixed backgrounds rather than in a quantum sum. But does this have any physical significance for quantum gravity theories?

Regge triangulations are in a strong sense lattice-like. They are co-ordinate free, meaning that the metrical information is all in the form of lengths between vertices, and the lattice connectivity (or in some generalisations, areas of triangles and/or angles). Having said that, the classical theory of Regge calculus is not applied with finite lattice spacing, except as an approximation; the limit of small edge lengths must be taken to recover general relativity and all its continuum symmetries. It has never been claimed that, with a finite lattice spacing, a space of Regge triangulations could approximate a reasonable space of spacetimes without violating local Lorentz symmetry (this will be discussed more precisely later).

Thus it would not be surprising if a quantum theory built on such Regge triangulations violated Lorentz symmetry, unless a continuum limit were taken. After all, the classical theory being quantised is Lorentz violating. However, in some approaches it is claimed that the problem can be overcome by a quantum sum over many triangulations. The program to which this issue is of most relevance is spin-foams, where fundamental "Planck-scale" discreteness is claimed. In other Regge calculus inspired approaches the lattice spacing is taken to zero. In the causal set approach, the histories are not lattice-like in this sense. They are claimed to be discrete *and* to have local Lorentz symmetry, in any situation in which a continuum approximation is valid [47, 48].

6.1. *A condition for symmetry preservation in discrete quantum gravity*

In theories like spin-foams, the spacetimes of GR (perhaps with additional fields) are effective descriptions of the fundamental discrete structures. When assessing such theories from the histories perspective, we should therefore apply the discussion of section 4, and the conditions on effective descriptions developed in section 4.1. We have the fundamental histories Ω and the effective histories Ω'. There should be a binary relation $\rightharpoonup$ such that $\gamma \rightharpoonup \mathcal{M}$ if the spacetime $\mathcal{M}$ is an effective description of the history γ. To apply the conditions from section 4.1 to the case of quantum gravity:

Physical conditions on emergent spacetime: $\rightharpoonup$ *is surjective;*

$$\forall \gamma \in \Omega \text{ and } \mathcal{M}_1, \mathcal{M}_2 \in \Omega', \quad \gamma \rightharpoonup \mathcal{M}_1 \text{ and} \gamma \rightharpoonup \mathcal{M}_2 \Rightarrow \mathcal{M}_1 \sim \mathcal{M}_2; \tag{10}$$

here, the relation of physical indistinguishability $\sim$ *must respect all symmetries of GR that we wish to preserve.*

To recap, the last clause follows if we accept the discussion of section 3 on symmetries. As before, the symmetries are not applied directly to the discrete structures themselves. These symmetries, such as local Lorentz symmetry, have their first definition at the level of the approximating continua. That is where they are properly applied and tested.

Journal of Physics: Conference Series **174** (2009) 012020 doi:10.1088/1742-6596/174/1/012020

This is an important and non-trivial condition, which would be difficult to formulate without the histories perspective given in earlier sections. It depends only on Ω and $\rightharpoonup$ for each theory (suitable definitions of Ω' and $\sim$ should be fixed by GR), and so the condition is in some sense only kinematical; we do not need to solve the theory or even know the full dynamics to test it.

In causal set theory this condition is known as the *hauptvermutung* or "fundamental conjecture" and justifying it is taken as a crucial task. In that case, a discrete/continuum correspondence can be defined that does indeed respect the symmetries of GR in the effective continuum description (*i.e.* it involves no special choice of co-ordinates, frames, foliation, *etc.*). As well as this, there is now compelling evidence for the *hauptvermutung*: timelike distances [49], dimension [50, 51, 52] and topology [53] can all be approximately reconstructed from a causal set.

In the spin-foam approach, no attempt has been made to justify a condition of the form of eqn.(10). There is no explicit proposal for $\rightharpoonup$. There is no discussion in the spin-foam literature explaining why the condition is safely to be ignored, leaving us to reconstruct the probable arguments. It may be due to disagreements on some statements made that led up to (10), for example in section 4.1. Disagreement there could lead one to refute that the condition is necessary in order to produce GR in a semi-classical limit. On the other hand, it may be due to some implicit confidence that such a condition could be satisfied by any reasonable discretisation of spacetime, with a good choice of $\rightharpoonup$.

In contrast, I conjecture that the condition must fail in spin-foam theories, for any choice of $\rightharpoonup$ of the expected form, if a continuum limit is not taken. By "expected" I mean that vertices will in some sense correspond to spacetime points, and so on for higher dimensional simplices, as suggested by the Regge calculus on which the spin-foams approach relies. The conjecture will not be proved in this paper. However, some reasons will be given to doubt that the physical conditions given above can be satisfied for spin-foams. After this, the symmetry arguments from section 4.1 will be briefly reviewed in the QG context to answer the other possible disagreement.

6.2. The problem with lattices

After the above discussion, we can reassess the question: is it significant that lattice-like histories violate Lorentz symmetry, when taken individually? First we should say what this means. Let us concentrate on Minkowski space, in which Lorentz symmetry is global, bearing in mind that this is a good approximation to the geometry of significant regions of the actual universe.

Let us first consider a fixed background case. The lattice can be considered as an abstract graph up to isomorphisms, perhaps with length-labelled edges. Consider a square (or diamond) lattice L with vertices v_L, and fields $\phi_L : v_L \longrightarrow \mathbb{R}$ on the lattice, and let each field configuration be a member of Ω. Now consider approximating these by real scalar fields on Minkowski space $\phi_{\mathbb{M}} \in \Omega'$. This kind of discretisation is very common, and the definition of $\rightharpoonup$ is simple. It can be written in a way that explicitly invokes a preferred co-ordinate frame in Minkowski space, but this is not necessary. Another form that does not reference a special frame can be written. The rough form of such an $\rightharpoonup$ is given in the following:

$\phi_L \rightharpoonup \phi_{\mathbb{M}}$ iff there exists some embedding $\iota : v_L \rightarrow \mathbb{M}$ (i.e. a map from vertices of the graph L to points of $\mathbb{M}$) such that $\forall x \in v_L$, $\phi_L(x) \approx \phi_{\mathbb{M}}(\iota(x))$, and such that the lattice distances "match" those in $\mathbb{M}$.

There is no need to be specific about the type of matching of distances to be used, as the following argument is not sensitive to these details. It could be exact or approximate. We want to discreteness scale to be "Planckian" in some sense, in accord with familiar arguments in QG; again we need not be specific. We could say, for example, that the embedding should include some Plankian order of vertices in a region of unit volume.

Such a discrete-continuum correspondence $\rightharpoonup$ must be used when the continuum limit is not taken, but it does not satisfy our physical conditions on emergent spacetime. Although there is no use of a frame selected *ab initio*, the embedding introduces a special lattice frame, and so the correspondence obviously depends on a frame. Some properties of the continuum histories are missing from the lattice histories.

Consider a field configuration $\phi_{\mathbb{M}1}$ in which that there is one plane wave travelling. The lattice can only support a limited bandwidth and so the above definition for $\rightharpoonup$ cannot distinguish between such a field configuration and another, $\phi_{\mathbb{M}2}$ in which there is a second wave with frequency vastly differing from the first. This is straightforward to see: in whatever embedding was used to show that $\phi_{\mathbb{M}1} \rightharpoonup \phi_L$, we add a wave with the lattice spacing as its wavelength to form $\phi_{\mathbb{M}2}$, and thus find that the same embedding can be used to show $\phi_{\mathbb{M}2} \rightharpoonup \phi_L$. But, in contradiction of our physical condition on emergent spacetime given in equation (10), it cannot be that $\phi_{\mathbb{M}1} \sim \phi_{\mathbb{M}2}$. Whatever the details of our approximation $\sim$ are, the presence or absence of a wave of arbitrary amplitude – whatever the frequency happens to be in any particular frame – is surely something that should be considered physically distinguishable, *if* $\sim$ is to be Lorentz invariant. It is an uncontroversial claim that such a lattice kinematics inevitably leads to Lorentz symmetry breaking; it can only work with a Lorentz violating condition for $\sim$.

In this case, symmetry can be restored by passing to a random lattice. In the background dependent case, if points are distributed according to a Poisson process this eliminates the lattice frame [48]. But the positions of the points cannot be allowed *the slightest* uncertainty. For, if we were to randomly move them in a ball-shaped neighbourhood, then the frame defined by these balls would re-introduce Lorentz breaking. And the Lorentzian analog of a neighbourhood of x, *i.e.* those points with Lorentzian distance ϵ of x, is of infinite volume for any non-zero ϵ in Minkowski space. In a theory in which lengths are dynamical, therefore, it is difficult to imagine this random lattice trick working (unless one uses an infinite valency lattice like a causal set).

It will be left to another place to fully explain how the approximation fails, and how similar the background independent case is. However, it is not hard to see that similar problems arise in that case. The above correspondence can be easily extended to fields over many lattices, and many spacetimes, instead of just one; replace $\mathbb{M}$ with $\mathcal{M}$ in the above condition.

None of the "rules" of fundamental discreteness were tacitly broken here. The discrete structure is a "co-ordinate free", abstract graph and requires no background manifold to live on. No background geometry is presupposed. However, a continuum spacetime must come into the discussion when we consider continuum approximations, and that is where the notion of embedding enters. Even then, the requirement is only that *some* embedding exists, which is a (possibly edge labelled) graph invariant and co-ordinate invariant requirement. Hence this conclusion cannot be evaded on the grounds that the lattice or the discrete/continuum approximation is not "background independent" enough. Given the need for a discrete/continuum correspondence $\rightharpoonup$ to exist, no *more* background independent realisation of the idea is possible.

These considerations, at least, cast doubt on the idea that spin-foams or any lattice-like structure based on finite-valency graphs can satisfy the condition of equation (10). But we should also re-examine the question whether such a condition is necessary in this context.

6.3. *Examining the condition*

If the above conjecture is correct, a spin-foam may have some properties corresponding to those of Minkowski space, but will lack some others that it would need to be Lorentz invariant. According to the claims, in a spin-foam theory it would then be impossible to measure for all of these properties in a classical situation. Despite this apparent problem, arguments have been made in favour of the recovery of local Lorentz invariance in spin-foam models. These are briefly given in this section.

In [42], the point is made that, in a semiclassical state of Loop Quantum Gravity representing Minkowski space, it is possible for two sharp measurements related by a boost to be non-commutative, just as measurements of components of angular momentum are non-commutative in standard quantum theory. In the histories language, this means that both cannot exactly correspond to properties of a history. These sharp measurements would not be classical, and so they are not directly relevant to the above problem. However, it has been argued[13] that some caveats in the discussion of classical measurements can be exploited to evade the problem raised in the previous section.

The argument is that, in quantum gravity, the strong idea of the "generic classical situation" put forward in section 3 need not apply. This undermines claim 5, and in turn, one of the physical requirements on effective descriptions in the argument above. Without generic classical situations, there is no reason why the discrete/continuum correspondence should respect the symmetries of GR. Properties of the histories that are lacking are simply not behaving classically.

Consider two coarse-grained measurements P and P' related by a boost. In a state supposed to represent Minkowski space, a measurement P could be measured in one Lorentz frame and considered part of a classical situation – but could carrying out both P and P' yet be disallowed in a classical situation? Well, this would be consistent as long as these boost-related properties are never both present in a realistic set of classical measurements. This is a view of Lorentz symmetry sometimes put forward for spin-foam quantum gravity. Thus is it claimed that a semi-classical, Lorentz invariant vacuum state for quantum gravity could still exist in such a theory, and give rise to physically acceptable classical situations. The claim is the following: it is not the symmetry which fails in this case, but classicality.

Let us imagine the earth's frame to be the one near which most measurements are carried out. In this scenario, for example, there may be no semi-classical state in the theory describing both the everyday events on and near the earth, and a photon of trans-Planckian frequency (with respect to the earth) moving on a classical background near the earth. The only "preferred frame" here, it is argued, is the one in which measurements are carried out. In a fuller treatment, that is merely a consequence of the configuration of the measurement apparatus.

There are a number of criticisms that could be made of this argument. Many reasons have already been given for the requirement of generic classical situations in section 3. As we claimed there, the existence of such generic classical situations is not usually doubted in any standard quantum theory. Any new theory that did not share this property would therefore be a radical departure; since one of the underpinnings of the classical approximation may have been compromised, this would necessitate a thorough evaluation of the plausibility of this new kind of quantum theory. As discussed before, this change may also be at odds with quasi-classical reasoning about quantum theory.

But the main problem would, again, be the interpretation of experiments designed to test the symmetry.

When we talk of Lorentz violation we are primarily interested in making predictions for the many relevant observations being undertaken. For example, a Lorentz invariant theory should return null results for photon time-of-flight experiments, (see *e.g.* [54]). These observations are a test of a classical symmetry: they use classical methods as discussed in section 3, and their interpretation does not directly invoke the quantum nature of spacetime. The connection between this phenomenological Lorentz invariance, and the Lorentz invariance of a quantum state, is destroyed if we do not demand generic classical situations, as has already been argued in the general case.

Say we were given this semi-classical state, and wanted to test its Lorentz invariance. Consider some set of measurements designed to do this, A, and a set A' in which all of those

[13] The argument is based on my understanding of some comments made by Carlo Rovelli and Daniele Oriti in private communication.

measurements have been Lorentz transformed (with the same transformation). We could carry out measurements A, or A', and we would of course find the same results (statistically) in either case. But in reality, we cannot control all of the measurements made (a case discussed in section 3). No-one doubts that the geometry of the universe looks semi-classical near the cosmological frame: it must be decohered by environmental effects, and this semi-classicality is confirmed by literally every experiment. Therefore, even if classically we are in Minkowski space (at least approximately and in some region), the quantum state has already lost the expected symmetry. What, then, do we predict for the current "Lorentz violation" tests? We could claim that far from the cosmic frame, it is not the symmetry but classicality which could fail, but that does not help us to predict anything. Without generic classical situations, *The Lorentz invariance of the state gives no reason to predict null results for these tests.* Thus this type of "Lorentz invariance" without the generic classical situation does not connect with the phenomenological criterion of Lorentz invariance currently under scrutiny. That should be the final criterion for Lorentz invariance.

Thus, claiming a Lorentz invariant quantum state without any further assumptions is not sufficient to rule out positive results for the real Lorentz violation tests. However, the physical conditions on emergent spacetimes given in section 6.1 do give a necessary condition for the recovery of classical symmetries by all possible tests of this type. If no spin-foam has the necessary properties to preserve Lorentz invariance in a classical situation, such a theory should be considered as giving rise to the type of Lorentz violating scenario currently being tested for.

More on the applying the histories approach to emergent geometry scenarios, including condensed matter analogs, is included in appendix Appendix A.

7. Conclusion

It is well-known that many classical intuitions fail in quantum mechanics. Despite this, we have discussed one idea that can be preserved: a correspondence between measurement results and properties of histories. From its beginning, this histories perspective has been especially useful in describing the emergence of classical behaviour in quantum systems, and this advantage has been applied to quantum gravity. The existence of many views on quantum mechanics is useful for progress in new theories, as well as for new insights on standard physics. When a result is brought to light by application of one perspective, as above, other approaches can sometimes be used to further examine it. However, the question arises in each such case: does switching to a different perspective actually show a loop-hole to the result, or only obfuscate the result?

Consideration of Feynman's histories approach has revealed something about the possible phenomenology of quantum gravity theories. Lorentz violation has been the most consistently pursued avenue for detecting QG effects, and it has been conjectured above that fundamentally discrete spin-foam theories cannot evade Lorentz violation, using the histories view. In particular, it was argued that this conclusion could not be evaded by invoking a quantum sum over lattices. To extend the work, more evidence could be given for the conjecture of section 6.1, that spin-foams cannot satisfy the physical condition (10) that is necessary to preserve Lorentz invariance. However, as mentioned already, since there is no current evidence that spin-foams *can* satisfy the condition, the onus is arguably on supporters of this view to provide evidence for it (or to undermine the condition (10) in some other way).

It would also be of use to study the properties of individual histories in plausible spin-foam path integrals, and to give more details of a correspondence relation $\rightharpoonup$ to continuum spacetimes. This would help to decide with more precision which metric properties are not well approximated (in histories that approximate a well-behaved spacetime near one frame). This would help to determine possible observational consequences.

The idea of deformed Lorentz symmetry [45] is also of possible relevance in this discussion. It may be possible to square the above conclusions of undeformed Lorentz violation with this idea.

No way of doing this is immediately apparent, and so this presents another avenue for study.

As already commented, the consequences of *the possibility* of a histories viewpoint cannot be ignored if one wishes to preserve standard quantum theory. In the converse case, one would lose the standard arguments for the emergence of the classical approximation; a new justification for a correspondence principle would have to be arrived at. This presents a great challenge for any "trans-quantum" theory. But even so, perhaps such radical steps are exactly what is necessary to reach the next level in fundamental theory [55, 56].

Acknowledgements
The author would like to thank Johannes Kofler for helpful comments on the classical limit of quantum mechanics, Harvey Brown for comments on gravity, and many participants at the Loops conference series for discussion of their views on Lorentz invariance, particularly Carlo Rovelli and Daniele Oriti. This work also benefitted from conversations with Fay Dowker and Rafael Sorkin.

References

[1] Markopoulou F and Smolin L 2007 *Class. Quant. Grav.* **24** 3813–3824 (*Preprint* arXiv:gr-qc/0702044)
[2] Markopoulou F 2003 *Class. Quant. Grav.* **20** 777–800 (*Preprint* arXiv:gr-qc/0203036)
[3] Feynman R P 1948 *Rev. Mod. Phys.* **20** 367–387
[4] Caves C M 1986 *Phys. Rev.* D **33** 1643–1665
[5] Sinha S and Sorkin R D 1991 *Found. Phys. Lett.* **4** 303–335
[6] Bell J 1987 *Speakable and unspeakable in quantum mechanics* (Cambridge, UK: Cambridge U. Press)
[7] Zinn-Justin J 2005 *Path integrals in quantum mechanics* (New York: Oxford U. Press)
[8] Feynman R P and Hibbs A R 1965 *Quantum mechanics and path integrals* (New York: McGraw-Hill)
[9] Sorkin Rafael D 1994 *Mod. Phys. Lett.* **A9** 3119–3128 (*Preprint* arXiv:gr-qc/9401003)
[10] Hartle J B 1995 *Proceedings of the Les Houches Summer School on Gravitation and Quantizations, Les Houches, France, 6 Jul - 1 Aug 1992* ed J Zinn-Justin and B Julia (North-Holland) (*Preprint* arXiv:gr-qc/9304006)
[11] Kofler J and Brukner C 2007 (*Preprint* arXiv:arXiv:0706.0668)
[12] Ashtekar A, Bombelli L and Corichi A 2005 *Phys. Rev.* D **72** 025008 (*Preprint* arXiv:gr-qc/0504052)
[13] Peres A 1995 *Quantum Theory: Concepts and Methods* (Kluwer Academic Publishers)
[14] Ambjorn J, Anagnostopoulos K N, Ichihara T, Jensen L and Watabiki Y 1998 *JHEP* **11** 022 (*Preprint* arXiv:hep-lat/9808027)
[15] Kofler J and Brukner C 2007 *Phys. Rev. Lett.* **99** 180403 (*Preprint* arXiv:quant-ph/0609079)
[16] Habib S, Shizume K and Zurek W H 1998 *Phys. Rev. Lett.* **80** 4361–4365 (*Preprint* arXiv:quant-ph/9803042)
[17] Wiebe N and Ballentine L E 2005 (*Preprint* arXiv:quant-ph/0503170)
[18] Schlosshauer M 2006 (*Preprint* arXiv:arXiv:quant-ph/0605249)
[19] Ballentine L E 1998 *Quantum Mechanics: A Modern Development* (World Scientific Publishing)
[20] Sorkin R D 1995 (*Preprint* arXiv:gr-qc/9507057)
[21] Kent A 1998 *Phys. Scripta* **T76** 78–84 (*Preprint* arXiv:gr-qc/9809026)
[22] Bohr N and Rosenfeld L 1933 *Mat. Fys. Medd. K. Da. Vidensk Selsk.* **12** translated and reprinted in [57]
[23] Wilson K G 1975 *Rev. Mod. Phys.* **47** 773
[24] Williams R M 2006 *J. Phys.: Conf. Ser.* **33** 38–48
[25] Ambjorn J, Jurkiewicz J and Loll R 2006 (*Preprint* arXiv:hep-th/0604212)
[26] Ambjorn J, Jurkiewicz J and Loll R 2005 (*Preprint* arXiv:hep-th/0509010)
[27] Ambjorn J, Jurkiewicz J and Loll R 2005 *Phys. Rev.* D **72** 064014 (*Preprint* arXiv:hep-th/0505154)
[28] Oriti D 2001 *Rep. Prog. Phys.* **64** 1489–1544 (*Preprint* arXiv:gr-qc/0106091)
[29] Perez A 2004 (*Preprint* arXiv:gr-qc/0409061)
[30] Bombelli L, Lee J H, Meyer D and Sorkin R 1987 *Phys. Rev. Lett.* **59** 521
[31] Henson J 2006 (*Preprint* arXiv:gr-qc/0601121)
[32] Dowker F 2006 *AIP Conf. Proc.* **861** 79–88
[33] Sorkin R D 1997 *Int. J. Theor. Phys.* **36** 2759–2781 (*Preprint* arXiv:gr-qc/9706002)
[34] Amelino-Camelia G 2004 (*Preprint* arXiv:gr-qc/0412136)
[35] Jacobson T, Liberati S and Mattingly D 2006 *Annals Phys.* **321** 150–196 (*Preprint* arXiv:astro-ph/0505267)
[36] Collins J, Perez A, Sudarsky D, Urrutia L and Vucetich H 2004 *Phys. Rev. Lett.* **93** 191301 (*Preprint* arXiv:gr-qc/0403053)

[37] Collins J, Perez A and Sudarsky D 2006 (*Preprint* arXiv:hep-th/0603002)
[38] Gambini R and Pullin J 1999 *Phys. Rev.* D **59** 124021 (*Preprint* arXiv:gr-qc/9809038)
[39] Sahlmann H and Thiemann T 2006 *Class. Quant. Grav.* **23** 867–908 (*Preprint* arXiv:gr-qc/0207030)
[40] Sahlmann H and Thiemann T 2006 *Class. Quant. Grav.* **23** 909–954 (*Preprint* arXiv:gr-qc/0207031)
[41] Bojowald M, Morales-Tecotl H A and Sahlmann H 2005 *Phys. Rev.* D **71** 084012 (*Preprint* arXiv:gr-qc/0411101)
[42] Rovelli C and Speziale S 2003 *Phys. Rev.* D **67** 064019 (*Preprint* arXiv:gr-qc/0205108)
[43] Livine E R and Oriti D 2004 *JHEP* **06** 050 (*Preprint* arXiv:gr-qc/0405085)
[44] Freidel L and Livine E R 2006 *Phys. Rev. Lett.* **96** 221301 (*Preprint* arXiv:hep-th/0512113)
[45] Amelino-Camelia G 2005 (*Preprint* arXiv:gr-qc/0506117)
[46] Sudarsky D 2005 *Int. J. Mod. Phys.* D **14** 2069–2094 (*Preprint* arXiv:gr-qc/0512013)
[47] Dowker F, Henson J and Sorkin R D 2004 *Mod. Phys. Lett.* A **19** 1829–1840 (*Preprint* arXiv:gr-qc/0311055)
[48] Bombelli L, Henson J and Sorkin R D 2006 (*Preprint* arXiv:gr-qc/0605006)
[49] Brightwell G and Gregory R 1991 *Phys. Rev. Lett.* **66** 260–263
[50] Meyer D 1988 Ph.D. thesis MIT
[51] Myrheim J *Preprint* CERN-TH-2538
[52] Reid D D 2003 *Phys. Rev.* D **67** 024034 (*Preprint* arXiv:gr-qc/0207103)
[53] Major S, Rideout D and Surya S 2007 *J. Math. Phys.* **48** 032501 (*Preprint* arXiv:gr-qc/0604124)
[54] Ellis J R, Mavromatos N E, Nanopoulos D V, Sakharov A S and Sarkisyan E K G 2006 *Astropart. Phys.* **25** 402–411 (*Preprint* arXiv:astro-ph/0510172)
[55] Hardy L 2007 *J. Phys.* A **40** 3081 (*Preprint* arXiv:gr-qc/0608043)
[56] Doring A and Isham C J 2007 (*Preprint* arXiv:quant-ph/0703060)
[57] A W J and Zurek W H 1983 *Quantum theory and measurement* (Princeton, NJ: Princeton U. Press)
[58] Barcelo C, Liberati S and Visser M 2005 *Living Rev. Rel.* **8** 12 (*Preprint* arXiv:gr-qc/0505065)
[59] Visser M and Weinfurtner S 2007 (*Preprint* arXiv:arXiv:0712.0427)
[60] Volovik G 2006 (*Preprint* arXiv:gr-qc/0612134)
[61] Konopka T, Markopoulou F and Smolin L 2006 (*Preprint* arXiv:hep-th/0611197)
[62] Konopka T, Markopoulou F and Severini S 2008 (*Preprint* arXiv:arXiv:0801.0861)
[63] Weinfurtner S, Liberati S and Visser M 2006 *J. Phys.: Conf. Ser.* **33** 373–385 (*Preprint* arXiv:gr-qc/0512127)
[64] Markopoulou F 2006 (*Preprint* arXiv:hep-th/0604120)
[65] Markopoulou F 2007 (*Preprint* arXiv:gr-qc/0703097)
[66] Dreyer O 2006 (*Preprint* arXiv:gr-qc/0604075)
[67] Dreyer O 2007 (*Preprint* arXiv:arXiv:0710.4350)
[68] Kribs D W and Markopoulou F 2005 (*Preprint* arXiv:gr-qc/0510052)
[69] Girelli F, Liberati S and Sindoni L 2008 (*Preprint* arXiv:0807.4910)
[70] Meschini D, Lehto M and Piilonen J 2005 *Stud. Hist. Philos. Mod. Phys.* **36** 435–464 (*Preprint* arXiv:gr-qc/0411053)
[71] Khoury J and Parikh M 2006 (*Preprint* arXiv:hep-th/0612117)
[72] Healey R 2007 *Gauging What's Real* (New York, NY: Oxford U. Press)

Appendix A. Emergent Geometry

We have already discussed the idea of considering spacetime as an approximation to some other more fundamental entity, and its place in some popular approaches to quantum gravity. This idea has been much influenced by condensed state matter analogs, in which relativistic field dynamics can appear as a hydrodynamical approximation to an underlying atomic dynamics [58, 59, 60]. This appendix reviews and comments on some ideas that have been inspired in this way, using the histories perspective developed above.

How far can we go in removing geometry from the fundamental level of our theory? Claim 6 clarifies what is possible and what is not. Following the discussion of section 4, it is indeed possible to have a history space Ω for quantum gravity that is not merely a reasonable set of Lorentzian manifolds Ω', or unsmooth generalisations thereof. However, there must be some way to approximate at least some of these histories with continuum spacetimes. We need a correspondence relation $\rightharpoonup$ between the history spaces that satisfies certain conditions. Examples include (a) causal sets, in which a reasonably explicit discrete/continuum correspondence has been given [31], and (b) the "quantum graphity" model [61, 62], which implicitly invokes a discrete/continuum correspondence when the occurrence of lattice-like graphs is said to be sign of emergent geometry (some form of graph-connectivity/distance correspondence is being presupposed). Not all histories or configurations must have geometrical approximations, but some must. This is as far from geometrical degrees of freedom as it is possible to get if the theory can be cast in a histories framework.

The effective degrees of freedom may have an unexpected relation to those at the underlying level. For instance, if the fundamental structures are graphs, it need not be true that the graph distance between two vertices is directly related to the length between two corresponding points in the approximating manifold [1]. This would amount to a counter-intuitive property of $\rightharpoonup$. However, if we agree with the discussion of section 4, we can conclude that all problems in relating fundamental and effective degrees of freedom can be expressed as problems with defining and interpreting $\rightharpoonup$; there is no further ambiguity introduced when carrying out the quantum sum. The properties of spacetime that we can measure (which are, of course, classical) are nothing but the common properties of some set of fundamental histories.

A histories perspective like this can help to explain some properties of known emergent geometry scenarios. Consider the condensed matter "analog models", in which relativistic fields emerge as an effective description, *e.g.* of density peturbations in a fluid [58, 59, 60]. In these scenarios, there is a preferred frame in the fundamental description: the average rest frame of the fluid. The atomic/continuum correspondence makes explicit use of this frame. The effective, Lorentzian description is not perfect, and it breaks down when considering high frequency modes in the field (with respect to that underlying rest frame) [63]. Why is this? Using the histories perspective, there is an obvious conjecture: the fundamental description, *e.g.* in terms of molecular worldlines, does not have any properties corresponding to field modes of smaller wavelength than the molecular separation. The effective description in terms of a mass density field cannot work at that scale. The discrete/continuum correspondence $\rightharpoonup$ picks out a time direction in the effective spacetime, and violates the physical conditions necessary for Lorentz symmetry to be maintained. This seems a compelling conjecture, as we have already seen how difficult it is to design discrete/continuum correspondences $\rightharpoonup$ that respect Lorentz invariance.

Some hope to recover GR in a similar way: as an effective description of some system with a fundamental cutoff and a preferred time direction (this includes [64, 65] discussed below). It has been argued that the symmetries, as measured by an observer who is internal to the system at the effective level, need not be affected by the fundamental preferred frame [66, 67]. But in the case of Lorentz symmetry, there is no example of this known in condensed state matter analogs – the effective symmetry always falls victim to the fundamental preferred frame at high energies. The above conjecture would explain this, and show how generic the problem is. The point is that this type of fundamental history space Ω and fundamental/effective correspondence $\rightharpoonup$ do not obey the physical conditions of sections 4.1 and 6.1. This would arguably be a stumbling block for any such "emergent gravity" theory, with Planckian discreteness and a preferred fundamental time direction, if local Lorentz symmetry is not to be broken at high energies.

While it is true that effective symmetries need not be present or meaningful on the fundamental level, that is not to say that properties at the fundamental level are completely irrelevant to the recovery of effective symmetries. We have already seen in detail how they relate, via $\rightharpoonup$, in sections 4 and 6.1. In all of the known analog models, however $\rightharpoonup$ would look in detail, it is clear that it links the preferred underlying time direction with a preferred time direction in the effective spacetime. According to the conjecture above, Lorentz violation could only be evaded if this troublesome property of $\rightharpoonup$ could be removed; *i.e* if the correspondence $\rightharpoonup$ could be made fully Lorentz invariant. In causal sets, inspired by a histories viewpoint, this property is "designed in" from the start. This conclusion cannot be evaded by a preference for some other viewpoint on emergence; it is only made less obvious.

Appendix A.1. No Geometry?

Recently, the idea of "emergent geometry" has been treated in an apparently novel way, giving rise to a new class of approaches to quantum gravity [68, 64, 65]. In the fundamental theory, it has been said, there need be no geometrical degrees of freedom. This "no geometry" approach is argued to go beyond even approaches such as spin foams or causal sets, where the discrete histories have properties that generalise, and sometimes approximate to, geometrical properties. These are still seen as geometrical degrees of freedom from this perspective.

New perspectives on the emergence of geometry are very useful. For example, they have given rise to exciting new ideas on the origin of gravity itself [67, 69]. However, the histories viewpoint is valid here: these theories with no fundamental geometry need not go beyond the normal quantum mechanical framework in any radical way (the most general structure so far employed in this context is a quantum mechanics of open systems). This new idea can therefore be discussed in the light of claim 6.

What is novel about the approaches described in [68, 64, 65], restated in the histories language? It is the approach to finding the correct theory, rather than some difference in the final method of recovering spacetime. In models like spin-foams or causal sets, the quantum histories have been designed with a relation to continuum Lorentzian geometry already in mind: for spin-foams, the foundation of the discrete/continuum approximation is in area information and connectivity, whereas it is in the causal relation in the latter case. In the new approach, a fundamental quantum theory is somehow picked, but without first specifying anything about $\rightharpoonup$. Indeed, without positing *a priori* the relation to continuum geometry, the histories basis is not yet known. Then, it is hoped, by exploiting properties of the quantum system it may be possible to pick out a suitable histories basis and $\rightharpoonup$ that will provide the theory with an interpretation as quantum gravity.

Thus we do not approach the formulation of the theory with pre-established geometrical notions in mind, and in this sense there is no geometry at the outset. From a histories perspective a different point might be emphasised: there are (hopefully) geometrical degrees of freedom, but we just don't identify them at the outset. If ever such a theory would be completed, it would be just as geometrical as a spin-foam theory (albeit that the discrete/continuum correspondence $\rightharpoonup$ might take an unexpected form). If any correspondence $\rightharpoonup$ exists, that implies that some fundamental histories will have some properties corresponding to continuum lengths, and so on. Without any such corresponding properties, geometry could not emerge as an effective description, as it must do to recover GR[14]. Once the theory was developed enough to define some correspondence $\rightharpoonup$, the problem of maintaining Lorentz invariance would again have to be faced.

In this connection it should be mentioned that, even in the classical theory, one can attempt to rewrite GR, in order to support a more radically relational standpoint in which spacetime has no intrinsic properties that do not derive from matter (see [71] and references therein). This viewpoint is philosophically interesting, and may be inspiring to new approaches to quantum gravity (although arguably experience in QED *etc.* speaks against writing out field degrees of freedom in this way [72]). But even when taking such positions, it must be possible to approximately reproduce the results of GR from the quantum gravity theory. That is all that is needed for claim 6 to be valid, unless a radical break with standard quantum theory is contemplated.

[14] This comment goes slightly beyond quantum histories approaches: there must be *something* in any new fundamental theory that *somehow* corresponds to geometry, or it will never incorporate GR. It is therefore pointless to give up on a theory as being "too geometrical" at the first sign of a property even vaguely analogous to, *e.g.*, length. This refutes the arguments made in [70] against current programs of emergent geometry.

Journal of Physics: Conference Series **174** (2009) 012021 doi:10.1088/1742-6596/174/1/012021

Concept of Time in Canonical Quantum Gravity and String Theory

Claus Kiefer

Institut für Theoretische Physik, Universität zu Köln, Zülpicher Straße 77, 50937 Köln, Germany.

Abstract. The accommodation of gravity into the quantum framework will entail changes for the concepts of space and time at the most fundamental level. Here I shall review the concept of time, but include also some remarks on the concept of space. After a general introduction, I shall discuss time in canonical quantum gravity and its application to cosmology. The most important feature is the fundamental timelessness of the theory. I shall conclude with a brief discussion of space and time in string theory.

1. The Problem of Time

There exist many reasons why a final quantum theory of gravity is not yet at our disposal. First, there is the persistent lack of experiments which could serve as a trustful guide. Second, there are many mathematical problems. An third, there are conceptual problems which have no counterparts in other quantum theories. Among them is the problem of time, which is the subject of my contribution. In the following, I shall give a brief introduction to this problem and present some ideas about the role that time plays in some of the current approaches to quantum gravity – canonical quantum gravity and string theory. A more detailed exposition can be found, for example, in my monograph [1]; the reader can find there also references to the original literature and other sources.

The problem of time in quantum gravity is directly related with the background independence of general relativity. Consider, first, ordinary quantum mechanics, which is governed by the Schr¨odinger equation,

$$\mathrm{i}\hbar\frac{\partial\psi}{\partial t} = \hat{H}\psi\ . \tag{1}$$

The parameter t in this equation is Newton's absolute time which serves here as a background structure par excellence. Does the situation change with the advent of Einstein's theory of special relativity? Not in this respect. Although space and time by themeself are no longer absolute, but are melted into a four-dimensional spacetime, this ensuing Minkowski spacetime is again an absolute structure. It exerts influence on all matter and fields, for example in the form of inertia, but cannot be acted upon by them. This is why Minkowski spacetime is a background structure. Standard quantum field theory, for example QED, is defined *on* this background. Moreover, the symmetry of this background, as given by the Poincaré group, is mandatory for the formulation of these theories.

The situation changes dramatically if general relativity is taken into account. Governed by the Einstein field equations,

$$R_{\mu\nu} - \frac{1}{2}g_{\mu\nu}R = \frac{8\pi G}{c^4}T_{\mu\nu}\ , \tag{2}$$

spacetime (as specified by its metric $g_{\mu\nu}$) is dynamical and no longer a background structure. The general movement of matter and fields generates gravitational waves which modify spacetime. If the gravitational field is to be quantized, one can no longer quantize it on a background. Everything must be quantized and, consequently, a quantum theory of gravity must contain new features; it can, in particular not contain any absolute time or spacetime.

In his closing speech at the Bern relativity meeting in 1955, Wolfgang Pauli seemed to have realized this problem clearly. He writes (in my translation, see [1] for the original reference):

> It seems to me ... that it [the main obstacle in quantizing gravity, C.K.] is not so much the linearity or non-linearity which forms the heart of the matter, but the very fact that here a more general group than the Lorentz group is present ... .

With "more general group" he means the diffeomorphism group of general relativity, which is the counterpart of background independence in the mathematical formalism.

Approaches to quantum gravity can roughly be divided into two classes. The first is the oldest and the most conservative. It is the attempt to search directly for a quantum theory of the gravitational field, without embarking on a unification of all forces. Since one mostly deals with general relativity, I shall call this class *quantum general relativity*. Its members are covariant gravity (e.g. perturbation theory and the path-integral approach) and canonical gravity. Depending on the chosen canonical variables, the latter can be subdivided further into geometrodynamics, loop quantum gravity, and others. The second approach is the attempt to solve the problem of quantum gravity within a unified quantum framework of all interactions. The only known example is string theory. In the following, I shall first study canonical quantum gravity, address then its application to cosmology, and finally turn to string theory.

2. Canonical Quantum Gravity

One can arrive at the central equations of canonical quantum gravity by various means. Perhaps the most physical one is to proceed analogously to Erwin Schr¨odinger when he arrived at his wave equation. In 1926, he started from the Hamilton–Jacobi equation of classical mechanics, from which he guessed a wave equation that would yield this equation in the limit of "geometrical optics" (or WKB limit, as we would say today). In [3] he writes (given here in English translation)

> *We know today, in fact, that our classical mechanics fails for very small dimensions of the path and for very great curvatures.* Perhaps this failure is in strict analogy with the failure of geometrical optics ... that becomes evident as soon as the obstacles or apertures are no longer great compared with the real, finite, wavelength. ... Then it becomes a question of searching for an undulatory mechanics, and the most obvious way is by an elaboration of the Hamiltonian analogy on the lines of undulatory optics.

Could one not proceed similarly for general relativity? Yes, this is possible. The first non-trivial step in this direction is, of course, to find the Hamilton–Jacobi version of the Einstein equations (2). This was achieved by Asher Peres in 1962 [4]. Instead of the ten Einstein field equations (2) one is now confronted with two *functional* equations for a functional $S[h_{ab}]$, where h_{ab} denotes the *three*-dimensional metric on a spacelike slice through spacetime. Because of their functional nature, these equations are fully equivalent to Einstein's equations. Restricting, for simplicity, to the vacuum case, they read (with $c = 1$)

$$\begin{aligned} 16\pi G\, G_{abcd}\frac{\delta S}{\delta h_{ab}}\frac{\delta S}{\delta h_{cd}} - \frac{\sqrt{h}}{16\pi G}({}^{(3)}R - 2\Lambda) &= 0\,, \\ D_a\frac{\delta S}{\delta h_{ab}} &= 0\,. \end{aligned} \tag{3}$$

The second equation guarantees that $S[h_{ab}]$ is invariant under coordinate transformations on the spatial section. In (3), G_{abcd} is an ultralocal function of the three-metric, which plays the role of a metric on the configuration space of all three-metrics; it is called "DeWitt metric". Furthermore, h is the determinant of the three-metric, ${}^{(3)}R$ the three-dimensional Ricci scalar, Λ the cosmological constant, and D_a denotes the three-dimensional covariant derivative. It is important to emphasize that these are purely spatial equations; there is, in particular, no analogue of the $\partial S/\partial t$-term in classical mechanics. The time t has disappeared! But where, then, does the classical spacetime come from? Spacetime can, in fact, be constructed by an equation analogously to $p = \partial S/\partial q$ in classical mechanics, which allows the construction of a particle trajectory. In this sense, the classical spacetime can be interpreted as a generalized trajectory in which a single q corresponds to a whole three-geometry.

In fact, there hold interesting and important "interconnection theorems" between the Hamilton–Jacobi formulation and the ensuing spacetime picture. For example, the validity of (3) on each spatial slice guarantees the validity of all ten Einstein equations on the resulting spacetime.

So far for the classical part. Adopting the path originally followed by Schr¨odinger, we shall look for a wave functional, $\Psi[h_{ab}]$, obeying a wave equation which in the WKB limit

$$\Psi[h_{ab}] = C[h_{ab}] \exp\left(\frac{\mathrm{i}}{\hbar} S[h_{ab}]\right) , \tag{4}$$

in which the metric-dependence of $C[h_{ab}]$ is neglected compared to the one of $S[h_{ab}]$, leads back to (3). In the vacuum case, this equation has the form

$$\hat{\mathcal{H}}_\perp \Psi \equiv \left(-2\kappa\hbar^2 G_{abcd} \frac{\delta^2}{\delta h_{ab}\delta h_{cd}} - (2\kappa)^{-1}\sqrt{h}({}^{(3)}R - 2\Lambda)\right)\Psi = 0 , \tag{5}$$

where $\kappa = 8\pi G$, which is called the "Wheeler–DeWitt equation". It generalizes the first equation in (3). The second equation is generalized by

$$\hat{\mathcal{H}}^a \Psi \equiv -2\nabla_b \frac{\hbar}{\mathrm{i}} \frac{\delta \Psi}{\delta h_{ab}} = 0 \tag{6}$$

and is called the quantum diffeomorphism (momentum) constraint. Equations (5) and (6) are quantum constraints for the wave functional and constitute the central equations of canonical quantum gravity (quantum geometrodynamics). Provided the linear structure of quantum theory is universally valid, these equations should hold – independent of their exact validity at the most fundamental level – at least approximately on length scales bigger than the Planck length. One may thus expect to get at least some reliable results from this framework.

What about the problem of time in the light of canonical quantum gravity? The solution to this problem seems to be straightforward – external time has disappeared! Instead of the Schr¨odinger equation (1) one now possesses an equation of the form $\hat{H}\psi = 0$. Spacetime has disappeared, too, and only space remains: the argument of the wave functional is the three-dimensional metric (plus non-gravitational fields, if present).

A closer inspection of (5) reveals that it is of a local hyperbolic form. This follows from the structure of the DeWitt metric. Despite the absence of an external time, one can thus define an *intrinsic* timelike variable from the structure of the kinetic term in (5). This variable is constructed solely from three-dimensional quantities (the three-metric and perhaps part of the non-gravitational degrees of freedom).

Since the presence of an external time is mandatory for the interpretational structure of ordinary quantum theory – in particular, for the use of a Hilbert-space structure and the conservation of probability *in* time –, the question arises what is the fate of this structure in a timeless world. This is also circumscribed by the "Hilbert-space problem" and the "problem of observables" [1]. A final solution is not yet available.

The standard approach to canonical quantum gravity starts with a $3+1$-decomposition of spacetime (assumed to be globally hyperbolic), that is, a decomposition of spacetime into a foliation of spacelike hypersurfaces. Consider a timelike vector field t^ν that points from a point at a hypersurface with $t =$ constant to a point at an infinitesimally neighbouring hypersurface with the same coordinate values,

$$\dot{X}^\nu \equiv t^\nu = N n^\nu + N^a X^\nu_{,a} \,, \tag{7}$$

see Figure 1. It is decomposed into its normal component (specified by the lapse function N) and

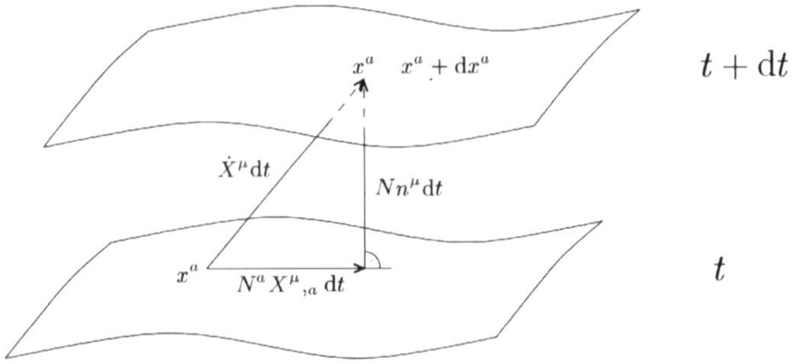

Figure 1. Geometric interpretation of lapse and shift

its tangential components (specified by the shift vector N^a). The four-dimensional line element can accordingly be decomposed as

$$\begin{aligned} \mathrm{d}s^2 &= g_{\mu\nu}\mathrm{d}x^\mu \mathrm{d}x^\nu = -N^2\mathrm{d}t^2 + h_{ab}(\mathrm{d}x^a + N^a\mathrm{d}t)(\mathrm{d}x^b + N^b\mathrm{d}t) & (8)\\ &= (h_{ab}N^aN^b - N^2)\mathrm{d}t^2 + 2h_{ab}N^a\mathrm{d}x^b\mathrm{d}t + h_{ab}\mathrm{d}x^a\mathrm{d}x^b \,. & (9)\end{aligned}$$

The configuration variable turns out to be the three-metric h_{ab}, and the conjugated momentum is a linear function of the extrinsic curvature. Rewriting the Einstein–Hilbert action in terms of these variables, one arrives at a form in which it is the sum of constraints only (up to surface terms). Upon quantization, then, these constraints become the Wheeler–DeWitt equation (5) and the momentum constraints (6).

We can now follow the path that originally led us to (5) and (6) in the opposite direction. That is, we can perform a semiclassical approximation for the wave functional. We write

$$|\Psi[h_{ab}]\rangle = C[h_{ab}]\mathrm{e}^{\mathrm{i}m_{\mathrm{P}}^2 S[h_{ab}]}|\psi[h_{ab}]\rangle \,, \tag{10}$$

where m_{P} is the Planck mass. This ansatz is close to the WKB form (4), but includes also non-gravitational fields. The notation chosen here is such that the "ket" in the state refers to the non-gravitational degrees of freedom for which we assume that a standard Hilbert-space structure is available. That is, we could instead also write $\psi[h_{ab}, \phi]$, where ϕ stands for the non-gravitational fields.

The semiclassical approximation scheme [1] then yields in the highest orders the Hamilton–Jacobi equations (3) for the functional S corrected by the expectation value of the Hamiltonian for the non-gravitational fields. In this sense the classical Einstein equations are recovered as approximate equations. The next order yields the following equations for $|\psi[h_{ab}]\rangle$:

$$\left(\hat{\mathcal{H}}^{\mathrm{m}}_\perp - \langle\psi|\hat{\mathcal{H}}^{\mathrm{m}}_\perp|\psi\rangle - \mathrm{i}G_{abcd}\frac{\delta S}{\delta h_{ab}}\frac{\delta}{\delta h_{cd}}\right)|\psi[h_{ab}]\rangle = 0 \,, \tag{11}$$

$$\left(\hat{\mathcal{H}}^{\mathrm{m}}_a - \langle\psi|\hat{\mathcal{H}}^{\mathrm{m}}_a|\psi\rangle - \frac{2}{\mathrm{i}}h_{ab}D_c\frac{\delta}{\delta h_{bc}}\right)|\psi[h_{ab}]\rangle = 0 \,, \tag{12}$$

where the subscript 'm' refers to the non-gravitational parts of the Hamiltonian and momentum constraints (not considered in (5) and (6)). One now evaluates $|\psi[h_{ab}]\rangle$ along a solution of the classical

Einstein equations, $h_{ab}(\mathbf{x},t)$, corresponding to a solution, $S[h_{ab}]$, of the Hamilton–Jacobi equations; this solution is obtained from

$$\dot{h}_{ab} = NG_{abcd}\frac{\delta S}{\delta h_{cd}} + 2D_{(a}N_{b)}\ . \tag{13}$$

Defining now, in this limit, a time parameter t by

$$\frac{\partial}{\partial t}|\psi(t)\rangle = \int \mathrm{d}^3x\, \dot{h}_{ab}(\mathbf{x},t)\, \frac{\delta}{\delta h_{ab}(\mathbf{x})}|\psi[h_{ab}]\rangle\ , \tag{14}$$

one obtains a functional Schr¨odinger equation for the quantized non-gravitational fields in the chosen external classical gravitational field,

$$\begin{aligned} \mathrm{i}\hbar\frac{\partial}{\partial t}|\psi(t)\rangle &= \hat{H}^{\mathrm{m}}|\psi(t)\rangle\ , \\ \hat{H}^{\mathrm{m}} &\equiv \int \mathrm{d}^3x \left\{N(\mathbf{x})\hat{\mathcal{H}}^{\mathrm{m}}_{\perp}(\mathbf{x}) + N^a(\mathbf{x})\hat{\mathcal{H}}^{\mathrm{m}}_a(\mathbf{x})\right\}\ . \end{aligned} \tag{15}$$

Here, $\hat{H}^{\mathrm{m}}$ denotes the full Hamiltonian of the non-gravitational fields in the Schr¨odinger picture, which parametrically depends on the (generally non-static) metric coefficients of the curved space–time background.[1] The emerging "WKB time" t controls the dynamics in this limit: time has been regained from timeless quantum gravity under well-defined special circumstances.

This recovery of time only works in a straightforward way if the ansatz (10) is used in which the dominating factor is of the complex form $\exp(\mathrm{i}S_0)$, where $S_0 = m_{\mathrm{P}}^2 S$. If instead one uses, for example, a superposition containing the terms $\exp(\mathrm{i}S_0)$ and $\exp(-\mathrm{i}S_0)$, the question arises how (15) with its t and its i emerges.

The answer is decoherence [5]. Consider an analogy from molecular physics: the emergence of chirality. Figure 2 shows the simplest possibility of a chiral molecule and the corresponding effective potential with respect to the transition from one configuration into the other.

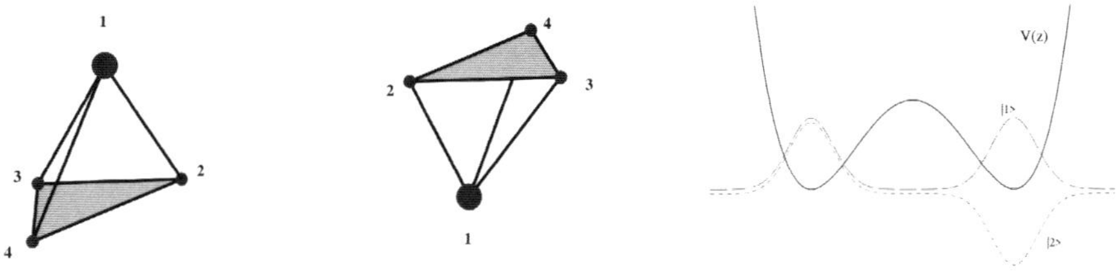

Figure 2. Simplest possible structure of an optically active molecule and a schematic picture of the effective potential for the inversion coordinate as well as of the two lowest-lying eigenstates.

Low-mass molecules (such as ammonia) indeed behave quantum mechanically, that is, they exhibit a superposition between the two configurations shown in Figure 2. For heavier mass this superposition is suppressed by the interaction with the environment, for example, by the scattering with light or air molecules. This is called decoherence. Since the Hamiltonian for such molecules is parity-invariant, the observation of one of the two definite molecular structures in Figure 2 constitutes a symmetry breaking – a chiral state has emerged.

In the case of the semiclassical approximation to quantum gravity, decoherence can lead to a similar symmetry breaking: although the full quantum state obeying the Wheeler–DeWitt equation is real, the actual observed semiclassical component is complex. Decoherence by interaction with small gravitational waves and tiny density fluctuations makes the $\exp(\mathrm{i}S_0)$- and the $\exp(-\mathrm{i}S_0)$-part of the

[1] There may, of course, be situations in which part of the gravitational fi eld (e.g. the gravitons) contribute to $\hat{H}^{\mathrm{m}}$ and part of the non-gravitational fi elds (e.g. macroscopic matter) contribute to S.

wave function dynamically independent [2, 5]. This justifies in retrospect the use of the complex ansatz (10).

If the functional Schr¨odinger equation can be recovered from full quantum gravity in an appropriate limit, the question arises whether one can go beyond this limit and calculate quantum gravitational correction terms. This can be done at least at a formal level, that is, at the level where one treats the functional derivatives like partial derivatives.

The next order in the Born–Oppenheimer approximation then gives corrections to the Hamiltonian for the non-gravitational fields,

$$\hat{H}^{\rm m} \to \hat{H}^{\rm m} + \frac{1}{m_{\rm P}^2}\,(\text{various terms}) \ . \tag{16}$$

The detailed form of these terms can be found in [1]. Future investigations should deal with a concrete application of these terms in cosmology, for example, in the search for quantum gravitational effects in the anisotropy spectrum of the Cosmic Background Radiation.

A simple example is the calculation of the quantum gravitational correction to the trace anomaly in de Sitter space. For a conformally coupled scalar field, the trace of the energy–momentum tensor, although being zero classically, is non-vanishing in the quantum theory; this "anomalous trace" is proportional to $\hbar$. It corresponds to the following expectation value, ε, of the Hamiltonian density,

$$\varepsilon = \frac{\hbar H_{\rm dS}^4}{1440\pi^2 c^3}\ , \tag{17}$$

where $H_{\rm dS}$ is the constant Hubble parameter of de Sitter space. It turns out that the ratio of the the first quantum gravitational correction $\delta\varepsilon$ to ε, as calculated from the Born–Oppenheimer expansion, reads

$$\frac{\delta\varepsilon}{\epsilon} \approx -\frac{1}{2160\pi}\left(\frac{t_{\rm P}}{H_{\rm dS}^{-1}}\right)^2\ , \tag{18}$$

where $t_{\rm P}$ denotes the Planck time. Numerically, the ratio (18) is, of course, small. Using values motivated by inflationary cosmology, one can assume that $H_{\rm dS}$ lies between 10^{13} and 10^{15} GeV, leading for the ratio (18) to values between roughly 10^{-16} and 10^{-22}. It is at present an open question whether these correction terms can become big enough to be observable.

At the end of this section I shall make some brief remarks on the situation in loop quantum gravity [6]. This is a particular case of the canonical theory and differs from geometrodynamics by the choice of variables. Since it contains the same timeless constraints, the situation is similar to the one discussed above. In addition, loop quantum gravity seems to give some new insights into space. At the kinematical level, it turns out that geometric operators can have a discrete spectrum of eigenvalues. Therefore, at least in an operational sense, three-dimensional space exhibits discrete features. One example is the quantization of area,[2]

$$\hat{A}(\mathcal{S})\Psi_S[A] = 8\pi\beta l_{\rm P}^2 \sum_{P\in S\cap\mathcal{S}} \sqrt{j_P(j_P+1)}\,\Psi_S[A] \tag{19}$$

where $l_{\rm P}$ is the Planck length and β is a free parameter in this approach.

3. Quantum Cosmology

Quantum cosmology is the application of quantum theory to the Universe as a whole. Applying canonical quantum gravity to cosmology, one encounters intriguing consequences of the concept of time.

[2] The sum is over all crossing points P of a spin network S with a surface $\mathcal{S}$, and j_P can assume integer and half-integer values.

As in classical cosmology, one starts with an ansatz for a highly symmetric metric. For a closed Friedmann–Lemaître universe with scale factor a, the metric can be chosen to be of the form

$$\mathrm{d}s^2 = -N^2(t)\mathrm{d}t^2 + a^2(t)\mathrm{d}\Omega_3^2\,. \tag{20}$$

As a representative for matter, we choose in addition a homogeneous massive scalar field ϕ. The Wheeler–DeWitt equation then reads (with units $2G/3\pi = 1$)

$$\frac{1}{2}\left(\frac{\hbar^2}{a^2}\frac{\partial}{\partial a}\left(a\frac{\partial}{\partial a}\right) - \frac{\hbar^2}{a^3}\frac{\partial^2}{\partial\phi^2} - a + \frac{\Lambda a^3}{3} + m^2a^3\phi^2\right)\psi(a,\phi) = 0\,. \tag{21}$$

The configuration space here is only two-dimensional (consisting of a and ϕ); such drastically reduced configuration spaces are called "minisuperspaces". The factor ordering (which is a general problem for the Wheeler–DeWitt equation) has been chosen in order to achieve covariance in minisuperspace.

The remaining part of this section is similar to the corresponding section in [2] and is therefore adapted from that source.

It is evident that equations such as (21) do not possess the mathematical problems of the full functional equation (5). One can thus focus attention on physical applications. One important application is the imposition of boundary conditions. Other important applications include the discussion of wave packets, the validity of the semiclassical approximation, the origin of classical behaviour and the arrow of time, and the possible quantum avoidance of classical singularities [1, 7].

Before picking out one particular model, I want to emphasize one important conceptual point which is relevant for the problem of time discussed above, see Figure 3.

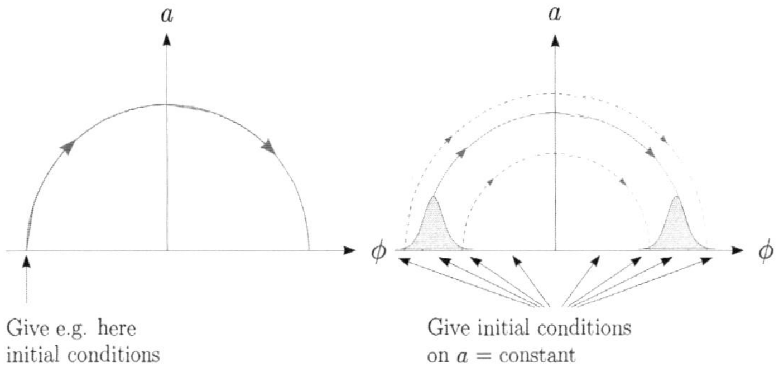

Figure 3. The classical and the quantum theory of gravity exhibit drastically different notions of determinism [1].

Consider a two-dimensional minisuperspace model with the variables a and ϕ as above. The figure on the left shows the classical trajectory in configuration space for a universe which is expanding and recollapsing. Classically, one can give initial conditions, for example, on the left end of the trajectory for small a and then determine the whole trajectory. In this sense, the recollapsing part of the trajectory is the deterministic successor of the expanding part. One could, of course, also start from the right end of the trajectory because there is no distinguished direction; but the important point is that a trajectory exists. Not so in the quantum theory where both the trajectory and the time parameter t are absent! If one wants to find a solution of the Wheeler–DeWitt equation which describes a wave packet following the classical trajectory, one has to specify two packets at the would-be ends of the classical trajectory, see the right figure. The reason is that (21) is a hyperbolic equation with respect to intrinsic time a, and the natural formulation of boundary conditions is to specify the wave function (and its derivative) at constant a. If one imposed only one of the two wave packets, the full solution would be a smeared-out wave function which does not resemble anything like a wave packet following the classical trajectory. In this sense, the "recollapsing" wave packet must be present "initially". This is a direct consequence of the concept of time in quantum gravity.

Let me now turn to a specific example [8]: a cosmological model with a "big brake". Classically, the model is characterized by an equation of state of the form $p = A/\rho$, where $A > 0$ ("anti-Chaplygin gas"). This can be realized by a scalar field ϕ with the following potential:

$$V(\phi) = V_0 \left(\sinh\left(\sqrt{3\kappa^2}|\phi|\right) - \frac{1}{\sinh\left(\sqrt{3\kappa^2}|\phi|\right)} \right) \; ; \; V_0 = \sqrt{A/4} \, . \tag{22}$$

This model universe develops a pressure singularity at the end of its evolution where it comes to an abrupt halt: $\dot{a}$ remains finite there but $\ddot{a}(t)$ tends to minus infinity; this is why it is called a "big brake". Since this model does not describe an accelerating universe, it is as such in conflict with present observations. However, it can easily be generalized in order to accommodate such an acceleration, without modifying the following discussion. The total lifetime of this universe would be

$$t_0 \approx 7 \times 10^2 \frac{1}{\sqrt{V_0 \left[\frac{\text{g}}{\text{cm}^3}\right]}} \text{ s} \, ,$$

which would be much bigger than the current age of our Universe for

$$V_0 \ll 2.6 \times 10^{-30} \, \frac{\text{g}}{\text{cm}^3} \, .$$

The classical trajectory in configuration space is shown in Figure 4. The big-brake singularity is at $\phi = 0$. In addition, there are the usual big-bang and big-crunch singularities at $a = 0$ and $\phi \to \pm\infty$.

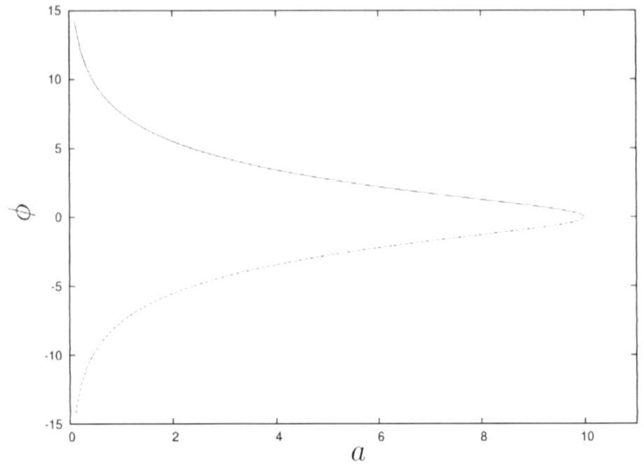

Figure 4. Classical trajectory in configuration space [8].

In the quantum theory, one encounters the following Wheeler–DeWitt equation:

$$\begin{aligned} &\frac{\hbar^2}{2}\left(\frac{\kappa^2}{6}\frac{\partial^2}{\partial\alpha^2} - \frac{\partial^2}{\partial\phi^2}\right)\Psi(\alpha,\phi) \\ &+ \; V_0 e^{6\alpha}\left(\sinh\left(\sqrt{3\kappa^2}|\phi|\right) - \frac{1}{\sinh\left(\sqrt{3\kappa^2}|\phi|\right)}\right)\Psi(\alpha,\phi) = 0 \, , \end{aligned} \tag{23}$$

where $\alpha = \ln a$. In order to study the behaviour near the region of the classical singularity, it is sufficient to study the limit of small ϕ. One can then use the approximate equation

$$\frac{\hbar^2}{2}\left(\frac{\kappa^2}{6}\frac{\partial^2}{\partial\alpha^2} - \frac{\partial^2}{\partial\phi^2}\right)\Psi(\alpha,\phi) - \frac{\tilde{V_0}}{|\phi|}e^{6\alpha}\Psi(\alpha,\phi) = 0\,, \tag{24}$$

where $\tilde{V_0} = V_0/3\kappa^2$. A crucial input is now the choice of boundary conditions. Firstly, we have to impose the condition that the wave function go to zero for large a; this is because the classical evolution stops at finite a. Secondly, we demand normalizability with respect to ϕ. The resulting solutions are then of the form

$$\Psi(\alpha,\phi) = \sum_{k=1}^{\infty} A(k)k^{-3/2}\mathrm{K}_0\left(\frac{1}{\sqrt{6}}\frac{V_\alpha}{\hbar^2 k\kappa}\right)\left(2\frac{V_\alpha}{k}|\phi|\right)e^{-\frac{V_\alpha}{k|\phi|}}\mathrm{L}^1_{k-1}\left(2\frac{V_\alpha}{k}|\phi|\right)\,, \tag{25}$$

where K_0 is a Bessel function, L^1_{k-1} denotes the Laguerre polynoms, and $V_\alpha \equiv \tilde{V_0}e^{6\alpha}$. Inspection of this solution shows that it *vanishes* at $\phi = 0$, that is, at the classical big-brake singularity. Therefore, this singularity is avoided in the quantum theory. In fact, the normalization condition with respect to ϕ also guarantees that the big-bang singularity is absent. One is thus left with a singularity-free quantum universe.

A wave-packet solution following the classical solution of Figure 4 and approaching zero when $\phi \to 0$ (that is, when approaching the region of the classical big-brake singularity), is shown in Figure 5.

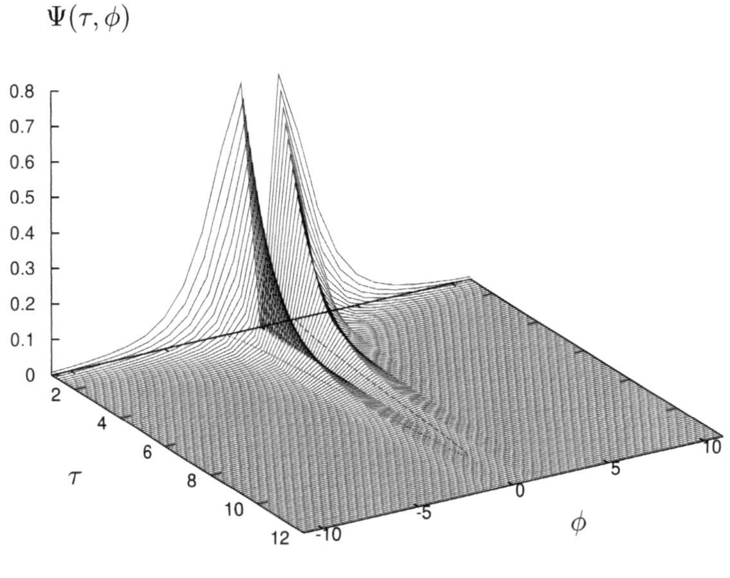

Figure 5. The wave packet for the big-brake model. The packet follows the classical trajectory but becomes zero at the classical singularity [8].

We have seen in the last section that a semiclassical time parameter t exists only under special circumstances. These are no longer fulfilled in the regions of the classical singularities where the timelessness of the fundamental quantum world takes over.

Journal of Physics: Conference Series **174** (2009) 012021 doi:10.1088/1742-6596/174/1/012021

4. String Theory

In contrast to the approaches discussed above, the ambition for string theory (or M-theory) is to construct a unified quantum theory of all interactions from the very beginning, see, for example, [9, 1] for an introduction. Quantum gravity then emerges in an appropriate limit where the various interactions can be conceptually separated from each other. Although string theory, too, does not yet exist in a final form, it possesses a couple of attractive features. Besides the unification of interactions, the inclusion of gravity is unavoidable; as Ed Witten has put it [10]: "String theory forces general relativity upon us." Necessary ingredients are gauge invariance, supersymmetry, but also the presence of higher dimensions from which so far no trace has been seen. The perturbation theory is probably finite to all orders, although the sum of all terms diverges. And there are only three fundamental dimensionful constants, c, $\hbar$, and the string length, $l_{\rm s}$, from which in principle all other scales can be derived.

What can be said about the concept of time (and space) in string theory [11]? A central ingredient of the formalism is the Euclidean path integral,

$$Z = \int \mathcal{D}X\mathcal{D}h\, \mathrm{e}^{-S/\hbar} , \tag{26}$$

where the sum is over the embedding variables, X, describing the embedding of the string into a higher-dimensional auxiliary spacetime, and the metric on the two-dimensional worldsheet, shortly denoted by h. Therefore, X and h are the quantum degrees of freedom. In addition, however, the action depends on external 'background' fields, such as the metric of the auxiliary embedding spacetime, the dilaton field, and others. They are at this stage not considered as quantum variables. Since there should be no background fields also in string theory at the most fundamental level, this cannot yet be the final stage. Expression (26) is at the heart of the string perturbation theory.

Demanding the absence of quantum anomalies, it follows that the background fields obey Einstein's equations up to order $l_{\rm s}^2$; these equations can be derived from an effective action. It also follows that the number of dimensions of the embedding spacetime is restricted to be 10 (or 11).

From an operational point of view, one can argue that there is a lower bound on the uncertainty, Δx, of a spacetime point. This is, of course, a consequence of the extended nature of the string. In heuristic gedanken experiments one can derive a generalized uncertainty relation of the form

$$\Delta x \geq \frac{\hbar}{\Delta p} + \frac{l_{\rm s}^2}{\hbar}\Delta p , \tag{27}$$

which expresses this fact.

Since string theory contains general relativity, the arguments leading to the timelessness of canonical quantum gravity apply here, too. Thus, for scales larger than the Planck length, the Wheeler–DeWitt equation should be approximately valid.

The "problem of time" is thus the same in string theory and canonical quantum gravity. New insights are obtained for the concept of space, although a final picture is not available. In this context, the following two approaches are relevant. First, motivated by M-theory one can devise *matrix models* which contain a finite number of degrees of freedom connected with a system of D0-branes for the description of M-theory; the fundamental scale is the 11-dimensional Planck scale. Second, the *AdS/CFT correspondence* plays a key role: non-perturbative string theory in a background spacetime which is asymptotically anti-de Sitter (AdS) is dual to a conformal field theory (CFT) defined in a flat spacetime of one less dimension. This is often considered as a mostly background-independent definition of string theory, since the background metric enters only through boundary conditions at infinity.

The AdS/CFT correspondence is also connected with the realization of the "holographic principle": it turns out that laws including gravity in space dimension $d = 3$ are equivalent to laws excluding gravity in $d = 2$. In a sense, space has here vanished, too, and it has been claimed that gravity is only an "illusion" [12]. This should, however, not be interpreted literally, since the holographic principle only refers to the equivalence of two theories in an operational sense and does not postulate the absence of gravity in our world.

Acknowledgments
I am grateful to the organizers of the DICE08 conference for inviting me to a stimulating meeting at a beautiful place.

References

[1] Kiefer C 2007 *Quantum Gravity* second edition (Oxford: Oxford University Press)
[2] Kiefer C 2008 Quantum geometrodynamics: whence, whither *preprint* arXiv:0812.0295, to appear in *Gen. Rel. Grav.*
[3] Schrödinger E 1926 *Annalen der Physik* **384** 489
[4] Peres A 1962 *Nuovo Cimento* **XXVI** 53
[5] Joos E, Zeh H D, Kiefer C, Giulini D, Kupsch J and Stamatescu I O 2003 *Decoherence and the appearance of a classical world in quantum theory* second edition (Berlin: Springer); see also `www.decoherence.de`
[6] Rovelli C 2004 *Quantum Gravity* (Cambridge: Cambridge University Press)
[7] Zeh H D 2007 *The physical basis of the direction of time* fi fth edition (Berlin: Springer)
[8] Kamenshchik A Y, Kiefer C and Sandhöfer B 2007 *Phys. Rev.* D **76** 064032
[9] Zwiebach B 2004 *A first course in string theory* (Cambridge: Cambridge University Press)
[10] Witten E 2005 *Nature* **438** 1085
[11] Horowitz G T 2005 *New. J. Phys.* **7** 201
[12] Maldacena J 2007 *Scientific American Reports* (April) 75

Source of the observed thermodynamic arrow

L. S. Schulman

Physics Department, Clarkson University, Potsdam, New York 13699-5820, USA

E-mail: schulman@clarkson.edu

Abstract. The puzzle of the thermodynamic arrow of time reduces to the question of how the universe could have had lower entropy in the past. I show that no special entropy lowering mechanism (or fluctuation) is necessary. As a consequence of expansion, at a particular epoch in the history of the universe a state that was near maximum entropy under the dominant short range forces becomes extremely unlikely, due to a switchover to newly dominant long range forces. This happened at about the time of decoupling, prior to which I make no statement about arrows. The role of cosmology in thermodynamics was first suggested by T. Gold.

The thermodynamic arrow of time, basically the second law of thermodynamics, is often said to be one of the mysteries of nature, standing in stark contrast to the near time-symmetry of other dynamical laws accepted in physics.

This article is less ambitious than other recent discussions of the problem, for example, Ref. [1]. I only want to explain the observed phenomenon in terms of other known and accepted pieces of physics. I will not go to the very early universe (not even the first millennium), nor will I invoke unobserved phenomena. What I do assert is that there is less mystery than is alleged. The essential features of this argument have already appeared both in my own writings [2, 3, 4] and in the writings of others [5].[1] In the present article I will review and extend these arguments.

As foreshadowed, the explanation will involve cosmology. Notwithstanding the asserted limited scope, the claim is far from trivial: it relates local thermodynamics to the universe at large, a connection that is neither obvious nor intuitive. This leads me to a secondary objective of this presentation, namely to trace the origin of these ideas to their proper source, namely Thomas Gold, who in at least some of the current literature is either ignored or misquoted.[2]

Around 1960, 30 years after the discovery of the expansion of the universe, Gold suggested [6] that the thermodynamic arrow was a consequence of that expansion. That this was a significant step can be seen in an earlier article by Wheeler and Feynman in which they start from a time-symmetric electrodynamics and recover the usual time-asymmetric theory by means of their "absorber theory" [7]. Their work deals in an essential way with the issue of irreversibility (for example in presenting a framework for Dirac's calculation of radiation reaction [8]) and in some

[1] See p. 135 in [2], p. 358 in [3], Sec. 7 in [4] and p. 705 in [5].

[2] I do not allude to Ref. [1]. The *mis*quotation is the claim that Gold talked about a universe with expansion and contraction, with reversal of the arrow at the midpoint. He did talk about a contracting universe and the idea that the arrow would be reversed in it. But both at once—not. Now I happen to like that idea, but whatever he thought of it, he was adamant that it should not be attributed to him (personal communication). I don't know the first of these careless attributions, but at least some of the misquotation comes from those who used earlier misquotations rather than going to the easily available primary source [6].

of their attempts to establish the absorber theory Wheeler and Feynman make reference to the universe at large. Nevertheless, the essential feature, that the universe is *expanding*, comes nowhere in their article. Ref. [7] even mentions discussion with Einstein. This was in the 1940's and expansion did not seem the route to solving this great puzzle. For context it is interesting to read the record of a conference convened by Gold and Bondi [9] in 1963. Some participants were sympathetic to the idea that expansion could play a role (e.g., [10]); many were not.

In any case, the arguments of Gold indicated ways in which large scale expansion could be felt locally. He reasoned [6] that any local arrow is imposed by an outside influence, with the ultimate outside influence being the expansion of the universe, manifested through outgoing wave electromagnetic boundary conditions. Although I found his thesis attractive, I felt [11] that a proper presentation of his argument required time-symmetric boundary conditions. In the present article a slightly different approach is taken, based on actual observation of earlier states of the universe, combined with known dynamical processes.

It was realized long ago by Boltzmann that the problem of justifying the arrow was not so much showing that the entropy of isolated systems increased (accomplished with controversy by his H-theorem), but explaining why there was low entropy in the past. His justification [12] was that there was an enormous "fluctuation." For most of the universe and for most of its history it is in equilibrium, but at some point in our past there was a fluctuation leading to our present situation. It is hard to argue with this assertion; it is also hard to believe. See [2], notes to Sec. 4.0, and many other discussions.

I now argue that what we already know about the history of the universe provides that low entropy, and it occurs by virtue of the expansion.[3]

Consider the era of recombination or decoupling. For that epoch, we have an excellent image of the universe, namely that provided by the cosmic background radiation. It shows, for the luminous matter, a tremendous level (1 part in 10^5) of uniformity. This is reasonable. Under short range forces, for example the usual intermolecular interactions, the most likely state, that of maximum entropy, is uniformity. Forces were short range prior to recombination because the system was basically a plasma. Although electromagnetic forces are long range, because of the presence of uncancelled positive and negative charges there was screening of the electromagnetic forces, making them effectively short range. Of course gravitational forces were also present, but for luminous matter they were negligible. For dark matter, it is generally felt that at that time there must already have been some level of non-uniformity, but that does not affect our argument. Similarly, there is no evidence for the presence of black holes. To summarize the significant observation: the luminous matter was in equilibrium under its dominant forces and that equilibrium was a uniform state.

With the advent of recombination, charged particles combined to form neutral objects, H atoms, and the universe became transparent. Atom-atom forces were also negligible since particle separations were on the order of mm. Gravity now became the dominant force. Under gravity, uniformity is *not* the most likely state.[4] On the contrary, matter tends to clump, in fact it

[3] A note of caution must be sounded, even when arguing from an early low entropy state. Once a particular time has been selected, symmetric dynamics *away from* that time, in either direction, will tend to increase (or not decrease) entropy. This is the first conundrum of the Boltzmann H-theorem. As pointed out in [11, 2], the use of *initial* conditions generally provides an arrow of time with little else required. Thus to make a non-circular argument, one can take these low entropy conditions as *two-time* boundary conditions. Then if entropy increases towards the middle the desired result is established. Such arguments appear in [2]. For the reasoning in the present article this is not a problem, since there is a substantial change in dominant dynamics at the time of decoupling. Moreover, I make no claim about having or not having an arrow at earlier epochs.

[4] As an alternative way of characterizing a state as "unlikely" I'll say that its entropy is low. Nevertheless, entropy is problematic when long range forces are involved. Binney and Tremaine ([13], p. 268) give a Newtonian-gravity argument showing that in a galaxy entropy can increase without limit through rearrangements of the stars. I would also say that there is a more fundamental issue here. In the usual gas-with-short-range forces,

seems to do so hierarchically. The uniform distribution that in an earlier era was the most likely configuration, now becomes unstable.

This then is the punch line: with no mechanism other than expansion, entropy has been lowered. Interestingly, it's not the state that changes, but the dominant forces. What was maximum entropy in one regime becomes low entropy in another. And this happens because of expansion, since it's expansion that induces the cooling, causing in turn the formation of H atoms, and finally leading to a world in which the newly dominant force—gravity—controls a state that is unlikely from a gravitational perspective. This then is Gold's thesis, although the detailed arguments are different.

Getting from this state of disequilibrium to our experienced arrow of time is qualitatively straightforward, although quantitatively challenging. Under gravitational forces there is a continuing emergence of ever greater levels of non-uniformity. Eventually this gives rise to star formation and the release of nuclear energy. This in turn leads to the negentropy flow that permits the preparation of local low entropy states. In our case, this negentropy flow takes the form of 6000 K photons that allow us to remain out of equilibrium on earth.

Establishing this picture quantitatively is the objective of the fields of structure formation and stellar evolution. Puzzles in the former area suggest that, as remarked above, a degree of structure was already present before recombination. Presumably, this would ultimately have led to the high degree of non-uniformity that we see today, but it would not have done so on the observed time scale (about 13 billion years). This presumption, however, applies to a fictional world sans expansion: in our world, expansion implies decoupling, the trigger for the relatively rapid development of non-uniformity.

The foregoing argument says nothing about a thermodynamic arrow *before* recombination. Perhaps there was none or perhaps it was there because of even earlier physical processes.[5] We offer no opinion on this issue, since this paper is devoted to explaining the thermodynamic arrow we now experience.[6]

Finally a comment about the term "adiabatic" when used in connection with the post-decoupling expansion of the universe. I mention this because in the discussion in [2] the expansion is characterized as rapid. Certainly, adiabatic, suggesting a slow process, is appropriate when considering the ability of the photon gas to keep up with the expansion. However, the clumping process, that which ultimately leads to stars, etc., is far slower than the expansion. In other

one *loses* information as the gas becomes homogeneous. Ergodicity is equivalent to a system's losing macroscopic observables, until the only one left is energy (or in some circumstances, momentum and angular momentum). With gravity, what may have been a homogeneous system begins to *acquire* features; portions of the "gas" separate and new observables are born. However, for the purposes of discussion in the present article, I will use the word "entropy" as a shorthand for the likelihood or unlikelihood of states. But the kind of calculation in which one assigns a particular numerical value of entropy to a certain state (as one does in conventional thermodynamics) is not at all contemplated. Where there is common ground is in the consequences of a system being in a likely or an unlikely state. As for the usual low entropy situation, a system that can be identified as of low likelihood will progress to states of higher likelihood. In our case, the low likelihood state—under gravity—is uniformity.

[5] A fascinating essay by Misner [14] suggests that time should be measured logarithmically in the early universe (the way one often thinks about low temperatures), so that there could have been a succession of eras with arrows and perhaps with heat deaths and revivals under mechanisms such as described here.

[6] Having or not having an arrow need not be a yes-or-no issue. For example, one would say that a gas *at* equilibrium has no arrow, while a gas released from a small volume surely does. This suggests that rate of change of entropy be a quantitative measure of an arrow. However, this too has pitfalls. For a metastable system described by stochastic dynamics the derivative may be small (fixed by the second largest eigenvalue of the matrix of transition probabilities [15]) while for any particular exemplar the apparent changes may be great (after a nucleation event) or seemingly nonexistent (in the absence of such an event). For the systems discussed here there are two differences from the scenario just mentioned. First, with long range forces entropy is problematic. Second, the nucleation events are *not* sudden and local, but may occur at large distance scales. As such, on smaller scales (where entropy may be defined in a local sense) there will be no perceivable arrow.

words, as the universe expands, it does not manage to reach the state of greatest likelihood associated with gravity-dominated dynamics (which may not exist).

In this article I have given qualitative arguments. To what extent can greater precision be obtained? This could be in the form of better estimates of inhomogeneity, or perhaps model systems exhibiting the features I postulate. Frankly, for my purposes I don't think this is necessary, and for two reasons. First, behind my assertions are decades of experience and calculations by many, many people. Gravity makes things clump; of course much more than that is known, time scales under various conditions and the like, but this is greater precision than I need. Second, I am not predicting inhomogeneity; I am observing it. The evident property of gravity to head toward such a configuration is essential to my argument, but for that argument it is enough simply to observe that uniformity and gravity are at odds.

Similar remarks apply to models, but discussing the phenomenon in phase transition terms is instructive. Consider a Hamiltonian with both short and long range interparticle forces. If there is short-range repulsion and it is bounded in magnitude, there may very well may be no stable state of the system. However, if the long-range force is much weaker than the other, there can be a long-lived (uniform density) metastable phase. The passage out of this metastable state can be considered a phase transition,[7] from something that looks like equilibrium under short-range forces to something for which the dynamics never comes to equilibrium. Note that I am not saying that the actual universe suddenly passed through such a transition. What I mean by "phase transition" is the dramatic dependence of system behavior on a parameter. In our case we could take the parameter to be the strength of the short range force. The lifetime of the metastable phase would then drop rapidly as the short range force weakened. Thus, if we considered the plasma (sans dark matter) that preceded recombination but in a *non*expanding universe, the lifetime for gravity to take over and cause clumping would be extremely long.[8] However, if the short range coupling constant is reduced (as it is, because of cooling and expansion), then that lifetime has the dramatic reduction that one associates with a phase transition.

I review what has or has not been explained here. I have *not* explained why the universe was in a metastable state prior to recombination, a state in which not only the inhomogeneity of the gravitational dominance was absent, but the ultimate inhomogeneity of black holes was also absent. What I *have* explained is how the state that is in fact observed leads to the second law of thermodynamics, as we now see it. My point is thus the connection or correlation of two observed phenomena. In one sentence I restate the explanation: Expansion-induced cooling shifted the strength-balance of long and short range forces, allowing nucleation of a matter distribution that is clumpy and appears to have no ultimate equilibrium. Or more loosely speaking, what *was* high entropy with respect to the dominant forces becomes low entropy by virtue of an expansion-induced change in which forces dominate.

As declared at the outset, the scope of the explanation offered here is more modest than that of other recent attempts. There is good reason to be conservative. The field of cosmology undergoes major revisions of its paradigms with impressive frequency.[9] A great deal of modern cosmology is concerned with properties of black holes, and for the topic at hand focuses particularly on their entropic properties. As far as I know there is no experimental or observational evidence

[7] This phase transition would not be well described using the conventional notion of a breakdown in analyticity. However, except for mean field models (e.g., the van der Waals gas), analyticity methods fail to describe metastable states, and more general concepts, relying on time-scale differences, are needed [15].

[8] Moreover, without expansion the dark matter inhomogeneities that developed beginning with matter dominance would require a much extended time scale before they affected the luminous matter, assuming it remained a plasma.

[9] Examples come to mind: adjusting Hubble's distance scale by a factor two, dark matter, inflation, dark energy. Although at least one cosmologist (www.hep.phys.soton.ac.uk/~sfk/cosmology.ppt) seems to think that Landau's adage about cosmologists, "often in error, never in doubt," has been abrogated, I don't share that opinion.

on this matter, nor is there a shred of evidence, beyond theoretical conviction, for the existence of radiation from black holes.

To summarize, the existence of the observed thermodynamic arrow of time has been traced to observed and non-controversial features of the not-too-early universe. Given the state observed at that time, coupled with the expansion of the universe, one obtains the contemporary arrow of time.

Acknowledgments

I am grateful to the organizers of the DICE2008 conference for the opportunity to present these ideas. At the conference itself I gave additional background with emphasis on the need to avoid "initial conditions prejudice," in which the natural structure of language can lure one into circular arguments. This part of my lecture has been omitted from the present article and is mostly covered in [2]. More recent work of mine on this is [16] in which the compatibility of interacting opposite-arrow regions is shown, as well as [17] in which macroscopic causality loses its status as an independent property of nature and is shown to be a consequence of general entropy increase, i.e., the thermodynamic arrow.

The article itself was significantly enhanced by extended discussions with Amos Ori and Marco Roncadelli. I am also grateful to Bernard Gaveau and Huw Price for helpful discussions. Much of this interaction took place in the "Advanced Study Group," *Time: quantum and statistical mechanics aspects*, of the Max Planck Institute for the Physics of Complex Systems, in Dresden. This work was partly supported by the United States National Science Foundation grant PHY 05 55313. I am also grateful to the Lewiner Institute for Theoretical Physics for partial support during a visit to the Technion.

References

[1] Wald R M 2005 Lecture given at the Seven Pines "Arrows of Time" meeting, 2004, arXiv:gr-qc/0507094
[2] Schulman L S 1997 *Time's Arrows and Quantum Measurement* (New York: Cambridge University Press)
[3] Schulman L S 2003 *Hyperbolic Differential Operators and Related Problems* ed Ancona V and Vaillant J (New York: Marcel Dekker) pp 355–370, proceedings of a June 2000 conference, in honor of Jean Vaillant. URL `arXiv:cond-mat/0009139 10 Sep 2000`
[4] Schulman L S 2005-6 *Time-related issues in statistical mechanics,* lectures given at the Institut de Physique Théorique, Direction des Sciences de la Matière, CEA-Saclay and at the Physics Dept., Technion (Haifa), Sec. 7, current URL `http://people.clarkson.edu/~schulman/`
[5] Penrose R 2004 *The Road to Reality* (London: Random House) Vintage Books Edition, New York, 2007.
[6] Gold T 1962 *Am. J. Phys.* **30** 403–410
[7] Wheeler J A and Feynman R P 1945 *Rev. Mod. Phys.* **17** 157–181. See also [18]
[8] Dirac P A M 1938 *Proc. Roy. Soc. London* **A 167** 148–169
[9] Gold T 1967 *The Nature of Time* (Ithaca, New York: Cornell Univ. Press) ed. with the assistance of D. L. Schumacher. Proc. conf. at Cornell University, 1963.
[10] Hoyle F and Narlikar J V 1964 *Proc. Roy. Soc. London* **277A** 1–23
[11] Schulman L S 1973 *Phys. Rev. D* **7** 2868–2874
[12] Boltzmann L 1995 *Lectures on Gas Theory* (New York: Dover Pub.) German original: Vorlesungen über Gastheorie, trans. S. G. Brush. Univ. of California Press, 1964.
[13] Binney J and Tremaine S 1987 *Galactic Dynamics* (Princeton, New Jersey: Princeton University Press)
[14] Misner C W 1969 *Phys. Rev.* **186** 1328–1333
[15] Gaveau B and Schulman L S 1998 *J. Math. Phys.* **39** 1517–1533
[16] Schulman L S 1999 *Phys. Rev. Lett.* **83** 5419–5422
[17] Schulman L S 2001 *Causality is an effect*, in *Time's Arrows, Quantum Measurement and Superluminal Behavior* ed D Mugnai, A Ranfagni and L S Schulman (Rome: Consiglio Nazionale delle Ricerche) pp 99–112
[18] Wheeler J A and Feynman R P 1949 *Rev. Mod. Phys.* **21** 425–433

Noncommutativity and quantum structure of spacetime

B P Dolan[1], **Kumar S Gupta**[2] **and A Stern**[3]

[1] Dept. of Mathematical Physics, National University of Ireland, Maynooth, Ireland
[2] Theory Division, Saha Institute of Nuclear Physics, 1/AF Bidhannagar, Kolkata, India
[3] Department of Physics, University of Alabama, Tuscaloosa, Alabama 35487, USA

E-mail: bdolan@thphys.nuim.ie, kumars.gupta@saha.ac.in, astern@bama.ua.edu

Abstract. Noncommutative geometry is a candidate for describing physics at the Planck scale. Although there are various proposals to formulate gravity theories in the noncommutative framework, not many exact solutions of noncommutative gravity theories are known. In this talk we describe a possible noncommutative analogue of the BTZ black hole, which is consistent with the geometric structure of the problem. The spacetime Poisson brackets that we obtain indicate the possibility of the quantization of time. We briefly discuss the issues of symmetries and particle statistics in such a spacetime.

1. Introduction

The description of physics at the Planck scale, where quantum aspects of gravity are expected to be manifest, remains an outstanding challenge in theoretical physics. It is plausible that strong quantum gravity effects associated with such high energy scales can affect the smooth structure of the spacetime manifold. There have been various proposals to replace the continuum picture of spacetime with a discrete atomistic structure at the Planck scale. In a related approach, it has been argued that Heisenberg's uncertainty principle together with Einstein's theory of general relativity leads to spacetime noncommutativity [1]. In this picture, the smooth manifold structure of the spacetime is replaced with a noncommutative algebra at high energies, and the associated dynamics is governed by a noncommutative generalization of gravity.

Most approaches to noncommutative gravity have involved deforming the commutative Einstein equations [2]. These approaches range from simply replacing point-wise products by Moyal star products in the Einstein-Hilbert action to the approach of Aschieri *et al.* [3] which preserves the diffeomorphism invariance of general relativity. In our approach [4], we write down Poisson brackets which are consistent with the geometry of some classical solution. In particular, we focus on the Banados-Teitelboim-Zanelli (BTZ) solution of $2+1$ gravity. The noncommutative counterpart of the BTZ solution is then obtained by 'quantization', with the deformation parameter being the noncommutativity parameter θ. The Poisson brackets and the resulting noncommutative algebra are not unique. In this regard, it may be desirable to impose the restriction that the isometry of the classical solution survives quantization and is implementable in any irreducible representation of the noncommutative algebra. In this talk we focus on the deformations with such additional constraints and show that they lead to spacetime noncommutativity where the time is quantized [5, 6].

Analysis of quantum field theory (QFT) on curved spacetime often reveals the features of the underlying geometry. It is thus desirable to have a formulation of QFT for a spacetime arising from noncommutative gravity. Here we take the preliminary steps for setting up QFT in such a background and briefly indicate the relevant issues.

2. Commutative BTZ

In this Section we shall briefly review certain features of the commutative BTZ solution that would be relevant for our analysis.

In terms of Schwarzschild-like coordinates (r,t,ϕ) the invariant measure for the BTZ black hole [7, 8] is expressed as

$$ds^2 = \left(M - \frac{r^2}{\ell^2} - \frac{J^2}{4r^2}\right)dt^2 + \left(-M + \frac{r^2}{\ell^2} + \frac{J^2}{4r^2}\right)^{-1} dr^2 + r^2\left(d\phi - \frac{J}{2r^2}dt\right)^2 , \tag{1}$$

$$0 \le r < \infty \;,\; -\infty < t < \infty \;,\; 0 \le \phi < 2\pi \;,$$

where M and J are the mass and spin, respectively, and $\Lambda = -1/\ell^2$ is the cosmological constant. For $0 < |J| < M\ell$, there are two horizons, the outer and inner horizons, corresponding respectively to $r = r_+$ and $r = r_-$, where

$$r_\pm^2 = \frac{M\ell^2}{2}\left\{1 \pm \left[1 - \left(\frac{J}{M\ell}\right)^2\right]^{\frac{1}{2}}\right\} . \tag{2}$$

The two horizons coincide in the extremal case $|J| = M\ell > 0$, while the inner one disappears for $J = 0$, $M > 0$. The metric is diagonal in the coordinates (χ_+, χ_-, r), where

$$\chi_\pm = \frac{r_\pm}{\ell}t - r_\mp\phi \,, \tag{3}$$

$$ds^2 = \frac{-(r^2 - r_+^2)d\chi_+^2 + (r^2 - r_-^2)d\chi_-^2}{r_+^2 - r_-^2} + \frac{\ell^2 r^2 dr^2}{(r^2 - r_+^2)(r^2 - r_-^2)} , \tag{4}$$

which shows that χ_+ is the time-like coordinate in the region I) $r \ge r_+$, r is the time-like coordinate in the region II) $r_- \le r \le r_+$, and χ_- is the time-like coordinate in the region III) $0 \le r \le r_-$.

It was shown that the manifold of the BTZ black hole solution is the quotient space of the universal covering space of AdS^3 by some elements of the group of isometries of AdS^3. The connected component of the latter is $SO(2,2)$. Say AdS^3 is spanned by coordinates (t_1, t_2, x_1, x_2) parameterizing R^4, satisfying

$$-t_1^2 - t_2^2 + x_1^2 + x_2^2 = -\ell^2 \;. \tag{5}$$

Alternatively, one can introduce 2×2 real unimodular matrices

$$g = \frac{1}{\ell}\begin{pmatrix} t_1 + x_1 & t_2 + x_2 \\ -t_2 + x_2 & t_1 - x_1 \end{pmatrix} , \qquad \det g = 1 \,, \tag{6}$$

belonging to the defining representation of $SL(2,R)$. The isometries correspond to the left and right actions on g,

$$g \to h_L g h_R \,, \qquad h_L, h_R \in SL(2,R) \;. \tag{7}$$

Since (h_L, h_R) and $(-h_L, -h_R)$ give the same action, the connected component of the isometry group for AdS^3 is $SL(2,R) \times SL(2,R)/Z_2 \approx SO(2,2)$.

The BTZ black-hole is obtained by discrete identification of points on the universal covering space of AdS^3. This insures periodicity in ϕ, $\phi \sim \phi + 2\pi$. The condition is

$$g \sim \tilde{h}_L g \tilde{h}_R \;, \tag{8}$$

where $(\tilde{h}_L, \tilde{h}_R)$ are certain elements of $SO(2,2)$. $\tilde{h}_L$ and $\tilde{h}_R$ can be expressed as diagonal $SL(2,R)$ matrices

$$\tilde{h}_L = \begin{pmatrix} e^{\pi(r_+ - r_-)/\ell} & \\ & e^{-\pi(r_+ - r_-)/\ell} \end{pmatrix} \;, \qquad \tilde{h}_R = \begin{pmatrix} e^{\pi(r_+ + r_-)/\ell} & \\ & e^{-\pi(r_+ + r_-)/\ell} \end{pmatrix} \;. \tag{9}$$

For $0 < |J| < M\ell$, the universal covering space of AdS^3 is covered by three types of coordinate patches which are bounded by the two horizons at $r = r_+$ and $r = r_-$. For all three coordinate patches, g can be decomposed according to

$$g = \begin{pmatrix} e^{\frac{1}{2\ell}(\chi_+ - \chi_-)} & \\ & e^{-\frac{1}{2\ell}(\chi_+ - \chi_-)} \end{pmatrix} g^{(0)}(r) \begin{pmatrix} e^{\frac{1}{2\ell}(\chi_+ + \chi_-)} & \\ & e^{-\frac{1}{2\ell}(\chi_+ + \chi_-)} \end{pmatrix} \;, \tag{10}$$

where $g^{(0)}(r)$ is an $SO(2)$ matrix which only depends on r and the coordinate patch. The periodicity condition for ϕ easily follows from (8). The identification (8) breaks the $SO(2,2)$ group of isometries to a two-dimensional subgroup $\mathcal{G}_{BTZ}$, consisting of only the diagonal matrices in $\{h_L\}$ and $\{h_R\}$. $\mathcal{G}_{BTZ}$ is the isometry group of the BTZ black hole, and from (10) is associated with translations in χ_+ and χ_-, or equivalently t and ϕ, on $r =$ constant surfaces.

3. Noncommutative BTZ

In this Section we shall find the nontrivial Poisson brackets among the $SL(2,R)$ matrix elements which describe a BTZ black hole.

For generic spin, $0 < |J| < M\ell$ (and $M > 0$), we shall search for Poisson brackets for the matrix elements of g which are polynomial of lowest order. They should be consistent with the quotienting (8), as well as with the unimodularity condition and, of course, with the Jacobi identity. For convenience we write the $SL(2,R)$ matrix as

$$g = \begin{pmatrix} \alpha & \beta \\ \gamma & \delta \end{pmatrix} \;, \qquad \alpha\delta - \beta\gamma = 1 \;. \tag{11}$$

Under the quotienting (8):

$$\begin{array}{rcl} \alpha & \sim & e^{\;2\pi r_+/\ell}\;\alpha \\ \beta & \sim & e^{-2\pi r_-/\ell}\;\beta \\ \gamma & \sim & e^{\;2\pi r_-/\ell}\;\gamma \\ \delta & \sim & e^{-2\pi r_+/\ell}\;\delta \;. \end{array} \tag{12}$$

All quadratic combinations of matrix elements scale differently, except for $\alpha\delta$ and $\beta\gamma$, which are invariant under (12). Lowest order polynomial expressions for the Poisson brackets of α, β, γ and δ which are preserved under (12) are quadratic and have the form:

$$\begin{array}{lll} \{\alpha,\beta\} = c_1\alpha\beta & \{\alpha,\gamma\} = c_2\alpha\gamma & \{\alpha,\delta\} = f_1(\alpha\delta,\beta\gamma) \\ \{\beta,\delta\} = c_3\beta\delta & \{\gamma,\delta\} = c_4\gamma\delta & \{\beta,\gamma\} = f_2(\alpha\delta,\beta\gamma) \end{array} \;, \tag{13}$$

where c_{1-4} are constants and $f_{1,2}$ are functions. They are constrained by

$$c_1 + c_2 \;=\; c_3 + c_4$$

$$\begin{aligned} f_1(\alpha\delta,\beta\gamma) &= (c_1+c_2)\beta\gamma \\ f_2(\alpha\delta,\beta\gamma) &= (c_2-c_4)\alpha\delta\ , \end{aligned} \tag{14}$$

after demanding that $\det g$ is a Casimir of the algebra. From (14) there are three independent constants c_{1-4}. Further restrictions on the constants come from the Jacobi identity, which leads to the following two possibilities:

$$\mathrm{A}:\quad c_2=c_4 \qquad \text{and} \qquad \mathrm{B}:\quad c_2=-c_1\ .$$

Both cases define two-parameter families of Poisson brackets. Say we call c_2 and c_3 the two independent parameters. The two cases are connected by an $SO(2,2)$ transformation. Case A goes to case B when

$$g=\begin{pmatrix}\alpha & \beta\\ \gamma & \delta\end{pmatrix}\rightarrow g'=\begin{pmatrix}\beta & -\alpha\\ \delta & -\gamma\end{pmatrix}=gh_R^{(0)}\ ,\qquad h_R^{(0)}=\begin{pmatrix} & -1\\ 1 & \end{pmatrix}\ , \tag{15}$$

along with

$$c_3\rightarrow c_2\ ,\qquad c_2\rightarrow c_3\ . \tag{16}$$

In terms of the embedding coordinates, this corresponds to $(t_1,t_2,x_1,x_2)\rightarrow(t_2,-t_1,x_2,-x_1)$.

There are three types of coordinate patches in the generic case of $M>0$ and $0<|J|<M\ell$, and their boundaries are the two horizons. Denote them again by: I) $r\geq r_+$, II) $r_-\leq r\leq r_+$ and III) $0\leq r\leq r_-$. The corresponding maps to $SL(2,R)$ are given by (10), with
I) $r\geq r_+$,

$$g^{(0)}(r)=g_I^{(0)}(r)=\frac{1}{\sqrt{r_+^2-r_-^2}}\begin{pmatrix}\sqrt{r^2-r_-^2} & \sqrt{r^2-r_+^2}\\ \sqrt{r^2-r_+^2} & \sqrt{r^2-r_-^2}\end{pmatrix}\ , \tag{17}$$

II) $r_-\leq r\leq r_+$,

$$g^{(0)}(r)=g_{II}^{(0)}(r)=\frac{1}{\sqrt{r_+^2-r_-^2}}\begin{pmatrix}\sqrt{r^2-r_-^2} & -\sqrt{r_+^2-r^2}\\ \sqrt{r_+^2-r^2} & \sqrt{r^2-r_-^2}\end{pmatrix}\ , \tag{18}$$

III) $0\leq r\leq r_-$,

$$g^{(0)}(r)=g_{III}^{(0)}(r)=\frac{1}{\sqrt{r_+^2-r_-^2}}\begin{pmatrix}\sqrt{r_-^2-r^2} & -\sqrt{r_+^2-r^2}\\ \sqrt{r_+^2-r^2} & -\sqrt{r_-^2-r^2}\end{pmatrix}\ . \tag{19}$$

Using the maps (17-19), we can write the Poisson brackets for the various cases in terms of the Schwarzschild-like coordinates (r,t,ϕ). The results are the same in all three coordinate patches. For the two-parameter families A one gets:

$$\begin{aligned} \{\phi,t\} &= \frac{\ell^3}{2}\frac{c_3-c_2}{r_+^2-r_-^2} \\ \{r,\phi\} &= -\frac{\ell r_+(c_3+c_2)}{2r}\frac{r^2-r_+^2}{r_+^2-r_-^2} \\ \{r,t\} &= -\frac{\ell^2 r_-(c_3+c_2)}{2r}\frac{r^2-r_+^2}{r_+^2-r_-^2}\ , \end{aligned} \tag{20}$$

for the two-parameter families B one gets:

$$\{\phi,t\} = \frac{\ell^3}{2}\frac{c_3-c_2}{r_+^2-r_-^2}$$

$$\begin{aligned}\{r,\phi\} &= -\frac{\ell r_-(c_2+c_3)}{2r}\frac{r^2-r_-^2}{r_+^2-r_-^2} \\ \{r,t\} &= -\frac{\ell^2 r_+(c_2+c_3)}{2r}\frac{r^2-r_-^2}{r_+^2-r_-^2}\ .\end{aligned} \tag{21}$$

These Poisson brackets are invariant under the action of the isometry group $\mathcal{G}_{BTZ}$ of the BTZ black hole. The first bracket agrees in both cases. The latter two brackets vanish at the outer horizon $r = r_+$ for case A, and at the inner horizon $r = r_-$ for case B. A central element of the Poisson algebra can be constructed out of the Schwarzschild coordinates for both cases. It is given by

$$\rho_\pm = (r^2 - r_\pm^2)\exp\left\{-\frac{2\kappa\chi_\pm}{\ell}\right\}\ , \qquad c_2 \neq c_3\ , \tag{22}$$

where the upper and lower signs correspond to case A and B, respectively,

$$\kappa = \frac{c_3+c_2}{c_3-c_2}\ , \tag{23}$$

and $\chi_\pm$ were defined in (3). The $\rho_\pm$ = constant surfaces define symplectic leaves which are topologically R^2 for generic values of the parameters (more specifically, $c_2 \neq \pm c_3$). We can coordinatize them by χ_+ and χ_-. One then has a trivial Poisson algebra in the coordinates $(\chi_+, \chi_-, \rho_\pm)$:

$$\{\chi_+,\chi_-\} = \frac{\ell^2}{2}(c_3-c_2)\ , \qquad \{\rho_\pm,\chi_+\} = \{\rho_\pm,\chi_-\} = 0\ . \tag{24}$$

The action of the $\mathcal{G}_{BTZ}$ transforms one symplectic leaf to another, except for the case $c_2 = -c_3$ in which we shall focus from now on.

In passing to the noncommutative theory, the operator associated with $\rho_\pm$ is central in the quantum algebra and proportional to the identity in any irreducible representation. Irreducible representations then select $\rho_\pm$ = constant surfaces and the isometry group $\mathcal{G}_{BTZ}$ maps between different irreducible representations, and thus cannot be implemented as inner transformations. However, for the special case of $c_2 = -c_3$, the parameter κ vanishes ($\kappa = 0$) and the irreducible representation is preserved under the action of $\mathcal{G}_{BTZ}$. In this case, the radial coordinate is in the center of the algebra, r = constant defines $R \times S^1$ symplectic leaves, and they are invariant under the action of $\mathcal{G}_{BTZ}$. The coordinates ϕ and t parametrizing any such surface are canonically conjugate:

$$\{\phi,t\} = \frac{c_3\ell^3}{r_+^2-r_-^2}\ , \qquad \{\phi_\pm, r\} = \{t,r\} = 0\ . \tag{25}$$

The Poisson brackets can be interpreted in terms of a twist [9] in the decomposition of g given in (10), where the twist is with respect to the first and third matrices. In passing to the noncommutative theory, we need to define a deformation of the commutative algebra generated by $t, e^{i\phi}$ and r. Call the corresponding quantum operators $\hat{t}, e^{i\hat{\phi}}$ and $\hat{r}$, respectively. Their commutation relations are

$$[e^{i\hat\phi},\hat t] = \theta e^{i\hat\phi}\ , \qquad [\hat r,\hat t] = [\hat r, e^{i\hat\phi}] = 0\ , \tag{26}$$

where from (25) the constant θ is linearly related to $\ell^3/(r_+^2 - r_-^2)$. There are now two central elements in the algebra: i) $\hat{r}$ and ii) $e^{-2\pi i\hat{t}/\theta}$. From i), irreducible representations select the $R \times S^1$ symplectic leaves. The action of $\mathcal{G}_{BTZ}$ does not take you out of any particular irreducible representation, and in this sense we can say that the isometry of the classical solution survives quantization.

With regard to the central element ii) $e^{-2\pi i\hat{t}/\theta}$, one can identify it with $e^{i\chi}$ in an irreducible representation. The spectrum of $\hat{t}$ is then discrete [5, 6]:

$$n\theta - \frac{\chi\theta}{2\pi}\ , \ \ n \in Z\ . \tag{27}$$

This implies that if the above description admits a Hamiltonian formulation, then the corresponding time-independent Hamiltonian operator would be conserved only modulo $2\pi/\theta$.

4. QFT on noncommutative BTZ

In this Section we give a very brief outline of the issues involved in setting up a QFT in the noncommutative BTZ spacetime.

In the commutative case, the steps involved in setting up a QFT include the mode expansion of the field and imposing commutation relations between the creation and annihilation operators depending on the field statistics and, finally, representing the operators of the theory in a suitable Hilbert space. It is assumed that the symmetries of the theory should be implemented in such a way that the statistics be superselected. Finally, the discrete spacetime symmetries should also be properly implemented in the quantum theory.

In the case of a Moyal plane, which is endowed with a star product, one way to implement the symmetries is by using the Drinfeld twist. In the presence of a finite noncommutative parameter, the coproduct for any continuous symmetry, e.g. Poincare symmetry, must be twisted so that it is compatible with the rule of star multiplication. This in turn implies that the flip operator, which defines the statistics, should be twisted as well. As a result, the algebra of the creation and annihilation operators differs from that in the commutative case. In addition, in the Moyal case, parity cannot be implemented as an automorphism and hence is not a good symmetry. These issues have been discussed extensively in the literature [10, 11, 12, 13].

In our formulation, the deformation of the BTZ black hole is described by a noncommutative cylinder. As we have seen, the spectrum of the time operator is quantized in such a case. In order to set up a consistent QFT for this noncommutative spacetime, the issues mentioned above must be addressed properly, which is presently under investigation.

References

[1] Doplicher S, Fredenhagen K and Roberts J E 1994 *Phys. Lett.* B **331** 39; *do.* 1995 *Commun. Math. Phys.* **172** 187 (*Preprint* arXiv:hep-th/0303037)
[2] Madore J and Mourad J 1994 *Int. J. Mod. Phys.* D **3** 221
[3] Aschieri P, Blohmann C, Dimitrijevic M, Meyer F, Schupp P and Wess J 2005 *Class. Quant. Grav.* **22** 3511; Aschieri P, Dimitrijevic M, Meyer F and Wess J 2006 *Class. Quant. Grav.* **23** 1883
[4] Dolan B, Gupta Kumar S and Stern A 2007 *Class. Quant. Grav.* **24** 1647 (*Preprint* arXiv:hep-th/0611233)
[5] Chaichian M, Demichev A, Presnajder P and Tureanu A 2001 *Eur. Phys. J.* C **20** 767; *do.* 2001 *Phys. Lett.* B **515** 426
[6] Balachandran A P, Govindarajan T R, Martins A G and Teotonio-Sobrinho P 2004 *JHEP* **0411** 068
[7] Banados M, Teitelboim C and Zanelli J 1992 *Phys. Rev. Lett.* **69** 1849; Banados M, Henneaux M, Teitelboim C and Zanelli J 1993 *Phys. Rev.* D **48** 1506
[8] Carlip S 1995 *Class. Quant. Grav.* **12** 2853
[9] Connes A and Landi G 2001 *Commun. Math. Phys.* **221** 141
[10] Oeckl R 2000 *Nucl. Phys.* B **581** 559 (*Preprint* arXiv:hep-th/0003018)
[11] Chaichian M, Nishijima K and Tureanu A 2003 *Phys. Lett.* **B568** 146 (*Preprint* arXiv:hep-th/0209008)
[12] Balachandran A P, Mangano G, Pinzul A and Vaidya S 2006 *Int. J. Mod. Phys.* **A21** 3111 (*Preprint* arXiv:hep-th/0508002)
[13] Balachandran A P, Govindarajan T R, Mangano G, Pinzul A, Qureshi B A and Vaidya S 2007 *Phys. Rev.* D **75** 045009 (*Preprint* arXiv:hep-th/0608179)

Quantum measurement and quantum gravity: many-worlds or collapse of the wavefunction?

T P Singh

Tata Institute of Fundamental Research, Homi Bhabha Road, Mumbai 400005, India

E-mail: `tpsingh@tifr.res.in`

Abstract. At present, there are two possible, and equally plausible, explanations for the physics of quantum measurement. The first explanation, known as the many-worlds interpretation, does not require any modification of quantum mechanics, and asserts that at the time of measurement the Universe splits into many branches, one branch for every possible alternative. The various branches do not interfere with each other because of decoherence, thus providing a picture broadly consistent with the observed Universe. The second explanation, which requires quantum mechanics to be modified from its presently known form, is that at the time of measurement the wavefunction collapses into one of the possible alternatives. The two explanations are mutually exclusive, and up until now, no theoretical reasoning has been put forward to choose one explanation over the other. In this article, we provide an argument which implies that the collapse interpretation is favored over the many-worlds interpretation. Our starting point is the assertion (which we justify) that there ought to exist a reformulation of quantum mechanics which does not refer to a classical spacetime manifold. The need for such a reformulation implies that quantum theory becomes nonlinear on the Planck mass/energy scale. Standard linear quantum mechanics is an approximation to this nonlinear theory, valid at energy scales much smaller than the Planck scale. Using ideas based on noncommutative differential geometry, we develop such a reformulation and derive a nonlinear Schrödinger equation, which can explain collapse of the wavefunction. We also obtain an expression for the lifetime of a quantum superposition. We suggest ideas for an experimental test of this model.

1. The quantum measurement problem

Suppose that in a quantum measurement one measures an observable $\hat{O}$ of a quantum system, and suppose that prior to the measurement the system is in a state $|\psi\rangle$ which can be expanded in a basis of orthonormal eigenstates $|\psi_n\rangle$ of the observable $\hat{O}$ as

$$|\psi\rangle = \sum_n a_n|\psi_n\rangle. \tag{1}$$

It is then known from experiment that, if the selected states of the measuring apparatus are in one-to-one correspondence with the basis $|\psi_n\rangle$, then after the measurement the quantum system is found to be in one of the eigenstates, say $|\psi_n\rangle$. Repeated measurements on identical copies of the quantum system show that the system is found to be in one or the other eigenstates $|\psi_n\rangle$, with the probability to be found in state $|\psi_n\rangle$ being given by $|a_n|^2$ (the Born probability rule).

The transition $|\psi\rangle \rightarrow |\psi_n\rangle$ that takes place during a quantum measurement cannot be described by the Schrödinger equation. This is of course because the Schrödinger equation is linear, and Schrödinger evolution will preserve the superposition expressed in Eq. (1). The observed transition $|\psi\rangle \rightarrow |\psi_n\rangle$, on the other hand, breaks linear superposition. The quantum measurement problem can be stated as follows: what is the correct physical description of this measurement process and of the observed result? This description should explain why the transition takes place in the first place, and why it obeys the Born probability rule. There are two possible explanations, which we elaborate on below.

1.1. *First explanation: the many-worlds interpretation*

According to the many-worlds interpretation of quantum mechanics, (originally due to Everett [1]), despite appearances, the transition $|\psi\rangle \rightarrow |\psi_n\rangle$ in fact does not take place at all during a quantum measurement, and superposition continues to be preserved after the measurement has taken place. Rather, it is assumed that during a quantum measurement the Universe splits into many branches, with a particular outcome, say $|\psi_n\rangle$, being realized in our branch of the Universe. Here, the term Universe is meant to refer also to the measuring apparatus, and the observer as well. It has also been asserted that the many-worlds interpretation is consistent with the Born probability rule [2]. (For another discussion on the probability rule in the many-worlds picture see [3].) A recent account of the many-worlds interpretation, including recent developments in its understanding, can be found in [4].

The different branches of the Universe do not interfere with each other because of the phenomenon of decoherence. The process of decoherence, which has been experimentally observed [5], destroys the interference amongst various alternatives in the quantum state of a macroscopic system consisting of a superposition of various alternatives [6]. Thus in the many-worlds picture the various branches of the Universe continue to remain superposed, as required by Schrödinger evolution, but do not interfere with each other, as a consequence of decoherence. In this picture, quantum mechanics does not have to be modified in order to explain a quantum measurement. This has been called 'economy of assumptions, and extravagance of Universes'.

The many-worlds interpretation may appear counter-intuitive, but it is difficult to find logical inconsistencies in the interpretation. The picture is in fact attractive because it can explain quantum measurement without having to change the laws of quantum mechanics. Its shortcoming perhaps is that it does not appear to be experimentally falsifiable (unless experimental proof can be found for the 'collapse' interpretation discussed next). Also, if the Universe splits into many branches, it is not clear how the obscure issue of 'splitting of the consciousness of an observer into many branches' is to be understood. Neither of these shortcomings however can by themselves rule out the possibility that the many-worlds interpretation could be the correct explanation of quantum measurement.

1.2. *Second explanation: collapse of the wavefunction*

The second explanation is that the transition $|\psi\rangle \rightarrow |\psi_n\rangle$ does indeed take place, and the Universe does not split into many branches. It is assumed that there is only one branch to the Universe, the one that we directly observe, and live in. The two explanations (many-worlds and collapse) are clearly mutually exclusive: one, and only one, out of the two explanations must be correct.

In order to explain quantum measurement by invoking collapse of the wavefunction, quantum mechanics and the Schrödinger equation must be modified. Quantum mechanics as we presently know it can only be an approximation to a more general theory, with the more general theory having the capacity to explain wavefunction collapse. This has been called 'economy of Universes, and extravagance of assumptions'.

For instance, it may be possible to explain quantum measurement by generalizing the Schrödinger equation to a nonlinear Schrödinger equation. The nonlinearity is assumed to become important in the measurement domain, but is negligible in the microscopic domain. In principle, the presence of the nonlinearity can result in breakdown of superposition, driving the quantum system to one particular alternative, in a manner consistent with the Born probability rule. This particular approach to collapse of the wavefunction will be the focus of the present paper. Other models of collapse are briefly reviewed in Section 4. Objections against a nonlinear quantum mechanics (such as superluminality) are briefly addressed in the Discussion section.

It is only fair to say that as of now, there is no universally accepted theory for collapse of the wavefunction, supported by experiment. Neither is there any experimental evidence that quantum mechanics has to be modified from its present form. Needless to say, this situation could change in the future.

1.3. Goal of the present paper

Up until now, there has been no experimental or theoretical motivation to favor the many-worlds interpretation of quantum measurement over the collapse interpretation, or vice-versa. Critics of the many-worlds interpretation could say its unfalsifiable, while adherents of this interpretation consider it an advantage that it requires no changes in the existing formulation of quantum theory. Proponents of the collapse model find it unphysical that unobservable parallel Universes are invoked so as to protect the prevailing structure of quantum theory, whereas critics label the collapse models as *ad hoc*; having been invented for the sole purpose of explaining quantum measurement, and not embedded in a broader theoretical framework.

The purpose of the present paper is to argue that there are additional theoretical reasons, hitherto unemphasized, which suggest that the collapse explanation is favored over the many-worlds interpretation. Our starting point, which we elaborate on in detail in Section 3, is that the notion of time which is used to describe time evolution in quantum theory is a classical notion. There ought to exist a reformulation of quantum mechanics which does not make a reference to this external classical notion of time. Furthermore, under appropriate circumstances, i.e. as and when one chooses to make reference to (a possibly available) external classical time, this reformulation should become equivalent to standard quantum mechanics.

The central thesis of our paper is that the requirement that there be such a reformulation of quantum mechanics leads to the conclusion that standard quantum theory is a limiting case of a more general, nonlinear, quantum theory. The nonlinearity becomes important in the measurement domain, and could cause the collapse of the wavefunction during a quantum measurement. This is why we say that the collapse based explanation is favored over the many-worlds interpretation. We arrive at this conclusion, not in an *ad hoc* fashion, but by addressing an entirely different incompleteness in the existing formulation of quantum theory, namely, its undesirable reference to an external classical time. Thus our conclusion is that removing the notion of classical time from quantum mechanics has an important and significant byproduct — one may be able to explain quantum measurement as a consequence of wavefunction collapse. The requirement that the aforementioned reformulation of quantum mechanics should exist is unavoidable; and nonlinearity in quantum mechanics is its inevitable consequence. The many-worlds interpretation is thus disfavored, not because it might appear to be counter-intuitive and discomforting, but because of compelling theoretical reasoning which extends outside and beyond the current formulation of quantum mechanics.

In Section 3 we will propose a tentative reformulation of quantum mechanics, borrowing ideas from noncommutative geometry. We will arrive at a nonlinear Schrödinger equation which generalizes the standard linear Schrödinger equation, and which can explain at least some significant aspects of the process of quantum measurement.

Before we present this reformulation, we will review in the next Section an illustrative toy-model for collapse induced by nonlinearity, which is due to Grigorenko [7]. This model will be of help to us in understanding the relation between wavefunction collapse and the nonlinear equation we arrive at in Section 3.

[A few other significant interpretations of quantum mechanics, not discussed in the present paper, have also been proposed in the literature. These include the works of Bohm [8] (and its generalization to nonlocal hidden variable theories), and those of Gell-Mann and Hartle [9], and Omnes [10]. These works present a different formulation of standard quantum mechanics, without implying that the experimental predictions of the new formulation differ from that of the standard theory.]

2. A toy model for nonlinear quantum mechanics and collapse of the wavefunction

2.1. Introduction

We propose a nonlinear Schrödinger equation which during a quantum measurement can dynamically induce the transition $|\psi\rangle \rightarrow |\psi_n\rangle$ with a probability $|a_n|^2$, when the initial state of the quantum system is given by Eq. (1). Since such a nonlinear equation must not violate what is known about quantum mechanics from experiments (including stringent bounds on nonlinearity in the microscopic domain), it must satisfy several conditions, which we enumerate below:

- The nonlinearity must become significant only during the onset of a quantum measurement, and not before that.
- The nonlinear equation must reduce to the standard linear Schrödinger equation when applied to the special case of microscopic systems.
- The nonlinear equation must reduce to the standard equation of motion of classical mechanics when applied to the special case of macroscopic systems.
- The Hamiltonian must have a non-Hermitean part, which is also nonlinear, so that the initial superposition of states can decay into one of the alternatives. However, the action of the full Hamiltonian on the states must be norm-preserving.
- Since in this scenario the outcome of a quantum measurement is deterministic but random, the nonlinear equation must contain one or more random variables. One or the other outcome of a measurement is realized, in consistency with the Born rule, depending on the relative values of the random variables.
- The predictions of the nonlinear equation must be experimentally testable, and must not contradict the experimentally verified features of standard quantum mechanics.
- The nonlinear equation must not be *ad hoc*, but must instead be a consequence of 'some other requirement'. In other words, there must be good theoretical reasons, independent of quantum measurement, such as those outlined in Section 1.3, for such a nonlinear generalization of quantum mechanics.

In this Section, we will work with a nonlinear Schrödinger equation which belongs to the following general class of norm-preserving nonlinear Schrödinger equations:

$$i\hbar\frac{d|\psi\rangle}{dt} = H|\psi\rangle + (1 - P_\psi)U|\psi\rangle \ . \tag{2}$$

Here, H is the Hermitian part of the Hamiltonian, as in standard quantum mechanics. $(1-P_\psi)U$ is the non-Hermitian part, $P_\psi = |\psi\rangle\langle\psi|$ is the projection operator, and U is an arbitrary nonlinear operator. This equation has been discussed by [11] and by [7]. We will make the crucial assumption (to be justified later, in Section 3) that the non-Hermitean part of the Hamiltonian becomes significant only when the mass of the system becomes comparable to or larger than Planck mass, $m_{Pl} = (\hbar c/G)^{1/2} \sim 10^{-5}$ g.

We use the term 'initial system' to refer to the quantum system $\mathcal{Q}$ on which a measurement is to be made by a classical apparatus $\mathcal{A}$, and the term 'final system' to refer jointly to $\mathcal{Q}$ and $\mathcal{A}$ after the initial system has interacted with $\mathcal{A}$. A quantum measurement will be thought of as an increase in the mass (equivalently, number of degrees of freedom) of the system, from the initial value $m_Q \ll m_{Pl}$ to the final value $m_Q + m_{\mathcal{A}} \gg m_{Pl}$. Clearly then, the non-Hermitian part of the Hamiltonian will play a crucial role in the transition from the initial system to the final system.

We assume that $\mathcal{A}$ measures an observable $\hat{O}$ of $\mathcal{Q}$, having a complete set of eigenstates $|\psi_n\rangle$. Let the quantum state of the initial system be given as $|\psi\rangle = \Sigma_n a_n|\psi_n\rangle$. The onset of measurement corresponds to mapping the state $|\psi\rangle$ to the entangled state $|\psi\rangle_F$ of the final system as

$$|\psi\rangle \rightarrow |\psi\rangle_F \equiv \sum_n a_n|\psi\rangle_{Fn} = \sum_n a_n|\psi_n\rangle|A_n\rangle \; , \tag{3}$$

where $|A_n\rangle$ is the state the measuring apparatus would result in, had the initial system been in the state $|\psi_n\rangle$. During a quantum measurement the non-Hermitian part of the Hamiltonian dominates over the Hermitian part, and governs the evolution of the state $|\psi\rangle_F$.

2.2. The toy model

As a useful and illustrative toy-model, we will consider a special case of the nonlinear equation (2) with the operator U given by

$$U = i\gamma\sum_n q_n|\psi\rangle_{Fn}\langle\psi|_{Fn} \; , \qquad H = 0 \; , \tag{4}$$

where the q_n are random real, positive constants [7]. γ is a constant which will be assumed to be zero before the onset of measurement, and nonzero during the measurement. This is consistent with the assumption, made above, that the non-Hermitean part of the Hamiltonian becomes significant only when the mass of the system becomes comparable to or larger than Planck mass. Thus by system here we mean the 'final system', which jointly refers to the measuring apparatus $\mathcal{A}$ and the initial quantum system $\mathcal{Q}$. The state $|\psi\rangle_{Fn}$ has been defined in (3) above. The remaining Hermitean part of the Hamiltonian has been set to zero for simplicity; because here we want to demonstrate how the non-Hermitean part is responsible for the decay of the superpositions initially present in the quantum system Q. Including the Hermitean part will not prevent the breakdown of superposition — it will only make the analysis more complicated.

[We are using Grigorenko's model to illustrate collapse induced by nonlinearity, because of its simplicity. However, the essential idea has been put forward much earlier, probably first by Bohm and Bub [12], who suggested a nonlinear Schrödinger equation to explain measurement, using hidden variables as random variables. More pertinent to our context is a 1976 paper by Pearle [13] who proposed a nonlinear Schrödinger equation with phases of states as random variables, to explain measurement. This insightful and highly readable paper already put forth, in spirit, the nonlinear mechanism proposed by us, except that a fundamental origin for the nonlinearity was not suggested there. We find it curious that subsequent focus shifted mainly to other dynamical reduction models (not involving the phase as a random variable) and Pearle's original suggestion was largely forgotten. In a sense, our present work revives Pearle's idea, bringing it in as a byproduct of considerations originating in quantum gravity.]

Let us now analyze Eq. (2), with the understanding that the state $|\psi\rangle$ is here $|\psi\rangle_F$. Substituting the form of the Hamiltonian given in (4) into the nonlinear Schrödinger equation (2) gives, after using the expansion for $|\psi\rangle_{Fn}$ given in (3),

$$da_n/dt = \gamma a_n(q_n - L) \; , \tag{5}$$

where $L = \Sigma_n q_n|a_n|^2$. Hence we can write

$$\frac{d}{\gamma dt}\left(\ln\frac{a_i}{a_j}\right) = q_i - q_j \; . \tag{6}$$

Since the evolution is norm-preserving, it is easily inferred that the system evolves to the state with the largest value of q_n. In a repeated measurement, different outcomes can be achieved by different sets of values of the random variables q_n. For instance, if the q_n each lie in the range $[0, \infty]$ and have a probability distribution

$$\omega(q_n) = |\langle\psi(t_0)|\psi_n\rangle|^2 \exp[|\langle\psi(t_0)|\psi_n\rangle|^2 q_n] \ , \tag{7}$$

then the probability that the system evolves to the state $|\psi_n\rangle$ can be shown to be $|\langle\psi(t_0)|\psi_n\rangle|^2$, as required by experiments [7].

More details and interesting features of this and related models can be found in Grigorenko's paper. For instance, the number of random variables can be less than the number of system states. Furthermore, an attractive candidate for a random variable is the phase of the initial quantum state $\psi(t_0)$. As explained by Grigorenko, as the initial phase varies randomly and uniformly in the range $[0, 2\pi]$, it is possible to have different outcomes in a repeated measurement, depending on the value of the initial phase, and in consistency with the Born probability rule. Grigorenko also discusses how to overcome the problem of superluminality in his model.

The implication of this model is that evolution during a quantum measurement is deterministic. Probabilities enter the picture because of the presence of one or more random variables in a nonlinear Schrödinger equation with a non-Hermitean part. Nonlinearity breaks superposition. The non-Hermitean part causes decay/growth of different components of the quantum state. The presence of the random variables ensures that in a repeated measurement, different outcomes are realized, and the Born probability rule can be recovered by associating suitable probability distributions with the random variables.

Nevertheless, this is a toy model, and the form of the Hamiltonian is decidedly *ad hoc.* In Section 3 we will show that a Hamiltonian with the same features as in this toy model arises from addressing a fundamental incompleteness of quantum mechanics — the presence of an external classical time in the theory. We will hence be able to provide a collapse based explanation of quantum measurement, for reasons having to do with quantum gravity. We also note that the toy model has a limitation that the form of the nonlinear operator can only be written down *after* choosing a basis. This limitation will not arise for the nonlinear equation derived in Section 3.

2.3. The Doebner-Goldin equation

In this brief digression we point out an important nonlinear Schrödinger equation which belongs to the general class (2). This is the Doebner-Goldin equation [14]

$$i\hbar\frac{\partial\psi}{\partial t} = -\frac{\hbar^2}{2m}\nabla^2\psi + iD\hbar\left(\nabla^2\psi + \frac{|\nabla\psi|^2}{|\psi|^2}\psi\right) . \tag{8}$$

Here, D is a real constant. The origin of the D-G equation has been discussed in detail in [14]; its generalizations are discussed by Goldin [15] and also in [16].

With some further assumptions, the D-G equation can be used to explain the collapse of the wavefunction [17], in the manner of the toy-model described above. The nonlinear Schrödinger equation that we arrive at in the next Section is very similar to the above D-G equation.

3. Quantum gravity suggests that quantum mechanics is nonlinear

3.1. Outline of the approach

We outline below the key steps in the development of our intended nonlinear theory:

- There ought to exist a reformulation of quantum mechanics which does not refer to a classical spacetime manifold. This provides a new path to quantum gravity.

- It then follows as a consequence that quantum theory as we know it is a limiting case of a nonlinear quantum theory.
- We propose the desired reformulation of quantum mechanics using ideas from noncommutative differential geometry.
- This has implications for the quantum measurement problem: we arrive at a nonlinear Schrödinger equation, with the nonlinearity becoming significant in the vicinity of the Planck mass scale.

A detailed discussion of the approach can be found in [16].

3.2. Why quantum mechanics without classical spacetime?

The concept of time that appears in quantum mechanics is a classical concept. It is part of a classical spacetime manifold. The overlying metric on this spacetime manifold is produced by classical matter fields. In our present Universe we take the presence of such classical matter fields as given. But the Universe could in principle be in a state in which there are no classical matter fields.

If only quantum matter fields are present in the Universe, then the metric produced by them will undergo quantum fluctuations. If the metric is undergoing quantum fluctuations, then one cannot assign physical significance to the underlying classical spacetime manifold. This is the essence of the Einstein hole argument [18, 19]. Hence it is necessary to have a fundamental reformulation of quantum mechanics which does not refer to a classical spacetime manifold.

[**The Einstein hole argument:** Consider a spacetime manifold $\mathcal{M}$ having matter fields, except in a hole H inside the manifold, which is devoid of matter fields. The only field present in the hole is the gravitational field, described by the metric $g_{\mu\nu}$. Consider an active diffeomorphism ϕ on $\mathcal{M}$, which is by definition identity outside the hole, and on its boundary, but different from identity inside the hole. Clearly, the stress-tensor $T_{\mu\nu}$ remains unchanged under this diffeomorphism, all over the spacetime. However, inside the hole, the metric changes under the diffeomorphism, say from $g_{\mu\nu}(p)$ at the point p to $\phi * g_{\mu\nu}(q)$ at the point q, where q is the point to which the point p is mapped, by the active diffeomorphism. Since $T_{\mu\nu}$ has not changed, it is natural to expect that the physical gravitational field has not changed either. This is achieved by appealing to general covariance, namely that the metrics $g_{\mu\nu}(p)$ and $\phi * g_{\mu\nu}(q)$ describe the same physical gravitational field.

However, Einstein realized that general covariance comes at a price. The fields $g_{\mu\nu}(p)$ and $\phi * g_{\mu\nu}(q)$ can be regarded as physically identical only if the points p and q are also regarded as physically identical! It is thus a consequence of general covariance that points on a spacetime manifold cease to have any physical meaning as separate, distinct points of a spacetime. The only way to restore a physical attribute to points of the spacetime manifold (i.e. consider them as events) is through the presence, on the manifold, of a specific dynamically determined metric tensor field — the metric tensor field serves to provide a label to the points.

In our present context, we hence see that if the metric tensor is undergoing quantum fluctuations, one can no longer regard the underlying classical spacetime manifold as having a physically meaningful status [20].]

As and when the Universe is in a state in which it is dominated by classical matter fields, a classical spacetime manifold, and a classical metric become available. Quantum mechanics could then be equivalently described in two ways: (i) either in the proposed reformulation, which continues to avoid making reference to the time which is now externally available, or, as we always do (ii) from the point of view of an observer in the classical Universe, as standard quantum mechanics wherein evolution is described with respect to an external time. The proposed reformulation should become equivalent to the standard formulation when an external time becomes available.

In such a reformulation there is no classical spacetime; however, we can envisage the concept of a 'quantum spacetime' and a 'quantum gravitational field' which is produced by quantum matter fields. The 'quantum gravitational field', like its classical counterpart, is assumed to act as a source for itself. This makes the quantum theory of gravity a nonlinear theory — this feature is central to the thesis of this paper. Such a theory is completely different from the Wheeler-DeWitt equation, wherein the quantum theory of gravity is linear by construction.

A nonlinear quantum gravity might appear counterintuitive because we are generally conditioned to building a quantum theory by 'quantizing' a classical theory. These rules of quantization are linear by definition, and further, they assume an external classical time as given. However, when such an external classical time is no longer available, we do not have such rules of quantization ready at hand. We are compelled to adopt a top-down approach, and then we see that it is very natural that 'quantum gravity' 'quantum gravitates', in precisely the same sense in which classical gravity acts as a source for itself — thus leading to a nonlinear quantum gravity. What this means, for instance, is that if we were to write an analog of the Wheeler-DeWitt equation for the nonlinear case, the Hamiltonian of the theory would depend on the quantum state.

As has been explained in detail in [16], this nonlinearity in the quantum gravity theory becomes significant only at the Planck mass/energy scale. At much lower energy scales, the theory is linear, to an excellent approximation. This has an analogy with classical general relativity — the classical theory is nonlinear only in the strong-field regime, but linear in the weak field Newtonian regime.

Next, we consider what the equation of motion of a quantum field or a quantum mechanical particle is, in such a 'quantum spacetime'. Again, as shown in detail in [16], if the mass of the particle is comparable to Planck mass, its quantum dynamics is influenced by its own quantum gravitational field, and the equation of motion is nonlinear. From the point of view of an external spacetime (as and when the latter becomes available) this equation of motion (in the nonrelativistic limit) will appear to be a nonlinear Schrödinger equation. When the particle's mass is much smaller than Planck mass, the nonlinear equation will reduce to the standard linear Schrödinger equation.

3.3. A reformulation based on noncommutative differential geometry

We suggest the concept of a noncommuting coordinate system, which 'covers' a noncommutative manifold, wherein commutation relations between coordinates are introduced on physical grounds. Commutation relations amongst momenta must also be introduced.

Our proposal is that basic laws are invariant under general coordinate transformations of noncommuting coordinates.This generalizes the standard concept of general covariance to noncommuting coordinates. This formulation should satisfy two important properties:

- Firstly, in the limit in which the system becomes macroscopic, the noncommutative spacetime should be indistinguishable from ordinary commutative spacetime, and the dynamics should reduce to classical dynamics.
- Secondly, if a dominant part of the system becomes macroscopic and classical, and a sub-dominant part remains quantum (as our Universe is) then seen from the viewpoint of the dominant part, the quantum dynamics of the sub-dominant part should be the same as the standard quantum dynamics known to us.

Our overall proposal for the reformulation of quantum mechanics in the language of noncommutative geometry can be stated as follows:

- I. Noncommutative special relativity gives the new formulation of relativistic quantum mechanics which does not refer to a classical spacetime. The standard quantum commutation relations of quantum mechanics are deduced from the spacetime and the momentum commutation relations.
- II. Noncommutative general relativity is quantum gravity.

In the next sub-section we suggest the reformulation of quantum mechanics in terms of noncommuting coordinate systems, and the recovery of standard quantum mechanics from this reformulation. We then show how inclusion of self-gravity leads to a nonlinear Schrödinger equation. Much work still remains to be done, in terms of making rigorous contact with noncommutative differential geometry, and arriving at the field equations for noncommutative general relativity. However the flow of ideas appears rather natural, and we arrive at an explanation for collapse-induced quantum measurement which can be subjected to an experimental test. We would like to emphasize once again that the ideas in this section were not developed with the *ad hoc* purpose of explaining quantum measurement, but have been concerned with an altogether different aspect of quantum mechanics — the unsatisfactory presence of an external classical time in the theory.

3.4. Quantum Minkowski spacetime

Consider a system of quantum mechanical particles having a total mass-energy much less than Planck mass m_{Pl}, and assume that no external classical spacetime manifold is available. Since Planck mass scales inversely with the gravitational constant, we are justified here in neglecting the gravitational field, and the resulting quantum spacetime produced by the system will be called a 'quantum Minkowski spacetime'.

To describe the dynamics using noncommutative geometry consider a particle with mass $m \ll m_{Pl}$ in a 2-d noncommutative spacetime with coordinates $(\hat{x}, \hat{t})$. On the quantum Minkowski spacetime we introduce the non-Hermitean flat metric

$$\hat{\eta}_{\mu\nu} = \begin{pmatrix} 1 & 1 \\ -1 & -1 \end{pmatrix} \tag{9}$$

and the corresponding noncommutative line-element

$$ds^2 = \hat{\eta}_{\mu\nu} d\hat{x}^\mu d\hat{x}^\nu = d\hat{t}^2 - d\hat{x}^2 + d\hat{t}d\hat{x} - d\hat{x}d\hat{t} \tag{10}$$

which is invariant under a generalized Lorentz transformation.

Noncommutative dynamics is constructed by formally defining a velocity $\hat{u}^i = d\hat{x}^i/ds$ which, from (10), satisfies the relation

$$1 = \hat{\eta}_{\mu\nu} \frac{d\hat{x}^\mu}{ds} \frac{d\hat{x}^\nu}{ds} = (\hat{u}^t)^2 - (\hat{u}^x)^2 + \hat{u}^t \hat{u}^x - \hat{u}^x \hat{u}^t \ . \tag{11}$$

We define a generalized momentum as $\hat{p}^i = m\hat{u}^i$ which hence satisfies

$$\hat{p}^\mu \hat{p}_\mu = m^2 \ . \tag{12}$$

Here, $\hat{p}_\mu = \hat{\eta}_{\mu\nu}\hat{p}^\mu$ is well-defined. Written explicitly, this equation becomes

$$(\hat{p}^t)^2 - (\hat{p}^x)^2 + \hat{p}^t \hat{p}^x - \hat{p}^x \hat{p}^t = m^2 \ . \tag{13}$$

Dynamics is constructed by introducing a *complex* action $S(\hat{x}, \hat{t})$ and by defining the momenta introduced above as gradients of this complex action. In analogy with classical mechanics this converts (13) into a (noncommutative) Hamilton-Jacobi equation which describes the dynamics.

When an external classical Universe with a classical manifold (x, t) becomes available (see below), one defines the generalized momentum (p^t, p^x) in terms of the complex action $S(x, t)$ as

$$p^t = -\frac{\partial S}{\partial t}, \qquad p^x = \frac{\partial S}{\partial x}, \tag{14}$$

and from (13) the following fundamental rule for relating noncommutative dynamics to standard quantum dynamics

$$(\hat{p}^t)^2 - (\hat{p}^x)^2 + \hat{p}^t\hat{p}^x - \hat{p}^x\hat{p}^t = (p^t)^2 - (p^x)^2 + i\hbar\frac{\partial p^\mu}{\partial x^\mu}. \tag{15}$$

A detailed justification for this key equation has been given in [16].

In terms of the complex action the right hand side of this equation can be written as

$$\left(\frac{\partial S}{\partial t}\right)^2 - \left(\frac{\partial S}{\partial x}\right)^2 - i\hbar\left(\frac{\partial^2 S}{\partial t^2} - \frac{\partial^2 S}{\partial x^2}\right) = m^2, \tag{16}$$

and from here, by defining a quantum state ψ in a natural manner: $\psi = e^{iS/\hbar}$, we arrive at the Klein-Gordon equation

$$-\hbar^2\left(\frac{\partial^2}{\partial t^2} - \frac{\partial^2}{\partial x^2}\right)\psi = m^2\psi. \tag{17}$$

In this manner we have arrived at standard quantum mechanics, starting from a formulation which did not make reference to a classical time.

The proposed commutation relations on the noncommutative spacetime are motivated by the standard quantum uncertainty relations which one would like to deduce from these commutation relations. We propose

$$[\hat{t}, \hat{x}] = iL_{Pl}^2 f^{-1}, \qquad [\hat{p}^t, \hat{p}^x] = iP_{Pl}^2 f. \tag{18}$$

The function f, of coordinates and momenta, should be chosen such that these commutation relations are consistent with the Casimir relation (13). In other words, the momenta should commute with the Casimir relation. In [16] the choice $f = 1$ was made; it turns out however that such a choice is not consistent with the Casimir relation. (I would like to thank Harold Steinacker for pointing this out.) In fact in two dimensions, it can be shown that the only consistent choice is $f = 0$ which clearly is not of interest. There does not appear to be any such restriction on f in three or higher dimensions, but the exact form of f remains to be discovered, and a detailed 'q-deformed calculus' for such commutation relations remains to be developed. Here, we will continue our heuristic discussion using the illustrative 2-d case which is adequate for addressing the measurement problem — we do not need to know the exact form of f for our present purpose. Also, we have restricted the discussion here to a single particle; the generalization to the multi-particle case, in four dimensions, is straightforward, and outlined in [16].

3.5. Including self-gravity

If the mass-energy of the particle is not negligible in comparison to Planck mass its self-gravity must be taken into account. The 'flat' metric (9) gets modified to the 'curved' metric

$$\hat{h}_{\mu\nu} = \begin{pmatrix} \hat{g}_{tt} & \hat{\theta} \\ -\hat{\theta} & -\hat{g}_{xx} \end{pmatrix}. \tag{19}$$

We have made the significant assumption that in addition to the standard symmetric metric $\hat{g}_{\mu\nu}$ the noncommutative 'curved' metric also has an antisymmetric component $\hat{\theta}_{\mu\nu}$. The component $\hat{\theta}_{\mu\nu}$ will play a central role in our explanation of quantum measurement.

With the introduction of the curved metric, Eqs. (10), (12), (15) and (16) are respectively replaced by the plausible equations

$$ds^2 = \hat{h}_{\mu\nu} d\hat{x}^\mu d\hat{x}^\nu = \hat{g}_{tt} d\hat{t}^2 - \hat{g}_{xx} d\hat{x}^2 + \hat{\theta}[d\hat{t}d\hat{x} - d\hat{x}d\hat{t}] \ , \tag{20}$$

$$\hat{h}_{\mu\nu}\hat{p}^\mu\hat{p}^\nu = m^2 \ , \tag{21}$$

$$\hat{g}_{tt}(\hat{p}^t)^2 - \hat{g}_{xx}(\hat{p}^x)^2 + \hat{\theta}\left(\hat{p}^t\hat{p}^x - \hat{p}^x\hat{p}^t\right) = m^2 \ , \tag{22}$$

$$g_{tt}(p^t)^2 - g_{xx}(p^x)^2 + i\hbar\theta\frac{\partial p^\mu}{\partial x^\mu} = m^2 \ . \tag{23}$$

These replacements have been made very much in the same spirit in which one goes from flat spacetime equations to curved spacetime equations in classical general relativity. Like in general relativity, the metric $\hat{h}_{\mu\nu}$ is assumed to be determined by the mass m via the quantum state $S(\hat{x}, \hat{t})$. The field equations are assumed to be covariant under general coordinate transformations of noncommuting coordinates. If these field equations could be determined, they would constitute the field equations of quantum gravity, in this approach.

In the macroscopic limit $m \gg m_{Pl}$ the antisymmetric component θ is assumed to go to zero; the noncommutative spacetime (20) is then indistinguishable from ordinary commutative spacetime, and Eq. (23) reduces to classical dynamics. In the microscopic limit $m \ll m_{Pl}$ we have that θ goes to one, g_{tt} and g_{xx} also go to one, and we recover standard quantum mechanics. Thus we see that when θ is different from zero and one, we get a new mechanics which is neither standard linear quantum mechanics, nor classical mechanics!

There are, however, issues which remain to be resolved. We have assumed θ to be real, which appears a reasonable choice, considering that it represents an additional component of the gravitational field, significant only in the mesoscopic domain. We do not know at the moment the explicit spacetime dependence of θ — we will assume as of now that θ depends only on the mass m, and on the quantum state $S(x, t)$.

3.6. A nonlinear Schrödinger equation

If we substitute the momenta in (23) in terms of the complex action using (14) and then substitute $\psi = e^{iS/\hbar}$ and take the nonrelativistic limit, the resulting effective Schrödinger equation is nonlinear [16]. It is very similar to the Doebner-Goldin equation discussed before — the latter arises when one classifies physically different quantum systems by considering unitary representations of the group of diffeomorphisms $Diff(R^3)$.

The simplest case is obtained when in (23) one approximates the diagonal metric components to unity, giving the nonlinear Schrödinger equation

$$i\hbar\frac{\partial\psi}{\partial t} = -\frac{\hbar^2}{2m}\frac{\partial^2\psi}{\partial x^2} + \frac{\hbar^2}{2m}(1-\theta)\left(\frac{\partial^2\psi}{\partial x^2} - [(\ln\psi)']^2\psi\right) + V(x)\psi \ . \tag{24}$$

We have generalized by including a potential $V(x)$. This equation bears a striking resemblance to the Doebner-Goldin equation (8); the two equations have been compared in detail in [16].

This equation can be rewritten as,

$$i\hbar\frac{\partial\psi}{\partial t} = -\frac{\hbar^2}{2m}\frac{\partial^2\psi}{\partial x^2} + \frac{\hbar^2}{2m}(1-\theta)\left(\frac{\partial^2[\ln\psi]}{\partial x^2}\right)\psi + V(x)\psi \ , \tag{25}$$

and then more usefully, by expanding the nonlinear term into real and imaginary parts, as

$$i\hbar\frac{\partial\psi}{\partial t} = -\frac{\hbar^2}{2m}\frac{\partial^2\psi}{\partial x^2} + V(x)\psi + \frac{\gamma(m)\hbar^2}{2m}q\frac{\partial^2(\ln R)}{\partial x^2}\psi + i\frac{\gamma(m)\hbar}{2m}q\frac{\partial^2\phi}{\partial x^2}\psi + V(x)\psi \ , \tag{26}$$

where $\gamma(m)q = (1-\theta)$ and $\psi = Re^{i\phi/\hbar}$. We have made the plausible assumption that $(1-\theta)$ can be written as a product of two positive terms — a part $\gamma(m)$ which does not depend on the state, and a part q which depends on the state, but not on the mass. We see in the second-last term of (26) the emergence of the non-Hermitean, nonlinear part which is of interest to us. Gravity is responsible for this term because θ is actually a function of m/m_{Pl}, and so is γ.

It may appear that the nonlinear equation we have derived is complicated, and does not possess the simplicity of the linear Schrödinger equation. However it is worth recalling that this equation is the nonrelativistic limit of the highly symmetric Eq. (21). The relativistic equation, when written in noncommuting coordinates, and in terms of the complex action, has a symmetric and simple form.

The above equation is thus similar to the nonlinear Schrödinger equation (2) reviewed in the previous section. However, our equation is not norm-preserving! It is norm-preserving if the probability density is defined as $|\psi|^{2/\theta}$, instead of $|\psi|^2$. We need not regard this circumstance as an implausible one, since in this mesoscopic domain (where θ is neither one nor zero) we would not know a priori what the exact definition of norm, in terms of the wavefunction, is [16].

One could ask for a reason for the presence of the non-Hermitean term in the Hamiltonian in Eq. (24). The answer is that the presence of such a term is generic; it is only in the small mass, linear, limit that this term is negligible. Further, so long as the evolution is norm-preserving, the presence of such a term cannot be regarded as objectionable.

In passing, we note that in terms of the complex action function S defined earlier as $\psi = e^{iS/\hbar}$ the nonlinear Schrödinger equation (24) is written as

$$\frac{\partial S}{\partial t} = -\frac{S'^2}{2m} + \frac{i\hbar}{2m}\theta(m)S'' + V(x) \; . \tag{27}$$

This equation is to be regarded as the nonrelativistic limit of Eq. (23). We easily see here that in the limit $\theta = 0$ classical mechanics is recovered; and that setting $\theta = 1$ gives the linear Schrödinger equation. The intermediate regime, where θ is neither one nor zero, is different from both classical and quantum mechanics. This regime is consistent both with classical and quantum mechanics, but will go undetected in experiments unless one examines properties of mesoscopic systems. Interestingly enough, if θ is different from one, the nonlinear equation can explain the collapse of the wavefunction, as we will now see.

3.7. Explaining quantum measurement

Prior to the onset of a quantum measurement, evolution is described by

$$i\hbar\frac{\partial\psi}{\partial t} = -\frac{\hbar^2}{2m}\frac{\partial^2\psi}{\partial x^2} + V(x)\psi \; , \tag{28}$$

thus preserving superposition. This is because we have $m \ll m_{Pl}$ and $\theta \to 1$.

We recall that the onset of measurement corresponds to mapping the state $|\psi\rangle$ to the state $|\psi\rangle_F$ of the final system as

$$|\psi\rangle \to |\psi\rangle_F \equiv \sum_n a_n|\psi\rangle_{Fn} = \sum_n a_n|\psi_n\rangle|A_n\rangle \; , \tag{29}$$

where $|A_n\rangle$ is the state the measuring apparatus would result in, had the initial system been in the state $|\psi_n\rangle$.

At the onset of measurement, evolution is described by the equation

$$i\hbar\frac{\partial\psi_F}{\partial t} = H_F\psi_F + \frac{\gamma(m)\hbar^2}{2m_F}q\frac{\partial^2(\ln R_F)}{\partial x^2}\psi_F + i\frac{\gamma(m)\hbar}{2m_F}q\frac{\partial^2\phi_F}{\partial x^2}\psi_F \; . \tag{30}$$

H_F is the standard linear, Hermitean, part of the Hamiltonian for the final system (including both quantum system and measuring apparatus). Also, $\gamma(m_F)q = (1-\theta_F)$ and $\psi_F = R_F e^{i\phi_F/\hbar}$. This equation should be compared with the nonlinear equation in the toy model.

The states $|\psi\rangle_{Fn}$ cannot evolve as a superposition because the evolution is now nonlinear. However, the initial state at the onset of measurement *is* a superposition of the $|\psi_{Fn}\rangle$; it is simply the entangled state $|\psi\rangle_F$ at the onset of measurement. This initial superposition must thus break down during further evolution, according to the law

$$i\hbar\frac{\partial a_n}{\partial t} = i\frac{\gamma(m_F)\hbar}{2m_F}q\frac{\partial^2\phi_F}{\partial x^2}a_n \ , \tag{31}$$

which follows after substituting the expansion (29) in (30). Here, we have assumed that the linear part of the Hamiltonian can be ignored, and that it is dominated by the nonlinear part — it is only the non-Hermitean part contained in the new terms which can cause growth/decay of the amplitude. Also, ϕ_F is the value of the phase of the state $|\psi\rangle_F$ at the onset of measurement. It can be shown from the above nonlinear equation that the phase evolves much more slowly, compared to the amplitude of the wavefunction. Note also that the q_n's are different for different states. This is because it is natural to assume that the component θ of the gravitational field, which is produced by the mass m, should depend on the quantum state.

We thus get, as before

$$\frac{d}{dt}\ln\frac{a_i}{a_j} = \frac{\gamma}{2m_F}(q_i - q_j)\phi_F'' \ , \tag{32}$$

and like for the toy model, only the state with the largest q survives. This is ensured because the a_n satisfy the condition $\Sigma|a_n|^2 = 1$ because of the initial conditions imposed on them. In addition, as noted above, $|\psi|^{2/\theta_F}$, and hence $(\Sigma|a_n|^2)^{1/\theta_F}$ is preserved during evolution. It is important to note the subtlety that there will be a θ_F associated with the state $|\psi\rangle_F$ and different θ_{Fn} associated with each of the states $|\psi\rangle_{Fn}$.

In order to recover the Born rule, the q_n must be random variables. Only further development in theory can determine if this is so, and what their probability distribution is. One could nonetheless, following Grigorenko, assign a probability distribution for the q_n so as to recover the Born rule. Once again, the phase of the initial quantum state $\psi(t_0)$ is an attractive candidate for the desired random variables.

From Eq. (31) we can define an important quantity, the lifetime τ_{sup} of a superposition. It can be read off from this equation to be

$$\tau_{sup} = \frac{2m}{(1-\theta)\phi_F''} \ . \tag{33}$$

The first inference we can draw is that, since θ is strictly equal to one in standard linear quantum mechanics, a quantum superposition has an infinite lifetime in the linear theory, as one would expect. However, the situation begins to change in an interesting manner as the value of the mass m approaches and exceeds m_{Pl}. Since we know that in this limit θ approaches zero, we can neglect θ, and the superposition lifetime will then essentially be given by

$$\tau_{sup} \approx \frac{m}{\phi_F''} \sim \frac{mL^2}{\phi} \ , \tag{34}$$

where L is the linear dimension of the system, and ϕ_F is the phase of the state $|\psi\rangle_F$ at the onset of measurement. For a macroscopic system we can get a numerical estimate of the life-time by noting that we are close to the classical limit, where the phase coincides with the classical action in the Hamilton-Jacobi equation. To leading order, the magnitude of the classical action is given by $S_{cl} = mc^2 t$, where t is the time over which we observe the classical trajectory; approximately, this could be taken to be the value of the phase ϕ, and τ_{sup} is then roughly given by

$$\tau_{sup} \sim \frac{1}{t}\left(\frac{L}{c}\right)^2 . \tag{35}$$

For a measuring apparatus, if we take the linear dimension to be say 1 cm, and the time of observation to be say 10^{-3} second, we get the superposition lifetime to be 10^{-18} second, which is an encouragingly small number. This could possibly explain why the wavefunction collapses so rapidly during a measurement.

We can get a very rough estimate of τ_{sup} for a mesoscopic system using (35), and continuing to ignore θ. Let us take $L \sim 10^{-3}$ cm and correspondingly $m \sim 10^{-9}$ g. Such a composite object has approximately 10^{15} particles, and we could take $\phi \sim N\hbar$ with $N \sim 10^{15}$. This gives $\tau_{sup} \sim 10^{-3}$ second.

Thus, in making a transition from a microscopic system which obeys linear quantum mechanics, to a macroscopic system such as a measuring apparatus, we find that the lifetime of a superposition changes from an astronomically large value to an immeasurably small value. We could thus be certain that there must exist intermediate, mesoscopic systems for which the lifetime of a superposition is an easily measurable number, say one millisecond. Unfortunately our present understanding of the approach described here is not good enough to say at what value of the mass m this will happen. What we can be certain about is that if the ideas described here are on the right track, then as experimentalists check for quantum superposition in larger and larger quantum systems, they will discover systems for which the lifetime of a superposition will become small enough to be measurable in the laboratory in principle. In practice, it remains to be seen whether or not decoherence will permit this lifetime to be measured.

3.8. Ideas for an experimental test of the model

The possibilities for an experimental test of this nonlinear model, and of its implications for quantum measurement, can be divided into three classes:

(i) Looking for breakdown of quantum superposition: This suggestion is in line with the kind of experiments already in progress — constructing larger and larger composite quantum objects (i.e. C_{60} and beyond) and checking whether or not linear superposition holds for such objects. Our prediction is that by the time the number of particles in the composite object reaches to about 10^{15} particles, the lifetime of the superposition will become small enough to be observable in the laboratory, and one will actually observe the decay of superposition in such a system. The greatest obstacle to the detection of such an effect, even if it is there, will come from the phenomenon of decoherence. In order to ascertain whether or not there is a breakdown of nonlinearity due to superposition, one will have to first ensure that the object is sufficiently well-isolated from its environment, so that decoherence can be avoided.

(ii) Difference between the predictions of the linear theory and the nonlinear theory: If one were to calculate expectation values of observables, using the nonlinear equation (24), the result will be different from what one will get from the linear theory, because now θ is nonzero. This aspect has been discussed in some detail in [16]. Care has to be taken to construct gauge-invariant observables for the nonlinear theory. It was shown in [16] that the ratio $\hbar/m$ is one such gauge-invariant observable — the nonlinear model predicts that the effective Planck's constant in the theory is $\hbar\theta(m)/m$. Thus, according to the nonlinear model, an experiment to measure the value of $\hbar/m$ for a mesoscopic system will give results consistent with theory only if one assumes Planck's constant to have an effective value $\hbar\theta(m)$.

(iii) Looking for a correlation between the absolute value of the initial phase and the outcome of a quantum measurement: If the random variable responsible for the outcome of a quantum measurement in the nonlinear Schrödinger equation (30) is indeed the absolute phase of the initial state $\psi(t_0)$, then the correlation between this phase and the outcome of a measurement should in principle be detectable by experiment. Conventional wisdom is that absolute phase cannot be measured and that only differences in phase are measurable. However, if the absolute value of the phase is going to decide the outcome of a measurement, it does not seem unnatural to expect that there might be a way to determine the value of this initial phase. In other words, while absolute phase is not observable in the linear theory, it probably does become observable in the nonlinear theory, and one should explore possible ways to measure it in the nonlinear theory, i.e. for mesoscopic systems.

4. Other models for collapse of the wavefunction

Various researchers have proposed modifications of the Schrödinger equation, with the purpose of explaining quantum measurement as a dynamically induced collapse. Most of these models also estimate the lifetime of a quantum superposition — the lifetime being large for microscopic systems, and small for macroscopic ones. While some models do not involve gravity, it is remarkable that many of the models assert that gravity is responsible for collapse. Below, we review some of the models very briefly, without really attempting to critically compare these models with each other, or with our own approach. (A comparison across various models of collapse will be reported elsewhere. The literature on the subject is large, and we confine ourselves to giving pointers to the literature. A nice, though somewhat older, review of collapse models has been given by Pearle [21]).

4.1. Models that do not involve gravity

Dynamical Reduction Models based on Nonlinear Stochasticity: The first suggestion that collapse could be explained as a dynamical reduction process using stochastic differential equations came from Pearle [22]. The Schrödinger equation was to be augmented by a stochastic term which could induce collapse. Significant development in this direction came from the work of Ghirardi, Rimini and Weber (GRW) [23]. This program, and its progress, has been reviewed in [24] and in [25]. There were two guiding principles for this dynamical reduction model known as Quantum Mechanics with Spontaneous Localization (QMSL) [24]:

1. The preferred 'basis' — the basis on which reductions take place — must be chosen in such a way as to guarantee a definite position in space to macroscopic objects.

2. The modified dynamics must have little impact on microscopic objects, but at the same time must reduce the superposition of different macroscopic states of macro-systems. There must then be an amplification mechanism when moving from the micro to the macro level.

The reduction is achieved by making the following set of assumptions:

1. Each particle of a system of n distinguishable particles experiences, with a mean rate λ_i, a sudden spontaneous localization process.

2. In the time interval between two successive spontaneous processes the system evolves according to the usual Schrödinger equation.

In their model, GRW introduced two new fundamental constants of nature, assumed to have definite numerical values, so as to reproduce observed features of the microscopic and macroscopic world. The first constant, $\lambda^{-1} \sim 10^{16}$ seconds, alluded to above, determines the rate of spontaneous localization (collapse) for a single particle. For a composite object of n particles, the collapse rate is $(\lambda n)^{-1}$ second. The second fundamental constant is a length scale $a \sim 10^{-5}$ cm which is related to the concept that a widely spaced wavefunction collapses to a length scale of about a during the localization.

A gravity based implementation of the QMSL model has been studied by Diosi [29] and discussed in [30].

The QMSL model has the limitation that it does not preserve symmetry of the wavefunction under particle exchange, and has been improved into what is known as the CSL (Continuous

Spontaneous Localization) model [26, 28]. In CSL a randomly fluctuating classical field couples with the particle number density operator of a quantum system to produce collapse towards its spatially localized eigenstates. The narrowing of the wavefunction amounts to an increase in the particle's energy, and actually amounts to a violation of energy conservation. This intriguing aspect of the collapse model has been discussed in [31] and an interesting suggestion for preserving energy conservation has been made therein.

An outstanding open question with regard to the dynamical reduction models is the origin of the random noise, or the randomly fluctuating classical scalar field, which induces collapse. We see herein the possibility of a connection, worth exploring further, with our proposal that the nonlinear Schrödinger equation has its origin in quantum gravity. Could the randomly fluctuating classical field of CSL be related to the random parameter $\theta(m)$ we have in our nonlinear equation?

Another collapse model is due to Adler (see [27] and related references therein), where an energy-driven stochastic Schrödinger equation is a phenomenological model for state-vector reduction. Collapse in this model is energy conserving and reproduces the Born probability rule. It is interesting to note that in this model, the terms that influence wavefunction reduction are directly related to mass-energy, and hence once again a connection with gravity is suggested.

4.2. Models that involve gravity

The gravitational field produced by a classical object obeys the laws of general relativity, or in the limiting case, those of Newtonian gravity. Since the position of the classical object is subject to tiny quantum uncertainties, the gravitational field and the curvature tensor produced by it are also subject to quantum fluctuations. Karolyhazy [32, 33] built an interesting and plausible model to explain how quantum superposition could be destroyed in macroscopic objects, as a result of these quantum fluctuations in the gravitational field of the object. The possible connection of such a model with the CSL approach has been discussed in [21]. Other models discussing the possible role of gravity in collapse are presented in [34, 35].

In spirit and concept, our work here comes closest to Penrose's idea that gravity is responsible for wavefunction reduction [36]. Penrose argues convincingly that the principle of general covariance and the principle of linear superposition in quantum mechanics are in direct conflict with each other. Our starting point has been essentially the same — we argued that because of general covariance, one cannot have a classical spacetime manifold coexisting with a Universe which has only quantum matter fields. This led us to conclude that if gravity cannot be neglected, quantum theory must be nonlinear — which is essentially what Penrose has argued: general covariance and linear superposition are incompatible with each other.

Penrose develops an estimate for the lifetime of the quantum superposition of two different positions of a macroscopic object, and demonstrates it to be of the order $\hbar/\Delta E$, where ΔE is the gravitational self-energy of the difference between the mass distributions of each of the two locations of the macroscopic object, see also Ref. [29]. An experimental test of Penrose's idea has been proposed in [37].

5. Discussion

In putting forth the two possible explanations of quantum measurement, we have assumed that the wavefunction describes an individual quantum system, and not a statistical ensemble of quantum systems. Also, we have ignored making any mention of the Copenhagen interpretation, which essentially states that upon measurement, the wavefunction collapses into one of the eigenstates, but the interpretation does not suggest any mechanism for the collapse.

We believe that our case for the necessity of a reformulation of quantum mechanics is robust. Equally robust is the inference that the standard linear quantum theory is a limiting case of a nonlinear quantum theory. However, only partial justification can be given for the use of noncommutative differential geometry for constructing such a reformulation. Having accepted to work in the framework of noncommutative geometry, we regard it as highly attractive that linear quantum theory, and its nonlinear generalization, are respectively the equations of motion in a noncommutative special relativity, and in its generalization to a noncommutative general relativity. Various issues here remain to be understood much better. These include: (i) the physical meaning associated with the noncommutative metric (9), (ii) the nature of commutation relations to be imposed on the 'curved' noncommutative metric $\hat{h}_{\mu\nu}$, (iii) the full development of the concept of general covariance in noncommutative geometry, and the related generalization of the concept of curvature (for a review of the current status of this aspect see for instance [38, 39]), (iv) the field equations, analogous to those in general relativity, for this metric, (v) the justification for retaining $\theta(m)$ in the description of the nonlinear Schrödinger equation (24), while ignoring the diagonal components $g_{\mu\nu}$.

In spite of these important issues which are yet to be addressed and resolved, we regard it as a natural consequence of the required reformulation that there is a new mechanics in the intermediate, mesoscopic domain; and that classical mechanics, as well as linear quantum mechanics, are its limiting cases. The limits are obtained by letting $\theta \to 0$ and $\theta \to 1$ respectively.

We re-emphasize that there is no reason to believe a priori that quantum mechanics continues to hold unchanged in the mesoscopic domain. And if there indeed are reasons to expect a departure from standard quantum theory in this domain (and these reasons are independent of the issue of quantum measurement) then experiments must be carried out to test the laboratory predictions of these new ideas.

There are very stringent experimental bounds on the presence of nonlinear terms in quantum mechanics in the atomic domain [40, 41, 42]. However, these bounds do not extend to the mesoscopic domain — thus for instance there are no bounds on nonlinear quantum mechanics when such a theory is applied to an object containing say 10^{15} particles.

Superluminality: It has also been pointed out that the presence of a nonlinearity in quantum mechanics can result in the possibility of superluminal communication [43, 44]. In our approach, the nonlinearity is a relic of a more fundamental description of the theory in terms of a noncommutative spacetime. We do not have at present a good understanding of the 'light-cone structure' in a noncommutative spacetime, and it is difficult to assess the possibility or otherwise of superluminal communication in a noncommutative geometry. More importantly, the nonlinearity we predict becomes significant only in the mesoscopic domain, and we are suggesting that mechanical laws here are different from both the classical and the quantum case. Thus in this domain the issue of superluminality needs to be addressed afresh. Once again, it needs to be stressed that the effects of nonlinearity could be severely masked by decoherence resulting from interaction with the environment, making the mesoscopic mechanics effectively indistinguishable from classical or quantum mechanics.

With regard to superluminality, it has also been pointed out by Doebner and Goldin [45] that by virtue of nonlinear gauge transformations, many nonlinear Schrödinger equations are physically equivalent to linear equations. So one could not possibly deduce superluminal communication for such nonlinear equations.

The preferred basis problem: We would like to suggest, following GRW, that a preferred basis is one in which positions of macroscopic objects are localized. This is consistent with the oft expressed view that "*all* quantum mechanical measurements consist of or are obtained from positional measurements made at various times" [45]. As stated by Feynman and Hibbs [46], p. 96:

"Indeed all measurements of quantum mechanical systems could be made to reduce eventually to position and time measurements. Because of this possibility a theory formulated in terms of position measurements is complete enough in principle to describe all phenomena."

The cosmological constant puzzle: Further investigations suggest that the introduction of the antisymmetric component θ of the gravitational field gives rise to a quantum-classical duality which might in turn explain why we observe a very tiny nonzero cosmological constant. These ideas have been discussed in [47].

Acknowledgments

For useful discussions and conversations during various stages of this work, it is a pleasure to thank Joy Christian, Patrick Das Gupta, Atri Deshamukhya, Avinash Dhar, Heinz-Dietrich Doebner, H. T. Elze, Thomas Filk, M. Genovese, G. C. Ghirardi, Gerald Goldin, T. R. Govindarajan, Kumar S. Gupta, Sashideep Gutti, Friedrich Hehl, Eric Joos, F. Karolyhazy, Aruna Kesavan, Romesh Kaul, Claus Kiefer, Kinjalk Lochan, Gautam Mandal, N. Mavromatos, Shiraz Minwalla, Ayan Mukhopadhyay, T. Padmanabhan, Aseem Paranjape, Roger Penrose, Rainer Plaga, L. Sriramkumar, H. Steinacker, Rakesh Tibrewala, A. V. Toporensky, Sandip Trivedi, C. S. Unnikrishnan, Cenalo Vaz, Albrecht von Müller and Spenta Wadia.

References

[1] Everett III H 1957 "Relative State" Formulation of Quantum Mechanics *Rev. Mod. Phys.* **29** 454
[2] Deutsch D and Wallace D 2007 Probability in the Everett Interpretation: State of Play http://pirsa.org/07090062
[3] Buniy R V, Hsu H D S and Zee A 2006 Discreteness and the origin of probability in quantum mechanics *Phys. Lett.* B **640** 219 (*Preprint* arXiv:hep-th/0606062)
[4] Tegmark M 2007 Many Lives in Many Worlds *Nature* **448** 23
[5] Brune M, Hagley H, Dreyer D, Maitre X, Maali A, Wunderlich C, Raimonde J M and Haroche S 1996 Observing the progressive decoherence of the "meter" in a quantum measurement *Phys. Rev. Lett.* **77** 4887
[6] Joos E, Zeh H-D , Kiefer C, Giulini D, Kupsch J and Stamatescu I-O 2005 *Decoherence and the Appearance of a Classical World in Quantum Theory* (New York: Springer) 2nd Edn.
[7] Grigorenko A N 1995 Measurement description by means of a nonlinear Schrödinger equation *J. Phys. A: Math. Gen.* **28** 1459
[8] Bohm D 1952 A suggested interpretation of the quantum theory in terms of "hidden" variables. I *Phys. Rev.* **85** 166
[9] Gell-Mann M and Hartle J B 1993 Classical equations for quantum systems *Phys. Rev.* D **47** 3345
[10] Omnes R 1992 Consistent interpretations of quantum mechanics *Rev. Mod. Phys.* **64** 339
[11] Gisin N 1981 A simple nonlinear dissipative quantum evolution equation *J. Phys. A: Math. Gen.* **14** 2259
[12] Bohm D and Bub J 1966 A proposed solution of the measurement problem in quantum mechanics by a hidden variable theory *Rev. Mod. Phys.* **38** 453
[13] Pearle P 1976 Reduction of the state-vector by a nonlinear Schrödinger equation *Phys. Rev.* D **13** 857
[14] Doebner H-D and Goldin G A 1992 On a general nonlinear Schrödinger equation admitting diffusion currents *Phys. Lett.* A **162** 397
[15] Goldin G A 2000 Perspectives on nonlinearity in quantum mechanics *Preprint* arXiv:quant-ph/0002013
[16] Singh T P 2006 Quantum mechanics without spacetime – a case for noncommutative geometry *Bulg. J. Phys.* **33** 217 (*Preprint* arXiv:gr-qc/0510042)
[17] Singh T P 2008 The inevitable nonlinearity of quantum gravity falsifies the many-worlds interpretation of quantum mechanics *Int. J. Mod. Phys.* D **17** 611 (*Preprint* arXiv:0705.2357)
[18] Einstein A 1914 Die formale Grundlage der allgemeinen Relativitätstheorie *Königlich Preussische Akademie der Wissenschaften (Berlin) Sitzungsberichte* 1030-1085
[19] Christian J 1998 Why the quantum must yield to gravity? *Physics meets Philosophy at the Planck Scale* ed C Callender and N Huggett (Cambridge, UK: Cambridge U. Press) (*Preprint* arXiv:gr-qc/9810078)
[20] Carlip S 2001 Quantum Gravity: A Progress Report *Rep. Prog. Phys.* **64** 885 (*Preprint* arXiv:gr-qc/0108040)
[21] Pearle P 1999 Collapse Models *Open Systems and Measurement in Relativistic Quantum Theory* ed F Petruccione and H P Breuer (Heidelberg: Springer Verlag) (*Preprint* arXiv:quant-ph/9901077)
[22] Pearle P 1979 Toward explaining why events occur *Int. J. Theor. Phys.* **18** 489
[23] Ghirardi G C, Rimini A and Weber T 1986 Unified dynamics for microscopic and macroscopic systems *Phys. Rev.* D **34** 470
[24] Bassi A and Ghirardi G C 2003 Dynamical reduction models *Phys. Rep.* **379** 257 (*Preprint* quant-ph/0302164)
[25] Bassi A 2007 Dynamical reduction models: present status and future developments *J. Phys.: Conf. Ser.* **67** 012013 (*Preprint* arXiv:quant-ph/0701014)

[26] Pearle P 1989 Combining stochastic dynamical state-vector reduction with spontaneous localization *Phys. Rev.* A **39** 2277
[27] Adler S 2003 Weisskopf-Wigner decay theory for the energy driven stochastic Schrödinger equation *Phys. Rev.* D **67** 025007
[28] Ghirardi G C, Pearle A and Rimini P 1990 Markov processes in Hilbert space and continuous spontaneous localization of systems of identical particles *Phys. Rev.* A **42** 78
[29] Diosi L 1989 Models for universal reduction of macroscopic quantum fluctuations *Phys. Rev.* D **40** 1165
[30] Ghirardi G C, Grassi R and Rimini A 1990 Continuous-spontaneous-reduction model involving gravity *Phys. Rev.* A **42** 1057
[31] Pearle P 2000 Wavefunction collapse and conservation laws *Found. Phys.* **30** 1145 (*Preprint* arXiv:quant-ph/0004067)
[32] Karolyhazy F 1966 Gravitation and quantum mechanics of macroscopic objects *Nuovo Cimento* A **42** 390
[33] Karolyhazy F, Frenkel A and Lukacs B 1986 On the possible role of gravity in the reduction of the wave function *Quantum concepts in space and time* ed R Penrose and C J Isham (Oxford: Oxford U. Press)
[34] Percival I C 1995 Quantum spacetime fluctuations and primary state difusion *Proc. Roy. Soc.* A **451** 503
[35] Fivel D I 1997 Dynamical reduction theory of Einstein-Podolsky-Rosen correlations and a possible origin of CP-violation *Phys. Rev.* A **56** 146
[36] Penrose R 1996 On gravity's role in quantum state reduction *Gen. Rel. Grav.* **28** 581
[37] Marshall W, Simon C, Penrose R and Bouwmeester D 2003 Towards quantum superpositions of a mirror *Phys. Rev. Lett.* **91** 130401
[38] Connes A 2000 Noncommutative geometry: Year 2000 *Preprint* math/0011193
[39] Madore J 1999 *An Introduction to Noncommutative Differential Geometry and its Physical Applications* (Cambridge, UK: Cambridge U. Press)
[40] Weinberg S 1989 Precision tests for quantum mechanics *Phys. Rev. Lett.* **62** 485
[41] Weinberg S 1989 Testing quantum mechanics *Ann. Phys.* **194** 336
[42] Bollinger J J, Heinzin D J, Itano W M, Gilbert S L and Wineland D J 1989 Test of the linearity of quantum mechanics by rf spectroscopy of the $^9Be^+$ ground state *Phys. Rev. Lett.* **63** 1031
[43] Gisin N 1989 Stochastic quantum dynamics and relativity *Helv. Phys. Acta* **62** 363
[44] Polchinski J 1991 Weinberg's nonlinear quantum mechanics and the Einstein-Podolsky-Rosen paradox *Phys. Rev. Lett.* **66** 397
[45] Doebner H-D and Goldin G 1996 Introducing nonlinear gauge transformations in a family of nonlinear Schrödinger equations *Phys. Rev.* A **54** 3764
[46] Feynman R P and Hibbs A R 1965 *Quantum Mechanics and Path Integrals* (New York: McGraw Hill)
[47] Singh T P 2008 Noncommutative gravity, a 'no-strings attached' quantum-classical duality, and the cosmological constant puzzle *Gen. Rel. Grav.* **40** 2037

Constrained quantum dynamics

Anna C T Gustavsson

Blackett Laboratory, Imperial College London, London SW7 2AZ, UK

E-mail: anna.gustavsson05@imperial.ac.uk

Abstract. We give an overview of the two different methods that have been introduced in order to describe the dynamics of constrained quantum systems; the symplectic formulation and the metric formulation. The symplectic method extends the work of Dirac on constrained classical systems to quantum systems, whereas the metric approach is purely a quantum mechanical method having no immediate classical counterpart. Two examples are provided that illustrate the nonlinear motion induced by the constraint.

1. Introduction

A question that has recently been asked is how does one implement unitary motion on a quantum system that is subject to a set of constraints? In classical mechanics, constrained evolutions are often constructed using an approach developed by Dirac [1, 2], which uses the symplectic geometry of the classical phase space. Recently, several authors [3, 4, 5] have worked on extending the classical theory of constraint to quantum systems, using the phase space formulation of quantum mechanics. There are, however, pros and cons in this approach: It is for example only applicable to systems subject to an even number of constraints. To address this and other issues, an alternative approach for treating constrained quantum systems was introduced in [6]. In this paper we review both these approaches and explore through examples the kind of nonlinear motion that results from constraining a unitary evolution in quantum mechanics.

The classical method uses the property that the classical phase space is a symplectic manifold endowed with a symplectic structure. The quantum mechanical phase space has the geometry of a Kähler manifold, with both a symplectic structure and a metric structure. As the method considered by [3, 4, 5] uses the symplectic geometry of the quantum phase space, it was given the name the *symplectic approach* to quantum constraints [5]. The alternative formalism introduced in [6] makes a novel use of the metric structure of the quantum state space, thus referred to as the *metric approach* to quantum constraints.

In this paper we will first review some of the main features of the phase space formulation of quantum mechanics. We will then highlight the key features of both the symplectic and metric approaches to implementing constrained unitary motion and discuss the resulting evolutions by looking at two examples.

2. The phase-space formulation of quantum mechanics

The phase space description of quantum mechanics is based on a projective Hilbert space formulation (see for example [7, 8, 9, 10, 11, 12, 13, 14, 15] and references cited therein). The idea is as follows. The expectation value of any observable in quantum mechanics is independent of the norm of the state vector

$$|x\rangle \sim \lambda|x\rangle \qquad \lambda \in \mathbb{C} - \{0\}. \tag{1}$$

A state in quantum mechanics can hence be viewed as a ray through the origin of the Hilbert space $\mathcal{H}^{n+1}$, and we can, without loss of any physical information about the system, consider quantum mechanics in the space of rays. The space of rays correspond to the complex projective space $\mathbb{P}^n$, where each state of the system is now given by a point. In real terms $\mathbb{P}^n$ is an even dimensional manifold Γ^{2n} equipped with a Riemannian structure given by the Fubini-Study metric g_{ab} and a compatible symplectic structure Ω_{ab}. The transition probability between two states in Γ is given by the associated geodesic distance between the two points. It is interesting to note that the probabilistic aspects of quantum theory are hence associated with the underlying metric geometry of the state space. In comparison the dynamical aspects of the theory are captured by underlying symplectic geometry. Physical observables are functions on Γ given by the expectation value of the corresponding Hilbert space operator at each point x. The Hamiltonian, for example, is given by

$$H(x) = \frac{\langle x|\hat{H}|x\rangle}{\langle x|x\rangle}. \tag{2}$$

The eigenstates of the Hamiltonian are given by the fixed points of H:

$$\nabla_a H(x) = 0, \tag{3}$$

and the corresponding eigenvalues are the values of $H(x)$ at each fixed point. Unitary evolution is governed by the Schrödinger equation, which takes the form of Hamilton's equations:

$$\frac{\mathrm{d}x^a}{\mathrm{d}t} = \Omega^{ab}\nabla_b H(x), \tag{4}$$

where Ω^{ab} is the inverse symplectic structure given by $\Omega_{ab}\Omega^{bc} = -\delta_a^c$. Here, and throughout the paper, there is an implied summation over repeated indices. We parameterise the state space by the "action-angle" parameterisation (cf. [16]) so that the equations of motion (4) take a particularly simple and familiar form. If we let the state $|x\rangle$ in energy basis be given by

$$|x\rangle = \sum_{i=1}^{n} \sqrt{p_i}\mathrm{e}^{-\mathrm{i}q_i}|E_i\rangle + \sqrt{1 - \sum_{i=1}^{n} p_i}|E_{n+1}\rangle, \tag{5}$$

then the expectation of the Hamiltonian operator $\hat{H} = \sum_j E_j|E_j\rangle\langle E_j|$ is given by

$$H(q_i, p_i) = E_{n+1} + \sum_{i=1}^{n} \omega_i p_i, \quad \text{where} \quad \omega_i = E_i - E_{n+1}. \tag{6}$$

The equations of motion then take the form

$$\dot{q}_i = \frac{\partial H(q_i, p_i)}{\partial p_i} \quad \text{and} \quad \dot{p}_i = -\frac{\partial H(q_i, p_i)}{\partial q_i}, \tag{7}$$

with solutions $q_i(t) = q_i(0) + \omega_i t$ and $p_i(t) = p_i(0)$.

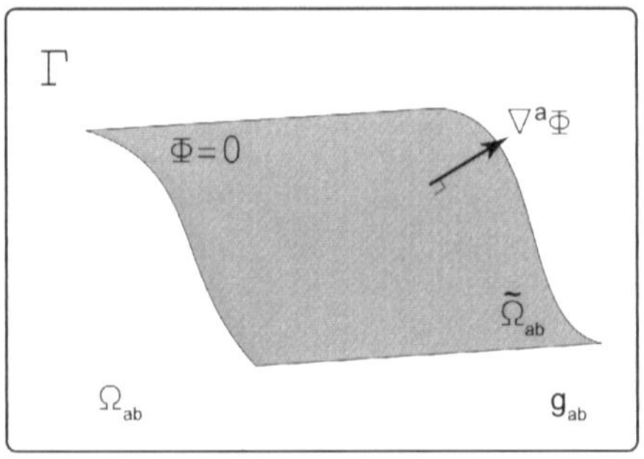

Figure 1. The constraints $\Phi^i = 0$ define a subspace of Γ. The evolution of the system can be constrained to this subspace using either the symplectic or the metric method. When using the symplectic approach the modified equations of motion take the same form as the original one, but with a modified symplectic structure $\tilde{\Omega}_{ab}$. In the metric approach the system is constrained by removing all components orthogonal to the surface from the vector field $\dot{x}^a = \Omega^{ab}\nabla_b H$.

3. Constrained dynamics

Let us now assume that the evolution of the quantum system is subject to one or more constraints. The type of constraints that we will be considering can be written in the form

$$\Phi^i(x) = 0, \qquad i = 1, 2, \ldots, N, \tag{8}$$

where N is the total number of constraints. Equation (8) defines a subspace Φ of Γ, given by the intersection of each of the N subspaces defined by $\Phi^i = 0$, onto which the evolution will be constrained, see figure 1.

We shall here consider two different types of constraints. The first type is given by the conservation of a set of observables, i.e.

$$\Phi^i(x) = \frac{\langle x|\hat{\Phi}^i|x\rangle}{\langle x|x\rangle}, \tag{9}$$

where $\hat{\Phi}^i$ is the Hilbert space operator corresponding to the observable Φ^i. The second type of constraint is when the evolution of the system is constrained to an algebraically well defined subspace of Γ that does not correspond to the conservation of any observable. An example of this is to constrain the motion of a pair of spin-$\frac{1}{2}$ particles that are initially disentangled to remain disentangled. There is no observable that corresponds to this property of the system, but as disentangled states form a subspace $Q = \mathbb{P}^1 \times \mathbb{P}^1$ [15, 17] of the overall state space $\mathbb{P}^3$ (see figure 2), we can constrain the system algebraically to remain on this subspace. We shall

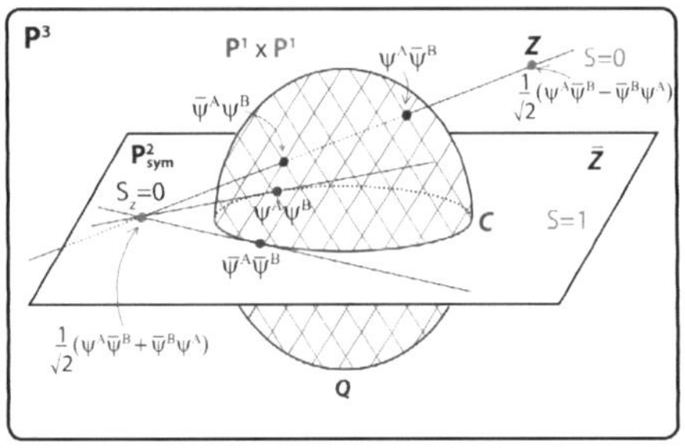

Figure 2. The state space of 2 spin-$\frac{1}{2}$ particles. The disentangled states form a subspace of the total state space given by $Q = \mathbb{P}^1 \times \mathbb{P}^1$, where $\mathbb{P}^1$ is the complex projective line. Further details of the state space structure of this system can be found in [15, 17].

return to this example and look at it in more detail in the next section where we give an outline of the symplectic approach to implementing quantum constraints.

3.1. Symplectic approach

Let us first consider the symplectic approach and follow the methodology introduced in [5]. In order to implement the constraints (8) the equations of motion are modified such that

$$\frac{\mathrm{d}x^a}{\mathrm{d}t} = \Omega^{ab}\nabla_b H(x) + \lambda_i \Omega^{ab}\nabla_b \Phi^i(x), \tag{10}$$

where λ_i are Lagrange multipliers associated with the constraints $\Phi^i(x)$. In some cases these Lagrange multipliers can be found explicitly by considering $\dot{\Phi}^i = 0$. By expanding $\dot{\Phi}^i$ using the chain rule $\dot{\Phi}^i = \dot{x}^a \nabla_a \Phi^i$ we find that the Lagrange multiplier is given by:

$$\lambda_i = \omega_{ji}\Omega^{ab}\nabla_a \Phi^j \nabla_b H, \tag{11}$$

where ω_{ij}, assuming it exists, is the inverse of

$$\omega^{ij} := \Omega^{ab}\nabla_a \Phi^i \nabla_b \Phi^j. \tag{12}$$

Note that when the constraints are given by conserved observables (9), then ω^{ij} correspond to the commutator between the two observables. By substituting the Lagrange multipliers back into (10) we find that the equations of motion can be rewritten in the form [5]

$$\frac{\mathrm{d}x^a}{\mathrm{d}t} = \tilde{\Omega}^{ab}\nabla_b H(x), \tag{13}$$

where $\tilde{\Omega}^{ab}$ is the induced symplectic structure on the constraint subspace $\Phi^i = 0$ given by

$$\tilde{\Omega}^{ab} = \Omega^{ab} + \Omega^{ac}\Omega^{bd}\omega_{ij}\nabla_c \Phi^i \nabla_d \Phi^j. \tag{14}$$

In other words, the equations of motion for the constrained system take the same form as for the unconstrained system. Only the symplectic structure is no longer the globally defined Ω_{ab}, but the symplectic structure induced locally on the subspace $\Phi^i = 0$ (see figure 1). A consequence of this is that we can no longer expect the resulting motion to be unitary, as the induced symplectic structure $\tilde{\Omega}_{ab}$ has a nonlinear term. In order to illustrate the kind of motion that could result from this kind of constraint we consider the example of two spin-$\frac{1}{2}$ particles constrained to remain disentangled, as mentioned above. Let us first note however that the symplectic approach is not applicable to all systems. In order to use it, one requirement is that N, the total number of constraints being imposed, must be even. This follows by construction as otherwise the symplectic geometry is not preserved on the constraint surface. Another condition that must be satisfied in order to find the lagrange multipliers is that the inverse of (12) must exist.

Example 1: Let us now consider the example of two spin-$\frac{1}{2}$ particles with Hamiltonian:

$$\hat{H} = -J\hat{\sigma}_1 \otimes \hat{\sigma}_2 - B(\hat{\sigma}_1^z \otimes \mathbf{1}_2 + \mathbf{1}_1 \otimes \hat{\sigma}_2^z), \tag{15}$$

where the subscripts 1 and 2 labels the two particles, J is the strength of the spin-interaction, B is the strength of the external magnetic field (orientated along the z-axis), and $\hat{\sigma}_z$ is the Pauli spin matrix in the z-direction. We further require the initial state to be disentangled, and we wish to constrain the system so that it remains so. In other words we want the state of the system to remain on the surface $Q = \mathbb{P}^1 \times \mathbb{P}^1$ as it evolves. To find the equations of motion we follow the method outlined above, first finding (12), taking its inverse, using it to evaluate the Lagrange multipliers and hence finding $\tilde{\Omega}_{ab}$ (please see [5] for full details of this calculation).

Since the complex projective line is isomorphic to the two-sphere $\mathbb{P}^1 \sim \mathbb{S}^2$, we can view Q as the product space of a pair of 2-spheres $\mathbb{S}_1^2 \times \mathbb{S}_2^2$. Changing coordinates into spherical coordinates

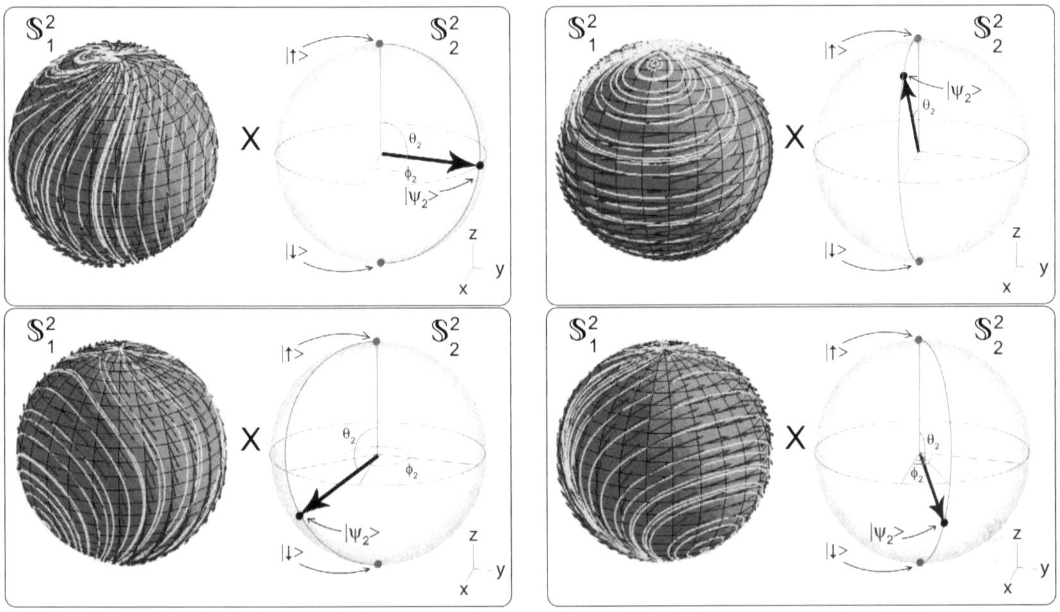

Figure 3. Four examples of 'snapshots' of the vector field and its integral curves as two spin-$\frac{1}{2}$ particles are constrained to remain disentangled so that their evolution lies on the product space $\mathbf{S}^2_1 \times \mathbf{S}^2_2$. The figures show the vector fields on $\mathbf{S}^2_1$ as the second particle evolves through the states ($\theta_2 = \frac{\pi}{2}, \phi_2 = \frac{\pi}{2}$) (top left), ($\theta_2 = \frac{\pi}{6}, \phi_2 = 0$) (top right), ($\theta_2 = \frac{2\pi}{3}, \phi_2 = \frac{5\pi}{3}$) (bottom left) and ($\theta_2 = \frac{3\pi}{4}, \phi_2 = \frac{\pi}{4}$) (bottom right), on $\mathbf{S}^2_2$.

(see [5]) we can visualise the evolution on the surface of the 2-spheres. The resulting equations of motion consist of four coupled non-linear differential equations:

$$\begin{aligned}
\dot{\theta}_1 &= \sin(\phi_1 - \phi_2)\sin\theta_2[(\omega_1 - \omega_2)\cos\theta_1 + \omega_2 - \omega_3] \\
\dot{\theta}_2 &= \sin(\phi_1 - \phi_2)\sin\theta_1[(\omega_2 - \omega_1)\cos\theta_2 - \omega_2 + \omega_3] \\
\dot{\phi}_1 &= \tfrac{1}{2}[-\omega_1 + (\omega_2 - \tfrac{\omega_1}{2})\cos\theta_2 + (\tfrac{3}{2}\omega_1 - \omega_2 - 2\omega_3)\cos\theta_1 + [\cos(\phi_1 - \phi_2)/(\sin\theta_1\sin\theta_2)] \\
&\quad \times \big(2(\omega_3 - \omega_2)\sin^2\theta_1\cos\theta_2 + (\omega_1 - \omega_2)(\cos^2\theta_1 - \cos^2\theta_2)\big)] \\
\dot{\phi}_2 &= \tfrac{1}{2}[-\omega_1 + (\omega_2 - \tfrac{\omega_1}{2})\cos\theta_1 + (\tfrac{3}{2}\omega_1 - \omega_2 - 2\omega_3)\cos\theta_2 + [\cos(\phi_1 - \phi_2)/(\sin\theta_1\sin\theta_2)] \\
&\quad \times \big(2(\omega_3 - \omega_2)\cos\theta_1\sin^2\theta_2 - (\omega_1 - \omega_2)(\cos^2\theta_1 - \cos^2\theta_2)\big)]\,.
\end{aligned} \tag{16}$$

It is difficult to solve these equations of motion in closed form. We will therefore explore the solution numerically by looking at various 'snapshots' of the vector field and associated trajectories as the system passes through certain points. In figure 3 we see four examples of this. The top left picture shows us the vector field on $\mathbb{S}^2_1$ as the evolution on $\mathbb{S}^2_2$ passes through the point $\theta_2 = \phi_2 = \frac{1}{2}\pi$. As the overall state of the system changes, we evolve on both spheres and as we change state on $\mathbb{S}^2_1$ the vector field on $\mathbb{S}^2_2$ changes and vice versa. To fully explore the solution numerically, we require an 'interactive simulation'; figure 3 merely gives a flavour for the often complicated trajectories that result from constraining the system to remain on Q. The strength of the formalism is that the solutions are nonetheless fully tractable, at least numerically.

3.2. Metric approach

In the previous section we found that there is a way to implement quantum constraints using the underlying symplectic geometry of the phase space. Let us now look at how some of the limitations of the symplectic formalism can be overcome by using the metric geometry of the phase space.

Another way of constraining the evolution of a system whose initial state lies on the surface $\Phi = 0$ is to remove all the components of the vector field normal to it [6] (see figure 1). This will force the system to remain on $\Phi = 0$. To implement this constraint we still use the method of Lagrange multipliers, but we substitute the symplectic structure in the second term of (10) with the metric structure. Hence the modified equations of motion become:

$$\frac{\mathrm{d}x^a}{\mathrm{d}t} = \Omega^{ab}\nabla_b H - \lambda_i g^{ab}\nabla_b \Phi^i, \tag{17}$$

where g^{ab} is the inverse of the Fubini-Study metric. Because we are now making use of the metric geometry of the constraint subspace, we no longer require that the total number of constraints N be even. The Lagrange multipliers λ_i can again be found explicitly in a similar way to the previous approach by considering $\dot{\Phi}^i = 0$. Following the method in [6], we hence find that

$$\lambda_i = M_{ij}\Omega^{ab}\nabla_a \Phi^j \nabla_b H, \tag{18}$$

where M_{ij} is the inverse of the matrix

$$M^{ij} := g^{ab}\nabla_a \Phi^i \nabla_b \Phi^j. \tag{19}$$

For the case when the constraints correspond to a family of conserved quantum observables, the matrix M^{ij} corresponds to the anticommutators between the observables. We hence require $\det(M^{ij}) \neq 0$ in order to find the equations of motion:

$$\dot{x}^a = \Omega^{ab}\nabla_b H - M_{ij}\Omega^{cd}\nabla_c \Phi^j \nabla_d H g^{ab}\nabla_b \Phi^i. \tag{20}$$

It is no longer in general the case that these can be rewritten in the form $\dot{x}^a = \tilde{\Omega}^{ab}\nabla_b H$, where $\tilde{\Omega}^{ab}$ is the inverse of a modified symplectic structure. We would, however, expect that systems that can be treated with both methods return the same result. It turns out that this is very hard to verify in general. In [6], a sufficient condition was derived under which the metric approach reduces to the symplectic approach with the modified equations of motion taking the form of (13), (see [6] for further details on this). We therefore take the view that the metric approach is the more general way of treating quantum constraints, but under certain conditions it reduces to an extension of the classical method to quantum systems as in the symplectic approach. We shall now look at one of the simplest examples of a system subject to only one constraint.

Example 2: Let us consider the example of a single spin-$\frac{1}{2}$ particle with Hamiltonian $\hat{H} = \hat{\sigma}_z$, where $\hat{\sigma}_z$ is a Pauli spin matrix, constrained such that

$$\Phi(x) = \frac{\langle x|\hat{\sigma}_x|x\rangle}{\langle x|x\rangle} \tag{21}$$

is conserved [6]. This example is interesting as we have an observable that does not commute with the Hamiltonian. Following the method outlined above, the equations of motion are obtained by considering (19), taking its inverse, finding the Lagrange multipliers (18) and hence the equations of motion (20). Just as in the previous example we look at the problem in spherical coordinates to more easily visualise the resulting evolution as trajectories on the surface of the 2-sphere. The modified equations of motion are given by [6]

$$\dot{\theta} = \tfrac{1}{2}\left(\frac{\sin(2\theta)\sin(2\phi)}{1-\sin^2\theta\cos^2\phi}\right), \quad \text{and,} \quad \dot{\phi} = \frac{2\cos^2\theta\cos^2\phi}{1-\sin^2\theta\cos^2\phi}. \tag{22}$$

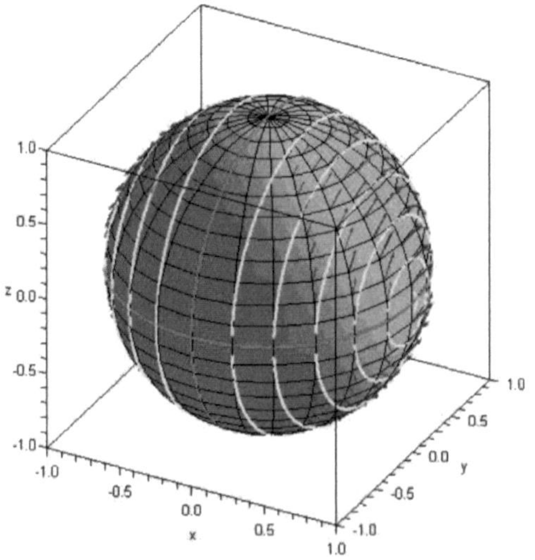

Figure 4. The vector field resulting from a single spin-$\frac{1}{2}$ particle with Hamiltonian $\hat{H} = \hat{\sigma}_z$ being constrained such that $A(x) = \frac{\langle x|\hat{\sigma}_x|x\rangle}{\langle x|x\rangle}$ is conserved [6]. The yellow lines are integral curves of the vector field and the red lines correspond to fixed points.

Figure 4 shows the resulting vector field and some of its integral curves. We recall that unitary unconstrained motion for this system corresponds to rigid rotation around the z-axis. As we are now forcing the system to remain on $\langle\hat{\sigma}_x\rangle = 0$, corresponding to circles around the x-axis, the resulting solution has a very interesting fixed point structure. Instead of only having two fixed points, the poles along the z-axis, as in the unconstrained case, the state space is now effectively partitioned into quarters of the sphere that do not interact with each other. Systems whose initial state is not a fixed point hence evolve along the half circular trajectories towards the fixed points along the equator.

Acknowledgments

ACTG would like to thank D C Brody, L P Hughston and D W Hook for many useful discussions, and for useful comments on an earlier version of this article. ACTG would also like to thank the organisers of the DICE2008 conference in Castiglioncello, Italy, 22-26 September 2008, for the opportunity to present this work there.

References

[1] Dirac P A M 1950 Generalized Hamiltonian dynamics *Canadian J. Math.* **2** 129-148
[2] Dirac P A M 1958 Generalised Hamiltonian dynamics *Proc. R. Soc. London* A **246** 326-332
[3] Burić N 2008 Hamiltonian quantum dynamics with separability constraints *Ann. Phys.* **323** 17-33
[4] Corichi A 2008 On the geometry of quantum constrained systems *Class. Quantum Grav.* **25** 135013
[5] Brody D C, Gustavsson A C T and Hughston L P 2008 Symplectic approach to quantum constraints *J. Phys.* A: Math. Theor. **41** 475301
[6] Brody D C, Gustavsson A C T and Hughston L P 2009 Metric approach to quantum constraints *Preprint* arXiv:quant-ph/0903.5261
[7] Strocchi F 1966 Complex coordinates and quantum mechanics *Rev. Mod. Phys.* **38** 36-40
[8] Cantoni V 1977 The Riemannian structure on the states of quantum-like systems *Comm. Math. Phys.* **56** 189-193

[9] T W B Kibble 1979 Geometrization of quantum mechanics *Commun. Math. Phys.* **65** 189-201
[10] Weinberg S 1989 Testing quantum mechanics *Ann. Phys.* **194** 336-386
[11] Cirelli R, Mania A and Pizzocchero L 1990 Quantum mechanics as an infinite-dimensional Hamiltonian system with uncertainty structure I, II *J. Math. Phys.* **31** 2891-2903
[12] Gibbons G W 1992 Typical states and density matrices *J. Geom. Phys.* **8** 147-162
[13] Hughston L P 1995 Geometric aspects of quantum mechanics *Twistor Theory* ed S Huggett (New York: Marcel Dekker) 59-79
[14] Ashtekar A and Schilling T A 1995 Geometry of quantum mechanics *CAM-94 Physics Meeting: AIP Conf. Proc.* **342** ed A Zapeda (Woodbury, New York: AIP Press) 471-478
[15] Brody D C and Hughston L P 2001 Geometric quantum mechanics *J. Geom. Phys.* **38** 19-53
[16] Oh P and Kim M 1994 Action angle variables for complex projective space and semiclassical exacness *Mod. Phys. Lett.* A **9** 3339-3346
[17] Brody D C, Gustavsson A C T and Hughston L P 2007 Entanglement of 3-qubit geometry *J. Phys.: Conf. Series* **67** 012044

Journal of Physics: Conference Series **174** (2009) 012026 doi:10.1088/1742-6596/174/1/012026

Center of mass in special and general relativity and its role in an effective description of spacetime

C Chryssomalakos[1], H Hernandez-Coronado[1] and E Okon[2]

[1] Institute for Nuclear Sciences, National Autonomous University of Mexico, Mexico City 045120, Mexico

[2] Department of Applied Mathematics, University of Waterloo, Ontario N2L 3G1, Canada

E-mail: hcoronado@nucleares.unam.mx, chryss@nucleares.unam.mx, eokon@uwaterloo.ca

Abstract. In this contribution, we suggest the approach that geometric concepts ought to be defined in terms of physical operations involving quantum matter. In this way it is expected that some (presumably nocive) idealizations lying deep within the roots of the notion of spacetime might be excluded. In particular, we consider that spacetime can be probed only with physical (and therefore extended) particles, which can be effectively described by coordinates that fail to commute by a term proportional to the spin of the particles.

1. Introduction

The only physical way that information can be extracted from spacetime is through matter which obeys the laws of quantum mechanics. However, some of the notions that general relativity is based on are incompatible with quantum mechanics. Then, the spacetime properties related to the notions referred to above cannot be accessed by physical means.

That is, on one hand, general relativity assumes that spacetime is a differential manifold and so the notions of *point* and *path* naturally arise. Moreover, physical meaning can be assigned to those notions, for example by postulating that pointlike test particles follow geodesics. On the other hand, paths cannot be associated to quantum particles, since a particle following a path has position and momentum simultaneously defined with infinite precision, and that violates the uncertainty principle. Furthermore, even if the concept of point is not forbidden by quantum mechanics, in general, quantum states are not localized – in spacetime – and therefore the concept of point becomes irrelevant.

Thus, while probing spacetime with classical pointlike test particles allows us to use paths and points, doing so with quantum test particles requires considering a generalization of points and paths. A first approach that can be taken in order to consider this, is to probe spacetime with extended classical distributions which can neither be localized nor associated to a single path. We expect that the above mentioned approach can teach us "something quantum" about spacetime.

In the present work we explore the possibility of an effective description of spacetime. That is, we consider the possibility of obtaining a picture of spacetime in terms of properties we are used to, such as points and paths, and yet, taking into account that test particles are quantum, and consequently are not precisely localized. In order to do so, extended test particles must be related to a single point, which should be selected according to its properties (the "center of mass", for instance).

The structure of the paper is as follows: in Section 2 we discuss what we think is the most suitable center of mass definition in Special Relativity and examine its properties following Pryce [1], in particular we find that its coordinates fail to commute, in the context of Poisson brackets. In Section 3 we comment on the possible implications of the noncommutativity found in Section 2, mention some further directions that can be pursued, and conclude with a comment on the center of mass definition in the General Relativity case.

2. Center of mass in special relativity

Among all the possible points that can be used for effectively describing the position of an extended object, a natural choice is the center of mass, nevertheless, the problem of defining the relativistic analogue of the Newtonian center of mass for a distribution is not fully resolved. A good review of the various proposals up to 1948 is given by Pryce [1]. In particular, for our purposes, we think that the most suitable one is such that its coordinates, X^μ, are defined as:

$$X^\mu = \frac{J^{\mu\nu} P_\nu}{m^2} + \alpha P^\mu, \tag{1}$$

where $\mu = 0, 1, 2, 3$, α parameterizes the center of mass worldline, P^μ and $J^{\mu\nu}$ are the total linear and angular momenta of the mass distribution, respectively, and $m^2 = P^\mu P_\mu$. If we think of the mass distribution as a set of pointlike particles, definition (1) is the *average of particle positions, weighted by their total energies in the center of momentum frame, i.e., the frame in which the total three momentum vanishes.* According to the above definition, the center of mass satisfies the following properties:

a. Its spatial components are part of a four vector, whose zeroth component is the time at which they are measured.
b. It is at rest in the "center of momentum" frame.
c. When no external forces act on the distribution, it moves with constant velocity P^μ/P^0.

By looking at the above properties, it could be concluded that nothing really gets modified by considering the extension of test particles while probing spacetime, however, it is interesting to note that the spatial components of the center of mass defined as in (1) satisfy the nontrivial commutation relations[1] (in the sense of Poisson brackets),

$$[X^i, X^j] = \epsilon^{ij}{}_k \left(\frac{S^k}{E^2} + \frac{\mathbf{S} \cdot \mathbf{P}}{m^2 E^2} P^k \right) , \tag{2}$$

with $\mathbf{S} = (S^{23}, S^{31}, S^{12})$ where $S^{\mu\nu} = J^{\mu\nu} - X^\mu P^\nu + X^\nu P^\mu$. The (classical) noncommutativity of the X's found in (2) follows from the fact that the system is extended in space and our interpretation is relativistic. In order to clarify the last statement, suppose the system is observed in the center of momentum frame, where it is found to have non zero angular momentum along the z-axis, and therefore non-zero S_z (since angular momentum in the center of momentum frame is spin by definition). Then upon quantization and according to the Heisenberg uncertainty principle, the above expression implies that the system's center of mass must be situated within a fuzzy region of the x-y plane for any point on the z-axis. This should be compared to the Newtonian case, where the center of mass can be located exactly since its coordinates commute. In other words, the behavior of extended systems reduces exactly to that of a pointlike particle

[1] It is precisely because of this fact that Pryce concluded that there was not a single fully satisfactory relativistic center of mass definition.

located at its center of mass (modulo the internal motion of the system), only in the Newtonian limit.

The relations (2) become relevant in the quantum version, where the center of mass coordinates are elevated to operators and the Poisson brackets to commutators. Their relevance in the quantum gravitational context resides in the aspiration of finding a quantum analogue to the classical geodesic motion of point particles. As discussed earlier, the hope is to assign some sort of effective path to a quantum particle, for which a mean position is needed because its state is, in general, spread out. Under the most favorable circumstances, given (2), the quantum particle, described by an effective point particle located at the "mean position", will feel some sort of average of the metric over a region whose extent is of the order of the r.h.s. of (2).

Notice also that the present noncommutativity differs from others in the literature, in the sense that it naturally arises from physical considerations, on one side, and in that it depends on the particle properties so that the physical interpretation is fundamentally different, on the other (for a more extensive discussion the reader is referred to [2]). It is also interesting to note that this same kind of noncommutativity appears in several different physical contexts (*e.g.*, [3], [4]).

Another feature of (1) worth-mentioning is that it forces us to consider spin, which is included in the total angular momentum $J^{\mu\nu}$. In the context of classical systems, as defined above, spin means *intrinsic angular momentum*, *i.e.*, the part of the angular momentum which does not depend on the origin of coordinates. If spin is not taken into account, the definition turns out to be nonassociative, that is, when computing the center of mass for, say, three mass distributions, the result depends on the order in which they are composed. This occurs even if the initial distributions have spin zero because the mass distribution resulting from such composition has in general non-zero spin.

Finally, note that the position coordinates of a single spinless particle commute, and expression (2) then establishes that (a quantum version of) it can be exactly localized, as in standard quantum mechanics. Notice likewise that it implies for an electron's center of mass to be found within a region whose area is of the order of (Compton wavelength)$^2 \sim 10^{-25}$ m^2. This is clearly a consequence of considering only positive energy states in our description, since in Dirac's theory for spin one-half particles – where negative energy states are admitted – the position operators are commutative. Invocation of antiparticles, miraculously, restores commutativity.

3. Further directions

Let us try to understand what we have learned from the previous discussion. The starting point was the assumption that the only way in which we can probe spacetime is through matter, which is quantum, and is in general described by a nonlocalized wavefunction. Intending to take account of the nonlocalized character of quantum states in the description of spacetime, classical extended matter was considered. In the Newtonian limit, classical extended matter can be described exactly as an effective point particle, located at the center of mass, with all of its classical properties. In the relativistic situation, its effective description naturally leads to noncommutative center of mass coordinates (in the sense of Poisson brackets). And finally, quantum mechanics implies the nonlocalization of the center of mass because of its noncommutative coordinates. This suggests that noncommutativity could be an important element for considering "quantum aspects" of spacetime.

An immediate next step in the program, should be to look for some sort of effective metric such that "effective quantum matter" follows effective geodesics (conveniently defined). Of course, it is not clear that such effective entities should exist.

Another direct way to proceed is to consider general spacetimes, that is, to consider General Relativity, where the situation is much more complicated. The authors know of at least three inequivalent center of mass definitions [5]-[7] that reduce to (1) when spacetime is flat. They have different properties and the choice of which should be used, depends on the specific physical situation that is dealt with. The differences between them stem from the fact that, in curved spacetimes, parallel transport is not trivial and then the comparison between vectors associated to different points of the mass-energy distribution is nontrivial. As a consequence, there is not only one way to define the total mass or total angular and linear momenta. Explicit expressions for the previous definitions are complicated and their commutators have not been computed. The only effort towards that direction, to the knowledge of the authors, is the work of Bonder and Sudarsky who are currently working on the computation of the Poisson brackets for the coordinates of their general relativistic center of mass definition [7].

Acknowledgments

CC and HH would like to thank the organizers of DICE 2008, in particular, Thomas Elze, for efficiently handling their requests and the high scientific level they helped to maintain. CC and HH also wish to acknowledge partial financial support from DGAPA-UNAM project IN 121306.

References

[1] Pryce M H L 1948 *Proc. Roy. Soc. Lon.* **195** 62-81
[2] Chryssomalakos C, Hernández H, Okon E and Vázquez-Montejo P 2007 *J. Phys.: Conf. Ser.* **68** 012003
[3] Casalbouni R 1976 *Nuovo Cimento* A **33** 389
[4] Bette A 1984 *J. Math. Phys.* **25** 2456
[5] Beiglböck W 1967 *Commun. Math. Phys.* **5** 106
[6] Dixon W G 1964 *Nuovo Cimento* **34** 317; *do.* 1970 *Proc. Roy. Soc. Lon.* A **314** 499
[7] Bonder Y and Sudarsky D *Private communication*

And what if gravity is intrinsically quantic?

Houri Ziaeepour

Mullard Space Science Laboratory (MSSL)-University College London, Holmbury St.Mary, Dorking RH5 6NT, Surrey, UK

E-mail: hz@mssl.ucl.ac.uk

Abstract. Since the early days of search for a quantum theory of gravity the attempts have been mostly concentrated on the quantization of an otherwise classical system. The two most contentious candidate theories of gravity, string theory and quantum loop gravity are based on a quantum field theory – the latter is a quantum field theory of connections on a $SU(2)$ group manifold and the former is a quantum field theory in two dimensional spaces. Here we argue that there is a very close relation between quantum mechanics (QM) and gravity. Without gravity, QM becomes ambiguous. We consider this observation as the evidence for an intrinsic relation between these fundamental laws of nature. We suggest a quantum role and definition for gravity in the context of a quantum Universe, and present a preliminary formulation for gravity in a system with a finite number of particles.

1. Introduction and motivations

Since the beginning of search for quantum theory of gravity all approaches have considered the quantization of a classical field theory. Nonetheless the idea of an intrinsic relation between gravity and QM is not new. The first evidence can be claimed to be the black hole entropy and its analogy with thermodynamics. However, it is well known that without considering Hawking radiation [1] and Bekenstein entropy limit [3], this analogy seems to be just a mathematical similarity. In fact the black hole entropy is a purely geometrical property related to the diffeomorphism symmetry and independent of the special case of the Einstein gravity Lagrangian [4, 2]. The discovery of similarity between Hawking radiation temperature and the equivalent temperature obtained from black hole entropy gave another dimension to this as a possible mediator between classical and quantum gravity theories. More recently it has been proved that in D-brane models [5] – a subclass of compactified string models – at low energies and weak couplings the direct counting of the states of an extremal or near extremal black hole lead to an entropy similar to the classical black holes [6]. These results enforce the relation between the classical black hole entropy and the quantum nature of gravity.

On the way to a quantum theory or a quantum connection for gravity various strategies have been taken. Apart from investigations of candidate models such as string theory, loop quantum gravity, causal set models, etc., in which the metric $g_{\mu\nu}$ or connection $\Gamma^{\rho}_{\mu\nu}$ are quantum fields, many authors tried to derive a gravity theory from a field theory in a curved space time. For instance T. Padmanabhan [7] has used Rindler metric – a frame with constant acceleration – as background, rather than Minkowski metric, and a somehow general Lagrangian for gravity. It includes a non-trivial surface term on space-like 3-surfaces. By postulating the relation between entropy, temperature, and surface gravity obtained from Einstein gravity, a dynamical equation similar to one from Einstein gravity has been obtained. This exercise shows explicitly that there is a close relation between the laws of black hole thermodynamics and Einstein gravity. However, black hole entropy is the Noether charge in the models with diffeomorphism symmetry [4]. Therefore it is not a surprise that in a given background the assumption of entropy relation obtained from Einstein gravity leads to the Einstein equation. This is analogue to gauge symmetry of electrodynamics in which the presence of a global conserved charge (surface gravity) and the application of the Gauss theorem permits to define a 2-form (connection) on a 3-surface. Then by using the gauge symmetry (diffeomorphism) and its conserved current, one can obtain the Maxwell equation and its corresponding Lagrangian.

Another popular approach is the construction of a class of theories generally called *emergent gravity models* in which gravity is considered to be a collective low energy/classical effect of a fundamental microscopic model in a flat Euclidean or Lorentzian background and very different from classical gravity [8]. In this case, gravity is considered to be a classical effect similar to condensation process in condensed matter physics. A curved space time with pseudo-Riemannian metric and diffeomorphism invariance have been found in a subset of these models notably called *analogue gravity* [9]. For the time being, in none of these models a tensor field theory with properties similar to what we know from Einstein gravity has been found. But there have been progresses in this direction. For instance, recently in a work by F. Girelli *et al.* [10] a Nordström scalar gravity model has been constructed from an N-scalar field theory on a Euclidean background.

In all the models mentioned above the background space is classical and a quantum field theory is defined on this space. The same line of thinking has been considered also in noncommutative field theory models with a quantum spacetime [11]. The main physical motivation for introducing noncommutative geometries is the fact that due to Heisenberg uncertainty, a very precise measurement of spacetime will create very large uncertainty in energy-momentum that can produce a black hole preventing further investigation of smaller distances. To solve this problem, coordinates are considered to have non-zero commutators. New progresses [12, 13] in this class of models show that a classical gravity model similar to Einstein gravity can emerge from these models under some conditions and simplifications. As the effect of a quantum spacetime in these models is measurable only at Planck scales, at present there is no direct way to verify them.

Here we want to consider the relation between classical gravity and quantum mechanics in a more intuitive way and see what physical properties join them and what make them inconsistent with each other.

There are a number of unsolved issues in gravity and in QM that have fundamental consequences for understanding these phenomena. Some of them are the following:

(i) Why is gravity a universal force?

(ii) Why is the Planck constant $\hbar$ universal?

(iii) Why is there no fundamental mass/length scale in QM?

The first question above is another way of asking the origin of the Equivalence Principle. This issue has been studied intensively in the framework of classical and quantum gravity models, and in grand unified theories. In fact in many candidate models of quantum gravity such as string theory, the Equivalence Principle is a low energy effect that breaks down at high energies. Nonetheless, at present the upper limit on the breakdown of equivalence strongly constrains some low energy quantum gravity models such as brane models inspired from D-brane compactification of strings. This rises the issue of what the most fundamental principle in gravity is? Its dependence only on energy-momentum or its geometrical origin as the metric of a pseudo-Riemannian space, or both? Breaking the Equivalence Principle violates the first condition. Therefore if we consider this principle as part of the *definition* of what we call gravity, models violating it can not be considered as a genuine quantum gravity model.

The second universality means that the amount of *uncertainty or randomness* in all physical systems is the same regardless of their mass, size, and couplings. Let's assume that the value of $\hbar$ for a particle of mass m is different by a factor of α. The Schrödinger-Klein-Gordon equation becomes:

$$(\alpha^2\hbar^2\Box - m^2)|\psi\rangle = 0 \Longleftrightarrow (\hbar^2\Box - m'^2)|\psi\rangle = 0 \,, \qquad m' = \frac{m}{\alpha} \,. \tag{1}$$

Therefore the non-universality of $\hbar$ can be removed by redefinition of the mass. But mass is the gravity charge! In the same way a different coupling G to gravity can be removed by redefinition of the mass which in its turn modifies the Schrödinger equation. Note that if mass is generated by interaction and is related to the vacuum expectation value of a field, this scaling can be performed on the vacuum expectation value. This operation is allowed in quantum mechanics because the energy reference is arbitrary. But considering gravity, scaling modifies the coupling of fields to gravity, except in conformal models where the scaling can be absorbed in the space volume. In fact it has been shown that Einstein and higher order gravity models are related to each others by conformal transformation if these models have a scalar field in their matter sector [14]. This is the case for all gauge models where the masses of particles are due to the vacuum expectation values of Higgs scalar fields. In this case additional terms from non-conformal matter can be considered as non-minimal interaction terms.

The origin of the close dependence of quantum mechanics on gravity is the third item above: QM lacks any fundamental energy or length scale. Therefore a theory as fundamental as QM, its extension, the quantum field theory, and thereby the Standard Model of particle physics need the presence of gravity for being a complete and applicable theory to Nature. On the other hand, gravity is supposed to be just an interaction and the general tendency is to see it with the same eye as the other interactions (except in emergent gravity models mentioned above). QM does not need any of the other known interactions, e.g. electromagnetism, to be meaningful. Thus we can conclude that gravity is fundamentally different and must have a much closer connection to QM.

Although emergent gravity models seem to have shown this connection, gravity at low energies remains classical and its connection to a quantum Universe appears only at the Planck scale. Nonetheless, with simple examples we can show that even at low energies the non-deterministic nature of QM is in conflict with a classical (deterministic) universal force like gravity.

2. Ambiguity of a classical universal force in a quantic Universe

Consider an empty Minkowski space. We are only interested in low energy scales, therefore we neglect vacuum fluctuations and all the phenomena related to quantum field theory. We add to this space a single particle of mass m. To remove the ambiguity due to black hole formation, we can consider a small but finite size for the particle larger than its Schwarzschild radius. In this case if the resolution of the measurement is much larger than the size of the particle it can be considered as point-like and the argument below is correct up to this approximation.

As there is no other observer/particle in this space, without loss of generality we can consider that the particle is at rest. Because we know the energy and momentum of this particle, according to the Heisenberg uncertainty rule we lose all information about its location, i.e. at any instant of time the particle can be at any point in the space. This has a number of consequences:

(i) A classical particle in a flat Minkowski space breaks the translation symmetry to a spherical symmetry and changes the flat metric to Schwarzschild.

(ii) A quantum particle as defined above does not break the translation symmetry. The wave function has the form of a free wave $\psi(x) \propto \exp(ip_\mu x^\mu)$ and the probability of detecting the particle in any finite volume approaches to zero (up to coarse-grain approximation mentioned before). To estimate the modification of the spacetime due to the presence of this particle, we can use the semiclassical Einstein equation:

$$R_{\mu\nu} - \frac{1}{2}g_{\mu\nu}R = \frac{8\pi G}{c^4}\langle\psi|\hat{T}_{\mu\nu}|\psi\rangle \sim \eta_{\mu\nu}\Lambda \ , \qquad \Lambda \to 0 \ . \tag{2}$$

The solution of (2) is simply a De Sitter metric which is conformally flat and as $\Lambda = m/V \to 0$, where V is the space volume, the metric becomes very close to the metric of the initial flat Minkowski space. Note that up to coarse-grain approximation this result is independent of the mass of the particle. Conclusion: In the context of QM a single massive particle does not change the spacetime by its gravity!

(iii) If we add a second particle only with gravitational interaction with the first one, we can use the Wheeler-DeWitt equation and the above metric as background to determine the wave function of the second particle. As the background metric is at most De-Sitter, otherwise flat, the wave function would be again a free wave but the wave number gradually decreases due to the expansion of the space time, and at coarse-grain limit it approaches its value in a flat spacetime. Moreover, the homogeneity of the spacetime does not change by the presence of a second particle either. This means that they do not interact gravitationally. In the same way we can add any number of particles and up to coarse-grain limit they do not feel each others! This paradoxical result is evidently inconsistent with the behaviour we know from classical gravity.

(iv) At present we do not have any experimental evidence that tells us what happens in such a setup.

The origin of the conflict between classical and quantum predictions is the fact that gravity and its dynamical equation are local, in contrast to QM which is intrinsically nonlocal. Note that a noncommutative spacetime and emergent gravity can not solve these ambiguities. The setup of this example satisfies the noncommutative relation simply because the location of the particle is completely uncertain in all directions. This sort of ambiguities should be added to the well known conflict between diffeomorphism invariance that leads to a null Hamiltonian and the lack of the concept of time evolution in quantum gravity [15].

These observations rise the idea that maybe we should think about gravity or whatever shows itself as gravity at low energies in another way, presumably in a quantic manner. Here we present some ideas about what the role of gravity can be, with properties we know in a quantum world, and how possibly a corresponding field theory can be constructed.

3. The role of gravity in a quantic Universe

Consider the phase space of a classical system and the Hilbert space of the same system when it is quantized. We can define quantization as the definition of an equivalence class in the phase space:

$$\mathcal{H} \equiv \Phi/\mathcal{S} \tag{3}$$

where Φ is the phase space and $\mathcal{S}$ is the set of surfaces with area $= \hbar$. For simplicity we neglect the spacetime indexes and consider the phase space like a 2-dimensional surface. If we have a system consisting of two non-interacting particles, their phase space or Hilbert space is the direct product of their respective single particle spaces[1]:

$$\Phi = \Phi_1 \otimes \Phi_2 \quad , \quad \mathcal{H} = \mathcal{H}_1 \otimes \mathcal{H}_2 \tag{4}$$

[1] Note that for simplifying the notation, we interchange the meaning of Hilbert space as the vector space to which all the state of a physical system belong and the orbit of the Lagrangian operator of the system for a given initial condition. The meaning should be clear from the context.

This means that particles don't *feel* each other. However, the assumption of no interaction between two particles is not realistic because we know that gravity is a general force and whatever the nature of the two particles, they interact through their gravity and therefore:

$$\mathcal{H} \neq \mathcal{H}_1 \otimes \mathcal{H}_2 \; . \tag{5}$$

Here we want to suggest a new definition for gravity in a quantum context:

Gravity between quantum particles presents the minimum deviation of their Hilbert space from direct products of single-particle Hilbert spaces.

This definition is consistent with the universality of gravity and with the semiclassical results for one particle – a single particle with known energy and momentum *does not* feel its own gravity! A direct consequence of this property is that such a particle does not make a black hole![2] By contrast, if the same particle is localized, then its energy-momentum would be completely uncertain similar to particles inside a black hole, which for a far observer can be at any place inside the horizon[3].

The presence of gravity becomes visible only if we have at least two particles in the system. A nontrivial projection of $\mathcal{H}_1 \otimes \mathcal{H}_2$ to itself defines the new Hilbert space $\mathcal{H}$. In fact when an interaction is present such a projection happens also in the phase space for both classical and quantum systems. What is more important in this model and different from others is that gravity mixes the equivalence classes:

$$\mathcal{H} = \frac{\Phi}{\mathcal{S}} \; , \quad \mathcal{S} \neq \mathcal{S}_1 \otimes \mathcal{S}_2 \; . \tag{6}$$

This means that the uncertainty surfaces for the two particles are not anymore separate but define an inseparable subspace of the total phase space Φ. This can solve the puzzle of universal $\hbar$. Assuming that this projection depends on the Newton coupling constant G or equivalently on the Planck mass, this provides the missing mass/length scale for QM. The extension of this construction to multi-particle systems is trivial and we do not present details here.

The nontrivial mixing of equivalence spaces representing the Heisenberg uncertainty relation means that the uncertainties of coordinate and momentum of two particles are no more independent. This is somehow similar to the noncommutative models. However, in the latter case the spacetime is considered as an independent entity with noncommutative coordinates. Here the nontrivial commutation relation is not an intrinsic but an induced property. It is possible that for large particle numbers the two models have some sort of relation. We leave the investigation of this eventuality to future works.

The nontrivial projection of the combined Hilbert space $\mathcal{H}_1 \otimes \mathcal{H}_2$ can be considered as a basis transformation with the condition that in the new coordinates the transformed particles can be considered as free and separable. This basis would play the role of a local inertial frame in the classical general relativity and should constrain the projection. However, at this level of the development of the model we can not prove that it always exists. Therefore we assume that its existence is granted.

We expect that in the classical limit the dynamic equation of this model approaches the Schrödinger-Klein-Gordon equation with minimal interaction to gravity:

$$(\hbar^2 \Box + \frac{R}{6} - m^2)|\psi\rangle = 0 \; . \tag{7}$$

[2] At first sight there is an ambiguity in this argument because in both classical and QM a particle can be composite. We assume that Eq. (5) is applied to fundamental particles. In addition, in the framework of this model the concept of a single particle can be even meaningless. See the end of this section for more details.

[3] Note that the Schwarzschild radius here is determined by the energy-momentum uncertainty and not by the invariant mass of the particle.

We consider the following transformation from one-particle coordinates to *free* coordinates X:

$$\hat{X}_i = \hat{x}_i \quad , \quad \hat{P}_i = \hat{p}_i + \sum_{j \neq i} f(\hat{x}_j, \hat{p}_j) \ , \tag{8}$$

where $f(\hat{x}_j, \hat{p}_j)$ is an arbitrary function. It is easy to show that commutators become:

$$[\hat{x}_i, \hat{x}_j] = [\hat{X}_i, \hat{X}_j] = 0 \quad i, j = 1, 2, \ldots \tag{9}$$

$$[\hat{p}_i, \hat{p}_j] = [\hat{P}_i, \hat{P}_j] = 0 \tag{10}$$

$$[\hat{x}_i, \hat{p}_i] = i\hbar \tag{11}$$

$$[\hat{X}_i, \hat{P}_i] = i\hbar \tag{12}$$

$$[\hat{x}_i, \hat{p}_j] = [\hat{X}_i, \hat{P}_j] = 0 \quad i \neq j \tag{13}$$

$$[\hat{x}_i, \hat{P}_j] = [\hat{x}_i, f(\hat{x}_i, \hat{p}_i)] \tag{14}$$

Note that in Eqs. (9)-(14) the indices refer to particles and the spacetime indices are neglected. We assume the following form for the function f:

$$f(\hat{x}_i, \hat{p}_i) = \frac{1}{\Lambda \hat{x}_i} \hat{p}_i \tag{15}$$

where Λ is an energy scale. Then:

$$\hat{P}_i = \hat{p}_i + \sum_{j \neq i} \frac{1}{\Lambda \hat{x}_j} \hat{p}_j \ , \tag{16}$$

$$[\hat{x}_i, \hat{P}_j] = [\hat{X}_i, \hat{P}_j] = \frac{i\hbar}{\Lambda \hat{x}_i} \ . \tag{17}$$

Finally, we apply (16) to the Schrödinger-Klein-Gordon equation (1):

$$\left\{ \sum_i \hat{p}_i^2 + 2 \sum_{j \neq i} \frac{1}{\Lambda \hat{x}_j} \hat{p}_i \hat{p}_j + \sum_{j,k \neq i} \frac{1}{\Lambda^2 \hat{x}_j \hat{x}_k} \hat{p}_j \hat{p}_k + m^2 \right\} |\psi\rangle = 0 \ . \tag{18}$$

The operator part of equation (18) has the same structure as (7) except that it contains discrete operators functioning on different particles.

In the formulation above we neglected spacetime indices. Apart from simplifying the notation, there is a deeper reason for this ignorance. First of all it proves that this formalism can be applied to any background spacetime. More importantly, the similarity of the role of species/particles index to spacetime index indicates that there is an interchangeable role between what is called *a particle* and *a point in the spacetime*. Therefore we can claim that in this model there is a natural unification or embedding of the spacetime with a group manifold determining the symmetries and variety of particles. This aspect is similar to string theories. The hope is that in the extension of this model to continuum, geometrical properties and symmetries determine the signature of spacetime metric, the emergence of a time as well as the dimension of space.

4. Outline

We suggested a modified structure of Hilbert space of multi-particle quantum system to present the effect of gravity. We showed that the redefinition of coordinate and momentum operators leads to a dynamical equation similar to the Schrödinger-Klein-Gordon equation for quantum mechanics in curved spacetime. A natural extension of the model to supergravity seems possible. Before being able to claim this model as a candidate model for quantum gravity we need to investigate a number of issues such as extension to infinite number of particles, in another word a field theory, the emergence of a time coordinate, and possible relation between this model and other candidate quantum gravity models.

Figure 1. The Universe, presented in this figure by the map of Cosmic Microwave Background anisotropy observed by the WMAP satellite, can be the large scale manifestation of an intrinsically indeterministic Universe based on the uncertainty rules of Quantum Mechanics, symbolically presented by the Schrödinger Cat à la da Vinci to remind us of Tuscany, the home country of da Vinci and the venue of the DICE2008 Workshop.

Acknowledgments
I would like to thank all the organizers of the DICE2008 Workshop, notably Prof. Thomas Elze, for their formidable organization and the choice of a wonderful place for the venue.

References

[1] Hawking S W 1971 *Phys. Rev. Lett.* **26** 1344
Bardeen J M, Carter B and Hawking S W 1973 *Commun. Math. Phys.* **31** 161
[2] Wald R M 1996 *Proceedings Symposium on Black Holes and Relativistic Stars in honor of S. Chandrasekhar, December 14-15, 1996* (*Preprint* gr-qc/9702022)
[3] Bekenstein J D 1973 *Phys. Rev.* D **7** 2333
[4] Wald R M 1993 *Phys. Rev.* D **48** 3427
[5] Horava P and Witten E 1996 *Nucl. Phys.* B **460** 506
[6] Strominger A and Vafa C 1996 *Phys. Lett.* B **379** 99
Callan C G and Maldacena J M 1996 *Nucl. Phys.* B **472** 591
Horowitz G T and Strominger A 1996 *Phys. Rev. Lett.* **77** 2368
[7] Padmanabhan T 2002 *Mod. Phys. Lett.* **17** 1147
[8] Hu B L 2005 *Int. J. Theor. Phys.* **44** 1785
[9] Barcelo C, Liberati S and Visser M 2005 *Living Rev. Rel.* **8** 12
[10] Girelli F, Liberati S and Sindoni L 2008 *Preprint* arXiv:0806.4239
[11] Doplicher S, Fredenhagen K and Roberts J E 1995 *Commun. Math. Phys.* **172** 187 (*Preprint* hep-th/0303037)
[12] Yang H S 2007 *Mod. Phys. Lett.* **22** 1119 (*Preprint* hep-th/0611174)
[13] Steinacker H 2007 *J. High Ener. Phys.* **712** 49
[14] Cotsakis S 2008 *Grav. Cosmol.* **14** 176 (*Prepint* gr-qc/0408095)
[15] Wald R M 1994 *Quantum field theory in curved spacetime and black hole thermodynamics* (Chicago: The University of Chicago Press)

Fourth International Workshop DICE2008 IOP Publishing
Journal of Physics: Conference Series **174** (2009) 012028 doi:10.1088/1742-6596/174/1/012028

Possibility of realizing Bussey's thought experiment on collapse of the wave function at the microscopic level

Kouichi Toyoshima, Takasi Endo and Yutaka Hirayoshi

Faculty of Science and Engineering, Saga University, Saga 840-8502, Japan

E-mail: toyo@cc.saga-u.ac.jp

Abstract. We discuss on feasibility of experiments to test if the wave function collapses at the level of microscopic events. The original idea is Bussey's 'thought' experiment introduced in 1984. Bussey's proposal is to collide two particles and then two pairs of scattered and unscattered waves of each particle are examined by a pair of Mach-Zehnder type interferometers (MZI). Different coincidence rates between different sets of combinations of MZI outputs are expected depending on the collapse/non collapse assumption. In our proposition to realize this idea, a pair of low-energy electron beams is collided nearly head-on in a nanometer-size region. Then each combination of two waves scattered at small angles is thrown into the MZI. Each MZI is composed of Moellenstedt biprisms as two arms and mirrors, and then brought together at single crystal films which act as half reflecting mirrors (beam splitters). Coincidence counting is done among four outputs of the beam splitters. Another method to measure the position correlation after the biprisms is also proposed.

1. Introduction

The problem of "wave function collapse" is still one of the most fundamental issues in quantum mechanics. Recently, apart from the theoretic and philosophical aspect of this issue, it may find a "practical" application as a measure to obtain final results in quantum computer, if it is realized. But the problem is still open and we have, as far as we know, no experimental test on how the wave function collapses.

More than 20 years ago Bussey [1] proposed a thought experiment to test if the wave function collapses at the microscopic level. As shown in Fig. 1, two beams are collided, the scattered and the unscattered beams enter two Mach-Zehnder type interferometers. Then coincidence counts are compared between different combinations of 4 output ports of interferometers. We denote the coincidence count between counter 1 and 3 by C_{13}, etc.

Here we repeat the introductory part of [1] with a little extension. If the wave function does not collapse, namely, if it is a superposition of scattered and unscattered waves

$$|\Psi >= \alpha|\mathrm{A}\rangle|\mathrm{B}\rangle + \beta|\mathrm{A}'\rangle|\mathrm{B}'\rangle \qquad (1)$$

then coincidence counts C_{13} and C_{14} are expressed as

$$C_{13} = |\tau|^4|\alpha + \gamma^2\beta|^2 \ , \quad C_{14} = |\gamma|^2|\tau|^4|\alpha + \beta|^2 \ .$$

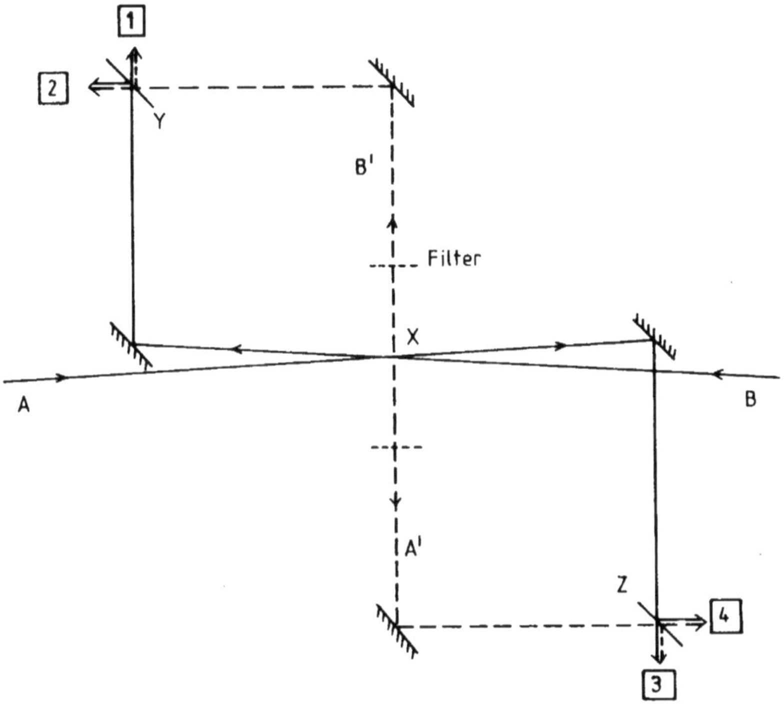

Figure 1. Bussey's double interferometer. Reproduction of Fig. 1 of [1].

But if it does collapse to one of these scattered or unscattered waves then:

$$|\Psi> \;=\; |\mathrm{A}\rangle|\mathrm{B}\rangle \quad \text{or} \quad |\mathrm{A}'\rangle|\mathrm{B}'\rangle \tag{2}$$

with weights $|\alpha|^2$ and $|\beta|^2$ respectively, these two coincidences are as follows:

$$C_{13} = |\tau|^4(|\alpha|^2 + |\gamma|^4|\beta|^2)\,, \quad C_{14} = |\gamma|^2|\tau|^4(|\alpha|^2 + |\beta|^2)\,,$$

where τ is the transmission coefficient of the half-mirror and γ is the ratio of the reflection coefficient ρ to transmission coefficient, $\rho = \gamma\tau$. Thus different sets of counts are expected depending on collapse/non-collapse postulates. In Bussey's paper [1] $\gamma = e^{i\pi/2}$ is assumed where a maximum contrast between non-collapse/collapse cases is obtained.

It is difficult to find particles for this purpose. Photons are not usable since they do not scatter each other in an ordinary situation. Charged particles seem to be suitable since they scatter each other by the Coulomb force. Low energy electrons from field emission sources seem to be adequate, since they have good temporal and spatial coherence, besides, they have relatively long wave lengths. On the other hand it is difficult to find a good half mirror for charged particles.

In our previous paper [2] we proposed an experiment to test a strange effect of disappearance of interference due to entanglement. In that article we examined the possibility of keeping coherence in the collision of field emission electron beams, whose geometry is similar to the arrangement in [1]. We will claim that two ways of extensions of that proposition could be the realization of Bussey's thought experiment.

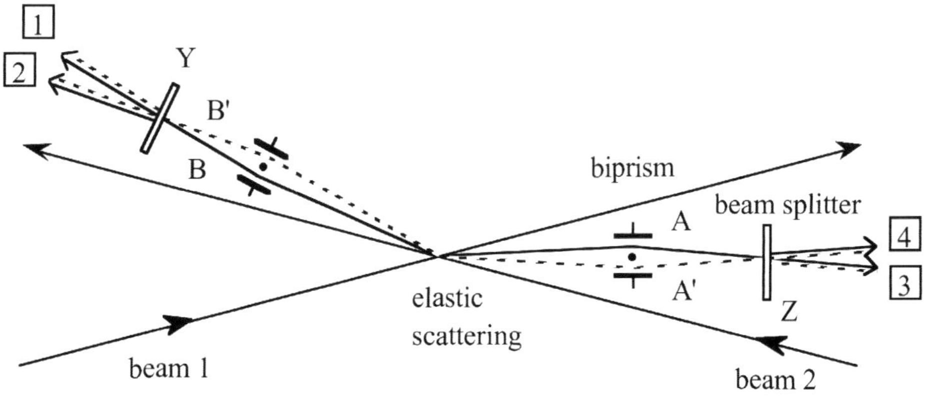

Figure 2. Test of wave function collapse by crystal beam splitters.

2. Proposed setup to test the wave function collapse

2.1. Use of thin crystal films as beam splitters

We propose a realistic version of Bussey's thought experiment by field emission electrons. As shown in Fig. 2, thin electron beams of the same energy (our assumption is 2.5 keV) are collided at a nearly head-on geometry. Elastically scattered electron beams which are nearly at 180 degree w.r.t. each other are introduced into electron biprisms on both sides. Instead of the unscattered and the scattered particles of [1], we use a combination of particles scattered at different angles and going through different sides of the central wires of the biprisms.

The two biprisms provide a combination of different paths of the "double interferometer" of Fig. 1. For half-reflecting mirrors (or beam splitters) we will put thin crystal films which partly reflect the electron beam by Bragg scattering. The two planes, in which the combinations of different paths are located, are better taken to be perpendicular to the Fig.'s plane in order to minimize the energy difference due to kinematics.

Table 1. Parameters for the proposed setup.

Item	Description or value
electron source	field emission type
beam energy	2.5 keV
beam crossing angle	7 degree
beam width at the colliding region	5 nm
size of the biprism (d in Fig. 3)	0.1 mm
biprism to beam splitter distance (L)	1.28 mm

Fabrication of a self-supporting Si crystal of 10 nm thickness is reported [3]. For Si crystal the mean free path for 2.5 keV electron is 3.2 nm, which means the attenuation to 4.4% after this thickness. Diffraction by (h, k, l)=(1, 1, 1) planes gives the smallest Bragg diffraction angle of 4.49 degrees. Phase shift of the diffracted wave to the transmitted one is a critical parameter, as shown in Section 1. We cannot find this data for this very material and condition. But according to the table of crystallography (Table 2.5A of [4]) the phase shift is 49.9 degree for 5 degree scattering angle of 10 keV electron from Si atom (compared to 8.9 degree to forward direction), with a tendency to increase when energy is decreased.

Parameters of the proposed setup are summarized in Table 1. Except for beam splitters (BS), these are the same as those in our previous article [2] to study the disappearance of interference due to entanglement. In [2] we have confirmed that the coherence of the scattered electron wave is not lost at the position where the BS is located by this set of parameters.

The diffraction angle and the size of the biprism (d in Fig. 3) define the biprism-to-BS distance (L in Fig. 3) to coincide the diffracted beam with the straight (transmission) beam from the other path of the biprism.

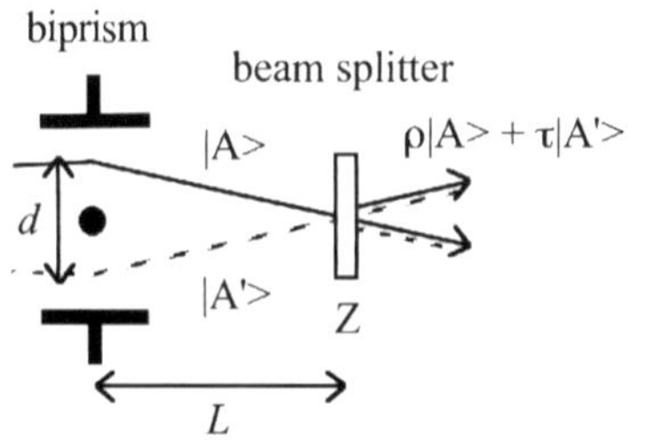

Figure 3. Geometrical parameters of the biprism.

2.2. *Observation of interference effects in coincidence counting*

Another method we propose to test the wave function collapse is to observe interference fringes in coincidence counting. This is rather a simple extension of the setup of [1].

As we have shown in [1], we cannot observe any interference fringe as long as we observe only one part of the entangled pair. Namely it looks like the wave function collapse, but it does not necessarily mean that, as described in the paper. But if we observe both parts of the entangled pair, namely we take a correlation measurement on positions of two electrons, we will get an interference fringe if the wave function does not collapse. Indeed, we have already proposed this measurement as a calibration of the experimental setup to prove that it is able to observe interference fringes in single counting.

If the wave function collapses at the collision event, no such fringe is seen in the coincidence experiment either. So this "calibration" is used as a test of collapse. If we find fringes there, we see that the wave function does not collapse. (If we find no fringe, to prove the collapse of the wave function we must find another method of calibration to confirm that this is not due to the poor precision of the apparatus.)

Fig. 4 shows the experimental arrangement. In place of beam splitters in Fig. 2 we put position sensitive detectors and measure a correlation of two electrons. (Like in the experiment with beam splitters, deflecting directions of biprisms should be taken to be perpendicular to the Fig.'s plane.) Practically the fringe at the crossing point immediately after the biprism is too narrow, so that it should be magnified by electron lenses as is employed by Tonomura [5].

If the wave function does not collapse, it is described as an entangled form. We rewrite (1) as a function of the spatial coordinates:

$$\langle \boldsymbol{r}_1, \boldsymbol{r}_2 | \Psi \rangle = \alpha \phi_{\mathrm{A}}(\boldsymbol{r}_1) \phi_{\mathrm{B}}(\boldsymbol{r}_2) + \beta \phi_{\mathrm{A}'}(\boldsymbol{r}_1) \phi_{\mathrm{B}'}(\boldsymbol{r}_2) \ .$$

Since the scattering amplitude is overwhelming at forward angle where we observe, we neglected the antisymmetrization. Let us describe the single particle states by plane waves, e.g. $\phi_{\mathrm{A}}(\boldsymbol{r}_1) =$

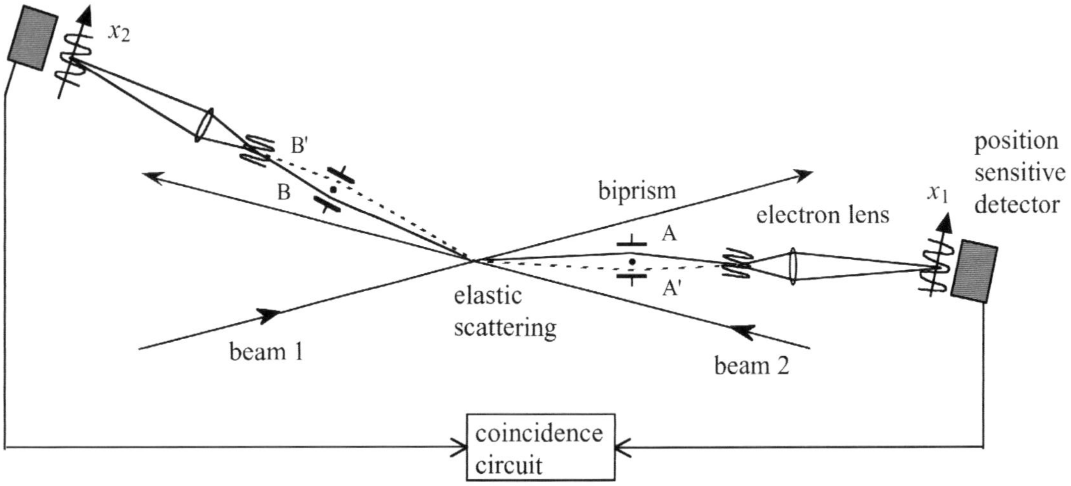

Figure 4. Test of wave function collapse by position correlation measurement.

$\exp(i\boldsymbol{k}_{\rm A}\boldsymbol{r}_1)$, etc. After the biprism each electron wave undergoes an inward deflection to get an additional momentum $\pm\boldsymbol{q}_{\rm A}/2$, where $\boldsymbol{q}_{\rm A} = g_{\rm A}(\boldsymbol{k}_{\rm A} - \boldsymbol{k}_{\rm A'})$. Here we defined $g_{\rm A}$ as a factor to express the strength of the biprism, which should be greater than 1 so that the biprism converges the incident beam. Similarly the electron in the state $\phi_{\rm B}(\boldsymbol{r}_2)$ turns to have a momentum $\boldsymbol{k}_{\rm B} - \boldsymbol{q}_{\rm B}/2$, and $\phi_{\rm B'}(\boldsymbol{r}_2)$ to have $\boldsymbol{k}_{\rm B'} + \boldsymbol{q}_{\rm B}/2$. Here $\boldsymbol{q}_{\rm B} = g_{\rm B}(\boldsymbol{k}_{\rm B} - \boldsymbol{k}_{\rm B'})$ where $g_{\rm B}$ is the same parameter for another biprism. By momentum conservation $\boldsymbol{k}_{\rm B} - \boldsymbol{k}_{\rm B'} = \boldsymbol{k}_{\rm A'} - \boldsymbol{k}_{\rm A} = \Delta\boldsymbol{k}$. Then the coincidence rate between the detectors after the biprisms is proportional to

$$|\langle \boldsymbol{r}_1, \boldsymbol{r}_2|\Psi\rangle|^2 = |\alpha|^2 + |\beta|^2 + 2|\alpha||\beta| \cos[\Delta\boldsymbol{k}(\boldsymbol{r}_1 - \boldsymbol{r}_2)(1-g) + \delta] \ .$$

Here we put $g_{\rm A} = g_{\rm B} = g > 1$ and δ comes from the phases of α and β. Then if we perform a coincidence experiment to take the correlation of positions of electrons along the direction of $\Delta\boldsymbol{k}$, namely x_1 and x_2 in Fig. 3, interference fringes at 45 degree with spacings $\sqrt{2}\pi/[|\Delta\boldsymbol{k}|(1-g)]$ should appear in the $x_1 - x_2$ plane. If the wave function collapses as in Eq. (2), no such fringe is observed.

Both methods are technically very challenging but we believe they are worth putting into practice.

References

[1] Bussey P J 1984 *Phys. Lett.* A **106** 407
[2] Toyoshima K, Endo T and Hirayoshi Y 2006 *Czech. J. Phys.* **56** 1361
[3] Utteridge S J, Sashina V A, Canney S A, Ford M J, Fang Z, Olivera D R, Vosb M and Weigold E 2000 *Applied Surface Science* **359** 162-163
[4] Cromer D T and Waber J T 1974 *International Tables for X-ray Crystalography IV* (Birmingham: Kynoch Press)
[5] Tonomura A 1987 *Rev. Mod. Phys.* **59** 639

Journal of Physics: Conference Series **174** (2009) 012029 doi:10.1088/1742-6596/174/1/012029

Virtual states and decoherence in a single impurity atom solid state quantum device

Enrico Prati

Laboratorio Nazionale Materiali e Dispositivi per la Microelettronica, CNR-INFM, Via Olivetti 2, I-20041 Agrate Brianza, Italy

E-mail: enrico.prati@mdm.infm.it

Abstract. When a single donor atom is embedded in solid state nanodevice, a rich variety of phenomena in the proximity of $T = 0$ emerges due to the long phase coherence times. Such phenomena are probed by means of charge sensing via tunneling currents. Some aspects connected to the transition from quantum to classical in quantum nanodevices are overviewed: the realization of transport through virtual states, and the definition of the temperature T of a microscopic/mesoscopic system as a key parameter required in the master equation when decoherence is considered. I also discuss the opportunity to study the transition from quantum to classical offered by single atom impurity devices on a time scale below the decoherence time, with a particular emphasis on the experimental test of the Leggett-Garg inequalities.

1. Introduction

Recent advances in solid state nanoelectronics have led to quantum devices where a single impurity atom is confined in a semiconductor nanostructure lattice [1, 2]. Due to their decananometric size, in the proximity of $T = 0$ such systems behave like a microscopic quantum laboratory where the confined magnetic donor atom provides discrete energy levels in the bandgap of the hosting semiconductor [3]. Consequently, it can be charged with one or two electrons which carry a one-half spin each. The number of electrons bound to the donor is controlled from 0 to 2 with a gate placed at a distance of few nanometers from the atom. The atom may interact with the leads, consisting of two electron reservoirs, each filled up to its Fermi energy, and with acoustic phonons. Even if single atom impurity quantum device are conceived for future quantum information processing applications[4, 5], they also represent an unique opportunity to study the interaction of quantum objects like the impurity atom and the bound electrons with their environment. The environment is constituted by the electrons in the reservoirs, the nuclei of the lattice (via spin-orbit) and the phonons. Here I discuss three relevant aspects of the transition from quantum to classical in such microscopic systems, connected to the investigation the decohering degrees of freedom. In particular I will refer to both recent and possible experiments. First, the spin-flip (relaxation) time T_1 is important for applications of spintronics involving classical information and it is encoded in the spin states $|\uparrow\rangle$ and $|\downarrow\rangle$, however, for quantum computing tasks and in order to study decoherence, the relevant timescale is the spin decoherence time T_2. T_2 is the lifetime for a coherent superposition of $|\uparrow\rangle$ and $|\downarrow\rangle$, generally much less than the relaxation time. I discuss the role of ergodicity and orthodicity to define both T_2 and the temperature of a single electron system in thermal and particle exchange with a mesoscopic environment made of few electrons. Indeed, the temperature T of a microscopic/mesoscopic system is a key parameter required in the master equation when decoherence is considered and it requires a generalization to small N (few units) systems. Second, like in other artificial quantum dots [6], forbidden transitions occurring at a time scale below the Heisemberg indetermination time are realized [7]. The virtual events are made possible by the energy levels provided by the single atom confined in the quantum well between the electron reservoirs. Such purely quantum events are detected as a sub-nanoampere classical current by means of Coulomb blockade stability diagrams. Finally, macrorealism is considered from the point of view of the single donor system. According to Leggett and Garg (1985) [8], macrorealism is defined by the simultaneous assumptions that: (A1) A macroscopic object, which has available to it two or more macroscopically distinct states, is at any given time in a definite one of those states. (A2) It is possible in principle to determine which of these states the system is in without any effect on the state itself, or on the subsequent system dynamics. Here I speculate that the Leggett-Garg inequalities, which quantify the compatibility of macrorealism with quantum mechanics, in analogy to Bell inequalities, can be experimentally fulfilled or violated by considering time correlations of observables like the number of electrons in the single donor quantum dot. I conclude with new perspectives opened by the control of single atoms in quantum dots in order to further proceed in the investigation of decoherence and the transition from quantum to classical at mesoscopic scale.

2. Decoherence time and temperature for small N ergodic electron systems

Both transverse relaxation times and thermodynamical quantities are ensamble related concepts. They are experimentally and theoretically studied as averages on a given ensamble. In this section I stress the role of ergodicity and orthodicity in order to define such quantities in a few particles system. One of the major ideas introduced the field of quantum mechanics after the '30 and after the Bell inequalities is decoherence [9]. The decoherence affects the evolution of an hyphotetical isolated system because of the coupling with the degrees of freedom of the environment, with a temperature dependent extra term of the master equation for the density matrix ρ. The evolution of the density matrix of a mixture is described by a master equation which takes into account the Von Neumann term which corresponds to $\dot{p} = \nabla V$, the relaxation term $\dot{p} = -\gamma p$ and the decoherence term which depends on the temperature T of the thermal bath. Such terms are explicitly written as

$$\dot{\rho} = -\frac{1}{\hbar}\left[H, \rho\right] - \gamma\left(x - x'\right)\left(\frac{\partial}{\partial x} - \frac{\partial}{\partial x'}\right)\rho - \frac{2m\gamma k_B T}{\hbar^2}\left(x - x'\right)\rho \tag{1}$$

The definition of temperature in statistical physics refers to the inverse of the partial derivative of the energy with respect to the entropy, at a fixed volume [10]. Such definition is a consequence of the observation that for two isolated systems in mutual thermal equilibrium such quantity is the same. Temperature, as the other thermodynamical quantities, is defined for a macroscopic system. Nanoscale systems like single impurity atoms in quantum devices [2, 11], defects in a crystal [12, 13], and quantum dots [14] may contain few electrons in exchange with a thermal bath consisting in a two or three dimensional electron system [15]. Here we consider a system constituted by a single electron bound to the donor which exchanges another electron from a refrigerated low dimensional electron system confined in the leads. Because of the weak eletron-phonon coupling at low temperature, such system is suitable to experimentally realize a mesoscopic condition with few electrons involved. On the other hand, relaxation time T_2 is generally defined on the average on a space ensamble, as the relaxation time of the transverse magnetization M_{xy} in the rotating frame when the $\mathbf{B}$ field is along z, with the following (Bloch) equation:

$$\frac{\partial M_{xy}}{\partial t} = -\frac{M_{xy}}{T_2} \tag{2}$$

Since the system under investigation contains an individual spin, the relaxation time definition requires a change of perspective. Such requirement is also needed to provide consistency to thermodynamics of individual/small number of particles. The fundamental hypothesis in the following is that the system is ergodic, in other words that averages on a space ensamble provide the same results of averages on the equivalent time ensamble, which is constituted by measurements on the same individual particle. Even is ergodicity is a very reasonable assumption of a space and time ensamble of electron spins in a Silicon single donor quantum dot, at the present time there is no experimental confirmation, based for example on a echo decay T_2 measurement. The fact that the thermal bath is made of few particles is quantified by considering a finite and small N, which is an uncommon condition for thermodynamics, generally derived in the limit of $N \to \infty$. The system here considered is described in the most general case by means of the grand partition function. It holds the further hypothesis that the quantum dot changes its occupation from 1 to 2 electrons only and viceversa. Since the grandcanonical ensamble is constituted of canonical ensambles, and canonical ensamble is natively orthodic, the system under investigation is always well defined in the limit of finite N.

Under the conditions that the system is ergodic and orthodic, one separates its Hamiltonian $H(N)$ of N electrons distributed between a single/double occupied island ($N_1 = 1 \leftrightarrow 2$) and the 2-dimensional system of N_2 electrons, by $H = H_1(N_1) + H_2(N_2)$. The quantum grand partition function for identical particles is:

$$Z(\Sigma, N, T) = \sum_{N_1=1}^{2} tr\left(e^{-\beta H_1}\right) tr\left(e^{-\beta H_2}\right) \tag{3}$$

where Σ is the total surface occupied by the confined electron(s) Σ_1 and the leads Σ_2, and $\beta = (k_B T)^{-1}$. One finds the Helmholtz potential $\Psi = -\beta^{-1} logZ$ for N_2 fermions of mass m, confined in a 2-dimensional system of area S, which is

$$\Psi(N_2, S) = \frac{N_2}{2}\mu(N_2, S) + \frac{1}{9}\frac{S}{\hbar^2} m k_B^2 T^2 \tag{4}$$

from which all the physics involved can be extracted, including the temperature T [16]. Temperature can be experimentally determined by means of the ratio between the occupation of the states corresponding to 1 and 2 electrons in the dot. Indeed the ratio between the occupations corresponds to the ratio between the average time spent in the states 1 and 2 respectively, or in other words, to the inverse of the probability of the tunneling on/off the island.

3. Virtual states in cotunneling processes

Beyond the transport in the sequential regime, which consists of transport to lowest order in the coupling to the reservoirs, higher-order tunneling processes become relevant as the resistance of the tunneling barriers reach the resistance $R_t = h/e^2 = 25.813k\Omega$. In the latter case, quantum fluctuations dominate transport because electrons tunnel via intermediate virtual states where first order tunneling would be suppressed. Thus, the intrinsic width of the energy levels of the quantum dot may include contributions also from tunneling via virtual states. Such higher-order tunneling events are called cotunneling processes. Here I compare the typical time scale involved in such processes and the decoherence time scale. Experimental results on cotunneling in semiconductor quantum dots have been reported [7, 17]. It is possible to observe the effect of the virtual state in a Silicon Nano-Field Effect Transistor (NanoFET) of the kind already

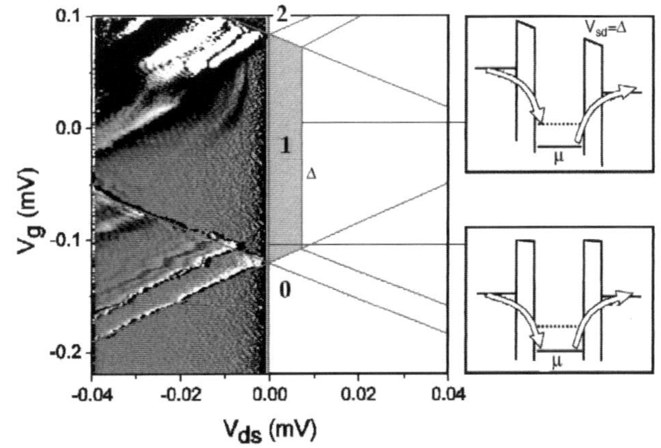

Figure 1.
Left: stability diagram of a single donor switching from 0 to 2 electrons as a function of the gate voltage (V_G). An inelastic cotunneling feature appears in the region 1 between the two peaks. Right: theoretical description of the stability diagram and schemes of the processes originating the $V_{DS} = \Delta$ (inelastic) and $V_{DS} = 0$ (elastic) tunneling.

described in Ref. [13]. Figure 1 shows the cotunneling in a single donor Coulomb blockade diamond in a NanoFET sample measured at 300 mK, compared to the theoretical prediction. On the left side, the experimental stability diagram of the donor quantum dot in shown. It represents the first derivative of the conductance G of the NanoFET below the threshold voltage where the peaks attributed to the single donors are visible. The numbers 0, 1 and 2 label the regions where sequential tunneling is forbidden due to the Coulomb blockade. Cotunneling appears in the forbidden region 1. On the right side, it is shown the scheme of the stability diagram and the processes involved in the elastic (up) and inelastic (down) cotunneling. Each V-shaped diagram corresponds to the ground state (GS) of the donor with 0 or 1 electrons, while the extra lines at Δ mV are originated by the first excited state (ES) energy level. The inelastic cotunneling occurs in correspondence of the ES level in a region where it does not participate to tunneling at first order.

In the case of such a double tunnel barrier, a second order process is possible for bias voltages below the Coulomb blockade, because the change in free energy for a process with tunneling in both junctions is negative even though the energy differences for independent tunneling at either junction are positive. It is possible to calculate the characteristic times for both the tunneling process τ_T (from the linewidth of the current peak) and for the decoherence since:

$$\tau_D = \tau_R \left(\frac{\hbar}{\Delta x \sqrt{2mk_BT}} \right)^2 \tag{5}$$

where τ_R is the relaxation time. For a Si NanoFET of 40 nm at 250 mK, $\tau_D = 100$ ns while $\tau_T = 1$ ps. It is interesting to notice that cotunneling happens at a time scale which must be within the limit τ_T imposed by the Heisemberg principle, but decoherence at the low temperature considered has a much longer characteristic time. The experiment of Figure 1 shows that the measurement of inelastic cotunneling is possible in terms of classical current of electrons. Decoherence must therefore have a time shorter than what predicted by the previous equation. Transport through virtual states imposes an experimental upper bound to the time scale for decoherence to have occurred. Indeed, according to decoherence, the classical output of the measurement is realized from the virtual quantum process involved on the decoherence timescale.

4. Probing the quantum to classical transition in donor quantum dots

The decoherence of quantum states has been considered and developed to introduce a time scale for which the fluctuations of the environment dominate on the phase coherence of a quantum state. Furthermore, how quantum mechanics does govern the macroscopic systems is an open field of investigation. Another open issue is indeed provided by macrorealism, which is based on the assumption that a macroscopic system is always in a definite state. The Leggett-Garg inequalities consist of an ultimate constraint, like the Bell inequalities, to discriminate whether macrorealistic representation of world is correct or not [8]. In other words, Leggett-Garg inequalities discriminate whether Nature consists of macroscopic objects with objective properties prior to and independent of measurements, or not . Such inequalities are based on the temporal correlations between two-valued quantities $Q(t)$ which can take the values of $+1$ or -1 and there is no experimental knowledge about their violation at the present time [18]. As discussed, nanotechnology is now capable to provide semiconductor samples small enough to isolate electrons and/or atoms in quantum dot. The energy discrete spectrum of confined electrons, the possible spin of nuclei, and the chance of driving the entanglement between the subsystems by means of suitable gates make semiconductor quantum dots (and possibly impurity atoms) the natural environment where the transition from quantum to classical physics can be resolved in the time domain. This is particularly true in Silicon quantum dots, for which coherence time is much longer than in GaAs quantum dots (in Silicon, while $T_2 = 20$ ns due to reversible frozen hyperfine field in free-induction decay, one has $T_2^* = 0.3$ ms [19] with Hahn echo experiments, which is a 4 order of magnitude longer time).

According to Leggett and Garg [8, 18], the statements A1 and A2 are patently in conflict with quantum mechanics. There are two observables which are of particular interest to realize such test. The first is the electron spin component along one axis, let's say z. For that, a spin to charge conversion is required. The second is the number of electrons in the impurity dot. A time resolved correlation in contrast with the Leggett-Garg would definitively prove the quantum nature of macroscopic objects and the lack of a breakdown of quantum mechanics at some time/space scale.

Isolated electrons in quantum dots possesses a dichotomic variable $Q(t)$ which can take only one of two discrete values which we label by convention $+1$ and -1, respectively. The electron occupation of the dot with 1 or 2 electrons in a measurement of $Q(t)$ which does always produce the value 1. For such system, if it is ideally isolated from its environment (which includes internal dissipative mechanisms), quantum mechanics predicts, independently of the density matrix of the system (or more accurately of the time ensemble formed by a sequence of experiments starting from the same initial conditions) 2-time correlations of the form $\langle Q(t)Q(t')\rangle = cos\omega(t-t')$, where ω is a tunneling frequency between the two states $Q = 1$. The measurements have to be performed under the condition of noninvasive measurability in order to grant the system not to be perturbed after the first measurement (not required for the second). Such issue can be solved by means of *ideal negative results* procedure: the measuring apparatus is arranged so that if $Q(t)$ is $+1$ it is triggered, while if $Q(t)$ is -1 nothing happens. Such situation can be realized by means of quantum point contacts or multiple quantum dots, where the 10-100 fA scale current fluctuates between bi-stable states according to the occupation of the site under investigation. To do that a fast time resolution is required and it can be achieved by ultrafast digital boards capable to reach the ns scale, which is required to grant coherence. In such case the observable is the number of electrons confined in the dot, whose number fluctuates according to the intrinsic probability amplitude (tunneling frequency) imposed by the geometry of the device. Suitably arranged (donor) quantum dots will provide in the future a ground to experimentally test the Leggett-Garg inequalities and to improve the knowledge about the transition from quantum world to the macroscopic scale.

5. Conclusions

To conclude, I examinated three aspects of the transition from quantum to classical from the point of view of single donor quantum dots. First, I pointed out the role of ergodicity and orthodicity for both small space and time ensambles in order the define the temperature which is required to parametrize the master equation of a system subject to decoherence. I reported the Helmholtz energy function of a two dimensional system constituted by a small N number of electrons. Second, it has been shown the inelastic cotunneling in a single donor quantum dot and the discrepancy between the tunneling time scale via the virtual states, and the decoherence time calculated as usual. Finally the use of single donor quantum dots has been proposed for the experimental verification of the Leggett-Garg inequalities, capable to eventually rule out macrorealism and to fix a breakdown of quantum mechanics. Single impurity quantum dots offer opportunities for the study of the transition from quantum to classical, complementary to the applications in nanoelectronics and the study of quantum device functionalities.

References

[1] Lansbergen G *et al* 2008 *Nature Physics* doi:10.1038/nphys994
[2] Prati E, Latempa R and Fanciulli M 2008 *Preprint* arXiv:0807.5026
[3] Taniguchi M *et al* 1976 *Sol. State Comm.* **20** 131
[4] Kane B E 1998 *Nature* **393** 133
[5] Cerletti V *et al* 2005 *Nanotechnology* **16** R27-R49
[6] Koppens F H L *et al* 2006 *Nature* **442** 766
[7] De Franceschi S, Sasaki S, Elzerman J M, van der Wiel W G, Tarucha S and Kouwenhoven L P 2001 *Phys. Rev. Lett.* **86** 878
[8] Leggett A J and Garg A 1985 *Phys. Rev. Lett.* **54** 857
[9] Zurek H 1991 *Physics Today* **44** 36-44
[10] Landau L and Lifshitz L 1980 *Statistical Physics – Course of Theoretical Physics Vol. 5* 3rd ed. (Oxford: Pergamon)
[11] Sellier H, Lansbergen G P, Caro J, Rogge S, Collaert N, Ferain I, Jurczak M and Biesemans S 2006 *Phys. Rev. Lett.* **97** 206805 (2006)
[12] Prati E *et al* 2007 *Phys. Lett.* A **370** 491493
[13] Prati E *et al* 2008 *J. Appl. Phys.* **103** 104503
[14] Sanquer M, Specht M, Ghenim L, Deleonibus S and Guegan G 2000 *Phys. Rev.* B **61** 7249
[15] Giazotto F *et al* 2006 *Rev. Mod. Phys.* **78** 217
[16] Prati E, in preparation
[17] Glattli D C, Pasquier C, Meirav U, Williams F I B, Jin Y and Etienne B 1991 *Z. Phys.* B **85** 375
[18] Leggett A J 2008 *Rep. Prog. Phys.* **71** 022001
[19] Chiba M and Hirai A 1972 *J. Phys. Soc. Japan* **33** 730

Fourth International Workshop DICE2008 IOP Publishing
Journal of Physics: Conference Series **174** (2009) 012030 doi:10.1088/1742-6596/174/1/012030

Purity sieve for models with factorizable interactions

Oleg Lychkovskiy[1,2]

[1] Institute for Theoretical and Experimental Physics, 117218, B. Cheremushkinskaya 25, Moscow, Russia

[2] Moscow Institute of Physics and Technology, 141700, Institutskii per. 9, Dolgoprudny, Moscow Region, Russia

E-mail: `lychkovskiy@itep.ru`

Abstract. Evolution of purity in case of factorizable interaction between an open system and an environment is investigated. We derive a simple expression for purity decrease at the first instants of evolution (when purity is close to unity), which appears to be quadratic in time. According to this expression, purity at very small times is maximal when the initial state of an open system coincide with one of the eigenstates of the interaction operator, no matter how strong the interaction is. However, it is widely known that eigenstates of the interaction Hamiltonian are not pointer states when, for example, interaction is weak (compared to self-Hamiltonian). Therefore the procedure of selecting pointer states by purity maximization (known as "purity sieve") should not rely on short-time purity behavior. We propose a modification of the purity sieve criterion which takes into account purity evolution at longer times. As an example of its applicability we recover known results for pointer states of a particle undergoing quantum Brownian motion; we point out that the criterion is not applicable for some other models, however. It is argued that the proposed modified purity sieve may be used for selecting pointer states of a particle undergoing decoherence through collisions.

1. Introduction

Purity sieve [1],[2] is one of the criteria for selecting pointer states of an open quantum system, i.e. the states most robust against decoherence. In this work purity sieve is investigated in the case of factorizable interaction between an open system and an environment. This case is relevant for the majority of models usually considered in the context of decoherence, for example for quantum Brownian motion model, spin-boson model and some spin-spin models. The paper is organized as follows. In the next Section we briefly review the conventional setting and the basic concepts of the decoherence program, outline the role of purity and remind the definition of the purity sieve. In Section 2 we analyze the short-time behavior of purity. Also a modified (with respect to canonical purity sieve) criterion for selecting pointer states is proposed in this Section. In Section 3 we introduce a model which has the following features: (1) it may be reduced to the "almost factorizable" form; (2) under certain assumptions decoherence in this model may be regarded as decoherence through collisions; (3) the particular case of this model is quantum Brownian motion. We use the proposed modified criterion to find pointer states for this model.

2. Some basics of decoherence

In this Section we give a very brief and incomplete sketch of the basic setting of the decoherence program. One may find a number of extensive reviews (e.g. [3]) and pedagogical introductions to the topic (e.g. [4]). In particular, a list of references to the original works is given in Ref. [3].

Consider a System (with Hilbert space $\mathcal{S}$) which interacts with an Environment (with Hilbert space $\mathcal{E}$). The total Hamiltonian of the combined closed system (with Hilbert space $\mathcal{H} = \mathcal{S} \otimes \mathcal{E}$) reads

$$\hat{H} = \hat{H}^{\mathcal{S}} + \hat{H}^{\mathcal{E}} + \hat{V} , \tag{1}$$

$\hat{H}^{\mathcal{S}}$, $\hat{H}^{\mathcal{E}}$ and $\hat{V}$ being self-Hamiltonians and interaction Hamiltonian respectively. The usage of superscripts $\mathcal{S}$ and $\mathcal{E}$ here and in what follows is self-explanatory.

The state of the combined system in $\mathcal{H}$ may be described by a state vector Ψ or by a density matrix $\hat{\rho}$. The states of the System $\mathcal{S}$ and the Environment $\mathcal{E}$ are described by the reduced density matrices

$$\hat{\rho}^{\mathcal{S}} \equiv \mathrm{tr}_{\mathcal{E}}\, \hat{\rho} , \quad \hat{\rho}^{\mathcal{E}} \equiv \mathrm{tr}_{\mathcal{S}}\, \hat{\rho} . \tag{2}$$

Density matrices evolve according to the master equations:

$$\frac{d}{dt}\hat{\rho} = -\mathrm{i}[\hat{H}, \hat{\rho}] , \tag{3}$$

$$\frac{d}{dt}\hat{\rho}^{\mathcal{S}} = -\mathrm{i}[\hat{H}^{\mathcal{S}}, \hat{\rho}^{\mathcal{S}}] - \mathrm{i}\ \mathrm{tr}_{\mathcal{E}}[\hat{V}, \hat{\rho}] . \tag{4}$$

If $\hat{\rho}$ is pure then, according to the Schmidt theorem, at any time $\hat{\rho}^{\mathcal{S}}$ and $\hat{\rho}^{\mathcal{E}}$ may be diagonalized with the same eigenvalues:

$$\hat{\rho}^{\mathcal{S}}(t) = \sum_i p_i(t) \hat{P}_i^{\mathcal{S}}(t) , \tag{5}$$

$$\hat{\rho}^{\mathcal{E}}(t) = \sum_i p_i(t) \hat{P}_i^{\mathcal{E}}(t) . \tag{6}$$

Here $\{\hat{P}_i^{\mathcal{S}}(t)\}$ and $\{\hat{P}_i^{\mathcal{E}}(t)\}$ are systems of mutually orthogonal projectors, and $\{p_i(t)\}$ is a set of non-negative numbers (probabilities).

An essential ingredient of the decoherence program is the following *Decoherence Hypothesis:* for a generic System, a sufficiently large Environment,for generic interaction and (almost) any initial conditions, the projectors $\{\hat{P}_i^{\mathcal{S}}(t)\}$ almost always belong to some set of fixed projectors, which correspond to the so-called *pointer states.*

In other words, the reduced density matrix is diagonal almost always in the basis of pointer states. Exceptional periods of time when the diagonalization basis substantially deviates from the pointer basis are short and rare. If the initial state of the System is far from pointer, the first moments of the evolution form such an exception: the System *decoheres* into the mixture of the pointer states during this period.

One of the main problems of the decoherence theory is to find pointer states, given a Hamiltonian $\hat{H}$ of the combined system $\mathcal{H}$. One of the helpful tools for this is *purity sieve* [1],[2]. In order to introduce purity let us first note that there is a simple method to obtain the probability $p_{\max}$, which is by definition the largest one among all p_i, and the corresponding projector $\hat{P}^{\mathcal{S}}_{\max}$, given the reduced density matrix $\hat{\rho}^{\mathcal{S}}$:

$$\hat{P}^{\mathcal{S}}_{\max} = \lim_{k\to\infty} \left[(\hat{\rho}^{\mathcal{S}})^k / \mathrm{tr}_{\mathcal{S}} (\hat{\rho}^{\mathcal{S}})^k \right] , \tag{7}$$

$$p_{\max} = \mathrm{tr}_{\mathcal{S}} \hat{P}^{\mathcal{S}}_{\max} \hat{\rho}^{\mathcal{S}} = \lim_{k\to\infty} \left[\mathrm{tr}_{\mathcal{S}} (\hat{\rho}^{\mathcal{S}})^{k+1} / \mathrm{tr}_{\mathcal{S}} (\hat{\rho}^{\mathcal{S}})^{k} \right] . \tag{8}$$

In practice, approximate expressions for $\hat{P}^{\mathcal{S}}_{\max}$ and $p_{\max}$ may be used:

$$\hat{P}^{\mathcal{S}}_{\max} \simeq (\hat{\rho}^{\mathcal{S}})^{k} / \mathrm{tr}_{\mathcal{S}} (\hat{\rho}^{\mathcal{S}})^{k} , \tag{9}$$

$$p_{\max} \simeq \mathrm{tr}_{\mathcal{S}} (\hat{\rho}^{\mathcal{S}})^{k+1} / \mathrm{tr}_{\mathcal{S}} (\hat{\rho}^{\mathcal{S}})^{k} . \tag{10}$$

Here the larger is k, the better is the approximation.

Assume now that at $t = 0$ the state of the System is pure, $\hat{\rho}^{\mathcal{S}}(0) = |\psi^{\mathcal{S}}\rangle\langle\psi^{\mathcal{S}}|$. Then $p_{\max}$ initially equals 1, and stays close to 1 during some time. In this case even $k = 1$ appears to be a good approximation, and one gets

$$p_{\max} \simeq \mathrm{tr}_{\mathcal{S}} (\hat{\rho}^{\mathcal{S}})^2. \tag{11}$$

The quantity on the r.h.s. of the above equation is called *purity*. We denote it by $\mathfrak{P}$:

$$\mathfrak{P} \equiv \mathrm{tr}_{\mathcal{S}} (\hat{\rho}^{\mathcal{S}})^2. \tag{12}$$

Sometimes $-\mathfrak{P}$ is called "linear entropy".

For the pure state $\hat{\rho}^{\mathcal{S}} = |\psi^{\mathcal{S}}\rangle\langle\psi^{\mathcal{S}}|$ purity equals 1, and it decreases in the course of evolution. The rate of the decrease differs for different initial states. It is natural to assume (in the spirit of the *Decoherence Hypothesis*) that for pointer states this rate should be small. The small rate of purity loss implies that the corresponding state is stable against decoherence. This reasoning suggests that in order to find pointer states one should minimize the purity decrease rate with respect to the initial pure states. This procedure is called *purity sieve*.

To summarize, purity sieve is a procedure for selecting pointer states of the System. It comes to the following two steps:

1. calculating purity $\mathfrak{P}$ as a functional of the initial state of the System $\psi^{\mathcal{S}}$, a functional of the initial state of the Environment $\hat{\rho}^{\mathcal{E}}$, and a function of time t, assuming the initial product state $\hat{\rho} = |\psi^{\mathcal{S}}\rangle\langle\psi^{\mathcal{S}}| \otimes \hat{\rho}^{\mathcal{E}}$;
2. maximizing $\mathfrak{P}$ with respect to $\psi^{\mathcal{S}}$ for some fixed (reasonable) t and $\hat{\rho}^{\mathcal{E}}$.

The states which provide the local maxima should be the pointer states. The procedure makes sense if the result weakly depends on $\hat{\rho}^{\mathcal{E}}$ and t, provided t is greater than some characteristic time, called *decoherence time*.

3. Purity sieve for factorizable interactions

Numerous models, investigated in the context of decoherence studies, have factorizable interaction Hamiltonian (see e.g. [2], [5]-[8]):

$$\hat{V} = \hat{S}\hat{E} . \tag{13}$$

Such form of interaction allows to simplify the master equation (4):

$$\frac{d}{dt}\hat{\rho}^{\mathcal{S}} = -\mathrm{i}[\hat{H}^{\mathcal{S}}, \hat{\rho}^{\mathcal{S}}] - \mathrm{i}\,[\hat{S}\,\mathrm{tr}_{\mathcal{E}}[\hat{E}\hat{\rho}] - \mathrm{h.c.}] . \tag{14}$$

Using this expression, we calculate the first two derivatives of purity at $t = 0$:

$$\dot{\mathfrak{P}} = 0 , \tag{15}$$

$$\ddot{\mathfrak{P}} = -4\delta\hat{E}^2\delta\hat{S}^2. \tag{16}$$

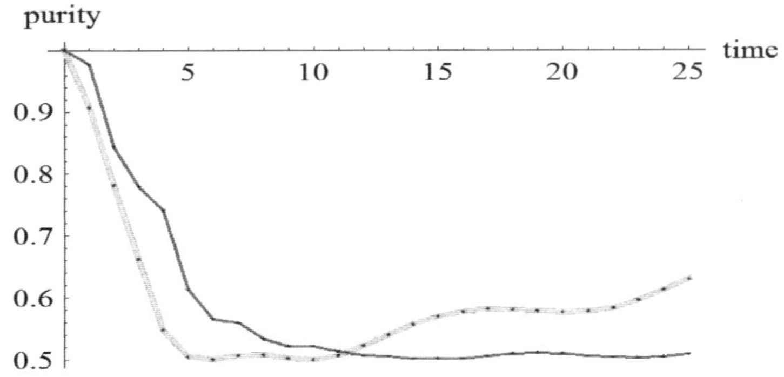

Figure 1. Purity as a function of time for the central spin model (20) and for two different initial states, $|0z\rangle$ (thick green line) and $|0x\rangle$ (thin red line). The parameters of the model are $N = 6,\ \omega = 1,\ \epsilon = 0.1$. The initial state of the environment is a mixture of all possible states with equal weights ($\hat{\rho}^{\mathcal{E}} = 1/2^N$).

Here the dispersions squared are calculated for the initial states of the System and the Environment:

$$\delta\hat{E}^2 \equiv \mathrm{tr}_{\mathcal{E}}\ \hat{E}^2\hat{\rho}^{\mathcal{E}} - (\mathrm{tr}_{\mathcal{E}}\ \hat{E}\hat{\rho}^{\mathcal{E}})^2\ , \tag{17}$$

$$\delta\hat{S}^2 \equiv \langle\psi^{\mathcal{S}}|\hat{S}^2|\psi^{\mathcal{S}}\rangle - \langle\psi^{\mathcal{S}}|\hat{S}|\psi^{\mathcal{S}}\rangle^2\ . \tag{18}$$

The first derivative of purity vanishes at $t = 0$ merely because the evolution starts from the state with maximal purity, $\mathfrak{P} = 1$.

The formula (16) for the second derivative is a result of the straightforward calculation; it appears to be less lengthy if one uses the interaction representation instead of the Schrödinger representation.

The short-time behavior of purity is quadratical in time and reads:

$$\mathfrak{P} \simeq 1 - 2\delta\hat{E}^2\delta\hat{S}^2t^2\ . \tag{19}$$

A noticeable fact is that in the case of factorizable interactions the second derivative at $t = 0$ also factorizes. As a consequence, as long as Eq. (19) holds, purity decrease is minimal for eigenstates of the interaction operator $\hat{S}$, which give $\delta\hat{S}^2 = 0$ and $\mathfrak{P} = 1 - O(t^3)$. Does this mean that purity sieve selects the eigenstates of $\hat{S}$ to be the pointer states of the System? In general, the answer is negative. For example, if the interaction is weak compared to the self-Hamiltonian of the System, pointer states are known to be the eigenstates of the self-Hamiltonian $\hat{H}^{\mathcal{S}}$ [9].

In order to clarify this subtle question, let us consider a simple central-spin model as an example:

$$\hat{H} = \frac{\omega}{2}\sigma_z + \sum_{i=1}^{N}\frac{\omega}{2}\sigma_z^i + \epsilon\sigma_x\sum_{i=1}^{N}\sigma_x^i. \tag{20}$$

Here σ_x and σ_z correspond to a central spin, σ_x^i and σ_z^i correspond to N bath spins, constants ω and ϵ characterize the strengths of self-Hamiltonian and interaction respectively. $\mathfrak{P}(t)$ is calculated numerically for two initial directions of the central spin, along $0z$ axis, $|0z\rangle$ (the eigenstate of the self-Hamiltonian $\hat{H}^{\mathcal{S}} = \frac{\omega}{2}\sigma_z$), and along $0x$ axis, $|0x\rangle$ (the eigenstate of the interaction operator $\hat{S} = \sigma_x$). The result for $N = 6,\ \omega = 1,\ \epsilon = 0.1$ is shown in Fig. 1. One can see that at first instants the eigenstate $|0x\rangle$ of the interaction operator 'wins' the competition, providing minimal purity decrease. However, at longer times the situation changes: purity evolution is no longer described by the quadratic law (19), and purity sieve selects another candidate, $|0z\rangle$, to be a pointer state.

Thus, we come to the following conclusion: maximizing the short-time expression (19) does not, in general, provide correct pointer states. However, it gives a hint how one can try to upgrade a selection criterion. The key quantity in Eq. (19) is the squared dispersion $\delta\hat{S}^2$ of the interaction operator. In order to maximize purity[1] we minimized $\delta\hat{S}^2$, calculated for the initial state $\psi^{\mathcal{S}}$. A straightforward generalization is to try to calculate the dispersion of the interaction operator for longer times, to average it and then – to minimize it. Thus, a modified criterion for selecting pointer states may be proposed, which comes to the minimization of

$$\int_0^t dt'\delta\hat{S}^2(t') \, , \tag{21}$$

where

$$\delta\hat{S}^2 \equiv \langle\psi^{\mathcal{S}}(t)|\hat{S}^2|\psi^{\mathcal{S}}(t)\rangle - \langle\psi^{\mathcal{S}}(t)|\hat{S}|\psi^{\mathcal{S}}(t)\rangle^2 \, , \tag{22}$$

and $\psi^{\mathcal{S}}(t)$ is an evolving state of the System (which is assumed to be approximately pure). It satisfies a Shrödinger-like equation

$$i\frac{d}{dt}\psi^{\mathcal{S}}(t) = (\hat{H}^{\mathcal{S}} + \kappa\hat{S})\psi^{\mathcal{S}}(t) \, , \quad \kappa \equiv \mathrm{tr}_{\mathcal{E}}\hat{E}\hat{\rho}^{\mathcal{E}} \, . \tag{23}$$

This criterion appears to work satisfactory in the case when the interaction operator is a coordinate operator: $\mathcal{S} = \hat{x}$. This may be verified by considering specific examples (see the next Section). On the other hand, one can easily check that it does not work, for example, in spin-spin model (20). The reason for such diverse behavior seems to be connected with the singularity or non-singularity of the interaction operator. However, this question is far from being clearly understood. It will be discussed elsewhere.

4. Purity sieve for quantum Brownian motion and for decoherence through collisions

In this Section we give an example of implementation of the the proposed modified selection criterion.

The following Hamiltonian is considered:

$$\hat{H} = \frac{p^2}{2M} + \frac{1}{2}M\Omega^2x^2 + \sum_{k=1}^{N}(\frac{p_k^2}{2m} + \frac{1}{2}m\omega^2x_k^2) + \sum_{k=1}^{N}U_1(x_k - x) + \sum_{k>j}^{N}U_2(x_k - x_j) \, . \tag{24}$$

It describes one particle of mass M and N particles with masses equal to m in oscillator potential wells. All particles interact with each other through potentials U_1 and U_2. If this potentials are collision potentials, the model describes decoherence through collisions. If the potentials are quadratic, this is a quantum Brownian motion (QBM) model.

Let us rewrite this Hamiltonian in an "almost factorizable" form. For this we use the following transformation of variables:

$\tilde{x} = x$
$\tilde{x}_k = x_k - x$
$\tilde{p} = p + \sum p_k$
$\tilde{p}_k = p_k$.

In new variables the Hamiltonian $\hat{H} = \hat{H}^{\mathcal{S}} + \hat{H}^{\mathcal{E}} + \hat{V}$ reads:

[1] The factor $\delta\hat{E}^2$ is expected to take some typical value for large environments; it should not affect the result.

$$\hat{H}^{\mathcal{S}} = \frac{\tilde{p}^2}{2M} + \frac{1}{2}(M\Omega^2 + Nm\omega^2)\tilde{x}^2 \ , \tag{25}$$

$$\hat{H}^{\mathcal{E}} = \sum_{k=1}^{N}(\frac{\tilde{p}_k^2}{2m} + \frac{(\sum \tilde{p}_k)^2}{2M} + \frac{1}{2}m\omega^2\tilde{x}_k^2) + \sum_{k=1}^{N} U_1(\tilde{x}_k) + \sum_{k>j}^{N} U_2(\tilde{x}_k - \tilde{x}_j) \ , \tag{26}$$

$$\hat{V} = -\frac{\tilde{p}\sum \tilde{p}_k}{M} + m\omega^2\tilde{x}\sum \tilde{x}_k \ . \tag{27}$$

One can see that the interaction Hamiltonian is a sum of two factorizable terms now. We could make some further assumptions in order to make one of the terms in (27) negligible and get rid of it. However, it appears to be unnecessary. Indeed, let us calculate the integral (21), which is subject to minimization, separately for both terms:

$$\int_0^T dt'\delta\tilde{p}^2(t') = \frac{1}{2}(M^2\tilde{\Omega}^2\delta x(0)^2 + \delta p(0)^2)T \ , \tag{28}$$

$$\int_0^T dt'\delta\tilde{x}^2(t') = \frac{1}{2}(\delta x(0)^2 + \delta p(0)^2/M^2\tilde{\Omega}^2)T \ , \tag{29}$$

where

$$\tilde{\Omega}^2 \equiv \Omega^2 + \frac{Nm}{M}\omega^2 \ , \tag{30}$$

and integration is performed over the oscillator period T. We see that due to the $p-x$ symmetry in the oscillator Hamiltonian, the two contributions appear to coincide up to a constant factor. One can prove that the interference term vanishes. Minimizing the obtained expression one gets the pointer states to be those with

$$\delta\tilde{x}\ \delta\tilde{p} = 1/2 \ , \tag{31}$$

$$\delta\tilde{x}^2 = (2M\tilde{\Omega})^{-1} \ . \tag{32}$$

Thus pointer states are minimum uncertainty states with the coordinate uncertainty defined by (32).

Previously this result was obtained for the QBM [2]. From the mathematical point of view calculations in [2] coincide with those in the present paper, although a different approach, based on the approximate master equation, was used in [2]. Approaches, which are also very different from the presented one, are usually used to study decoherence through collisions (see, e.g. [7],[10]). However, these studies normally predict the pointer states to be phase-space localized states (as demanded by the quantum-to-classical correspondence principle), which agrees with our result.

Although our result for the model (24) seems to be in general reasonable, one should be cautious when applying it to specific forms of Hamiltonian with certain functions U_1 and U_2. For example, if the interaction vanishes, $U_1 = 0$, the very conception of pointer states makes no sense, but our selection criterion does not 'feel' it and yet gives minimum-uncertainty states as pointer states. As it was mentioned above, the applicability conditions of the proposed criterion are not entirely explored.

Acknowledgments

The author is grateful to organizers of the Fourth International Workshop DICE2008 for hospitality. The work was financially supported by the Dynasty Foundation scholarship, RF President grant NSh-4568.2008.2, RFBR grants 07-02-00830-a and RFBR-08-02-00494-a.

Journal of Physics: Conference Series **174** (2009) 012030 doi:10.1088/1742-6596/174/1/012030

References

[1] Zurek W H 1993 *Prog. Theor. Phys.* **89** 281
[2] Zurek W H, Habib S and Paz J P 1993 *Phys. Rev. Lett.* **70** 1187
[3] Zurek W H 2003 *Rev. Mod. Phys.* **75** 715
[4] Schlosshauer M 2007 *Decoherence and the Quantum-to-Classical Transition* (Berlin: Springer)
[5] Zurek W H 1982 *Phys. Rev.* D **26** 1862
[6] Caldeira A O and Leggett A J 1983 *Physica* A **121** 587
[7] Paz J P, Habib S and Zurek W H 1993 *Phys. Rev.* D **47** 488
[8] Cucchietti F M, Paz J P and Zurek W H 2005 *Phys. Rev.* A **72** 052113
[9] Paz J P and Zurek W H 1999 *Phys. Rev. Lett.* **82** 5181
[10] Joos E and Zeh H D 1985 *Z. Phys.* B **59** 223

Journal of Physics: Conference Series **174** (2009) 012031 doi:10.1088/1742-6596/174/1/012031

A foundation of quantum mechanics based on focusing and symmetry

Inge S Helland

Department of Mathematics; University of Oslo, POBox 1053 Blindern, 0316 Oslo, Norway

E-mail: ingeh@math.uio.no

Abstract. A new foundation of quantum mechanics for systems symmetric under a compact symmetry group is proposed. The foundation is given by a link to classical statistics and coupled to the concept of parameter. A vector ϕ of parameters is called an inaccessible c-variable if experiments can be provided for the single parameter, but no experiment can be provided for ϕ. This is related to the concept of complementarity in quantum mechanics, but more generally to contrafactual parameters. Using these concepts and some weak assumption, the Hilbert space of quantum mechanics is constructed. The complete set of axioms of quantum mechanics is provided by proving Born's formula under weak assumptions.

1. Introduction

Since Birkhoff and von Neumann (1936) there have been several attempts to find a new basis for quantum mechanics. It is stressed by Zeilinger (1999) that both special and general relativity are based on firm foundational principles, but that one can question whether the same can be said about quantum theory. Of course one has the ordinary Hilbert space formulation, but the question is how this can be motivated as the only possible basis, and even more important, how logical this basis is. A very recent development is due to Doering and Isham (2007): They propose a new, abstract foundation of quantum theory based upon topos theory, a variant of category theory.

In this paper, and the related paper Helland (2008) a new foundational aspects of quantum theory is proposed related to an extension of mathematical statistics. A point of departure is that sometimes we may translate languages and learn from other cultures. More precisely, what we will discuss and compare in this paper is:

Quantum theory: Most of the literature is based upon a framework where states are represented by unit vectors in an abstract complex vector space.

Mathematical statistics: Most of the literature is based on a framework given by parametric classes of probability measures.

Each basis here has lead to an extremely rich and varied culture. Predictions done within each culture have been verified empirically. I will claim that one can find links between these completely different logical bases, and try to sketch these links explicitly.

2. Basic concept of mathematical statistics

Mathematical statistics is based upon parametric classes of probability measures. The parameters λ are fundamental, and provide the links to the real world being investigated. More precisely, the question asked in any investigation is: What is the value of λ? The tool used in this investigation is given by the likelihood: $p^{\lambda}(y)$ - the probability distribution of the data y as a function of λ. There is an extensive methodology with this point of departure. The simplest case is when we have a discrete parameter λ. Then the possible answer is always of the form $\lambda = \lambda_k$. More generally, the answer can be expressed in terms of estimates, of confidence intervals or of tests of hypotheses.

Consider two simple examples.

Focusing I.

Imagine that a medical patient can be given one of two exclusive treatments and that he has
- expected lifetime λ^1 if he gets the treatment 1;
- expected lifetime λ^2 if he gets the treatment 2.

In any case, he receives the treatment, or the treatment begins, at a certain time t.

Crucial point: We can estimate from data either λ^1 or λ^2, but we can never estimate the c-variable $\phi = (\lambda^1, \lambda^2)$.

A natural generalization of this example is to what is called counterfactual parameters, where only one component of $\phi = (\lambda^1, \lambda^2, ..., \lambda^m)$ can be estimated.

Focusing II.

An apparatus for a very special length measurement is so sensitive that it is destroyed after one single measurement. Let μ be the length which is to be measured.

Assume furthermore that the measurement uncertainty σ only can be estimated by destroying the whole apparatus.

Let $\phi = (\mu, \sigma)$. Then it is impossible to estimate the whole c-variable ϕ, only μ or σ can be estimated.

In general, the choice of experimental question is essential.

Extensions of the statistical language considered in this paper:

Focusing: Often one can first introduce a conceptually defined variable ϕ, a vector-variable which as a rule does not take a value. Call this a c-variable ϕ. The parameters of statistical models λ are chosen as complementary functions of ϕ. *Questions to nature are asked in terms of one such* λ*:* What is the value of λ^a? If this question can be approached through estimation from data, we say that λ^a is *accessible*.

Symmetry: For certain given statistical models there is a symmetry group G acting upon the c-variable space Φ, and hence an induced group acting upon each parameter of the model.

These extensions will later be coupled to quantum mechanics.

3. More on symmetry and on the construction

Let the group G act upon the c-variable space Φ. For covenience, the action of a group element g is written to the right of the c-variable or parameter. Note that it makes sense to talk about ϕg even when ϕ is an inaccessible c-variable. In the example above with two or more expected lifetimes, a joint time unit shift may well make sense.

The (right) invariant measure ρ is defined by $\rho(Ag) = \rho(A)$ for all A and g.

Journal of Physics: Conference Series **174** (2009) 012031 doi:10.1088/1742-6596/174/1/012031

A subparameter $\lambda = \lambda(\phi)$ is called *permissible* if $\lambda(\phi_1) = \lambda(\phi_2)$ implies $\lambda(\phi_1 g) = \lambda(\phi_2 g)$ for all g. Then λ transforms under G by

$$\lambda(\phi) \rightarrow (\lambda g)(\phi) = \lambda(\phi g).$$

For any subparameter λ^a there is a maximal subgroup G^a of G under which λ^a is permissible.

3.1. Construction of the quantum space I.

Let λ^a be maximal discrete parameters such that $\phi = (\lambda^1, ..., \lambda^m)$ is an inaccessible c-variable. We first have to define the concept of *state:*

Definition.

The state consists of two elements:

a) A focused question: Given the focus a, a maximal accessible parameter, what is the value of this parameter λ^a?

b) An answer: $\lambda^a = \lambda_k$.

Below we will show how this can be connected to the quantummechanical formalism.

Start of construction:

Let $\mathbf{L}^a$ consist of all functions in $\mathrm{L}^2(\Phi, \rho)$ which are of the form $f(\phi) = \tilde{f}(\lambda^a(\phi))$ for some function $\tilde{f}$.

Let the operator S^a on $\mathbf{L}^a$ be defined by multiplying:

$$S^a f(\phi) = \lambda^a(\phi) f(\phi).$$

Then the eigenfunctions of S^a are of the form $u_k = I(\phi : \lambda^a(\phi) = \lambda_k)$.

3.2. Construction of the quantum space II.

Let W be an arbitrary unitary operator on $\mathrm{L}^2(\Phi, \rho)$. Then we can transform:

$$\mathbf{H}^a = W\mathbf{L}^a, \quad v_k = W u_k, \quad T^a = W S^a W^\dagger.$$

Assumption 1.

For each pair of focused parameter λ^a and λ^b there is a group element $g_{ab} \in G$ such that

$$\lambda^b(\phi) = \lambda^a(\phi g_{ab}).$$

Assumption 2.

The reduced groups $G^a, G^b, ...$ generate the whole group G.

A standard result from group representation theory says:

Theorem 1.

For a compact group, every irreducible unitary representation $V(g)$ can be written as $V(g) = WU(g)W^\dagger$ for some W, with $U(g)$ being a subrepresentation of the right regular representation, i.e., the one given by $U(g)f(\phi) = f(\phi g)$.

In this paper we will assume that the group G is compact. This covers the spin 1/2 case and many other cases.

Theorem 2.
Under these assumptions one can find an invariant representation of the group G with invariant space $\mathbf{H}^a$; that is $V(g)\mathbf{H}^a = \mathbf{H}^a$ for all $g \in G$. From this, from Theorem 1 and from a suitable choice of W, we get $\mathbf{H}^a = \mathbf{H}^b = \mathbf{H}^c = ...$, and this can be taken as the Hilbert space $\mathbf{H}$.

Theorem 2 is proved in detail in Helland (2008).

3.3. Construction of the quantum space III.
First we establish the connection to the state concept defined above.

Theorem 3.
With $u_k^a = I(\phi : \lambda^a(\phi) = \lambda_k)$ and $v_k^a = Wu_k^a$, the state function v_k^a can be given a unique interpretation as follows: 1) There is a question: What is the value of $\lambda^a(\phi)$? 2) An answer: $\lambda^a(\phi) = \lambda_k$.

Theorem 4.
For each choice of focused question a there is an operator $T^a = WS^aW^\dagger$ on the Hilbert space $\mathbf{H}$ which corresponds to a perfect experiment with parameter λ^a. This operator has eigenvectors v_k^a with eigenvalues λ_k.

3.4. Density operator and effect
Let prior or posterior probability π_k over the parameter λ^a be given. A density operator is then defined by:

$$\sigma = \sum_k \pi_k v_k^a v_k^{a\dagger}.$$

The statistical interpretation of the density operator is straghtforward: It describes a mixed state, a probability distribution according to $\{\pi_k\}$ over the pure states given by v_k^a.

A more modern concept is that of an effect (Busch, 2003). Let a statistical experiment with likelihood $p_j^b(y)$ and parameter λ^b be given. Then the corresponding effect is defined by:

$$\mathcal{E} = \sum_j p_j^b(y) v_j^b v_j^{b\dagger}.$$

The statistical interpretation can be connected to the state concept defined above: The state v_j^b is interpreted as above as a focused question together with an answer: $\lambda^b = \lambda_j^b$. Then the likelihood corresponding to this parametric value is $p_j^b(y)$. Thus, given an observation y, the effect uniquely determines the state, the implied parametric value and the ensuing statistical likelihood. As elsewhere in this paper the parameter, and also the observation are assumed to be discrete, so $0 \le p_j^b(y) \le 1$ always.

4. Born's formula
Born's formula is fundamental in linking the differently focused experiments:

$$\mathrm{P}(\lambda^b = \lambda_j | \lambda^a = \lambda_k) = |v_j^{b\dagger} v_k^a|^2.$$

This is proved under weak assumptions by Deutsch, Wallace, Saunders, Zurek and others.

An alternative proof under weak assumptions can be based upon the concept of effects and the Busch-Gleason theorem. The Busch-Gleason is a recent variant of the Gleason theorem - due to Busch (2003) - which has the advantage that it is also valid for dimension 2. It is formulated in terms of set functions for effects. This proof, carried out in Helland (2008), also rests on an assumption, related to randomization between experiments. More precisely, if $\pi(\mathcal{E}^r) = \sum p_j^{b_r}(y)P(\lambda^{b_r} = \lambda_j|\lambda^a = \lambda_k)$ is the set function of experiment $\mathcal{E}^r$, corresponding to parameter b_r, then it is assumed that $\pi(\frac{1}{n}\sum_r \mathcal{E}^r) = \frac{1}{n}\sum_r \pi(\mathcal{E}^r)$. The fact that Born's formula can be proved under different, rather weak conditions, serves as an argument for the fact that it must be true in extensive generalivy in quantum theory. Besides, there are empirical evidences for the formula.

4.1. Consequence of Born's formula

a) Let the state of the system be given by $v = v_k^a$, and let $\lambda = \lambda^b$ be an arbitrary parameter with corresponding operator $T = \sum_j \lambda_j v_j^b v_j^{b\dagger}$.

Then the expected result of a perfect measurement will be:

$$\mathrm{E}_v(\lambda) = \sum_j \lambda_j \mathrm{P}(\lambda^b = \lambda_j|\lambda^a = \lambda_k) = v^\dagger T v.$$

This is a basic postulate of conventional quantum mechanics.

b) Let an experiment have focus on a parameter λ^b and likelihood $p_j(y)$ for $\lambda^b = \lambda_j$. Define an operatorvalued measure $\mathcal{M}$ by

$$\mathcal{M}(dy) = \sum_j p_j(y) v_j^b v_j^{b\dagger} dy.$$

Assume the the initial state is given by $\lambda^a = \lambda_k$. Then the probability distribution for the result of experiment b is given by:

$$\mathrm{P}(dy|\lambda^a = \lambda_k) = v_k^{a\dagger}\mathcal{M}(dy)v_k^a.$$

More generally let the initial probability distribution be determined by the density matrix σ. Then $\mathrm{P}(dy) = \mathrm{tr}(\sigma\mathcal{M}(dy))$, a standard formula.

c) *Collapse of the wave packet.* Assume an initial state v_k^a, and assume that a perfect measurement of λ^b has been done without knowing that value. Then the state is described by a density matrix $\sum_j |v_k^{a\dagger} v_j^b|^2 v_j^b v_j^{b\dagger}$.

After measurement $\lambda^b = \lambda_j$, the state then changes to the vector v_j^b. From the previous statement, this happens with probability $|v_k^{a\dagger} v_j^b|^2$.

Note that the reason why this collapse of the wave packet is unproblematic in this approach is that we never assume any ontological interpretation of the state vector. The state represents just a question together with an ideal answer.

5. Discussion

There are several themes that can be further discussed in this setting. One is the action at a distance and Bell's inequality, and others are related to the socalled 'paradoxes' in quantum mechanics. Furthermore, the theory can be extended to continuous parameters, and in this connection the Planck constant can be defined in a natural way. Finally, time development can be discussed and the Schrödinger's equation can be derived. All these themes are further discussed in Helland (2008).

One aspect of the present paper is that it proposes a link between classical statistics and quantum mechanics. In an independent paper such a link has also recently been argued for by Wetterich (2008).

References

[1] Birkhoff G and von Neumann J 1936 *Ann. Math.* **37** 823
[2] Zeilinger A 1999 *Found. Phys.* **29** 631
[3] Doering A and Isham CJ 2007 *Preprints* arXiv:quant-ph/0703060, arXiv:0703062, arXiv:0703064 and arXiv:0703066
[4] Helland I S 2008 *Found. Phys.* **38** 818
[5] Deutsch D 1999 *Proc. Roy. Soc. London* A **455** 3129
[6] Wallace D 2002 *Preprint* arXiv:quant-ph/0211104
[7] Saunders S 2004 *Proc. Roy. Soc. London* A **460** 1
[8] Zurek W H 2005 *Physical Review* A **71** 052105-1
[9] Busch P 2003 *Phys. Rev. Lett.* **91** 120403-1
[10] Wetterich C 2008 *Preprint* arXiv:quant-ph/0811.0927

Journal of Physics: Conference Series **174** (2009) 012032 doi:10.1088/1742-6596/174/1/012032

Symmetries of nonrelativistic phase space and the structure of quark-lepton generation

Piotr Żenczykowski

Division of Theoretical Physics, Institute of Nuclear Physics, Polish Academy of Sciences, Radzikowskiego 152, 31-342 Kraków, Poland

E-mail: piotr.zenczykowski@ifj.edu.pl

Abstract. According to the Hamiltonian formalism, nonrelativistic phase space may be considered as an arena of physics, with momentum and position treated as independent variables. Invariance of $\mathbf{x}^2 + \mathbf{p}^2$ constitutes then a natural generalization of ordinary rotational invariance. We consider Dirac-like linearization of this form, with position and momentum satisfying standard commutation relations. This leads to the identification of a quantum-level structure from which some phase space properties might emerge. Genuine rotations and reflections in phase space are tied to the existence of new quantum numbers, unrelated to ordinary 3D space. Their properties allow their identification with the internal quantum numbers characterising the structure of a single quark-lepton generation in the Standard Model. In particular, the algebraic structure of the Harari-Shupe preon model of fundamental particles is reproduced exactly and without invoking any subparticles. Analysis of the Clifford algebra of nonrelativistic phase space singles out an element which might be associated with the concept of lepton mass. This element is transformed into a corresponding element for a single coloured quark, leading to a generalization of the concept of mass and a different starting point for the discussion of quark unobservability.

'I do not believe that a real understanding of the nature of elementary particles can ever be achieved without a simultaneous deeper understanding of the nature of spacetime'

Roger Penrose - [1]

1. Elementary particles and space

In our search for the underlying components of matter, we have identified several species of fundamental elementary fermions. According to the Standard Model, there are three generations of such particles, each generation composed of eight objects: two leptons and two triplets of quarks. Interactions of these particles proceed through an exchange of various gauge bosons. The particles themselves (and the systems composed thereof) differ in their properties, with the differences corresponding to different eigenvalues of various quantum numbers.

Two groups of such quantum numbers may be identified. The first one is composed of the so-called 'spatial' quantum numbers, while the other one comprises 'internal' quantum numbers.

The spatial quantum numbers of a particle are those standardly written in the form 'J^{PC}'. Here J denotes the spin of the particle, P denotes its parity, and C is the so-called charge-conjugation parity. They are all connected with the properties of spacetime: spin is related to ordinary rotations, parity - to reflections in our 3D space, while C-parity - being related to complex conjugation - is clearly connected with time reflection. It is important to notice that all these spatial quantum numbers are nonrelativistic in origin. In particular, the emergence of antiparticles is not a relativistic phenomenon as the Dirac equation might suggest: they appear also when the strictly nonrelativistic Schrödinger equation is linearized [2].

As far as the internal quantum numbers are concerned, many physicists believe that these quantum numbers should also be connected with some 'classical arena' which would constitute some kind of an extension of spacetime. In view of the nonrelativistic nature of all spatial quantum numbers it seems then natural to expect that the minimal extension of the concept of space needed for an understanding of the internal quantum numbers should be nonrelativistic as well.

2. Space, time and quantum

It might be argued that the reasoning of the preceding section, based on an expected analogy between spatial and internal quantum numbers, is not sufficient to justify the adoption of a nonrelativistic approach well enough. After all, the standard form of the theory of special relativity involves transformations which mix time with space, thereby undermining the very concept of absolute simultaneity.

In fact, however, the connection between space and time is more subtle than the standard form of special relativity would suggest: the latter form emerges only when the Einstein radiolocation prescription for the synchronization of distant clocks is adopted. Yet, distant clocks may be synchronised in various ways, reflecting the presence of gauge freedom related to the impossibility of measuring the one-way speed of light. Absolute simultaneity may then be achieved with the help of a suitable gauge [3].

A different argument in favour of a nonrelativistic approach may also be given. Namely, although relativistic field theory does unite special relativity and quantum physics, this marriage of quantum and relativistic ideas is in the opinion of many physicists somewhat uneasy [4]. The actual wording takes various forms, such as e.g. 'The construction of a fully objective theory of state-vector reduction which is consistent with the spirit of relativity is a profound challenge, since 'simultaneity' is a concept (...) foreign to relativity' [5]. Considerations of this and similar types lead to a widely advocated idea that space and time are emergent phenomena, absent at the underlying quantum level.

We conclude therefore that it is well justified to adopt an approach in which one does not *start* from mixing time with space. Yet, if the internal quantum numbers are to be connected with the properties of the macroscopic 'space', the ordinary 3D space clearly has to be somehow extended into a broader 'arena'.

3. The arena

3.1. Max Born: a hint

The issue of a possible relation between some particle properties, such as mass, and the surrounding 'emergent' space was of significant concern already to Max Born, over half a century ago. In his 1949 paper [6] he writes: 'I think that the assumption of the observability of the 4-dimensional distance of two events inside atomic dimensions (no clocks or measuring rods) is an extrapolation...' He then continues with the discussion of a difference between the position and momentum spaces for elementary particles. First, he notes that the concept of mass appears in the relation $p^2 = m^2$, and that different observed elementary particles correspond to different discrete values of m^2, thus rendering p^2 observable. Then, he stresses that x^2, the corresponding invariant in coordinate space (with x^2 of atomic dimensions), seems to be 'no observable at all'.

At the same time, he points out that the laws of nature such as

$$\dot{x}_k = \frac{\partial H}{\partial p_k}, \quad \dot{p}_k = -\frac{\partial H}{\partial x_k},$$
$$[x_k, p_l] = i\hbar\delta_{kl},$$
$$L_{kl} = x_k p_l - x_l p_k \tag{1}$$

are invariant under 'reciprocity' transformations:

$$x_k \to p_k, \qquad p_k \to -x_k. \tag{2}$$

Noting that the reciprocity symmetry somehow does not apply to elementary particles, he concludes: 'This lack of symmetry seems to me very strange and rather improbable'.

3.2. Phase space as the arena

In view of the preceding arguments and with quantum mechanics 'living in phase space', it seems natural to start from mixing the 3D space of positions with the 3D space of momenta, and to look for any additional quantum numbers that might possibly emerge in such a broader scheme. In other words, instead of implicitly identifying the ordinary 3D space of positions with the nonrelativistic arena on which physical processes take place, I propose that it is the nonrelativistic phase space that might and should be viewed as such an arena.

It should be stressed that the procedure of mixing the spaces of positions and momenta leads to such a generalization of the concept of ordinary space which could be termed 'minimal'. In particular, no additional dimensions – 'hidden' from our sight – are introduced in this way. We just have to recognize that this macroscopic arena is around us, fully visible.

In the following, I will infer the existence of some internal quantum numbers from the properties of this generalized arena. In a more fundamental approach, however, one has to reverse the arrow of implications and – starting from the underlying quantum structure, as manifested in the quantum numbers of elementary particles – actually build the 'Emergent Phase Space' (EPS).

3.3. Phase space and the Standard Model

The simplest phase-space generalization of the 3D concepts of reflection and rotation requires a fully symmetric treatment of the two O(3) invariants, $\mathbf{x}^2$ and $\mathbf{p}^2$, which is achieved by adding them together:

$$\mathbf{x}^2 + \mathbf{p}^2. \tag{3}$$

This generalization leads to $O(6)$, which goes beyond both Born's reciprocity and the familiar symmetries of 3D space. Treating $\mathbf{x}^2$ and $\mathbf{p}^2$ as operators, we now require their commutators to be form invariant. As is well known from the case of the 3D harmonic oscillator, the original $O(6)$ symmetry is then reduced to $U(1) \otimes SU(3)$. The $U(1)$ factor describes (in particular) Born's reciprocity transformations and their squares: 3D reflections. The $SU(3)$ factor takes care of the standard rotations (among other transformations).

The appearance of the $U(1) \otimes SU(3)$ group from the first principles and the presence of the same group in the Standard Model (SM) raise the possibility that the SM internal symmetry group is actually related to phase space symmetries. A confirmation of this suggestion seems to require the construction of the SM gauge prescription from and/or upon the underlying quantum structure. The gauge structure would have to appear alongside the emerging (phase) space. Consequently, I think, it lies beyond our reach at the moment. In the following, I shall show, however, that the structure of quantum numbers obtained at the quantum level of the phase-space-related approach exactly parallels that observed in the real world.

4. Linearization of $\mathbf{x}^2 + \mathbf{p}^2$

In order to reach the quantum level, we linearize $\mathbf{x}^2 + \mathbf{p}^2$ à la Dirac. Using anticommuting A_k and B_k ($k = 1, 2, 3$), with explicite representation

$$\begin{aligned} A_k &= \sigma_k \otimes \sigma_0 \otimes \sigma_1 \\ B_k &= \sigma_0 \otimes \sigma_k \otimes \sigma_2 \\ B_7 &= \sigma_0 \otimes \sigma_0 \otimes \sigma_3 \end{aligned} \tag{4}$$

(B_7 is the seventh anticommuting element of the relevant Clifford algebra), one finds

$$(\mathbf{A} \cdot \mathbf{p} + \mathbf{B} \cdot \mathbf{x})(\mathbf{A} \cdot \mathbf{p} + \mathbf{B} \cdot \mathbf{x}) = (\mathbf{p}^2 + \mathbf{x}^2) + \sum_1^3 \sigma_k \otimes \sigma_k \otimes \sigma_3. \tag{5}$$

The first term on the r.h.s., denoted below by R, appears here because all six elements A_k and B_l anticommute among themselves. The second term, denoted by R^σ, is due to the fact that x_k and p_k do not commute. These two terms sum up to a total $R^{tot} = R + R^\sigma$.

The $SU(4)/SO(6)$ generators are constructed as antisymmetric bilinears of A_k, B_l. In particular, the generator of standard rotation has the explicit form

$$S_k = \frac{1}{2}(\sigma_k \otimes \sigma_0 + \sigma_0 \otimes \sigma_k) \otimes \sigma_0 \tag{6}$$

and corresponds to simultaneous (same size and sense) rotations in the momentum and position subspaces.

5. Eigenvalues of R^{tot}

5.1. Gell-Mann-Nishijima relation

We now introduce operator Y, to be identified shortly with the (weak) hypercharge:

$$Y \equiv \frac{1}{3} R^\sigma B_7 = \frac{1}{3} \sum_1^3 \sigma_k \otimes \sigma_k \otimes \sigma_0 \equiv \sum_1^3 Y_k. \tag{7}$$

Since the 'partial hypercharges' Y_k commute, they may be simultaneously diagonalized. One then gets the pattern shown in Table 1 (as the matrices are 8×8, this pattern is obtained twice).

Table 1. Decomposition of eigenvalue of Y into eigenvalues of its components

colour	0	1	2	3
Y	-1	$+\frac{1}{3}$	$+\frac{1}{3}$	$+\frac{1}{3}$
Y_1	$-\frac{1}{3}$	$-\frac{1}{3}$	$+\frac{1}{3}$	$+\frac{1}{3}$
Y_2	$-\frac{1}{3}$	$+\frac{1}{3}$	$-\frac{1}{3}$	$+\frac{1}{3}$
Y_3	$-\frac{1}{3}$	$+\frac{1}{3}$	$+\frac{1}{3}$	$-\frac{1}{3}$

In [7] a conjecture was put forward that the electric charge Q is just an appropriately normalized operator $R^{tot} B_7$, evaluated for the lowest level of R, i.e.:

$$Q = \frac{1}{6}(R_{lowest} + R^\sigma) B_7 = I_3 + \frac{Y}{2} \tag{8}$$

where, with $R_{lowest} = (\mathbf{p}^2 + \mathbf{x}^2)_{lowest} = 3$, the eigenvalues of $I_3 = B_7/2$ are $\pm 1/2$. The above equation, *derived* here in a phase-space-related approach, is known as the Gell-Mann-Nishijima relation (with I_3 known as weak isospin), and constitutes a law of nature. With the help of Table 1, it yields the charges of all eight leptons and quarks from a single SM generation.

5.2. *Harari-Shupe rishons*

With the growing number of fundamental fermions, the problem of understanding why they group into generations composed of eight particles has been addressed by many physicists. The most widely cited proposal (over 320 citations) is due to Haim Harari and Michael Shupe [8]. The Harari-Shupe model describes the structure of a single SM generation with the help of a composite model. It builds all eight fermions of a single generation from only two spin-1/2 'preons' V and T (or 'rishons' as Harari dubbed them), of charges 0 and +1/3 respectively. This is shown in Table 2, where total charges and hypercharges of particles are also listed. Note that rishons obey strange statistics, i.e. it is the states with ordered rishons (e.g. VTT, TVT, and TTV) which are to correspond to the three colours of $SU(3)$ and which, consequently, are deemed different.

The rishon model, though algebraically very economical, has several drawbacks, however. These include, among others: the issue of preon confinement at extremely small distance scales (when confronted with the uncertainty principle), the apparent absence of spin-3/2 fundamental particles, and the lack of explanation as to why the ordering of the three rishons is important and leads to $SU(3)$.

Table 2. Rishon structure of leptons and quarks with weak isospin $I_3 = +1/2$

	ν_e	u_R	u_G	u_B	e^+	$\bar{d}_R$	$\bar{d}_G$	$\bar{d}_B$
	VVV	VTT	TVT	TTV	TTT	TVV	VTV	VVT
Q	0	$+\frac{2}{3}$	$+\frac{2}{3}$	$+\frac{2}{3}$	+1	$+\frac{1}{3}$	$+\frac{1}{3}$	$+\frac{1}{3}$
Y	−1	$+\frac{1}{3}$	$+\frac{1}{3}$	$+\frac{1}{3}$	+1	$-\frac{1}{3}$	$-\frac{1}{3}$	$-\frac{1}{3}$

5.3. *Preonless resolution of problems*

A comparison of Tables 1, 2 (using Eq. (8)) shows that the phase-space approach reproduces the main structure of the Harari-Shupe model exactly. In fact, it does not only that: it also solves the three problems of the rishon model.

Namely, the phase-space approach explains the structure of charge eigenvalues without assuming *any* subparticle components of quarks and leptons. Therefore, there is no problem of 'where are spin-3/2 fundamental fermions', and there is no problem with preon confinement. Furthermore, the strange statistics of rishons and its connection with SU(3) are naturally explained.

One can readily understand the meaning of the 'ordered rishon structure' (such as VTT) in phase-space terms. Thus, the position of rishon corresponds to one of three directions in our macroscopic 3D space: VTT corresponds therefore to the partial hypercharge eigenvalue of $-1/3$ in direction (x, p_x) and to the same eigenvalue of $+1/3$ in both remaining directions, (y, p_y) and (z, p_z). Now, any discussion of rotations requires three directions, not just one. Hence the concept of spin simply cannot be applied to a single rishon. This is in line with Heisenberg's opinion [9] concerning the idea of dividing matter again and again: '...the antinomy of the smallest dimensions is solved in particle physics in a very subtle manner, of which neither Kant nor the ancient philosophers could have thought: The word 'dividing' loses its meaning'.

6. Transformations in phase space

6.1. Genuine $SO(6)$ transformations

In order to see the relation between quarks and leptons, consider [10, 11] a transformation generated by F^{σ}_{-2}, one of six 'genuine' $SU(4)/SO(6)$ generators $F^{\sigma}_{\pm n}$ $(n = 1, 2, 3)$:

$$F^{\sigma}_{-n} = \frac{1}{2}\left(\sigma_0 \otimes \sigma_n - \sigma_n \otimes \sigma_0\right)\sigma_3 \tag{9}$$

$$F^{\sigma}_{+n} = \frac{1}{2}\epsilon_{nkl}\sigma_k \otimes \sigma_l \otimes \sigma_3. \tag{10}$$

.

Under F^{σ}_{-2}-generated transformations one obtains:

$$\begin{array}{ll} A'_k = A_1 \cos\phi - A_3 \sin\phi & B'_1 = B_1 \cos\phi + B_3 \sin\phi \\ A'_2 = A_2 & B'_2 = B_2 \\ A'_3 = A_3 \cos\phi + A_1 \sin\phi & B'_3 = B_3 \cos\phi - B_1 \sin\phi \end{array} \tag{11}$$

i.e. **A** and **B** rotate in opposite senses. For $\phi = \pm\pi/2$ one then gets:

$$Y = Y_1 + Y_2 + Y_3 \to Y' = -Y_3 + Y_2 - Y_1 \tag{12}$$

and consultation of Table 1 shows that lepton and quark # 2 are exchanged, while the remaining two quarks are left untouched. The same result is obtained when the analogous transformation generated by F^{σ}_{+2} is considered.

6.2. Rotations in phase space

When the corresponding transformations in phase space are considered, one gets (for F^{σ}_{-2}-generated rotations):

$$[x'_k, x'_l] = [p'_k, p'_l] = 0 \tag{13}$$

$$[x'_k, p'_l] = i\Delta_{kl} \tag{14}$$

with

$$\Delta = \begin{bmatrix} \cos 2\phi & 0 & \sin 2\phi \\ 0 & 1 & 0 \\ -\sin 2\phi & 0 & \cos 2\phi \end{bmatrix}. \tag{15}$$

For the case of lepton-quark # 2 interchange ($\phi = \pm\pi/2$), one then obtains (for both F^{σ}_{-2}- and F^{σ}_{+2}- generated rotations):

$$\text{(quark)} \quad \Delta = \begin{bmatrix} -1 & 0 & 0 \\ 0 & +1 & 0 \\ 0 & 0 & -1 \end{bmatrix} \quad \leftrightarrow \quad \Delta = \begin{bmatrix} 1 & 0 & 0 \\ 0 & 1 & 0 \\ 0 & 0 & 1 \end{bmatrix} \quad \text{(lepton)} \tag{16}$$

It may be said therefore that quark is a lepton rotated in phase space.

Taking into account the remaining two types of genuine $SO(6)$ transformations, we get the following four sets of generalized commutation relations:

$$\begin{array}{llll} \text{lepton} & \text{quark 1} & \text{quark 2} & \text{quark 3} \\ [x_1, p_1] = i & [x_1, p_1] = i & [p_1, x_1] = i & [p_1, x_1] = i \\ [x_2, p_2] = i & [p_2, x_2] = i & [x_2, p_2] = i & [p_2, x_2] = i \\ [x_3, p_3] = i & [p_3, x_3] = i & [p_3, x_3] = i & [x_3, p_3] = i. \end{array} \tag{17}$$

6.3. Reflections in phase space

When a reflection in phase space (e.g.. $p'_k = p_k$, $x'_k = -x_k$, $i \to i$) is performed, the number of sets is doubled from four to eight, with i in Eqs (17) changed to $-i$ (this corresponds to doublets of weak isospin and is different from charge conjugation). Thus, we obtain 8 disjoint sectors, corresponding to 8 particles of a single generation of the Standard Model.

The fact that the three additional sets of commutation relations in Eq. (17) are not rotationally invariant is in my opinion an asset of the approach. Namely, the only condition that quarks must really fulfill is that it is the systems composed thereof (i.e. mesons, baryons) that must be covariant under rotations. Thus, my conjecture is that quarks must conspire (see example in the next section).

7. Clifford algebra and mass

The $U(1) \otimes SU(3)$ structure of (a half of) the Clifford algebra of nonrelativistic phase space is shown in Table 3. Here the even elements (linear combinations of the $SU(4)/SO(6)$ generators and the unit element) with $I_3 = +1/2$ (hence superscript '+') are given on the left, while the odd elements (linear combinations of products of an odd number of A_m and B_n) with left and right eigenvalues of $I_{3l} = +1/2$ and $I_{3r} = -1/2$ are shown on the right. In the columns marked Y_l and Y_r, the left and right eigenvalues of Y are given. A more detailed explanation of entries in this table may be found in [12].

Table 3. Classification of Clifford algebra elements according to their $U(1) \otimes SU(3)$ properties

$U(1)$	$SU(3)$	element	Y_l	Y_r	$U(1)$	$SU(3)$	element	Y_l	Y_r
-2	3^*	H^+_{m0}	-1	$+\frac{1}{3}$	$+1$	3^*	$U^\dagger_k$	$+\frac{1}{3}$	$+\frac{1}{3}$
$+2$	3	H^+_{0m}	$+\frac{1}{3}$	-1	-1	3	V_k	$+\frac{1}{3}$	-1
0	8	F^+_a	$+\frac{1}{3}$	$+\frac{1}{3}$	-1	3	W_k	-1	$-\frac{1}{3}$
0	1	Y^+_{-1}	-1	-1	$+1$	6	$G_{\{kl\}}$	$+\frac{1}{3}$	$+\frac{1}{3}$
0	1	$Y^+_{+\frac{1}{3}}$	$+\frac{1}{3}$	$+\frac{1}{3}$	-3	1	G_0	-1	-1

The algebraic counterpart of the lepton mass should be odd (just like the odd A_m is associated with p_m) and is identified with the only odd $Y = -1$ (and $SO(3)$ scalar) element in Table 3, i.e. with G_0. The $F_{\pm 2}$-generated transformation from the lepton to the quark sector changes G_0 into $G_{\{22\}}$, which is a member of the $SU(3)$ sextet, and is not rotationally invariant. However, the sum of the three quark mass terms, i.e. $G_{\{kk\}}$, is rotationally invariant [12], just as the idea of quark conspiracy suggests.

8. Summary

The phase-space approach provides a possible theoretical explanation of the structure of a single generation of the Standard Model. The symmetry obtained involves $U(1) \otimes SU(3)$ and resembles the full $U(1) \otimes SU(3) \otimes SU(2)_L$ symmetry group of the Standard Model quite closely (in fact, the $SU(2)$ partners of $I_3 = B_7/2$, i.e. $I_{1,2} = \sigma_0 \otimes \sigma_0 \otimes \sigma_{1,2}$, automatically do not commute with the 3D reflections, thus suggesting parity violation). We have derived the Gell-Mann-Nishijima relation and reproduced the structure of the Harari-Shupe model (visualized in Fig. 1), while evading its main problems related to the introduction of 'confined preons'. The existence of eight particles in a single SM generation has been related to the $2^3 = 8$ possible sets of $[x_k, p_k] = \pm i$

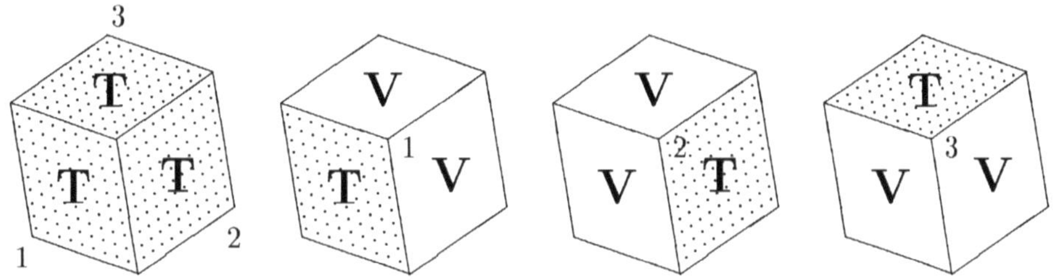

Figure 1. Four of the eight 'DICE1979' corresponding to the Harari-Shupe model. All dice are identical when their rotations are admitted. Each corner corresponds to one particle of a single SM generation. Each face corresponds to a rishon. The three dice to the right (coloured quarks) show the leftmost die (lepton) rotated by π around axes $1, 2, 3$, so that different corners (as marked) present themselves to the reader.

commutation relations, with the $\pm$ sign adopted independently for each direction in our 3D space.

The proposed aproach obviously raises many questions. The phase-space-related modification of the way in which the imaginary unit enters into our theories brings in the question of whether it is possible to extend the concept of the arena further, so that parity violation in weak interactions (together with the emergence of three quark-lepton generations and the related appearance of the CP-violating i) could be described in a more realistic way. Then, there are other important questions such as the issue of the emergence of points, the construction of composite systems, etc. I hope to be able to address some of them in the future.

Acknowledgments

This work has been partially supported by the Polish Ministry of Science and Higher Education research project No N N202 248135.

References

[1] Penrose R 1968 Structure of Spacetime *Batelle Rencontres: 1967 Lectures in Mathematics and Physics* ed C M DeWitt and J A Wheeler (New York: Benjamin) 121
[2] Horzela A and Kapuścik E 2003 *Electromagn. Phenom.* **3** 63
[3] Mansouri R and Sexl R 1977 *Gen. Rel. Grav.* **8** 497
Selleri F 2005 *Found. Phys. Lett.* **18** 325
[4] Bell J S 1981 Quantum mechanics for cosmologists *Quantum Gravity 2* ed C J Isham, R Penrose and D W Sciama (Oxford: Clarendon Press) p 611, reprinted in J S Bell 1987 *Speakable and unspeakable in quantum mechanics* (Oxford: Clarendon Press) 117-138
[5] Penrose R 1989 *The Emperor's New Mind* (Oxford: Oxford University Press) Section 8
[6] Born M 1949 *Rev. Mod. Phys.* **21** 463
[7] Żenczykowski P 2007 *Acta Phys. Pol.* B **38** 2053
[8] Harari H 1979 *Phys. Lett.* B **86** 83
Shupe M A 1979 *Phys. Lett.* B **86** 87
[9] Heisenberg W 1976 *Physics Today* **29** 32
[10] Żenczykowski P 2007 *Acta Phys. Pol.* B **38** 2631
[11] Żenczykowski P 2008 *Phys. Lett.* B **660** 567
[12] Żenczykowski P 2009 *J. Phys.* A **42** 045204

Journal of Physics: Conference Series **174** (2009) 012033 doi:10.1088/1742-6596/174/1/012033

On the Ricci flow and emergent quantum mechanics

José M. Isidro[1], **J.L.G. Santander**[2] **and P. Fernández de Córdoba**[3]

Grupo de Modelización Interdisciplinar, Instituto de Matemática Pura y Aplicada, Universidad Politécnica de Valencia, Valencia 46022, Spain

E-mail: [1]`joissan@mat.upv.es`, [2]`jlgonzalez@mat.upv.es`, [3]`pfernandez@mat.upv.es`

Abstract. The Ricci flow equation of a conformally flat Riemannian metric on a closed 2–dimensional configuration space is analysed. It turns out to be equivalent to the classical Hamilton–Jacobi equation for a point particle subject to a potential function that is proportional to the Ricci scalar curvature of configuration space. This allows one to obtain Schrödinger quantum mechanics from Perelman's action functional: the quantum–mechanical wavefunction is the exponential of i times the conformal factor of the metric on configuration space. We explore links with the recently discussed *emergent quantum mechanics*.

1. A conformally flat configuration space

Let us consider a 2–dimensional, closed Riemannian manifold M. In isothermal coordinates x, y the metric reads

$$g_{ij} = \mathrm{e}^{-f}\delta_{ij}, \tag{1}$$

where $f = f(x, y)$ is a function, hereafter referred to as *conformal factor*. Our conventions are $g = |\det g_{ij}|$ and $R_{im} = g^{-1/2}\partial_n\left(\Gamma_{im}^{n} g^{1/2}\right) - \partial_i\partial_m\left(\ln g^{1/2}\right) - \Gamma_{is}^{r}\Gamma_{mr}^{s}$ for the Ricci tensor, $\Gamma_{ij}^{m} = g^{mh}\left(\partial_i g_{jh} + \partial_j g_{hi} - \partial_h g_{ij}\right)/2$ being the Christoffel symbols. The volume element on M equals

$$\sqrt{g}\,\mathrm{d}x\mathrm{d}y = \mathrm{e}^{-f}\mathrm{d}x\mathrm{d}y. \tag{2}$$

Given an arbitrary function $\varphi(x, y)$ on M, we have the following expressions for the Laplacian $\nabla^2\varphi$ and the squared gradient $(\nabla\varphi)^2$:

$$\nabla^2\varphi := \frac{1}{\sqrt{g}}\partial_m\left(\sqrt{g}g^{mn}\partial_n\varphi\right) = \mathrm{e}^{f}\left(\partial_x^2\varphi + \partial_y^2\varphi\right) =: \mathrm{e}^{f}D^2\varphi, \tag{3}$$

$$(\nabla\varphi)^2 := g^{mn}\partial_m\varphi\partial_n\varphi = \mathrm{e}^{f}\left[(\partial_x\varphi)^2 + (\partial_y\varphi)^2\right] =: \mathrm{e}^{f}\left(D\varphi\right)^2, \tag{4}$$

where $D^2\varphi$ and $(D\varphi)^2$ stand for the flat–space values of the Laplacian and the squared gradient, respectively. The Ricci tensor reads

$$R_{ij} = \frac{1}{2}D^2 f\,\delta_{ij} = \frac{1}{2}\mathrm{e}^{-f}\nabla^2 f\,\delta_{ij}. \tag{5}$$

From here we obtain the Ricci scalar

$$R = \mathrm{e}^{f}D^2 f = \nabla^2 f. \tag{6}$$

A manifold M such as that considered here is in fact a compact Riemann surface without boundary. Physically it will play the role of a configuration space for a certain point mechanics on M. It may be worthwhile to remember that any Riemannian metric on a 2–dimensional manifold can be written in the form (1). This is often referred to as the property that any 2–dimensional metric is always conformally flat. This nice property of the 2–dimensional case no longer holds in $n \geq 3$ dimensions. We will however restrict our attention to $n = 2$.

2. A crash course in Ricci flow

For an introduction to the Ricci flow and its applications, a good reference is [2]. More technical are Perelman's original article and references therein [1]. Perelman's functional $\mathcal{F}[\varphi, g_{ij}]$ on the manifold M is defined as

$$\mathcal{F}[\varphi, g_{ij}] := \int_M \mathrm{d}x\mathrm{d}y\sqrt{g}\,\mathrm{e}^{-\varphi}\left[(\nabla\varphi)^2 + R(g_{ij})\right], \tag{7}$$

where g_{ij} is a metric on M and φ a real function on M. We will take the above equation as our starting point. It may be regarded physically as providing an action functional, on configuration space M, for the two independent fields g_{ij} and φ. Now some aspects of the functional $\mathcal{F}[\varphi, g_{ij}]$ are worth mentioning. Setting $\varphi = 0$ identically we have the Einstein–Hilbert functional for gravity on M. Admittedly Einstein–Hilbert gravity, being a boundary term in $n = 2$ dimensions, is trivial in $n = 2$ dimensions. However the generalisation of 2–dimensional gravity provided by the functional $\mathcal{F}[\varphi, g_{ij}]$ when $\varphi \neq 0$ is interesting. Indeed, Perelman's functional arises in string theory as the low–energy effective action of the bosonic string [3]. Already these two properties suffice to justify our interest in the functional (7). If this were not enough, the mathematical applications of Ricci–flow theory are impressive [1, 2, 4], although they will not be dealt with here. Instead we will concentrate our attention on the relation between the Ricci flow and quantum mechanics on M.

We first compute the Euler–Lagrange extremals corresponding to the fields g_{ij} and φ. Next we set the equations of motion so obtained equal to the first–order time derivatives of g_{ij} and φ, respectively. This results in the two evolution equations

$$\frac{\partial g_{ij}}{\partial t} = -2\left(R_{ij} + \nabla_i\nabla_j\varphi\right), \quad \frac{\partial\varphi}{\partial t} = -\nabla^2\varphi - R. \tag{8}$$

We stress that the right–hand sides of (8), once equated to zero, are the Euler–Lagrange equations of motion corresponding to (7), and that the time derivatives on the left–hand sides have been put in by hand. In fact time is not a coordinate on configuration space M, but an external parameter. The two equations (8) are referred to as the *gradient flow* of $\mathcal{F}$, since they provide a set of time–evolution equations. Via a time–dependent diffeomorphism, one can show that the set (8) are equivalent to

$$\frac{\partial g_{ij}}{\partial t} = -2R_{ij}, \quad \frac{\partial\varphi}{\partial t} = -\nabla^2\varphi + (\nabla\varphi)^2 - R. \tag{9}$$

We will use (9) rather than (8).

As already remarked, one advantage of having a 2–dimensional configuration space is that all metrics on it are conformal, so we can substitute (1) throughout. By (2) and (6), we can express $\mathcal{F}[\varphi, g_{ij}]$ as

$$\mathcal{F}[\varphi, f] := \mathcal{F}[\varphi, g_{ij}(f)] = \int_M \mathrm{d}x\mathrm{d}y\,\mathrm{e}^{-\varphi-f}\left[(\nabla\varphi)^2 + \nabla^2 f\right]. \tag{10}$$

In order to understand the physical meaning of the flow eqns. (9), let us analyse them in more detail. Using (1) and (5) we see that the first flow equation,

$$\frac{\partial g_{ij}}{\partial t} = -2R_{ij}, \tag{11}$$

is equivalent to

$$\frac{\partial f}{\partial t} = \nabla^2 f. \tag{12}$$

This is the usual heat equation, with the important difference that the Laplacian operator ∇^2 is given by (3): indeed M is not flat but only *conformally* flat. So conformal metrics on the (curved) manifold M evolve in time according to the heat equation wih respect to the corresponding (curved) Laplacian. The second flow equation in (9) will be the subject of our attention in what follows.

So far, the conformal factor f and the scalar φ have been considered as independent fields. Setting now $\varphi = f$ in (10) we obtain

$$\mathcal{F}[f] := \mathcal{F}[\varphi = f, f] = \int_M \mathrm{d}x\mathrm{d}y\, \mathrm{e}^{-2f}\left[(\nabla f)^2 + \nabla^2 f\right]. \tag{13}$$

After setting $\varphi = f$ we appear to have a contradiction, since we have two different flow equations in (9) for just one field f. That there is in fact no contradiction can be seen as follows. In (9) we have two independent flow equations for the two independent fields f and φ. Equating the latter two fields implies that the two flow equations must reduce to just one. This can be achieved by substituting one of the two flow eqns. (9) into the remaining one. By (6) and (12) we have $R = \partial f/\partial t$, which substituted into the second flow equation of (9) leads to

$$\frac{\partial f}{\partial t} = \frac{1}{2}(\nabla f)^2 - \frac{1}{2}\nabla^2 f. \tag{14}$$

We will later on find it useful to distinguish notationally between the time–independent conformal factor f, as it stands in the functional (13), and the time–dependent conformal factor as it stands in the flow equation (14). We therefore rewrite (14) as

$$\frac{\partial \tilde{f}}{\partial t} = \frac{1}{2}\left(\nabla \tilde{f}\right)^2 - \frac{1}{2}\nabla^2 \tilde{f}, \tag{15}$$

where a tilde on top of a field indicates that it is a time–dependent quantity.

3. A correspondence between conformal metrics and mechanical actions

In what follows we will regard the manifold M as the configuration space of a mechanical system, to be identified presently. We will establish a 1–to–1 correspondence between conformally flat metrics on configuration space M, and (classical or quantum) mechanical systems on M. Let us consider classical mechanics in the first place. We recall that, for a point particle of mass m subject to a time–independent potential U, the Hamilton–Jacobi equation for the time–dependent action $\tilde{S}$ reads

$$\frac{\partial \tilde{S}}{\partial t} + \frac{1}{2m}\left(\nabla \tilde{S}\right)^2 + U = 0. \tag{16}$$

It is well known that, separating the time variable as per

$$\tilde{S} = S - Et, \tag{17}$$

Journal of Physics: Conference Series **174** (2009) 012033 doi:10.1088/1742-6596/174/1/012033

with S the so–called reduced action, one obtains

$$\frac{1}{2m}\left(\nabla S\right)^2 + U = E. \tag{18}$$

Eqn. (17) suggests separating variables in (15) as per

$$\tilde{f} = f + Et, \tag{19}$$

where the sign of the time variable is reversed[1] with respect to (17). Substituting (19) into (15) leads to

$$\frac{1}{2}\left(\nabla f\right)^2 - \frac{1}{2}\nabla^2 f = E. \tag{20}$$

Comparing (20) with (18) we conclude that, picking a value of the mass $m = 1$, the following identifications can be made:

$$S = f, \qquad U = -\frac{1}{2}\nabla^2 f = -\frac{1}{2}R. \tag{21}$$

So the potential U is proportional to the scalar Ricci curvature of the configuration space M, while the reduced action S equals the conformal factor f. This concludes the first half of our dictionary: to construct a classical mechanics starting from a conformal metric on M.

Conversely, if we are given a classical mechanics as determined by an arbitrary potential function U on M, and we are required to construct a conformal metric on M, then the solution is given by the function f satisfying the Poisson equation $-2U = \nabla^2 f$, where the Laplacian is computed with respect to the unknown function f.

Although we have so far considered the *classical* mechanics associated with a given conformal factor, one can immediately construct the corresponding *quantum* mechanics, by means of the Schrödinger equation for the potential U. We can therefore restate our result as follows: we have established a 1–to–1 correspondence between conformally flat metrics on configuration space, and quantum–mechanical systems on that same space.

4. Schrödinger's functional from Perelman's functional

Let us summarise our results. We have considered a conformally flat Riemannian metric on a closed 2–dimensional manifold M, and regarded the latter as the configuration space of a classical mechanical system. We have formulated a dictionary between such conformal metrics, on the one hand, and quantum mechanics on the same space, on the other. This dictionary has a nice geometrical interpretation: the reduced mechanical action S equals the conformal factor f, and the potential function U is (proportional to) the Ricci curvature of M. It is interesting to observe that the Ricci scalar as a potential function has also arisen in the context of Bohmian mechanics [5, 6].

The previous correspondence can be exploited further: we will exchange a conformally flat metric for a wavefunction satisfying the Schrödinger equation for the potential U. We first observe that the Schrödinger equation itself can be obtained as the extremal of the action functional

$$\mathcal{S}[\psi, \psi^*] := \int_M \mathrm{d}x\mathrm{d}y\sqrt{g}\left(\mathrm{i}\psi^*\frac{\partial\psi}{\partial t} - \frac{1}{2m}\nabla\psi^*\nabla\psi - U\psi^*\psi\right). \tag{22}$$

Pick $m = 1$ as before, and substitute

$$\psi = \mathrm{e}^{\mathrm{i}\tilde{f}} \tag{23}$$

[1] This time reversal is imposed on us by the time–flow eqn. (15), with respect to which time is reversed in the mechanical model. This is just a rewording of (part of) section 6.4 of ref. [2], where a corresponding heat flow is run *backwards* in time.

into (22) to obtain $-\partial_t \tilde{f} - \frac{1}{2}(\nabla \tilde{f})^2 + \frac{1}{2}\nabla^2 \tilde{f}$ within the integrand. As before, let us consider the stationary case, where $\partial_t \tilde{f} = 0$ and $\tilde{f}$ becomes f. Then (22) turns into

$$\mathcal{S}[f] := \mathcal{S}\left[\psi = \mathrm{e}^{\mathrm{i}f}\right] = \frac{1}{2}\int_M \mathrm{d}x\mathrm{d}y\,\mathrm{e}^{-f}\left[-(\nabla f)^2 + \nabla^2 f\right]. \tag{24}$$

Comparing the functionals (13) and (24) we arrive at the following interesting relation:

$$\mathcal{F}\left[f/2\right] = \mathcal{S}[f] + \frac{3}{2}\mathcal{K}[f], \tag{25}$$

where

$$\mathcal{K}[f] := \frac{1}{2}\int_M \mathrm{d}x\mathrm{d}y\,\mathrm{e}^{-f}\,(\nabla f)^2 \tag{26}$$

is the kinetic energy functional on M. In $\mathcal{F}\left[f/2\right]$ above, the Laplacian $\nabla^2 f$ and the squared gradient $(\nabla f)^2$ are computed with respect to the conformal factor f, even if the functional $\mathcal{F}$ is evaluated at $f/2$. Thus, on a compact Riemann surface without boundary, the Schrödinger functional $\mathcal{S}[f]$ turns out to be a close cousin of the Perelman functional $\mathcal{F}[f/2]$. Altogether we have proved that Schrödinger quantum mechanics on a 2–dimensional, compact configuration space arises from Perelman's functional.

5. Discussion and perspectives

The Ricci flow has provided many far–reaching insights into long–standing problems in topology and geometry [1, 2]. Recent works [7, 8, 9, 10, 11, 12, 13, 14] have shed light on applications of conformal symmetry and the Ricci flow to foundational issues in quantum mechanics. In this contribution we have established a 1–to–1 correspondence between conformally flat metrics on configuration space, and quantum mechanics on that same space. This correspondence has been used to prove that Schrödinger quantum mechanics in two space dimensions arises from Perelman's functional on a compact Riemann surface.

We have worked in the 2–dimensional case for simplicity. Now Perelman's functional arises in string theory as the low–energy effective action of the bosonic string [3]. In view of these facts it is very tempting to try and interpret quantum mechanics itself, in any number of dimensions, possibly also noncompact, as some kind of effective, low–energy approximation to some more fundamental theory. Related ideas have been put forward in the literature [15, 16, 17, 18], where standard quantum mechanics has been argued to *emerge* from an underlying deterministic theory. Basically, in *emergent* quantum mechanics, one starts from a *deterministic* model and, via some dissipative mechanism that implements information loss, one ends up with a *probabilistic* theory. Several mechanisms implementing information loss have been proposed in the literature. Thus, in ref. [15], dissipation is effected by an attractor on phase space, which produces a lock–in of classical trajectories around some fixed point; instead, in ref. [18] dissipation arises as a coarse–graining of classical information via a probability distribution function on phase space. A somewhat different dissipative mechanism, based on the Ricci flow equation (11), has been put forward in ref. [19].

Some features of emergent quantum mechanics are present in our picture. Most notable among them is the presence of dissipation, or information loss: as remarked above, this underlies the passage from a classical description to a quantum description. Indeed, in our setup, the classical description is provided by the conformal factor f of the metric g_{ij} on configuration space, while the quantum description is given by the wavefunction $\psi = \mathrm{e}^{\mathrm{i}f}$. The latter contains less information than the former, as there exist different conformal factors f giving rise to just one quantum wavefunction ψ. This situation is analogous to that described in [15, 16], in which quantum states arise as equivalence classes of classical states: different classical states may fall into one and the same quantum equivalence class. Beyond the trivial case of any two conformal factors f_1 and f_2 differing by 2π times an integer, there is the more interesting case of f_1 and f_2 satisfying $\nabla_1^2 f_1 = \nabla_2^2 f_2$, where the subindices 1, 2 refer to the fact that the corresponding Laplacians are computed with respect to the conformal factors f_1 and f_2, respectively. If M is such that the Laplace–like equation $\nabla_1^2 f_1 - \nabla_2^2 f_2 = 0$ admits nontrivial solutions, then any two such f_1 and f_2 (different classical states) fall into the same quantum state, as both f_1 and f_2 give rise to the same potential function $-2U = \nabla_1^2 f_1 = \nabla_2^2 f_2$.

Another feature of emergent quantum mechanics that is present in our picture is the following. In refs. [15, 16, 17] it has been established that to every quantum system there corresponds at least one deterministic system which, upon prequantisation, gives back the original quantum system. In our setup this existence theorem is realised alternatively as follows. Let a quantum system possessing the potential function V be given on the configuration space M, the latter satisfying the same requirements as above. Let us consider the Poisson equation on M, $\nabla_V^2 f_V = -2V$, where f_V is some unknown conformal factor, to be determined as the solution to this Poisson equation, and ∇_V^2 is the corresponding Laplacian. We claim that the deterministic system, the prequantisation of which gives back the original quantum system with the potential function V, is described by the following data: configuration space M, with classical states being conformal factors f_V and mechanics described by the action functional (13). The lock–in mechanism (in the terminology of refs. [15, 16]) is the choice of one particular conformal factor, with respect to which the Laplacian is computed, out of all possible solutions to the Poisson equation on M, $\nabla_V^2 f_V = -2V$. The problem thus becomes topological–geometrical in nature, as the lock–in mechanism has been translated into a problem concerning the geometry and topology of configuration space M, namely, whether or not the Poisson equation possesses solutions on M, and how many.

As a perspective for future work, it would be an interesting question to ask whether or not Ricci–flow techniques could be used to implement some kind of renormalisation–group flow from an underlying deterministic model to an emergent quantum mechanics. This would be in line with the view that quantum mechanics is an infrared phenomenon [20, 21], and also with models of spacetime whereby classical gravity has been argued to arise by some process of *thermalisation* of some underlying quantum theory. We hope to address these issues in the future.

Acknowledgments
It is a great pleasure to thank H.-T. Elze for interesting technical discussions. J.M.I. also thanks Max–Planck–Institut für Gravitationsphysik, Albert–Einstein–Institut (Potsdam, Germany) for hospitality extended over a long period of time.

References

[1] Perelman G 2002 The entropy formula for the Ricci flow and its geometric applications *Preprint* arXiv:math/0211159
[2] Topping P 2006 *Lectures on the Ricci Flow* (London Mathematical Society Lecture Notes Series **325** (Cambridge, UK: Cambridge U. Press)
[3] Deligne P *et al.* eds 1999 *Quantum Fields and Strings: a Course for Mathematicians* vol. 2 (Providence: American Mathematical Society)
[4] Carroll R 2006 *Fluctuations, Information, Gravity, and the Quantum Potential* (Berlin: Springer)
[5] Koch B 2008 Geometrical interpretation of the quantum Klein–Gordon equation *Preprint* arXiv:0801.4635
[6] Koch B 2008 Relativistic Bohmian mechanics from scalar gravity *Preprint* arXiv:0810.2786
[7] Bertoldi G, Faraggi A and Matone M 2000 Equivalence principle, higher dimensional Mobius group and the hidden antisymmetric tensor of quantum mechanics *Class. Quant. Grav.* **17** 3965 (*Preprint* arXiv:hep-th/9909201)
[8] Matone M 2005 The cocycle of the quantum HJ equation and the stress tensor of CFT *Braz. J. Phys.* **35** 316 (*Preprint* arXiv:hep-th/0502134)
[9] Carroll R 2007 Some remarks on Ricci flow and the quantum potential *Preprint* arXiv:math-ph/0703065

[10] Carroll R 2007 Ricci flow and quantum theory *Preprint* arXiv:0710.4351
[11] Carroll R 2007 Remarks on Weyl geometry and quantum mechanics *Preprint* arXiv:0705.3921
[12] Carroll R 2007 Remarks on the Friedman equations *Preprint* arXiv:0712.3251
[13] Carroll R 2008 More on the quantum potential *Preprint* arXiv:0807.1320
[14] Carroll R 2008 Remarks on Fisher information *Preprint* arXiv:0807.4158
[15] 't Hooft G 2007 The mathematical basis for deterministic quantum mechanics *J. Phys.: Conf. Ser.* **67** 012015 (*Preprint* arXiv:quant-ph/0604008)
[16] 't Hooft G 2007 Emergent quantum mechanics and emergent symmetries *Preprint* arXiv:0707.4568
[17] Elze H-T 2008 Note on the existence theorem in "Emergent quantum mechanics and emergent symmetries" *J. Phys. A: Math. Theor.* **41** 304020 (*Preprint* arXiv:0710.2765
[18] Elze H-T 2009 *Int. J. Quant. Inf.* **7** 83 (*Preprint* arXiv:0806.3408)
[19] Isidro J M, Santander J.L.G. and Fernández de Córdoba P 2008 Ricci flow, quantum mechanics and gravity *Preprint* arXiv:0808.2351
[20] Elze H-T 2007 *Int. J. Quant. Inf.* **5** 215 (*Preprint* arXiv:hep-th/0605154)
[21] Elze H-T 2007 *J. Phys.: Conf. Ser.* **67** 012016 (*Preprint* arXiv:0704.2559)

Journal of Physics: Conference Series **174** (2009) 012034 doi:10.1088/1742-6596/174/1/012034

Can quantum mechanics be an emergent phenomenon?

Massimo Blasone[a], **Petr Jizba**[b,c] **and Fabio Scardigli**[d,e]

[a] INFN, Gruppo Collegato di Salerno; DMI, Università di Salerno, Fisciano (SA) - 84084 Italy
[b] ITP, Freie Universität Berlin, Arnimallee 14 D-14195 Berlin, Germany;
[c] FNSPE, Czech Technical University in Prague, Břehová 7, 115 19 Praha 1, Czech Republic
[d] Leung Center for Cosmology and Particle Astrophysics (LeCosPA),
Department of Physics, National Taiwan University, Taipei 106, Taiwan;
[e] Yukawa Institute for Theoretical Physics, Kyoto University, Kyoto 606-8502, Japan

E-mail: blasone@sa.infn.it, jizba@physik.fu-berlin.de, fabio@phys.ntu.edu.tw

Abstract. We raise the issue whether conventional quantum mechanics, which is not a hidden variable theory in the usual Jauch-Piron's sense, might nevertheless be a hidden variable theory in the sense recently conjectured by G. 't Hooft in his pre-quantization scheme. We find that quantum mechanics might indeed have a fully deterministic underpinning by showing that Born's rule naturally emerges (i.e., it is not postulated) when 't Hooft's Hamiltonian for be-ables is combined with Koopmann–von Neumann operatorial formulation of classical physics.

1. Introduction

Although quantum mechanics (QM) has been the undisputed basis for most of progress in fundamental physics during the last 90 years or so, the extension of the current theoretical frontier to Planck's scale physics, and recent enlargements of our experimental capabilities, may make the 21st century the period in which possible limits of quantum theory will be subjected to a thorough scrutiny. The basic premise of this paper is that QM is actually not a complete ontological system, but in fact it represents a very accurate low-energy approximation to a deeper level of dynamics. But what exactly the "deeper level dynamics" should be? There is a growing interest in this *bone of contention* which is partially fueled by the belief that in order to make a convincing synthesis of QM and general relativity (GR) a new conceptual paradigm is needed to tackle physics at very small space-time scales. There is, however, a fundamental discord in how this should be achieved. One way of thinking maintains that at high energies the rules of GR should be changed/modified while the rules of quantum physics should be kept untouched. Such a view is typically justified by arguing that QM is the most precise theory ever, being accurate to about one part in 10^{11} (for the latest precision tests of QED see, e.g. Ref. [1]). As pointed by R. Penrose this view is deceptive [2]: in fact GR has now been experimentally tested by measuring the orbit-period slowdown in the Hulse-Taylor binary pulsar PSR 1913 + 16 with accuracy to one part in 10^{14} – and this precision is apparently limited only by the accuracy of clocks on Earth [3]. Another way of thinking holds that it is rather QM that should be replaced at high energy scales with a more fundamental paradigm, while GR may, or may not, be modified at such high energies. Along this line, motivated by the black-hole thermodynamics,

G. 't Hooft has hinted [4] that the fundamental rules should be, at very high energy scale – perhaps at the Planck scale – deterministic.

There is a long standing suspicion towards determinism that might underlie QM which mainly steams from Bell's inequality (BI) [5] and its various generalizations [6]. While deterministic theories always satisfy Bell's inequality, QM systems evidently experimentally violate BI [7, 8, 9]. This is then interpreted as a proof of the non-existence of a deterministic underpinning for QM. At this stage it should be stressed that BI's are not a result about QM. They simply state that *usual* deterministic theories obeying the *usual* (Kolmogorovian) probability theory inevitably satisfy certain inequalities between the mean values of certain (suitably chosen) observed quantities. By "usual deterministic theories" one means theories that are *realistic*, i.e. the system has an intrinsic existence independent of observation, and *local*, i.e. sufficiently separated measurements should not influence each other. Though these assumptions about the underlying deterministic nature of QM are certainly intuitively plausible, they are too much rooted in our everyday laboratory-scale experience. Planck's scale ($\sim 10^{19}$ GeV, i.e. $\sim 10^{-35}$m) deterministic dynamics (whatever this means) may, after all, look non-local to a human observer whose observational scale is at best few TeV ($\sim 10^{-19}$m). In fact, according to 't Hooft [10], the enormous amount of information that would be lost in the process of "coarse graining" from 10^{19}GeV to 10^{3}GeV would lead to formation of equivalence classes, in the sense that many distinct states of the Planckian-scale dynamics would go into a single state of the observational-scale dynamics. For example, in Ref. [10] is proposed that, when the dynamical equations at be-ables level describe the configuration-space of a chaotic system, the equivalence classes could be related to the stable orbits of such system (e.g., limit cycles). The mechanism responsible for clustering of trajectories to equivalence classes is identified by 't Hooft as information loss: after a while one cannot retrace back the initial conditions of a given trajectory, one can only say at what attractive trajectory it will end up. Explicit examples of be-able systems that give rise, at a macroscopic level, to a genuine quantum behavior can be found in Refs. [4, 10, 11, 12]. Applications of the outlined scenario have been given, e.g., in Refs. [11, 12, 13, 14, 15]. For other approaches see also Refs. [16, 17, 18]

The passage to quantum theory without conventional quantization procedure is called by 't Hooft *pre-quantization*. In what follows we wish to use the pre-quantization concept to tackle the issue of Born's rule from first principles, i.e. without need to postulate it. The paper is organized as follows: In Section 2 we present some fundamentals of 't Hooft pre-quantization method. An extension of this is presented in the second half of Section 2. There we put forward some ideas concerning the dependance of 't Hooft's loss-of-information constraint on observer's energy scale. A connection with Koopman–von Neumann (KvN) operatorial formulation of classical physics is discussed in Section 3. There we demonstrate in some detail that Born's rule is, at the primordial energy scale, merely a convenient mathematical instrument for doing classical statistical physics, while other, equivalent means are also available. It is only at low-energy (observational) scales, where the operatorial and deterministic modes of description substantially depart, that Born's rule starts to have its own independent existence. Various remarks and speculations are postponed till the concluding Section 4.

2. Essentials of 't Hooft's pre-quantization method

In this section we briefly outline 't Hooft's continuous-time pre-quantization method [10]. A discrete-time version, which employs cellular automata, is discussed, e.g. in Refs. [4, 12, 19], and will not be followed here. We start with the assumption that the dynamics at the primordial deterministic level is described by the Hamiltonian

$$H = \sum_i p_i f_i(\mathbf{q}) + g(\mathbf{q})\,. \tag{1}$$

Journal of Physics: Conference Series **174** (2009) 012034 doi:10.1088/1742-6596/174/1/012034

Here f_i and g are functions of $\mathbf{q} = \{q_1, \ldots, q_n\}$. One should note that the equation of motion for q_i, i.e.

$$\dot{q}_i \;=\; f_i(\mathbf{q})\,, \tag{2}$$

is autonomous, since the p_i variables are decoupled. The above system is obviously deterministic, however its Hamiltonian is not bounded from below. It is also worthy of noting that

$$q_i(t+\Delta t) \;=\; q_i(t) + f_i(\mathbf{q})\Delta t + \frac{1}{2} f_k(\mathbf{q}) \frac{\partial^2 H}{\partial p_i \partial q_k}(\Delta t)^2 \;+\; \cdots \;=\; F_i(\mathbf{q}(t), \Delta t)\,, \tag{3}$$

where F_i is some function of $\mathbf{q}(t)$ and Δt but not $\mathbf{p}$. Since (3) holds for any Δt we get the Poisson bracket

$$\{q_i(t'), q_k(t)\} = 0 \quad \text{for} \quad \forall\; t, t'\,. \tag{4}$$

Because of the autonomous character of Eq.(2) one can define a formal Hilbert space $\mathcal{H}$ spanned by the states $\{|\mathbf{q}\rangle\}$, and associate with p_i the operator $\hat{p}_i = -i\partial/\partial q_i$. It is not difficult to see that the generator of time translations (i.e., the "Hamiltonian" operator), of the form $\hat{H} = \sum_i \hat{p}_i f_i(\hat{\mathbf{q}}) + g(\hat{\mathbf{q}})$ generates precisely the deterministic evolution (2). Indeed, we first observe that because $\hat{H}$ is generator of time translations then in Heisenberg's picture

$$\hat{q}_i(t+\Delta t) = e^{i\Delta t \hat{H}} \hat{q}_i(t) e^{-i\Delta t \hat{H}}\,, \tag{5}$$

which for infinitesimal Δt implies

$$\hat{q}_i(t+\Delta t) - \hat{q}_i(t) = i\Delta t[\hat{H}, q_i(t)] \quad \Rightarrow \quad \dot{\hat{q}}_i = f_i(\hat{\mathbf{q}})\,. \tag{6}$$

On the other hand for arbitrary finite Δt we have from (5)

$$\hat{q}_i(t+\Delta t) \;=\; \sum_{n=0}^{\infty} \frac{1}{n!}[\hat{H}, [\hat{H}, [\cdots [\hat{H}, \hat{q}_i(t)]] \cdots]] \;=\; \tilde{F}_i(\hat{\mathbf{q}}(t), \Delta t)\,. \tag{7}$$

In Eq.(7) $\hat{H}$ appears in the generic term of the sum n times. On the other hand $\tilde{F}_i$ is some function of $\hat{\mathbf{q}}(t)$ and Δt but not $\hat{\mathbf{p}}$, which immediately implies that

$$[\hat{q}_i(t), \hat{q}_j(t')] \;=\; 0\,, \tag{8}$$

for any t and t' (this in turn gives $F_i = \tilde{F}_i$). Result (8) shows that the Heisenberg equation of motion for $\hat{q}_i(t)$ in the $\mathbf{q}$-representation is identical with the c-number dynamical Eq.(2). This is because $\hat{q}_i(t+\Delta t)$ and $\hat{q}_i(t)$ commute, and hence $\hat{q}_i(t+\Delta t)$, $\hat{q}_i(t)$, $f_i(\hat{\mathbf{q}})$ and also $\mathrm{d}\hat{q}_i(t)/\mathrm{d}t$ can be simultaneously diagonalized. In this diagonal basis we get back the c-numbered autonomous Eq.(2). In other words, operators $\hat{q}_i$ evolve deterministically even after "quantization". This evolution is only between base vectors. From the Schrödinger-picture point of view this means that the *state vector* evolves smoothly from one *base vector* to another (in Schrödinger picture base vectors are time independent and fixed). So at each instant the state vector coincides with some specific base vector. Because of this, there is *no* non-trivial linear superposition of the state vector in terms of base vectors and hence *no* interference phenomenon shows up when measurement of $\mathbf{q}$-variable is performed. Dynamical variables fulfilling Eq.(8) were first considered by Bell [5] who called them *be-ables* as opposed to observed dynamical variables which are in QM called *observables.*

Let us now come back to Eq.(1). One may immediately notice that H is unbounded from below. This fact should not disturb us too much since actual dynamics of be-ables is described by Eq.(2). The H was merely introduced to set up the parallel operatorial formulation of the be-able dynamics. In the following section we will see that this Hamiltonian has yet another useful role, namely it helps to formulate a classical statistical mechanics for be-ables. On the observational (i.e. emergent) level the Hamiltonians are key objects providing both the equations of motions and the energy of the system. The *emergent* Hamiltonians must be bounded from below. The concept of lower bound is, in 't Hooft's proposal, just an emergent property formed during the coarse graining of the be-able degrees of freedom down to the observational ones. We can devise a simple toy model mechanism showing how a lower bound for H may develop. To this end consider $\rho(\hat{q})$ to be some positive function of $\hat{q}_i$ (but not $\hat{\mathbf{p}}$) with $[\hat{\rho}, \hat{H}] = 0$. One then defines the splitting

$$\begin{aligned} \hat{H} &= \hat{H}_+ - \hat{H}_-, \\ \hat{H}_+ &= \left(\hat{\rho} + \hat{H}\right)^2 \frac{\hat{\rho}^{-1}}{4}, \quad \hat{H}_- = \left(\hat{\rho} - \hat{H}\right)^2 \frac{\hat{\rho}^{-1}}{4}, \end{aligned} \tag{9}$$

where $\hat{H}_+$ and $\hat{H}_-$ are positive-definite operators satisfying

$$[\hat{H}_+, \hat{H}_-] = [\hat{\rho}, \hat{H}_\pm] = 0. \tag{10}$$

At this stage we introduce the "coarse-graining" operator $\hat{\Phi}$ that describes the loss of information occurring during the passage from the be-able to observational scale. One possible choice is [1]

$$\hat{\Phi}_E = (1 - e^{-(E_p - E)/E})\hat{H}_-, \tag{11}$$

where E refers to the observer's energy scale, while E_p is the be-able energy scale, which we take to be Planck's scale. The operator $\hat{\Phi}_E$ is then implemented as a constraint on the Hilbert space $\mathcal{H}$. So at each observational energy scale E, the observed physical states $|\psi\rangle_{phys}$ are given by the condition

$$\hat{\Phi}_E|\psi\rangle_{phys} = 0. \tag{12}$$

This equation identifies the states that are not affected by the coarse graining, i.e., states that are still distinguishable at the observational scale E. The c-number factor $(1 - e^{-(E_p-E)/E})$ is *not* irrelevant and one cannot use directly $\hat{H}_-$ instead of $\hat{\Phi}_E$. In fact the constraint (12) is, according to Dirac's classification of constraints [20], a first class primary constraint because $[\hat{\Phi}_E, \hat{\Phi}_E] = 0$ and $[\hat{\Phi}_E, \hat{H}] = 0$. First-class constraints generate gauge transformation and hence not only restrict the full Hilbert space (from $\mathcal{H}$ to $\mathcal{H}_c$) but also produce equivalence classes of states, which are in general non local (e.g., points with a space-like separation are identified). Two states belong to the same class if they can be transformed into each other by a gauge transformation with the generator $\hat{\Phi}_E$. Let $\mathcal{G}_E$ be the one parameter group of these gauge transformations generated by $\hat{\Phi}_E$. The equivalence classes built out of such gauge transformations represent at each scale E the physical states (i.e., observables). If $\mathcal{O}_E$ denotes the space spanned by the observables, we can identify $\mathcal{O}_E$ with a quotient space

$$\mathcal{O}_E = \mathcal{H}_c/\mathcal{G}_E. \tag{13}$$

The quotient space $\mathcal{O}_E$ (its structure and dimensionality) depends on the energy scale E. In particular at the level of be-ables where $E = E_p$ the constraint $\hat{\Phi}_E$ is identically zero and the

[1] This form might be inspired by the entropy considerations reported in [11]. We shall discuss this issue elsewhere.

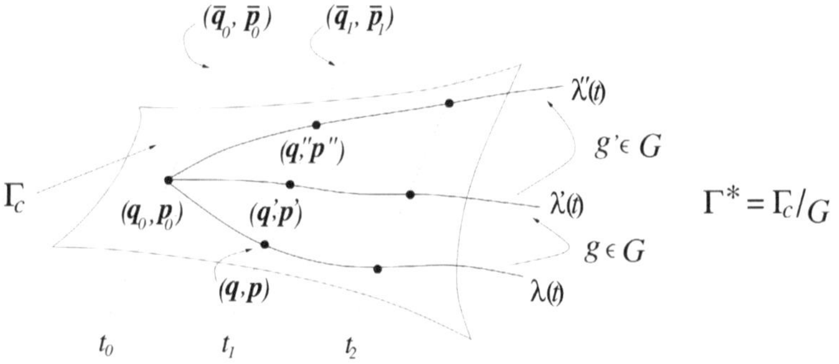

Figure 1. Pictorial representation of the gauge freedom inherent in the evolution $dq_i/dt = \{q_i, H_e\}$. Here $(\mathbf{q}, \mathbf{p})$, $(\mathbf{q}', \mathbf{p}')$, $(\mathbf{q}", \mathbf{p}")$, ... describe the same physical state $(\bar{\mathbf{q}}_1, \bar{\mathbf{p}}_1)$. g are elements of the gauge group generated by Φ_E. For simplicity we do not explicitly write down the sub-index E.

space of observables is directly the Hilbert space $\mathcal{H}$. On the other hand, when $E \ll E_p$, e.g. at scales available to a human observer, one has $\hat{\Phi}_E = \hat{H}_-$. The latter is the constraint originally considered by 't Hooft [4, 10]. In such a case the resulting physical state space, i.e. the space of observables, has the energy eigenvalues that are trivially bounded from below owing to

$$\hat{H}|\psi\rangle_{phys} = \hat{H}_+|\psi\rangle_{phys} = \hat{\rho}|\psi\rangle_{phys} \,. \tag{14}$$

Thus, in the Schrödinger picture the equation of motion

$$\frac{\mathrm{d}}{\mathrm{d}t}|\psi_t\rangle_{phys} = -i\hat{H}_+|\psi_t\rangle_{phys} \,, \tag{15}$$

has only positive frequencies on physical states. Above coarse-graining procedure was so far implemented in an operatorial mode of description of the be-able dynamics. As a result we have obtained on observational energy scales a Schrödinger equation with a "well behaved" (i.e., bounded from below) emergent Hamiltonian. Constraining procedure can be, in parallel, also applied to a deterministic mode of description of the be-able dynamics. This can be done through the Dirac–Bergmann algorithm [21]. Here one implements the constraint Φ_E directly on the phase space Γ spanned by $\mathbf{p}$ and $\mathbf{q}$. This results in the constrained hyper surface Γ_c. Because Φ_E Poisson-commutes with itself and with H, it is a first class primary constraint [21]. Similarly as in the operatorial case, this means that Φ_E not only restricts the phase space but also provides classes of equivalence for possible evolutions. This is because Φ_E is unable to determine the Lagrange multiplier in the "extended" Hamiltonian $H_e(\mathbf{q}, \mathbf{p}) = H(\mathbf{q}, \mathbf{p}) + \lambda(t)\Phi_E(\mathbf{q})$. In particular, the self-consistency equation

$$0 = \frac{\mathrm{d}\Phi_E}{\mathrm{d}t} = \{\Phi_E, H_e\} = \{\Phi_E, H\} + \lambda(t)\{\Phi_E, \Phi_E\} \,, \tag{16}$$

is trivially fulfilled without giving any information on $\lambda(t)$. This leads to the class of equivalent dynamics described by the equation $dq_i/dt = \{q_i, H_e\}$. Each of the possible trajectories in the class is parameterized by different $\lambda(t)$, see Fig.1. Two points belong to the same class if they can be transformed into each other by gauge transformation generated by Φ_E. For infinitesimal transformations this means that for two equivalent dynamics with slightly different $\lambda(t)$

$$\delta\mathbf{q}(t) = \mathbf{q}_{\lambda(t)} - \mathbf{q}_{\lambda'(t)} = \varepsilon\{\mathbf{q}(t), \Phi_E\} \,, \tag{17}$$

with $\varepsilon = t(\lambda - \lambda')$. Gauge classes can be identified by choosing a single representative element from each class. This is done by a gauge-fixing condition χ such that $\{\Phi_E, \chi\} \neq 0$. The emergent physical space (space of classically observed degrees of freedom at scale E) is a quotient space $\Gamma_E^* = \Gamma_c/\mathcal{G}_E$. It should be noted that the identification of the equivalent classes is a non-local procedure, since at a given instant one generally identifies also points with a space-like separation. This is, in spirit, deeply different from the quotioning procedure used in gauge field theories. There the quotient space is obtained from the configuration space that in itself is non-physical (it contains huge amount of physically equivalent configurations). The physical space emerges only after one identifies each equivalence class with a distinct physical configuration. In contrast, in our case we start with the configuration space (space of be-able dynamics) that is *a priori physical* and the identification of equivalent classes is done only because we wish to find an "effective" (or emergent) description at some energy scale E that is much lower than the be-ables scale.

3. Koopman–von Neumann formulation and 't Hooft's be-able dynamics

Let us come back to 't Hooft's Hamiltonian (1) and try to understand its role a bit more. There are three immediate issues that should be addressed. These are: the issue of hermiticity, the issue of the canonical momenta p_i, and finally the issue of the ensuing state-vectors and related Born's rule. The hermiticity is not a real problem at the energy scale E_P. There the dynamics is driven by Eq.(2) and the Hamiltonian is only a formal tool that allows to generate the corresponding evolution equations. Should one wish to have hermitian Hamiltonian from the very scratch, one may compensate for the non-hermitian ordering between $\hat{p}_i$ and $f_i(\hat{\mathbf{q}})$ by adding the (non dynamical) function $g(\hat{\mathbf{q}})$ to $\hat{H}$. This should be done in such a way that $g^\dagger(\hat{\mathbf{q}}) - g(\hat{\mathbf{q}}) = \sum_i [\hat{p}_i, f_i(\hat{\mathbf{q}})]$, so that $\hat{H}^\dagger = \hat{H}$. In the particular case when $g(\hat{\mathbf{q}})$ is anti-hermitian the Hamiltonian is not only hermitian but also Weyl ordered. Indeed, in this case we have

$$\begin{aligned} \hat{H} &= \sum_i \hat{p}_i f_i(\hat{\mathbf{q}}) + g(\hat{\mathbf{q}}) = \tfrac{1}{2}\sum_i \left(\hat{p}_i f_i(\hat{\mathbf{q}}) + f_i(\hat{\mathbf{q}})\hat{p}_i\right) + \tfrac{1}{2}\sum_i [\hat{p}_i, f_i(\hat{\mathbf{q}})] + g(\hat{\mathbf{q}}) \\ &= \tfrac{1}{2}\sum_i \left(\hat{p}_i f_i(\hat{\mathbf{q}}) + f_i(\hat{\mathbf{q}})\hat{p}_i\right) = W\left(\sum_i \hat{p}_i f_i(\hat{\mathbf{q}})\right). \end{aligned} \tag{18}$$

In this connection it is interesting to notice that the Weyl form (18) also enters in the Pontryagin's approach [22] to quantization of non-Hamiltonian systems described by equations (2). The hermiticity itself is, however, an important concept at low energies, where the emergent Hamiltonian is directly interpreted as the energy of the system. However, in the existent models of 't Hooft's constraining conditions the emergent Hamiltonian is hermitian. The canonical momentum p_i is at the be-able level just a dummy variable which merely allows to set up a Hamiltonian formalism. In the previous section we have seen that it also allows to formulate an operatorial version of the be-able dynamics. It should be noticed that, while at the Planck energy E_P the Poisson bracket (or alternatively the commutator) of the be-able position with its *kinetic* momentum vanishes, i.e. $\{q_i(t), p_j^{kin}(t)\} \propto \{q_i(t), \dot{q}_j(t)\} = 0$, on the contrary the corresponding Poisson bracket with canonical momentum is the canonical one, $\{q_i(t), p_j(t)\} = 1$. Since the be-able dynamics (2) is formulated on the tangent rather than cotangent bundle, it is the kinetic rather than canonical momenta that is physically a more relevant object at E_P. The third and more pressing issue is the Born rule. We now show that the Born rule is closely related to Koopman–von Neumann's (KvN) operatorial formulation of classical physics [23] [24]. To this end we can ask ourselves how an observer "living" in the be-able world would do statistical physics on systems described by Eq.(2). In fact, the simplest way is to go through the Hamiltonian (1). In this case Koopman and von Neumann found a simple recipe for defining the probability density function. Following KvN we define a "wave function" $\psi(\mathbf{p}, \mathbf{q}, t)$ that evolves in time with the Liouville operator, i.e., the Hamilton field (summation over i is understood)

$$\hat{\mathcal{H}} = -i\partial_{p_i}H(\mathbf{p},\mathbf{q})\partial_{q_i} + i\partial_{q_i}H(\mathbf{p},\mathbf{q})\partial_{p_i}\,, \tag{19}$$

according to the equation

$$i\frac{\partial}{\partial t}\psi = \hat{\mathcal{H}}\psi\,. \tag{20}$$

The vectors ψ are complex wave functions on the phase space $\Gamma = (\mathbf{p},\mathbf{q})$, with the normalization $\int d\mathbf{p}d\mathbf{q}\,|\psi(\mathbf{p},\mathbf{q})|^2 = 1$. The corresponding complex conjugation turns Eq.(20) to

$$i\frac{\partial}{\partial t}\psi^* = \hat{\mathcal{H}}\psi^*\,. \tag{21}$$

By multiplying (20) with ψ^* and (21) with ψ, and adding them together we obtain

$$i\frac{\partial}{\partial t}\varrho = \hat{\mathcal{H}}\varrho \quad\Leftrightarrow\quad \frac{\partial}{\partial t}\varrho = (-\partial_{p_i}H\partial_{q_i} + \partial_{q_i}H\partial_{p_i})\varrho\,, \tag{22}$$

where we have defined $\varrho(\mathbf{p},\mathbf{q}) = \psi^*(\mathbf{p},\mathbf{q})\psi(\mathbf{p},\mathbf{q})$. The second equation is nothing but the well-known Liouville equation for the probability density function of a (classical) statistical system whose dynamics is driven by a (classical) Hamiltonian H. If one defines the scalar product between two wave functions ψ_1 and ψ_2 as $\langle\psi_1|\psi_2\rangle = \int d\mathbf{p}d\mathbf{q}\,\psi_1^*\psi_2$, then it is easy to show that $\langle\psi_1|\hat{\mathcal{H}}\psi_2\rangle = \langle\hat{\mathcal{H}}\psi_1|\psi_2\rangle$, i.e., the Liouvillian $\hat{\mathcal{H}}$ is a self-adjoint operator, which implies that the norm of any state is conserved during the evolution. The latter is consistent with the interpretation of $\varrho = \psi^*\psi$ as a probability density function on the phase space Γ. Now, since the $\mathbf{p}$ variables are in 't Hooft's proposal only dummy variables (true degrees of freedom are be-ables, i.e., variables $\mathbf{q}$) the relevant density matrix is the reduced density matrix $\tilde{\varrho}(\mathbf{q}) = \int d\mathbf{p}\,\varrho(\mathbf{p},\mathbf{q})$. By using 't Hooft's Hamiltonian (1) and integrating (22) over $\mathbf{p}$ we obtain

$$\begin{aligned}\frac{\partial}{\partial t}\tilde{\varrho}(\mathbf{q}) &= -f_i(\mathbf{q})\partial_{q_i}\tilde{\varrho}(\mathbf{q}) - [\partial_{q_i}f_i(\mathbf{q})]\tilde{\varrho}(\mathbf{q}) = -\partial_{q_i}(f_i(\mathbf{q})\tilde{\varrho}(\mathbf{q}))\\ &= -i\hat{p}_i f_i(\mathbf{q})\tilde{\varrho}(\mathbf{q}) = -i\hat{H}(\mathbf{p},\mathbf{q})\tilde{\varrho}(\mathbf{q})\,,\end{aligned} \tag{23}$$

where $\hat{H}$ is 't Hooft's Hamiltonian without the g term. If now we define

$$\tilde{\psi}(\mathbf{q}) = \int_{-\infty}^{\infty} d\mathbf{p}\,\psi(\mathbf{p},\mathbf{q})\,, \tag{24}$$

we can use Eq.(20) (respectively Eq.(21)) to compute the evolution of $\tilde{\psi}(\mathbf{q})$ (respectively $\tilde{\psi}^*(\mathbf{q})$) under the 't Hooft's Hamiltonian $\hat{H}$, so that we obtain

$$i\frac{\partial}{\partial t}\tilde{\psi}(\mathbf{q}) = \hat{H}(\mathbf{p},\mathbf{q})\tilde{\psi}(\mathbf{q}) \quad\text{and}\quad i\frac{\partial}{\partial t}\tilde{\psi}^*(\mathbf{q}) = \hat{H}(\mathbf{p},\mathbf{q})\tilde{\psi}^*(\mathbf{q})\,. \tag{25}$$

Multiplying the first by $\tilde{\psi}^*(\mathbf{q})$ and the second by $\tilde{\psi}(\mathbf{q})$, summing together, and taking advantage of the particular form of 't Hooft's Hamiltonian, we get

$$\frac{\partial}{\partial t}(\tilde{\psi}^*(\mathbf{q})\tilde{\psi}(\mathbf{q})) = -i\hat{H}(\mathbf{p},\mathbf{q})(\tilde{\psi}^*(\mathbf{q})\tilde{\psi}(\mathbf{q})) \tag{26}$$

which, confronted with (23), bring us to the identification $\tilde{\varrho}(\mathbf{q}) = \tilde{\psi}^*(\mathbf{q})\tilde{\psi}(\mathbf{q})$

This is precisely the (configuration-space) Born rule. While in the previous section we have seen that the be-able dynamics can be equivalently described both by classical or (quantum-like) operatorial means, this section reinforces the picture by showing that this parallelism can be extended also to statistical physics. The parallel mode of description of course breaks when constraint (12) is imposed. We thus see that 't Hooft's Hamiltonian has a privileged role among classical Hamiltonians. In fact, as shown in Ref. [25], there are no other Hamiltonian systems with the peculiar property that their full quantum evolution coincides with the classical one.

We should finally note that the above canonical scenario has also a path-integral counterpart. In this case the KvN operatorial version of the deterministic system (1) is taken over by the classical path integral of Gozzi *et al* [25, 26]. In fact, because both $\tilde{\psi}(\mathbf{q})$ and $\tilde{\varrho}(\mathbf{q}) = |\tilde{\psi}(\mathbf{q})|^2$ satisfy the same equation of the motion – Schrödinger's equation, the solutions can be written through the same evolution kernel, i.e.

$$\tilde{\varrho}(\mathbf{q},t) = \int d\mathbf{q}' K(\mathbf{q},t;\mathbf{q}',t')\tilde{\varrho}(\mathbf{q}',t')\,, \quad \text{and} \quad \tilde{\psi}(\mathbf{q},t) = \int d\mathbf{q}' K(\mathbf{q},t;\mathbf{q}',t')\tilde{\psi}(\mathbf{q}',t')\,. \tag{27}$$

Although both above kernels represent formally identical expressions, they are qualitatively different. The first kernel in (27) corresponds to a transition *probability*, while the second corresponds to a transition *amplitude*. The fact that the same kernel propagates both $\tilde{\psi}$ and $|\tilde{\psi}|^2$ was, in the KvN framework, firstly recognized and discussed in Refs. [26]. The corresponding discussion for 't Hooft Hamiltonian system was done in Ref. [25]. The evolution kernel itself has the path integral representation

$$K(\mathbf{q},t;\mathbf{q}',t') = \mathcal{N}\int_{\xi(t)=\xi_1}^{\xi(t')=\xi_2} \mathcal{D}\xi \, \exp\left[i\int_t^{t'} dt \, L(\xi,\dot{\xi})\right] = \mathcal{N}\int_{\mathbf{q}(t)=\mathbf{q}}^{\mathbf{q}(t')=\mathbf{q}'} \mathcal{D}\mathbf{q} \prod_i \delta[\dot{q}_i - f_i(\mathbf{q})]\,. \tag{28}$$

Here we have defined $2N$ configuration-space coordinates as

$$\xi_i \;=\; p_i, \quad i = 1,\ldots,N \qquad \text{and} \qquad \xi_i \;=\; q_i, \quad i = N+1,\ldots,2N\,, \tag{29}$$

and the Lagrangian

$$L(\xi,\dot{\xi}) \;=\; \tfrac{1}{2}\xi^i\omega_{ij}\dot{\xi}^j - H(\xi)\,, \tag{30}$$

(ω is the $2N\times 2N$ symplectic matrix). The factor $\mathcal{N}$ is a normalization factor, and the measure can be explicitly written as

$$\mathcal{N}\int_{\xi(t)=\xi_1}^{\xi(t')=\xi_2} \mathcal{D}\xi \;=\; \mathcal{N}\int_{\mathbf{q}(t)=\mathbf{q}_1}^{\mathbf{q}(t')=\mathbf{q}_2} \mathcal{D}\mathbf{q}\mathcal{D}\mathbf{p}\,. \tag{31}$$

The kernel representation (28) corresponds to the classical path integral of Gozzi *et al* [26]. The constraining procedure can be now applied directly to the path integral (28) either through Dirac–Bergmann [21] or Faddeev–Jackiw procedure [25, 27, 28]. The actual constraining mechanism is however different when applied to path integrals representing transition amplitudes or to path integrals representing transition probabilities. In particular, for transition probabilities one should rewrite the path integral in the Feynman–Vernon closed-time-path form [29], and then apply the constraint to each time branch independently. This lead to a different emergent kernel than in the case of transition amplitudes. So, similarly as in the operatorial framework, in the path-integral formulation the constraint (12) removes the dichotomy of classical-quantal formulation of the be-able dynamics. It should be stressed that the symplectic structure in both kernels will be, however, still preserved, yet now with an effective Hamiltonian and with different degrees of freedom [28].

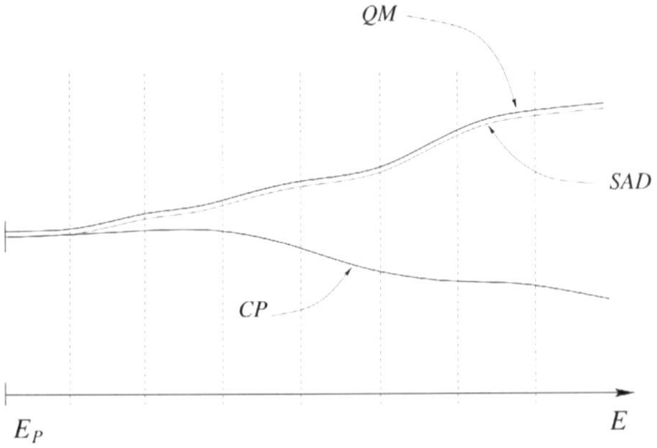

Figure 2. Schematic representation of the emergent quantum description. At the primordial level E_P both classical and quantal descriptions are equivalent. Also, both the state ψ and the deterministic density matrix ϱ are described via the same evolution kernel K. As the energy scale lowers, the information is continuously lost in the process of coarse graining and the observer sees only effective dynamics, i.e. equivalence classes of primordial dynamics. The dashed lines remind that constraint (12) must be applied at all energy scales. Operatorial and deterministic description are coarse-grained together and at low energies are perceived as QM and SAD, respectively. On the statistical level SAD is described via a density matrix with a symplectic evolution kernel. The latter is what the observer perceives as CP.

4. Conclusions and outlooks

In this paper we have demonstrated that when KvN operatorial formulation of classical physics is combined with 't Hooft's Hamiltonian for be-ables, one naturally arrives at the Born's rule without postulating it. This result substantiates 't Hooft's pre-quantization scheme and affirmatively answer to the question whether QM can be a hidden variable theory. It should be stressed that the be-able dynamics is fully local (see Eq.(1)). The non-locality that provides the loophole in Bell inequalities is realized in 't Hooft's pre-quantization by identifying the physical space of observables Γ^*_E with a quotient space $\Gamma_c/\mathcal{G}_E$ that has a non-local structure with respect to the physical space of be-ables Γ.

The picture one can draw from our discussion is the following (see Fig.2): at the primordial scale E_P the dynamics is purely deterministic and it can be equivalently tackled with classical (description on the level of single trajectories) and operatorial (description on the level of evolution in the Hilbert space of states) means. As one lowers the energies, the typical length and time scales are getting larger, and the corresponding description is getting coarse-grained. This leads to a huge information loss and brings one to an effective level of description. A typical observer at $E \ll E_P$ has then at his/her disposal the operatorial mode of description – the Schrödinger equation (15) with the effective Hamiltonian $\hat{H}_+$, which he/she calls *Quantum Mechanics*. On a parallel footing, there is also available to him/her a non-operatorial, non-local, coarse-grained description of the deterministic system that propagates at energy E_P. This non-local imprint of the be-able dynamics[2] (SAD) has, however, the statistical density matrix that is described with the standard classical Liouville equation, or equivalently with a classical kernel for transition probabilities. The kernel itself can be represented via path integral with the Lagrangian that exhibits an explicit symplectic structure. It is thus natural for our observer to define as *Classical Physics* (CP) the physics experienced at the macroscopic *statistical* level. This emergent macroscopic description exhibits a symplectic structure which is then identified

[2] We denote the non-local imprint of the be-able dynamics as SAD to stress its formal connection with Einstein's "spooky action at distance".

as a typical signature of CP. In this view, one should not be surprised that BI's are violated at an observational scale. This is because our understanding of CP (on which Bell's inequalities are ultimately based) is not derived from SAD dynamics alone (where BI's could be easily violated) but only from its statistical behavior.

Acknowledgments

We are particularly grateful to G. 't Hooft and H. Kleinert for inspiring conversations. M.B.acknowledges partial financial support from MIUR and INFN. F.S. thanks Japan Society for the Promotion of Science for support under the fellowship P06782, National Taiwan University and LeCosPA for support under the research contract 097510/52313, and ITP Freie Universität Berlin for warm hospitality. P.J. would like to acknowledge financial supports from the Ministry of Education of the Czech Republic (research plan no. MSM 6840770039) and from the Deutsche Forschungs Gemeinschaft (grant Kl256/47).

References

[1] Amsler C *et al* 2008 *Phys. Lett.* B **667** 1
[2] Hawking S and Penrose R 2000 *The Nature of Space and Time* (New York: Princeton U. Press)
[3] Weisberg J M and Taylor J H 2004 *Preprint* arXiv:astro-ph/0407149
[4] 't Hooft G 1988 *J. Stat. Phys.* **53** 323; *do.* 1996 *Class. Quant. Grav.* **13** 1023; *do.* 1999 *Class. Quant. Grav.* **16** 3263; *do.* 2002 *Preprint* arXiv:quant-ph/0212095
[5] Bell J S 1987 *Speakable and Unspeakable in Quantum Mechanics* (Cambridge, UK: Cambridge U. Press)
[6] see, e.g., Clauser J F, Horne M A, Shimony A and Holt R A 1969 *Phys. Rev. Lett.* **26** 880; Greenberger D M, Horne M, Shimony A and Zeilinger A 1990 *Am. J. Phys.* **58** 1131
[7] Aspect A, Dalibard J and Roger G 1982 *Phys. Rev. Lett.* **49** 1804; Aspect A, Grangier P and Roger G 1982 *Phys. Rev. Lett.* **49** 91
[8] Tittel W, Brendel J, Zbinden H, and Gisin N 1998 *Phys. Rev. Lett.* **81** 3563
[9] Rowe M A *et al* 2001 *Nature* **409** 791
[10] 't Hooft G 2003 *Int. J. Theor. Phys.* **42** 355; *do.* 1999 *in 37th International School of Subnuclear Physics, Erice* ed A Zichichi (London: World Scientific) (*Preprint* arXiv:hep-th/0003005)
[11] Blasone M, Jizba P and Vitiello G 2001 *Phys. Lett.* A **287** 205
[12] Blasone M, Jizba P and Vitiello G 2003 *J. Phys. Soc. Jap. Suppl.* **72** 50; Blasone M and Jizba P 2002 *Can. J. Phys.* **80** 645; Blasone M, Celeghini E, Jizba P and Vitiello G 2003 *Phys. Lett.* A **310** 393
[13] Scardigli F 2007 *Found. of Phys.* **37** 1278
[14] Elze H T 2005 *Braz. J. Phys.* **35**(2A) 343 (*Preprint* arXiv:hep-th/0503069)
[15] Elze H T 2009 *Int. J. Qu. Info. (IJQI)* **7** 83–96 (*Preprint* arXiv:0806.3408)
[16] Wetterich C 2004 in *Decoherence and Entropy in Complex Systems* ed H-T Elze, Lecture Notes in Physics **633** (Berlin: Springer)
[17] Adler S L 2005 *Quantum Mechanics as an Emergent Phenomenon* (Cambridge, UK: Cambridge U. Press)
[18] Smolin L 2006 *Preprint* arXiv:quant-ph/0609109
[19] Blasone M and Jizba P 2007 *J. Phys.: Conf. Ser.* **67** 012046
[20] see, e.g., Sundermeyer K 1982 *Constrained Dynamics with Applications to Yang-Mills theory, General Relativity, Classical Spin, Dual String Model* (Berlin: Springer)
[21] Dirac P A M 1950 *Can. J. Math.* **2** 129; *do.* 1958 *Proc. R. Soc. London.* A **246** 326; *do.* 1965 *Lectures in Quantum Mechanics* (New York: Yeshiva University); Bergmann P G 1949 *Phys. Rev.* **75** 680
[22] Pontryagin L S, Boltanskij V G, Gamkrelidze R V and Miscenko E F 1962 *The Mathematical Theory of Optimal Processes* (New York: Wiley)
[23] Koopman B O 1931 *Proc. Natl. Acad. Sci. U.S.A.* **17** 315; von Neumann J 1932 *Ann. Math.* **33** 587; *do.* 1932 *Ann. Math.* **33** 789
[24] The KvN formalism has been considered, in connection with 't Hooft deterministic QM, also in: Elze H T 2006 *J. Phys.: Conf. Ser.* **33** (2006) 399, although in a different context from the one of the present article
[25] Blasone M, Jizba P and Kleinert H 2005 *Phys. Rev.* A **71** 052507
[26] Gozzi E 1988 *Phys. Lett.* B **201** 525; Gozzi E, Reuter M and Thacker W D 1989 *Phys. Rev.* D **40** 3363; Mauro D 2002 *Int. J. Mod. Phys.* A **17** 1301
[27] Faddeev L D and Jackiw R 1988 *Phys. Rev. Lett.* **60** 1692
[28] Blasone M, Jizba P and Kleinert H 2005 *Ann. Phys.* **320** 468
[29] Feynman R P and Vernon F L 1963 *Ann. Phys.* **24** 118

Born's rule for momentum measurement from prequantum field theory

Andrei Khrennikov
International Center for Mathematical Modeling
in Physics and Cognitive Sciences,
University of Växjö, S-35195, Sweden

E-mail: Andrei.Khrennikov@vxu.se

Abstract. Recently a new attempt to go beyond quantum mechanics (QM) was done in the form of so called prequantum classical statistical field theory (PCSFT). In this approach QM is considered as incomplete theory – in total accordance with Einstein [1]. PCSFT is (one of possible) completions of QM. Hidden variables are given by classical fields. (It is again in accordance with Einstein.) These prequantum fields fluctuate on time and space scales which are essentially finer than scales of QM, cf. 't Hooft's attempt to go beyond QM [2, 3]. However, opposite to Einstein, PCSFT does not assign results of quantum observations directly to hidden variables. Bohr's idea that the whole experimental arrangement should be taken into account is basic for PCSFT-based measurement theory. This theory for measurements of momentum of quantum particles is presented in this paper.

1. Introduction

Since the first days of creation of quantum mechanics (QM), its formalism and especially its interpretation has been inducing the permanent feeling of unsatisfaction. Especially disturbing is Bohr's postulate on completeness of QM, i.e., impossibility to go beyond it and to develop a finer description of micro-world than given by QM-formalism. Scientific curiosity and belief that we can proceed infinitely deeply in mysteries of Nature induced the strong feeling of protest against Bohr's no-go principle. We remind that completeness of QM is a purely philosophic principle. All later attempts to justify it with help of mathematical theorems, so called no-go theorems – von Neumann, Bell, Kochen-Specker, etc. did not satisfy people who are more or less well educated in questions of quantum foundations, see, e.g., [4] for detailed analysis of this problem. As a consequence of this unsatisfactory situation in quantum foundations, various attempts to go beyond QM have been done during the last 80 years. We can mention de Broglie's theory of double solution which was later elaborated in Bohmian mechanics, stochastic electrodynamics, semiclassical model for quantum optics, Nelson's stochastic QM and its generalization by Davidson and, recently, 't Hooft's model, see, e.g., [5]–[10], [2, 3]; see also recent paper of T. Elze on justification of 't Hooft's approach [11].

Recently [12]–[15] a new attempt to go beyond QM was done in the form of the so-called prequantum classical statistical field theory (PCSFT). Hidden variables are given by classical fields. These prequantum fields fluctuate on time and space scales which are essentially finer than scales of QM, cf. 't Hooft's attempt to go beyond QM [2, 3]. However, opposite to Einstein, PCSFT does not assign results of quantum observations directly to hidden variables. Bohr's idea that the whole experimental arrangement should be taken into account is basic for PCSFT-based measurement theory. This theory for measurements of momentum of quantum particles in presented in this paper. Thus our theory combines peacefully ideas of both Einstein and Bohr. On the one hand, we deny Bohr's philosophic principle on completeness of QM. On the other hand, we deny naive Einsteinian realism – assigning results of all possible observations to a hidden variable.

We remind that the *semiclassical model* provides a simple formalism for calculation of probabilities of photon detection, see, e.g., [5]–[7]. The semiclassical model is a purely classical field model. In particular, such a fundamental element of the quantum mechanical description as noncommutativity is eliminated. Quantum observables are presented by classical random variables. However, the semiclassical model is applicable only to the electromagnetic field. Moreover, not all probability distributions of QM can be reproduced, see, e.g., Scully and Zubairy [5] for discussions.

We shall also consider a model with prequantum field variables. Comparing the semiclassical model with our model (which we call *prequantum classical statistical field theory* – PCSFT), we can say:

a) PCSFT is based on a different mathematical model of random field;

b) it is applicable not only to the electromagnetic field, but for any type of field;

c) it reproduces Born's rule in the general framework, hence it reproduces any quantum probability distribution.

In PCSFT the basic objects are not particles, but classical fields, therefore by b) any type of quantum particle is described by the corresponding prequantum random field. For example, we can consider the electron field or the neutron field.

The foundations of PCSFT were given in the author's papers [12]–[15]. It was shown that quantum averages can be considered as averages with respect to ensembles of classical prequantum fields. However, measurement theory based on PCSFT has not yet been developed. The absence of the corresponding measurement theory induced the impression that PCSFT is a kind of ontic theory which does not have a direct relation with quantum measurements. In our paper measurement theory based on PCSFT will be presented. The basic idea is that a so-called quantum state (e.g., a pure state) is simply a label for an ensemble of classical random fields. We describe the process of interaction of detectors with a random field.

In our approach the probability of selection of a field is proportional to its power (its L_2-norm). We call this approach *power field-signal detection theory – PFSDT*. The power of signal (classical prequantum field) given by its L_2-norm plays the crucial role in PFSDT. It will be shown that in PFSDT *Born's rule* is a consequence of the well known *Bayes formula* for conditional probability.

In our measurement theory everything – fields, detectors – si classical, the output probabilities are, nevertheless, the same as in QM. In this paper we restrict our considerations to the case of *momentum measurement*. We consider a scalar complex field. Generalization to vector fields is evident.

2. Classical random fields: the momentum representation

We consider the configuration space of complex random fields:

$$Z = L_2(\mathbf{R}^3) \ ,$$

and the space of square integrable complex fields $\phi : \mathbf{R}^3 \to \mathbf{C}$. It is endowed with the norm

$$\|\phi\|^2 = \int_{\mathbf{R}^3} |\phi(x)|^2 dx \ .$$

A random field is a Z-valued random variable $\phi(\omega) \in Z$. Here ω is a chance parameter. We denote by μ the probability distribution of the random variable $\phi(\omega)$. It is a probability measure on Z.[1] Of course, Z has infinite dimension, but mathematical theory of such measures is well developed. We shall proceed on the physics level of rigorousness.

We are interested in the momentum representation of the (classical) random field. The position and momentum representations are connected through the Fourier transform:

$$\phi \to \tilde{\phi} , \quad \tilde{\phi}(p) = \mathcal{F}(\phi)(p) ,$$

and the configuration space is represented as $Z = L_2(\mathbf{R}_p^3)$. Thus a random field can be considered as an L_2-function of two variables, momentum and chance: $\tilde{\phi}(p, \omega)$.

We shall consider only random fields with zero mean value: $E\tilde{\phi}(\omega) = 0$. This equality simply means that for any function $f \in L_2(\mathbf{R}_p^3)$:

$$E \int_{\mathbf{R}^3} \overline{f(x)} \tilde{\phi}(p, \omega) dp = 0 .$$

Covariance of a random field is defined as

$$C_\mu(f, g) = E\Big(\int_{\mathbf{R}^3} \overline{f(x)} \tilde{\phi}(p, \omega) dx \Big) \Big(\int_{\mathbf{R}^3} g(x) \overline{\tilde{\phi}(p, \omega)} dp \Big) .$$

The corresponding operator is denoted by C_μ. We recall that the covariance operator has all main features of von Neumann's density operator (self-adjoint, positively defined, trace class), apart from normalization. Dispersion is given by

$$\sigma^2(\mu) = E\|\tilde{\phi}(\omega)\|^2 (= E\|\phi(\omega)\|^2) .$$

We shall consider dispersion as a parameter, say α, of our model. We remind the following useful equality:

$$\alpha = \mathrm{TrC}_\mu .$$

3. Random field-signals and momentum measurement

We denote by $\mathcal{E}_\mu$ an ensemble of fields represented by a probability measure μ on Z. It is the probability distribution of a random field $\tilde{\phi}(\omega)$. (We use the same symbol μ to denote the probability distribution of the random field both in the position and momentum representations.) The $\mathcal{E}_\mu$ gives realizations of the corresponding random field.

Let p be the momentum observable. This observable is considered as an observable on fields. Thus

$$p(\omega) = p(\tilde{\phi}(\omega))$$

is a random variable. It takes its values in $\mathbf{R}$. Our aim is to find the probability distribution of this random variable from the probability distribution of the random field. In Section 3 we shall present a model of detection and on the basis of this model we shall find the probability distribution of $p(\omega)$.

In measurement theory for momentum we consider a random field as a *random signal* interacting with detectors. Suppose that the momentum observable p is given by a measurement device M_p. For example, for any point p_0, we can consider the observable p_{p_0} given by a detector $M_p(p_0)$ corresponding to selection of the momentum $p = p_0$. For any (sufficiently regular) set $I \subset \mathbf{R}_p^3$ we can consider the observable p_I given by a collection of detectors $M_p(I)$ located in the domain $I \subset \mathbf{R}_p^3$ (e.g. a cube).

[1] To be completely rigorous mathematically, we should consider a Kolmogorov probability space $(Z, F, \mathbf{P})$, where Z is the space of chance parameters, F is a σ-algebra of its subsets and $\mathbf{P}$ is a probability. Then $\phi : Z \to Z$ is a measurable function. Its probability distribution is given by $\mu(U) = \mathbf{P}(\omega \in Z : \phi(\omega) \in U)$ for a Borel subset U of Z.

4. Power field-signal detection theory – PFSDT

In our model of detection the measurement process over a random field consists of two steps:

a) selection of a field $\tilde{\phi} \in \mathcal{E}_\mu$;
b) measurement on this field: $p(\tilde{\phi})$.

We assume that measurement devices (detectors) are sensitive to the *power* of the (classical) field-signal. Let $p_0 \in \mathbf{R}^3_p$ and $I \subset \mathbf{R}^3_p$. For a field-signal ϕ, we introduce its momentum power at the point p_0 as

$$\pi_2^{(p)}(p_0, \tilde{\phi}) = |\tilde{\phi}(p_0)|^2 \ .$$

It induces the momentum power of this signal in a domain I as

$$\pi_2^{(p)}(I, \tilde{\phi}) = \int_I |\tilde{\phi}(p)|^2 dp \ .$$

The total power (or simply power) of momentum is given by

$$\pi_2^{(p)}(\tilde{\phi}) = \int_{\mathbf{R}^3} |\tilde{\phi}(p)|^2 dp = \|\phi\|^2 = \int_{\mathbf{R}^3} |\phi(x)|^2 dx.$$

We now formulate the fundamental feature of the class of detectors under consideration, namely, *sensitivity to the power of a field-signal* in the form of two postulates:

Postulate 1. *The probability P_μ to select a fixed field $\tilde{\phi}(p)$ from the random field-signal $\tilde{\phi}(p, \omega)$ (the ensemble $\mathcal{E}_\mu$) is proportional to the power of the field $\tilde{\phi}(p)$:*

$$dP_\mu(\tilde{\phi}) = K_\mu \pi_2^{(p)}(\tilde{\phi}) d\mu(\tilde{\phi}). \tag{1}$$

The coefficient of proportionality K_μ can be found from the normalization of probability:

$$K_\mu = \frac{1}{\int_Z \pi_2^{(p)}(\tilde{\phi}) d\mu(\tilde{\phi})} = \frac{1}{\sigma^2(\mu)} = \frac{1}{\alpha} \ .$$

Thus, we get

$$dP_\mu(\tilde{\phi}) = \frac{\pi_2^{(p)}(\tilde{\phi})}{\alpha} d\mu(\tilde{\phi}) \ . \tag{2}$$

The selection step a) of a signal from a random field for the momentum measurement was formalized by Postulate 1. This postulate is intuitively attractive: more powerful signals are selected more often. The measurement step b) is formalized as follows.

Postulate 2. *The probability $P(p = p_0|\phi)$ to get the result $p = p_0$ for the fixed field-signal $\tilde{\phi}(p, \omega)$ is proportional to the momentum power $\pi_2^{(p)}(p_0, \tilde{\phi})$ of this field at the point p_0. The coefficient of proportion does not depend on p_0.*

Hence, the coefficient of proportion $k^{(p)}(p_0|\tilde{\phi}) \equiv k^{(p)}_{\tilde{\phi}}$, thus

$$k^{(p)}_{\tilde{\phi}} = \frac{1}{||\tilde{\phi}||^2} = \frac{1}{||\phi||^2} \ ,$$

and hence:

$$dP(p| \ \tilde{\phi}) = \frac{|\tilde{\phi}(p)|^2}{||\tilde{\phi}||^2} dp \ .$$

Now we would like to find the probability $P_\mu^{(p)}(p = p_0)$ to get the result $p = p_0$ for momentum detection for a random field with the probability distribution μ. By using the classical *Bayes' formula* for composition of probability and conditional probability we obtain:

$$P_\mu^{(p)}(p = p_0) = \int_Z P(p|\ \tilde{\phi}) dP_\mu(\tilde{\phi}) = \frac{1}{\alpha}\int_Z |\tilde{\phi}(p_0)|^2 d\mu(\tilde{\phi}) \ , \tag{3}$$

and hence

$$P_\mu^{(p)}(p \in I) = \frac{1}{\alpha}\int_Z \int_I |\tilde{\phi}(p)|^2 dp d\mu(\tilde{\phi}) \ . \tag{4}$$

4.1. Ideal and real detectors

We developed measurement theory for detectors taking into account the power of the field-signal. The field $\delta(p - p_0)$ describes a point-like detector. This is an ideal measurement. The real detector is described by a field $u_{p_0}(p) \equiv u(p - p_0)$ giving a continuous approximation of $\delta(p - p_0)$. The corresponding power function is given by

$$\pi_2^{(p)}(u_{p_0}, \tilde{\phi}) = \Big| \int \overline{u(p - p_0)} \tilde{\phi}(p) dp \Big|^2 .$$

5. Coupling between PFSDT and QM

To find coupling between the presented theory of detection of field-signals (PFSDT) and the quantum formalism, we introduce projectors (in the space of classical fields):

$$\hat{I}\tilde{\phi}(p) \equiv \chi_I(p)\tilde{\phi}(p) \ ,$$

where $\chi_I(p)$ is the characteristic function of a Borel set $I \subset \mathbf{R}_p^3$ (e.g. a cube).

Theorem. *The probability measure* $P_\mu^{(p)}$ *can be represented in the operator form:*

$$P_\mu^{(p)}(I) = \mathrm{Tr}(\hat{\rho}\hat{\mathrm{I}}) \ , \tag{5}$$

where

$$\hat{\rho} = C_\mu / \alpha \ . \tag{6}$$

Proof. We have:

$$P_\mu^{(p)}(I) = \frac{1}{\alpha}\int_Z \langle \hat{I}\tilde{\phi}, \tilde{\phi}\rangle d\mu(\tilde{\phi}) = \mathrm{Tr}(\hat{\rho}\hat{\mathrm{I}}) \ .$$

We remark that the operator $\hat{\rho}$ has all properties of the von Neumann density operator. This theorem motivates the following correspondence between classical random fields and von Neumann's density operators (quantum states):

Any classical random field induces a quantum state by mapping the field in the density operator given by (6) and vice versa.

The probability distribution $P_\mu^{(p)}$ on $\mathbf{R}_p^3$ given by PFSDT coincides with the probability distribution given by QM. We remark that the correspondence between random fields and quantum states is not one-to-one: a random field is not determined uniquely by its covariance operator. However, if one restricts considerations to Gaussian random fields only, then the correspondence will become one-to-one.

Let Ψ be a normalized vector – a "pure state" of QM. We consider the momentum representation of the L_2-space. Hence, this pure state is given by a square summable function

$\tilde{\Psi}(p)$. We consider the Gaussian measure $\mu_{\tilde{\Psi}}$ on Z with zero mean value and covariance operator $C_{\tilde{\Psi}} = \tilde{\Psi} \otimes \tilde{\Psi}$. Formally we can write:

$$d\mu_{\tilde{\Psi}}(\tilde{\phi}) = \exp\left(-\left|\int_{-\infty}^{+\infty} \overline{\tilde{\Psi}(p)}\tilde{\phi}(p)dp\right|^2\right)\prod_p d\tilde{\phi}(p) .$$

Then we find easily that

$$P_{\tilde{\Psi}}^{(p)}(p) \equiv P_{\mu_{\tilde{\Psi}}}^{(p)}(p) = |\tilde{\Psi}(p)|^2, \tag{7}$$

thus

$$P_{\tilde{\Psi}}^{(p)}(I) = \int_I |\tilde{\Psi}(p)|^2 dp . \tag{8}$$

This is nothing else than *Born's rule.*

We find the mean value of the momentum p with respect to the probability measure $P_\mu^{(p)}$. We restrict our considerations to one dimensional case. We have:

$$\langle p \rangle_{P_\mu^{(p)}} = \int_{-\infty}^{+\infty} p dP_\mu^{(p)}(p) = \mathrm{Tr}(\hat{\rho}\hat{\mathrm{p}}) ,$$

where $\hat{p}$ is the momentum operator in Schrödinger's representation of QM.

Thus PFSDT (a measurement model for classical random fields) produces the same probability distributions and averages as the conventional QM model. PFSDT provides a possibility to go beyond QM. Each "quantum particle" in an ensemble of systems prepared for the momentum measurement can be considered as a realization of a classical random field.

6. Coupling between PFSDT and PCSFT

As was pointed out in the Introduction, in a series of papers [12]–[15] we developed a random field model PCSFT – prequantum classical statistical field model – reproducing (in particular) quantum averages. The correspondence between random fields and density operators was given by (6). Thus the considered model of detection of field-signals – PFSDT – borrowed this correspondence from PCSFT. In PCSFT we also considered classical physical variables defined as functionals of classical fields $\phi \to f(\phi)$. The correspondence rule between PCSFT-variables and quantum observables was given in [12]–[15] by

$$\hat{A} = \frac{1}{2}\frac{\delta^2 f}{\delta\phi^2}(0) .$$

The second derivative of the classical functional $f(\phi)$ at the point $\phi = 0$ gives the corresponding quantum observable $\hat{A}$. We remind that the second derivative is always a symmetric operator (Hessian). This map is noninjective as well: different classical observables having the same second derivative at $\phi = 0$ are mapped into the same quantum observable. However, if the class of classical observables is restricted to quadratic forms of classical fields, then this map is on-to-one.

PCSFT gives the following rule for computation of averages:

$$\langle f \rangle_\mu = Ef(\phi(\omega)) = \int_Z f(\phi)\mu(\phi) .$$

Let us consider the classical PCSFT-variable inducing the momentum operator:

$$f_p(\tilde{\phi}) = \int_{-\infty}^{+\infty} p|\tilde{\phi}|^2 dp .$$

Of course,

$$\hat{p} = \frac{1}{2}\frac{\delta^2 f_p}{\delta\tilde{\phi}^2}(0) \ ,$$

where $\hat{p}$ is the momentum operator. Since it was shown in [12]–[15] that for quadratic forms PCSFT-average coincides with QM-average and in the previous Section we have shown that PFSDT-average also coincides with QM-average, we have

$$\langle p \rangle_{P_\mu^{(p)}} = \langle f_p \rangle_\mu \ .$$

But we can also show this directly:

$$\langle p \rangle_{P_\mu^{(p)}} = \int_Z \Big(\int_{-\infty}^{+\infty} p|\tilde{\phi}(p)|^2 dp \Big) d\mu(\tilde{\phi}) = \langle f_p \rangle_\mu .$$

6.1. Discussion on methodology of measurements

Our model PFSDT – PCSFT completed by measurement theory – is a model with hidden variables, but it is not a conventional model. On the one hand, we have a classical prequantum (field-type) model that reproduces statistical predictions of QM. On the other hand, the methodology of measurement in PFSDT does not differ so much from the Copenhagen methodology.

Our basic objects are classical fields. The momentum p measured by a detector is not generated deterministically by the classical field $\tilde{\phi}(p)$. Any field is distributed over a domain in $\mathbf{R}_p^3$ (in general over the whole space). The concrete value $p = p_0$ is produced by a measurement device M_p. The main difference from Copenhagen is that in PFSDT such a methodology of measurement is not based on a special nonclassical mathematics. Moreover, one could proceed to reality, without modification of classical views.

Appendix

Considerations of Section 4 were performed on the physics level of rigorousness. For $p_0 \in \mathbf{R}_p^3$ the map $\tilde{\phi} \to \tilde{\phi}(p_0)$ is not a continuous linear functional on L_2. Therefore, for an arbitrary measure μ on Z the integral (3) is not well defined. Nevertheless, the integral $\int_I |\tilde{\phi}(p)|^2 dp$ is well defined for any Borel set (e.g. a cube) and, moreover, we have:

$$\int_Z \int_I |\tilde{\phi}(p)|^2 dp d\mu(\tilde{\phi}) \leq \int_Z ||\tilde{\phi}||^2 d\mu(\tilde{\phi}) < \infty.$$

Thus we obtain the following.

Proposition *For any probability distribution μ of a random field $\tilde{\phi}(p,\omega)$, the $P_\mu^{(p)}$ given by (4) is a probability measure on $\mathbf{R}^3$.*

Acknowledgments

I am thankful for discussions on attempts to go beyond QM to L. Accardi, S. Albeverio, M. Appleby, L. E. Ballentine, V. Belavkin, L. Hardy, I. Bengtsson, M. Bozejko, C. Fuchs, S. Gudder, K. Gustafson, I. Helland, A. Holevo, R. Hudson, S. Kozyrev, E. Loubenets, M. Manko, V. Manko, O, Manko, D. Petz, R. Schack, A. Caticha, D. Aerts, S. Aerts, B. Coecke, A. Grib, P. Lahti, A. Aspect, D. Mermin, G. Adenier, W. De Baere, W. De Muynck, K. Hess and W. Philipp, R. Gill, D. Greenberger, J-A. Larsson, A. Peres, I. Pitowsky, M. Scully, I. Volovich, G. 't Hooft, H-T. Elze, A. Legett, M. Davidson, E. Nelson, L. De La Pena, A. M. Cetto, K. Svozil, M. S. Zubairy, R. Penrose, M. Genovese, S. Kulik, A. Holevo, M. D'Ariano, A. Plotnitsky, G. Jaeger, I. Bengtson, J.-A. Larsson, K. Gustafson.

Journal of Physics: Conference Series **174** (2009) 012035 doi:10.1088/1742-6596/174/1/012035

[1] Einstein A 1993 *The Collected Papers of Albert Einstein* (Princeton: Princeton U. Press)
[2] 't Hooft G 2001 Quantum Mechanics and Determinism *Preprint* hep-th/0105105
[3] 't Hooft G 2002 Determinism beneath Quantum Mechanics *Preprint* quant-ph/0212095
[4] Khrennikov A 2007 *J. Russian Laser Research*, **28** 244-254
[5] Scully M O and Zubairy M S 1997 *Quantum optics* (Cambridge: Cambridge U. Press)
[6] Louisell W H 1973 *Quantum Statistical Properties of Radiation* (New York: J. Wiley)
[7] Mandel L and Wolf E 1995 *Optical Coherence and Quantum Optics* (Cambridge: Cambridge U. Press)
[8] De la Pena L and Cetto A M 1996 *The Quantum Dice: An Introduction to Stochastic Electrodynamics* (Dordrecht: Kluwer)
[9] Nelson E 1985 *Quantum Fluctuation* (Princeton: Princeton University Press)
[10] Davidson M 1979 *J. Math. Phys.* **20** 1865-1870
[11] Elze H-T 2009 *Int. J. Qu. Info. (IJQI)* **7** 83-96
[12] Khrennikov A 2005 *J. Phys. A: Math. Gen.* **38** 9051-9073
[13] Khrennikov A 2006 *Phys. Lett.* A **357** 171-176
[14] Khrennikov A 2006 *Nuovo Cimento* B **121** 1005-1021
[15] Khrennikov A 2006 *Found. Phys. Lett.* **18** 637-650

Journal of Physics: Conference Series **174** (2009) 012036 doi:10.1088/1742-6596/174/1/012036

On Schrödinger's quest for new physics for life

Alfredo Iorio

Institute of Particle and Nuclear Physics, Charles University of Prague, V Holešovickach 2, 18000 Prague 8, Czech Republic

E-mail: `Alfredo.Iorio@mff.cuni.cz`

Abstract. Two recent investigations are reviewed: quantum effects for DNA aggregates and scars formation on virus capsids. The possibility that scars could explain certain data recently obtained by Sundquist's group in electron cryotomography of immature HIV-1 virions is also briefly addressed. Furthermore, a bottom-up reflection is presented on the need to invent new physics to pave the way to a rigorous physical theory of biological phenomena. Our experience in the two researches presented here and our personal interpretation of Schrödinger's vision are behind the latter request.

1. Introduction

It seems to us that the frontier of physics is rapidly moving to the biological side not only in the obvious way that the highly detailed knowledge of the biological data requires increasingly sophisticated physics for their explanation, but also (and even more intriguingly for the theoretical physicist's mind) in the sense that the time might be mature for a profound reformulation of biology that makes it a science similar to physics: few principles, mathematical derivations, explanation and prediction of phenomena (and back). Nonetheless, the problem is very difficult if faced with the proper rigor and we have no answer to propose, a part from the conviction that some important changes are necessary on the physical approach to the problems of biology and that a "physics tailored for life phenomena" needs to be invented. On this we comment only in the last section where we arrive by taking a feet-on-the-ground road that makes two stops prior to getting there: the study of (relativistic) quantum effects in DNA condensation [1] and the role of certain topological defects (scars, see, e.g., [2]) for virus capsids shape-changes [3]. This way we have the chance to face important biological problems on the table and to try to learn on the battle-field what is that is missing from the big picture.

2. Quantum fields holding together DNA aggregates

2.1. The story

DNA molecules in aqueous solution ionize and become highly charged anions which strongly repel [4]-[12] (for reviews see, e.g., [13, 14]). When specific cations are added (i) DNA attracts and *binds* them to make a new structure, the DNA-cations complex, and (ii) when about 90 per cent of the DNA negative charge is screened the *like-sign* DNA-cations strands *attract and collapse* (see, e.g., [13, 14, 15]) to form finite-size aggregates whose shape is either rod-like or spheroidal or, most commonly, toroidal [16].

Long ago Oosawa and Manning (OM) explained the counterions condensation as a phase transition within the classical Poisson-Boltzmann (PB) mean field electrostatic theory [17].

Despite important advancements, the second part of the puzzle still presents many open questions: (i) it is not understood why the aggregate does not grow forever; (ii) there is no general consensus on the necessity to go beyond the PB classical theory, as proposed in [18] (see also [5]) and as opposed in [7] (see also [14, 19]); (iii) whichever approach is used (the zero frequency Casimir/van der Waals interaction [20] (see also [21]); the "electrostatic zipper" model of [7]; correlations of thermal Gaussian fluctuations of the number density of counterions [5, 18]; the Wigner crystal approach [11, 22]) the paradigm is that the interactions are *classical* [13]. The lack of appreciation of quantum effects for this phenomenon was probably due to the lack of calculations based on the appropriate *codimension two* (lines in three dimensions) Casimir-like technique[1] [25] and with cations not directly participating in the interaction but playing the passive role of screening the electrostatic repulsion.

2.2. A new model based on relativistic quantum fields

In [1] we singled-out the zero-point interaction due to the disturbances induced in the *quantum electric vacuum* by the presence of the (nearly) neutral DNA-cations complexes. We modeled the N anions (the DNA strands) as infinite lines (finite length effects are unimportant) all parallel to the z-axis and located at $\vec{l}_i$ in the $x-y$ plane with certain coefficients $\nu_i(z_i)$ carrying information on the charge structure of the DNA and taken to be $\nu_i(z_i) = \nu =$ constant, $\forall i = 1, ..., N$. The cations screen the total charge and set the length scale. The effective model for the electric potential $\Phi(\vec{x})$ in this set-up is given by [1]

$$\left[-\partial_z^2 - \nabla_\perp^2 + \mu^2 + \lambda \sum_{i=1}^{N} \delta^{(2)}(\vec{x}_\perp - \vec{l}_i)\right] \Phi(\vec{x}) = J \,, \tag{1}$$

which is a *modified* Debye-Hückel (DH) equation, with $\mu^2 = k^2\kappa^2$, k the cations valency, $\kappa^{-1} = (\epsilon k_B T/(8\pi e^2 n_0))^{1/2}$ the Debye screening length, ϵ the dielectric constant of water, n_0 fixes the zero of the potential, $\lambda = 4\pi\nu|q|^2/\epsilon k_B T$, $J = -(1/\epsilon)4\pi|q|\nu\sum_{i=1}^N \delta^{(2)}(\vec{x}_\perp - \vec{l}_i)$, $T = 300$K, and with Eq.(1) the limit for weak Φ of a PB equation modified to include the DNA strands in the charge (Boltzmann) distribution [1]

$$\rho_{\rm DNA}(\vec{x}, T) = -n^0_{\rm DNA}(\vec{x})|q| \exp\left(\frac{|q|\Phi(\vec{x})}{k_B T}\right) \tag{2}$$

where $q < 0$ is the charge of the DNA strand with $n^0_{\rm DNA}(\vec{x}) = \sum_{i=1}^N \nu_i(z_i)\delta^2(\vec{x}_\perp - \vec{l}_i)$ and N the number of strands.

Eq.(1) is only a starting point, as our concern are time dependent fluctuations, $\Phi(\vec{x}) \to \Phi(\vec{x}) + \phi(\vec{x}, t)$, that we treat as quantum by considering the associated action (we set $\hbar = c = 1$, c speed of light in the medium)

$$\bar{\mathcal{A}}(\phi) = \int d^4x \frac{1}{2}\phi \left(-\partial_t^2 - \partial_z^2 - \nabla_\perp^2 + \mu^2 + \lambda \sum_{i=1}^{N} \delta^{(2)}(\vec{x}_\perp - \vec{l}_i)\right)\phi \,, \tag{3}$$

which we infer from the action for Φ that gives Eq.(1) as equation of motion, $\mathcal{A}(\Phi) = \int d^4x \left(\frac{1}{2}\Phi[-\partial_z^2 - \nabla_\perp^2 + \mu^2 + \lambda\sum_{i=1}^N \delta^{(2)}(\vec{x}_\perp - \vec{l}_i)]\Phi + J\Phi\right)$, to be of the order of $\hbar$. Note that in $\bar{\mathcal{A}}(\phi)$ the term with the coupling to J is zero because $\int d^4x J\phi = \int d^3x J \int_0^\tau dt\phi = 0$.

These fluctuations are disturbances of the quantum vacuum induced by the delta-functions and, since they are the electric field, they travel at the speed of light in the

[1] The work in [20, 6] is based on the Lifshitz computation for the Casimir effect in presence of dielectric media [23] (see also [24] and references therein) i.e. a codimension one (surfaces in three dimensions) calculation.

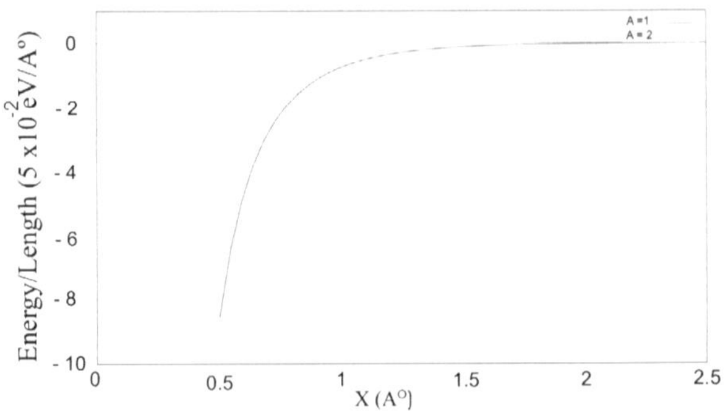

Figure 1. Two-body interaction energy/length $\mathcal{E}(x)$ against lattice size x for $a = 1$ (upper curve) and $a = 2$ (lower curve)

medium. We considered the generating functional $Z[\Phi, \phi] = \int [D\Phi] e^{i\mathcal{A}(\Phi)} \int [D\phi] e^{i\bar{\mathcal{A}}(\phi)} = \int [D\Phi] e^{-(\mathcal{A}(\Phi) + \text{corrections})}$ where we Wick rotate on the time direction $t \to it$, and by identifying the effective action as $\mathcal{A}_{\text{eff}}(\Phi) = \mathcal{A}(\Phi) + \mathcal{E}\tau$ one sees that [1]

$$\mathcal{E} = \frac{1}{2} \int_{-\infty}^{+\infty} \frac{dp}{2\pi} \int_0^{+\infty} dE \rho(E) \sqrt{E + p^2} \sim \frac{1}{2} \sum_k \omega_k \tag{4}$$

where $\rho(E)$ is the density of states, i.e. $\mathcal{E}$ is the zero point "Casimir-like" energy of the system, that is found to be [1], [25] (see also [26])

$$\mathcal{E} = \frac{\hbar c}{8\pi} \int_0^{\infty} dE \ln \left[\det \left(\delta_{ij} - \frac{K_0(\sqrt{E + \mu^2}\, l_{ij})}{\ln(\sqrt{E + \mu^2}/M)} (1 - \delta_{ij}) \right) \right] , \tag{5}$$

where we reintroduced $\hbar$ and c, $K_0(x)$ is the modified Bessel function of the second kind of order zero, μ is the scale parameter introduced earlier, $l_{ij} = |\vec{l_i} - \vec{l_j}|$ are the relative distances and M is a further mass scale parameter originated by the codimension two (for stability it satisfies $M < \mu$). We are not considering charge density or positional fluctuations of counterions, hence we do not have slow-moving fluctuations, as, e.g., the authors of [6] that had to consider only zero frequency (i.e. classical) contribution to the usual Casimir/van der Waals effect. We have here a universal mechanism of interaction that (i) only becomes important when the OM condensation has taken place, (ii) whose temperature dependence is via the length scale μ and (iii) with a free (but highly constrained in units of μ) parameter M.

2.3. New results

The two-body interaction indeed is attractive as shown in Fig. 1. Our focus needs be on configurations as close as possible to real cases [16], such as that shown in Fig. 2, where the relative distances are $l_{ij} = c_{ij} x$ with x the basic lattice distance and c_{ij} taking the symmetry of the given arrangement into account. The units are $\mu^{-1} \sim \mathcal{O}(10)$ Å, for distances, and $\mathcal{E} \sim 5 \times 10^{-2}$ eV/Å, for lineal energy density. $M = e^{-1/a}$ (we choose $a = 1$ and $a = 2$). Thermal fluctuations, as computed, e.g., in [5], give for the two-body interaction a maximum value of $\mathcal{E} \sim 5 \times 10^{-3}$ eV/Å at 10 Å. At this distance our two-body interaction is 5×10^{-2} eV/Å (for $a = 1$) or 2×10^{-1} eV/Å (for $a = 2$), i.e. between one and two orders of magnitude stronger. For the many-body case, the case of importance for the aggregates, this factor grows enormously [1] but we cannot trust our approximations for x too close to the singular value of the logarithm. It is clear, though, that at the distances of relevance this quantum energy is stronger (or much stronger) than thermal energy. Thus we can conclude that it is relativistic quantum field zero-point energy that holds together DNA aggregates! This is nice but...

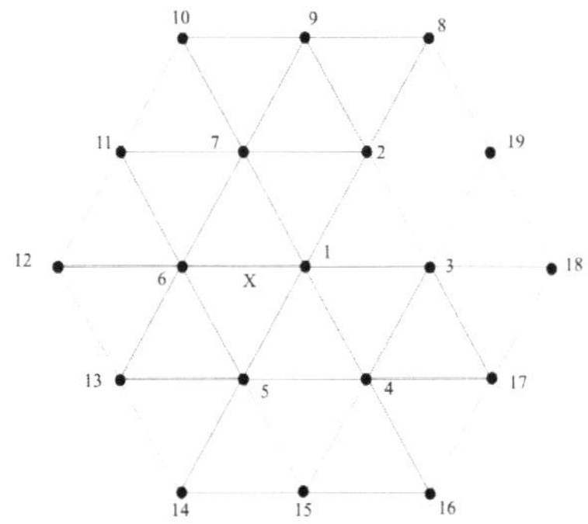

Figure 2. Configuration used to compute the interaction of 19 strands.

2.4. Effects of the lack of a general theory of biological physics

... important pieces are missing here due to the lack of a proper "general frame": (i) where is, in the physics of the problem, that those filaments are the carriers of genetic information? (ii) is the description of this phenomenon complicated because we are not using the most appropriate approach? and is that meaning that we need new variables or a different coarse-graining or that we need to take into account that DNA compaction is just part of a series of "moves" made with a "purpose" (e.g., to pack the DNA in the bacteriophage's head or into the sperm cell and that is, in turn, just one step of virus assembly or of fecundation, and the latter one step in the "life" cycle of the virus or in the reproduction of eukaryotic cell organisms, and so on to higher and higher levels of complexity) (iii) how can we translate into physical language the "purpose", or, in other words, the clear fact that all these moves are done by "using the laws of physics for a goal"? is it matter of changing the definition of physical rules or what? finally, (iv) is quantum physics used to full strength, i.e. as pivotal for life itself [27] (see also [28])?

3. Scars on viruses: the general conjecture and the case of immature HIV-1

3.1. The two stories: virus structure and Thomson problem

A virus is a piece of DNA or RNA surrounded by a protein coat, the capsid, sometimes cased into a lipidic membrane, the envelop [29]. Capsid's shapes can be: helical, icosahedral or complex (sphero-cylindrical, conical, tubular or more complicated shapes) [29]. Viruses may change their shape (polymorphism) as an important step in their "life"-cycle. For instance, HIV-1 is only infective when its capsid has changed from spherical to conical (maturation) [30].

The theory of icosahedral capsids was proposed by Crick and Watson [31] and later established by Caspar and Klug (CK) [32]: those capsids all have $60T$ proteins arranged into 12 pentamers and $10(T-1)$ hexamers, where $T = n^2 + m^2 + nm = 1, 3, 4, 7, \ldots$. These numbers descend from the Poincarè-Euler theorem[2]

$$\sum_N (6-p)N_p = 6\chi \,, \tag{6}$$

applied to the sphere, $\chi = 2$, and from the CK "quasi-equivalence" principle [32], see also [3]. The CK theory is nowadays an established paradigm among virologists [29] and various modifications/generalizations have been proposed by physicists and mathematicians [3, 33, 34, 35, 36] (for a review see [37]), but none made its way to virology textbooks.

An intimately related physical set-up is that of N electrons on the surface of a sphere. To find their minimum energy configurations means to solve the so-called "Thomson problem"

[2] N_p is the number of p-gons used to tile a surface, e.g., N_5 pentagons, N_6 hexagons, etc. The resulting polyhedron P has $V_P = 1/3\sum_N N_p p$ vertices, $E_P = 1/2\sum_N N_p p$ edges, and $F_P = \sum_N N_p$ faces, giving for the Euler characteristic $\chi = V_P - E_P + F_P$ the expression in Eq. (6).

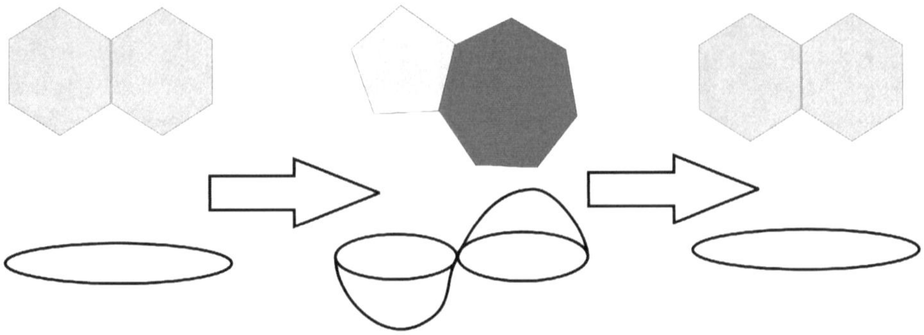

Figure 3. Idealized scar formation-annihilation mechanism.

[38], unsolved in general. For $N < \mathcal{O}(500)$ (and of the form $N = 10T + 2$) the icosahedron is the solution, as shown in [39], a work inspired by the CK theory of virus capsids. For bigger N configurations that present "scars" are the solution as shown in [40] and for spherical elastic materials (non-Coulomb potentials) in [41]. A scar is a lineal pattern of pentagons-heptagons that initiate at the pentagonal vertices of the icosahedron, whose length (i.e. number of pentagons and heptagons in it) and geometrical arrangement depend on N (see [2] for an approach based on spontaneous symmetry breaking of the icosahedral group). Scars have been experimentally found (in spherical crystals of mutually repelling polystyrene beads self-assembled on water droplets in oil [42]) but still need a deep understanding in terms of phase transitions. Nonetheless, it is easy to see that they are allowed by the theorem (6) written as

$$(6-5)\;N_5 + (6-6)\;N_6 + (6-7)\;N_7 = 12\;. \tag{7}$$

This means that N_6 can be arbitrary (hence also the required $N_6 = 10(T-1)$) and $N_5 - N_7 = 12$, hence, starting off with the 12 pentagons of the icosahedron one can add *pairs* pentagon-heptagon, but not a pentagon or a heptagon separately. Geometrically, we are saying that a unit sphere has curvature $R_{\rm sphere} = +1$ and each polygon contributes to this curvature with $R_p = (6-p)/12$: a hexagon with $R_6 = 0$, a pentagon with $R_5 = +1/12$, a heptagon with $R_7 = -1/12$. The total energy is then the sum of bending energy and stretching energy, $\mathcal{E}_t = \mathcal{E}_b + \mathcal{E}_s$ that compete for the minimization of the total energy [41].

3.2. Scar formation-annihilation mechanism for viruses

Below the threshold for the scar production $N_{\rm scar}$, $\mathcal{E}_b$ is provided by the 12 pentagons, while the hexagons contribute to $\mathcal{E}_s$ only. One way to imagine the transition at $N > N_{\rm scar}$ is to think of a pair 6-6, with zero total and local curvature and zero bending energy, *converted into* a pair 5-7 (see Fig. 3), again with zero total curvature but with nonzero local curvature hence with nonzero bending energy given by $2E_b$, where E_b is the energy necessary to convert a 6 into a 5 or into a 7. In [3] we proposed that scars can *appear on virus capsids* at an intermediate stage of their evolution towards a non-spherical shape and can actually *drive* such shape-change. There E_b should be related to the *conformational switching energy* [43] (for HIV-1 see, e.g., [44]).

The (idealized) shape-change mechanism of [3] is (see Fig. 3): i) proteins first make a CK icosahedron; ii) when, e.g., N reaches $N_{\rm scar}$, they form scars: $6-6 \to 5-7$; iii) the capsid changes shape via the release of the bending energy into stretching energy at the location of the scar with the consequent "annihilation" of the 5-7 pair: $5-7 \to 6-6$. The resulting capsid has the CK morphological units but not the spherical shape.

An example given in [3] is that of scars created only near the 10 inner vertices via a mechanism that respects a C_5 rotation symmetry around the north pole-south pole axis. In Fig. 4 the vertices where the scars are formed are indicated with •, while the other two are indicated with ○. If we require that this mechanism is area preserving, i.e. the total number of proteins before and after is always $60T$ the final capsid obtained is the spherocylinder of Fig. 5 (a shape taken by certain bacteriophages' head) with the 12 pentamers at the vertices and the $10(T-1)$ hexamers distributed differently with respect to the intermediate icosadeltahedron. In general a variety of final capsid shapes could be obtained via this mechanism. For the C_5 symmetric example given, if, for instance, the scars carry a bigger bending energy the final shape could be a backy-tube capsid; if the orientation of the scars in the previous setting is such that C shrinks, hence B becomes longer (see Fig. 5) then a disk-like shape is obtained; etc..

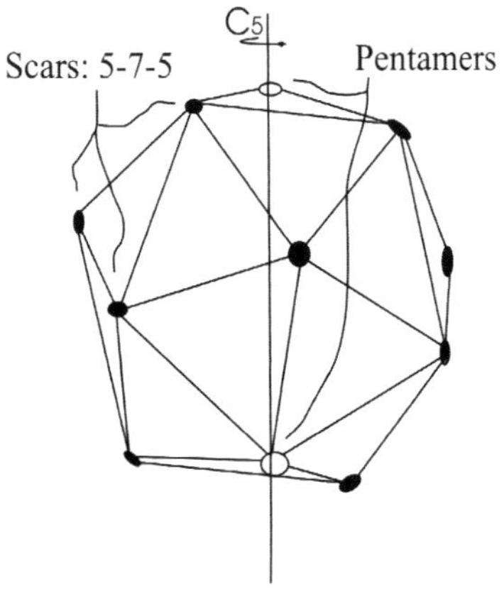

Figure 4. The intermediate spherical capsid.

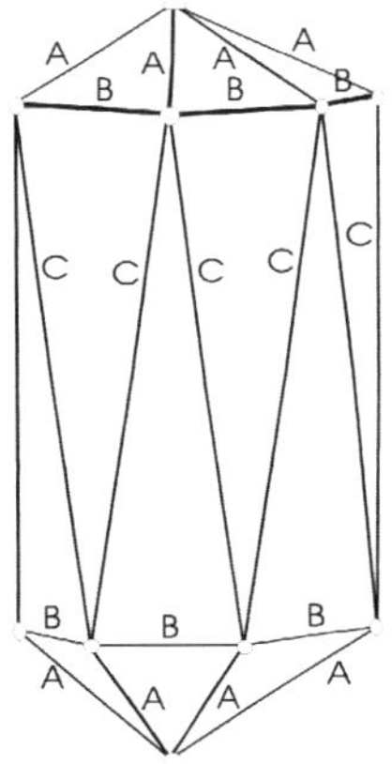

Figure 5. The final spherocylindrical capsid.

3.3. Are scars at work on the HIV-1 immature virions? The data of Sundquist's group.
This is just an idealized model and to make full contact with real viruses one has to be ready to adjust it and, as scars are supposed to disappear, it is also a delicate matter to find the right experimental set-ups for spotting the scars. A case that seems particularly promising is the important case (for human health) of the HIV-1. There the number of Gag polyproteins (essentially made of the three proteins MA, CA, NC) that make up the immature quasi-spherical capsid is estimated to be between 2500 [45] and 5000 [46] (even 11000 is reported [46]), i.e. well above the $N_{\rm scar}$ of the Thomson problem. Indeed, the impressive data of Sundquist's group taken by electron cryotomography of the immature capsid [45] (see also [30]) reveal the structure of the CA shell that is seen to have an hexameric texture and large areas with an apparent lack of symmetry ("disordered" areas in the terminology of [45]). The role of these a-symmetric areas in the maturation process, the apparent lack of pentameric structures, the mechanism that allows for closure of the spherical capsid or the absence of such closure, and other aspects are important issues to be understood.

We claim here that in those areas of the capsid of the immature HIV-1 virions the scar formation-annihilation mechanism we discussed might be at work, although, since it takes place within a many-layer structure comprising the rest of the Gag and the lipidic envelop, it does not give the effect described above in the idealized case. What it presumably does instead is: (i) to stretch the structure, hence not allowing the perfect spherical shape; (ii) in combination with the proteolytic cleavage is responsible for the dramatic reduction of the number of proteins (the generally accepted number for the mature capsid is around 1500); (iii) it might even explain the apparent absence of pentamers if one supposes that when the annihilation takes places it involves also the "primitive" pentamers, i.e. what in the model are the untouched (by the annihilation) vertices of the intermediate icosahedron.

Of course we need to explore those assertions in a quantitative fashion within the model, and we have work in progress in that direction. In the meanwhile, a faster way is to probe those claims directly in experiments, ideally with the help of Sundquist's laboratory.

3.4. Effects of the lack of a general theory of biological physics

Again in this case, as for the DNA compaction, it is clear that the appropriate general frame is lacking. With some adjustment (e.g., "DNA packing" goes to "capsid shape-changes", etc) all questions of Subsection 2.4 apply here, except (i). One important point here is that viruses are at the *border* between dead matter (where standard physics applies) and living matter (where it appears that new physics is necessary), hence we deem them to be the best candidates to understand how to build-up the missing frame. For instance, invariably all models (including ours) on virus structure deal with (free) energy minimization (see [47] for a review), just like, for instance, an elastic ball. May be this is a place to start looking for changes.

4. Biology as physics

So, what is it this new physics that we claim to be necessary for a true understanding of living matter? The vast majority among physicists and biologists says that there is nothing to be done: the physics for living matter is just the same as for dead matter (elastic balls, charged wires, classical machines at the nanometer scale, etc.) it is "just more complicated to apply". Even taking this conservative view, it seems to us that (to say it with a metaphor) it is like trying to solve a (difficult) geometry problem with spherical symmetry insisting in using cartesian coordinates: results are more difficult to get, their interpretation becomes obscure and, most importantly, we might miss global information insisting with a description that is good locally but not globally (see also the metaphor of the mechanical engineer bumping into an electric motor of [27]). The conservative approach to biophysics is nowadays achieving impressive results, and this approach is just what we have been using in our own work reviewed in the previous two Sections, but already at this level of the analysis (i.e. by just focusing on the problems to be solved and not on the general problems of method) we pointed out in Subsections 2.4 and 3.4 that the proper frame is missing and this is not just our own impression.

There are some calls to theoretical physics from the biology frontier to help systematize biology in a rigorous manner (see, e.g., [48]). We physicists could partially answer these calls by, for instance, fitting a given biological problem into a given established physical theory and use the logical consistency of the latter. This is an exciting and worth thing to do and when it is possible to do that there is plenty of cross-fertilizations between the two fields. In a way our proposal of the scar creation-annihilation mechanism for viruses (and the whole "relationship" between the Thomson problem, or elastic theory of spherical membranes [41] or the Landau-Ginzburg theory of phase transitions [49] on the physics side and virus structure on the biology side) is an example. Then the application of physical theories to biological problems might make *that* sector of biology just like *that* physical theory, hence the frame in that case would be set without mayor changes in the methods. This, though, would not solve the problem in general but would work only for certain particular features of certain particular biological entities and it is not certain at all that this would work all the time (and why should it?).

We have now identified what is missing: a general theory of physical biology (GTPB), namely, *the* conceptual frame that should *always* work, for all biological problems, no matter the complexity, and that "just" needs to be applied case by case. It must give the correct description of the phenomenon and, via mathematical elaborations, must produce predictions. This is surely a tremendous task. Let us now see whether this is accomplishable with building blocks we have already or whether we need to change something on the physics side as well.

In 1943 such a change of physics to move the frontier of knowledge from dead to living matter was deemed necessary by Schrödinger in his famous lectures delivered at Trinity College in Dublin [27]. Two were the main messages there: the carrier of genetic information needs to be an aperiodic solid and the thermodynamics of living matter is based on the order from order paradigm rather than the order from disorder of dead matter. Both considerations were rooted into quantum mechanics.

The first message was a clear prediction of the structure and role of DNA, discovered only ten years later. The second message is still an open question that found some partial answers in the work of Prigogine and others who invented Non-Equilibrium Thermodynamics (NET) (see, e.g., [50] for a textbook chapter). Since living organisms are far from equilibrium open systems, NET is surely one of the building blocks we need for GTPB, but it cannot be the end of the story because, similar considerations apply to other non-living systems (financial markets, traffic jams, weather systems, etc, i.e. what is sometimes called a "complex system") and this is not what we are looking for. We need a tailored physics for life, or, in other words, what we might call a "principle of distinction" at work *exclusively* for living matter, and this is still not there. Another point is that the role of quantum mechanics as pivotal for life phenomena [27, 28] needs to be fully exploited.

To build up the GTPB is the most exciting and challenging problem ahead, but it is very difficult. For the time being, we can solve the many challenging problems of standard biophysics, investigate better the role of quantum physics for life and build up a "GTPV": a general theory of physics for viruses.

Acknowledgments
It is my pleasure to thank Jiři Horejši and the Institute of Particle and Nuclear Physics of Charles University for supporting my incursions into biological physics, Samik Sen for lending his own time to the DNA project and, finally, Siddhartha Sen for joyfully joining in from the very beginning.

References

[1] Iorio A, Sen S and Sen S 2008 Do quantum effects hold together DNA condensates? *Preprint* arXiv:0806.3130v1; *do.* Quantum-vacuum-driven DNA condensation *Preprint* in preparation
[2] Iorio A and Sen S 2006 *Phys. Rev.* B **74** 052102; *do.* 2007 *Phys. Rev.* B **75** 099901 (E)
[3] Iorio A and Sen S 2008 *Cent. Eur. J. Biol.* **3(4)** 380; *do.* 2007 Virus Structure: From Crick and Watson to a new conjecture *Preprint* arXiv:0707.3690v1, invited talk by A Iorio at Fourth International Summer School and Workshop on Nuclear Physics Methods and Accelerators in Biology and Medicine - Prague, 8-19 July 2007
[4] Kholodenko A L and Beyerlein A L 1995 *Phys. Rev. Lett.* **74** 4679
[5] Grønbech-Jensen N, Mashl R J, Bruinsma R F and Gelbart W M 1997 *Phys. Rev. Lett.* **78** 2477; Ha B-Y and Liu A 1997 *Phys. Rev. Lett.* **79** 1289
[6] Podgornik R and Parsegian V A 1998 *Phys. Rev. Lett.* **80** 1560
[7] Kornyshev A A and Leikin S 1998 *Proc. Natl. Acad. Sci. USA* **95** 13579; *do.* 1999 *Phys. Rev. Lett.* **82** 4138
[8] Borukhov I, Bruinsma R F, Gelbart W M and Liu A J 2001 *Phys. Rev. Lett.* **86** 2182
[9] Angelini T E, Liang H, Wriggers W and Wong G C L 2003 *Proc. Natl. Acad. Sci. USA* **100** 8634
[10] O'Shaughnessy B and Yang Q 2005 *Phys. Rev. Lett.* **94** 048302; Zribi O V, Kyung H, Golestanian R, Liverpool T B and Wong G C L 2006 *Phys. Rev.* E **73** 031911; Trizac E and Tellez G 2006 *Phys. Rev. Lett.* **96** 038302
[11] Grason G M and Bruinsma R F 2006 *Phys. Rev. Lett.* **97** 027802; *do.* 2007 *Phys. Rev.* E **76** 021924
[12] Qiu X, Andresen K, Kwok L W, Lamb J S, Park H Y and Pollack L 2007 *Phys. Rev. Lett.* **99** 038104
[13] Gelbart W M, Bruinsma R F, Pincus P A and Parsegian V A 2000 *Phys. Today* **53** 38
[14] Kornyshev A A, Lee D J, Leikin S and Wynveen A 2007 *Rev. Mod. Phys.* **79** 943
[15] Levin Y 2002 *Rept. Prog. Phys.* **65** 1577; Grosberg A Yu, Nguyen T T and Shklovskii B I 2002 *Rev. Mod. Phys.* **74** 329
[16] Hud N V and Vilfan I D 2005 *Annu. Rev. Biophys. Biomol. Struct.* **34** 295
[17] Oosawa F 1968 *Biopolymers* **6** 1633; Manning G S 1969 *J. Chem. Phys.* **51** 924

[18] Oosawa F 1971 *Polyelectrolytes* (New York, NY: Dekker)
[19] Cherstvy A G, Kornyshev A A and Leikin S 2002 *J. Phys. Chem.* B **106** 13362; Cherstvy A G 2005 *J. Phys.: Condens. Matter* **17** 1363
[20] Ninham B W and Parsegian V A 1970 *Biophys. J.* **10** 646
[21] Mitchell D J, Ninham B W and Richmond P 1973 *Biophys. J.* **13** 370
[22] Rouzina I and Bloomfield V A 1996 *J. Phys. Chem.* **100** 9977
[23] Lifshitz E M 1955 *Zh. Eksp. Teor. Fiz.* **29** 94 [Eng. transl.: 1956 *Soviet. Phys. JETP* **2** 73]; Schwinger J, DeRaad Jr. L L amd Milton K A 1978 *Ann. Phys.* **115** 1
[24] Milonni P W 1993 *The Quantum Vacuum: an introduction to quantum electrodynamics* (San Diego, CA: Academic Press)
[25] Scardicchio A 2005 *Phys. Rev.* D **72** 065004
[26] Jaffe R L and Scardicchio A 2005 *J. High Energy Phys.* JHEP0506(2005)006
[27] Schrödinger E 1992 *What is Life?* (Cambridge, UK: Cambridge U. Press)
[28] Penrose R 1989 *The emperor's new mind* (Oxford, UK: Oxford U. Press); Frölich H 1975 *Phys. Lett.* A **51** 21; Davies P C W 2004 *BioSystems* **78** 69
[29] Chiu W, Burnett R M and Garcea R L Eds 1997 *Structural Biology of Viruses* (New York, NY: Oxford U. Press)
[30] Ganser-Pornillos B K, Yeager M and Sundquist W I 2008 *Curr. Opin. Struct. Bio.* **18** 203
[31] Crick F and Watson J D 1956 *Nature* **177** 473
[32] Caspar D and Klug A 1963 *Cold Spring Harb Symp. Quant Biol.* **27** 1
[33] Twarock R 2004 *J. Theor. Biol.* **226** 477
[34] Bruinsma R F, Gelbart W M, Reguera D, Rudnick J and Zandi R 2003 *Phys. Rev. Lett.* **24** 248101
[35] Zandi R, Reguera D, Bruinsma R F, Gelbart W M and Rudnick J 2004 *Proc. Natl. Acad. Sci. USA* **101** 15556
[36] Nguyen T T, Bruinsma R F and Gelbart W M 2005 *Phys. Rev.* E **72** 051923
[37] Zlotnick A 2005 *J. Mol. Recognit.* **18** 479
[38] Thomson J J 1904 *Phil. Mag.* **7** 237
[39] Altschuler E L, Williams T J, Ratner E R, Tipton R, Stong R, Dowla F and Wooten F 1997 *Phys. Rev. Lett.* **78** 2681
[40] Perez-Garrido A, Dodgson M J W, Moore M A, Ortuño M and Diaz-Sanchez A 1997 *Phys. Rev. Lett.* **79** 1417
[41] Bowick M J, Nelson D R and Travesset A 2000 *Phys. Rev.* B **62** 8738; Bowick M J, Cacciuto A, Nelson D R and Travesset A 2002 *Phys. Rev. Lett.* **89** 185502; *do.* 2007 *Phys. Rev.* B **73** 024115
[42] Bausch A R, Bowick M J, Cacciuto A, Dinsmore A D, Hsu M F, Nelson D R, Nikolaides M G, Travesset A and Weitz D A 2003 *Science* **299** 1716
[43] Harrison S 2001 *Curr. Opin. Struct. Biol.* **11** 195
[44] Gross I, Hohenberg H, Wilk T, Wiegers K, Grättinger M, Müller B, Fuller S and Kräusslich H G 2000 *EMBO J.* **19** 103
[45] Wright E R, Schooler J B, Ding H J, Kieffer C, Fillmore C, Sundquist W I and Jensen G J 2007 *EMBO J.* **26** 2218
[46] Briggs J A, Simon M N, Gross I, Kräusslich H G, Fuller S D, Vogt V M and Johnson M C 2004 *Nat. Struct. Mol. Biol.* **11** 672
[47] Bruinsma R F 2006 *Eur. Phys. J.* E **19** 303
[48] Macklem P T 2008 *J. Appl. Physiol.* **104** 1844; Phillips R and Quake S R 2006 *Phys. Today* **59** 38
[49] Guerin T and Bruinsma R 2007 *Phys. Rev.* E **76** 061911
[50] Volkenshtein M V 1983 *Biophysics* (Moscow, RU: Mir Publishers) Ch. 9

Noise stabilization effects in models of interdisciplinary physics

B Spagnolo[1], G Augello[1], P Caldara[1], A Fiasconaro[1], A La Cognata[1], N Pizzolato[1], D Valenti[1], A A Dubkov[2] and A L Pankratov[3]
[1]Dipartimento di Fisica e Tecnologie Relative, Group of Interdisciplinary Physics[1], Università di Palermo,
Viale delle Scienze, pad. 18, I-90128 Palermo, Italy
[2]Radiophysical Department, N. Novgorod State University,
Gagarin ave.23, Nizhniy Novgorod, 603950 Russia
[3]Institute for Physics of Microstructures of RAS, GSP-105, 603950 Nizhny Novgorod, Russia
E-mail: spagnolo@unipa.it

Abstract. Metastability is a generic feature of many nonlinear systems, and the problem of the lifetime of metastable states involves fundamental aspects of nonequilibrium statistical mechanics. The investigation of noise-induced phenomena in far from equilibrium systems is one of the approaches used to understand the behaviour of physical and biological complex systems. The enhancement of the lifetime of metastable states through the noise enhanced stability effect and the role played by the resonant activation phenomenon will be discussed in models of interdisciplinary physics: (i) polymer translocation dynamics; (ii) transient regime of FitzHugh-Nagumo model; (iii) market stability in a nonlinear Heston model; (iv) dynamics of Josephson junctions; (v) metastability in a quantum bistable system.

1. Introduction

Relaxation in many natural systems proceeds through metastable states, often observed in condensed matter physics, in cosmology, biology and high-energy physics. In spite of such ubiquity, the microscopic understanding of metastability still raises fundamental questions. This is because we are dealing with a typical out of equilibrium statistical dynamical phenomenon. A physical quantity which characterize the stability of a metastable state is the mean first passage time, which gives a measure of the lifetime of the metastable state. The dependence of the lifetime on the noise intensity for metastable and unstable systems was revealed to have resonance character. Typical examples are the resonant activation (RA) phenomenon [1]-[3] and the noise enhanced stability (NES) [4]. This resonance-like phenomenon shows that the noise can modify the stability of the system in a counterintuitive way: the system remains in the metastable state for a longer time than in the deterministic case [4]-[12]. In this paper we will shortly review the enhancement of the lifetime of metastable states and the role played by the resonant activation phenomenon in models of interdisciplinary physics: (i) polymer translocation dynamics; (ii) transient regime of FitzHugh-Nagumo model; (iii) market stability in a nonlinear Heston model; (iv) dynamics of Josephson junctions; (v) metastability in a quantum bistable system.

[1] Electronic address: http://gip.dft.unipa.it

2. Polymer translocation dynamics

The study of the dynamics of translocation of individual polymers across nanometer-scale pores is essential to help the understanding of how biological systems work. In a fundamental work on polymer translocation [13], it was demonstrated that the crossing time of single-stranded DNA molecules, forced by an applied electric field to move through a $2.6-nm$ diameter ion channel, linearly increases with the polymer length. By using a similar experimental setup, Meller *et al.* [14] have also shown that the time-length relationship depends on specific interactions between the DNA and the nanopore walls. Very interesting experimental results by Han *et al.* [15] show longer crossing times for shorter DNA molecules, suggesting the existence of a quasi-equilibrium state of the polymer after being trapped for a time. This work suggests that, in same cases, a special condition for the polymer travelling across a nanometer-scale channel can be realized to make it temporarily trapped into a metastable state. In this framework, our investigations on the dynamics of polymers under the influence of a noisy environment show the presence of the noise enhanced stability (NES) effect [4]-[12] for short polymers escaping from a potential barrier. We also show how thermal fluctuations influence the motion of short chain molecules, by inducing two different regimes of translocation in the transport dynamics, depending on the length of the polymer.

2.1. Model and dynamics simulations

The polymer is modeled by a semi-flexible chain of N beads connected by harmonic springs according the Rouse model [16]. The van der Waals interactions between all beads are kept into account by introducing a Lennard-Jones potential. We neglect any hydrodynamic effect and volume effect. In order to confer a suitable stiffness to the chain, a bending recoil torque is included in the model, with a rest angle $\theta_0 = 0$ between two consecutive bonds. The potential energy of the modeled chain molecule is

$$U = U_{\rm Har} + U_{\rm Bend} + U_{\rm LJ}, \tag{1}$$

where $U_{\rm Har} = \sum_{i=1}^{N-1} K_{\rm r}(r_{i,i+1} - d)^2$ represents the energy required to extend the bond between two consecutive beads, $U_{\rm Bend} = \sum_{i=2}^{N-1} K_\theta(\theta_{i-1,i+1} - \theta_0)^2$ the work required to bend the chain, and $U_{\rm LJ} = 4\epsilon_{\rm LJ}\sum_{i,j(i\neq j)}\left[\left(\frac{\sigma}{r_{ij}}\right)^{12} - \left(\frac{\sigma}{r_{ij}}\right)^{6}\right]$ the Lennard-Jones potential. Here $K_{\rm r}$ is the elastic constant, r_{ij} the distance between particles i and j, K_θ the bending modulus, $\epsilon_{\rm LJ}$ the LJ energy depth and σ the monomer diameter. The polymer motion is modeled in a two-dimensional domain as a stochastic process of diffusion in the presence of a potential barrier having the form

$$U_{\rm Ext}(x) = ax^2 - bx^3, \tag{2}$$

with parameters $a = 3\cdot 10^{-3}$ and $b = 2\cdot 10^{-4}$. The drift of the $i^{\rm th}$ monomer of the chain molecule is described by the following overdamped Langevin equations

$$\gamma\frac{dx}{dt} = -\frac{\partial U}{\partial x} - \frac{\partial U_{\rm Ext}(x)}{\partial x} + \sqrt{D}\xi_{\rm x}(t), \tag{3}$$

$$\gamma\frac{dy}{dt} = -\frac{\partial U}{\partial y} + \sqrt{D}\xi_{\rm y}(t), \tag{4}$$

where U is the interaction potential, defined by Eq. (1), $\xi_x(t)$ and $\xi_y(t)$ are white Gaussian noise modeling the temperature fluctuations, with the usual statistical properties, namely $\langle\xi_k(t)\rangle = 0$ and $\langle\xi_k(t)\xi_l(t+\tau)\rangle = D\delta_{(k,l)}\delta(\tau)$ for $(k, l = x, y)$ and γ is the friction coefficient. In this work we set the γ parameter equal to one. Our simulation time t_s is scaled with the friction parameter as $t_s = t/\gamma$, therefore we use arbitrary units for all computed translocation times. We set both $K_{\rm r}$ and K_θ equal to 10, $\epsilon_{\rm LJ}$ equal to 0.1 and σ equal to 3, in arbitrary units. The inter-beads rest length d is chosen equal to 5, while the number of monomers N ranges from 2 to 40 units.

2.2. Results and conclusions

First results concern the case of harmonic potential forces U_{Har}, the cubic potential of Eq. (2) and initial unstable conditions. We find the noise enhanced stability (NES) effect for a short polymer escaping from a potential barrier [16]. In Fig. 1a we show how the MFPT changes with the noise intensity at different polymer lengths. Assuming identical unstable initial conditions, we find that the NES effect is more relevant for shorter chains (lower number of monomers). The maximum of the MFPT rapidly decreases at increasing polymer length and its position shifts towards higher levels of noise. The dynamics of shorter chains is much more similar to that of

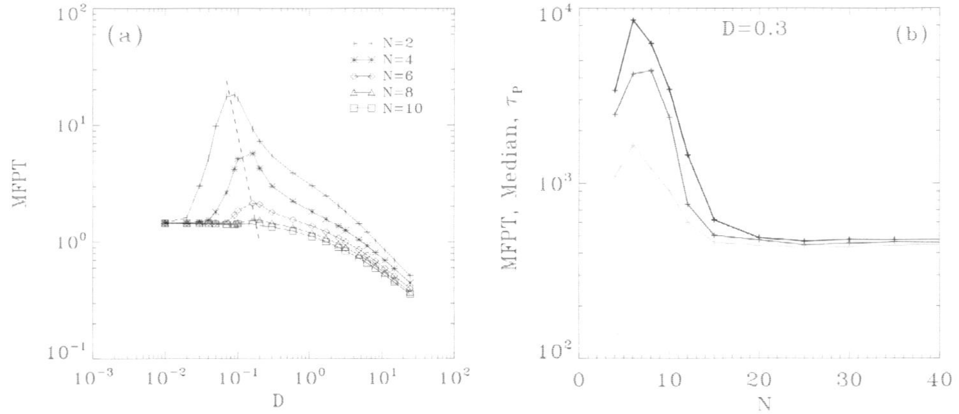

Figure 1. (a) Mean first passage time vs. noise intensity for five different numbers N of monomers in the polymer. The initial unstable condition $x_0 = 1.3$ is the same for all the monomers. The dashed line indicate the shift of the maximum of the MFPT. (b) Mean first passage time (blue line), median of the crossing time distribution (red line) and most probable translocation time (green line) vs. polymer length (number of beads) for noise intensity $D = 0.3$.

a single particle escaping from the potential well. As a consequence shorter chains show more pronounced NES effect, with higher maximum values of the MFPT. On the other hand, longer polymers need higher noise intensity values to move the center of mass back into the metastable state, starting from initial unstable states.

Now we consider all the potential energies reported in Eq. (1) and the polymer initially placed in the metastable state [17]. The different dynamics of translocation of short and long polymers is investigated as a function of the noise intensity. The results are reported in Fig. 1b. We report the MFPT, the median and the mode (τ_p) of the polymer translocation time distribution as a function of the chain length, for a noise intensity $D = 0.3$. Two different regimes of translocation are clearly present: for $N \leq 15$ all characteristic times show a maximum at $N \simeq 7$, while for $N \geq 15$ a common plateau is observed. This maximum is more evident in the MFPT than in τ_p, but the height of both peaks is rapidly reduced when the noise intensity increases. This maximum occurs at a polymer length corresponding to the persistence length of the chain L_p, which represents the length over which the correlations in the direction of the tangent to the polymer are lost. Experimental and theoretical studies on the translocation time, taken by a chain molecule to cross a membrane, show a strong dependence of this time with the length of the polymer. In this work we propose a possible reason of this dependence. After been trapped for a while, the molecule leaves the top of the barrier exhibiting an enhancement of the mean crossing time as a function of the temperature fluctuations. This is a typical signature of the presence of a metastable state in the system investigated. Our simulations demonstrate that longer polymers can travel the same channel distance faster than shorter chains. This result is in agreement with the experimental measurements of DNA mobility carried out by Han *et al.* [15], but a deeper analysis on the role of initial conditions on the translocation time is necessary.

3. FitzHugh-Nagumo model with external noise

In this section we study the dynamics of a neuron considered as an open system perturbed by periodic deterministic forcing and random fluctuations of environmental parameters. We analyze the stochastic dynamics of a neuron described by the FitzHugh-Nagumo (FHN) model

$$\dot{x} = x - \frac{x^3}{3} - y + Asin(\omega t + \varphi_0) + \zeta_x, \quad (5)$$

$$\dot{y} = \epsilon(x + I), \quad (6)$$

where x is the membrane potential, characterized by a fast dynamics, and y is the recovery variable, which exhibits a slow dynamics responsible for the refractory behaviour of the neuron. In Eq. (5), A and ω are respectively amplitude and frequency of the external forcing, $\zeta_x(t)$ is a self-correlated noise described by the Ornstein-Uhlenbeck process [18]

$$\frac{d\zeta_x(t)}{dt} = -\frac{1}{\tau}\zeta_x(t) + \frac{1}{\tau}\xi_x(t), \quad (7)$$

with τ correlation time of the colored noise, $\xi_x(t)$ Gaussian white noise with zero mean and correlation function $< \xi_x(t)\xi_x(t') >= \sigma_x \delta(t - t')$, and σ_x noise intensity. A self-correlated noise is more suitable than white noise to represent the random fluctuations in real systems, where the noise spectrum is characterized by the presence of a cut-off. The correlation function of the noise source is $\langle \zeta_x(t)\zeta_x(t')\rangle = \frac{\sigma_x}{2\tau} e^{-|t-t'|/\tau}$.

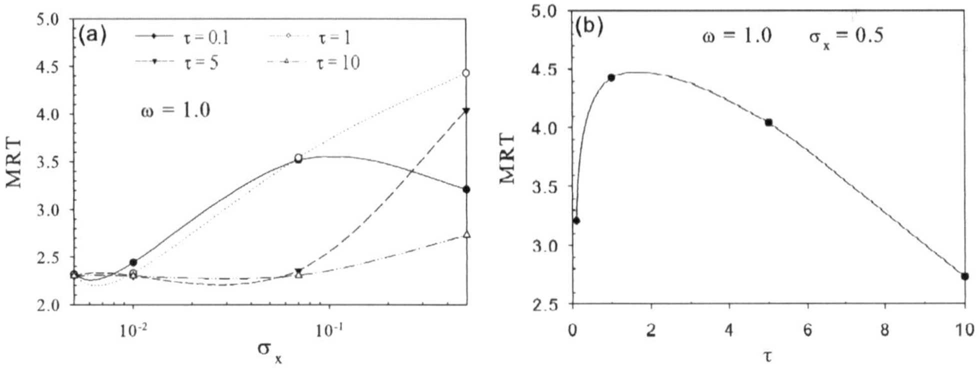

Figure 2. (a) Nonmonotonic behaviour (with a maximum) of MRT as a function of the noise intensity σ_x, for $\omega = 1.0$: in the presence of colored noise this effect disappears as correlation time increases. (b) Nonmonotonic behaviour of the MRT as a function of the correlation time τ, for $\omega = 1.0$ and $\sigma_x = 0.5$. The values of the parameters are: $I = 1.1$, $\epsilon = 0.05$, $\varphi_0 = 0$, $A = 0.5$. The initial state is (x_0, y_0), with $x_0 = -I$, $y_0 = -I + I^3/3$.

In a previous work the system response has been studied, under the influence of a periodic driving signal, as a function of the white noise intensity [19]. The authors showed the occurrence of resonant activation (RA) and noise enhanced stability (NES). The presence of random fluctuations produces modifications in the resonant activation (RA) phenomenon observed in deterministic regime and causes noise enhanced stability to appear in the minimum of RA. NES effect and modifications in RA are connected with the appearance of neuron excitability even if A and ω take values in the deterministic no-firing region. This indicates that noise, because of its cooperative action with the periodic force, can induce neuron excitability, also when, in deterministic regime, the firing is prohibited due to the parameter values.

In order to study the dynamics of the FHN system we introduce the mean response time defined as $< T >= 1/N \sum_{i=1}^{N} T_i$, with T_i the *first response time*, that is, the time for which the first spike occurs. We analyze the behaviour of the mean response time (MRT) of the neuron, in the presence of periodical stimulus, as a function of the noise intensity. The results, shown in Fig. 2a, evidence the presence of a noise enhanced stability (NES) phenomenon. In particular, MRT shows a nonmonotonic behavior as a function of σ_x (see panel a): starting from $\sigma_x = 0.005$, for intermediate levels of noise intensity an increase of MRT appears. For higher noise intensities a decrease of MRT is observed. We note that the strong dependence of MRT on the noise intensity is suppressed as correlation time increases. Finally, we study MRT as a function of the correlation time τ. The results, shown in Fig. 2b, indicate the presence of a nonmonotonic behaviour of MRT as a function of τ for a fixed value of noise intensity ($\sigma_x = 0.5$), confirming that high values of the correlation time cause the suppression of the noise effects on MRT.

4. A market model with stochastic volatility

Econophysics is a new interdisciplinary research field appeared and developed during last years. Theories and methods, originally developed by physicists in statistical physics to study complex systems, are applied to analyse financial market data [20, 21]. A big amount of work has been addressed to the empirical investigation of different phenomena to find universal laws underlying the dynamics of financial market. During last decades, a considerable effort has been devoted to the construction of models to reproduce and predict the behavior of real market, where the stock option evolution is determined by many traders interacting with each other and using different strategy to increase their own profit. Arbitrariness and nonlinearity present in the system lead to regard the financial market as a complex system where the randomness of the human behavior can be modelled by using stochastic processes. A common approach, used to describe the evolution of stock prices, is based on the Heston model. In this context, empirical verification with stocks and options showed good agreement with real data [22]. A further investigation on the Heston model has been recently performed [23]-[27]. This model consists of two coupled stochastic differential equations which represent the log-normal geometric Brownian motion stock process and the Cox-Ingersoll-Ross (CIR) mean-reverting stochastic process [24]. By using the *log* of the price, $x(t) = ln\ p(t)$, a generalization of the Heston model has been proposed to take into account feedback effects responsible for price fluctuations. Here, a cubic nonlinearity is introduced to consider the different dynamical regimes by the motion of a fictitious "Brownian particle" moving in an effective potential with a metastable state. This nonlinear Heston (NLH) model is defined by the following equations

$$dx(t) = -\left(\frac{\partial U}{\partial x} + \frac{v(t)}{2}\right) dt + \sqrt{v(t)} dW_1(t) \quad (8)$$

$$dv(t) = a(b - v(t))dt + c\sqrt{v(t)} dW_2(t), \quad (9)$$

where $U(x) = 2x^3 + 3x^2$ is the *effective* cubic potential with a metastable state at $x_{me} = 0$, a maximum at $x_m = -1$, and a cross point between the potential and the x axes at $x_c = -1.5$. $W_1(t)$ and $W_2(t)$ are uncorrelated Wiener processes with the usual statistical properties $< dW_i >= 0$, $< dW_i(t) dW_j(t') >= dt\ \delta(t - t')\ \delta_{i,j}\ (i, j = 1, 2)$. Here, v is the CIR process, which is defined by three parameters, b, a and c. They represent respectively the long term variance, the rate of mean reversion to the long term variance, and the volatility of variance, often called the *volatility of volatility*. The stochastic volatility $v(t)$ is characterized by exponential autocorrelation and *volatility clustering* [21, 26], that is alternating calm with burst

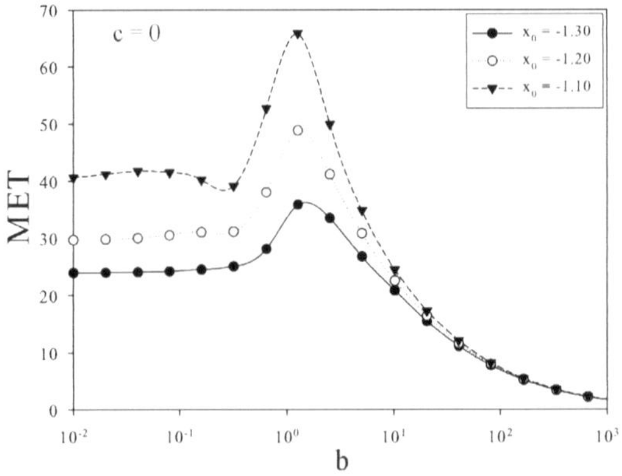

Figure 3. Mean escape time (MET) for 3 different unstable starting positions, when only the reverting term is present: $a = 10^{-2}$, $c = 0$. The curves are averaged over 10^5 escape events.

periods of volatility. The average exit time of the system from the stable to the unstable domain of the metastable potential $U(x)$ may be prolonged by imposing external noise: this phenomenon is named noise enhanced stability (NES). The NES effect and its different dynamical regimes can be explained considering the barrier "*seen*" by the Brownian particle starting at the initial position x_0, that is $\Delta U_{in} = U(x_{max}) - U(x_0)$, and by comparing it with the height of the barrier ΔU characterizing the metastable state [6, 9]. By investigating the mean escape time (MET), as a function of the model parameters a, b and c, we find the parameter region where a nonmonotonic behavior of MET is observable in our NLH model with stochastic volatility $v(t)$ [26, 27]. This behaviour is similar to that observed for MET *versus* v in the NES effect with constant volatility v. We investigated the dynamics of the system for $c = 0$. In this case, after an exponential transient, the volatility reaches the asymptotic value b, and the NES effect is observable as a function of b. This case corresponds to the usual parametric constant volatility regime. The results are reported in Fig. 3, where MET *versus* b is plotted for three different starting unstable initial positions. The simulations were performed considering the initial positions of the process $x(t)$ in the unstable region $[x_c, x_m]$ and using an absorbing barrier at $x = -6.0$. When the "Brownian particle" $x(t)$ hits the barrier, the escape time is registered and another simulation is started, placing $x(t)$ at the same starting position x_0 and using for the volatility the value taken on at the hitting time. The nonmonotonic behavior, which is more evident for starting positions near the maximum of the potential, is always present. After the maximum, when the values of b are much greater than the potential barrier height, the exponential behavior is recovered. The results of our simulations show that the NES effect can be observed as a function of the volatility reverting level b, the effect being modulated by the parameter $(ab)/c$. The phenomenon disappears if the noise term is predominant in comparison with the reverting term: the effect is no more observable because the parameter c pushes the system towards a too noisy region. When the noise term is coupled to the reverting term, we observe the NES effect as a function of the parameter c. Finally we note that the effect disappears if b is so high as to saturate the system [27].

5. Transient dynamics of Josephson junctions

Josephson junctions (JJs) are the basis components of a great variety of supeconducting devices. They are employed in High Temperature Superconductors (HTSs), such as nanoscale Superconducting Quantum Interference Devices (SQUIDs) for the detection of very weak magnetic fields, and in low temperatures solid state superconducting quantum bits (qubits), such as charge, flux and phase qubits [28]-[31]. In JJs, working both at high and low temperatures, the environment strongly affects the behavior of the system. In the low temperature superconductive devices it is very difficult to avoid the influence of environment that constitutes mainly a decoherence source for the system. Therefore, the study of transient dynamics of JJs in the presence of noise sources is very interesting for the understanding of the interaction between these systems and environment. The dynamics of a JJ is studied considering a fictitious Brownian particle moving in a washboard potential [32] and the behavior of the current-voltage characteristic of a JJ is strictly related to the lifetime of the superconductive metastable state of the particle. The decay of the particle from the metastable state, in fact, depends on the fluctuations of the voltage across the junction. Recently noise induced effects were experimentally observed in underdamped Josephson junctions [3, 12], and the switching to resistive state of an annular Josephson junction due to thermal activation was analyzed [33].

We present the study of the transient dynamics of short overdamped and long JJs under the influence of fluctuating bias current and oscillating potential. We numerically calculate the lifetime of the superconductive metastable state also called the mean switching time (MST) to the resistive state for short JJ (SJJ) and long JJ (LJJ). We demonstrate the presence of noise induced effects such as resonant activation (RA) [1, 2, 3, 7, 8] and noise enhanced stability (NES) [4]-[12]. We analyze both the effects of thermal and correlated noise sources. In SJJ and LJJ we consider white noise, accounting for the thermal fluctuations, and correlated (colored) noise separately. Moreover, in LJJ we present an analysis considering together the effects of white and colored noise. For given values of frequency of the driving signal and suitable noise intensity, we find maxima of the lifetime of the superconductive state. This is an interesting feature for the study of the coherence time of these devices. Our results hold for low temperature superconducting devices when we consider only the effects of the colored noise, and they can be extended to high temperatures, when both colored and white noise come into play.

5.1. Short overdamped Josephson junctions

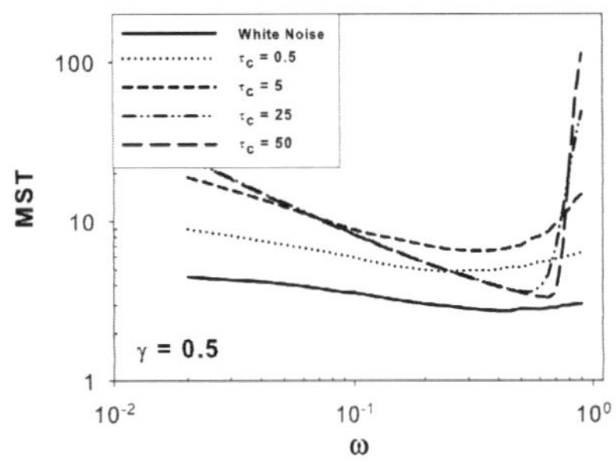

Figure 4. MST vs ω for white noise and different τ_c, γ=0.5, i_0=0.8 and A=0.7.

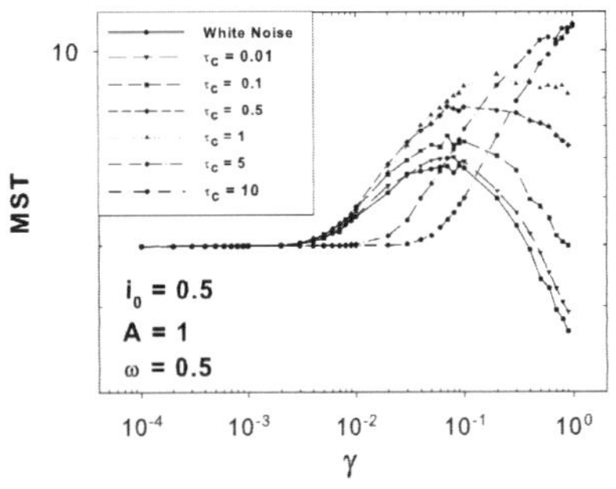

Figure 5. MST vs γ for white noise and different values of τ_c, ω=0.5, A=1, i_0=0.5.

The study of SJJs is performed in the framework of the resistive shunted junction (RSJ) model formalism [34], with a fluctuating current term. The overdamped Langevin equation to study the dynamics is [35]

$$\frac{d\phi}{dt} = -\omega_c \frac{dU(\phi)}{d\phi} - \omega_c \zeta(t), \tag{10}$$

where ϕ is the phase difference of the wave functions in the ground state between left and right superconductive sides of the junction. The characteristic frequency of the Josephson junction is $\omega_c = 2eR_N I_c/\hbar$, where e is the electron charge, R_N^{-1} is the normal conductivity, I_c is the critical current and $\hbar = h/2\pi$ with h the Plank constant. In Eq. (10) the time is normalized to the inverse of the characteristic frequency of the junction ω_c. In our analysis $\zeta(t)$ is a colored noise source, generated by an Ornstein-Uhlenbeck (OU) process

$$d\zeta(t) = -\frac{1}{\tau_c}\zeta(t)dt + \frac{\sqrt{\gamma}}{\tau_c}dW(t), \tag{11}$$

where γ is the noise intensity and $W(t)$ is the Wiener process with the usual statistical properties: $\langle dW(t)\rangle = 0$, and $\langle dW(t)dW(t')\rangle = \delta(t-t')dt$. The potential profile $U(\phi)$ of Eq. (10) is given by

$$U(\phi) = 1 - cos\phi - i(t)\phi, \tag{12}$$

where $i(t) = i_0 + f(t)$, $i_0 = i_b/I_c$ is the constant dimensionless bias current and $f(t) = Asin\omega t$ is the driving current with dimensionless amplitude $A = i_s/I_c$ and frequency ω (i_b and i_s represent the bias current and the driving current amplitude respectively). We investigate the dynamics of the Brownian particle in the presence of a time-dependent nonlinear periodic potential, by solving numerically Eq. (10), for $\omega_c=1$. To study the lifetime of the superconductive state we take, as initial condition for the particle, the minimum of the potential profile, corresponding to the condition $\phi_0 = arcsin(i_0)$ and we calculate the time spent by the particle to reach the next maximum. In Fig. 4 we report the behavior of MST vs ω, for the noise intensity $\gamma = 0.5$, and different values of τ_c. The non-monotonic behavior of the curves shows that the RA phenomenon, already found in the presence of white noise [8, 35], appears also with colored noise. The values of MST around the minimum are influenced by the variation of τ_c. Moreover, we find, in a wide range of frequency ($0.3 < \omega < 0.8$) (see Fig. 4), a non-monotonic behavior of MST as a function of the correlation time.

In Fig. 5, we show the behavior of MST as a function of the noise intensity for different values of the correlation time. We note the appearance of noise enhanced stability (NES), a phenomenon already found in short JJs in the presence of white noise [8, 35]. Here we observe a range of values of correlation time, namely $0.01 < \tau_c < 1$, in which the curves present a non-monotonic behavior. For $\tau_c = 5, 10$, the non-monotonic behavior disappears.

5.2. Long Josephson junctions

The investigation of LJJs is developed in the framework of the sine-Gordon model [34, 36]. The dynamics of a LJJ is described by the motion of a phase string in a one dimensional washboard potential. The transient dynamics is represented by a nonlinear partial differential equation with a stochastic term [37, 38]

$$\beta\frac{\partial^2\phi}{\partial t^2} + \frac{\partial\phi}{\partial t} + \frac{\partial^2\phi}{\partial x^2} = i(x) - \sin\phi + i_f(x,t), \tag{13}$$

where $\beta = \omega_c RC$, with R and C the LJJ equivalent resistance and capacitance, respectively. Here $i(x)$ and $i_f(x,t)$ are the bias current density and the fluctuating current density normalized to the critical current density I_c of the junction, respectively. In Eq. (13) time is normalized to the inverse of the characteristic frequency, $\omega_c = 2eRI_c/\hbar$, of the junction. Analogously length is normalized to the Josephson penetration depth (λ_J). The boundary conditions for Eq. (13) are:

$$\frac{\partial\phi(0,t)}{\partial x} = \frac{\partial\phi(L,t)}{\partial x} = 0. \tag{14}$$

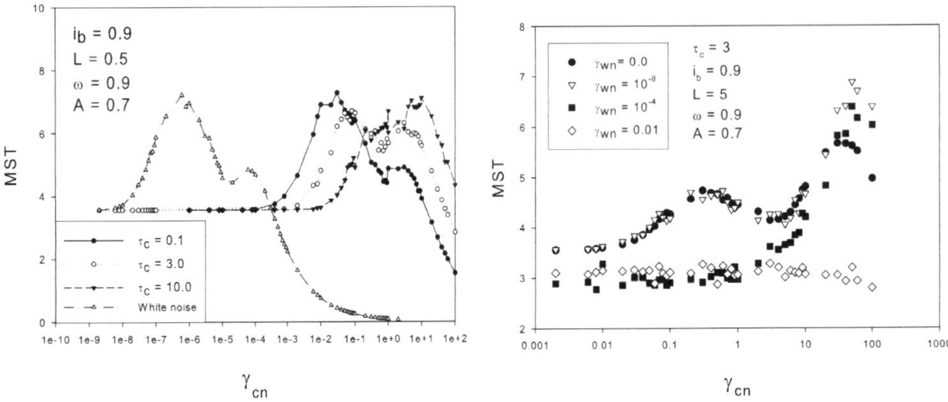

Figure 6. Left panel: MST vs γ_{cn} with $i_b = 0.9$ and L=0.5. Right panel: MST vs γ_{cn} with $i_b = 0.9$ and L=5, for different values of white noise intensity.

where $L = l/\lambda_J$ with l the length of the junction. As initial position of the phase string in the washboard potential, we consider the minimum, represented by the condition $\phi(x,0) = arcsin(i_b)$. We studied, first, the effects of a fluctuating current signal described by a correlated noise and, then, the effects of both colored and white noise on the system, by inserting in Eq. (13) a fluctuating term given by the sum of a correlated noisy current signal, $i_{cn}(x,t)$, and a thermal current signal, $i_{wn}(x,t)$, namely $i_f(x,t) = i_{cn}(x,t) + i_{wn}(x,t)$. In the presence of colored noise we found the noise induced effects such as RA and NES. In particular in left panel of Fig. 6 the non-monotonic behavior of MST as a function of the noise intensity γ_{cn} indicates the presence of the NES effect. The maxima of the curves depend by different values of τ_c. The effects of both colored and white noise are presented in right panel of Fig. 6, where the curves of MST vs γ_{cn} are calculated for different white noise intensities γ_{wn}, with $L = 5$. We note that when γ_{wn} is greater than γ_{cn} the effects of colored noise disappear. If we consider γ_{wn} suitably lower than γ_{cn}, the effects of colored noise become evident. In fact, when γ_{cn} is at least five order of magnitude higher than γ_{wn}, the curve of MST vs γ_{cn} (empty triangles) matches with good agreement that (black circles) obtained in the absence of white noise (see right panel of Fig. 6 for $\gamma_{wn}=10^{-8}$).

6. Metastability in a quantum system

Quantum tunneling effect occurs often in condensed matter physics, examples are JJs, hetero-nanostructures, etc.. The tunneling effect plays an important role in the nonlinear relaxation time from a metastable state in an open quantum system, interacting with a thermal bath. Symmetrical and asymmetric bistable systems are good quantum model systems for analysis of the "superconducting quantum bits" and decoherence phenomena. To obtain very long coherence times in the presence of interaction between the qubit and the noisy environment is one of the greatest challenges of physics. The influence of the environment in quantum tunneling has been in the focus of intense research over the last years [39]-[42]. The environment is commonly described as an ensemble of harmonic oscillators (thermal bath) at thermal equilibrium at temperature T, with a bilinear coupling between the quantum system and the thermal bath. By this kind of coupling between system and environment the quantum mechanical analogue of the generalized Langevin equation can be derived [39]. Time-dependent driving fields, such as laser beams, have most interesting implications for quantum systems. These time-dependent fields give rise to interesting effects, such as the coherent destruction of tunneling [43], the effect of quantum stochastic resonance [44], and the control and reduction of decoherence in open quantum systems [45]. In this work we analyze a time-dependent asynmmetric bistable potential by using the approach of the Feynman-Vernon functional [46] in discrete variable representation (DVR) [47, 48].

6.1. The model

The starting point of our investigation is the Caldeira-Leggett model [39] for a quantum particle with mass M interacting with a thermal bath. The particle, with position operator $\mathbf{q}$ and momentum operator $\mathbf{p}$ is moving in an asymmetric one-dimensional double-well potential $V_0(\mathbf{q})$ in the presence of a time-dependent external force, $A\sin(\Omega t)$. The unperturbed Hamiltonian of the system is

$$\mathbf{H}_u n(t) = \frac{\mathbf{p}^2}{2M} + V_0(\mathbf{q}) - A\sin(\Omega t), \tag{15}$$

and the potential is

$$V_0(\mathbf{q}) = \frac{M^2\omega_0^4}{64\Delta U}\mathbf{q}^4 - \frac{M\omega_0^4}{4}\mathbf{q}^2 - \mathbf{q}\epsilon, \tag{16}$$

where ϵ is the asymmetry parameter. The environment is modelled [39] as an ensemble of harmonic oscillators

$$\mathbf{H}_B = \sum_{j=1}^{\mathcal{N}} \frac{1}{2}\left[\frac{\mathbf{p}_j^2}{m_j} + m_j\omega_j^2\left(\mathbf{x}_j - \frac{c_j}{m_j\omega_j^2}\mathbf{q}\right)^2\right]. \tag{17}$$

The full Hamiltonian (system + bath) is $\mathbf{H}(t) = \mathbf{H}_u n(t) + \mathbf{H}_B$. As $\mathcal{N} \to \infty$, in Eq. (17), we have a continuous spectral density. We use an Ohmic spectral density with an exponential cut-off $J(\omega) = \eta\omega\exp\left(\frac{\omega}{\omega_c}\right)$. Because of the bilinear coupling between the coordinate $\mathbf{q}$ of the system and the coordinate $\mathbf{x}$ of thermal bath we can consider this as a quantum analogue of the multiplicative noise in classical systems.

By following the Feynman-Vernon approach [46] the reduced density operator can be written as

$$\rho(q_f, q_f'; t) = \int dq_0 \int dq_0' K(q_f, q_f', t; q_0, q_0', t_0)\rho_S(q_0, q_0', t_0), \tag{18}$$

where the propagator K is given by

$$K(q_f, q_f', t; q_0, q_0', t_0) = \int_{q(t_0)=q_0}^{q(t)=q_f} \mathcal{D}q \int_{q'(t_0)=q_0'}^{q'(t)=q_f'} \mathcal{D}q' \mathcal{A}[q]\mathcal{A}^*[q']\mathcal{F}_{FV}[q, q'], \tag{19}$$

and $\mathcal{A}[q] = \exp\left(\frac{iS_S[q]}{\hbar}\right)$ with $S_S[q]$ being the classical action functional. In Eq. (19) $\mathcal{F}_{FV}[q, q'] = \exp\left(\frac{-\phi_{FV}[q,q']}{\hbar}\right)$ is the Feynman Vernon influence functional, which depends on the bath correlation function [41].

6.2. Discrete variable representation

By using the discrete variable representation (DVR) [47], with the discrete eigenbasis of the position operator $\mathbf{q}$, and the generalized non-interacting cluster approximation [48] it is possible to obtain a generalized Master Equation (GME)

$$\rho_{\mu\mu}(t) = \sum_{\nu=1}^{M} \int_{t_0}^{t} dt' H_{\mu\nu}(t - t')\rho_{\nu\nu}(t') + I_\mu(t - t_0), \quad \mu = 1, \ldots, M \ , \tag{20}$$

neglecting all the intercluster interactions and taking into account the paths that start in an off-diagonal element (see Ref. [48] for details). Since the inhomogeneities $I_\mu(t, t_0)$ do not contribute to the long-time dynamics because decay exponentially with time on a rather short time scale, this term can be neglected. We assume furthermore that the characteristic memory time τ_{mem} of the kernels of Eq. (20) is the smallest time scale of the problem (*Markovian limit*). This means that we can substitute the argument of $\rho_{\nu\nu}(t')$ under the integral by the time t and draw $\rho_{\nu\nu}(t)$ in front of the integral. We then obtain the *Markovian approximated generalized master equation*

$$\dot{\rho}_{\mu\mu}(t) = \sum_{\nu=1}^{M} \Gamma_{\mu\nu}(t)\rho_{\nu\nu}(t), \tag{21}$$

with the time-dependent rate coefficients

$$\Gamma_{\mu\nu}(t) = \int_0^\infty d\tau H_{\mu\nu}(t, t-\tau). \tag{22}$$

Since the diagonal elements $\rho_{\mu\mu}(t)$ obey Eq. (21), the long-time dynamics in this regime is ruled by a single exponential decay. Thus the Eq. (21) is a set of coupled ordinary first-order differential equations, which can be decoupled via a diagonalization procedure. The diagonalized rate matrix reads

$$\sum_{\kappa_1,\kappa_2=1}^{M} (S^{-1})_{\mu\kappa_1}\Gamma_{\kappa_1\kappa_2}S_{\kappa_2\nu} = \Lambda_\mu\delta_{\mu\nu}, \tag{23}$$

where $S_{\mu\nu}$ denotes the element of the transformation matrix and Λ_μ the eigenvalues of the rate matrix. The general solution of the Markov approximated GME is

$$\rho_{\mu\mu}(t) = \sum_{\nu,\kappa=1}^{M} S_{\mu\nu}(S^{-1})_{\mu\kappa}e^{\Lambda_\nu(t-t_0)}\rho_{\kappa\kappa}(t_0). \tag{24}$$

Since the diagonal elements of the rate matrix are the negative sum of the matrix elements of the corresponding columns, one eigenvalue equals to zero, i.e., $\Lambda_1 = 0$. Therefore,

$$\rho_{\mu\mu}(t) = \rho_{\mu\mu}^\infty + \sum_{\nu=2}^{M}\sum_{\kappa=1}^{M} S_{\mu\nu}(S^{-1})_{\mu\kappa}e^{\Lambda_\nu(t-t_0)}\rho_{\kappa\kappa}(t_0). \tag{25}$$

with $\rho_{\mu\mu}^\infty = \sum_{\kappa=1}^{M} S_{\mu,1}(S^{-1})_{1,\kappa}\rho_{\kappa\kappa}(t_0)$ being the asymptotic population of the DVR-state $\mid q_\mu >$. The rate which determines the dynamics on the largest time-scale is the smallest non-zero absolute value of the real part of the eigenvalues of the rate matrix, i.e. *the quantum relaxation rate*

$$\Gamma \equiv \min\{\mid Re\Lambda_\nu \mid; \nu = 2, \ldots, M\}. \tag{26}$$

6.3. Results

In Fig. 7 we report the first four global states $< q \mid \psi_1 >, \ldots, < q \mid \psi_4 >$ for barrier height $E_B = \Delta U/\hbar\omega_0 = 1.4$ and asymmetry parameter $\varepsilon = 0.23$. The corresponding potential profile has a metastable state on the left well. The quantum relaxation rate Γ as a function of the asymmetry parameter is reported in Fig. 8 for different temperatures. There are many overlapping regions where is visible a nonmonotonic behaviour of Γ as a function of the temperature. We can distinguish two different nonmonotonic behaviours: one with a maximum, reported in Fig. 9, and the other one with a minimum. This last one corresponds to the noise enhanced stability effect revealed in classical metastable systems, that is a noise enhanced stability quantum effect.

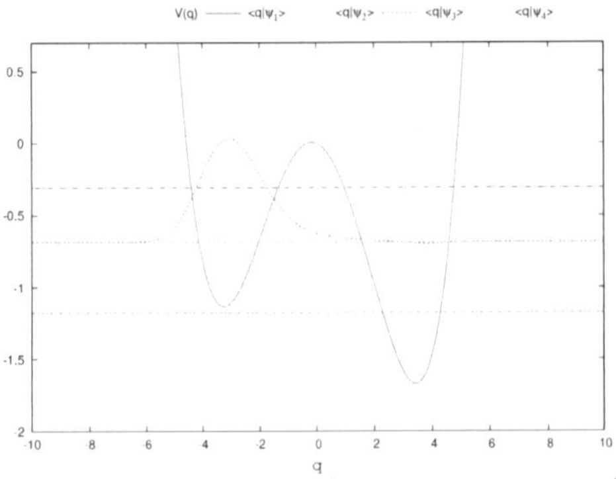

Figure 7. The first four global states $< q \mid \psi_1 >, \ldots, < q \mid \psi_4 >$ for barrier height $E_B = \Delta U/\hbar\omega_0 = 1.4$ and bias $\varepsilon = 0.23$.

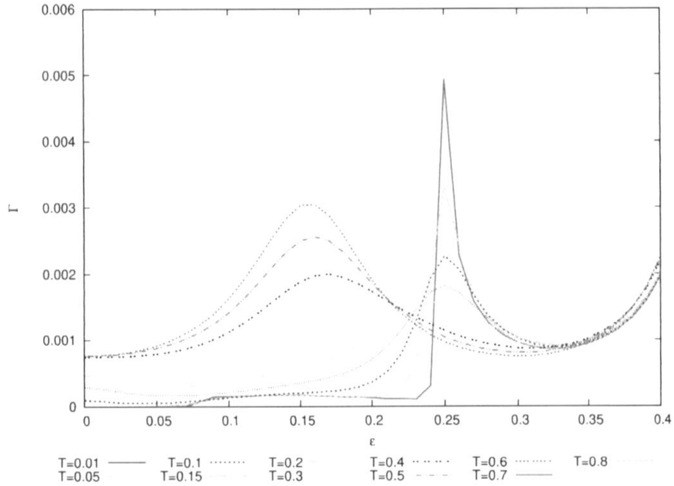

Figure 8. Quantum relaxation rate Γ as a function of the asymmetry parameter ε for different temperatures T. The barrier height is $E_B = 1.4$ and the number of energy levels is $M = 4$.The bath parameters are $\eta = 0.1$ and $\omega_c = 10.0$.

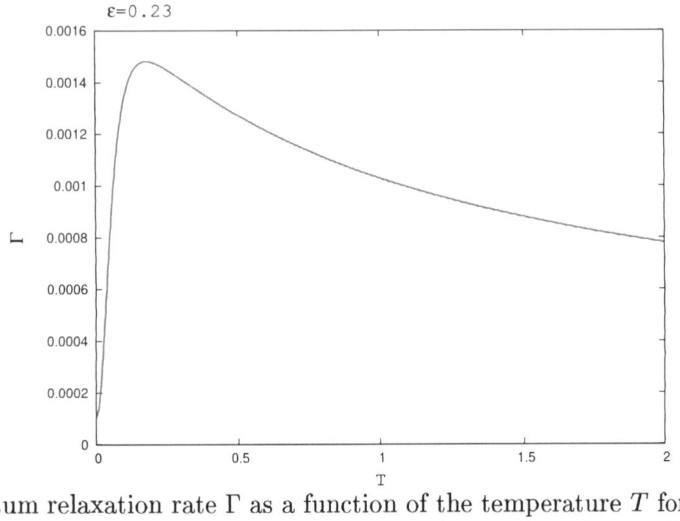

Figure 9. Quantum relaxation rate Γ as a function of the temperature T for fixed asymmetric parameter $\epsilon = 0.23$. The barrier height is $E_B = 1.4$ and the number of energy levels is $M = 4$.The bath parameters are $\eta = 0.1$ and $\omega_c = 10.0$.

Acknowledgments
This work was supported by MIUR.

References
[1] Doering C R and Gadoua J A 1992 *Phys. Rev. Lett.* **69** 2318
[2] Mantegna R N and Spagnolo B 2000 *Phys. Rev. Lett.* **84** 3025
[3] Yu Y and Han S 2003 *Phys. Rev. Lett.* **91** 127003
[4] Mantegna R N and Spagnolo B 1996 *Phys. Rev. Lett.* **76** 563
[5] Mielke A 2000 *Phys. Rev. Lett.* **84** 818.
[6] Agudov N V and Spagnolo B 2001 *Phys. Rev.* E **64**, 035102(R)
[7] Dubkov A A, Agudov N V and Spagnolo B 2004 *Phys. Rev.* E **69** 061103
[8] Pankratov A L and Spagnolo B 2004 *Phys. Rev. Lett.* **93** 177001
[9] Fiasconaro A, Spagnolo B and Boccaletti S 2005 *Phys. Rev.* E **72** 061110(5)
[10] Hurtado P I, Marro J and Garrido P L 2006 *Phys. Rev.* E **74** 050101(R)
[11] Spagnolo B, Dubkov A A, Pankratov A L, Pankratova E V, Fiasconaro A and Ochab-Marcinek A 2007 *Acta Physica Polonica* B **38** 1925
[12] Sun G *et al* 2007 *Phys. Rev.* E **75** 021107(4)
[13] Kasianowicz J J, Brandin E, Branton D and Deamer D W 1996 *Proc. Natl. Acad. Sci. USA* **93** 13770
[14] Meller A and Branton D 2002 *Electrophoresis* **23** 2583.
[15] Han J, Turner S W and Craighead H G 1999 *Phys. Rev. Lett.* **83** 1688
[16] Pizzolato N, Fiasconaro A and Spagnolo B 2008 *Int. J. Bif. Chaos* **18** 2871
[17] Pizzolato N, Fiasconaro A and Spagnolo B 2009 *J. Stat. Mech.* **P01011**
[18] Valenti D, Augello G and Spagnolo B 2008 *Eur. Phys. J.* B **65** 443
[19] Pankratova E V, Polovinkin A V and Spagnolo B 2005 *Phys. Lett. A* **344** 43
[20] Mantegna R N and Stanley H E 2000 *An Introduction to Econophysics: Correlations and Complexity in Finance* (Cambridge, UK: Cambridge U. Press)
[21] Bouchaud J P and Potters M 2004 *Theory of Financial Risks and Derivative Pricing* (Cambridge, UK: Cambridge U. Press)
[22] Drăgulescu A A and Yakovenko V M 2002 *Quantitative Finance* **2** 443
[23] Miccichè S, Bonanno G, Lillo F and Mantegna R N 2002 *Physica* A **314** 756
[24] Silva A C, Prange R E and Yakovenko V M 2004 *Physica* A **344** 227
[25] Bonanno G and Spagnolo B 2005 *Fluc. Noise Lett.* **5**, L325
[26] Bonanno G, Valenti D and Spagnolo B 2006 *Eur. Phys. J.* B **53** 405
[27] Bonanno G, Valenti D and Spagnolo B 2007 *Phys. Rev.* E **75** 016106
[28] Wu C H *et al* 2008 *Nanotechnology* **19** 315304
[29] Nakamura Y *et al* 1999 *Nature* **398** 786
[30] Friedman J R *et al* 2000 *Nature* **406** 43
[31] Martinis J M *et al* 2002 *Phys. Rev. Lett.* **89** 117901
[32] Yu Y *et al* 2002 *Science* **296** 889
[33] Gulevich D and Kusmartsev F2006 *Physica* C **435** 87
[34] Barone A and Paternò G 1982 *Physics and Application of the Josephson Effect* (New York: Wiley) 122
[35] Gordeeva A V *et al* 2008 *Int. J. Bif. Chaos* **18** 2823.
[36] Gulevich D R *et al* 2008 *Phys. Rev. Lett.* **101** 127002
[37] Fedorov K G and Pankratov A 2007 *Phys. Rev.* B **76** 024504
[38] Fedorov K G, Pankratov A L and Spagnolo B 2008 *Int. J. Bif. Chaos* **18** 2855
[39] Caldeira A O and Leggett A L 1981 *Phys. Rev. Lett.* **46** 211
[40] Leggett A J, Chakravarty S, Dorsey A T, Fisher M, Garg A and Zwerger W 1987 *Rev. Mod. Phys.* **59** 1
[41] Weiss U 1999 *Quantum Dissipative Systems* (Singapore: World Scientific)
[42] Grifoni M and Hänggi P 1998 *Phys. Rep.* **304** 230
[43] Grossmann F, Dittrich T, Jung P and Hänggi P 1991 *Phys. Rev. Lett.* **67** 516
[44] Löfstedt R and Coppersmith S. N. 1994 *Phys. Rev. Lett.* **72** 1947.
[45] Viola L, Knill E and Lloyd S 1999 *Phys. Rev. Lett.* **82** 2417
[46] Feynman R P and Vernon F L, Jr. 1963 *Ann. Phys. (N.Y.)* **24** 118
[47] Harris D O, Engerholm G G and Gwinn W D 1965 *J. Chem. Phys.* **43** 1515
[48] Thorwart M, Grifoni M and Hänggi P 2001 *Annals of Physics* **293** 15

Quantum-classical interactions and measurement: a consistent description using statistical ensembles on configuration space

M Reginatto[1] and M J W Hall[2]

[1]Physikalisch-Technische Bundesanstalt, Bundesallee 100, 38116 Braunschweig, Germany
[2]Theoretical Physics, Research School of Physics and Engineering, Australian National University, Canberra ACT 0200, Australia

E-mail: marcel.reginatto@ptb.de

Abstract. We provide an overview of a canonical formalism that describes mixed quantum-classical systems in terms of statistical ensembles on configuration space, and discuss applications to measurement theory. It is shown that the formalism allows a general description of the measurement of a quantum system by a classical apparatus without running into inconsistencies. An example of classical and quantum particles interacting gravitationally is also given.

1. Introduction

There are a number of good reasons for modeling the interaction of quantum and classical systems. Some of them are practical: it is sometimes convenient for physical or computational reasons to model part of a physical system classically, in which case it is of interest to have a formalism that is free of inconsistencies and allows for well defined approximation schemes. But there are other reasons which stem from open problems in the foundations of physics. A good example can be found in measurement theory: in the standard Copenhagen interpretation of quantum mechanics, the measuring apparatus must be described in classical terms [1, 2] and this implies a coupling of some sort between a quantum system and the apparatus which is treated as a classical system. A measurement performed according to these principles will be discussed in Section 4. Gravity provides another example: since there is no full quantum theory yet, one would like to know to what extent the quantization of gravity is forced upon us by consistency arguments alone [3], and one way to do this is to investigate how far we can get with a mixed system in which the gravitational field remains classical while matter is assumed to consist of quantum fields. This is a difficult problem that is beyond the scope of the paper. We will consider here a much simpler problem which may however be relevant to this issue, that of measurements where quantum and classical non-relativistic particles interact gravitationally; this is described in Section 5.

There are, of course, a number of obstacles that have to be dealt with when modeling the interaction of quantum and classical systems. First of all, quantum mechanics and classical mechanics are formulated using very different mathematical structures, and it is necessary to to find a common mathematical framework before progress can be made. Furthermore, there are difficult conceptual issues, i.e. uncertainty principle, superposition, etc. What aspects of quantum and classical mechanics should be preserved in the description of a mixed system?

In the following Sections we give an overview of a solution to this problem based on a canonical formalism that describes statistical ensembles on configuration space [4]. The formalism can be applied to both quantum and classical systems and allows a general and consistent description of interactions between them. This approach overcomes difficulties arising in previous attempts; in particular, the correct equations of motion for the quantum and classical sectors are recovered in the limit of no interaction, conservation of probability and energy are satisfied, uncertainty relations hold for conjugate quantum variables, and the formalism allows a back reaction of the quantum system on the classical system. A discussion of other approaches falls outside the scope of this paper, but see Ref. [4] and the references quoted therein, and Ref. [5].

2. Ensembles in configuration space: the basic idea

We start from the assumption that, as seems to be implied by quantum mechanics, the configuration of a physical system is an inherently statistical concept. The system is then described by an ensemble of configurations with probability density P, where $P \geq 0$ and $\int dx\, P(x,t) = 1$. To derive equations of motion we introduce an *ensemble Hamiltonian* $\tilde{H}[P,S]$ where S is canonically conjugate to P. The equations of motion take the form

$$\frac{\partial P}{\partial t} = \left\{P, \tilde{H}\right\}_{PB} = \frac{\delta \tilde{H}}{\delta S}\,, \qquad \frac{\partial S}{\partial t} = \left\{S, \tilde{H}\right\}_{PB} = -\frac{\delta \tilde{H}}{\delta P}\,, \tag{1}$$

where $\{A, B\}_{PB}$ is the Poisson bracket of the fields A and B.

The following ensemble Hamiltonians are of interest in that they lead to equations that describe the evolution of quantum and classical non-relativistic systems:

$$\begin{aligned} \tilde{H}_C[P,S] &= \int dx\, P\left[\frac{|\nabla S|^2}{2m} + V(x)\right]\,, \\ \tilde{H}_Q[P,S] &= \tilde{H}_C[P,S] + \frac{\hbar^2}{4}\int dx\; P\frac{|\nabla \log P|^2}{2m}\,. \end{aligned} \tag{2}$$

For example, the equations of motion derived from $\tilde{H}_Q[P,S]$ are given by

$$\frac{\partial P}{\partial t} + \nabla.\left(P\frac{\nabla S}{m}\right) = 0\,, \quad \frac{\partial S}{\partial t} + \frac{|\nabla S|^2}{2m} + V + \frac{\hbar^2}{2m}\frac{\nabla^2 P^{1/2}}{P^{1/2}} = 0\,, \tag{3}$$

while the equations of motion derived from $\tilde{H}_C[P,S]$ are the same as Eq. (3) but with $\hbar = 0$. The first equation in Eq. (3) is a continuity equation, the second equation is the classical Hamilton-Jacobi equation when $\hbar = 0$ and a modified Hamilton-Jacobi equation when $\hbar \neq 0$. Defining $\psi := \sqrt{P}\, e^{iS/\hbar}$, Eq. (3) takes the form

$$i\hbar\frac{\partial \psi}{\partial t} = \frac{-\hbar^2}{2m}\nabla^2\psi + V\psi\,, \tag{4}$$

which is the usual form of the Schrödinger equation. Therefore, in this formalism, quantum and classical particles are treated on an equal footing, with differences being primarily due to the different forms of the respective ensemble Hamiltonians. Hence we have a framework that enables us to extend the formalism to mixed quantum-classical systems in a natural way, as discussed in the next Section.

We end this Section with some remarks concerning the interpretation of the canonically conjugate variables. While the interpretation of P is straightforward, there are some subtle issues concerning the physical interpretation of S. To maintain full generality, S should not be regarded as a "momentum potential". In particular, for an ensemble of classical particles with uncertainty described by the probability P, it will not be assumed that the momentum of a member of the ensemble is a well-defined quantity proportional to the gradient of S, as it is done in the usual deterministic interpretation of the Hamilton–Jacobi equation. This avoids forcing a similar deterministic interpretation in the quantum and quantum-classical cases. A deterministic picture can be recovered for classical ensembles precisely in those cases in which trajectories are operationally defined [4].

It is however possible to define local energy and momentum densities in terms of S. If $\tilde{H}[\lambda P, S] = \lambda \tilde{H}[P, S]$ (which is the case for the ensemble Hamiltonians that we consider in this article), then

$$\tilde{H} = \int dx\, P \frac{\delta \tilde{H}}{\delta P} = - \int dx\, P \frac{\partial S}{\partial t} = -\langle \partial S/\partial t \rangle\ , \tag{5}$$

which shows that $\partial S/\partial t$ is a local energy density. Furthermore, $\int dx\, P \nabla S$ is the canonical infinitesimal generator of translations, since

$$\begin{aligned} \delta P(x) &= \delta \mathbf{x} \cdot \left\{ P, \int dx\, P \nabla S \right\}_{PB} = -\delta \mathbf{x} \cdot \nabla P\ , \\ \delta S(x) &= \delta \mathbf{x} \cdot \left\{ S, \int dx\, P \nabla S \right\}_{PB} = -\delta \mathbf{x} \cdot \nabla S\ , \end{aligned} \tag{6}$$

under action of the generator, and therefore $P\nabla S$ can be considered a local momentum density. These results are generally valid; i.e. they hold true for classical, quantum and mixed systems.

3. Interaction of quantum and classical systems: a configuration space description

The formalism of configuration-space ensembles allows a general and consistent description of interactions between quantum and classical ensembles. We consider non-relativistic systems as an example here. A mixed quantum-classical ensemble Hamiltonian on a configuration space with coordinates q, x is given by [4]:

$$\tilde{H}_{QC}[P, S] = \int dq\, dx\, P \left[\frac{|\nabla_x S|^2}{2M} + \frac{|\nabla_q S|^2}{2m} \right] + \int dq\, dx\, P \left[\frac{\hbar^2}{4} \frac{|\nabla_q \log P|^2}{2m} + V(q, x, t) \right]\ . \tag{7}$$

Here q denotes the configuration space coordinate of a quantum particle of mass m and x that of a classical particle of mass M, and $V(q, x, t)$ is a potential energy function describing the quantum-classical interaction. The equations of motion for P and S derived from $\tilde{H}_{QC}$ are

$$\begin{aligned} \frac{\partial P}{\partial t} &= -\nabla_q \cdot \left(P \frac{\nabla_q S}{m} \right) - \nabla_x \cdot \left(P \frac{\nabla_x S}{M} \right)\ , \\ \frac{\partial S}{\partial t} &= -\frac{|\nabla_q S|^2}{2m} - \frac{|\nabla_x S|^2}{2M} - V + \frac{\hbar^2}{2m} \frac{\nabla_q^2 P^{1/2}}{P^{1/2}}\ . \end{aligned} \tag{8}$$

The approach based on configuration space ensembles overcomes difficulties arising in previous approaches. We collect here a number of properties which are proven in Refs. [4, 5]:

1. Probability and energy are conserved.
2. There is backreaction of the quantum system on the classical system.
3. The correct quantum and classical equations of motion are recovered in the limit of no interaction.

4. The equations satisfy Galilean invariance for potentials of the form $V(x, q, t) = V(|x - q|)$ in Eq. (8).

5. The two "minimal requirements for a quantum-classical formulation" specified by Salcedo [6, 7] are satisfied: a Lie bracket may be defined on the set of observables, and this Lie bracket is equivalent to a Poisson bracket for classical observables and to $(i\hbar)^{-1}$ times the quantum commutator for quantum observables.

6. "Configuration separability" is satisfied: the classical configuration is invariant under any canonical transformation applied to the quantum component, and vice versa.

7. The "definite benchmark ... for an acceptable quantum-classical hybrid system" specified by Peres and Terno [8] is satisfied: the expectation values for the position and momentum observables of linearly-coupled quantum and classical oscillators obey the classical equations of motion.

8. Generalized Ehrenfest relations are satisfied.

We are not aware of any other formulation that satisfies all of these properties. In particular, requirements 5 is highly non-trivial [6, 7, 9]; it therefore provides a critical test for any formulation of mixed quantum-classical systems.

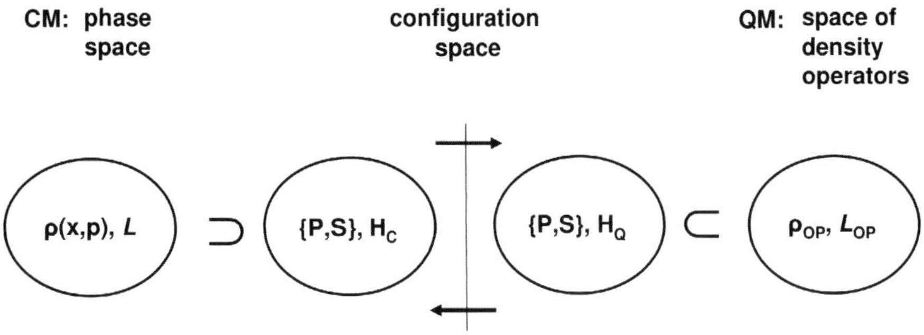

Figure 1. Relationship between phase space (classical mechanics), configuration space (classical, quantum mechanics) and the space of density operators (quantum mechanics).

We end this Section with some observations that concern the relation between the theory of ensembles in configuration space and the more familiar formulations of statistical theories. States are usually described by density operators in the case of quantum mechanics and by phase space distributions in the case of classical mechanics (the important case of a classical point particle corresponds to a special phase space distribution, i.e. a delta function, and hence it can be handled with the same formalism that is used for classical statistical theories). Fig. 1 shows the relation between phase space (classical mechanics), configuration space (classical, quantum mechanics) and the space of density operators (quantum mechanics). One of the reasons why the formalism of ensembles in configuration space is able to provide a consistent description of mixed quantum-classical systems has to do with the way in which these spaces are related. What is important here is that classical ensembles in configuration space are not as general as ensembles in phase space: in general, ensembles in phase space have to be described by *mixtures* of classical ensembles in configuration space. This is analogous to the situation in quantum mechanics, where density matrices describe mixtures of pure states (wave functions). One therefore has the following relations:

a. Point particles $\subset$ Configuration space ensembles $\subset$ Phase space ensembles (CM).

b. Configuration space ensembles $\subset$ Density operators (QM).

c. Mixtures of configuration space ensembles $=$ Phase space ensembles (CM).

d. Mixtures of configuration space ensembles = Density operators (QM).

In our formulation, mixed quantum-classical states are described in a unified way in terms of ensembles in configuration space, ensuring mathematical compatibility of the quantum and classical sectors. Furthermore, in this unified approach interactions between quantum and classical systems can be easily incorporated. By contrast, in other formulations attempts are made to bring together spaces that are mathematically incompatible, e.g. the quantum Hilbert space of wave functions is join either the space of point particles or the space of distributions in phase space of classical mechanics, and this leads to unsurmountable difficulties.

4. Measurements of position

As an illustration of the application of the formalism to measurement theory, we consider a measurement of the position of a quantum particle carried out by a classical apparatus.

To interpret the outcome it will be helpful to introduce some additional concepts which we now define. Consider the conditional probability $P(q|x) = P(q,x)/P(x)$. The conditional wave function is defined by

$$\psi(q|x) := \sqrt{P(q|x)}\, e^{iS(q,x)/\hbar}, \qquad |\psi_x\rangle := \int dq\, \psi(q|x)\, |q\rangle \; . \tag{9}$$

We can also introduce a conditional density operator,

$$\rho_{Q|C} := \int dx\, P(x)\, |\psi_x\rangle\langle\psi_x| \; . \tag{10}$$

It is important to keep in mind that $\psi(q|x)$ and $\rho_{Q|C}$ do *not* satisfy linear Schrödinger and Liouville equations, nor unitary invariance properties. These quantities only contain *partial* information.

To model a measurement of position [4], introduce the ensemble Hamiltonian

$$\tilde{H}_{\text{position}} = \tilde{H}_{QC} + \kappa(t) \int dq\, dx\, P\, q.\nabla_x S \; . \tag{11}$$

For an interaction over a short time period $[0,T]$ such that $\tilde{H}_{QC}$ can be ignored during the interaction,

$$\frac{\partial P}{\partial t} = -\kappa(t)\, q.\nabla_x P \; , \qquad \frac{\partial S}{\partial t} = -\kappa(t)\, q.\nabla_x S \; , \tag{12}$$

which, with $K = \int_0^T dt\, \kappa(t)$, integrates to:

$$P(q,x,T) = P(q, x - Kq, 0) \; , \qquad S(q,x,T) = S(q, x - Kq, 0) \; . \tag{13}$$

Consider the case where the initial position of the pointer is sharply defined, $P(q,x,0) = \delta(x - x_0) P_Q(q)$. Then, after the measurement, $\Delta x \neq 0$, with probability $P_Q(q)$ of finding $x = x_0 - Kq$. In other words, uncertainty is transferred from the quantum ensemble to the classical ensemble. The conditional density for the quantum component is diagonal in the position basis:

$$\rho_{Q|C} = \int dq\, P_Q(q)\, |q\rangle\langle q| \tag{14}$$

and thus 'decoheres' with respect to position.

This simple example shows that it is possible to describe measurements via quantum-classical interactions, as required for example by the Copenhagen interpretation, without running into inconsistencies. The main features are: (i) the measuring apparatus is described classically, as is required for the unambiguous communication and comparison of physical results; (ii) information about quantum ensembles is obtained via an appropriate interaction with an ensemble of classical measuring apparatuses which correlates the classical configuration with a corresponding quantum property, and (iii) there is a conditional decoherence of the quantum ensemble relative to the classical ensemble, which depends upon the nature of the quantum-classical interaction.

For a more detailed analysis of this example, as well as an example of a measurement of spin, see Ref. [4].

5. Quantum and classical particles interacting gravitationally

In this Section, we consider the scattering of two non-relativistic particles, one is a classical particle of mass M (the projectile) and the other one is a quantum particle of mass m (the target). The interaction is assumed to be caused by the gravitational attraction between the two particles. This scattering experiment constitutes a measurement of the quantum system by a classical apparatus, in the sense that information about the state of the quantum particle can be inferred from measurements of the position of the classical particle made after the interaction has taken place.

An interesting case is the one in which the quantum system is prepared in such a way that the initial amplitude for the quantum particle (i.e. as $t \to -\infty$ when the two particles are very far from each other so that the interaction term can be neglected) has two peaks of equal magnitude, A and B, that are well separated. Assume further that the classical particle is launched in the direction of peak A. What happens when the classical particle scatters from the quantum particle? A "naive approach" suggests three possible mutually exclusive outcomes: (a) the quantum particle is found at the location of peak A and the classical particle comes very close to the quantum particle of mass m: the scattering is very strong; (b) the quantum particle is found at the location of peak B and the classical particle never comes very close to the quantum particle of mass m: the scattering is very weak; (c) as in semiclassical gravity [10], the classical particle 'sees' a mass $m/2$ at the location of peak A and it 'sees' a mass $m/2$ at the location of peak B: the scattering is about one half of what one would calculate under assumption (a). Each of these possibilities seems unrealistic.

The source of difficulties is, clearly, the use of the "naive approach": it is impossible to derive any reasonable conclusion without introducing a concrete model for the mixed quantum-classical system, and we now proceed to do this.

The ensemble Hamiltonian will be of the form given in Eq. (7) with interaction term

$$V(q,x) = -G\frac{mM}{|x-q|} \ . \tag{15}$$

The ensemble Hamiltonian that describes the system is then

$$\tilde{H}_{QC}[P,S] = \int dq\, dx\, P\left[\frac{|\nabla_q S|^2}{2m} + \frac{|\nabla_x S|^2}{2M} - G\frac{mM}{|x-q|} + \frac{\hbar^2}{4}\frac{|\nabla_q \log P|^2}{2m}\right] \ . \tag{16}$$

It is convenient to introduce center-of-mass and relative coordinates $\overline{x}$ and r, and the total and relative masses M_T and μ. Then

$$\begin{aligned} \tilde{H}_{QC} \quad = \quad & \underbrace{\int d\overline{x}dr\, P\left[\frac{|\nabla_{\overline{x}} S|^2}{2M_T} + \frac{\hbar^2 m}{4(m+M)}\frac{|\nabla_{\overline{x}} \log P|^2}{2M_T}\right]}_{\text{(i) quantum-like term: free center-of-mass motion}} \\ & + \underbrace{\int d\overline{x}dr\, P\left[\frac{|\nabla_r S|^2}{2\mu} + \frac{\hbar^2 M}{4(m+M)}\frac{|\nabla_r \log P|^2}{2\mu} - G\frac{\mu M_T}{|r|}\right]}_{\text{(ii) quantum-like term: relative-motion in a potential}} \end{aligned} \tag{17}$$

$$- \underbrace{\frac{2\mu}{M_T} \int d\overline{x} dr \, P \left[\frac{\hbar^2 M}{4(m+M)} \frac{(\nabla_{\overline{x}} \log P \cdot \nabla_r \log P)}{2\mu} \right]}_{\text{(iii) interaction term}} .$$

The qualitative features of the solution can be determined without carrying out a detailed calculation. To interpret this expression, compare it to the ensemble Hamiltonian of a purely quantum system; i.e. Eq. (2), second line. We have a sum of three terms: (i) a quantum-like term corresponding to free centre-of-mass motion but with a rescaled Planck constant:

$$\hbar_{\overline{X}} := [m/(m+M)]^{1/2} \, \hbar \, ; \tag{18}$$

(ii) a quantum-like term corresponding to relative motion in a potential $V(r) = -G\mu M_T/|r|$ but with a rescaled Planck constant:

$$\hbar_R := [M/(m+M)]^{1/2} \, \hbar \, ; \tag{19}$$

and (iii) an intrinsic interaction term.

The predictions of the mixed quantum-classical system described here differ therefore substantially from the outcomes predicted using the "naive approach". In particular, one expects a solution with *qualitative* features that resemble those of a purely quantum system, however with important modifications induced by the (iii) term and by the rescaling of the Planck constant in (i) and (ii). In particular, the classical and quantum configuration spaces will be correlated once the particles interact, as in the example of position measurement discussed previously. This will lead to a final state that is similar to an entangled state. Cases (a) and (b) of the "naive approach" are "either-or" outcomes that are clearly not in agreement with this observation. Furthermore, the theory that we have used is fundamentally different from a theory built along the lines of standard semiclassical gravity. Therefore, case (c) of the "naive approach" is also excluded. A more extensive discussion of this system can be found in Appendix B of Ref. [3].

More generally, it is possible in principle to model the interaction of a classical gravitational field with quantized matter fields [3, 4, 11]. This leads to a theory that is fundamentally different from semiclassical gravity.

6. Concluding remarks

It is seen that a consistent description of quantum-classical interactions is possible, based on a canonical formalism for ensembles in configuration space. Furthermore, this is the only description which satisfies all of properties 1-8 in Section 3.

In the standard Copenhagen interpretation of quantum mechanics, it is assumed that the measuring apparatus is described classically and that information about a quantum system is obtained via an appropriate interaction between the quantum system and the classical apparatus. In this article, we have shown that the formalism of configuration-space ensembles allows a general and consistent description of the measurement of a quantum system by a classical apparatus. Using two examples, we have shown how this formalism is implemented "in practice".

The models described in this article provide illustrations of some qualitative features that one expects to be present in all such solutions: for example, uncertainty is transferred from the quantum system to the classical measuring apparatus, and there is decoherence of the quantum component. Concepts used to describe measurement in quantum theory will have their counterparts in this formalism, although perhaps with limited validity.

References

[1] Bohr M 1958 *Atomic Physics and Human Knowledge* (New York: Wiley)
[2] Heisenberg W 1958 *Physics and Philosphy* (London: Allen and Unwin)
[3] Albers M, Kiefer C and Reginatto M 2008 *Phys. Rev.* D **78** 064051
[4] Hall M J W and Reginatto M 2005 *Phys. Rev.* A **72** 062109
[5] Hall M J W 2008 *Phys. Rev.* A **78** 042104
[6] Caro J and Salcedo L L 1999 *Phys. Rev.* A **60** 842
[7] Salcedo L L 2007 *J. Chem. Phys.* **126** 057101
[8] Peres A and Terno D 2001 *Phys. Rev.* A **63** 022101
[9] Agostini F, Caprara S and Ciccoti G 2007 *Europhys. Lett.* **78** 30001
[10] Page D N and Geilker C D 1981 *Phys. Rev. Lett.* **47** 979
[11] Reginatto M 2005 *Braz. J. Phys.* **35** 476

Journal of Physics: Conference Series **174** (2009) 012039 doi:10.1088/1742-6596/174/1/012039

Bohmian mechanics and the emergence of classicality

A Matzkin

Laboratoire de Spectrométrie Physique (CNRS Unité 5588), Université Joseph-Fourier Grenoble-1, BP 87, 38402 Saint-Martin d'Hères, France

E-mail: `Alexandre.Matzkin@ujf-grenoble.fr`

Abstract. Bohmian mechanics is endowed with an ontological package that supposedly allows to solve the main interpretational problems of quantum mechanics. We are concerned in this work by the emergence of classicality from the quantum mechanical substrate. We will argue that although being superficially attractive, the de Broglie-Bohm interpretation does not shed new light on the quantum-to-classical transition. This is due to nature of the dynamical law of Bohmian mechanics by which the particles follow the streamlines of the probability flow. As a consequence, Bohmian trajectories can be highly non-classical even when the wavefunction propagates along classical trajectories, as happens in semiclassical systems. In order to account for classical dynamics, Bohmian mechanics needs non-spreading and non-interfering wave packets: this is achieved for practical purposes by having recourse to decoherence and dense measurements. However one then faces the usual fundamental problems associated with the meaning of reduced density matrices. Moreover the specific assets of the de Broglie-Bohm interpretation – in particular the existence of point-like particles pursuing well-defined trajectories – would play no role in accounting for the emergence of classical dynamics.

1. Introduction

The de Broglie-Bohm theory of motion (henceforth coined 'BM' for Bohmian mechanics) is generally regarded as the main alternative interpretation to standard quantum mechanics. The aim is to give an objective account of the quantum processes by introducing a well-defined ontology obeying causal dynamical laws. The program is undoubtedly attractive when compared to the plague of interpretational problems affecting the formalism of standard quantum mechanics; these problems arise because the theoretical entities of the formalism do not refer unambiguously to objects and properties of the observable universe [1]. On the other hand, according to Bohmian mechanics the quantum world is made up of waves and particles pursuing deterministic trajectories. Waves and particles are taken to be real, allowing to unify the classical and quantum descriptions of nature: "*there is no need for a break or 'cut' in the way we regard reality between quantum and classical levels*" [2].

The point we want to examine in this work is whether regarding the emergence of classical mechanics, BM lives up to its expectations. We will argue it doesn't, because by construction the quantum trajectories cannot approximate those of classical mechanics. This is particularly striking in semiclassical systems in which the wavefunction is propagated along classical trajectories whereas the probability flow – that provides the structure for Bohmian trajectories – is highly non-classical. Only an ad-hoc mechanism producing localized wave packets while transforming superpositions into mixtures would lead to classical trajectories. Such a mechanism is well-known – decoherence –, but relying on decoherence to obtain approximate classical dynamics is a non-starter for any realist interpretation (except, with some provisions, for the many-worlds interpretation), that is for any interpretation wishing quantum mechanics to be something more than an instrumentalist book of recipes. Moreover adopting decoherence harms the attractive parallel that the de Broglie Bohm interpretation makes between the statistical structure of quantum and classical mechanics.

We will first briefly recall the main ingredients and rationales for introducing Bohmian mechanics. We will then give examples of the dynamical quantum-classical correspondence which is well-known from the field of quantum chaos, and contrast this dynamical correspondence with the dynamics obtained from the BM laws of motion. We will display the dynamical mismatch [3] impeding the emergence of classical dynamics from BM for closed systems. Finally, we will argue why decoherence cannot constitute a viable solution for BM in order to recover classicality.

2. Why Bohmian mechanics?

The de Broglie-Bohm interpretation [4, 5] consists in an ontological and dynamical package aimed at solving the interpretational problems of quantum mechanics. It starts from the standard quantum mechanical theoretical terms but reinterprets them in a specific way. Hence although at best BM can only match the predictions of quantum mechanics (which it does in the non-relativistic domain), its alleged advantage is epistemological – BM supposedly describes what is happening in a quantum system 'in reality'. According to BM, a quantum system is composed of a particle and a wave. The particle, which lives in physical space-time, is what makes detectors click. The wave appears as a field having a privileged representation in position configuration space; the particle is guided by the density current of the field. From the polar decomposition of the wavefunction ψ in the position representation,

$$\psi(\mathbf{x},t) = R_{\psi}^{1/2}(\mathbf{x},t)\exp(iS_{\psi}(\mathbf{x},t)/\hbar) \ , \tag{1}$$

the Schrödinger equation yields the coupled equations

$$\frac{\partial R_{\psi}(\mathbf{x},t)}{\partial t} + \frac{1}{m}\nabla\cdot(R_{\psi}(\mathbf{x},t)\nabla S_{\psi}(\mathbf{x},t)) = 0 \tag{2}$$

and

$$\frac{\partial S_{\psi}(\mathbf{x},t)}{\partial t} + \frac{(\nabla S_{\psi}(\mathbf{x},t))^2}{2m} + V(\mathbf{x},t) + Q_{\psi}(\mathbf{x},t) = 0 \ , \tag{3}$$

where $V(\mathbf{x},t)$ is the usual potential and $Q_{\psi}(\mathbf{x},t)$ is a term known as the quantum potential given by

$$Q_{\psi}(\mathbf{x},t) \equiv -\frac{\hbar^2}{2m}\frac{\nabla^2 R_{\psi}^{1/2}}{R_{\psi}^{1/2}} \ . \tag{4}$$

The momentum and the velocity of the particle are introduced via a configuration space field defined from the polar phase function through

$$\mathbf{p}_{\psi}(\mathbf{x},t) = m\mathbf{v}_{\psi}(\mathbf{r},t) = \nabla S_{\psi}(\mathbf{x},t) \tag{5}$$

allowing to obtain the particle's equation of motion in a pseudo-Newtonian form

$$\frac{d\mathbf{p}_{\psi}}{dt} = -\nabla(V(\mathbf{x},t) + Q_{\psi}(\mathbf{x},t)) \ . \tag{6}$$

Introducing a particle to account for quantum phenomena puts quantum and classical mechanics partially on the same footing, thereby allowing to transfer to the quantum domain

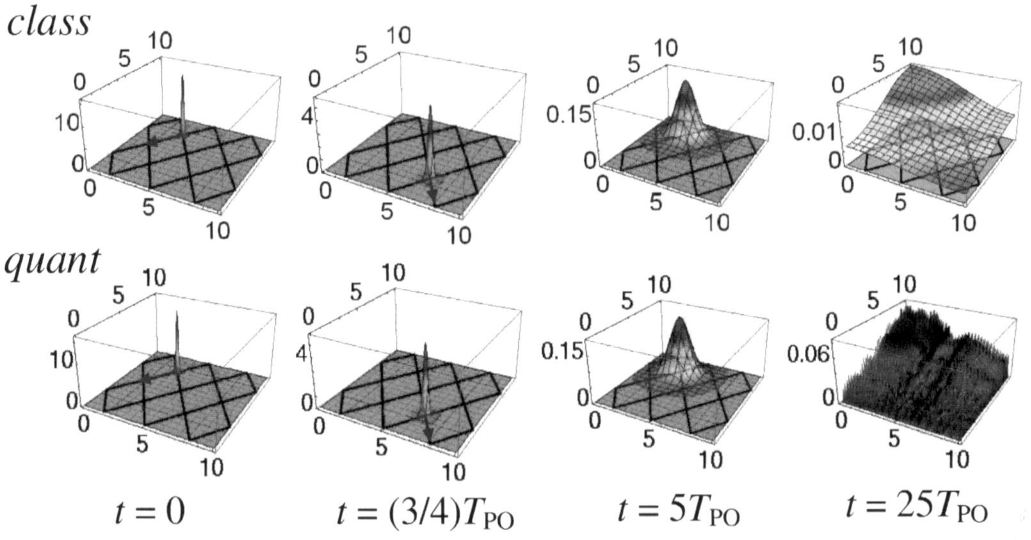

Figure 1. Dynamics of a Gaussian statistical distribution in a classical (top) and a quantum (bottom) square billiard with sides of length $L = 10$ (arbitrary units). At $t = 0$ the maximum of the distribution sits on the periodic orbit (the black line at the bottom of each plot) and is launched in the direction of the arrow. T_{PO} is the period of the orbit.

the intuitive epistemological categories employed to understand the classical domain. The non-classical, counter-intuitive aspects are attributed to the wave; some properties of the wave remain thus mysterious (their non-locality, the fate of the empty waves). Our interest here is focused on the particle dynamics. From this perspective, we see that the defining equations of BM are similar to those of classical mechanics in the Hamilton-Jacobi formalism: the Hamilton-Jacobi equation is indeed recovered by setting $Q = 0$ and deleting the ψ indices in Eq. (3). However from a structural point of view, this analogy is superficial. The equations of classical mechanics arise from the flows in phase-space obeying the canonical equations of motion for any choice of canonical coordinates. Any statistical distribution can be decomposed into elementary phase-space elements obeying these equations, i.e. the dynamics of the distribution depends on the elementary phase-space dynamics. In the de Broglie-Bohm theory on the other hand, the dynamics of the particle depends on the wavefunction (this is reflected in our notation with the indices labeled by ψ). The law of motion for an *individual* particle depends on the *statistical distribution* R_ψ to which the particle belongs. As can be seen from Eqs. (2) and (5), the dynamics of the particle is determined by the direction of the probability flow in configuration space, the local current density $\mathbf{j}_\psi$ taking the form

$$\mathbf{j}_\psi(\mathbf{x}, t) = R_\psi(\mathbf{x}, t)\mathbf{v}_\psi(\mathbf{x}, t) \; . \tag{7}$$

3. The quantum-classical correspondence and the dynamical mismatch

The investigations of the quantum-classical correspondence, which have their origins in the early days of quantum mechanics, were revived more recently in the context of quantum chaos [6]. It is today well-established that several types of quantum systems – known generically as semiclassical systems – display the manifestations of properties belonging to the classical analog of these systems. This is due to the fact that the wavefunction propagates essentially along the trajectories of the corresponding classical system; indeed in these cases the semiclassical

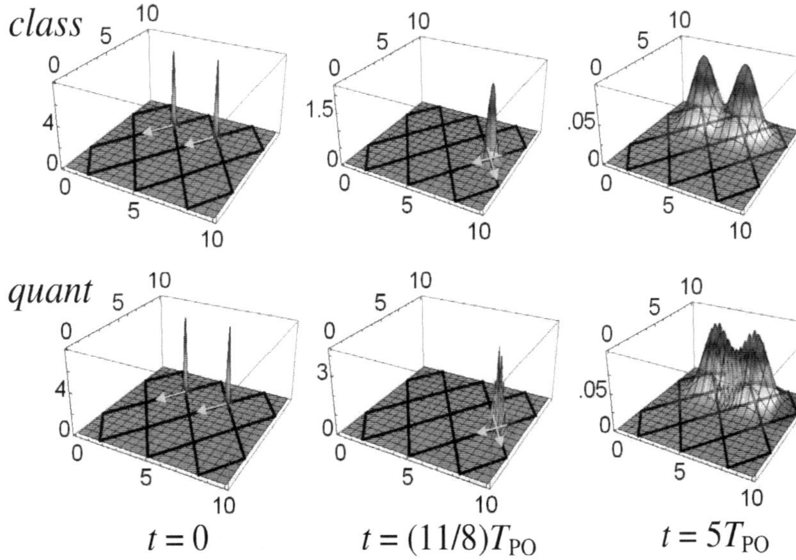

Figure 2. Same as Fig. 1 for an initial state composed of two Gaussian distributions localized on different points of the orbit launched in the direction of the arrows. The central plots show the situation at $t = (11/8)T_{PO}$ when the distributions cross.

approximation to the path integral propagator, given by [7]

$$K(\mathbf{x}_0, \mathbf{x}, t) = \sum_k \frac{1}{2i\pi\hbar} \left| \det \frac{\partial^2 \mathcal{S}_k}{\partial \mathbf{x} \partial \mathbf{x}_0} \right|^{1/2} \exp\left(i\phi_k + i\mathcal{S}_k(\mathbf{x}_0, \mathbf{x}, t)/\hbar\right) , \tag{8}$$

becomes excellent. Here the sum runs on all the classical trajectories k connecting $\mathbf{x}_0$ to $\mathbf{x}$ in the time t. $\mathcal{S}_k$ is the classical action for the kth trajectory and the determinant is the inverse of the Jacobi field familiar from the classical calculus of variations, reflecting the local density of the paths; ϕ_k is a phase accounting for reflections and conjugate points encountered along the kth trajectory.

Illustrative examples are given in Figs. 1-3. Figs. 1 and 2 show wave packet propagation in a square billiard compared with the evolution of a classical distribution of particles. A square billiard is a particularly simple and tractable non-trivial system for which the semiclassical approximation is exact [8]. When an initial wave packet is placed on a classical periodic orbit, the wave packet follows that orbit and spreads as a function of time (Fig. 1). This is precisely the behaviour of a statistical distribution of particles in classical mechanics, the quantum and classical probabilities spreading at the same rate as expected from Eq. (8). The difference between the classical and quantum evolutions is to be seen in the interferences: when the part of the wavefunction carried by two classical trajectories k and k' reach the same point $\mathbf{x}$ (which is more likely to happen as the wavefunction spreads), they both contribute to the propagator (8) yielding interferences in the probability distribution (averaging over the interferences allows to recover the classical distribution). This is even more the case when one chooses an initial distribution localized at two spatially different points (Fig. 2).

Fig. 3 shows features of a hydrogen atom in a magnetic field. This system – an electron subjected to competing Coulomb and magnetic fields – has been widely investigated as a prototype of quantum chaos [9]: the dynamics of a *classical* electron can be regular, mixed phase-space, or chaotic depending on the relative intensities of the fields (see Fig. 3(a)). The

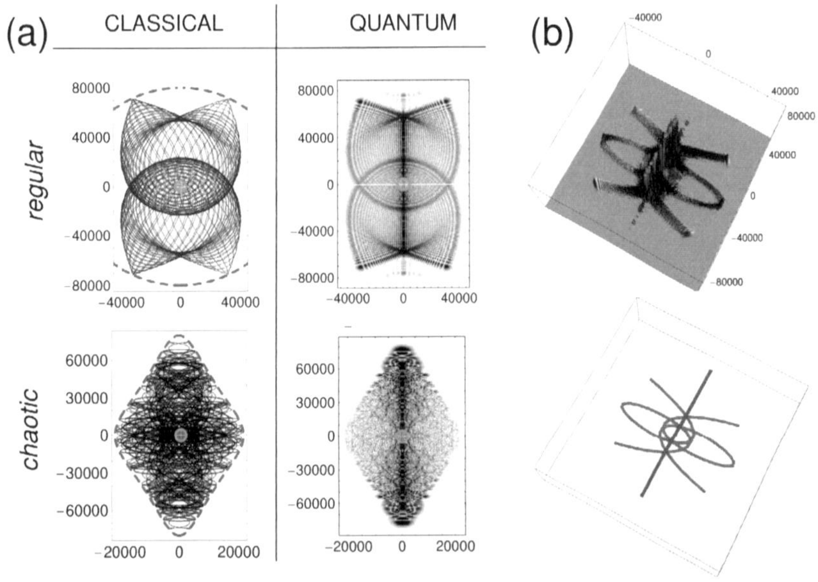

Figure 3. The hydrogen atom in a magnetic field and its classical counterpart (a charged particle in Coulomb and magnetic fields). The axial symmetry reduces the non-trivial dimensions to a 2D problem. (a) Left panel: a single classical trajectory is launched from the nucleus (shown in yellow) and is left to evolve when the relative Coulomb and magnetic field intensities and electron energy are set so that the dynamical regime is *regular* (top) and *chaotic* (bottom). Right panel: typical Hamiltonian eigenstates computed by solving numerically the Schrödinger equation when the relative field intensities and electron energy are set to be the same as in the classical system. (b) An energy eigenstate (top) displaying a localisation of the probability distribution. Most of the probability actually lies on two classical periodic orbits existing when the field and energy parameters are the same as in the quantum system; the two orbits are shown on the bottom plot in red and blue.

quantum system's behaviour is similar to these trajectories, as can be seen by computing and plotting energy eigenstates for the same field parameters and electron energy employed in the computation of the classical trajectories. Fig. 3(b) shows an energy eigenstate for which the probability is highly concentrated on two classical periodic orbits existing at the same energy and field parameters. From a more quantitative stance, the quantum autocorrelation function following electron excitation (which is an experimentally observable quantity) displays sharp peaks appearing at times that coincide with the periods of the trajectories that a classical electron would follow if the classical autocorrelation function were computed; the statistical distribution of the energy levels of the quantum system has also been shown to depend on average phase-space properties of the classical system.

Fig. 4 shows Bohmian trajectories for the square billiard. According to Eqs. (5)-(7) the particle dynamics depends on the probability flow. In the square billiard, the probability flow is localized for short times along the periodic orbit and the particle follows that motion (Fig. 4(a)). For longer times however, as the wavefunction spreads, the local probability flow results from multiple interferences and a given trajectory will sense very fine details of the probability flow. Even for short times, i.e. when the wave packet is localized along the classical periodic orbit, interferences make Bohmian trajectories deviate from the classical ones: Fig. 4(b) shows Bohmian trajectories for the initial double Gaussian. When the wave packets of Fig. 2

cross, the net current density vanishes locally and its direction is reversed: the Bohmian particle is turned back, retracing its path and reaching the initial point at the period of the periodic orbit. The classical path is recovered only on a statistical basis but individual Bohmian trajectories will not follow the periodic orbit, although according to Eq. (8) each of the two bits of the initial wavefunction appears as being propagated by following the classical trajectory.

In the hydrogen atom in a magnetic field, the vortex structure is considerably more complex. The Bohmian trajectories have nothing in common with the classical ones [3]; moreover there is no relationship between the regularity or chaoticity of the Bohmian trajectories and the classical dynamical regime (actually one can easily build quantum states leading to highly chaotic Bohmian trajectories even for quantum systems whose classical counterpart is perfectly integrable, therefore displaying only regular trajectories). The quantum-classical correspondence in terms of the dynamical regimes, leading to the striking classical features visible in the quantum properties as illustrated in Fig. 3, is therefore spoiled. This is to be expected, given that Bohmian trajectories depend on very fine details of the local probability current density, whereas the classical manifestations affect the wave propagation and are visible in the large-scale quantum structures.

The dynamical mismatch between Bohmian trajectories and classical motion has serious implications concerning the empirical acceptability of the de Broglie-Bohm theory as describing the *real* behaviour of the quantum world. The usual hand-waving argument – quantum and classical particle motions belong to different domain, the former approaching the latter in some limit as $Q_\psi \to 0$ – is clearly not applicable: it seems indeed untenable to explain the appearance of classical structure in the wave and in the statistical distribution of the particles while upholding that the real dynamics of the particle remains highly non-classical, without any type of correspondence with the classical dynamics that should emerge at some point (in the classical limit) despite the fact that observable properties of these quantum systems display the manifestation of the underlying classical dynamics. The most cogent answer that has been given [5, 10] is that although Bohmian dynamics is irremediably non-classical in closed quantum systems, the particle motion changes when the system interacts with the environment: classical behaviour is then obtained from open-system Bohmian dynamics. However there are certain obstacles that stand in the way in order to arrive at this conclusion.

4. Decoherence and the classical limit

According to Bohmian mechanics the particle is guided by the wave. In order to recover classical mechanics, the wave must therefore follow classical trajectories. But as we have seen above, this necessary condition is not sufficient: the wave packet must in addition not interfere, banishing spreading (which produces self-interference in a bounded system) and multiply-localized evolving wave packets. The only physical phenomenon that can account for these effects is decoherence, which, with an appropriate timescale, will transform the interfering wave packets in improper mixtures. But BM can hardly rely on decoherence to achieve classical dynamics for at least two reasons:

(i) By tracing out the environmental variables, one is applying the projection postulate and averaging over the "measurements" performed by the environment, but the total wavefunction remains entangled [12]. From a de Broglie-Bohm dynamics point of view, this total wavefunction – which is the one guiding the particle – has several branches corresponding to the simultaneous couplings with the environmental states. Each interaction of the system with the environment leads to only one branch of the wavefunction being active (the one containing the particle); the other branches are the so-called 'empty waves' – they still exist and can potentially become active (and interfere) again. The collapse is therefore only apparent and Bohmian mechanics (nor standard quantum mechanics for

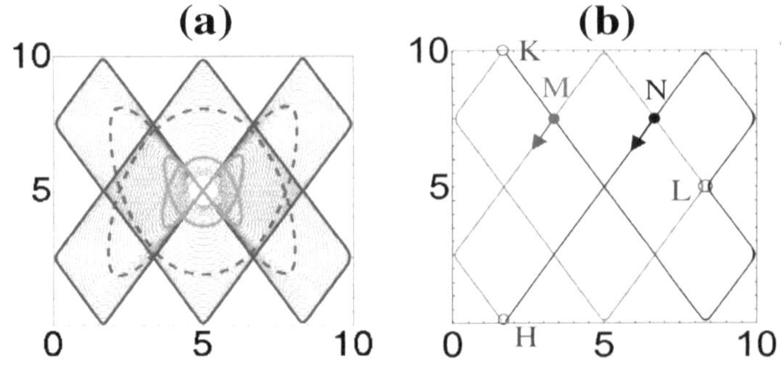

Figure 4. Bohmian trajectories in the square billiard. (a) At $t = 0$ the particle sits at the maximum of the distribution shown in Fig. 1. For short times the particle travels with the wave packet and thus follows the classical periodic orbit: the blue thick line shows the *Bohmian* trajectory in the time interval $0 < t < T_{PO}$. As the wave packet spreads the Bohmian trajectory slows down while spiralling towards the central fixed point: the red-dashed and yellow lines show the trajectory in the time intervals $15T_{PO} < t < 16T_{PO}$ and $30T_{PO} < t < 31T_{PO}$ respectively. (b) Two Bohmian trajectories are shown for short-times ($0 < t < T_{PO}$), corresponding to the initially doubly localized quantum state shown in Fig. 2. The red line is the trajectory when the particle lied initially on the maximum of the Gaussian at M: it starts by following the classical periodic orbit, but turns back upon reaching H, retraces its path, is deflected at L and reaches M again at $t \approx T_{PO}$. The blue line corresponds to the trajectory when the particle is initially on top of the other Gaussian at N. Taken individually, none of these trajectories is classical.

that matter) does not provide a dynamical theory accounting for the fate of the empty waves. There is therefore no guarantee that the branches of the wavefunction will remain dynamically independent of each other, unless one discards the empty waves 'by hand' (or one subscribes to the many worlds interpretation, a move that most supporters of BM would find dreadful). Alternatively, it is possible to define Bohmian trajectories directly from the reduced density matrices [11]: even in the absence of interferences, these 'reduced' trajectories are non-classical (they actually represent a weighted average of the possible classical motions).

(ii) By itself, decoherence converts, at least for practical purposes, a pure state into a mixed one but does not necessarily impede spreading. Hence a specific interaction with the measurement apparatus must be postulated so that the wave packet follows the classical motion without spreading. Formally this can be achieved via a continuous dense measurement. From a general standpoint however, there is no compelling reason to uphold that achieving the classical limit requires non-spreading wave packets [13]; in particular we have seen that in the square billiard the average quantum amplitude spreads exactly like the corresponding classical distribution. But in the de Broglie-Bohm theory, the statistical distributions, despite being reminiscent of the classical ones, cannot spread if classical motion is to be recovered. Ultimately the wave, referring to an individual physical phenomenon, takes precedence over the particle aspect (with its statistical distribution); the particle ends up not playing any role in accounting for the emergence of classical dynamics.

What we can conclude from these two remarks is that by relying on decoherence and non-spreading wave packets in order to achieve the emergence of classicality, Bohmian mechanics does not employ its specific interpretational assets, but uses instead practical recipes available to other interpretations. Not only do these recipes not need the ontological propositions of BM, but they face interpretational problems regarding the status of improper mixtures [14]. At any rate it is not necessary to postulate a point-like particle as part of the quantum mechanical reality if classical mechanics is identified with a non-spreading wave packet. The unifying appeal of the de Broglie-Bohm proposal, attractive because of its superficial similarity with classical concepts, is therefore lost when making the quantum to classical transition. Bohmian mechanics is of high interest and thought-challenging, but it does not seem to shed new light regarding the emergence of classical mechanics.

References

[1] Matzkin A 2002 Realism and the wavefunction *Eur. J. Phys.* **23** 285
[2] Bohm D and Hiley B J 1985 Unbroken quantum realism from microscopic to macroscopic levels *Phys. Rev. Lett.* **55** 2511
[3] Matzkin A and Nurock V 2008 Classical and Bohmian trajectories in semiclassical systems: Mismatch in dynamics, mismatch in reality? *Studies in History and Philosophy of Science* B **39** 17
[4] Holland P R 1993 *The Quantum Theory of Motion* (Cambridge, UK: Cambridge U. Press)
[5] Bohm D and Hiley B J 1993 *The Undivided Universe: an ontological interpretation of quantum theory* (London: Routledge)
[6] Brack M and Badhuri R 2003 *Semiclassical Physics* (Boulder CO, USA: Westview Press)
[7] Schulman L S 1981 *Techniques and applications of Path Integration* (New York: Wiley)
[8] Matzkin A 2008 Bohmian mechanics, the quantum-classical correspondence and the classical limit: the case of the square billiard *Preprint* arXiv:0806.3240v2
[9] Friedrich H and Wintgen D 1989 The hydrogen atom in a uniform magnetic field — An example of chaos *Phys. Rep.* **183** 37
[10] Appleby D M 1999 Bohmian Trajectories Post-Decoherence *Found. Phys.* **29** 1885
[11] Sanz A S and Borondo F 2007 A quantum trajectory description of decoherence *Eur. Phys. J.* D **44** 319
[12] Zeh H D 2003 in *Decoherence and the Appearance of a Classical World in Quantum Theory* (Berlin: Springer) 7-40
[13] Wiebe N and Ballentine L E 2005 Quantum mechanics of Hyperion *Phys. Rev.* A **72** 022109
[14] Adler S L 2003 Why decoherence has not solved the measurement problem *Studies in History and Philosophy of Science* B **34** 135

Fourth International Workshop DICE2008 IOP Publishing
Journal of Physics: Conference Series **174** (2009) 012040 doi:10.1088/1742-6596/174/1/012040

Quantum phases due to cosmic gravity and the spin-statistics connection

Aditya Gilra
Tata Institute of Fundamental Research, Mumbai, India
E-mail: agilra@tifr.res.in

Abstract. Recently, it was proposed that the spin-statistics connection arises due to a quantum dynamical phase involving cosmic gravity (Unnikrishnan C S, gr-qc/0406043). There it was assumed that the gravitational quantum phase accumulated on the wavefunctions of each of two identical particles undergoing scattering in the presence of cosmic matter with the direction of their momentum changing by some angle is equivalent to the particles forward scattering with the entire universe rotating oppositely by the same angle. However, if one considers the gravitational phase accumulated by an electron in orbit to be analogous to that acquired by a scattering particle, the accumulated phase inferred from the fine-structure splitting in an atom is too small to explain the spin statistics connection. Though, the desirability of a dynamical proof of the spin-statistics connection remains.

1. Spin-statistics connection

The spin-statistics connection (SSC) in quantum mechanics is simply stated as: Integer spin particles follow Bose-Einstein statistics, while half-odd integer spin particles follow Fermi-Dirac statistics. Operationally, for two identical spin-s particles scattering off each other, the total scattering amplitude is the sum / difference of the direct and the particles-exchanged scattering amplitudes depending on whether the spin s is integer / half-odd integer ($\hbar = 1$). For present purposes, we will consider both spins perpendicular to the plane of scattering. The two indistinguishable processes (direct and particles-exchanged) that lead to the same final configuration are shown in figure 1.

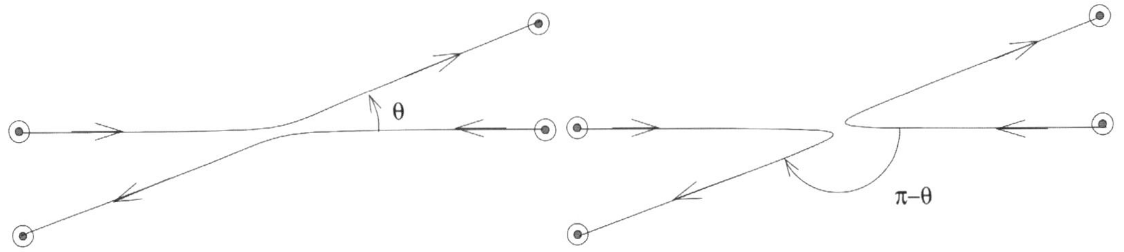

(a) process 1: each particle rotates ccw by θ (b) process 2: each particle rotates ccw by $\theta - \pi$

Figure 1. Scattering of two identical particles into the same final configuration by two different processes (spins perpendicular to plane of scattering)

2. Dynamical proof

The ad-hoc SSC rule in quantum mechanics is to add/subtract the amplitudes of these two processes depending on whether s is integer / half-odd integer. Proofs of the SSC are kinematical and well-established ones require quantum field theory. C S Unnikrishnan in [1] has attempted a dynamical explanation of the SSC involving cosmic gravity. Taking spin as a matter current, just as magnetic moments and effects are due to charge currents, it was argued that spin-dependent effects be due to matter currents / gravitomagnetism. Further since the SSC holds for all known scatterings/interactions, a dynamical proof should perforce require gravity, as it is the only known universal force. Further, it would involve cosmic gravity since it is large (as a homogeneous potential) and pervasive.

2.1. Cosmic gravitational phases

This dynamical cosmic gravity based explanation involves the relative phase between the two scattering amplitudes of the processes in figures 1(a) and 1(b). These processes take place in the presence of existent cosmic matter of approximately critical density which defines a preferred frame, namely the comoving frame of the Robertson-Walker universe. In figure 1(a), the momentum of each particle rotates with respect to the cosmic frame through angle θ (positive anti-clockwise), while in figure 1(b), the momentum of each particle rotates through $\theta - \pi$.

The question here is what is the phase picked up by a spin s quantum particle as its momentum changes direction with respect to the cosmic frame. The crucial assumption made in [1] is that the scattering of identical particles by some angle θ is equivalent in physical effects to forward scattering while the entire universe rotates by angle $-\theta$. Thus the cosmic / lab frame phase on each particle is equivalent to that calculated in the center-of-mass frame rotating with the particle's momentum say at angular frequency ω during scattering after which the particles continue in their initial direction in this frame. Or it can be calculated in a frame comoving with the particle but keeping one axis aligned with the momentum i.e. rotating at ω with respect to cosmic frame. In these latter two frames, the cosmic matter rotates in the opposite direction and this constitutes a matter current causing a gravitomagnetic field. It is as if for the process in figure 1(a), the universe effectively rotates by $-\theta$, while for the process in figure 1(b) by $-\theta + \pi$.

To calculate the phase in a frame rotating at ω with respect to cosmic matter, one transforms the critical density Robertson-Walker metric ($c = 1$ throughout)

$$ds^2 = -dt^2 + a(t)(dx'^2 + dy'^2 + dz'^2) \tag{1}$$

by

$$x' = x\cos\omega t - y\sin\omega t\ , \quad y' = x\sin\omega t + y\cos\omega t \tag{2}$$

to get off-diagonal metric components ($a(t_{now}) = 1$)

$$g_{0i} = (-y\omega, x\omega, 0) \tag{3}$$

which act as a gravitomagnetic vector potential

$$\mathbf{A} = (\frac{1}{2}y\omega, -\frac{1}{2}x\omega, 0), \tag{4}$$

the curl of which gives a gravitomagnetic field

$$\mathbf{B} = (0, 0, -\omega). \tag{5}$$

The gravitomagnetic moment of a spin is $\mu_g = -\mathbf{s}$. Similar to a magnetic field interacting with a magnetic dipole, the spin-gravitomagnetic interaction energy is

$$H_{int} = -\mu_g.\mathbf{B} = -\mathbf{s}.\omega. \tag{6}$$

Thus the phase picked up by each scattered particle in figure 1(a) in this rotating frame is ($\hbar = 1$ throughout)

$$\exp(-i \int H_{int} dt) = \exp(is\theta) \tag{7}$$

and the product wavefunction of the two particles picks up a phase $\exp(i2s\theta)$. For the figure 1(b), a rotating frame is taken which rotates with the momentum of a particle in this process (opposite sense) to give the phase on the product wavefunction as $\exp(i2s(-\pi + \theta))$. These calculated phases are used directly in the cosmic/lab frame. Thus these phases appear in the scattering amplitudes of the two processes which are then simply added, yielding a relative phase between the two amplitudes $\exp(-i2\pi s)$ which is just ± 1 depending on s being integer or half-odd integer. Hence one gets the operational spin-statistics connection from the spin-gravitomagnetic interaction.

2.2. Relation to fine structure splitting

Now in analogy with the scattering case, if we assume that in an atom, an electron going around the nucleus is equivalent to the universe rotating at the rate of the orbital frequency (as seen from a frame comoving with the electron and one axis aligned with its momentum), then the electron should pick up the afore-mentioned gravitomagnetic phase too. Here the only spin-dependent phase on an electron in an atom can be inferred from the experimentally measured fine structure splitting of the energy levels of spin up and spin down electrons in an atom (due to spin-orbit coupling). There are two parts to the fine structure splitting, the first is due to the proton's electric field and the second which is opposite and half (taking electron g-factor = 2) of the first, is a Thomas correction term related to Thomas precession. There is no role for gravity in the first, while the magnitude of the second is only $-s(1-\gamma)\omega$ ($\gamma = \frac{1}{\sqrt{1-v^2}}$, v the speed of the electron). For non-relativistic ($\gamma \approx 1$) quantum mechanics, this is very small compared to the $-s\omega$ required for the above explanation of SSC to work.

A caveat remains that this is a particle in a bound orbit, while the SSC concerns scattering particles, so one is unsure if the gravitational phase accumulated on it is analogous to that on a scattering particle. A gravity-based dynamical explanation of the SSC needs to be formulated consistently such that in the scattering case, the phase gained is $-s\omega$, while in the orbit case, there is only a simple Thomas correction phase $-s(1-\gamma)\omega$ (despite the particles in both cases changing their momentum direction at rate ω). But if the orbit and scattering phases are analogous then given the fine structure splitting, the phase calculated in the rotating center-of-mass frame on a scattering particle cannot be the same as that accumulated in the cosmic / lab frame. Thus the fine-structure splitting acts as a consistency check for any dynamical explanation of the spin statistics connection. Further, one notes that for the fine structure splitting, a gravitomagnetic calculation in a frame which is Fermi-Walker transported along with the electron i.e. only undergoes Thomas precession at angular velocity $(1-\gamma)\omega$ (generally considered the rest frame of the electron) gives the correct phase as required for fine structure splitting in the cosmic / lab frame. Thomas precession has also been considered in relation with gravitomagnetism in [2].

3. Desirability of dynamical explanation

A dynamical explanation of the SSC is desirable due to the following reasons and benefits. It will provide a positive proof of the SSC unlike present proofs which are negative proofs showing the impossibility of wrong statistics. Considering spin as a matter current, one expects the spin dependent effects in dynamics to be related to matter currents via gravitomagnetism (analogous to electrodynamics, as one expects in any relativistic theory of gravity). A dynamical proof of this sort would not require quantization of spin and would work for fractional spin and statistics too and further would generalize easily to any number of dimensions.

Acknowledgments

I would like to thank my PhD advisor Prof C S Unnikrishnan for introducing me to this problem and for detailed and patient discussions on the same.

References

[1] Unnikrishnan C S 2004 Spin-statistics connection and the gravity of the universe: the cosmic connection *Preprint* gr-qc/0406043
[2] Unnikrishnan C S 2001 *Mod. Phys. Lett.* A **16** 429–40

Journal of Physics: Conference Series **174** (2009) 012041 doi:10.1088/1742-6596/174/1/012041

Random Hamiltonian in thermal equilibrium

Dorje C Brody, David C P Ellis and Darryl D Holm

Department of Mathematics, Imperial College London, London SW7 2AZ, UK

E-mail: `david.ellis102@imperial.ac.uk`

Abstract. A framework for the investigation of disordered quantum systems in thermal equilibrium is proposed. The approach is based on a dynamical model—which consists of a combination of a double-bracket gradient flow and a uniform Brownian fluctuation—that 'equilibrates' the Hamiltonian into a canonical distribution. The resulting equilibrium state is used to calculate quenched and annealed averages of quantum observables.

1. Introduction

In the conventional treatment of quantum statistical mechanics there is a natural division between (a) the system under study, which is treated quantum mechanically and whose states are subject to thermal fluctuations, and (b) the Hamiltonian of the system, which is treated essentially classically and is held fixed. For some quantum systems, however, the Hamiltonian itself may fluctuate for one reason or another. Questions that interest us in this connection, in particular, are: "How can a randomly fluctuating Hamiltonian approach its equilibrium state?" and "What is the form of the equilibrium distribution, and how do we calculate observable expectation values in equilibrium?" The former is a question of a *dynamical* nature, whereas the latter is a question of a *static* nature. The purpose of the present paper is to propose an approach to address these questions. Specifically, we shall derive a dynamical model having the property that a given initial Hamiltonian evolves randomly—but isospectrally—in such a way that the associated density function on the space of isospectral Hamiltonians approaches a steady state distribution given by the *canonical ensemble*. Furthermore, we apply the resulting equilibrium state to calculate thermal expectation values of other observables, leading to new physical predictions in some limiting (quenched and annealed) cases.

2. Approach to equilibrium

In classical statistical mechanics the notion of a *gradient flow* plays an important role in describing the approach to equilibrium: A system immersed in a heat bath naturally tends to release its energy into the environment and thus approach its minimum energy state, and this tendency is characterised by a Hamiltonian gradient flow. An equilibrium state is attained when this flow is on average counterbalanced by thermal noise due to a random interaction with the bath. Here the magnitude of the noise is determined by the temperature of the bath. Accordingly, the idea we are going to introduce here is a gradient flow equation on the space of Hamiltonians with the property that the eigenstates of an arbitrary initial Hamiltonian H_0 at time $t = 0$ tend toward alignment with those of a reference Hamiltonian, denoted by G. Thus, G plays the role of the 'fixed' Hamiltonian in conventional quantum statistical mechanics. The

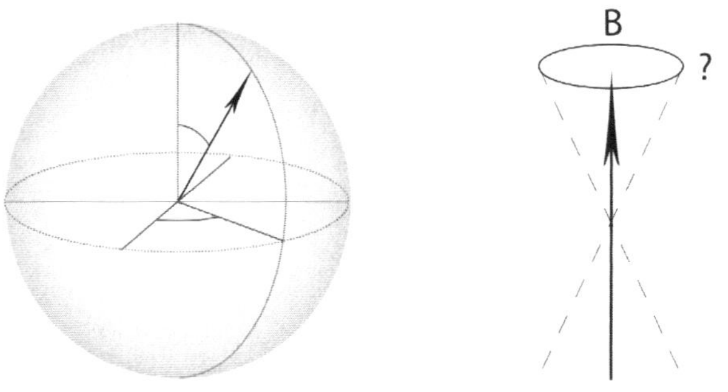

Figure 1. *Spin in fluctuating magnetic field.* The space of pure states of a spin-$\frac{1}{2}$ particle, in external magnetic field $\mathbf{B}$, is the surface of the Bloch sphere. In statistical theory of quantum mechanics the state is represented by statistical distributions of the pure states, whereas the magnetic field $\mathbf{B}$ that specifies the Hamiltonian is held fixed. What happens if the direction of the field $\mathbf{B}$ is itself subject to a small fluctuation?

eigenstates of H_t thus evolve toward those of G under the flow. By introducing a suitable noise term, we are able to characterise the approach to an equilibrium distribution.

The dynamical model for characterising approach to equilibrium is given by

$$\frac{\mathrm{d}H_t}{\mathrm{d}t} = -\lambda\,[H_t,[H_t,G]] + [H_t,\omega_t], \tag{1}$$

where $\lambda \in \mathbb{R}_+$. Here $\{\omega_t\}$ denotes a skew-symmetric matrix of independent white noise terms, and $[H_t,\omega_t]$ is the Lie bracket of these with H_t (see also [1]). Hence $[H_t,\omega_t]$ is symmetric and linear in both H_t and ω_t. The term $[H_t,[H_t,G]]$ gives rise to the aforementioned gradient flow in the space of Hamiltonians. The Hermitian matrix G plays the role of the 'Hamiltonian of the Hamiltonians' in the sense that G determines the motion in the space of Hamiltonians. In particular, we can regard the linear function $\mathrm{tr}(HG)$ on the space of the totality of Hermitian matrices H as representing the 'energy' function defined on that space.

There is an invariant measure (stationary solution) associated with the evolutionary equation (1). This is given by the canonical density:

$$\rho(H) \propto \exp\big(-\lambda\,\mathrm{tr}(HG)\big), \tag{2}$$

which can be used as a new basis for studying quantum statistical mechanics. In units $\hbar = 1$ the coupling λ has dimension $[\mathrm{Energy}]^{-1}$, and has the interpretation of representing the inverse temperature for the 'Hamiltonian bath' (not to be confused with the thermal bath in which the system, and possibly also the apparatus determining the Hamiltonian, is immersed). The canonical density (2) can also be derived by entropy maximisation subject to the constraint that the energy $\mathrm{tr}(HG)$ has a definite expectation value.

3. The gradient flow: double bracket equation

Let us begin by examining properties of the gradient term in (1). Specifically, consider the following dynamical equation for Hermitian matrices:

$$\frac{\mathrm{d}H_t}{\mathrm{d}t} = -\lambda\,[H_t,[H_t,G]]\,. \tag{3}$$

Note that in terms of the Hermitian matrix $X = \mathrm{i}[H,G]$ the double-bracket evolution (3) can be rewritten as

$$\frac{\mathrm{d}H}{\mathrm{d}t} = \mathrm{i}\lambda[H,X], \tag{4}$$

which formally is just the Heisenberg equation of motion. However, owing to the H-dependence of X the evolution is nonunitary. The Hamiltonians H_0 and G are both assumed nondegenerate. The flow induced by (3) satisfies the following properties: (i) the evolution is isospectral, i.e. the eigenvalues of H_0 are preserved; and (ii) the evolution gives the 'alignment' $\lim_{t\to\infty}[H_t, G] = 0$.

We remark that the double bracket flow was first introduced in the context of magnetism (Landau-Lifshitz equation) [2]. In its modern form it was introduced by Brockett [3] and has been successfully applied to many areas, such as optimal control, linear programming, sorting algorithms, and dissipative systems (see references cited in [4]).

In the case of a 2×2 Hamiltonian the gradient flow equation (3) can be solved straightforwardly. In this case we express the Hamiltonian in terms of the Pauli matrices:

$$H_t = \tfrac{1}{2}u_t\mathbb{1} + \tfrac{1}{2}\nu\,\boldsymbol{\sigma}\cdot\mathbf{n}_t, \tag{5}$$

where $\mathbf{n}_t = (\mathrm{x}_t, \mathrm{y}_t, \mathrm{z}_t)$. Similarly for the reference Hamiltonian G we write

$$G = \tfrac{1}{2}v\mathbb{1} + \tfrac{1}{2}\mu\,\boldsymbol{\sigma}\cdot\mathbf{g} \tag{6}$$

for a unit vector $\mathbf{g}$. A calculation shows that the solution to (3) reads

$$H_t = \tfrac{1}{2}\begin{pmatrix} u_0 - \nu\tanh(\omega t - c_0) & \nu\,\mathrm{sech}(\omega t - c_0)\,\mathrm{e}^{-\mathrm{i}\phi_0} \\ \nu\,\mathrm{sech}(\omega t - c_0)\,\mathrm{e}^{\mathrm{i}\phi_0} & u_0 + \nu\tanh(\omega t - c_0) \end{pmatrix}, \tag{7}$$

where $\omega = \lambda\nu\mu$, and $c_0 = \tanh^{-1}(\cos\theta_0)$ and ϕ_0 are initial values [4]. Furthermore, the eigenvalues of H_t are time-independent, and we have

$$\lim_{t\to\infty} H_t = \tfrac{1}{2}\begin{pmatrix} u_0 - \nu & 0 \\ 0 & u_0 + \nu \end{pmatrix}. \tag{8}$$

Thus, the Hamiltonian is asymptotically diagonalised in the G-basis. Observe that $\mathrm{tr}H_t$ and $\det H_t$ are conserved quantities. Therefore, the flow induced by (3) for fixed initial values u_0 and $|\mathbf{n}_0|$ is confined to a two-sphere $\mathcal{L}$, which is isomorphic to the state space of a two-level system. It follows that in two dimensions we have the equivalence of the Schrödinger and Heisenberg pictures, even though the dynamical equation is not unitary. Since u_0 and $|\mathbf{n}_0|$ are constant, we fix these and focus our attention on $\mathcal{L}$ parameterised by the dynamical coordinates (θ_t, ϕ_t).

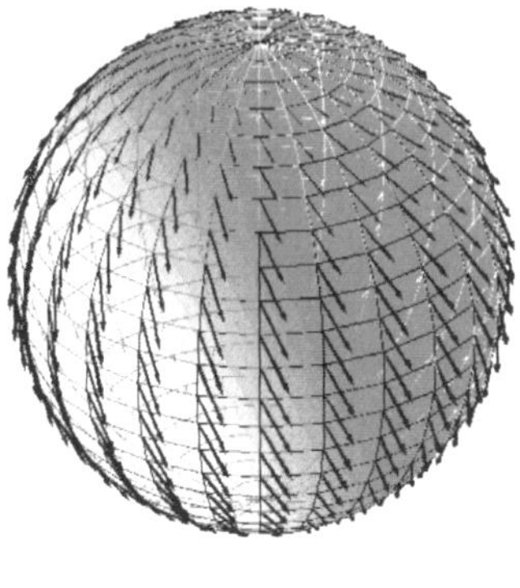

Figure 2. *Gradient flow with unitary motion on the sphere* $\mathcal{L}$. The vector field generated by the unitarily-modified gradient-flow (9) is plotted. The first term in (9) generates a rotation around the G-axis (which is chosen to be the z axis here), while the second term generates geodesic flows toward the south pole. The axis $\mathbf{n}_0$ of the initial Hamiltonian H_0 spirals around the G-axis $\mathbf{g}$ and is asymptotically aligned with the latter (south pole in this example).

We remark that the dynamical equation (3) can be modified to include a unitary term:

$$\frac{\mathrm{d}H_t}{\mathrm{d}t} = -\mathrm{i}[H_t, G] - \lambda\,[H_t, [H_t, G]] \tag{9}$$

without greatly affecting its physical characteristics. In the 2×2 example, the only change occurs in the phase so that instead of $\phi_t = \phi_0$ we have $\phi_t = \phi_0 + \mu t$.

4. Elements of stochastic differential geometry

We now wish to introduce a Brownian term into the deterministic flow (3). Specifically, we consider a uniform Brownian field on the isospectral subspace of the space of Hermitian matrices (the sphere $\mathcal{L}$ in the 2×2 case). Before we proceed, however, it will be useful to recall how stochastic motions can be defined on a manifold. The basic process we consider is the Wiener process $\{W_t\}$ defined on a filtered probability space $(\Omega, \mathcal{F}, \mathbb{P})$. Here Ω is the sample space, $\mathcal{F}$ is a σ-field on Ω, and $\mathbb{P}$ is the probability measure. The filtration of $\mathcal{F}$ determines the causal structure of $(\Omega, \mathcal{F}, \mathbb{P})$. This is given by a parameterised family $\{\mathcal{F}_t\}_{0 \leq t < \infty}$ of nested σ-subfields satisfying $\mathcal{F}_s \subset \mathcal{F}_t \subset \mathcal{F}$ for any $s \leq t < \infty$. We say that $\{W_t\}$ is a Wiener process if: (a) $W_0 = 0$; and (b) $\{W_t\}$ is Gaussian such that $W_{t+h} - W_t$ has mean zero and variance $|h|$ (see [5]). A process $\{\sigma_t\}$ is said to be adapted to the filtration $\{\mathcal{F}_t\}$ generated by $\{W_t\}$ if its random value at time t is determined by the history of $\{W_t\}$ up to that time.

If $\{\sigma_t\}$ is $\mathcal{F}_t$-adapted, then the stochastic integral $M_t = \int_0^t \sigma_s \mathrm{d}W_s$ exists, provided that $\{\sigma_t\}$ is almost surely square-integrable. If the variance of $\{M_t\}$ exists, then $\{M_t\}$ satisfies the *martingale* conditions $\mathbb{E}[|M_t|] < \infty$ and $\mathbb{E}[M_t|\mathcal{F}_s] = M_s$, where $\mathbb{E}[-]$ denotes expectation with respect to the measure $\mathbb{P}$. The latter condition implies that given the history of the Wiener process up to time s the expectation of M_t for $t \geq s$ is given by its value at s.

A general *Ito process* is defined by an integral of the form

$$x_t = x_0 + \int_0^t \mu_s \mathrm{d}s + \int_0^t \sigma_s \mathrm{d}W_s, \tag{10}$$

where $\{\mu_t\}$ and $\{\sigma_t\}$ are called the drift and the volatility of $\{x_t\}$. A convenient way of expressing (10) is to write $\mathrm{d}x_t = \mu_t \mathrm{d}t + \sigma_t \mathrm{d}W_t$, and to regard the initial condition x_0 as implicit. In the special case $\mu_t = \mu(x_t)$ and $\sigma_t = \sigma(x_t)$, where $\mu(x)$ and $\sigma(x)$ are prescribed functions, the process x_t is said to be a diffusion.

This analysis can be generalised to the case of a diffusion $\{x_t\}$ taking values on a manifold $\mathfrak{M}$, driven by an m-dimensional Wiener process $\{W_t^i\}_{i=1,\ldots,m}$. Let ∇_a be a torsion-free connection on $\mathfrak{M}$ such that for any vector field ξ^a its covariant derivative in local coordinates is

$$\delta_{\mathbf{b}}^b \delta_a^{\mathbf{a}} (\nabla_b \xi^a) = \frac{\partial \xi^{\mathbf{a}}}{\partial x^{\mathbf{b}}} + \Gamma^{\mathbf{a}}_{\mathbf{bc}} \xi^{\mathbf{c}}, \tag{11}$$

where $\delta_a^{\mathbf{a}}$ is the standard coordinate basis in a given coordinate patch. Suppose we have an Ito process taking values in $\mathfrak{M}$. Let $x_t^{\mathbf{a}}$ denote the coordinates of the process in a particular patch. Then writing $h^{\mathbf{ab}} = \sigma_i^{\mathbf{a}} \sigma^{\mathbf{b}i}$ we define the drift process $\mu^{\mathbf{a}}$ by

$$\mu^{\mathbf{a}} \mathrm{d}t = \mathrm{d}x^{\mathbf{a}} + \tfrac{1}{2} \Gamma^{\mathbf{a}}_{\mathbf{bc}} h^{\mathbf{bc}} \mathrm{d}t - \sigma_i^{\mathbf{a}} \mathrm{d}W_t^i. \tag{12}$$

Alternatively, we can write the *covariant Ito differential* as $\mathrm{d}x^a = \delta_{\mathbf{a}}^a (\mathrm{d}x^{\mathbf{a}} + \frac{1}{2} \Gamma^{\mathbf{a}}_{\mathbf{bc}} h^{\mathbf{bc}} \mathrm{d}t)$, where $\delta_{\mathbf{a}}^a$ is the dual coordinate basis. Then (12) can be represented as

$$\mathrm{d}x^a = \mu^a \mathrm{d}t + \sigma_i^a \mathrm{d}W_t^i. \tag{13}$$

If $\mu^a(x)$ and $\sigma_i^a(x)$ are $m+1$ vector fields on $\mathfrak{M}$, then the general diffusion process on $\mathfrak{M}$ is governed by a stochastic differential equation $\mathrm{d}x^a = \mu^a(x)\mathrm{d}t + \sigma_i^a(x)\mathrm{d}W_t^i$, where $\mathrm{d}x^a$ is the covariant Ito differential associated with the given connection (see Hughston [6]).

For the characterisation of the diffusion process it suffices to specify a connection on $\mathfrak{M}$, and a metric is not required. The quadratic relation $\mathrm{d}x^a \mathrm{d}x^b = h^{ab} \mathrm{d}t$, where $h^{ab} = \sigma_i^a \sigma^{bi}$, follows from the Ito identities $\mathrm{d}t^2 = 0$, $\mathrm{d}t\mathrm{d}W_t^i = 0$, and $\mathrm{d}W_t^i \mathrm{d}W_t^j = \delta^{ij} \mathrm{d}t$. Then for any smooth function $\phi(x)$ on $\mathfrak{M}$ we define the associated process $\phi_t = \phi(x_t)$, and Ito's formula takes the form

$$\begin{aligned} \mathrm{d}\phi_t &= (\nabla_a \phi)\mathrm{d}x^a + \tfrac{1}{2}(\nabla_a \nabla_b \phi)\mathrm{d}x^a \mathrm{d}x^b \\ &= \left(\mu^a \nabla_a \phi + \tfrac{1}{2} h^{ab} \nabla_a \nabla_b \phi\right) \mathrm{d}t + \sigma_i^a \nabla_a \phi \mathrm{d}W_t^i. \end{aligned} \tag{14}$$

The probability law for x_t is characterised by a density function $\rho(x,t)$ on $\mathfrak{M}$ that satisfies the Fokker-Planck equation

$$\frac{\partial \rho}{\partial t} = -\nabla_a(\mu^a \rho) + \tfrac{1}{2}\nabla_a \nabla_b (h^{ab}\rho). \tag{15}$$

The diffusion is said to be nondegenerate if h^{ab} is of maximal rank. If g_{ab} is a Riemannian metric on $\mathfrak{M}$ and ∇_a is the associated Levi-Civita connection, then if $h^{ab} = \sigma^2 g^{ab}$, the process x_t is a Brownian motion with drift on $\mathfrak{M}$, with volatility parameter σ.

5. Diffusion model for thermalisation

Consider a stochastic differential equation of the form

$$\mathrm{d}x^a = \mu^a \mathrm{d}t + \kappa \sigma_i^a \mathrm{d}W_t^i \tag{16}$$

on a real manifold $\mathfrak{M}$. Here κ is a constant, the drift μ^a is a vector field on $\mathfrak{M}$, and the vectors $\{\sigma_i^a\}$ constitute an orthonormal basis in the tangent space of $\mathfrak{M}$. In this case the associated Fokker-Planck equation reads

$$\frac{\partial}{\partial t}\rho_t(x) = -\nabla_a(\mu^a \rho_t) + \tfrac{1}{2}\kappa^2 \nabla^2 \rho_t. \tag{17}$$

For our model we require that the drift vector μ^a represent the double-bracket gradient flow (3). This is achieved by choosing

$$\mu^a = -\tfrac{1}{2}\kappa^2 \lambda \nabla^a G, \tag{18}$$

where $G(x)$ is a function on $\mathfrak{M}$ given by $\mathrm{tr}(HG)$. Then it follows that there exists a unique stationary solution to (17), given by the canonical density

$$\rho(x) = \frac{\exp(-\lambda G(x))}{\int_{\mathfrak{M}} \exp(-\lambda G(x))\mathrm{d}V}. \tag{19}$$

If $\mathfrak{M}$ is the space of pure states, then we have a model for thermalisation of quantum states introduced by Brody & Hughston [7].

6. Two-dimensional case in more detail

To illustrate these results in more explicit terms we consider a system consisting of a single spin-$\frac{1}{2}$ particle immersed in an external magnetic field. The Hamiltonian is then $H = -\mathbf{B}\cdot\mathbf{S}$, where $\mathbf{B}$ denotes the field and $\mathbf{S}$ the spin vector. The direction of the field $\mathbf{B}$, however, is subject to fluctuations around its stable direction, specified by G (directed along the z-axis). For the dynamical equation we obtain [4]:

$$\begin{cases} \mathrm{d}\theta_{\mathrm{t}} = \omega \sin\theta_{\mathrm{t}}\mathrm{dt} + \sqrt{2\nu}(\mathrm{dW}_{\mathrm{t}}^1 + \mathrm{dW}_{\mathrm{t}}^2) \\ \mathrm{d}\phi_t = -\frac{1}{\sin\theta_t}\sqrt{2\nu}(\mathrm{d}W_t^1 - \mathrm{d}W_t^2). \end{cases} \tag{20}$$

The associated Fokker-Planck equation reads

$$\dot{\rho} = -\omega(\cos\theta + \sin\theta\,\partial_\theta)\rho + 2\nu\left(\partial_\theta^2 + \tfrac{1}{\sin^2\theta}\,\partial_\phi^2\right)\rho, \tag{21}$$

where $\partial_\theta = \partial/\partial\theta$ and $\partial_\phi = \partial/\partial\phi$. The asymptotic solution is the canonical density function:

$$\rho(\theta,\phi) = \frac{\lambda\mu}{2\pi\sinh(\frac{1}{2}\lambda\mu)}\exp\left(-\tfrac{1}{2}\lambda\mu\cos\theta\right). \tag{22}$$

Direct substitution shows that (22) is the stationary solution to (21).

It follows from (22) and the use of the volume element $\mathrm{d}V = \frac{1}{4}\sin\theta\mathrm{d}\theta\mathrm{d}\phi$ that the equilibrium mean Hamiltonian is

$$\langle H\rangle = \tfrac{1}{2}\begin{pmatrix} u_0 + \nu\langle\cos\theta\rangle_\lambda & 0 \\ 0 & u_0 - \nu\langle\cos\theta\rangle_\lambda, \end{pmatrix}, \tag{23}$$

where

$$\langle\cos\theta\rangle_\lambda = \frac{2}{\lambda\mu} - \frac{1}{\tanh(\frac{1}{2}\lambda\mu)}. \tag{24}$$

We may regard the parameter λ as representing the 'inverse temperature' for the Hamiltonian. If the noise level is high ($\lambda \ll 1$), then the direction of the external field $\mathbf{B}$ on the average lies close to the xy-plane so that $\langle\cos\theta\rangle_\lambda \simeq 0$. If the noise level is low $\lambda \gg 1$, then the field $\mathbf{B}$ on the average is parallel to the z-axis and we have $\langle\cos\theta\rangle_\lambda \simeq -1$. We plot $\langle\cos\theta\rangle_\lambda$ as a function of $\tau = 1/\lambda$ (see Fig. 3).

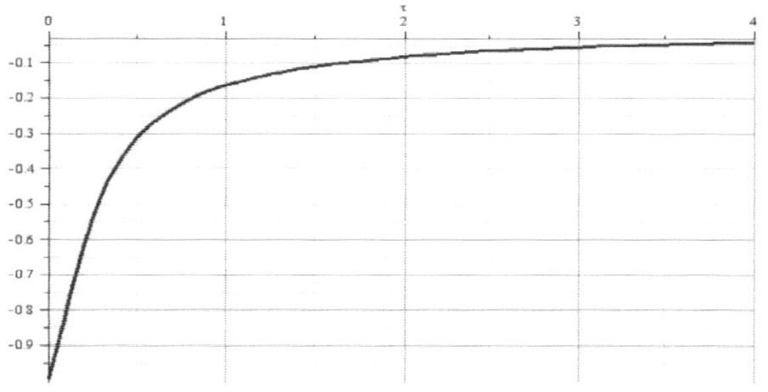

Figure 3. The plot of the expectation $\langle\cos\theta\rangle_\lambda$ as a function of the inverse Hamiltonian temperature $\tau = 1/\lambda$. For $\tau \ll 1$ we have $\langle\cos\theta\rangle_\lambda \simeq -1$, whereas for $\tau \gg 1$ we find $\langle\cos\theta\rangle_\lambda \simeq 0$.

7. Quantum statistical mechanics of disordered systems

Now we consider how the statistical theory of Hamiltonians presented above can be applied to quantum statistical mechanics, when the system and the apparatus specifying the Hamiltonian are both immersed in a heat bath with inverse temperature β. In this context it is natural to borrow ideas from the spin glass literature. We may take the averaged Hamiltonian $\langle H\rangle_\lambda$ as the starting point of the analysis—this gives the analogue of an *annealed* average:

$$\langle O\rangle_A = \frac{\mathrm{tr}\left(O\mathrm{e}^{-\beta\langle H\rangle_\lambda}\right)}{\mathrm{tr}\left(\mathrm{e}^{-\beta\langle H\rangle_\lambda}\right)} \tag{25}$$

of an observable O. Such an averaging, however, will change the eigenvalues of H.

Alternatively, we may use the 'unaveraged' Hamiltonian to compute the expectation of an observable O, and then take its average:

$$\langle O\rangle_Q = \left\langle \frac{\mathrm{tr}(O\mathrm{e}^{-\beta H})}{\mathrm{tr}(\mathrm{e}^{-\beta H})} \right\rangle_\lambda. \tag{26}$$

This gives the analogue of a *quenched* average. The canonical quenched average of the Hamiltonian $G = \sigma_z$ is

$$\langle G\rangle_Q = \tfrac{1}{2}\mu\tanh\left(\tfrac{1}{2}\beta\nu\right)\left(\frac{1}{\tanh(\frac{1}{2}\lambda\mu)} - \frac{2}{\lambda\mu}\right), \tag{27}$$

whereas the canonical annealed average of G is

$$\langle G\rangle_A = \tfrac{1}{2}\mu\tanh\left[\tfrac{1}{2}\beta\nu\left(\frac{1}{\tanh(\frac{1}{2}\lambda\mu)} - \frac{2}{\lambda\mu}\right)\right]. \tag{28}$$

These results suggest a new line of studies on the extended quantum statistical mechanics of disordered systems. The plot below shows the annealed (blue) and quenched (red) averages of σ_z, as a function of the bath temperature $T = 1/\beta$, for fixed λ such that $\lambda^{-1} = 0.1$.

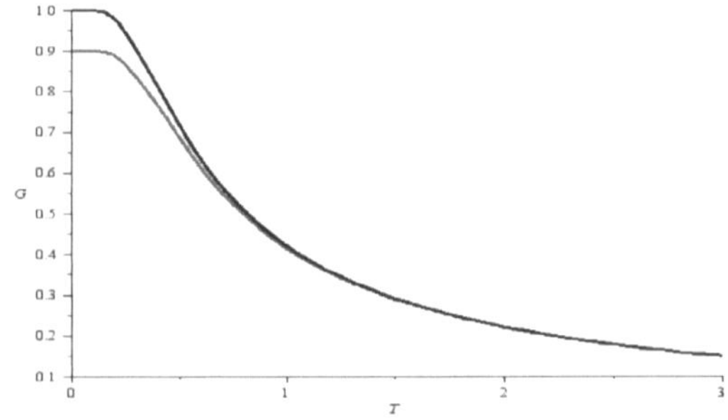

Figure 4. *Quenched and annealed averages of G.* The functions $\langle G\rangle_Q$ and $\langle G\rangle_A$ are plotted against the temperature $T = \beta^{-1}$, where we set $\lambda^{-1} = 0.1$, $v = 0$, $\nu = 1$, and $\mu = 2$ so that $G = \sigma_z$. The 'quenched magnetisation' $\langle\sigma_z\rangle_Q$ does not attain the maximum value 1.0 at zero temperature unless $\lambda^{-1} = 0$.

8. Examination of higher-dimensional cases

The geometry of higher-dimensional Hermitian matrices is somewhat more intricate than the two-dimensional case examined above. The space of $N\times N$ Hermitian matrices has the structure of the product $\mathbb{R}^N \times \mathbb{CP}^N \times \mathbb{CP}^{N-1} \times \cdots \times \mathbb{CP}^1$, where $\mathbb{CP}^k$ denotes the complex projective k-space. This space is considerably larger than the space $\mathbb{CP}^N$ of pure states upon which $N\times N$ Hermitian matrices act, and as a consequence the equivalence of the Schrödinger and Heisenberg pictures for a nonunitary motion would in general be lost.

A 3×3 Hermitian matrix H can be written as $H = E_1|E_1\rangle\langle E_1| + E_2|E_2\rangle\langle E_2| + E_3|E_3\rangle\langle E_3|$, where $\{E_i\}$ are the eigenvalues and $\{|E_i\rangle\}$ are the associated eigenstates. The degrees of freedom for the energy eigenvalues correspond to the open space $\mathbb{R}^3$; the remaining degrees of freedom corresponding to $\mathbb{CP}^2\times\mathbb{CP}^1$ are encoded in the specification of the energy eigenstates. Writing $\{|g_i\rangle\}$ for the eigenstates of G, we can express $|E_1\rangle$ in the form:

$$|E_1\rangle = \sin\tfrac{1}{2}\vartheta\cos\tfrac{1}{2}\varphi|g_1\rangle + \sin\tfrac{1}{2}\vartheta\sin\tfrac{1}{2}\varphi\mathrm{e}^{\mathrm{i}\xi}|g_2\rangle + \cos\tfrac{1}{2}\vartheta\mathrm{e}^{\mathrm{i}\eta}|g_3\rangle, \tag{29}$$

which determines a point in $\mathbb{CP}^2$. There is a $\mathbb{CP}^1$ worth degrees of freedom left for $|E_2\rangle$:

$$\begin{aligned} |E_2\rangle &= \left(\cos\tfrac{1}{2}\alpha\sin\tfrac{1}{2}\vartheta\cos\tfrac{1}{2}\varphi - \sin\tfrac{1}{2}\alpha\sin\tfrac{1}{2}\varphi\mathrm{e}^{\mathrm{i}\beta}\right)|g_1\rangle \\ &\quad + \left(\cos\tfrac{1}{2}\alpha\cos\tfrac{1}{2}\vartheta\sin\tfrac{1}{2}\varphi\mathrm{e}^{\mathrm{i}\xi} + \sin\tfrac{1}{2}\alpha\cos\tfrac{1}{2}\varphi\mathrm{e}^{\mathrm{i}(\beta+\xi)}\right)|g_2\rangle + \cos\tfrac{1}{2}\alpha\sin\tfrac{1}{2}\vartheta\mathrm{e}^{\mathrm{i}\eta}|g_3\rangle. \end{aligned} \tag{30}$$

The specifications of $|E_1\rangle$ and $|E_2\rangle$ leave no further freedom left for $|E_3\rangle$ and we have

$$\begin{aligned} |E_3\rangle &= \left(\sin\tfrac{1}{2}\alpha\cos\tfrac{1}{2}\vartheta\cos\tfrac{1}{2}\varphi + \cos\tfrac{1}{2}\alpha\sin\tfrac{1}{2}\varphi\mathrm{e}^{\mathrm{i}\beta}\right)|g_1\rangle \\ &\quad + \left(\sin\tfrac{1}{2}\alpha\cos\tfrac{1}{2}\vartheta\sin\tfrac{1}{2}\varphi\mathrm{e}^{\mathrm{i}\xi} - \cos\tfrac{1}{2}\alpha\cos\tfrac{1}{2}\varphi\mathrm{e}^{\mathrm{i}(\beta+\xi)}\right)|g_2\rangle + \sin\tfrac{1}{2}\alpha\sin\tfrac{1}{2}\vartheta\mathrm{e}^{\mathrm{i}\eta}|g_3\rangle. \end{aligned} \tag{31}$$

In this manner we obtain the nine parameters required for the specification of an arbitrary 3×3 Hermitian matrix.

It should be remarked that the foregoing procedure is merely an example of how one might parameterise a generic $N \times N$ Hermitian matrix in a systematic manner; the scheme is somewhat unconventional in that a generic 2×2 Hermitian matrix in this parameterisation is written

$$H = \omega_+ \begin{pmatrix} 1 & 0 \\ 0 & 1 \end{pmatrix} + \omega_- \begin{pmatrix} \cos\theta & \sin\theta\, \mathrm{e}^{-\mathrm{i}\phi} \\ \sin\theta\, \mathrm{e}^{\mathrm{i}\phi} & -\cos\theta \end{pmatrix}, \tag{32}$$

where $\omega_\pm = \frac{1}{2}(E_1 \pm E_2)$. This set of coordinates is nevertheless convenient because it isolates invariant quantities $\omega_\pm$ from the coordinates (θ, ϕ) of the isospectral submanifolds.

For our application in quantum statistical mechanics of disordered systems we are required to determine the partition function

$$Z(\lambda) = \int \mathrm{e}^{-\lambda\, \mathrm{tr}(HG)} \mathrm{d}V. \tag{33}$$

In the case of a 3×3 Hamiltonian, since the dynamical equation (1) preserves the three eigenvalues of H, the dynamical motion stays on the isospectral submanifold spanned by the six angular variables $\vartheta, \varphi, \alpha \in [0, \pi]$ and $\xi, \eta, \beta \in [0, 2\pi]$, with volume element

$$\mathrm{d}V = \tfrac{1}{128} \sin\alpha \sin\vartheta (1 - \cos\vartheta) \sin\varphi \, \mathrm{d}\alpha \, \mathrm{d}\beta \, \mathrm{d}\vartheta \, \mathrm{d}\varphi \, \mathrm{d}\xi \, \mathrm{d}\eta. \tag{34}$$

Since we are working in the G-basis, the calculation of $\mathrm{tr}(HG)$ is straightforward, and the integral (33) reduces to an expression analogous to the integral representation for a Bessel function.

The choice of parameterisation adapted here for a generic $N \times N$ Hermitian matrix need not be the most adequate for our purpose. An alternative approach is to write $H = U^\dagger E U$, where E is a diagonal matrix with eigenvalues $\{E_i\}$, and U is a unitary matrix. Then the calculation of the partition function becomes semi-Gaussian, and we are left with an integration over the invariant Haar measure [8]. Ideas from matrix theory or random matrices might prove useful in the statistical analysis of disordered quantum systems introduced here.

Acknowledgments
The authors thank R. Brockett and E. J. Brody for comments and stimulating discussions. DDH thanks the Royal Society of London for partial support by its Wolfson Merit Award.

References
[1] Brockett R W 1937 *Systems and Control in the Twenty-First Century ed* C Byrnes *et al* (Boston: Birkhäuser) 75-101
[2] Landau L D and Lifshitz E M 1935 *Phys. Z. Sowietunion* **8** 153-169
[3] Brockett R W 1991 *Lin. Alg. Appl.* **146** 79-91
[4] Brody D C, Ellis D C P, and Holm D D 2008 *J. Phys.* A **41** 502002
[5] Hida T 1980 *Brownian Motion* (Berlin, Germany: Springer)
[6] Hughston L P 1996 *Proc. Roy. Soc. London* A **452** 953-979
[7] Brody D C and Hughston L P 1999 "Thermalisation of quantum states" *J. Math. Phys.* **40** 12-18
[8] Sugiura M 1990 *Unitary Representations and Hamonic Analysis* (Amsterdam: North-Holland)

Fourth International Workshop DICE2008 IOP Publishing
Journal of Physics: Conference Series **174** (2009) 012042 doi:10.1088/1742-6596/174/1/012042

Complex nonlinear relations in classical and quantum physics

Dieter Schuch

Institut für Theoretische Physik, Goethe–Universität, Max-von-Laue-Str. 1, D-60438 Frankfurt am Main, Germany

E-mail: schuch@em.uni-frankfurt.de

Abstract. An aim of this paper is to show that the major difference between classical and quantum physics is not only that in the latter case complex quantities are a conceptional element, but also that real and imaginary parts, or phase and amplitude, respectively, are not independent of each other but coupled via some conservation laws. This also holds for systems where the Hamiltonian is no longer a constant of motion but a so-called Ermakov invariant with the dimension of an action still exists. In particular, this is valid for certain dissipative systems where a transition between a formal Hamiltonian level and a physical nonlinear level can be achieved via a non-unitary transformation.

1. Introduction

The major difference between classical physics and quantum physics is the introduction of **complex quantities** as **essential** entities to describe the physical world. The Nobel Laureate C.N. Yang described this with the following words [1]:"Complex numbers became a conceptual element of the very foundation of pyhsics: the fundamental equations of matrix mechanics and of wave mechanics ", the commutator of position and momentum and the time-dependent Schrödinger equation (SE), "both explicitly contain the imaginary unit $i = \sqrt{-1}$. It is to be emphasized that the very meaning of these equations would be totally destroyed if one tries to get rid of i by writing" these equations "in terms of real and imaginary parts."

One aim of this article is to show that the reason for this qualitative change in the concept of the description of a physical system is **not only** the use of **complex quantities, but also** the fact that **real and imaginary parts** of these quantities, or **phase and amplitude** (in a polar representation), are **not independent** of each other but **uniquely coupled** via some **conservation laws**.

Another point to be emphasized is that, apart from the Hamiltonian that plays a central role in classical as well as in quantum mechanics (and does not only represent the energy of a given physical system but also determines its dynamics), there are other **physical quantities** that are **still conserved** - even if the **Hamiltonian is not** - and are also intimately connected with the classical and quantum time-evolution of the system. A quantity of this kind is the so-called **Ermakov invariant** that essentially has the dimension of an **action** and is - unlike the Hamiltonian - still a **conserved** quantity, e.g. for a harmonic oscillator (HO) with **time-dependent frequency** $\omega(t)$ or certain d**issipative systems** with linear velocity dependent friction forces.

In Section 2, it will be shown how this invariant that was originally obtained in a more mathematical context by V.P. Ermakov in 1880 [2], i.e., long before quantum mechancis existed, emerges from the analysis of exact analytic Gaussian wave packet (WP) solutions of the time-dependent SE. It will also be shown that this invariant is related to a complex nonlinear Riccati equation that can be treated in different ways, leading to different physical aspects of the same (quantum) system. The quantities involved in this formalism are also related to the time-dependent Green's function and the representation of canonical transformations in quantum mechanics.

In Section 3 the discussion will focus on certain effective descriptions of dissipative systems with irreversible dynamics possessing an exact Ermakov invariant. In particular, two approaches will be considered. One is based on an explicitly time-dependent Hamiltonian and, after canonical quantization, yields an explicitly time-dependent linear SE. The other approach is based on an irreversible Fokker-Planck-type equation for the probability density that can be separated into two (complex conjugate) nonlinear Schrödinger equations (NLSE). It will be shown that both approaches are physically equivalent and can be transformed into each other. The price one must pay for this is a **non-unitary** transformation.

Finally, in Section 4, some conclusions will be drawn and perspectives for further developments mentioned.

2. Dynamical invariant for systems with time-reversible dynamics

Amongst the few quantum mechanical problems that can be solved exactly with analytical solutions, the free motion (potential $V = 0$) and the HO ($V = \frac{m}{2}\omega^2 x^2$) are particularly important, because in these cases also the time-dependent version of the SE has exact analytic solutions in the form of Gaussian WPs. This still holds, when the frequency of the HO is time-dependent.

In the following we consider the HO (in one dimension) with possibly time-dependent frequency, $\omega = \omega(t)$, in which case the Hamiltonian is no longer a dynamical invariant ($\frac{d}{dt}H \neq 0$). So we start from the corresponding time-dependent SE

$$i\hbar\frac{\partial}{\partial t}\Psi(x,t) = \left\{-\frac{\hbar^2}{2m}\frac{\partial^2}{\partial x^2} + \frac{m}{2}\omega^2(t)x^2\right\}\Psi(x,t) \tag{1}$$

with exact analytic Gaussian WP solution written in the form

$$\Psi(x,t) = N(t)\exp\left\{i\left[y(t)\tilde{x}^2 + \frac{<p>}{\hbar}\tilde{x} + K(t)\right]\right\} \tag{2}$$

where $\tilde{x} = x - <x> = x - \eta(t)$ with $<x> = \int_{-\infty}^{+\infty}\Psi^* x\Psi dx = \eta(t)$ = classical trajectory, $<p> = m\dot{\eta}$ = classical momentum; $N(t)$ (normalizing factor) and $K(t)$ are purely time-dependent terms which are not relevant for the following. The complex, possibly time-dependent, quantity $y(t) = y_R(t) + iy_I(t)$ is connected with the WP width, or the position uncertainty, via $y_I = \frac{1}{4<\tilde{x}^2>}$ with $<\tilde{x}^2> = <x^2> - <x>^2$.

By inserting this WP (2) into the SE (1) it is obtained that the two parameters that characterize the Gaussian function, namely the maximum, located at $x = <x> = \eta(t)$ and the width, contained in $y_I(t)$, fulfil the classical equation of motion

$$\ddot{\eta} + \omega^2(t)\eta = 0\ , \tag{3}$$

(where overdots denote time derivatives) and the complex nonlinear Riccati equation

$$\frac{2\hbar}{m}\dot{y} + \left(\frac{2\hbar}{m}y\right)^2 + \omega^2(t) = 0\ , \tag{4}$$

Journal of Physics: Conference Series **174** (2009) 012042 doi:10.1088/1742-6596/174/1/012042

respectively.

Using the imaginary part of this equation, the real part y_R can be expressed in terms of y_I. It turns out to be advantageous to introduce a new variable via

$$\frac{2\hbar}{m} y_I = \frac{1}{\alpha^2(t)} , \tag{5}$$

where $\alpha(t)$ is directly proportional to the WP width, or position uncertainty, i.e., $\alpha = \left(\frac{m}{2\hbar y_I}\right)^{1/2} = \left(\frac{2m<\tilde{x}^2(t)>}{\hbar}\right)^{1/2}$, showing that the real part of $y(t)$ just describes the relative change in time of the WP width according to

$$\frac{2\hbar}{m} y_R = \frac{\dot{\alpha}}{\alpha} = \frac{1}{2} \frac{\frac{d}{dt} < \tilde{x}^2 >}{< \tilde{x}^2 >} . \tag{6}$$

With the help of Eqs. (5) and (6), the complex nonlinear Riccati equation can be rewritten as a real nonlinear equation of the form

$$\ddot{\alpha} + \omega^2(t)\alpha = \frac{1}{\alpha^3} . \tag{7}$$

It has been shown by Ermakov in 1880 [2] that the system of differential equations (3) and (7), coupled via the possibly time-dependent frequency ω, leads, via elimination of this quantity, to a dynamical invariant of the form

$$I_L = \frac{1}{2} \left[(\dot{\eta}\alpha - \eta\dot{\alpha})^2 + \left(\frac{\eta}{\alpha}\right)^2 \right] = \text{const} . \tag{8}$$

This invariant, which has (up to a constant factor of mass m) the dimension of an **action**, not only depends on the classical variables $\eta(t)$ and $\dot{\eta}(t)$, but also on the quantum uncertainty, connected with $\alpha(t)$, and its change in time, $\dot{\alpha}(t)$. The Ermakov invariant has been rediscovered by several authors in the 20th century [3], also in a quantum mechanical context [4].

Additional interesting insights into the relation between η and α can be obtained by introducing a new complex variable $\lambda(t)$ and linearizing the Riccati equation (4) with the ansatz

$$\left(\frac{2\hbar}{m} y\right) = \frac{\dot{\lambda}}{\lambda} , \tag{9}$$

thus turning it into the complex linear equation

$$\ddot{\lambda} + \omega^2(t)\lambda = 0 , \tag{10}$$

looking exactly like Eq. (3) for $\eta(t)$, which is not just accidental.

For the following we choose to write $\lambda(t)$ in polar coordinates as well as in cartesian coordinates, i.e., $\lambda = \alpha e^{i\varphi} = u + iz$. The choice of the symbol α for the absolute value of λ is also not coincidental, as will now be shown. Writing relation (9) in polar coordinates yields

$$\frac{\dot{\lambda}}{\lambda} = \frac{\dot{\alpha}}{\alpha} + i\dot{\varphi} = \left(\frac{2\hbar}{m} y\right) . \tag{11}$$

The real part already looks identical to the one given in Eq. (6). So the absolute value of λ would be (up to a constant factor) identical with the square root of the position uncertainty $< \tilde{x}^2 >$, if

$$\dot{\varphi} = \frac{1}{\alpha^2} \tag{12}$$

is fulfilled. This can easily be verified by inserting $\left(\frac{2\hbar}{m}y_R\right)$ and $\left(\frac{2\hbar}{m}y_I\right)$, as given in Eq. (11), into the imaginary part of the Riccati equation, leading to $\dot{\varphi}\alpha^2 = k = constant$. This conservation law couples amplitude and phase of the complex quantity $\lambda(t)$ and formally reminds of the conservation of angular momentum for the two-dimensional motion, e.g., in a central force potential. However, it should be kept in mind that $\lambda(t)$ moves in the **complex** plane, not in a real physical plane (for further details see [5]).

Rewriting Eq. (12) in cartesian coordinates leads to $\dot{z}u - \dot{u}z = 1$ which shows that, in this case, real and imaginary parts of λ are also not independent of each other but coupled by this conservation law.

To show the physical meaning of the cartesian components of λ it can be used that the WP solution (2) of the SE can also be obtained from an initial WP $\Psi(x', t' = 0)$ and a time-dependent Green's function, also called time-propagator or Feynman kernel. For an initial Gaussian WP, the Green's function is also Gaussian. It can be shown (for details see [6]) that the dynamics is entirely determined by two parameters, which are identical with the cartesian components u and z of λ. Comparing the WP obtained in this way with the one given by Eq. (2) shows, in particular, that the imaginary part z is, according to $z = \frac{m}{\alpha_0 p_0}\eta(t)$, directly proportional to the classical trajectory $\eta(t)$, thus explaining the formal similarity between Eqs. (3) and (10). The meaning of u can be obtained by replacing η and $\dot{\eta}$ in the Ermakov invariant by z and $\dot{z}$ and using that $\left(\frac{z}{\alpha}\right)^2 = \sin^2\varphi$. So, for I_L to be constant, $(\dot{z}\alpha - z\dot{\alpha})^2 = \cos^2\varphi = \left(\frac{u}{\alpha}\right)^2$ must be fulfilled, leading (up to a $\pm$ sign) to

$$u = \dot{z}\alpha^2 - z\dot{\alpha}\alpha = \alpha^2\left(\dot{z} - \frac{\dot{\alpha}}{\alpha}z\right), \qquad (13)$$

which shows that, for constant width, i.e., $\dot{\alpha} = 0$, the real part of λ is proportional to the classical velocity or momentum, whereas, for $\dot{\alpha} \neq 0$, an additional contribution proportional to the classical trajectory $\eta(t)$ occurs.

Finally it should be mentioned that the Green's function also must fulfil the SE. Inserting it into (1) shows that in the particular case of a Gaussian WP, the constant k occurring in $\dot{\varphi}\alpha^2 = k$ (see above) is equal to 1, thus leading to (12).

So far, we have obtained from our complex equations of motion the two conservation laws, namely, the conservation of "angular momentum" in the complex plane, coupling real and imaginary parts of λ, or amplitude and phase, respectively. The other conserved quantity with dimension of an action, the Ermakov invariant I_L, couples the amplitude of λ with its cartesian components. In the following, it shall be shown that also these two invariants are not independent of each other.

In classical mechanics, the time-evolution of a Hamiltonian system is described by canonical transformations in phase space that can be represented by a group of 2×2 matrices with determinant equal to 1, the so-called two-dimensional real symplectic group $Sp(2, R)$.

It has been shown in [7] how to obtain the representation of this group of linear canonical transformations in time independent quantum mechanics. Recently [8], this has been extended to include time-dependent quantum systems.

In this case, the relation connecting initial position x' and momentum p' with their values x and p at a later time t is given by

$$\begin{pmatrix} x \\ \frac{p}{m} \end{pmatrix} = \begin{pmatrix} \dot{z} & -\dot{u} \\ -z & u \end{pmatrix} \begin{pmatrix} \frac{x'}{\alpha_0} \\ -\frac{\alpha_0 p'}{m} \end{pmatrix} = \mathbf{M} \begin{pmatrix} \frac{x'}{\alpha_0} \\ -\frac{\alpha_0 p'}{m} \end{pmatrix}, \qquad (14)$$

where u and z are again the cartesian components of $\lambda(t)$, and also the initial uncertainty α_0 must be considered (for details see [8]). The determinant of the transformation matrix $\mathbf{M}$ is obviously $\dot{\hat{z}}\hat{u} - \dot{\hat{u}}\hat{z} = 1$, which is just our conservation of angular momentum (in the complex plane).

Expressing $z, \dot{z}$ and u in terms of $\eta, \dot{\eta}, \alpha$ and $\dot{\alpha}$ with the additional relation

$$\dot{u} = \left(\frac{m}{\alpha_0 p_0}\right)\left[\dot{\eta}\dot{\alpha}\alpha - \eta\left(\dot{\alpha}^2 + \frac{1}{\alpha^2}\right)\right] , \tag{15}$$

which follows straightfotwardly from (13), using (3) and (7), the matrix $\mathbf{M}$ can be rewritten as

$$\mathbf{M} = \left(\frac{m}{\alpha_0 p_0}\right)\begin{pmatrix} \dot{\eta} & -\dot{\eta}\dot{\alpha}\alpha + \eta\left(\dot{\alpha}^2 + \frac{1}{\alpha^2}\right) \\ -\eta & \dot{\eta}\alpha^2 - \eta\dot{\alpha}\alpha \end{pmatrix} , \tag{16}$$

now with the determinant in the form $\det(\mathbf{M}) = 2\left(\frac{m}{\alpha_0 p_0}\right)^2 I_L = 1$.

So, essentially up to constant factors, our two invariants are not only related to the same complex quantity λ in its different coordinate representations but, moreover, are even essentially identical - which was not at all obvious before.

Because of the certainly important role that the Ermakov invariant plays for the dynamics of quantum systems, it should be interesting to investigate dynamical systems that do not possess a constant of motion in the form of a Hamiltonian but still possess an Ermkov invariant. Two examples of this kind have been found [10]. They represent dissipative systems with irreversible dynamics and will be discussed in the following section.

3. Dynamical invariant for systems with irreversible dissipative dynamics

As the conventional approach [10] to describe dissipation by coupling the system of interest to an environment of very many external degrees of freedom often leads to cumbersome or, for certain nonlinear or chaotic systems even impossible calculations, alternative approaches try to find a way of including the effect of the environment without having to take into account all the details of its many degrees of freedom.

In the context of such an effective description, explicitly time-dependent Hamiltonians [11], as well as nonlinear modifications of the Schrödinger equation (NLSE) are discussed in the literature [12, 13]. Since our focal point is the Ermakov invariant, we will concentrate on those approaches for which an exact invariant of Ermakov type exists. As has been pointed out by Nassar [9], this is only the case for the explicitly time-dependent Hamiltonian of Caldirola and Kanai (CK) [11] and for a NLSE with logarithmic nonlinearity [13]. In both cases exact WP solutions exist and the equations of motion for the corresponding maxima and widths shall be discussed.

The approach of CK starts with the an explicitly time-dependent classical Lagrangian. With the corresponding canonical momentum, $\hat{p} = \frac{\partial}{\partial \dot{x}} L_{CK} = m\dot{x}e^{\gamma t} = pe^{\gamma t}$, the Hamiltonian

$$\hat{H}_{CK} = \frac{1}{2m} e^{-\gamma t}\, \hat{p}^2 + e^{\gamma t} V(x) \tag{17}$$

can be formulated which yields the proper equations of motion including a linear velocity dependent friction force with friction coefficient γ, but it neither represents the energy of the system, nor is it a constant of motion. It is important to realize that the canonical momentum $\hat{p}$ is not only different from the physical (kinetic) momentum $p = m\dot{x}$, but, in particular, the transition from the *physical* variables x and $p = m\dot{x}$ to the *canonical* variables $\hat{x} = x$ and $\hat{p} = pe^{\gamma t}$ represents a **non-canonical** transformation.

The transition to quantum mechanics is achieved by replacing, as usual, the *canonical* momentum with a differential operator according to $\hat{p} \rightarrow \hat{p}_{op} = \frac{\hbar}{i}\frac{\partial}{\partial x}$, leading to the corresponding (explicitly time-dependent) Hamiltonian operator and, thus, to the modified SE. For the systems considered in this work, this equation also possesses exact Gaussian WP solutions. The equation of motion for the WP maximum is just the one for the classical trajectory, in our notation

$$\ddot{\eta} + \gamma\dot{\eta} + \omega^2\eta = 0. \tag{18}$$

The modified complex Riccati-type equation (connected with the WP-width) can again, with the definition $\frac{2\hbar}{m}\hat{y}_I = \frac{1}{\alpha_{CK}^2(t)} = \frac{\hbar}{2m<\tilde{x}^2>_{CK}}$, be transformed into the real Ermakov equation

$$\ddot{\alpha}_{CK} + \gamma\dot{\alpha}_{CK} + \omega^2\alpha_{CK} = \frac{e^{-2\gamma t}}{\alpha_{CK}^3} \ . \tag{19}$$

This equation, together with Eq. (18), forms the required system of equations of motion that possesses an exact Ermakov-type invariant, here given in the form

$$I_{CK} = \frac{1}{2}\left[e^{2\gamma t}\left(\dot{\eta}\;\alpha_{CK} - \eta\;\dot{\alpha}_{CK}\right)^2 + \left(\frac{\eta}{\alpha_{CK}}\right)^2\right] = \text{const} \ . \tag{20}$$

The case of free motion ($V = 0$) is obtained in the limit $\omega \to 0$. It should be mentioned that, in this model - compared to the free motion without dissipation - the friction force leads to a slower spreading of the WP.

The other approach starts already on the quantum mechanical level and introduces a time-symmetry-breaking diffusion term into the continuity equation for the probability density $\varrho(x,t) = \Psi^*(x,t)\Psi(x,t)$, thus turning it into an irreversible Fokker–Planck-type equation. Following a method developed by Madelung and Mrowka [14], the (real) Fokker–Planck-type equation can be separated into two complex equations for the wave function $\Psi(x,t)$ and its complex conjugate, Ψ^*, if the separation condition

$$-D\frac{\frac{\partial^2}{\partial x^2}\varrho}{\varrho} = \gamma\left(\ell n\varrho - <\ell n\varrho>\right) \tag{21}$$

with diffusion coefficient D is fulfilled. This leads to the NLSE

$$i\hbar\frac{\partial}{\partial t}\Psi_{NL}(x,t) = \left\{-\frac{\hbar^2}{2m}\frac{\partial^2}{\partial x^2} + V(x) + \gamma\frac{\hbar}{i}\left(\ell n\Psi_{NL} - <\ell n\Psi_{NL}>\right)\right\}\Psi_{NL}(x,t) \tag{22}$$

with a complex logarithmic nonlinearity (for details see [13]). The real part of this nonlinearity is identical with an approach by Kostin [12]. This part depends on the phase of the WP and is responsible for the motion of the WP maximum. The imaginary part that corresponds to the diffusion term in the Fokker-Planck equation has no influence on the WP maximum but strongly influences the dynamics of the WP width.

The equation of motion for the WP maximum is identical with Eq. (18); the equation of motion for the width can be obtained from the modified Riccati equation which can, again using the definition (5) for α_{NL}, be transformed into the Ermakov-type equation

$$\ddot{\alpha}_{NL} + \left(\omega^2(t) - \frac{\gamma^2}{4}\right)\alpha_{NL} = \frac{1}{\alpha_{NL}^3} \ . \tag{23}$$

Also in this case, the pair of Eqs. (18) and (23) leads to an exact dynamical invariant of Ermakov type,

$$I_{NL} = \frac{1}{2}e^{\gamma t}\left[\left(\dot{\eta}\;\alpha_{NL} - \left(\dot{\alpha}_{NL} - \frac{\gamma}{2}\alpha_{NL}\right)\eta\right)^2 + \left(\frac{\eta}{\alpha_{NL}}\right)^2\right] = \text{const} \ . \tag{24}$$

Further details concerning the linearization of the Riccati equation, corresponding Green's functions, etc. can be found in [6].

Also in this approach the results for the free motion are obtained in the limit $\omega \to 0$. It should be mentioned that for the NLSE, in contrast with the CK (and also Kostin's) model, the dissipation accelerates the spreading of the WP (for details, see [13]). This can be understood if the relation between these two models is analysed.

Again, the action plays an important role. So Schrödinger introduced in his first communication [15] the wave function Ψ via the action S and the corresponding Hamilton-Jacobi equation by defining $S = \frac{\hbar}{i} \ln \Psi$. The logarithmic NLSE (22) can be transformed, following Schrödinger's arguments, exactly into the form of a Hamilton-Jacobi equation if a new action $\hat{S} = e^{\gamma t} S$ is introduced. The corresponding Hamiltonian $\hat{H} = e^{\gamma t} H$ is identical with the one of CK. The important point, however, is that, due to the newly defined action $\hat{S}$, also the wave function in this transformed (canonical) system has a different meaning from the usual one. In particular, the wave function $\hat{\Psi}$ in the transformed system (corresponding to the CK-ansatz) is connected with the wave function in the physical system (corresponding to the usual SE and the NLSE) via the **non-unitary** transformation

$$\ell n \hat{\Psi}_{CK} = e^{\gamma t} \ell n \Psi_{NL} \ . \tag{25}$$

Therefore, α_{CK} and α_{NL} are conencted via $\alpha_{CK} = e^{-\gamma t/2} \alpha_{NL}$, which also explains the different spreading behaviours of the free motion WP in the two approaches (deceleration vs. acceleration). Inserting this last relation into Eq. (19) for α_{CK} turns it exactly into Eq. (23) for α_{NL} and, hence, the invariant I_{CK} (Eq. (20)) is identical with I_{NL} (Eq. (24)), $I_{CK} \equiv I_{NL}$.

4. Conclusions

It has been shown for some model systems with time-dependent Hamiltonian and/or dissipative irreversible dynamics that, even if in these cases, where the Hamiltonian (if it at all exists) is no longer a constant of motion, there still exists another invariant (Ermakov) that has the dimension of an **action**. This invariant connencts the classical motion describing the dynamics of the WP maximum that follows the classical trajectory ($\eta(t)$) with the quantum dynamics describing the change of the WP width ($\alpha(t)$) that is proportional to the position uncertainty. The dynamics of this quantum aspect can be derived from a complex nonlinear Riccati equation that can not only be transformed into a real nonlinear Ermakov equation, but, also be linearized to a complex Newotian equation for a quantity $\lambda(t)$, which moves in the complex plane under the restriction of conserved "angular momentum". This conservation law couples real and imaginary parts, or phase and amplitude of λ, respectively. It has also been shown that his conservation law is (up to a constant factor) identical with the Ermakov invariant.

Going beyond isolated systems and considering systems coupled to some environment, thus exhibiting irreversible dissipative dynamics, the Hamiltonian, or energy of the system, is definitely no longer a constant of motion but there are still effective quantum mechanical descriptions of these systems where an exact Ermakov invariant exists.

The consequences are different in the various approaches. One attempt uses a NLSE where the nonlinearity (in general) destroys the superposition principle. So, coupling to an environment (heat bath, observer, etc.) avoids Schrödinger cat; however, operators and wave functions have the same physical meaning as in ordinary linear quantum mechancis. The other method is based on an explicity time-dependent Hamiltonian where the corresponding SE is linear, so the canonical formalism is conserved but operators and wave functions cannot be interpreted directly in the usual physical sense. Nevertheless, both approaches can, unambiguously, be transfomed into each other; the price one must pay is that of a **non-unitary** transformation (related to a non-canonical transformation in the classical case).

Further studies should go beyond quadratic Hamiltonians (e.g., Coulomb problem, etc.). In time-independent quantum mechancis, similar relations between phase and amplitude of the wave function exist with respect to changes in space as they exist for λ with respect to changes in time [16]. The physical meaning and consequences of the corresponding conservation laws and invariants shall also be investigated further.

References

[1] Yang C N 1987 *Schrödinger – Centenary Celebration of a Polymath* (Cambridge, UK: Cambridge U. Press) 41-52
[2] Ermakov V P 1880 *Univ. Izv. Kiev* **20** 1
[3] Pinney E 1950 *Proc. Amer. Math. Soc.* **1** 681
[4] Milne W E 1930 *Phys. Rev.* **35** 863; Lewis H R 1967 *Phys. Rev. Lett.* **18** 510
[5] Schuch D 1989 *Int. J. Quantum Chem.: Quantum Chem. Symp.* **23** 59; *do.* 1992 *Int. J. Quantum Chem.* **42** 663
[6] Schuch D and Moshinsky M 2006 *Phys. Rev.* A **73** 062111
[7] Moshinsky M and Smirnov Y F 1996 *The Harmonic Oscillator in Modern Physics* (Amsterdam: Harwood Acad. Publishers) Ch. 35
[8] Schuch D and Moshinsky M 2008 *SIGMA* **4** 056
[9] Nassar A B 1986 *J. Math. Phys.* **27** 755; *do.* 1986 *J. Math. Phys.* **27** 2949
[10] Caldeira A O and Leggett A J 1983 *Ann. Phys. (NY)* **149** 374; *ibid* 1983 **153** 445E
[11] Caldirola P 1941 *Nuovo Cimento* **18** 393; Kanai E 1948 *Progr. Theor. Phys.* **3** 440
[12] Kostin M D 1972 *J. Chem. Phys.* **57** 3589
[13] Schuch D, Chung K-M and Hartmann H 1983 *J. Math. Phys.* **24** 1652; *ibid* 1984 **25** 3086; Schuch D and Chung K-M 1986 *Int. J. Quantum Chem.* **29** 1561
[14] Madelung E 1950 *Die Mathematischen Hilfsmittel des Physikers* (Berlin: Springer) 432; Mrowka B 1951 *Z. Phys.* **130** 164
[15] Schrödinger E 1926 *Ann. d. Phys.* **79** 361
[16] Reinisch G 1994 *Physica* A **206** 229

Properties of some nonlinear Schrödinger equations motivated through information theory

Liew Ding Yuan and Rajesh R Parwani
Department of Physics, National University of Singapore, Kent Ridge, Singapore
E-mail: parwani@nus.edu.sg

Abstract. We update our understanding of nonlinear Schrödinger equations motivated through information theory. In particular we show that a $q-$deformation of the basic nonlinear equation leads to a perturbative increase in the energy of a system, thus favouring the simplest $q = 1$ case. Furthermore the energy minimisation criterion is shown to be equivalent, at leading order, to an uncertainty maximisation argument. The special value $\eta = 1/4$ for the interpolation parameter, where leading order energy shifts vanish, implies the preservation of existing supersymmetry in nonlinearised supersymmetric quantum mechanics. Physically, η might be encoding relativistic effects.

1. Introduction

We have previously argued, see for example [1], that the maximum uncertainty (entropy) method [2], used widely in several disciplines [3], is also a plausible framework to investigate potential nonlinear generalisations of Schrödinger's equation [4].

We consider here for simplicity a single particle in one space dimension. The basic nonlinear equation of Ref. [4] is

$$i\hbar\frac{\partial\psi}{\partial t} = -\frac{\hbar^2}{2m}\frac{\partial^2\psi}{\partial x^2} + V(x)\psi + F(p)\psi\,, \tag{1}$$

with $F(p) = Q_{nl} - Q$ where:

$$Q_{nl} = \frac{\mathcal{E}}{\eta^4}\left[\ln\frac{p}{(1-\eta)p+\eta p_+} + 1 - \frac{(1-\eta)p}{(1-\eta)p+\eta p_+} - \frac{\eta p_-}{(1-\eta)p_- + \eta p}\right], \tag{2}$$

$$Q = -\frac{\hbar^2}{2m}\frac{1}{\sqrt{p}}\frac{\partial^2\sqrt{p}}{\partial x^2}, \tag{3}$$

and $p(x) = \psi^\star(x)\psi(x)$ and $p_\pm(x) = p(x \pm \eta L)$. The nonlinearity length scale L and the energy parameter $\mathcal{E}$ are constrained through the relation $\mathcal{E}L^2 = \hbar^2/4m$. The dimensionless parameter $0 < \eta \leq 1$ plays the role of an interpolating parameter and also regularises potential singularities.

The nonlinear equation preserves a number of important properties of the linear Schrödinger equation, such as the conservation of probability, existence of the usual plane wave solutions and invariance under a scaling of the wavefunction, $\psi \to \lambda\psi$.

2. Physical meaning of the η parameter

In Ref. [5] nonlinear Dirac equations were constructed using an axiomatic approach [6]. In taking the non-relativistic limit of those equations [7] it was found that the potential singularities in the naive limit were regularised if one kept sub-leading relativistic corrections coming from the lower components of the Dirac spinor. This suggests an interpretation in which the η parameter in the nonlinear Schrödinger equation (1) is encoding relativistic effects present already in the linear quantum theory, while the nonlinearity is parametrised by the scale L.

Alternatively, one may proceed as follows. In Ref. [8] we computed the perturbative effect of the nonlinearity (1) on the energy spectrum of the linear theory with smooth external potentials $V(x)$. For an unperturbed state with quantum number n, with N nodes in the wavefunction, the result is

$$\delta E \approx \frac{\hbar^2 |L| \pi}{6m} \sqrt{\eta(1-\eta)} \; (1-4\eta) \sum_{p=1}^{N} C_{np}^2 + O(L/a)^2 , \tag{4}$$

where a is the characteristic length scale of the linear theory. The coefficients C_{np} depend on the slope of the unperturbed wavefunctions near the nodes, and hence on the form of the potential. Notice however that the dependence of the energy shift on η is universal, that is, independent of $V(x)$. (For states without nodes the energy shift is of lower order, $O(L/a)^2$).

It was suggested in [1] that η be fixed by minimising the energy of the system. The result (4) indicates that this happens when $\eta \approx 0.80$. Now, to leading order, the energy shift (4) is given by $\langle F(p) \rangle$, evaluated using the unperturbed states. Assuming integration over an infinite domain, this reduces after a simple shift of variables [4] to $I_{NL} - I_F$, where I_F is the Fisher information measure that leads to the linear Schrödinger equation [9, 10] and $I_{NL} = -I_{KL}$ is the information measure that leads to the nonlinear generalisation (1); I_{KL} is the Kullback-Liebler measure.

Thus minimising the energy shift is equivalent to minimising the information content of the nonlinear theory relative to that of the linear theory, and this is achieved for excited states at the universal value $\eta \sim 0.8$.

An interesting aspect of the result (4) is that it vanishes for values $\eta = 0, 1/4, 1$. The $\eta = 0$ case is obvious because then the nonlinearity is switched off. The $\eta = 1$ case is amusing because there the nonlinearity exists but the intermediate perturbative calculations are ill-defined if one starts with $\eta = 1$; thus $\eta \to 1$ may only to taken after the calculations. We will see in the next section that the vanishing of the energy shift at $\eta = 1$ is likely to be an accident, valid only at leading order.

So what about the $\eta = 1/4$ case? This is quite puzzling and we still do not understand if it indicates some hidden symmetry of the theory at that value. However we see that at that special value the nonlinearity can help preserve, to leading order, an existing supersymmetry of the linear theory: Consider one dimensional supersymmetric quantum mechanics [11] with two partner Hamiltonians H_1, H_2. Except for the ground state of H_1 the two Hamiltonians are isospectral but with differing wavefunctions. For example, the first excited state of H_1 will have the same energy as the ground state of H_2; the corresponding wavefunction of H_1 will have one node while that for H_2 will be nodeless. Thus if we nonlinearise both Hamiltonians as in (1), we see that at $\eta = 1/4$, all energy shifts, whether for excited or ground states will be of order $(L/a)^2$ and so to leading order the pair of Hamiltonians will have their original supersymmetry preserved.

3. Robustness of results

The nonlinearity in Ref. [4] was obtained using, arguably, the simplest generalisation of the Fisher information measure which preserved various desirable properties of the former. How sensitive are the above results to the particular choice of I_{KL}? We can study this issue by considering deformations of the Kullback-Liebler measure.

Consider therefore the $q-$deformed measure ($q > 0$),

$$M(\eta, q) \equiv \frac{-1}{q\eta^4} \int dx\ dt\ p(x) \left[\ln_q \frac{p(x)}{(1-\eta)p(x) + \eta p(x+\eta L)}\right], \tag{5}$$

with

$$\ln_q y \equiv \frac{y^{q-1} - 1}{q-1}. \tag{6}$$

This measure reduces, when $q = 1$, to the measure $-I_{KL}$ used in the derivation of (1) and retains for general q the various desirable properties of I_{KL}. Using the more general (5) leads to another nonlinear Schrödinger equation which can be used to calculate leading order energy shifts as in Ref. [8]. For unperturbed states with nodes in a symmetric potential the correction may be written

$$\delta E(q) \approx \frac{\hbar^2 |L|}{4m} J(\eta, q) \sum_{p=1}^{N} C_{np}^2 + O(L/a)^2, \tag{7}$$

where

$$J(\eta,q) \equiv \int_{-\infty}^{\infty} \frac{dy y^2}{\eta^4 q} \Big[\ln_q \frac{y^2}{(1-\eta)y^2 + \eta(y+\eta)^2} + \left(\frac{y^2}{(1-\eta)y^2 + \eta(y+\eta)^2}\right)^{q-1} - $$
$$ -(1-\eta)\left(\frac{y^2}{(1-\eta)y^2 + \eta(y+\eta)^2}\right)^{q} - \eta\left(\frac{(y-\eta)^2}{(1-\eta)(y-\eta)^2 + \eta y^2}\right)^{q} \Big].$$

For integer values of q one may evaluate the function $J(\eta, q)$ explicitly but this becomes time consuming for large q values. Thus we evaluated it numerically for a range of real $0 < q < 1$. The results are as follows.

For fixed $q > 1$ and $0 < \eta < 1$: J has one local maximum at η_{max} where it is positive, one global minimum at $\eta_{min} > \eta_{max}$ where it is negative, and it becomes zero at three points $0 = \eta_0 < \eta_1 < \eta_2 < 1$. η_{max} is between zero and η_1 while η_{min} is between η_1 and η_2. J diverges to positive infinity as $\eta \to 1$. The values of η_1, η_2 decrease as q increases. The minimum value of J also increases (becomes less negative) as q increases. In the limit $q \to 1$, $\eta_2 \to 1$ and thus J vanishes at $\eta = 1$ as we saw in the last section.

For fixed $0 < q < 1$ and $0 < \eta < 1$: J has one global maximum at η_{max} where it is positive, one global minimum at $\eta_{min} > \eta_{max}$ where it is negative, and it becomes zero at two points $0 = \eta_0 < \eta_1 < 1$. η_{max} is between zero and η_1 while η_{min} is between η_1 and 1. J is negative at $\eta = 1$. The value of η_1 decreases as q increases. The minimum value of J decreases (becomes more negative) as q increases.

The results of the last two paragraphs show that $J(\eta, q)$ reaches its minimum (which is negative) for $q = 1$ and for $\eta = 0.8$; see Fig. 1. Thus using the motivation of the last section, minimising the energy of the system, or equivalently minimising our information about the system, leads to the $q = 1$ nonlinear equation (1) as the preferred choice, at least when analysing static properties of states with definite parity.

So far we have focused on states with nodes. For states without nodes the energy shifts are $O(L^2)$. For the simple harmonic oscillator we have explicitly

$$\delta E(q) = \eta^2(\eta^2(q+1)(q+2) - 4\eta(q+1) + 2)\frac{\hbar^2 L^2}{8ma^4} + O(L^4). \tag{8}$$

This has a global minimum at $q \sim 1.1$ and $\eta \sim 0.76$, very close to the values for the states with nodes.

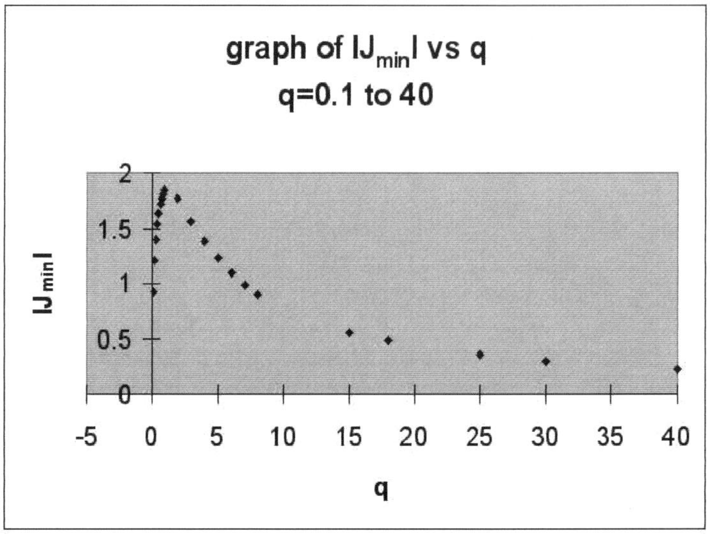

Figure 1. The magnitude of the minimum value of $J(\eta, q)$ for each q is plotted

4. Other results and outlook

Some non-perturbative aspects of the nonlinear Schrödinger equation (1) were studied in [12]. In [13] we used Eq. (1) to study the issue of singularity avoidance in quantum cosmology and further work in this direction is in progress. We have also used information theoretic ideas to construct integrable nonlinear Schrödinger equations in Ref. [14]. An application of the nonlinear Dirac equations of Ref. [5] is in Ref. [15].

Acknowledgment

R.P. thanks the organisers of the DICE workshop for their hospitality and the opportunity to present our results in a wonderful environment.

[1] Parwani R 2007 *Theoretical and Mathematical Physics* **152(1)** 101216
[2] Jaynes E T 2004 *Probability Theory, The Logic of Science* (Cambridge, UK: Cambridge U. Press)
Balian R 2005 *Studies in Hist. and Phil. of Mod. Phys.* **36** 323
[3] Buck B and Macaulay V A 1991 *Maximum Entropy in Action* (New York: Oxford U. Press)
Kapur J N and Kesavan H K 1992 *Entropy Optimization Principles with Applications* (New York: Academic Press)
[4] Parwani R 2005 *Ann. Phys.* **315** 419
[5] Ng W K and Parwani R 2007 *Preprint* arXiv:0707.1553
[6] Parwani R 2006 *AIP Conf. Proc.* **810** 387-92
Parwani R 2006 *Int. J. Theor. Phys.* **45** 1901-13
[7] Ng W K and Parwani R 2008 *Preprint* arXiv:0807.1877
[8] Parwani R and Tabia G 2007 *J. Phys. A: Math. Theor.* **40** 5621-35
[9] Parwani R 2005 *J. Phys. A: Math. Gen.* **38** 6231
[10] Reginatto M 1998 *Phys. Rev.* A **58** 1775
Frieden B R 1989 *J. Mod. Opt* **35** 1297

Frieden B R 1989 *Am. J. Phys.* **57** 1004
[11] Cooper F, Khare A and Sukhatme U 1995 *Phys. Rep.* **251** 267-385
[12] Parwani R and Tan H S 2007 *Phys. Lett.* A **363** 197-201
Nguyen L H, Tan H S and Parwani R 2008 *Preprint* arXiv:0801.0183
[13] Nguyen L H and Parwani R 2008 *Preprint* arXiv:0805.4294
[14] Parwani R and Pashaev O K 2008 *J. Phys. A: Math. Theor.* **41** 235207
[15] Ng W K and Parwani R *Preprint* arXiv:0805.3015

Journal of Physics: Conference Series **174** (2009) 012044 doi:10.1088/1742-6596/174/1/012044

Matrix models, emergent gravity and gauge theory

Harold Steinacker
Fakultät für Physik, Universität Wien
Boltzmanngasse 5, A-1090 Wien, Austria

E-mail: harold.steinacker@univie.ac.at

Abstract. Matrix models of Yang-Mills type describe noncommutative spaces, which are considered as models for dynamical space-time. Gravity arises on these branes through an intrinsically noncommutative mechanism. We review recent progress in the understanding of this emergent gravity. The metric is not fundamental but arises effectively in the semi-classical limit, along with nonabelian gauge fields. This leads to a mechanism for protecting certain geometries from quantum corrections, including flat space-time.

1. Background and motivation

Quantum field theory and general relativity (GR) provide the basis of our present understanding of fundamental matter and interactions. In spite of the success of these two theories, there is up to now no satisfactory way to reconcile them in a consistent quantum theory. In particular, quantum mechanics combined with GR strongly suggests a "foam-like" or quantum structure at the Planck scale 10^{-33} cm, where quantum fluctuations of space-time are expected to be important. While some kind of quantum structure of space-time indeed arises e.g. in string theory, a satisfactory understanding is still missing. The cosmological constant problem should be seen in the same context: the observed tiny (or zero) value of the cosmological constant is in striking contradiction with quantum mechanical expectations, which are off by a factor of order 10^{120}. Reconciling quantum mechanics with gravity is therefore of utmost importance in theoretical physics.

In view of these problems, it is natural to consider noncommutative (NC) or quantum spaces as models for space-time. For fixed backgrounds, considerable work has been done in this context, leading to NC field theory [4]. Recently, it was understood that gravity emerges naturally from NC gauge theory, without having to introduce an explicit dynamical metric. Earlier forms of this idea [6, 5] can be cast in concise form for matrix models of Yang-Mills type [8], which describe dynamical quantum spaces. We discuss basic results of this approach. The IKKT model [10] is singled out as a prime candidate for a quantum theory of space-time and matter.

2. The quantization of Poisson manifolds

Space-time in GR is modeled by a 4-dimensional manifold $\mathcal{M}$ with metric $G_{\mu\nu}(x)$. The basic assumption of the present approach is that space-time carries an additional Poisson structure $\{x^\mu, x^\nu\} = \theta^{\mu\nu}(x)$ (which will be related to the metric in (23)), more precisely that space-time is the *quantization* $\mathcal{M}_\theta$ of such a Poisson manifold.

In principle, a Poisson structure breaks (local) Lorentz invariance, which may seem incompatible with observation. However, it turns out that $\theta^{\mu\nu}$ does not enter explicitly the effective action of the models discussed here, to leading order in an expansion in $\theta^{\mu\nu}$. If we assume that the scale of noncommutativity Λ_{NC} defined by $\det\theta^{\mu\nu} = \Lambda_{NC}^8$ is at or near the Planck scale, it is then quite conceivable that its presence through higher-order terms in the effective action has not been detected up to now.

It is well-known that a Poisson-manifold $(\mathcal{M}, \theta^{\mu\nu}(x))$ can be quantized [12]. This means that there exists a quantization map

$$\begin{array}{rcl} \mathcal{C}(\mathcal{M}) & \to & \mathcal{A} \subset L(\mathcal{H}) \\ f(x) & \mapsto & \hat{f}(X) \end{array} \tag{1}$$

such that $i\{f,g\} \mapsto [\hat{f},\hat{g}] + O(\theta^2)$. Here $\mathcal{C}(\mathcal{M})$ denotes a suitable space of functions on $\mathcal{M}$, and $\mathcal{A}$ is interpreted as quantized algebra of functions on $\mathcal{M}$. The quantization of the coordinate functions x^μ will be denoted by X^μ. The matrix model will indeed provide preferred coordinates $X^\mu \sim x^\mu$, where $\sim$ indicates the leading contribution in a semi-classical expansion in powers of $\theta^{\mu\nu}$. These x^μ are not observable and thus not in conflict with any experimental constraints. The integral is replaced by the trace, more precisely by the volume of the symplectic volume form

$$\begin{array}{rcl} (2\pi)^2\,\mathrm{Tr}\hat{f} & \sim & \int d^4x\, \rho(x)\, f \\ \rho(x) & = & (\det\theta^{-1}_{\mu\nu})^{1/2} \end{array} \tag{2}$$

assuming that $\theta^{\mu\nu}$ is non-degenerate. This is essentially the Bohr-Sommerfeld quantization law.

A simple example of a Poisson manifold is given by $\mathbb{R}^4$ together with a constant antisymmetric matrix $\{x^\mu, x^\nu\} = \bar{\theta}^{\mu\nu}$. Its quantization gives the Moyal-Weyl quantum plane $\mathbb{R}^4_\theta$, where

$$[\bar{X}^\mu, \bar{X}^\nu] = i\bar{\theta}^{\mu\nu}\,. \tag{3}$$

This is formally the (doubled) Heisenberg algebra, i.e. the usual quantum-mechanical phase space. However, we need to consider the quantization of generic Poisson manifold here. We only consider the *semi-classical* or geometrical limit of such a quantum space in this paper. This means that the space is described in terms of functions on $\mathcal{M}$ using (1), keeping only the Poisson bracket on the rhs of (1) and dropping all higher-order terms in θ. Accordingly, we will always replace $[\hat{f}(X), \hat{g}(X)] \to i\{f(x), g(x)\}$ and $[X^\mu, X^\nu] \to i\theta^{\mu\nu}(x)$. In particular,

$$[X^\mu, f(X)] \sim i\theta^{\mu\nu}(x)\frac{\partial}{\partial x^\nu}f(x). \tag{4}$$

3. Yang-Mills matrix models and their effective geometry

Consider a matrix model with action

$$S_{YM} = -\mathrm{Tr}[X^a, X^b][X^{a'}, X^{b'}]\eta_{aa'}\eta_{bb'}, \tag{5}$$

for hermitian matrices or operators X^a, $a = 1, ..., D$ acting on some Hilbert space $\mathcal{H}$. Here

$$\eta_{ab'} = \mathrm{diag}(1,1,...,1) \quad \text{or} \quad \eta_{ab} = \mathrm{diag}(-1,1,...,1) \tag{6}$$

in the Euclidean resp. Minkowski case. The above action is invariant under the following gauge symmetry

$$X^\mu \to U^{-1}X^\mu U, \qquad U \in U(\mathcal{H}). \tag{7}$$

The equations of motion are

$$[X^a, [X^b, X^{a'}]]\eta_{aa'} = 0. \tag{8}$$

The matrix model should be considered as background-independent, since no geometry whatsoever is present a priori; rather, the geometry will arise dynamically. In particular, the model admits as solutions 4 -dimensional noncommutative spaces $\mathcal{M}_\theta \subset \mathbb{R}^D$, interpreted as space-time embedded in D dimensions. To see this, we split the matrices as

$$X^a = (X^\mu, \phi^i), \qquad \mu = 1, ..., 4, \ \ i = 1, ..., D-4 \tag{9}$$

where the "scalar fields" $\phi^i = \phi^i(X^\mu)$ are assumed to be functions of X^μ. The prototype of such a solution is a flat embedding of a 4-dimensional quantum space

$$\begin{aligned} [X^\mu, X^\nu] &= i\theta^{\mu\nu}, \qquad && \mu,\nu = 1, ..., 4, \\ \phi^i &= 0, && i = 1, ..., D-4 \end{aligned} \tag{10}$$

where X^μ generates a 4-dimensional NC space $\mathcal{M}_\theta$; for example, $\mathbb{R}^4_\theta \subset \mathbb{R}^D$ is realized by $[X^\mu, X^\nu] = i\theta^{\mu\nu}\mathbb{1}$.

In general, $\phi^i(X) \sim \phi^i(x)$ will be nontrivial. One could interpret $\phi^i(x)$ as scalar fields on $\mathbb{R}^4_\theta$; however, it is more appropriate to interpret $\phi^i(x)$ as purely geometrical degrees of freedom, defining the embedding of a submanifold $\mathcal{M} \subset \mathbb{R}^D$. In the semi-classical limit, suitable "optimally localized states" of $\langle X^a \rangle \sim x^a$ will then be located on $\mathcal{M} \subset \mathbb{R}^D$. This $\mathcal{M}$ carries the induced metric

$$g_{\mu\nu}(x) = \eta_{\mu\nu} + \partial_\mu\phi^i\partial_\nu\phi^j\delta_{ij} \ = \ \partial_\mu x^a \partial_\nu x^b \eta_{ab} \tag{11}$$

via pull-back of η_{ab}. Note that $g_{\mu\nu}(x)$ is *not* the metric responsible for gravity, and will enter the action only implicitly. Rather, it turns out that all fields arising from the matrix model will only live on the brane $\mathcal{M}$, and couple to the effective metric $G_{\mu\nu}$ given below (17). As opposed to standard braneworld-scenarios, there really is no higher-dimensional "bulk" which could carry physical degrees of freedom here, not even gravitons.

Effective metric and Poisson structure Expressing the ϕ^i in terms of X^μ, we obtain

$$[\phi^i, f(X^\mu)] \ \sim \ i\theta^{\mu\nu}\partial_\mu\phi^i\partial_\nu f \tag{12}$$

in the semi-classical limit. This involves only the components $\mu = 1, ..., 4$ of the antisymmetric "tensor" $[X^a, X^b] \sim i\theta^{ab}(x)$, which has rank 4 in the semi-classical limit. Here the derivations

$$-i[X^\mu, .] \ \sim \ \theta^{\mu\nu}\partial_\nu \tag{13}$$

span the 4-dimensional tangent space of $\mathcal{M} \subset \mathbb{R}^D$, and define a preferred frame. We can interpret

$$[X^\mu, X^\nu] \sim i\theta^{\mu\nu}(x) \tag{14}$$

as Poisson structure on $\mathcal{M}$, noting that the Jacobi identity is trivially satisfied. This is the Poisson structure on $\mathcal{M}$ whose quantization is given by the matrices X^μ, $\mu = 1, ..., 4$, interpreted as quantization of the coordinate functions x^μ on $\mathcal{M}$. In particular, the rank of $\theta^{\mu\nu}$ coincides with the dimension of $\mathcal{M}$. Its inverse $\theta^{-1}_{\mu\nu}(x)$ defines a symplectic form on $\mathcal{M}$.

We can now extract the semi-classical limit of the matrix model and its physical interpretation. To understand the effective geometry on $\mathcal{M}$, consider a (test-) particle on $\mathcal{M}$, modeled by an additional scalar field φ. Due to gauge invariance, the only reasonable kinetic term is

$$S[\varphi] \ \equiv \ -\mathrm{Tr}[X^a, \varphi][X^b, \varphi]\eta_{ab} = -\mathrm{Tr}\left([X^\mu, \varphi][X^\nu, \varphi]\eta_{\mu\nu} + [\phi^i, \varphi][\phi^j, \varphi]\delta_{ij}\right) \tag{15}$$

(for example, φ could be an $su(n)$ component of ϕ^i). In the semi-classical limit, this becomes

$$S[\varphi] \sim \frac{1}{(2\pi)^2}\int d^4x\,|G_{\mu\nu}|^{1/2}\,G^{\mu\nu}\partial_\mu\varphi\partial_\nu\varphi \tag{16}$$

which has the correct covariant form, where[1] [2]

$$\boxed{G^{\mu\nu}(x) = e^{-\sigma}\,\theta^{\mu\mu'}(x)\theta^{\nu\nu'}(x)g_{\mu'\nu'}(x).} \tag{17}$$

Here $g_{\mu\nu}(x)$ is the metric (11) induced on $\mathcal{M} \subset \mathbb{R}^D$ via pull-back of η_{ab}, and

$$\rho = (\det\theta^{-1}_{\mu\nu})^{1/2}, \qquad e^{-\sigma} = \rho\,|g_{\mu\nu}|^{-\frac{1}{2}}, \tag{18}$$

$$\eta(x) = \frac{1}{4}e^{\sigma}\,G^{\mu\nu}g_{\mu\nu}. \tag{19}$$

Therefore the kinetic term on $\mathcal{M}_\theta$ is governed by the metric $G_{\mu\nu}(x)$, which depends on the Poisson tensor $\theta^{\mu\nu}$ and the embedding metric $g_{\mu\nu}$. We note that

$$|G_{\mu\nu}(x)| = |g_{\mu\nu}(x)|, \tag{20}$$

hence $G_{\mu\nu}$ is unimodular for trivially embedded branes (in the preferred matrix coordinates x^μ) [8]. Since $\theta^{\mu\nu}$ does not enter the Riemannian volume at all, this leads to a very interesting mechanism for stabilizing flat space, which may hold the key for the cosmological constant problem as discussed below.

4. Covariant equations of motion

We are now in a position to rewrite the basic matrix e.o.m. (8) in a covariant way. For the scalar fields ϕ^i, the e.o.m. is

$$\begin{aligned} 0 &= [X^a,[X^b,\phi^i]]\eta_{ab} = [X^\mu,[X^\nu,\phi^i]]\eta_{\mu\nu} + [\phi^i,[\phi^j,\phi^i]]\,\delta_{ij} \\ &= i[X^\mu,\theta^{\nu\eta}\partial_\eta\phi^i]\eta_{\mu\nu} + i[\phi^i,\theta^{\nu\eta}\partial_\nu\phi^j\partial_\eta\phi^i]\,\delta_{ij} \\ &\sim -\theta^{\mu\rho}\partial_\rho(\theta^{\nu\eta}\partial_\eta\phi^i)\eta_{\mu\nu} - \theta^{\mu\rho}\partial_\mu\phi^i\partial_\rho(\theta^{\nu\eta}\partial_\nu\phi^j\partial_\eta\phi^i)\,\delta_{ij} \\ &= e^\sigma(\Gamma^\eta\partial_\eta\phi^i - G^{\rho\eta}\partial_\rho\partial_\eta\phi^i) = -e^\sigma\Delta_G\phi^i\,; \end{aligned} \tag{21}$$

for a detailed derivation see [2]. The same computation gives

$$\Delta_G x^\mu = 0 = \Gamma^\mu, \tag{22}$$

consistent with the ambiguity of the splitting $X^a = (X^\mu, \phi^i)$ into coordinates and scalar fields. Together with (21) this implies [2]

$$\boxed{G^{\gamma\eta}\,\nabla_\gamma(e^\sigma\theta^{-1}_{\eta\nu}) = e^{-\sigma}\,G_{\mu\nu}\,\theta^{\mu\gamma}\,\partial_\gamma\eta} \tag{23}$$

Here ∇ denotes the Levi-Civita connection with respect to $G_{\mu\nu}$. Remarkably, this equation is also a consequence of a "matrix" Noether theorem due to the symmetry $X^a \to X^a + c^a\mathbb{1}$, as shown in [1]. This means that equation (23) is protected from quantum corrections, and should be taken serious at the quantum level. It has the structure of covariant Maxwell equations, and gives the relation between the noncommutativity $\theta^{\mu\nu}(x)$ and the effective metric $G^{\mu\nu}$. For

[1] $G^{\mu\nu}$ corresponds to $\tilde{G}^{\mu\nu}$ in [1, 2].

given asymptotical behavior, $\theta^{\mu\nu}(x)$ should therefore be completely determined (apart from gravitational waves, see below) by the scalar functions $\eta(x)$ and e^{σ}.

We note that (21) and (22) have a simple interpretation: the embedding $\mathcal{M} \subset \mathbb{R}^D$ is harmonic w.r.t. $G_{\mu\nu}$. In particular, the dynamical matrices $X^\mu \sim x^\mu$ define harmonic coordinates for on-shell geometries, which in general relativity would be interpreted as gauge condition. However, (21) and (22) individually might be subject to quantum corrections as indicated below.

The main message is that the kinetic term in the matrix model necessarily involves the metric $G_{\mu\nu}(x)$. This also holds for gauge fields and fermions, possibly up to density factors. Therefore $G_{\mu\nu}$ must be interpreted as gravitational metric. It is dynamical, because it depends on the embedding fields ϕ^i and the Poisson structure $\theta^{\mu\nu}$, both of which are dynamical. Using a standard embedding theorem [11], it follows that $G_{\mu\nu}$ can describe in principle the most general metric in 4 dimensions for the case $D \geq 10$ (ignoring the e.o.m. for now). Therefore the matrix model provides a non-perturbative definition of a theory of space-time and gravity, which naturally includes also gauge fields and matter. The quantization can also be performed relatively easily in a perturbative manner, taking advantage of an alternative interpretation as noncommutative gauge theory: the dynamical Poisson structure can be parametrized in terms of a $u(1)$ gauge field as indicated below. This leads to the hope that this type of matrix model may serve as a model for quantum gravity, quite possibly more accessible than other, more conventional approaches.

5. Dynamical emergent gravity

Before making contact with real physics, we need to understand the dynamics of emergent gravity. Notice that we have not encountered anything like the Einstein-Hilbert action, and in fact it seems impossible to simply add such a term in the matrix model. This is of course extremely interesting: either the model will fail hopelessly, or it will provide a very remarkable and predictive new mechanism for gravity, free of theoretical prejudice.

We can give at least 2 arguments which suggest that the dynamics of the geometry resp. gravity in the matrix model should be quite close to GR:

(i) the dynamics of $\theta^{\mu\nu}(x)$ as given by the matrix model indeed implies $R_{\mu\nu} = 0$ at least for linearized metric fluctuations (30) around flat Moyal-Weyl space without matter. Moreover, it provides the 2 physical degrees of freedom for gravitons (30). This is a remarkable observation, essentially due to Rivelles [5]; see [1] for more details.

(ii) as soon as the matrix model is quantized, the effective action will contain an induced Einstein-Hilbert action. This is discussed below in more detail.

5.1. Quantization and induced gravity

The quantization of the matrix model (5) is defined by

$$Z = \int dX^a e^{-S_{YM}[X]}, \tag{24}$$

which is easily generalized to include fermions, notably in the IKKT model [10] for $D = 10$. As shown in [15], this integral is well-defined for finite N (in $D \geq 3$ dimensions), apart from the flat directions corresponding to $X^a \to X^a + c^a \mathbb{1}$ which could be handled using standard methods. This provides a non-perturbative definition of the model at the quantum level. To gain some insight, we consider a perturbative treatment around a given background as discussed above. For simplicity, consider the quantization of an additional scalar field φ coupled to the matrix model as in (15), which upon integration leads to an effective action

$$e^{-\Gamma_\varphi} = \int d\varphi \, e^{-S[\varphi]}, \quad \text{where} \quad \Gamma_\varphi = \frac{1}{2}\mathrm{Tr}\log\Delta_G\,. \tag{25}$$

A standard argument using the heat kernel expansion of Δ_G gives

$$\Gamma_\varphi = \frac{1}{16\pi^2}\int d^4x\sqrt{|G_{\mu\nu}|}\,\left(c_1\Lambda_1^4 + c_4 R[G]\,\Lambda_4^2 + O(\log\Lambda)\right) , \qquad (26)$$

which is also the general structure of the 1-loop effective action for the geometrical sector due to other fields. The coefficients c_i as well as the effective cutoffs Λ_i depend on the detailed field content of the model, and can be obtained essentially from Seeley-de Witt coefficients[2]. This is essentially the mechanism of induced gravity [14].

The action (26) suggests to relate the cutoff Λ_4^2 with the gravitational constant $\frac{1}{G}$. On the other hand, Λ_4 is expected to be the scale of $N=4$ SUSY breaking, since only $N=4$ supersymmetric models do not induce this term [7, 9]; this is related to the lack of UV/IR mixing in these models as discussed below. This suggests that in order to have a finite gravitational coupling constant, the model should have $N=4$ supersymmetry above a certain scale. This is realized in the IKKT model, where $D=10$. The scale Λ_1 may be different from Λ_4.

Note that a term $\int d^4x\sqrt{G}\,\Lambda^4$ is usually interpreted as cosmological constant, and its scaling with Λ^4 usually presents a major problem. This is nothing but the cosmological constant problem, which is arguably the biggest challenge to our understanding of gravity. We claim that this problem may be resolved here. To see this, recall $|G_{\mu\nu}| = |g_{\mu\nu}|$ (20), independent of $\theta^{\mu\nu}(x)$. Moreover,

$$\delta\int d^4x\sqrt{G} \sim \int d^4x\sqrt{g}g^{\mu\nu}\delta g_{\mu\nu} \sim \int d^4x\sqrt{g}\delta\phi^i\Delta_g\phi^j\delta_{ij}$$

vanishes for harmonic embeddings $\Delta_g\phi^i = 0$. Therefore for such backgrounds, this "would-be cosmological constant term" is irrelevant and does not enter the equations of motion for the geometry. For example, flat (Moyal-Weyl) space is a solution even at one loop[3], *without fine-tuning* Λ. Therefore minimally or harmonically embedded branes (w.r.t. $g_{\mu\nu}$) are *protected from the cosmological constant problem* [2]; the term $\int d^4x\sqrt{G}\,\Lambda^4$ is present but simply does not imply any cosmological constant. This is due to the particular parametrization of the geometry in terms of $\theta^{\mu\nu}$ and ϕ^i rather than a fundamental metric. Notice that $g_{\mu\nu}$ does not necessarily (but quite possibly) coincide with the effective metric $G_{\mu\nu}$. This makes the full analysis of the gravitational sector in the matrix model quite non-trivial, and more work is required. Once more realistic solutions are found which satisfy this stability condition, the present model would arguably be in much better agreement with observation than GR without fine-tuning.

6. Gauge fields

We now relate the above discussion with noncommutative gauge theory. The main result is that the matrix model naturally includes $su(n)$ gauge theory coupled to gravity. If the same model is (improperly) viewed as $u(n)$ gauge theory on $\mathbb{R}^4_\theta$, then the trace-$u(1)$ sector is afflicted by the notorious UV/IR mixing. This is nothing but a reflection of the induced gravitational action.

6.1. Would-be $u(1)$ gauge fields

Recall that a particular solution of the matrix equations of motion (8) is given by the generators $\bar X^\mu$ of the Moyal-Weyl quantum plane $\mathbb{R}^4_\theta$. Its effective geometry is indeed flat, given by

$$\bar G^{\mu\nu} = \bar\rho\,\bar\theta^{\mu\mu'}\,\bar\theta^{\nu\nu'}g_{\mu'\nu'} \qquad \bar\rho = (\det\bar\theta^{\mu\nu})^{-1/2} \equiv \Lambda_{NC}^4 \qquad (27)$$

[2] However, the contributions from fermions and the "would-be $U(1)$ fields" are non-standard as discussed in [9].

[3] And presumably to all loops; this can also be seen from the point of view of $U(1)$ gauge theory discussed below.

assuming trivial embedding. Fluctuations of X^a can be parametrized in terms of $u(1)$ gauge fields A_μ and scalar fields ϕ^i on $\mathbb{R}^4_\theta$, using

$$\begin{aligned} X^a &= (\bar{X}^\mu + \mathcal{A}^\mu, \phi^i) \\ [\bar{X}^\mu + \mathcal{A}^\mu, f] &= i\bar{\theta}^{\mu\nu}(\frac{\partial}{\partial \bar{x}^\nu} + i[A_\nu, .])f \equiv i\bar{\theta}^{\mu\nu} D_\nu f \end{aligned} \tag{28}$$

where $\mathcal{A}^a \sim \mathcal{A}^a(\bar{x}) \equiv -\bar{\theta}^{ab} A_b(\bar{x})$. The actions (5) could then be written as

$$S_{YM} = \int d^4\bar{x} \left(\bar{\rho}^{-1} \bar{G}^{\mu\mu'} \bar{G}^{\nu\nu'} F_{\mu\nu} F_{\mu'\nu'} + 2\bar{G}^{\mu\nu} D_\mu \phi^i D_\nu \phi^i \delta_{ij} - \bar{\rho}[\phi^i, \phi^j][\phi^{i'}, \phi^{j'}]\delta_{ij}\delta_{i'j'} + \bar{G}^{\mu\nu} g_{\mu\nu}\right)$$

where $F_{\mu\nu} = \partial_\mu A_\nu - \partial_\nu A_\mu + i[A_\mu, A_\nu]$ is the $u(1)$ field strength. This might suggest an interpretation as noncommutative $u(1)$ gauge theory on $\mathbb{R}^4_{\bar{\theta}}$. However, $A_\mu(\bar{x})$ is completely absorbed in the metric $G^{\mu\nu}(x)$ resp. $\theta^{\mu\nu}(x)$ in the proper geometrical interpretation, via

$$\theta^{\mu\nu}(x) = \bar{\theta}^{\mu\nu}(\bar{x}) - \bar{\theta}^{\mu\mu'}\bar{\theta}^{\nu\nu'} F_{\mu'\nu'}. \tag{29}$$

This explains why these would-be $u(1)$ gauge fields cannot be disentangled e.g. from the $su(n)$ components: all fields couple to the metric $G_{\mu\nu}$, which is a function of $F_{\mu\nu}$ via (29). In particular, an induced gravitational action (26) arises upon quantization, which does *not* simply renormalize the tree-level action. This is precisely the "strange" UV/IR contribution [16] to the effective action for the would-be $u(1)$ gauge fields in the IR limit [3].

Expanding the effective metric $G_{\mu\nu} = \bar{G}_{\mu\nu} + h_{\mu\nu}$ around the flat Moyal case to leading order in $F_{\mu\nu}$ leads to an expression for the linearized metric fluctuations

$$h_{\mu\nu} = -\bar{G}_{\nu\nu'}\bar{\theta}^{\nu'\rho}\bar{F}_{\rho\mu} - \bar{G}_{\mu\mu'}\bar{\theta}^{\mu'\rho}\bar{F}_{\rho\nu} \tag{30}$$

Remarkably, the equations of motion of the matrix model imply that the vacuum geometries are Ricci-flat [5],

$$R_{\mu\nu}[\bar{G} + h] = 0 \quad + O(\theta^2), \qquad \text{while} \quad R_{\mu\nu\rho\sigma} \neq 0. \tag{31}$$

6.2. Nonabelian gauge fields

Finally we show how $su(n)$-valued gauge fields arise in the same matrix model, on a suitable background. To avoid confusion we denote such a "nonabelian" background with

$$Y^a = \begin{pmatrix} Y^\mu \\ Y^i \end{pmatrix} = \begin{cases} X^\mu \otimes \mathbb{1}_n, & a = \mu = 1, ..., 4, \\ \phi^i \otimes \mathbb{1}_n, & a = 4 + i, \ i = 1, ..., D - 4. \end{cases} \tag{32}$$

This can be understood as n copies of the brane configurations considered above. We want to understand general fluctuations around this new background. Since the $u(1)$ components describe the geometry, we expect to find $su(n)$-valued gauge fields as well as scalar fields in the adjoint. It turns out that the following gives an appropriate parametrization of these general fluctuations:

$$\begin{pmatrix} Y^\mu \\ Y^i \end{pmatrix} = \begin{pmatrix} X^\mu \otimes \mathbb{1}_n + \mathcal{A}^\mu \\ \phi^i \otimes \mathbb{1}_n + \Phi^i + \mathcal{A}^\rho \partial_\rho(\phi^i \otimes \mathbb{1}_n + \Phi^i) \end{pmatrix} \tag{33}$$

where

$$\begin{aligned} \mathcal{A}^\mu &= \mathcal{A}^\mu_\alpha \otimes \lambda^\alpha = -\theta^{\mu\nu} A_{\nu,\alpha} \otimes \lambda^\alpha, \\ \Phi^i &= \Phi^i_\alpha \otimes \lambda^\alpha \end{aligned} \tag{34}$$

parametrize the $su(n)$-valued gauge fields resp. scalar fields, and λ^α denotes the generators of $su(n)$. This amounts to the leading term in a Seiberg-Witten map [13], relating noncommutative and commutative $su(n)$ gauge fields.

One can now show that the semi-classical limit of the matrix model action (5) for these $su(n)$-valued gauge fields A_μ on general 4-dimensional $\mathcal{M}_\theta \subset \mathbb{R}^D$ is given by

$$S_{YM}[\mathcal{A}] \sim \int d^4x \sqrt{|G_{\mu\nu}|}\, e^\sigma\, G^{\mu\mu'} G^{\nu\nu'}\, \mathrm{tr}(F_{\mu\nu}\, F_{\mu'\nu'}) + 2\int \eta(x)\ \mathrm{tr}\, F \wedge F\ . \tag{35}$$

where $F_{\mu\nu} = \partial_\mu A_\nu - \partial_\nu A_\mu + i[A_\mu, A_\nu]$ is the $su(n)$-valued field strength. This was shown in [8] for $D = 4$ using a direct but long computation, and in [1] for the general case via the equations of motion. A similar result applies for higher-dimensional branes. Remarkably, the corresponding Yang-Mills equations of motion are a direct consequence of a matrix Noether theorem corresponding to the symmetry $X^a \to X^a + c^a \mathbb{1}$ of the matrix model [1].

Finally, fermions are naturally included in these models [10, 2] and turn out to couple to the same metric $G_{\mu\nu}$, albeit possibly (depending on the geometry) with a non-standard spin connection; see [9] for results on the $D = 4$ case.

We conclude that Yang-Mills matrix models are strong candidates for a unified theory of fundamental interactions, and promise advantages over GR for quantization and the cosmological constant problem. A simple and intrinsically noncommutative mechanism for gravity has been identified. However, more work is required to obtain a thorough understanding and judgment.

Acknowledgments
This work was supported by FWF project P20017.

References
[1] Steinacker H 2008 Covariant Field Equations, Gauge Fields and Conservation Laws from Yang-Mills Matrix Models *Preprint* arXiv:0812.3761
[2] Steinacker H 2009 *Nucl. Phys.* B **810** (2009) 1 (*Preprint* arXiv:0806.2032)
[3] Grosse H, Steinacker H and Wohlgenannt M 2008 *JHEP* **0804** 023 (*Preprint* arXiv:0802.0973)
[4] for basic reviews see: Douglas M R and Nekrasov N A 2001 *Rev. Mod. Phys.* **73** 977 (*Preprint* arXiv:hep-th/0106048); Szabo R J 2003 *Phys. Rept.* **378** 207 (2003) (*Preprint* arXiv:hep-th/0109162)
[5] Rivelles V O 2003 *Phys. Lett.* B **558** 191 (*Preprint* arXiv:hep-th/0212262)
[6] Yang H S 2007 *Mod. Phys. Lett.* A **22** 1119 (*Preprint* arXiv:hep-th/0611174)
[7] Matusis A, Susskind L and Toumbas N 2000 *JHEP* **0012** 002 (*Preprint* arXiv:hep-th/0002075)
[8] Steinacker H 2007 *JHEP* **12** 049 (*Preprint* arXiv:0708.2426)
[9] Klammer D and Steinacker H 2008 *JHEP* **0808** 074 (*Preprint* arXiv:0805.1157)
[10] Ishibashi N, Kawai H, Kitazawa Y and Tsuchiya A 1997 *Nucl. Phys.* B **498** 467 (*Preprint* arXiv:hep-th/9612115)
[11] Clarke C J S 1970 *Proc. Roy. Soc. London* A **314** 417
[12] Kontsevich M 2003 *Lett. Math. Phys.* **66** 157 (*Preprint* arXiv:q-alg/9709040)
[13] Seiberg N and Witten E 1999 *JHEP* **9909** 032 (*Preprint* hep-th/9908142)
[14] Sakharov A D 1968 *Sov. Phys. Dokl.* **12** 1040 [*do.* 1967 *Dokl. Akad. Nauk Ser. Fiz.* **177**]
[15] Austing P and Wheater J F 2001 *JHEP* **0104** 019 (*Preprint* arXiv:hep-th/0103159)
[16] Minwalla S, Van Raamsdonk M and Seiberg N 2000 *JHEP* **0002** 020 (*Preprint* arXiv:hep-th/9912072)

Fourth International Workshop DICE2008 IOP Publishing
Journal of Physics: Conference Series **174** (2009) 012045 doi:10.1088/1742-6596/174/1/012045

The emergence of gravity as a retro-causal post-inflation macro-quantum-coherent holographic vacuum Higgs-Goldstone field

Jack Sarfatti[1] **and Creon Levit**[2]

[1] Internet Science Education Project
[2] NASA Ames Research Center

E-mail: `adastra1@mac.com` `creon.levit@nasa.gov`

Abstract.
We present a model for the origin of gravity, dark energy and dark matter: Dark energy and dark matter are residual pre-inflation false vacuum random zero point energy ($w = -1$) of large-scale negative, and short-scale positive pressure, respectively, corresponding to the "zero point" (incoherent) component of a superfluid (supersolid) ground state. Gravity, in contrast, arises from the 2nd order topological defects in the post-inflation virtual "condensate" (coherent) component. We predict, as a consequence, that the LHC will never detect exotic real on-mass-shell particles that can explain dark matter $\Omega_{\mathrm{DM}} \approx 0.23$. We also point out that the future holographic dark energy de Sitter horizon is a total absorber (in the sense of retro-causal Wheeler-Feynman action-at-a-distance electrodynamics) because it is an infinite redshift surface for static detectors. Therefore, the advanced Hawking-Unruh thermal radiation from the future de Sitter horizon is a candidate for the negative pressure dark vacuum energy.

1. Gravity from topological singularities in the quantum vacuum

We consider the possibility that the Einstein-Cartan 1-forms consistent with 1915 General Relativity (GR) are local macro-quantum emergent *supersolid* [1] c-number fields. We mean this in the same sense that v (the locally irrotational superflow 3D Galilean relativity group velocity 1-form in superfluid ^{4}He) is emergent, with quantized circulation. The single-valuedness of the associated S^1 macroquantum coherent Higgs-Goldstone order parameter $\Psi = |\Psi|e^{i\Theta}$ emerges from an effective spontaneous broken [2] non-electromagnetic $U(1) = O(2)$ ground state gauge symmetry[1].

$$v = \frac{1}{2\pi}\frac{h}{m}\mathrm{d}\Theta \tag{1}$$

$$\oint v = \frac{1}{2\pi}\frac{h}{m}\oint d\Theta = n\frac{h}{m} \tag{2}$$

$$n = \pm1, \pm2, ... \tag{3}$$

[1] This corresponds to Hagen Kleinert's multi-valued singular phase transformations (discussed elsewhere in these proceedings).

In analogy to the above, we use a phenomenological model for the moment of inflation with eight macroquantum coherent relative phase 0-forms Θ^I and Φ^I that form two Lorentz group 4-vectors with magnitudes Θ and Φ, respectively. These magnitudes, in turn, are the phases of a S^2 vacuum order parameter manifold. This S^2 fiber bundle over real spacetime supports stable point monopole topological defects (simultaneous nodes of the three corresponding Higgs fields) [3].

These GeoMetroDynamic (GMD) point monopoles are the lattice points in spacelike slices of Hagen Kleinert's "world crystal lattice" [4]. They correspond to a non-trivial 2nd homotopy group of emergent effective post-inflation field $O(3)$ mappings of surrounding non-bounding 2-cycles $S^2{}_{3D}$ in 3D physical space to the vacuum manifold S^2.

The inhomogeneities in all eight phases Θ^I and Φ^J form the emergent GMD tetrad field A^I. For details, see equations 14-20, below.

In the world hologram conjecture [5][6], with total hologram screen area A, the mean separation ΔL of the lattice points is for our pocket universe $c/H_0 = \sqrt{N}L_P = 10^{29}$ cm on the cosmic landscape of the megaverse, given by[7]:

$$\Delta L = \left(L_P{}^2\sqrt{A}\right)^{1/3} = \left(L_P{}^2\frac{c}{H_0}\right)^{1/3} \approx 10^{-13}\text{cm} \tag{4}$$

$$N = \frac{A}{4L_P{}^2} = \frac{1}{4\Lambda L_P{}^2} = \frac{A^{(3/2)}}{(\Delta L)^3} = \frac{V}{(\Delta L)^3} \tag{5}$$

$$A = \partial V \tag{6}$$

where ∂ is the quasi-boundary operator because the surrounding future light cone surface is a non-bounding de Rham 2-cycle. This 2-cycle encloses N GMD point monopole defects in the three effective real $O(3)$ Higgs field macroquantum coherent Penrose-Onsager off-diagonal-long-range-order post-inflation vacuum parameters[8].

Wrapping once around $S^2{}_{3D}$ through solid angle 4π wraps an integer N times round the vacuum manifold S^2. This is analogous to the global quantized circulation vortices in superfluid ^{4}He that are the stable topological defects in the first homotopy group for an $O(2)$ mapping with only a single relative Goldstone phase Θ and two real Higgs scalars Ψ_1, Ψ_2 (instead of the three Ψ_1, Ψ_2,Ψ_3 with two relative Goldstone phases Θ, Φ over 3D spacelike slices of physical spacetime in our toy model).

This wrapping integer N is the explanation for the Bekenstein bit quantized areas of null black hole event horizons and observer-dependent cosmological (e.g. dark energy future de Sitter) horizons[2] of area A that obey Hawking's entropy and temperature formulas

$$S = k_B A/4L_P{}^2 = Nk_B \tag{7}$$

$$T = \frac{\partial E}{\partial S} = \frac{hc}{k_B\sqrt{A}} = \frac{hc\sqrt{\Lambda}}{k_B} = \frac{hc}{k_B L_P\sqrt{N}} \tag{8}$$

$$\oiint 2d\Theta \wedge d\Phi = 4\pi N = A/4L_P{}^2 = 10^{124} \text{ Bekenstein BITs,} \tag{9}$$

where the double integral around the vacuum manifold is induced by a single wrap around the future asymptotic de Sitter horizon. The de Sitter horizon is a surrounding (but non-bounding [9]) closed 2-cycle at lightlike conformal infinity. It is also a stretched thermal horizon for comoving observers in the accelerating Hubble flow [5].

[2] In this paper we use the term "de Sitter horizon" informally. A more precise description would be "future cosmological event horizon".

Journal of Physics: Conference Series **174** (2009) 012045 doi:10.1088/1742-6596/174/1/012045

The remaining six Goldstone phase angles form the Calabi-Yau space of string theory - the same field Gennady Shipov calls the "oriented point" [10]. Einstein's 1915 curvature field is simply the local gauge field from the 4-parameter translation universal spacetime symmetry group $T(4)$ for all matter fields (i.e., strong equivalence principle) with the constraint of zero torsion.

Locally gauging the 10-parameter Poincare group $P(10)$ of Einstein's 1905 special relativity gives the Einstein-Cartan theory of (*dislocation* defect) torsion[4] in addition to (*disclination* defect) curvature [3] of the symmetric Levi-Civita connection. Indeed, this local gauge field model can be reinterpreted in terms of the eight multi-valued Goldstone phases of the coherent post-inflation vacuum field. The Calabi-Yau space seems to be simply the torsion field in disguise.

2. Dark energy from the future

The future de Sitter event horizon world hologram is "our past light cone at the end of time"[11]. It can be pictured as a pixelated spherical shell of area $NL_P{}^2$ infinitely far from our detectors (in proper time) on their *future* light cone, with thickness L_P and duration L_P/c. This shell, or "screen", has 4D volume $NL_P{}^4$ with dark energy density hc/(4DVolume Hologram Screen). This screen projects the voxels of our accelerating expanding 3D space hologram image back from the future - indeed, back to the moment of inflation 13.7 billion years ago in what Igor Novikov calls a "globally self-consistent" strange loop in time.

To summarize: The area of an observer's future de Sitter horizon holographically determines the dark energy density seen by that observer.

For a static local-non-inertial-frame (LNIF) observer (with covariant acceleration $g = c^2/\sqrt{N}L_P = cH_0 \approx 10^{-9}$ m/sec^2 relative to the $\Lambda = 0$, $k = 0$ spatially flat post-inflation background Friedman metric) the Hawking-Unruh temperature of the future de Sitter horizon is proportional to his acceleration. This is in similar to a static outside observer adiabatically approaching the horizon of a black hole – being "slowly lowered down on a cable" – who measures a temperature which approaches the Planck temperature $hc/L_P k_B$ as he approaches the horizon.

Of course, the locally coincident geodesic observer relative to the Friedman metric sees no heat radiation - only $w = -1$ positive dark zero point energy density vibrations of equal but opposite negative pressure per large space dimension, as required by the Einstein Equivalence Principle (EEP).

In contrast the NASA WMAP isotropic black body radiation in the comoving Friedman frame is coming along our past light cone from the surface of last contact 380,000 years after the post-inflation reheating of the Big Bang, and its energy density weakens $a(t)^{-4}$ as space expands because it has $w = +1/3$ ratio of pressure to energy density[4].

[3] There are two classes of defects: The monopole defects which form the "atoms" of the supersolid world crystal lattice, and the disclinations and dislocations in this lattice, which account for, respectively, the curvature and torsion of spacetime.

[4] There is a lack of consensus and clarity in the literature on who sees what. The Unruh effect in globally flat Minkowski spacetime is: A covariantly proper accelerating local detector (not on a timelike geodesic, which by definition has zero covariant proper 4-acceleration) sees thermal equilibrium blackbody radiation whose temperature is proportional to its covariant 4-acceleration magnitude. In contrast, a momentarily coincident non-accelerating detector sees only zero point vacuum fluctuations instead of the thermal radiation. That is, some of the vacuum fluctuation energy is converted into thermal radiation in the rest frame of the intrinsically accelerating detector.

Static detectors outside the event horizon of an ideal Schwarzschild black hole are covariantly properly accelerating in order to "stand still" at a fixed Schwarzschild radial coordinate r. This is in accord with the actual Pound-Rebka Harvard Tower experiment showing the gravity redshift using the nuclear Mossbauer effect. Furthermore, the comoving detectors in the Robertson-Walker representation at constant χ are analogous to the previous case at constant r [11]:

$$ds^2 = -c^2dt^2 + R(t)^2[d\chi^2 + S_k^2(\chi)d\psi^2],$$

We propose that the dark energy zero point vacuum fluctuations measured by non-rotating covariantly unaccelerated Local Inertial Frame (LIF) detectors (on timelike geodesics relative to the physical spacetime multi-valued connection local gauge field $\Gamma^{\mu}{}_{\nu\lambda}$ that forms curved and torsioned spacetime) appear as advanced Wheeler-Feynman quasi-thermal blackbody "Unruh radiation". It comes back from the future de Sitter horizon "perfect absorber" with temperature that has order of magnitude

$$T = \frac{\mathrm{hg}}{\mathrm{ck}_B} \to \frac{\mathrm{hc}}{\mathrm{k}_B \mathrm{L}_P \sqrt{N}} \approx \frac{\mathrm{hc}}{\mathrm{k}_B \mathrm{L}_P 10^{62}} \tag{10}$$

for covariantly accelerated Local Non-Inertial Frame (LNIF) detectors off timelike geodesics.

For example, a static observer outside the event horizon of a non-rotating black hole must covariantly accelerate away from the black hole radially with

$$g = -\frac{GM}{r^2} \frac{1}{\sqrt{1 - \frac{2GM}{c^2 r}}} \sim T \tag{11}$$

in order to stay at fixed r in the curved spacetime outside the black hole. They need to fire their rocket engines in order to remain static. These static LNIF observers see the event horizon as a "stretched membrane" with Unruh temperature T. Coincident LIF observers do not see this at all. This is an example of what Leonard Susskind calls "horizon complementarity" [5], in analogy with Niels Bohr's quantum complementarity of wave-particle duality from the non-commutativity of the Lie algebra of observable operators on qubit Hilbert space fibers over classical field configuration space.

The comoving observers that see an approximately isotropic WMAP Cosmic Microwave Background (CMB) in our accelerating expanding universe are analogous to the static LNIF observers in the Schwarzschild model where now there is a universal acceleration $g \approx 10^{-9}$ m/sec^2, which is the same order of magnitude of the anomalous Sun-centered radial accelerations of the two NASA Pioneer space probes beyond the orbit of Jupiter. This is a curious coincidence that possibly has deeper significance, although it is surprising to find Hubble's parameter H_0 appearing on such a short scale local metric field. In contrast LIF detectors see this advanced quasi-thermal Unruh Wheeler-Feynman radiation as zero point vacuum fluctuation energy density

$$\rho_{\mathrm{DE}} = \text{string tension} \times \text{vacuum curvature} = \frac{\text{string tension}}{\text{area of future cosmic horizon}} \tag{12}$$

$$= \frac{\mathrm{hc}}{\mathrm{NL}_P{}^4} = \frac{\mathrm{hc}}{(10^{-2}\mathrm{cm})^4} = 0.73 \times 10^{-29} \text{grams/cc}.$$

It is as if there is an effective high frequency cutoff at $c/10^{-2}$cm $= 3 \times 10^{12}$ Hz for the $w = -1$ zero point dark energy virtual photon vibrations with critical wavelength equal to the geometric mean of the future de Sitter horizon scale and the Planck scale. The world hologram model posits that the number N of interior 3D voxels N of size ΔL *equals* the number of 2D pixels of size L_P on the world hologram future de Sitter horizon.

However, e.g., Davies & Davis [12] write "the response of a particle detector travelling along a geodesic in a de Sitter invariant vacuum state; the detector behaves as if immersed in a bath of thermal radiation" Of course,"geodesic" depends on choice of the local GCT frame invariant gravity tetrad field. Thus, the detector on a geodesic in the de Sitter gravity tetrad field is actually properly accelerating with respect to the geodesic in the zero cosmological constant Robertson-Walker gravity tetrad field. This is the point of view we take in this paper and it agrees operationally with how the dark energy data is actually interpreted.

Journal of Physics: Conference Series **174** (2009) 012045 doi:10.1088/1742-6596/174/1/012045

3. Calabi-Yau from torsion. Brane theory from the 1970s

The four tetrad 1-form fields e^I are the General Coordinate Invariant (GCI) gravitational fields in Einstein's 1915 GR. They form a single 4-vector under the 6-parameter homogeneous Lorentz group $SO(1,3)$. The non-trivial curvilinear 4D General Coordinate Transformations (GCT) connect covariantly accelerating coincident LNIFs with g-forces on its rest detectors. A non-gravity force is required to create a translational covariant acceleration. Conservation of angular momentum maintains a rotating LNIF in the absence of friction in deep space once the external torque is removed. The Lorentz group transformations connect coincident covariantly non-accelerating LIFs with vanishing g-forces. The tetrad field components $e^{\mu}{}_I$ and their inverses $e^I{}_{\mu}$ connect locally coincident LIFs with LNIFs. The LNIF curvilinear metric field is $g_{\mu\nu}$. The coincident LIF Center Of Mass (COM) metric η_{IJ} is that of Minkowski space-time of Einstein's 1905 SR. The Strong Equivalence Principle (SEP) implies for the absolute differential local frame invariant ds

$$ds^2 = g_{\mu\nu}(\text{LNIF})dx^{\mu}dx^{\nu} = \eta_{IJ}(\text{LIF})e^I e^J \tag{13}$$

The multi-valued Goldstone phase transformations in our toy model form a 4x4 M-Matrix of non-closed 1-forms where the non-trivial parts of the four curvature-only tetrad 1-forms A^I and the six non-trivial torsion field spin connection 1-forms ϖ^{IJ} = -ϖ^{JI} are the diagonals and antisymmetrized off-diagonal M-Matrix elements.

$$M^{IJ} = d\Theta^I \wedge \Phi^J - \Theta^I \wedge d\Phi^J \tag{14}$$

$$dM^{IJ} = -2d\Theta^I \wedge d\Phi^J \tag{15}$$

$$d^2 = 0 \tag{16}$$

$$A^I = diag(M^{IJ}) \tag{17}$$

$$e^I = I^I + A^I = e^I{}_{\mu}e^{\mu}{}_{\text{LNIF}} = e^{\mu}{}_I e^I{}_{\text{LIF}} \tag{18}$$

$$I^{\mu}{}_I = \delta^{\mu}{}_I \tag{19}$$

$$\varpi^{IJ} = M^{[I,J]} \tag{20}$$

The Einstein 1915 zero torsion field i.e. $\varpi^{IJ} = 0$, curvature field 2-form is

$$R^{IJ} = D\omega^{IJ} = d\omega^{IJ} + \omega^I{}_K \wedge \omega^{KJ} \tag{21}$$

Where the torsion field 2-form in Einstein-Cartan theory beyond 1915 GR would be

$$\varpi^I = De^I = \varpi^I{}_K \wedge e^K \tag{22}$$

The 1915 GR Einstein-Hilbert pure gravity field action density is the 0-form

$$L_G = \epsilon_{IJKL}R^{IJ} \wedge e^K \wedge e^L \tag{23}$$

$$S_G = \int L_G d^4x. \tag{24}$$

$$\frac{\delta S_G}{\delta e^I} = 0 \tag{25}$$

is the pre-Feynman action principle. It is the critical point pure gravity vacuum classical field equation in the absence of matter field sources, that in the usual tensor notation is

$$R_{\mu\nu} = 0. \tag{26}$$

Adding the on-mass-shell matter fields of real quanta plus their off-mass-shell virtual quanta, which contribute to the cosmological scalar $\Lambda_{\rm zpf}$ field, replaces the above classical vacuum curvature field equation. The vanishing functional derivative of the *total* action S with respect to the 4 GCT invariant tetrad 1-forms[5] i.e:

$$\frac{\delta S}{\delta e^I} = 0 \tag{27}$$

$$S = \mathrm{S}_G + \mathrm{S}_M + \mathrm{S}_{\rm zpf} \tag{28}$$

$$\mathrm{R}_{\mu\nu} - \left(\frac{1}{2}R + \Lambda_{\rm zpf}\right)\mathrm{g}_{\mu\nu} = -\frac{8\pi G}{\mathrm{c}^4}\mathrm{T}_{\mu\nu} \tag{29}$$

We can no longer assume the zero torsion field limit ($\varpi^{\rm IJ} = 0$) of the Bianchi identity of 1915 GR. And so we dream that the final theory will use a more general connection than that of Levi-Civita. This more general connection is induced by locally gauging a more general spacetime symmetry group (e.g. the Poincare[13], de Sitter or perhaps the conformal group) instead of gauging $T(4)$ as is usually done in GR.

We use ";" and "|" to denote covariant differentiation with respect to the LC connection and the more general connection, respectively, and so rewrite the divergence of the Einstein equation as:

$$\mathrm{G}_{\mu\nu}{}^{|\nu} + \frac{\partial\Lambda_{\rm zpf}}{\partial \mathrm{x}^\nu} g^\nu_\mu + \frac{8\pi G}{\mathrm{c}^4}\mathrm{T}_{\mu\nu}{}^{|\nu} = 0 \tag{30}$$

This more general divergence (flow) equation with its additional stress-energy currents suggests new channels inter-connecting the incoherent random vacuum zero point fluctuations to the smooth coherent (generalized) curvature field. However, it must be supplemented by a similar torsion field equation that comes from the vanishing functional derivative of the total action S with respect to the 6 dynamically independent spin connection ϖ^{IJ} 1-forms. The ϖ^{IJ} form the compensating gauge field induced from localizing the 6-parameter homogeneous Lorentz group $SO(1,3)$ that also acts equally on all matter field actions, i.e.,

$$\frac{\delta S}{\delta \varpi^{\rm IJ}} = 0 \tag{31}$$

The extra-dimensional brane M-Theory, when mature, should predict a renormalization group flow to larger gravity coupling strength in the short-distance limit. This, when combined with the world hologram conjecture, suggests Abdus Salam's 1973 bi-metric f-gravity [14] formula

$$G = \mathrm{G}_{\rm Newton}(1 + \alpha e^{-r/\Delta L}) \tag{32}$$

$$\alpha \gg 1$$

has been rediscovered by brane theorists [15] looking at the Calabi-Yau (torsion) field.

4. Conclusions

In this model there is no quantum gravity in the usual sense of starting with a classical field and quantizing it. Rather, we go the opposite way in the spirit, though not the letter, of Sakharov's 1967 proposal [16]. What hitherto was called the classical gravity field is seen to be really an emergent effective *macro-quantum coherent* c-number post-inflation vacuum field.

[5] The tetrad 1-forms are the compensating gauge fields from localizing the global space-time universal translation symmetry group acting equally on all matter field actions.

We claim the residual random negative-zero-point-pressure advanced virtual bosons *back-from-the-future* manifest as the anti-gravitating universally repulsive dark energy. This is because the future de Sitter horizon for a co-moving observer in our universe is a Wheeler-Feynman perfect absorber – an infinite red shift surface – just like a black hole event horizon is for a static LNIF observer.

In contrast, we claim the universally attracting dark matter comes from residual positive-zero-point-pressure virtual fermion-antifermion pairs. In this picture, looking for real on-mass-shell dark matter particles in the LHC or in underground WIMP detectors is like looking for the motion of the Earth through the mechanical aether of Galilean relativity using Michelson and Morley's Victorian interferometer.

This model pocket universe, relative to our Earth-bound detectors, is created nonlocally and self-consistently by what Igor Novikov and Kip Thorne call "a loop in time" [17]. Advanced Hawking radiation is emitted from the future cosmic ("de Sitter") horizon at all times and is blue-shifted as it travels into the past. However, only the advanced radiation emitted at time $t_{trigger} \approx 4$ Gyr arrives back at $t = 0$, where it ignites the big bang (see figure 1.1 in [11]).

Advanced information (Hawking radiation) flows from the infinite future proper time "Omega" S^2 horizon of the observer's world line (shown in the lower "conformal time" (τ) diagram in [11] Fig 1.1) down the null cosmic event horizon to its intersection with the null particle horizon (at $t_{trigger} \approx$ 4Gyr, $\tau \approx$ 32Gyr). It flows back along the particle horizon, winding up at the initial "Alpha" moment of inflation completing what John Cramer calls a "transaction"[18].

Furthermore, looking at Fig 5.1 of [11] we see that the future cosmic de Sitter horizon area $A_c \sim N(t)$, where t is the proper cosmic Robertson-Walker time (corresponding to spacelike hypersurfaces of maximal CMB isotropy), rises from "zero" (1 Bit) to de Sitter asymptote $N(t \to \inf) \approx 10^{124}$ Bits rather quickly. The advanced dark energy thermal Hawking radiation reaching us now backward-through-time along our future light cone is very close to the asymptotic value.

The final entropy of our (retro)-causal[18][19] universe is $Nk_B \approx 10^{124}$ Bekenstein bits, as mandated by the area of the future de Sitter horizon, whereas the initial entropy of the universe is exactly $k_B = 1$ bit, as mandated by the (Planck) area of the initial singularity.

This shows very clearly why the cosmological arrow of time is aligned with the thermodynamic arrow of time, solving Roger Penrose's main objection [20] to inflationary cosmology: why the early universe has relatively low entropy. We need both retrocausality and the world hologram principle to properly understand the Arrow of Time of the Second Law of Thermodynamics.

5. Background related reading

The works of John Wheeler and Richard Feynman [21], Fred Hoyle and Jayant Narlikar [22], James Woodward [23], Michael Ibison [24], Robert Becker [25] and John Cramer [18] on advanced electromagnetic waves, radiation reaction and vacuum fluctuations, Leonard Susskind [5] and Jack Ng's work on the world hologram [7], Gennady Shipov on torsion fields [10], and finally Hagen Kleinert's work on singular multi-valued phase transformations as local gauge transformations [4] are essential background reading. We have also found Chapter 2 of Rovelli's lectures on quantum gravity [26] very clear regarding the use of Cartan forms in gravity theory.

During the preparation of this paper we be became aware of a publication entitled: "Is dark energy from cosmic Hawking radiation?"[27] that relates $w = -1$ dark energy to observer-dependent Hawking radiation. However, those authors do not clearly specify which horizon they are referring to. They omit the (key) notion of retro-causality. We claim retrocausality is necessarily implied when the correct cosmic horizon - the *future lightlike* de Sitter horizon – is specified. Tamara Davis's thesis [11] (especially her figures 1.1 and 5.1) clarifies much of the rampant confusion over this subtle issue.

References

[1] Sarfatti J 1969 *Phys. Lett.* A **30** 300
[2] Sarfatti J and Stoneham M 1967 *Proc. Phys. Soc.* **91** 214
[3] Thouless D 1998 *Topological Quantum Numbers in Nonrelativistic Physics* (River Edge, NJ: World Scientific)
[4] Kleinert H 2008 *Multivalued Fields in Condensed Matter, Electromagnetism, and Gravitation* (River Edge, NJ: World Scientific)
[5] Susskind L and Lindesay J 2005 *An Introduction to Black Holes, Information and the String Theory Revolution* (River Edge, NJ: World Scientific)
[6] Hogan C 2008 *Phys. Rev.* D **77** 104031
[7] Ng Y 2008 *preprint* arXiv:0808.1261v1
[8] Yang C 1962 *Rev. Mod. Phys.* **34** 694
[9] Kiehn R 2008 *Non-equilibrium Thermodynamics from the Perspective of Continuous Topological Evolution: Vol 1 Non-Equilibrium Systems and Irreversible Processes* (http:www.lulu.com/kiehn)
[10] Shipov G 1998 *A Theory of Physical Vacuum* (Moscow: Russian Academy of Natural Sciences)
[11] Davis T 2004 *Fundamental Aspects of the Expansion of the Universe and Cosmic Horizons* Ph.D dissertation (University of New South Wales) (*Preprint* arXiv:astro-ph/0402278v1)
[12] Davies P and Davis T 2003 *preprint* arXiv:astro-ph/0310522v1
[13] Kibble T 1961 *J. Math. Phys.* **2** 212
[14] Salam A and Strathdee J 1977 *Phys. Rev.* D **16** 2668
[15] Esposito-Farese G 2005 *Proc. 33rd SLAC Summer Inst. on Particle Physics: Gravity In The Quantum World And The Cosmos* ed Hewett J *et al* (Menlo Park, CA: SLAC) *preprint* http://www.slac.stanford.edu/econf/C0507252/proceedings.htm
[16] Sakharov A 1968 *Sov. Phys. Doklady* **12** 1040
[17] Hawking S, Thorne K, Novikov I, Ferris T, Lightman A and Price R 2002 *The Future of Spacetime* (New York: W. W. Norton)
[18] Cramer J 1986 *Rev. Mod. Phys.* **58** 647
[19] Price H 2008 *preprint* arXiv:0802.3230v1
[20] Penrose R 2005 *The Road to Reality: A Complete Guide to the Laws of the Universe* (New York: Alfred A. Knopf)
[21] Wheeler J and Feynman R 1945 *Rev. Mod. Phys.* **17** 157; *do.* 1949 *Rev. Mod. Phys.* **21** 425
[22] Hoyle F and Narlikar J 1995 *Rev. Mod. Phys.* **67** 113
[23] Woodward J 1996 *Found. Phys. Lett.* **9** 1
[24] Ibison M 2006 *AAAS Conf. Proc. on Reverse Causation* (*preprint* arXiv:0705.0083v1)
[25] Becker R 2003 *Proc. Int. High Frequency Gravitational Wave Working Group* ed P Murad and R M L Baker Jr. (McLean, VA: Mitre Corp.)
[26] Rovelli C 2007 *Quantum Gravity* (Cambridge UK: Cambridge U. Press)
[27] Lee L *et al* 2008 *preprint* arXiv:0803.1987v4

Journal of Physics: Conference Series **174** (2009) 012046 doi:10.1088/1742-6596/174/1/012046

Path integrals on causal sets

Steven Johnston

Blackett Laboratory, Imperial College London, Prince Consort Road, London, SW7 2AZ, UK.

E-mail: `steven.johnston02@imperial.ac.uk`

Abstract. We describe a quantum mechanical model for particle propagation on a causal set. The model involves calculating a particle propagator by summing amplitudes assigned to trajectories within the causal set. This "discrete path integral" is calculated using a matrix geometric series. Amplitudes are given which, when the causal set is generated by sprinkling points into 1+1 or 3+1 Minkowski spacetime, ensure the particle propagator agrees, in a suitable sense, with the retarded causal propagator for the Klein-Gordon equation.

1. Introduction

Causal set theory provides a model for discrete spacetime. There are a number of motivations for postulating that spacetime has a discrete structure (see, for example, [1, 2]). One of the main motivations is that the discreteness could regularise the divergences present in quantum field theory. Before such an aim can be achieved, however, a version of quantum field theory suitable for the discrete spacetime background must be developed.

In this work we perform quantum mechanical path integrals (or rather path *sums*) on a causal set. Due to the discrete nature of a causal set these can be defined as actual sums over all particle trajectories (which move forwards in time) leading from one spacetime event to another. A particle propagator is defined by summing amplitudes assigned to particle trajectories which move forwards in time. We present amplitudes which ensure that, for causal sets generated by sprinklings into 1+1 and 3+1 dimensional Minkowski spacetime, this propagator gives (in a suitable sense) the Klein-Gordon retarded causal propagator from quantum field theory. This may provide a route to defining quantum field theory on a causal set.

In this work we take the viewpoint that matter is modelled as point particles (rather than fields or potentials) propagating according to quantum mechanical laws. This is in keeping with the propagator approach to quantum field theory pioneered by Feynman (See [3]). A particle description rather than a field description is also, perhaps, more in keeping with the "discrete" causal set philosophy.

The work presented here and at DICE2008 is a summary of the results obtained in [4].

1.1. Definitions

A *causal set* (or *causet*) is a locally finite partially ordered set. This means it is a pair $(C, \preceq)$ with a set C and a partial order relation $\preceq$ defined on C that is

- Reflexive: $x \preceq x$,
- Antisymmetric: $x \preceq y \preceq x$ implies $x = y$,
- Transitive: $x \preceq y \preceq z$ implies $x \preceq z$,

- Locally finite: $|\{z \in C | x \preceq z \preceq y\}| < \infty$

for all $x, y, z \in C$. Here $|A|$ denotes the cardinality of a set A. We write $x \prec y$ to mean $x \preceq y$ and $x \neq y$.

The set C represents the set of spacetime events and the partial order $\preceq$ represents the causal order between pairs of events. If $x \preceq y$ we say "x precedes y" or "x is to the causal past of y".

A *chain* in a causal set C is a subset in which any two elements are related. If the chain is finite it is a sequence of distinct elements $v_{i_0}, \ldots, v_{i_n}$ each of which precedes the next element in the sequence: $v_{i_0} \prec v_{i_1} \prec v_{i_2} \prec \ldots \prec v_{i_n}$. The length of the chain is n.

A *link* (or *covering relation*) between $u, v \in C$ is a relation $u \prec v$ such that there exists no $w \in C$ with $u \prec w \prec v$. We say u and v are *nearest neighbours* (or v *covers* u) and write $u \prec\!* \, v$.

A *path* is a subset $P \subset C$ which is a maximal (or saturated) chain. This means P is a chain with no element $w \in C - P$ such that $u \prec w \prec v$ for some $u, v \in P$. If the path is finite it is a sequence of distinct elements $v_{i_0}, \ldots, v_{i_n}$ each of which is linked to the next element in the sequence: $v_{i_0} \prec\!* \, v_{i_1} \prec\!* \, v_{i_2} \prec\!* \, \ldots \prec\!* \, v_{i_n}$. The length of the path is n, the number of links.

1.2. Example

We'll consider, as an example, a particular 5-element causal set $(\mathcal{C}, \preceq)$ with

$$\mathcal{C} = \{v_1, v_2, v_3, v_4, v_5\}, \tag{1}$$

and relations

$$\begin{aligned} & v_1 \preceq v_1, v_1 \preceq v_2, v_1 \preceq v_3, v_1 \preceq v_4, v_1 \preceq v_5 \\ & v_2 \preceq v_2, v_2 \preceq v_3, v_2 \preceq v_5, \\ & v_3 \preceq v_3, v_3 \preceq v_5, \\ & v_4 \preceq v_4, v_4 \preceq v_5, \\ & v_5 \preceq v_5. \end{aligned} \tag{2}$$

We can represent a causal set as a directed graph in which the causal set elements are drawn as vertices and relations between distinct pairs of elements are drawn as directed lines between vertices. The positioning of the vertices in the page is of no importance—the graph merely serves as a visualisation of the causal relations. Figure 1 represents our example causal set.

2. Adjacency matrices

A useful way to represent a finite causal set is by its *adjacency matrix*. We now define two types of adjacency matrices that will be used to compute the path integrals in Section 3. We'll look at finite causal sets as this allows us to work with finite square matrices. The generalisation to arbitrary causal sets is given in [4, Section 4].

Before defining the adjacency matrices we must label the elements in our causal set. For a finite causal set C with p elements we choose a labelling $v_1, \ldots, v_p$ for the elements of C. There are then two $p \times p$ adjacency matrices of interest[1]: C and L.

The *causal matrix* is defined by

$$C_{ij} := \begin{cases} 1 & \text{if } v_i \prec v_j \\ 0 & \text{otherwise.} \end{cases} \tag{3}$$

The *link* matrix is defined by

$$L_{ij} := \begin{cases} 1 & \text{if } v_i \prec\!* \, v_j \\ 0 & \text{otherwise.} \end{cases} \tag{4}$$

[1] In [4] we write A_C instead of C and A_R instead of L

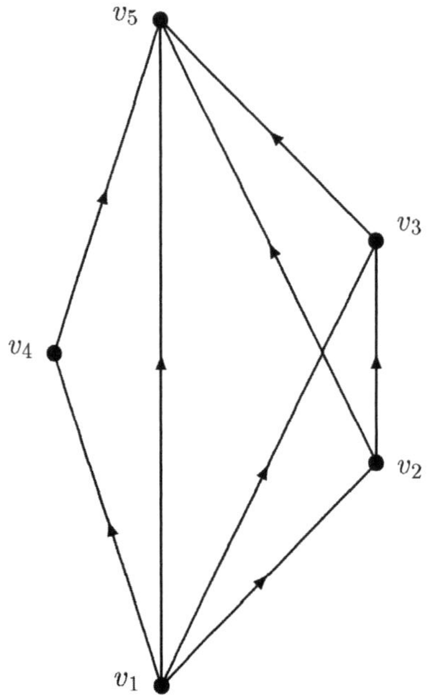

Figure 1. A directed graph representing our example causal set.

These are both zero on the main diagonal and, from the definition of a causal set, the labelling can always be chosen to ensure they are both strictly upper triangular matrices (in which case the labelling is called a *natural labelling*).

Powers of these matrices have the following useful properties[5]:

$$(C^n)_{ij} = \text{The number of chains of length } n \text{ from } v_i \text{ to } v_j, \tag{5}$$

$$(L^n)_{ij} = \text{The number of paths of length } n \text{ from } v_i \text{ to } v_j. \tag{6}$$

2.1. Example

For our example causal set $(\mathcal{C}, \preceq)$ we have

$$C = \begin{pmatrix} 0 & 1 & 1 & 1 & 1 \\ 0 & 0 & 1 & 0 & 1 \\ 0 & 0 & 0 & 0 & 1 \\ 0 & 0 & 0 & 0 & 1 \\ 0 & 0 & 0 & 0 & 0 \end{pmatrix}, \quad L = \begin{pmatrix} 0 & 1 & 0 & 1 & 0 \\ 0 & 0 & 1 & 0 & 0 \\ 0 & 0 & 0 & 0 & 1 \\ 0 & 0 & 0 & 0 & 1 \\ 0 & 0 & 0 & 0 & 0 \end{pmatrix}. \tag{7}$$

3. Path integrals on causal sets

A classical model for massive particle propagation is the swerves model described in [6]. The massless generalisation of these models were discussed by Sorkin[7] and Philpott[8] at DICE2008. These models are, however, not quantum mechanical.

In the continuum the propagation of particles is described by a propagator $K(x, y)$ which can be interpreted as the quantum mechanical amplitude (or just amplitude) for a particle to travel between two spacetime points x and y (see, for example, [3, chapter 6]). The path integral approach to quantum mechanics [9, 10] allows the propagator to be calculated by assigning amplitudes to each possible particle trajectory from x to y. Summing these amplitudes over all possible trajectories from x to y gives the propagator $K(x, y)$. As we shall show, in the discrete spacetime of a causal set this sum over trajectories can be simply performed.

To define path integrals on a causal set we have to make two choices: which trajectories to sum over and what amplitudes to assign to each trajectory. In this paper we'll consider two types of trajectories within the causal set: either all possible *chains* between two elements or all possible *paths* between two elements. The amplitude assigned to each trajectory is then defined in terms of two basic amplitudes: a and b.

The constant a is the amplitude for the particle to 'hop' once along the trajectory from one element to the next. The constant b is the amplitude for the particle to 'stop' at one element of the trajectory (the initial and final elements are not regarded as stops). The amplitude for the whole trajectory is then the product of the amplitudes for each hop and each stop. For a chain or path of length n (so there are n hops and $n-1$ intermediate stops) this compound amplitude is therefore $a^n b^{n-1}$.

We'll deal only with finite causal sets but the extension to arbitrary causal sets is straightforward because any causal set is locally finite. For a causal set with p elements we define a $p \times p$ matrix Φ.

If we sum over chains we define

$$\Phi := aC. \tag{8}$$

If we sum over paths we define

$$\Phi := aL. \tag{9}$$

The total amplitude to go from v_i to v_j along a trajectory of any length is then K_{ij} where K is the $p \times p$ matrix

$$K := \Phi + b\Phi^2 + b^2\Phi^3 + \ldots = \sum_{n=1}^{\infty} b^{n-1}\Phi^n, \tag{10}$$

where matrix multiplication is used.

Each term in this sum is the contribution to the total amplitude from chains or paths of a particular length. As an example,

$$b(\Phi^2)_{ij} = \sum_{k=1}^{p} b\Phi_{ik}\Phi_{kj}, \tag{11}$$

is the sum (over all intermediate positions v_k) of the amplitudes for all length two trajectories v_i to v_k to v_j.

Since the causal set is finite and the trajectories move only forwards in time the sum in (10) terminates and we have

$$K = \Phi(I - b\Phi)^{-1}, \tag{12}$$

where I is the $p \times p$ identity matrix. This matrix inverse is simple to perform if the causal set has been labelled to ensure Φ is strictly upper triangular. In this case the upper triangular matrix $I - b\Phi$ can be inverted using elementary row operations.

3.1. Example

If we sum over chains in our example causal set $(\mathcal{C}, \preceq)$ we have

$$\Phi = aC = \begin{pmatrix} 0 & a & a & a & a \\ 0 & 0 & a & 0 & a \\ 0 & 0 & 0 & 0 & a \\ 0 & 0 & 0 & 0 & a \\ 0 & 0 & 0 & 0 & 0 \end{pmatrix}, \tag{13}$$

$$K = \Phi(I - b\Phi)^{-1} = \begin{pmatrix} 0 & a & a+a^2b & a & a+3a^2b+a^3b^2 \\ 0 & 0 & a & 0 & a+a^2b \\ 0 & 0 & 0 & 0 & a \\ 0 & 0 & 0 & 0 & a \\ 0 & 0 & 0 & 0 & 0 \end{pmatrix}. \tag{14}$$

Each term K_{ij} is equal to the sum of the amplitudes for all chains from v_i to v_j. For example $K_{15} = a + 3a^2b + a^3b^2$ is the total amplitude for 1 chain of length 1, 3 chains of length 2 and 1 chain of length 3 (see Figure 2).

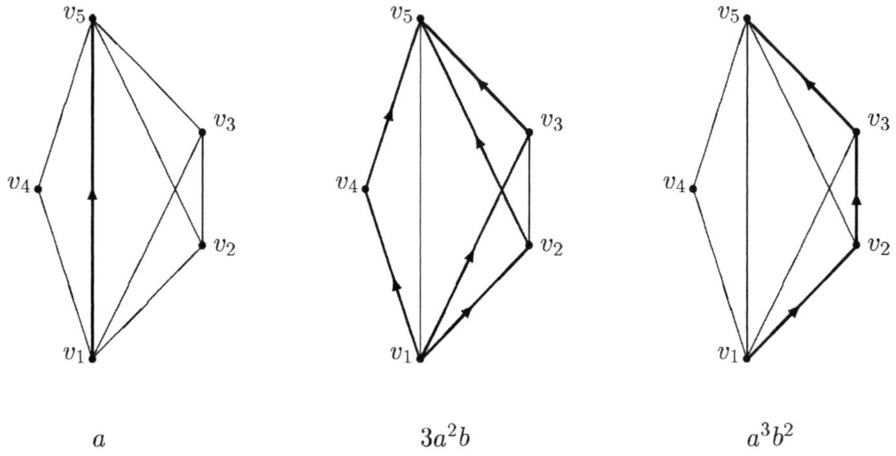

Figure 2. Contributions to K_{15} for chains of different lengths in our example causal set.

4. Sprinklings and expected values

A *sprinkling* is a way to generate a causal set from a d-dimensional causal Lorentzian manifold (M, g). To perform a sprinkling, points are placed at random within M using a Poisson process with sprinkling density ρ. This means that the points are randomly distributed throughout M with n points being sprinkled into a region of d-volume V with probability

$$\text{Prob}(n \text{ points in region of } d\text{-volume } V) = \frac{(\rho V)^n}{n!} e^{-\rho V}. \tag{15}$$

This ensures the expected number of points in a region of d-volume V is equal to ρV. The sprinkling generates a causal set whose elements are the sprinkled points. The causal set's partial order relation is "read off" from the manifold's causal relation restricted to the sprinkled points. If the Lorentzian manifold satisfies a strong enough causality condition (e.g. if it's strongly causal) then we expect the manifold to well-approximate the sprinkled causal set.

Causal sets generated by sprinklings into a Lorentzian manifold will not be identical because the particular points that are sprinkled are chosen randomly. The value of the propagator calculated from one sprinkled causal set therefore depends in detail on the particular causal set that is generated. This randomness in the sprinkling process means we are interested in the *expected value* (and variance) for the propagator calculated by averaging over all possible sprinkled causal sets.

We shall look at sprinklings into d-dimensional Minkowski spacetime. First we fix two points x and y in Minkowski spacetime. We then sprinkle a causal set. There is zero probability that the sprinkled causal set will contain x and y so we then add x and y to the sprinkled causal set. We then calculate the propagator from x to the y by summing over chains or paths in the sprinkled causal set. If we average the value of the propagator calculated in this way over all sprinkled causal sets (with a fixed sprinkling density ρ) we define two functions on Minkowski spacetime.

The function $K_C(x-y,\rho)$ is the average of the propagator which sums over chains. The function $K_P(x-y,\rho)$ is the average of the propagator which sums over paths. Both functions depend only on the difference $x-y$ because of the translation invariance of Minkowski spacetime. In [4] integral equations satisfied by K_C and K_P are derived for sprinklings into d-dimensional Minkowski spacetime. These equations are used to determine suitable values for a and b (given here by equations (23) and (25)) for sprinklings into 1+1 and 3+1 dimensional Minkowski spacetime. Before looking at these values we first define the continuum propagators we seek to reproduce.

5. Minkowski spacetime propagators

In d-dimensional Minkowski spacetime the causal relations are particularly simple. For $x=(x_0,\vec{x})$, $y=(y_0,\vec{y})$ we have:

$$x \preceq y \iff x_0 \leq y_0 \quad \text{and} \quad (x_0-y_0)^2 \geq (\vec{x}-\vec{y})^2. \tag{16}$$

The scalar propagators in d-dimensional Minkowski spacetime quantum field theory are Green's functions of the Klein-Gordon equation:

$$(\Box + m^2)K_m^{(d)}(x-y) = \delta^d(x-y). \tag{17}$$

Here m is the mass of the particle, δ^d is the d-dimensional Dirac delta function and we choose units with $\hbar = c = 1$. The d'Alembertian is given by

$$\Box := \frac{\partial^2}{\partial {x_0}^2} - \frac{\partial^2}{\partial {x_1}^2} - \frac{\partial^2}{\partial {x_2}^2} - \ldots - \frac{\partial^2}{\partial {x_{d-1}}^2}. \tag{18}$$

We can solve (17) using a d-dimensional Fourier transform to get

$$K_m^{(d)}(x-y) := -\frac{1}{(2\pi)^d}\int \mathrm{d}^d p \frac{e^{-ip(x-y)}}{p_0^2 - \vec{p}^2 - m^2}. \tag{19}$$

The integrand in this expression contains poles so the propagator in Minkowski spacetime is only uniquely defined if we specify a contour of integration (or equivalently boundary conditions for the solution of (17)). The *retarded causal propagator* is the Green's function obtained by avoiding the poles at $p_0 = \pm\sqrt{\vec{p}^2+m^2}$ by two small semi-circles in the upper-half p_0 complex plane. This is equivalent to

$$K_m^{(d)}(x-y) := \lim_{\epsilon\to 0+} -\frac{1}{(2\pi)^d}\int \mathrm{d}^d p \frac{e^{-ip(x-y)}}{(p_0+i\epsilon)^2 - \vec{p}^2 - m^2}. \tag{20}$$

This can be calculated explicitly in different dimensions. We consider the expressions in $d=2$ and $d=4$ dimensions.

In 1+1 dimensional Minkowski spacetime we have:

$$K_m^{(2)}(x-y) = \begin{cases} \frac{1}{2}J_0(m\tau_{xy}) & \text{if } x \preceq y \\ 0 & \text{otherwise,} \end{cases} \tag{21}$$

Journal of Physics: Conference Series **174** (2009) 012046 doi:10.1088/1742-6596/174/1/012046

Here $\tau_{xy} = \sqrt{(x_0 - y_0)^2 - (x_1 - y_1)^2}$ is the proper time from x to y and J_0 is the zeroth-order Bessel function of the first kind.

In 3+1 dimensional Minkowski spacetime[11]:

$$K_m^{(4)}(x-y) = \begin{cases} \frac{1}{2\pi}\delta(\tau_{xy}^2) - \frac{m}{4\pi}\frac{J_1(m\tau_{xy})}{\tau_{xy}} & \text{if } x \preceq y \\ 0 & \text{otherwise,} \end{cases} \tag{22}$$

Here $\tau_{xy} = \sqrt{(x_0 - y_0)^2 - (\vec{x} - \vec{y})^2}$ is the proper time from x to y and J_1 is the first-order Bessel function of the first kind.

6. Results

It can be shown[4] that, for 1+1 dimensional Minkowski spacetime, if we sum over *chains* with

$$a = \frac{1}{2}, \qquad b = -\frac{m^2}{\rho}, \tag{23}$$

we get

$$K_C(x-y,\rho) = K_m^{(2)}(x-y) \tag{24}$$

In 3+1 dimensional Minkowski spacetime if we sum over *paths* and take the infinite density limit with

$$a = \frac{\sqrt{\rho}}{2\pi\sqrt{6}}, \qquad b = -\frac{m^2}{\rho}, \tag{25}$$

we get that

$$\lim_{\rho\to\infty} K_P(x-y,\rho) = K_m^{(4)}(x-y) \tag{26}$$

The physically interesting result is the 3+1 case. Although we get the exact continuum propagator in the infinite density limit there are two types of deviations that will occur when we calculate the propagator on a single sprinkled causal set.

Firstly, the value of $K_P(x-y,\rho)$ for a sprinkled causal set with finite sprinkling density ρ differ from the continuum propagator because the sprinkling density is finite. For high density sprinkling, however, $K_P(x-y,\rho)$ will be close to the continuum propagator.

Secondly, the value of the propagator calculated on a *single* sprinkled causal set will deviate from the *expected value* $K_P(x-y,\rho)$ by random fluctuations. These fluctuations will depend on the particular causal set that is sprinkled but, as discussed in [4], numerical simulations indicate that their size reduces as the sprinkling density is increased.

Modelling spacetime as a single causal set generated by sprinkling into 3+1 dimensional Minkowski spacetime[2] (with a Planckian sprinkling density, say) would, in light of these differences, give deviations from continuum behaviour for particle propagation. These deviations are welcomed provided they are small enough to have gone unnoticed so far. If quantum field theory on a causal set can be developed it is just such deviations that could provide experimental evidence for spacetime discreteness.

7. Conclusions and further work

A path integral model for particle propagation on a causal set spacetime has been presented. For causal sets generated by sprinkling into 1+1 and 3+1 dimensional Minkowski spacetime the propagator calculated on the causal set gives, in a suitable sense, the scalar retarded causal propagator. It is hoped that the propagator derived here could be used to define quantum field

[2] Ultimately the aim is to define suitable causal set dynamics which would generate causal set spacetimes without reference to sprinkling into a background continuum spacetime.

theory on a causal set but, before this is possible, a number of difficult steps must first be completed. We finish by outlining some ideas for one possible route to a quantum field theory on a causal set:

(i) In standard quantum field theory the *Feynman propagator*, rather than the retarded or advanced propagators[3], is used in calculations. We are left with two possibilities:
  (a) Re-express quantum field theory without using the Feynman propagator and instead using only the retarded and advanced propagators.
  (b) Develop a model which gives the Feynman propagator on a causal set. This may involve summing amplitudes assigned to trajectories which repeatedly reverse their time direction. Unfortunately, without further restriction, this would allow infinitely many such trajectories from one element to another (if the causal set corresponds to a connected graph) which could cause the sum-over-trajectories to diverge.

(ii) The work presented here deals only with scalar particles. For a realistic model of matter one must be able to model spin 1/2 and spin 1 particles. This may be achieved by including extra internal degrees of freedom associated with the particles on the causal set. In the case of spin degrees of freedom these would have to relate somehow to the spacetime geometry.

(iii) We are interested in describing interacting particles. The graph-based language of Feynman diagrams (or similar diagrammatic methods) seems very well suited to a causal set spacetime.

Acknowledgments

The author is grateful to the DICE2008 organisers for hosting an enjoyable conference and giving an opportunity to present this work. This work was supported by a STFC studentship.

References

[1] Bombelli L, Lee J, Meyer D and Sorkin R 1987 *Phys. Rev. Lett.* **59** 521
[2] Henson J 2006 The causal set approach to quantum gravity *Preprint* arXiv:gr-qc/0601121v2
[3] Bjorken J D and Drell S D 1964 *Relativistic Quantum Mechanics* (McGraw-Hill)
[4] Johnston S 2008 *Class. Quantum Grav.* 25 202001
[5] Stanley R P 1986 *Enumerative Combinatorics: Vol I* (Wadsworth and Brooks/Cole)
[6] Dowker F, Henson J and Sorkin R D 2003 Quantum Gravity Phenomenology, Lorentz Invariance and Discreteness *preprint* arXiv:gr-qc/0311055v3
[7] Sorkin R D *Direct transmission of light along causal links* Talk given at DICE2008, Castiglioncello, Italy, September 22-26 2008
[8] Philpott L *Massless particle diffusion in causal set theory* Talk given at DICE2008, Castiglioncello, Italy, September 22-26 2008
[9] Feynman R P and Hibbs A R 1965 *Quantum Mechanics and Path Integrals* (McGraw-Hill)
[10] Feynman R P 1948 *Rev. Mod. Phys.* **20** 367
[11] De Jager E M 1967 *SIAM Journ. Appl. Math.* **15** No. 4. 944-963

[3] If $K^{(d)}(x-y)$ denotes the retarded propagator the advanced propagator is given by $K^{(d)}(y-x)$. On the causal set if the retarded propgator is the matrix K the advanced propagator is its transpose K^T obtained by summing over trajectories that only travel *backwards* in time.

Emergent non-commutative matter fields from group field theory models of quantum spacetime

Daniele Oriti
Institute for Theoretical Physics, Utrecht University,
Leuvenlaan 4, 3584 TD, Utrecht, The Netherlands, EU
and
Albert Einstein Institute,
Am Muehlenberg 4, Golm, Germany, EU

E-mail: daniele.oriti@aei.mpg.de

Abstract. We offer a perspective on some recent results obtained in the context of the group field theory approach to quantum gravity, on top of reviewing them briefly. These concern a natural mechanism for the emergence of non-commutative field theories for matter directly from the GFT action, in both 3 and 4 dimensions and in both Riemannian and Lorentzian signatures. As such they represent an important step, we argue, in bridging the gap between a quantum, discrete picture of a pre-geometric spacetime and the effective continuum geometric physics of gravity and matter, using ideas and tools from field theory and condensed matter analog gravity models, applied directly at the GFT level.

1. Introduction

In this contribution we offer a perspective on some recent results [1, 2] obtained in the context of the group field theory (GFT) approach to quantum gravity [3, 4], on top of reviewing them briefly. These concern a natural mechanism for the emergence of non-commutative field theories for matter directly from the GFT action, in both 3 and 4 dimensions and in both Riemannian and Lorentzian signatures. The interest of such results is manifold. First, they show a straightforward link between spin foam/loop quantum gravity models, via GFTs, and non-commutative geometry. Second, they are, in principle, a crucial step forward in the attempt to relate this class of models with effective frameworks, like Deformed Special Relativity, that form the basis of much current Quantum Gravity Phenomenology [5]. On top of all this, in this contribution we argue that they can be naturally understood within a certain interpretative framework for GFTs and for the relation between its quantum microscopic spacetime structures and continuum spacetime physics. The perspective we present is not, of course, the only possible consistent one, nor the one shared by all the researchers working in this specific area. It is, however, a scenario that is consistent with what we presently know about GFTs, and with a recent proposal about the emergence of continuum physics from them, as put forward by the author. Moreover, it is a scenario that makes these results all the more exciting, which is, we believe, an added benefit. It is based on three main points: 1) GFTs are the most complete and fundamental definition of quantum gravity models based on spin network or simplicial gravity structures; 2) continuum spacetime physics is to be looked for in the collective "many-particle"physics of GFT quanta; 3) therefore, instead of usual LQG, spin foam or simplicial gravity techniques, one could try to adopt a condensed matter perspective on quantum spacetime, and see what condensed matter ideas and tools give when applied to the quantum spacetime system directly at the GFT level; 4) in particular, this shift in perspective and tools can be very useful for bridging the gap between a microscopic, quantum, discrete picture of a pre-geometric spacetime and the effective continuum geometric physics of gravity and matter we are accustomed and have phenomenological access to. We have exposed the above points and drawn some possible consequences of this perspective in [6]. Here we point out how the results on the emergence of non-commutative field theories from GFTs can fit in it.

1.1. Group field theory formulation of quantum gravity

Group Field Theories [3, 4] are quantum field theories on (possibly, extensions of) group manifolds characterized by a peculiar non-local coupling of fields, designed to produce, in their perturbative expansion around the vacuum, Feynman diagrams that can be put in 1 to 1 correspondence with d-dimensional simplicial complexes. The fundamental field is interpreted as a second quantized (d-1)-simplex or as a (second quantized) elementary spin network vertex. The Feynman amplitudes can be understood as simplicial gravity path integrals (although not necessarily written in terms of some gravity action) or, dually, as spin foam models, i.e. sum over histories of spin networks states. GFTs are thus at the same time a generalization of matrix models for 2d quantum gravity to higher dimensions, and a new version of the simplicial quantum gravity approaches, and a complete definition of the covariant dynamics of spin networks in loop quantum gravity. Let us be a bit more specific.

In their simplest type of models, the field is a $\mathbb{C}$-valued function of d group elements $\phi(g_1,..,g_d)$, for a group G (configuration space) being for example the $SO(d-1,1)$ Lorentz group. Each argument corresponds to one of the boundary (d-2)-faces of the (d-1)-simplex represented by the field. Additional symmetry requirements can be imposed. Typically, one imposes invariance under diagonal action of G: $\phi(g_1,...,g_d) = \phi(g_1 g,...,g_d g)$. Fields and action can then be expanded in modes, i.e. in group representations. Both group and representation variables can be given a straightforward interpretation (justified by their role in the corresponding amplitudes) as (pre-) geometric data, i.e. data defining a discrete geometry for the simplicial complex associated with each Feynman diagram. The details of course depend on the specific model. A GFT model is defined by a choice of action:

$$S = \frac{1}{2}\int dg_i d\tilde{g}_i\, \phi(g_i)\mathcal{K}(g_i\tilde{g}_i^{-1})\phi(\tilde{g}_i) + \frac{\lambda}{(d+1)!}\int dg_{ij}\, \phi(g_{1j})...\phi(g_{d+1j})\,\mathcal{V}(g_{ij}g_{ji}^{-1}),$$

i.e. of a kinetic and interaction functions $\mathcal{K}$ and $\mathcal{V}$. The interaction term describes the interaction of (d-1)-simplices to form a d-simplex by gluing them along their (d-2)-faces (arguments of the fields); this gives the mentioned non-local combinatorial pairing of fields. The nature of the interaction is specified by the choice of function $\mathcal{V}$. The kinetic term involves two fields each representing a given (d-1)-simplex seen from one of the two d-simplices (interaction vertices) sharing it, so that the choice of kinetic functions $\mathcal{K}$ specifies how the geometric degrees of freedom corresponding to their d (d-2)-faces are propagated from one vertex to the next. The quantum theory is defined in terms of the expansion in Feynman diagrams of the partition function:

$$Z = \int \mathcal{D}\phi\, e^{-S[\phi]} = \sum_{\Gamma} \frac{\lambda^N}{sym[\Gamma]} Z(\Gamma),$$

where N is the number of interaction vertices in the Feynman graph Γ, $sym[\Gamma]$ is a symmetry factor for the graph and $Z(\Gamma)$ the corresponding Feynman amplitude. By construction ([3, 4] the Feynman diagram Γ is a collection of 2-cells (faces), edges and vertices, i.e. a 2-complex, topologically dual to a d-dimensional simplicial complex. The resulting complexes/triangulations can have arbitrary topology, each corresponding to a particular *interaction process* of the fundamental building blocks of space, i.e. (D-1)-simplices. In representation space each Feynman diagram is a spin foam (a 2-complex with faces f labelled by group representations), and each Feynman amplitude defines a spin foam model. Because of the geometric interpretation for the (group-theoretic) variables these Feynman amplitudes define a sum-over-histories for discrete quantum gravity, i.e. a sum over simplicial geometries, within a sum over simplicial topologies.

Several GFT models have been and are currently studied, and much more is known about GFTs in general. For all this, we refer to the literature [3, 4, 7, 8].

1.2. From GFT to continuum spacetime physics?

GFTs are thus a (tentative) class of models for quantum spacetime, and thus *not defined on any spacetime*, in which its microscopic structure is described by elementary building blocks given by spin network vertices or simplices, labelled by pre-geometric data. This description is valid and useful in the regime in which the perturbative expansion of the GFT partition function is valid, i.e. small coupling constant and few GFT quanta involved in the physical process one is considering. In this approximation a discrete spacetime *emerges* as a Feynman diagram from this "pre-spacetime" theory.

This is one of the intriguing aspects of GFTs. It is also the origin of the main open issue: how to recover a continuum spacetime, its continuum geometry and all continuum physics, including usual General Relativity and quantum field theories for matter, from such a different description of the pre-geometric regime? This is the outstanding problem faced by *all* current discrete approaches to quantum gravity[9].

The problem of the continuum involves actually several intertwined questions. Some of them are conceptual: where does a concept of geometry originate from in a non-geometric or pre-geometric framework? where to look for notions and properties of space and time in models defined in absence of any spacetime structure? Some are purely kinematical: should we expect the pre-geometric GFT data to give rise to the continuum geometry? or could an emergent macroscopic continuum geometry be totally unrelated to these microscopic and pre-geometric data? what is the correct approximation scheme to obtain a continuum spacetime from the discrete GFT structures? Some are instead dynamical: what is the physical regime, the physical conditions, in which the continuum approximation is viable, justified and/or useful? what is the physical dynamical process leading to this regime from the pre-geometric one, and how to describe it?

We have discussed in [6] our personal perspective on this problem, and the possibility that GFTs could be the right setting for tackling it, and how. We summarize it here, briefly. From the GFT point of view, the crucial issue is whether we expect the continuum approximation (as opposed to large scale, semi-classical or other a priori distinct approximations) to involve very large numbers of GFT quanta or not[1]. We opt for a positive answer, as naive reasoning would suggest (one would expect a generic continuum spacetime to be formed by zillions of Planck size building blocks, rather than few macroscopic ones). If this is the case, then we are dealing, from the GFT point of view, with a many-particle system whose microscopic theory is given by some fundamental GFT action and we are interested in its collective dynamics and states in some thermodynamic approximation. This simple thought alone suggests us to look for ideas and techniques from statistical field theory and condensed matter theory, and to try to apply/reformulate/re-interpret them in a GFT context.This also immediately suggests that the GFT formalism is the most natural setting for studying this dynamics even coming from the pure loop quantum gravity perspective or from simplicial quantum gravity. In the first case, in fact, GFTs offer a second quantized formalism for the same quantum geometric structures,

[1] In the full theory; the same question in a symmetry reduced context, for example, may have a different answer.

Journal of Physics: Conference Series **174** (2009) 012047 doi:10.1088/1742-6596/174/1/012047

and quantum field theory is indeed what comes natural in condensed matter when dealing with large collections of particles/atoms. In the second case, GFTs offer an alternative non-perturbative definition of the quantum dynamics of simplicial gravity, re-interprets the usual sum over discrete geometries as a perturbative expansion around the no-spacetime vacuum, and in doing so suggest the possibility of different vacuum states and a different reformulation of the same dynamics that is better suited for studying the dynamics of (combinatorially) complicated simplicial geometries. See [6] for more details.

2. Condensed matter perspective and GFT implementation

So, what can we learn from condensed matter models in our quest for bridging the gap between microscopic quantum and discrete pre-geometry in absence of spacetime, and macroscopic continuum spacetime and associated physics, with matter fields, geometry, gauge interactions and all that?

Condensed matter theory comes in help in two main ways, one very general, one more specific.

The first: condensed matter theory is exactly the area of physics concerned with the general issue of understanding the physical behaviour of large assemblies of microscopic constituents, in their collective behaviour, of elucidating their emergent properties, and of developing the appropriate notions and mathematical tools for describing the different layers of collective organization that coexists in a given phase, as well as the transition from one phase to another. Therefore, once we realize that a classical continuum spacetime is the result of the collective properties and dynamics of large numbers of GFT quanta, it is natural to look at condensed matter theory for insights.

One could be a bit more specific (and speculative) and conjecture that continuum spacetime arises as a *fluid phase* of this large assembly of GFT quanta, maybe after a process analogous (mathematically) to condensation in GFT momentum space (Bose condensation, for example). This conjecture is also argued for in [6]. Continuum geometrodynamics, be it classical (General Relativity) or quantum could then arise from the effective GFT hydrodynamics in this fluid (or condensed) phase. This is a second conjecture put forward in [6]. Again, in order to put these ideas to test, one has to turn to condensed matter theory and to statistical field theory, i.e. we should develop statistical group field theory, and study that special condensed matter system that is a quantum spacetime. A corresponding research programme has been lied down tentatively in [6], and it involves developing first for GFTs and then applying several techniques, common in statistical field theory and condensed matter. One of them is the renormalization group, and in fact one way to study the phase structure of GFTs as well as the behaviour of GFT systems at different scales would certainly be in terms of perturbative and non-perturbative renormalization group transformations. No systematic study of GFT renormalization has been performed until now, but work is in progress from various groups. In particular, a systematic treatment of perturbative GFT renormalization for the 3d Boulatov model is under way, and its first results can be found in [10]. Another avenue would be to develop an Hamiltonian statistical mechanics for GFTs, and in turn this involves several steps: a canonical/hamiltonian GFT formalism, which made highly non-trivial by the non-local nature of these models, a precise understanding of the GFT analogue of the notion of energy, temperature, pressure, etc, and a re-interpretation of any such notion in quantum gravity terms. With these tools at hand one should analyze the phase structure of various GFT models and prove the existence of an appropriate fluid phase in some of them. Then, the hydrodynamic approximation in this phase should be developed and the final goal would be to show the emergence of General Relativity from this GFT hydrodynamic description. It is too early to report on progress on this front.

There is a second way, however, in which condensed matter theory comes into the game. It provides specific examples of systems in which the collective behaviour of the microscopic constituents in some hydrodynamic approximation gives rise to effective emergent geometries as well as matter fields. Thus it gives further support, by means of explicit examples, to the idea that continuum geometry and gravity may emerge naturally from fundamental systems which do not have a geometric or gravitational nature per se, at least in their fundamental formulation (GFTs are of this type). These are the so-called analog gravity models [11].

The emergence of gravity and (generically) curved geometries in analog condensed matter models takes place in the hydrodynamic regime, i.e. usually at the level of a (modified) Gross-Pitaevskii equation or of the corresponding Lagrangian/Hamiltonian, which in turn is usually obtained in the mean field theory approximation (or refined version of the same) around some background configuration of the fluid under study. The background configurations that have proven more interesting are those identifying Bose-Einstein condensates or fermionic superfluids [11]. However, the emergence of effective metric fields is more general and not confined to these rather peculiar systems. What happens is that the collective parameters describing the fluid and its dynamics in these background configurations (e.g. the density and velocity of the fluid in the laboratory frame) can be recast as the component functions of an *effective metric field.*

This would be only cosmetics if not for two further results. 1) In some very special cases and in some particular approximation the hydrodynamic equations governing the dynamic of the effective metrics, when recast in geometric terms, can also be seen to reproduce known geometrodynamic theories, at least in part, ranging from Newtonian gravity to (almost) GR; see [12]. This means that it is possible to reproduce also gravitational dynamics as emergent from systems that are not geometric in nature or form. 2) The effective dynamics of perturbations (quasi-particles, themselves collective excitations of the fundamental constituents of the fluid) around the same background configurations turns out to be given by matter field theories in curved spacetimes, whose geometry is indeed the one identified by the effective metrics obtained from the collective background parameters of the fluid.

The first type of results is so far limited to special systems, peculiar approximations, and ultimately not fully satisfactory, in the sense that it has not been possible yet to reproduce, say, the Einstein-Hilbert dynamics in any, however idealized, condensed matter system [12]. This seems to suggest that the ideas and techniques developed in this context and to this aim are certainly useful and interesting, but need to be applied to some very peculiar system representing (quantum) spacetime, if one wants to explain in this way the emergence of geometry and General Relativity in the real world. Therefore, these results certainly encourage and support the idea of applying similar ideas and techniques in Group Field Theory, but they do not provide a sure guidance on how exactly this should be done.

The second type of results, however, is much more general and applies to a very large class of systems and approximations, including systems as common as ordinary fluids (e.g. water) in everyday physical conditions [11]. It is this type of results that we focus on in this contribution, because we are able to obtain similar results in a GFT context.

The general scheme of what goes on in all these condensed matter systems (including BEC, superfluids, etc) concerning the emergence of matter field theories on effective metric spacetimes is well captured by the following "meta-model"described in [11]. Consider a system described by a single scalar field $\phi(x)$, living on a flat metric spacetime of trivial topology (a good approximation for the quantum geometry of any lab in any research institute on Earth). The field ϕ can be the effective order parameter for a BEC, the collective field encoding the velocity and density of an ordinary fluid in the hydrodynamic approximation, or whatever else. Assume that its dynamics is encoded in a lagrangian $L(\phi, \partial_\mu \phi)$ depending on the field and its partial derivatives. Let us expand the field around some classical solution ϕ_0 of the equations of motion, as: $\phi(x) = \phi_0(x) + \phi_1(x)$. Next we expand the lagrangian itself to obtain an effective action for the perturbation field $\phi_1(x)$ (we focus our attention to the kinetic term only); we obtain, generically, an effective Klein-Gordon operator on a curved metric:

$$\begin{aligned} S(\phi) &= S(\phi_0) + \frac{1}{2}\int \left[\frac{\partial^2 L}{\partial(\partial_\mu\phi)\partial(\partial_\nu\phi)}\,|_{\phi_0}\; \partial_\mu\phi_1\partial_\nu\phi_1 + \left(\frac{\partial^2 L}{\partial\phi\partial\phi} - \partial_\mu\frac{\partial^2 L}{\partial(\partial_\mu\phi)\partial\phi}\right)|_{\phi_0}\right] + (....) \\ &= S(\phi_0) + \frac{1}{2}\int \sqrt{-g}\,\left[\phi_1\Box_{\phi_0}\phi_1 - V(\phi_0)\phi_1^2\right] + (\text{.... } interactions\text{ }), \end{aligned} \quad (1)$$

with the operator $\Box_{\phi_0} = g^{\mu\nu}\partial_\mu\partial_\nu$, for the effective (inverse) metric $\sqrt{-g}\,g^{\mu\nu} = \frac{\partial^2 L}{\partial(\partial_\mu\phi)\partial(\partial_\nu\phi)}\,|_{\phi_0}$. It is the possible to invert to obtain the effective metric and from this the other tensors characterizing the effective spacetime geometry. Notice that both the "fundamental"field and the quasi-particle one live on a 4-dimensional spacetime of trivial topology, although endowed with a different metric in general. We refer to the literature for further details and applications of the above general result. The main point to notice here is that one generically obtains an effective spacetime geometry to which the quasi-particles couple, depending only in its precise functional form on the fundamental Lagrangian L and on the classical solution ϕ_0 chosen; they do not couple to the initial (laboratory) flat background metric.

This is the type of mechanism we want to reproduce in a Group Field Theory context. Assuming that a given GFT model (Lagrangian) describes the microscopic dynamics of a *discrete quantum spacetime*, and that some solution of the corresponding fundamental equations can be interpreted as identifying a given quantum spacetime configuration, 1) can we obtain an effective macroscopic *continuum* field theory for matter fields from it? and if so, 2) what is the effective spacetime and geometry that these emergent matter fields see?

Answering these questions means, as we have tried to emphasize, making an important step towards bridging the gap between our fundamental discrete models of spacetime, and the usual continuum description of spacetime. It also means getting closer to possible quantum gravity phenomenology, and to experimental falsifiability. Let us also notice that, while the correspondence between classical solutions of the fundamental equations and effective geometries is a priori unexpected in condensed matter systems, which are non-geometric in nature, so that they are referred to as *analog* gravity models, in the GFT case the situation is different. We have here models which are non-geometric and far from usual geometrodynamics in their formalism, but which at the same time are expected to encode quantum geometric information and indeed to determine, in particular in their classical solutions, a (quantum and therefore classical) geometry for spacetime [3], also at the continuum level, the issue being how they exactly do so. We are, in other words, far beyond a pure analogy. What we find is that it is possible to apply the same procedure to GFT models and that one can obtain rather straightforwardly effective continuum field theories for matter fields. The effective matter field theories that we obtain most easily from GFTs are quantum field theories on non-commutative spaces of Lie algebra type.

3. Results on emergent effective matter

Let us now introduce and review briefly the results obtained so far.

3.1. Non-commutative field theories as group field theories

First of all we introduce the relevant class of non-commutative field theories. The basic point is the duality between Lie algebra and corresponding Lie group re-interpreted as the non-commutative version of the usual duality between coordinate and momentum space. More precisely, if we have a non-commutative spacetime of Lie algebra type $[X_\mu, X_\nu] = C^\lambda_{\mu\nu}X_\lambda$, the corresponding momentum space is naturally identified with the corresponding Lie group, on which non-commutative coordinates X_μ act as (Lie) derivatives. From this perspective, we understand the origin of the spacetime non-commutativity to be the curvature of the corresponding momentum space, a sort of Planck scale "co-gravity"[18]. The link with GFTs is then obvious: in momentum space the field theory on such non-commutative spacetime will be given, by definition, by some sort of group field theory. The task will then be to derive the relevant field theories for matter from interesting GFT models of quantum spacetime.

In 3 spacetime dimensions the results obtained concern a euclidean non-commutative spacetime given by the $\mathfrak{su}(2)$ Lie algebra , i.e. whose spacetime coordinates are identified with the $\mathfrak{su}(2)$ generators with $[X_i, X_j] = i\frac{1}{\kappa}\epsilon_{ijk}X_k$. Momenta are instead identified with group elements SU2) [13], acquiring a non-commutative addition property following the group composition law. A scalar field theory in momentum space is then given by a group field theory of the type:

$$S[\psi] = \frac{1}{2}\int_{\mathrm{SU}(2)} dg\psi(g)\mathcal{K}(g)\psi(g^{-1}) - \frac{\lambda}{3!}\int [dg]^3\, \psi(g_1)\psi(g_2)\psi(g_3)\delta(g_1 g_2 g_3), \tag{2}$$

in the 3-valent case, where the integration measure is the Haar measure on the group. The non-commutative Fourier transform [13] relates then functions on the group and non-commutative functions in the enveloping algebra of $\mathfrak{su}(2)$. It is based on the non-commutative plane waves $e_g = e^{ik_i X_i} \in \widehat{C}_\kappa(\mathfrak{su}(2))$, where k_i are local coordinates on the group manifold labeling the group element (and thus the momentum) g and bounded as $\mid k \mid \leq \kappa$ (so that it is natural to identify κ with a Planckian maximal mass/momentum scale). It reads:

$$\widehat{\phi}(X) = \int_{\mathrm{SU}(2)} dg\, e_{g(k_i)}\, \phi(g(k)), \quad X \in \mathfrak{su}(2), \quad \widehat{\phi}(X) \in \mathcal{U}(\mathfrak{su}(2)). \tag{3}$$

Strictly speaking, this construction works on SO(3), but can be extended to SU(2) in several ways [14]. Using this, the action above can be rewritten in configuration space. We will see below some simple example of such action, derived straightforwardly from a group field theory model for 3d quantum gravity. We notice that one can define a further map from elements of the enveloping algebra of $\mathfrak{su}(2)$ to functions on $\mathbb{R}^3$ (isomorphic to the same Lie algebra as a vector space) endowed with a non-commutative star product again reflecting the non-commutative composition of momenta following the rules of group multiplication. See [13, 14, 15] for details. We also notice that the Feynman amplitudes of the above scalar field action (with simple kinetic terms) can be derived from the Ponzano-Regge spin foam model coupled to point particles [16], in turn obtainable from an extended GFT formalism [17]. We will see that the GFT construction to be presented allows to bypass completely the spin foam formulation of the coupled theory.

The 4-dimensional non-commutative spacetime that is of most direct relevance for Quantum Gravity phenomenology is so-called κ-Minkowski [18]. We recall here the main features of such space and of the non-commutative field theory defined on it, referring for further details to [2] and references therein. κ-Minkowski space-time can be identified with the Lie algebra $\mathfrak{an}_3$, which is a subalgebra of $\mathfrak{so}(4,1)$. Indeed, if $J_{\mu\nu}$ are the generators of $\mathfrak{so}(4,1)$, the $\mathfrak{an}_3$ generators are:

$$X_0 = \frac{1}{\kappa}J_{40}, \quad X_k = \frac{1}{\kappa}(J_{4k} + J_{0k}), \quad k = 1, ..., 3\ , \tag{4}$$

which, once identified with the coordinates of our non-commutative space, characterize it with the commutation relations:

$$[X_0, X_k] = -\frac{i}{\kappa}X_k, \qquad [X_k, X_l] = 0, \quad k, l = 1, ..., 3. \tag{5}$$

Using this, we can then define non-commutative plane waves with the AN_3 group elements as $h(k_\mu) = h(k_0, k_i) \equiv e^{ik_0 X_0}e^{ik_i X_i}$, thus identifying the coordinates on the group k_μ as the wave-vector (in turn related to the momentum). From here, a non-commutative addition of wave-vectors follows from the group multiplication of the corresponding plane waves.

Crucial for our construction, the Iwasawa decomposition relates SO(4,1) and AN_3 as [19]:

$$SO(4,1) = AN_3\, SO(3,1) \cup AN_3\mathcal{M}\, SO(3,1), \tag{6}$$

where the two sets are disjoint and $\mathcal{M}$ is the diagonal matrix with entries $(-1,1,1,1,-1)$ in the fundamental 5d representation of SO(4,1). Since De Sitter space-time dS_4 can be defined as the coset $SO(4,1)/SO(3,1)$, an arbitrary point v on it can be uniquely obtained as:

$$v = (-)^\epsilon h(k_\mu).v^{(0)} = h(k_\mu)\mathcal{M}^\epsilon.v^{(0)}, \qquad \epsilon = 0 \text{ or } 1, \quad h \in AN_3, \tag{7}$$

where we have taken a reference space-like vector $v^{(0)} \equiv (0,0,0,1) \in \mathbb{R}^4$, such that its little group is the Lorentz group SO(3,1) and the action of SO(4,1) on it sweeps the whole De Sitter space, and defined the vector $v \equiv h(k_\mu).v^{(0)}$ with coordinates:

$$v_0 = -\sinh\frac{k_0}{\kappa} + \frac{\mathbf{k}^2}{2\kappa^2}e^{k_0/\kappa} \quad v_i = -\frac{k_i}{\kappa} \quad v_4 = \cosh\frac{k_0}{\kappa} - \frac{\mathbf{k}^2}{2\kappa^2}e^{k_0/\kappa}. \tag{8}$$

The sign $(-)^\epsilon$ corresponds to the two components of the Iwasawa decomposition. We then introduce the set $AN_3^c \equiv AN_3 \cup AN_3\mathcal{M}$, such that the Iwasawa decomposition reads $SO(4,1) = AN_3^c\, SO(3,1)$ and that AN_3^c is isomorphic to the full de Sitter space. Actually, one can check that AN_3^c is itself a group. A crucial point is that the component v_4 of the above vector is left invariant by the action of the Lorentz group SO(3,1). This suggests to use this function of the "momentum" k_μ as a new (deformed) invariant energy-momentum (dispersion) relation, in the construction of a deformed version of particle dynamics and field theory on κ-Minkowski spacetime. This is the basis for much current QG phenomenology [5].

Finally, we will need an integration measure on AN_3 in order to define a Fourier transform. The group AN_3 is provided with two invariant Haar measures: $\int dh_L = \int d^4k_\mu$, $\int dh_R = \int e^{+3k_0/\kappa}\, d^4k_\mu$, which are respectively invariant under the left and right action of the group. The left invariant measure can be obtained from the 5d parametrization used above as:

$$\kappa^4 \int \delta(v_A v^A - 1)\theta(v_0 + v_4)\, d^5v_A = \int d^4k_\mu = \int dh_L, \tag{9}$$

so it is the natural measure on AN_3 inherited from the Haar measure on SO(4,1). However, this measure is not Lorentz invariant, due to the restriction $v_+ > 0$. To get a Lorentz invariant measure, we write the same measure as a measure on $AN_3^c \equiv AN_3 \cup AN_3\mathcal{M} \sim dS$:

$$\int dh_L \equiv \int_{AN_3} dh_L^+ + \int_{AN_3\mathcal{M}} dh_L^- = \int \delta(v_A v^A - 1)d^5v. \tag{10}$$

Another way to obtain a Lorentz invariant measure is to consider the elliptic de Sitter space $dS/\mathbb{Z}_2$ where we identify $v_A \leftrightarrow -v_A$, which amounts to identifying the group elements $h(k_\mu) \leftrightarrow h(k_\mu)\mathcal{M}$. This space is indeed isomorphic to AN_3 as a manifold. One way to achieve nicely this restriction at the field theory level is to consider only fields on De Sitter space (or on AN_3^c) which are however invariant under the parity transformation $v_A \leftrightarrow -v_A$ [15].

For the free real scalar field $\phi : G \to \mathbb{R}$, we define the action

$$\mathcal{S}(\phi) = \int dh\, \phi(h)\, \mathcal{K}(h)\, \phi(h), \quad \forall h \in G, \tag{11}$$

where dh is the left invariant measure. We then interpret $G = AN_3, AN_3^c$ as the momentum space. We demand $\mathcal{K}(h)$ to be a function on G invariant under the Lorentz transformations, which suggests to use some function $\mathcal{K}(h) = f(v_4(h))$. Two common choices are

$$\mathcal{K}_1(h) = (\kappa^2 - \pi_4(h)) - m^2, \quad \mathcal{K}_2(h) = \kappa^2 - (\pi_4(h))^2 - m^2, \quad \pi_4 = \kappa v_4. \tag{12}$$

The above action is then Lorentz invariant if we choose a Lorentz invariant measure, i.e. dh_L in the case of either generic fields on AN_3^c or symmetric fields on AN_3.

Finally, the following generalized Fourier transform relates functions on the group $\mathcal{C}(G)$ and elements of the enveloping algebra $\mathcal{U}(\mathfrak{an}_3)$, i.e. non-commutative fields on the non-commutative spacetime $\mathfrak{an}_3$, i.e. on κ-Minkowski. For $G = \mathrm{AN}_3^c, \mathrm{AN}_3$, respectively:

$$\widehat{\phi}(X) = \int_{\mathrm{AN}_3} dh_L^+\, h(k_\mu)\, \phi^+(k) + \int_{\mathrm{AN}_3\mathcal{M}} dh_L^-\, h(k_\mu)\, \phi^-(k), \quad X \in \mathfrak{an}_3, \;\; \widehat{\phi}(X) \in \mathcal{U}(\mathfrak{an}_3)$$

$$\widehat{\phi}(X) = \int_{\mathrm{AN}_3} dh_L\, h(k_\mu)\, \phi(k), \quad X \in \mathfrak{an}_3, \;\; \widehat{\phi}(X) \in \mathcal{U}(\mathfrak{an}_3)$$

where we used the non-abelian plane-wave $h(k_\mu)$ [13, 15]. The group field theory action on G can now be rewritten as a non-commutative field theory on κ-Minkowski (in the AN_3 case)

$$\mathcal{S}(\phi) = \int dh_L\, \phi(h)\, \mathcal{K}(h)\, \phi(h) = \int d^4X \left(\partial_\mu \widehat{\phi}(X) \partial^\mu \widehat{\phi}(X) + m^2 \widehat{\phi}^2(X)\right). \qquad (13)$$

The Poincaré symmetries are naturally deformed in order to be consistent with the non-trivial commutation relations of the κ-Minkowski coordinates [15].

3.2. 3d case

The group field theory we start from is the Boulatov model for 3d quantum gravity [3]. We consider a real field $\phi : \mathrm{SU}(2)^3 \to \mathbb{R}$ invariant under the diagonal right action of SU(2):

$$\phi(g_1, g_2, g_3) = \phi(g_1 g, g_2 g, g_3 g), \quad \forall g \in \mathrm{SU}(2). \qquad (14)$$

The action for this 3d group field theory involves a trivial propagator and the tetrahedral vertex:

$$\begin{aligned} S_{3d}[\phi] &= \frac{1}{2} \int [dg]^3 \phi(g_1, g_2, g_3)\phi(g_3, g_2, g_1) \\ &\quad - \frac{\lambda}{4!} \int [dg]^6 \phi(g_1, g_2, g_3)\phi(g_3, g_4, g_5)\phi(g_5, g_2, g_6)\phi(g_6, g_4, g_1). \end{aligned} \qquad (15)$$

The Feynman diagrams of the theory are then, by construction, 3d triangulations, while the corresponding Feynman amplitudes are given by the Ponzano-Regge spin foam model [3].

Now [1] we look at two-dimensional variations of the ϕ-field around classical solutions of the corresponding equations of motion:

$$\phi(g_3, g_2, g_1) = \frac{\lambda}{3!} \int dg_4 dg_5 dg_6 \phi(g_3, g_4, g_5)\phi(g_5, g_2, g_6)\phi(g_6, g_4, g_1). \qquad (16)$$

Calling $\phi^{(0)}$ a generic solution to this equation, we look at field variations $\delta\phi(g_1, g_2, g_3) \equiv \psi(g_1 g_3^{-1})$ which do not depend on the group element g_2. We consider a specific class of classical solutions, named "flat" solutions (they can be interpreted as quantum flat space on some a priori non-trivial topology):

$$\phi^{(0)}(g_1, g_2, g_3) = \sqrt{\frac{3!}{\lambda}} \int dg\; \delta(g_1 g) F(g_2 g) \delta(g_3 g), \quad F : G \to \mathbb{R}. \qquad (17)$$

As shown in [1], this ansatz gives solutions to the field equations as soon as $\int F^2 = 1$.

This leads to an effective action for the 2d variations ψ:

$$\begin{aligned} S_{eff}[\psi] &= \frac{1}{2}\int \psi(g)\mathcal{K}(g)\psi(g^{-1}) - \frac{\mu}{3!}\int [dg]^3\psi(g_1)\psi(g_2)\psi(g_3)\delta(g_1g_2g_3) \\ &-\frac{\lambda}{4!}\int [dg]^4\psi(g_1)..\psi(g_4)\delta(g_1..g_4), \end{aligned} \tag{18}$$

with the kinetic term and the 3-valent coupling given in term of F:

$$\mathcal{K}(g) = 1-2\left(\int F\right)^2 - \int dhF(h)F(hg), \qquad \frac{\mu}{3!} = \sqrt{\frac{\lambda}{3!}}\int F.$$

with $F(g)$ assumed to be invariant under conjugation $F(g) = F(hgh^{-1})$.

Such an action defines a non-commutative quantum field theory invariant under the quantum double of SU(2) (a quantum deformation of the Poincaré group) [16, 13, 1, 20, 14].

Being an invariant function, F can be expanded in group characters:

$$F(g) = \sum_{j\in\mathbb{N}/2} F_j\chi_j(g), \qquad F_0 = \int F, \quad F_j = \int dg\, F(g)\chi_j(g), \tag{19}$$

where the F_j's are the Fourier coefficients of the Peter-Weyl decomposition in irreducible representations of SU(2), labelled by $j \in \mathbb{N}/2$. The kinetic term reads then:

$$\mathcal{K}(g) = 1 - 3F_0^2 - \sum_{j\geq 0}\frac{F_j^2}{d_j}\chi_j(g) = \sum_{j\geq 0} F_j^2\left(1-\frac{\chi_j(g)}{d_j}\right) - 2F_0^2 \equiv Q^2(g) - M^2. \tag{20}$$

It is easy to check that $Q^2(g) \geq 0$ with $Q(\mathbb{I}) = 0$. We interpret this term as the generalized "Laplacian" of the theory while the 0-mode F_0 defines the mass $M^2 \equiv 2F_0^2$.

If we choose the simple solution (other choices will give more complicated kinetic terms)

$$F(g) = a + b\chi_1(g), \qquad \int F = a^2 + b^2 = 1, \tag{21}$$

we obtain

$$\mathcal{K}(g) = \frac{4}{3}(1-a^2)\vec{p}^{\,2} - 2a^2. \tag{22}$$

3.3. 4d case

Let us consider a general 4d GFT related to topological BF quantum field theories, i.e. whose Feynman expansion leads to amplitudes that can be interpreted as discrete BF path integrals for gauge group $\mathcal{G}$. This is given by the following action:

$$\begin{aligned} S_{4d} &= \frac{1}{2}\int [dg]^4\phi(g_1,g_2,g_3,g_4)\phi(g_4,g_3,g_2,g_1) \\ &-\frac{\lambda}{5!}\int [dg]^{10}\phi(g_1,g_2,g_3,g_4)\phi(g_4,g_5,g_6,g_7)\phi(g_7,g_3,g_8,g_9)\phi(g_9,g_6,g_2,g_{10})\phi(g_{10},g_8,g_5,g_1), \end{aligned} \tag{23}$$

where the field is again required to be gauge-invariant, $\phi(g_1,g_2,g_3,g_4) = \phi(g_1g,g_2g,g_3g,g_4g)$ for any $g \in \mathcal{G}$. The relevant group for our construction will be SO(4,1), which requires some regularization to avoid divergencies due to its non-compact nature.

We generalize to 4d the "flat solution" ansatz of the 3d group field theory as [1]:

$$\phi^{(0)}(g_i) \equiv \sqrt[3]{\frac{4!}{\lambda}} \int dg\, \delta(g_1 g) F(g_2 g) \widetilde{F}(g_3 g) \delta(g_4 g), \tag{24}$$

with $(\int F\widetilde{F})^3 = 1$. The effective action around such background is [2]:

$$\begin{aligned} S_{eff}[\psi] &= \frac{1}{2}\int \psi(g)\psi(g^{-1})\mathcal{K}(g) \\ &- \sqrt[3]{\frac{\lambda}{4!}} \int F \int \widetilde{F} \int \psi(g_1)..\psi(g_3)\, \delta(g_1..g_3) \left[\int F \int \widetilde{F} + \int dh F(hg_3)\widetilde{F}(h)\right] \\ &- \left(\sqrt[3]{\frac{\lambda}{4!}}\right)^2 \int F \int \widetilde{F} \int \psi(g_1)..\psi(g_4)\, \delta(g_1..g_4) - \frac{\lambda}{5!}\int \psi(g_1)..\psi(g_5)\, \delta(g_1..g_5), \end{aligned} \tag{25}$$

with the kinetic operator given by:

$$\mathcal{K}(g) = \left[1 - 2\left(\int F \int \widetilde{F}\right)^2 \int F\widetilde{F} - 2\int F \int \widetilde{F} \int dh F(hg)\widetilde{F}(h) \int dh F(h)\widetilde{F}(hg)\right]. \tag{26}$$

A simpler special case of the classical solution above is obtained choosing $\tilde{F}(g) = \delta(g)$ while keeping F arbitrary but with $F(\mathbb{I}) = 1$. Calling $c \equiv \int F$, the effective action becomes:

$$\begin{aligned} S_{eff}[\psi] &= \frac{1}{2}\int \psi(g)\psi(g^{-1}) \left[1 - 2c^2 - 2cF(g)F(g^{-1})\right] \\ &- c\left(\sqrt[3]{\frac{\lambda}{4!}}\right) \int \psi(g_1)..\psi(g_3)\, \delta(g_1..g_3) \left[c + F(g_3)\right] \\ &- c\left(\sqrt[3]{\frac{\lambda}{4!}}\right)^2 \int \psi(g_1)..\psi(g_4)\, \delta(g_1..g_4) - \frac{\lambda}{5!}\int \psi(g_1)..\psi(g_5)\, \delta(g_1..g_5). \end{aligned} \tag{27}$$

In order to make contact with deformed special relativity, we now specialize this construction to one that gives an effective field theory based on the momentum group manifold AN_3.

We start then, as anticipated, from the group field theory describing SO(4, 1) BF-theory.

From the quantum gravity perspective, there are several reasons of interest in this model: 1) the McDowell-Mansouri formulation (as well as related ones [21]) defines 4d gravity with cosmological constant as a BF-theory for SO(4, 1) plus a potential term which breaks the gauge symmetry from SO(4, 1) down to the Lorentz group SO(3, 1); this suggests to try to define Quantum Gravity in the spin foam context as a perturbation of a topological spin foam model for $SO(4,1)$ BF theory. These ideas could also be implemented directly at the GFT level, and the starting point would necessarily be a GFT for $SO(4,1)$ of the type we use here. 2) we expect [3, 4] any classical solution of this GFT model to represent quantum De Sitter space on some given topology, and such configurations would be physically relevant also in the non-topological case. 3) the spinfoam/GFT model for SO(4, 1) BF-theory seems the correct arena to build a spin foam model for 4d quantum gravity plus particles on De Sitter space [22], treating them as topological curvature defects for an $SO(4,1)$ connection, similarly to the 3d case [16].

Following the above procedure we naturally obtain an effective field theory living on SO(4, 1). We want then to obtain from it an effective theory on $AN_3^{(c)}$. We choose:

$$F(g) = \alpha(v_4(g) + a)\vartheta(g), \qquad \tilde{F}(g) = \delta(g). \tag{28}$$

The function v_4 is defined as matrix element of g in the fundamental (non-unitary) five-dimensional representation of $SO(4,1)$, $v_4(g) = \langle v^{(0)}|g|v^{(0)}\rangle$, where $v^{(0)} = (0,0,0,0,1)$ is, as previously, the vector invariant under the $SO(3,1)$ Lorentz subgroup. $\vartheta(g)$ is a cut-off function providing a regularization of F, so that it becomes an L^1 function. Assuming that $\vartheta(\mathbb{I}) = 1$, we require $\alpha = (a+1)^{-1}$ in order for the normalization condition to be satisfied.

Then we can derive the effective action around such classical solutions for 2d field variations:

$$\begin{aligned} S_{eff}[\psi] &= \frac{1}{2}\int \psi(g)\psi(g^{-1})\left[1 - 2c^2 - \frac{2c\vartheta^2(g)(a+v_4(g))^2}{(a+1)^2}\right] \\ &\quad -c\left(\frac{\lambda}{4!}\right)^{\frac{1}{3}}\int \psi(g_1)..\psi(g_3)\,\delta(g_1..g_3)\,[c+F(g_3)] \\ &\quad -c\left(\frac{\lambda}{4!}\right)^{\frac{2}{3}}\int \psi(g_1)..\psi(g_4)\,\delta(g_1..g_4) - \frac{\lambda}{5!}\int \psi(g_1)..\psi(g_5)\,\delta(g_1..g_5), \end{aligned} \tag{29}$$

where $c = \int F$. Thus the last issue to address in order to properly define this action is to compute the integral of F; this can be done, and we refer to [2] for details.

We recognize the correct kinetic term for a DSR field theory. However, the effective matter field is still defined on a $SO(4,1)$ momentum manifold. The only remaining issue is therefore to understand the "localization" process of the field ψ to AN_3^c.

The kinetic term does not show any dependence on the Lorentz sector. This suggests that the SO(3,1) degrees of freedom are non-dynamical and that the restriction of the field ψ to AN_3^c group elements defines the complete dynamics of the theory. This would be trivially true if not for the fact that the interaction term depend also on the Lorentz degrees of freedom. One way to make this manifest is, for example, to assume that the perturbation field ψ has a product structure $\psi(g) = \tilde{\psi}(h)\Psi(\Lambda)$. The only contribution to the kinetic term from the Lorentz sector is a constant multiplicative term $\int_{SO(3,1)} d\Lambda\Psi(\Lambda)\Psi(\Lambda)$. Therefore we get an exactly DSR-like and κ-Poincaré invariant free field theory. On the other hand, the vertex term couples Lorentz and AN_3 degrees of freedom; thus the κ-Poincaré symmetry is broken and the pure DSR-like form lost. The above also shows that, if we were to choose the dependence of the perturbation field on the Lorentz sector to be trivial, i.e. $\Psi(\Lambda) \equiv 1$, and thus to *start* from a perturbation field defined only on the AN_3 subgroup, we would indeed obtain a DSR field theory, but with an interaction term that would be more complicated that a simple polinomial interaction, due to to the integrations over the Lorentz group. Of course, it would be a possible DSR field theory nevertheless. Still, because of the form of the kinetic term, we believe a reduction to the AN_3 sector to happen dynamically, or that a proper canonical analysis would show that the SO(3,1) modes are pure gauge and can thus drop from the action altogether. Anyway, notice that a restricted theory obtained from the above and living on AN_3^c only is dynamically stable. In fact, if we consider only excitations of the field in AN_3^c, we will never obtain excitations in SO(3,1) due to momentum conservation $\delta(g_1..g_n)$ since AN_3^c is a subgroup.

Then, restricting ourselves to group elements $h_i \in \mathrm{AN}_3^c$, we have the field theory:

$$\begin{aligned} S_{final}[\psi] &= \frac{1}{2}\int \psi(h)\psi(h^{-1})\left[1 - 2c^2 - 2cv_4(h)^2\vartheta(h)^2\right] \\ &\quad -c\left(\frac{\lambda}{4!}\right)^{\frac{1}{3}}\int \psi(h_1)..\psi(h_3)\delta(h_1..h_3)\,[c+v_4(h_3)\vartheta(h_3)] \\ &\quad -c\left(\frac{\lambda}{4!}\right)^{\frac{2}{3}}\int \psi(h_1)..\psi(h_4)\delta(h_1..h_4) - \frac{\lambda}{5!}\int \psi(h_1)..\psi(h_5)\delta(h_1..h_5), \end{aligned} \tag{30}$$

with implicit left-invariant measure on AN_3^c. We have thus derived a DSR scalar field theory with a κ-deformed Poincaré symmetry from the GFT for SO(4,1) topological BF-theory.

For other possible strategies leading to the same result, and for more details on the above one, see [2]. In particular, notice that we could have directly started from a BF-like GFT action for the group AN_3. Following the same procedure, and with appropriate regularization, we would have obtained easily a DSR field theory of the type we want. What would be less clear, in this case, and this is why we have not focused on this simplified setting, is the link between the initial theory and known classical or quantum formulations of gravity.

4. Conclusions

In this contribution, we have briefly reviewed recent results concerning the relation between non-commutative matter field theories, characterized by a configuration space with a Lie algebra structure and a momentum space being a group manifold, and a symmetry group given by quantum deformations of the Poincaré group, and group field theories, which are candidate models for the microscopic description of quantum spacetime and of its dynamics, merging the insights of canonical loop quantum gravity, covariant spin foam models and simplicial gravity approaches. In particular, we have seen how the former emerge naturally from the later, in a way reminiscent of the emergence of field theories for quasi-particles from hydrodynamics in condensed matter theory and analog gravity models. We have motivated this line of research as a step forward in a research programme trying to bridge the gap between group field theories as fundamental descriptions of a quantum, discrete spacetime at the Planck scale, and the continuum spacetime physics of gravity and matter at macroscopic scales, using ideas and methods from quantum and statistical field theory and condensed matter physics, applied directly at the GFT level. To be sure: 1) there are many technical and physical aspects of the procedure used to derive non-commutative matter field theories from GFTs that need further analysis; 2) one important difference with respect to analog gravity models should be understood: while quasi-particle dynamics on effective metrics is there obtained from the hydrodynamics of the fundamental system, in our GFT context we have applied a similar procedure directly to the fundamental microscopic field theory; one could speculate that this is why we have obtained *non-commutative* field theories for the emergent matter, as opposed to ordinary field theries; 3) the same results have motivations and relevance that are independent from the perspective we have adopted here to present them. These relate to formal aspects of the relation between non-commutative geometry and group field theory, and to the importance of these non-commutative models (and of the general DSR idea) to Quantum Gravity phenomenology [2]. For all this, we refer to the literature [14, 13, 15, 5]. However, we believe that these interesting results fit well in the more general scheme we have outlined, on the one hand supporting the picture of GFTs as fundamental theories of quantum gravity, rather than just auxiliary mathematical tools, and on the other hand acquiring, from this more general perspective, an additional reason of interest.

Acknowledgments

We thank Florian Girelli and Etera Livine, for the enjoyable collaboration, and for many discussions on non-commutative field theories, DSR and group field theory, and the organizers of DICE2008 for a stimulating conference and for the opportunity to present our work and ideas.

References

[1] Fairbairn W, Livine E 2007 *Class. Quant. Grav.* **24** 5277 (*Preprint* arXiv:gr-qc/0702125)
[2] Girelli F, Livine E, Oriti D 2009 4d deformed special relativity from group field theories *Preprint* arXiv:0903.3475
[3] Oriti D 2006 *Preprint* arXiv:gr-qc/0607032; Oriti D 2007 *Mathematical and Physical Aspects of Quantum Gravity* ed B Fauser, J Tolksdorf and E Zeidler (Basel: Birkhaeuser) (*Preprint* arXiv:gr-qc/0512103)
[4] Freidel L 2005 *Int. J. Theor. Phys.* **44** 1769 (*Preprint* arXiv:hep-th/0505016)
[5] Amelino-Camelia G 2005 *Lect. Notes Phys.* **669** 59-100 (*Preprint* arXiv:gr-qc/0412136)
[6] Oriti D 2007 *Proceedings of Science Preprint* arXiv:0710.3276

[7] Rovelli C 2004 *Quantum Gravity* (Cambridge, UK: Cambridge U. Press)
[8] Perez A 2003 *Class. Quant. Grav.* **20** R43 (*Preprint* arXiv:gr-qc/0301113)
[9] Oriti D *ed* 2009 *Approaches to Quantum Gravity* (Cambridge, UK: Cambridge U. Press)
[10] Freidel L, Gurau R and Oriti D *Preprint* GFT renormalization - 3d case: power counting of divergences, to appear
[11] Barcelo C, Liberati S and Visser M 2005 *Living Rev. Rel.* **8** 12 (*Preprint* arXiv:gr-qc/0505065)
[12] Volovik G E 2001 *Phys. Rep.* **351** 195 (*Preprint* arXiv:gr-qc/0005091); *do.* 2006 *Preprint* gr-qc/0612134
[13] Freidel L and Majid S 2006 *Preprint* arXiv:hep-th/0601004
[14] Livine E 2008 *Preprint* arXiv:0811.1462
[15] Freidel L, Kowalski-Glikman J and Nowak S 2007 *Preprint* arXiv:0706.3658
[16] Freidel L and Livine E 2006 *Class. Quant. Grav.* **23** 2021-2062 (*Preprint* arXiv:hep-th/0502106)
[17] Oriti D and Ryan J 2006 *Class. Quant. Grav.* **23** 6543 (*Preprint* arXiv:gr-qc/0602010)
[18] Majid S 2000 *Foundations of quantum group theory* (Cambridge, UK: Cambridge U. Press)
[19] Vilenkin N and Klimyk A *Representations of Lie Groups* (Berlin: Springer)
[20] Joung E, Mourad J and Noui K 2008 *Preprint* arXiv:0806.4121
[21] Freidel L and Starodubtsev A 2005 *Preprint* arXiv:hep-th/0501191
[22] Kowalski-Glikman J and Starodubtsev A 2008 *Phys. Rev.* D **78** 084039 (*Preprint* arXiv:0808.2613)

Fourth International Workshop DICE2008 IOP Publishing
Journal of Physics: Conference Series **174** (2009) 012048 doi:10.1088/1742-6596/174/1/012048

Massless particle diffusion in causal set theory

Lydia Philpott[1], **Fay Dowker**[1] **and Rafael Sorkin**[2]
[1] Blackett Laboratory, Imperial College London, London SW7 2AZ, UK
[2] Perimeter Institute for Theoretical Physics, Waterloo ON, Canada
E-mail: l.philpott06@imperial.ac.uk

Abstract. Particles propagating in a fundamentally discrete spacetime will not follow the smooth geodesics predicted by general relativity. In causal set theory we can construct a simple Lorentz invariant model for the 'swerving' of massive particles caused by discreteness. On macroscopic scales these swerves can be approximated by a diffusion equation in phase space. For massless particles a model is less simple to construct, but a diffusion equation can still be obtained. We find that massless particles will undergo a diffusion and drift in momentum magnitude. This effect will be most significant over cosmic distances. The diffusion and drift distorts a blackbody spectrum, and thus the perfect blackbody nature of the cosmic microwave background allows us to place tight constraints on the values of the diffusion and drift constants.

1. Introduction

Is spacetime a discrete or continuous structure? Causal set theory, a discrete approach to quantum gravity, gives a concrete framework in which to investigate this question. The hypothesised discreteness scale lies well outside the current possibilities of direct observations but this does not mean the question is unanswerable. If spacetime is discrete the signature of that discreteness may be visible to us on macroscopic scales, much like the macroscopic Brownian motion of pollen in water reveals the presence of microscopic molecules. One obvious place to look for evidence of an underlying discreteness is in the behaviour of particles. The macroscopic behaviour of massive particles on an underlying discrete spacetime was first addressed in [1]. Here the case of massless particles is investigated. Although there is as yet no underlying model for the propagation of massless particles on a causal set, the Lorentz invariance and discreteness of causal set theory leads to a general diffusion equation in the momentum of the particles. In Sections 2 and 3 the diffusion equation in, respectively, affine time and 'cosmic' time is discussed. Bounds on the free parameters are obtained in Section 4 by considering the blackbody nature of the cosmic microwave background. The results presented here are a summary of work that appears in [2].

For comprehensive reviews of causal set theory see for example [3, 4, 5]. In detail, a causal set is a set C endowed with a binary relation $\preceq$ satisfying:

(i) transitivity: if $x \preceq y$ and $y \preceq z$ then $x \preceq z$, $\forall x, y, z \in C$;

(ii) reflexivity: $x \preceq x$, $\forall x \in C$;

(iii) acyclicity: if $x \preceq y$ and $y \preceq x$ then $x = y$, $\forall x, y \in C$;

(iv) local finiteness: $\forall x, z \in C$ the set $\{y \mid x \preceq y \prec z\}$ of elements is finite.

The continuum Lorentzian manifold we observe is assumed to be an approximation in some sense to an underlying causal set, where the partial order in the causal set gives rise to the causal order on the manifold and the number of elements in a region determines the volume of that region. Defining what exactly is meant by an approximation and identifying the continuum manifold or manifolds that approximate a causal set is not a simple task. To investigate the properties of causal sets it is helpful to work backwards: begin with a manifold and from it obtain a causal set for which the manifold is an approximation. This is done by a process known as sprinkling. Given a Lorentzian manifold, (M, g), points are selected from M randomly via a Poisson process in which the probability measure is equal to the spacetime volume measure in some fundamental units. The selected points are the elements of a causal set once they have been endowed with the partial order induced by the spacetime causal order. The number of points chosen from any region of the manifold will be approximately equal to the volume of the region (in fundamental units) up to Poisson fluctuations.

Although the very simple structure of a causal set is theoretically sufficient to determine all the information about the geometry of a spacetime, things become complicated when we begin to consider matter. As we know mass and spacetime curvature are intimately related it would seem that the presence of matter should be encoded in the very structure of the causal set. Perhaps particles can be described by patterns of causal relations within the causal set. Alternatively we could think of representing matter by some sort of field, assigning a number or set of numbers to each element in our causal set. The phenomenology discussed in this article is only one of many directions from which the problem of particles on a causal set is being tackled. For work on QFT on a causal set see [6] and Johnston's paper within this volume. The propagation of fields on a causal set is discussed in [7]. Ultimately the search for an understanding of particles in causal set theory cannot be separated from the search for a quantum causal set dynamics.

We can begin to gain some intuition into the behaviour of massive particles in a discrete spacetime by considering simple classical point particle models. Examples of such models for massive particles are discussed in detail in [1, 2] so we will only briefly give one example here. It is first necessary to give some causal set definitions. Let C be a causal set.

- A *chain* is a totally ordered subset of C.
- A *longest chain* between two elements is a chain whose length is maximal amongst chains between those endpoints, and is the closest approximation we have on a causal set to a timelike geodesic. There may be more than one longest chain between two elements. For two causal set elements a and b the length of a longest chain between a and b will be denoted $d(a, b)$. The special case $d(a, b) = 1$ is called a *link*.

Given a causal set C, consider a particle trajectory to consist of a chain of elements $\ldots e_{n-2} \prec e_{n-1} \prec e_n \ldots$. The trajectory is constructed iteratively, where the trajectory's past determines its future, but only a certain number of elements n_f (the 'forgetting number') into the past is relevant. Given a partial particle trajectory $\ldots e_{n-1}, e_n$ the next element e_{n+1} is chosen such that

- $d(e_n, e_{n+1}) = n_f$,
- $d(e_{n-1}, e_{n+1}) = 2n_f$,

(see Figure 1). These requirements do not guarantee the existence of a unique e_{n+1}, but as there will almost certainly be finitely many such e_{n+1} we can construct a trajectory by choosing an element at random from those eligible. The trajectory constructed by this method will be approximately geodesic over all $\{e_{n-1} : e_{n+1}\}$ segments, but on shorter scales it will 'swerve' a little.

Motivated by these swerves that appear in point particle models Dowker et al. [1] found that the macroscopic behaviour of massive particles on an underlying discrete spacetime can

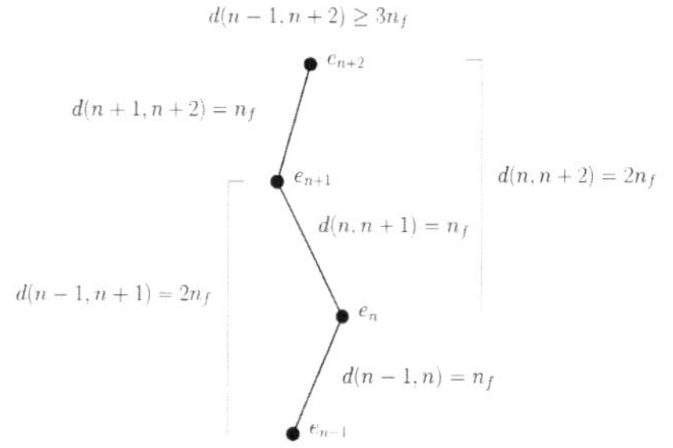

Figure 1. One method of constructing a trajectory in a causal set

be approximated by a continuum diffusion equation (much like the standard 1-dimensional diffusion equation can be derived from a discrete random walk process). For massless particles constructing a point particle model is somewhat more tricky. If we consider a causal set constructed by sprinkling into Minkowski spacetime, for any given element p there will almost surely be no element q sprinkled on the future lightcone of p. If we, therefore, attempt to construct a simple model with a massless particle jumping from one element to another causally related element the final trajectory will not approximate a null geodesic. Hopefully in the future a more complicated algorithm (perhaps involving jumps both forward and backward in time, or jumps only along links) or the study of massless fields on a causal set will allow a concrete model to be constructed. Although models can be constructed for massive particles the derived diffusion equation did not rely on a particular underlying model, and indeed is a general equation that should arise from any underlying method of massive particle propagation that is Lorentz invariant and translation invariant. Following the same method a general diffusion equation for massless particles can be derived without the need for any specific underlying model.

2. A diffusion equation for massless particles

A diffusion equation for massless particles can be obtained by following the prescription given by Sorkin in [8]. A process that undergoes stochastic evolution on a manifold of states, $\mathcal{M}$, in some time parameter, t, can be described by a current and a continuity equation:

$$J^A = -\partial_B \left(K^{AB} \rho \right) + v^A \rho \tag{1}$$

$$\frac{\partial \rho}{\partial t} = -\partial_A J^A. \tag{2}$$

Here ρ is the probability density for the system, a scalar density in $\mathcal{M}$. The coefficients K^{AB} and v^A depend on the process under investigation. The state space of a massless particle in Minkowski spacetime is $\mathbb{M}^4 \times \mathbb{H}^3_0$, where $\mathbb{H}^3_0$ denotes the 'light cone' in momentum space defined by $p_\mu p^\mu = 0$. The diffusion process takes place in affine time λ defined by $dx^\mu = p^\mu d\lambda$. By requiring the process be Lorentz invariant and translation invariant one can determine K^{AB} and v^A and thus the diffusion equation. Full details of the derivation can be found in [2]. The diffusion equation is found to be:

$$\begin{aligned} \frac{\partial \rho_\lambda}{\partial \lambda} &= -p^\mu \frac{\partial \rho_\lambda}{\partial x^\mu} + k_1 \frac{\partial}{\partial E} \left(E^3 \frac{\partial}{\partial E} \left(\frac{\rho_\lambda}{E} \right) \right) \\ &\quad - k_2 \frac{\partial}{\partial E} \left(E \rho_\lambda \right) \end{aligned} \tag{3}$$

where $E = p$ is the energy of the particle and k_1 and k_2 are two free parameters, a diffusion and drift coefficient respectively. This equation shows that a photon will undergo a diffusion in energy, but no change in direction: it will continue to propagate on the lightcone, at the speed of light. As it is expressed in terms of an unobservable affine time, this equation does not allow us to make direct connections to observations. In order to determine how such a diffusion would appear to us, and to place constraints on the diffusion and drift coefficients, it is necessary to know how a distribution would evolve in time in the rest frame of a detector, which time will be referred to as *cosmic time*.

3. Momentum diffusion for massless particles in cosmic time

Reexpressing 3 in terms of cosmic time, t, one obtains [2]:

$$\frac{\partial \rho_t}{\partial t} = -\frac{p^i}{E}\partial_i \rho_t - (k_1 + k_2)\frac{\partial \rho_t}{\partial E} + k_1 E \frac{\partial^2 \rho_t}{\partial E^2}\,, \tag{4}$$

where $E = p$ is the energy and note that the scalar density ρ_t is not the same as ρ_λ. If a distribution of photons is initially peaked at some energy the distribution will spread over time at a rate given by the diffusion parameter k_1. There will also be a drift of the spectrum to higher or lower energies depending on the magnitude and sign of k_2. The next step is to place bounds on the values of k_1 and k_2.

4. Bounding the constants k_1 and k_2

In developing a phenomenological model, one aims to provide both a model for currently unexplained observations and suggestions for new observations that might be made to test a theory. But before proposing new observations, one should of course constrain one's model as tightly as possible, based on what is already known. This model has two parameters: a positive diffusion constant k_1 and a 'drift' constant k_2, which may be either positive or negative. To place the strongest bounds on the values of these parameters, it seems sensible to look at photons that have been travelling for a very long time in the frame of the detector and thus have had the maximum possibility to experience any underlying discreteness. The cosmic microwave background (CMB) seems an ideal testing ground in this sense. Not only are its photons the "oldest" we can observe, but its spectrum has been determined with great precision. Most of the photons in the CMB have been "free streaming" for approximately 13.7 billion years, or on the order of 10^{60} Planck times. When the universe became transparent at recombination, they would have had a blackbody spectrum with a temperature $3000\,K$ (see for example [9]). Current observations yield a temperature of $2.728 \pm 0.004\,K$ and measure the spectrum to be Planckian (blackbody) over the $2 - 21cm^{-1}$ frequency range to within a weighted rms deviation of only 50 parts per million (ppm) of the peak brightness [10]. Since a diffusion as described above would have distorted the energy distribution, the fact that the CMB photons have travelled so far but remained so perfectly thermal allows the parameters to be constrained very tightly. The derivation of the massless cosmic time equation assumes spacetime to be Minkowskian, and thus to constrain k_1 and k_2 a simplified model of the CMB is considered in which the expansion of the universe is ignored and the CMB temperature remains constant at $2.728K$. This model is discussed in detail in Section 5 of [2]. A model incorporating the expansion of the universe is discussed in Section 6 of [2] and shown to give bounds of the same order of magnitude as the simplified model. To place constraints on k_1 and k_2 an initially blackbody spectrum at a temperature of $2.728K$ was evolved numerically according to 4 for a range of values of k_1 and k_2, and for a time equal to the time since the surface of last scattering. The rms deviation between the resulting final spectrum and a best-fit blackbody spectrum was then calculated and compared to the deviation of 50 parts per million (ppm) of the peak brightness allowed by observations. To obtain concrete bounds a bound on k_1 was determined by setting $k_2 = 0$ and

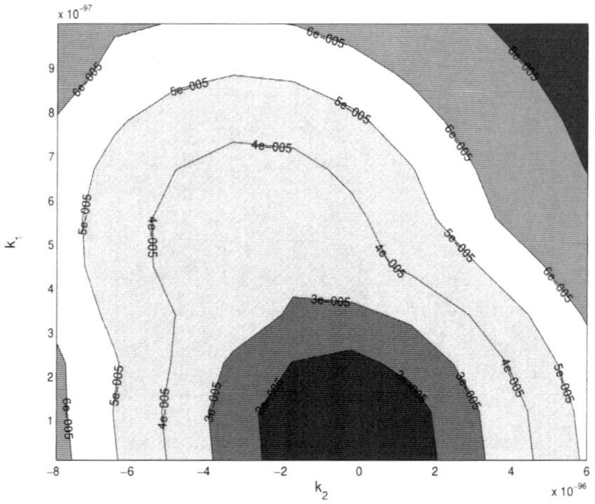

Figure 2. The rms deviation of the simulated spectrum from Planckian as a proportion of the spectrum peak, varying both k_1 and k_2. Values of k_1 and k_2 within the $5e{-}05$ contour give a spectrum that is Planckian to within 50ppm of the peak.

a bound on k_2 was determined by setting $k_1 = 0$. The resulting bounds in SI units are:

$$k_1 < 3 \times 10^{-44} kgm^2 s^{-3} \quad (5)$$

$$-1 \times 10^{-43} < k_2 < 1 \times 10^{-43} kgm^2 s^{-3}. \quad (6)$$

The deviation from Planckian varying values of both k_1 and k_2 can be seen in Figure 2.

5. Conclusion

Given an underlying discrete spacetime, particle momenta would be subject to stochastic variations, and if these variations were small, their effects would be describable on large scales as a diffusion in momentum space. This model gives us an opportunity to seek observational evidence for spacetime discreteness. The Lorentz invariance of the model discussed here is particularly important as the majority of quantum gravity phenomenological models involve violations or deformations of Lorentz invariance, which to date have not been detected. The parameters of our model are constrained by the blackbody character of the CMB radiation. This is by no means the only means of constraining the parameters: one could, for example, consider the broadening or shifting of spectral lines. The time for which the photons in the CMB have been travelling and the accuracy to which the spectrum is measured, however, suggest the bounds set here will be among the most stringent. To this point the particles have been considered to be classical and spinless, how an underlying discreteness would affect the polarization of photons remains an open question.

Acknowledgments

The authors would like to thank the organisers of DICE2008 for the opportunity to present this paper. LP acknowledges the support of a TEC doctoral scholarship.

References

[1] Dowker F, Henson J and Sorkin R D 2004 *Mod. Phys. Lett.* A **19** 1829–1840 (*Preprint* arxiv:gr-qc/0311055)
[2] Philpott L, Dowker F and Sorkin R 2008 *Preprint* arxiv:0810.5591

[3] Henson J 2009 *Approaches to Quantum Gravity: Towards a New Understanding of Space, Time and Matter* ed D Oriti (Cambridge, UK: Cambridge U. Press) (*Preprint* arxiv:gr-qc/0601121)
[4] Dowker F 2006 *Contemp. Phys.* **47** 1–9
[5] Sorkin R D 2005 *Lectures on Quantum Gravity, Proceedings of the Valdivia Summer School, Valdivia, Chile, January 2002* ed A Gomberoff and D Marolf (Plenum) (*Preprint* arxiv:gr-qc/0309009)
[6] Johnston S 2008 *Class. Quant. Grav.* **25** 202001 (*Preprint* arxiv:0806.3083)
[7] Dowker F, Henson J and Sorkin R D The coherence of light from distant sources in a discrete universe, in preparation
[8] Sorkin R 1986 *Ann. Phys.* **168** 119–147
[9] Kolb E W and Turner M S 1990 *The Early Universe* (Addison-Wesley)
[10] Fixsen D J *et al* 1996 *Astrophys. J.* **473** 576 (*Preprint* arxiv:astro-ph/9605054)

Journal of Physics: Conference Series **174** (2009) 012049 doi:10.1088/1742-6596/174/1/012049

A result in 2d causal set theory: the emergence of spacetime.

Graham Brightwell[1], Joe Henson[2] and Sumati Surya[3]

[1]London School of Economics, London, UK,
[2]Perimeter Insitute, Waterloo, Canada,
[3]Raman Research Institute, Bangalore, India

E-mail: ssurya@rri.res.in

Abstract. A 2d model of causal set quantum gravity is constructed using a continuum-inspired dynamics. Apart from a restriction to causal set dimension and topology, the model is fully dynamical and includes all relevant 2d conformally flat degrees of freedom. Surprisingly, in the large N limit the partition function is dominated by causal sets that resemble 2d Minkowski spacetime. Thus, in this model the "entropy problem" of causal set theory is overcome and a sensible low energy limit is obtained.

One of the biggest challenges of causal set theory [1, 2] is to construct a viable quantum dynamics which gives rise to manifoldlike causal sets in the classical limit. A causal set, or a locally finite partially ordered set (Fig (a)) is in general devoid of continuum characteristics; the order relation maps to the causal relation only for a causal set which admits a *continuum*

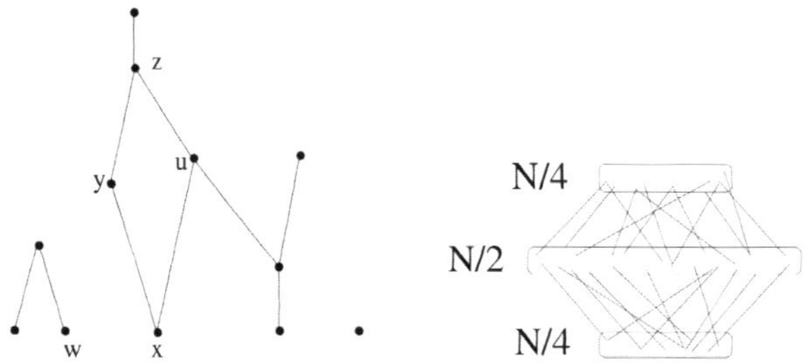

(a) (b)

Figure 1. (a) A Hasse diagram of a poset of 12 elements, with a relation $\prec$ which is (i) acyclic: $x \not\prec x$, (ii) transitive: $x \prec y$ and $y \prec z \Rightarrow x \prec z$, and (iii) finite: the set $\{u|x \prec u, u \prec z\}$ for every (x, z) is of finite cardinality. If $x \prec y$ and there exists no p such that $x \prec p \prec y$, then x and y are said to be *linked*. (b) A typical large N-element causal set has three levels with approximately $N/2$ elements in the middle layer and $N/4$ each in the remaining two layers.

approximation. Indeed, in the limit of large N most N-element causal sets are non-manifoldlike, being "squat" with only 3-levels or "moments of time" (Figure (b)) [3]. Manifoldlike behaviour is thus far from being an obvious consequence of the theory and any viable dynamics must suppress this "entropic" dominance of squat causal sets.

There are two ways to proceed in constructing a dynamics for causal sets. One is a top-down approach in which the continuum is used as a "prop" to obtain a discretised dynamics. This is the method adopted by most other approaches to quantum gravity. The other, bottom-up approach, is an attempt to start from a new, fundamental set of principles, and to obtain in the appropriate limits, the dynamics familiar to us from continuum physics [1]. These approaches are not delinked, since the former can suggest what the fundamental principles adopted by the latter should be. An important example is the principle of general covariance. Its discrete version, label invariance, is a natural (and necessary) principle to adopt for causal set dynamics.

The bottom-up approach has been successfully used to construct a class of classical sequential growth dynamics based on the principles of general covariance or label invariance and an appropriate "Bell" causality condition [4, 5]. Starting with a single element, the Markovian stochastic dynamics "grows" a causal set to the future by adding a new element at every stage, which does not lie to the past of any of the existing elements. In a large class of these dynamics, the entropy from squat causal sets is suppressed in the large N limit. However, manifoldlike causal sets do not seem to appear in any generic way in these theories. While slow progress is being made in constructing a viable quantum dynamics from causal sets, it is a worthwhile excercise to examine whether there are further lessons to be learnt from the top-down approach.

A standard assumption, adopted by all other approaches to quantising gravity but not required by causal set theory, is to restrict to spacetime dimension. In addition, as is again often the practice, one can restrict the topology to some fixed manifold M. The continuum partition function is

$$Z_M = \int [dg] e^{iS_{EH}(g)} \tag{1}$$

where g is the spacetime metric and $S_{EH}(g)$ the Einstein-Hilbert action. In a top-down approach the idea is to mimic this structure by first discretising the spacetimes (M, g) and then the dynamics. This is the method followed by the Regge-triangulation type approaches to quantisation [6, 7] both in the Lorentzian and Euclidean contexts.

In causal set theory, one has a concrete prescription for discretising (M, g) via a *Poisson sprinkling* at some density V_c^{-1}. The probability that a volume v region of (M, g) contains n elements from such a sprinkling is given by

$$P_v(n) = \frac{1}{n!} e^{-\frac{v}{V_c}} \left(\frac{v}{V_c}\right)^n. \tag{2}$$

The causal set C is then constructed by adding in the causal relations induced from (M, g) to these sprinkled elements. This choice of a random discretisation maintains a covariant number to volume correspondence $\langle n \rangle = \frac{V}{V_c}$. Along with the causal relation, it thus provides a discrete version of Malament's continuum result paraphrased as "causal structure + local volume element= (strongly causal) spacetime" [8]. Conversely, a causal set C is said to be approximated by (M, g) if it can arise from such a discretisation with high probability. Thus, the path integral over spacetimes can be replaced by a sum over causal sets, each of which is approximated by a spacetime in the class $\{(M, g)\}$.

On the other hand, because of the randomness, a nearest neighbour based discretisation of the Einstein-Hilbert action is not possible. Every element in a causal set that is approximated by Minkowski spacetime, for example, has infinite "nearest neighbours" which lie inside the future and past unit hyperboloids (Figure 2). Thus, any prescription for a discretised action must necessarily take the randomness of the lattice into consideration. One approach, adopted

[1] Rafael Sorkin calls the first an "opportunistic" route and the second a "principled" route.

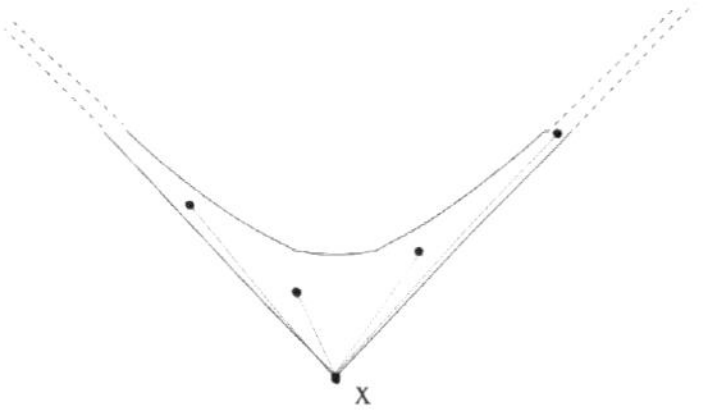

Figure 2. The number of elements linked to x in 2d Minkowski spacetime is infinite .

in [9] constructs the action S_{EH} from small Alexandrov intervals with minimal "fluctuations". Another approach is to construct the Ricci scalar using the so-called Synge function from the Dalembertian operator [10].

If we were to stop at this stage, the discretisation would provide only a spacetime regularisation of the partition function since each causal set in the sum has a spacetime approximation. While this in itself is a reasonable direction to take, it sidesteps the question of whether a theory based more fundamentally on causal sets can give rise to manifoldlike behaviour in the large N limit. In order to address this question one would have to generalise the sum to include causal sets that are not manifoldlike, but which nevertheless share some property of the continuum class of spacetimes $\{(M, g)\}$. This property may in general be difficult to define, but we will present a situation in which it is indeed natural [13].

We start with the class of 2d conformally flat spacetimes $\{(M, g)\}$ with $M \simeq \mathbb{R}^2$. Every $g_{ab}\, dx^a dx^b = \Omega^2(u, v)\, \eta_{ab}\, dx^a dx^b$ where $\eta_{ab}\, dx^a dx^b = -du\, dv$ is a fiducial flat metric on M, (u, v) being lightcone coordinates, and $\Omega^2(u, v)$ a conformal factor [13]. The continuum action is

$$S_{EH} = \frac{1}{16\pi G} \int_M R dV - \frac{1}{8\pi G} \int_{\partial M} k dS - \frac{1}{8\pi G} \Lambda \int_M dV, \tag{3}$$

where Λ is the cosmological constant. It is natural in the causal set context to impose in addition, the so-called unimodular constraint thus fixing the four volume $V \equiv \int_M dV$. This means restricting the sum over causal sets to those of a fixed cardinality[2]. The most natural choice for fixed volume spacetimes in the Lorentzian context are the interval spacetimes (I, g) (Figure 3) for which the action is

$$S_{EH}(V) = \frac{1}{16\pi G} \int_I R dV - \frac{1}{8\pi G} \int_{\partial I} k dS - \sum_j \frac{1}{8\pi G} \theta_j - \frac{1}{8\pi G} \Lambda V, \tag{4}$$

where θ_j are the boost parameters associated with the four joints in ∂I. Using the Gauss Bonnet theorem on a Lorentzian interval [14] $S_{EH}(V)$ simplifies to a constant $\Phi = \frac{1}{16\pi G}(2\pi - 2\Lambda V)$, so that

$$Z(V) = e^{i\Phi} \times \int [dg] \tag{5}$$

A causal set that is approximated by a 2d conformally flat spacetime has a particularly simple and elegant characterisation as a *2D order*. Let C be an N-element causal set obtained from (I, g), and let U and V be the set of induced lightcone coordinates u and v, respectively, for the embedded elements of C. The probability that any two of the u coordinates coincide is zero (since the spacetime volume between two elements on a lightcone is zero) which implies that U

[2] This constraint is "mild" in that it does not modify the field equations [11]. It is an important ingredient in obtaining the prediction for the size of the fluctuation in the cosmological constant [12].

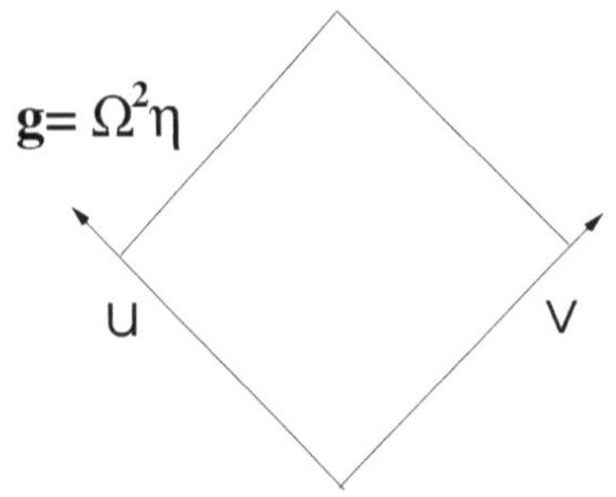

Figure 3. The interval (I, g) can be obtained by cutting out an Alexandrov interval (I', η) in 2d Minkowski spacetime and replacing it with (I, g), so that $\partial I = \partial I'$.

is a *total order*, i.e., every pair of distinct elements in U is related. Similarly, V is a total order. Moreover, for any pair of elements $e_1 = (u_1, v_1)$, $e_2 = (u_2, v_2)$ in C

$$e_1 \prec e_2 \quad \Leftrightarrow \quad u_1 < u_2 \quad \text{and} \quad v_1 < v_2. \tag{6}$$

The "intersection" of the two linear orders $U \cap V$ is a 2D order and gives a labelling of C (Figure 4(a)). However, the class of 2D orders also include those that are non-manifoldlike (Figure 4(b)).

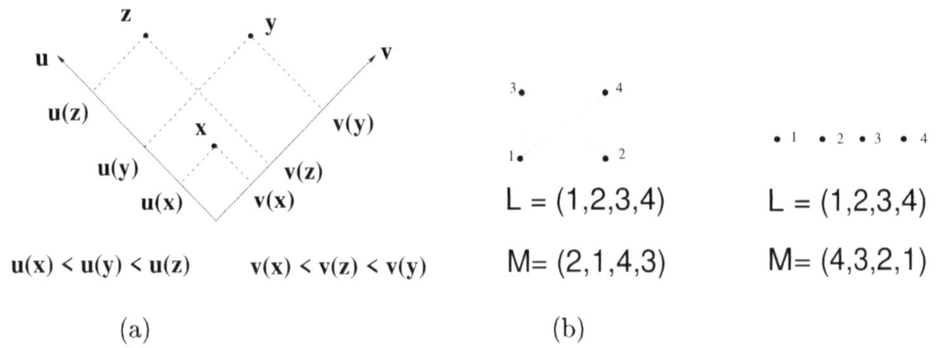

Figure 4. (a) An embedding of 3 points into a portion of 2d Minkowski spacetime. (b) Examples of 2D orders $L \cap M$: $e_i \prec e_j$ in $L \cap M$, iff $e_i \prec e_j$ in L and M.

As shown in [13] 2D orders moreover do not contain non-trivial 1-cycles and hence cannot be approximated by 2d spacetimes not in the conformally flat class [13]. Thus, extending the partition function to a sum over all unlabelled 2D orders would include not only all causal sets that are approximated by conformally flat spacetimes, but also those that are non-manifoldlike, yet in the same topological class as 2d conformally flat spacetimes.

The simplest route to discretising the action (5) then gives us a partition function for 2d causal set quantum gravity

$$Z(N) = e^{i\Phi} \times \sum_{\text{2Dorders}} 1 = e^{i\Phi} \times \; U(N) \tag{7}$$

where $U(N)$ is the uniform distribution over the space of unlabelled 2D orders. At first sight this oversimplification suggests that the entropy problem has been revived in a 2D order guise. However, quite surpringly, the results of El-Zahar and Sauer indicate the contrary [15].

In order to understand that result, we visit another distribution on the space of labelled 2D orders, the *random 2D order* $P(N)$. Starting from a labelled N-element set, pick two linear orderings L and M randomly and independently, and construct the 2D order $P(N) = L \cap M$. Even though there are $N!$ ways of choosing each, the set of distinct labelled 2D orders is of cardinality $< N!^2/2$ [16], thus giving rise to a non-trivial distribution $P(N)$. $P(N)$ has a special place in causal set theory, since one can show that it is precisely the distribution obtained via a Poisson sprinkling into the flat interval spacetime (I, η).

It was shown by El-Zahar and Sauer that in the large N limit, all label-invariant properties of the two distributions $U(N)$ and $P(N)$ coincide [15, 16]. Interpreted in our language, this implies that in the large N limit, the partition function $Z(N)$ is dominated by causal sets that are approximated by the 2d Minkowski interval. In this sense, Minkowski spacetime is a *prediction* of the model.

Starting from a model of causal set quantum gravity in which non-manifoldlike excitations are included in a fully non-perturbative setting the results of [15, 16] tell us that the continuum is emergent in the large N limit. Moreover, it can be shown that deviations of $U(N)$ from $P(N)$ are of $O(1/N)$ and hence suitably suppressed at large N. The natural interpretation of the $N \to \infty$ limit in causal set theory is the infinite volume limit, with the sprinkling density V_c fixed, so that full Minkowski spacetime is the prediction of the model[3].

We end with some general remarks. We have sidestepped the issues of quantum interpretation in quantum gravity by the use of a partition function in which the complex part of the measure appears as an overall phase. Thus, the standard thermodynamic interpretation suffices for our purpose. Introducing relative phases into the partition function will force us to confront these questions. A promising candidate is the anhomomorphic logic based approach [17], and it would be of interest to apply it to such a simplified model. Even within a thermodynamic context, it is useful to see if a more intrinsic order theoretic choice for the partition function will give rise to a similar result. We have also ignored matter contributions, so that our model has no simple relationship to standard continuum 2D quantum gravity. Introducing a scalar field as in [10] would be another way of generalising our model. Indeed, the relative simplicity of the class of 2D orders should make it possible in future to use it to test a variety of conceptual as well as technical questions in causal set quantum gravity.

References

[1] Bombelli L, Lee J H, Meyer D and Sorkin R D 1987 *Phys. Rev. Lett.* **59** 521
[2] Sorkin R D 2002 *Lectures on quantum gravity – School on Quantum Gravity, Valdivia, Chile, 4-14 Jan 2002* (Published in Valdivia 2002) 305-327 (*Preprint* arXiv:gr-qc/0309009)
[3] Kleitman D and Rothschild B L 1975 *Trans. Am. Math. Soc.* **205** 205
[4] Rideout D P and Sorkin R D 2000 *Phys. Rev.* D **61** 024002
[5] Varadarajan M and Rideout D P 2006 *Phys. Rev.* D **73** 104021 (*Preprint* arXiv:gr-qc/0504066)
[6] Regge T 1961 *Nuovo Cimento* **19** 558; Sorkin R D 1975 *Phys. Rev.* D **12** 385; Hartle J B and Sorkin R D 1981 *Gen. Rel. Grav.* **13** 541
[7] David F 1993 *Preprint* arXiv:hep-th/9303127; *do.* 1990 *Lectures at 8th Jerusalem Winter School for Theoretical Physics, Two-Dimensional Gravity and Random Surfaces, Jerusalem, Israel, Dec 27 - Jan 4, 1991* published in Jerusalem Gravity 1990:0125-141 (QC178:J4:1990)
[8] Malament D 1977 *J. Math. Phys* **18** 1399
[9] Sverdlov R and Bombelli L 2008 *Preprint* arXiv:gr-qc/0801.0240
[10] Sorkin R D 2006 Does locality fail at intermediate length scales? *Approaches to Quantum Gravity - Toward a new understanding of space and time* ed D Oriti (Cambridge, UK: Cambridge U. Press); Henson J 2006 The causal set approach to quantum gravity *Approaches to Quantum Gravity - Toward a new understanding of space and time* ed D Oriti (Cambridge, UK: Cambridge U. Press)
[11] Unruh W G 1989 *Phys. Rev.* D **40** 1048; Sorkin R D 1994 *Int. J. Theor. Phys.* **33** 523

[3] Keeping the spacetime volume V fixed instead gives us the (unphysical) continuum limit of the model.

[12] Sorkin R D 1997 *Int. J. Theor. Phys.* **36** 2759; Ahmed M, Dodelson S, Greene P B and Sorkin R D 2004 *Phys. Rev.* D **69** 103523
[13] Brightwell G, Henson J and Surya S 2008 *Class. Quant. Grav.* **25** 105025 (*Preprint* arXiv:gr-qc/0706.0375)
[14] Birman G S and Nomizu K 1984 *Mich. Math. J.* **31** 77-81
[15] El-Zahar M H and Sauer N W 1988 *Order* **5** 239
[16] Winkler P 1991 *Order* **7** 329
[17] Sorkin R D 2007 *J. Phys.: Conf. Ser.* **67** 012018 (*Preprint* arXiv:quant-ph/0703276)

Quantum tetrahedra: a toy model for classical limit of the quantum gravity

Daniel R. Terno

Centre for Quantum Computer Technology, Faculty of Science, Macquarie University, Sydney NSW 2113, Australia

E-mail: `dterno@physics.mq.edu.au`

Abstract. We present an asymptotically optimal generalized measurement for the joint extraction of classical information from a quantum tetrahedron for set of non-commuting observables. The optimal uncertainty in dihedral angles is shown to scale as an inverse of the surface area. However, clustering of many small tetrahedra leads to a faster convergence to a classical geometry.

1. Introduction

One of the major results of loop quantum gravity (LQG) [1, 2] is a quantization of space: the spectrum of geometric operators representing area or volume is discrete. Basis states in the kinematical Hilbert space of LQG are represented by spin networks, which are finite, directed, labeled graphs. Their edges are labeled by SU(2) spins, and SU(2)-invariant tensors (intertwiners) label the vertices. A spin label attached to a link determines the area of a surface that is intersected by it, while an intertwiner is associated with the volume of a spatial region that contains the vertex.

The simplest state that represents a finite volume of space involves a four-valent spin network vertex. A dual picture associates vertices with tetrahedra and edges with two-dimensional surfaces, so an *atom of space* can be represented as a quantum tetrahedron. Its triangular areas fix the spin labels and the volume determines the intertwiner. The same mathematical structure results from a formal quantization of a classical tetrahedron [3].

Geometry of a classical tetrahedron is determined by six parameters. Six geometric variables are associated with observables on the Hilbert space $\mathcal{H}_0$ of a quantized tetrahedron. However, the maximal set of commuting observables contains only five operators, which makes a retrieved classical description intrinsically fuzzy. A study of the emergence of sharp values for a set of non-commuting observables a useful toy model for the emergence of classical space [4]. Techniques of quantum information theory play a central role in our analysis.

2. Classical and quantum tetrahedra

There are several sets of six numbers that determine tetrahedron's shape [3, 6]. We use four facial areas and two independent angles between them, as a shape-defining set. This choice gives a natural way to compare classical values with the estimates obtained from a quantum tetrahedron.

We label the outward normals to the faces as $\mathbf{J}_i$, and following the convention take their lengths to be twice the triangular areas, $J_i = 2A_i$. Being a closed surface a tetrahedron satisfies the closure condition

$$\sum_{i=1}^{4} \mathbf{J}_i = 0. \tag{1}$$

There is a number of useful relations between areas, angles and volume [6]. Angles between the triangular faces $\mathbf{J}_i$ are the (inner) dihedral angles, which are related to the outer dihedral angles θ_{ij},

$$J_{ij} \equiv \mathbf{J}_i \cdot \mathbf{J}_j = J_i J_j \cos\theta_{ij} \tag{2}$$

as $\theta_{ij}^{\text{in}} = \pi - \theta_{ij}$. The volume can be expressed in terms of the area vectors,

$$V^2 = -\frac{1}{36}\epsilon_{abc} J_1^a J_2^b J_3^c = -\frac{1}{36}\mathbf{J}_1 \cdot \mathbf{J}_2 \times \mathbf{J}_3. \tag{3}$$

In the following we take J_1,J_2, J_3, J_4, θ_{12} and θ_{23} to form the shape-defining set.

In the quantized problem the four normals are identified with the generators of SU(2). For later use we note that the Casimir operator asymptotically equals to $J \equiv \sqrt{j(j+1)} \simeq j + \frac{1}{2}$.

The closure constraint Eq. (1) restricts the Hilbert space to its SU(2) invariant subspace (of the total zero spin). Its basis can be constructed in two steps, e.g., first by coupling spins 1 with 2, $\hat{\mathbf{L}}_{12} = \hat{\mathbf{J}}_1 + \hat{\mathbf{J}}_2$, so $\hat{J}_{12} = \hat{\mathbf{J}}_1 \cdot \hat{\mathbf{J}}_2$, and 3 with 4, and then forming the singlets from the intermediate pairs

$$|l\rangle = \sum_m \frac{(-1)^{l-m}}{\sqrt{2j+1}} |l, m\rangle_{12} |l, -m\rangle_{34}, \tag{4}$$

where we suppressed the labels $j_1, \ldots, j_4$. An alternative basis is defined by the eigenvalues of $\hat{\mathbf{L}}_{23}^2$. The two sets of basis vectors are related through $6j$-symbols,

$$\langle l|l'\rangle = (-1)^{\sum j_i}\sqrt{d_l d_{l'}} \begin{Bmatrix} j_1 & j_2 & l \\ j_3 & j_4 & l' \end{Bmatrix}. \tag{5}$$

The commutator of $\hat{J}_{12}$ and $\hat{J}_{23}$ is related to the volume through [7]

$$[\hat{J}_{12}, \hat{J}_{23}] = -i\epsilon_{abc}\hat{J}_1^a \hat{J}_2^b \hat{J}_3^c \equiv i\hat{U}. \tag{6}$$

In agreement with LQG results its absolute value operator $|\hat{U}|$ is identified with the quantized classical squared volume $36V^2$.

There are no self-adjoint angle operators $\hat{\theta}$ on finite-dimensional representations of SU(2) [8], so a classical angular parameter θ is estimated by using a positive operator-valued measure (POVM). Following [3, 5] we restrict ourselves to the case where four areas have well-defined values. Then the angles are easily identified through the quantum analog of Eq. (2), where the operators $\hat{J}_{ij}$ are extracted from

$$\hat{\mathbf{L}}_{ik}^2 = \hat{\mathbf{J}}_i^2 + \hat{\mathbf{J}}_k^2 + 2\hat{J}_{ik}. \tag{7}$$

To simplify the following formulas we consider the case of four equal areas: $j_1 = \ldots = j_4 = j$. Accordingly, the estimate of the classical angle θ_{12} is given by

$$z_{ik}(\rho) \equiv \cos\theta_{ik}(\rho) = \frac{\langle \hat{J}_{ik}\rangle}{j(j+1)}. \tag{8}$$

An obvious spread estimator of a classical random variable z is the standard deviation

$$(\Delta z)^2 \equiv \Delta^2 z = \langle z^2\rangle - \langle z\rangle^2. \tag{9}$$

In quantum theory a pair of non-commutative self-adjoint operators satisfies the Schrödinger-Robertson relation [8, 9]. In the case of $\hat{J}_{12}$ and $\hat{J}_{23}$ it is

$$\Delta^2(\hat{J}_{12})\Delta^2(\hat{J}_{23}) \geq \tfrac{1}{4}|\langle\hat{U}\rangle|^2 + \sigma^2(\hat{J}_{12}, \hat{J}_{23}), \tag{10}$$

where

$$\Delta^2\hat{X} \equiv \langle\hat{X}^2\rangle - \langle\hat{X}\rangle^2, \qquad \sigma(\hat{X},\hat{Y}) \equiv \langle\hat{X}\hat{Y} + \hat{Y}\hat{X}\rangle/2 - \langle\hat{X}\rangle\langle\hat{Y}\rangle. \tag{11}$$

A good semiclassical state that describes a tetrahedron with fixed triangular areas should have small angular and volume uncertainties,

$$\Delta z_{ik}/\langle z_{ik}\rangle \to 0, \qquad \Delta V/\langle V\rangle \to 0 \tag{12}$$

across the range of the angle and volume expectations that correspond to the classical values.

When Eq. (8) is used to obtain the dihedral angles, then is easy to see that for such a state

$$\Delta z_{12}\Delta z_{23} = \frac{\Delta(\hat{J}_{12})}{J^2}\frac{\Delta(\hat{J}_{23})}{J^2} \geq \frac{|\langle\hat{U}\rangle|}{2J^4} \sim \frac{1}{J} \tag{13}$$

Rovelli and Speziale [5] constructed a family of such states. We discuss them in Sec. 4.

3. The optimal convergence rate

Minimization of $\Delta z_{12}\Delta z_{23}$ does not necessarily produce the best semiclassical states. For example, the eigenstates of either of $\hat{J}_{ik}$ result in $\Delta z_{12}\Delta z_{23} \equiv 0$. Moreover, fixing the expectation values z_{12} and z_{23} does not fix the expectation of $\hat{U}$. The volume can be recovered by classical calculation from the determined values of θ_{ik}, but the physical significance of the states with $36V_{\rm class}^2 \neq \langle\hat{U}\rangle$ is not clear.

To study the asymptotics let us introduce some rescaled quantities. We set $\langle\hat{J}_{12}\rangle = \ell_* K^2$, $\langle\hat{J}_{23}\rangle = k_* K^2$ and $J = j_* K$. Thus the shape

$$z_{12} = \ell_*/j_*, \qquad z_{23} = k_*/j_*, \tag{14}$$

is fixed even if the size goes to infinity, $K \to \infty$.

If the goal is to minimize the left hand side of Eq. (13), then not fixing the volume expectation leaves a larger parameter space, and one can expect a smaller product of variances. Appearance of the eigenvalues of $\hat{U}$ in pairs $\pm u$ makes it appealing to expect the optimal states to have a zero expectation of the squared volume.

Nevertheless, even the unconstrained minimization gives $\Delta z_{12}\Delta z_{23} \propto 1/K$. The rest of this Section deals with derivation of this bound. We introduce another parametrization of a tetrahedron which allows to map the unconstrained search of the minimal uncertainty states to the problem of optimal direction transmission [8, 10].

The four vectors of equal length that satisfy the closure condition (1) can be represented as

$$\mathbf{J}_{1,2} = J(\mp\sin\theta, 0, \cos\theta), \qquad \mathbf{J}_{3,4} = J(\pm\sin\theta\cos\phi, \pm\sin\theta\cos\phi, -\cos\theta). \tag{15}$$

Using the relations for the dihedral angles we find that

$$2\theta = \theta_{12} = \pi - \theta_{12}^{\rm in}, \qquad \cos\phi = \frac{\cos\theta_{23} + \cos\theta}{\sin^2\theta}. \tag{16}$$

For example, a regular tetrahedron is parameterized by $\cos\theta = 1/\sqrt{3}$ and $\phi = \pi/2$.

Eq. (15) maps a problem of finding the minimum of $\Delta z_{12}\Delta z_{23}$ to the task of finding the optimal direction transmission protocol. Given a spatial direction $\mathbf{n}_0(\theta_0,\phi_0)$ there is an encoding and decoding scheme that results in the optimal estimate $\mathbf{n}(\theta,\phi)$. The protocol is optimal with respect to the *fidelity* and related error measures [8, 10]. Fidelity is defined as

$$F=\frac{1+\langle\cos\chi\rangle}{2}, \tag{17}$$

where $\cos\chi=\mathbf{n}_0\cdot\mathbf{n}$, and the average is taken over the resulting probability distribution $p(\theta_0,\phi_0;\theta,\phi)$. SU(2) covariance properties allow to reduce the problem to sending a single fiducial direction. A complementary quantity $D=1-F$ is the mean square error of the measurement, if the error is defined as $\sin^2\chi/2=|\mathbf{n}_0-\mathbf{n}|^2$.

There are many different versions of this communication task. They differ in physical systems that serve as information carriers (see [8, 10] and the references therein). However, in all of them the optimal measurement is unbiased, i.e.,

$$\langle\theta\rangle=\theta_0, \qquad \langle\phi\rangle=\phi_0, \tag{18}$$

and the fidelity asymptotically approaches unity as

$$D\propto 1/d, \qquad d\to\infty, \tag{19}$$

where $d=\dim\mathcal{H}$ is the dimension of the available Hilbert space, and a pre-factor depends on the set-up details. If the information carrier is a single spin-j particle, then $d=d_j=2j+1$.

In our case the Hilbert space is the intertwiner space, and the assumption $j_1=\ldots=j_4=j$ gives

$$d=2d_j. \tag{20}$$

To compare this result with the product of variances $\Delta z_{12}\Delta z_{23}$ we first express the deviation angle χ in terms of $\delta\theta=\theta-\theta_0$ and $\delta\phi=\phi-\phi_0$. Assuming $|\delta\theta|<\theta_0$, it is easy to derive the asymptotic result

$$\chi^2\simeq\sin^2\theta_0\delta\phi^2+\delta\theta^2. \tag{21}$$

Estimating the individual variances, and taking into account that $D\simeq\chi^2/4$, we see that the uncertainty of the optimal shape transmission

$$\Delta(z_{12})\Delta(z_{23})=c(k_*,\ell_*)D\propto\frac{1}{d}=\frac{1}{2j_*K}. \tag{22}$$

Hence even if we disregard the physical meaning of the resulting states, the uncertainty still behaves as $1/J$.

4. Joint POVM and its optimality

Expressions like Eq. (13) refer to an ensemble of identically prepared systems, where in half of the cases one measures $\hat{J}_{12}$ and in the other half $\hat{J}_{23}$. Since classical dihedral angle variables J_{12} and J_{23} are defined simultaneously, the emergence of classicality is properly described [8, 9] by convergence of the joint probability distribution $p(J_{12},J_{23})$ to their sharp classical profiles. This is achieved by a POVM that is described in this Section.

A generic pure state of a tetrahedron is given by

$$|\psi\rangle=\sum_{l=0}^{2j}c_l|l\rangle, \qquad c_l\equiv|c_l|e^{i\phi_l}. \tag{23}$$

For example, if the sextet of the classical parameters of a tetrahedron is completed by

$$j_{12} = \langle \hat{J}_{12} \rangle, \qquad j_{23} = \langle \hat{J}_{23} \rangle, \tag{24}$$

then the amplitude of the states proposed in [5] has a Gaussian profile centered around l_0, $j_{12} = \frac{1}{2}l_0(l_0+1) - \frac{1}{2}j_1(j_1+1) - \frac{1}{2}j_2(j_2+1)$.

The phase is determined with the help of an auxiliary tetrahedron that is made from the four edges $j_1 + \frac{1}{2}, \ldots, l_4 + \frac{1}{2}$ and $j_{12} + \frac{1}{2}$ and $j_{23} + \frac{1}{2}$. It equals to

$$\phi_l = \phi_0 l, \tag{25}$$

where ϕ_0 is dihedral angle between the two faces of an auxiliary tetrahedron that share a link $j_{12} + \frac{1}{2}$.

Partial traces of an arbitrary state on $\mathcal{H}_{12} = \mathbb{C}^{d_1} \otimes \mathbb{C}^{d_2}$

$$\rho_{12} = \sum_l \frac{|c_l|^2}{2l+1} \sum_{m=-l}^{l} |l,m\rangle\langle l,m|, \tag{26}$$

and on $\mathcal{H}_{23}$ (a similar expression) are diagonal, a feature we use below.

To construct a joint measurement that results in z_{12} and z_{23} we use a "commutative spin observable". A POVM that is used to identify the directions is built from the normalized SU(2) coherent states,

$$\hat{E}_{\theta\phi} = \frac{2j+1}{4\pi} |\theta,\phi\rangle\langle\theta,\phi|, \qquad \int \hat{E}_{\theta\phi} d\Omega_{\theta\phi} = \mathbb{1}, \tag{27}$$

where

$$|\theta,\phi\rangle = \sum_{m=-j}^{j} D^{(j)}(\phi,\theta,\psi=0)_{mj} |j,m\rangle, \tag{28}$$

and $D^{(j)}$ is a spin-j rotation matrix.

A commutative angular momentum observable is a collection of (statistical) moment operators that correspond to this POVM. The first moment operator is used to calculate the expectation values,

$$\hat{\mathbf{M}}^{(1)} = (j+1) \int \mathbf{n} \hat{E}_{\theta\phi} d\Omega_{\theta\phi}, \tag{29}$$

where $\mathbf{n} = (\sin\theta\cos\phi, \sin\theta\sin\phi, \cos\theta)$ is a unit vector. The measurement is unbiased, i. e., for any state ρ

$$\mathrm{tr}(\hat{\mathbf{M}}^{(1)}\rho) = \mathrm{tr}(\hat{\mathbf{J}}\rho). \tag{30}$$

The statistical moments for $\mathbf{J}_{ik}$ are most conveniently obtained through the integration over the original angles,

$$\hat{M}_{ik}^{(1)} = (j+1)^2 \int (\mathbf{n}_1 \cdot \mathbf{n}_2) \hat{E}_{\theta_i\phi_i} \hat{E}_{\theta_k\phi_k} d\Omega_i d\Omega_k \tag{31}$$

and

$$\hat{M}_{ik}^{(2)} = (j+1)^4 \int (\mathbf{n}_1 \cdot \mathbf{n}_2)^2 \hat{E}_{\theta_i\phi_i} \hat{E}_{\theta_k\phi_k} d\Omega_i d\Omega_k. \tag{32}$$

Their matrix elements are spelled out in [4]. These operators have a simple overall structure. In particular, in the corresponding (l,m) basis they have a form

$$\hat{M}_{lm,l'm'}^{(i)} = \hat{M}_{lm,l'm}^{(i)} \delta_{mm'}. \tag{33}$$

Unlike the sharp projective measurements of $\mathbf{J}_{12}$ and $\mathbf{J}_{23}$, the above construction allows a simultaneous estimate of the angles θ_{12} and θ_{23}.

It follows from Eq. (26) that we are interested in the asymptotic behavior of their expectation values on the states

$$\rho_l = \mathbb{1}_l/(2l+1), \tag{34}$$

where $j = j_* K$, $l = l_* K$, and K goes to infinity. The operator $\hat{M}^{(1)}_{12}$ is not an unbiased estimator: for a generic $|\psi\rangle \in \mathcal{H}_{12}$ the expectation $\langle\psi|\hat{M}^{(1)}_{12}|\psi\rangle$ is different from $\langle\psi|\hat{J}_{12}|\psi\rangle$. However, it is possible to show that

$$\mathrm{tr}(\hat{M}^{(1)}_{12}\rho_l) = \mathrm{tr}(\hat{J}_{12}\rho_l) = j_{12} = \tfrac{1}{2}l(l+1) - j(j+1). \tag{35}$$

The asymptotics of Δz_{12} was investigated both analytically and numerically. It was found that for ρ_l

$$\lim_{K\to\infty} \frac{\sigma^2_{j_{12}}(M)}{\langle J_{12}\rangle^2} \equiv \frac{\langle\hat{M}^{(2)}_{12}\rangle - j^2_{12}}{j^2_{12}} \propto \frac{1}{K}. \tag{36}$$

A simple error analysis shows that if the state is peaked on $\langle\psi|\hat{J}_{12}|\psi\rangle \simeq j_{12}$ the variance of the unsharply measured J_{12} is the sum of the sharp variance $\Delta^2(\hat{J}_{12}) = \langle\psi|\hat{J}^2_{12}|\psi\rangle - \langle\psi|\hat{J}_{12}|\psi\rangle^2$ and the measurement unsharpness $\langle j_{12}|\hat{M}^{(2)}_{12}|j_{12}\rangle^2$,

$$\Delta^2(J_{12}) \simeq \Delta^2(\hat{J}_{12}) + \sigma^2_{j_{12}}(M). \tag{37}$$

Hence the result of Eq. (36) guaranties that if a state is such that $\langle\hat{J}_{12}\rangle = \ell_* K^2$, $\langle\hat{J}_{23}\rangle = k_* K^2$ and

$$\Delta(\hat{J}_{12})\Delta(\hat{J}_{23}) \sim K^3, \tag{38}$$

as the states of [5] are, then the estimate obtained from the joint POVM asymptotically behaves as

$$\Delta z_{12}\Delta z_{23} \sim 1/K, \tag{39}$$

5. Fast convergence to the classical limit

While we established that the uncertainties in the shape of an atom of space scale inversely with its surface area, it is interesting to investigate convergence of geometry to its classical value for more complicated structures. Our goal is to check the intuitive assumption that "many small tetrahedra approach the classicality faster than just a scaled up single tetrahedron".

First we subdivide a single classical tetrahedron through a number of iterative steps that are described below. *Subdivision* is an extensively studied technique in computer aided geometric design and visualisation, as well as in numerical analysis, particularly in computational fluid dynamics (see, e.g., [12, 13] and references therein). Given this refined triangulation we set up a refined spin network as a dual of the new triangulation, and label its edges according to the triangular areas they pierce.

At every step the most direct approach results in dividing a tetrahedron into eight descendants. The four tetrahedra are obtained by cutting off the corners of the parent tetrahedron at the edge midpoints, as shown on Fig 1. They are obviously similar to their parent. Each of its faces is now composed of the outer faces of three child tetrahedra and one of the faces of an octahedron. The remaining octahedron can be split into two pyramids, each of which is separated into two tetrahedra. This splitting has a number of drawbacks [13] and we use it only at the last iteration, to produce a four-valent spin network. In all other steps we use a different refinement rule that consists in subdividing an octahedron into six child octahedra and eight tetrahedra. It is done by connecting the edge midpoints of each face (Fig.1) and by connecting all edge midpoints to the barycenter of the parent octahedron (Fig. 2). Even for an arbitrary initial tetrahedron its barycenter $b = \frac{1}{4}(v_1+v_2+v_3+v_4)$, coincides with the barycenter of the child octahedron, which ensures that the eighth second generation tetrahedra are similar to the initial one with the scale factor $1/4 = (1/2)^2$. This similarity can be established either by elementary geometry, using Fig. 3 as an aid, or by finding out the explicit transformation law.

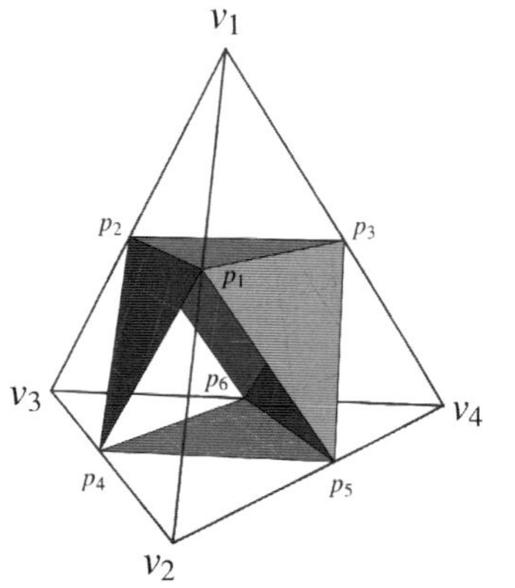

Figure 1. The first stage of the regular subdivision: four child tetrahedra and an octahedron. The faces $(p_1p_4p_5)$ and $(p_2p_4p_6)$ are transparent.

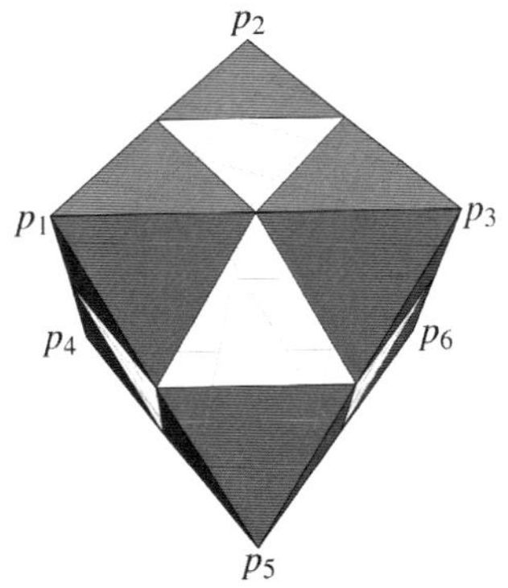

Figure 2. All the octahedron's edges are divided into equal halves. Connecting the points results in $4\times 8 = 32$ triangles. The white triangles are the faces of the second-generation tetrahedra. Each face contributes one tetrahedron, and each of the octahedron's vertices contributes a child octahedron.

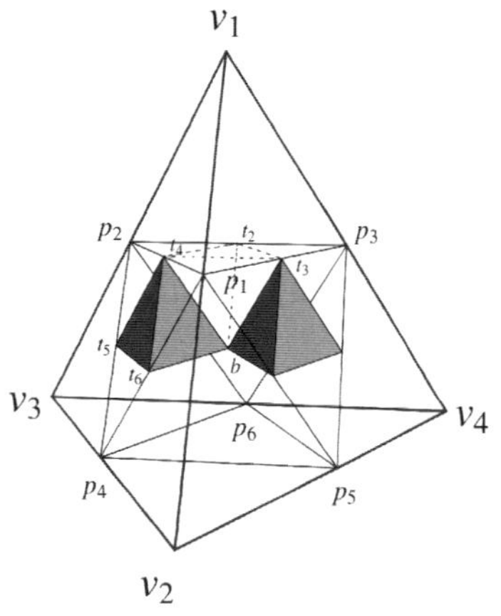

Figure 3. The barycenter b is a common vertex of all the second generation tetrahedra. Two of them are shown as solid bodies, and one as a dotted outline. For a general initial tetrahedron the octahedron's diagonals (p_1p_6), (p_2p_5), and (p_3p_4) do not intersect at the barycenter.

If at some stage the triangulation consists of T tetrahedra and O octahedra, then one refitment step results in $T \mapsto 4T + 8O, O \mapsto T + 6O$. Consequently, after n subdivisions

$$T_n = \frac{1}{3}\left(2^{3n} + 2^{n+1}\right) \sim \frac{2^{3n}}{3}, \qquad O_n = \frac{1}{3}\left(2^{3n-1} - 2^{n-1}\right) \sim \frac{2^{3n}}{6}. \tag{40}$$

Hence the volume fraction of the tetrahedra that are similar to the initial one asymptotically reaches $\frac{1}{3}$. Since the final iteration is followed by a subdivision of the octahedra into four tetrahedra, the n step iterative procedure gives us T_n tetrahedra that are similar to the original one, and $T'_n = 4O_n$ tetrahedra of other classes.

Assume that the number of steps is such that the surface areas of the small tetrahedra T (and two out of four faces of the tetrahedra T') still satisfy $2A_n = J_n \approx j_* \gg 1$. Hence $2A = 2^{2n} j_* \equiv K j_*$. To establish our point in the simplest possible way we focus only on the tetrahedra T. The dimensionality of a (SU(2) invariant) Hilbert space that is associated with a single such tetrahedron is $d_* = 2j_*$, and their total number is $K^{3/2}/3$. The total Hilbert space dimension is

$$d(K) = (2j_*)^{\frac{1}{3}K^{3/2}} = (2j_*)^{\frac{1}{3}\left(\frac{2A}{j_*}\right)^{3/2}}. \tag{41}$$

As a result, if the shape is encoded in the information-theoretical optimal way, the shape uncertainty decreases super-exponentially, as

$$\Delta(z_{12})\Delta(z_{23}) \propto 1/d(K). \tag{42}$$

This result ignores the expectation of the volume operator, and its usefulness is mainly in setting the upper bound on the convergence to classicality.

We can use the same construction to show a modest improvement even when the total measured state is given by

$$|\Psi\rangle = \otimes_a^{\frac{1}{3}K^{3/2}} |\psi\rangle_a, \tag{43}$$

and the state of each $|\psi\rangle$ is given by Eq. (23). If the (commuting!) J_{12} and J_{23} are estimated for all the small tetrahedra independently by separately applying a POVM of Sec. III to each tetrahedron, then the statistical averaging over the entire sample leads to

$$\Delta(z_{12})\Delta(z_{23}) \simeq \frac{3c}{2j_*}\frac{1}{K^{3/2}}, \tag{44}$$

where a constant c is determined by the asymptotics of a single tetrahedron. As a result, the uncertainty goes to zero faster than the uncertainty of Eq. (22)

6. Conclusions

We constructed a SU(2)-invariant positive operator valued measure that simultaneously extracts two classical parameters that are associated with non-commutative observables. Mapping the semiclassical problem into a quantum-informational task, we showed that for a single tetrahedron the fuzziness of geometry is reduced only as an inverse of the area. However, a judicious choice of more complicated states speeds-up the convergence. It still remains to be seen weather an exponential convergence to the classical limit is possible.

References

[1] Rovelli C 2004 *Quantum Gravity* (Cambridge, UK: Cambridge U. Press); Thiemann T 2007 *Modern Canoninical Quantum General Relativity* (Cambridge, UK: Cambridge U. Press)
[2] Dittrich B and Thiemann T 2007 *Preprint* arXiv:0708.1721; Rovelli C 2007 *Preprint* arXiv:0708.2481.
[3] Barbieri A 1998 *Nucl. Phys.* B **518** 714; Baez J C and Barrett J W 1999 *Adv. Theor. Math. Phys.* **3** 815
[4] Terno D R 2008 *Preprint* arXiv:0808.2533
[5] Rovelli C and Speziale S 2006 *Class. Quant. Grav.* **23** 5861
[6] Jackson F and Weisstein E M *Tetrahedron* http://mathworld.wolfram.com/Tetrahedron.html; Coxeter H S M 1963 *Regular Polytopes* (New York: Macmillan)
[7] Ashtekar A and Lewandowski J 1995 *J. Geom. Phys.* **17** 191
[8] Holevo A S 1982 *Probabilistic and Statistical Aspects of Quantum Theory* (Amsterdam: North-Holland);
[9] Busch P M, Grabowski M and Lahti P J 1995 *Operational Quantum Physics* (Berlin: Springer)

[10] Bagan E, Baig M, Brey A, Muñoz-Tapia R and Tarrach R 2000 *Phys. Rev. Lett.* **85** 5230; Bartlett S D, Rudolph T and Spekkens R W 2007 *Rev. Mod. Phys.* **79** 555
[11] Peres A and Terno D R 2004 *Rev. Mod. Phys.* **76** 93
[12] Greiner G and Grosso R 2000 *The Visual Computer* **16** 357
[13] Schaefer S, Hakenberg J and Warren J 2004 in *Eurographics Symposium on Geometry Processing* 151

Journal of Physics: Conference Series **174** (2009) 012051 doi:10.1088/1742-6596/174/1/012051

Matter in toy dynamical geometries

Tomasz Konopka
ITP, Utrecht University, Utrecht 3584 CE, The Netherlands
E-mail: `t.j.konopka@uu.nl`

Abstract. One of the objectives of theories describing quantum dynamical geometry is to compute expectation values of geometrical observables. The results of such computations can be affected by whether or not matter is taken into account. It is thus important to understand to what extent and to what effect matter can affect dynamical geometries. Using a simple model, it is shown that matter can effectively mold a geometry into an isotropic configuration. Implications for "atomistic" models of quantum geometry are briefly discussed.

1. Introduction

In the context of quantum gravity (see e.g. [1] for an overview of different approaches), much attention is devoted to studies of pure-gravity systems in which all matter degrees of freedom and their interactions are switched off. Whereas the use of this assumption is understandable given that most proposals for quantum theory of gravity are difficult to study even without the inclusion of additional fields, it is known that the presence of matter can alter the behavior of a gravitational system [2]. The purpose of this paper is to illustrate the effect of matter on geometry in a simple setting. In particular, the purpose is to show that matter can induce isotropy in a model that does not assume it a-priori - an effect not easily seen from standard treatments of matter in dynamical geometries [2].

Physical theories are often formulated using the Feynman path integral and many candidates for a theory of quantum gravity [1] are also defined within this framework. These candidates propose to study path integrals (partition functions) formally written as

$$Z_\phi = \int \mathcal{D}[g] \int \mathcal{D}[\phi]\, e^{-E(g,\,\phi)} \tag{1}$$

where integrals denote summations over geometries g and matter degrees of freedom ϕ, and $E(g, \phi)$ is a weighting function that depends on both the geometry and the matter content (this function can be interpreted as the action or the energy of the system, according to the setting). The weighting function is often assumed to be of the form $E(g,\, \phi) = E_g(g) + E_\phi(g,\, \phi)$ so that the contribution from the matter fields in $E_\phi(g,\, \phi)$ is separated from the pure-geometry part $E_g(g)$. Furthermore, it is often assumed that if the matter contribution $E_\phi(g,\, \phi)$ is set to zero, the resultant, simpler, path integral

$$Z_{\phi=0} = \int \mathcal{D}[g]\, e^{-E_g(g)} \tag{2}$$

has many of the same properties as the original one (1). That is, it is assumed that expectation values $\langle \mathcal{O} \rangle_\phi$ and $\langle \mathcal{O} \rangle_{\phi=0}$ of a geometric observable $\mathcal{O}$ computed in the two models have similar properties. It is this assumption that motivates the study of "pure-geometry" systems even when the real universe is known to contain matter fields of various types.

The purpose of this paper is to demonstrate that geometric observables can in fact be different depending on whether or not matter is included in the calculation. The strategy to demonstrate this will be to compute expectation values explicitly in the setting of a two dimensional rectangular box in which the sides lengths L_x and L_y are variable but the total area $L_xL_y = A$ is fixed. This system was already introduced in [3] but it is explored here more fully. An important property of this system is that it contains a single configuration where $L_x = L_y$ and the box is a square but it contains a multitude of other configurations in which $L_x \neq L_y$; the geometric observable $R = L_x/L_y$ differentiates the square configuration from the non-square ones. The main result will be that the expectation value for this ratio in a system without matter must be different from one, but that it can be one when matter is present.

The rest of the paper is organized as follows. Sec. 2 describes the flexible box system and defines its sum over geometries. The expectation value of the observable in the absence of matter is also computed. Sec. 3 discusses the effect of adding particles. It is shown through numerical studies that the expectation value of the observable $R = L_x/L_y$ in the presence of particles can be close to unity. This indicates that matter can select a homogenous and isotropic geometry from the ensemble. In Sec. 4 this discussion is extended to the case of fields and it shown in what circumstances the Casimir energy of fields has a similar effect. A discussion of applications to current research programs on "atomistic" models[1] of spacetime are presented in Sec. 5.

2. Toy dynamical geometries

2.1. Regularized sum over geometries

Formal expressions like (1) and (2) do not specify the space of geometries to be integrated over or how exactly to carry out the integrals. To overcome this, several approaches (e.g. [3–5]) propose to study systems like (1) in a discretized setting. In those approaches, one replaces the integral over geometries by a finite sum over a set $\mathcal{C}$ of geometry configurations. The same strategy is adopted here. Since the purpose of this paper is to study effects of matter on dynamical geometries and not to propose a sensible way or regularizing the path integral for quantum gravity, the set $\mathcal{C}$ is chosen to make this specific application possible.

Consider configurations spaces $\mathcal{C}_A$ corresponding to sets describing rectangular boxes having sides of length L_x and L_y and area A. To make each of these sets $\mathcal{C}_A$ finite, the possible lengths L_x and L_y must be restricted to a discrete or semi-discrete spectrum characterized by at least one length scale, ℓ, designating a smallest allowed length. If the allowed lengths are assumed to be equally spaced when they are small, a concrete length spectrum that can be used is

$$L \in \begin{cases} \mathbb{N}\ell & \text{if } L \leq \lambda \\ \mathbb{R} & \text{if } L > \lambda. \end{cases} \tag{3}$$

The minimal value and next larger values for L are ℓ and integer multiples of ℓ, respectively, up to a scale λ. Above this scale, L can take any real number value. It will be convenient to choose $\lambda = \sqrt{A}$. With this convention, the sets $\mathcal{C}_A$ can be written out explicitly as

$$\mathcal{C}_A = \left\{ \left(\ell, \frac{A}{\ell}\right), \left(2\ell, \frac{A}{2\ell}\right), \ldots, \left(\sqrt{A}, \sqrt{A}\right), \ldots, \left(\frac{A}{\ell}, \ell\right) \right\}. \tag{4}$$

Note that these sets are countable and well defined for all real $A > \ell^2$.

[1] The term "atomistic" model is due to H-T. Elze.

Journal of Physics: Conference Series **174** (2009) 012051 doi:10.1088/1742-6596/174/1/012051

2.2. Partition function and geometric observables

After defining the configuration space of a dynamical box with area A, one must specify how the configurations should be weighted in the ensemble defining the partition function (1). In the case where they are all equally weighted, the partition function of the system is

$$Z = \sum_{p=1}^{\lambda/\ell} (1+1) = \frac{2\lambda}{\ell} = \frac{2\sqrt{A}}{\ell}\,. \tag{5}$$

The first term shown in the sum represents values of L_x in the discrete part of the length spectrum, i.e. the first half of the configurations listed in (4). The second term is the analogous sum over discrete values of ${L_y}^2$. In the end, Z turns out to be equal to the size of the set $\mathcal{C}_A$; the thermodynamic properties of the empty box system are determined entirely by entropy.

Expectation values of observables describing the box geometry can now be computed. For concreteness, consider the ratio $R = L_x/L_y$. The expectation value for R should be close to unity if the box is likely to be square, and very different from unity if the box is likely to be long and narrow. One finds

$$\langle R\rangle = \frac{1}{AZ}\left[\sum_{p=1}^{\lambda/\ell}\left((p\ell)^2 + \left(\frac{A}{p\ell}\right)^2\right)\right] \to \frac{\pi^2\sqrt{A}}{12\,\ell}\,. \tag{6}$$

The last expression denotes the evaluation of $\langle R\rangle$ in the limit $A\to\infty$. Since the result is divergent in this limit, (also when $\ell\to 0$), the calculation suggests that the box should be expected to be long and narrow in that limit. Curiously, note that the expectation value for the inverse ratio, $\langle L_y/L_x\rangle$, is given by exactly the same diverging expression; this is a consequence of the $L_x \leftrightarrow L_y$ symmetry of the system and implies that these ratios provide measures for the degree to which the two sides of the box are different and cannot be used to infer which length, L_x or L_y, is larger.

Another useful expectation value is

$$\langle R^2\rangle = \frac{1}{A^2Z}\left[\sum_{p=1}^{\lambda/\ell}\left((p\ell)^4 + \left(\frac{A}{p\ell}\right)^4\right)\right] \to \frac{\pi^4 A^{3/2}}{180\,\ell^3}; \tag{7}$$

the last expression again stands for the $A\to\infty$ limit. This quantity can be used to compute the fluctuations $(\Delta R)^2 = \langle R^2\rangle - \langle R\rangle^2$ of the ratio R. In the limit $A\to\infty$, the contribution to $(\Delta R)^2$ from the first term, $\langle R^2\rangle$, is much larger than that from the second, $\langle R\rangle^2$, and thus the fluctuations diverge as $A^{3/2}$.

3. Particles in dynamical geometries

3.1. Particles as standing waves

Suppose that the flexible box is populated with quantum mechanical particles of mass m. The dynamics of each particle is described by the Schrödinger equation in two dimensions with a potential that is zero inside the box and infinity everywhere else. After setting $\hbar = c = 1$, the energy spectrum for each particle is

$$E = \frac{1}{2m}\left(k_x^2 + k_y^2\right), \tag{8}$$

[2] The given prescription for evaluating Z carries a risk of over-counting whenever the two configurations corresponding to $p = \lambda/\ell$ are in fact equivalent. This problem can be averted by considering A such that λ/ℓ is not an integer. In any case, the error involved becomes negligible when the area is large.

where k_x and k_y are the momenta in the x and y directions, respectively. The allowed values for these momenta $(i = x, y)$ are

$$k_i = \frac{(2-\sigma)\pi n_i}{L_i} \tag{9}$$

where σ depends on the type of boundary conditions,

$$\sigma = \begin{cases} 1 & \text{Dirichlet} \\ 0 & \text{Periodic,} \end{cases} \tag{10}$$

and $n_i = \sigma, 1, 2, \ldots$ are quantum numbers whose range also depends on the boundary conditions.

For a given box configuration (L_x, L_y), the partition function for a single particle is

$$z(L_x, L_y) = \sum_{n_x=\sigma}^{\infty} \exp\left(-\frac{\beta(2-\sigma)^2\pi^2 n_x^2}{2mL_x^2}\right) \sum_{n_y=\sigma}^{\infty} \exp\left(-\frac{\beta(2-\sigma)^2\pi^2 n_y^2}{2mL_y^2}\right) \tag{11}$$

where $\beta = 1/kT$ is the inverse temperature. For a system of N identical Maxwell-Boltzmann particles, their combined contribution is

$$z_N(L_x, L_y) = \left(z(L_x, L_y)\right)^N / N! \tag{12}$$

The total partition for the particles as well as the box can then be written as

$$Z = 2\sum_{p=1}^{\lambda/\ell} z_N\left(p\ell, \frac{A}{p\ell}\right) \tag{13}$$

where the factor of two in front has the same origin as in (5).

It is important to emphasize here that z and z_N depend explicitly on the lengths of the sides of the box and the parameter σ. In particular, it can be shown that the largest term in Z is due to configurations with $L_x = L_y$ when $\sigma = 1$, or to configurations with $L_x \gg L_y$ and $L_y \gg L_x$ when $\sigma = 0$. These features form the basis for understanding the results described next.

3.2. Evaluation of observables

The expectation value for R can now be computed, analogously to (6), using

$$\langle R\rangle = \frac{1}{ZA}\sum_{p=1}^{\lambda/\ell}\left((p\ell)^2 + \left(\frac{A}{p\ell}\right)^2\right) z_N\left(p\ell, \frac{A}{p\ell}\right). \tag{14}$$

This can be evaluated numerically given certain values for the parameters. To make for a realistic situation, the mass of the particles and the temperature of the system should be much smaller than the inverse minimal length, $m, kT \ll 1/\ell$. Thus $\beta(2-\sigma)^2\pi^2/8m \gg 1/\ell^2$. After fixing this ratio in terms of ℓ, which can then be set to unity without loss of generality, the remaining free variables are A and N. In the following, the number of particles N is expressed in the form $N = N_0 + cA^\alpha$ with N_0 and α positive dimensionless constants, and c a positive constant with dimension depending on the value of α.

Numerical results for $\langle R\rangle$ and $(\Delta R)^2$ with $\ell = 1$, $\beta(2-\sigma)^2\pi^2/8m = 10$ are shown in Fig. 1. Curves corresponding to a fixed number of particles ($N = 1$, $\sigma = 0, 1$) show that the expectation value $\langle R\rangle$ and its fluctuations rise as the area of the boxes is increased. The asymptotic slopes match the behavior expected from the empty box calculations in (6) and (7). Curves corresponding to systems wherein the number of particles scales with area ($N = 0.01A$, $\sigma = 0, 1$)

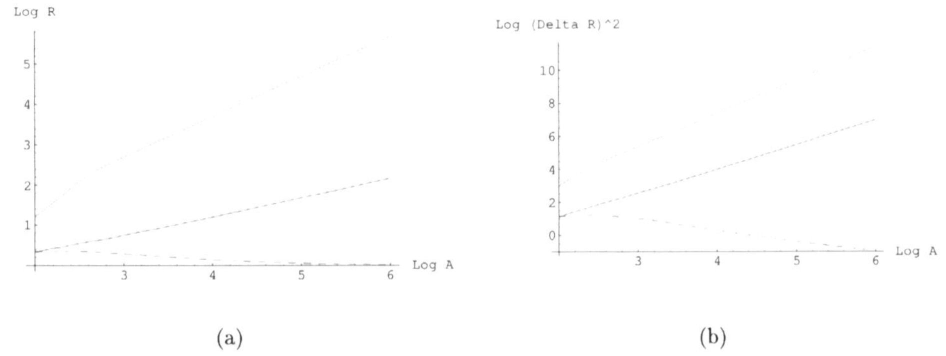

Figure 1. Geometric observables in toy geometries. In both plots, the lines represent (top to bottom) systems with $(N, \sigma) = (0.01A, 0)$, $(1, 0)$, $(1, 1)$, $(0.01A, 1)$.

exhibit more interesting behavior. When Dirichlet boundary conditions are imposed ($\sigma = 1$), the asymptotic behavior for large A is $\langle R \rangle \to 1$ and $\langle (\Delta R)^2 \rangle \to 0$. Thus, in this case, the ensemble of particles mold the dynamical geometry into a stable isotropic configuration. The situation is different for periodic boundary conditions ($\sigma = 0$); there both the expectation value for R and its fluctations diverge faster than in the empty box situations.

The most interesting result of these plots is the emergence of isotropy, $\langle R \rangle \to 1$, in the case of Dirichlet boundary conditions. It can be understood qualitatively as follows. The isotropic configuration contributes the most out of all the terms composing the partition function $z(L_x, L_y)$ of a single particle. This advantage is then raised to a power cA^α. Since the number of non-isotropic configurations grows at most polynomially with A (it grows as $\sqrt{A}$), all of those configurations contribute negligibly to the overall partition function in the large area limit. The difference in growth rates between the polynomial and the exponential components also explains why the approach to perfect isotropy occurs rather quickly.

4. Casimir energy in dynamical geometries

4.1. Regularization of Casimir energy

Another way of implementing matter is using fields instead of particles. The expectation value of the Hamiltonian operator of a field is in general given by a sum of two terms: one term proportional to the number of particles in some state (zero for the vacuum), and one other term that is divergent even in the absence of physical particles. This latter term is called the Casimir energy and it is usually attributed to quantum vacuum fluctuations [7]. It is (setting $\hbar = c = 1$)

$$E_C = \frac{\eta}{2} \sum_i \omega_i. \tag{15}$$

The factor η is positive for bosonic fields and negative for fermionic fields. It is different from one if a field (such as a vector field) can have multiple polarizations. The summation is over all field mode energies.

In a flat $2+1$ dimensional background, relativistic field modes have energies ω given by $\omega^2 = m^2 + k_x^2 + k_y^2$ in terms of momenta $k_{x,y}$ and mass m. The momenta are again constrained by boundary conditions given in (9) and (10). Thus the Casimir energy is

$$E_0 = \frac{1}{2}\eta\pi(2-\sigma)\sum_{n_x=\sigma}^{\infty}\sum_{n_y=\sigma}^{\infty}\left(\widetilde{m}_\sigma^2 + \frac{n_x^2}{L_x^2} + \frac{n_y^2}{L_y^2}\right)^{1/2}, \tag{16}$$

where $\widetilde{m}_\sigma = m/(2-\sigma)\pi$. This quantity is ultra-violet divergent; much work on the Casimir energy revolves around extracting its physically relevant components [7]. One technique to do this is to define a modified version of the summation (16) which includes a cutoff function D_ℓ multiplying each summand as follows:

$$E_{0,\ell} = \frac{1}{2}\eta\sum_{n_x=\sigma}^{\infty}\sum_{n_y=\sigma}^{\infty}\omega_{n_x,n_y}\, D_\ell\left(\frac{n_x}{L_x}, \frac{n_y}{L_y}\right). \tag{17}$$

The cutoff function D_ℓ must be chosen to satisfy certain criteria [9]. First, it must reduce to unity when $\ell \to 0$ so that $E_{0,\ell} \to E_0$ in this limit. Second, for $\ell \neq 0$, the function should fall off quickly when its arguments become large as to make $E_{0,\ell}$ finite. Third, it should not have any singularities or branch cuts in the complex plane. Lastly, it is also convenient to choose D_ℓ so that it is symmetric in its arguments, $n_x/L_x \leftrightarrow n_y/L_y$.

To evaluate (17), one applies the Abel-Plana formula

$$\sum_{n=0}^{\infty} F(n) = \int_0^\infty F(t)\,dt + \frac{1}{2}F(0) + i\int_0^\infty \frac{dt}{e^{2\pi t}-1}\left(F(it) - F(-it)\right) \tag{18}$$

twice in order to exchange the two summations over n_x and n_y into integrals. The resultant $E_{0,\ell}$ can be written (see [7–9] for more details) in the form

$$E_{0,\ell} = \frac{1}{2}\eta\,\pi\,(2-\sigma)\,(E_{\ell,A} + E_{\ell,B} + E_{AR}). \tag{19}$$

The first two terms in the parenthesis are

$$\begin{aligned} E_{\ell,A} &= (L_x L_y)\int_0^\infty du_x \int_0^\infty du_y\, \sqrt{\widetilde{m}_\sigma^2 + u_x^2 + u_y^2}\; D_\ell\,(u_x, u_y) \propto \frac{L_x L_y}{\ell^3} \\ E_{\ell,B} &= \frac{1}{2}\,(1-2\sigma)\,(L_x + L_y)\int_0^\infty du\, \sqrt{\widetilde{m}_\sigma^2 + u^2}\; D_\ell\,(u,0) \propto (1-2\sigma)\,\frac{L_x + L_y}{\ell^2}. \end{aligned} \tag{20}$$

These terms are divergent when $\ell \to 0$ and thus parametrize the ultra-violet behavior of E_0. When $\ell \neq 0$ and $m \ll 1$, their leading behavior is as shown. The precise coefficients are dependent on the details of the cutoff function. Interestingly, the terms are proportional to the area and perimeter of the two-dimensional box, and the latter has different sign for the two considered boundary conditions.

The last term in (19), E_{AR}, consists of the remaining contributions obtained from the expansions using the Abel-Plana formula. For the present discussion, the important features of E_{AR} are that it is finite in the $\ell \to 0$ limit and that its magnitude decreases with area [6–8].

The Casimir energy is an active subject of research [6, 7] because it is a purely quantum effect that leads to a macroscopic effect in the form of the Casimir effect/force. In the standard description of the Casimir effect, it is the finite term E_{AR} that is physically interesting. A clear explanation of why the divergent terms can be discarded when describing situations relevant for experiments testing the Casimir force is given in works on cavities with a piston [9]. In short, the reason is that the divergent contributions from inside and outside a cavity cancel from formulae describing measurable quantities (the force on the piston).

In the context considered in this paper, the field is defined only within the boundaries of the box, and thus a cancellation of divergent terms from inside and outside cannot occur. The analysis of the effect of Casimir energy on the statistical properties of the dynamical box must therefore either include the regularized divergent terms, or discard them in an ad-hoc manner.

4.2. Casimir energy and expected geometry
The partition function of the flexible box with field is

$$Z = 2\sum_{p=1}^{\sqrt{A}/\ell} z_\phi(\beta, p) \tag{21}$$

where $z_\phi(\beta, p)$ stands for the partition function of a field theory at inverse temperature β in a box configuration labelled by p. $z_\phi(\beta, p)$ should have contributions from the Casimir energy and also from the real particle present in a thermal state. However, in the low temperature regime where the average thermal energy is smaller than the mass of the particles, it is reasonable to approximate z_ϕ by the contribution from the Casimir energy alone. Thus $z_\phi(\beta, p) \sim e^{-\beta E_0(p)}$ with E_0 given by (19) for each box configuration labelled by p.

Consider first the option of including the divergent but regularized terms of E_0 in the calculations. Since the area of the boxes is kept fixed, the term $E_{\ell,A}$ contributing to E_0 is the same for all the box configurations and so can be factored out and essentially ignored. The other terms depend on the geometry of the box and thus must all be taken into consideration. Being divergent in the $\ell \to 0$ limit, the term $E_{\ell,B}$ proportional to the perimeter of the box is dominant over the term E_{AR}. The latter term can therefore be ignored. The partition function thus becomes

$$Z = 2\sum_{p=1}^{\lambda/\ell} \exp\left(-\beta\eta\pi(2-\sigma)(1-2\sigma)\frac{1}{\ell^2}\left(p\ell + \frac{A}{p\ell}\right)\right). \tag{22}$$

Expectation values can be computed numerically similarly as done in the previous section. The results are that the square configuration become dominant if the parameters σ and η are chosen as either $\sigma = 1$ and $\eta < 0$, or as $\sigma = 0$ and $\eta > 0$. That is, the square configuration is preferred if either the field is fermionic with Dirichlet boundary conditions or if the field is bosonic with periodic boundary conditions.

Next, consider the option of discarding the regularized divergent terms of E_0 and using only the finite contribution E_{AR}. Since E_{AR} is decreasing with A, it's effect would be insignificant in the partition function for the dynamical box. The statistical properties of the dynamical box should thus be expected to match those described in Sec. 2. In other words, the finite component of the Casimir energy is not sufficient for shaping the dynamical geometry into a particular preferred configuration.

5. Discussion
Any successful theory of quantum dynamical geometry must eventually include coupling to the matter content observed in the universe. A-priori, average properties of dynamical geometries should be expected to be different when computed in theories with and without matter. Consequently, it is important to understand in detail the way in which matter may affect geometric observables. This issue was addressed in this paper using a toy model in which dynamical geometry was described by a two-dimensional "flexible" rectangular box of fixed area A. The expectation value of the observable $R = L_x/L_y$ was shown to be very different depending on whether matter was present or not, and depending on the boundary conditions imposed on the matter. In particular, in the presence of a finite density of particles and Dirichlet boundary conditions, it was shown that $\langle R\rangle$ can approach unity and the fluctuations $(\Delta R)^2$ can vanish in the large area limit. This scenario can thus be argued to give rise to a classical, stable, and isotropic geometry. A similar effect can arise also in the case of quantum fields if the divergent but regularized Casimir energy is included in the analysis.

The purpose of the calculations presented in this paper is to exemplify *that* and illustrate *how* matter can affect a dynamical geometry. One technical assumption that was used regarded the form of the spectrum for the lengths L_x and L_y. Although the spectrum was chosen only for convenience and not according to a deep physical principle, it should be possible to formulate precisely the conditions that must be satisfied by the physical length spectrum for the results presented to continue to hold. This kind of analysis, however, goes beyond the scope of this paper. Nonetheless, it is worth pointing out that the spectrum used is a natural one from the point of view of "atomistic" models of spacetime models in which large-scale geometry is thought of being composed of a large number of elementary building blocks.

A natural application of the presented ideas is in the context of "atomistic" models of spacetime. In this context, conventional approaches to studying matter on dynamical geometries (see e.g. [2]) have limited use because the notion of geometry is only emergent, but the general mechanism described in this paper can be applied. Consider for example the class of models called graphity [3, 4]: the basic version of these models does not generate extended manifold-like graphs, but modified versions that include a homogeneity requirement do [3]. It has been suggested that matter degrees of freedom might provide such a homogeneity requirement via a mechanism similar to the one described here. Similarly, it is possible that matter might play an important role in other models of emergent spacetime such as causal sets [10, 11] or group field theory [12].

Another application of the ideas presented here is in cosmology: instead of putting homogeneity and isotropy as inputs in an ansatz for a cosmological spacetime solution, it may be interesting to see them arising from the statistical mechanics of matter fields.

Acknowledgments

I would like to thank the organizers of DICE2008 for a diverse and extremely stimulating workshop. I have also benefitted from discussions with J. Ambjorn and B. Z. Foster.

References

[1] Oriti D ed 2008 *Approaches to Quantum Gravity* (Cambridge,UK: Cambridge U. Press)
[2] Birrell N D and Davies P C 1982 *Quantum fields in curved space* (Cambridge,UK: Cambridge U. Press)
[3] Konopka T 2008 *Phys. Rev.* D **78** 044032 (*Preprint* arXiv:hep-th/0805.2283)
[4] Konopka T, Markopoulou F and Severini S 2008 *Phys. Rev.* D **77** 104029 (*Preprint* arXiv:hep-th/0801.0861)
[5] Ambjorn J, Jurkiewicz J and Loll R 2000 *Phys. Rev. Lett.* **85** 924 (*Preprint* arXiv:hep-th/0002050)
[6] Milton K A 2004 *J. Phys.* A **37** R209 (*Preprint* arXiv:hep-th/0406024
[7] Bordag M, Mohideen U and Mostepanenko VM 2001 *Phys. Rep.* **353** 1 (*Preprint* arXiv:quant-ph/0106045)
[8] Ambjorn J and Wolfram S 1983 *Ann. Phys.* **147** 1
[9] Cavalcanti R M 2004 *Phys. Rev.* D **69** 065015 (*Preprint* arXiv:quant-ph/0310184)
[10] Rideout D and Wallden P 2008 *Preprint* arXiv:gr-qc/0810.1768
[11] Johnston S 2008 *Class. Quant. Grav.* **25** 202001 (*Preprint* arXiv:hep-th/0806.3083)
[12] Oriti D 2007 *Preprint* arXiv:gr-qc/0710.3276.

The emergence of spacetime

B.G. Sidharth

Dipartimento di Matematica e Informatica
University di Udine, Via Delle Scienze 206, Udine 33100, ITALY

E-mail: birlasc@gmail.com

Abstract. We briefly review two concepts of time - the usual time associated with "being" and more recent ideas, answering to the description of "becoming". The approximation involved in the former is examined. Finally we argue that it is (unpredictable) fluctuations that underlie time.

1. Introduction

Twentieth century physics inherited the Newtonian concept of space and time. Even with the advent of Special Relativity some of these ideas remained though time became less absolute and the concept of simultaneity got modified. In particular time was reversible. With the advent of Quantum Mechanics, there was a carry over of these concepts, even though there were new inputs like the effect of observation. So the description is that of a time reversible universe, in contrast to real life experience. It may be mentioned however that all this is strictly speaking valid for a single particle universe or at best an isolated system of a few particles. In statistical mechanics we encounter a different scenario where time is irreversible. Indeed the arrow of time is associated with an increase of entropy.

All this can be brought out by the simple example of a container with two compartments. Compartment A is filled with a swarm of molecules. In contrast compartment B is empty. A tiny hole is then bored into the dividing wall. Molecules from A gradually leak into compartment B, till both compartments have more or less an equal number of molecules. Molecules would still be crossing over from A to B and vice versa, but this state of equilibrium persists. Here with the passage of time the empty compartment B has got filled up with molecules. There is however no reversibility as in classical and Quantum Theory in the sense that we do not get back a situation, even in principle where the compartment B gets emptied out and all the molecules in B fill up compartment A.

There have been a number of arguments for and against the arrow of time in the context of modern theories of physics [1, 2, 3, 4, 5]. Nevertheless the fact that there should not be an arrow of time even within the context of Quantum Theory has been discussed [1]. Indeed as Penrose prophesized, [6] "... the tiny fact of an almost completely hidden time-asymmetry seems genuinely to be present in the K^0-decay. It is hard to believe that nature is not, so to speak, trying to tell something through the results of this delicate and beautiful experiment." It may be observed that the time arrows actually encountered are:

i) The thermodynamic arrow of time, which represents an evolution towards a greater entropy.
ii) The electromagnetic arrow of time, which becomes apparent in the retarded nature of

electromagnetic radiation chosen due to considerations of causality.
iii) The CP violating arrow of time.
iv) The cosmological arrow of time which appears in the expansion of the universe or the formation of structures.
These arrows are generally sought to be explained in terms of the usual time symmetric dynamics, but with the additional input of time asymmetric boundary conditions, for instance in the above example of a container A with molecules being in contact with an empty container B, we start ab initio with an asymmetric situation. The initial asymmetry in the universe is provided by the Big Bang, as another example [7, 8]. This is also the content of the Ehrenfest's so called double urn model [9] – it is essentially the above two compartment argument. In any case this is how an asymmetric outcome is understood to arise from symmetric considerations–the arrow of time from time reversible laws.

2. Prigogine's analysis

Prigogine had analyzed the above situation quite clearly. He observed [10] that according to the classical point of view, nature would be an automaton. However, today we discover everywhere instabilities, bifurcations, evolution. This demands a different formulation of the laws of nature to include probability and time symmetry breaking. He argued that the difficulties in the classical formulation come from too narrow a point of view concerning the fundamental laws of dynamics (classical or quantum). The classical model has been a model of integrable systems (in the sense of Poincaré). It is this circumstance that leads to determinism and time reversibility. He argued that when we leave this model and consider a class of non-integrable systems, the difficulties are overcome. Further, this approach unifies dynamics, thermodynamics and probability theory.

Prigogine goes on to point out that integrable systems are the exception, rather than being the rule. He observes, "...classically people were thinking in terms of points but here we have to speak in terms of ensembles. We cannot any more make a physics of points but we have to make a physics of distributions. This means that we have a statistical description. That also means that we have to give up classical determinism." Other new elements are also introduced into the theory. It must be mentioned that even in Quantum Theory, be it the Schrödinger equation or the Dirac equation, the wave function still follows a deterministic theory as in classical mechanics. The indeterminism of Quantum Mechanics comes in through a different mechanism, that is an observation disturbs the system and there is at that very instant a non causal collapse of the wave function into one of its eigen states.

3. Microscopic considerations

In the preceding section, the analysis was based on notions of macroscopic mechanics. However, Salecker and Wigner [11, 12] argued that the macroscopic concepts of space and time have no operational meaning for micro systems. That is spacetime descriptions may be valid only for macroscopic systems. To quote Wigner, "... the inherent limitations on the accuracy of a clock of given weight and size, which should run for a period of a certain length, are quite severe. In fact, the result in summary is that a clock is an essentially non-microscopic object. In particular, what we vaguely call an atomic clock, a single atom which ticks off its periods, is surely an idealization which is in conflict with fundamental concepts of measurability."

Zimmerman and others [13, 14] suggested that space and time arise from the properties of microscopic particles in a thermodynamic fashion in the sense that the former are a result of interactions among many microscopic systems, without a detailed description of the interactions amongst these systems in spatio-temporal terms.

In this context there have been more recent efforts in Quantum Gravity approaches. While some resemble Hawking's Euclidean Quantum Gravity in that the super position principle is applied for the several possible ways the entire universe could evolve in time (cf. refs. [15]

for a readable account), a more recent variant approach has been that of causal dynamical triangulation of Loll, Ambjorn and others. Here in the spirit of the usual triangulation, spacetime develops from the micro level in a self organizational scheme [16, 17, 18]. It must be mentioned that these recent approaches are based on three pillars which are, discreteness rather than continuum, emergence (as opposed to a reductionist approach) and causality (except in Hawking's approach). While this is reasonable from a physical point of view, particularly the requirement of causality does not go far enough. It already hides an implicit arrow of time. In any case these approaches cannot as yet be termed as being successful or the last word [19].

In this spirit the author has argued (cf. [20] and several references therein) that physical concepts of space and time or spacetime arise outside the Compton wavelength within which these concepts no longer hold [21]. Indeed this is already implicit in Wigner's analysis referred to. Within the Compton wavelength, we have unphysical phenomena like Zitterbewegung and a breakdown of causality- only an average over the Compton scale restores physics (cf. also [20, 22]). Within the Compton scale time can be modelled by a double Weiner process corresponding to a one dimensional Random Walk. It must be reiterated that within the Compton wavelength there is no causal physics [23]. Indeed it can be argued that Special Relativity becomes operational outside the Compton wavelength (cf. ref. [21]). In this sense we are justified in considering time, rather than spacetime.

To appreciate all this let us consider the motion of a particle with position given by $x(t)$, subject to random correction given by, as in the usual theory, (cf. [24, 25, 26]),

$$|\Delta x| = \sqrt{< \Delta x^2 >} \approx \nu \sqrt{\Delta t},$$

$$\nu = \hbar / m, \nu \approx lv \tag{1}$$

where ν is the so called diffusion constant and is related to the mean free path l as above. We can then proceed to deduce the Fokker-Planck equation as follows (cf. ref. [24] for details):
We first define the forward and backward velocities corresponding to having time going forward and backward (or positive or negative time increments) in the usual manner,

$$\frac{d_+}{dt} x(t) = \mathbf{b}_+ , \frac{d_-}{dt} x(t) = \mathbf{b}_- \tag{2}$$

This leads to the Fokker-Planck equations

$$\partial \rho / \partial t + div(\rho \mathbf{b}_+) = V \Delta \rho,$$

$$\partial \rho / \partial t + div(\rho \mathbf{b}_-) = -U \Delta \rho \tag{3}$$

defining

$$V = \frac{\mathbf{b}_+ + \mathbf{b}_-}{2} \quad ; U = \frac{\mathbf{b}_+ - \mathbf{b}_-}{2} \tag{4}$$

We get on addition and subtraction of the equations in (3) the equations

$$\partial \rho / \partial t + div(\rho V) = 0 \tag{5}$$

$$U = \nu \nabla ln \rho \tag{6}$$

It must be mentioned that V and U are the statistical averages of the respective velocities. We can then introduce the definitions

$$V = 2\nu \nabla S \tag{7}$$

$$V - \imath U = -2\imath \nu \nabla (ln \psi) \tag{8}$$

We next observe the decomposition of the Schrödinger wave function as

$$\psi = \sqrt{\rho} e^{\imath S/\hbar}$$

leads to the well known Hamilton-Jacobi type equation

$$\frac{\partial S}{\partial t} = -\frac{1}{2m}(\partial S)^2 + \bar{\mathbf{V}} + Q, \tag{9}$$

where

$$Q = \frac{\hbar^2}{2m} \frac{\nabla^2 \sqrt{\rho}}{\sqrt{\rho}}$$

From (7) and (8) we can finally deduce the usual Schrödinger equation or (9) [27].

We note that in this formulation three conditions are assumed, conditions whose import has not been clear. These are [24]:
(1) The current velocity is irrotational.
(2) In spite of the fact that the particle is subject to random alterations in its motion there exists a conserved energy, defined in terms of its probability distribution.
(3) The diffusion constant is inversely proportional to the inertial mass of the particle, with the constant of proportionality being a universal constant $\hbar$ (cf. equation (1))
We note that the complex feature above disappears if the fractal or non-differentiable character is not present, (that is, the forward and backward time derivatives(4) are equal): What distinguishes Quantum Mechanics is the adhoc feature, the diffusion constant ν of (1) in Nelson's theory and the "Quantum potential" Q of (9) which appears in Bohm's theory [28] as well, though with a different meaning.

Interestingly from the Uncertainty Principle,

$$m \Delta x \frac{\Delta x}{\Delta t} \sim \hbar$$

we get back equation (1) of Brownian motion.

We would like to emphasize that we have arrived at the Quantum Mechanical Schrödinger equation from Classical considerations of diffusion, though with some new assumptions. In the above, effectively we have introduced a complex velocity $V - \imath U$ which alternatively means that the real coordinate x goes into a complex coordinate

$$x \to x + \imath x' \tag{10}$$

That is, in this non relativistic development either we use forward and backward time derivatives and the usual space coordinate as in (4), or we use the derivative with respect to the usual time coordinate but introduce complex space coordinates as in (10). Already, we can get a glimpse of the special relativistic hyperbolic geometry with real space and imaginary time coordinates (or vice versa).

We now try to generalize this complex coordinate to three dimensions. Then we encounter a surprise - we end up with not three, but four dimensions,

$$(1, \imath) \to (I, \sigma),$$

where I is the unit 2×2 matrix and σs are the Pauli matrices. We get the special relativistic Lorentz invariant metric at the same time. (In this sense, as noted by Sachs [29], Hamilton who made this generalization would have hit upon Special Relativity, if he had identified the new fourth coordinate with time).

That is,

$$x + \imath y \to Ix_1 + \imath x_2 + jx_3 + kx_4,$$

where $(\imath, j, k)$ now represent the Pauli matrices; and, further,

$$x_1^2 + x_2^2 + x_3^2 - x_4^2$$

is invariant. Before proceeding further, we remark that special relativistic time emerges above from the generalization of the complex one dimensional space coordinate to three dimensions.

While the usual Minkowski four vector transforms as the basis of the four dimensional representation of the Poincaré group, the two dimensional representation of the same group, given by the right hand side in terms of Pauli matrices, obeys the quaternionic algebra of the second rank spinors (cf. ref. [30, 31, 29] for details).

In the above context let us try to see what the time of the usual theory is: In the stochastic approach, we deal with a double Wiener process which leads to a complex velocity $V - \imath U$. As noted it is this complex velocity that leads to Quantum Theory from the usual diffusion theory.

To see this in a simple way, let us write the usual diffusion equation as

$$\Delta x \cdot \Delta x = \frac{h}{m} \Delta t \equiv \nu \Delta t \tag{11}$$

We saw that equation (11) can be rewritten as the usual Quantum Mechanical relation,

$$m \frac{\Delta x}{\Delta t} \cdot \Delta x = h = \Delta p \cdot \Delta x \tag{12}$$

We are dealing here, with phenomena within the Compton or De Broglie wavelength.

We now treat the diffusion constant ν to be very small, but non vanishing. That is, we consider the semi classical case. It is well known that in this situation we can use the WKB approximation [32]. Whence the right hand side of the wave function,

$$\psi = \sqrt{\rho} e^{\imath/\hbar S}$$

goes over to, in the one dimensional case, for simplicity,

$$(p_x)^{-\frac{1}{2}} e^{\frac{1}{\hbar}} \int p(x) dx$$

so that we have, on comparison,

$$\rho = \frac{1}{p_x} \tag{13}$$

ρ being the probability density. In this case the condition $U \approx 0$, that is, the velocity potential becoming real, implies

$$\nu \cdot \nabla ln(\sqrt{\rho}) \approx 0 \tag{14}$$

This semiclassical analysis suggests that $\sqrt{\rho}$ is a slowly varying function of x, in fact each of the factors on the left side of (14) would be $\sim 0(h)$, so that the left side is $\sim 0(h^2)$ (which is being neglected). Then from (13) we conclude that p_x is independent of x, or is a slowly varying function of x. The equation of continuity now gives

$$\frac{\partial \rho}{\partial t} + \vec{\nabla}(\rho \vec{v}) = \frac{\partial \rho}{\partial t} = 0$$

That is the probability density ρ is independent or nearly so, not only of x but also of t. We are thus in a stationary and homogenous scenario. This is strictly speaking, possible only in a single particle Universe, or for a completely isolated particle, without any effect of the environment. Under these circumstances we have the various conservation laws and the time reversible theory, all this taken over into Quantum Mechanics as well. This is an approximation valid for small, incremental changes, as indeed is implicit in the concept of a differentiable spacetime manifold.

To put it simply, if dt is the change or time interval in the usual time, then N such intervals would imply a passage of time of magnitude Ndt, whereas in our approach, if $d\tau$ is the basic interval, then the time passage would be $\sqrt{N}d\tau$.

We recognize that time is statistical and depends on the number of constituents. There could be collisions amongst N particles, or equivalently, as we will see below the fluctuational creation of such particles from the background Dark Energy. It could then be possible to ascribe a general minimum time interval τ for all the N particles, which would be for example the mean free time between the collisions or the fluctuational creation of particles. On the other hand let τ_0 represent the corresponding quantity, but this time associated with each individual event rather than the entire assembly. We would then have (cf. also ref.[14])

$$\tau = \lambda\tau_0, \tag{15}$$

where λ represents the statistical dispersion effect and is given by

$$\lambda = \left[\frac{1}{N}\sum_{\imath=1}^{N}(\theta_\imath)^2 - \left(\frac{1}{N}\sum_{\imath=1}^{N}\theta_\imath\right)^2\right]^{1/2} \tag{16}$$

$\theta_\imath$ representing the coefficient to be multiplied into each individual τ_0, in order to get the corresponding interval for the $\imath$th particle. As N is large and the $\theta_\imath$ are all of the order 1, using (16) in (15) we get

$$\tau = 1/\sqrt{N}\tau_0 \tag{17}$$

As there are N such events, the time elapsed T would be $N\tau$ which from (17) is given by

$$T = \sqrt{N}\tau_0 \tag{18}$$

The relevance of all this is the following. The author's 1997 model was one in which particles were fluctuationally created from a background Dark Energy within the Compton time τ_0. It may be mentioned that this model lead to a cosmology in which the universe was accelerating with a small cosmological constant, besides many other consistent but otherwise inexplicable astrophysical relations being deduced from the theory (cf. refs. [33, 28, 34, 20] and several references therein). As is known all this was confirmed in 1998 itself through the observations of distant supernovae. In this model given n particles at any time, $\sqrt{n}$ particles would be fluctuationally created in the time τ_0. That is we would have,

$$\frac{dn}{dt} \approx \frac{\sqrt{n}}{\tau_0}$$

because the interval τ_0 being small, we can approximate with derivates. Whence on integration from time $t = 0$ to $t = T$, we get,

$$T = \sqrt{N}\tau_0 \tag{19}$$

In (19), at time T there would be N particles in the universe. Not only is (19) consistent, because there are N of the order 10^{80} particles in the universe while T is of the order 10^{17} seconds and τ_0 a typical Compton time is of the order 10^{-23}seconds, but as can be seen (19) is identical to

(18) though it is obtained by a different route. The above is very much in the spirit of a one dimensional Random Walk or two Weiner process encountered earlier. In this case time would flip forward and backward like steps to the right and steps to the left at random. The nett dispersion or time elapsed after N such stages would be exactly as in equations (18) or (19). In the above cosmological model this would mean that particles would be created and destroyed at random, from and into the background Dark Energy, but the nett result is given by (19). In this formulation, there is no inbuilt causality.

4. Remarks

Time is associated with change. In a changeless universe, there would be no time. It must be mentioned here that various theories of time, already imply time or more generally change (cf. also [35]). The question is, what type of a change do we consider? We have argued that the time which we usually use is based on an incremental change - it is almost as if there were no change. For example the law of conservation of energy is based on a time translation symmetry - an infinitesimal translation in time leaves everything unchanged [36]. Clearly this is only an approximation which assumes that there is no change in a very short interval.

In fact, time is essentially an ordering or sequencing of events. The key here is the way in which this ordering is done so that causality and other laws of physics hold or emerge rather than being inputs. If we consider the universe as a sequence of instantaneous space slices, to start with, then a random sequence would represent a lawless, and literally chaotic universe.

On the contrary we have seen that at the micro Compton scale, there is no causality and there is the chaotic Zitterbewegung. This means that if it were possible for a creature or a measuring device to be so small as to be within the Compton scale, then such a creature or device would perceive a lawless, chaotic universe. However physics, emerges once averages over the unphysical Compton scale are taken. In this sense the universe that is perceived and measured is a macroscopic universe. At the micro level, as pointed out by Wigner and Sackler, we can no longer extrapolate these macro concepts. In a sense, this is connected with the Copenhagen debate on the role of macroscopic measuring devices in obtaining information about microscopic systems.

At the macro scale however we have two different situations. One, which we encountered in the first section can further be exemplified with the example of a porcelain plate that falls to the ground and breaks into many pieces. Here a highly ordered system, namely the porcelain plate becomes a highly disordered system. As long as we do not specify the exact shapes and sizes of the shattered pieces in advance, this can always happen - it provides an arrow of time with increasing entropy. This time, furthermore is irreversible. The shattered pieces then combining to form the plate or equivalently the shattered pieces, which describe a very definite prescribed shape and size would be an impossibility and would represent the reversal of time. In any case, the connection between time and probability is brought about. If the probability of something happening is (in advance) zero or nearly so, time does not "evolve" to such a situation. Time's arrow or flow is in the direction of non vanishing probabilities.

There is another change that we had considered in the previous section - this is the fluctuational creation of particles. This gives rise to a macroscopic time, through a Brownian process, leading to the correct age of the universe as exemplified in (19). In contrast to the change which our usual time represents, this latter Brownian change is no longer incremental. This picture is a far cry from the smoothly flowing time of usual theory. The contrast is between "becoming" and "being". Moreover, this time is based on, not just the local particle, but a whole assembly of particles, as brought out by (16), for example. Our contention is that it is this irreversible change of fluctuations that represents our actual time. This is in the spirit of Bergson's future being creation (or invention) [37].

In any case, the above discussion shows the breakdown of the concept of point time or instants of time. Indeed as Wigner (loc.cit) notes, this concept has no meaning at least within the time interval thrown up by the Uncertainty Principle, and perhaps even beyond. Finally, it may be mentioned that the probabilistic feature of time discussed above, brings in to play via observer participation, two poorly understood concepts as of now: information and consciousness.

References

[1] Sidharth B G 2000 *Chaos, Solitons and Fractals* **11** 1045-1046
[2] Coveney P V 1988 *Nature* **333** 409-415
[3] Ranada F A and Tiemblo A 2008 *Found.Phys* **38** 458-469
[4] Gell'man M and Hartle J B 1996 Time Symmetry and Asymmetry in Quantum Mechanics and Quantum Cosmology *Physical Origins of Time Asymmetry* eds J J Halliwell, J Perez-Mercader and W H Zurek (Cambridge, UK: Cambridge U. Press)
[5] Lebowitz J L 1996 Times Arrow and Boltzmann Entropy *Physical Origins of Time Asymmetry* eds J J Halliwell, J Perez-Mercader and W H Zurek (Cambridge, UK: Cambridge U. Press)
[6] Hawking S and Israel W 1979 *General Relativity: An Einstein Centenary Survey* (Cambridge, UK: Cambridge U. Press)
[7] Hu B L 1996 Fluctuation, Dissipation and Irreversibility in Cosmology *Physical Origins of Time Asymmetry* eds J J Halliwell, J Perez-Mercader and W H Zurek (Cambridge, UK: Cambridge U. Press)
[8] Hawking S W 1996 The No Boundary Condition and the Arrow of Time *Physical Origins of Time Asymmetry* eds J J Halliwell, J Perez-Mercader and W H Zurek (Cambridge, UK: Cambridge U. Press)
[9] Wheeler J A 1996 Time Today in Physical Origin *Physical Origins of Time Asymmetry* eds J J Halliwell, J Perez-Mercader and W H Zurek (Cambridge, UK: Cambridge U. Press)
[10] Prigogine I 2008 *A Century of Ideas* ed B G Sidharth (Springer, Netherlands)
[11] Wigner E P 1952 *Rev. Mod. Phys.* **29** 255
[12] Salecker H and Wigner E P 1958 *Phys. Rev.* **109** 571-577
[13] Zimmerson E J 1962 *Am. J. Phys* **30** 97
[14] Karimov A R 2008 *New Advances in Physics* **2** Number 1 103-107
[15] Halliwell J 1991 *Scientific American* December 1991
[16] Cho A 2004 *Phys. Rev.* September 28; also http://focus.aps.org/story/v14/st13
[17] Ambjorn J, Gorlich A, Jurkiewicz J and Loll R 2008 *Phys. Rev. Lett.* **100** 091304 (*Preprint* arXiv:0712.2485)
[18] Loll R 2008 *Class. Qu. Grav.* **25** 114006 (*Preprint* arXiv:0711.0273)
[19] Smolin L 2005 *The Trouble with Physics* (New York: Basic Books)
[20] Sidharth B G 2008 *The Thermodynamic Universe* (Singapore: World Scientific)
[21] Sidharth B G 2008 *Found. Phys.* **38** 695-706
[22] Sidharth B G Revisiting Zitterbewegung *Int. J. Th. Phys.*, in press (*Preprint* arXiv:0806.0985)
[23] Weinberg S 1972 *Gravitation and Cosmology* (New York: John Wiley & Sons) 61ff
[24] Smolin L 1986 *Quantum Concepts in Space and Time* eds R Penrose and C J Isham (Oxford: OUP) 147-181
[25] Reif F 1965 *Fundamentals of Statistical and Thermal Physics* (Singapore: McGraw-Hill Book Co.)
[26] Sidharth B G 2000 *Chaos, Solitons and Fractals* **12**(1) 173-178
[27] Nottale L 1993 *Fractal Space-Time and Microphysics: Towards a Theory of Scale Relativity* (Singapore: World Scientific) 312
[28] Sidharth B G 2001 *Chaotic Universe: From the Planck to the Hubble Scale* (New York: Nova Science)
[29] Sachs M 1982 *General Relativity and Matter* (D. Reidel Publishing Company, Holland) 45ff
[30] Sidharth B G 2003 *Found. Phys. Lett.* **16**(1) 91-97
[31] Shirokov Yu M 1958 *Sov. Phys. JETP* **6**(33) No.5 929-935
[32] Schiff L I 1968 *Quantum Mechanics* (London: McGraw Hill)
[33] Sidharth B G 1998 *Int. J. Mod. Phys.* A **13**(15) 2599
[34] Sidharth B G 2005 *The Universe of Fluctuations* (Springer, Netherlands)
[35] Aiello M, Castagnino M and Lombardi O 2008 *Found. Phys* **38** 257-292
[36] Roman P 1965 *Advanced Quantum Theory* (Reading, Mass.: Addison-Wesley) 31
[37] Prigogine I 1985 *Order Out of Chaos* (London: Flamingo, Harper Collins) xix

Journal of Physics: Conference Series **174** (2009) 012053 doi:10.1088/1742-6596/174/1/012053

Noncommutative quantum cosmology

C Bastos[1,2], O Bertolami[1,2], N C Dias[3,4] and J N Prata[3,4]

[1] Departamento de Física, Instituto Superior Técnico, Avenida Rovisco Pais 1, 1049-001 Lisboa, Portugal
[2] Instituto de Plasmas e Fusão Nuclear, Instituto Superior Técnico, Avenida Rovisco Pais 1, 1049-001 Lisboa, Portugal
[3] Departamento de Matemática, Universidade Lusófona de Humanidades e Tecnologias, Avenida Campo Grande, 376, 1749-024 Lisboa, Portugal
[4] Grupo de Física Matemática, Universidade de Lisboa, Avenida Prof. Gama Pinto 2, 1649-003, Lisboa, Portugal

E-mail: cbastos@fisica.ist.utl.pt, orfeu@cosmos.ist.utl.pt, ncdias@mail.telepac.pt, joao.prata@mail.telepac.pt

Abstract.
We present a phase-space noncommutative extension of Quantum Cosmology in the context of a Kantowski-Sachs (KS) minisuperspace model. We obtain the Wheeler-DeWitt (WDW) equation for the noncommutative system through the ADM formalism and a suitable Seiberg-Witten map. The resulting WDW equation explicitly depends on the phase-space noncommutative parameters, θ and η. Numerical solutions of the noncommutative WDW equation are found and, interestingly, also bounds on the values of the noncommutative parameters. Moreover, we conclude that the noncommutativity in the momenta sector leads to a damped wave function implying that this type of noncommutativity can be relevant for a selection of possible initial states for the universe.

1. Introduction

Noncommutative space-time is believed to be a fundamental ingredient of quantum gravity [1] and arises in String Theory/M-Theory, where a noncommutative effective low-energy gauge theory when describing the excitations of open strings in the presence of a Neveu-Schwarz constant background field [2]. These ideas have sparked the interest in noncommutative field theories (QFT)[see e.g. [3] and Refs. therein]. In what concerns the one-particle sector of QFT, several noncommutative versions of quantum mechanics have been discussed [4, 5, 6, 7, 8, 9, 10]. The recent measurement of the first two quantum states of the gravitational quantum well for ultra cold neutrons has led one to consider noncommutative extensions of the gravitational quantum well [6, 8]. In that work, it is concluded that momentum noncommutativity should be considered for the noncommutative problem to differ from the commutative one.

The most general canonical noncommutative quantum mechanics (NCQM) is characterized by an extension of the Heisenberg-Weyl algebra. In general, time is assumed as a commutative parameter and a $2d$-dimensional phase-space of operators with noncommuting position and momentum variables is considered. This extended Heisenberg-Weyl algebra is given by:

$$[\hat{q}_i, \hat{q}_j] = i\theta_{ij} \ , \ [\hat{q}_i, \hat{p}_j] = i\hbar\delta_{ij} \ , \ [\hat{p}_i, \hat{p}_j] = i\eta_{ij} \ , \ i,j = 1, ..., d \qquad (1)$$

where η_{ij} and θ_{ij} are antisymmetric real constant $(d \times d)$ matrices and δ_{ij} is the identity matrix. One can relate this extended algebra with the standard Heisenberg-Weyl algebra:

$$\left[\hat{R}_i, \hat{R}_j\right] = 0 \ , \ \left[\hat{R}_i, \hat{\Pi}_j\right] = i\hbar\delta_{ij} \ , \ \left[\hat{\Pi}_i, \hat{\Pi}_j\right] = 0 \ , \ i,j = 1, ..., d \ , \tag{2}$$

by a class of linear (non-canonical) transformations:

$$\hat{q}_i = \hat{q}_i\left(\hat{R}_j, \hat{\Pi}_j\right) \ , \ \hat{p}_i = \hat{p}_i\left(\hat{R}_j, \hat{\Pi}_j\right) \tag{3}$$

which are often referred to as the Seiberg-Witten (SW) map [2]. With these transformations, one is able to convert a noncommutative system into a modified commutative one, which is dependent on the noncommutative parameters and of the particular SW map. The usual Schrödinger equation with a modified η, θ-dependent Hamiltonian determines the dynamics of the system and the wave functions in the Hilbert space correspond to the states of the system. However, it should be stressed that the physical properties of the system such as expectation values, probabilities and eigenvalues of operators are independent of the chosen SW map [10].

In this contribution we report on the results of a study of the phase space noncommutative KS minisuperspace model [11]. The KS minisuperspace model has been previously examined in the context of noncommutativity in the configuration space [14, 15]. We regard our more general approach fairly natural given that noncommutativity is presumably a relevant feature of quantum gravity and it is expected that its effects ought be important in the early universe. To address this question we implement the canonical quantization procedure in the context of the phase space noncommutative KS minisuperspace model in order to obtain the corresponding WDW equation of the problem (see e.g. Refs. [12, 13] for general discussions).

Aiming to solve the noncommutative WDW equation, one implements a constraint that commutes with the Hamiltonian (and is associated with a constant of motion of the classical problem) and that allows for turning a second order partial differential equation into an second order ordinary differential equation that is then solved numerically. By examining the resulting physical solutions one is able to restrict the set of possible values of the noncommutative parameters.

Furthermore, we conclude that momentum space noncommutativity leads to a richer structure of states for the early universe. In this case the fundamental solutions of the WDW equation (which are featureless oscillations for both the commutative and configuration noncommutative cases) display a damping behaviour. This suggests a criterion for selection of states for the early universe.

2. The cosmological model

One considers a cosmological model described by the KS metric. In the Misner parametrization, the line element can be written as [16]

$$ds^2 = -N^2dt^2 + e^{2\sqrt{3}\beta}dr^2 + e^{-2\sqrt{3}\beta}e^{-2\sqrt{3}\Omega}(d\theta^2 + \sin^2\theta d\varphi^2) \ , \tag{4}$$

where β and Ω are the scale factors and N is the lapse function. The presence of at least two scale factors is necessary in order to impose a noncommutative algebra. The Hamiltonian for this metric is obtained via the ADM formalism [16]:

$$H = N\mathcal{H} = Ne^{\sqrt{3}\beta+2\sqrt{3}\Omega}\left[-\frac{P_\Omega^2}{24} + \frac{P_\beta^2}{24} - 2e^{-2\sqrt{3}\Omega}\right] \ , \tag{5}$$

where P_Ω and P_β are the canonical momenta conjugate to Ω and β, respectively. Given that the results are all gauge independent, one chooses $N = 24e^{-\sqrt{3}\beta-2\sqrt{3}\Omega}$. This choice corresponds to a particular gauge choice, which is only motivated by technical simplicity. At quantum level, the treatment is manifestly covariant as the lapse function does not enter at all in the formalism. In the next sections, this will be shown explicitly. The classical and the quantum formulations will be considered separately.

2.1. *The classical model*

In the classical picture, the equations of motion for the four variables Ω, β, P_Ω and P_β can be obtained from the Poisson's bracket algebra. The commutative and the configuration space noncommuativity cases were discussed previously in Refs. [14, 15]. Analytical solutions to classical equations of motion were obtained for the two cases.

Here, one considers the most general noncommutative extension that can be obtained by imposing a noncommutative relation between the two scale factors, Ω and β, and between the two canonical momenta, P_Ω and P_β:

$$\{\Omega, P_\Omega\} = 1 \ , \ \{\beta, P_\beta\} = 1 \ , \ \{\Omega, \beta\} = \theta \ , \ \{P_\Omega, P_\beta\} = \eta \ . \tag{6}$$

In the constrained hypersurface

$$\mathcal{H} \approx 0 \ , \tag{7}$$

the classical equations of motion for the noncommutative system are given by

$$\begin{aligned} &\dot{\Omega} = -2P_\Omega \ , \quad (a) \\ &\dot{P_\Omega} = 2\eta P_\beta - 96\sqrt{3}e^{-2\sqrt{3}\Omega} \ , \quad (b) \\ &\dot{\beta} = 2P_\beta - 96\sqrt{3}\theta e^{-2\sqrt{3}\Omega} \ , \quad (c) \\ &\dot{P_\beta} = 2\eta P_\Omega \ . \quad (d) \end{aligned} \tag{8}$$

Due to the entanglement among the four variables, an analytical solution is hard to find. However, it is possible to obtain a numerical solution to this system as well as a constant of motion from Eqs. (8a) and (8d):

$$\dot{P_\beta} = -\eta(-2P_\Omega) = -\eta\dot{\Omega} \Rightarrow P_\beta + \eta\Omega = C \ ; \tag{9}$$

this will play an important role in solving the phase space noncommutative WDW equation.

2.2. *The quantum model*

Here and henceforth, one assumes a system of units where $c = \hbar = G = 1$, that is the Planck units. Regarding the noncommutative parameters, θ and η, as an intrinsic feature of quantum gravity, they are expected to be of order one in Planck units.

By canonical quantization of the classical Hamiltonian constraint, Eq. (7), one obtains the commutative WDW equation for the wave function of the universe. For the simplest ordering of operators,

$$\exp\left(\sqrt{3}\hat{\beta} + 2\sqrt{3}\hat{\Omega}\right)\left[-\hat{P}_\Omega^2 + \hat{P}_\beta^2 - 48e^{-2\sqrt{3}\hat{\Omega}}\right]\psi(\Omega, \beta) = 0 \ . \tag{10}$$

where $\hat{P}_\Omega = -i\frac{\partial}{\partial\Omega}$, $\hat{P}_\beta = -i\frac{\partial}{\partial\beta}$ are the fundamental momentum conjugate operators to $\hat{\Omega} = \Omega$ and $\hat{\beta} = \beta$, respectively. Notice that, as it is usual in Quantum Cosmology, Eq.(10) depends on the prescribed factor ordering. One chooses the simplest factor ordering, which has already been studied in Refs. [14, 15]. This allows for a direct comparison with the results in the literature.

For the commutative case, the solutions of Eq. (10) are [14],

$$\psi_\nu^\pm(\Omega, \beta) = e^{\pm i\nu\sqrt{3}\beta} K_{i\nu}(4e^{-\sqrt{3}\Omega}) \ , \tag{11}$$

where $K_{i\nu}$ are modified Bessel functions.

One requires that the coordinates and the canonical momenta do not commute. Thus, the extended Heisenberg algebra is,

$$\left[\hat{\Omega}, \hat{\beta}\right] = i\theta \ , \ \left[\hat{P}_{\Omega}, \hat{P}_{\beta}\right] = i\eta \ , \ \left[\hat{\Omega}, \hat{P}_{\Omega}\right] = \left[\hat{\beta}, \hat{P}_{\beta}\right] = i \ . \tag{12}$$

To obtain a representation of this algebra (12), one transforms it into the standard Heisenberg algebra through a SW map [11]:

$$\begin{aligned} \hat{\Omega} &= \lambda\hat{\Omega}_c - \frac{\theta}{2\lambda}\hat{P}_{\beta_c} \ , \ \hat{\beta} = \lambda\hat{\beta}_c + \frac{\theta}{2\lambda}\hat{P}_{\Omega_c} \ , \\ \hat{P}_{\Omega} &= \mu\hat{P}_{\Omega_c} + \frac{\eta}{2\mu}\hat{\beta}_c \ , \ \hat{P}_{\beta} = \mu\hat{P}_{\beta_c} - \frac{\eta}{2\mu}\hat{\Omega}_c \ , \end{aligned} \tag{13}$$

where the index c denotes commutative variables, i.e. variables for which $\left[\hat{\Omega}_c, \hat{\beta}_c\right] = \left[\hat{P}_{\Omega_c}, \hat{P}_{\beta_c}\right] = 0$ and $\left[\hat{\Omega}_c, \hat{P}_{\Omega_c}\right] = \left[\hat{\beta}_c, \hat{P}_{\beta_c}\right] = i$. This transformation can be inverted only if:

$$\xi \equiv \theta\eta < 1. \tag{14}$$

In this case, the inverse transformation reads:

$$\begin{aligned} \hat{\Omega}_c &= \frac{1}{\sqrt{1-\xi}}\left(\mu\hat{\Omega} + \frac{\theta}{2\lambda}\hat{P}_{\beta}\right) \ , \ \hat{\beta}_c = \frac{1}{\sqrt{1-\xi}}\left(\mu\hat{\beta} - \frac{\theta}{2\lambda}\hat{P}_{\Omega}\right) \ , \\ \hat{P}_{\Omega_c} &= \frac{1}{\sqrt{1-\xi}}\left(\lambda\hat{P}_{\Omega} - \frac{\eta}{2\mu}\hat{\beta}\right) \ , \ \hat{P}_{\beta_c} = \frac{1}{\sqrt{1-\xi}}\left(\lambda\hat{P}_{\beta} + \frac{\eta}{2\mu}\hat{\Omega}\right) \ . \end{aligned} \tag{15}$$

Substituting the noncommutative variables, expressed in terms of the commutative ones, into the commutation relations (12), one obtains a relation between the dimensionless constants λ and μ:

$$(\lambda\mu)^2 - \lambda\mu + \frac{\xi}{4} = 0 \Leftrightarrow \lambda\mu = \frac{1+\sqrt{1-\xi}}{2} \ . \tag{16}$$

Hence, through the transformation Eq. (13), one may regard Eq. (12) as an algebra of operators acting on the usual Hilbert space $L^2(\mathbb{R}^2)$. In this representation, the WDW Eq. (10) is deformed into a modified second order partial differential equation, which exhibits an explicit dependence on the noncommutative parameters:

$$\left[(-i\mu\frac{\partial}{\partial\Omega_c} + \frac{\eta}{2\mu}\beta_c)^2 - (-i\mu\frac{\partial}{\partial\beta_c} - \frac{\eta}{2\mu}\Omega_c)^2 + 48\exp\left[-2\sqrt{3}(\lambda\Omega_c + i\frac{\theta}{2\lambda}\frac{\partial}{\partial\beta_c})\right]\right]\psi(\Omega_c, \beta_c) = 0 \ . \tag{17}$$

This equation is fairly complex and cannot be solved analytically. However, the noncommutative quantum version of the constant of motion Eq. (9):

$$\hat{C} = \hat{P}_{\beta} + \eta\hat{\Omega} = \sqrt{1-\xi}\left(\mu\hat{P}_{\beta_c} + \frac{\eta}{2\mu}\hat{\Omega}_c\right) \tag{18}$$

commutes with the noncommutative Hamiltonian constraint Eq. (17), for the chosen operator ordering. This allows one to transform the partial differential Eq. (17) into an ordinary differential equation, which can be then solved numerically [11]. The results are presented in the next section.

Journal of Physics: Conference Series **174** (2009) 012053 doi:10.1088/1742-6596/174/1/012053

3. Solutions

We consider now Eq. (17) in detail. Defining $\hat{A} = \frac{\hat{C}}{\sqrt{1-\xi}}$ from Eq. (18), it then follows that:

$$\mu \hat{P}_{\beta_c} + \frac{\eta}{2\mu}\hat{\Omega}_c = \hat{A} \ . \tag{19}$$

As already referred to, the noncommutative WDW Eq. (17) is very complex and no analytical solution is likely to be found. The strategy is to solve it numerically transforming it into an ordinary differential equation. First, one verifies that the constant of motion, Eq. (9), commutes with the Hamiltonian in the constrained space of states:

$$\left[\hat{P}_\beta + \eta\hat{\Omega}, \hat{H}\right] = \left[\hat{P}_\beta + \eta\hat{\Omega}, -\hat{P}_\Omega^2 + \hat{P}_\beta^2 - 48e^{-2\sqrt{3}\hat{\Omega}}\right] = 0 \ . \tag{20}$$

Then, one looks for solutions of Eq. (17) that are simultaneous eigenstates of the Hamiltonian and of the constraint Eq. (19). Let $\psi_a(\Omega_c, \beta_c)$ be an eigenstate of the operator Eq. (19) with a real eigenvalue, $a \in \mathbb{R}$, then:

$$\left(-i\mu\frac{\partial}{\partial\beta_c} + \frac{\eta}{2\mu}\Omega_c\right)\psi_a(\Omega_c, \beta_c) = a\psi_a(\Omega_c, \beta_c) \ . \tag{21}$$

This equation admits the solution:

$$\psi_a(\Omega_c, \beta_c) = \Re(\Omega_c)\exp\left[\frac{i}{\mu}\left(a - \frac{\eta}{2\mu}\Omega_c\right)\beta_c\right] . \tag{22}$$

Substituting the wave function (22) into Eq. (17) one obtains

$$\mu^2\left[\frac{\Re''}{\Re} - i\frac{\eta}{\mu^2}\frac{\Re'}{\Re}\beta_c - \frac{\eta^2}{4\mu^4}\beta_c^2\right] + i\eta\left[\frac{\Re'}{\Re} - i\frac{\eta}{2\mu^2}\beta_c\right]\beta_c - \frac{\eta^2}{4\mu^2}\beta_c^2 + \left(a - \frac{\eta}{2\mu}\Omega_c\right)^2 -$$
$$-\frac{\eta}{\mu}\left(a - \frac{\eta}{2\mu}\Omega_c\right)\Omega_c + \frac{\eta^2}{4\mu^2}\Omega_c^2 - 48\exp\left[-2\sqrt{3}\left(\lambda\Omega_c - \frac{\theta}{2\lambda\mu}\left(a - \frac{\eta}{2\mu}\Omega_c\right)\right)\right] = 0 \ , \tag{23}$$

where $\Re' \equiv \frac{d\Re}{d\Omega_c}$. After some algebraic manipulations, one finally gets

$$\mu^2\Re'' + \left(\eta\frac{\Omega_c}{\mu} - a\right)^2\Re - 48\exp\left[-2\sqrt{3}\frac{\Omega_c}{\mu} + \frac{\sqrt{3}\theta}{\lambda\mu}a\right]\Re = 0 \ . \tag{24}$$

Performing the following change of variables,

$$z = \frac{\Omega_c}{\mu} \ \rightarrow \ \frac{d}{dz} = \mu\frac{d}{d\Omega_c} \ , \tag{25}$$

it finally yields for $\phi(z) \equiv \Re(\Omega_c(z))$

$$\phi''(z) + (\eta z - a)^2\phi(z) - 48\exp[-2\sqrt{3}z + \frac{\sqrt{3}\theta}{\lambda\mu}a]\phi(z) = 0 \ . \tag{26}$$

This second order ordinary differential equation can be solved numerically. Its solutions depend on the eigenvalue a and on the noncommutative parameters θ and η.

Our problem involves with four initial conditions: $\Omega(0)$, $\beta(0)$, $P_\Omega(0)$ and $P_{\beta(0)}$. With the exception of $\beta(0)$, all the others are related to each other due to the constraint Eq. (7). Thus, if

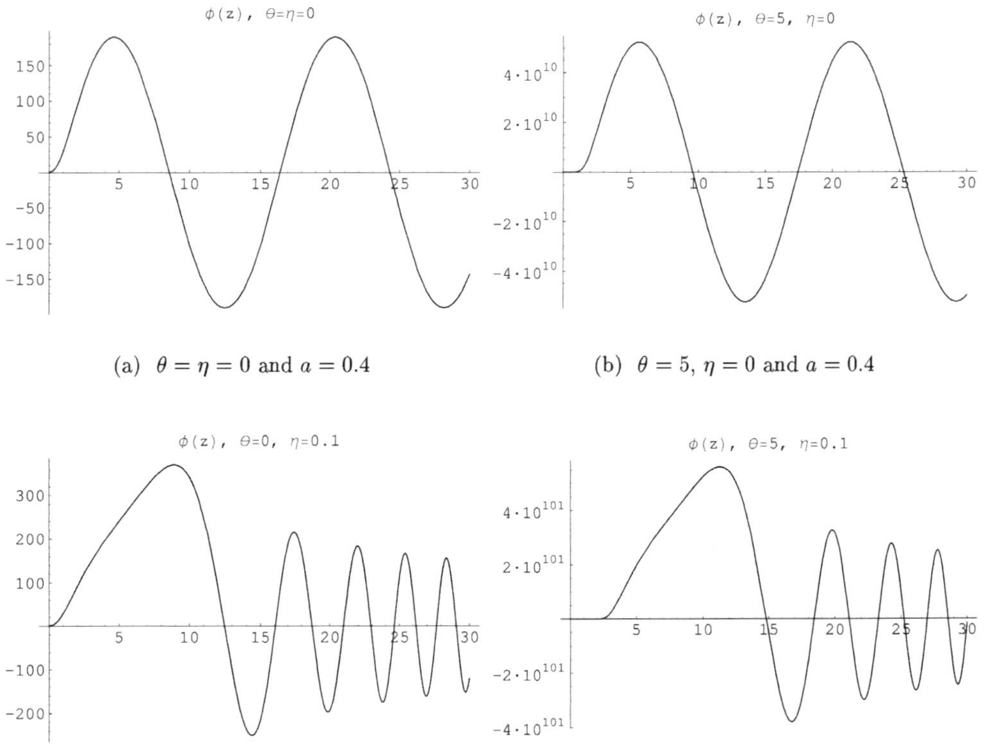

(a) $\theta = \eta = 0$ and $a = 0.4$

(b) $\theta = 5$, $\eta = 0$ and $a = 0.4$

(c) $\theta = 0$, $\eta = 0.1$ and $a = 0.565$

(d) $\theta = 5$, $\eta = 0.1$ and $a = 0.799$

Figure 1. Representation of the numerical solutions of Eq. (26) for different values to the noncommutative parameters. In the four plots $P_\beta(0) = 0.4$ and $\Omega(0) = 1.65$.

we choose some numerical values for $P_{\beta(0)}$ and $P_\Omega(0)$, one immediately obtains a value for $\Omega(0)$. $\beta(0)$ is an independent initial condition.

In Fig. 1 we present the numerical solutions of Eq. (26) for particular sets of values for a, θ and η. The eigenvalue a was taken to be $a = \frac{C}{\sqrt{1-\theta\eta}}$ and is determined through Eq. (9) from the classical values $P_\beta(0)$ and $\Omega(0)$ used to generate the solutions of Eqs. (8). These classical values are fairly typical, and once again they were borrowed from the previous studies of the noncommutative KS cosmological model [15], so as to allow for comparison with previous results. However, one finds that under variations of the relevant parameters, the qualitative behaviour of the obtained wave function is not significantly changed.

Indeed after a thorough analysis of the results, one observes that the qualitative features of the solutions displayed in Fig. 1 is unchanged for a rather broad range of values for θ, η and a [11]. The choice $\theta = 5$ is fairly typical in what concerns the properties of the wave function. Furthermore, it is consistent with the point of view that the noncommutative parameters should be of order one close to the fundamental quantum gravity scale. The summary of the main results of Ref. [11] is the following:

(i) For $\theta = 5$, the wave function has a damping behaviour for η in the range $0.05 < \eta < 0.12$;
- For $\eta_c > 0.12$ the wave function blows up, suggesting that it is an upper limit for momenta noncommutativity;

(ii) For $\theta > \eta$, varying θ affects the numerical values of $\phi(z)$, but its qualitative features remain unchanged.
   - The lower limit for η which exhibits a damping behaviour is around $\eta \sim 0.05$ for all $\theta > \eta$. Clearly, higher η values (c.f. Fig. 1) have a great influence on the wave function;
   - For $\eta = 0$ the wave function is essentially oscillatory;
   - For $0 < \eta < 0.05$, the wave function is actually amplified instead of exhibiting a damping behaviour;

(iii) For $\eta > \theta$, the damping behaviour of the wave function is harder to observe as the wave function does not blow up only for certain values of θ, in particular when $\eta \in [1, 2]$;
   - For $\eta \geq 3$ there are no possible ranges for θ for which the wave function is well defined;

(iv) For large z the qualitative behaviour of the wave function is analogous to the one depicted in Fig. 1 for $z \leq 30$.

The criterion to determine bounds for the noncommutative parameters is based on the existence of well defined smooth solutions of the WDW equation. As discussed, solutions do not exist for arbitrary values of θ and η. Notice that it is the noncommutativity introduced in the momenta space that affects the behaviour of the wave function. It is interesting to observe that this kind of noncommutativity turned oscillatory functions of the commutative and noncommutative in configuration space cases into damped wave functions. This novel feature is a welcome property as it introduces structure into the wave function, suggesting a natural selection of states for the quantum cosmological model and thus for the set of initial conditions of the classical cosmological model.

4. Conclusions

In this contribution we have reviewed the effects of a phase space noncommutativity on the minisuperspace KS quantum cosmological model. Through the analysis of the noncommutative system, one finds a classical constant of motion that allows for a numerical solution of the noncommutative WDW equation. One clearly sees that the quantum model is affected by the noncommutativity on the momenta sector, that is the wave function presents a damping behaviour for growing values of the Ω variable. Thus, the wave function is more peaked for small values of Ω, which is a rather interesting and a new portrait of the quantum behaviour for the very early universe.

Acknowledgments

The work of CB is supported by Fundação para a Ciência e a Tecnologia (FCT) under the fellowship SFRH/BD/24058/2005. The work of OB is partially supported by the FCT project No. POCTI/FP/63916/2005. The work of NCD and JNP was partially supported by Grant No. POCTI/0208/2003 and PTDC/MAT/69635/2006 of the FCT.

References

[1] Connes A 1996 *Commun. Math. Phys.* **182** 155
[2] Seiberg N and Witten E 1999 *JHEP* **9909** 032 and Refs. therein
[3] Szabo R 2003 *Phys. Rep.* **378** 207
[4] Gamboa J *et al* 2001 *Mod. Phys. Lett.* A **16** 2075; Gamboa J, Loewe M and Rojas J C 2001 *Phys. Rev.* D **64** 067901
[5] Zhang J-Z 2004 *Phys. Rev. Lett.* **93** 043002; Zhang J-Z 2004 *Phys. Lett.* B **584** 204
[6] Bertolami O, Rosa J G, Aragão C, Castorina P and Zappalà D 2005 *Phys. Rev.* D **72** 025010
[7] Acatrinei C 2005 *Mod. Phys. Lett.* A **20** 1437
[8] Bertolami O, Rosa J G, Aragão C, Castorina P and Zappalà D 2006 *Mod. Phys. Lett.* A **21** 795
[9] Bastos C and Bertolami O 2008 *Phys. Lett.* A **372** 5556

[10] Bastos C, Bertolami O, Dias N C and Prata J N 2008 *J. Math. Phys.* **49** 072101
[11] Bastos C, Bertolami O, Dias N C and Prata J N 2008 *Phys. Rev.* D **78** 023516
[12] Hartle J B and Hawking S W 1983 *Phys. Rev.* D **28** 2960
[13] Bertolami O and Mourão J M 1991 *Class. Quantum Grav.* **8** 1271
[14] Garcia-Compeán H, Obrégon O and Ramírez C 2002 *Phys. Rev. Lett.* **88** 161301
[15] Barbosa G D and Pinto-Neto N 2004 *Phys. Rev.* D **70** 103512
[16] Ryan M and Shepley L 1975 *Homogeneous Relativistic Cosmologies* (Princeton: Princeton U. Press)

The emergence of probabilities in anhomomorphic logic

Yousef Ghazi-Tabatabai[1] and Petros Wallden[2]
[1]Blackett Laboratory, Imperial College, London, SW7 2AZ, U.K.

E-mail: yousef.ghazi05@imperial.ac.uk, petros.wallden@gmail.com
[2]Raman Research Institute, Theoretical Physics Group
Sadashivanagar, Bangalore - 560 080, India

Abstract.
Anhomomorphic logic is a new interpretation of Quantum Theory (due to R. Sorkin). It is a *histories* formulation (c.f. consistent histories, quantum measure theory). In this approach, reality is a co-event, which is essentially an assignment of a truth value {True, False} to each question. The way this assignment is done mimics classical physics in as much as possible, allowing however for sufficient flexibility to accommodate quantum 'paradoxes', as is shown by the analysis of Kochen-Specker theorem. In this contribution, after briefly reviewing the approach, we will examine how probabilistic predictions can arise. The Cournot principle and the use of approximate preclusions will play a crucial role. Facing similar problems in interpreting probability as in classical probability theory, we will resort to the *weak* form of Cournot principle, where possible realities will be preclusive co-events and the quantum measure is used to obtain predictions. Examples considered, includes the fair coin and the double slit pattern arguably one of the most important paradigms for quantum theory.

1. Introduction

Quantum theory challenges the picture we had in classical physics about what reality is. In order to retain the classical picture alternative formulations, such as hidden variables, were proposed, facing however serious problems due to the severe restrictions from observations that result from Bell's inequalities [1] and the Kochen-Specker theorem [2]. Quantum theory (as it stands now) requires some external (to the system) observer, in order to "make sense". However, for the needs of quantum cosmology for example, where we have a truly closed system, we do lack an interpretation. Consistent histories [3, 4, 5], can be seen as an attempt to deal with this, that also removed in some way the very special role that time has in standard quantum theory. The latter is another desired feature, if one is interested in building a quantum theory of gravity where time is in same footing with space. Consistent histories however, failed to provide a fully satisfactory interpretation of the quantum theory of closed systems due to the context dependence (or else the dependence of predictions on the consistent set realized, which arises due to the fact that there exist many incompatible consistent sets). As a development came a novel interpretation which retains essentially the same mathematical structure (history space

Journal of Physics: Conference Series **174** (2009) 012054 doi:10.1088/1742-6596/174/1/012054

and decoherence functional) the so-called, *anhomomorphic logic* [1] [6, 7]. In the next section we introduce this approach briefly, before we come back to the core of this contribution, which is how probabilistic predictions are dealt within this approach[2].

2. Introducing anhomomorphic logic

2.1. Classical physics, histories and logic

Let us first consider classical physics, in which we use Histories and (classical) Logic in order to ask questions about nature.

We have a set Ω (call it *History Space*) of all possible histories h_i. We can think of each h_i as a trajectory (or more generally a full specification of the particle in every moment of time). In classical physics one and only one of these histories is actually realized. However, in stochastic physics (non-deterministic) we do not know which one is realized, but have a probability measure on Ω[3]. All possible questions about a system can now be rephrased in terms of asking whether the real history belongs to some subset of Ω. So for example if we want to ask whether the particle is at the interval Δ at time t, we ask if the realized history h belongs to the subset $\Delta_t := \{h_i | h_i(t) \in \Delta\} \subset \Omega$.

Associated with Ω is its power set $\mathcal{U}$ (set of subsets of Ω), that has a Boolean algebra structure with intersection as multiplication and symmetric difference ($A\triangle B := (A \cup B) \setminus (A \cap B)$) as addition.

We also have the set of truth values $\mathcal{T}$ (e.g. {True, False}) which also has a Boolean algebra structure (that of $\mathbb{Z}_2$, identifying 'True' with 1 and 'False' with 0). Finally we have the possible valuations ϕ_i. A valuation, ϕ_i is an assignment of a truth value to each question, i.e. in other words a map from $\mathcal{U}$ to $\mathcal{T}$. We moreover require that this valuation respects the Boolean structures of $\mathcal{U}$ and $\mathcal{T}$ by being a homomorphism:

$$\begin{aligned} \phi(A\triangle B) &= \phi(A) + \phi(B) \\ \phi(A \cap B) &= \phi(A)\phi(B) \end{aligned} \tag{1}$$

It can be shown that if one requires the maps ϕ to be homomorphic there is a one to one correspondence between these homomorphic maps and single histories, i.e. each homomorphism corresponds to a characteristic map of a history:

$$\phi_i | \phi_i(A) = 1 \text{ iff } h_i \in A \tag{2}$$

2.2. Consistent histories and quantum measure

In quantum theory the above picture does not hold. The only fully developed histories formulation (at present) is the consistent histories approach (also known as decoherent histories) Ref. [3, 4, 5].[4] The basic elements are histories of the system and to each pair of histories a complex number is assigned by the means of the *decoherence functional* and it corresponds to the interference between these histories. It is defined as:

[1] Also known with different names as for example "co-event interpretation".

[2] More details can be found in Ref. [8].

[3] Histories that have measure zero, are never realized, and thus are not even possible realities. This will be of use later.

[4] Mathematically this is very close to the well known sum-over-histories formalism of Feynman, but the interpretation is quite different.

$$\begin{aligned} D(A,B) &= D^*(B,A) \\ D(A \sqcup B, C) &= D(A,C) + D(B,C) \\ D(A,A) &\geq 0 \\ \sum_{h,h'} D(h,h') &= 1 \end{aligned} \tag{3}$$

The diagonal elements correspond to *candidate* probabilities and under certain circumstances[5] we can interpret this (real) number as the probability of this history (or in general coarse grained history i.e. subset of histories) actually occuring (this probability is conditional on the classical domain, or consistent set, that is actually realized). For more details the reader is referred to the original references [3, 4, 5].

This candidate probability interpretation led R. Sorkin to think of the diagnoal elements of a decoherence functional as a *quantum measure*, as they mimic some of the properties of a classical measure, and can be thought of as the first step in a chain of generalizations of classical measure theory (see Refs. [9, 10]). We will use the notation $\mu(A) := D(A,A)$ for the quantum measure. While it is normalized to unity and is positive-definite, it fails to obey the "additivity of disjoint regions of the sample space", a necessary requirement to have a probability measure,

$$\mu(A \sqcup B) \neq \mu(A) + \mu(B) \tag{4}$$

This is due to interference (c.f. double slit). Note though that the quantum measure *does* obey the weaker

$$\mu(A \sqcup B \sqcup C) = \mu(A \sqcup B) + \mu(A \sqcup C) + \mu(B \sqcup C) - \mu(A) - \mu(B) - \mu(C) \tag{5}$$

Which means that in quantum theory, there is no fundamentally new three paths interference. In particular, it can be shown that any quantum measure that obeys Eq. (5) can arise from some decoherence functional (as defined above, and thus the equivalence between quantum measures and decoherence functionals.).

Due to the difference of the quantum measure compared to classical measure, the picture of classical histories and logic analyzed in the previous section, cannot be retained. The most striking problem, comes essentially from the Kochen-Specker theorem, and is discussed in Ref. [11]. A simplified version form of this problem comes from the three slit experiment, where we have $\mu(A \sqcup B) = \mu(B \sqcup C) = 0$ but $\mu(A \sqcup C) \neq 0$. The first two imply that all the set $\{A, B, C\}$ is impossible (since it is covered by measure zero sets), however one subset, $\{A, C\}$, *is* possible.

2.3. Anhomomorphic logic

We can see now that in order to be compatible with quantum theory, one needs to alter something in the classical picture. Anhomomrphic logic is the approach where we maintain the same set of possible questions (namely $\mathcal{U}$), same possible truth values ($\mathcal{T} = \{\text{True, False}\}$) but we change the allowed valuation maps, ϕ_i, by weakening the requirement the map is a homomorphism. These maps (no longer homomorphic) are called *co-events*. This approach was initiated by R. Sorkin in Refs. [6, 7]. As already stated, F. Dowker and Y. Ghazi-Tabatabai

[5] If it belongs to a partition that the elements pairwise have zero interference.

Journal of Physics: Conference Series **174** (2009) 012054 doi:10.1088/1742-6596/174/1/012054
in Ref. [11] showed how the suggested approach evades Kochen-Specker theorem[6] and thus is a good candidate for a realistic interpretation of quantum theory.

While we want to weaken the requirement to be homomorphism we need it to maintain sufficient structure to be able to make deductions (deductive logic), retain some sense of reality and in the same time be able to accommodate the paradoxes of quantum theory. There are three conditions we require.

(i) We retain the preservation of multiplication under ϕ but no longer require the preservation addition in general,

$$\begin{aligned} \phi(A \cap B) &= \phi(A)\phi(B) \\ \phi(A \triangle B) &\neq \phi(A) + \phi(B) \end{aligned} \tag{6}$$

This is called *multiplicative* co-event[7]. It has the following desirable properties.

(a)

$$\phi(A) = 1 \text{ and } A \subset B \Rightarrow \phi(B) = 1 \tag{7}$$

This is the basic inference rule ("modus ponens"), and can be used to make deductive proofs[8]. However we *cannot* use proofs by contradiction (excluded middle), since

$$\phi(A) = 0 \nRightarrow \phi(\Omega \setminus A) = 1 \tag{8}$$

This is the case in intuitionistic logic also (see also constructive mathematics for mathematics using only deductive proofs). Note though, that the following *does* hold:

$$\phi(A) = 1 \Rightarrow \phi(\Omega \setminus A) = 0 \tag{9}$$

(b) It can be shown, that each multiplicative co-event[9] ϕ_A corresponds to a subset A called *dual* of the co-event[10], such that

$$\phi_A(B) = 1 \text{ if and only if } A \subseteq B \tag{10}$$

This is similar to the case at classical physics, where the duals were single element subsets (corresponding to characteristic maps). For convenience we will occasionally identify the co-events with its dual.

(ii) We require that every subset of measure zero is always mapped to zero (i.e. it is always false).

$$\mu(A) = 0 \Rightarrow \phi(A) = 0 \tag{11}$$

Co-events obeying equaiton 11 are called *preclusive.* This requirement is same as in classical physics, where the histories that are (or belong to a set) of measure zero cannot occur, in other words are not possible realities.

[6] See also Ref. [12] for related work.

[7] Earlier alternative definitions were use as in Ref. [6] but this turned out to be the most satisfactory due to several reasons.

[8] e.g. "I am physicist"-True, along with "Physicists are humans"-True, implies that "I am human"-True.

[9] At least for finite dimensional history space Ω; however, these considerations can be generalized.

[10] In the literature, the dual is also referred occasionally as the "support".

Journal of Physics: Conference Series **174** (2009) 012054 doi:10.1088/1742-6596/174/1/012054

(iii) We require that the co-event (or actually its dual) is *primitive*. This means the following: We (partially) order all possible duals (which are simply subsets of Ω) with respect to set inclusion. Then from all the (multiplicative) preclusive co-events we choose only those that are minimal with respect to this order, i.e. the finest grained (smaller) duals. Remember that in classical physics, these are simply single element subsets (h_i), so by requiring primitivity we come as close as possible to the picture we have in classical physics.

We therefore get to the point where we identify these co-events, which are (multiplicative), preclusive and primitive, as possible realities (PPC). The multiplicativity, allows us to view the relevant duals as what actually happens, and thus we have a very similar picture with classical physics, where the difference is that reality is no longer a fine-grained description but rather a coarser-grained one[11]. A very interesting thing to point out here, is that the logic that arises, depends on the dynamics (and initial condition), via the use of the quantum measure (the zero sets) and it is *not* fixed a-priori.

Before moving to the main text of this contribution, we shall stress one more interesting point. Classical physics corresponds to homomorphisms while the "quantum" nature is encoded in the parts of the map that are anhomomorphic. Thus we could say that a classical domain arises if we consider some coarse graining (i.e. some subalgebra of $\mathcal{U}$) such that the induced map on this subalgebra, is a homomorphism for ANY of the allowed (PPC) co-events. What is most interesting, is that with this notion of a classical domain there exists a unique finest grained classical domain that all the others arise as coarse-grainings. This is in striking difference with consistent histories where the main problem was the existence of many incompatible classical domains. The reader is once again refered to Ref. [8] for details.

3. Probabilities in anhomomorphic logic

3.1. Probabilities in classical closed systems: Cournot Principle

In the analysis we have done so far, we have argued about what a possible reality is (a PPC co-event), but have omitted what is probable. In most everyday life cases, the predictions we get (particularly from quantum theory) are probabilistic. However here we shall recall that in consistent histories we have a single closed system, and in such cases the very concept of probability (in classical physics) is not easily and unambiguously defined. For example a probabilistic statement about the state of the full universe cannot be testable, since either outcome (finding universe possessing the property in question or not) would not falsify our initial assertion.

There was a big philosophical debate by the founders of probability theory on how one is expect to understand a probabilistic statement[12]. The standpoint which we shall adopt (which suits best the case of a single closed system such as the universe) is the use of the *Cournot* Principle, and is closely related with experimentally falsifying a theory :

(Strong) Cournot Principle: In a repeated trial, an event A of small measure (ϵ arbitrarily chosen), does *not* occur.

Note that in a repeated trial, distributions of outcomes that differ drastically from the probability distribution of the single trial, would have very small measure and thus they do not occur. However, this picture is problematic in classical physics. To see this, we can consider the case of tossing a fair coin. The probability measure of having all heads in N trials, goes as

[11] We can view this, as either finer grained questions are un-physical, or physics at finer grained scales has some (classically) contradictory properties.

[12] See for example Kolmogorov in the "Grundbegriffe".

Journal of Physics: Conference Series **174** (2009) 012054 doi:10.1088/1742-6596/174/1/012054

$1/2^N$, while the probability of getting heads less than (say) 60% is $\sum_{k=1}^{0.6N}\binom{N}{k}$ which is much greater than $1/2^N$. We could thus claim that an outcome of all heads is impossible, while an outcome with 50% heads is possible. However, we note, that in any actual realization (series of outcomes), we will get a sequence of results (e.g. $htt tht thht\cdots$) that has probability of occurring $1/2^N$ exactly the same with the one we get with only heads. This probability was very small ($< \epsilon$) and thus this outcome was prohibited. In this example, we note that *any* possible outcome has measure less than ϵ and if we take the (strong) Cournot seriously that would come to a direct contradiction with reality (since something actually happens). Such considerations lead us to a milder version of the Cournot principle:

> **Weak Cournot Principle**: In a repeated trial, an event A *singled out in advance*, of small measure (ϵ arbitrarily chosen), will *not* occur.

This means that if we ask in advance: "Is the outcome 50% heads possible" we will get the answer yes. However, if we ask in advance "will the sequence *hththththt* occur" we will get the answer no, since the measure for this outcome is small. We should stress here that there is a split between the ontology and predictability of the theory. According to weak Cournot, any outcome that doesn't have identically zero measure, is possible. However if we have a theory and we ask a question that has small ($< \epsilon$) non-zero measure, if this outcome occurs, we falsify our theory, even if our theory did not exclude this event from occuring. For example, if N coin tosses give N heads, we falsify our assertion that the coin was fair, even if it is conceivable that it was a fair coin. To sum up, in strong Cournot, no approximately zero measure set is a possible reality, while in weak Cournot, approximately zero measure sets are possible realities. In order to make scientific predictions, we need to preselect questions and we may need to falsify our theory even if the experimental outcome is (ontologically speaking) a possible reality.

Coming back to the case of a closed *single* system, note that while we cannot make a probabilistic statement for the universe itself, we can instead make probabilistic predictions for subsystems that 'look identical'. In the above example we made statements about the distribution of outcomes of tosses of coins (subsystems). These statements, if we view them as statements about coins, correspond to non-trivial probabilities. However, they correspond to 'almost trivial' (i.e. almost zero or almost one) probabilities for the full system of N coins. The reader is refered to Ref. [8] for further details.

3.2. Quantum theory

One of the things that stopped us from taking the strong Cournot seriously, was the fact that we had an example that all Ω was covered by histories (possible outcomes, h_i's[13]) each of which was precluded because it had measure $\mu(h_i) < \epsilon$. This resembles the Kochen Specker theorem. Anhomomorphic logic, by considering as possible realities subsets of Ω other than singletons (fine-grained histories h_i), had evaded this contradiction. One would hope that it may be possible to take strong Cournot seriously in the quantum case, by defining Approximate PPC (APPC) co-events by replacing the condition '$\mu(A) = 0$' with the condition '$\mu(A) < \epsilon$' in the definition of preclusion. (c.f. with Eq.11). We hope that by making this change we can encode all the 'useful' information in the quantum measure in our set of APPC co-events; then the quantum measure itself would no longer be needed for predictions. However this leads to contradictions with observation. For if the dual of a co-event contains two histories giving different answers to a particular question, then the co-event itself will answer our question in a manner inconsistent with Boolean logic; even when this question relates to an experimentlaly

[13] Note that by h_i here we mean a sequence of single outcomes, as for example *hththththt* in the coin case.

Journal of Physics: Conference Series **174** (2009) 012054 doi:10.1088/1742-6596/174/1/012054

observable outcome (see Ref. [8] for further details). For example, returning to our repeated trial of a coin, assume we have a APPC co-event that has in its dual one history that has a heads outcome in the first coin toss, and a second history that has a tails outcome. Then by the definition of anhomomorphic logic, we would get NO to the question "did the first coin toss result in heads?" and also NO to the question "did the first coin toss result in tails". But we can experimentally verify that exactly one of the two outcomes 'heads', 'tails' will actually occur if we are to throw our coin. Such co-events (that are possible for APPC co-events), would not be possible had we stick with *exactly* preclusive co-events (see Ref. [8] for further details). We are then pushed back to the use of the weak Cournot.

In quantum theory when we use weak Cournot, the possible realities are still the PPC co-events (c.f. in classical physics, possible realities were fine grained histories, that had non-zero measure, even if it was arbitrarily small). To deduce probabilistic predictions we need to use once again the quantum measure. We select a question corresponding to a subset A of the history space Ω. If $\mu(A) < \epsilon$, we say that this result is not possible, and this is the prediction we make. If this outcome actually arises, we then say that our initial assumption about what was the quantum measure (which includes initial conditions and dynamics) has been falsified.

Note, the following difference of the quantum case compared to the classical one. In classical physics the measure that is used to derive predictions, is on the space of possible realities, which is the history space. In quantum theory, the quantum measure is again on the history space, however this is no longer the space of possible realities, since the possible realities are the PPC co-events. The reader is refered to Ref. [8] for further details.

3.3. The double-slit example

Finally, let us show how the quantum measure can be used along with weak Cournot to get predictions that reproduce the double slit pattern. For simplicity we will consider a discrete screen, with 5 slots ($i \in \{0, \pm1, \pm2\}$)[14]. We also have two slits (s_1, s_2). Each fine grained history (for a single repetition), consists from the particle crossing one slit and hitting the screen at on slot. The measure is given:

$$\begin{aligned}
|(s_j, i)| &= 0.1 \\
|(s_1, i)| \sqcup (s_2, i)| &= \begin{cases} 0.3 & \text{if } i = \pm 2, 0 \\ 0.05 & i = \pm 1 \end{cases} \\
|(s_k, i) \sqcup (s_l, j)| &= |(s_k, i)| + |(s_l, j)| \\
&= 0.2 \text{ if } i \neq j
\end{aligned}$$

If we have many independent repetitions we can simply use the product measure. Now assume that we have a closed system with 10 particles that are to cross this double slit, and we assume that anything that has measure less than $\epsilon = 10^{-3}$ is precluded, meaning that if such outcome comes we falsify our initial theory (according to the weak Cournot). We can easily see that the question "Is the distribution uniform" consists of all the permutations where we have 2 particles at each slot, and this number is $\binom{10}{2}\binom{8}{2}\binom{6}{2}\binom{4}{2} = 113400$ which is much larger than the number of permutations for getting the double slit pattern (i.e. 3 particles at $\pm 2, 0$ slots and 1 at either 1 or -1). The latter is 4800. However the measure for each fine grained history of 10 particles that are uniformly distributed is only $\approx 5 \times 10^{-9}$. The total measure of all histories that are of uniform distribution is then $113400 \times 5 \times 10^{-9} \approx 5 \times 10^{-4}$. This is less than ϵ, so the

[14] Slots 0 and ± 2 are the bright fringes and ± 1 the dark fringes of the pattern.

uniform distribution is precluded. In contrast while there are much fewer combinations that give the double slit distribution, the measure for each of them is $\approx 10^{-6}$ and thus the total measure for getting a distribution 'like' the double slit pattern is $4800 \times 10^{-6} \approx 5 \times 10^{-3}$, which is greater than ϵ and thus the double slit distribution is not ruled out by our theory[15]. Here, in this simplified model, we see how predictions of the type of double slit pattern (i.e. ruling out distributions such as the uniform one) can arise in anhomomorphic logic, when we use the quantum measure of many repetitions of the system along with the weak Cournot principle.

4. Summary and Conclusions

In this contribution, we reviewed the anhomomorphic logic approach to quantum theory, which is a development of the consistent histories approach. Reality is no longer a single fine grained history, but a primitive preclusive multiplicative co-event, which can also be viewed as a coarse grained history. The Kochen Specker theorem is evaded, as is the problem of many incompatible classical domains faced by consistent histories. The core of this contribution was how to deal with probabilistic predictions in this formalism. We resorted to the Cournot principle to give meaning to probabilistic statements and essentially in a frequentist's view on probability (rather than propensity). The use of strong Cournot principle was ruled out (alas for different reasons than in classical physics). The weak Cournot, introduces a split of ontology and predictions. In classical physics this leads to:

(a) Ontology: Fine grained histories are the possible realities. In other words one and only one history is actually realized.
(b) Predictions: The (classical) measure on Ω is used, in order to make predictions, that are of the type "if A is realized and $\mu_c(A) \leq \epsilon$ then our initial assumption is rejected".

In quantum theory the picture is similar:

(a) Ontology: Possible realities are the multiplicative, primitive and (exactly) preclusive co-events. If we view the dual picture (the duals), we can say that what is realized is a coarse grained set of histories or else a non-trivial subset of Ω.
(b) Predictions: The *quantum* measure on Ω is used, in order to make predictions, that are of the type "if A is realized and $\mu(A) \leq \epsilon$ then our initial assumption is rejected". Note however, the fact that the quantum measure is on the history space, which in the quantum case, is no longer the space of possible realities.

Acknowledgments

The authors would like to thank Fay Dowker, Rafael Sorkin and Sumati Surya for many discussions on Anhomomorphic Logic. PW thanks the organizers for carrying out this very interesting conference and giving him the opportunity to give this talk. Partial support from the Royal Society International Joint Project 2006-R2 is acknowledged and YGT was supported by an STFC studentship.

[15] Note, however, that if we asked more detailed question, such as "the first particle will hit slot 0, the second slot 2, etc" even if the distribution was correct the measure would be very small ($\approx 10^{-6}$ here) and thus ruled out by our predictions. The "paradox" here, as in classical physics, would be that the eventual outcome of the experiment, would be one of those ruled out histories, and the contradiction is evaded by the crucial concept of *pre*-selected questions, in the definition of weak Cournot.

References

[1] Clauser J F, Horne M A, Shimony A and Holt R A 1969 *Phys. Rev. Lett.* **23** 880
[2] Kochen S and Specker E 1967 *J. Math. Mech.* **17** 5983
[3] Griffiths R B 1984 *J. Stat. Phys.* **36** 219
[4] Omnès R 1988 *J. Stat. Phys.* **53** 893
[5] Gell-Mann M and Hartle J 1990 *Complexity, Entropy and the Physics of Information, SFI Studies in the Science of Complexity, Vol. VIII* ed W Zurek (Reading, Mass.: Addison-Wesley)
[6] Sorkin R D 2007 *J. Phys.* A **40** 3207
[7] Sorkin R D 2007 *J. Phys.: Conf. Ser.* **67** 012018
[8] Ghazi-Tabatabai Y and Wallden P 2009 *Preprint* arXiv:0901.3675v1
[9] Sorkin R D 1994 *Mod. Phys. Lett.* A **9** 3119
[10] Sorkin R D 1997 *Quantum Classical Correspondence: Proceedings of the 4th Drexel Symposium on Quantum Nonintegrability – Philadelphia, September 8-11, 1994* ed D H Feng and B-L Hu (Cambridge, Mass.: International Press) 229-251 (*Preprint* arXiv:gr-qc/9507057)
[11] Dowker F and Ghazi-Tabatabai Y 2008 *J. Phys.* A **41** 205306
[12] Surya S and Wallden P 2008 *Preprint* arXiv:0809.1951v2

Arithmetic for the unification of quantum mechanics and general relativity

Victor Korotkikh

School of Computing Sciences, Faculty of Business and Informatics, Central Queensland University, Mackay, Queensland, 4740, Australia

E-mail: v.korotkikh@cqu.edu.au

Abstract. In the paper we bring attention to description of complex systems in terms of self-organization processes of prime integer relations. Revealed through the unity of two equivalent forms, arithmetical and geometrical, the description may have the potential for the unification of quantum mechanics and general relativity. Remarkably, based on integers and controlled by arithmetic only such processes can define nonlocal correlations between parts of a complex system and the geometry of their spacetimes.

1. Introduction

In the paper we consider the unification of quantum mechanics and general relativity from a new perspective: arithmetic. It is given by description of complex systems in terms of self-organization processes of prime integer relations. The description is realized through the unity of two equivalent forms, arithmetical and geometrical. In sections 2 and 3 we recognize features of quantum mechanics and general relativity in the arithmetical and geometrical forms accordingly. Quantization as a natural aspect of the description and one large symmetry of the processes in terms of conservation of "quantum" numbers are considered in section 4. In section 5 we propose a context to draw a parallel between string theory and our description. In Conclusions possible implications of the results are discussed.

2. Arithmetical form and features of quantum mechanics

In the description a complex system is defined by a set of global quantities. They remain invariant under the transitions between the states of the system. It turns out that to preserve the system the parts have to be correlated through hierarchical structures built in accordance with self-organization processes of prime integer relations [1,2,3]. Features of quantum mechanics [4,5] can be recognized in the arithmetical form. In particular, the correlation structures provide the statistical information about the system without reference to the distances between parts in space and time.

Let us consider elementary parts $P_1, ..., P_N, N \geq 2$ as the initial building blocks in the formation of a complex system. Elementary part P_i is given in its local reference frame specified by two parameters $\varepsilon_i > 0$ and $\delta_i > 0, i = 1, ..., N$. The global quantities of the system are defined through the geometrical form, where the parameters are demanded to be the same $\varepsilon = \varepsilon_i,\ \delta = \delta_i, i = 1, ..., N$. As a result the parameters ε and δ become constants and can not be changed unless simultaneously in all reference frames.

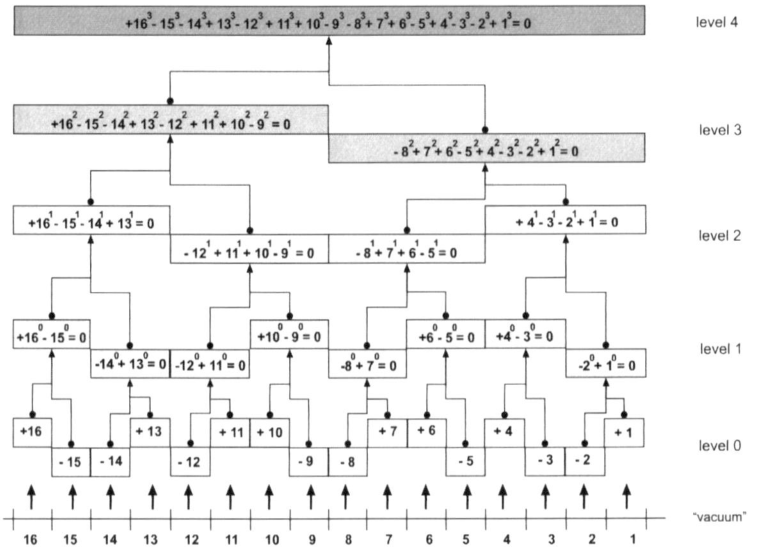

Figure 1. As an example a hierarchical structure of prime integer relations built by a self-organization process is shown. The process starts when integers $16, 13, 11, 10, 7, 6, 4, 1$ in the positive "state" and integers $15, 14, 12, 9, 8, 5, 3, 2$ in the negative "state" are generated from the "vacuum" to appear at level 0. Controlled by arithmetic only the process forms prime integer relations from level 1 to level 4. The hierarchical structure of prime integer relations determines a correlation structure of a complex system. Under the correlation structure the space changes $\Delta X_1, ..., \Delta X_{16}$ of the elementary parts $P_1, ..., P_{16}$, as its initial building blocks, are encoded by the sequence $+1-1-1+1-1+1+1-1-1+1+1-1+1-1-1+1$.

Let I be an integer alphabet and $I_N = \{\bar{x} = x_1...x_N, x_i \in I, i = 1, ..., N\}$ be the set of sequences of length N. The state of elementary part P_i in its local reference frame is characterized by a space coordinate $x_i\delta$, where $x_i \in I, i = 1, ..., N$. The state of all elementary parts is given by $x_1\delta, ..., x_N\delta$ and is referred to the sequence $\bar{x} = x_1...x_N \in I_N$ it is encoded by.

For a complex system let $I_N(Q_1, ..., Q_k)$ be the set of states of the elementary parts with the transitions leaving the global quantities $Q_1, ..., Q_k$ of the system unchanged. Under the transitions the system is preserved as a whole, because the quantities it is defined by remain the same. Let n be the number of states in $I_N(Q_1, ..., Q_k)$ enumerated as $\bar{x}_1 = x_{11}...x_{1N}, ..., \bar{x}_n = x_{n1}...x_{nN}$. For a pair of states $\bar{x}_i = x_{i1}...x_{iN}, \bar{x}_j = x_{j1}...x_{jN} \in I_N(Q_1, ..., Q_k)$ with the changes $\Delta\bar{x}_{ij} = \Delta x_{ij1}...\Delta x_{ijN} = (x_{i1} - x_{j1})...(x_{iN} - x_{jN})$ of the elementary parts we are interested to specify the correlations between the parts preserving the quantities of the system. The matrix $\mathcal{M}(I_N(Q_1, ..., Q_k)) = \{\Delta\bar{x}_{ij}, i, j = 1, ..., n\}$ is used to arrange the changes between the states of $I_N(Q_1, ..., Q_k)$. It is shown [1] that $Q_1, ..., Q_k$ quantities of a complex system remain invariant, if and only if the correlations between the parts are defined by a system of k Diophantine equations

$$(M+N)^{k-1}\Delta x_{ij1} + ... + (M+1)^{k-1}\Delta x_{ijN} = 0$$

$$. \quad . \quad . \quad . \quad . \quad . \quad . \quad .$$

$$(M+N)^{0}\Delta x_{ij1} + ... + (M+1)^{0}\Delta x_{ijN} = 0 \qquad (1)$$

and an inequality

$$(M+N)^{k}\Delta x_{ij1} + ... + (M+1)^{k}\Delta x_{ijN} \neq 0$$

with the coefficients of the matrix $\mathcal{M}(I_N(Q_1, ..., Q_k))$ as solutions and M as an integer. Notably, the equations (1) and the inequality do not include the constants ε and δ. The coefficients of the system could become the entries of the Vandermonde matrix, once the number of the equations would be N.

By using the Diophantine equations (1) a certain type of integer relations can be identified. Remarkably, their analysis reveals hierarchical structures, which can be interpreted as a result of self-organization processes of prime integer relations [1]. Each of the processes has integers $\{(M+N-l+1), l=1,...,N\}$ as the basic building blocks and starts when $|\Delta x_{ijl}|$ of integers $(M+N-l+1), l=1,...,N$ in the "state" $sign(\Delta x_{ijl})$, i.e., positive or negative, are generated at level 0 to form prime integer relations at level 1. The process follows the single organizing principle: *at each level the powers of the integers in the prime integer relations should be increased by one, so that through emerging arithmetical interdependencies the prime integer relations could self-organize to make up prime integer relations at the higher level* (Figure 1).

By analogy with prime numbers a prime integer relation cannot be divided but into integer relations built at the level by processes [1]. It is an undivided whole, complete system. For example, an integer relation $+4^0 - 3^0 + 2^0 - 1^0 = 0$ is not prime, because it can be divided into two integer relations $+4^0 - 3^0 = 0$ and $+2^0 - 1^0 = 0$, which in their turn are prime as they cannot be further divided and are built by some processes. In other words, a prime integer relation cannot be formed upon removal of any of its components, all prime integer relations it is made of are necessary and sufficient.

Therefore, in the arithmetical form a complex system is given by correlation structures defined by hierarchical structures of prime integer relations. As each of the correlation structures is ready to exercise its own scenario and there is no mechanism specifying which of them is going to take place, an intrinsic uncertainty about the complex system exists. At the same time, the information about the correlation structures can be used to evaluate the probability of an observable to take each of the measurement outcomes. Thus, the arithmetical form of the description provides the statistical information about a complex system. In particular, let $\mathcal{X}_i$ be the set of all possible values defined by the Diophantine equations (1) for the space change ΔX_i of the elementary part $P_i, i=1,...,N$ as its observable. For each possible value $\chi \in \mathcal{X}_i$ let $\mu_i(\chi)$ be the number of times the correlation structures provide the condition $\Delta X_i = \chi$. To predict the result of a measurement this information can be characterized statistically. The probability $p(\Delta X_i = \chi)$ that the observable ΔX_i takes the value of χ is given by

$$p(\Delta X_i = \chi) = \frac{\mu_i(\chi)}{\sum_{\chi' \in \mathcal{X}_i} \mu_i(\chi')}.$$

Remarkably, through the Diophantine equations (1) the arithmetical form reveals nonlocal correlations without reference to the distances between parts in space and time. According to the description parts of a complex system may be far apart in space and time and yet remain interconnected with instantaneous effect on each other. Namely, if a correlation structure of a system is selected and some parts are specified, then through the prime integer relations in control of the correlation structure the other parts are simultaneously defined.

3. Geometrical form and features of general relativity

The geometrical form is nothing but the self-organization processes isomorphically represented by the transformations of two-dimensional patterns [1]. Under the isomorphism the hierarchical structures of prime integer relations defining the correlation structures of a complex system become equivalently represented by hierarchical structures of geometrical patterns determining the dynamics of the system. Features of general relativity [6] can be recognized in the geometrical form. In particular, the formation of a complex system can be realized through the transformation of the spacetimes of the parts to make them shaped and connected exactly right to match the boundary curves of the geometrical patterns. The force holding the parts together as one system is completely determined by the geometry of the spacetimes.

Let us consider a hierarchical structure of geometrical patterns (Figure 2), which for given ε and δ is isomorphic to the hierarchical structure of prime integer relations (Figure 1). The dynamics of the elementary parts $P_1, ..., P_{16}$ as the initial building blocks in the formation of the correlation structure is given by a piecewise constant function $\Psi_1^{[0]}(t)$, $t_0 \leq t \leq t_{16}$ at level 0. It is defined when a curve $\gamma_{i0}, i = 1, ..., 16$, represented by the graph of the function $\Psi_1^{[0]}(t)$, $t_{i-1} < t \leq t_i$, $t_i = \varepsilon i$, $i = 1, ..., 16$, is used to specify that the space coordinate $X_{i0} = X_i$ of the elementary part $P_{i0} = P_i$ changes by $\Delta X_{i0} = \Delta X_i = \Psi_1^{[0]}(t_i) = \pm\delta$, while the time coordinate T_{i0} changes by

$$\Delta T_{i0} = Length(\gamma_{i0}) = \int_{t_{i-1}}^{t_i} \sqrt{1 + (\frac{d\Psi_1^{[0]}}{dt})^2 dt} = \varepsilon.$$

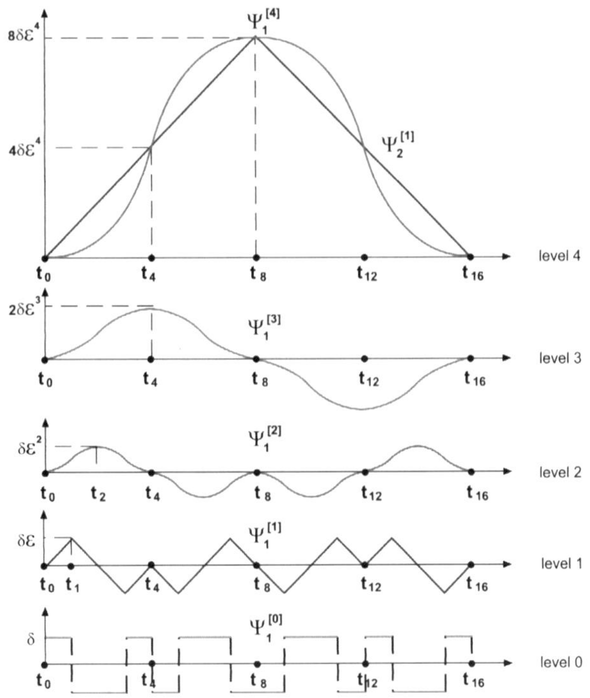

Figure 2. A hierarchical structure of geometrical patterns corresponding to the hierarchical structure of prime integer relations. The geometrical pattern of elementary part $P_{ik}, i = 1, ..., 16$, $k = 1, ..., 4$ at level k is the region bounded by its boundary curve, i.e., the graph of the function $\Psi_1^{[k]}(t), t \in [t_{i-1}, t_i]$, the vertical lines $t = t_{i-1}$ and $t = t_i$ and the t-axis. Importantly, the geometrization makes possible to do physics using prime integer relations. Determined by the boundary curves geometrical patterns allow to use the fundamental theorem of calculus in connecting quantities of elementary parts at different levels. All geometrical patterns corresponding to prime integer relations are symmetrical and their symmetries are interconnected by the self-organization process.

Due to the isomorphism, at one side the prime integer relations of level $k = 1, ..., 3$ form the prime integer relations of level $k+1$ (Figure 1) providing the correlations for the elementary parts to become the elementary parts of more complex structures. At the other side under the integration of the function $\Psi_1^{[k]}$ the geometrical patterns of level k transform into the geometrical patterns of level $k+1$ to determine the dynamics of the elementary parts (Figure 2). In particular, as elementary part $P_{ik}, i = 1, ..., 16$ becomes elementary part $P_{i,k+1}$ of higher level, the boundary curve of its geometrical pattern specifying the dynamics transforms from $\Psi_1^{[k]}(t)$ to $\Psi_1^{[k+1]}(t)$, $t_{i-1} \leq t \leq t_i$. As a result of this transition the space and time coordinates of the elementary part are determined to change interdependently in accordance with $\Psi_1^{[k+1]}(t)$. In fact, at level $k = 1, ..., 4$ the boundary curve $\gamma_{ik}, i = 1, ..., 16$, as the graph of the function $\Psi_1^{[k]}(t)$, $t_{i-1} \leq t \leq t_i$, determines the motion of the elementary part P_{ik} specifying the space coordinate $X_{ik}(t) = \Psi_1^{[k]}(t),\ t_{i-1} \leq t \leq t_i$ interdependently with the time coordinate

$$T_{ik}(t) = \int_{t_{i-1}}^{t} \sqrt{1 + (\frac{d\Psi_1^{[k]}(t')}{dt'})^2} dt' = \int_{t_{i-1}}^{t} \sqrt{1 + (\frac{dX_{ik}(t')}{dt'})^2} dt',\ t_{i-1} \leq t \leq t_i. \quad (2)$$

Under the correlations the change of the space position of the elementary part P_{ik} during the time

$$\Delta T_{ik} = Length(\gamma_{ik}) = \int_{t_{i-1}}^{t_i} \sqrt{1 + (\frac{d\Psi_1^{[k]}(t)}{dt})^2} dt = \int_{t_{i-1}}^{t_i} \sqrt{1 + (\frac{dX_{ik}(t)}{dt})^2} dt \quad (3)$$

is

$$\Delta X_{ik} = \Psi_1^{[k]}(t_i) - \Psi_1^{[k]}(t_{i-1}). \quad (4)$$

The condition (2) can be expressed in the differential form suggesting an invariant

$$(\frac{dT_{ik}(t)}{dt})^2 - (\frac{dX_{ik}(t)}{dt})^2$$

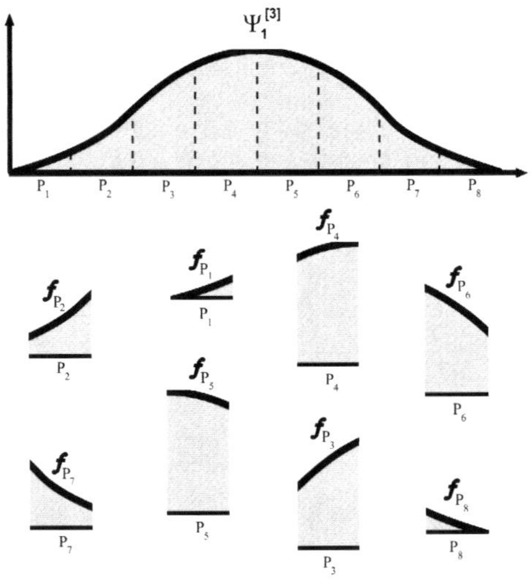

Figure 3. The geometrical pattern fully defines the geometrical patterns of the elementary parts. Thus, the spacetimes of the elementary parts, like the pieces of a puzzle, have to make exactly the boundary curve of the geometrical pattern. Since an elementary part can be described by a one-dimensional curve, the situation may be also seen through interactions of the curves in accordance with the prime integer relation $+16^2 - 15^2 - 14^2 + 13^2 - 12^2 + 11^2 + 10^2 - 9^2 = 0$ composing the boundary curve of the geometrical pattern.

It appears that the boundary curve of an elementary part functions like its spacetime. Thus from this perspective to form a complex system the spacetimes of the elementary parts have to be transformed to match the boundary curves of the geometrical patterns (Figure 3). Unless this condition is not met the corresponding prime integer relations are not in place to provide the correlation structures for the system.

The situation can be seen in terms of a force holding elementary parts together as one system. Since the condition for an elementary part to be held in a system is determined by the geometry of its spacetime, the force acting on the part can be also defined by the geometry of the spacetime. Because the spacetimes of the elementary parts have to geometrically reproduce the prime integer relations in control of the correlation structures, then it is arithmetic that through the prime integer relations dictates how the boundary curves of the geometrical patterns have to be curved and thus determines the spacetimes of the elementary parts and the force. In this context *the spacetimes of elementary parts and the force acting on them can be viewed as properties of arithmetic expressed through prime integer relations and their processes.* Therefore, in our description spacetime turns out to be a manifestation of the correlations between parts and does not have the absolute character much in line with general relativity.

For example, it can be shown that the elementary parts $P_1, ..., P_8$ have the same dynamics at level 1 and level 2. But at level 3 as they become the elementary parts of the correlation structure controlled by the prime integer relation $+16^2 - 15^2 - 14^2 + 13^2 - 12^2 + 11^2 + 10^2 - 9^2 = 0$ (Figure 3), their dynamics appear out to be different. By calculating the function $\Psi_1^{[3]}(t), t \in [t_0, t_8]$ (Figure 2) when $\varepsilon = 1, \delta = 1$ we can find the boundary curves of the elementary parts as

$$f_{P_1}(t) = +(1/3!)t^3,\ f_{P_2}(t) = -(1/3!)t^3 + t^2 - t + 1/3,$$
$$f_{P_3}(t) = -(1/3!)t^3 + t^2 - t + 1/3,\ f_{P_4}(t) = +(1/3!)t^3 - 2t^2 + 8t - 26/3,$$
$$f_{P_5}(t) = -(1/3!)t^3 + 2t^2 - 8t + 38/3,\ f_{P_6}(t) = +(1/3!)t^3 - 3t^2 + 17t - 87/3,$$
$$f_{P_7}(t) = +(1/3!)t^3 - 3t^2 + 17t - 87/3,\ f_{P_8}(t) = -(1/3!)t^3 + 4t^2 - 32t + 256/3, \qquad (5)$$

where for simplicity the index specifying the level has been omitted. From (3),(4) and (5) for the elementary parts P_1, P_4, P_5, P_8 we can obtain $\Delta T_{i3} \approx 1.02$, $|\Delta X_{i3}| = 1/6$, $i = 1, 4, 5, 8$, while for the elementary parts P_2, P_3, P_6, P_7 - $\Delta T_{i3} \approx 1.30$, $|\Delta X_{i3}| = 5/6$, $i = 2, 3, 6, 7$. The spacetimes of the elementary parts have to be mutually self-consistent to geometrically reproduce the prime integer relation in control of the correlation structure. And it turns out that to meet this condition the spacetimes have to be different. Namely, in the reference frame of the elementary part P_1 its space position during the time $\Delta T_{13} \approx 1.02$ changes by $\Delta X_{13} = 1/6$, while in the reference frame of the elementary part P_2 its space position during the time $\Delta T_{23} \approx 1.30$ changes by $\Delta X_{23} = 5/6$.

The notion of simultaneity in our description finds the following meaning. Under the prime integer relation the realization of the correlations between the elementary parts happens simultaneously. But for the parts the effect of the correlations is not the same and for each part it is appropriately determined by the prime integer relation. In particular, the correlations manifest in different times passed in the reference frames of the elementary parts. We may say that the clock in the reference frame of the elementary part P_2 runs faster than the clock in the reference frame of the elementary part P_1.

4. Quantization and one large symmetry of the processes

In this section we consider that quantization as a natural aspect of the description plays an important role in expressing one large symmetry of the processes in terms of conservation of "quantum" numbers.

The concept of symmetry plays a profound role in understanding of nature. In our description arithmetic through the self-organization processes of prime integer relations reveals one large symmetry. In particular, the self-organization processes can be associated with a certain type of symmetry transformations. Under such symmetry transformations global quantities of a complex system remain unchanged due to the correlation structures defined by self-organization processes of prime integer relations [1]. The symmetry of the processes can be identified as the geometrical patterns of the prime integer relations appear symmetrical while their transformations integrate the symmetries together (Figure 2).

Remarkably, the symmetry of the geometrical pattern of a prime integer relation is global and applies to a corresponding correlation structure as a whole. At the same time, it is also possible to consider the situation from the perspective of the elementary parts. In our description an elementary part P_{ik} at level k can be characterized by the boundary curve of the geometrical pattern, which at level k is a polynomial $f_{P_{ik}} = a_{i1}t^k + a_{i2}t^{k-1} + ... + a_{i,k+1}$ of degree k (Figure 3) and thus can specify the elementary part by a set of "quantum" numbers $QN(P_{ik}) = (a_{i1}, a_{i2}, ..., a_{i,k+1})$. The symmetry of the geometrical pattern can be expressed by the conservation of "quantum" numbers of the elementary parts and represented through their symmetrical patterns.

For example, the correlation structure considered above is defined by a process reaching level 3, three global quantities of the system remain invariant. From (5) we can find that the first, second and third "quantum" numbers of the elementary parts are conserved, i.e., their sums equal zero. For the first "quantum" numbers we have $+1/3! - 1/3! - 1/3! + 1/3! - 1/3! + 1/3! + 1/3! - 1/3! = 0$, the second "quantum" numbers: $0 + 1 + 1 - 2 + 2 - 3 - 3 + 4 = 0$ and the third "quantum" numbers: $0 - 1 - 1 + 8 - 8 + 17 + 17 - 32 = 0$.

To obtain symmetrical patterns of the elementary parts we consider the geometrical pattern in a reference frame with the origin at the center of the geometrical pattern. By the change of variable $t = t' + 4$ we make the transition and get new polynomials

$$g_{P_1}(t') = +(1/3!)t'^3 + 2t'^2 + 8t' + 32/3, \; g_{P_2}(t') = -(1/3!)t'^3 - t'^2 - t' + 17/3,$$

$$g_{P_3}(t') = -(1/3!)t'^3 - t'^2 - t' + 17/3, \; g_{P_4}(t') = +(1/3!)t'^3 + 2,$$

$$g_{P_5}(t') = -(1/3!)t'^3 + 2, \; g_{P_6}(t') = +(1/3!)t'^3 - t'^2 + t' + 5/3,$$

$$g_{P_7}(t') = +(1/3!)t'^3 - t'^2 + t' + 5/3, \; g_{P_8}(t') = -(1/3!)t'^3 + 2t'^2 - t' + 32/3.$$

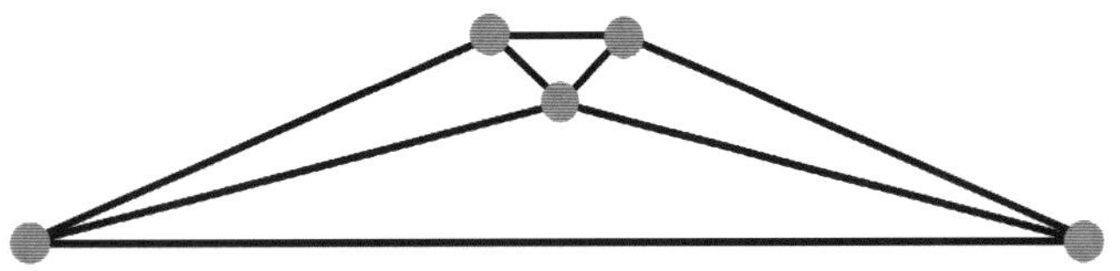

Figure 4. The figure shows the symmetrical pattern of the elementary parts obtained by using their second and third "quantum" numbers and rotated for a better presentation.

As before, the first, second and third "quantum" numbers of the elementary parts are conserved, i.e., the first "quantum" numbers: $+1/3! - 1/3! - 1/3! + 1/3! - 1/3! + 1/3! + 1/3! - 1/3! = 0$, the second "quantum" numbers: $+2 - 1 - 1 + 0 + 0 - 1 - 1 + 2 = 0$ and the third "quantum" numbers: $+8 - 1 - 1 + 0 + 0 + 1 + 1 - 8 = 0$. By using the conserved "quantum" numbers symmetrical patterns of the elementary parts can be produced (Figure 4).

5. Correlation structures as scattering processes

The conservation of "quantum" numbers suggests to represent a correlation structure of a system as a scattering process. In the representation some of the elementary parts play the role of the initial particles and the other elementary parts play the role of the final ones. The initial particles collide and emerge after the scattering process as the final particles.

In Figure 5 the diagrammatic language of S-matrix theory is used to represent the above correlation structure as a scattering process, where the first, second and third "quantum" numbers of the particles are conserved. In addition, there is one more conservation. Let E_i be the quantity of elementary part $P_i, i = 1, ..., 8$ specified by the area of its geometrical pattern, we call it the "energy" of the elementary part. We can find that in the scattering process the "energy" of the particles is conserved, i.e., the sum of the "energies" of the initial particles equals the sum of the "energies" of the final $E_1 + E_4 + E_6 + E_7 = E_2 + E_3 + E_5 + E_8$ (Figure 3).

Although in the scattering process the particles can be described as one-dimensional curves (Figure 3), in the representation they are shown point-like (Figure 5). This association of point-like particle with one-dimensional curve makes us to recall the origin of string theory when in the context of S-matrix theory [7] it was found that strongly interacting elementary particles, viewed before as points, could be actually described as one-dimensional curves [8].

However, string theory and our description are far different starting with the character and origin of the curves. In particular, in characterizing elementary particles string theory sees the curves as vibrations of fundamental strings going no much further about their explanation. In

Journal of Physics: Conference Series **174** (2009) 012055 doi:10.1088/1742-6596/174/1/012055

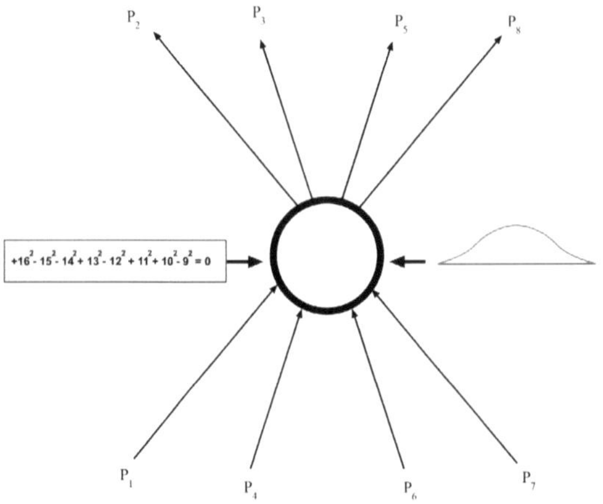

Figure 5. The representation of the correlation structure as a scattering process, where elementary parts P_1, P_4, P_6, P_7 play the role of the initial particles and elementary parts P_2, P_3, P_5, P_8 play the role of the final particles. The interaction of the particles is controlled by the prime integer relation in order to compose the boundary curve of the geometrical pattern. In our description particles can be described by curves as constituents of the boundary curves of geometrical patterns and thus corresponding prime integer relations. Therefore through this link particles can be unified by the structure of the prime integer relations.

the description the curves are the constituents from which the boundary curves of the geometrical patterns are made of. At the same time the geometrical patterns serve to express the prime integer relations geometrically. A prime integer relation is made of and determined by integers. Therefore, in our description the basic building blocks are the integers. Moreover, principles underlying the behavior and defining properties of the curves are fully explained as through the self-organization processes of prime integer relations they are completely determined by arithmetic.

6. Conclusions

In the paper we have brought attention to description of complex systems in terms of self-organization processes of prime integer relations as it may have the potential for the unification of quantum mechanics and general relativity. The processes are fully controlled by arithmetic and the possibility that arithmetic can be involved in the unification matters suggests significant implications. In particular, it is unlikely the unification could be delivered by a master equation. Probably, the perspective should be changed to consider things through a new stage, i.e., the hierarchical network, the structure built by the totality of the processes. Remarkably, there is only one "force" on this stage: arithmetic. As processes take place they can encode systems to form and become more complex, yet without reference to the distances in space and time. It is through the geometry of the processes spacetimes and other quantities of the systems emerge. And it is promising that features turned out to be separately described by quantum mechanics and general relativity can be recognized in the processes.

7. References

[1] Korotkikh V 1999 *A Mathematical Structure for Emergent Computation* (Dordrecht: Kluwer Academic Publishers)
[2] Korotkikh V 2005 *Braz. J. Phys.* **35** 2B 509
[3] Korotkikh V and Korotkkh G 2006 *Int. J. Complex Syst.* 1751 (*Preprint* arXiv:nlin/0606023)
[4] Feynman R P 1985 *QED: The Strange Theory of Light and Matter* (Princeton, N.J.: Princeton U. Press)
[5] Gisin N 2005 Can relativity be considered complete? From Newtonian nonlocality to quantum nonlocality and beyond *Preprint* arXiv:quant-ph/0512168
[6] Einstein A 1960 *Relativity: The Special and the General Theory - A Popular Exposition* (London: Routledge)
[7] Chew G F 1968 *Science* **161** 762
[8] Vecchia P D and Schwimmer A 2007 The beginning of string theory: a historical sketch *Preprint* arXiv:phyics.hist-ph/0708.3940

Fourth International Workshop DICE2008 IOP Publishing
Journal of Physics: Conference Series **174** (2009) 012056 doi:10.1088/1742-6596/174/1/012056

Krein space quantization of linear gravity in de Sitter spacetime and generalization to QFT

Farrin Payandeh
Payame Noor University, Tehran 19395-4697, IRAN
E-mail: payandeh@aut.ac.ir

Abstract. It has been shown that the negative-norm states or negative-frequency solutions of the field equation are indispensable for a fully covariant quantization (Krein space quantization) of the minimally coupled free scalar field in de Sitter spacetime. The new method of quantization has been applied for the free boson and spinor fields in Minkowski spacetime, and the basic structure of quantum field theory is reformulated in Krein space. It is seen that the presence of unphysical negative-frequency states plays the role of an automatic renormalization tool for the theory.

1. Introduction

One of the greatest challenges of physics today is the achievement of a proper theory of quantized gravitational fields. In other words a theory, which quantizes the gravitational fields without any anomaly, is sought for the past seven decades without a thorough success. The element of time, which has two completely different concepts in quantum mechanics as opposed to general relativity, is the most outstanding problem of this theory. By introducing the background field method this problem has been resolved, although it could not have been applied to the very early moments of evaluation of the universe (the Planck scale) where the perturbation of metric is in the same order as the background metric itself. In the background field method, the theory of quantum gravity is not renormalized. Three different opinions have been stated for explanation of this anomaly. The first view sees the problem as an inherent problem of general relativity and the second, as an intrinsic problem of quantum mechanics. The third group, believe that both, quantum mechanics and general relativity are problematic and an alternative theory has to be replaced.

Combination of quantum theory and special relativity results in appearance of singularity in quantum field theory (QFT). To cope with this problem regularization and renormalization procedures, which are not related to fundamental concepts of quantum mechanic and/or special relativity, have been successfully utilized to remove divergences for certain problems. This procedure, however, cannot be extended to the quantum general relativity, and the theory remains divergent. We believe that the root of this anomaly lies in the QFT.

The origion of this work goes back to the quantization of massless minimally coupled scalar field in de Sitter spacetime which plays an important role in the inflationary model as well as in the linear quantum gravity. It has been shown that negative frequency or norm states necessarily appear in a covariant quantization of the free minimally coupled scalar field in de Sitter spacetime [1]. As proved by Allen [2], a covariant quantization of minimally coupled scalar field could not be constructed by positive norm states alone. To cope with this problem the Krein space quantization has been utilized [3]. Preserving the covariance principle and ignoring the positivity condition, the field quantization has been performed in Krein space, a combined Hilbert and anti-Hilbert space. This process resembles Gupta-Bleuler quantization of vector field in electrodynamics in Minkowski spacetime. Mathematically, for the minimally coupled scalar field in de Sitter spacetime [1], it has been proven that the use of the two sets of solutions (positive and negative frequency states) is unavoidable for preservation of (i) causality (locality), (ii) covariance, and (iii) elimination of the infrared divergence. This causal approach was generalized further to the calculation of the graviton propagator in de Sitter spacetime, and to the one-loop effective action for the scalar field in a general curved spacetime [4]. A fully covariant quantization is obtained for linear gravity (the traceless rank-2 "massless" tensor field)in dS spacetime [5]. In this process the ultraviolet and infrared divergences have been automatically eliminated [4]. The new method of quantization has been extended to free boson and spinor fields in Minkowski spacetime and the one-loop approximation of Møller scattering has been performed [6]. The role of "unphysical" states appears in the method as a natural renormalization tool. In this paper, after a brief review of Krein space quantization we show that how the singularities appear in the standard QFT and how they disappear in the Krein QFT.

2. Krein QFT calculation in Minkowski spacetime

As we know, in the quantum level the classical field must be replaced by the field operator and we should be careful in writing the product of fields for physical quantities. Moreover, in quantization of a point particle, AB in the Hamiltonian is replaced by the following relation in the quantum level

$$AB = BA \text{ (classical theory)} \longrightarrow \frac{1}{2}[\hat{A}\hat{B} + \hat{B}\hat{A}] \quad \text{(quantum theory)}.$$

The above discussion is equivalent to define the field operator by the following relation

$$\hat{\phi}(x) = \frac{1}{\sqrt{2}}\int d^3\vec{k}[\hat{a}(\vec{k})u_p(k,x) + \hat{a}^\dagger(\vec{k})u_p^*(k,x) + \hat{b}(\vec{k})u_n(k,x) + \hat{b}^\dagger(\vec{k})u_n^*(k,x)],$$

where the coefficients $\hat{a}(\vec{k})$ and $\hat{b}(\vec{k})$ are operators acting on both Hilbert and anti-Hilbert spaces, respectively. Similar to the classical theory, these operators are independent of each other. However, in the standard QFT they are complex conjugate with together and satisfy the following commutation relations

$$[a(\vec{k}), a^\dagger(\vec{k}')] = \delta^3(\vec{k} - \vec{k}'), \quad [b(\vec{k}), b^\dagger(\vec{k}')] = -\delta^3(\vec{k} - \vec{k}'), \quad [a(\vec{k}), b(\vec{k}')] = 0.$$

The operators $a^\dagger(\vec{k})$ and $a(\vec{k})$ create and destroy respectively the mode $u_p(k,x)$ with positive energy $(k^0 = \omega_{\vec{k}})$, which may be considered as the operators of creation and annihilation of a particle and the operators $b^\dagger(\vec{k})$ and $b(\vec{k})$ create and destroy respectively the mode $u_n(k,x)$ with negative energy $(-k^0 = -\omega_{\vec{k}})$, which may be considered as the operators of creation and annihilation of an "anti-particle" in the inverse time direction. Due to the above commutation relations the two sets of modes do not affect on each other and in the standard QFT, the negative energy states are eliminated in the quantum field operators, which is the origin of the anomaly or divergence. On the contrary, the anomaly disappears by taking these states into account.

Generalizing to the Minkowski spacetime, $\eta^{\mu\nu} = diag(1, -1, -1, -1)$, the Krein space method has been successfully implemented in boson field quantization [6], resulting in both positive and negative norms for the auxiliary negative-frequency states. This approach has been applied to the study of the free spinor field, $(s = \frac{1}{2})$, too. In so doing, the auxiliary negative-frequency states, with positive norms, have been utilized, the modes of which do not interact with the physical states or the real physical world. Since the norms of the unphysical states are positive, the Krein space quantization (which deals with negative norms) should be extended to the " generalized krein space quantization". Naturally these modes can not be affected by the physical boundary conditions. Following this scheme, the normal ordering procedure is rendered useless because the ultraviolet divergence in the stress tensor disappears and the vacuum energy remains convergent [1]. The results of the new approach are as the following [6] (K and T stand for quantities in Krein space and the time ordered product, respectively):

(i) Klein-Gordon field

$$(\partial^2+m^2)\phi(x)=0\ ,\ \phi_K(x)=\frac{1}{\sqrt{2}}\ [\phi_p(x)+\phi_n(x)]\ ,\ iG_T(x,x')=<0|T\phi_K(x)\phi_K(x')|0>$$

$$u_p(k,x)=\frac{e^{-ik.x}}{\sqrt{(2\pi)^3 2\omega_{\vec{k}}}}\ ,\ u_n(k,x)=\frac{e^{ik.x}}{\sqrt{(2\pi)^3 2\omega_{\vec{k}}}}\ ,\ \omega_{\vec{k}}=\sqrt{\vec{k}.\vec{k}+m^2}$$

$$\phi_K(x)=\frac{1}{\sqrt{2}}\int d^3\vec{k}\ [(a_{\vec{k}}+b^{\dagger}_{\vec{k}})u_p(k,x)+(a^{\dagger}_{\vec{k}}+b_{\vec{k}})u_n(k,x)]$$

$$[a_{\vec{k}},a_{\vec{k}'}{}^{\dagger}]=+\delta^3(\vec{k}-\vec{k}')\quad ,\quad [b_{\vec{k}},b_{\vec{k}'}{}^{\dagger}]=-\delta^3(\vec{k}-\vec{k}')$$

$$<1_{\vec{k}'}|1_{\vec{k}}>=+\delta^3(\vec{k}-\vec{k}')\quad ,\quad <\bar{1}_{\vec{k}'}|\bar{1}_{\vec{k}}>=-\delta^3(\vec{k}-\vec{k}')$$

$$H=\int d^3\vec{k}\ \omega_{\vec{k}}\ (a_{\vec{k}}{}^{\dagger}a_{\vec{k}}+b_{\vec{k}}{}^{\dagger}b_{\vec{k}}+a_{\vec{k}}{}^{\dagger}b_{\vec{k}}{}^{\dagger}+a_{\vec{k}}b_{\vec{k}})\quad ,\quad <0\mid H\mid 0>=0$$

(ii) Maxwell field

$$\partial^2 A_\mu=0\ ,\ A^K_\mu(x)=\frac{1}{\sqrt{2}}[A^p_\mu(x)+A^n_\mu(x)]\ ,\ D^T_{\mu\nu}(x,x')=-\eta_{\mu\nu}G_T(x,x')$$

$$A^K_\mu(x)=\frac{1}{\sqrt{2}}\int d^3\vec{k}\sum_{\lambda=0}^{3}\epsilon^\lambda_\mu(\vec{k})[(a^\lambda_{\vec{k}}+b^{\lambda\dagger}_{\vec{k}})\ u_p(k,x)+(a^{\lambda\dagger}_{\vec{k}}+b^\lambda_{\vec{k}})u_n(k,x)]$$

$$[a^\lambda_{\vec{k}},a^{\lambda'\dagger}_{\vec{k}'}]=-\eta^{\lambda\lambda'}\delta^3(\vec{k}-\vec{k}')\quad ,\quad [b^\lambda_{\vec{k}},b^{\lambda'\dagger}_{\vec{k}'}]=+\eta^{\lambda\lambda'}\delta^3(\vec{k}-\vec{k}')$$

$$|one\ \ physical\ \ state>=|1^\lambda_{\vec{k},a}>\propto a^{\lambda\dagger}_{\vec{k}}|0>\qquad \lambda=1,2$$

$$|one\ \ unphysical\ \ state>=|\bar{1}^\lambda_{\vec{k},a}>\propto a^{\lambda\dagger}_{\vec{k}}|0>\qquad \lambda=0,3$$

$$|one\ \ unphysical\ \ state>=|\bar{1}^\lambda_{\vec{k},b}>\propto b^{\lambda\dagger}_{\vec{k}}|0>\qquad \lambda=0,1,2,3$$

$$H=\int d^3\vec{k}\ \omega_{\vec{k}}\sum_{\lambda=0}^{3}(-\eta_{\lambda\lambda})(a^{\lambda\dagger}_{\vec{k}}a^\lambda_{\vec{k}}+b^{\lambda\dagger}_{\vec{k}}b^\lambda_{\vec{k}}+a^{\lambda\dagger}_{\vec{k}}b^{\lambda\dagger}_{\vec{k}}+a^\lambda_{\vec{k}}b^\lambda_{\vec{k}})\ ,\ <0\mid H\mid 0>=0$$

Four of the above states have negative norms, and two of them are unphysical positive norm states.

(iii) Dirac field (generalized Krein space)

$$(i\ \partial\!\!\!/-m)\psi(x)=0\ ,\ \psi(x)=\frac{1}{\sqrt{2}}\ [\psi_p(x)+\psi_n(x)]\ ,\ S_T(x,x')=(i\ \partial\!\!\!/+m)G_T(x,x')$$

$$\mathcal{U}^s(k,x)=\sqrt{\frac{m}{(2\pi)^3\omega_{\vec{k}}}}\ u^s(\vec{k})e^{-ik.x}\quad ,\quad \mathcal{V}^s(k,x)=\sqrt{\frac{m}{(2\pi)^3\omega_{\vec{k}}}}\ v^s(\vec{k})e^{ik.x}$$

$$\psi_K(x) = \frac{1}{\sqrt{2}} \int d^3\vec{k} \sum_{s=1,2} [(b_{\vec{k}s} + c^\dagger_{\vec{k}s})\mathcal{U}^s(k,x) + (d^\dagger_{\vec{k}s} + a_{\vec{k}s})\mathcal{V}^s(k,x)]$$

$$\{b_{\vec{k}s}, b^\dagger_{\vec{k}'s'}\} = \{d_{\vec{k}s}, d^\dagger_{\vec{k}'s'}\} = \delta_{ss'}\ \delta^3(\vec{k}-\vec{k}')\ ,\ \{a_{\vec{k}s}, a^\dagger_{\vec{k}'s'}\} = \{c_{\vec{k}s}, c^\dagger_{\vec{k}'s'}\} = \delta_{ss'}\ \delta^3(\vec{k}-\vec{k}')$$

$$< 1^b_{\vec{k},s} \mid 1^d_{\vec{k}',s} >= \delta_{bd}\ \delta^3(\vec{k}-\vec{k}')\quad ,\quad < \bar{1}^a_{\vec{k},s} \mid \bar{1}^c_{\vec{k}',s} >= \delta_{ac}\ \delta^3(\vec{k}-\vec{k}')$$

$$H = \int d^3\vec{k}\ \omega_{\vec{k}} \sum_{s=1,2} [b^\dagger_{\vec{k}s} b_{\vec{k},s} + b^\dagger_{\vec{k}s} c^\dagger_{\vec{k}s} + c_{\vec{k},s} b_{\vec{k},s} + c_{\vec{k},s} c^\dagger_{\vec{k}s} - d_{\vec{k},s} d^\dagger_{\vec{k}s} - d_{\vec{k},s} a_{\vec{k}s} - a^\dagger_{\vec{k}s} d^\dagger_{\vec{k}s} - a^\dagger_{\vec{k}s} a_{\vec{k}s}]$$

where $< 0 \mid H \mid 0 >= 0$. $b^\dagger_{\vec{k}s}$, $d^\dagger_{\vec{k}s}$ are respectively the creation operators of physical one-particle and one-antiparticle states with positive energy, and $c^\dagger_{\vec{k}s}$, $a^\dagger_{\vec{k}s}$ are called the creation operators of unphysical one-particle and one-antiparticle states respectively with negative energy.

Applying the Krein space method in quantum electrodynamics (QED), Wick's theorem and Feynman rules could be expressed in Krein space. So that, the basic structure of QFT could be reformulated, resulting in a theory without divergence [7].

3. Divergence free propagator in Krein space

The origin of divergences in standard quantum field theory lies in the singularity of the Green's function at short relative distances (ultraviolet divergence) or in the large relative distances (infrared divergence):

$$G^P_F(x,x') = -\frac{1}{8\pi}\delta(\sigma_0) + \frac{m^2}{8\pi}\ \theta(\sigma_0)[\frac{J_1(\sqrt{2m^2\sigma_0}) - iN_1(\sqrt{2m^2\sigma_0})}{\sqrt{2m^2\sigma_0}}] - \frac{im^2}{4\pi^2}\ \theta(-\sigma_0)\frac{K_1(\sqrt{2m^2(-\sigma_0)})}{\sqrt{2m^2(-\sigma_0)}}$$

The divergence appears in the imaginary part of the Feynman propagator, and the real part is convergent. Consideration of negative frequency states, which are also solutions of the field equation conveniently removes singularity of the Green function with exception of delta function singularity:

$$G_T(x,x') = \frac{1}{2}[G^p_F(x,x') + (G^p_F(x,x'))^*] = \Re G^p_F(x,x') = \int \frac{d^4k}{(2\pi)^4} e^{-ik.(x-x')} PP\frac{1}{k^2-m^2}$$

or

$$G_T(x-x') \equiv G_T(x,x') = -\frac{1}{8\pi}\delta(\sigma_0) + \frac{m^2}{8\pi}\theta(\sigma_0)\frac{J_1(\sqrt{2m^2\sigma_0})}{\sqrt{2m^2\sigma_0}},\qquad \sigma_0 \geq 0$$

The latter singularity can be eliminated by the quantum metric fluctuations [8]. It has been shown that the combination of the above methods (*i.e.* Krein space quantization and quantum metric fluctuations) totally removes the divergences of QFT [9].

$$< G_T(x-x') >= -\frac{1}{8\pi}\sqrt{\frac{\pi}{2 < \sigma_1^2 >}} \exp(-\frac{\sigma_0^2}{2 < \sigma_1^2 >}) + \frac{m^2}{8\pi}\ \theta(\sigma_0)\frac{J_1(\sqrt{2m^2\sigma_0})}{\sqrt{2m^2\sigma_0}}$$

where $2\sigma_0 = \eta_{\mu\nu}(x^\mu - x'^\mu)(x^\nu - x'^\nu)$, and $< \sigma_1^2 >$ is related to the density of gravitons. When $\sigma_0 = 0$, due to the metric quantum fluctuation $< \sigma_1^2 > \neq 0$, and we have

$$< G_T(0) >= -\frac{1}{8\pi}\sqrt{\frac{\pi}{2 < \sigma_1^2 >}} + \frac{m^2}{16\pi}$$

The time-ordered propagator of spinor field in generalized Krein space also reads:

$$S_T(x,x') = \frac{1}{8\pi}\, i\gamma^\mu(x_\mu - x'_\mu)\{\sqrt{\frac{\pi}{2<\sigma_1^2>}}\, e^{-\frac{\sigma_0^2}{2<\sigma_1^2>}}[\frac{\sigma_0}{<\sigma_1^2>} + m^2\frac{J_1(\sqrt{2m^2\sigma_0})}{\sqrt{2m^2\sigma_0}}]$$
$$+\frac{m}{2\sqrt{2}}\,\theta(\sigma_0)[\sqrt{2m^2\sigma_0}\; J_0(\sqrt{2m^2\sigma_0}) - 2J_1(\sqrt{2m^2\sigma_0})]\}$$
$$+\frac{m}{8\pi}\,[-\sqrt{\frac{\pi}{2<\sigma_1^2>}}e^{-\frac{\sigma_0^2}{2<\sigma_1^2>}} + m^2\theta(\sigma_0)\frac{J_1(\sqrt{2m^2\sigma_0})}{\sqrt{2m^2\sigma_0}}].$$

The most interesting result of this construction is the convergence of the Green's function at large distances, which means that the infrared divergence is gauge dependent [10, 11]. Presence of the "unphysical" states plays the role of an automatic renormalization tool for the theory. The physical interpretation of the unphysical negative energy states is not yet clear and any further progress calls for more investigations [12, 13, 14, 15]. However, in the case of spinor field one can interpret as the unphysical particle and antiparticle running in the inverse of the associated time direction to physical ones!

4. Evaluating divergent integrals of QED in Krein space

The Feynman propagator for spinor field in Krein space is

$$S_F(x,x') = \frac{1}{2}\left(S_F^p(x,x') + \gamma^5\gamma^0[S_F^p(x,x')]^\dagger\gamma^0\gamma^5\right),$$

$$S_F(x,x') = \frac{1}{2}(i\not{\partial} + m)[G_F^p(x,x') + G_F^{p*}(x,x')] = (i\not{\partial} + m)G_F(x,x'),$$

For the propagator in the momentum space we obtain

$$\tilde{S}_F(k) = \tilde{S}_F^p(k) - \gamma^0[\tilde{S}_F^p(k)]^\dagger\gamma^0,$$

$$\frac{1}{\not{k} - m + i\epsilon} - \frac{1}{\not{k} - m - i\epsilon} = -i2\pi(\not{k} + m)\delta(k^2 - m^2).$$

where the above relations are written without considering the fluctuations of metric. Since continuing the calculations with $S_T(x,x')$ is difficult, so we content ourself to work with $\tilde{S}_F(k)$.

In the previous section we obtained that the free fields in Krein space are free of any divergence. For the case of interaction field we briefly recall the problem of singularity or anomaly in the standard QED. We would like to show that the singularities appeared in the standard QFT, disappear in the Krein QFT method. So that, we present the following integral which appears in the standard QFT

$$I(n,l,M^2) = \int_L \frac{d^4k}{(2\pi)^4}\frac{(k^2)^n}{(k^2 - M^2 + i\epsilon)^l} = \frac{1}{(l-1)!}\frac{\partial^{l-1}}{(\partial M^2)^{l-1}}\int_L \frac{d^4k}{(2\pi)^4}\frac{(k^2)^n}{k^2 - M^2 + i\epsilon}$$

$$= -\frac{1}{16\pi^2}\frac{\Gamma(n+2)}{\Gamma(l)}\frac{\partial^{l-1}}{(\partial M^2)^{l-1}}\int_0^\infty ds s^{-2-n}e^{-iM^2s}(i)^n,\quad n \geq 0, l \geq 1,\ M^2 \neq 0,$$

where Γ is the Gamma function. For $2n + 4 \geq 2l$ the integral is divergent. We can also see very simply that the real part is always convergent

$$2\Re I^p = I^p + (I^p)^* = -\frac{1}{16\pi^2}\frac{\Gamma(n+2)}{\Gamma(l)}\frac{\partial^{l-1}}{(\partial M^2)^{l-1}}(M^2)^{n+1}$$

$$\int_0^\infty \frac{ds}{s^{2+n}}\begin{cases} 2(-1)^j\cos s & \text{for } n=2j \\ 2(-1)^j \sin s & \text{for } n=2j+1 \end{cases} \quad M^2\neq 0.$$

In the standard QED the divergent quantities are found in the self-energy graph, the vacuum polarization graph and the vertex graph. The self-energy graph in the standard QED is given by the following integral

$$\Sigma^p(q) = -ie^2\int_0^1 dx\gamma_\mu(\not{q} - \not{q}x + m)\gamma^\mu \int_L \frac{d^4k}{(2\pi)^4}\frac{1}{[k^2-m^2+q^2x(1-x)]^2},$$

In Krein QED it must be replaced by the following expression, which is convergent

$$\Sigma(q) = [\Sigma^p(q)] - \gamma^0[\Sigma^p(q)]^\dagger\gamma^0 = -ie^2\int_0^1 dx\gamma_\mu(\not{q} - \not{q}x + m)\gamma^\mu[I(0,2,M_1^2) + I(0,2,M_1^2)^*)]$$

where $M_1^2 = m^2 - q^2x(1-x)$. The vacuum polarization graph is given by the following integral

$$\Pi^p_{\mu\nu}(q) = 4ie^2\int_0^1 dx\int_L \frac{d^4k}{(2\pi)^4}\Big(\frac{2q_\mu q_\nu}{[k^2-m^2+q^2x(1-x)]^2}$$
$$-\frac{2x(1-x)(q_\mu q_\nu - g_{\mu\nu}q^2)}{[k^2-m^2+q^2x(1-x)]^2} - \frac{g_{\mu\nu}}{k^2-m^2+q^2x(1-x)}\Big).$$

The contributions of the first and third terms in the integrand cancel. If we calculate the new expression in Krein space

$$\Pi_{\mu\nu}(q) = [\Pi^p_{\mu\nu}(q)] - \gamma^0[\Pi^p_{\mu\nu}(q)]^\dagger\gamma^0 = -ie^2\int_0^1 dx 2x(1-x)[q_\mu q_\nu - g_{\mu\nu}q^2](I(0,2,M_1^2) + I(0,2,M_1^2)^*).$$

we can see very simply that it is convergent.

5. Conclusion

In standard quantum field theory, to eliminate the divergences that appear in the physical quantities, the normal ordering (renormalization) procedure has been adopted for free (interacting) fields. However, the divergences seem to disappear once the requirement of the positivity of norm and energy is relaxed. The addition of the new unphysical states, thus, leads us to the Krein space quantization. Once again it is found that the theory is automatically renormalized. We hope to apply the new method successfully to reformulate QED and QCD in Krein space.

The physical meaning of the unphysical negative energy states is not yet clear. However, in the case of spinor field one can interpret as the unphysical particle and antiparticle running in the inverse time direction! This case will need a careful consideration. But the most important question is that: *Does this quantization only regularize the theory without changing the physical content of the theory?*

Acknowlegements: I am grateful and thank Prof. M.V. Takook and Prof. M. Mehrafarin for very useful discussions as well as the DICE2008 organizers.

Journal of Physics: Conference Series **174** (2009) 012056 doi:10.1088/1742-6596/174/1/012056

References

[1] De Bievre S and Renaud S 1998 *Phys. Rev.* D **57** 6230; Gazeau J P, Renaud J and Takook M V 2000 *Class. Quant. Grav.* **17** 1415

[2] Allen B 1985 *Phys. Rev.* D **32** 3136

[3] Mintchev M 1979 *J. Phys. A: Math. Gen.* **13** 1841

[4] Takook M V 1999 Covariant Two-point Function for Linear Gravity in de Sitter Space *Preprint* gr-qc/0001052; Garidi T Gazeau J P and Takook M V 2003 *J. Math. Phys.* **44** 3838; Takook M V 2005 *Int. J. Mod. Phys.* E **14** 219; Takook M V 2001 *Mod. Phys. Lett.* A **16** 1691

[5] Behroozi S, Rouhani S, Takook M V and Tanhayi M R 2006 *Phys. Rev.* D **74** 124014; Dehghani M, Rouhani S, Takook M V and Tanhayi M R 2008 *Phys. Rev.* D **77** 064028; Rouhani S, Takook M V and Tanhayi M R 2009 Linear Weyl Gravity in de Sitter Universe *Preprint* gr-qc/0903.2670

[6] Payandeh F, Mehrafarin M, Rouhani S and Takook M V 2008 *UJP* **53** 1203

[7] Payandeh F, Mehrafarin M and Takook M V 2009 *Science in China Series G: Physics, Mechanics and Astronomy* **52** 212

[8] Ford H L 1997 Quantum Field Theory in Curved Spacetime *Preprint* gr-qc/9707062

[9] Rouhani S and Takook M V 2006 A Naturally Renormalized Quantum Field Theory *Preprint* gr-qc/0607027

[10] Takook M V 2001 *Mod. Phys. Lett.* A **16** 1691

[11] Gazeau J P, Renaud J and Takook M V 2000 *Class. Quant. Grav.* **17** 141

[12] Ramirez A and Mielinik B 2003 *Rev. Mex. Fis.* **49S2** 130

[13] Helfer A D 1998 The Physics of Negative Energy Densities *Preprint* hep-th/9811081

[14] Visser M and Barcelo C 2000 Energy Conditions and their Cosmological Implications *Preprint* gr-qc/0001099

[15] Blinnikov S 2000 *Surv. High Energy. Phys.* **15** 37

Two-point function of massless spin-2 field in de Sitter space-time

A. Ch. Shirazi
Plasma Physics Research Center, Islamic Azad University ,P.O.BOX 14835-157, Tehran, IRAN
E-mail: at_shirazi@physics.org

Abstract. Some problems arise in covariant quantization of massless spin-2 tensor field in linear approximation (linear quantum gravity) in de Sitter space-time. This quantum field breaks de Sitter invariance and its two-point function has logarithmic divergence for largely separated points. This divergence contributes to the tree level scattering amplitude for scalar field in de Sitter space-time. For covariant quantization of massless minimally coupled scalar field in de Sitter space-time, which is similar to the massless spin-2 field, Krein space quantization has been successfully suggested. In this paper, we consider the two-point function for linear gravity and its divergences in Krein space procedure.

1. Introduction

In order to obtain a unified theory of all the fundamental forces, general relativity and quantum theory should be unified. Unfortunately, some problems arise while constructing quantum gravity. First, microcausality condition implies that the commutator of two fields at spacelike separated points is required to vanish but whether or not two such points are "spacelike" depends on metric which-in the case of quantum gravity-is it self an operator. Second, the notion of time becomes unclear since whether or not a particular direction is timelike is metric dependent and is hence subject to quantum effects [1]. Existence approaches to quantum gravity tackle these problems in different ways. One of these techniques is "weak field" limit or linear approximation. In this scheme the full metric tensor is decomposed into a background field and a small perturbation

$$g_{\mu\nu} = g^b_{\mu\nu} + h_{\mu\nu}. \tag{1}$$

The causal structure of space time is described by background metric $g^b_{\mu\nu}$ and $h_{\mu\nu}$ will be quantized. Quantization of $h_{\mu\nu}$ results in a massless spin-2 particle.

The de Sitter space-time (dS) is a maximally symmetric solution of the vacuum Einstein field equations with positive cosmological constant . The recent observational data are strongly in favor of a positive acceleration of the present universe. Therefore, in a first approximation, the background space-time might be considered as a de Sitter space-time. In usual quantization methods for this massless rank-2 tensor field two problems arise. The first is that the theory is not fully covariant under the de Sitter group and the second that the two-point function has a logarithmic divergence for largely separated points [2, 3]. Tsamis and Woodard have considered a quantum field operator in flat coordinates that only covers one-half of de Sitter hyperboloid and found a quantum field that breaks de Sitter invariance [4].

In order to preserve dS covariance, Gupta-Bleuler method has been successfully suggested for the massless minimally coupled scalar field [7]. Following this procedure, we consider this method for spin-2 case and for the graviton propagator. This paper is organized as follows. In section 2 we obtain the equation of linear gravity in de Sitter space-time. In section 3 we briefly consider the quantization of this field. In section 4, the two-point function is considered. A summary is given in section 5. Appendix A, contains details of the calculation in section 2.

2. Linear quantum gravity

The vacuum solution of Einstein field equations in de Sitter space-time with positive cosmological constant $\Lambda = 3H^2$ is

$$R_{\mu\nu} - \frac{1}{2}g_{\mu\nu}R + 3H^2 g_{\mu\nu} = 0. \tag{2}$$

The field equation for the linearized gravity which is obtained from (2) by the linear approximation $g_{\mu\nu} = g^b_{\mu\nu} + h_{\mu\nu}$, is [Appendix A]

$$\frac{1}{2}\Big[\Box h_{\mu\nu} + \nabla_\mu\nabla_\nu h - \nabla_\mu\nabla_\lambda h^\lambda_\nu - \nabla_\nu\nabla_\lambda h^\lambda_\mu\Big]$$

$$+\frac{1}{2}g^b_{\mu\nu}\Big(\nabla_\lambda\nabla_\rho h^{\lambda\rho} - \Box h\Big) - H^2\Big(\frac{1}{2}g^b_{\mu\nu}h + h_{\mu\nu}\Big) = 0, \tag{3}$$

where $h = h^\mu_\mu$ and $\Box$ is the Laplace-Beltrami operator on de Sitter space and is given by

$$\Box = \nabla_\mu\nabla^\mu = \frac{1}{\sqrt{g^b}}\partial^\nu\Big(\sqrt{g^b}g^b_{\nu\mu}\partial_\mu\Big). \tag{4}$$

The ∇_μ is the covariant derivative and the indices are raised and lowered by the background metric. Equation (2.2) is invariant under the gauge transformation

$$h_{\mu\nu} \to h'_{\mu\nu} = h_{\mu\nu} + \nabla_\mu\varepsilon_\nu + \nabla_\nu\varepsilon_\mu. \tag{5}$$

One can choose a general family of gauge conditions

$$\nabla^\mu h_{\mu\nu} = \zeta\nabla_\nu h \tag{6}$$

where ζ is a constant. As mentioned before, in weak field the causal structure of space time is described by background metric $g^b_{\mu\nu}$ and $h_{\mu\nu}$ will be quantized. In the next section we briefly consider the quantization of this field.

3. Krein space quantization

The de Sitter space-time is made identical to the four dimensional one-sheeted hyperboloid

$$X_H = \{x \in R^5;\;\; x^2 = \eta_{\alpha\beta}x^\alpha x^\beta = -H^2,\;\; \alpha,\beta = 0,1,2,3,4\}, \tag{7}$$

where $\eta_{\alpha\beta} = diag\{1,-1,-1,-1,-1\}$. The de Sitter metric is

$$ds^2 = \eta_{\alpha\beta}dx^\alpha dx^\beta = g^{dS}_{\mu\nu}dX^\mu dX^\nu, \mu = 0,1,2,3, \tag{8}$$

where X^μ are the 4 space-time coordinates in dS hyperboloid. Any geometrical object in this space can be written in terms of the four local coordinates X^μ (intrinsic) or in terms of the five global coordinates x^α (ambient space). We now work in 5-dimensional ambient space notation for simplicity of the calculations. The tensor field $K_{\alpha\beta}(x)$ in this notation is locally related to the tensor field $h_{\alpha\beta}(X)$ through the relation

$$h_{\mu\nu}(X) = \frac{\partial x^\alpha}{\partial X^\mu}\frac{\partial x^\beta}{\partial X^\nu}K_{\alpha\beta}(x(X)). \tag{9}$$

In ambient space notation the traceless part of $h_{\alpha\beta}$ can be constructed from a projector tensor $D_{\alpha\beta}(x,\partial)$ and a massless minimally coupled (m.m.c) scalar field φ [5]

$$K_{\alpha\beta}(x) = D_{\alpha\beta}(x,\partial)\varphi, \tag{10}$$

$$\Box\varphi = 0. \tag{11}$$

The canonical quantization method is solely based on the presence of positive norm states. This norm for scalar field is defined by the Klein-Gordon inner product. The positivity condition comes from the concept of probability in usual quantum mechanics. However,(m.m.c) scalar field, φ, cannot be quantized in a fully covariant way through the usual method [6]. In this case, the space generated by positive modes is not de Sitter invariant. In other words, this set is not invariant under the action of the dS group: by applying the action of the dS group negative norm states φ^*,$\langle\phi_k^*,\phi_k^*\rangle = -1$, appear [7]. Therefore, the appearance of negative norm modes is the price to pay in order to obtain a fully covariant theory. Thus, the quantization will be done in a Krein space that is defined as a direct sum of an Hilbert and an anti-Hilbert space

$$H = H_+ \oplus H_+^*. \tag{12}$$

The existence of these non-physical states leads to a Gupta-Bleuler type field quantization and field operator is defined through

$$\varphi(X) = \frac{1}{\sqrt{2}}[\varphi_p(X) + \varphi_n(X)], \tag{13}$$

where

$$\varphi_p(X) = \sum_k a_k\phi_k(X) + H.C., \quad \varphi_n(X) = \sum_k b_k\phi_k^*(X) + H.C. \ . \tag{14}$$

The Fock space is built over a Krein space; $a_k^\dagger$ create physical states and $b_k^\dagger$ create unphysical states

$$a_k^\dagger|0\rangle = |physical\ \ state\rangle, b_k^\dagger|0\rangle = |unphysical\ \ states\rangle \tag{15}$$

$$\langle physical\ \ state|physical\ \ state\rangle = 1 \tag{16}$$

$$\langle unphysical\ \ state|unphysical\ \ state\rangle = -1 \tag{17}$$

$$\langle physical\ \ state|unphysical\ \ state\rangle = 0. \tag{18}$$

Vacuum is defined by

$$a_k|0\rangle = b_k|0\rangle = 0, \tag{19}$$

and the only non vanishing commutation relations between the operators are

$$[a_k, a_k^\dagger] = 1, [b_k, b_k^\dagger] = -1. \tag{20}$$

Different aspects of Krein space quantization is discussed in [8].

4. Two-point function

The graviton two-pint function which represents correlation of vacuum fluctuation in the gravitational field is defined by

$$G^{\mu\nu\rho'\sigma'}(x,x') = \langle 0|h^{\mu\nu}(x)h^{\rho'\sigma'}(x')|0\rangle. \tag{21}$$

This function has been studied by many authors using various gauges and different methods. It has been found that the graviton propagator (in the usual linear approximation for gravitational field) exhibits a pathological behavior for large separated points [2, 3].

However, recently Hawking, Hertog and Turok, have shown that the logarithmic growth of the physical two-point function in spatially-flat de Sitter space-time is a gauge artefact [9] .Also Higouchi and Kouris have considered the graviton two-point function in some covariant and non covariant gauges and obtained that this growth will not be reflected in the two-point function of a local gauge invariant quantity [10, 11]. The two-point function in the ambient space is defined through [12]

$$W_{\alpha\beta\alpha'\beta'}(x,x') = \langle \Omega, K_{\alpha\beta}(x)K_{\alpha'\beta'}(x')\Omega\rangle. \tag{22}$$

The traceless-part of this function can be written in the form [5]

$$W_{\alpha\beta\alpha'\beta'}(x,x') = \triangle_{\alpha\beta\alpha'\beta'}(x,x')W_{mc}(x,x'), \tag{23}$$

where $\triangle_{\alpha\beta\alpha'\beta'}(x,x')$ is a projector tensor and $W_{mc}(x,x')$ is the two-point function of the massless minimally coupled scalar field .

In usual quantization procedure W_{mc} is obtained in terms of an invariant quantity z [13]

$$W^p_{mc} = \frac{H^2}{8\pi^2}[\frac{1}{1-z} - \ln(1-z) + f_{\alpha\beta}(\eta,\eta') + const.], \quad z = -H^2 x.x', \tag{24}$$

which breaks dS invariance because of the last term.

By adopting Krein QFT framework, the Wightman two-point function reads

$$W(x,x') = \langle 0|\varphi(x)\varphi(x')|0\rangle = \frac{1}{2}\langle 0|\varphi_p(x)\varphi_p(x')|0\rangle + \frac{1}{2}\langle 0|\varphi_n(x)\varphi_n(x')|0\rangle$$

$$-\frac{1}{2}\langle 0|\varphi_p(x)\varphi_n(x')|0\rangle - \frac{1}{2}\langle 0|\varphi_n(x)\varphi_p(x')|0\rangle. \tag{25}$$

Condition (3.18) implies that the last two terms vanish. Therefore, we have

$$W(x,x') = \frac{1}{2}[W_p(x,x') + W_n(x,x')]. \tag{26}$$

where for the the rest of this section $W(x,x')$ denotes the two-point function in Krein space. This function for m.m.c. scalar case is [14]

$$W_{mc}(x,x') = \frac{H^2 i}{8\pi}\varepsilon(x^0,x'^0)[\delta(1-z) + \Theta(z-1)] \tag{27}$$

where $\Theta(z-1)$ is the step function and $\varepsilon(x^0,x'^0)$ is the sign function. This function is covariant and convergent in the large separated points limit. Therefore, the traceless-part of the graviton two-point function will become covariant and convergent .

5. Conclusion

We discussed Krein space quantization method for quantizing the linear gravity in dS space-time. In this scheme, quantization in indefinite inner product space, the logarithmic divergence of the traceless-part of the graviton two-point function disappears. Therefore, logarithmic behavior of the tree-level scattering amplitude of the scalar field with one graviton exchange that comes from the transverse-traceless part of the graviton two-point function disappears in this construction and the theory is automatically renormalized [15].

Although our result is just valid for traceless part, Higuchi and Kouris have shown that the pure trace part of a covariant graviton two-point function whose growth with distance was suggested to be physical [2, 3] gives a pure gauge contribution, when combined with another part [16].

Acknowledgments

We are grateful to M.V. Takook for very useful discussions.

Appendix A.

As mentioned before, the full metric in linear limit is decomposed into a background field and a small perturbation

$$g_{\mu\nu} = g^b_{\mu\nu} + h_{\mu\nu}, \qquad g^{\mu\nu} = g^{\mu\nu}_b - h^{\mu\nu}. \tag{A.1}$$

In linear limit in $h_{\mu\nu}$ equation (2) equals to

$$R^{(1)}_{\mu\nu} - (1/2)g^b_{\mu\nu}R^{(1)} - (1/2)h_{\mu\nu}R^{(0)} + \Lambda h_{\mu\nu} = 0 \tag{A.2}$$

where $R^{(0)} = 4\Lambda = 12H^2$ and $R^{(0)}_{\mu\nu}$ are Ricci scalar and Ricci tensor in dS space-time. $R^{(1)}$ and $R^{(1)}_{\mu\nu}$ are Ricci scalar and Ricci tensor in first order of $h_{\mu\nu}$ that are obtained through applying (A.1) into their relations:

$$R_{\mu\nu} = \partial_\nu\Gamma^\lambda_{\mu\lambda} - \partial_\lambda\Gamma^\lambda_{\mu\nu} + \Gamma^\rho_{\mu\lambda}\Gamma^\lambda_{\nu\rho} - \Gamma^\rho_{\mu\nu}\Gamma^\lambda_{\rho\lambda}, \tag{A.3}$$

$$\Gamma^\nu_{\mu\lambda} = \frac{1}{2}g^{\nu\sigma}(\nabla_\mu g_{\lambda\sigma} + \nabla_\lambda g_{\sigma\mu} - \nabla_\sigma g_{\mu\lambda}. \tag{A.4}$$

Therefore in firt order their relations read:

$$R^{(1)}_{\mu\nu} = \frac{1}{2}[\nabla^\lambda\nabla_\lambda h_{\mu\nu} + \nabla_\nu\nabla_\mu h - \nabla_\lambda\nabla_\mu h^\lambda_\nu - \nabla_\lambda\nabla_\nu h^\lambda_\mu], \tag{A.5}$$

$$R^{(1)} = g^{\mu\nu}_b R^{(1)}_{\mu\nu} - h^{\mu\nu}R^{(0)}_{\mu\nu} = -\nabla_\mu\nabla_\nu h^{\mu\nu} + \Box h + 3H^2 h, \tag{A.6}$$

$$h = h^\mu_\mu. \tag{A.7}$$

By applying these relations in (A.2) we obtain the linear field equation in dS space-time (2.3).

References

[1] Isham C J 1996 *Class. Quant. Grav.* **13** A5-9
[2] Allen B and Turyn M 1987 *Nucl. Phys.* B **292** 813-52
[3] Antoniadis I and Mottola E 1991 *J. Math. Phys.* **32** 1037
[4] Tsamis N C and Woodard R P 1992 *Phys. Lett.* B **292** 269
[5] Dehghani M, Rouhani S, Takook M V and Tanhayi M R 2008 *Phys. Rev.* D **77** 064028
[6] Allen B and Follacci A 1987 *Phys. Rev.* D **35** 3771
[7] Gazeau J P, Renaud J and Takook M V 2000 *Class. Quant. Grav.* **17** 1415
[8] Garidi T, Huguet E and Renaud J 2005 *J. Phys.* A **38** 245-56
[9] Hawking S W, Hertog T and Turok N 2000 *Phys. Rev.* D **62** 063502

[10] Higuchi A and Kouris S S 2000 *Class. Quant. Grav.* **17** 3077
[11] Higuchi A and Kouris S S 2001 *Class. Quant. Grav.* **18** 4317
[12] Garidi C, Gazeau J P and Takook M V 2003 *J. Math. Phys.* **44** 3838
[13] Follacci A 1991 *J. Math. Phys.* **32** 2828
[14] Takook M V 2001 *Mod. Phys. Lett.* A **16** 1691
[15] Rouhani S and Takook M V 2004 *Euro. Phys. Lett.* **68** 1
[16] Higuchi A and Kouris S S 2001 *Class. Quant. Grav.* **18** 2933

Journal of Physics: Conference Series **174** (2009) 012058 doi:10.1088/1742-6596/174/1/012058

Gravitational renormalization of quantum field theory: a "conservative" approach

Roberto Casadio
Dipartimento di Fisica, Università di Bologna, and I.N.F.N., Sezione di Bologna, via Irnerio 46, 40126 Bologna, Italy
E-mail: casadio@bo.infn.it

Abstract. We propose general guidelines in order to incorporate the geometrical description of gravity in quantum field theory and address the problem of UV divergences non-perturbatively. In our aproach, each virtual particle in a Feynman graph should be described by a modified propagator and move in the space-time generated by the other particles in the same graph according to Einstein's (semiclassical) equations.

1. Introduction

Pauli, long ago [1], suggested that gravity could act as a regulator for the ultraviolet (UV) divergences that plague quantum field theory (QFT) by providing a natural cut-off at the Planck scale. Later on, classical divergences in the self-mass of point-like particles were indeed shown to be cured by gravity [2], and the general idea has since then resurfaced the literature many times (see, *e.g.*, Refs. [3, 4, 5, 6, 7, 8]). In spite of that, Pauli's ambition has never been fulfilled.

As it happens, QFT is successfully used to describe particle physics in flat [9] (or curved but still fixed [10]) space-time where standard renormalization techniques allow one to obtain testable results, notwithstanding the presence of ubiquitous singularities stemming from the very foundations of the theory, that is the causal structure of (free) propagators. We have thus grown accustomed to the idea that the parameters in a Lagrangian have no direct physical meaning and infinite contributions may be subtracted to make sense of mathematically diverging integrals. The modern approach to renormalization [11] views the occurrence of such infinities as a measure of our theoretical ignorance of nature and every Lagrangian should, in turn, be considered as an effective (low energy) description doomed to fail at some UV energy scale Λ [12]. Moreover, gravitational corrections to the Standard Model amplitudes to a given order in the (inverse of the) Planck mass $m_{\rm p}$ are negligibly small at experimentally accessible energies [13]. These facts briefly elucidate the main theoretical reason that makes it so difficult to use gravity as a regulator: if it is to provide a natural solution to the problem of UV divergences, gravity must be treated non-perturbatively [7].

In the QFT community, gravity is mainly viewed as a spin-2 field which also happens to describe distances and angles (to some extent). As such, the most advanced strategy to deal with it is the background field method for functional integrals [14, 9], according to which one expands the Einstein-Hilbert Lagrangian (or a generalisation thereof) around all of the fields' classical values, including the classical background metric. The latter is reserved the role of defining the causal structure of space-time, whereas the quantum mechanical part yields the graviton propagator and matter couplings (of order m_p^{-2}). The effect of gravity on matter fields can then be analysed perturbatively by computing the relevant Feynman graphs [15]. A notorious consequence of this approach is that, by simple power counting, pure gravity is seen to be non-renormalizable, a "text-book" statement [16], yet occasionally debated. For example, Ref. [7] suggested that perturbative expansions are usually performed in the wrong variables and that Einstein gravity would appear manifestly renormalizable if one were able to resum logarithmic-like series [5]. In the physically more interesting case with matter, non-perturbative results can be obtained in just a very few cases, one of particular interest being the correction to the self-mass of a scalar particle, which becomes finite once all ladder-like graphs containing gravitons are added [4]. A remarkable approach was developed in Refs. [17], in which a tree-level effective action for gravity at the energy scale μ is derived within the background field method but without specifying the background metric *a priori*. The latter is instead, *a posteriori* and self-consistently, equated to the quantum expectation value determined by the effective action at that scale. This method does not involve cumbersome loop contributions and hints that gravity might be *non-perturbatively* renormalizable [18], with a non-Gaussian UV fixed point, thus realising the *asymptotic safety* conjectured several decades ago by Weinberg [19].

Based on the idea that QFT is an effective approach [12], different attempts have taken a shortcut and addressed the effects of gravity on the propagation of matter field directly, *e.g.*, by employing modified dispersion relations or uncertainty principles at very high (*trans-Planckian*) energy [20]. Some works have postulated such modifications, whereas others tried to derive them from (effective) descriptions of quantum gravity (see, *e.g.*, Refs. [21]). It is in fact common wisdom that, for energies of the order of m_p or larger, the machinery of QFT fails and one will need a more fundamental quantum theory of gravity, such as String Theory [22] or Loop Quantum Gravity [23]. Quite interestingly, both approaches hint at space-time non-commutativity [24] as an effective implementation of gravity as a regulator, with the scale of non-commutativity of the order of the Planck length ℓ_p. A new feature which, in turn, follows from space-time non-commutativity is the IR/UV mixing, whereby physics in the infrared (IR) is affected by UV quantities [25]. This feature gives us hope of probing (indirectly) such an extreme energy realm in future experiments or even using available data of very large scale (cosmological) structures.

In Ref. [26], we proposed yet a different strategy to incorporate gravity in the body of QFT. Instead of proposing a new, or relying on an available, fundamental theory of quantum gravity, we tried to define modified propagators in a very "conservative" (minimal) way inspired by the simple semiclassical perspective in which gravity is described by Einstein's geometrical theory and matter by perturbative QFT. In this approach, gravity is therefore not viewed as a spin-2 field, but rather as the causal structure of space-time (or the manifestation thereof) at all loops in QFT, a property the background field method instead reserves to the classical part of the metric only. The modified propagators for matter fields should therefore take into consideration the presence of each and every source, classical or virtual, in a given process mathematically described by Feynman's diagrams. Of course, philosophical perspectives aside, the relevant question is whether this idea leads to different (or the same) phenomenological predictions with respect to the other approaches to UV physics currently available. However, we are in a fairly premature stage to assess that. In fact, even realising the relatively simple guidelines which we review here poses serious technical problems, and a very preliminary attempt, based on several further working assumptions, can also be found in the second part of Ref. [26].

We shall use units with $c = \hbar = 1$ and the Newton constant $G = \ell_p/m_p$.

2. Geometrical gravity in QFT

In order to make contact with the physics, let us note that one needs to consider two basic energy scales, one related to phenomenology and one of theoretical origin, namely:

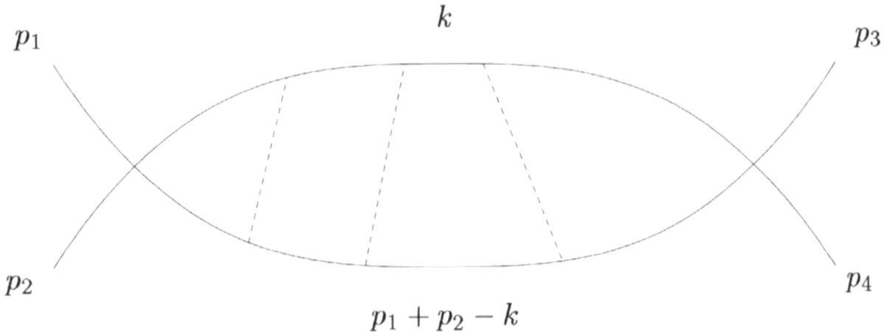

Figure 1. Usual one-loop correction to the four-point function in $\lambda\,\phi^4$ (solid lines) and graviton exchanges (dashed lines).

a) the highest energy presently available in experiments, say $E_{\rm exp} \simeq 1\,$TeV, and

b) the Planck energy $m_{\rm p} \simeq 10^{16}\,$TeV.

It is well assessed that, for energies up to $E_{\rm exp}$, the Standard Model of particle physics (without gravity) and renormalization techniques yield results in very good agreement with the data. Further, finite, albeit experimentally negligible, quantum gravitational corrections can be obtained by employing the effective QFT approach [13] (which also yields some – but not all – of the general relativistic corrections to the Newtonian potential). At the opposite end of the spectrum, for energies of the order of $m_{\rm p}$ or larger, QFT presumably breaks down and one needs a new quantum theory which includes gravity in a fundamental manner, like String Theory [22] or Loop Quantum Gravity [23].

In any case, we expect that gravitational corrections to QFT amplitudes play an increasingly important role for larger and larger energy scale $\mu > E_{\rm exp}$, and that it should be possible to describe such effects in perturbative QFT directly (at least in the regime $E_{\rm exp} \lesssim \mu \lesssim m_{\rm p}$). We call this window the realm of "semiclassical gravity", and that is the range where our proposal is more likely to shed some new light [1].

2.1. Semiclassical gravity

At intermediate energies $E_{\rm exp} \lesssim \mu \ll m_{\rm p}$, we expect that a semiclassical picture holds in which the space-time can be reliably described as a classical manifold with a metric tensor $g_{\alpha\beta}$ that responds to the presence of (quantum) matter sources according to the well-known equation [10]

$$R_{\alpha\beta} - \frac{1}{2}\,R\,g_{\alpha\beta} = \frac{\ell_{\rm p}}{m_{\rm p}}\,\langle \hat{T}_{\alpha\beta} \rangle \ , \tag{1}$$

where $R_{\alpha\beta}$ (R) is the Ricci tensor (scalar) and $\langle \hat{T}_{\alpha\beta} \rangle$ the expectation value of the matter stress tensor obtained from QFT on that background. All the same, if one takes Eq. (1) at face value, the way perturbative terms are computed in QFT appears questionable, since loops of virtual particles are included in Feyman's diagrams whose four-momentum $|k^2| = |k_\alpha\,k^\alpha|$ formally goes all the way to infinity (*i.e.*, to $m_{\rm p}$ and beyond) but are still described by the (free) propagators computed on a fixed (possibly flat) background.

For example, let us consider the graph for scalar particles with self-interaction $\lambda\,\phi^4$, represented by solid lines in Fig. 1, which is a pictorial representation of the integral

$$\Gamma^{(4)}(p) \simeq \int \frac{k^3\,{\rm d}k}{(2\,\pi)^4}\,\tilde{G}_{\rm F}(k)\,\tilde{G}_{\rm F}(p-k) \ , \tag{2}$$

[1] We actually attempted at pushing our predictions even further and address the very problem of UV divergences in Ref. [26].

where $\tilde{G}_{\rm F}$ is the momenutm-space Feynman propagator in four dimensions,

$$\tilde{G}_{\rm F}(p) = \frac{1}{p^2 + i\,\epsilon} \ . \tag{3}$$

Although the external momenta p_i ($i = 1, \ldots, 4$) are taken within the range of experiments (that is, $|p_i^2| \lesssim E_{\rm exp}^2$ in the laboratory frame), the two virtual particles in the loop have unconstrained momenta k and $p_1 + p_2 - k$, respectively. One might therefore wonder if it is at all consistent to describe those two particles using the above flat-space propagator. The common QFT approach to this problem would result in adding gravity in the form of graviton exchanges (the dashed lines in Fig. 1) and estimate deviations from purely flat-space results. This procedure is however likely to miss non-perturbative contributions that the UV physics might induce into the IR. For sure, it will not render finite diverging integrals, such as the one in Eq. (2), unless one is able to resum an infinite number of perturbative terms.

The interplay among propagators, UV divergences and the causal structure of space-time can be better appreciated by noting that, in any approach in which the space-time structure is a fixed background, the short distance behaviour of QFT (in four dimensions) is described by the Hadamard form of the propagators [6]

$$G(x, x') = \frac{U(x, x')}{\sigma} + V(x, x')\, \ln(\sigma) + W(x, x') \ , \tag{4}$$

where U, V and W are regular functions and $2\,\sigma$ is the square of the geodesic distance between x and x'. For instance, in Minkowski space-time, one has

$$2\,\sigma = (x - x')^2 \ , \tag{5}$$

and the propagator contains divergences for $\sigma \to 0$ (*i.e.*, along the light cone and for $x \to x'$). Calculations based on the use of propagators in QFT therefore (implicitly) rely on the formalism of distribution theory and UV divergences appear as a consequence when one tries to compute (mathematically) ill-defined quantities, such as the four-point function in Eq. (2). One can devise mathematical workarounds for this problem, but what matters here is that, if only the relation (5) is modified (like in QFT on a curved space-time), the divergences for $\sigma \to 0$ will remain. Nonetheless, a few partial results suggest that deeper modifications of the causal structure might occur at the quantum level. For example, it was shown that the divergence on the light-cone disappears (with a smearing at large momenta of the form considered in Ref. [27]) if graviton fluctuations are in a coherent state [6] [2].

It seems appropriate to us to tackle this problem by pushing further the validity of the semiclassical Einstein equations. We shall hence assume that virtual particles propagate in a background compatible with Eq. (1) at the scale $\mu \sim k = \sqrt{|k^2|}$ and their propagators be correspondingly adjusted [3]. As we mentioned before, our underlying viewpoint is not that gravity is just a field (although with a very complicated dynamics), but the geometrical perspective according to which gravity *is* the space-time and, in particular, the causal structure obeyed by all (other) fields. Let us remark again that this view is partly incorporated into the background field method, whereby the metric is split into two parts,

$$g_{\mu\nu} = \eta_{\mu\nu} + h_{\mu\nu} \ . \tag{6}$$

The classical part $\eta_{\mu\nu}$ possesses the expected symmetries of General Relativity and determines the causal structure for all (other) classical and quantum fields, whereas $h_{\mu\nu}$ is just another

[2] With the further inclusion of negative norm states, all UV divergences were claimed to be cured in Ref. [28].

quantum field which acts on the matter fields via usual (although complicated) interaction terms, hence in a non-geometrical way. One could actually view our approach as a step backward, since the gravitational field is not explicitly quantised [29] (there is no analogue of the above $h_{\mu\nu}$), and it is in fact not even defined separately (*i.e.*, in the absence of matter [3]).

2.2. Gravity in propagators and transition amplitudes

We can now formulate the basic prescriptions for defining a "gravitationally renormalised" QFT:

A1) perturbative QFT defined by Feynman diagrams is a viable approach to particle physics for energies μ below a cut-off $\Lambda \gg E_{\text{exp}}$;

A2) in a (one-particle irreducible) Feynman diagram with N internal lines, each virtual particle is described by a Feynman propagator

$$G(x, y) = G^{(\Lambda)}_{\{x_i\}}(x, y) \tag{7}$$

corresponding to the space-time generated by the other $N - 1$ virtual particles in the same graph with coordinate positions x_i ($i = 1, \ldots, N - 1$) and constrained according to **A1**;

A3) Standard Model results are recovered at low energy, $\mu \lesssim E_{\text{exp}} \ll m_{\text{p}}$.

Several comments on the above guidelines are in order. First of all, we explicitly introduced a cut-off in **A1**, having in mind that our approach is not meant to be the final theory of everything, but should rather be regarded as a computational recipe. A second, essential, simplification was introduced in **A2**, in that each virtual particle is treated like a test particle in the space-time generated by the other particles, its own gravitational backreaction thus being neglected [4]. Another consequence of **A2** is that integration over positions inside loops can now be viewed as also purporting a (quantum mechanical) superposition of (virtual) metrics, and there is hope that this can smear the usual divergences of (3) out (as was shown in Ref. [6] for particular gravitational states). It also should not go unnoticed that we did not mention a Lagrangian (or action) from which the modified propagators satisfying **A2** could be obtained. In this respect, our proposal follows the philosophy of Ref. [15], which gives the Lagrangian a secondary role with respect to Feynman's rules for computing perturbative amplitudes. However, the symmetries of a system are far more readable if a Lagrangian is available [12] and it would be interesting to find out whether an action principle can be devised to streamline the derivation and show which symmetries are preserved or broken. The latter kind of analysis can also be performed perturbatively, although, as is well known for the Slavnov-Taylor identities of (non-Abelian) Yang-Mills theory, that task requires a lot more effort. A final observation is that the Standard Model of particle physics (without gravity) is a rigid theory and it is very likely that a generic modification of the sort we are proposing here has hazardous effects in the range of presently available data, thus compromising **A3**. One should therefore check very careful that none of the assessed predictions of the Standard Model is lost in our approach.

We cannot ignore the technical fact that the N-body problem in General Relativity is extremely complicated, to say the least, already for $N = 2$. To the general guidelines, we therefore add two working assumptions:

W1) starting from the coordinate-space propagator in Eq. (7), it is possible to define a momentum-space propagator $\tilde{G}_{\{k_i\}}(p)$;

[3] This is somewhat reminiscent of the "relational mechanics" approach to gravity (see, *e.g.*, Ref. [30] and References therein).

[4] Let us note in passing that this somewhat parallels a perturbative result of non-commutative QFT, according to which there is no tree-level correction to the commutative case [24].

W2) one can approximate the momentum-space propagator for each virtual particle

$$\tilde{G}^{(\Lambda)}_{\{k_i\}}(p) \simeq \tilde{G}^{(\Lambda)}_{q}(p) \ , \tag{8}$$

where $q \simeq \sqrt{|\sum k_i|^2}$ is the total momentum of the remaining $N-1$ particles.

The latter is a "mean field" assumption devised to deal with graphs containing more than two virtual particles. The approximate equality in Eq. (8) may thus be replaced with other expressions of choice, the key point being that the problem reduces to studying the propagator for a test particle in a background generated by an "average" source.

For example, the "gravitationally renormalized" analogue of the four-point amplitude in Eq. (2) is obtained by replacing each particle's propagator with the new expression (8),

$$\Gamma^{(4)}_{\rm GR}(p;\Lambda) \simeq \int^{\Lambda} \frac{k^3\,{\rm d}k}{(2\,\pi)^4}\,\tilde{G}^{(\Lambda)}_{(p-k)}(k)\,\tilde{G}^{(\Lambda)}_{(k)}(p-k) \ . \tag{9}$$

Provided $\tilde{G}_{(k)}(q)$ falls off fast enough at large k, one can therefore hope to obtain finite transition amplitudes even in the limit $\Lambda \to \infty$. Of course, in order to obtain explicit expressions, one first needs to solve for the geometry produced by virtual particles and then obtain the momentum-space form of the propagator (an early attempt can be found in the second part of Ref. [26]).

3. Final remarks

Inspired by the observation that a semiclassical description of gravity should be possible in processes that involve energies below the Planck scale, we formulated general guidelines that can be employed to adjust QFT in order to include gravitational contributions. Such guidelines were listed in the form of general prescriptions (**A1-A3**) and more specific working assumptions (**W1** and **W2**) that formalise our approach to include gravity within the Standard Model of particle physics in a geometrical and non-perturbative way. All of them are of course debatable and subject to possible refinements.

Preliminary results, reported in Ref. [26], showed that one might indeed expect significant UV modifications from the dependence of the propagators on the momenta of virtual particles. However, further working assumptions were used therein, whose impact must be clarified. And, of course, a realistic QFT should be analysed before the final word can be spoken on UV divergences and that old idea of Pauli.

References

[1] See, *e.g.*, in Pauli W 1985 (Letter of Heisenberg to Peierls 1930) *Scientific Correspondence* Vol II ed von Meyenn K (Berlin: Springer) 15 ; Klein O 1956 *Helv. Phys. Acta. Suppl.* **4** 58
[2] Arnowitt R, Deser S and Misner C W 1960 *Phys. Rev. Lett.* **4** 375
[3] Deser S 1957 *Rev. Mod Phys.* **29** 417
[4] DeWitt B S 1964 *Phys. Rev. Lett.* **13** 114
[5] Isham C J, Salam A and Strathdee J 1971 *Phys. Rev.* D **3** 1805
[6] Ford L H 1997 Quantum field theory in curved spacetime *Preprint* arXiv:gr-qc/9707062
[7] Woodard R P 1998 Particles as bound states in their own potentials *Preprint* arXiv:gr-qc/9803096
[8] Robinson S P and Wilczek F 2006 *Phys. Rev. Lett.* **96** 23160
[9] Peskin M E and Schroeder D W 1995 *An introduction to quantum field theory* (Reading: Perseus Books)
[10] Birrell N D and Davies P C W 1982 *Quantum fields in curved space* (Cambridge, UK: Cambridge U. Press)
[11] Wilson K G and Kogut J B 1974 *Phys. Rep.* **12** 75
[12] Weinberg S 1997 What is quantum field theory, and what did we think it was? *Preprint* arXiv:hep-th/9702027
[13] Donoghue J F 1994 *Phys. Rev. Lett.* **72** 2996
[14] DeWitt B S 1967 *Phys. Rev.* **162** 1195
[15] Veltman M J G 1964 *Diagrammatica: The Path to Feynman rules* (Cambridge, UK: Cambridge U. Press)

[16] Shomer A 2007 A pedagogical explanation for the non-renormalizability of gravity *Preprint* arXiv:hep-th/0709.3555
[17] Reuter M 1998 *Phys. Rev.* D **57** 971; Souma W 1999 *Prog. Theor. Phys.* **102** 181
[18] Reuter M and Weyer H 2008 Background Independence and Asymptotic Safety in Conformally Reduced Gravity *Preprint* arXiv:hep-th/0801.3287
[19] Weinberg S 1979 Ultraviolet divergences in quantum theories of gravitation *General relativity: an Einstein centenary survey* ed S Hawking and W Israel (Cambridge, UK: Cambridge U. Press)
[20] Unruh W G 1995 *Phys. Rev.* D **51** 2827; Corley S and Jacobson T 1996 *Phys. Rev.* D **54** 1568
[21] Maggiore M 1993 *Phys. Lett.* B **304** 65; Scardigli F 1999 *Phys. Lett.* B **452** 39; Capozziello S, Lambiase G and Scarpetta G 2000 *Int. J. Theor. Phys.* **39** 15; Alberghi G L, Casadio R and Tronconi A 2006 *Phys. Rev.* D **74** 103501; Amelino-Camelia G 2008 Quantum Gravity Phenomenology *Preprint* arXiv:gr-qc/0806.0339
[22] Becker K, Becker M and Schwarz J H 2007 *String theory and M-theory: a modern introduction* (Cambridge, UK: Cambridge U. Press)
[23] Rovelli C 1998 *Living Rev. Rel.* **1** 1
[24] Szabo R J 2003 *Phys. Rep.* **378** 207
[25] Minwalla S, Van Raamsdonk M and Seiberg N 2000 *JHEP* **0002** 020; Matusis A, Susskind L and Toumbas N 2000 *JHEP* **0012** 002
[26] Casadio R 2008 Gravitational renormalization of quantum field theory *Preprint* arXiv:hep-th/0806.0501
[27] Smailagic A and Spallucci E 2003 *J. Phys.* A **36** L517; *do.* 2004 *J. Phys.* A **37** 7169
[28] Rouhani S and Takook M V 2006 A naturally renormalized quantum field theory *Preprint* arXiv:gr-qc/0607027
[29] Kauffmann S K Orthodox quantization of Einstein's gravity: might its unrenormalizability be technically fathomable and physically innocuous?, unpublished
[30] Anderson E 2008 Triangleland. I. Classical dynamics with exchange of relative angular momentum *Preprint* arXiv:gr-qc/0809.1168
[31] Nicolini P 2008 Noncommutative Black Holes, The Final Appeal To Quantum Gravity: A Review *Preprint* arXiv:hep-th/0807.1939

Journal of Physics: Conference Series **174** (2009) 012059 doi:10.1088/1742-6596/174/1/012059

A window to quantum gravity phenomena in the emergence of the seeds of cosmic structure

Daniel Sudarsky[†] **and Adolfo De Unanue**[††]

Instituto de Ciencias Nucleares, Universidad Nacional Autónoma de México, A. Postal 70-543, México D.F. 04510, México

E-mail: [†]`sudarsky@nucleares.unam.mx`, [††]`adolfo@nucleares.unam.mx`

Abstract. Inflationary cosmology has, in the last few years, received a strong dose of support from observations. The fact that the fluctuation spectrum can be extracted from the inflationary scenario through an analysis that involves quantum field theory in curved space-time, and that it coincides with the observational data has lead to a certain complacency in the community, which prevents the critical analysis of the obscure spots in the derivation. We argue here briefly, as we have discussed in more detail elsewhere, that there is something important missing in our understanding of the origin of the seeds of Cosmic Structure, as is evidenced by the fact that in the standard accounts the inhomogeneity and anisotropy of our universe seems to emerge from an exactly homogeneous and isotropic initial state through processes that do not break those symmetries. This article gives a very brief recount of the problems faced by the arguments based on established physics. The conclusion is that we need some new physics to be able to fully address the problem. The article then exposes one avenue that has been used to address the central issue and elaborates on the degree to which, the new approach makes different predictions from the standard analyses. The approach is inspired on Penrose's proposals that Quantum Gravity might lead to a real, dynamical collapse of the wave function, a process that we argued has the properties needed to extract us from the theoretical impasse described above.

1. Introduction

One of the big problems facing the search for a unification of Quantum Theory and Gravitation is the almost complete absence of experimental data to be used as guidance by theorists. In fact there is up to this date no single man made experiment in which the realms of Quantum Mechanics and Gravitation intersect, i.e. are simultaneously needed to account for the results. The famous COW experiments [1] and the recent "cold neutron" experiments [2] that are some times exhibited as examples of tests of the interface of gravity and quantum mechanics, can not be taken as such when viewed from within the relativistic paradigm, as they can be fully accounted for in terms of physics within a single inertial reference frame, and thus Einsteinian gravity can not be said to be playing any role (for a more detailed discussion of this point see for instance [3]). There exists however **one single situation** offered to us by nature, which satisfies the two criteria of, being observationally accessible and requiring, for a complete understanding, both general relativity and quantum physics. That is: the origin of the seeds of cosmic structure.

Among the most important achievements in observational cosmology we have the precision measurements of the anisotropies in the CMB[4]. These together with an extensive set of observational studies of large scale matter distribution, led to a very satisfactory picture of the evolution of structure our universe based on a detailed understanding of the physics behind it. In fact it is nowadays widely accepted that the origin of structure in of our universe has its natural explanation within the context of the inflationary scenarios: Inflation takes relatively arbitrary initial conditions presumably emerging from a Planck Era and leads to an *almost* de-Sitter phase of accelerated expansion that essentially erases all memories of the initial conditions. At this point inflation has lead to a featureless universe, which seems to lack even a small degree of inhomogeneity and anisotropy that is necessary to lead to the subsequent structure formation. At this point quantum mechanics is thought to provide this essential ingredient: The quantum fluctuations of the inflaton field, which has been put by inflation in its "ground state". These are thought to provide the seeds of the anisotropies and inhomogeneities that eventually evolve into the structure we first see in the CMB and eventually into the obvious features of our universe such as galaxy clusters, galaxies, stars, etc. The remarkable fact is that the calculations based on the above scheme seem to lead naturally to the correct spectrum of these primordial fluctuations.

There is however a serious hole in this seemingly blemish-less picture: The description of our Universe– or the relevant part thereof- starts[1] with an initial condition which is homogeneous and isotropic both in the background space-time and in the quantum state that is supposed to describe the "fluctuations", and it is quite easy to see that the subsequent evolution through dynamics that do not break these symmetries can only lead to an equally homogeneous and anisotropic universe.

In fact many arguments have been put forward in order to deal with this issue, that is often phrased in terms of the Quantum to Classical transition –without focusing on the required concomitant breakdown of homogeneity and isotropy in the state– the most popular ones associated with the notion of decoherence [16]. These the alternatives have been critically discussed in[5, 6]. One of the main obstacles is that in order to justify any explanation based on decoherence, one has to argue that certain degrees of freedom must be traced over because they are unobservable, and this in turn can only be justified by relying on the limitations we humans currently have, in making certain measurements. The problem is that in so doing we would be using our existence as input, but what cosmology is all about is understanding the evolution of the universe and its structure including the emergence of the conditions that make humans possible. In other words, in order to understand the emergence of an inhabitable universe, we would be relying on the existence and limitations of such inhabitants. Therefore any explanation based purely on standard decoherence becomes circular by definition. There are further problems in each of the specific proposals, and we direct the reader to the above references for extended discussion of this issues. There are other cosmologists that have acknowledged that there is problem here, and that quantum mechanics as we know it needs modifications to be applicable to the cosmology [17], with one of them explicitly stating that decoherence does not offer a complete and satisfactory resolution to this problem [18].

Moreover, if we were to think in terms of first principles, we would start by acknowledging that the correct description of the problem at hand would involve a full theory of quantum gravity coupled to a theory of all the matter quantum fields, and that there, the issue would be whether we start with a quantum state that is homogeneous and isotropic or not?. Even if these notions do not make sense within that level of description, a fair question is whether or not, the inhomogeneities and anisotropies we are interested on, can be traced to aspects of

[1] Here we refer to the era relevant to the starting point of the analysis that leads to the "fluctuation spectrum". In the standard view of inflation, the relevant region of the universe starts with a Planck regime containing large fluctuations of essentially all relevant quantities, but then, a large number of inflationary e-folds leads to an homogeneous and isotropic universe which is in fact the starting point of the analysis that takes us to the primordial fluctuation spectrum. One might wish, instead, to regard such fluctuation spectrum as a remnant of the earlier anisotropic and inhomogeneous conditions but then one ends up giving up any pretense that one can explain its origin and account for its specific form.

the description that have no contra-part in the approximation we are using. Recall that such description involves the separation of background *vs.* fluctuations and thus must be viewed only as an approximation, that allows us to separate the nonlinearities in the system–as well as those aspects that are inherent to quantum gravity– from the linear part of problem represented by the fluctuations, which are treated in terms of linear quantum field theory. In this sense, we might be tempted to ignore the problem and view it as something inherent to such approximation. This would be fine, but we should recognize then that we could not argue that we understand the origin of the CMB spectrum, if we view the asymmetries it embodies as arising from some aspect of the theory we do not know, rely on, or touch upon. In fact, in the treatment which we describe next, the proposal is to bring up one particular element or aspect, that we view as part of the quantum gravity realm, to the forefront of the treatment, in order to modify –in a minimalistic way– the semiclassical treatment, that, as we said, we find lacking, and provide a setting in which the obscure issues can at least be focus on. It is of course not at all clear that the problem we are discussing should be related to quantum gravity, but the later is the only sphere which is now believed capable of leading to a radical change in the paradigm of fundamental physics which is of course what we are considering here.

The approach taken in [5] is influenced by Penrose's suggestion that quantum gravity might play a role in triggering a real dynamical collapse of the wave function of systems [12]. His proposals would have a system collapsing whenever the gravitational interaction energy between two alternative realizations that appear as superposed in a wave function of a system reaches a threshold naturally identified with M_{Planck}. We will show that these ideas can, in principle, be investigated in the present context , and that they could lead to observable effects. In fact the very early universe can be seen as a case for which there exists already a wealth of empirical information and one which, as we have argued can not be fully understood without involving some New Physics, with features that would seem to be quite close to those of Penrose's proposals. There are of course alternatives settings in which modifications of quantum theory, in principle unrelated to Quantum Gravity, could play the role of the New Physics[19] that we have argued is needed in order to account for the seeds of cosmic structure, but we will limit ourselves here to a rather generic setting motivated by the first set of ideas.

2. Quantum gravity and its effective semiclassical description

Before we present the treatment we are proposing, which should be consider as being of phenomenological nature it is worthwhile to see how would it fit within the context of a fundamental theory such as a quantum theory of gravity.

The first thing one should note is that the notions of space-time are likely to change dramatically when considered in a fully quantum theory of gravitation[7]. A fundamental theory of quantum gravity (with or without matter) is naturally expected to be a timeless theory[2], but General Relativity is certainly not.

In fact we have a good example of this arising in the past and current attempts to apply the canonical quantization procedure to the theory of General Relativity: In all such schemes one ends with a timeless theory in which the wave functionals $\Psi[h_{ab}]$ depend on (see [10]) a spatial metric h_{ab} associated with a slice Σ of space-time (its canonical conjugate momentum π^{ab} is a closely related to the extrinsic curvature of such slice K_{ab}), or, equivalent variables such as triads and connections as in the modern incarnation of the program known as Loop Quantum Gravity (see for example [11]).

The recovery of the usual space-time notions –that in the appropriate limit should be GR– is thus expected to proceed trough a rather indirect way which involves the identification of one

[2] There are of course approaches that do otherwise like String Theory (see for instance [8]) or the Causal Sets program [9].

of the variables of the formalism with something playing the role of a physical clock[13], and in fact this seems to be more easily achieved in a theory involving not only the gravitational degrees of freedom, but also some suitable matter fields [14], for instance a scalar field ϕ, where one would have a wave functional $\Psi[h_{ab}, \phi]$. The idea is then that upon the identification of ϕ as a physical clock variable, one would be able to talk about the probability that the space-time metric and its conjugate variable take such and such value when the clock takes a given value, and from such information one would presumably be able to estimate the most likely space-time, correlations and so forth. Examples of application of such approach can be seen in [15]. These effective descriptions can be expected quite generally to incorporate some degree of breakdown in unitary evolution [13].

Our point here is that the recovering of the usual space-time picture is expected, to be a complex procedure even if we have a complete theory of quantum gravity in interaction with all matter fields. It should be said that in the LQG program, even the simpler spatial notions as distance volumes and to a lesser degree areas, turn out to be also a rather involved procedure.

The recovery of the standard evolution of physical degrees of freedom in space-time can be expected to involve an even more cumbersome procedure including suitable approximations and averaging. It is thus not unnatural to consider that those might include what we would call "jumps", and in general the sort of general behavior that would look as an effective collapse of "the wave-function" as seen from the stand point of the effective description. Needless is to say that we have at this time no hope of being able to describe the above procedure in any detail, among other reasons because we do not have at the moment a fully satisfactory and workable theory of quantum gravity.

On the other hand the effective description is expected to lead in the appropriate limit to General Relativity as the description of space-time, and in the corresponding appropriate limit, to quantum field theory for the description of matter fields. We assume that the situation of interest (the inflationary regime of the early universe) lies in a region where both these descriptions are approximately valid, but where some modifications tied to the fact that these picture is only an effective one, need to be incorporated in a seemingly *ad hoc* manner. Of course we would be complete loss if we did not have any other guidance as to the nature of these modifications, but here is where we look in the opposite direction and guide ourselves on some empirical facts: The inflationary account of the origin of the seeds of cosmic structure works "almost well" and its defects might be dealt with by the introduction of one such additional feature to the picture. What is required is the assumption of the existence of a process that can take a symmetric (i.e H&I) state of a close system (the Universe), into a state with small departures from a symmetric state while the standard evolution would have preserved those symmetries. As we will see, a quantum mechanical collapse of a wave function seems to have the required characteristics, except that it is usually assumed to be associated with the interaction of the quantum mechanical system with an external classical "apparatus" or "observer". It is clear that in the situation at hand we can not call upon any such feature so we will assume that the feature in question appears as a self induced collapse, along the lines that have been suggested, based on quite different arguments, as a likely feature of quantum gravity by R. Penrose.

These observations and ideas lead us to consider, situations where a quantum treatment of other fields would be appropriate but an effective classical treatment of gravitation would be justified. That is the realm of semi-classical gravity that we will assume to be valid in our context except at those instants where would break down in association with the jumps or collapses of the state of the quantum field that we considered to be part of the effective description of underlying fundamental quantum theory containing gravitation.

In accordance with the ideas above we will use a semi-classical description of gravitation in interaction with quantum fields as reflected in the semi-classical Einstein's equation $R_{ab} - (1/2)g_{ab}R = 8\pi G\langle T_{ab}\rangle$ whereas the other fields are treated in the standard quantum field theory (in curved space-time) fashion. As indicated this could not hold when a quantum gravity induced collapse of the wave function occurs, at which time, the excitation of the fundamental quantum gravitational degrees of freedom must be taken into account, with the corresponding breakdown of the semi-classical approximation. The possible breakdown of the semi-classical approximation is formally represented by the presence of a term Q_{ab} in the semi-classical Einstein's equation which is supposed to become nonzero **only** during the collapse of the quantum mechanical wave function of the matter fields. Thus we write

$$R_{ab} - (1/2)g_{ab}R + Q_{ab} = 8\pi G \langle T_{ab} \rangle \tag{1}$$

Thus, we consider the development of the state of the universe during the time at which the seeds of structure emerge to be initially described by a H.& I. state for the gravitational and matter D.O.F. At some time, η_c, the quantum state of the matter fields reaches a stage whereby the corresponding state for the gravitational D.O.F. is forbidden, and a quantum collapse of the matter field wave function is triggered. This new state of the matter fields does no longer need to share the symmetries of the initial state, and by its connection to the gravitational D.O.F. now accurately described by Einstein's semi-classical equation leads to a geometry that is no longer homogeneous and isotropic.

That is the approach that is taken in the work presented here, where the intent will be to emphasize the phenomenology potential and the lessons that can be extracted from it in order to make it clear that the proposal can be viewed as a viable path to investigate aspects of the physical world that have for long seemed absolutely beyond reach.

3. The quantum origin of the seeds of cosmic structure

Next we give a short description of the analysis of the origin of the primordial cosmological inhomogeneities and anisotropies based on the ideas outlined above. The staring point is as usual the action of a scalar field coupled to gravity.

$$S = \int d^4x \sqrt{-g}[\frac{1}{16\pi G}R[g] - 1/2\nabla_a\phi\nabla_b\phi g^{ab} - V(\phi)], \tag{2}$$

where ϕ stands for the inflaton and for its potential V. One then splits both, metric and scalar field into a spatially homogeneous "background" part and an inhomogeneous part "fluctuation", i.e. the scalar field is written $\phi = \phi_0 + \delta\phi$, while the perturbed metric can, (after appropriate gauge fixing and by focusing on the scalar perturbation) be written

$$ds^2 = a(\eta)^2 \left[-(1+2\Psi)d\eta^2 + (1-2\Psi)\delta_{ij}dx^i dx^j\right], \tag{3}$$

where Ψ is the relevant perturbation called the "Newtonian potential".

The background solution corresponds to the standard inflationary cosmology during the inflationary era has a scale factor $a(\eta) = -\frac{1}{H_I\eta}$, with $H_I^2 \approx (8\pi/3)GV$while the scalar ϕ_0 field in the slow roll regime so $\dot{\phi}_0 = -(a^3/3\dot{a})V'$.

The perturbation of the scalar field leads to a perturbation of the energy momentum tensor, and thus Einstein's equations at lowest order lead to

$$\nabla^2\Psi = 4\pi G\dot{\phi}_0\delta\dot{\phi} \equiv s\delta\dot{\phi} \tag{4}$$

where $s = 4\pi G\dot{\phi}_0$.

We must now write the quantum theory of the rescaled the field $y = a\delta\phi$. For definiteness we consider the field in a box of side L, and write the field and momentum operators as

$$\hat{y}(\eta, \vec{x}) = \frac{1}{L^3}\sum_{\vec{k}} e^{i\vec{k}\cdot\vec{x}}\hat{y}_k(\eta), \qquad \hat{\pi}^{(y)}(\eta, \vec{x}) = \frac{1}{L^3}\sum_{\vec{k}} e^{i\vec{k}\cdot\vec{x}}\hat{\pi}_k(\eta), \tag{5}$$

where the sum is over the wave vectors $\vec{k}$ satisfying $k_i = 2\pi n_i/L$ for $i = 1,2,3$ with n_i integers, then, we write the fields in terms of the annihilation and creation operators $\hat{y}_k(\eta) \equiv y_k(\eta)\hat{a}_k + \bar{y}_k(\eta)\hat{a}^\dagger_{-k}$ and $\hat{\pi}_k(\eta) \equiv g_k(\eta)\hat{a}_k + \bar{g}_k(\eta)\hat{a}^\dagger_{-k}$ with

$$y_k^{(\pm)}(\eta) = \frac{1}{\sqrt{2k}}\left(1 \pm \frac{i}{\eta k}\right)\exp(\pm ik\eta), \qquad g_k^{\pm}(\eta) = \pm i\sqrt{\frac{k}{2}}\exp(\pm ik\eta). \tag{6}$$

Given that we are interested in considering a kind of self induced collapse which operates in close analogy with a "measurement" which normally involves self adjoint operators, we work find it convenient with the real and imaginary components of the fields and thus we write $\hat{y}_k(\eta) = \hat{y}_k{}^R(\eta) + i\hat{y}_k{}^I(\eta)$ and $\hat{\pi}_k(\eta) = \hat{\pi}_k{}^R(\eta) + i\hat{\pi}_k{}^I(\eta)$ where the operators $\hat{y}_k^{R,I}(\eta)$ and $\hat{\pi}_k^{R,I}(\eta)$ are hermitian. Let $|\Xi\rangle$ be any state in the Fock space of $\hat{y}$,and assign to each such state the following quantity: $d_k^{R,I} = \langle\hat{a}_k^{R,I}\rangle_\Xi$. The expectation values of the modes of the fundamental field operators are then expressible as

$$\langle\hat{y}_k{}^{R,I}\rangle_\Xi = \sqrt{2}\Re(y_k d_k^{R,I}), \qquad \langle\hat{\pi}_k^{(y)\,R,I}\rangle_\Xi = \sqrt{2}\Re(g_k d_k^{R,I}). \tag{7}$$

For the vacuum state $|0\rangle$ we have of course: $\langle\hat{y}_k{}^{R,I}\rangle_0 = 0, \langle\hat{\pi}_k^{(y)\,R,I}\rangle_0 = 0$, while their corresponding uncertainties are

$$(\Delta\hat{y}_k{}^{R,I})_0^2 = (1/2)|y_k|^2(\hbar L^3), \qquad (\Delta\hat{\pi}_k{}^{R,I})_0^2 = (1/2)|g_k|^2(\hbar L^3). \tag{8}$$

The collapse: Next we provide a simple specification of what we mean by "the collapse of the wave function" by stating the form collapsed state in terms of its collapse time. We assume the collapse to be analogous to some sort of imprecise measurement of the operators $\hat{y}_k^{R,I}(\eta)$ and $\hat{\pi}_k^{R,I}(\eta)$. In order to describe is the state $|\Theta\rangle$ after the collapse we must specify $d_k^{R,I} = \langle\Theta|\hat{a}_k^{R,I}|\Theta\rangle$. This is done by making the following assumption about the state $|\Theta\rangle$ after collapse:

$$\langle\hat{y}_k^{R,I}(\eta_k^c)\rangle_\Theta = x_{k,1}^{R,I}\sqrt{(\Delta\hat{y}_k^{R,I})_0^2} = x_{k,1}^{R,I}|y_k(\eta_k^c)|\sqrt{\hbar L^3/2}, \tag{9a}$$

$$\langle\hat{\pi}_k^{(y)\,R,I}(\eta_k^c)\rangle_\Theta = x_{k,2}^{R,I}\sqrt{(\Delta\hat{\pi}_k^{(y)R,I})_0^2} = x_{k,2}^{R,I}|g_k(\eta_k^c)|\sqrt{\hbar L^3/2}, \tag{9b}$$

where $x_{k,1}, x_{k,2}$ are selected randomly from within a Gaussian distribution centered at zero with spread one. We note that our universe, corresponds to a single realization of the random variables, and thus each of the quantities $x^{R,I}{}_{k,1,2}$ has a single specific value. Later, we will see how to make relatively specific predictions, despite these features.

The connection to gravitational sector is at the semi-classical level so Eq.(4) turns into

$$\nabla^2\Psi = s\langle\delta\dot{\phi}\rangle. \tag{10}$$

We note that before the collapse, the expectation value on the right hand side is zero. Next we determine what happens after the collapse: To this end, we need to solve the equations (9) for d_k and then substitute this in (7) and then in the Fourier transform of Eq.(10) and obtain

$$\Psi_k(\eta) = \frac{-s}{k^2}\langle\delta\dot{\phi}_k\rangle_\Theta = \frac{-s}{k^2}\sqrt{\hbar L^3}k\frac{1}{2a}F(k), \tag{11}$$

where

$$F(k) = (1/2)[A_k(x_{k,1}^R + ix_{k,1}^I) + B_k(x_{k,2}^R + ix_{k,2}^I)], \tag{12}$$

with

$$A_k = \frac{\sqrt{1+z_k^2}}{z_k}\sin(\Delta_k), \qquad B_k = \cos(\Delta_k) + (1/z_k)\sin(\Delta_k), \tag{13}$$

and where $\Delta_k = k\eta - z_k$ with $z_k = \eta_k^c k$.

Turning to the observational quantities we recall that the quantity that is measured is $\frac{\Delta T}{T}$ as a function of (θ,φ), the coordinates on the celestial two-sphere which is expressed as $\sum_{lm}\alpha_{lm}Y_{l,m}(\theta,\varphi)$. The angular variations of the temperature are then identified with the corresponding variations in the "Newtonian Potential" Ψ, by the understanding that they are the result of gravitational red-shift in the CMB photon frequency ν so $\frac{\delta T}{T} = \frac{\delta\nu}{\nu} = \frac{\delta(\sqrt{g_{00}})}{\sqrt{g_{00}}} \approx \Psi$ (we are ignoring, for simplicity the complications of the late time physics such as reheating or acoustic oscillations). Thus, the quantity of interest is the "Newtonian potential" on the surface of last scattering: $\Psi(\eta_D, \vec{x}_D)$, from where one extracts $a_{lm} = \int \Psi(\eta_D,\vec{x}_D)Y_{lm}^* d^2\Omega$. To evaluate the expected value for the quantity of interest we use (11) and (12) to write

$$\Psi(\eta,\vec{x}) = \sum_{\vec{k}} \frac{s}{k^2}\sqrt{\frac{\hbar k}{L^3}\frac{1}{2a}}F(\vec{k})e^{i\vec{k}\cdot\vec{x}}, \tag{14}$$

then, after some algebra we obtain

$$\alpha_{lm} = s\sqrt{\frac{\hbar}{L^3}\frac{1}{2a}}\sum_{\vec{k}}\frac{U(k)\sqrt{k}}{k^2}F(\vec{k})4\pi i^l j_l((|\vec{k}|R_D)Y_{lm}(\hat{k}), \tag{15}$$

where $\hat{k}$ indicates the direction of the vector $\vec{k}$. It is in this expression that the justification for the use of statistics becomes clear. The quantity we want to evaluate is the result of the combined contributions of an ensemble of collapsing harmonic oscillators each one contributing with a complex number to the sum, leading to what is in effect a bi-dimensional random walk whose total displacement corresponds to the observational quantity. We can not of curse evaluate such total displacement but only its most likely value We do so and then take the continuum limit and which after rescaling the variable of integration to $x = kR_D$, becomes

$$|\alpha_{lm}|^2_{M.L.} = \frac{s^2\hbar}{2\pi a^2}\int \frac{C(x/R_D)}{x^4} j_l^2(x)x^3 dx, \tag{16}$$

where

$$C(k) \equiv 1 + (2/z_k^2)\sin(\Delta_k)^2 + (1/z_k)\sin(2\Delta_k). \tag{17}$$

In the exponential expansion regime where μ vanishes and in the limit $z_k \to -\infty$ where $C = 1$, we find:

$$|\alpha_{lm}|^2_{M.L.} = \frac{s^2\hbar}{2a^2}\frac{1}{l(l+1)}. \tag{18}$$

which has the standard functional result. However we must consider the effect of the finite value of times of collapse η_k^c codified in the function $C(k)$. We note is that in order to get a reasonable spectrum there is a single simple option: That z_k be essentially independent of k that is the time of collapse of the different modes should depend on the mode's frequency according to $\eta_k^c = z/k$. There are of course other possible schemes of collapse and we have investigated the most natural ones and their corresponding effects on the primordial fluctuation spectrum, with results that to a large extent confirm that the above conclusion is rather robust (see figure 1, section 5 and specifically [21] for a deeper discussion). Thus we can conclude that the above pattern of times of collapse seems to be implied by the data (as far as our preliminary analysis has shown so far). In our view such conclusion represents one important and relevant piece of information about whatever the mechanism of collapse is.

Journal of Physics: Conference Series **174** (2009) 012059 doi:10.1088/1742-6596/174/1/012059

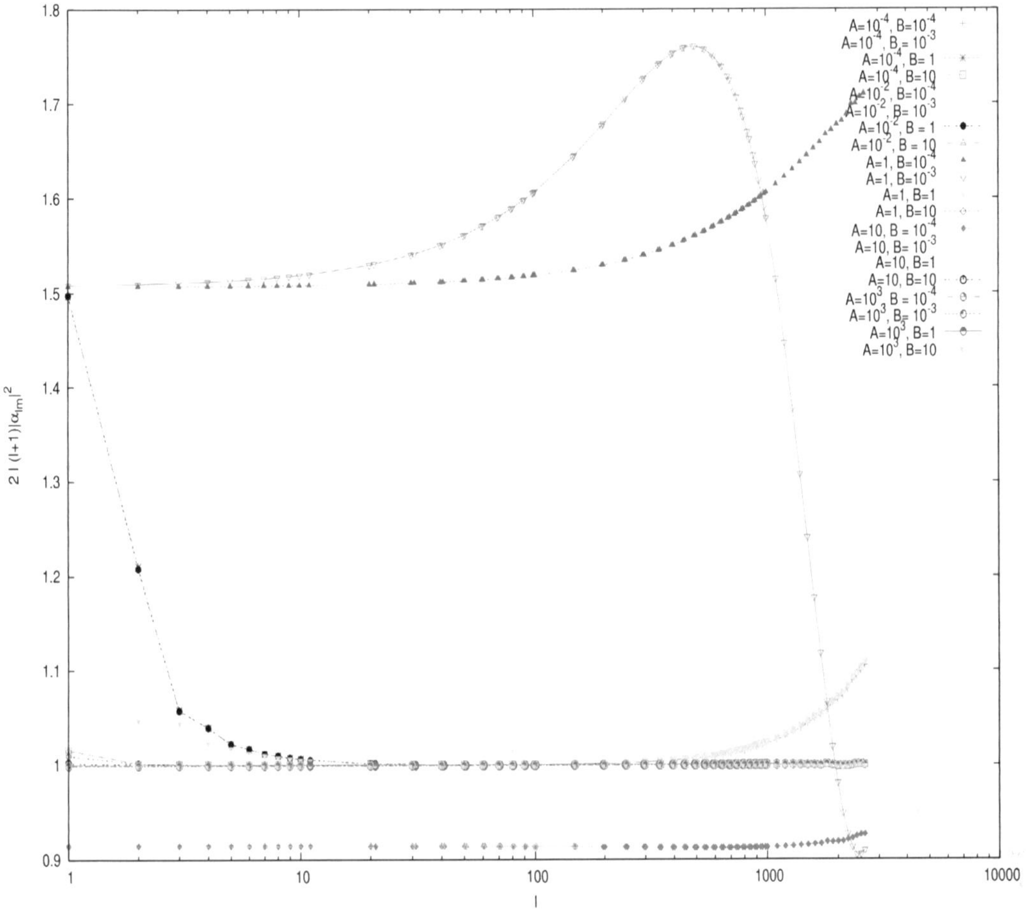

Figure 1. Semi-log plot of $|\alpha_{lm}|^2(C_1(k))$ assuming a small departure from the result "z_k independent of k". This small departure is represented by $\tilde{z}_x = A + Bx$, where $x = kR_D$, with R_D standing for the radius of last scattering surface. Here, we are showing different values of (A, B), representing how robust is the scheme of collapse under this circumstances. The abscissa ranges from $l = 0$ to $l = 2600$. See section 5 for more details about the meaning of the variables A and B.

4. A version of 'Penrose's mechanism' for collapse in the cosmological setting

Based on the analysis of the inadequacies of Quantum Mechanics as a complete theory of nature and the places from where solutions can arise, R. Penrose has argued that the collapse of quantum mechanical wave functions is an actual dynamical process, independent of observation, and that the fundamental physics is related to quantum gravity. More precisely, according to this suggestion, the collapse into one of several coexisting quantum mechanical alternatives would take place when the gravitational interaction energy between the alternatives exceeds a certain threshold. We have considered a naive realization of Penrose's ideas appropriate for the present setting to be as follows: Each mode would collapse by the action of the gravitational interaction between it's own possible realizations. In our case, one could estimate the interaction energy $E_I(k, \eta)$ by considering two representatives of the possible collapsed states on opposite sides of the Gaussian associated with the vacuum. We interpret Ψ, literally as the Newtonian potential and consequently , $\rho = a^{-2}\dot{\phi}_0\delta\dot{\phi}$ should be identified with matter density. Then the gravitational interaction energy between alternatives should be:

$$E_I(\eta) = \int \Psi^{(1)}\rho^{(2)}dV = (a/L^3)\dot{\phi}_0\Sigma_k\Psi_k^{(1)}(\eta)\delta\dot{\phi}_k^{(2)}(\eta), \tag{19}$$

where $(1), (2)$ refer to the two different realizations chosen. Recalling that $\Psi_k = (-s/k^2)\delta\dot{\phi}_k$, we find

$$E_I(\eta) = -4\pi G(a/L^3)\dot{\phi}_0^2\Sigma_k(1/k^2)\delta\dot{\phi}_k^{(1)}(\eta)\delta\dot{\phi}_k^{(2)}(\eta) \approx \Sigma_k(\pi\hbar G/ak)(\dot{\phi}_0)^2. \tag{20}$$

Where we have used equation (8), to estimate $\delta\dot{\phi}_k^{(1)}(\eta)\delta\dot{\phi}_k^{(2)}(\eta)$ by $| < \delta\dot{\phi}_k > |^2 = \hbar k L^3(1/2a)^2$.

This result can be interpreted as the sum of the contributions of each mode to the interaction energy of different alternatives. We view each mode's collapse as occurring independently although at this point it is rather unclear if this can be fully justified, and thus, the collapse of mode k would occur when this energy $E_I(k,\eta) = (\pi\hbar G/ak)(\dot{\phi}_0)^2 = \frac{\pi\hbar G}{9H_I^2}(a/k)(V')^2$ reaches the value of the Planck Mass M_p. Thus the condition determining the time of collapse η_k^c of the mode k becomes,

$$z_k = \eta_k^c k = \frac{\pi}{9}(\hbar V')^2(H_I M_p)^{-3} = \frac{\epsilon}{8\sqrt{6\pi}}(\tilde{V})^{1/2} \equiv z^c, \tag{21}$$

which is independent of k, and thus, as we saw in the previous section leads to a roughly scale invariant spectrum of fluctuations in accordance with observations.

5. Further phenomenological analysis

The scheme of collapse we have considered in section 3, and which will be referred as "the symmetric scheme" or scheme No 1, is evidently far from unique and other similarly natural schemes can be considered here we will briefly discuss two alternatives: "the momentum preferred scheme" or scheme No 2 and the "Wigner functional scheme" or scheme No 3. The first corresponds to the assumption that it is only the momentum conjugate mean value that changes during the collapse according to equation 9 while the field's expectation value maintains its initial value during the collapse, namely zero. The scheme No 3 corresponds to the assumption that after the collapse the expectation values of field and momentum modes, follow the correlations in the corresponding uncertainties that existed in the of the pre-collapse state, namely:

$$\left\langle \hat{y}_k^{(R,I)}(\eta_k^c)\right\rangle_\Omega = x_k^{(R,I)}\Lambda_k\cos\Theta_k\,, \quad \left\langle \hat{\pi}_k^{(R,I)}(\eta_k^c)\right\rangle_\Omega = x_k^{(R,I)}\Lambda_k k\sin\Theta_k, \tag{22}$$

where Λ_k is given by the major semi-axis of the ellipse characterizing the bi dimensional Gaussian function (the ellipse corresponds to the boundary of the region in "phase space" where the Wigner function has a magnitude larger than 1/2 its maximum value), and Θ_k is the angle between that axis and the $y_k^{R,I}$ axis.

The subsequent analysis proceeds in the same fashion as that presented in section 3 and the result has the same mathematical expression as in equation (16), with the sole exception being the exact expression of the function $C(k)$, which for the scheme no. 2 takes the form:

$$C_2(k) = 1 + \sin^2\Delta_k\left(1 - \frac{1}{z_k^2}\right) - \frac{1}{z_k}\sin(2\Delta_k), \tag{23}$$

and in the scheme no. 3 is:

$$C_3(k) = \frac{32z_k^2}{\mathcal{A}(1+5z_k^2-\mathcal{A})}\times$$
$$\left\{ [\mathcal{A}-1+3z_k^2]\left(\cos\Delta_k - \frac{\sin\Delta_k}{z_k}\right)^2 + \sin^2\Delta_k\left[\mathcal{A}-3z_k^2-7\right] + 8z_k\cos\Delta_k\sin\Delta_k\right\}. \tag{24}$$

with $\mathcal{A} = \sqrt{1 + 10z_k^2 + 9z_k^4}$. In [21] it was shown that, despite the fact that the expression for C_3 looks by far more complicated that C_2, their dependence in z_k is very similar, except for the amplitude of the oscillations.

These in turn lead to particular forms of the primordial spectrum (i.e the spectrum which emerges from inflation and has not yet been modified to include the late time physics such as the acoustic oscillations responsible for the famous peaks). At the approximation level we are working here the spectra would be all identical to the standard ($n = 1$) scale invariant Harrison-Zel'dovich (HZ) spectrum corresponding to a flat line in the graph of $2l(l+1)|\alpha_{lm}|^2$ vs. l, if we assume that z_k is independent of k (corresponding to a time of collapse of mode k given by $\eta_k^c = z/k$). Therefore, at this level the different collapse schemes are not distinguishable.

We thus were let to considered the sensitivity of the resulting spectrum to small deviations of the "z_k independent of k pattern" by studying a linear departure from the k independent z_k characterized by $\tilde{z}_x$ as $\tilde{z}_x = A + Bx$ ($x = kR_D$ where R_D stands for the radius of the surface of s last scattering) in order to examine the robustness of the various collapse schemes in as far as predicting the observational spectrum.

Some of the results of these analysis can be seen in the graphs 2, 3 in which we see that each collapse scheme leads to a particular pattern of modifications of the spectrum, clearly showing the potential of the present approach to teach us something about the effective collapse, or alternatively to account for possible deviations if anything of this sort were to be detected in future observations.

One of the most important predictions of the scheme, is the absence of tensor modes, or at least their very strong suppression. This can be understood by considering the semi-classical version of Einstein's equation and its role in describing the manner in which the inhomogeneities and anisotropies in the metric arise in our scheme. As indicated in the introduction, the metric is taken to be an effective description of the gravitational D.O.F., in the classical regime, and not as the fundamental D.O.F. susceptible to be described at the quantum level. It is thus the matter degrees of freedom (which in the present context are represented by the inflaton field) the ones that are described quantum mechanically and which, as a result of an hypothetical fundamental aspect of gravitation at the quantum level, would be subject to an effective quantum collapse (the reader should recall that our point of view is that gravitation at the quantum level will be drastically different from standard quantum theories, and that, in particular, it will not involve universal unitary evolution). This leads to a nontrivial value for $\langle T_{ab} \rangle$, which leads to the appearance of the metric fluctuations. The point is that the energy momentum tensor contains linear and quadratic terms in the expectation values of the quantum matter field fluctuations, which are the source terms determining the geometric perturbations. In the case of the scalar perturbations, there are first order contributions to the perturbed energy momentum tensor, which are proportional to $\dot{\phi}_0 \langle \delta\dot{\phi} \rangle$, while there are no similar first order terms that would appear as source of the tensor perturbations (i.e. of the gravitational waves). In the usual treatment, and besides its conceptual shortcomings, no such natural suppression of the tensor modes can be envisaged. At the time of the writing of this article, the tensor modes had not been detected, in contrast with the scalar modes.

6. Discussion

We have presented the first steps in the proposal involving the introduction of novel aspect (a self induce collapse of the wave function) of physics in the description of the emergence from the quantum uncertainties in the state of the inflaton field, of the seeds of structure in our universe. We have argued that such novel aspect is likely to be associated to the connection of a fundamental theory of quantum gravity, with the effective description in terms of the equations of semi-classical general relativity. We find it quite remarkable that in doing so, we are able to obtain a relatively satisfactory picture. We do not know what exactly is the physics of collapse

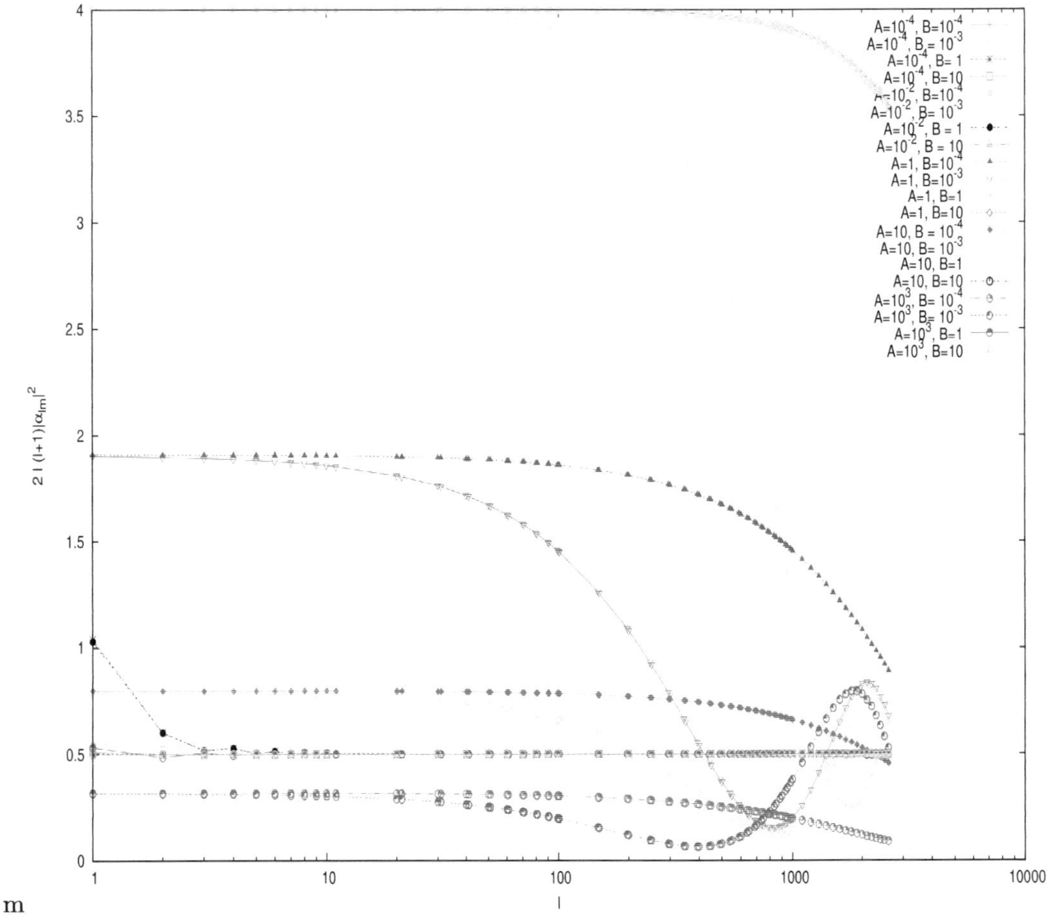

Figure 2. Semi-log plot of $|\alpha_{lm}|^2(C_2(k))$ for different values of (A, B), representing how robust is the scheme of collapse when it departs from z_k constant.The abscissa ranges from $l = 0$ to $l = 2600$

but we were nevertheless able to obtain some constraints on it (about the time of collapse of the different modes), and shown that a simplistic extrapolation of Penrose's ideas satisfy this constraint. We have not investigated the possible connection of our proposal with other more developed schemes involving similar non unitary modifications of quantum theory such as the various schemes considered by colleagues participating in this meeting and others. Te reason for that is that we found it better to try to extract some information about what would be needed for the scheme to work in the cosmological case on which we have centered our interest, and would hope to be able to explore the connections of our proposal, with such schemes an their compatibility of their implications with the conclusions extracted in this initial analysis.

We have reviewed the serious shortcoming of the inflationary account of the origin of cosmic structure, and have given a brief account of the proposals to deal with them which were first reported in [5]. These lines of inquiry have lead to the recognition that something else seems to be needed for the whole picture to work and that it could be pointing towards an actual manifestation quantum gravity. We have shown that not only the issues are susceptible of scientific investigation based on observations, but also that a simple account of what is needed, seems to be provided by the extrapolation of Penrose's ideas to the cosmological setting.

Journal of Physics: Conference Series **174** (2009) 012059 doi:10.1088/1742-6596/174/1/012059

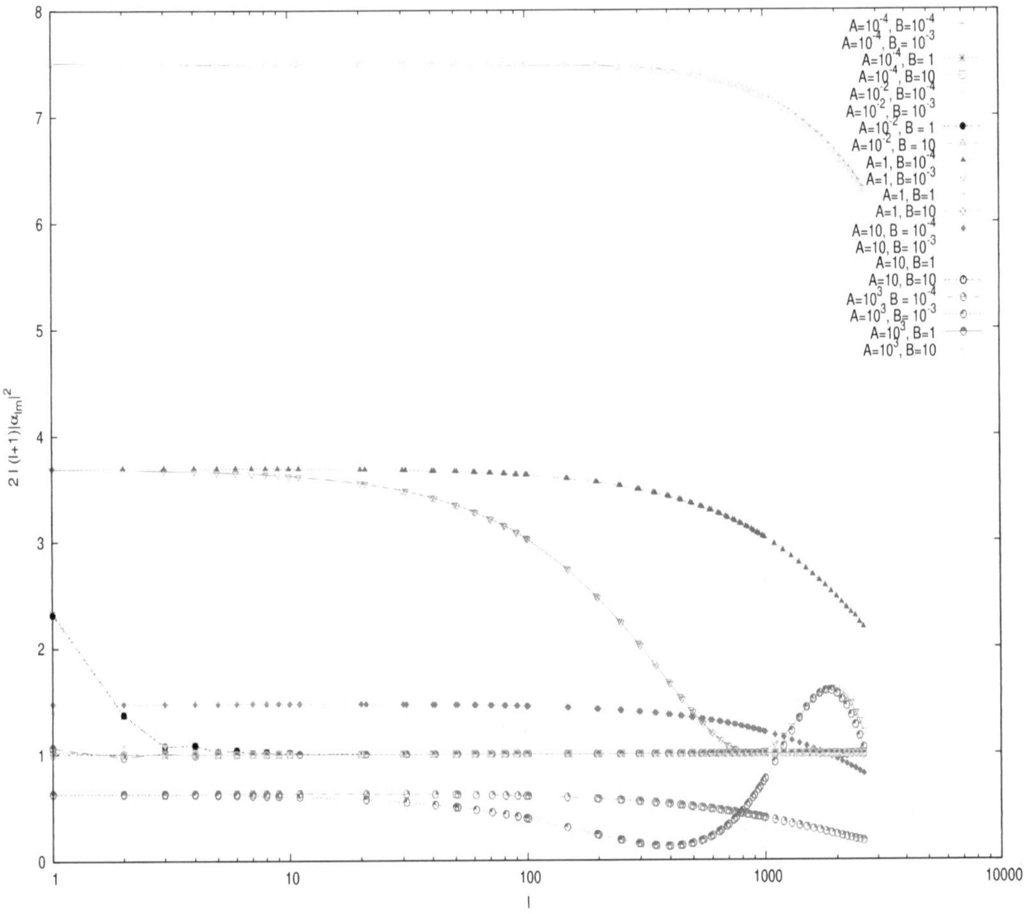

Figure 3. Semi-log plot of $|\alpha_{lm}|^2(C_3(k))$ for different values of (A, B), representing how robust is the scheme of collapse when it departs from z_k constant. The abscissa is l up to $l = 2600$.

Interestingly the scheme does in fact lead to some deviations from the standard picture where the metric and scalar field perturbations are quantized. For instance, as discussed in the last section, one is lead to expect no excitation of the tensor modes, something that we can expected to be able confront with relatively precise data in the near future.

We also find new avenues to address the fine tuning problem that affects most inflationary models, because one can follow in more detail the objects that give rise to the anisotropies and inhomogeneities, and by having the possibility to consider independently the issues relative to formation of the perturbation, and their evolution through the reheating era (for a more extended discussion of this point see [20]).

Other aspects that can, in principle, be tested, were discussed in the last section. The noteworthy fact is that what initially could have been thought to be essentially a philosophical problem, leads instead to truly physical issues.

Our main point is however that in our search for physical manifestations of new physics tied to quantum aspects of gravitation, we might have been ignoring what could be the most dramatic such occurrence: The cosmic structure of the Universe itself.

Acknowledgments
It is a pleasure to acknowledge very helpful conversations with J. Garriga, E. Verdaguer and A. Perez. This work was supported in part by DGAPA-UNAM IN108103 grant.

References
[1] Colella R, Overhauser A W and Werner S 1975 *Phys. Rev. Lett.* **34** 1472-1474
[2] Abele H, Bässler S and Westphal A 2003 *Quantum Gravity From Theory to Experimental Search* ed D Giulini, C Kiefer, and C Lämmerzahl (Berlin: Springer Verlag)
[3] Chryssomalakos C and Sudarsky D 2003 *Gen. Relat. Grav.* **35** 605-617
[4] Lange A E *et al* 2001 *Phys. Rev.* D **63** 042001;
Hinshaw G *et. al.* 2003 *Astrophys. J. Supp.* **148** 135;
Gorski K M *et al* 1996 *Astrophys. J.* **464** L11;
Bennett C L *et al* 2003 *Astrophys. J. Suppl.* **148** 1
[5] Perez A, Sahlmann H and Sudarsky D 2006 *Class. Quant. Grav.* **23** 2317-54 (*Preprint* arXiv:gr-qc/0508100)
[6] Perez A and Sudarsky A *Preprint in preparation* Shortcomings in the Understanding of Why Cosmological Perturbations Look Classical
[7] Corichi A, Ryan M and Sudarsky D 2002 *Mod. Phys. Lett.* A **17**, 555;
Sudarsky D 2008 *Int. J. Mod. Physics* D **17** 425 (*Preprint* arXiv:gr-qc/0712.3242)
[8] Polchinski J 1998 *String Theory, vol. 1: "An introduction to the Bosonic String"* and *String Theory, vol. 2: "Superstring Theory and Beyond"* (Cambridge, UK: Cambridge U. Press)
[9] Bombelli L, Lee J H, Meyer D and Sorkin R 1987 *Phys. Rev. Lett.* **59** 521;
Sorkin R 2002 in "Valdivia 2002, Lectures on quantum gravity", pp 305-327 (*Preprint* arXiv:gr-qc/0309009)
[10] Dewitt B S 1967 *Phys. Rev.* **160** 1113 (1967)
Wheeler J A, 1968 *Battelle Rencontres: 1967 Lectures in Mathematics and Physics* ed C DeWitt and J A Wheeler (New York: W.A. Benjamin)
[11] Rovelli C 1998 *Living Rev. Rel.* **1** 1 (*Preprint* arXiv:gr-qc/9710008);
Ashtekar A 2001 Quantum geometry and gravity: Recent advance *Preprint* arXiv:gr-qc/0112038;
Thiemann T 2001 Introduction to modern canonical quantum general relativity *Preprint* arXiv:gr-qc/0110034
[12] Penrose R 1989 *The Emperor's New Mind* (Oxford U. Press);
Penrose R 1996 On Gravity's Role in Quantum State Reduction *Gen. Rel. Grav.* **28** 581-600, reprinted in *Physics meets philosophy at the Planck scale* ed C Callender pp 290–304
[13] Gambini R and Pullin J 2004 *Phys. Rev. Lett.* **93** 240401;
Gambini R, Porto R A and Pullin J 2004 *Phys. Rev.* D **70** 124001 (*Preprint* arXiv:gr-qc/0408050)
[14] Hartle J B, Hawking S W and Hertog T 2008 *Phys. Rev.* D **77** 123537
[15] Gambini R, Porto R and Pullin J 2004 *New J. Phys.* **6** 45 (*Preprint* arXiv:gr-qc/0402118)
[16] Halliwell J J 1989 *Phys. Rev.* D **39** 2912;
Kiefer C 2000 *Nucl. Phys. Proc. Suppl.* **88** 255;
Polarski D and Starobinsky A A 1996 Semiclasicallity and decoherence of Cosmological perturbations *Preprint* arXiv:gr-qc/9504030;
Zurek W H 1990, Environment Induced Superselection In Cosmology *Moscow 1990, Proceedings, Quantum gravity* (QC178:S4:1990) 456-72, see High Energy Physics Index 30 (1992) No. 624;
Laflamme R and Matacz A 1993 *Int. J. Mod. Phys.* D **2** 171;
Castagnino M and Lombardi O 2003 *Int. J. Theor. Phys.* **42** 1281;
Lombardo F C and Lopez Nacir D 2005 *Phys. Rev.* D **72** 063506;
Martin J 2005 *Lect. Notes Phys.* **669** 199;
Kiefer C, Lohmar I, Polarski D and Starobinsky A A 2006 *Preprint* arXiv:astro-ph/0610700
[17] Padmanabhan T 1996 *Cosmology and Astrophysics Through Problems* (Cambridge, UK: Cambridge U. Press);
Hartle J B 2005 Generalized Quantum mechanics for Quantum Gravity *Preprint* arXiv:gr-qc/0510126
[18] Mukhanov V 2005 *Physical Foundations of Cosmology* (Cambridge, UK: Cambridge U. Press) 348
[19] Ghirardi G C, Rimini A and Weber T 1986 *Phys. Rev.* D **34** 470;
Bassi A 2007 *J. Phys.: Conf. Ser.* **67** 012013 (*Preprint* arXiv:quant-ph/0701014v2);
Bassi A and Ghirardi G C 2003 Dynamical Reduction Models *Preprint* arXiv:quant-ph/0302164v2
[20] Sudarsky D 2007 *J. Phys.: Conf. Ser.* **68** 012029 (*Preprint* arXiv:gr-qc/0612005)
[21] De Unanue A and Sudarsky D 2008 *Phys. Rev.* D **78** 043510 (*Preprint* arXiv:gr-qc/0801.4702)

Fourth International Workshop DICE2008 IOP Publishing
Journal of Physics: Conference Series **174** (2009) 012060 doi:10.1088/1742-6596/174/1/012060

Coupling dark energy with Standard Model states

M C Bento[1,2], A E Bernardini[3] and O Bertolami[1,4]

[1] Departamento de Física, Instituto Superior Técnico, Avenida Rovisco Pais 1, 1049-001 Lisboa, Portugal.
[2] Centro de Física Teórica de Partículas, Instituto Superior Técnico, Avenida Rovisco Pais 1, 1049-001 Lisboa, Portugal.
[3] Departamento de Física, Universidade Federal de São Carlos, PO Box 676, 13595-905, São Carlos, SP, Brasil.
[4] Instituto de Plasmas e Fusão Nuclear, Instituto Superior Técnico, Avenida Rovisco Pais 1, 1049-001 Lisboa, Portugal.

E-mail: orfeu@cosmos.ist.utl.pt

Abstract. In this contribution one examines the coupling of dark energy to the gauge fields, to neutrinos, and to the Higgs field. In the first case, one shows how a putative evolution of the fundamental couplings of strong and weak interactions via coupling to dark energy through a generalized Bekenstein-type model may cause deviations on the statistical nuclear decay Rutherford-Soddy law. Existing bounds for the weak interaction exclude any significant deviation. For neutrinos, a perturbative approach is developed which allows for considering viable varying mass neutrino models coupled to any quintessence-type field. The generalized Chaplygin model is considered as an example. For the coupling with the Higgs field one obtains an interesting cosmological solution which includes the unification of dark energy and dark matter.

1. Introduction

The nature of the dark energy and its connection with the Standard Model (SM) states is one of the most intriguing issues related with the negative pressure component required to understand the accelerated expansion of the universe.

A putative time variation of the electromagnetic coupling on cosmological time scales from the observation of absorption spectra of quasars [1] was recently reported and has led to a revival of interest on ideas about the variation of fundamental couplings (see Ref. [2] for a review). This variation is fairly natural in unification models with extra dimensions. Moreover, if there is a variation of the electromagnetic coupling, it is rather natural to expected variations of other gauge or Yukawa couplings as well. It is therefore of great interest to confirm these observations and to propose realistic particle physics models consistent with a cosmological variation of the fundamental couplings. One starts analyzing how a putative variation of the strong and weak couplings due to coupling to dark energy may affect the nuclear decay Rutherford-Soddy law [3]. In order to do that one considers the available bounds on the variation of strong and weak couplings and a generalized version of the Bekenstein model [4].

Given that observations suggest that the variation of the electromagnetic coupling is a late event in the history of the universe it is natural to associate it to the observed late accelerated expansion of the universe [5]. This late time acceleration of the expansion can be driven by a cosmological constant (see [6] and references therein), by a scalar field with a suitable potential, quintessence [7], or by a fluid with an exotic equation of state like the generalized Chaplygin gas (GCG) [8]. For the analysis of the implications that dark energy may have on the Rutherford-Soddy law one considers the variation of the strong and weak couplings in the context of an exponential potential quintessence model and of the GCG model.

The GCG model considers an exotic perfect fluid described by the equation of state [8]

$$p = -\frac{A}{\rho_{ch}^{\alpha}} , \tag{1}$$

where A is a positive constant and α is a constant in the range $0 \leq \alpha \leq 1$. For $\alpha = 1$, the equation of state is reduced to the Chaplygin gas scenario. The covariant conservation of the energy-momentum tensor for an homogeneous and isotropic spacetime implies that

$$\rho_{ch} = \rho_{ch0}\left[A_s + \frac{(1-A_s)}{a^{3(1+\alpha)}}\right]^{\frac{1}{1+\alpha}} , \tag{2}$$

where $A_s = A/\rho_{ch0}^{(1+\alpha)}$, ρ_{ch0} is the present energy density of the GCG and a the scale factor of the universe. Hence, one can see that at early times the energy density behaves as matter while at late times it behaves like a cosmological constant. This dual role is at the core of the surprising properties of the GCG model. Moreover, this dependence with the scale factor indicates that the GCG model can be interpreted as an entangled admixture of dark matter and dark energy.

This model has been thoroughly studied from the observational point of view and it is shown to be compatible with the Cosmic Microwave Background Radiation (CMBR) peak location and amplitudes [9, 10], with SNe Ia data [11, 12, 13], gravitational lensing statistics [14], cosmic topology [15], gamma-ray bursts [16] and variation of the electromagnetic coupling [17]. The issue of structure formation and its difficulties [18] have been recently addressed [19]. Most recent analysis based on CMBR data indicates that $\alpha < 0.25$ and $A_s > 0.93$ [10].

One also studies the coupling of dark energy to neutrinos in the context of the so-called mass varying neutrino (MaVaN) models [20, 21, 22, 23, 24]. This possibility is particularly interesting since the coupling of neutrinos to the dark energy scalar field component may lead to a number of significant phenomenological consequences. If the neutrino mass m_{ν} is generated by the dynamical value of a cosmologically active scalar field ϕ it would be an evolving quantity.

Hence one considers the possibility that neutrino masses arise from an interaction with the scalar field that drives the accelerated expansion of the universe. One considers the MaVaN contribution to the energy conservation equation of the cosmic fluid as a perturbation that iteratively modifies the background fluid equation of state [24].

Dark matter is most often not considered in the formulation of MaVaN models; however, the possibility of treating dark energy and dark matter in a unified scheme naturally offers this possibility. The GCG is particularly relevant in this respect, as already mentioned, it is consistent with known observational constraints. Our analysis considers the coupling of MaVaN's to the underlying scalar field of the GCG model. Since the neutrino contribution is perturbative, one obtains a small deviation from the stability condition characteristic of the unperturbed GCG equation of state [24].

Finally, one also considers a possible coupling between the Higgs boson and dark energy [25]. In addition to unraveling the mass generation for fermions and electroweak gauge bosons, the Higgs boson may also be a portal to new physics hitherto hidden in a SM singlet sector [26].

The most important consequences of this extension, both cosmological and phenomenological, have been analyzed. The phenomenological implications can arise in two ways: mixing of the singlet with the Higgs boson and the possibility of invisible decay of the Higgs boson into two singlet bosons [26, 27, 28, 29, 30]. A coupling of the form $\Phi H^{\dagger}H$, where H is the Higgs doublet and Φ is the scalar singlet field, will generate a mixing between the physical Higgs boson and the singlet after spontaneous electroweak symmetry breaking. In addition, a quartic coupling like $\Phi\Phi H^\dagger H$ results in the possibility of Higgs boson decay into a pair of singlets and, in the case of a non-zero vacuum expectation value of the singlet, a mixing can also be induced. The invisible decay of the Higgs boson could be detected through the weak gauge boson fusion at the Large Hadron Collider (LHC) [31]

From the cosmological point of view, an ultra-light scalar singlet is stable enough to be a good dark matter candidate, most often referred to as phion [28, 30, 32, 33, 35, 37]. Bounds on the phion-Higgs coupling can be obtained from the constraints on the phion relic abundance.

One repeats here the analysis of Ref. [25] on the consequences of identifying the scalar singlet not only with the dark matter, but also with the dark energy field, responsible for the recent stage of accelerated expansion of the universe. It is shown that it is possible to obtain, under conditions, an unified picture of dark matter and dark energy, where dark energy is the zero-mode classical field rolling the usual quintessence potential and the dark matter candidate is the quantum excitation (particle) of the field, which is generated in the universe due to its coupling to the Higgs boson.

2. Coupling to gauge fields

As an example of the coupling of dark energy to gauge fields, one considers the impact that this putative coupling might have on the well established experimental fact that the nuclear decay rate of any nuclei is described by the statistical Rutherford-Soddy law:

$$\frac{dN}{dt} = -\lambda N \quad , \tag{3}$$

where $\lambda = \frac{1}{\tau_0}$, τ_0 being the nuclei's lifetime. For a constant λ, which reflects the fact that the rate of transformation of an element is constant under all conditions [38], one obtains the well known exponential decay law:

$$N(t) = N_i e^{-\lambda(t-t_i)} \quad , \tag{4}$$

where N_i is the nuclei's number counting at t_i. This statistical law is a consequence of radiative processes of electromagnetic, strong and weak nature that take place within the nuclei. Thus, τ_0 is related to the amplitude of the relevant decay process being therefore a function of order γ of the coupling constant α_g:

$$\tau_0 = A\alpha_g^\gamma \quad , \tag{5}$$

with $\alpha_S(E) = g_s^2(E)/\hbar c$ for the strong interactions, g_s being the strong coupling constant, $\alpha_W = G_F m_p^2 c/\hbar^3$ the weak coupling constant, G_F being Fermi's constant and m_p the proton mass. Parameter A is a constant related to a specific process, suitable integration over the phase space, binding energy, quark masses, etc.

On the other hand, one expects that if dark energy couples with the whole gauge sector of the Standard Model, this coupling can be modelled by a generalization of the so-called Bekenstein model [4]

$$\mathcal{L}_{gauge} = -\frac{1}{16\pi} f(\phi) F^a_{\mu\nu} F^{\mu\nu a} \; , \tag{6}$$

where $f(\phi)$ is an arbitrary function of the dark energy field, ϕ, and $F^a_{\mu\nu}$ the gauge field strength. Given that the variation of the gauge couplings is presumably small (c.f. below), one expands this function to first order

$$f(\phi) = \frac{1}{\alpha_i}\left[1 + \zeta\left(\frac{\phi - \phi_i}{M}\right)\right] \; , \tag{7}$$

where α_i is the initial value of the gauge structure constant, ζ is a constant, ϕ_i is the initial value of the quintessence field and M a characteristic mass scale of the dark energy model. We point out that a model with a quadratic variation is discussed in Ref. [39]. For the linear model it follows that

$$[f(\phi)]^{-1} \simeq \alpha_i \left[1 - \zeta\left(\frac{\phi - \phi_i}{M}\right)\right] . \tag{8}$$

Thus, the gauge coupling evolution is given by

$$\alpha_g = [f(\phi)]^{-1} \simeq \alpha_i \left[1 - \zeta\left(\frac{\phi - \phi_i}{M}\right)\right] , \tag{9}$$

hence, for its variation, one obtains

$$\frac{\Delta\alpha_g}{\alpha_g} = \zeta\left(\frac{\phi - \phi_i}{M}\right) . \tag{10}$$

For the electromagnetic interaction one should take into account the Equivalence Principle limits, which implies [40]

$$\zeta \leq 7 \times 10^{-4} . \tag{11}$$

Of course, this bound does not apply to short range interactions. However, as will be seen, this parameter must be constrained in order to ensure that strong and weak gauge coupling do not change significantly so to ensure the experimental standing of the Rutherford-Soddy law.

In what follows one considers only strong and weak decays given that the effects of a putative variation of the electromagnetic coupling have been the subject of various studies (see e.g. Ref. [41] and references therein). Thus, for strong and weak decays, keeping quark masses unchanged and disregarding the running of the couplings, one can write for τ:

$$\tau_{DE} = Af(\phi)^{-\gamma} , \tag{12}$$

and since $\zeta << 1$

$$\tau_{DE} \simeq \tau_0 \left[1 - \gamma\zeta_{S,W}\left(\frac{\phi_0 - \phi_i}{M}\right)\right] , \tag{13}$$

depending on whether the process is driven by strong or weak interactions, where ϕ_0 refers to the value of the dark energy scalar field at present. Introducing this expression into the rate of change of atoms

$$\frac{dN}{dt} \simeq -\frac{N}{\tau_0}\left[1 + \gamma\zeta_{S,W}\left(\frac{\phi_0 - \phi_i}{M}\right)\right] , \tag{14}$$

from which follows that [3]

$$N(t) \simeq N_i \exp\left[-\lambda(\Delta t + \gamma\zeta_{S,W} I)\right] , \tag{15}$$

where

$$I = \int_{t_i}^{t_0} \left(\frac{\phi_0 - \phi_i}{M}\right) dt , \tag{16}$$

and $\Delta t = t_0 - t_i$.

One considers now the most stringent bounds on the variation of the gauge couplings (latest bounds on the variation of the electromagnetic coupling can be found in [41] and references therein). In what concerns the strong interaction, considerations on the stability of two-nucleon systems yield the bound [2]

$$\left|\frac{\Delta\alpha_S}{\alpha_S}\right| < 4 \times 10^{-2} . \tag{17}$$

On its hand, for weak interactions, considerations on the β-decay rates of ^{187}Re lead to the bound [42, 43]

$$\frac{\Delta\alpha_W}{\alpha_W} < 3 \times 10^{-7} \ . \tag{18}$$

In order to make quantitative predictions concerning the level of validity of the statistical Rutherford-Soddy decay law, one examines two observationally viable dark energy models.

For simplicity, one starts considering the exponential potential $V(\phi) = V_o e^{-\beta\phi/M_P}$, in the case that the contribution of the scalar field is dominant, there is a family of solutions [44]

$$\phi(t) = \phi_0 + \frac{2}{\beta}\ln(tM_P) \ , \tag{19}$$

$$\phi_0 = \frac{2}{\beta}\ln\left(\frac{V_o\beta^2}{2M_P^4(6-\beta^2)}\right) \ , \tag{20}$$

$$\rho_\phi \propto \frac{1}{a^{\beta^2}} \qquad a \propto t^{2/\beta^2} \ . \tag{21}$$

in terms of the scale factor of the universe. For $\beta < \sqrt{6}$, ϕ_0 can always be chosen to be zero by redefining the origin of ϕ, in which case $V_0 = \frac{2}{\beta^2}(\frac{6}{\beta^2} - 1)M_P^4$. The parameters of this model can be set to satisfy all phenomenological constraints (see e.g. last reference in [7]).

One also considers the GCG model. Following Refs. [12, 24], one describes the GCG through a real scalar field. One starts with the Lagrangian density

$$\mathcal{L} = \frac{1}{2}\dot{\phi}^2 - V(\phi) \ , \tag{22}$$

where the potential for a flat, homogeneous and isotropic universe has the following form

$$V = \frac{1}{2}A_s^{\frac{1}{1+\alpha}}\rho_0\left[\cosh(\beta\phi)^{\frac{2}{\alpha+1}} + \cosh(\beta\phi)^{-\frac{2\alpha}{\alpha+1}}\right] \ , \tag{23}$$

where $\beta = 3(\alpha+1)/2$.

For the energy density of the field one has

$$\rho_\phi = \tfrac{1}{2}\dot{\phi}^2 + V(\phi) = \left[A + \frac{B}{a^{3(1+\alpha)}}\right]^{\frac{1}{1+\alpha}} = \rho_{ch} \ . \tag{24}$$

The Friedmann equation can be written as

$$H^2 = \frac{8\pi G}{3}\rho_\phi \tag{25}$$

and thus

$$H(t) = H_0\Omega_\phi^{1/2} \tag{26}$$

where Ω_ϕ can be written as

$$\Omega_\phi = \left[A_s + \frac{(1-A_s)}{a^{3(1+\alpha)}}\right]^{\frac{1}{3(1+\alpha)}} \ . \tag{27}$$

Using Eq. (1) to obtain the pressure and integrating one obtains

$$\phi(a) = -\frac{1}{2\beta}\ln\left[\frac{\sqrt{1-A_s(1-a^{2\beta})} - \sqrt{1-A_s}}{\sqrt{1-A_s(1-a^{2\beta})} + \sqrt{1-A_s}}\right] \ . \tag{28}$$

The integral I, Eq. (16), has been computed for the quintessence model with an exponential potential (Model I) and for the GCG model (Model II) and the implications for the variation of the Rutherford-Soddy decay law for strong and weak interactions have been examined. For Model I one has set $\beta^2 = 2$ and that at present the scalar field contribution to the energy density is 70% of the critical density. For Model II, one has considered the values $\alpha = 0.2$, $A_s = 0.95$, and used the relationship $1 + z = 1/a$ to express the integral I in terms of the red-shift.

For the strong interaction decay one considers the $^{238}_{92}U$ nucleus for which $\tau_0 = 4.468 \times 10^9$ years. For the weak interaction decay, oone uses data from the β^- decay of the ^{187}Re for which $\tau_0 = 4.558 \times 10^9$ years. Using bounds (17), (18), and the integration interval equal to τ_0 we compute $\zeta_{S,W}$ and the respective relative deviation from the Rutherford-Soddy law N/N_{RS} for the case of weak and strong interactions. Our results are shown in the following Table for $\gamma = 2$:

	I (yrs)	$\mid \zeta_S \mid$	$\mid \zeta_W \mid$	$\mid \frac{N}{N_{RS}} \mid_S -1 \mid$	$\mid \frac{N}{N_{RS}} \mid_W -1 \mid$
Model I	1.22×10^9	7.1×10^{-2}	5.24×10^{-7}	3.8×10^{-2}	2.79×10^{-7}
Model II	1.08×10^9	2.27×10^{-1}	1.7×10^{-6}	1.04×10^{-1}	8.1×10^{-7}

3. Coupling to neutrinos

The existence of a cosmological neutrino background is a firm prediction of the cosmological standard model, hence any hint about this component of the universe energy density is quite relevant.

The neutrino energy density and pressure are expressed through a Fermi-Dirac distribution function without a chemical potential, $f(q)$, where $q \equiv \frac{|\boldsymbol{p}|}{T_{\nu 0}}$, and sub-index 0 denotes present values. Assuming a flat Friedman-Robertson-Walker cosmology with $a_0 = 1$, one has

$$\rho_\nu(a,\phi) = \frac{T_{\nu 0}^4}{\pi^2 a^4} \int_0^\infty dq\, q^2 \left(q^2 + \frac{m^2(\phi)\, a^2}{T_{\nu 0}^2}\right)^{1/2} f(q), \tag{29}$$

$$p_\nu(a,\phi) = \frac{T_{\nu 0}^4}{3\pi^2 a^4} \int_0^\infty dq\, q^4 \left(q^2 + \frac{m^2(\phi)\, a^2}{T_{\nu 0}^2}\right)^{1/2} f(q).$$

By observing that

$$m_\nu(\phi) \frac{\partial \rho_\nu(a,\phi)}{\partial m_\nu(\phi)} = \left(\rho_\nu(a,\phi) - 3p_\nu(a,\phi)\right), \tag{30}$$

and from Eq. (30), one obtains the energy-momentum conservation for the neutrino fluid

$$\dot\rho_\nu(a,\phi) + 3H(\rho_\nu(a,\phi) + p_\nu(a,\phi)) = \dot\phi \frac{dm_\nu(\phi)}{d\phi} \frac{\partial \rho_\nu(a,\phi)}{\partial m_\nu(\phi)}, \tag{31}$$

where $H = \dot a/a$ is the expansion rate of the universe and the *dot* denotes differentiation with respect to cosmic time.

The coupling between cosmological neutrinos and the scalar field as specified in Eq. (30) is restricted to times when neutrinos are non-relativistic (NR), i. e. $\frac{\partial \rho_\nu(a,\phi)}{\partial m_\nu(\phi)} \simeq n_\nu(a) \propto a^{-3}$ [21, 23, 45]. On the other hand, as long as neutrinos are relativistic ($T_\nu(a) = T_{\nu 0}/a >>$

$m_\nu(\phi(a))$), the decoupled fluids evolve adiabatically since the strength of the coupling is suppressed by the relativistic increase of pressure ($\rho_\nu \sim 3p_\nu$).

Treating the system of NR neutrinos and the scalar field as a single unified fluid (UF) which adiabatically expands with energy density $\rho_{UF} = \rho_\nu + \rho_\phi$ and pressure $p_{UF} = p_\nu + p_\phi$ leads to

$$\dot{\rho}_{UF} + 3H(\rho_{UF} + p_{UF}) = 0 \quad \Rightarrow \quad \dot{\rho}_\phi + 3H(\rho_\phi + p_\phi) = -\dot{\phi}\frac{dm_\nu}{d\phi}\frac{\partial \rho_\nu}{\partial m_\nu}, \tag{32}$$

where the last step is derived from the substitution of Eq. (31) into Eq. (32).

It is well known that the relative contribution of the energy densities components of the universe with respect to the one of the dark energy is on its own a problem. The assumptions proposed in Ref. [21] and subsequently developed elsewhere [23, 45, 46, 48] introduce a stationary condition (SC) which allows circumventing the coincidence problem for cosmological neutrinos, by imposing that the dark energy is always diluted at the same rate as the neutrino fluid, that is,

$$\frac{dV(\phi)}{d\phi} = -\frac{dm_\nu}{d\phi}\frac{\partial \rho_\nu}{\partial m_\nu}. \tag{33}$$

This condition introduces a constraint on the neutrino mass since it promotes it into a dynamical quantity, as indicated in Eq. (32). In this context, the main feature of the scenario of Ref. [21] is that, in what concerns to dark sector, it is equivalent to a cosmological constant-like equation of state and an energy density that is as a function of the neutrino mass [24].

At our approach, the effect of the coupling of the neutrino fluid to the scalar field fluid is quantified by a linear perturbation $\epsilon\phi$ ($|\epsilon| << 1$) such that $\phi \rightarrow \varphi \approx (1+\epsilon)\phi$. It then follows a novel equation for energy conservation

$$\ddot{\varphi} + 3H\dot{\varphi} + \frac{dV(\varphi)}{d\varphi} = -\frac{dm_\nu}{d\varphi}\frac{\partial \rho_\nu}{\partial m_\nu}. \tag{34}$$

After some straightforward manipulation one obtains for the value of the coefficient of the perturbation [24, 26]

$$\epsilon \simeq \frac{-\frac{dm_\nu}{d\phi}\frac{\partial \rho_\nu}{\partial m_\nu}}{\left[\phi^2 \frac{d}{d\phi}\left(\frac{1}{\phi}\frac{dV(\phi)}{d\phi}\right)\right]}, \tag{35}$$

for which the condition $|\epsilon| << 1$ is required. Upon fulfilling all known phenomenological requirements, the above result allows for addressing a wide class of scalar field potentials and related equations of state for various candidates for the dark sector (dark energy and dark matter), which through the SC would be incompatible with realistic neutrino mass generation models.

In order to verify under which conditions Eq. (33) agrees with the proposed perturbative approach for a given background equation of state, the coefficient of the linear perturbation should be given by

$$\epsilon \simeq \frac{\frac{dV(\phi)}{d\phi}}{\left[\phi^2 \frac{d}{d\phi}\left(\frac{1}{\phi}\frac{dV(\phi)}{d\phi}\right)\right]} \qquad (|\epsilon| << 1). \tag{36}$$

This means that one must search for a neutrino mass dependence on the scale factor for which the above condition is satisfied. Thus, once one sets the equation of state for the dark sector, there will be a period at late times for which the SC and the perturbative approach match. In particular, this feature can be reproduced by the GCG equation of state.

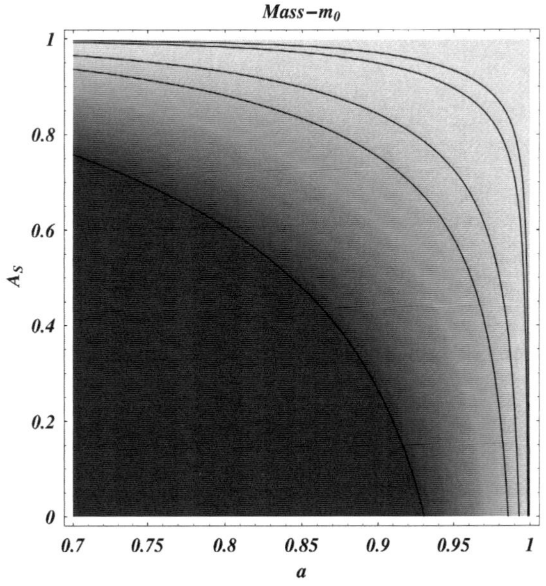

Figure 1. Present-day values of the neutrino mass m_0 and the corresponding values of a_{NR} for which the transition between the NR and UR regimes takes place in the GCG scenario with $\alpha = 1/2$ and variable A_s. The increasing *graylevel* corresponds to increasing values of m_0, for the boundary values $m_0 = 0.05\,eV$, $0.1\,eV$, $0.5\,eV$, $1\,eV$, $5\,eV$, respectively.

Given a potential, the explicit dependence of m_ϕ on the scale factor can be immediately obtained from Eq. (33). Furthermore, it is necessary to determine for which values of the scale factor the neutrino-scalar field coupling becomes important. For convenience one sets the value of $a = a_{NR}$ for which $\rho_{\nu,NR} = \rho_{\nu,UR}$ holds, usually established by the condition of $m_\nu \gtrsim T_\nu$, that parameterizes the transition between the NR and the ultra-relativistic (UR) regime. In fact, this takes place when

$$m_\nu(a) = m_0(\phi_0/\phi(a))^n = \chi \frac{T_{\nu,0}}{a} \tag{37}$$

where χ is a numerical factor estimated to be about $\chi \simeq 94$ considering that $\rho_\nu/\rho_{\text{Crit}} = m_0\,[eV]/(94\,h^2\,[eV])$, where h is the value of the Hubble constant in terms of $100\,km\,s^{-1}\,Mpc^{-1}$. Such a correspondence between a_{NR} and m_0 is illustrated in the Fig. 1 for the case of $\alpha = 1/2$. Considering the whole set of parameters that characterize the background fluid, one notices that it is rather difficult to see that the maximal value assumed by the ϵ parameter corresponds to its present-day value.

One observes that the interval of parameters A_s, m_0 and eventually α, for which our approximation can be applied ($\epsilon < 1$), is valid for $a > a_{NR}$ and severally constrained by the condition $a_{NR} \lesssim 1$. For values of A_s ($0.7 \lesssim A_s \lesssim 1$) [9, 13] one finds that $0.1 < \epsilon \lesssim 1$. Just under quite special circumstances the usual SC and the perturbative contribution of MaVaN's match. In the original MaVaN scenario [21], the SC corresponds to the adiabatic solution ($H^2 \ll \mathrm{d}^2V/\mathrm{d}\phi^2$) of the scalar field equation of motion. In this case, the kinetic energy terms of the scalar field can be safely neglected. The consistency of our perturbative scenario with the stationary condition can be achieved only when the kinetic energy contribution is not relevant at late times.

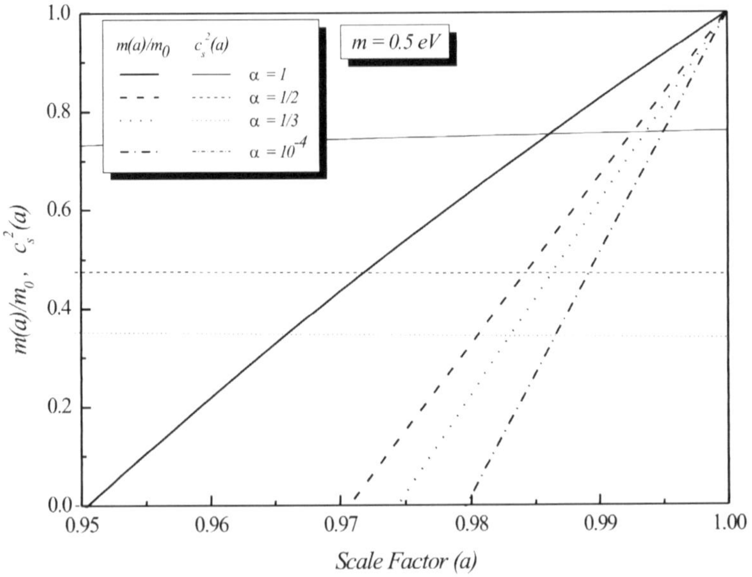

Figure 2. Model independent perturbative modification on squared speed of sound c_s^2 as a function of the scale factor for the neutrino-GCG coupled fluid in comparison with the adiabatic GCG fluid for $A_s = 0.7$ with $\alpha = 1, 1/2,\ 1/3,\ 10^{-4}$, for a present-day value of neutrino mass, $m_0 = 0.5\,eV$.

For $A_s = 0$, the GCG behaves always as matter, whereas for $A_s = 1$, it behaves always as a cosmological constant. Consequently, it is natural that the relevance of the kinetic energy term at present times is suppressed when the parameter A_s gets close to unity, which further ensures the agreement between the perturbative approach and the SC analysis.

Fig. 2 illustrates the results for an increasing neutrino mass with the scale factor for a set of phenomenologically consistent parameters in the context of the GCG model. Interestingly, for $m_0 = 0.5\,eV$, a fairly typical value, one can see that stable MaVaN perturbations correspond to a well defined effective squared speed of sound,

$$c_s^2 \simeq \frac{dp_\phi}{d(\rho_\phi + \rho_\nu)} > 0. \tag{38}$$

The greater the m_0 values, the more important are the corrections to the squared speed of sound, up to the limit where the perturbative approach breaks down. However, one finds that as far as the perturbative approach is concerned, our model does not run into stability problems in the NR neutrino regime. In opposition, in the SC treatment, where neutrinos are just coupled to dark energy, cosmic expansion in combination with gravitational drag due to cold dark matter have a major impact on the stability of MaVaN models. Usually, for a general fluid for which the equation of state is known, the dominant behaviour on c_s^2 arises from the dark sector component and not by the neutrino component. For the models where the SC (cf. Eq. (33)) implies in a cosmological constant-type equation of state, $p_\phi = -\rho_\phi$, one inevitably obtains $c_s^2 = -1$.

Thus, the perturbative approach is in agreement with the assumption that the coupling between neutrinos and dark energy (and/or dark matter) is weak. It is found that the stability condition related to the squared speed of sound of the coupled fluid is predominantly governed by dark energy equation of state. This is rather similar to the dynamics of the weakly coupled *cosmon* fields [47]. Actually, such a troublesome behaviour should have already been observed as the SC implies that $c_s^2 = -1$ from the very start, and the role of recovering the stability is relegated to the neutrino contribution [48]. The loosening of the stationary constraint Eq. (33) emerges from the dynamical dependence on φ, more concretely due to a kinetic energy component [24]. The knowledge of the background fluid equation of state for the dark sector (the GCG in our example), and the criterion for the applicability of the perturbative approach, do allow to overcome the c_s^2 negative problem, independently from the neutrino mass dependence set by the SC.

4. Coupling to the Higgs field

In order to study the coupling of dark energy to the Higgs field one considers first the coupling of a very light scalar to the Higgs field. One examines a model where the fields are already in their minimum energy configuration (contrary to the usual quintessence models, where the fields are displaced from their minimum). As an example one adopts the minimal spontaneously broken hidden sector model of Schabinger and Wells [29] involving the usual Higgs doublet H and a further singlet complex scalar field Φ with a potential

$$V(\Phi, H) = -m_H^2|H|^2 - m_\Phi^2|\Phi|^2 + \lambda|H|^4 + \rho|\Phi|^4 + \eta|H|^2|\Phi|^2. \tag{39}$$

The fields develop non-vanishing vacuum expectation values, $\langle|H|^2\rangle = v^2/2$ and $\langle|\Phi|^2\rangle = \xi^2/2$, and in unitary gauge, the physical fields $h = H - v/\sqrt{2}$ (the Higgs boson field) and $\phi = \Phi - \xi/\sqrt{2}$ mix through a non-diagonal mass matrix due to a non-vanishing η. The mixing angle ω between the two scalars is given by

$$\tan\omega = \frac{\eta v \xi}{(\rho\xi^2 - \lambda v^2) \pm \sqrt{(\rho\xi^2 - \lambda v^2)^2 + \eta^2 v^2 \xi^2}} \,. \tag{40}$$

In order to derive bounds on the mixing ω one considers fifth force constraints [49]. The mixing of the SM Higgs field with a light scalar induces a long range Yukawa-like interaction due to the exchange of an ultra-light singlet field with strength $g_e \sin\omega$ for electrons and $g_N \sin\omega$ for nucleons, where $g_e = 2.9 \times 10^{-6}$ is the electron Yukawa coupling and $g_N = 2.1 \times 10^{-3}$ is the nucleon Yukawa coupling [50].

Thus, a non-relativistic test body of mass M placed in the gravitational field of the Earth (with mass M_E) at a distance r from its center undergoes an acceleration given by:

$$a = a_{gr} + a_\phi \,, \tag{41}$$

where

$$a_{gr} = \frac{M_E}{M_{Pl}^2 r^2} \,, \tag{42}$$

is the usual Newtonian acceleration, M_{Pl} is the Planck mass and

$$a_\phi = \frac{\omega^2}{Mr^2}[g_N^2 N_N^E N_N^t + g_N g_e(N_N^E N_e^t + N_e^E N_N^t) + g_e^2 N_e^E N_e^t] \tag{43}$$

is the extra acceleration due to a new force arising from ϕ exchange, as long as $m_\Phi r \ll 1$; $N_{(N,e)}^{(E,t)}$ is the number of nucleons or electrons in the Earth and in the test body.

With these definitions one finds that the difference in acceleration between two test bodies of distinct compositions is given by

$$\varepsilon = 2\frac{|a_1 - a_2|}{|a_1 + a_2|} = \frac{M_{Pl}}{\bar{m}}\omega^2 g_N g_e \Delta f_p \tag{44}$$

where $\bar{m}$ is the average nucleon mass and Δf_p is the difference in isotopic composition of the two test bodies:

$$\Delta f_p = \frac{N_p^{(1)}}{N_p^{(1)} + N_n^{(1)}} - \frac{N_p^{(2)}}{N_p^{(2)} + N_n^{(2)}} \tag{45}$$

with $N_{(p,n)}$ being the number of protons or neutrons in the test body.

For typical isotopic differences of the order of $\Delta f_p = \mathcal{O}(10^{-1})$ and using the experimental result $\varepsilon < \mathcal{O}(10^{-13})$ [51, 52] one gets an estimate for upper bound on the mixing angle

$$\omega < \mathcal{O}(10^{-20}) \ . \tag{46}$$

Hence the mixing of an ultra-light scalar singlet is severely constrained by fifth force-type experiments. However, one should point out that this is expected in the discussed model, since in the limit $\xi \ll v$ the mixing angle is given by

$$|\omega| \simeq \frac{\eta}{2\lambda}\frac{\xi}{v} \simeq \eta\frac{m_\phi}{m_H} \tag{47}$$

where one assumes quartic couplings of $\mathcal{O}(1)$. In order for this limit to apply, the Compton length of the light scalar should be of the order of the Earth radius, implying in a mass of the order of 10^{-14} eV, to be compared with a Higgs mass of the order of 100 GeV. This leads to $\omega \simeq \eta \times 10^{-25}$. Therefore, one concludes that there are no limits on the mixing constant η, which implies that it is possible that the Higgs may have a large invisible width into very light particles.

It is also possible that these light scalars are dark matter candidates. Since in this case there is no symmetry preventing ϕ decay, one must check whether in fact they can survive till present. The light scalar decays through its mixing with the Higgs boson. If $m_\phi < 2m_e$, the dominant decay rate is into photons dominated by a top quark triangle ($g_{\phi\bar{t}t} = \omega g_{h\bar{t}t}$), which can be readily estimated:

$$\Gamma(\phi \to \gamma\gamma) \simeq \omega^2 \left(\frac{\alpha}{\pi}\right)^2 \frac{G_F m_\phi^3}{36\sqrt{2}\pi} \tag{48}$$

which implies that

$$\Gamma(\phi \to \gamma\gamma)^{-1} \simeq \frac{10^{-3}}{\omega^2 (m_\phi/1\ \mathrm{eV})^3} t_U \tag{49}$$

where $t_U \simeq 10^{17}s$ is the age of the universe. Hence, in order for the light scalar to survive till present one must require

$$\omega^2 \left(\frac{m_\phi}{1\ eV}\right)^3 < 10^{-3} \tag{50}$$

which is easily satisfied for the considered case.

Hence one should consider the bounds discussed in Ref. [36], whose reasoning shows that the scalar ϕ cannot be a dark matter candidate. In order to do that one considers the cosmological evolution of the light scalar field. If η is sufficiently small, the singlet field decouples early in the thermal history of the universe and is diluted by subsequent entropy production. In Ref. [34], one has considered out-of-equilibrium singlet production via inflaton decay in the context of $N = 1$ supergravity inflationary models (see. e.g. [36] and references therein).

On the other hand, for certain values of the coupling η, it is possible that Φ particles are in thermal equilibrium with ordinary matter. In order to determine whether this is the case, one makes the usual comparison between the thermalization rate Γ_{th} and the expansion rate of the universe, H. In Ref. [34] it was found that if $\eta > 10^{-10}$ the singlets can be brought into thermal equilibrium right after the electroweak phase transition, being therefore as abundant as photons.

Since one is interested in a stable ultra-light singlet field, it is a major concern avoiding its overproduction if it decouples while relativistic. In fact, in this case, there is an analogue of Lee-Weinberg limit for neutrinos (see e.g. [53]):

$$\Omega_\phi h^2 \simeq 0.08 \left(\frac{m_\phi}{1 \text{ eV}}\right) \quad , \tag{51}$$

which shows that as $m_\phi \simeq 10^{-14}$ eV, the thermally produced ultra-light scalar ϕ particles cannot be a dark matter candidate, as for it is required that $\Omega_\phi h^2 \simeq 0.1$. Of course, this model requires a considerable fine-tuning to generate the different scalar mass scales and it should be embedded in a more encompassing model, such as the Minimal Supersymmetric Standard Model with the addition of one singlet chiral superfield [54].

In what follows one considers the case where the Higgs is coupled to a quintessence field and pay special attention to the fact that it is not settled in its minimum. It is in this respect that a realistic dark energy candidate differs from the generic dark matter candidate modelled by a scalar Φ considered above .

The most natural potential coupling the Higgs field to the quintessence field is of the "hybrid inflation" type [55]:

$$V(\Phi, H) = U(\Phi) + \lambda\left(|H|^2 - \frac{v^2}{2}\right)^2 + \lambda_1 \Phi^2 |H|^2 \tag{52}$$

where $U(\Phi)$ is the usual quintessential-type potential which can, for instance, be of the form of an exponential or an inverse power law in the singlet field [56, 57]. The second term is the SM Higgs potential and the last term gives rise to Higgs-quintessence coupling, where λ_1 is a dimensionless coupling constant.

Working in the unitary gauge and denoting $\tilde{h}$ as the Higgs field, one can expand the fields around their classical, homogeneous vacuum configurations:

$$\Phi(\vec{x}, t) = \phi_c(t) + \varphi(\vec{x}, t); \;\; \tilde{h}(\vec{x}, t) = \frac{1}{\sqrt{2}}(\tilde{h}_c(t) + h(\vec{x}, t)) \tag{53}$$

with $\langle \Phi(\vec{x}, t)\rangle = \phi_c(t)$, $\langle \tilde{h}(\vec{x}, t)\rangle = \tilde{h}_c(t)/\sqrt{2}$ and the physical fields have zero vacuum expectation values, $\langle \varphi(\vec{x}, t)\rangle = \langle h(\vec{x}, t))\rangle = 0$.

The classical field configurations obey the equations of motion:

$$\begin{aligned} \ddot{\phi}_c + 3H\dot{\phi}_c + \left.\frac{dU}{d\Phi}\right|_{\Phi=\phi_c} + \lambda_1 \phi_c \tilde{h}_c^2 &= 0 \\ \ddot{\tilde{h}}_c + 3H\dot{\tilde{h}}_c + \lambda(\tilde{h}_c^2 - v^2)\tilde{h}_c + 2\lambda_1 \phi_c^2 \tilde{h}_c &= 0. \end{aligned} \tag{54}$$

Since this is a model of dynamical dark energy, the classical quintessence field has not relaxed to the minimum of its potential which causes the Higgs field to be displaced from the usual vacuum expectation value ($\tilde{h}_c \neq v$, where $v = 246$ GeV). This results in a time-varying weak scale or G_F, for which there are very strong bounds [58, 59] (see also the bound Eq. (18)).

In order to guarantee that $\tilde{h}_c = v$, one considers the following *ad hoc* modifications to the last term of the potential Eq. (52), representing the interaction between the Higgs and quintessence fields:

$$\begin{aligned} V(\Phi, H)_{int} &= \lambda_1 \Phi^2 (|H| - v/\sqrt{2})^2 \qquad (A) \\ V(\Phi, H)_{int} &= \lambda_1 (\Phi - \phi_c)^2 |H|^2 \qquad (B) \\ V(\Phi, H)_{int} &= \lambda_1 (\Phi - \phi_c)^2 (|H| - v/\sqrt{2})^2 \qquad (C) \end{aligned} \tag{55}$$

For case (A) the equations of motion become:

$$
\begin{aligned}
\ddot{\phi}_c + 3H\dot{\phi}_c + \left.\frac{dU}{d\Phi}\right|_{\Phi=\phi_c} + \lambda_1\phi_c(\tilde{h}_c - v)^2 &= 0 \\
\ddot{\tilde{h}}_c + 3H\dot{\tilde{h}}_c + \lambda(\tilde{h}_c^2 - v^2)\tilde{h}_c + \sqrt{2}\lambda_1\phi_c^2(\tilde{h}_c - v) &= 0
\end{aligned}
\tag{56}
$$

and hence $\tilde{h}_c = v$ is a solution in which case the classical quintessence field obeys the usual equation:

$$
\ddot{\phi}_c + 3H\dot{\phi}_c + \left.\frac{dU}{d\Phi}\right|_{\Phi=\phi_c} = 0. \tag{57}
$$

It is interesting to notice that this vacuum state for the Φ field does not arise from a mechanism of spontaneous symmetry breaking, but is instead a consequence of initial conditions for the slow roll evolution of the quintessence field. For most of the models ϕ_c evolves essentially as a logarithm of the cosmic time, so that its typical rate of change is the Hubble time:

$$
\frac{\dot{\phi}_c}{\phi_c} \sim H\ . \tag{58}
$$

Thus, ϕ_c evolves in a quasi-stationary regime set by the Hubble scale.

The Lagrangian density for the physical fields is given by

$$
\mathcal{L} = \frac{1}{2}\left[(\dot{\phi}_c + \dot{\varphi})^2 - (\vec{\nabla}\varphi)^2\right] + \frac{1}{2}(\partial_\mu h)^2 - V(\phi_c, v, \varphi, h) \tag{59}
$$

where

$$
\begin{aligned}
V(\phi_c, v, \varphi, h) &= U(\phi_c + \varphi) + \lambda\left(\frac{1}{2}(h+v)^2 - v^2/2\right)^2 + \\
&\quad \tfrac{1}{2}\lambda_1(\phi_c + \varphi)^2 h^2.
\end{aligned}
\tag{60}
$$

Writing

$$
U(\phi_c + \varphi) = U(\phi_c) + U'(\phi_c)\varphi + \tfrac{1}{2}U''(\phi_c)\varphi^2 + \tfrac{1}{6}U'''(\phi_c)\varphi^3 + \cdots \tag{61}
$$

and using the classical equation of motion to cancel the tadpole in Eq. (61) with the $\dot{\phi}_c\dot{\varphi}$ term in Eq. (59) one finally obtains:

$$
\begin{aligned}
\mathcal{L} &= \left[\tfrac{1}{2}\dot{\phi}_c^2 - U(\phi_c)\right] + \tfrac{1}{2}(\partial_\mu\varphi)^2 + \tfrac{1}{2}(\partial_\mu h)^2 - \tfrac{1}{2}m_\varphi^2\varphi^2 - \\
&\quad \lambda(h^2/2 + vh)^2 - \tfrac{1}{2}\lambda_1(\phi_c + \varphi)^2 h^2 - \tfrac{1}{6}g\varphi^3 + \mathcal{O}(\varphi^4)
\end{aligned}
\tag{62}
$$

where $m_\varphi^2 = U''(\phi_c)$ and $g = U'''(\phi_c)$.

The term in brackets in Eq. (62) reflects the classical zero mode contribution of the scalar field to the energy density of the universe which causes its accelerated expansion. The other terms express the behaviour of the quantum excitations (particles) in the universe. In particular, the Higgs-quintessence interaction results in a new, time-dependent contribution to the Higgs boson mass given by:

$$
m_h^2 = 2\lambda v^2 + \lambda_1\phi_c^2\ . \tag{63}
$$

Since in usual quintessence models $\phi_c = \mathcal{O}(M_{Pl})$ one expects a very large deviation from the Standard Model value of the Higgs mass unless the coupling constant λ_1 is very tiny,

$\lambda_1 = \mathcal{O}(v^2/M_{Pl}^2)$ practically closing the Higgs portal. Notice, however, that unitarity in longitudinal gauge boson scattering is not violated since the relevant couplings are still small even if the Higgs boson mass is large.

In case (B) the only difference arises in the interaction term, which becomes

$$\mathcal{L}_{int} = -\tfrac{1}{2}\lambda_1\varphi^2(v+h)^2 \tag{64}$$

and there is no new contribution to the Higgs mass. Interestingly, in this case the coupling generates a new contribution to the mass of the quintessence particle φ:

$$m_\varphi^2 = U''(\phi_c) + \lambda_1 v^2 \tag{65}$$

that, contrary to usual quintessence models, does not have to be small, since it does not affect the evolution of $\phi_c(t)$ given by Eq. (57). Furthermore, since the original quintessence field is not coupled to ordinary matter and there is no mixing to the Higgs boson, there are no bounds arising from fifth force constraints. Notice that in strict terms, this choice renders an ill defined field theory as the φ mass will vary on a cosmological scale. However, for models where the acceleration of the expansion is transient, it is possible that $U''(\phi_c)$ vanishes around present time (see e.g. Refs. [60, 61, 62]). Under this condition, this case also introduces a trilinear coupling $\varphi^2 h$ which, in addition to generating an invisible Higgs boson decay, can bring the quintessence particle into thermal equilibrium in the early universe. The analysis of the contribution arising from this quintessence particle to the dark matter component follows that of Ref. [33, 36, 37]. An interesting situation occurs for the natural case of $\lambda_1 = \mathcal{O}(0.04)$, which implies $m_\varphi \simeq 50$ GeV. In this case, the φ particle can be brought into thermal equilibrium through the Higgs coupling in the early universe and, for a Higgs mass $m_H \simeq 140$ GeV it decouples while non-relativistic, generating the dark matter density in the observed range.

Therefore, in this unified picture, dark energy is the zero-mode classical field $\phi_c(t)$ rolling down the usual quintessence potential and the dark matter candidate is the quantum excitation (particle) φ, which is produced in the universe due to its coupling to the Higgs boson.

Finally, in case (C) there are no new contributions to either to the Higgs or to the quintessence particle. Only a quartic interaction $\varphi^2 h^2$ is generated and the quintessence particle turns out not to be a good dark matter candidate.

5. Conclusions

In this contribution, the implications of coupling dark energy to SM fields, namely gauge fields, neutrinos and the Higgs field, are examined. One has first considered a possible coupling of dark energy to gauge fields and the impact it might have on the nuclear decay Rutherford-Soddy law. From the available bounds on the variation of the strong and weak gauge couplings it has been estimated, the deviations from the Rutherford-Soddy law. It has been considered a linear model for the variation of the couplings. One finds that the stringent bound on the variation of the weak coupling arising from ^{187}Re implies in significant constraints on any variation of the Rutherford-Soddy law for weak interactions, actually at the level 3×10^{-7} for the quintessence field with an exponential potential and at 8×10^{-7} level for the GCG model. For strong interactions, the bounds are much less stringent and are about 4×10^{-2} for the quintessence model with exponential potential and of order 10^{-1} for the GCG model. Stronger bounds on the deviation of the nuclear decay law for strong interactions could be obtained if variations of the strong coupling were tighter. Furthermore, it is interesting that somewhat different deviations are found for distinct dark energy models. This may be relevant to distinguish them given that most often they are degenerate with respect to the observable cosmological parameters.

In what concerns the coupling of neutrinos to dark energy, one finds that models with increasing neutrino mass with the scale factor is phenomenologically consistent in the context of the GCG model. For $m_0 = 0.5\,eV$, a fairly typical value, one can see that stable MaVaN perturbations correspond to a well defined effective squared speed of sound. The greater the m_0 values are, the more important are the corrections to the squared speed of sound, up to the limit where the perturbative approach breaks down [24].

As for the coupling of dark energy to the Higgs field, the implications of a scenario where the Higgs boson is coupled to a SM singlet field responsible for the accelerated expansion of the universe were examined. The most natural "hybrid-like" potential is disfavoured since it leads to a time-variation of the weak scale. A modification of the potential can however, give origin to the quite interesting possibility where the classical zero-mode component of the singlet field corresponds to the dark energy particle while its excitation plays the role of dark matter. In order to make this excitation consistent with the cosmological density requirement $\Omega_\varphi h^2 \simeq 0.1$ one gets a coupling to the Higgs field $\lambda_1 = \mathcal{O}(10^{-2})$, which implies $m_\varphi = \mathcal{O}(10)$ GeV. Quite interestingly, this scenario can be be scrutinized in the forthcoming generation of accelerators through the invisible decay of the Higgs boson [25].

Acknowledgments
O.B would like to thank R. Rosenfeld for collaboration on the work reported on Section 4. He would also like to thank T. Elze for setting up so successfully DICE2008 and for creating such a nice atmosphere for discussion and fruitful exchange of ideas.

References

[1] Webb J K *et al* 1999 *Phys. Rev. Lett.* **82** 884; *do.* 2001 *Phys. Rev. Lett.* **87** 091301; Murphy M T, Webb J K, Flambaum V V and Curran S J 2003 *Astrophys. Space Sci* **283** 577
[2] Uzan J P 2003 *Rev. Mod. Phys.* **75** 403
[3] Bento M C and Bertolami O 2009 *Preprint* arXiv:0901.1818
[4] Bekenstein J D 1982 *Phys. Rev.* D **25** 1527
[5] Perlmutter S J *et al* (Supernova Cosmology Project) 1999 *Ap. J.* **517** 565; Riess A G *et al* (High-Z Supernova Search Team) 1998 *Astron. J.* **116** 1009; Garnavich P M *et al* 1998 *Ap. J.* **509** 74; Tonry J L *et al* (Supernova Search Team Collaboration) 2003 *Ap. J.* **594** 1.
[6] Bento M C and Bertolami O 1999, *Gen. Rel. Gravitation* **31** 1461; Bento M C, Bertolami O and Silva P T 2001 *Phys. Lett.* B **498** 62
[7] Bronstein M 1993 *Phys. Zeit. Sowjet Union* **3** 73; Bertolami O 1986 *Il Nuovo Cimento* **93** B 36; *do.* 1986 *Fortsch. Phys.* **34** 829; Ozer M and Taha M O 1987 *Nucl. Phys.* B **287** 776; Ratra B and Peebles P J E 1988 *Phys. Rev.* D **37** 3406; *do.* 1988 *Ap. J. Lett.* **325** 117; Wetterich C 1988 *Nucl. Phys.* B **302** 668; Caldwell R R, Dave R and Steinhardt P J 1998 *Phys. Rev. Lett.* **80** 1582; Zlatev I, Wang L and Steinhardt P J 1999 *Phys. Rev. Lett.* **82** 986; Uzan J P 1999 *Phys. Rev.* D **59** 123510; Bertolami O and Martins P J 2000 *Phys. Rev.* D **61** 064007; Fujii Y 2000 *Phys. Rev.* D **61** 023504; Sen A A, Sen S and Sethi S 2001 *Phys. Rev.* D **63** 107501; Bento M C, Bertolami O and Santos N C 2002 *Phys. Rev.* D **65** 067301
[8] Kamenshchik A, Moshella U and Pasquier V 2001 *Phys. Lett.* B **511** 265; Bilić N, Tupper G B and Viollier R D 2002 *Phys. Lett.* B **535** 17; Bento M C, Bertolami O and Sen A A 2002 *Phys. Rev.* D **66** 043507
[9] Bento M C, Bertolami O and Sen A A 2003 *Phys. Rev.* D **67** 063003; *do.* 2003 *Phys. Lett.* B **575** 172; *do.* 2003 *Gen. Relativity and Gravitation* **35** 2063
[10] Barreiro T, Bertolami O and Torres P 2008 *Phys. Rev.* D **78** 043530
[11] Alcaniz J S, Jain D and Dev A 2003 *Phys. Rev.* D **67** 043514
[12] Bertolami O, Sen A A, Sen S and Silva P T 2004 *Mon. Not. Roy. Astron. Soc.* **353** 329
[13] Bento M C, Bertolami O, Sen A A and Santos N C 2005 *Phys. Rev.* D **71** 063501
[14] Silva P T and Bertolami O 2003 *Ap. J.* **599** 829; Dev A, Jain D, Upadhyaya D D and Alcaniz J S 2004 *Astron. Astrophys.* **417** 847
[15] Bento M C, Bertolami O, Rebouças M J and Silva P T 2006 *Phys. Rev.* D **73** 043504
[16] Bertolami O and Silva P T 2006 *Mon. Not. Roy. Astron. Soc.* **365** 1149
[17] Bento M C, Bertolami O and Torres P 2007 *Phys. Lett.* B **648** 14
[18] Sandvik H, Tegmark M, Zaldarriaga M and Waga I 2004 *Phys. Rev.* D **69** 123524
[19] Bento M C, Bertolami O and Sen A A 2004 *Phys. Rev.* D **70** 083519
[20] Gu P, Wang X and Zhang X 2003 *Phys. Rev.* D **68** 087301
[21] Fardon R, Nelson A E and Weiner N 2004 *JCAP* **0410** 005

[22] Lesgourgues J and Pastor S 2006 *Phys. Rep.* **429** 307
[23] Bjaelde O E *et al* 2008 *JCAP* **0801** 026
[24] Bernardini A E and Bertolami O 2008 *Phys. Rev.* D **77** 083506; *do.* 2008 *Phys. Lett.* B **662** 97
[25] Bertolami O and Rosenfeld R 2008 *Int. J. Mod. Phys.* A **23** 4817
[26] Patt B and Wilczek F 2006 *Preprint* arXiv:hep-ph/0605188
[27] van der Bij J J 2006 *Phys. Lett.* B **636** 56
[28] Binoth T and van der Bij J J 1997 *Z. Phys.* C **75** 17
[29] Schabinger R and Wells J D 2005 *Phys. Rev.* D **72** 093007
[30] Barger V, Langacker P, McCaskey M, Ramsey-Musolf M J and Shaughnessy G 2008 *Phys. Rev.* D **77** 035005
[31] Eboli O J P and Zeppenfeld D 2000 *Phys. Lett.* B **495** 147
[32] Silveira V and Zee A 1985 *Phys. Lett.* B **161** 136
[33] McDonald J 1994 *Phys. Rev.* D **50** 3637
[34] Bento M C, Bertolami O, Rosenfeld R and Teodoro L 2000 *Phys. Rev.* D **62** 041302
[35] Holz D E and Zee A 2001 *Phys. Lett.* B **517** 239
[36] Bento M C, Bertolami O and Rosenfeld R 2001 *Phys. Lett.* B **518** 276
[37] Burgess C P, Pospelov M and ter Veldhuis T 2001 *Nucl. Phys.* B **619** 709
[38] Rutherford S E, Chadwick J and Ellis C 1930 *Radiations from Radioactive Substances* (Cambridge, UK: Cambridge U. Press)
[39] Olive K A and Pospelov M 2008 *Phys. Rev.* D **77** 043524
[40] Olive K A and Pospelov M 2002 *Phys. Rev.* D **65** 085044
[41] Bento M C, Bertolami O and Santos N M C 2004 *Phys. Rev.* D **70** 107304
[42] Olive K A, Pospelov M, Qian Y-Z, Manhes G, Vangioni-Flam E, Coc A and Casse M 2004 *Phys. Rev.* D **69** 027701; Y. Fujii and A. Iwamoto 2003 *Phys. Rev. Lett.* **91** 261101
[43] Dent T, Stern S and Wetterich C 2008 *Preprint* arXiv:0809.4628
[44] Ferreira P G and Joyce M 1998 *Phys. Rev.* D **58** 023508
[45] Peccei R D 2005 *Phys. Rev.* D **71** 023527
[46] Brookfield A W, van de Bruck C, Mota D F and Tocchini-Valentini D 2006 *Phys. Rev.* D **73** 083515
[47] Wetterich C 1995 *Astron. Astrophys.* **301** 321
[48] Takahashi R and Tanimoto M 2006 *JHEP* **0605** 021
[49] Dvali G and Zaldarriaga M 2002 *Phys. Rev. Lett.* **88** 091303
[50] Cheng T P 1988 *Phys. Rev.* D **38** 2869
[51] Baessler S *et al* 1999 *Phys. Rev. Lett.* **83** 3585
[52] Adelberger A G 2001 *Class. Quant. Gravity* **18** 2397
[53] Kolb E W and Turner M S 1990 *The Early Universe* (Reading, MA.: Addison-Wesley)
[54] Balazs C, Carena M, Freitas A and Wagner C E M 2007 *JHEP* **06** 066
[55] Linde A D 1994 *Phys. Rev.* D **49** 748
[56] Peebles P J E and Ratra B 2003 *Rev. Mod. Phys.* **75** 559
[57] Padmanabhan T 2003 *Phys. Rep.* **380** 235
[58] Wetterich C 2003 *JCAP* **0310** 002
[59] Uzan J P 2003 *Rev. Mod. Phys.* **75** 403
[60] Albrecht A and Skordis C 2000 *Phys. Rev. Lett.* **84** 2076
[61] Barrow J D, Bean R and Magueijo J 2000 *Mon. Not. R. Ast. Soc.* **316** L41
[62] Bento M C, Bertolami O and Santos N C 2002 *Phys. Rev.* D **65** 067301

Emergent gravity and entanglement entropy of black holes

Mariano Cadoni

Dipartimento di Fisica, Università di Cagliari, and INFN sezione di Cagliari, Cittadella Universitaria 09042 Monserrato, ITALY

E-mail: mariano.cadoni@ca.infn.it

Abstract. When the gravitational interaction emerges from some underlying quantum field theory, black hole entropy should be completely explained in terms of the entanglement entropy (EE) of the quantum fields. We show that this is the case of the anti-de Sitter black hole in two spacetime dimensions. In this case two-dimensional (2D) Newton constant is wholly induced by a conformal field theory and an exact formula for the EE can be derived. In the large black hole mass limit the leading term of this formula reproduces exactly the Bekenstein-Hawking entropy of the hole, whereas the subleading term behaves logarithmically. This subleading term has the universal form of the entanglement entropy of physical systems described by effective conformal fields theories (e.g. one-dimensional statistical models at the critical point).

1. Introduction

The origin of the Bekenstein-Hawking (BH) black hole entropy has already been a rather controversial issue. The usual statistical paradigm should explain the BH entropy in terms of a microstate gas and an underlying unitary theory of quantum gravity (QG). However, the opposite point of view, which does not necessarily assume an underlying unitary theory and thinks of the BH entropy as fundamental feature of QG, cannot be completely excluded.

The issue is very difficult to settle mainly because we know very few about the emergence of the classical spacetime structure out of a microscopic theory of gravity. In this context the notion of entanglement (geometric) entropy obtained by tracing fundamental degrees of freedom over the inaccessible region of classical black hole geometry is a key concept for tackling the problem.

It is an old idea, first proposed by 't Hooft, that the BH black hole entropy could be understood in terms of the quantum entanglement of matter in the black hole geometry [1, 2, 3, 4, 5, 6, 7, 8, 9, 10]. The main support to this conjecture comes from the fact that both BH and entanglement entropy (EE) scale as the area of the boundary. However, the naive scaling arguments run easily into both conceptual and technical problems.

The usual statistical interpretation of the BH entropy is conceptually different from the EE, which measures the observer's lack of information about the quantum state of the system in inaccessible regions of spacetime. Moreover, the EE depends both on the number of species of matter fields localized in the black hole geometry and on the value of the UV cutoff arising owing to the presence of a sharp boundary between the accessible and inaccessible regions of the spacetime. On the other hand, the BH entropy should be universal, independent from both matter fields and UV cutoff. Some conceptual problems look somehow milder using Sakharov's induced gravity approach [11, 12, 13], but the fundamental issue is still there.

In recent years there has been considerable improvement concerning the understanding of EE in field theory and condensed matter systems. Exact formulas have been derived for the EE entropy of 2D conformal field theory (CFT) [14, 15]. Moreover, EE has been shown to play a key role for understanding quantum phase transitions (e.g for spin chains and quantum liquids) [16, 17, 18, 19, 20]. Entanglement entropy turned out to be an useful concept also for investigating general features of quantum field theory (QFT) and its relationship with gravity. In particular, the AdS/CFT correspondence has been used to calculate the EE of CFTs with gravitational duals, c-theorems have been shown to hold also for the EE and approximate formulas have been derived for the EE of generic QFTs in any dimension [21, 22, 23, 24, 25, 26].

The key question we will address in this note is the following: Can we use this new ideas and improved understanding of EE to solve old problems related with the EE of black holes? We will show that the answer to this question is yes, at least for 2D AdS black hole and most likely for three-dimensional (3D) Bañados-Teitelboim-Zanelli (BTZ) black hole.

2. Entanglement entropy for conformal field theory in two dimensions

The case in which the EE in QFT is best understood is that of 2D CFT. Let us consider a 2D spacetime with a compact spacelike dimension of length Λ. When only a spacelike slice Q (of length Σ) of our universe is accessible for measurement we loose information about the degrees of freedom (DOF) localized in the complementary region P and we have to trace over these DOF. The entanglement entropy originated in this way is given by the Von Neumann entropy $S_{ent} = -Tr_Q \hat{\rho}_Q \ln \hat{\rho}_Q$. The reduced density matrix, $\hat{\rho}_Q = Tr_P \hat{\rho}$, is obtained by tracing the density matrix $\hat{\rho}$ over states in the region P. The resulting EE for the ground state of the 2D CFT is given by [14, 15].

$$S_{ent} = \frac{c+\bar{c}}{6} \ln \left(\frac{\Lambda}{\epsilon\pi} \sin \frac{\pi\Sigma}{\Lambda} \right), \tag{1}$$

where c and $\bar{c}$ are the central charges of the 2D CFT. Owing to the presence of a sharp boundary separating the region P from the region Q, the entanglement entropy diverges. It can be regularized by introducing the ultraviolet cutoff ϵ. Eq. (1) gives the EE for a CFT on a spacetime with the topology of a cylinder (the timelike direction is noncompact).

The expression (1) for the EE as some interesting features:
a) Subadditivity.
b) Invariance under the transformation which exchanges the inside and outside regions

$$\Sigma \to \Lambda - \Sigma. \tag{2}$$

c) S_{ent} is not a monotonic function of Σ, but increases and reaches its maximum for $\Sigma = \Lambda/2$ and then decreases as Σ increases further.

For $\Lambda \gg \Sigma$ the compact spacelike dimension becomes infinite and Eq. (1) becomes the EE for a CFT at zero temperature on the plane [14, 15],

$$S_{ent} = \frac{c+\bar{c}}{6} \ln \left(\frac{\Sigma}{\epsilon} \right). \tag{3}$$

If we consider a CFT at finite temperature $T = 1/\beta$, the 2D spacetime is a cylinder with compact timelike dimension of length β and noncompact spacelike dimension. The EE in this case is given by [15]

$$S_{ent} = \frac{c+\bar{c}}{6} \ln \left(\frac{\beta}{\epsilon\pi} \sinh \frac{\pi\Sigma}{\beta} \right). \tag{4}$$

3. Two-dimensional AdS black holes and the dual CFT

In the simplest case 2D AdS black holes are classical solutions of a 2D dilaton-gravity theory with action $A = \int d^2x\sqrt{g}\Phi(R+2/L^2)$, where the AdS length L is related to cosmological constant of the AdS spacetime, $\lambda = 1/L^2$, and Φ is a scalar called dilaton. The 2D AdS black hole solutions can be written as follows, [28],

$$ds^2 = -\frac{1}{L^2}\left(r^2 - {r_h}^2\right)dt^2 + L^2\left(r^2 - r_h^2\right)^{-1}dr^2, \quad \Phi = \frac{r}{GL}, \tag{5}$$

where G is the dimensionless 2D Newton constant and r_h is the horizon radius.

The black hole mass, temperature and Bekenstein-Hawking entropy turn out to be given by [28]

$$M = \frac{{r_h}^2}{2GL^3}, \quad T = \frac{r_h}{2\pi L^2}, \quad S_{BH} = \frac{2\pi r_h}{GL}. \tag{6}$$

The 2D gravity theory has a dual description in terms of a chiral CFT with with central charge [29, 30, 31, 32]

$$c = \frac{12}{G}. \tag{7}$$

This AdS/CFT correspondence can be used to give a microscopical meaning to the thermodynamical entropy of the 2D AdS black hole (5). The BH entropy (6) can be also reproduced by counting states in the dual CFT [29, 30, 31, 32].

4. Entanglement entropy of the 2D black hole

It has been observed that in two spacetime dimensions black hole entropy can be ascribed to quantum entanglement if 2D Newton constant is wholly induced by quantum fluctuations of matter [4, 33, 12, 13]. This is exactly what equation (7) above, and more in general the AdS/CFT correspondence) tells us. It follows that in the semiclassical approximation the black hole EE should be identified with the entanglement entropy of the CFT vacuum in the black hole geometry (5). The black hole exterior has to be identified with the observable region Q and the black hole interior with the region P where the CFT degrees of freedom are traced over.

Two obstacles prevent direct application of the formulas (1),(3) for the EE of 2D CFTs. First, Eqs. (1),(3) hold only for flat spacetimes. Second, in the CFT calculation we have to use spacelike slices, whereas in the black hole case, owing to the coordinate singularities at $r = r_h$ and $r = \infty$ there is no global notion of spacelike coordinate. We cannot give a direct meaning to both Σ and $(\Lambda - \Sigma)$.

The second difficulty can be circumvented by using an appropriate regularization procedure, the first by using a generalized formula that holds in a curved spacetime [4],

$$S_{ent} = -\frac{c}{6}\left(\rho|_{boundary} - \ln\frac{\epsilon}{\Lambda}\right), \tag{8}$$

where c is the central charge (7), Λ is an IR cutoff and $\exp(2\rho)$ is the conformal factor of the metric in the coordinate system used to define the CFT vacuum. In this coordinate system the black hole metric (5) takes the form [28]

$$ds^2 = \frac{{r_h}^2}{L^2}\frac{(-dt^2 + d\sigma^2)}{\sinh^2(\frac{r_h\sigma}{L^2})}. \tag{9}$$

Notice that because we are dealing with a chiral CFT, in Eq. (8) we have contributions from only one sector (e.g. right movers) of the CFT. In Ref. [31, 32] it has been shown that the 2D

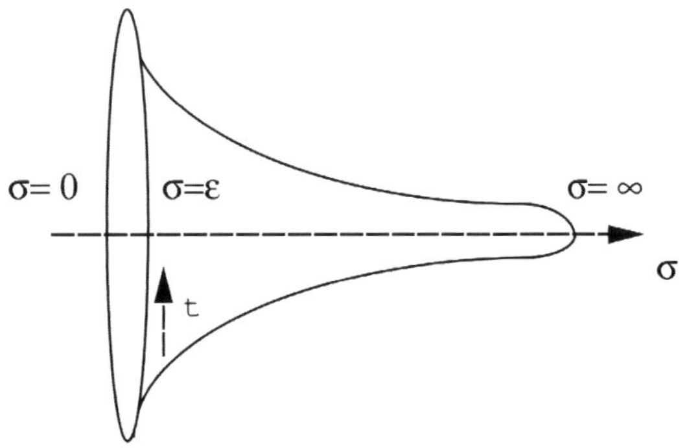

Figure 1. Regularized Euclidean instanton corresponding to the 2D AdS black hole in the coordinate system (t, σ) covering only the black hole exterior. The Euclidean time is periodic. The point $\sigma = \infty$ correspond to the black hole horizon. $\sigma = 0$ corresponds to the asymptotic timelike boundary of AdS_2.

AdS black hole is dual to an open string with appropriate boundary conditions. These boundary conditions are such that only one sector of the 2D CFT is present.

Working in the Euclidean, the 2D manifold has only a boundary at $\sigma = 0$ (corresponding to $r = \infty$), the timelike conformal boundary of AdS_2. The conformal factor (hence the EE) blows up on this boundary. To remove the singularity we consider a regularized boundary at $\sigma = \epsilon$. This plays the role of a UV cutoff for the coordinate σ, which is the natural spacelike coordinate of the dual CFT. On the other hand ϵ is an IR cutoff for the coordinate r, which is the natural spacelike coordinate for the AdS_2 black hole. We have the regularized Euclidean shown in figure (1). We get now from Eqs. (8) and (9)

$$S_{ent}^{(bh)} = \frac{c}{6} \ln \left(\frac{L^2}{\epsilon r_h} \sinh \frac{\epsilon r_h}{L^2} \right). \tag{10}$$

Notice that the regularized boundary is at finite proper distance from the horizon. ϵ acts also as IR regulator, making the presence of the IR cutoff Λ in Eq. (8) redundant.

The EE still depends on the UV cutoff ϵ. The AdS/CFT correspondence enable us to identify ϵ as the UV cutoff of the CFT: $\epsilon \propto L$. The proportionality factor can be determined by requiring that the analytical continuation of Eq. (11) below is invariant under the transformation (2). This will be discussed later in this section, where we will show that this requirement fixes $\epsilon = \pi L$, so that Eq. (10) becomes,

$$S_{ent}^{(bh)} = \frac{c}{6} \ln \left(\frac{L}{\pi r_h} \sinh \frac{\pi r_h}{L} \right), \tag{11}$$

which has to be identified with the EE of the 2D AdS black hole. The entanglement entropy (11) has the expected behavior as a function of the horizon radius r_h or, equivalently, of the

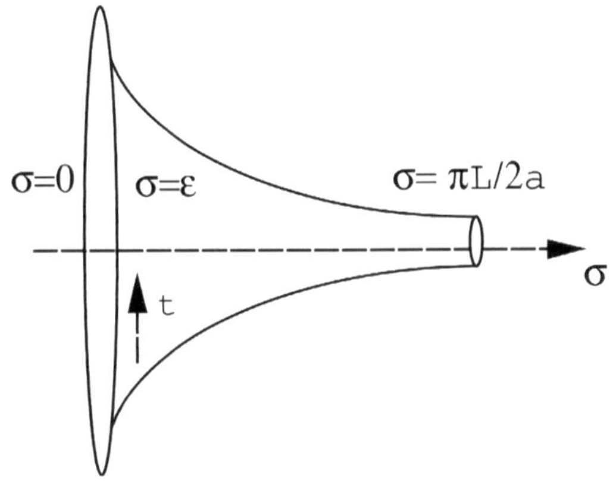

Figure 2. Regularized Euclidean instanton corresponding to the 2D AdS solution with negative mass. The Euclidean time is periodic. The point $\sigma = \pi L^2/2r_h$ corresponds to the black hole singularity at $r = 0$. $\sigma = 0$ corresponds to the asymptotic timelike boundary of AdS_2.

black hole mass M. $S_{ent}^{(bh)}$ becomes zero in the AdS ground state, $r_h = 0$ ($M = 0$), whereas it grows monotonically for $r_h > 0$ ($M > 0$).

The black hole EE (11) is analogue to the EE of a CFT at finite temperature given by Eq. (4) and differs from Eq. (1) by the exchange of hyperbolic with trigonometric functions. We can obtain the second from the first by analytic continuation $r_h \to r_h$ of our formula (10), i.e. by considering a solution with negative mass. The corresponding (regularized) Euclidean instanton is shown in Fig. (2). Repeating the calculation and redefining appropriately the UV cutoff ϵ, we get now

$$S_{ent} = \frac{c}{6} \ln\left(\frac{\Lambda}{\pi\epsilon} \sin\frac{\pi r_h}{\Lambda}\right). \tag{12}$$

The entanglement entropy of the 2D CFT in the curved background given by the AdS solution with negative mass has exactly the form given by Eq. (1) with the horizon radius r_h playing the role of Σ. The requirement that equation (12) is the analytic continuation of Eq. (10) fixes, as previously anticipated, the proportionality factor between ϵ and L in the calculations leading to Eq. (11).

4.1. Large black hole mass behavior

Let us now consider the large black hole mass behavior $r_h >> L$ of Eq. (11) corresponding to a macroscopic black hole:

$$S_{ent}^{(bh)} = \frac{2\pi r_h}{GL} - \frac{2}{G}\ln\frac{r_h}{L} + O(1) = S_{BH} - \frac{2}{G}\ln S_{BH} + O(1). \tag{13}$$

The leading term in the large mass expansion of the black hole entanglement entropy reproduces exactly the Bekenstein-Hawking entropy. In this regime the thermal length $\beta = 1/T \ll L$, thermal correlations dominate and the EE becomes Gibbs entropy. The subleading term has the universally predicted $\ln S_{BH}$ behavior for the quantum corrections to the Bekenstein-Hawking result. Although the logarithmic behavior of quantum corrections to the BH entropy seems to be a rather universal result, [34, 35, 36, 37, 38, 39, 40, 41, 42, 43], there is no general agreement about the value of the prefactor of this term. Equation (13) fixes the prefactor of $\ln S_{BH}$ in terms of the 2D Newton constant. The subleading correction has also the universal form shared by other physical system described by 2D CFTs, such as 1D statistical models near to the critical point or free scalars fields (with r_h playing the role of a correlation length), [15, 22].

For $\beta \gg L$ quantum correlation dominate. But we expect our description to break down at $\beta \sim L$ because we reach the limit of validity of our semiclassical approximation. We also expect in this regime a phase transition analogue to the Hawking-Page transition that takes place in 4D.

5. Entanglement entropy of the BTZ black hole

There are several arguments indicating that our derivation could be extended to black holes in 3D AdS spacetime, i.e. to the BTZ black hole. Pure gravity in three spacetime dimensions has the same degrees of freedom of 2D dilaton gravity, that is only global DOF and no propagating gravitons. Moreover, 2D AdS dilaton gravity can be considered as a circular symmetric dimensional reduction of 3D AdS gravity with the dilaton parametrizing the radius of the transverse circle. The 2D AdS black hole discussed in this paper can be obtained as dimensional reduction of the BTZ black hole with vanishing angular momentum J. The thermodynamical parameters, mass, temperature and entropy, characterizing the $J = 0$ BTZ black hole match exactly those pertinent to the 2D AdS black hole, given by Eq. (6), once one expresses the 3D Newton constant in terms of the 2D Newton constant G and of the de Sitter length L.

In the case of 2D AdS gravity the dual theory has the form of a chiral CFT whereas in the 3D case, owing to different boundary condition, we have a dual boundary 2D CFT with both left and right movers. An other important difference is that in the 3D case the AdS/CFT correspondence must be realized necessarily in a holographic way. The CFT does not live in the bulk 3D space but in the 2D, $r \to \infty$, timelike boundary of the spacetime. Thus we need to make use of the AdS/CFT correspondence to relate tracing of CFT DOF in the boundary to tracing of gravitational DOF in the bulk. These features make the method for calculating the EE used in this paper, difficult to use in the 3D case. However, preliminary calculations that make use of the IR/UV relation for the AdS/CFT correspondence indicate the validity of results similar to those described in this paper, also for the BTZ black hole [44].

6. Conclusions

We have found a simple and general formula for the EE of the 2D AdS black hole. The picture that emerges is intriguing but to a large extend expected: the black hole Entanglement entropy reduces to the BH entropy only for macroscopic black holes when thermal correlations dominate. In general the EE codifies information about the full quantum gravity description.

In particular the EE could be used to gain information about quantum gravity phase transitions (e.g. the Hawking-Page phase transition) and about the emergence of the spacetime description from the fundamental theory, in the same way as EE is used to gain information about quantum phase transitions in condensed matter systems. In the full quantum regime, $\beta \sim L$, the underlying microscopic theory has to be used.

The next step should be the generalization of our approach to higher-dimensional gravity theory. Work in progress indicates that this is possible at least for the 3D case.

References

[1] 't Hooft G 1985 *Nucl. Phys.* B **256** 727
[2] Bombelli L, Koul R K, Lee J H and Sorkin R D 1986 *Phys. Rev.* D **34** 373
[3] Frolov V P and Novikov I 1993 *Phys. Rev.* D **48** 4545 (*Preprint* arXiv:gr-qc/9309001)
[4] Fiola T M, Preskill J, Strominger A and Trivedi S P 1994 *Phys. Rev.* D **50** 3987 (*Preprint* arXiv:hep-th/9403137)
[5] Belgiorno F and Liberati S 1996 *Phys. Rev.* D **53** 3172 (*Preprint* arXiv:gr-qc/9503022)
[6] Hawking S, Maldacena J M and Strominger A 2001 *JHEP* **0105** 001 (*Preprint* arXiv:hep-th/0002145)
[7] Maldacena J M 2003 *JHEP* **0304** 021 (*Preprint* arXiv:hep-th/0106112)
[8] Brustein R, Einhorn M B and Yarom A 2006 *JHEP* **0601** 098 (*Preprint* arXiv:hep-th/0508217)
[9] Emparan R 2006 *JHEP* **0606** 012 (*Preprint* arXiv:hep-th/0603081)
[10] Valtancoli P 2006 *Preprint* arXiv:hep-th/0612049
[11] Jacobson T 1994 *Preprint* arXiv:gr-qc/9404039
[12] Frolov V P, Fursaev D V and Zelnikov A I 1997 *Nucl. Phys.* B **486** 339 (*Preprint* arXiv:hep-th/9607104)
[13] Frolov V P and Fursaev D V 1997 *Phys. Rev.* D **56** 2212 (*Preprint* arXiv:hep-th/9703178)
[14] Holzhey C, Larsen F and Wilczek F 1994 *Nucl. Phys.* B **424** 443 (*Preprint* arXiv:hep-th/9403108)
[15] Calabrese P and Cardy J L 2004 *J. Stat. Mech.* **0406** P002 (*Preprint* arXiv:hep-th/0405152)
[16] Vidal G, Latorre J I, Rico E and Kitaev A 2003 *Phys. Rev. Lett.* **90** 227902 (*Preprint* arXiv:quant-ph/0211074)
[17] Its A R, Jin B Q and Korepin V E 2005 *J. Phys.* A **38** 2975 (*Preprint* arXiv:quant-ph/0409027)
[18] Kitaev A and Preskill J 2006 *Phys. Rev. Lett.* **96** 110404 (*Preprint* arXiv:hep-th/0510092)
[19] Latorre J I, Lutken C A, Rico E and Vidal G 2005 *Phys. Rev.* A **71** 034301 (*Preprint* arXiv:quant-ph/0404120)
[20] Korepin V E 2003 *Phys. Rev. Lett.* **92** 964021
[21] Casini H and Huerta M 2004 *Phys. Lett.* B **600** 142 (*Preprint* arXiv:hep-th/0405111)
[22] Fursaev D V 2006 *Phys. Rev.* D **73** 124025 (*Preprint* arXiv:hep-th/0602134)
[23] Solodukhin S N 2006 *Phys. Rev. Lett.* **97** 201601 (*Preprint* arXiv:hep-th/0606205)
[24] Ryu S and Takayanagi T 2006 *Phys. Rev. Lett.* **96** 181602 (*Preprint* arXiv:hep-th/0603001)
[25] Ryu S and Takayanagi T 2006 *JHEP* **0608** 045 (*Preprint* arXiv:hep-th/0605073)
[26] Hubeny V E, Rangamani M and Takayanagi T 2007 *JHEP* **0707** 062 (*Preprint* arXiv:hep-th/0705.0016)
[27] Cadoni M 2007 *Phys. Lett.* B **653** 434 (*Preprint* arXiv:hep-th/0704.0140)
[28] Cadoni M and Mignemi S 1995 *Phys. Rev.* D **51** 4319 (*Preprint* arXiv:hep-th/9410041)
[29] Cadoni M and Mignemi S 1999 *Phys. Rev.* D **59** 081501 (*Preprint* arXiv:hep-th/9810251)
[30] Cadoni M and Mignemi S 1999 *Nucl. Phys.* B **557** 165 (*Preprint* arXiv:hep-th/9902040)
[31] Cadoni M and Cavaglia M 2001 *Phys. Lett.* B **499** 315 (*Preprint* arXiv:hep-th/0005179)
[32] Cadoni M and Cavaglia M 2001 *Phys. Rev.* D **63** 084024 (*Preprint* arXiv:hep-th/0008084)
[33] Susskind L and Uglum J 1994 *Phys. Rev.* D **50** 2700 (*Preprint* arXiv:hep-th/9401070)
[34] Fursaev D V 1995 *Phys. Rev.* D **51** 5352 (*Preprint* arXiv:hep-th/9412161)
[35] Mann R B and Solodukhin S N 1998 *Nucl. Phys.* B **523** 293 (*Preprint* arXiv:hep-th/9709064)
[36] Kaul R K and Majumdar P 2000 *Phys. Rev. Lett.* **84** 5255 (*Preprint* arXiv:gr-qc/0002040
[37] Carlip S 2000 *Class. Quant. Grav.* **17** 4175 (*Preprint* arXiv:gr-qc/0005017)
[38] Ghosh A and Mitra P 1994 *Phys. Rev. Lett.* **73** 2521 (*Preprint* arXiv:hep-th/9406210)
[39] Mukherji S and Pal S S 2002 *JHEP* **0205** 026 (*Preprint* arXiv:hep-th/0205164)
[40] Setare M R 2003 *Phys. Lett.* B **573** 173 (*Preprint* arXiv:hep-th/0311106)
[41] Domagala M and Lewandowski J 2004 *Class. Quant. Grav.* **21** 5233 (*Preprint* arXiv:gr-qc/0407051)
[42] Medved A J M 2005 *Class. Quant. Grav.* **22** 133 (*Preprint* arXiv:gr-qc/0406044)
[43] Grumiller D 2005 *Preprint* arXiv:hep-th/0506175
[44] Cadoni M and Melis M, in preparation

Multipartite entanglement in neutrino oscillations

Massimo Blasone† [1,2], **Fabio Dell'Anno**[1,2,3], **Silvio De Siena**[1,2,3] **and Fabrizio Illuminati**[1,2,3,4]

[1] Dipartimento di Matematica e Informatica, Università degli Studi di Salerno, Via Ponte don Melillo, I-84084 Fisciano (SA), Italy
[2] INFN Sezione di Napoli, Gruppo collegato di Salerno, Baronissi (SA), Italy
[3] CNR-INFM Coherentia, Napoli, and CNISM, Unità di Salerno, Italy
[4] ISI Foundation for Scientific Interchange, Viale Settimio Severo 65, I-10133 Torino, Italy

E-mail: †blasone@sa.infn.it

Abstract. Particle mixing is related to multi-mode entanglement of single-particle states. The occupation number of both flavor eigenstates and mass eigenstates can be used to define a multiqubit space. In such a framework, flavor neutrino states can be interpreted as multipartite mode-entangled states. By using two different entanglement measures, we analyze the behavior of multipartite entanglement in the phenomenon of neutrino oscillations.

1. Introduction

Quantum entanglement is a crucial physical resource in quantum information and communication science [1]. It has been mainly investigated in systems of condensed matter, atomic physics, and quantum optics. In the domain of particle physics, entanglement has been considered mainly in relation to two-body decay, annihilation, and creation processes, see for instance Refs. [2]. Very recently, multipartite entanglement has been studied in connection with the phenomenon of particle mixing [3, 4]. Such a phenomenon, associated with a mismatch between flavor and mass of the particle, appears in several instances: quarks, neutrinos, and the neutral K-meson system [5, 6]. Particle mixing is at the basis of important effects as neutrino oscillations and CP violation [7]. Flavor mixing for the case of three generations is described by the Pontecorvo-Maki-Nakagawa-Sakata (PMNS) in the lepton instance [8, 9]. The matrix elements represent the transition probabilities from one lepton to another. For example, flavor mixing of neutrinos for three generations is described by the 3×3 unitary mixing matrix $\mathbf{U}(\tilde{\theta}, \delta)$ [5],

$$\mathbf{U}(\tilde{\theta}, \delta) = \begin{pmatrix} c_{12}c_{13} & s_{12}c_{13} & s_{13}e^{-i\delta} \\ -s_{12}c_{23} - c_{12}s_{23}s_{13}e^{i\delta} & c_{12}c_{23} - s_{12}s_{23}s_{13}e^{i\delta} & s_{23}c_{13} \\ s_{12}s_{23} - c_{12}c_{23}s_{13}e^{i\delta} & -c_{12}s_{23} - s_{12}c_{23}s_{13}e^{i\delta} & c_{23}c_{13} \end{pmatrix}, \tag{1}$$

where $(\tilde{\theta}, \delta) \equiv (\theta_{12}, \theta_{13}, \theta_{23}; \delta)$ and $c_{ij} \equiv \cos\theta_{ij}$, $s_{ij} \equiv \sin\theta_{ij}$. The parameters θ_{ij} are the mixing angles, and δ is the phase responsible for CP violation. The three-flavor neutrino states are defined as

$$|\underline{\nu}^{(f)}\rangle = \mathbf{U}(\tilde{\theta}, \delta)\,|\underline{\nu}^{(m)}\rangle \tag{2}$$

where $|\underline{\nu}^{(f)}\rangle = (|\nu_e\rangle, |\nu_\mu\rangle, |\nu_\tau\rangle)^T$ are the states with definite flavor and $|\underline{\nu}^{(m)}\rangle = (|\nu_1\rangle, |\nu_2\rangle, |\nu_3\rangle)^T$ those with definite mass. From Eq. (2), we see that each flavor state is given by a superposition of mass eigenstates, i.e. $|\nu_\alpha\rangle = U_{\alpha 1}|\nu_1\rangle + U_{\alpha 2}|\nu_2\rangle + U_{\alpha 3}|\nu_3\rangle$. Let us recall that both $\{|\nu_\alpha\rangle\}$ and $\{|\nu_i\rangle\}$ are orthonormal, i.e. $\langle\nu_\alpha|\nu_\beta\rangle = \delta_{\alpha,\beta}$ and $\langle\nu_i|\nu_j\rangle = \delta_{i,j}$. Therefore, one can interpret the label i as denoting a quantum mode, and can establish the following correspondence with three-qubit states:

$$|\nu_1\rangle \equiv |1\rangle_1|0\rangle_2|0\rangle_3 \equiv |100\rangle, \quad |\nu_2\rangle \equiv |0\rangle_1|1\rangle_2|0\rangle_3 \equiv |010\rangle, \quad |\nu_3\rangle \equiv |0\rangle_1|0\rangle_2|1\rangle_3 \equiv |001\rangle, \tag{3}$$

where $|\rangle_i$ denotes states in the Hilbert space for neutrinos with mass m_i. Then, the occupation number allows to interpret the flavor states as constituted by entangled superpositions of the mass eigenstates. Quantum entanglement emerges as a direct consequence of the superposition principle. It is important to remark that the Fock space associated with the neutrino mass eigenstates is physically well defined. In fact, at least in principle, the mass eigenstates can be produced or detected in experiments performing extremely precise kinematical measurements [10]. In this framework, as discussed in Ref. [3], the quantum mechanical state (2) is entangled in the field modes, although being a single-particle state.

Mode entanglement defined for single-photon states of the radiation field or associated with systems of identical particles has been discussed in Ref. [11]. In the particular instance of Eq. (2), the multipartite flavor states can be seen as a generalized class of W states. These are multipartite entangled states that occur in several physical systems and can be engineered with pure quantum optical elements [12]. The concept of mode entanglement in single-particle states has been widely discussed and is by now well established [11, 13]. Successful experimental realizations using single-photon states have been reported as well [14]. Moreover, remarkably, the nonlocality of single-photon states has been experimentally demonstrated [15], verifying a theoretical prediction [16]. Furthermore, the existing schemes to probe nonlocality in single-particle states have been generalized to include massive particles of arbitrary type [17].

In the dynamical regime, flavor mixing (and neutrino mass differences) generates the phenomenon of neutrino oscillations. The mass eigenstates $|\nu_j\rangle$ have definite masses m_j and definite energies E_j. Their propagation can be described by plane wave solutions of the form $|\nu_j(t)\rangle = e^{-iE_j t}|\nu_j\rangle$. The time evolution of the flavor neutrino states Eq.(2) is given by:

$$|\underline{\nu}^{(f)}(t)\rangle = \widetilde{\mathbf{U}}(t)|\underline{\nu}^{(f)}\rangle\,, \qquad \widetilde{\mathbf{U}}(t) \equiv \mathbf{U}(\tilde{\theta},\delta)\,\mathbf{U}_0(t)\,\mathbf{U}(\tilde{\theta},\delta)^{-1}\,, \tag{4}$$

where $|\underline{\nu}^{(f)}\rangle$ are the flavor states at $t=0$, $\mathbf{U}_0(t) = diag(e^{-iE_1 t}, e^{-iE_2 t}, e^{-iE_3 t})$, and $\widetilde{\mathbf{U}}(t=0) = \mathrm{I}$.

Flavor neutrino states are well defined in the context of Quantum Field Theory (QFT), where they are obtained as eigenstates of the flavor neutrino charges [19]. In the relativistic limit, the exact QFT flavor states reduce to the usual Pontecorvo flavor states Eq.(2): flavor modes are thus legitimate and physically well-defined individual entities and mode entanglement can be defined and studied in analogy with the static case of Ref.[3]. We can thus establish the following correspondence with three-qubit states:

$$|\nu_e\rangle \equiv |1\rangle_{\nu_e}|0\rangle_{\nu_\mu}|0\rangle_{\nu_\tau}, \quad |\nu_\mu\rangle \equiv |0\rangle_{\nu_e}|1\rangle_{\nu_\mu}|0\rangle_{\nu_\tau}, \quad |\nu_\tau\rangle \equiv |0\rangle_{\nu_e}|0\rangle_{\nu_\mu}|1\rangle_{\nu_\tau}. \tag{5}$$

States $|0\rangle_{\nu_\alpha}$ and $|1\rangle_{\nu_\alpha}$ correspond, respectively, to the absence and the presence of a neutrino in mode α. Entanglement is thus established among flavor modes, in a single-particle setting. Eq. (4) can then be recast as

$$|\nu_\alpha(t)\rangle = \widetilde{\mathbf{U}}_{\alpha e}(t)|1\rangle_{\nu_e}|0\rangle_{\nu_\mu}|0\rangle_{\nu_\tau} + \widetilde{\mathbf{U}}_{\alpha\mu}(t)|0\rangle_{\nu_e}|1\rangle_{\nu_\mu}|0\rangle_{\nu_\tau} + \widetilde{\mathbf{U}}_{\alpha\tau}(t)|0\rangle_{\nu_e}|0\rangle_{\nu_\mu}|1\rangle_{\nu_\tau}\,, \tag{6}$$

with the normalization condition $\sum_\beta |\widetilde{\mathbf{U}}_{\alpha\beta}(t)|^2 = 1$ $(\alpha,\beta = e,\mu,\tau)$. The time-evolved states $|\underline{\nu}^{(f)}(t)\rangle$ are entangled superpositions of the three flavor eigenstates with time-dependent

Journal of Physics: Conference Series **174** (2009) 012062 doi:10.1088/1742-6596/174/1/012062

coefficients. Thus, flavor oscillations can be related to multi-mode (flavor) entanglement of single-particle states [4].

It is important to remark difference between Refs. [3] and [4]: In Ref. [3], the multipartite entanglement, associated with the multiqubit space of *mass modes*, has been analyzed in connection with the "decoherence" effects induced by free propagation of oscillating neutrinos. In Ref. [4], on the other hand, the entanglement is quantified with respect to the multiqubit space associated with *flavor modes*, and it arises as a consequence of the non-trivial time evolution of the flavor states.

In this paper, we extend the results of Ref. [4]. We compute the multipartite entanglement associated to three flavor neutrino oscillations, by using two different global measures of entanglement: the von Neumann entropy, based on the entropies related to all possible bipartitions of the system, and the K-component geometric measure, introduced in Ref. [20]. Finally, we compare the results corresponding to the two entanglement measures.

2. Neutrino oscillations

In this section, we briefly review the phenomenon of neutrino oscillations. At time t the probability associated with the transition $\nu_\alpha \to \nu_\beta$ is

$$P_{\nu_\alpha \to \nu_\beta}(t) \,=\, |\langle \nu_\beta | \nu_\alpha(t)\rangle|^2 \,=\, |\widetilde{\mathbf{U}}_{\alpha\beta}(t)|^2 \,, \tag{7}$$

where $\alpha, \beta = e, \mu, \tau$. The transition probability $P_{\nu_\alpha \to \nu_\beta}(t)$ is a function of the energy differences $\Delta E_{jk} = E_j - E_k$ $(j, k = 1, 2, 3)$ and of the mixing angles. Since detectable neutrinos are ultra-relativistic, the standard adopted approximation is $\Delta E_{jk} \simeq \frac{\Delta m^2_{jk}}{2E}$, where $\Delta m^2_{jk} = m^2_j - m^2_k$ and $E = |\vec{p}|$ is the energy of a massless neutrino (all massive neutrinos are assumed to have the same momentum $\vec{p}$). The mixing angles θ^{PMNS}_{ij}, and the squared mass differences are fixed at the experimental values reported in Ref. [21]:

$$\sin^2\theta^{PMNS}_{12} \,=\, 0.314\,, \quad \sin^2\theta^{PMNS}_{13} \,=\, 0.8 \times 10^{-2}\,, \quad \sin^2\theta^{PMNS}_{23} \,=\, 0.45\,, \tag{8}$$

$$\begin{aligned} &\Delta m^2_{21} \,=\, \delta m^2\,, \qquad \Delta m^2_{31} \,=\, \Delta m^2 + \frac{\delta m^2}{2}\,, \qquad \Delta m^2_{32} \,=\, \Delta m^2 - \frac{\delta m^2}{2}\,, \\ &\delta m^2 \,=\, 7.92 \times 10^{-5}\, eV^2\,, \qquad \Delta m^2 \,=\, 2.6 \times 10^{-3}\, eV^2\,. \end{aligned} \tag{9}$$

In the forthcoming analysis, we will put to zero the CP-violating phase δ. In Fig. 2 we show the behavior of the transition probabilities $P_{\nu_e \to \nu_\alpha}$ $(\alpha = e, \mu, \tau)$ as a function of the scaled, dimensionless time $T = \frac{2Et}{\Delta m^2_{12}}$.

3. Multipartite entanglement in neutrino oscillations

In this section we characterize the multipartite flavor entanglement possessed by the state (6).

Let us first introduce the entanglement measures which will be exploited to this aim. Bipartite entanglement of pure states is unambiguously quantified by the von Neumann entropy or by any other monotonic function of the former [22]. Moving from the two- to the three-flavor scenario, multipartite entanglement measures have been introduced in terms of functions of bipartite measures [23]. Adopting an approach similar to that of Refs. [23], we define as proper measures of multipartite entanglement the von Neumann entropies associated to every possible bipartition of the system, see Ref. [3]. Let $\rho^{(\alpha)} = |\nu_\alpha(t)\rangle\langle\nu_\alpha(t)|$ be the density operator corresponding to the pure state $|\nu_\alpha(t)\rangle$, Eq. (6). We denote by $\rho^{(\alpha)}_{i,j} \,=\, Tr_k[\rho^{(\alpha)}]$ the density matrix reduced with

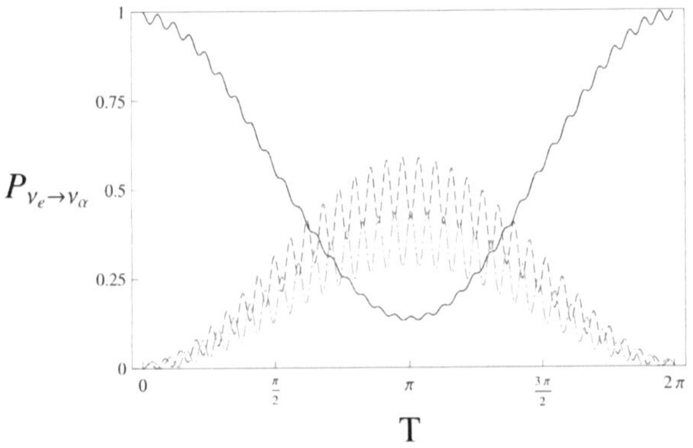

Figure 1. Transition probabilities $P_{\nu_e\to\nu_\alpha}$ as functions of the scaled time T. Parameters θ_{ij} and Δm^2_{ij} are fixed at the experimental values [21]. Curves correspond to $P_{\nu_e\to\nu_e}$ (full), $P_{\nu_e\to\nu_\mu}$ (dashed), and $P_{\nu_e\to\nu_\tau}$ (dot-dashed). The phase δ is fixed to zero.

respect to flavor k, with $i,j,k = e,\mu,\tau$ and $i \neq j \neq k$. The von Neumann entropy associated with such a bipartition will be given by

$$E^{(i,j;k)}_{vN\,\alpha} = -Tr_{i,j}[\rho^{(\alpha)}_{i,j} \log_2 \rho^{(\alpha)}_{i,j}]. \tag{10}$$

At last, we define the average von Neumann entropy

$$\langle E^{(2:1)}_{vN} \rangle = \frac{1}{3} \sum_k E^{(i,j;k)}_{vN}, \tag{11}$$

where the sum is intended over all the possible bipartitions of the system.
An alternative characterization of multipartite entanglement is given in Refs. [20, 24]. Given the pure state $|\nu_\alpha(t)\rangle$ belonging to a 3-dimensional Hilbert space, the geometric measure of entanglement is defined as:

$$E^{(K)}_G(|\nu_\alpha(t)\rangle) \equiv E^{(K)}_{G\,\alpha} = -\frac{1}{3}\log_2 \max_{|\Phi^{(K)}\rangle} \left|\langle\Phi^{(K)}|\nu_\alpha(t)\rangle\right|^2, \tag{12}$$

where the maximum is taken with respect to all K-separable pure states $|\Phi^{(K)}\rangle$, with $K = 1, 2$. This measure is intrinsically geometric because it coincides with the distance (in the Hilbert-Schmidt norm) between a given pure state and a set of K-separable reference pure states. For $K = 3$, the reference state $|\Phi^{(3)}\rangle$ is the fully product state:

$$|\Phi^{(3)}\rangle = \prod_k (\cos\xi_k|0\rangle_{\nu_k} + \sin\xi_k|1\rangle_{\nu_k}), \tag{13}$$

where ξ_k $(k = e,\mu,\tau)$ are real parameters. For $K = 2$, the reference state $|\Phi^{(2)}\rangle$ can be written in the normalized form:

$$\begin{aligned} |\Phi^{(2)}\rangle = {} & (\cos\delta_1|0\rangle_{\nu_i}|0\rangle_{\nu_j} + \sin\delta_1\cos\delta_2|0\rangle_{\nu_i}|1\rangle_{\nu_j} + \sin\delta_1\sin\delta_2\cos\delta_3|1\rangle_{\nu_i}|0\rangle_{\nu_j} + \\ & \sin\delta_1\sin\delta_2\sin\delta_3|1\rangle_{\nu_i}|1\rangle_{\nu_j}) \otimes (\cos\delta_4|0\rangle_{\nu_k} + \sin\delta_5|1\rangle_{\nu_k}), \end{aligned} \tag{14}$$

where $i,j,k = e,\mu,\tau$ and $i \neq j \neq k$, and δ_k are real parameters. In Fig. 3, panels (a) and (b), we plot the von Neumann entropy $E^{(i,j;k)}_{vN\,e}$ and the geometric component $E^{(2)}_{G\,e}(i,j;k)$ as functions of the scaled time T. Both the entanglement measures exhibit similar behavior. In fact, the curves have the same ordering with respect to each other, and share the same maxima and minima. Next we compare the behavior of the entanglement measures of Fig. 3 with that of the transition probabilities of Fig. 2. We observe that there is a strong correlations between the degree of mixing and the entanglement content of a given bipartition. In order to quantify the

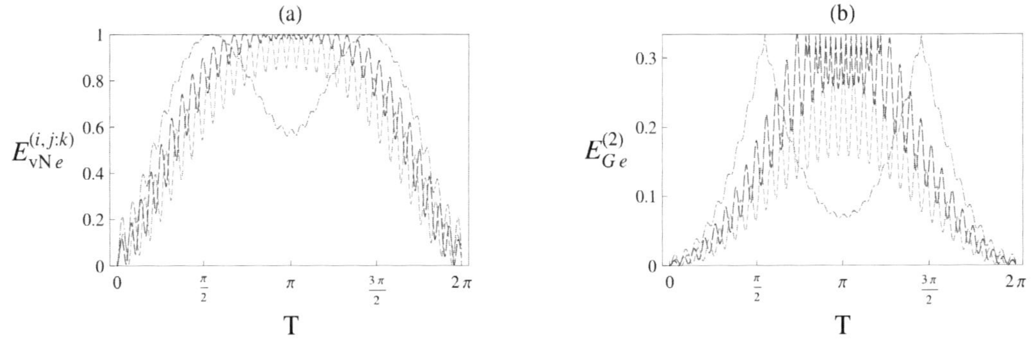

Figure 2. The von Neumann entropy $E_{vN\,e}^{(i,j;k)}$ (a) and the geometric component $E_{G\,e}^{(2)}(i,j;k)$ (b) as functions of the scaled time T. The plot style associated to each bipartition $(i,j;k)$ is: $i=1$, $j=2$, and $k=3$ (dashed line); $i=1$, $j=3$, and $k=2$ (long dashed line); $i=2$, $j=3$, and $k=1$ (dot-dashed line). Parameters θ_{ij} and Δm_{ij}^2 are fixed at the experimental values [21]. The phase δ is fixed to zero.

global entanglement of the state $|\nu_e(t)\rangle$, we compute the average von Neumann entropy $E_{vN\,e}^{(2:1)}$ and the geometric component $E_{G\,e}^{(3)}$. These quantities are plotted in Fig. 3, and exhibit a very similar behavior.

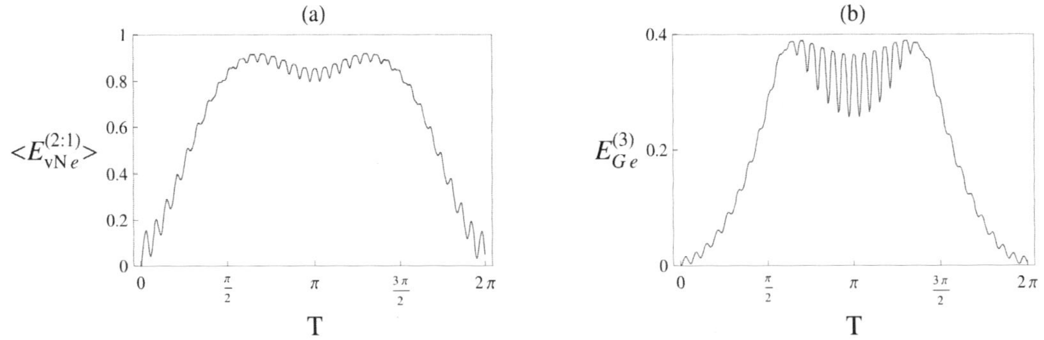

Figure 3. The average von Neumann entropy $\langle E_{vN\,e}^{(2:1)}\rangle$ (a) and the geometric component $E_{G\,e}^{(3)}$ (b) as functions of the scaled time T. Parameters θ_{ij} and Δm_{ij}^2 are fixed at the experimental values [21]. The phase δ is fixed to zero.

4. Conclusions

In conclusion, by generalizing the results of a previous paper [4], we have analyzed the amount and distribution of entanglement in the phenomenon of neutrino oscillations. The phenomenon of flavor oscillations can be interpreted in terms of the dynamical behavior of quantum correlations associated with time-evolved flavor states. In Ref.[4], we suggest the possibility of using entangled states of oscillating neutrinos as a resource for quantum information protocols, and propose experimental schemes for transferring it to spatially separated modes of stable leptonic particles.

Acknowledgments
We acknowledge partial financial support from MIUR, INFN, INFM and CNISM.

References

[1] Nielsen M A and Chuang I L 2000, *Quantum Information and Quantum Computation* (Cambridge, UK: Cambridge U. Press)
[2] Lee T D and Yang C N 1960 *Internal report, Argonne National Laboratory (unpublished)*; Inglis D R 1961 *Rev. Mod. Phys.* **33** 1; Day T B 1961 *Phys. Rev.* **121** 1204; Lipkin H J 1968 *Phys. Rev.* **176** 1715; Zralek M 1998 *Acta Phys. Polon.* B **29** 3925; Bertlmann R A 2006 *Lect. Notes Phys.* **689** 1; Li J L and Qiao C F *Preprint* arXiv:0708.0291
[3] Blasone M, Dell'Anno F, De Siena S, Di Mauro M and Illuminati F 2008 *Phys. Rev.* D **77** 096002
[4] Blasone M, Dell'Anno F, De Siena S and Illuminati F *Preprint* arXiv:0707.4476
[5] Cheng T and Li L 1989 *Gauge Theory of Elementary Particle Physics* (Clarendon Press)
[6] Eidelman S *et al* [Particle Data Group] 2004 *Phys. Lett.* B **592** 1
[7] Pontecorvo B 1967 *Zh. Eksp. Teor. Fiz.* **53** 1717 [1968 *Sov. Phys. JETP* **26** 984]
[8] Maki Z, Nakagawa M and Sakata S 1962 *Prog. Theor. Phys.* **28** 870
[9] Okun L and Pontecorvo B 1957 *Zh. Eksp. Teor. Fiz.* **32** 1587; Pontecorvo B 1958 *Sov. Phys. JETP* **7** 172
[10] Kayser B 1981 *Phys. Rev.* D **24** 110
[11] Zanardi P 2002 *Phys. Rev.* A **65** 042101
[12] Dell'Anno F, De Siena S and Illuminati F 2006 *Phys. Rep.* **428** 53
[13] Björk G, Jonsson P and Sánchez-Soto L L 2001 *Phys. Rev.* A **64** 042106; van Enk S J 2005 *Phys. Rev.* A **72** 064306; van Enk S J 2006 *Phys. Rev.* A **74** 026302; Terra Cunha M O, Dunningham J A and Vedral V 2007 *Proc. of the Royal Soc.* A **463** 2277
[14] Lombardi E, Sciarrino F, Popescu S and De Martini F 2002 *Phys. Rev. Lett.* **88** 070402; Lvovsky A I, Hansen H, Aichele T, Benson O, Mlynek J and Schiller S 2001 *Phys. Rev. Lett.* **87** 050402; Babichev S A, Appel J and Lvovsky A I 2004 *Phys. Rev. Lett.* **92** 193601
[15] Hessmo B, Usachev P, Heydari H and Björk G 2004 *Phys. Rev. Lett.* **92** 180401
[16] Tan S M, Walls D F and Collett M J 1991 *Phys. Rev. Lett.* **66** 252; Hardy L 1994 *Phys. Rev. Lett.* **73** 2279
[17] Dunningham J and Vedral V 2007 *Phys. Rev. Lett.* **99** 180404
[18] Casas J A and Ibarra A 2001 *Nucl. Phys.* B **618** 171
[19] Blasone M and Vitiello G 1995 *Ann. Phys. (N.Y.)* **244** 283; Blasone M, Jizba P and Vitiello G 2001 *Phys. Lett.* B **517** 471
[20] Blasone M, Dell'Anno F, De Siena S and Illuminati F 2008 *Phys. Rev.* A **77** 062304
[21] Fogli G L, Lisi E, Marrone A, Melchiorri A, Palazzo A, Serra P, Silk J and Slosar A 2007 *Phys. Rev.* D **75** 053001
[22] Amico L, Fazio R, Osterloh A and Vedral V 2008 *Rev. Mod. Phys.* **80** 517; Horodecki R, Horodecki P, Horodecki M and Horodecki K *Preprint* arXiv:quant-ph/0702225, *Rev. Mod. Phys.*, in press
[23] Meyer D A and Wallach N R 2002 *J. Math. Phys.* **43** 4273; Brennen G K 2003 *Quantum Inf. Comp.* **3** 619; Scott A J 2004 *Phys. Rev.* A **69** 052330; de Oliveira T R, Rigolin G and de Oliveira M C 2006 *Phys. Rev.* A **73** 010305(R); Facchi P, Florio G and Pascazio S 2006 *Phys. Rev.* A **74** 042331
[24] Wei T C and Goldbart P M 2003 *Phys. Rev.* A **68** 042307

Journal of Physics: Conference Series **174** (2009) 012063 doi:10.1088/1742-6596/174/1/012063

Particle mixing as possible explanation of the dark energy conundrum

Antonio Capolupo and Giuseppe Vitiello

Dipartimento di Matematica e Informatica, Università di Salerno and Istituto Nazionale di Fisica Nucleare, Gruppo Collegato di Salerno, 84100 Salerno, Italy

E-mail: capolupo@sa.infn.it

Abstract. The vacuum condensate due to neutrino and quark mixing behaves as a perfect fluid and, at the present epoch, as a cosmological constant. The very small breaking of the Lorentz invariance constrains today the value of the dark energy.

1. Introduction

Recent astrophysical data [1]-[7] represent observational evidence of the fact that the universe today observed is flat and it is undergoing an accelerated expansion. The hypothesis is that such an expansion is driven by a fluid called *dark energy* with negative pressure, nearly homogeneously distributed and making up to $\sim 70\%$ of the energy content of the universe.

Even if it seems clear how dark energy works, its nature remains an unsolved problem. Therefore, the understanding of the dark energy has become one of the main issue of modern physics. Its determination should provide the gravity vacuum state [8], should clarify the mechanisms which led from the early universe to the today observed large scale structures [9, 10], and to represent the theoretical solution to the observational data.

A lot of models have been proposed in order to explain the dark energy [11]-[19]. However, no comprehensive scheme is available, which may fit the observations and frame them into a fundamental theory.

In this paper we report recent results according to which the particle mixing phenomenon could explain the dark energy component of the universe [20]. Responsible of such a component is the non-perturbative vacuum structure associated with the particle mixing [21]-[23].

The estimated value of the cosmological constant is imposed by the small breaking of the Lorentz invariance of the flavor vacuum at the present epoch [20].

The paper is organized as follows. In Section II we present the QFT formalism for mixed fields [24]-[31] (for a detailed review see [29]). We introduce the particle mixing vacuum condensate and we show that it is homogeneous and isotropic [20]. In Section III we exhibit the mixing contributions to the dark energy at the present epoch and Section IV is devoted to conclusions.

2. Particle mixing vacuum condensate

The mixing transformations among three generations of Dirac fields are: $\Psi_f(x) = \mathcal{U}\,\Psi_m(x)$, where $\mathcal{U}$ is the CKM matrix and $\Psi_m^T = (\psi_1, \psi_2, \psi_3)$ are the fields with definite masses $m_1 \neq m_2 \neq m_3$. The mixing relations can be written as $\psi_\sigma^\alpha(x) \equiv G_\theta^{-1}(t)\, \psi_i^\alpha(x)\, G_\theta(t)$, where

$(\sigma, i) = (A,1),(B,2),(C,3)$; A, B, C denote lepton (e,μ,τ) or flavor (d,s,b) indices. $G_\theta(t)$ is the mixing generator [20, 28, 29]. The flavor annihilators, at each time, are defined as: $\alpha^r_{\mathbf{k},\sigma}(t) \equiv G^{-1}_\theta(t)\; \alpha^r_{\mathbf{k},i}(t)\; G_\theta(t)$, and $\beta^r_{\mathbf{k},\sigma}(t) \equiv G^{-1}_\theta(t)\; \beta^r_{\mathbf{k},i}(t)\; G_\theta(t)$. They annihilate the flavor vacuum $|0(t)\rangle_f = G^{-1}_\theta(t)\;|0\rangle_m$, where $|0\rangle_m$ is the vacuum annihilated by $\alpha^r_{\mathbf{k},i}$ and $\beta^r_{\mathbf{k},i}$, $i = 1,2,3$, $r = 1,2$. In the infinite volume limit $|0(t)\rangle_f$ turns out to be unitarily inequivalent to the vacuum $|0\rangle_m$ [24], [27]. Moreover, $|0(t)\rangle_f$ is a coherent condensate of particles. In the reference frame such that $\mathbf{k} = (0,0,|\mathbf{k}|)$, the numbers of condensed particles are:

$$\mathcal{N}^{\mathbf{k}}_1 \;=\; {}_f\langle 0(t)|N^{\mathbf{k},r}_{\alpha_1}|0(t)\rangle_f \;=\; {}_f\langle 0(t)|N^{\mathbf{k},r}_{\beta_1}|0(t)\rangle_f \;=\; s^2_{12}c^2_{13}\,|V^{\mathbf{k}}_{12}|^2 + s^2_{13}\,|V^{\mathbf{k}}_{13}|^2\,, \tag{1}$$

and similar relations for $\mathcal{N}^{\mathbf{k}}_2$, $\mathcal{N}^{\mathbf{k}}_3$. In Eq.(1), $N^{\mathbf{k},r}_{\alpha_i} = \alpha^{r\dagger}_{\mathbf{k},i}\alpha^r_{\mathbf{k},i}$, $N^{\mathbf{k},r}_{\beta_i} = \beta^{r\dagger}_{\mathbf{k},i}\beta^r_{\mathbf{k},i}$ with $i = 1,2,3$, and $V^{\mathbf{k}}_{ij}$ are the Bogoliubov coefficients entering the mixing transformations (see Refs.[20, 28, 29]).

As shown in Ref.[20], the vacuum condensate due to neutrino and quark mixing behaves as a perfect fluid. Indeed considering the Minkowski metric, the energy-momentum tensor density $\mathcal{T}_{\mu\nu}(x)$ for the fermion fields ψ_i, $i = 1,2,3$ is [32] : $\mathcal{T}_{\mu\nu}(x) := \frac{i}{2} : \left(\bar{\Psi}_m(x)\gamma_\mu \overleftrightarrow{\partial}_\nu \Psi_m(x)\right) :$. Then the energy momentum tensor density of the vacuum condensate is given by

$$\mathcal{T}^{cond}_{\mu\nu}(x) = {}_f\langle 0(t)| : \mathcal{T}_{\mu\nu}(x) : |0(t)\rangle_f\,, \tag{2}$$

for which the off-diagonal components are zero:

$$\int d^3x\; {}_f\langle 0(t)|\mathcal{T}_{0j}(x)|0(t)\rangle_f \;=\; \int d^3x\; {}_f\langle 0(t)|\mathcal{T}_{j\,l}(x)|0(t)\rangle_f \;=\; 0\,, \tag{3}$$

where $j \neq 0$, and $j \neq l$, respectively. Thus $\mathcal{T}^{cond}_{\mu\nu}$ is homogeneous and isotropic and can be written as

$$\mathcal{T}^{cond}_{\mu\nu} \;=\; diag(\mathcal{T}^{cond}_{00}\,, \mathcal{T}^{cond}_{11}\,, \mathcal{T}^{cond}_{22}\,, \mathcal{T}^{cond}_{33}\,)\,, \tag{4}$$

where

$$\int d^3x\; \mathcal{T}^{cond}_{00}(x) \;=\; \int d^3x\; {}_f\langle 0(t)| : \mathcal{T}_{00}(x) : |0(t)\rangle_f \;=\; 4\sum_i \int d^3\mathbf{k}\,\omega_{k,i}\,\mathcal{N}^{\mathbf{k}}_i\,, \tag{5}$$

$$\int d^3x\; \mathcal{T}^{cond}_{jj}(x) \;=\; \int d^3x\; {}_f\langle 0(t)| : \mathcal{T}_{jj}(x) : |0(t)\rangle_f \;=\; 4\sum_i \int d^3\mathbf{k}\,\frac{k_j k_j}{\omega_{k,i}}\,\mathcal{N}^{\mathbf{k}}_i\,, \tag{6}$$

(no summation on j is intended) and $\mathcal{N}^{\mathbf{k}}_i$ are the numbers of particles condensed in the vacuum (see Eq.(1)).

3. Particle mixing and dark energy

In Ref.[20] it has been shown that the energy density due to the vacuum condensate arising from particle mixing gives a contribution to the vacuum energy which evolves dynamically (see also Refs.[21], [22]). Moreover, it has been remarked that the negligible breaking of the Lorentz invariance of the vacuum is responsible of the very small value of the dark energy at the present epoch [20]. In this Section we report the main results obtained in Ref.[20].

Let us write the energy-momentum tensor density $\mathcal{T}_{\mu\nu}(x)$ for the fermion fields ψ_i, $i = 1,2,3$ as [33]

$$\begin{aligned} :\mathcal{T}_{\mu\nu}(x): \;&=\; :\Sigma_{\mu\nu}(x): + :\mathcal{V}_{\mu\nu}(x): \\ &=\; :\Big\{\frac{i}{2}\left(\bar{\Psi}_m(x)\gamma_\mu \overleftrightarrow{\partial}_\nu \Psi_m(x)\right) - \eta_{\mu\nu}\left[\frac{i}{2}\bar{\Psi}_m(x)\gamma^\alpha \overleftrightarrow{\partial}_\alpha \Psi_m(x)\right] \\ &+\; \eta_{\mu\nu}\left[\bar{\Psi}_m(x)\,\mathsf{M}_d\,\Psi_m(x)\right]\Big\}:\,, \end{aligned} \tag{7}$$

where

$$:\Sigma_{\mu\nu}(x): \;=\; :\left\{\frac{i}{2}\left(\bar{\Psi}_m(x)\gamma_\mu \overleftrightarrow{\partial}_\nu \Psi_m(x)\right) - \eta_{\mu\nu}\left[\frac{i}{2}\bar{\Psi}_m(x)\gamma^\alpha \overleftrightarrow{\partial}_\alpha \Psi_m(x)\right]\right\}:, \tag{8}$$

$$:\mathcal{V}_{\mu\nu}(x): \;=\; \eta_{\mu\nu}:\left[\bar{\Psi}_m(x)\,\mathsf{M}_d\,\Psi_m(x)\right]:, \tag{9}$$

$\mathsf{M}_d = diag(m_1, m_2, m_3)$, $\Psi_m = (\psi_1, \psi_2, \psi_3)^T$ and $\eta_{\mu\nu} = diag(1, -1, -1, -1)$.

In any epochs the contributions of the particle mixing to the vacuum energy density ρ_{mix} and to the vacuum pressure p_{mix} are given respectively by the $(0,0)$ and the (j,j) components of the energy momentum tensor density of the vacuum condensate:

$$\rho_{mix} \equiv \frac{1}{V}\eta^{00}\int d^3x\, \mathcal{T}^{cond}_{00}(x)\,, \tag{10}$$

$$p_{mix} \equiv -\frac{1}{V}\eta^{jj}\int d^3x\, \mathcal{T}^{cond}_{jj}(x)\,, \tag{11}$$

where no summation on the index j is intended. By using Eqs.(5) and (6) we have:

$$\rho_{mix} = \frac{2}{\pi}\sum_i \int dk\, k^2\, \omega_{k,i}\, \mathcal{N}_i^{\mathbf{k}}, \tag{12}$$

$$p_{mix} = \frac{2}{3\pi}\sum_i \int dk\, k^2\, \frac{k^2}{\omega_{k,i}}\, \mathcal{N}_i^{\mathbf{k}}. \tag{13}$$

We note that the use of the identity $\omega_{k,i} = \frac{k^2}{\omega_{k,i}} + \frac{m_i^2}{\omega_{k,i}}$ enables to write ρ_{mix} as

$$\rho_{mix} = \Sigma_{mix} + \mathcal{V}_{mix} \tag{14}$$

where the kinetic term Σ_{mix} and the potential term $\mathcal{V}_{mix}$ are given by

$$\Sigma_{mix} = \frac{2}{\pi}\sum_i \int_0^K dk\, k^2\, \frac{k^2}{\omega_{k,i}}\, \mathcal{N}_i^{\mathbf{k}}, \qquad \mathcal{V}_{mix} = \frac{2}{\pi}\sum_i \int_0^K dk\, k^2\, \frac{m_i^2}{\omega_{k,i}}\, \mathcal{N}_i^{\mathbf{k}}. \tag{15}$$

However, in the present epoch, the very small breaking of the Lorentz invariance [7], imposes that the vacuum expectation values of $\mathcal{T}_{\mu\nu}(x)$ are space-time independent. This implies that the kinematical part $\Sigma^{cond}_{\mu\nu}$ of $\mathcal{T}^{cond}_{\mu\nu}$ is today very small [20]:

$$\Sigma^{cond}_{\mu\nu} = {}_f\langle 0(t)|:\Sigma_{\mu\nu}(x):|0(t)\rangle_f \simeq 0 \tag{16}$$

and $\mathcal{T}^{cond}_{\mu\nu}$ is given by:

$$\mathcal{T}^{cond}_{\mu\nu} \simeq {}_f\langle 0(t)|:\mathcal{V}_{\mu\nu}(x):|0(t)\rangle_f = \eta_{\mu\nu}\, {}_f\langle 0(t)|:\bar{\Psi}_m(x)\,\mathsf{M}_d\,\Psi_m(x):|0(t)\rangle_f\,. \tag{17}$$

By using Eqs.(4) and (17), at the present epoch, we obtain:

$$\frac{1}{V}\int d^3x\, \mathcal{T}^{cond}_{\mu\nu}(x) = diag(\rho_{mix}, p_{mix}, p_{mix}, p_{mix})$$

$$= \eta_{\mu\nu}\sum_i m_i \int \frac{d^3x}{(2\pi)^3}\, {}_f\langle 0|:\bar{\psi}_i(x)\psi_i(x):|0\rangle_f\,. \tag{18}$$

The today state equation is then given by: $\rho_{mix} \simeq -p_{mix}$, that is, the adiabatic index is

$$w_{mix} = p_{mix}/\rho_{mix} \simeq -1\,. \tag{19}$$

This means that the vacuum condensate coming from particle mixing, today, mimics the behavior of the cosmological constant [21], [22]. From Eq.(18), introducing the cut-off on the momenta K, we derive ρ_{mix} [20]:

$$\rho_{mix} \simeq \frac{2}{\pi}\sum_i \int_0^K dk\, k^2\, \frac{m_i^2}{\omega_{k,i}}\, \mathcal{N}_i^{\mathbf{k}}\,. \tag{20}$$

The integral (20) diverges in K as $m_i^4 \log(2K/m_j)$, with $i,j = 1,2,3$ [22]. However, the value close to -1 of w_{mix} at the present epoch constrains the value of K and consequently the value of ρ_{mix} [20]. Indeed, by comparing Eqs.(12) and (20), we note that the condition (16) for the $(0,0)$ component of $\mathcal{T}_{\mu\nu}$ imposes $\Sigma_{mix} \ll \mathcal{V}_{mix}$, where Σ_{mix} and $\mathcal{V}_{mix}$ are given in Eq.(15). Such a condition is satisfied when K in Eq.(15) is $K \ll \sqrt[3]{m_1 m_2 m_3}$. For this reason here we do not need to consider the regularization problem of the ultraviolet divergence of the integral (20).

We now show that particular values of w_{mix}, for neutrinos and for quarks, lead to contributions of the dark energy that are in agreement with its estimated upped bound. To do that we derive an expression of w_{mix} as function of the cut-off K [20].

Let us consider the adiabatic expansion of a sphere of volume V. Since the pressure p, at which the sphere expands, does work, the total energy, $E = \rho V$, varies: $dE = -p\, dV$. That is $\rho\, dV + V\, d\rho = -p\, dV$, that can be expressed as $d[(\rho + p)V] = 0$, and then

$$\rho + p = \frac{const}{V}\,. \tag{21}$$

Thus if the volume is very large ($V \to \infty$), we have $\rho \simeq -p$ and $w = p/\rho \simeq -1$.

Being $\rho = \Sigma + \mathcal{V}$ where Σ and $\mathcal{V}$ are the kinetic and the potential terms respectively, and taking into account Eq.(21), for a volume $V \to \infty$, we obtain $\rho = \Sigma + \mathcal{V} \simeq -p$. Taking into account the condition $\Sigma \ll \mathcal{V}$ imposed by the negligible breaking of the Lorentz invariance at the present epoch, we have $\rho \simeq \mathcal{V} \simeq -p$. Then, from the equality: $\rho = \frac{const}{V} - p = \Sigma + \mathcal{V}$, we have $\Sigma \simeq \frac{const}{V} \simeq 0$, for $V \to \infty$. This implies that, at the present epoch, ρ can be written as $\rho = \Sigma - p \simeq -p$. Then, in the case of the flavor vacuum condensate, w_{mix} is given by:

$$w_{mix} = \frac{p_{mix}}{\Sigma_{mix} - p_{mix}}\,. \tag{22}$$

Since today $\Sigma_{mix} \sim 0$, then $w_{mix} \sim -1$. Eq.(22) gives an expression of w_{mix} as function of K (since Σ_{mix} and p_{mix} are function of K).

By using different values of w_{mix} close to -1, we compute the contributions given to the dark energy by the particle mixing condensates, at the present epoch. We found the following results [20]:

Neutrino mixing condensate contribution:

$\rho_{mix}^{\nu} \sim 10^{-47} GeV^4$ for $-0.98 \leq w_{mix}^{\nu} \leq -0.97$. Such contributions are compatible with the estimated upper bound of the dark energy and w_{mix}^{ν} is in agreement with the constraint on the dark energy state equation [7].

For $w_{mix}^{\nu} < -0.98$ we have negligible contributions of ρ_{mix}^{ν}. The results we found are dependent on the neutrino mass values one uses. (We have considered values of the neutrino masses such that the experimental values of squared mass difference [34] are satisfied, as for example: $m_1 = 4.6 \times 10^{-3} eV$, $m_2 = 1 \times 10^{-2} eV$, $m_3 = 5 \times 10^{-2} eV$).

Quark mixing condensate contribution:

A contribution compatible with the estimated upper bound of the dark energy: $\rho_{mix}^{q} = 1.5 \times 10^{-47} GeV^4$ is found for $w_{mix}^{q} = -1$ [20]. Very small deviations from the value $w_{mix}^{q} = -1$ give rise to contributions of ρ_{mix}^{q} that are beyond the accepted upper bound of the dark energy.

4. Conclusions and discussion

In this report we have shown that the vacuum condensate from particle mixing provides a contribution to the dark energy which is compatible with the estimated value of the cosmological constant. Such value is imposed by the small breaking of the Lorenz of the flavor vacuum at the present epoch.

Acknowledgements

Support from INFN and Miur is acknowledged.

References

[1] Perlmutter S *et al* 1999 *ApJ* **517** 565;
Knop R A *et al* 2003 *ApJ* **598** 102
[2] Riess A G *et al* 1998 *AJ* **116** 1009;
Tonry J L *et al* 2003 *ApJ* **594** 1
[3] de Bernardis P *et al* 2000 *Nature* **404** 955
[4] Stompor R *et al* 2001 *ApJ* **561** L7
[5] Spergel D N *et al* 2003 *ApJS* **148** 175
[6] Riess A G *et al* 2004 *ApJ* **607** 665
[7] Hinshaw G *et al* 2008 *Preprint* arXiv:astro-ph/0803.0732v1;
Hill R *et al* 2008 *Preprint* arXiv:astro-ph/0803.0570v1;
Gold B *et al* 2008 *Preprint* arXiv:astro-ph/0803.0715v1;
Wright E *et al* 2008 *Preprint* arXiv:astro-ph/0803.0577v1;
Nolta M *et al* 2008 *Preprint* arXiv:astro-ph/0803.0593v1;
Dunkley J *et al* 2008 *Preprint* arXiv:astro-ph/0803.0586v1;
Komatsu E *et al* 2008 *Preprint* arXiv: astro-ph/0803.0547v1.
[8] Weinberg S 1989 *Rev. Mod. Phys.* **61** 1
[9] Guth A 1981 *Phys. Rev.* D **23** 347; *do.* 1982 *Phys. Lett.* B **108** 389
[10] A.D. Linde A D 1982 *Phys. Lett.* B **108** 389; *do.* 1982 *Phys. Lett.* B **114** 431; *do.* 1983 *Phys. Lett.* B **129** 177; *do.* 1990 *Phys. Lett.* B **238** 160
[11] Sahni V and Starobinski A 2000 *Int. J. Mod. Phys.* D **9** 373
[12] Sahni V 2004 *Lect. Notes Phys.* **653** 141;
Carroll S M 2001 *Living Rev. Rel.* **4** 1;
Peebles P J E and Ratra B 2003 *Rev. Mod. Phys.* **75**;
Copeland E J, Sami M and Tsujikawa S 2006 *Int. J. Mod. Phys.* D **15** 1753
[13] Padmanabhan T 2003 *Phys. Rep.* **380** 235
[14] Kamenshchik A, Moschella U and Pasquier V 2001 *Phys. Lett.* B **511** 265
[15] Padmanabhan T 2002 *Phys. Rev.* D **66** 021301
[16] Bassett B A, Kunz M, Parkinson D and Ungarelli C 2003 *Phys. Rev.* D **68** 043504
[17] Freese K and Lewis M 2002 *Phys. Lett.* B **540** 1
[18] Dvali G R, Gabadadze G and Porrati M 2000 *Phys. Lett.* B **485** 208
[19] Nojiri S and Odintsov S D 2003 *Phys. Rev.* D **68** 123512;
Carroll S M, Duvvuri V, Trodden M and Turner M S 2004 *Phys. Rev.* D **70** 043528;
Allemandi G, Borowiec A and Francaviglia M 2004 *Phys. Rev.* D **70** 103503
[20] Capolupo A, Capozziello S and Vitiello G 2009 *Phys. Lett.* A **373** 601-610
[21] Capolupo A, Capozziello S and Vitiello G 2007 *Phys. Lett.* A **363** 53; *do.* 2007 *J. Phys.: Conf. Ser.* **67** 012032
[22] Capolupo A, Capozziello S and Vitiello G 2008 *Int. J. Mod. Phys.* A **23** 4979-4990
[23] Blasone M, Capolupo A, Capozziello S, Carloni S and Vitiello G 2004 *Phys. Lett.* A **323** 182;
Blasone M, Capolupo A, Capozziello S and Vitiello G 2008 *Nucl. Instrum. Meth.* A **588** 272
[24] Blasone M and Vitiello G 1995 *Annals Phys.* **244** 283
[25] Fujii K, Habe C and Yabuki T 1999 *Phys. Rev.* D **59** 113003; *do.* 2001 *Phys. Rev.* D **64** 013011
[26] Ji C R and Mishchenko Y 2001 *Phys. Rev.* D **64** 076004; *do.* 2002 *Phys. Rev.* D **65** 096015
[27] Hannabuss K C and Latimer D C 2000 *J. Phys.* A **33** 1369; *do.* 2003 *J. Phys.* A **36** L69
[28] Blasone M, Capolupo A and Vitiello G 2002 *Phys. Rev.* D **66** 025033
[29] Capolupo A 2004 Ph.D. Thesis *Preprint* arXiv:hep-th/0408228
[30] Blasone M, Capolupo A, Romei O and Vitiello G 2001 *Phys. Rev.* D **63** 125015
Capolupo A, Ji C R, Mishchenko Y and Vitiello G 2004 *Phys. Lett.* B **594** 135

[31] Blasone M, Capolupo A, Terranova F and Vitiello G 2005 *Phys. Rev.* D **72** 013003;
Blasone M, Capolupo A and Vitiello G 2005 *Acta Phys. Polon.* B **36** 3245;
Blasone M, Capolupo A, Ji C R and Vitiello G 2006 *Preprint* arXiv:hep-ph/0611106
[32] Itzykson C and Zuber J B 1980 *Quantum Field Theory* (New York: McGraw-Hill);
Schweber S 1961 *An Introduction Relativistic Quantum Field Theory* (Harper and Row)
[33] Leite Lopez J 1981 *Gauge Field Theories* (Pergamon Press)
[34] Altarelli G 2007 *Preprint* arXiv:hep-ph/0711.0161

Multipartite geometric entanglement in finite size XY model

Massimo Blasone[1,2], Fabio Dell'Anno[1,2,3], Silvio De Siena[1,2,3], Salvatore Marco Giampaolo[1,2,3] and Fabrizio Illuminati[1,2,3,4]

[1] Dipartimento di Matematica e Informatica, Università degli Studi di Salerno, Via Ponte don Melillo, I-84084 Fisciano (SA), Italy
[2] INFN Sezione di Napoli, Gruppo collegato di Salerno, Baronissi (SA), Italy
[3] CNR-INFM Coherentia, Napoli, Italy
[4] ISI Foundation for Scientific Interchange, Viale Settimio Severo 65, I-10133 Torino, Italy

E-mail: blasone@sa.infn.it

Abstract. We investigate the behavior of the multipartite entanglement in the finite size XY model by means of the hierarchical geometric measure of entanglement. By selecting specific components of the hierarchy, we study both global entanglement and genuinely multipartite entanglement.

1. Introduction

Quantum entanglement is one of the most peculiar features of quantum mechanics and is a fundamental resource in quantum information science [1]. In recent years, quantum correlations have been extensively investigated in the context of quantum many-body systems [2]. The behavior of quantum entanglement has been analyzed [3, 4, 5] in connection with quantum phase transitions and collective phenomena in strongly correlated systems [6, 7], with particular attention on quantum spin chains with Heisenberg interaction [8, 9]. Indeed, spin chains represent scalable qubit system suitable for quantum information tasks [10]. Recently, in order to observe finite size effects, the entanglement has been examined in finite spin chains [11].

In this paper, we exploit the hierarchical geometric measure of entanglement, introduced in Ref. [12], in order to characterize the multipartite entanglement in the finite size XY model. This measure can be used to distinguish the global multipartite entanglement from genuinely multipartite entanglement [12]. The hierarchical measure is a generalization of the Wei and Goldbart geometric measure [13]. The latter is defined as the Euclidean distance of a given multipartite state to the nearest fully separable state. The geometric measure can be considered as one of the most reliable quantifiers of multi-particle entanglement [14]: It exhibits interesting connections with other measures [13, 15] and can be efficiently estimated by quantitative entanglement witnesses amenable of experimental verification [16, 17]. The geometric measure can be extended by the convex roof procedure to the case of mixed states, and is a proper multipartite entanglement monotone. Remarkably, the geometric measure can be effectively exploited to quantify the entanglement of two distinct multipartite bound entangled states [18] and to study the behavior of global entanglement at the approach of quantum phase transitions

[3, 4, 5]. Generalized versions of this measure have been proposed [19]. Furthermore, procedures for the analytical determination of the geometric measure have been presented [12, 20].

In section 2, we review the hierarchical geometric measure of entanglement. In section 3, we resume the XY model: we introduce the Heisenberg Hamiltonian and we recall the essential formalism and the main mathematical tools necessary for the development of the discussion, including the lowest energy states of the system. In section 4, we introduce the hierarchical geometric components useful for the characterization of global multipartite entanglement and of genuine multipartite entanglement. Finally, in section 5 we present our results.

2. Hierarchical geometric measure of entanglement

Let us review the hierarchical geometric measure of entanglement, introduced in Ref. [12]. We consider a N-qubit system, corresponding to a tensor-product state space $\mathcal{H}^{d_N}$ of dimension $d_N = 2^N$. Let us introduce an integer K, $2 \leq K \leq N$, and the ordered sequence of integers $\{M_1, M_2, \ldots, M_K\}$, where $M_1 \leq M_2 \leq \ldots \leq M_K$, and $\sum_{s=1}^{K} M_s = N$. Let us consider the K-partition of the system in K subsystems described by the sets $\{Q_s\}_{s=1}^{K}$. Each set Q_s is composed of M_s elementary parties, i.e. $Q_s = \{i_1^{(s)}, i_2^{(s)}, \ldots, i_{M_s}^{(s)}\}$, where $i_j^{(s)} \in \{1, \ldots, N\}$ is a discrete index labeling the N elementary parties, and $Q_s \bigcap Q_{s'} = \emptyset$ for $s \neq s'$. Given a generic K-partition $Q_1|Q_2|\ldots|Q_K$ of the N-qubit system, any K-separable state associated to such a partition is defined as the tensor product of K M_s-qubit pure states $|\Phi_s^{(Q_s)}\rangle$. Each state $|\Phi^{(Q_s)}\rangle$ belongs to the Hilbert space $\mathcal{H}^{d_{Q_s}}$ of dimension $d_{Q_s} = 2^{M_s}$. A K-separable state can then be written as

$$\bigotimes_{s=1}^{K} |\Phi_s^{(Q_s)}\rangle \,. \tag{1}$$

Correspondingly, the Hilbert space $\mathcal{H}^{d_N}$ is decomposed in the tensor product $\bigotimes_{s=1}^{K} \mathcal{H}^{d_{Q_s}}$. Varying the integers M_s, one obtains different K-partitions $Q_1|Q_2|\ldots|Q_K$ and, correspondingly, different possible K-separable states. We denote by $\mathbf{S}_K$ the set of all K-separable states, defined as

$$\mathbf{S}_K = \bigcup_{\{Q_1,\ldots,Q_K\}} S_K(Q_1|\ldots|Q_K) \,, \tag{2}$$

where $S_K(Q_1|Q_2|\ldots|Q_K)$ is the set of all the K-separable states associated to a fixed K-partition. We can now define the geometric measures of entanglement with respect to K-separable pure states for an arbitrary N-qubit pure state $|\Psi^{(N)}\rangle$ as:

$$E_G^{(K)}(Q_1|\ldots|Q_K) = -\frac{1}{N} \log_2 \Lambda_K^2(Q_1|\ldots|Q_K) \,, \tag{3}$$

where the squared overlap is defined as

$$\Lambda_K^2(Q_1|\ldots|Q_K) = \max_{|\varphi\rangle \in S_K(Q_1|\ldots|Q_K)} \left|\langle\varphi|\Psi^{(N)}\rangle\right|^2 . \tag{4}$$

3. XY model

In this section we briefly review the anisotropic XY model described by the following spin 1/2 Hamiltonian:

$$H_{xy} = -\sum_{i=1}^{N} \left[(1+\gamma) S_i^x S_{i+1}^x + (1-\gamma) S_i^y S_{i+1}^y\right] - h \sum_{i=1}^{N} S_j^z \,, \tag{5}$$

where N is the number of sites, and the operators S_i^α, with $\alpha = x, y, z$, can be expressed in terms of the Pauli matrices σ_i^α, i.e. $S_i^\alpha = 2^{-1}\sigma_i^\alpha$, and satisfy the commutation relation

$[S_k^\alpha, S_h^\beta] = i\delta_{k,h}\epsilon_{\alpha\beta\gamma}S_k^\gamma$. Here we assume the periodic boundary condition $S_{N+1}^\alpha \equiv S_1^\alpha$. By following a standard procedure, the Hamiltonian H_{xy} can be written in diagonal form in terms of spinless fermion operators [7, 8, 9]. Indeed, the spin operators of Eq. (5) are mapped to the fermion operators c_i, $c_i^\dagger$ by means of the Jordan-Wigner transformation, i.e. $S_k^- = \prod_{j=1}^{k-1}(1 - 2c_j^\dagger c_j)c_k^\dagger$, $S_k^+ = \left[S_k^-\right]^\dagger$, where $S_i^\pm$ are the lowering and raising spin operators. The spin states at site i, parallel $|\uparrow\rangle_i$, or antiparallel $|\downarrow\rangle_i$, to the z axis, correspond to the fermion vacuum $|0\rangle_i$, or to the one fermion state $|1\rangle_i$. It can be easily shown that, in the XY model, the parity of the number of fermions is conserved, i.e. the operator $\hat{\Omega} = \prod_{i=1}^N(1 - 2c_i^\dagger c_i)$ commutes with the Hamiltonian. Let us consider the simple instance of $N = 3$; the lowest energy states take the forms:

$$|\Psi_e^{(3)}\rangle = \left\{\cos\theta_{\frac{\pi}{3}} + \frac{1}{\sqrt{3}}\sin\theta_{\frac{\pi}{3}}(c_1^\dagger c_2^\dagger + c_1^\dagger c_3^\dagger + c_2^\dagger c_3^\dagger)\right\}|0\rangle\rangle, \tag{6}$$

$$|\Psi_o^{(3)}\rangle = \left\{\frac{1}{\sqrt{3}}\cos\theta_{\frac{2\pi}{3}}(c_1^\dagger + c_2^\dagger + c_3^\dagger) + \sin\theta_{\frac{2\pi}{3}}c_1^\dagger c_2^\dagger c_3^\dagger\right\}|0\rangle\rangle, \tag{7}$$

where e and o denote the even and odd fermion sectors, respectively, and $|0\rangle\rangle = |0\rangle_1|0\rangle_2|0\rangle_3$ is the fermion vacuum. The parameter θ_q is defined by the equation [7]:

$$\tan 2\theta_q = \frac{\gamma\sin q}{\cos q - h}. \tag{8}$$

The energies E as functions of the magnetic field h, at fixed anisotropy parameter γ, associated with the states (6) and (7), are plotted in Fig. 1. We observe that, at given $\gamma = 0.1, 0.4, 0.7$,

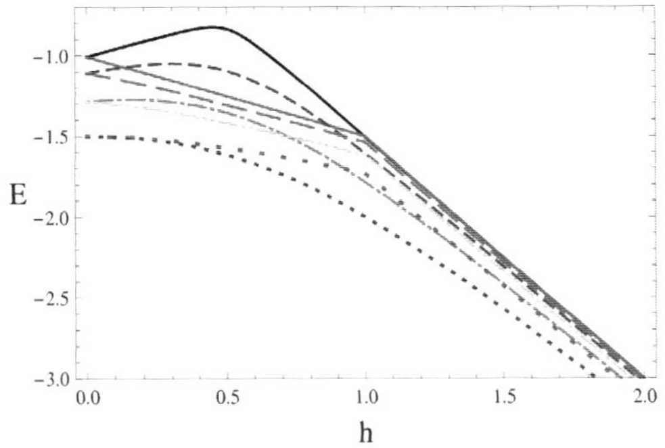

Figure 1. Energy E as a function of the magnetic field h, associated with the lowest energy states with definite parity $|\Psi_e^{(3)}\rangle$ and $|\Psi_o^{(3)}\rangle$. The anisotropy parameter is fixed at the values: a) $\gamma = 0.1$ (e black full line, o gray full line); b) $\gamma = 0.4$ (e dashed line, o long-dashed line); c) $\gamma = 0.7$ (e dot-dashed line, double-dot-dashed line); d) $\gamma = 1$ (e dotted line, o long-dotted line).

the energy curves corresponding to different parity (e and o) cross each other for nonvanishing h; for $\gamma = 1$ (Ising model), the curves intersect only at $h = 0$. We define the ground state $|\Psi_g^{(3)}\rangle$ of the system as follows:

$$|\Psi_g^{(3)}\rangle = \min_{\langle H_{XY}\rangle}\left\{|\Psi_e^{(3)}\rangle, |\Psi_o^{(3)}\rangle\right\}, \tag{9}$$

where $\langle H_{XY}\rangle = \langle\Psi_\alpha^{(3)}|H_{XY}|\Psi_\alpha^{(3)}\rangle$ $(\alpha = e, o)$.

4. Reference states and geometric components

In this section, we introduce the reference states and the geometric components used to characterize the multipartite entanglement for $N = 3$. Following Ref. [13], the reference state $|\varphi^{(3)}\rangle$ associated with the global geometric measure of entanglement is the fully separable state defined as:

$$|\varphi^{(3)}\rangle = \bigotimes_{i=1}^{3} \left[x_i + y_i S_i^-\right] | \uparrow\uparrow\uparrow\rangle \, , \tag{10}$$

where $x_j = \cos\delta_j$ and $y_j = \sin\delta_j$, and δ_i are real parameters. In terms of the fermion operators, Eq. (10) reads:

$$|\varphi^{(3)}\rangle = \bigotimes_{i=1}^{3} \left[x_i + y_i \, c_i^\dagger\right] |0\rangle\rangle \, , \tag{11}$$

The geometric component associated with the reference state (10), i.e. $E_G^{(3)}$, corresponds to a global measure of entanglement. According to the approach of Ref. [12], a complete characterization of multipartite geometric entanglement in the XY spin model would require the calculation of a further geometric component, i.e. $E_G^{(2)}(Q_1|Q_2)$. The geometric components associated to all the possible bipartitions of the system provide a quantification of genuine multipartite entanglement. In fact, if there exists a biseparable state that completely overlaps with the considered state, then the system is factorized in two subsystems, and the genuine multipartite entanglement is zero. Specifically, we will consider the bipartition (12|3), as, due to the translational invariance of the system, any other bipartition is equivalent to this one.

Therefore, we will consider the following reference states:

$$|\varphi^{(2)}\rangle = \left[x_1 + y_1 x_2 S_1^- + y_1 y_2 x_3 S_2^- + y_1 y_2 y_3 S_1^- S_2^-\right] \otimes \left[x_4 + y_4 S_3^-\right] | \uparrow\uparrow\uparrow\rangle \, , \tag{12}$$

where the operators S_i^- can be expressed in terms of the fermion operators c_i, $c_i^\dagger$ through the Jordan-Wigner transformations. Thus, in the next section, we will compute the following geometric components: the Wei and Golbart geometric measure $E_G^{(3)} = E_G^{(3)}(1|2|3)$, i.e. the minimum distance from the state (10); the component $E_G^{(2)}(12|3)$ associated with the bipartition (12|3), i.e. the minimum distances from the state (12).

5. Multipartite geometric entanglement: results

In this section we analyze the behavior of the multipartite entanglement in finite cyclic spin chains with $N = 3$ sites. In order to quantify the content of global entanglement, we compute the geometric component $E_G^{(3)}$, which is plotted in Fig. 2 as a function of the external magnetic field h and for different values of the anisotropy parameter γ, specifically $\gamma = 0.1, 0.4, 0.7, 1$ (panel a, b, c, and d, respectively). The quantity $E_G^{(3)}$ is plotted for $|\Psi_e^{(3)}\rangle$, $|\Psi_o^{(3)}\rangle$, and $|\Psi_g^{(3)}\rangle$, i.e Eqs. (6), (7), and (9), respectively.

We see that as h increases, $E_G^{(3)}(|\Psi_e^{(3)}\rangle)$ goes to zero. In fact, the ground state associated with a strong magnetic field tends to the state $| \uparrow\uparrow\uparrow\rangle$ (corresponding to the fermion vacuum), exhibiting a complete alignment of all the spins. On the contrary, for increasing h, $E_G^{(3)}(|\Psi_o^{(3)}\rangle)$ tends to an asymptotic saturation value as the state $|\Psi_o^{(3)}\rangle$ goes to the W state $|W\rangle = \frac{1}{\sqrt{3}}(| \downarrow\uparrow\uparrow\rangle + | \uparrow\downarrow\uparrow\rangle + | \uparrow\uparrow\downarrow\rangle)$. Let us recall that, in the thermodynamic limit $N \to \infty$, at the critical field $h_c = 1$ the system undergoes a quantum phase transition. Moreover, at the magnetic field $h_f = (1 - \gamma^2)^{1/2}$ the system can be in a factorized (fully product) state [3, 21]. For $h = h_f$, the energy is degenerate, i.e. the lowest energy states, associated with the two distinct parity sectors, possess the same energy. Thus, a superposition of these states can be chosen as the ground state of the system in a way that the state is fully factorized. From a physical point of view, a (small) external perturbation can be used to induce a controlled relaxation of the system towards a specific state.

In Fig. 2, we see that, for $\gamma = 0.1, 0.4, 0.7$, $E_G^{(3)}(|\Psi_g^{(3)}\rangle)$ exhibits a parity jump in correspondence of the magnetic field h_f. In fact, for $h < h_f$, the ground state coincides with the state $|\Psi_o^{(3)}\rangle$ (odd fermion sector), while for $h > h_f$ there is a jump to the even sector. For $\gamma = 1$, no parity jump can be observed, as the lowest energy state is even for any $h > 0$.

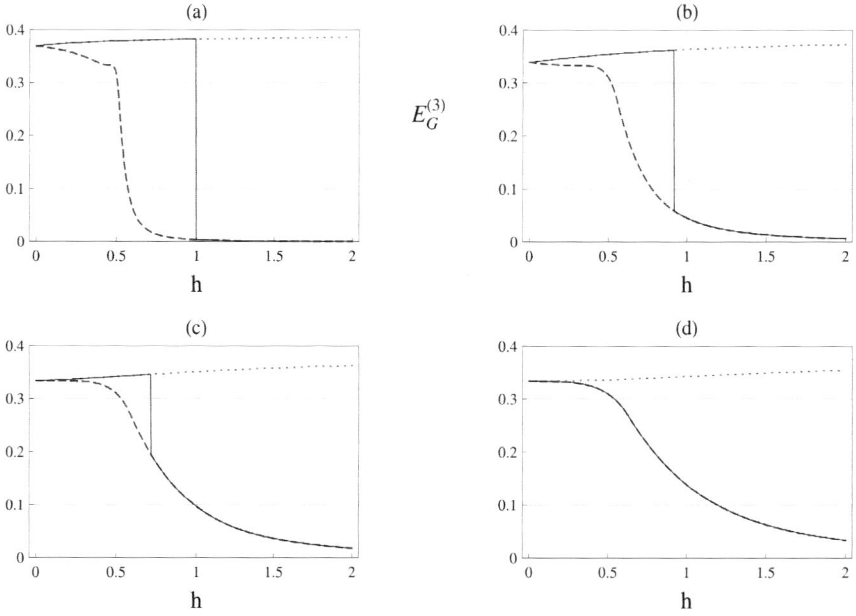

Figure 2. Geometric measure $E_G^{(3)}$ as a function of the magnetic field h associated with the lowest energy states $|\Psi_e^{(3)}\rangle$ (dashed line), $|\Psi_o^{(3)}\rangle$ (dotted line), and $|\Psi_g^{(3)}\rangle$ (full line). The anisotropy parameter is fixed at the values: (a) $\gamma = 0.1$, (b) $\gamma = 0.4$, (c) $\gamma = 0.7$, (d) $\gamma = 1$.

In order to analyze the behavior of genuine multipartite entanglement, we consider the geometric component $E_G^{(2)}(12|3)$, which is plotted in Fig. 3. We observe an overall suppression of the entanglement for any h, more marked for $h < h_f$. For $\gamma = 0.1, 0.4, 0.7$, the $E_G^{(2)}(|\Psi_e^{(3)}\rangle)$ exhibits an absolute maximum; for $\gamma = 1$ it is a decreasing function of h. $E_G^{(2)}(|\Psi_o^{(3)}\rangle)$ is a decreasing function of h for any γ. Looking at $E_G^{(2)}(|\Psi_g^{(3)}\rangle)$, we see that the parity jump is still present for $\gamma = 0.1, 0.4, 0.7$ as well as for the fully-separable component $E_G^{(3)}$; for $\gamma = 1$, the lowest energy state is even for any $h > 0$.

6. Conclusions

In conclusion, we have applied the hierarchical geometric measure of entanglement to a simple quantum spin chain, made by three spins only. This application shows that such a measure provides a complete characterization of the multipartite entanglement of the system under consideration. In fact, it allows to distinguish between the global and the genuine multipartite entanglement.

Such a kind of analysis can be extended to larger spin chains, which can be used for realizations of information protocols, such as teleportation. Obviously, when a large number of

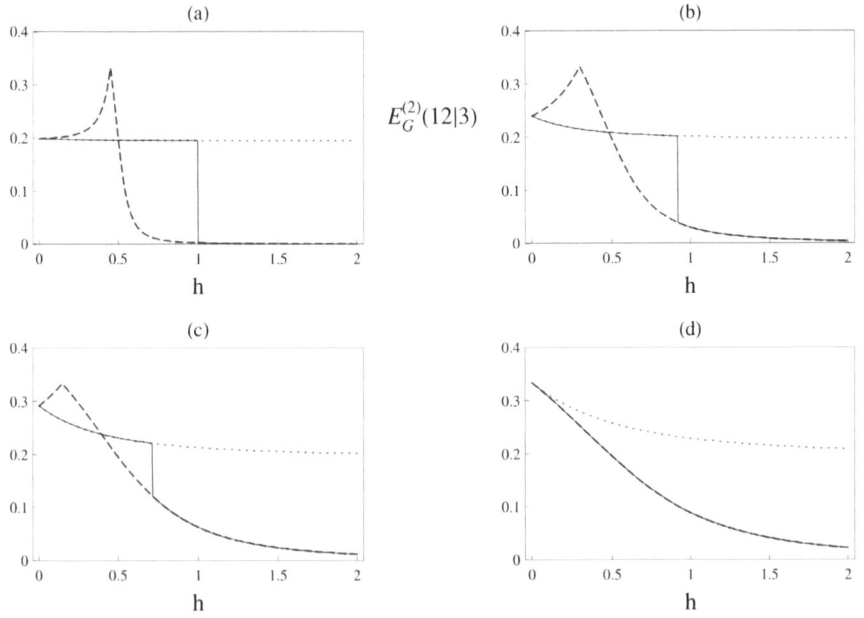

Figure 3. Geometric measure $E_G^{(2)}(12|3)$ as a function of the magnetic field h associated with the lowest energy states $|\Psi_e^{(3)}\rangle$ (dashed line), $|\Psi_o^{(3)}\rangle$ (dotted line), and $|\Psi_g^{(3)}\rangle$ (full line). The anisotropy parameter is fixed at the values: (a) $\gamma = 0.1$, (b) $\gamma = 0.4$, (c) $\gamma = 0.7$, (d) $\gamma = 1$.

spin sites is considered, the number of possible multipartions grows considerably and application of the present approach becomes a very demanding task.

Another interesting direction of investigation is to look at the various components of multipartite entanglement in connection with a quantum phase transition, by extending the results of Ref.[3].

In both the above outlined investigations, it will be interesting to consider the dependence of multipartite entanglement (and its decay rate) on suitably defined scales of distance.

References

[1] Nielsen M A and Chuang I L 2000, *Quantum Information and Quantum Computation* (Cambridge, UK: Cambridge U. Press)

[2] Osborne T J and Nielsen 2002 *Phys. Rev.* A **66** 032110; Osterloh A *et al* 2002 *Nature* **416** 608; Vidal G, Latoree J I, Rico E and Kitaev A 2003 *Phys. Rev. Lett.* **90** 227902; Roscilde T *et al* 2004 *Phys. Rev. Lett.* **93** 167203

[3] Wei T-C, Das D, Mukhopadyay S, Vishveshwara S and Goldbart P M 2005 *Phys. Rev.* A **71** 060305(R)

[4] Orús R 2008 *Phys. Rev. Lett.* **100** 130502

[5] Orús R, Dusuel S and Vidal J 2008 *Phys. Rev. Lett.* **101** 025701

[6] Sachdev A 1999 *Quantum Phase Transitions* (Cambridge, UK: Cambridge U. Press)

[7] Takahashi M 1999 *Thermodynamics of One-Dimensional Solvable Models* (Cambridge, UK: Cambridge U. Press)

[8] Lieb E H, Schultz T and Mattis D 1961 *Ann. Phys. (N.Y.)* **16** 417

[9] Barouch E, McCoy B M and Dresden M 1970 *Phys. Rev.* A **2** 1075; Barouch E and McCoy B M 1971 *Phys. Rev.* A **3** 786

[10] Di Vincenzo D P *et al* 2000 *Nature* **408** 339; Benjamin S C and Bose S 2003 *Phys. Rev. Lett.* **90** 247901

[11] Rossignoli R, Canosa N and Matera J M 2008 *Phys. Rev.* A **77** 052322; De Pasquale A, Costantini G, Facchi P, Florio G, Pascazio S and Yuasa K 2008 *Eur. Phys. J. Special Topics* **160** 127

[12] Blasone M, Dell'Anno F, De Siena S and Illuminati F 2008 *Phys. Rev.* A **77** 062304

[13] Wei T-C and Goldbart P M 2003 *Phys. Rev.* A **68** 042307
[14] Eisert J and Gross D 2006 in *Lectures on quantum information* eds D Bruss and G Leuchs (Weinheim: Wiley-VCH)
[15] Cavalcanti D 2006 *Phys. Rev.* A **73** 044302
[16] Gühne O, Reimpell M and Werner R F 2007 *Phys. Rev. Lett.* **98** 110502
[17] Eisert J, Brandão F G S L and Audenaert K M R 2007 *New J. Phys.* **9** 46
[18] Wei T-C, Altepeter J B, Goldbart P M and Munro W J 2004 *Phys. Rev.* A **70** 022322
[19] Cao Y and Wang A M 2007 *J. Phys. A: Math. Theor.* **40** 3507; Ostapchuk D, Passante G, Kobes R and Kunstatter G *Preprint* arXiv:0707.4020
[20] Tamaryan L, Park D and Tamaryan S 2008 *Phys. Rev.* A **77** 022325; Tamaryan L, Park D, Son J-W, Tamaryan S 2008 *Phys. Rev.* A **78** 032304
[21] Giampaolo S M, Adesso G and Illuminati F 2008 *Phys. Rev. Lett.* **100** 197201

Fourth International Workshop DICE2008 IOP Publishing
Journal of Physics: Conference Series **174** (2009) 012065 doi:10.1088/1742-6596/174/1/012065

VIP experiment

S. BARTALUCCI[2], S. BERTOLUCCI[2], M. BRAGADIREANU[5], M. CATITTI[2], M. CARGNELLI[1], C. CURCEANU (PETRASCU)[2,5], S. DI MATTEO[2], J.-P. EGGER[6], C. GUARALDO[2], M. ILIESCU[2], T. ISHIWATARI[1], M. LAUBENSTEIN[3], J. MARTON[1], E. MILOTTI[4], D. PIETREANU[2], T. PONTA[5], D. L. SIRGHI[2,5], F. SIRGHI[2,5], L. SPERANDIO[2], O. VAZQUEZ DOCE[2], E. WIDMANN[1], J. ZMESKAL[1]

[1] "Stefan Meyer" Institute for Subatomic Physics, Boltzmanngasse 3, A-1090 Vienna, Austria
[2] Laboratori Nazionali di Frascati dell'INFN, CP 13, Via E. Fermi 40, I-00044, Frascati, Italy
[3] Laboratori Nazionali del Gran Sasso, S.S. 17/bis, I-67010 Assergi (AQ), Italy
[4] Dipartimento di Fisica, Università di Trieste and INFN– Sezione di Trieste, Via Valerio, 2, I-34127 Trieste, Italy
[5] "Horia Hulubei" National Institute of Physics and Nuclear Engineering, Str. Atomistilor no. 407, P.O. Box MG-6, Bucharest - Magurele, Romania
[6] Institut de Physique de l'Université, CH–2000 Neuchâtel, Switzerland

E-mail: `dorel.pietreanu@lnf.infn.it`

Abstract. The VIP (Violation of the Pauli Exclusion Principle) experiment is dedicated to check the validity of one of the basic principles of modern physics. This investigation is done searching for anomalous X-rays emitted by copper atoms in a conductor: any detection of these anomalous X-rays would mark a Pauli forbidden transition. VIP is currently taking data at the Gran Sasso underground laboratories, and its scientific goal is to improve by three-four orders of magnitude the previous limit on the probability of Pauli violating transitions, bringing it into the $10^{-29\div-30}$ region. The new experimental results and future plans are presented.

1. Introduction

The Pauli Exclusion Principle (PEP), direct consequence of the spin-statistics connection [1], has been tested in the last years by several experiments [2, 3, 4, 5, 6]. Regarding the violation of PEP for electrons, in 1988 Ramberg and Snow (RS) [7] performed a dedicated experiment, searching for anomalous X-ray transitions due to a small possible violation of PEP in a copper conductor. They established an upper limit on the probability for an electron circulating in the conductor to form a mixed symmetry state with the already present copper electrons as $\beta^2/2 \leq 1.7 \times 10^{-26}$.

2. The Experiment

The idea of the experiment was first suggested by George Snow [7] and was put forward in a paper of Greenberg and Mohapatra [8]: if there are electron states with wavefunctions with mixed symmetry, and if some electrons sit in the ground states of some atoms while others are free, then the free electrons may be captured in the ground atomic states, even though the ground (1S) states may already be filled with electron pairs. Such a radiative capture process is accompanied by the emission of X-rays, and because of the anomalous atomic electron configuration these X-rays differ from the characteristic X-rays emitted by the atom. The detection of these anomalous X-rays would indicate that non-Paulian atoms exist and would demonstrate a violation of the Pauli Exclusion Principle. One might argue that if such mixed symmetry states exist, any anomalous pair in a block of material should already fill the ground states of the atoms in the block and therefore no detection should be expected: George Snow suggested that one could reestablish a non-equilibrium situation simply by injecting fresh electrons with an external source, such as a current source. The current source substitutes the radioactive β-source in the experiment of Goldhaber and Scharff-Goldhaber [9] (as reinterpreted by Reines and Sobel[10]), which was actually the first search for anomalous X-rays.

VIP is a dedicated experiment for the measurements of the probability of the Pauli Exclusion Principle violation for electrons which uses the same methods like the Ramberg and Snow experiment, with a much better soft X-ray detector in a low-background experimental area - the INFN Gran Sasso underground laboratory. The detector consist of Charge-Coupled Devices (CCD), caracterized by the excellent background rejection capability, based on pattern recognition and good energy resolution (320 eV FWHM at 8 keV in the present measurement) [11].

The experimental method consists in the introduction of new electrons into a copper strip, by circulating a current, and in the search for X-rays resulting from the 2p $\rightarrow$ 1s anomalous radiative transition that occurs if one of the new electrons is captured by a copper atom and cascades down to the 1s state already filled by two electrons of opposite spin. The energy of the 2p$\rightarrow$1s transition, calculated with a multiconfiguration Dirac-Fock method with an estimated error smaller than 10 eV [12], would differ from the normal K_α transition by about 300 eV (7.729 keV instead of 8.04 keV), providing an unambiguous signal of the PEP violation. The measurement alternates periods without current in the copper strip, in order to evaluate the X-ray background in conditions where no PEP-violating transitions are expected to occur, to periods in which current is flown in the conductor, thus providing new electrons, which may possibly violate PEP. The rather straightforward analysis consists in the evaluation of the statistical significance of the normalized subtraction of the two spectra in the energy region of the PEP-violating transition.

3. Results

3.1. First Experimental Results

The first results were obtained in the period 21 November - 13 December 2005, at the Frascati National Laboratories of INFN, Italy, before moving and installing the apparatus at Gran Sasso National Laboratories(LNGS). Two types of measurements were performed:

- 14510 minutes (about 10 days) of measurements with a 40 A current circulating in the copper target;
- 14510 minutes of measurements without circulating current,

where CCD's were read-out every 10 minutes. The resulting calibrated in energy X-ray spectra are shown in figure 1. These spectra include data from 14 CCD's out of 16, because of noise problems in the remaining 2.

Both spectra, apart of the continuous background component generated by the cosmic ray background and natural radioactivity, display clear Cu K_α and K_β lines due to X-ray fluorescence also caused by the cosmic ray background and natural radioactivity. No other lines are present and this reflects the careful choice of the materials used in the setup, as for example the high purity copper and high purity aluminium.

The subtracted spectrum is shown in Figure 2 a) (whole energy scale) and b) (a zoom on the region of interest). Notice that the subtracted spectrum is normalized to zero within statistical

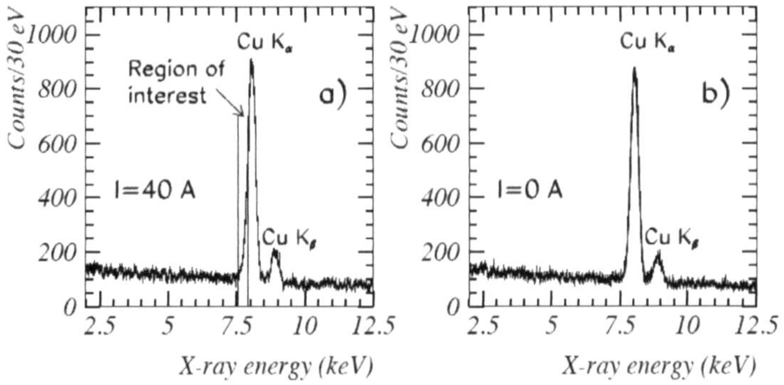

Figure 1. Energy spectra in the Frascati measurement: a) energy spectrum for the measurement with current (I=40 A); b) energy spectrum for the measurement without current.

error, and is structureless.

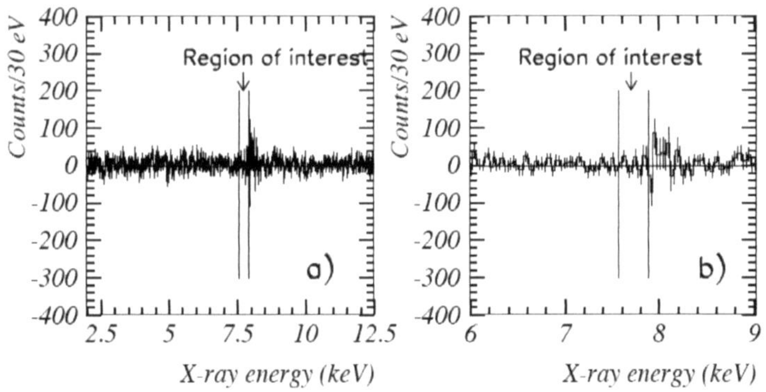

Figure 2. Subtracted energy spectra in the Frascati measurement, current minus no-current, giving the limit on PEP violation for electrons: a) whole energy range; b) expanded view in the region of interest (7.564 - 7.894 keV). No evidence for a peak in the region of interest is found.

The notation of Ignatiev and Kuzmin [13] , which has been incorporated in the paper of Greenberg and Mohapatra [8], was used: even though the model of Ignatiev and Kuzmin has been later shown to be incompatible with quantum field theory [14], the parameter β that measures the degree of PEP violation in the Ignatiev and Kuzmin model has stuck and is still found in the literature, also because it is easy to show that it is related to the parameter q of quon theory, by the relation [15]: $(1+q)/2 = \beta^2/2$ (in quon theory, $-1 \leq q \leq 1$, where $q = -1$ corresponds to fermions and $q = 1$ corresponds to bosons, so that here q must be close to -1 and $(1+q)/2$ must be very small, because we are dealing with electrons). Moreover, this parametrization was used for an easy comparison of our results with the RS experiment, which utilized the same formalism. To determine the experimental limit on $\beta^2/2$ from our data, using the same arguments of Ramberg and Snow [7, 16], we found [16]:

$$\frac{\beta^2}{2} \leq 4.5 \times 10^{-28} \text{ at } 99.7\% \text{ CL.}$$

3.2. Preliminary LNGS updated results

In order to reduce the background, the apparatus is currently installed in the LNGS underground laboratory, to reduce cosmicrays interactions, while the effects of natural radioactivity are moderated by a massive shield built by low-activity materials. The results presented here are based on one year of data acquisiton, in the above-mentioned conditions. Our data consist of two periods of data acquisition:

- 236005 minutes of measurements with a 40 A current circulating in the copper target;
- 172685 minutes of measurements without circulating current,

Performing the analysis as described in the previous section, from the subtracted spectra we find the new preliminary value for the violation parameter:

$$\frac{\beta^2}{2} \leq 6 \times 10^{-29} \text{ at } 99.7\% \text{ CL.}$$

improving the limit obtained by Ramberg and Snow by a factor about 250. Today this is the best value ever reached on the probability of PEP violation for the electrons.

4. Conclusions and Perspectives

A new measurement of the PEP violation limit for electrons is being performed by the VIP experiment. The search of a tiny violation is based on a measurement of PEP violating X-ray transitions in copper, under a circulating 40A current. A new limit for the PEP violation for electrons was established: 4.5×10^{-28}, lowering by about two orders of magnitude the previous one [7]. We have installed VIP at LNGS in spring 2006 and started the data-taking in reduced background condition. After one year of data taking the LNGS preliminary limit for the PEP violation for electrons was lowered to 6.0×10^{-29}, about three orders of magnitude better than the RS experiment. Data acquisition will continue at LNGS with the goal of bringing the limit on PEP violation for electrons into the 10^{-30} region, which is of particular interest [17] for all those theories related to possible PEP violation coming from new physics.

References

[1] Pauli W., *Phys. Rev.*, **58** (1940) 716.
[2] Bernabei R. et al., *Phys. Lett. B*, **408** (1997) 439.
[3] Borexino Collaboration (Back H. O. *et al.*), *Eur. Phys. J. C*, **37** (2004) 421.
[4] R. C. Hilborn and C. L. Yuca, Phys. Rev. Lett. **76** (1996) 2844.
[5] NEMO Collaboration, Nucl. Phys. **B87** (Proc. Suppl.) (2000) 510.
[6] E. Nolte et al., J. Phys. G: Nucl. Part. Phys. **17** (1991) S355.
[7] E. Ramberg and G. Snow, *Phys. Lett. B* **238**, 438 (1990).
[8] G. W. Greenberg and R. N. Mohapatra, *Phys. Rev. Lett.* **59**, 2507 (1987).
[9] M. Goldhaber and G. Scharff-Goldhaber, *Phys. Rev.* **73**, 1472 (1948).
[10] F. Reines and H. W. Sobel, *Phys. Rev. Lett.* **32**, 954 (1974).
[11] T. Ishiwatari et al., *Phys. Lett. B* **593**, 48 (2004); G. Beer et al., *Phys. Rev. Lett.* **94**, 212302 (2005).
[12] S. Di Matteo and L. Sperandio, VIP Note, IR-04, April 26, 2006 (the energy shift has been computed by P. Indelicato - private communication).
[13] A. Yu. Ignatiev and V. A. Kuzmin, *Yad. Fiz.* **46**, 786 (1987); see also ICTP preprint IC/87/13 (1987) and A. Yu Ignatiev, arXiv: hep-ph/0509258.
[14] A. B. Govorkov, *Phys. Lett.* **A137**, 7 (1989).
[15] O.W. Greenberg, *Phys. Rev.* **D43**, 4111 (1991).
[16] VIP Collab. (S. Bartalucci *et al.*), New experimental limit on the Pauli Exclusion Principle violation by electrons, to appear in *Phys. Lett. B*, also preprint arXiv:quant-ph/0605047.
[17] Duck I. and Sudarshan E. C. G., Am. J. Phys., 66 (1998) 284.

Naked singularity in a modified gravity theory

Remo Garattini

Università degli Studi di Bergamo, Facoltà di Ingegneria, Viale Marconi 5
24044 Dalmine (Bergamo) ITALY
INFN - sezione di Milano, Via Celoria 16, Milan, ITALY

E-mail: remo.garattini@unibg.it

Abstract. The cosmological constant induced by quantum fluctuation of the graviton on a given background is considered as a tool for building a spectrum of different geometries. In particular, we apply the method to the Schwarzschild background with positive and negative mass parameter. In this way, we put on the same level of comparison the related naked singularity $(-M)$ and the positive mass wormhole. We discuss how to extract information in the context of a $f(R)$ theory. We use the Wheeler-De Witt equation as a basic equation to perform such an analysis regarded as a Sturm-Liouville problem . The application of the same procedure used for the ordinary theory, namely $f(R) = R$, reveals that to this approximation level, it is not possible to classify the Schwarzschild and its naked partner into a geometry spectrum.

1. Introduction

Almost ten years ago, observation about type I supernovae data revealed that the Universe is in an acceleration phase [1]. Since then, no satisfactory explanation has been given. Indeed if the Friedmann-Robertson-Walker model of the universe, based on the Einstein's field equations is correct, the explanation of such an expansion should be due to approximately a 76% of what is known as *Dark Energy*. Dark Energy is based on the following equation of state $P = -\rho$ (where P and ρ are the pressure pf the fluid and the energy density, respectively). Dark Energy changes into *Phantom Energy* when $P < -\rho$. Nevertheless, neither Dark Energy nor Phantom Energy models appear to be satisfactory. In this scenario, the idea that General Relativity could be modified in something more general has been considered in recent years. In particular, the Einstein-Hilbert action $(\kappa = 8\pi G)$

$$S = \frac{1}{2\kappa}\int d^4x\sqrt{-g}R + S^{matter} \tag{1}$$

is replaced by[2]

$$S = \frac{1}{2\kappa}\int d^4x\sqrt{-g}f(R) + S^{matter}. \tag{2}$$

It is clear that other more complicated choices could be done in place of $f(R)$ [4]. In particular, one could consider $f\left(R, R_{\mu\nu}R^{\mu\nu}, R_{\alpha\beta\gamma\delta}R^{\alpha\beta\gamma\delta}, \ldots\right)$ or $f(R,G)$ where G is the Gauss-Bonnet invariant or any combination of these quantities [1]. Note that in principle one could consider

[1] For a recent review on $f(R)$, see Refs[3, 4], while a recent review on the problem of $f(G)$ and $f(R,G)$ can be found in[5].

Journal of Physics: Conference Series **174** (2009) 012066 doi:10.1088/1742-6596/174/1/012066

the replacement $R - 2\Lambda_c \to f(R)$ in such a way to avoid the use of the cosmological constant Λ_c. It is well known indeed, that there exists a factor of 120 orders of magnitude of discrepancy between the observed and the computed value. This huge disagreement is known as *cosmological constant problem*. It is important to remark that the cosmological constant plays an alternative rôle with respect to Dark Energy and modified gravity theories in explaining the acceleration of the Universe. Nevertheless, nothing forbids to consider them together in the context of the Wheeler-DeWitt equation (WDW)[6]

$$\mathcal{H}\Psi=0, \tag{3}$$

where [2]

$$\mathcal{H} = (2\kappa)\, G_{ijkl}\pi^{ij}\pi^{kl} - \frac{\sqrt{g}}{2\kappa}\left({}^3R - 2\Lambda_c\right), \tag{4}$$

${}^3R$ is the scalar curvature in three dimensions and G_{ijkl} is called the super-metric. $\mathcal{H}$ represents the time-time component of the Einstein's field equations without matter fields. It represents a constraint at the classical level and the invariance under *time* reparametrization. One can formally re-write the WDW equation as an eigenvalue problem

$$\frac{1}{V}\frac{\int \mathcal{D}\left[g_{ij}\right]\Psi^*\left[g_{ij}\right]\int_\Sigma d^3x\hat{\Lambda}_\Sigma\Psi\left[g_{ij}\right]}{\int \mathcal{D}\left[g_{ij}\right]\Psi^*\left[g_{ij}\right]\Psi\left[g_{ij}\right]} = \frac{1}{V}\frac{\left\langle\Psi\left|\int_\Sigma d^3x\hat{\Lambda}_\Sigma\right|\Psi\right\rangle}{\langle\Psi|\Psi\rangle} = -\frac{\Lambda_c}{\kappa}, \tag{5}$$

where

$$\hat{\Lambda}_\Sigma = (2\kappa)\, G_{ijkl}\pi^{ij}\pi^{kl} - \frac{\sqrt{g}}{2\kappa}{}^3R \tag{6}$$

and

$$V = \int_\Sigma d^3x\sqrt{g}. \tag{7}$$

It is clear that, what we interpret as an eigenvalue is an induced cosmological constant and, as pointed out in Ref.[8], we can use such an eigenvalue evaluated to one loop in different backgrounds as a tool to compute a geometrical "*spectrum*" based on Zero Point Energy (ZPE). In particular, in Ref.[8],we have computed the induced cosmological constant generated by a naked singularity associated to the Schwarzschild metric. The reason for such a choice is simple: if the Schwarzschild solution is

$$ds^2 = -\left(1 - \frac{2MG}{r}\right)dt^2 + \left(1 - \frac{2MG}{r}\right)^{-1}dr^2 + r^2\left(d\theta^2 + \sin^2\theta d\phi^2\right), \tag{8}$$

replacing M with $-M = \bar{M}$, we obtain a naked singularity with the following line element

$$ds^2 = -\left(1 + \frac{2\bar{M}G}{r}\right)dt^2 + \left(1 + \frac{2\bar{M}G}{r}\right)^{-1}dr^2 + r^2\left(d\theta^2 + \sin^2\theta d\phi^2\right). \tag{9}$$

It is immediate to see there is no horizon protecting the singularity. An immediate consequence of a negative Schwarzschild mass is that if one were to place two bodies initially at rest, one with a negative mass and the other with a positive mass, both will accelerate in the same direction going from the negative mass to the positive one. Furthermore, if the two masses are of the same magnitude, they will uniformly accelerate forever [3]. Another reason to consider a naked singularity described by the line element 9 is that it naturally represents a form of "*Dark Energy*" and therefore it deserves attention. In this paper we would like to extend the analysis

[2] See Ref.[7] for more details.

[3] See Ref.[9], for negative mass analysis in General Relativity. See also the problem of the *Cosmic Censorship Conjecture* postulated by R. Penrose[10]

Journal of Physics: Conference Series **174** (2009) 012066 doi:10.1088/1742-6596/174/1/012066

of Ref.[8] to $f(R)$ theories. Although the subject of this investigation is quite delicate, because as far as we know the subject of $f(R)$ theories combined with naked singularities has not been considered even at the classical level, it seems quite reasonable to apply the scheme of Eq.(5) to the negative Schwarzschild mass. Even in this case, we exclude the contribution of matter fields and the final result will be due only to quantum fluctuations. In practice, we desire to compute

$$\Delta\Lambda_c = \Lambda_c^S - \Lambda_c^N \geq (\leq)\ 0, \tag{10}$$

where $\Lambda_c^{S,N}$ are the induced cosmological constant computed in the different backgrounds. Moreover, the Schwarzschild solution for both masses, namely $\pm M$ is asymptotically flat. Therefore we are comparing backgrounds with the same asymptotically behavior. Nevertheless, in Eq.(3), surface terms never come into play because $\mathcal{H}$ as well as Λ_c/κ are energy densities and surface terms are related to the energy (e.g. ADM mass) and not to the energy density. We want to point up that even in the case of $f(R)$ theories, we are neither discussing the problem of forming the naked singularity nor a transition during a gravitational collapse. Rather the singularity is considered already existing. The semi-classical procedure followed in this work relies heavily on the formalism outlined in Refs.[8, 15].

2. Positive and negative Schwarzschild mass in a $f(R) = R$ theory

The Schwarzschild background is simply described by Eq.(8). In terms of the induced cosmological constant of Eq.(5), we get

$$\frac{\Lambda_{0,c}(\mu_0, r)}{8\pi G} = \frac{1}{64\pi^2}\sum_{i=1}^{2}\left(\frac{3MG}{r^3}\right)^2 \ln\left(\left|\frac{4r^3\mu_0^2}{3MG\sqrt{e}}\right|\right), \tag{11}$$

where we have removed the ultraviolet divergence with the help of a zeta function regularization and applied a Renormalization Group equation in order to avoid a dependence on the mass scale μ.

We know that an extremum appears, maximizing the induced cosmological constant for

$$\frac{3MG\sqrt{e}}{4r^3\mu_0^2} = \frac{1}{\sqrt{e}} \tag{12}$$

and leading to

$$\frac{\Lambda_{0,c}(\mu_0, r)}{8\pi G} = \frac{\mu_0^4}{4e^2\pi^2} \tag{13}$$

or [4]

$$\frac{\Lambda_{0,c}(\mu_0, r)}{8\pi G} = \left(\frac{3MG}{r^3}\right)^2 \frac{1}{64\pi^2} \qquad r \in \left[r_t, \frac{5}{4}r_t\right]. \tag{14}$$

Therefore, it appears that there exists a bound for $\Lambda_{0,c}$

$$\frac{9}{256\pi^2 r_t^4} \leq \frac{\Lambda_{0,c}(\mu_0, r)}{8\pi G} \leq \frac{225}{4096\pi^2 r_t^4}. \tag{15}$$

When we consider the naked Schwarzschild metric, we obtain an induced cosmological constant of the form

$$\frac{\Lambda_{0,c}^{naked}(\mu_0, r)}{8\pi G}$$

[4] See Refs.[7, 8] for technical details concerning the reasons of why $r \in \left[r_t, \frac{5}{4}r_t\right]$.

$$= \frac{1}{64\pi^2}\left[\left(\frac{15\bar{M}G}{r^3}\right)^2 \ln\left(\left|\frac{4r^3\mu_0^2}{15\bar{M}G\sqrt{e}}\right|\right) + \left(\frac{9\bar{M}G}{r^3}\right)^2 \ln\left(\left|\frac{4r^3\mu_0^2}{9\bar{M}G\sqrt{e}}\right|\right)\right]. \tag{16}$$

In order to find an extremum, it is convenient to define the following dimensionless quantity

$$\frac{9\bar{M}G\sqrt{e}}{4r^3\mu_0^2} = x, \tag{17}$$

then Eq.(16) becomes

$$\frac{\Lambda_{0,c}^{naked}(\mu_0, r)}{8\pi G} = -\frac{\mu_0^4}{4e\pi^2}\left[x^2 \ln x + \frac{25}{9}x^2 \ln\left(\frac{5x}{3}\right)\right]. \tag{18}$$

We find a solution when

$$\bar{x} = \frac{1}{\sqrt{e}}\left(\frac{3}{5}\right)^{\frac{25}{34}} \simeq 0.417 \tag{19}$$

corresponding to a value of

$$\frac{\Lambda_{0,c}^{naked}(\mu_0, r)}{8\pi G} = \frac{\mu_0^4}{4e^2\pi^2}\frac{17}{75}5^{\left(\frac{9}{17}\right)}3^{\left(\frac{8}{17}\right)} \simeq 0.328\frac{\mu_0^4}{4e^2\pi^2} = 0.328\frac{\Lambda_{0,c}(\mu_0, r)}{8\pi G}. \tag{20}$$

This means that

$$\frac{\Lambda_{0,c}^{naked}(\mu_0, r)}{\Lambda_{0,c}(\mu_0, r)} = 0.328 < 1. \tag{21}$$

A comment to this inequality is in order. Eq.(21) shows that the Schwarzschild naked singularity has a lower value of ZPE compared to the positive Schwarzschild mass. This means that, even if the order of magnitude is practically the same, the naked singularity is less favored with respect to the Schwarzschild wormhole. We now try to apply the same method to a modified gravity theory of the form $f(R)$.

3. Positive and negative Schwarzschild mass one loop energy for a generic $f(R)$ theory

In this section, we report the main steps discussed in Ref.[15] for a $f(R)$ theory in connection with the Sturm-Liouville problem of Eq.(5). Although a $f(R)$ theory does not need a cosmological constant, rather it should explain it, we shall consider the following Lagrangian density describing a generic $f(R)$ theory of gravity

$$\mathcal{L} = \sqrt{-g}\left(f(R) - 2\Lambda\right), \qquad with\ f'' \neq 0, \tag{22}$$

where $f(R)$ is an arbitrary smooth function of the scalar curvature and primes denote differentiation with respect to the scalar curvature. A cosmological term is added also in this case for the sake of generality, because in any case, Eq.(22) represents the most general lagrangian to examine. Obviously $f'' = 0$ corresponds to GR. The generalized Hamiltonian density for the $f(R)$ theory assumes the form [5]

$$\mathcal{H} = f'(R)\left[(2\kappa)G_{ijkl}\pi^{ij}\pi^{kl} - \frac{\sqrt{g}}{2\kappa}\left({}^{(3)}R - 2\Lambda_c\right)\right]$$

$$+2(2\kappa)\left[G_{ijkl}\pi^{ij}\pi^{kl} + \frac{\pi^2}{4}\right](f'(R) - 1) + \frac{1}{2\kappa}\left[V(\mathcal{P}) + 2g^{ij}\left(\sqrt{g}f'(R)\right)_{|ij}\right]. \tag{23}$$

[5] See Refs.[16, 15] for technical details.

where

$$\mathcal{P} = -\,6\sqrt{g}f'(R) \tag{24}$$

and

$$V(\mathcal{P}) = \sqrt{g}\left[Rf'(R) - f(R)\right]. \tag{25}$$

When $f(R) = R$, $V(\mathcal{P}) = 0$ as it should be. By imposing the Hamiltonian constraint and integrating over the hypersurface Σ, we obtain

$$\int_\Sigma d^3x \left\{ \left[(2\kappa)\, G_{ijkl}\pi^{ij}\pi^{kl} - \frac{\sqrt{g}}{2\kappa}{}^{(3)}R\right] + (2\kappa)\left[G_{ijkl}\pi^{ij}\pi^{kl} + \frac{\pi^2}{4}\right]\frac{2\left(f'(R)-1\right)}{f'(R)} + \frac{V(\mathcal{P})}{2\kappa f'(R)}\right\} = -\frac{\Lambda_c}{\kappa}\int_\Sigma d^3x\sqrt{g}, \tag{26}$$

where we have assumed that $f'(R) \neq 0$ and we have dropped a divergence form term. Eq.(26) can be cast in the form of Eq.(5), by formally repeating the same procedure. Thus one gets

$$\frac{1}{V}\frac{\left\langle \Psi \left| \int_\Sigma d^3x \left[\hat{\Lambda}^{(2)}_\Sigma\right]\right| \Psi\right\rangle}{\langle \Psi|\Psi\rangle} + \frac{2\kappa}{V}\frac{2\left(f'(R)-1\right)}{f'(R)}\frac{\left\langle \Psi \left| \int_\Sigma d^3x \left[G_{ijkl}\pi^{ij}\pi^{kl} + \pi^2/4\right]\right| \Psi\right\rangle}{\langle \Psi|\Psi\rangle} + \frac{1}{V}\frac{\left\langle \Psi \left| \int_\Sigma d^3x V(\mathcal{P})/\left(2\kappa f'(R)\right)\right| \Psi\right\rangle}{\langle \Psi|\Psi\rangle} = -\frac{\Lambda_c}{\kappa}. \tag{27}$$

From Eq.(27), we can define a "*modified*" $\hat{\Lambda}^{(2)}_\Sigma$ operator which includes $f'(R)$. Thus, we obtain

$$\frac{\left\langle \Psi \left| \int_\Sigma d^3x \left[\hat{\Lambda}^{(2)}_{\Sigma,f(R)}\right]\right| \Psi\right\rangle}{\langle \Psi|\Psi\rangle} + \frac{\kappa}{V}\frac{\left(f'(R)-1\right)}{f'(R)}\frac{\left\langle \Psi \left| \int_\Sigma d^3x \left[\pi^2\right]\right| \Psi\right\rangle}{\langle \Psi|\Psi\rangle} + \frac{1}{V}\frac{\left\langle \Psi \left| \int_\Sigma d^3x \frac{V(\mathcal{P})}{2\kappa f'(R)}\right| \Psi\right\rangle}{\langle \Psi|\Psi\rangle} = -\frac{\Lambda_c}{\kappa}, \tag{28}$$

where

$$\hat{\Lambda}^{(2)}_{\Sigma,f(R)} = (2\kappa)\, h(R)\, G_{ijkl}\pi^{ij}\pi^{kl} - \frac{\sqrt{g}}{2\kappa}\,{}^3R^{lin}, \tag{29}$$

with

$$h(R) = 1 + \frac{2\left[f'(R)-1\right]}{f'(R)} \tag{30}$$

and where ${}^3R^{lin}$ is the linearized scalar curvature. Note that when $f(R) = R$, consistently it is $h(R) = 1$. From Eq.(28), we redefine Λ_c

$$\Lambda'_c = \Lambda_c + \frac{1}{2V}\frac{\left\langle \Psi \left| \int_\Sigma d^3x \frac{V(\mathcal{P})}{f'(R)}\right| \Psi\right\rangle}{\langle \Psi|\Psi\rangle} = \Lambda_c + \frac{1}{2V}\int_\Sigma d^3x\sqrt{g}\frac{Rf'(R) - f(R)}{f'(R)}, \tag{31}$$

where we have explicitly used the definition of $V(\mathcal{P})$. In the same spirit of the previous section, we find that by replacing $\Lambda_0(\mu_0, r)$ with $\Lambda'_0(\mu_0, r)$, the TT tensors one loop contribution for a $f(R)$ theory is given by Eq.(11) and the extremum is given therefore by [6,7]

$$\Lambda'_0(\mu_0, \bar{x}) = \frac{G\mu_0^4}{2\pi e^2}, \tag{32}$$

[6] For a complete derivation of the effective action for a $f(R)$ theory, see Ref.[17].

[7] By a canonical decomposition of the gauge part ξ_a into a transverse part ξ^T_a with $\nabla^a\xi^T_a = 0$ and a longitudinal part $\xi^{\|}_a$ with $\xi^{\|}_a = \nabla_a\psi$, it is possible to show that most of the contribution comes from the longitudinal part (scalar). Evidence against scalar perturbation contribution in a Schwarzschild background has been discussed in Ref.[18].

with $\bar{x}$ expressed by Eq.(12) . In terms of $\Lambda_0(\mu_0, \bar{x})$, we find

$$\frac{1}{\sqrt{h(R)}}\left[\Lambda_0(\mu_0, \bar{x}) + \frac{1}{2V}\int_\Sigma d^3x\sqrt{g}\frac{Rf'(R) - f(R)}{f'(R)}\right] = \frac{G\mu_0^4}{2\pi e^2} \tag{33}$$

and isolating $\Lambda_0(\mu_0, \bar{x})$, we obtain

$$\Lambda_0(\mu_0, \bar{x}) = \sqrt{h(R)}\frac{G\mu_0^4}{2\pi e^2} - \frac{1}{2V}\int_\Sigma d^3x\sqrt{g}\frac{Rf'(R) - f(R)}{f'(R)}. \tag{34}$$

Note that $\Lambda_0(\mu_0, \bar{x})$ can be set to zero when

$$\sqrt{h(R)}\frac{G\mu_0^4}{2\pi e^2} = \frac{1}{2V}\int_\Sigma d^3x\sqrt{g}\frac{Rf'(R) - f(R)}{f'(R)}. \tag{35}$$

Let us see what happens when

$$f(R) = \exp(-\alpha R). \tag{36}$$

This choice is simply suggested by the regularity of the function at every scale. In this case, Eq.(35) becomes

$$\sqrt{\frac{3\alpha\exp(-\alpha R) + 2}{\alpha\exp(-\alpha R)}}\frac{G\mu_0^4}{\pi e^2} = \frac{1}{\alpha V}\int_\Sigma d^3x\sqrt{g}(1 + \alpha R). \tag{37}$$

For Schwarzschild, it is $R = 0$, and by setting $\alpha = G$, we have the relation

$$\mu_0^4 = \frac{\pi e^2}{G}\sqrt{\frac{1}{(3G + 2)G}}. \tag{38}$$

It is clear that the passage to the naked singularity is straightforward, at least at formal level. The result is identical to Eq.(32) with the replacement

$$\Lambda_0'(\mu_0, \bar{x}) \rightarrow \Lambda_0^{(naked)\prime}(\mu_0, \bar{x}), \tag{39}$$

which means that also in Eq.(33) we have to replace $\Lambda_0(\mu_0, \bar{x})$ with $\Lambda_0^{naked}(\mu_0, \bar{x})$. However, also in this case we have the freedom to choose the r.h.s of Eq.(33) in such a way to cancel $\Lambda_0^{naked}(\mu_0, \bar{x})$. This means that also a naked Schwarzschild singularity predicts a vanishing cosmological constant. Let us see the consequences on the renormalization point μ_0. If we further proceed and we fix the form of $f(R)$ like in Eq.(36), we get the relation

$$\mu_0^{(naked)4} = 0.328\mu_0^4 \qquad \longrightarrow \qquad \mu_0^{naked} = 0.757\mu_0. \tag{40}$$

As we can see, it seems that a generic $f(R)$ theory cannot be more different by the ordinary case, namely when $f(R) = R$. This means that in this approach and at this level of approximation we cannot discriminate the different geometries. The situation is more marked, if we had chosen as a boundary condition for the naked singularity a privileged point $r_0 = 2MG$, namely a *"fictitious throat"*. We use the term fictitious because there is no throat at all. Nevertheless, if one wishes to fix such a boundary condition, there should not be difference at all between positive and negative Schwarzschild mass. Therefore, it appears that we are in a position where we cannot build a spectrum of geometries *including* a naked singularity. On the other hand, the impact of this approach on the cosmological constant problem deserves further investigation.

Journal of Physics: Conference Series **174** (2009) 012066 doi:10.1088/1742-6596/174/1/012066

References

[1] Riess A G *et al* 1998 *Astron. J.* **116** 1009 (*Preprint* arXiv:astro-ph/9805201); Perlmutter S *et al* 1998 *Nature* **391** 5 (*Preprint* arXiv:astro-ph/9712212); Riess A G *et al* 1998 1999 *Astron. J.* **118** 2668 (*Preprint* arXiv:astro-ph/9907038); Perlmutter S *et al* 1999 *Astrophys. J.* **517** 565 (*Preprint* arXiv:astro-ph/9812133; Riess A G *et al* 2001 *Astrophys. J.* **560** 49(*Preprint* arXiv:astro-ph/0104455); Tonry J L *et al* 2003 *Astrophys. J.* **594** 1 (*Preprint* arXiv:astro-ph/0305008); Knop R *et al* 2003 *Astrophys. J.* **598** 102 (*Preprint* arXiv:astro-ph/0309368); Riess A G *et al* 2004 *Astrophys. J.* **607** 665 (*Preprint* arXiv:astro-ph/0402512; Barris B *et al* 2004 *Astrophys. J.* **602** 571 (*Preprint* arXiv:astro-ph/0310843)

[2] Capozziello S, Carloni S and Troisi A 2003 *Preprint* arXiv:astro-ph/0303041; Carroll S M, Duvvuri V, Trodden M and Turner M S 2004 *Phys. Rev.* D **70**, 043528

[3] Faraoni V 2008 $f(R)$ gravity: successes and challenges *Preprint* arXiv:gr-qc/0810.2602

[4] Capozziello S and Francaviglia M 2007 Extended Theories of Gravity and their Cosmological and Astrophysical Applications *Preprint* arXiv:astro-ph/0706.1146

[5] Nojiri S and Odintsov S D 2007 *Int. J. Geom. Meth. Mod. Phys.* **4** 115 (*Preprint* arXiv:hep-th/0601213)

[6] DeWitt B S 1967 *Phys. Rev.* **160** 1113

[7] Garattini R 2006 *J. Phys.* A **39** 6393 (*Preprint* arXiv:gr-qc/0510061); *do.* 2006 *J. Phys.: Conf. Ser.* **33** 215 (*Preprint* arXiv:gr-qc/0510062)

[8] Garattini R 2008 The induced Cosmological Constant as a tool for exploring geometries *Preprint* arXiv:gr-qc/0812.4037

[9] Bondi H 1957 *Rev. Mod. Phys.* **29** 423

[10] Penrose R 1969 *Riv. Nuovo Cim.* **252** 1; *do.* 2002 *Gen. Rel. Grav.* **34** 1141

[11] Gibbons G W, Hartnoll S A and Ishibashi A 2005 *Prog. Theor. Phys.* **113** 963 (*Preprint* arXiv:hep-th/0409307)

[12] Gleiser R J and Dotti G 2006 *Class. Quant. Grav.* **23** 5063 (*Preprint* arXiv:gr-qc/0604021)

[13] Dotti G, Gleiser R J, Ranea-Sandoval I F and Vucetich H 2008 *Class. Quant. Grav.* **25** 245012 (*Preprint* arXiv:gr-qc/0805.4306); Dotti G, Gleiser R J and Pullin J 2007 *Phys. Lett.* B **644** 289 (*Preprint* arXiv:gr-qc/0607052

[14] Perez-Mercader J and Odintsov S D 1992 *Int. J. Mod. Phys.* D **1** 401; Cherednikov I O 2002 *Acta Phys. Slov.* **52** 221; *do.* 2004 *Acta Phys. Polon.* B **35** 1607; Bordag M, Mohideen U and Mostepanenko V M 2001 *Phys. Rep.* **353** 1. Inclusion of non-perturbative effects beyond one-loop in de Sitter Quantum Gravity have been discussed in Falkenberg S and Odintsov S D 1998 *Int. J. Mod. Phys.* A **13** 607 (*Preprint* arXiv:hep-th/9612019)

[15] Capozziello S and Garattini R 2007 *Class. Quant. Grav.* **24** 1627 (*Preprint* arXiv:gr-qc/0702075)

[16] Querella L 1999 Variational Principles and Cosmological Models in Higher-Order Gravity - Ph.D. Thesis (*Preprint* arXiv:gr-qc/9902044)

[17] Cognola G, Elizalde E, Nojiri S, Odintsov S D and Zerbini S 2005 *JCAP* **0502** 10 (*Preprint* arXiv:hep-th/0501096)

[18] Garattini R 2004 *TSPU Vestnik* **44 N7** 72 (*Preprint* arXiv:gr-qc/0409016)

[19] Abdalla M C B, Nojiri S and Odintsov S D 2005 *Class. Quant. Grav.* **22** L35 (*Preprint* arXiv:hep-th/0409177)

Journal of Physics: Conference Series **174** (2009) 012067 doi:10.1088/1742-6596/174/1/012067

Towards the solution of cosmological constant and zero point energy problems through metric reversal symmetry

Recai Erdem

Department of Physics, İzmir Institute of Technology
Gülbahçe Köyü, Urla, İzmir 35430, Turkey

E-mail: recaierdem@iyte.edu.tr

Abstract. In this talk I review my studies on metric reversal symmetry and their further implications. The talk is mainly concentrated on the relevance of the metric reversal symmetry to the solutions of the cosmological constant and zero point energies. However the use of metric reversal symmetry to hide higher Kaluza-Klein modes at the scales larger than the size of extra dimensions is also discussed, and speculations on its possible relevance to Pauli-Villars and Lee-Wick model are also briefly mentioned.

1. Introduction

The (old) Cosmological constant problem [1] is a fundamental problem of physics that exists independent of the expansion of the universe being in an accelerated fashion [2] or not although the observation of the expansion rate of the universe made this problem more popular because a positive cosmological constant in Einstein field equations is the most standard explanation for the accelerated expansion of the universe. The problem can be summarized as follows: There are many contributions to the cosmological constant from quantum physics and high energy physics. These contributions must be extremely fine tuned so that they cancel or are hidden from gravity either exactly or almost exactly. These contributions should cancel exactly if the accelerated expansion of the universe is attributed to alternative mechanisms such as quintessence, phantom fields, or modified gravity [3, 4] while the cancellation of these contributions must be extremely fine tuned or be hidden from gravity in a fine tuned manner if the accelerated expansion of the universe is attributed to a positive cosmological constant [5]. The (dark) energy density [3, 4] corresponding to the observed value of the expansion rate of the universe is $\sim 10^{-3}$ eV [6] if the accelerated expansion of the universe is attributed to cosmological constant or other dark energy sources while the theoretically expected contribution to the energy density of cosmological constant due to the renormalized zero point energy of an electron is $\sim 10^{33}$ times, the contribution due to the OCD vacuum is $\sim 10^{44}$ times, the contribution due to the vacuum expectation value of the Higgs field is $\sim 10^{55}$ times the observed value. This is the old cosmological constant problem. There are other cosmological constant problems as well; cosmic coincidence (i.e. the energy densities of matter and dark energies being of the same order of magnitude) and if cosmological constant vary with time, etc.. In this talk what I mean by the cosmological constant problem is the old cosmological problem and it is one of the main topics of this talk.

There are many schemes that intend to solve the (old) cosmological constant problem such as symmetries, anthropic principle, string theory landscape, diluting through extra dimensions, adjustment mechanisms etc, and numerous models in each scheme [1, 7]. However none have been wholly successful up to now. In this talk I consider a symmetry that may be called metric reversal symmetry as a possible tool to make some progress in the direction of the solution of the (old) cosmological constant problem [8, 9, 10, 12], and to give an alternative mechanism to the Kaluza-Klein modes at length scales larger than the size of the extra dimension(s) [13]. The main approach in this study is to implement the metric reversal symmetry to forbid (or eliminate) a cosmological constant and then to attribute the accelerated expansion of the universe to other mechanism such as quintessence, phantom field, or modified gravity. However a non-vanishing cosmological constant may be accommodated into this scheme by breaking the metric reversal symmetry while setting the cosmological constant to zero seems me more aesthetic in this scheme. In the first part of the talk I consider the implementation of the symmetry in the context of classical field theory [8, 9, 10]. In the second part I consider extension of the symmetry to quantum domain by studying the contribution of the quantum zero modes to the cosmological constant problem [11, 12]. In the last part I present a scheme that uses metric reversal symmetry to hide higher Kaluza-Klein modes at the present length scales accessible to experiments [13]. I also briefly speculate on further implications of the symmetry in connection to Pauli-Villars regularization [14], and Lee-Wick model [15].

2. Metric reversal symmetry and the cosmological constant problem

I name the multiplication of the infinitesimal line element by minus one as metric reversal, that is,

$$ds^2 \;=\; g_{AB}dx^A\,dx^B \;\rightarrow\; -\,ds^2 \tag{1}$$

There are two ways to realize (1) given by [8, 7, 16]

$$x^A \rightarrow i\,x^A \;\;,\;\;\; g_{AB} \rightarrow g_{AB} \;\;. \tag{2}$$

and [17, 9, 18]

$$x^A \rightarrow x^A \;\;,\;\;\; g_{AB} \rightarrow -g_{AB} \;\;. \tag{3}$$

The transformations (2) and (3) may be regarded as the first and the second realizations of the metric reversal symmetry, respectively. In fact these two realizations are wholly equivalent at the level of gravity while become somewhat different when matter is included. For example the transformation rule of gauge fields under these two realizations are different.

The infinitesimal volume element, dV and the scalar curvature, R transform under the metric reversal as

$$dV \;=\; \sqrt{(-1)^S g}d^Dx \rightarrow (-1)^{\frac{D}{2}}dV \;\;\;,\;\;\; R \rightarrow -R \tag{4}$$

where S denotes the number of spatial dimensions and D the (total) dimension of the space. The requirement of gravitational action

$$S_R = \frac{1}{16\pi\, G}\int \sqrt{(-1)^S g}\, R\, d^Dx \tag{5}$$

be invariant under metric reversal symmetry constraints the number of dimensions by

$$D = 2(2n+1) \;\;\;,\;\;\; n = 0, 1, 2, 3, ... \;\;. \tag{6}$$

On the other hand the metric reversal symmetry in $D = 2(2n+1)$ dimensions forbids a cosmological constant term

$$S_C = \frac{1}{8\pi G}\int \sqrt{g}\,\Lambda\, d^D x. \tag{7}$$

One may reach a similar conclusion by requiring covariance of Einstein equations under metric reversal while in this case there is no constraint on the dimension of the space. However the formulation of the symmetry by action functional is more promising for the extension of the analysis to quantum domain. Moreover the use of an extra dimensional setting enables the implementation of the symmetry through extra dimensional reflections. For example one may take the metric and the extra dimensional reflection, respectively, as

$$ds^2 = \cos(kx_5)\, g_{AB}dx^A dx^B \;, \quad A,B = 0,1,2,...,5 \tag{8}$$
$$ds^2 \to -ds^2 \quad \text{as} \quad kx_5 \to \pi - kx_5 \;, \quad g_{AB} \to g_{AB} \tag{9}$$

In this way the orientation reversal induced through the infinitesimal volume element may be obtained in a more standard setting because in this way the orientation change in the space takes place only when one travels through the extra dimensions while there is no orientation change as one travels through the usual 4-dimensional space. This is the approach I adopt in this study for the second realization of the symmetry Eq.(3). The source of the first realization may be attributed to some mechanism at the level of quantum gravity as given in [19].

It is evident that the metric reversal symmetry forbids a bulk cosmological constant (CC). Further one may form models where both a possible brane CC or a CC that may be induced by the part of the cosmological constant that only depends on extra dimensions is forbidden. One may refer to references [8, 9, 10] for more detail and some specific models of this type. However the metric reversal symmetry in its simple form seems to be a classical symmetry that can not be extended into quantum domain. This is due to the complication brought by the inclusion of the matter into theory. This can be seen either by showing the failure of the symmetry to be compatible with correct boundary conditions necessary for quantum mechanics as shown in [16] or by studying the transformation properties of the classical fields under the symmetry. In the latter approach as shown in [12] one observes that either both the classical matter Lagrangian and the vacuum expectation value of energy-momentum are allowed or both are forbidden simultaneously because the energy-momentum tensor follows from the matter Lagrangian. However one can extend the symmetry to quantum domain if one realizes the symmetry through extra dimensional reflections and allows mixture of different Kaluza-Klein modes since some of the modes are even and the others are odd under the extra dimensional reflections. These main ideas were put into a concrete form in [12]. The resulting model does not have zero point energy problem of quantum fields. After diagonalization of the kinetic terms the result is a spectrum of fields consisting of the usual fields plus ghost-like fields. This is essentially Linde's model [20, 21]. This scheme automatically follows from metric reversal symmetry while Linde's model was put forward in an ad hoc way. In this way it is easy to see that the problem with the incorrect boundary conditions under the symmetry transformation is removed because the transformed boundary conditions belong to the ghost-like copy of the universe and vice versa. To see the situation more clearly let us summarize the work of [12] below.

3. Metric reversal symmetry and zero point energy problem: extension of the symmetry to the quantum domain

Consider a space consisting of the sum of 2(2n+1) and 2(2m+1) (e.g 6 and 2) dimensional subspaces with the metric

$$ds^2 \;=\; g_{AB}dx^A\, dx^B \,+\, g_{A'B'}dx^{A'}\, dx^{B'}$$

$$
\begin{aligned}
&= \Omega_z(z)[g_{\mu\nu}(x)\,dx^\mu dx^\nu + \tilde{g}_{ab}(y)\,dy^a dy^b] + \Omega_y(y)\tilde{g}_{A'B'}(z)\,dz^{A'}dz^{B'} \quad (10)\\
&\Omega_y(y) = \cos k|y|) \;,\quad \Omega_z(z) = \cos k'|z| \quad (11)\\
A,B &= 0,1,2,3,5,....N \;,\quad N = 2(2n+1)\;,\quad A',B' = 1',2',,....N' \;,\quad N' = 2(2m+1)\\
&\mu\nu = 0,1,2,3\;,\quad a,b = 1,2,...,N-4\;,\quad n,m = 0,1,2,3......\;.
\end{aligned}
$$

The usual four dimensional space is embedded in the first space $g_{AB}dx^A\,dx^B$. I take the action be invariant under both realizations of the metric reversal symmetry, that is,

$$ds^2 \to -ds^2 \quad \text{as} \quad x^A \to i\,x^A,\; x^{A'} \to i\,x^{A'},\; g_{AB} \to g_{AB},\; g_{A'B'} \to g_{A'B'} \quad (12)$$
$$\Rightarrow\; \Omega_z \to \Omega_z,\; \Omega_y \to \Omega_y,\; g_{\mu\nu} \to g_{\mu\nu},\; \tilde{g}_{ab} \to \tilde{g}_{ab},\; \tilde{g}_{A'B'} \to \tilde{g}_{A'B'} \quad (13)$$

and

$$ds^2 \to -ds^2 \quad \text{as} \quad ky \to \pi - ky\;,\; k'z \to \pi - k'z\;,\; x^A \to x^A,\; x^{A'} \to x^{A'} \quad (14)$$
$$\Rightarrow\; \Omega_z \to -\Omega_z,\; \Omega_y \to -\Omega_y,\; g_{\mu\nu} \to g_{\mu\nu},\; \tilde{g}_{ab} \to \tilde{g}_{ab},\; \tilde{g}_{A'B'} \to \tilde{g}_{A'B'}\;. \quad (15)$$

The requirements of the homogeneity and isotropy of the 4-dimensional space together with the equations (12-15) set $g_{\mu\nu}$ to the Minkowski metric $\eta_{\mu\nu} = diag(1,-1,-1,-1)$.

To see the essentials of the scheme consider the zero point energy induced by the kinetic Lagrangian of a scalar field

$$
\begin{aligned}
S_{\phi k} &= \int dV\,\mathcal{L}_{\phi k} \quad (16)\\
dV &= \sqrt{(-1)^S g}\sqrt{(-1)^{S'}g'}\,d^Dx\,d^Dx'\\
\mathcal{L}_{\phi k} &= \frac{1}{2}g^{AB}\partial_A\phi\partial_B\phi \quad (17)
\end{aligned}
$$

After writing ϕ in terms of its Kaluza-Klein modes and integrating over extra dimensions one obtains

$$
\begin{aligned}
S_{\phi k} =\;& \frac{1}{8}(LL')^2\int d^4x\,\{4\partial_\mu[\,\phi_{1,2}(x) + \phi_{1,0}(x)]\,\partial_\nu(\,\phi_{0,0}(x)\,)\\
&+4\partial_\mu[\,\phi_{0,2}(x) + \phi_{0,0}(x) + \phi_{2,2}(x) + \phi_{2,0}(x)]\,\partial_\nu(\,\phi_{1,0}(x)\,)\\
&+4\eta^{\mu\nu}\sum_{r=1,s=1}^{\infty}\partial_\mu[\,\phi_{|r-1|,|s-2|}(x) + \phi_{|r-1|,s+2}(x)\\
&+2\phi_{|r-1|,s}(x) + \phi_{r+1,|s-2|}(x) + \phi_{r+1,s+2}(x) + 2\phi_{r+1,s}(x)\,]\partial_\nu(\,\phi_{r,s}(x)\,)\\
&-4k^2\sum_{r=1,s=0} r[\,(|r-1|)(\,\phi_{|r-1|,|s-2|}(x) + \phi_{|r-1|,s+2}(x) + 2\,\phi_{|r-1|,s}(x)\,)\\
&+(r+1)(\,\phi_{r+1,|s-2|}(x) + \phi_{r+1,s+2}(x) + 2\,\phi_{r+1,s}(x)\,) - \phi_{r+1,s}(x)\,)\,]\phi_{r,s}(x)\\
&-4\frac{1}{2}k'^2\sum_{r=0,s=1} s\,[\,(|s-3|)\phi_{r,|s-3|}(x) + (s+3)\,\phi_{r,s+3}(x)\\
&+3(|s-1|)\,\phi_{r,|s-1|}(x) + 3(s+1)(\,\phi_{r,s+1}(x)\,]\,\phi_{r,s}(x)\}\;. \quad (18)
\end{aligned}
$$

One observes that Eq.(18) contains only off-diagonal mixing of Kaluza-Klein modes. So the corresponding 4-dimensional energy momentum tensor

$$T^\nu_\mu = \int dV\,\Omega_z^{-1}\{g^{\nu\tau}\partial_\tau\phi\partial_\mu\phi - \frac{1}{2}\delta^\nu_\mu\,g^{\rho\tau}\partial_\rho\phi\partial_\tau\phi\} \quad (19)$$

also consists of off-diagonally coupled Kaluza-Klein modes. Hence the vacuum expectation value of T_μ^ν consists of the following type of terms

$$< 0|T_\mu^\nu|0 > \; \propto \; < 0|\, a_{n,m} a_{r,s}^\dagger |0 > = 0 \,, \quad < 0|\, a_{r,s}^\dagger a_{r,s} |0 > = 0 \qquad n \neq r \quad \text{and/or} \quad m \neq s \tag{20}$$

(because $a_{r,s}|0> = 0$, and $[\, a_{n,m}, a_{r,s}^\dagger \,] = 0$ for $n \neq r$ and/or $m \neq s$) where $a_{n,m}$, $a_{n,m}^\dagger$ are the creation and annihilation operators in the expansion of $\phi_{n,m}$. Similar results apply to the extra dimensional piece of the kinetic term and bulk mass terms for ϕ, and to other fields such as fermions and gauge bosons. Any other power of ϕ also necessarily contains at least one pair of off-diagonal mixings of Kaluza-Klein modes due to the symmetry. The details about these points may be found in [12]. Therefore this scheme eliminates any non-zero contribution to the cosmological constant due to quantum zero modes.

One can see the situation in another way as well by diagonalizing the action. To see the main points in the derivation consider the second term in (18)

$$\begin{aligned} S_{\phi k} &= \frac{1}{2}(LL')^2 \int d^4x \, 2\eta^{\mu\nu}\partial_\mu\phi_{1,0}\partial_\nu\phi_{0,0} \\ &= \frac{1}{2}(LL')^2 \int d^4x \, [\eta^{\mu\nu}\partial_\mu\phi_1\partial_\nu\phi_1 \, - \, \eta^{\mu\nu}\partial_\mu\phi_2\partial_\nu\phi_2 \,] \end{aligned} \tag{21}$$

The other terms in the kinetic terms and in bulk mass terms may be diagonalized in similar way. In other words this scheme corresponds to a scheme where the field spectrum consists of the usual particles and their ghost-like copies with a possible interaction between these two sets through gravity, and trilinear and higher order terms (e.g. ϕ^4) in the Lagrangian. This is essentially a variant of Linde's model. However the usual and the ghost universe picture emerges automatically as a consequence of the metric reversal symmetry in this scheme while Linde's model is formulated in an ad hoc way.

The gravitational action in this space can not be taken simply as the Einstein-Hilbert action given by (5) because it cancels out since the scalar curvature consists of two parts one due to the $2(2n+1)$ dimensional subspace (that is even under $kz \to \pi - kz$) and the other due to the $2(2m+1)$ dimensional subspace (that is even under $ky \to \pi - ky$) while the volume element is odd under both. Instead one may adopt an R^2 or $R\phi$ type of term for the gravitational Lagrangian. It is shown in [12] that one may take a R^2 term for the gravitational sector and it reduces to the usual Einstein-Hilbert term after integration over extra dimensions.

One may allow the emergence of the Robertson-Walker metric from the Minkowski metric in (21) by breaking of the first realization of the symmetry given in (12,13) so that the accelerated expansion of the universe is accommodated while the second realization of the metric reversal symmetry given by (14,15) remains unbroken to insure the vanishing of the zero point energies. In this scheme the accelerated expansion of the universe is either should be attributed to other mechanism such as quintessence, phantom, modified gravity etc. or one may break the metric reversal through the introduction of a classical term (other than zero point energies) that breaks the first realization of the metric reversal symmetry. For example one may introduce a classical Lagrangian, that breaks the symmetry, given by

$$\mathcal{L}_{cl} \;=\; \alpha\, v_{1,0} v_{0,1} \cos k\, y \, \cos k'\, z \tag{22}$$

where $\alpha << 1$ is a constant that reflects that $\mathcal{L}_{cl}$ is small since it corresponds to the breaking of the $x^A \to i\, x^A$, $x^{A'} \to i\, x^{A'}$ symmetries separately by a small amount, and $v_{1,0}$, $v_{0,1}$ are some constants.

4. Finite number of Kaluza-Klein modes through metric reversal symmetry

The significance of metric reversal symmetry is not restricted to the cosmological constant problem. For example one may employ the metric reversal symmetry to hide some of the Kaluza-Klein modes at usual length scales even when they have already been excited. This may give an alternative way to explain the absence of the Kaluza-Klein modes at current accelerator energies while the usual prescription is to take the Kaluza-Klein modes be too heavy to be excited at current energies. In this section I give a simple scheme to hide Kaluza-Klein modes at the current length scales accessible in experiments by using metric reversal symmetry. The details may be found in [13].

To illustrate the scheme consider the free fermions in the following 5-dimensional metric

$$ds^2 = \cos k\, z\, [\, g_{\mu\nu}(x)\, dx^\mu dx^\nu - dz^2\,] \qquad \mu,\nu = 0,1,2,3 \tag{23}$$

where the extra dimension is taken to be compact and have the size L, and $k = \frac{2\pi}{L}$. The action for (free) fermionic fields for this space is

$$\begin{aligned} S_f &= \int (\cos kz)^{\frac{5}{2}}\, \mathcal{L}_f\, d^4x\, dz \\ &= \int (\cos kz)^2\, i\bar{\chi}\gamma^a(\,\partial_a + \frac{1}{16}\tan kz\,[\,\gamma_a\,,\,\gamma_5]\,)\chi\, d^4x\, dz \,+\, H.C. \\ &\{\gamma^a,\gamma^b\} = 2\eta^{ab}\ ,\ (\eta^{ab}) = diag(1,-1,-1,-1,-1) \end{aligned} \tag{24}$$

where $H.C.$ stands for Hermitian conjugate, and the second term is the spin connection term. The Kaluza-Klein decomposition of χ is

$$\chi = \sum\, [a_n \cos n\frac{kz}{2} + b_n \sin n\frac{kz}{2}]\chi_n(x) \tag{25}$$

I take the action (24) be invariant under an extra dimensional reflection realization of metric reversal symmetry given by

$$k\,z \rightarrow \pi + k\,z. \tag{26}$$

This transformation is essentially equivalent to (9) while (9) is more restrictive in constraining the conformal factors and more applicable to orbifold constructions. I also impose the action be invariant under the first realization of the metric reversal symmetry or its subgroup given by

$$x^a \rightarrow -x^a \qquad a = 0,1,2,3,4 \tag{27}$$

where the Kaluza-Klein modes χ_n are required to transform as

$$\chi_n(x) \rightarrow \epsilon^{\lambda_n}\, \mathcal{CPT}\, \chi_n(-x) \quad , \quad \lambda_n = \frac{i}{2}(-1)^{\frac{n}{2}} \tag{28}$$

where ϵ is some constant, and $\mathcal{CPT}$ denotes the usual 4-dimensional CPT operator (acting on the spinor part of the field). I impose χ satisfy anti-periodic boundary conditions i.e. $\chi(z=0) = -\chi(z=L)$. This sets n in (25,28) be odd.

The requirement of the invariance of the action (24) under (26), (27) (and (28)) results in the following expression for the 4-dimensional part of S_f

$$\begin{aligned} &\sum_{r,s=0}^{\infty}(\int d^4x\, [i\bar{\chi}_{(2|r|+1)}\gamma^\mu\partial_\mu\chi_{(2|s|+1)} \int dz\, (\cos kz)^2 \\ &\times [\, f^*_{2|r|+1}g_{2|s|+1} + (\cos\frac{2|r|+1}{2}kz + \sin\frac{2|r|+1}{2}kz)(\cos\frac{2|s|+1}{2}kz - \sin\frac{2|s|+1}{2}kz) \end{aligned}$$

$$+g^*_{2|r|+1}f_{2|s|+1})(\cos\frac{2|r|+1}{2}kz - \sin\frac{2|r|+1}{2}kz)(\cos\frac{2|s|+1}{2}kz + \sin\frac{2|s|+1}{2}kz)] + H.C.$$

$$= \sum_{r,s=0}^{\infty}(\int d^4x\,[2\,i\bar{\chi}_{(2|r|+1)}\gamma^\mu\partial_\mu\chi_{(2|s|+1)}\int dz\,(\cos kz)^2\,(f^*_{2|r|+1}g_{2|s|+1} + g^*_{2|r|+1}f_{2|s|+1})$$

$$\times[\cos\frac{2|r|+1}{2}kz\,\cos\frac{2|s|+1}{2}kz\; - \;\sin\frac{2|r|+1}{2}kz\,\sin\frac{2|s|+1}{2}kz\,] + H.C.$$

$$= \frac{1}{2}\sum_{r,s=0}^{\infty}(f^*_{2|r|+1}g_{2|s|+1} + g^*_{2|r|+1}f_{2|s|+1})\int d^4x\,i\bar{\chi}_{(2|r|+1)}\gamma^\mu\partial_\mu\chi_{(2|s|+1)}$$

$$\times\int_0^L dz\,[\,\cos\,(|r|+|s|-1)kz\,] + H.C. \tag{29}$$

where $2r+1 = 4l+1$, $2s+1 = 4p+3$ (l,p=0,1,2,....) or vice versa, , and $f_n = \frac{1}{2}(a_n+b_n)$, $g_n = \frac{1}{2}(a_n-b_n)$. Because of the periodicity of cosine function the terms in (29) give non-zero contributions after integration over z only if the arguments of cosines are zero. This is possible only when

$$|r|+|s|-1=0 \quad\Rightarrow\quad r=0\;,\;s=1 \quad \text{or} \quad s=1\;,\;r=0 \tag{30}$$

The result of z integration in (8) is

$$\frac{L}{2}\int d^4x\;\;[\;\;i\bar{\chi}_1\gamma^\mu\partial_\mu\chi_0\;+\;i\bar{\chi}_0\gamma^\mu\partial_\mu\chi_1\,] + H.C. \tag{31}$$

where the term $(f^*_1g_0 + g^*_1f_0)$ is absorbed into redefinition of χ_1, χ_0. The diagonalization of (31) gives

$$\frac{1}{2}L\int d^4x[i\bar{\psi}\gamma^\mu\partial_\mu\psi\;-\;i\bar{\tilde{\psi}}\gamma^\mu\partial_\mu\tilde{\psi}] + H.C. \tag{32}$$

$$\psi\;=\;\frac{1}{\sqrt{2}}(\chi_1\;+\;\chi_0)\;\;,\;\;\tilde{\psi}\;=\;\frac{1}{\sqrt{2}}(\chi_1\;-\;\chi_0) \tag{33}$$

Hence there are one usual fermion and one ghost fermion in the spectrum at the large length scales where the extra dimension can not be detected. All other Kaluza-Klein modes will be hidden at these length scales even when they are produced in high energy experiments.

5. Conclusion

I have reviewed the relevance of extra dimensional metric reversal symmetry to the solutions of cosmological constant, zero point energy problems, and to a scheme where higher Kaluza-Klein modes may be hidden at current length scales accessible to experiments even when they have already been produced. The results seem to be encouraging for the solutions of these problems in a quite natural framework. Moreover the implementation of metric reversal symmetry (or orientation reversal symmetry) through the use of extra dimensions makes the symmetry plausible. The emergence of a ghost field for each usual field due to the mixture of Kaluza-Klein modes of the same field implies that this framework may also be relevant to Pauli-Villars regularization and Lee-Wick model. These points will be studied in my future work.

Acknowledgments

This work was supported in part by Scientific and Technical Research Council of Turkey under grant no. 107T235.

References

[1] Weinberg S 1989 The Cosmological Constant Problem *Rev. Mod. Phys.* **61** 1

[2] Linder E V 2008 Mapping the Cosmological Expansion *Rep. Prog. Phys.* **71** 056901 (*Preprint* arXiv:0801.2968) and references therein

[3] Copeland E J, Sami M and Tsujikawa S 2006 Dynamics of dark energy *Int. J. Mod. Phys.* D **15** 1753 (*Preprint* hep-th/0603057) and references therein

[4] Frieman J, Turner M and Huterer D 2008 Dark Energy and the Accelerating Universe *Ann. Rev. Astron. Astrophys.* **46** 385 (*Preprint* arXiv:0803.0982) and references therein

[5] Carroll S 2001 The Cosmological Constant *Living Rev. Rel.* **4** 1 (*Preprint* astro-ph/0004075)

[6] Yao W-M *et al* 2006 Review of Particle Physics *J. Phys. G: Nucl. Part. Phys.* **33** 1

[7] Nobbenhuis S 2006 Categorizing Different Approaches to the Cosmological Constant Problem *Found. Phys.* **36** 613 (*Preprint* gr-qc/0411093)

[8] Erdem R 2005 A symmetry for vanishing cosmological constant in an extra dimensional toy model *Phys. Lett.* B **621** 11 (*Preprint* hep-th/0410063)

[9] Erdem R 2006 A symmetry for vanishing cosmological constant: Another realization *Phys. Lett.* B **639** 348 (*Preprint* gr-qc/0603080)

[10] Erdem R 2007 A symmetry for vanishing cosmological constant *J. Phys.* A **40**, 6945 (*Preprint* gr-qc/0611111)

[11] Zeldovich Ya B 1968 The Cosmological Constant and the Theory of Elementary Particles *Sov. Phys. Usp.* **11** 381

[12] Erdem R 2008 A way to get rid of cosmological constant and zero-point energy problems of quantum fields through metric reversal symmetry *J. Phys.* A **41** 235401 (2008) (*Preprint* ArXiv:0712.2989)

[13] Erdem R 2008 Finite number of Kaluza-Klein modes, all with zero masses *Preprint* arXiv:0809.2647

[14] Pauli W and Villars F 1949 On the Invariant Regularization in Quantum Field Theory *Rev. Mod. Phys.* **21** 434

[15] Lee T D and Wick G C 1970 Finite Theory of Quantum Electrodynamics *Phys. Rev.* D **2** 1033

[16] 't Hooft G and Nobbenhuis S 2006 Invariance under complex transformations, and its relevance to the cosmological constant problem *Class. Quant. Grav.* **23** 3819 (*Preprint* gr-qc/0602076)

[17] Bonelli G and Boyarsky A M 2000 Six dimensional topological gravity and the cosmological constant problem *Phys. Lett.* B **490** 147 (*Preprint* hep-th/0004058)

[18] Duff M J and Kalkkinen J 2006 Signature reversal invariance *Nucl. Phys.* B **758** 161 (*Preprint* hep-th/0605273)

[19] Guendelman E I and Kaganovich A B 2008 Transition to zero cosmological constant and phantom dark energy as solutions involving change of orientation of spacetime manifold *Class. Quantum. Grav.* **25** 235015 (*Preprint* arXiv:0804.1278)

[20] Linde A D 1988 The Universe Multiplication and the Cosmological Constant Problem *Phys. Lett.* B **200** 272
Linde A D 2003 *Inflation, Quantum Cosmology and the Anthropic Principle (Science and Ultimate Reality)* ed J D Barrow *et al* (Cambridge, UK: Cambridge U. Press) 426-458

[21] Kaplan D E and Sundrum R 2006 A Symmetry for the cosmological constant *J. High Energy Phys. JHEP* JHEP07(2006)042 (*Preprint* hep-th/0505265)
Moffat J W 2005 Charge Conjugation Invariance of the Vacuum and the Cosmological Constant Problem *Phys. Lett.* B **627** 9 (*Preprint* hep-th/0507020)

On the existence of time in non-singular spacetimes

E. Minguzzi
Dipartimento di Matematica Applicata, Università degli Studi di Firenze, Via S. Marta 3, I-50139 Firenze, Italy
E-mail: ettore.minguzzi@unifi.it

Abstract. I show that a physically reasonable spacetime (i.e. a chronological spacetime in which the null convergence and the null genericity conditions hold) which is null geodesically complete admits a time function.

1. Introduction

In this paper I shall consider the problem of justifying the existence of a global time in non-singular spacetimes. This work is based on the results obtained in [1, 2]. Recently I gave similar accounts in [3, 4].

The problem of deducing the existence of a global time from physical arguments is an old one. Geroch and Horowitz [5] regarded that of the existence of a global time as one of the most important open issues in the field of global general relativity. Little progress has been made on this problem since their 1979 report, indeed, the absence of compelling arguments for the globality of time could lead one to suspect that maybe such arguments do not exist, and thus that any attempt to prove the contrary is condemned to failure. Nevertheless, in this work I shall show that null geodesically complete spacetime which are physically reasonable should admit a time function. Unfortunately, this result can not be considered a proof of the existence of a global time because our Universe could indeed be geodesically null incomplete. Although this result does not guarantee the existence of a global time, at least it shows that its absence implies, without any assumption on trapped surfaces or diverging cosmological flows, the singularity of spacetime.

The basic idea is to improve a previous result by Hawking (1966 Adams prize essay, see [6]) which states that a chronological spacetime in which every inextendible lightlike geodesic has a pair of conjugate point is strongly causal. As we shall see this theorem holds with 'strongly causal' replaced with 'stably causal'. This fact is non-trivial because as observed by Carter [7] and Penrose [8, 9] there are infinite causality levels between strong causality and stable causality.

However, let us follow the logical development starting from the definition of time function.

2. Time function and stable causality

As it is customary in general relativity I shall write $p < q$ if there is a causal curve joining p to q, and $p \le q$ if $p < q$ or $p = q$. We shall also define $J^+ = \{(p,q) : p \le q\}$. A time function is a function $t : M \to \mathbb{R}$ which is continuous and such that if $p < q$ then $t(p) < t(q)$. In other words, a time function is a continuous function all over the spacetime which increases on every causal curve. If the relation J^+ has the antisymmetric property, $p \le q$ and $q \le p$ implies $p = q$, we say that $\leq$ (or J^+) is a partial order. A time function can be regarded as a total (linear) ordering which extends the usual causal ordering. Indeed, note that given two events $p, q \in M$ it is not always possible to 'decide' if $p \leq q$ or $q \leq p$, because they are not always comparable, but given a time function t one can introduce the following relation on M, $R_t^+ = \{(x, z) : t(x) \leq t(z)\}$, which is indeed a total ordering. Moreover, it extends J^+ in the sense that $J^+ \subset R_t^+$.

A spacetime is stably causal if it is possible to open the light cones all over the space time while preserving causality. The causal vectors for the old causal structure must be timelike for the new one, this means that at a given point the new light cone can not be tangent to the old one but should instead contain it strictly. There is a theorem by Hawking which shows the equivalence between stable causality and the existence of a time function [10]. In one direction Hawking's proof was not rigorous and it was subsequently amended in [11, 12], where the authors prove that whenever a time function exists it can be chosen smooth with a timelike gradient. The goal is thus to justify stable causality on physical grounds.

We shall make a minimal causality requirement namely that of chronology: there are no closed timelike curves. This assumption can be further weakened to the condition that the chronological violating set has an open component with compact boundary [1], nevertheless I shall not consider this generalization here.

Thus we have to justify in some way stable causality starting from chronology. We are going to assume, as Hawking and Penrose did in their 1970 singularity theorem [6], that the spacetime satisfies the (i) *null convergence condition*, namely at every point and for every null vector n, $R_{\mu\nu}n^\mu n^\nu \geq 0$, which is basically a positivity energy requirement, and the (ii) *null genericity condition*, namely every inextendible lightlike geodesic admits at some point a tangent vector n such that $n^c n^d n_{[a} R_{b]cd[e} n_{f]} \neq 0$. The null genericity condition basically states that the spacetime has no particular symmetry, so that following a lightlike geodesic one will ultimately find some curvature of the spacetime and hence a curvature tensor whose components are not 'aligned' with the lightlike tangent vector n. The conditions of chronology, null convergence, and null genericity are often regarded as physically reasonable if quantum mechanical effects are not taken into account.

Now, Hawking and Penrose [6] proved that if a lightlike geodesic is null complete and the null convergence condition and the null genericity conditions are satisfied then it has a pair of conjugate points. A standard result [13] then states that the are two chronologically related events on the lightlike geodesic. In mathematical terms, the lightlike geodesic is not a *lightlike line*, these curves being defined as inextendible achronal causal curves.

3. Absence of lightlike lines and transitivity of $\bar{J}^+$

According to the previous analysis it suffices to prove the claim

Claim: Chronological spacetimes without lightlike lines are stably causal.

to establish the existence of a time function in null geodesically complete spacetimes. I proved this claim in [1] and again, with a largely independent argument, in [2]. The first step is to establish that the absence of lightlike lines implies the transitivity of the closure of the causal relation $\bar{J}^+$. Let $(x, y) \in \bar{J}^+$ and $(y, z) \in \bar{J}^+$; one has to prove $(x, z) \in \bar{J}^+$. This fact follows from the application of the limit curve theorem to both sequences of limit curves, taking as limit point the middle point y. There are four cases depending on whether the two limit curves connect respectively the pairs (x, y) and (y, z). There is one possibly dangerous case, illustrated in figure 1, in which none of the two sequences converges to a connecting causal curve. By the limit curve theorem there is an inextendible causal curve passing through y, and since there are no lightlike lines, this causal curve has two chronologically related events denoted $\bar{x}$ and $\bar{z}$. From that it is possible to construct, using the fact that I^+ is open, a sequence of causal curves whose

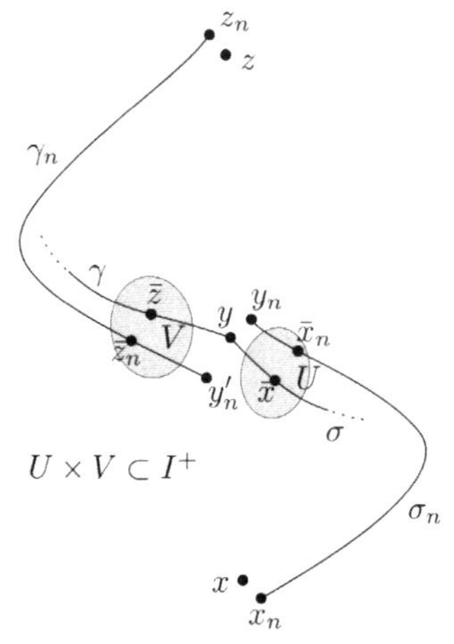

Figure 1. The potentially dangerous case in the proof of the transitivity of $\bar{J}^+$.

endpoints converge to (x, z). Also it is not hard to prove that under chronology the absence of lightlike lines implies that $\bar{J}^+$ is antisymmetric, thus we have a closed and transitive relation, which contains J^+ which is antisymmetric. The smallest closed and and transitive relation which contains J^+ is called [14] K^+ and it has been long conjectured that the antisymmetry of K^+, a property called K-causality, is equivalent to stable causality. Our argument above implies that $K^+ = \bar{J}^+$ and that this relation is antisymmetric, thus K-causality holds. In [2] I have proved that indeed K-causality coincides with stable causality, thus with this last implication the proof is finished.

The proof of the conjectured equivalence between K-causality and stable causality has also other interesting consequences. For instance, by using this result I proved [2] that the relation "the spacetime is strongly causal and $\bar{J}^+$ is transitive" is a new causality property which stays between stable causality and causal continuity.

Acknowledgments

This work has been partially supported by GNFM of INDAM.

References

[1] Minguzzi E 2008 Chronological spacetimes without lightlike lines are stably causal *Commun. Math. Phys.*, in press *Preprint* arXiv:0806.0153
[2] Minguzzi E 2008 *K*-causality coincides with stable causality *Commun. Math. Phys.*, in press *Preprint* arXiv:0809.1214
[3] Minguzzi E 2008 On the global existence of time. Essay submitted to the FQXi contest on the 'Nature of Time' http://www.fqxi.org/community/forum/topic/271. *Preprint* arXiv:0901.0904
[4] Minguzzi E 2008 Chronological null complete spacetimes admit a global time *Proceedings of the conference 'Spanish Relativity Meeting' ERE 2008, Salamanca, September 15 - 19* to appear *Preprint* arXiv:0807.0103
[5] Geroch R and Horowitz G T 1979 Global structure of spacetimes *General relativity: An Einstein centennary survey* (Cambridge, UK: Cambridge U. Press) 212–292
[6] Hawking S W and Penrose R 1970 *Proc. Roy. Soc. Lond.* A **314** 529–548
[7] Carter B 1971 *Gen. Relativ. Gravit.* **1** 349–391
[8] Penrose R 1972 *Techniques of Differential Topology in Relativity* Cbms-Nsf Regional Conference Series in Applied Mathematics (Philadelphia: SIAM)
[9] Minguzzi E 2008 *Class. Quantum Grav.* **25** 015009
[10] Hawking S W 1968 *Proc. Roy. Soc. London* A **308** 433–435
[11] Bernal A N and Sánchez M 2003 *Commun. Math. Phys.* **243** 461–470
[12] Bernal A N and Sánchez M 2005 *Commun. Math. Phys.* **257** 43–50

[13] O'Neill B 1983 *Semi-Riemannian Geometry* (San Diego: Academic Press)
[14] Sorkin R D and Woolgar E 1996 *Class. Quantum Grav.* **13** 1971–1993

Nature of tracing dimension, imaginary order of freedom and our observed time passing at constant speed [1]

S Nagao

Business Development and Licensing Department, Nippon Boehringer Ingelheim Co., Ltd., ThinkPark Tower, 2-1-1, Osaki, Shinagawa, Tokyo 141-6017, Japan

E-mail: snagao@lilac.plala.or.jp

Abstract. Nature of the time and requirements to work as a time dimension are investigated. A potential scenario of the development of the universe is conceptually investigated starting from energy as vibration in multiple dimensions. A model is proposed, in which the Big Bang is a phase transition of energy from vibration in 4-dimensional space to energy distribution in 3-D surface of a 4-D sphere. The Time which we observe passing at a constant speed is not such a reference frame which we unintentionally believe to be the time, but the radius dimension of the 4-D sphere. The feature of the Dark Matter and the mystery of the Dark Energy are naturally explained from the model.

1. Introduction

Currently there is a big mystery that the Supernovae data seem to imply the universe is expanding in an accelerating manner [1]-[5], for which successful explanation is missing. The general relativity of Einstein can nicely express various observations of the universe. However, there is a question if the Einstein tensor would really represent the curvature of spacetime. It may express only a gravitational field in the four dimensions of the space and the time. Furthermore, there would be a possibility that the gravitational force would directly work in a four-dimensional space at a certain instance of another time dimension, which is different from the time we observe.

We unintentionally perceive the time without a clear definition. In order to try to understand the development of the universe, natures of the time should be identified and a time dimension should be defined accordingly.

2. Nature and definition of the time

2.1. Tracing dimension

In a one-dimensional world, there is no nature of movement. There exists only a selection of value, which is not a movement. In order to express a movement in one dimension, another reference dimension, which traces the first one by changing its own value to express value change of the first dimension, is essential. For expressing a movement in plural dimensions, an additional

[1] This paper is a private work independent from the company.

dimension to trace it is also required. Let's define the "Tracing Dimension" as a dimension to trace other dimensions for expressing their movement by taking a moving single value. The term movement here is a value change in a dimension by a value variance of another dimension.

To work as a Tracing Dimension, any single value of the Tracing Dimension should correspond to a single value of other dimensions for a movement. If respective values of two (or plural) dimensions vary on one-to-one correspondence, either dimension can be a Tracing Dimension to the other.

There is another important aspect of the tracing for a movement. Neither the speed of tracing nor the direction of the tracing has any meaning. Only the relativity of the values of the object dimensions to the values of the Tracing Dimension is the essence of the movement. For instance, the speed of a movement in X-dimension with the tracing by T-dimension, dX/dT, is independent from dT/dt even if another tracing dimension t is virtually introduced.

2.2. Imaginary order of freedom

The first-order of freedom means that there is a chance or freedom to select a value, whereas there is no chance to select a value in the zero-order of freedom but only a constant value is taken. In addition let's define the "Imaginary-order of Freedom" as that there is no chance of selecting a value but the value is changing in one direction, that is, it takes a moving single value by the tracing. In case a dimension has the Imaginary-order of Freedom by a tracing, let's call it as an imaginary dimension.

An explanatory example of movements is shown in the Figure 1. Value changes of objects such as E1, E2, and E3 in three dimensions of X, Y and t are given. With tracing by t-dimension, there are plural values of objects in Y-dimension. However, all the objects show a common value in X-dimension, which moves in a single direction with the tracing by t-dimension. Thus, X-dimension has the Imaginary-order of Freedom in the tracing, while Y-dimension has the first-order of freedom. As values of an imaginary dimension and the Tracing Dimension vary on one-to-one correspondence, an imaginary dimension can work as another Tracing Dimension. In the case of the Figure 1, X-dimension can also work as a Tracing Dimension to express movements of the objects in Y-dimension.

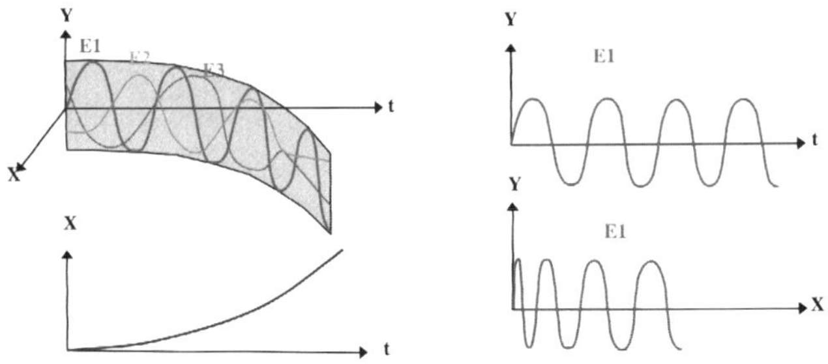

Figure 1. First order and imaginary order of freedom

2.3. Nature and definition of the time

Movement is, in general, shown or interpreted in reference to the time. The time is exactly a Tracing Dimension to express a movement in other dimensions. We can define that a time is a Tracing Dimension to express movements of objects in the space dimensions. The time is not only one absolute dimension for any movements, but can be relative to different movements. If a certain dimension shows the Imaginary-order of freedom with tracing by the original Tracing Dimension, it can also work as a time for expressing movements in other dimensions. From the meaning of tracing, a Tracing Dimension itself can be also regarded as an imaginary dimension of space reversely.

When one dimension works as a Tracing Dimension, the remaining dimensions to be expressed by it can be defined as a space dimension. There is no fundamental difference between a space dimension and a time dimension other than the degree of freedom. It can be summarized that a time is a dimension to trace other dimensions for expressing movements and is also an imaginary space dimension.

3. Gravitational force

Einstein's general relativity [6, 7] is fully accepted and meets very nicely various observations [8]-[12]. However, there are several questions, which are not perfectly explained, as follows:

- The Einstein equation primarily represents the gravitational field. Although Einstein insisted that the curvature of spacetime is the cause to bring about such gravity [13, 7], there is still missing a theoretical explanation to authorize the curvature of spacetime. The Einstein tensor may only represent the curvature of the gravitational field.
- Einstein insisted and it is well accepted that any interaction even gravity can not transmit faster than the light speed [14]. From this aspect, there arises a serious issue of the causality. For instance, an area of the universe, which expanded to the opposite direction from our area, has no longer any interaction with our area as the distance is expanding faster than the light speed.
- From the special relativity, there is no privileged frame but any inertial coordinates are equivalent [14]. However, the static mass of a substance is common by any inertial coordinate even if its velocity and kinetic energy are described differently by respective coordinates. The relativistic mass represents a mass for acceleration; however the static mass is believed to be the fundamental one corresponding to its intrinsic energy. For this explanation, to which coordinate is it stationary for giving the static mass? A special coordinate is considered as static in reality.

There would be a possibility that the time dimension described in the general relativity might be a space-like dimension having the imaginary order of freedom with tracing by another dimension.

As a model for consideration, let's think of a world consisting of four spatial dimensions and one original tracing dimension. An energy distribution is uniformly expanding on the 3-dimensional surface of a 4-dimensional sphere, a rough image of which is illustrated in the Figure 2. Because of the gravitational attraction working between all parts of the energy distribution, the expansion is getting slower and then will turn to shrink. In order to express this process, a time dimension to trace other dimensions is essential. With the tracing by this original time dimension, the radius dimension (referred to as the "S"-dimension) takes a single value, which is moving by tracing. Therefore, the S-dimension has the imaginary order of freedom, and can work as another tracing dimension to express energy distribution in the 3-dimensional surface space.

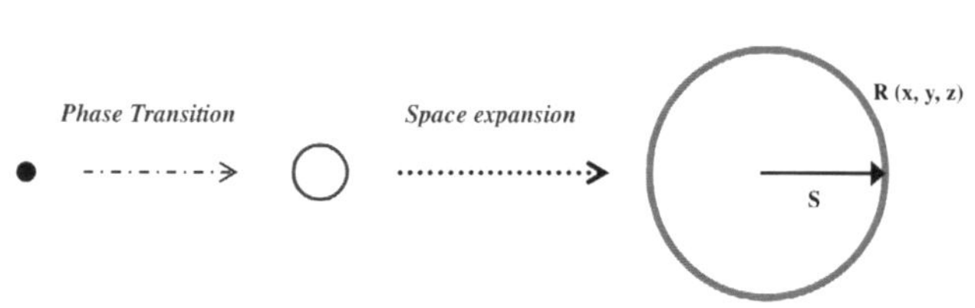

Figure 2. Expansion of the energy distribution of the space

There would be a possibility that by the original tracing time the gravitational force works in a completely time-independent manner corresponding to a distance in the 4-dimensional space. Gravitational field is spread not only in the 3-D surface space but also in the S-dimension. If we feel the S-dimension, which is an imaginary space dimension, as the time which we observe passing at a constant speed (referred to as the "Time"), the gravitational field affects not only the 3-D space but also the Time dimension. Taking these aspects into consideration, the gravity may be described by the Newtonian way from the original time dimension but from the observed Time dimension it might be described by the Einstein field equation.

Because any energy exists only in the 3-D surface membrane of the 4-D sphere, gravitational interaction between two massive objects should not be transmitted in any medium due to the lack of energy or substance inside the membrane toward the center of the universe, subject that the force is depending on the linear distance between the two objects in the 4-D sphere. Therefore there would be no need to consider that the gravitational interaction should propagate slower than the speed of light, although other interactions and any substances including the light have the limit of transmission as they exist only in the 3-D surface brane.

Another important factor of gravity is that the gravity works with almost no limitation of the distance. Not only between very distant galaxies but even nearly zero distance which causes gravitational collapse [15] the gravity works. It may also suggest the absence of the medium for gravitational interaction.

I propose that the gravitation would work in accordance with the Newton's equation depending on the linear distance in the 4-D sphere at a certain instance of the original tracing time dimension. However, the mass is not the so-called static mass but that corresponding to the total energy including kinetic and any kind of energy. As all energies are distributed in the 3-D surface, we would observe the surface as the space and the imaginary space dimension S as the Time which we observe passing constantly. From our observation based on the Time, the gravitational interaction would be described by the Einstein field equation or its analog.

4. Development of the universe

As a potential model for the development of the universe, I propose the following.

- The initial state of the universe is vibrations in multiple dimensions, which render energy.
- If an energy vibration has a boundary condition and shows a steady state due to its inside gravity, it can be treated as an object having a mass corresponding to its total energy.
- The gravitational force works between any pieces of energy based on the linear distance in the multiple dimensions.
- At the Big Bang, the energy distribution expanded in 4 space dimensions with tracing by one original time dimension. Other hidden dimensions remained compactified, vibrations in which are treated as a mass or energy.
- The energy distribution is in 3-D surface of a 4-D sphere. The phase transition of energy from vibration in 4-D dimensions to an energy distribution in the 3-D space released a tremendous energy, which firstly converted to a kinetic energy of extremely rapid expansion and to gravitational potential due to the expansion subsequently.
- Let's define the space of the universe as the area where energy is spread.
- Let's refer to the energy vibration in the 3-D surface as the "Spacia", which is the energy of the space itself.
- The Spacia has fluctuations in the radius S-dimension and in the 3-D surface space.
- The fluctuation in the S-dimension is kept in a steady state, which renders additional energy to the Spacia.

- The fluctuations in the 3-D surface space are firstly in a steady state, but immediately pulse waves, which are no longer in a steady state due to the expansion of the Spacia, emerge from them.
- The pulse wave is primarily the light.
- Due to high energy of the light and extremely short size, boundary conditions come out due to its inside gravity, by which other quantum particles are created. Remaining lights after completion of particle creation are now observed as the cosmic microwave background radiation (CMB).

The key features of this model are that the space is the area where energy is spread, and that any substances including the light are waves of the space energy called the Spacia. The propagation speed of the light is constant at a certain time, depending on the energy density of the Spacia.

5. Expansion of the universe and the time passing

This model implies that the expansion of the universe is slowing down due to the gravitational attraction between all areas of the universe, and the universe will turn to shrink. However, by the S-dimension, which is an imaginary space dimension, the universe is expanding at a constant speed ($dR/dS = 2\pi$, where R is the circumference of the surface with the radius of S), and will shrink at the same speed after the expansion stops.

I here strongly insist that our observed time passing, the Time, should be the S-dimension but not the original tracing dimension, the time. The relativity between the 3-D space and the Time (= S), which can be represented by dR/dT (R is the whole length of the space or its unit area) is independent from dT/dt. Even how slowly or rapidly the Time traces the space, the relativity is the same. This should be the real nature of the Time, which seems passing at a constant speed. If we see only the remaining Spacia after completion of particle creation or the size of universe in the 3-D space, the 3 space dimensions are also imaginary in degree of freedom for the expansion of the universe. This signifies that the size of the 3-D space (defined as "R") can also act as a tracing dimension for the Time (or S). The proceeding speed of the Time by the R is also a constant ($dT/dR = 1/2\pi$). This is exactly the nature that the Time is observed as passing at a constant speed. We can conclude with a very important notice as follows: "The fact that the space is expanding at a constant speed by the Time is identical to the fact that the Time is passing at a constant speed."

Gross outline illustrations of the expansion of the universe based on the original time and the observed Time are shown in the Figure 3 and the Figure 4, respectively.

For calculating the gravitational force of a part of universe having its mass = m, the counterpart can be calculated as the gravity center of the whole universe having the mass of M. Based on the time, the force (F) is given by the Newton's gravitational equation.

$$F = GMm/S^2 \tag{1}$$

The acceleration by the force is GM/S^2, which is unique to any part of the universe. Put the expansion speed of S by the time, $dS/dt = V$, $T = T_0$ at the beginning of the Big Bang, and $V = V_0$ at T_0, then the total energy of the area gives the following equation.

$$\frac{1}{2} \cdot mV^2 + \int_0^S FdS = \frac{1}{2} \cdot mV_0^2 + \int_0^{S_0} FdS \tag{2}$$

From the equations (1) and (2), the expansion speed $V = dS/dt$ is given as follows:

$$V = \pm\sqrt{2GM * (1/S - K)} \qquad (K = 1/S_0 - V_0^2/2GM) \tag{3}$$

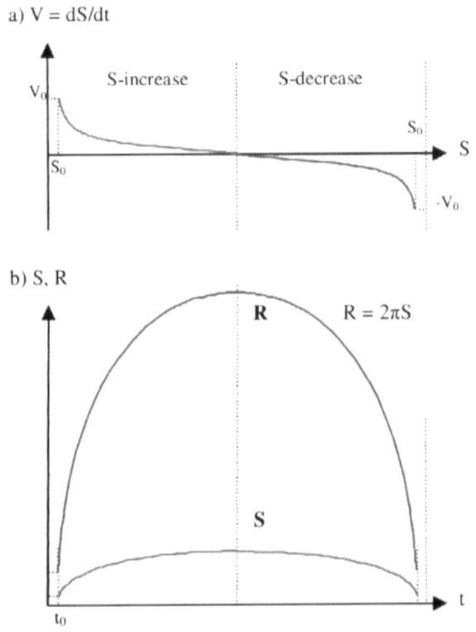

Figure 3. Expansion of the universe based on the original time (t): a) Speed of expanding S-radius, b) Size of the space (R) and the radius (S)

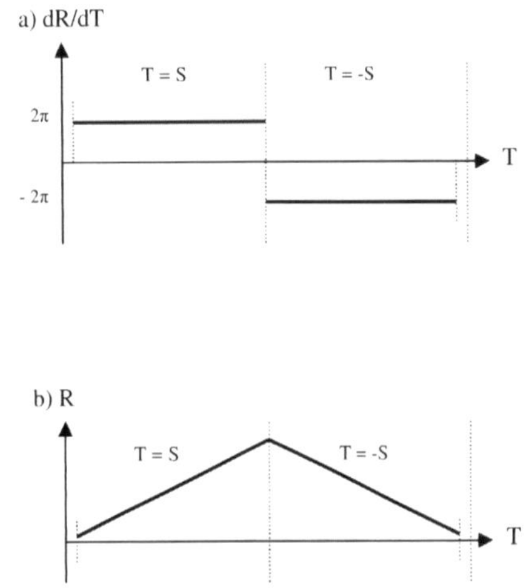

Figure 4. Expansion of the universe based on the detectable Time (T): a) Speed of space expansion, b) Size of the space

On the other hand, based on the Time, the space expansion can be characterized as follows. Firstly, in order to conform the Time passing direction to the time passing, let's define the Time as $-S$ for the shrinking process of the universe.

$dT = dS \qquad (dS/dt > 0)$
$dT = -dS \qquad (dS/dt < 0)$

The expansion speed of the 3-D surface by the Time, defined as dR/dT, is given as follows:

$dR/dT = dR/dS = 2\pi \qquad (dS/dt > 0)$
$dR/dT = -dR/dS = -2\pi \qquad (dS/dt < 0)$

6. Key features of the time

If the Time is the S-dimension, the arrow of the Time is a 4-dimensional position vector in the S and 3-D space dimensions. By a spherical coordinate with the radius of S, the Time arrow has a 3-dimensional space direction in addition to the direction in the S-dimension, shown as $(T, \theta_1, \theta_2, \theta_3)$. By a Cartesian coordinate system, any place has its own specific static space-Time coordinate different from that of other places, as shown in the Figure 5.

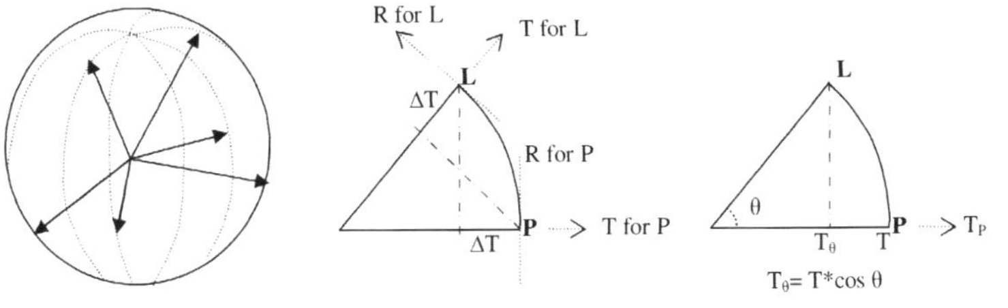

Figure 5. Time arrows and relative space-time Cartesian coordinates

In the Figure 5, the Time for L (T_L) at L is same as the Time for P (T_P) at P in magnitude. Any place in the universe has the common Time. However, T_P at L is smaller with the factor of $\cos\theta$ than T_P at P (or T_L at L). By T_P, P is the most advanced in the Time, but by T_L, L is the most advanced in the Time. Any location of the space is the most advanced in the Time than other locations by its own Time. The area of the universe, which expanded to the opposite direction from our area, has been moving to the past after the Big Bang by the Time of our area, while the both areas share the same Time for the whole universe as well as the same original tracing time.

We can see only the current figure of our selves, but can not see the past figure of us. On the other hand, we can see only the past figure of a distant galaxy but can not see the current figure of it. This phenomenon is derived from two aspects. One is that the light transmits only in the Spacia. Between the past position in the 4-D sphere of a static object and the current position of it, the Spacia does not exist. The second factor is that from a distant place, light takes a time to reach us and furthermore the distant location is behind in the Time by our Time.

7. Light speed and the supernovae data for space expansion

From observations of supernovae Ia, it is suggested that the universe is expanding in an accelerating manner. Let's check a time factor.

According to the proposed model in this article, the speed of light should depend on the density of the medium, the Spacia. Let's analogize the light speed from the speed of sound in the air. The speed of sound is given in general by

$c = \sqrt{C/\rho}$, where C is a coefficient of stiffness, ρ is the density of the air. Further, the equation is led to

$c = \sqrt{\frac{\gamma \cdot \kappa \cdot T}{m}}$, where γ is the adiabatic index, κ is the bulk modulus, T is the absolute temperature in kelvins, m is the mass of a single molecule in kilograms. The c_{air} is proportional to square root of the absolute temperature shown by

$c_{air} = 331.3m \cdot s^{-1}\sqrt{\frac{T}{273.15}} = 331.3m \cdot s^{-1}\sqrt{1 + \frac{\theta}{273.15}}$, where θ is the temperature in degrees Celsius [16, 17].

Speculated by analogy with the speed of sound, the light transmission velocity in the Spacia might be proportional to square root of the energy density or temperature of the Spacia. Therefore the velocity by the time, dC/dt, might be inversely proportional to square root of $S^3 (= T^3)$. Using the Time instead of the time, dC/dT = dC/dS, might be shown by an equation similar to the equation (4).

$$dC/dT = dC/dS = \frac{K_1}{\sqrt{S^3 * (1/S - K_2)}} \quad (4)$$

As a reference illustration, a graph of a related formula is shown in the Figure 6. Except for the early period and the neighbor period from dS/dt = 0, the light speed might be within a rather narrow variability.

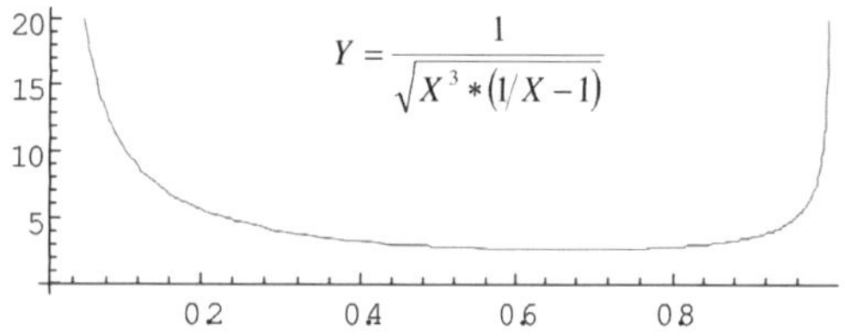

Figure 6. A graph of a similar formula of the equation (4) for light speed

Currently, the traveled time period of the light of supernovae is calculated to divide the luminosity distance by the constant light speed throughout the whole time period [3]-[5]. However, the light speed is getting slower by the Time (also by the time) passing. Therefore, the traveled time of the light of a very distant supernova is calculated longer than the real value. The distance is not proportional to the Time (time) back. For getting Hubble Plots, the Time (time) back for a red-shift value should be adjusted by the change of the light speed by the Time (time). The real expansion by the Time would be at a constant speed as proposed in this article.

8. Discussion

From the model of the time and universe proposed in this article, the Dark Matter would be interpreted as the remaining energy of the Spacia excluding photons and all other quantum particles. In other words, the Dark Matter would be the energy corresponding to vibrations of the Spacia in the S (= the Time) and hidden dimensions, which can be interpreted as the energy of vacuum of the space. The Dark Energy would be an imaginary energy, which does not exit but is introduced wrongly due to the lack of adjustment of the changing speed of the light in the analysis of supernovae data.

The scenario for the development of the universe described in this article is somewhat hypothetic. However, the nature and definition of the time discussed in the second chapter should be rather theoretical. I hope professional physicists would mathematically investigate the transformation of the Newtonian equation based on the original tracing dimension to a new one, like the Einstein field equation, based on the S-dimension as our observed Time.

References

[1] Perlmutter S *et al* 1997 *Astrophys. J.* **483** 565
[2] Perlmutter S *et al* 1998 *Nature* **391** 51
[3] Perlmutter S *et al* 1999 *Astrophys. J.* **517** 565
[4] Schmidt B *et al* 1998 *Astrophys. J.* **507** 46
[5] Perlmutter S and Schmidt B 2003 Measuring cosmology with Supernovae *Preprint* astro-ph/0303428
[6] Einstein A 1915 *Sitzungsberichte der Preussischen Akademie der Wissenschaften zu Berlin* 844-847
[7] Einstein A 1916 The Foundation of the General Theory of Relativity *Annalen der Physik* **49** (http://www.alberteinstein.info/gallery/gtext3.html)
[8] Pound R V and Rebka G A 1959 *Physical Review Letters* **3** 439
[9] Pound R V and Rebka G A 1960 *Physical Review Letters* **4** 337
[10] Pound R V and Snider J L 1964 *Physical Review Letters* **13** 539
[11] Hafele J and Keating R 1972 *Science* **177** 166
[12] Hafele J and Keating R 1972 *Science* **177** 168
[13] Einstein A and Grossmann M 1913 *Zeitschrift fuer Mathematik und Physik* 225-261
[14] Einstein A 1905 *Annalen der Physik* **17** 891
[15] Oppenheimer J R and Volkoff G M 1939 *Physical Review* **55** 374-381
[16] Speed of sound *Wikipedia* (http://en.wikipedia.org/wiki/Speed_of_sound) (accessed Nov. 23, 2008)
[17] Everest F 2001 *The Master Handbook of Acoustics* (New York: McGraw-Hill) pp 262-263

Fourth International Workshop DICE2008 IOP Publishing
Journal of Physics: Conference Series **174** (2009) 012070 doi:10.1088/1742-6596/174/1/012070

Event-driven power-law relaxation in weak turbulence: a liquid crystals mesoscopic experiment bridging quantum dots and the integration theory for the brain

Ludovico Silvestri[1], Leone Fronzoni[1], Angelo Gemignani[2,3], Paolo Grigolini[1,4], Danilo Menicucci[3,5] and Paolo Allegrini[1,6]

[1]Dipartimento di Fisica "E. Fermi"-Università di Pisa and INFM CRS-SOFT, Largo Pontecorvo 3, 56127 Pisa, Italy
[2]Department of Physiological Sciences, University of Pisa, Via San Zeno 31, 56127 Pisa, Italy
[3]Extreme Physiology Centre, Scuola Superiore SantAnna, Piazza Martiri della Libertà 7, 56127, Pisa, Italy
[4]Center for Nonlinear Science, University of North Texas, P.O. Box 311427, Denton, Texas 76203-1427, U.S.A.
[5]Istituto di Fisiologia Clinica del Consiglio Nazionale delle Ricerche, Via Moruzzi 1, 56124 Pisa, Italy
[6]Interdepartmental Research Center "E.Piaggio" Faculty of Engineering - University of Pisa Via Diotisalvi 2, 56126 Pisa, Italy

E-mail: `paolo.allegrini@df.unipi.it`

Abstract. We illustrate a theoretical approach to derive generalized master equations, in both quantum and classical physics, compatible with the occurrence of crucial events. We show how to derive out of it a new Fluctuation-Dissipation Theorem (FDT), called phenomenological FDT. We fail deriving from the same approach another FDT form, called dynamic FDT, which is proved by us to drive the response of liquid crystals to external perturbations. We argue that this experimental effect is incompatible with the adoption of a density approach, in the same way as the decoherence theory fails explaining the wave function collapse, if this corresponds to crucial events. We also argue that the same perspective applies to the brain, according to the tenets of the integration theory of consciousness.

1. Introduction

In the last decade there has been an increasing interest in the fluorescent intermittence of quantum dots [1]. The search for a theory of this phenomenon is a very active field of research and the authors of a recent review on the subject reached the conclusion that there does not exist yet a theory accounting for all the properties revealed by the experiment [1]. The phenomenon is reminiscent of the well known quantum jumps of the Delmeth experiment [2], the main difference being that the latter concerns a single atom in a trap, with a Poisson intermittence [3], while the blinking quantum dot phenomenon is produced by a collection of atoms, and the intermittent effect is non-Poisson. The time distance between two consecutive transitions from the light-on (off) to the light-off (on) condition is given by an inverse power law distribution density with a power index $\mu < 2$. A careful analysis of the experimental data reveals that the process is renewal [4].

Recently, the authors of [5] developed a theoretical model generating intermittence qualitatively similar to that produced by blinking quantum dots, suggesting that the phenomenon may correspond to the global properties produced by the cooperation among many units, each of which, in the absence of cooperation, may be described very well by using the Poisson statistics of quantum jumps. When the cooperation parameter reaches the critical value corresponding to the phase transition from the uncorrelated to the correlated condition, the signal generated by the sum of all individual contributions exhibits the typical intermittent behavior of blinking quantum dots.

In this paper we prove that the global luminescence of liquid crystals fits this property, thereby making the physics of liquid crystals an investigation ground especially suited for the experimental assessment of the emergence of cooperation-induced macroscopic effects. In this condition a new fluctuation-dissipation theorem applies [6, 7], which seems to conflict with the Liouville or Liouville-like treatment that are currently adopted by the advocates of decoherence theory to bypass the adoption of the wave function collapse [8]. We are convinced that this condition applies to the brain as well, thereby making liquid crystals suitable for illustrating the brain dynamics and its response to external stimuli. We believe that the connection between liquid crystal dynamics and the brain can be established on the basis of the integration theory of consciousness [9]. In fact, according to this theory consciousness corresponds to the capacity of integrating information. However, the different information units are not totally independent but they rather interact as the single units of the model of Ref. [5] and the liquid crystal defect formation and annihilation, which are the subject of the experiment illustrated in this paper.

The outline of this paper is as follows. In Section 2 we derive the structure of a master equation driven by crucial events and we show that it generates a linear response which implies the times of crucial event occurrence to be independent of the external perturbation. We stress the conflict between the trajectory and the density approach in processes dominated by crucial events, including the quantum case. In Section 3 we illustrate the experimental results on the liquid crystal response to perturbation. Section 4 is devoted to concluding remarks.

2. Crucial events and time convoluted master equations

Let us consider the master equation

$$\frac{d\mathbf{p}}{dt} = \mathbf{K}\mathbf{p}(t), \tag{1}$$

where $\mathbf{K}$ is a N by N matrix and $\mathbf{p}(t)$ is a N-dimensional vector. The range of application of this master equation is very extended, insofar as it can be used in both classical and quantum statistical mechanics. For instance, the celebrated Lindblad equation [10]

$$\frac{d}{dt}\rho = L\rho \tag{2}$$

can be written in the same form as Eq. (1) by interpreting the density matrix $\rho(t)$ as a vector and the Lindblad super-operator L as a matrix. To appreciate the importance of the Lindblad equation for the fundamental issue of the wave function collapse, let us consider the toy Hamiltonian [8]

$$H_T = G\Sigma_z\sigma_x + H_\sigma, \tag{3}$$

where

$$H_\sigma = -\frac{V}{\sqrt{2}}\sigma_z + g\sigma_x\eta + H_B. \tag{4}$$

The authors of [8] interpreted the Pauli matrix Σ as the pointer of a measurement device that has to establish whether the system of interest is the up or down state, and focused their attention on the intriguing case where the system is linear superposition of the two eigen-states of its z-component, Σ_z, with the same quantum mechanical probability. The interaction with the fast fluctuating dipole σ generates the relaxation of the density matrix $\rho_\Sigma(t)$ described by the Lindblad equation (2) with $\rho(t) = \rho_\Sigma(t)$ and

$$L\rho_\Sigma(t) = 2\frac{G^2}{\hbar^2\gamma}\left[\Sigma_z\rho_\Sigma(t)\Sigma_z - \rho_\Sigma(t)\right], \tag{5}$$

where $1/\gamma$ is the relaxation time of the autocorrelation function of σ_x. The matrix ρ_Σ is turned into a statistical density matrix yielding $\langle\Sigma_z\rangle = 0$ and this is interpreted as a wave function collapse into either the up or down state.

The interesting result of Ref. [8] has to do with the case where the fluctuation $\sigma_x(t)$ departs from the exponential condition. In this case, the predictions of de-choerence theory [11] and the occurrence of real wave function collapses generate different results, which are in principle accessible to experimental control. In other words, it is possible to see the difference between the case where the dynamics of σ_x is described by a non-Markovian de-coherence equation and the case when real wave function collapses exist, generating the same non-exponential correlation function as the de-coherence theory [8]. In the former case the pointer relaxation is an inverse power law and in the latter case is an exponential relaxation [8]. This conflict may be a consequence of the theoretical difficulties generated by the adoption of a Liouville approach to describe dynamic processes driven by the crucial events hereby defined [12].

To make further progresses on the intriguing issue of wave function collapses it is important to establish under which condition a generalized master equation is compatible with the existence of crucial events. To illustrate all this in a concise but clear way, let us go back to the master equation (1). Without loss of generality, let us set $\Delta t = 1$, where Δt is the integration time step. This allows us to replace t with the integer number n. Following the Continuous Time Random Walk (CTRW) prescription [13] we make the assumption that there exists a stochastic connection between the discrete time n and the physical (continuous) time t. Thus, we replace Eq. (1) with

$$\mathbf{p}(n+1) = (\mathbf{1} - \mathbf{K})\mathbf{p}(n). \tag{6}$$

At a generic time t we have

$$\mathbf{p}(t) = \sum_{n=0}^{\infty}\int_0^t dt'\psi_n(t')\Psi(t-t')\,(1-\mathbf{K})^n\,\mathbf{p}(0). \tag{7}$$

We now explain the physical meaning of Eq. (7) and at the same time we define the concept of crucial event. The time distance between the occurrence of two events (not necessarily crucial) is given by the waiting time distribution density $\psi(\tau)$. We set

$$\psi(\tau) = (\mu - 1)\frac{T^{\mu-1}}{(\tau+T)^\mu}. \tag{8}$$

We assume that $\tau(n) \equiv t(n+1) - t(n)$ is derived from Eq. (8). The state $\mathbf{p}(n)$ lasts from $t(n)$ to $t(n+1)$. The event occurring at time $t(n)$ is an abrupt change from the state $\mathbf{p}(n-1)$ to the state $\mathbf{p}(n)$. We define this event *crucial* if

$$1 < \mu < 3. \tag{9}$$

The reason for calling these events crucial is given by the work of Ref. [14] proving that the emerging power spectrum is $P(f) \propto 1/f^\eta$, with $\eta = 3 - \mu$, namely, it is divergent for $f \to 0$ in the whole range of Eq. (9). Note that $\psi_n(t)$ is the probability that n events occurred, the last of which at time t. The function $\Psi(t)$ denotes the probability that no event occurs up to time t. The Laplace transforms of the functions $\psi_n(t)$ and $\Psi(t)$ are expressed in terms of the Laplace transform of $\psi(t)$, $\hat{\psi}(u)$. It is straightforward to prove that the Laplace transform of $\mathbf{p}(t)$, $\hat{\mathbf{p}}(u)$, reads

$$\hat{\mathbf{p}}(u) = \frac{1}{u + \hat{\Phi}(u)\mathbf{K}}\mathbf{p}(0), \tag{10}$$

and that this is the Laplace transform of the generalized master equation

$$\frac{d\mathbf{p}}{dt} = \int_0^t dt'\Phi(t-t')\mathbf{K}\mathbf{p}(t'). \tag{11}$$

Note that

$$\hat{\Phi}(u) = \frac{u\hat{\psi}(u)}{(1-\hat{\psi}(u))}, \tag{12}$$

namely, the typical memory kernel emerging from the CTRW theory [13]. Using this approach it is possible, in principle, to create quantum mechanical generalized master equation that are compatible with the occurrence of crucial events [15].

This is an interesting result that, however, does not settle the issue of making quantum mechanical collapses compatible with quantum mechanics. In fact, although the generalized master equation of Eq. (11) is formally similar to those derived from the quantum Liouville picture by means of a contraction procedure, the memory kernel $\Phi(t)$ is not a correlation function [16]. The authors of Ref. [16] point out that the adoption of the projection procedure may generate the time convoluted structure of Eq. (11) with $\Phi(t)$ corresponding to a stationary correlation function. In the case of systems driven by crucial events the projection procedure should be extended so as to properly take into account the non-stationary nature of these processes. This is a theoretical problem very far from a successful solution.

Furthermore, we have to address the intriguing issue of the response to external perturbations. This can be taken into account by assuming that the matrix $\mathbf{K}$ depends on time. Sokolov has recently made a theoretical proposal that in the general case of Eq. (1) would read

$$\frac{d\mathbf{p}}{dt} = \int_0^t dt'\Phi(t-t')\mathbf{K}(t)\mathbf{p}(t'). \tag{13}$$

In the special case where $N = 2$,

$$\mathbf{K(t)} = \begin{pmatrix} -w_{12}(t) & +w_{21}(t) \\ +w_{12}(t) & -w_{21}(t) \end{pmatrix}, \tag{14}$$

$$w_{12}(t) = \frac{1}{2}(1+\epsilon\xi_P(t)) \tag{15}$$

and

$$w_{21}(t) = \frac{1}{2}(1-\epsilon\xi_P(t)), \tag{16}$$

we obtain, for $\Pi(t) \equiv p_1(t) - p_2(t)$

$$\frac{d}{dt}\Pi(t) = \epsilon\xi_P P(t) - \int_0 t dt'\Phi(t-t')\Pi(t'), \tag{17}$$

where

$$P(t) \equiv \sum_{n=1}^{\infty} \psi_n(t), \tag{18}$$

namely, the number of event per unit of time produced under the condition that all the realizations have an event at $t = 0$. Note that in the specific case that we are illustrating as an example for the prescription of Eq. (13), the system of interest generates a fluctuation $\xi_S(t)$, which is a sequence of 1 and -1, and

$$\Pi(t) = \langle \xi_S(t) \rangle . \tag{19}$$

It is straightforward to prove that Eq. (17) yields

$$\langle \xi_S(t) \rangle = \epsilon \int_0^t dt' P(t') \Psi(t - t') \xi_P(t'), \tag{20}$$

which fits the general structure of the linear response theory [17].

It is well known that Kubo established a close connection between the linear response theory and the Fluctuation Dissipation Theorem (FDT) of first kind [18]. Thus, the interesting result of Eq. (20) raises the important issue of assessing the proper FDT for a process driven by critical events. The typical form of the linear response theory is [17]

$$\langle \xi_S(t) \rangle = \epsilon \int_0^t dt' \chi(t, t') \xi_P(t'), \tag{21}$$

where $\chi(t, t')$ is called linear response function. In the case of ordinary non-equilibrium statistical physics, namely, when the regression to equilibrium can be expressed by means of stationary correlation functions,

$$\chi(t, t') = -\frac{d\Psi(t, t')}{dt'}, \tag{22}$$

where $\Psi(t, t')$ is a stationary correlation function, thereby yielding $\Psi(t, t') = \Psi(t - t')$. When the system is driven by crucial events we have that for $\mu < 2$ the equilibrium condition is never reached [19] , whereas when $2 < \mu < 3$ equilibrium is possible, in principle, but it takes a virtually infinite time time for the system to regress to equilibrium, thereby implying the ergodicity breakdown discussed by Lee [20, 21]. The authors of Refs. [6, 7] noticed that in this case

$$\frac{d\Psi(t, t')}{dt} \neq -\frac{d\Psi(t, t')}{dt'}. \tag{23}$$

In fact,

$$\frac{d\Psi(t, t')}{dt} = -\psi(t, t'), \tag{24}$$

where $\psi(t, t')$ is the probability that an event occurs at t given that the waiting time process begins at an earlier time t', subsequent, however, to the preparation time, and

$$\frac{d\Psi(t, t')}{dt'} = P(t') \Psi(t - t'). \tag{25}$$

We thus see that the interesting proposal of Sokolov corresponds to the choice of the derivative with respect to t'. We refer to the choice of Eq. (24) as dynamic FDT and to the choice of Eq. (25) as phenomenological FDT.

In Section 3 we prove that the response to perturbation of a sample of nematic liquid crystals, driven in a weak turbulent regime, corresponds to the dynamic FDT, thereby generating the important question of whether or not an elegant derivation from the generalized master equation exists in this case, for instance, the choice

$$\frac{d\mathbf{p}}{dt} = \int_0^t dt' \Phi(t - t') \mathbf{K}(t') \mathbf{p}(t'). \tag{26}$$

It is straightforward to prove that this choice yields

$$\langle \xi_S(t) \rangle = \epsilon \int_0^t dt' \psi(t-t') \xi_P(t'), \tag{27}$$

a result that seems to conflict with the non-stationary nature of the dominion of Eq. (9).

3. The experiment

Our setup consists in a planar-aligned nematic liquid crystal (methoxybenzylidene butylaniline, or MBBA) sandwiched between two transparent glass plates coated with Indium Oxide to allow electrical conductivity. We study the conducting regime of electrohydrominamic convection (EHC) [23] by applying an AC electric potential difference between the two plates with frequency of 60Hz. Sample dimensions are: Width $d = 10 \pm 1$ μm and lateral dimensions $A = 1.0 \pm 0.1$ cm^2, yielding aspect ratio $s = 1000 \pm 100$.

The sample is maintained at 18.000 ± 0.004^oC temperature. This precision is achieved using a double thermostat, where a high-precision resistance thermistor has the role of warming up the sample, cooled down by a thermal bath at a lower temperature (8.00 ± 0.01^oC in our setup), using a second thermostat. The lamp under the sample is stabilized through a photodiode-driven feedback circuit. A CCD captures the sample images magnified by a polarizer microscope and sends it to a PC equipped with a digital video-acquisition card which transforms digital images into 8-bit grayscale numerical matrices for subsequent offline analysis.

Above a threshold voltage $V_0 \approx 20$V a convective instability emerges, giving rise to the so-called Williams domains [24], visualized as a regular pattern of light *vs.* darkness stripes. As the voltage is further increased, several defects start to appear in the regular pattern with a specific dynamical behavior: It is possible to see births, deaths, drifts, and oscillations of defects through the observed region. A review of this phenomenon can be found in [23]. We note that the flcutuations of transmitted light intensity in EHC are mainly due to the creation and annihilation of such defects, correspounding to a sudden increase (light "on") or decrease (light "off") of the light transmittivity and, consequently, to a dichotomous signal.

The control parameter $\epsilon \equiv V/V_0 - 1$, called *reduced voltage*, is defined in such a way that $\epsilon > 0$ corresponds to the region where these dynamics occur. The time-scales (inverses of drift velocities, of oscillation frequencies, of death-birth rates) decrease with increasing ϵ, visually showing a less spacially-correlated turbulent pattern. For $0 < \epsilon < \epsilon_1 = 0.21 \pm 0.01$ we have spacial inhomogeneity in the pattern and the system does not equilibrate. We see lines of defects separating different zones of regular stripes with different orientations. Further increasing ϵ these regular domains become smaller and shorter living, till $\epsilon = \epsilon_1$, which signal the transition to another regime, $\epsilon_1 < \epsilon < \epsilon_2$ or Dynamic Scattering Mode 1 [23], where inhomogeneity disappears, and a stationary correlation function $\Phi_\xi(t)$ can be defined.

In all the abovementioned regimes, the fluctuations of transmitted light intensity show a $1/f^\eta$ power spectrum, thereby leading to the intriguing conjecture that the dynamic of the system is driven by crucial events. The proof of such conjecture is to be searched in the linear response properties of the sample, finding out whether it follows eq. 21 or not and, most important, if the response function proposed by the authors of [6], namely, the condition illustrated by Eq. (24), is suitable to reproduce experimental observation. The $1/f$ noise eventually disappears at $\epsilon = 0.9 \pm 0.1$, well below the final transition to fully developed turbulence (Dynamic Scattering Mode 2 [23]), which is thus beyond the scope of this study, focusing on the anomalous response properties of a system driven by crucial events.

In Fig. 1 we show the indexes μ stemming from the $1/f^\eta$ spectra through the prescription $\mu = 3 - \eta$ [14]. Spectral analysis was performed on recorded 1 hour's length signals of transmitted light intensity, with averages on moving windows of duration $327.68s$, corresponding to 2^{13} points at $0.04s$ sampling time, overlapping for half of the window lengths. We also evaluated

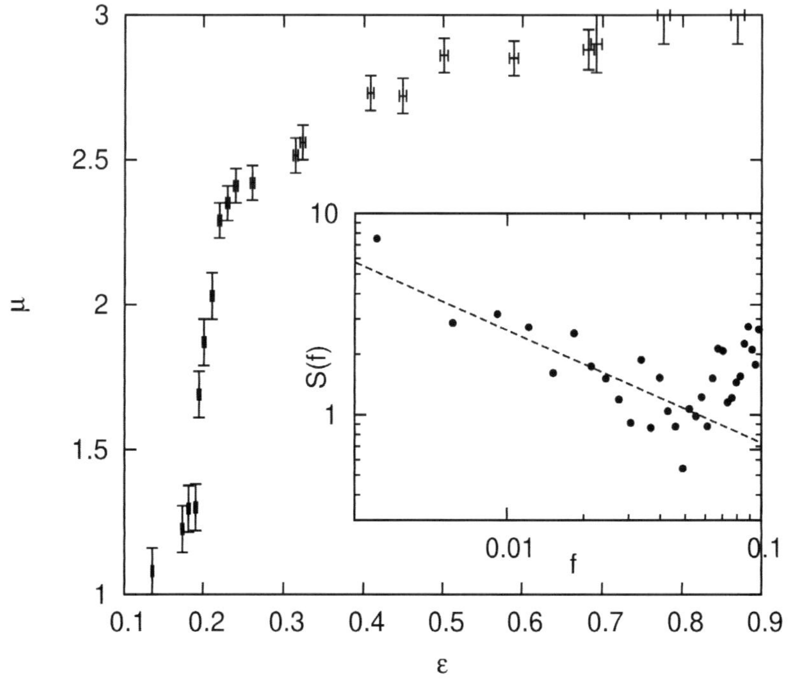

Figure 1. Main figure: points denote μ as a function of the reduced voltage ϵ. Inset: points denote power spectrum for $\epsilon = 0.260 \pm 0.003$; line is a fit yielding $\mu = 2.44$

μ using other indipendent techniques, like the Difffusion Entropy (DE) [26] and the Detrended Fluctuation Analysis (DFA) [27] giving the same results with greater accuracy. In Fig. 1 error bars are reported according to this higher precision, but the discussion of DE and DFA will be omitted here. We note that μ as a function of ϵ shows a fast change around $\mu = 2$, signaling the transition from the non-ergodic regime of [19] to the non-ergodic regime of [20, 21]. In the inset of Fig. 1 we show how the resulting measure is consistent with the empirical power spectrum.

To study the response properties of the system and finding then by what theoretical picture they are better described we study the relaxation of transmitted light intensity after an abrupt change in the applied voltage. For this purpose, we modulate the voltage amplitude with a square wave of period $P = 100s$. At the beginning the system is below threshold for a period $P_r = 10$ s in order to quench the system and delete any memory of spacial correlation. Then for a period $P_\epsilon = P - P_r = 90$ s the system is left to reach its stationary light transmittivity. Averages were performed on 200 periods to establish a connection with a Gibbs-like averages, that in a non-ergodic regime are not equivalent to temporal averages. The resulting light transmitted intensity is a linear function of $\langle \xi_S(t) \rangle$ and one can so manipulate it with linear transformation in order to have a signal that decays to zero, from a positive value.

As pointed out in [25], since the relaxation process starts from a non equilibrium condition, another term must be added to that coming from FDT, to take account of the Onsager-like regression to equilibrium. As the authors of [25] have shown, this term turn out to be the $O(t) \propto 1/t^{\mu-1}$. From this point on we refer to $O(t)$ as the Onsager-like regression, and to $L(t)$ as the term coming from FDT.

The relaxation in the case $\mu > 2$ are shown in Fig. 2 for two values of the control parameter,

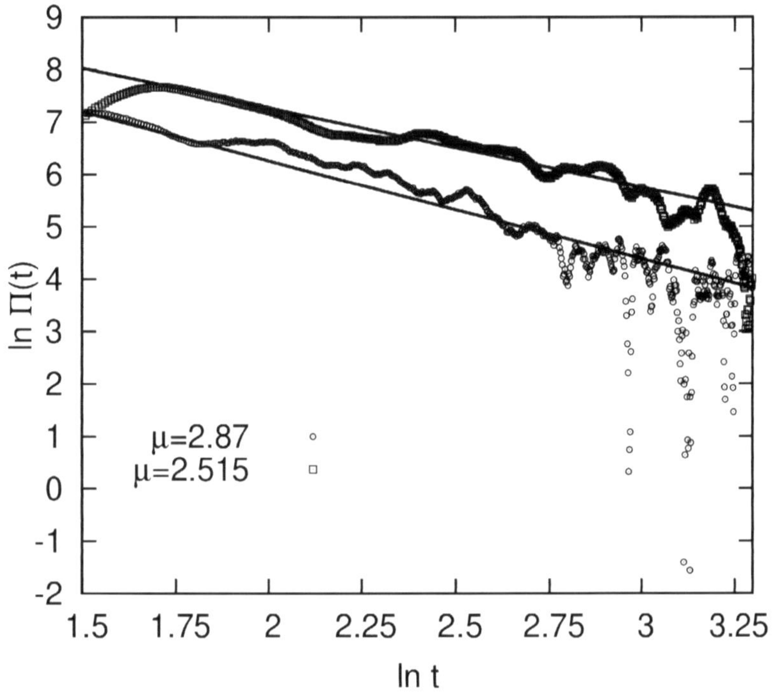

Figure 2. Main figure: Relaxation of $\Pi(t)$ for $\mu = 2.515$ and $\mu = 2.87$ (upper and lower set of points and curves, respectively). Points represent experimental observations, solid lines represent the theoretical power-index relaxation $1/t^{\mu-1}$.

$\epsilon = 0.324$ (corresponding to $\mu = 2.515$) and $\epsilon = 0.513$ ($\mu = 2.87$). In both case we compare the experimental results to the theoretical prediction $O(t)$, finding a good agreement.

The case $\mu = 1.225 (< 2)$ is shown in Fig. 3, where the full line corresponds to the complete theoretical prediction $O(t) + L(t)$. The dominant factor $O(t)$ can be subtracted to the experimental relaxation, and compared to the dynamical contribution $L(t)$ given by FDT. In the present case ($\mu < 2$), this last term turns out to be $\propto 1/t^{2-\mu}$ [6]. As shown in the insert of Fig. 3, this procedure unravels a contribution to relaxation proportional to $1/t^{2-\mu}$. Notice that a sequence of crucial events with $\mu < 2$ makes the Kolmogorov-Sinai entropy increase as $L^{\mu-1}$, generating a randomness rate per unit of time decreasing as $1/L^{2-\mu}$. This shows that the EHC fluctuations, requiring the EDFDT of Ref. [6], are a manifestation of this weakly chaotic condition.

4. Conclusion

In Section 3 we found that the response to external perturbation obeys the dynamic FDT and that the dynamic FDT seems to be incompatible with the adoption of a Liouville, or Liouville-like, approach. We thus conclude that the theory of weak turbulence for liquid crystals shares the same fundamental difficulty as the de-coherence theory [11] concerning the crucial issue of quantum mechanical collapses, insofar as the occurrence of crucial events seems to be incompatible with the Liouville picture. In the case of quantum mechanics it is well known [8] that the Lindblad equation, which corresponds to Poisson physics [3], and thus to processes with no crucial events, there is no conflict between the occurrence of wave function collapses and

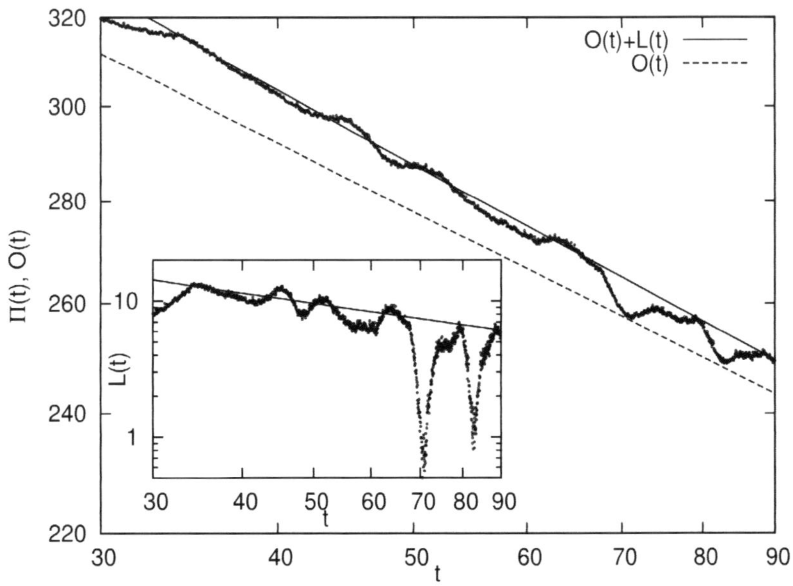

Figure 3. Main figure: Relaxation of $\Pi(t)$ for $\mu = 1.225$. Points represent experimental $\Pi(t)$. Solid line represent the theoretical sum of two power-index relaxations, respectively $L(t) \propto 1/t^{2-\mu}$ and $O(t) \propto 1/t^{\mu-1}$; dashed line represent relaxation $O(t)$ only. Inset: points denote experimental $L(t) = \Pi(t) - O(t)$ (from main figure), solid line represents an inverse-power law $\propto 1/t^{2-\mu}$.

the adoption of the standard quantum mechanical prescriptions. When we move from ordinary statistical physics, with no crucial events, to anomalous statistical physics, with crucial events, the smooth connection between quantum and classical physics established by the de-coherence theory is broken [8]. We refer the interested reader to the review paper of Ref. [22] for further examples of this kind. In this article, proving that the response of liquid crystals to perturbation obeys the theoretical prescription of Eq. (24), we afford another example of possible conflict between theories resting on the conventional Liouville approach and the *experimental* results. In the earlier work [22] experiment means numerical simulation. In this article we have established experimentally the validity of the dynamic FDT, and, consequently, through the arguments of Section 2, a possible conflict between the real experimental results and the adoption of the density approach.

We are convinced that one of the most important consequences of these results is in the field of neuro-sciences. According to Penrose [28] to understand consciousness it is necessary to adopt both the U and the R principle. The U principle is the unitary quantum mechanical time evolution and the R principle refers to the randomness associated to wave function collapses. Although this may be compatible with the vision emerging from our theoretical and experimental results, we are inclined to believe with the authors of Ref. [29] that the adoption of the quantum computation paradigm is irrelevant to explaining how the brain produces thought. We believe that our experimental approach may afford directions on how to analyze the brain function so as to generate further support to the theoretical perspective suggested by Edelman and Tononi [30]. The emergence of consciousness is determined by the interaction of many components generating a phase-transition condition, and, according to the perspective established by this and related articles [5, 6], crucial events. These crucial events generate a response to external perturbation, departing from the ordinary non-equilibrium statistical physics, and a careful analysis of the experimental data thus produced is expected to shed light into the system's complexity. According to the integrated information theory [31] the key ingredient of consciousness is the amount of information generated by a complex of elements, and the dynamic FDT, determining the liquid crystal response to perturbation, is the key ingredient for the transport of information from one to another complex system [32]. We think that this suggests a promising research direction to shed light into the consciousness phenomenon.

Acknowledgments

P. G. thanks ARO and WELCH for financial support of this work through Grants No. W911NF-08-1-077 and No. B-1577, respectively.

References

[1] Frantsuzov P, Kuno M, Jankó B and Marcus R A 2008 *Nature Physics* **4** 519
[2] Dehmelt H G 1975 *Bull. Am. Phys. Soc.* **20** 60
[3] Cook J and Kimble H J 1985*Phys. Rev. Lett.* **54** 1023
[4] Bianco S, Grigolini P and Paradisi P 2005 *J. Chem. Phys.* **123** 174704
[5] Bianco S, Geneston E, Grigolini P and Ignaccolo M 2008 *Physica* A **387** 1387
[6] Allegrini P, Bologna M, Grigolini P and West B J 2007 *Phys. Rev. Lett.* **99** 010603
[7] Aquino G, Grigolini P and West B J 2007 *Europhys. Lett.* **80** 10002
[8] Bologna M, Grigolini P, Pala M G and Palatella L 2003 *Chaos, Solitons and Fractals* **17** 601
[9] Tononi G 2004 *BMC Neuroscience* **5** 42
[10] Linblad G 1976 *Commun. Math. Phys.* **48** 119
[11] Joos E, Zeh H D, Kiefer C, Giulini D, Kupsch J and Stamatescu I-O 2003 *Decoherence and the Appearance of a Classical World in Quantum Theory* second edition (Berlin: Springer)
[12] Bologna M and Grigolini P submitted to *J. Stat. Mech*
[13] Montroll E W and Weiss G H 1965 *J. Math. Phys.* **6** 167
[14] Lukovic M and Grigolini P 2008 *J. Chem. Phys.* **129** 184102
[15] Budini A 2005 *Phys. Rev.* E **72** 056106
[16] Cakir R, Krokhin A and Grigolini P 2007 *Chaos, Solitons and Fractals* **34** 19
[17] Kubo R 1957 *Journal of the Physical Society of Japan* **12** 570
[18] Kubo R, Hashitsume N and Toda M 1985 *Statistical Physics Vol. 2* (Berlin: Springer)
[19] Margolin G and Barkai E 2006 *J. Stat. Phys.* **122** 137
[20] Lee M H 2007 *Phys. Rev. Lett.* **98** 110403
[21] Lee M H 2008 *Acta Physica Polonica* B **39** 1035
[22] Grigolini P 2006 *Advances Chem. Physics* **133** 357
[23] Tóth-Katona T and Gleeson J T 2004, *Phys. Rev.* E **69** 016302
[24] Kai S, Chizumi N and Kohno M 1989, *Phys. Rev.* A **40** 6554
[25] Silvestri L, Fronzoni L, Grigolini P and Allegrini P 2009 *Phys. Rev. Lett.* **102** 014502
[26] Scafetta N, Hamilton P and Grigolini P 2001 *Fractals* **9** 193; Grigolini P, Palatella L and Raffaelli G 2001, *Fractals* **9** 439
[27] Peng C K *et al* 1994 *Phys. Rev.* E **49** 1685
[28] Penrose R 1994 *Shadows of the mind* (Oxford: Oxford U. Press)
[29] Litt A, Eliasmith C, Kroon F W, Weinstein S and Thagard P 2006 *Cognitive Science* **30**, 593
[30] Edelman G M and Tononi G 2000 *A Universe of Consciousness: How Matter Becomes Imagination* (New York: Basic Books)
[31] Tononi G 2008 *Biological Bulletin* **215** 216
[32] West B J, Geneston E L and Grigolini P 2008 *Phys. Rep.* **468** 1

Quantum mechanics in q-deformed calculus

A Lavagno[1,3], **G Gervino**[2,3]

[1]Dipartimento di Fisica, Politecnico di Torino, I-10129 Torino, Italy
[2]Dipartimento di Fisica, Università di Torino, I-10126 Torino, Italy
[3]Istituto Nazionale di Fisica Nucleare (INFN), Sezione di Torino, I-10126 Torino, Italy

E-mail: andrea.lavagno@polito.it; gervino@to.infn.it

Abstract. Starting on the basis of q-deformed calculus and q-symmetric oscillator algebra, we introduce a generalized Schrödinger equation which admits factorized time-space solutions and the free plane wave functions can be expressed in terms of the so-called basic-hypergeometric functions. In this framework, q-deformed adjoint and q-hermitian operator properties occur in a natural way in order to satisfy the fundamental quantum mechanics assumptions.

1. Introduction

In describing complex systems, quantum algebra and quantum groups have been the subject of intensive research in several physical fields such as cosmic strings and black holes [1], conformal quantum mechanics [2], nuclear and high energy physics [3, 4, 5], fractional quantum Hall effect and high-T_c superconductors [6]. From the seminal work of Biedenharn [7] and MacFarlane [8], it was clear that the q-calculus, originally introduced by Jackson [9] in the study of the basic hypergeometric series [10, 11], plays a central role in the representation of the quantum groups with a deep physical meaning [12, 13, 14, 15, 16].

It is remarkable to observe that the q-calculus is very well suited for to describe fractal and multifractal systems. As soon as the system exhibits a discrete-scale invariance, the natural tool is provided by Jackson q-derivative and q-integral, which constitute the natural generalization of the regular derivative and integral for discretely self-similar systems [17]. Furthermore, it has been shown that a complete formulation of generalized thermostatistics of q-bosons and q-fermions can be established using basic numbers and employing the q-calculus based on the Jackson Derivative (JD). The thermodynamic functions such as entropy, pressure, internal energy, specific heat etc. of such deformed systems have been studied and compared with standard bosons and fermions [18, 19, 20].

In the framework of the deformed Heisenberg algebra, a q-deformed Schrödinger equations have been proposed [21, 22]. Although the proposed quantum dynamics is related on the noncummutative differential structure on configuration space, we believe that a fully consistent q-deformed formalism of the quantum dynamics, based on the properties of the q-calculus, has been still lacking.

In this paper, working in the context of the $q \leftrightarrow q^{-1}$ symmetric algebra, we study a generalized deformed quantum mechanics consistently with the prescriptions of the q-calculus. At this scope, we introduce a scalar product in a q-deformed square-integrable wave space and an associate definition of q-adjoint and q-Hermitian operators. In this framework we introduce a q-deformed Schrödinger equation which satisfies the basic assumptions of the quantum mechanics.

2. Deformed oscillator algebra and q-calculus

Let us briefly review the basic properties of q-oscillator algebra of creation and annihilation operators introduced by Biedenharn and McFarlane [7, 8] and the main features of q-calculus and basic-deformed elementary functions useful in the present investigation.

The symmetric deformed algebra is determined by the following commutation relations for a, $a^\dagger$ and the number operator N, thus (for simplicity we omit the particle index)

$$[a, a] = [a^\dagger, a^\dagger] = 0 \ , \quad a a^\dagger - q a^\dagger a = q^{-N} \ , \tag{1}$$

$$[N, a^\dagger] = a^\dagger \ , \quad [N, a] = -a \ . \tag{2}$$

The q-Fock space spanned by the orthornormalized eigenstates $|n\rangle$ is constructed according to

$$|n\rangle = \frac{(a^\dagger)^n}{\sqrt{[n]!}}|0\rangle \ , \quad a|0\rangle = 0 \ , \tag{3}$$

where the q-basic factorial is defined as

$$[n]! = [n][n-1]\cdots[1] \ , \tag{4}$$

and the so-called $q \leftrightarrow q^{-1}$ symmetric basic number $[x]$ is defined in terms of the q-deformation parameter

$$[x] = \frac{q^x - q^{-x}}{q - q^{-1}} \ . \tag{5}$$

In the limit $q \to 1$, the basic number $[x]$ reduces to the ordinary number x and all the above relations reduce to the standard boson relations.

The actions of a, $a^\dagger$ on the Fock state $|n\rangle$ are given by

$$a^\dagger|n\rangle = [n+1]^{1/2}|n+1\rangle \ , \tag{6}$$

$$a|n\rangle = [n]^{1/2}|n-1\rangle \ , \tag{7}$$

$$N|n\rangle = n|n\rangle \ . \tag{8}$$

¿From the above relations, it follows that $a^\dagger a = [N]$, $a a^\dagger = [N+1]$.

The transformation from Fock observables to the configuration space (Bargmann holomorphic representation) may be accomplished by the replacement [13]

$$a^\dagger \to x \ , \quad a \to \mathcal{D}_x^{(q)} \ , \tag{9}$$

where $\mathcal{D}_x^{(q)}$ is the JD [9]

$$\mathcal{D}_x^{(q)} = \frac{D_x - (D_x)^{-1}}{(q - q^{-1})\, x} \ , \tag{10}$$

and $D_x = q^{x\,\partial_x}$ is the dilatation operators. Its action on an arbitrary real function $f(x)$ is given by

$$\mathcal{D}_x^{(q)}\, f(x) = \frac{f(q\,x) - f(q^{-1}x)}{(q - q^{-1})\, x} \ . \tag{11}$$

In contrast to the usual derivative, which measures the rate of change of the function in terms of an incremental translation of its argument, the JD measures its rate of change with respect to a dilatation of its argument by a factor of q.

The JD satisfies some simple proprieties which will be useful in the following. For instance, its action on a monomial $f(x) = x^n$, where $n \geq 0$, is given by

$$\mathcal{D}_x^{(q)}\,(a\,x^n) = a\,[n]\,x^{n-1} \ , \tag{12}$$

where a is a constant.

Moreover, it is easy to verify the following q-version of the Leibnitz rule

$$\begin{aligned} \mathcal{D}_x^{(q)}\big(f(x)\,g(x)\big) &= \mathcal{D}_x^{(q)}\,f(x)\,g(q^{-1}x) + f(q\,x)\,\mathcal{D}_x^{(q)}\,g(x)\;, \\ &= \mathcal{D}_x^{(q)}\,f(x)\,g(q\,x) + f(q^{-1}x)\,\mathcal{D}_x^{(q)}\,g(x)\;. \end{aligned} \tag{13}$$

In addition the following property holds

$$\mathcal{D}_{ax}\,f(x) = \frac{1}{a}\,\mathcal{D}_x^{(q)} f(x)\;. \tag{14}$$

Consistently with the q-deformed theory, the standard integral must be generalized to the q-integral defined, for $0 < q < 1$ in the interval $[0, a]$, as [11]

$$\int_0^a f(x)\,d_qx = a\,(q^{-1} - q)\sum_{n=0}^{\infty} q^{2n+1}\,f(q^{2n+1}\,a)\,, \tag{15}$$

while in the interval $[0, \infty)$

$$\int_0^\infty f(x)\,d_qx = (q^{-1} - q)\sum_{n=-\infty}^{\infty} q^{2n+1}\,f(q^{2n+1})\,. \tag{16}$$

The indefinite q-integral is defined as

$$\int f(x)\,d_qx = (q^{-1} - q)\sum_{n=0}^{\infty} q^{2n+1}\,x\,f(q^{2n+1}\,x) + \text{constant}\,. \tag{17}$$

One can easily see that the q-integral approaches the Riemann integral as $q \to 1$ and also that q-differentiation and q-integration are inverse to each other, thus

$$\mathcal{D}_x^{(q)}\int_0^x f(t)\,d_qt = f(x)\,, \qquad \int_0^\alpha \mathcal{D}_x^{(q)} f(x)\,d_qx = f(a) - f(0)\,, \tag{18}$$

where the second identity occurs when the function $f(x)$ is q-regular at zero, i.e.

$$\lim_{n\to\infty} f(xq^n) = f(0)\,. \tag{19}$$

By using the deformed Leibnitz rule of Eq.(13), an analogue formula for integration by parts may easily be deduced as

$$\int_0^a f(qx)\,\mathcal{D}_x^{(q)} g(x)\,d_qx = f(x)\,g(x)|_{x=0}^{x=a} - \int_0^a (\mathcal{D}_x^{(q)} f(x))\,g(q^{-1}x)\,d_qx\,, \tag{20}$$

$$\int_0^a f(q^{-1}x)\,\mathcal{D}_x^{(q)} g(x)\,d_qx = f(x)\,g(x)|_{x=0}^{x=a} - \int_0^a (\mathcal{D}_x^{(q)} f(x))\,g(qx)\,d_qx\,. \tag{21}$$

For the following developments it appears very relevant to observe that the above integration by parts can be also expressed, for example, as

$$\int_0^a f(x)\,\mathcal{D}_x^{(q)} g(x)\,d_qx = f(qx)\,g(x)|_{x=0}^{x=a} - \int_0^a (\mathcal{D}_x^{(q)} f(qx))\,g(qx)\,d_qx\,, \tag{22}$$

or, equivalently,

$$\int_0^a f(x)\,\mathcal{D}_x^{(q)} g(x)\,d_qx = f(q^{-1}x)\,g(x)|_{x=0}^{x=a} - \int_0^a (\mathcal{D}_x^{(q)} f(q^{-1}x))\,g(q^{-1}x)\,d_qx\,, \tag{23}$$

where we have rewritten the q-Leibniz rule as

$$\mathcal{D}_x^{(q)}\Big(f(qx)\,g(x)\Big) = \mathcal{D}_x^{(q)}\,f(qx)\,g(qx) + f(x)\,\mathcal{D}_x^{(q)}\,g(x)\,, \tag{24}$$

$$\mathcal{D}_x^{(q)}\Big(f(q^{-1}x)\,g(x)\Big) = \mathcal{D}_x^{(q)}\,f(q^{-1}x)\,g(q^{-1}x) + f(x)\,\mathcal{D}_x^{(q)}\,g(x)\,. \tag{25}$$

From the above relations, as also pointed out in Ref.[17], it appears evident that JD derivative and q-calculus provides a custom made formalism in which to express scaling relations. When x is taken as the distance from a critical point, JD thus quantifies the discrete self-similarity of the function $f(x)$ in the vicinity of the critical point and can be identified with the generator of fractal and multifractal sets with discrete dilatation symmetries.

Let us now introduce the following q-deformed exponential function defined by the series

$$\mathrm{E}_q(x) = \sum_{k=0}^{\infty} \frac{x^k}{[k]_q!} = 1 + x + \frac{x^2}{[2]_q!} + \frac{x^3}{[3]_q!} + \cdots\,. \tag{26}$$

The function (26), reducing to the ordinary exponential function in the $q \to 1$ limit, defines the q-exponential, well known in the literature since a long time ago, originally introduced in the study of basic hypergeometric series [10, 11].

Among many properties, it is important to recall the following relation [11]

$$\mathcal{D}_x^{(q)}\mathrm{E}_q(\alpha\,x) = a\,\mathrm{E}_q(\alpha\,x)\ , \tag{27}$$

and its dual

$$\int_0^x \mathrm{E}_q(\alpha\,y)\,d_qy = \frac{1}{\alpha}\Big[\mathrm{E}_q(\alpha\,x) - 1\Big]\,, \tag{28}$$

where α is a real number different from zero. It is important to point out that Eqs.(27) and (28) are two important properties of the q-exponential which turns out to be not true if we employ the ordinary derivative or integral.

Beside to the q-deformed exponential, it is natural to introduce the q-deformed trigonometric functions as [11]

$$\mathrm{E}_q(ix) = C_q(x) + i\,S_q(x)\,. \tag{29}$$

As a consequence, we have

$$S_q(x) = \sum_{n=0}^{\infty} (-1)^n\,\frac{x^{2n+1}}{[2n+1]_q!}\,, \tag{30}$$

$$C_q(x) = \sum_{n=0}^{\infty} (-1)^n\,\frac{x^{2n}}{[2n]_q!}\,. \tag{31}$$

Furthermore,

$$\mathcal{D}_x^{(q)}S_q(\alpha x) = \alpha\,C_q(\alpha x)\,, \tag{32}$$

$$\mathcal{D}_x^{(q)}C_q(\alpha x) = -\alpha\,S_q(\alpha x)\,, \tag{33}$$

therefore, one can also easily see that $S_q(\alpha x)$ and $C_q(\alpha x)$ are the linearly independent solutions of the q-differential equation

$$(\mathcal{D}_x^{(q)})^2 u(x) + \alpha^2 u(x) = 0\,. \tag{34}$$

The above introduced basic-trigonometric functions can be expressed in terms of the basic-hypergeometric function and of the so-called Hahn-Exton q-Bessel function [10].

3. Deformed operator properties in a q-square-integrable space

Starting from the properties of the previous Section, we are ready to introduce the main ingredients to develop a consistent deformed quantum mechanics.

Let L_q^2 be the q-square-integrable space of all complex functions defined in $(-\infty, \infty)$ such that

$$\|\psi_q\| = \left(\int_{-\infty}^{\infty} |\psi_q(x)|^2 \, d_q x\right)^{1/2} < \infty \,. \tag{35}$$

The space L_q^2 defined above it is a linear space. If ψ_q and φ_q are two q-square-integrable functions, any linear combinations $\alpha\psi_q + \beta\varphi_q$, where α and β are arbitrarily chosen complex numbers, are also q-square-integrable functions. Moreover, L_q^2 is a separable Hilbert space with the inner (scalar) product

$$\langle \varphi, \psi \rangle_q = \int \varphi_q^\star(x) \, \psi_q(x) \, d_q x \,. \tag{36}$$

This is linear with respect to ψ_q, the norm of a function ψ_q is a real, non-negative number: $\langle \psi, \psi \rangle_q \geq 0$ and

$$\langle \psi, \varphi \rangle_q = \langle \varphi, \psi \rangle_q^\dagger \,. \tag{37}$$

Analogously to the undeformed case, it is easy to see that from the above properties of the q-scalar product follows the deformed Schwarz inequality

$$|\langle \varphi, \psi \rangle_q|_q^2 \leq \langle \varphi, \varphi \rangle_q \, \langle \psi, \psi \rangle_q \,. \tag{38}$$

Consistently with the above definitions, the q-adjoint of an operator $\hat{A}_q$ is defined by means of the relation

$$\langle \psi, \hat{A}_q^\dagger \varphi \rangle_q = \langle \varphi, \hat{A}_q \psi \rangle_q^\dagger \,, \tag{39}$$

and, by definition, a linear operator is q-Hermitian if it is its own q-adjoint. More explicitly, an operator $\hat{A}_q$ is q-Hermitian if for any two states φ_q and ψ_q we have

$$\langle \varphi, \hat{A}_q \psi \rangle_q = \langle \hat{A}_q \varphi, \psi \rangle_q \,. \tag{40}$$

On the basis of the results of the previous Section, it appear natural to associate the operators corresponding to the position coordinate and the momentum of a particle as follows

$$\hat{x}_q = x \,, \quad \hat{p}_q = -i\hbar \mathcal{D}_x^{(q)}. \tag{41}$$

At this point, it becomes a crucial question if the introduced deformed momentum operator is q-Hermitian. At this scope, we have to consider the following product

$$\langle \varphi, \mathcal{D}_x^{(q)} \psi \rangle_q = \int \varphi_q^\star(x) \, \mathcal{D}_x^{(q)} \psi_q(x) \,. \tag{42}$$

By considering the integration by parts of Eq.(13) and imposing that the function $\varphi_q(x)$ and $\psi_q(x)$ go to zero in the limit of $x \to \pm\infty$, we have

$$\begin{aligned}
\langle \varphi, \mathcal{D}_x^{(q)} \psi \rangle_q &= -\int \mathcal{D}_x^{(q)} \varphi_q^\star(qx) \, \psi_q(qx) \, d_q x \\
&= -\int \mathcal{D}_{qx} \varphi_q^\star(qx) \, \psi_q(qx) \, d_q(qx) \\
&= -\int \mathcal{D}_x^{(q)} \varphi_q^\star(y) \, \psi_q(y) \, d_q y \,,
\end{aligned} \tag{43}$$

where in the second equivalence we have used the property (14) and in the last equivalence we have changed the integration variable $y = qx$. Therefore, we have that the momentum operator, introduced in Eq.(41), is a q-Hermitian operator. Equivalently,

$$\langle\varphi, \hat{p}_q\psi\rangle_q = \langle\hat{p}_q\varphi, \psi\rangle_q \,. \tag{44}$$

Let us remark that this important property is a peculiarity of the $q \leftrightarrow q^{-1}$ symmetric q-framework defined in Eq.s (1), (5) and (9). Conversely, it is known that, in the non-symmetric framework, the momentum operator defined in term of the JD, like in Eq.(41), is not Hermitian [15, 20, 23].

4. Deformed Schrödinger equation

On the basis of the above properties, we have now the recipe to generalize the definition of observables and to introduce a deformed Schrödinger equation in the framework of q-deformed theory. Consistently with the standard quantum mechanics, we have two fundamental postulates: 1) with the dynamical variable $A(x,p)$ is associate the linear operator $\hat{A}_q(x, -i\hbar\mathcal{D}_x^{(q)})$; 2) the mean value of this dynamical variable, when the system is in the (normalized) state ψ_q, is

$$\langle\hat{A}\rangle_q = \int \psi_q^\star\, \hat{A}_q\, \psi_q\, d_qx \,. \tag{45}$$

Observables are real quantities, hence the expectation value (45) must be real for any state ψ_q:

$$\int \psi_q^\star\, \hat{A}_q\, \psi_q\, d_qx = \int (\hat{A}_q^\dagger\, \psi_q^\star)\, \psi_q\, d_qx \,, \tag{46}$$

therefore, on the basis of Eq.(40), observables must be represented by q-Hermitian operators.

If we require there is a state ψ_q for which the result of measuring the observable A is unique, in other words that the fluctuations

$$(\Delta A_q)^2 = \int \psi_q^\star\, (\hat{A}_q - \langle\hat{A}\rangle_q)^2\, \psi_q\, d_qx \,, \tag{47}$$

must vanish, we obtain the following q-eigenvalue equation of a q-Hermitian operator A_q with eigenvalue a

$$\hat{A}_q\, \psi_q = a\, \psi_q \,. \tag{48}$$

As a consequence, the eigenvalues of a q-Hermitian operator are real because $\langle\hat{A}\rangle_q$ is real for any state; in particular for an eigenstate with the eigenvalue a for which $\langle\hat{A}\rangle_q = a$.

Furthermore, as in the undeformed case, two eigenfunctions $\psi_{q,1}$ and $\psi_{q,2}$ of the q-Hermitian operator $\hat{A}_q$, corresponding to different eigenvalues a_1 and a_2, are orthogonal. We can always normalize the eigenfunction, therefore we can chose all the eigenvalues of a q-Hermitian operator orthonormal, i.e.

$$\int \psi_{q,n}^\star\, \psi_{q,m}\, d_qx = \delta_{n,m} \,. \tag{49}$$

Consequently, two eigenfunctions $\psi_{q,1}$ and $\psi_{q,2}$ belonging to different eigenvalues are linearly independent.

It easy to see that, adapting step by step the undeformed case to the introduced q-deformed framework, the totality of the linearly independent eigenfunctions $\{\psi_{q,n}\}$ of q-Hermitian operator $\hat{A}_q$ form a complete (orthonormal) set in the space of the previously introduced basic square-integrable wave functions. In other words, if ψ_q is any state of a system, then it can be expanded in terms of the eigenfunctions (with a discrete spectrum) of the corresponding q-Hermitian operator $\hat{A}_q$ associate to the observable:

$$\psi_q = \sum_n c_{q,n}\,\psi_{q,n}\,, \tag{50}$$

where

$$c_{q,n} = \int \psi^\star_{q,n}\,\psi_q\,d_qx\,. \tag{51}$$

The above expansion allows us, as usual, to write the expectation value of $\hat{A}_q$ in the normed state ψ_q as

$$\langle\hat{A}\rangle_q = \int \psi^\star_q\,\hat{A}_q\,\psi_q\;d_qx = \sum_n |c_{q,n}|^2_q\;a_n\,, \tag{52}$$

where $\{a_n\}$ are the set of eigenvalues (assumed, for simplicity, discrete and non-degenerate) and the normalization condition of the wave function can be written in the form

$$\sum_n |c_{q,n}|^2_q = 1\,. \tag{53}$$

Following the prescriptions at the beginning of this Section, we are able to introduce the q-Hamiltonian operator as

$$\hat{H}_q = -\frac{\hbar^2}{2m}\,\mathcal{D}^{(q)^2}_x + V_q(x)\,. \tag{54}$$

The above Hamiltonian is q-Hermitian and the time development of a system will be given by the following generalized Schrödinger equation

$$i\hbar\frac{\partial\psi_q(x,t)}{\partial t} = \hat{H}_q\,\psi_q(x,t)\,. \tag{55}$$

Important features of the above equation is that implies a consistent conservation in time of the probability densities. In fact, by taking the complex conjugation of Eq.(55), summing and integrating term by term the two equations, we get

$$i\hbar\frac{\partial}{\partial t}\int \psi^\star_q\,\psi_q\,d_qx = \int\left[\psi^\star_q(\hat{H}_q\psi_q) - (\hat{H}_q\psi^\star_q)\psi_q\right]\,d_qx = 0\,, \tag{56}$$

where the last equivalence follows from the fact that the operator Hamiltonian is q-Hermitian.

The Eq.(55) admits factorized solution $\psi_q(x,t) = \phi(t)\,\varphi_q(x)$, where $\phi(t)$ satisfies to the equation

$$i\hbar\frac{d\phi(t)}{dt} = E\phi(t)\,, \tag{57}$$

with the standard (undeformed) solution

$$\phi(t) = \exp\left(-\frac{i}{\hbar}\,E\,t\right)\,, \tag{58}$$

while $\varphi_q(x)$ is the solution of time-independent q-Schrödinger equation

$$\hat{H}_q\,\varphi_q(x) = E\varphi_q(x)\,. \tag{59}$$

In one dimensional case, for a free particle ($V_q = 0$) described by the wave function $\varphi^f_q(x)$, Eq.(59) becomes

$$\mathcal{D}^{(q)^2}_x\,\varphi^f_q(x) + k^2\varphi^f_q(x) = 0\,, \tag{60}$$

where $k = \sqrt{2mE/\hbar^2}$. The previous equation is equivalent to Eq.(34) of Section II and the solution can be written in term of the basic-exponential [11]

$$\varphi_q^f(x) = N_q \, \mathrm{E}_q(ikx) \,, \tag{61}$$

where N_q is a normalization constant. Equivalently, it can be expressed in term of a linear combination of the basic-trigonometric functions $S_q(kx)$ and $C_q(kx)$, introduced in Eq.s (30) and (31). The above equation generalizes the plane wave function in the framework of the q-calculus.

5. Conclusions

On the basis of the symmetric q-calculus, originally introduced in the framework of the basic-hypergeometric functions, we have introduced a generalized linear Schrödinger equation which satisfies the basic quantum mechanics assumptions, under an appropriate generalization of the operators properties in the space of q-square-integral functions. Although a complete description of the introduced quantum dynamical equations lies out the scope of this paper, the results derived here can represent a starting point for future investigations into a full consistent deformed quantum mechanics in the framework of the q-calculus.

References

[1] Strominger A 1993 *Phys. Rev. Lett.* **71** 3397
[2] Youm D 2000 *Phys. Rev.* D **62** 095009
[3] Bonatsos D and Daskaloyannis D 1999 *Progr. Part. Nucl. Phys.* **43** 537
[4] Lavagno A 2002 *Phys. Lett.* A **301** 13
[5] Lavagno A 2002 *Physica* A **305** 238; Alberico W M *et al* 2008 *Physica* A **387** 467
[6] Wilczek F 1990 *Fractional Statistics and Anyon Superconductivity* (Singapore: World Scientific)
[7] Biedenharn L 1989 *J. Phys. A: Math. Gen.* **22** L873
[8] MacFarlane A 1989 *J. Phys. A: Math. Gen.* **22** 4581
[9] Jackson F H 1909 *Am. J. Math.* **38** 26; *do.* 1909 *Mess. Math.* **38** 57
[10] Gasper G and Rahman M 1990 *Basic hypergeometric series* Encyclopedia of mathematics and its applications (Cambridge, UK: Cambridge U. Press)
[11] Exton H 1983 *q-Hypergeometric functions and applications* (Chichester: Ellis Horwood)
[12] Celeghini E *et al* 1995 *Ann. Phys.* **241** 50
[13] Finkelstein R 1996 *J. Math. Phys.* **37** 983; *do.* 1996 *J. Math. Phys.* **37** 2628
[14] Lavagno A and Quarati P 2001 Phys. Lett. B **498** 47; *do.* 2000 Nucl. Phys. B [PS] **87** 209
[15] Lavagno A, Scarfone A M and Swamy N P 2006 *Eur. Phys. J.* C **47** 253
[16] Lavagno A, Scarfone A M and Swamy N P 2007 *J. Phys. A: Math. Theor.* **40** 8635
[17] Erzan A and Eckmann J-P 1997 *Phys. Rev. Lett.* **78** 3245
[18] Lavagno A and Swamy N P 2000 *Phys. Rev.* E **61** 1218
[19] Lavagno A and Swamy N P 2002 *Phys. Rev.* E **65** 036101
[20] Lavagno A 2008 *J. Phys. A: Math. Theor.* **41** 244014
[21] Zhang-Zu J 1998 *Phys. Lett.* B **440** 66; 2000 *Phys. Lett.* B **477** 361
[22] Micu M 1999 *J. Phys.* A **32** 7765
[23] Fichtmüller M, Lorek A and Wess J 1996 *Z. Phys.* C **71** 533

The propagator for the step potential using the path decomposition expansion

James M Yearsley
Blackett Laboratory, Imperial College, London SW7 2BZ, UK
E-mail: james.yearsley@imperial.ac.uk

Abstract. We present a direct path integral derivation of the propagator in the presence of a step potential. The derivation makes use of the Path Decomposition Expansion (PDX), and also of the definition of the propagator as a limit of lattice paths.

1. Introduction

The propagator for the step potential, $V(x) = V_0\Theta(-x)$, is a simple model for scattering, and a useful ingredient in arrival time problems [1]. Expressions for this propagator are known [2], but existing derivations are complex, and give little insight into the structure. Here we show how this propagator may be obtained in a natural way via the Path Decomposition Expansion (PDX), in which the propagator is factored across a surface of constant time [3] and the Brownian motion definition of the path integral. [1]

This propagator may be viewed as a conditional probability density for a random walk on the real line. We then make this integral over paths into a concrete object by defining it as the continuum limit of a discrete sum of lattice paths, $u(0,T|0,0)$. We consider a rectangular lattice with spacing in the time direction of ϵ, and spacing in the x direction of η, and consider propagation for a time $T = 2\epsilon n$, so we have $2n$ steps in our paths. The Euclidean propagator $\overline{g}$ is then defined as the continuum limit of $u/2\eta$ where we take $\epsilon, \eta \to 0$, $n \to \infty$, keeping $\epsilon/\eta^2 = m$ and $T = 2\epsilon n$ fixed. That is,

$$\overline{g}(0,T|0,0) := \lim_{\eta,\epsilon\to 0}(2\eta)^{-1}u(0,T|0,0). \tag{1}$$

2. The Path Decomposition Expansion (PDX)

The PDX is a very useful tool for evaluating discontinuous potentials, such as the step potential. A typical path from $x_0 < 0$ to x_1 may cross $x = 0$ many times, but the set of paths may be partitioned according to their *first* and *last* crossing time. We therefore split every path into three parts: (A) a restricted part that starts at x_0 and does not cross $x = 0$, but that ends on $x = 0$ at time t_1, (B) an unrestricted part from $x = 0$ to $x = 0$ that may cross $x = 0$ many times and, (C) a further restricted part from $x = 0$ to x_1 that does not re-cross $x = 0$, see Fig. 1. The

[1] This contribution is based on [4].

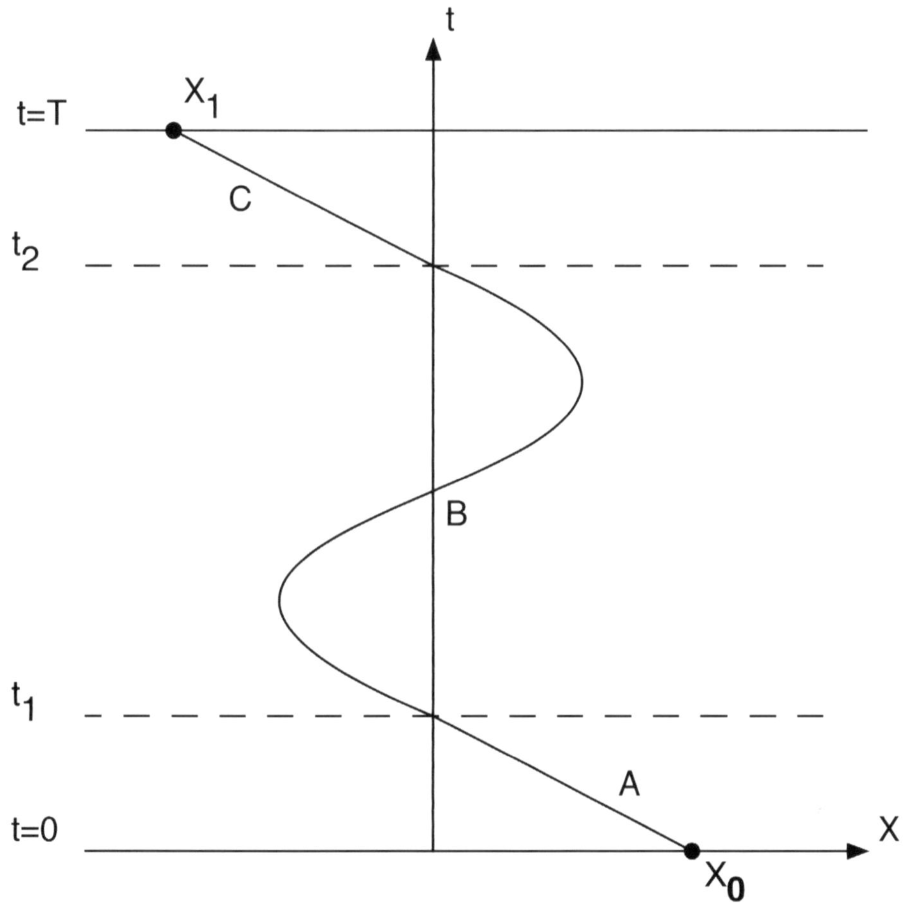

Figure 1. A typical path from x_0 to x_1.

propagator can therefore be written as [3];

$$\begin{aligned} g(x_1, T|x_0, 0) &= \frac{\hbar^2}{4m^2}\int_0^T dt_2 \int_0^{t_2} dt_1 \, \frac{\partial g_r}{\partial x}(x_1, t|x, t_2)|_{x=0} \, g(0, t_2; 0, t_1) \\ &\times \frac{\partial g_r}{\partial x}(x, t_1; x_0, 0)|_{x=0}. \end{aligned}$$

Where t_1 and t_2 are the first and last crossing times and the g_r are the restricted propagators for the regions $x < 0$ and $x > 0$, where the potential is constant. They are therefore easily computed by the method of images. Note that when x_1 and x_0 are both positive, there is an additional contribution from paths which never cross $x = 0$ [3]. The calculation of the propagator therefore reduces to the much simpler task of computing $g(0, T|0, 0)$, the propagator along the *edge* of a step potential.

3. Lattice paths

We first switch to working with the Euclidean propagator $\overline{g}$ by means of a Wick rotation and we specialise immediately to the case of $x_0 = x_1 = 0$. That is, we wish to calculate

$$\overline{g}(0,T|0,0) = \int_{x(0)=0}^{x(T)=0} \mathcal{D}x \; e^{-S_E/\hbar} \tag{2}$$

where S_E is the Euclidean action given by

$$S_E = \int_0^T dt \left(\frac{m\dot{x}^2}{2} + V_0\Theta(x) \right). \tag{3}$$

On the lattice the conditional probability $u(0,T|0,0)$ is given by a sum of paths, each weighted by $e^{-V_0\tau}$, where τ is the length of time spent in $x < 0$. This may be written as

$$u(0,T|0,0) = \frac{1}{2^{2n}} \sum_{k=0}^{n} n_k \; e^{-2k\epsilon V_0}, \tag{4}$$

where n_k is the number of paths spending a time $2k\epsilon < T$ in the region $x < 0$. Expresions for these n_k are known and are in fact independent of k and are equal to the Catalan numbers [4, 6].

$$C_n = \frac{1}{n+1}\binom{2n}{n} \tag{5}$$

where $2n$ is the total number of time steps. This is the central observation that allows this propagator to be computed in closed form.

The sum can therefore be performed and, using the very useful asymptotic form for the C_n

$$C_n \approx \frac{2^{2n}}{\sqrt{\pi} n^{3/2}}, \tag{6}$$

we find

$$u(0,T|0,0) \approx \frac{1}{\sqrt{\pi} 2\epsilon n^{3/2}} \left(\frac{1 - e^{-2\epsilon(n+1)V_0}}{V_0} \right).$$

Taking the continuum limit and Wick rotating back to real time then yields the final result,

$$g(0,T|0,0) = -i \left(\frac{m}{2\pi i} \right)^{1/2} \frac{(1 - e^{-iV_0 T})}{V_0 T^{3/2}}, \tag{7}$$

which is the expression for the propagator along the edge of a step potential. The full propagator from x_0 to x_1 may now be obtained using the PDX [4].

4. Summary

The PDX, together with some elementary combinatorics, allows for a simple derivation of the propagator for this potential. A similar method allows for the computation of the propagator for a delta function potential, by counting the number of crossings of the origin [4]. It is interesting to note that the method described reduces the computation of a path integral to a problem in combinatorics, and this suggests that other propagators may be computed in this way.

Acknowledgments

The author would like to thank the organisers of DICE2008, especially Hans-Thomas Elze, for the opportunity to to take part in this conference. He is also indebted to Jonathan J Halliwell for his support, and for many useful discussions.

References

[1] Halliwell J J 2008 *Phys. Rev.* A **77** 062103
[2] de Carvalho T O 1993 *Phys. Rev.* A **47** 2562
[3] Auerbach A and Kivelson S 1985 *Nucl. Phys.* B **257** 799
[4] Yearsley J M 2008 *J. Phys.* A **41** 285301
[5] See, for example, Hartle J B 1988 *Phys. Rev.* D **37** 2818
[6] A full proof of this can be found in many textbooks on combinatorics, see for example, Stanley RP 1999 *Enumerative Combinatorics vol 2* (Cambridge, UK: Cambridge U. Press)

AUTHOR INDEX

Albrecht, Andreas 58
Allegrini, Paolo 702
Arecchi, F Tito 108
Augello, G 442
Bartalucci, S 671
Bastos, C 571
Bento, M C 629
Bernardini, A E 629
Bertolami, O 571, 629
Bertolucci, S 671
Blasone, Massimo 415, 652, 664
Bombelli, Luca 268
Bragadireanu, M 671
Brida, Giorgio 175
Brightwell, Graham 540
Brodutch, Aharon 29
Brody, Dorje C 475
Cadoni, Mariano 645
Caldara, P 442
Capolupo, Antonio 658
Cargnelli, M 671
Casadio, Roberto 609
Catitti, M 671
Chryssomalakos, C 365
Curceanu, C 671
De Córdoba, P Fernández 408
De Siena, Silvio 652, 664
De Unanue, Adolfo 616
Degiovanni, Ivo Pietro 175
Dell'Anno, Fabio 652, 664
Di Matteo, S 671
Dias, N C 571
Diósi, Lajos 16
Doce, O Vasquez 671
Dolan, B P 332
Dowker, Fay 534
Dubkov, A A 442
Egger, J-P 671
Ellis, David C P 475
Elze, Hans-Thomas 99
Endo, Takasi 376
Erdem, Recai 682
Fiasconaro, A 442
Freeman, Walter J 125
Fronzoni, Leone 702
Gambini, Rodolfo 22
Garattini, Remo 675
Gemignani, Angelo 702
Genovese, Marco 175
Gervino, G 712
Ghazi-Tabatabai, Yousef 579
Ghirardi, Giancarlo 159
Giampaolo, Salvatore Marco 664
Gilra, Aditya 471
Gramegna, Marco 175
Grigolini, Paolo 702
Guaraldo, C 671
Gupta, Kumar S 332
Gustavsson, Anna C T 357
Hall, M J W 455
Halliwell, J J 39
Helland, Inge S 394
Henson, Joe 293, 540
Hernandez-Coronado, H 365
Hirayoshi, Yutaka 376
Holm, Darryl D 475
Hu, B L 187
Iliescu, M 671
Illuminati, Fabrizio 652, 664
Iorio, Alfredo 433
Ishiwatari, T 671
Isidro, José M 408
Jizba, Petr 415
Johnston, Steven 512
Khrennikov, Andrei 425
Kiefer, Claus 316
Kleinert, H 149
Konopka, Tomasz 555
Korotkikh, Victor 588
La Cognata, A 442
Laubenstein, M 671
Lavagno, A 712
Levit, Creon 504
Lychkovskiy, Oleg 387
Marton, J 671
Matzkin, A 463
Mavromatos, Nick E 203
Menicucci, Danilo 702
Milotti, E 671
Minguzzi, E 690
Nagao, S 694
Okon, E 365
Oriti, Daniele 520
Pankratov, A L 442
Parwani, Rajesh R 491
Payandeh, Farrin 596
Penrose, Roger 1
Philpott, Lydia 534
Piacentini, Fabrizio 175
Pietreanu, D 671
Pizzolato, N 442
Ponta, T 671
Prata, J N 571
Prati, Enrico 381
Pullin, Jorge 22
Reginatto, M 455
Rideout, D 237
Santander, J L G 408
Sarfatti, Jack 504
Scardigli, Fabio 415

AUTHOR INDEX

Schettini, Valentina....175
Schuch, Dieter....483
Schulman, L S....327
Shirazi, A C H....603
Sidharth, B G....563
Silvestri, Ludovico....702
Singh, T P....338
Sirghi, D L....671
Sirghi, F....671
Sorkin, Rafael....258, 534
Spagnolo, B....442
Sperandio, L....671
Steinacker, Harold....496
Stern, A....332
Sudarsky, Daniel....616
Surya, Sumati....540
Sverdlov, Roman....268
Terno, Daniel R....546
Toyoshima, Kouichi....376
Traina, Paolo....175
Vaidman, Lev....29
Valenti, D....442
Vitiello, Giuseppe....125, 658
Volovik, G E....69
Wallden, P....237, 579
Wetterich, C....79
Widmann, E....671
Yearsley, James M....720
Yuan, Liew Ding....491
Zenczykowski, Piotr....400
Ziaeepour, Houri....369
Zmeskal, J....671

Institute of Physics
Dirac House, Temple Back
Bristol BS1 6BE UK